Dietary Reference Intakes: Recommended Intakes for Individuals: Minerals

Life Stage Group	Calcium (mg/day)	Chromium (μg/day)	Copper (μg/day)	Fluoride (mg/day)	Iodine (μg/day)	Iron (mg/day)	Magnesium (mg/day)	Manganese (mg/day)	Molybdenum (μg/day)	Phosphorous (mg/day)	Selenium (μg/day)	Zinc (mg/day)	Sodium (g/day)	Chloride (g/day)	Potassium (g/day)
Infants															
0–6 mo	200*	0.2*	200*	0.01*	110*	0.27*	30*	0.003*	2*	100*	15*	2*	0.12*	0.18*	0.4*
6–12 mo	260*	5.5*	220*	0.5*	130*	**11**	75*	0.6*	3*	275*	20*	**3**	0.37*	0.57*	0.7*
Children															
1–3 y	**700**	11*	**340**	0.7*	**90**	**7**	**80**	1.2*	**17**	**460**	**20**	**3**	1.0*	1.5*	3.0*
4–8 y	**1000**	15*	**440**	1*	**90**	**10**	**130**	1.5*	**22**	**500**	**30**	**5**	1.2*	1.9*	3.8*
Males															
9–13 y	**1,300**	25*	**700**	2*	**120**	**8**	**240**	1.9*	**34**	**1,250**	**40**	**8**	1.5*	2.3*	4.5*
14–18 y	**1,300**	35*	**890**	3*	**150**	**11**	**410**	2.2*	**43**	**1,250**	**55**	**11**	1.5*	2.3*	4.7*
19–30 y	**1,000**	35*	**900**	4*	**150**	**8**	**400**	2.3*	**45**	**700**	**55**	**11**	1.5*	2.3*	4.7*
31–50 y	**1,000**	35*	**900**	4*	**150**	**8**	**420**	2.3*	**45**	**700**	**55**	**11**	1.5*	2.3*	4.7*
51–70 y	**1,000**	30*	**900**	4*	**150**	**8**	**420**	2.3*	**45**	**700**	**55**	**11**	1.3*	2.0*	4.7*
> 70 y	**1,200**	30*	**900**	4*	**150**	**8**	**420**	2.3*	**45**	**700**	**55**	**11**	1.2*	1.8*	4.7*
Females															
9–13 y	**1,300**	21*	**700**	2*	**120**	**8**	**240**	1.6*	**34**	**1,250**	**40**	**8**	1.5*	2.3*	4.5*
14–18 y	**1,300**	24*	**890**	3*	**150**	**15**	**360**	1.6*	**43**	**1,250**	**55**	**9**	1.5*	2.3*	4.7*
19–30 y	**1,000**	25*	**900**	3*	**150**	**18**	**310**	1.8*	**45**	**700**	**55**	**8**	1.5*	2.3*	4.7*
31–50 y	**1,000**	25*	**900**	3*	**150**	**18**	**320**	1.8*	**45**	**700**	**55**	**8**	1.5*	2.3*	4.7*
51–70 y	**1,200**	20*	**900**	3*	**150**	**8**	**320**	1.8*	**45**	**700**	**55**	**8**	1.3*	2.0*	4.7*
> 70 y	**1,200**	20*	**900**	3*	**150**	**8**	**320**	1.8*	**45**	**700**	**55**	**8**	1.2*	1.8*	4.7*
Pregnancy															
14–18 y	**1,300**	29*	**1,000**	3*	**220**	**27**	**400**	2.0*	**50**	**1,250**	**60**	**12**	1.5*	2.3*	4.7*
19–30 y	**1,000**	30*	**1,000**	3*	**220**	**27**	**350**	2.0*	**50**	**700**	**60**	**11**	1.5*	2.3*	4.7*
31–50 y	**1,000**	30*	**1,000**	3*	**220**	**27**	**360**	2.0*	**50**	**700**	**60**	**11**	1.5*	2.3*	4.7*
Lactation															
14–18 y	**1,300**	44*	**1,300**	3*	**290**	**10**	**360**	2.6*	**50**	**1,250**	**70**	**13**	1.5*	2.3*	5.1*
19–30 y	**1,000**	45*	**1,300**	3*	**290**	**9**	**310**	2.6*	**50**	**700**	**70**	**12**	1.5*	2.3*	5.1*
31–50 y	**1,000**	45*	**1,300**	3*	**290**	**9**	**320**	2.6*	**50**	**700**	**70**	**12**	1.5*	2.3*	5.1*

Note: This table (taken from the DRI reports, see www.nap.edu) presents Recommended Dietary Allowances (RDAs) in **bold** type and Adequate Intakes (AIs) in ordinary type followed by an asterisk (*). RDAs and AIs may both be used as goals for individual intakes. RDAs are set up to meet the needs of almost all (97–98%) healthy individuals in a group. It is calculated from an EAR. If sufficient scientific evidence is not available to establish an EAR, and thus calculate an RDA, an AI is usually developed. For healthy breastfed infants, the AI is the mean intake. The AI for all other life stage and gender groups is believed to cover the needs of all healthy individuals in the groups, but lack of data or uncertainty in the data prevent being able to specify with confidence the percentage of individuals covered by this intake.

Source: Dietary Reference Intake Tables: The Complete Set. Institute of Medicine, National Academy of Sciences. Available online at www.nap.edu. Reprinted with permission from *Dietary Reference Intakes: The Essential Guide to Nutrient Requirements*, 2006, by the National Academy of Sciences, Washington, D.C. Institute of Medicine, Food and Nutrition Board Dietary Reference Intakes for Calcium and Vitamin D (2011), National Academies Press, Washington DC, 2011.

Acceptable Macronutrient Distribution Ranges (AMDR) for Healthy Diets as a Percent of Energy

Age	*Carbohydrate*	*Added Sugars*	*Total Fat*	*Linoleic Acid*	*α- Linolenic Acid*	*Protein*
1–3 y	45–65	≤25	30–40	5–10	0.6–1.2	5–20
4–18 y	45–65	≤25	25–35	5–10	0.6–1.2	10–30
≥19 y	45–65	≤25	20–35	5–10	0.6–1.2	10–35

Source: Institute of Medicine, Food and Nutrition Board. "Dietary Reference Intakes for Energy, Carbohydrate, Fiber, Fat, Fatty Acids, Cholesterol, Protein, and Amino Acids." Washington, D.C.: National Academies Press, 2002, 2005.

Dietary Reference Intakes: Recommended Intakes for Individuals: Carbohydrates, Fiber, Fat, Fatty Acids, Protein, and Water

Life Stage Group	*Carbohydrate (g/day)*	*Fiber (g/day)*	*Fat (g/day)*	*Linoleic Acid (g/day)*	*α-Linolenic Acid (g/day)*	*Protein*		*Water[b] (L/day)*
						(g/kg/day)[a]	*(g/day)*	
Infants								
0–6 mo	60*	ND	31*	4.4*†	0.5*‡	1.52*	9.1*	0.7*
6–12 mo	95*	ND	30*	4.6*†	0.5*‡	1.50	11.0	0.8*
Children								
1–3 y	130	19*	ND	7*	0.7*	1.10	13	1.3*
4–8 y	130	25*	ND	10*	0.9*	0.95	19	1.7*
Males								
9–13 y	130	31*	ND	12*	1.2*	0.95	34	2.4*
14–18 y	130	38*	ND	16*	1.6*	0.85	52	3.3*
19–30 y	130	38*	ND	17*	1.6*	0.80	56	3.7*
31–50 y	130	38*	ND	17*	1.6*	0.80	56	3.7*
51–70 y	130	30*	ND	14*	1.6*	0.80	56	3.7*
>70 y	130	30*	ND	14*	1.6*	0.80	56	3.7*
Females								
9–13 y	130	26*	ND	10*	1.0*	0.95	34	
14–18 y	130	26*	ND	11*	1.1*	0.85	46	2.1*
19–30 y	130	25*	ND	12*	1.1*	0.80	46	2.3*
31–50 y	130	25*	ND	12*	1.1*	0.80	46	2.7*
51–70 y	130	21*	ND	11*	1.1*	0.80	46	2.7*
>70 y	130	21*	ND	11*	1.1*	0.80	46	2.7*
Pregnancy	175	28*	ND	13*	1.4*	1.10	71	3.0*
Lactation	210	29*	ND	13*	1.3*	1.10	71	3.8*

ND = not determined. *Values are AI (Adequate Intakes), † Refers to all ω-6 polyunsaturated fatty acids,
‡Refers to all ω-3 polyunsaturated fatty acids.

Source: Institute of Medicine, Food and Nutrition Board. "Dietary Reference Intakes for Energy, Carbohydrate, Fiber, Fat, Fatty Acids, Cholesterol, Protein, and Amino Acids" (2002/2005); "Dietary Reference Intakes for Water, Potassium, Sodium, Chloride, and Sulfate" (2005). Washington, D.C.: National Academies Press.

[a] Based on g protein per kg of body weight for the reference body weight, e.g., for adults 0.8 g/kg body weight for the reference body weight.

[b] Total water includes all water contained in food, beverages, and drinking water.

Nutrition:

Science and Applications

Third Edition

NUTRITION:

Science and Applications

THIRD EDITION

Lori A. Smolin, Ph.D.
University of Connecticut

Mary B. Grosvenor, M.S., R.D.

WILEY

VP & PUBLISHER	Kaye Pace
EXECUTIVE EDITOR	Kevin Witt
PROJECT EDITOR	Lorraina Raccuia
MARKETING MANAGER	Clay Stone
SENIOR PRODUCT DESIGNER	Linda Muriello
ASSISTANT CONTENT EDITOR	Lauren Morris
EDITORIAL ASSISTANT	Grace Bagley
DESIGN DIRECTOR	Harry Nolan
CONTENT MANAGER	Juanita Thompson
SENIOR MEDIA SPECIALIST	Svetlana Barskaya
SENIOR PRODUCTION EDITOR	Patricia McFadden
SENIOR PHOTO EDITOR	MaryAnn Price
TEXT DESIGN	Wendy Lai
COVER DESIGN	Wendy Lai

Cover Photo: Damian Davies/Getty Images
Chapter Outline Icons: iStockphoto

This book was set in Minion Pro Regular 10/12 pt by MPS Limited, Chennai, India and printed and bound by Courier Companies.

This book is printed on acid-free paper. ♾

Founded in 1807, John Wiley & Sons, Inc. has been a valued source of knowledge and understanding for more than 200 years, helping people around the world meet their needs and fulfill their aspirations. Our company is built on a foundation of principles that include responsibility to the communities we serve and where we live and work. In 2008, we launched a Corporate Citizenship Initiative, a global effort to address the environmental, social, economic, and ethical challenges we face in our business. Among the issues we are addressing are carbon impact, paper specifications and procurement, ethical conduct within our business and among our vendors, and community and charitable support. For more information, please visit our website: www.wiley.com/go/citizenship.

Library of Congress Cataloging in Publication Data:

ISBN-13: 978-1-118-28826-9
BRV ISBN: 978-1-118-34292-3

Printed in the United States of America

10 9 8 7 6 5 4

About the Authors

Lori A. Smolin received a bachelor of science degree from Cornell University, where she studied human nutrition and food science. She received a doctorate from the University of Wisconsin at Madison, where her doctoral research focused on B vitamins, homocysteine accumulation, and genetic defects in homocysteine metabolism. She completed postdoctoral training both at the Harbor–UCLA Medical Center, where she studied human obesity, and at the University of California–San Diego, where she studied genetic defects in amino acid metabolism. She has published articles in these areas in peer-reviewed journals. Dr. Smolin is currently at the University of Connecticut, where she has taught both in the Department of Nutritional Science and in the Department of Molecular and Cell Biology. Courses she has taught include introductory nutrition, life cycle nutrition, food preparation, nutritional biochemistry, general biochemistry, and introductory biology.

Mary B. Grosvenor holds a bachelor of arts in English and a master of science in Nutrition Science, affording her an ideal background for nutrition writing. She is a registered dietitian and has worked in clinical as well as research nutrition, in hospitals and communities large and small in the western United States. She teaches at the community college level and has published articles in peer-reviewed journals in nutritional assessment and nutrition and cancer. Her training and experience provide practical insights into the application and presentation of the science in this text.

Dedication

To my family for their love, support, and humorous outlook on life and to the country of Scotland, where I was fortunate to live during the writing of this edition. The breathtaking countryside and friendly people provided both inspiration for and diversion from long hours of writing.

LAS

To my sons, though grown and gone, for continuing to remind me what is important in life. To my husband, Peter, for his patience as well as his editorial talents.

MBG

Preface

Courtesy USDA

Jeffrey Coolidge/Getty Images, Inc.

Charles D. Winters/Photo Researchers, Inc.

Nutrition: Science and Applications, Third Edition, is intended as an introductory text for a science-oriented nutrition course. The material is appropriate for a college student at any level, freshman to senior, taking this course to fulfill a science requirement. The clear, concise writing style—reinforced visually with colorful, engaging illustrations—makes the science accessible. The strong metabolism coverage, clinical flavor, and critical-thinking approach to understanding science and nutrition research make this a text that will also prepare nutrition majors and other science majors for their future studies and careers. These students will discover that this text ties together information that they have studied in chemistry, physiology, biology, and biochemistry courses.

This up-to-date text includes the most recent recommendations from the DRIs, Dietary Guidelines, and MyPlate, and is extensively referenced from current literature. Recent concerns in nutrition science, such as nutrigenomics, the dual epidemics of obesity and diabetes, controlling world hunger, the risks and benefits of genetically modified foods, and the nutritional impact of dietary supplements, are discussed. The examples used throughout the text reflect the diverse ethnic and cultural mix of the American population.

Critical Thinking Enhances Problem-Solving Skills

Nutrition: Science and Applications takes a critical-thinking approach to understanding and applying human nutrition. Like other introductory texts, it offers students the basics of nutrition by exploring the nutrients, their functions in the body, and sources in the diet. But its unique critical-thinking approach gives students an understanding of the "whys" and "hows" behind nutrition processes and explores the issues that surround nutrition controversies. Within each chapter, separate Critical Thinking exercises introduce nutrition-related problems and use a series of thought-provoking questions to lead students through the logic needed to find solutions and to make healthy food and nutrition decisions. "Applications" at the end of each chapter then ask students to use this same process of logical scientific inquiry, along with the information in the chapter, to analyze their own diets and modify them to promote health and to reduce the risk of nutrient deficiencies and nutrition-related chronic diseases. "Think critically" questions also accompany many of the illustrations in the text and appear at the end of special features such as the Debate, Off the Label, and Science Applied. These are designed to promote more in-depth thought and focus student attention on the information in visuals or discussed in the features. This critical-thinking approach gives students the tools they need to bring nutrition out of the classroom and apply the logic of science to their own nutrition concerns—both as consumers and as future scientists and health professionals.

Jeffrey Coolidge/Getty Images, Inc.

Integrated Metabolism Reinforces Understanding

Nutrition: Science and Applications is distinctive in its integrated approach to the presentation of nutrient metabolism. Metabolism is one of the most challenging topics for students, but is critical in understanding how nutrients function and impact human health. The text includes a comprehensive discussion of metabolism as it applies to each of the energy-yielding nutrients, shows how the micronutrients are involved, and then ties it all together in discussions of energy balance and fueling physical activity. The presentation differs from that in other texts, however, in integrating discussions of metabolism throughout appropriate chapters. This approach makes metabolism more manageable and memorable for students because it presents material in smaller segments and highlights its relevance to the nutrient being discussed. It also reinforces understanding of metabolic processes by revisiting key concepts with each nutrient and adding relevant new information. *Nutrition: Science and Applications* introduces a simple overview of metabolism in Chapter 3 and then builds on this base with more complex discussions in Chapters 4 through 10. For example, the discussion of carbohydrate metabolism in Chapter 4 presents the basics of glucose metabolism. This information is reviewed and augmented in chapters on lipids, proteins, micronutrients, energy balance, and exercise. Each discussion of metabolism is highlighted by the metabolism icon. To tie the concepts together, the illustrations use the same basic diagram with new information superimposed over familiar portions to demonstrate how each nutrient fits into the process. The nutrients and steps of metabolism are also color coded for easier recognition. Students or instructors who want a slightly more in-depth summary of metabolism will find it online in Focus on Metabolism.

Integration of Health and Disease Relationships Holds Interest

© Mike Goldwater/Alamy Limited

Can I help my mom manage her blood cholesterol?

Why have I gained 10 pounds?

What should I eat to reduce my risk of cancer?

How can I change my diet to better support my athletic training?

These are some of the questions students want answered when they enroll in nutrition classes. To answer these and other health-related questions and to fuel student interest continuously, discussions of the relationships among nutrition, health, and disease are integrated throughout the text. The integration helps students recognize that a nutrient's function in the body is related to its role in health and disease. For example, just as discussing goiters in the section on iodine is logical and piques interest, so is discussing diabetes in sections on carbohydrates, osteoporosis with calcium, and hypertension with sodium. Covering nutrition-related chronic conditions with the topic or nutrient most related to the issue continuously reinforces the applicability of nutrition science to the student's lives, and also helps them appreciate how and why their diet affects their health.

The Total Diet, Not Individual Foods, Is the Focus

Nutrition: Science and Applications presents the message that each food choice makes up only a small part of your total diet and that it is the overall dietary pattern that determines the healthfulness of your diet. Each of the macronutrient chapters begins with a section that discusses the role of that nutrient in the diet—factors that affect our intake and how different food sources of each nutrient may make very different contributions to the diet. For example, choosing whole-wheat bread gives you a more nutrient-dense source of carbohydrate than choosing a slice of chocolate cake. However, this does not mean you should never have chocolate cake. The text emphasizes that there are no "bad" foods as long as the sum of food choices over a period of days or weeks makes up a healthy overall diet. To reinforce this, these chapters end with a discussion of how to meet your need for that specific nutrient while taking into consideration other dietary recommendations that promote health.

© Spaces Images/SuperStock

Distinctive Features

This text includes a number of features that both spark student interest and help them learn the basics of nutrition.

Chapter Outline

This brief outline of the chapter's content gives students and instructors an overview of the major topics presented in the chapter.

Case Study

Each chapter begins with a short case study. These health- and disease-oriented vignettes help spur student curiosity and provide a taste of some of the concepts to be explained in the chapter. For example, the case study for Chapter 6 describes the challenges a college student faces when trying to meet his nutrient needs on a vegan diet. The Chapter 10 case study recounts the story of an athlete who experienced dehydration during an Olympic marathon. The case study in Chapter 9 discusses a baby diagnosed with vitamin D deficiency. These intriguing and entertaining stories link the material in the chapter with everyday health and disease issues. The chapter content then helps students understand the issues associated with each case.

© Luca Manieri/iStockphoto

Proteins and Amino Acids

6

CHAPTER OUTLINE

6.1 Protein in Our Food
Sources of Protein in Our Diet
Nutrients That Accompany Protein Sources

6.2 Protein Molecules
Amino Acid Structure
Protein Structure

6.3 Protein in the Digestive Tract
Protein Digestion
Amino Acid Absorption
Why Undigested Protein Can Cause Food Allergies

6.4 Amino Acid Functions in the Body
Proteins Are Continually Broken Down and Resynthesized
How Amino Acids Are Used to Synthesize Proteins
Synthesis of Nonprotein Molecules
How Amino Acids Provide Energy

6.5 Functions of Body Proteins
How Proteins Provide Structure
How Proteins Facilitate and Regulate Body Processes

6.6 Protein, Amino Acids, and Health
Protein Deficiency
Protein Excess
Protein and Amino Acid Intolerances

6.7 Meeting Recommendations for Protein Intake
How Protein Requirements Are Determined
Protein Recommendations
Translating Recommendations into Healthy Diets
How to Meet Nutrient Needs with a Vegetarian Diet

CASE STUDY

Elliot thinks that it will be better for his health and the health of the planet if he stops eating animal products. He is a college student who eats some of his meals at the student union and does some of his own cooking. He doesn't know much about nutrition, but he knows that the union offers vegetarian choices at every meal. What he doesn't realize is that, although the vegetarian meals served at the student union do not contain meat, they may contain eggs and dairy products. He had planned to eliminate these from his diet as well as meat.

In starting his new eating plan, Elliot thought breakfast would be easy, but as he poured himself a bowl of cereal, he realized that milk was an animal product. He switched to toast, but the butter he had in the refrigerator was an animal product so he had to settle for jam alone. He likes cream in his coffee, but when he read the label on his nondairy coffee creamer it said it contained milk protein. For lunch he usually goes out with friends for a burger or a sandwich. The burger restaurant didn't have any vegetarian options, so he tried the sandwich shop. His only choice there was to order the veggie and cheese sub, without the cheese. This meal wasn't very filling, so on the way home he bought a big bag of chips—at least these didn't contain any animal products.

For dinner that night the cafeteria offered vegetarian lasagna, but it was full of cheese. To stick with his animal-product-free diet he just had some pasta with marinara sauce. Ice cream for dessert was out so he opted for a slice of apple pie once he was assured it didn't contain any animal products. **By the end of the day he was frustrated, hungry, and felt his diet had no variety and was actually less healthy than what he ate before he tried to become a vegetarian.**

© Chris Schmidt/iStockphoto

Outcome

Outcome, appearing at the end of each chapter, completes the case study stories begun in the chapter introduction. For example, the outcome at the end of Chapter 6 describes how the student is able to improve his vegan diet once he understands complementary proteins and recognizes the variety of healthy ethnic vegan choices that are available to him. The outcome in Chapter 9 describes how the vitamin D deficiency is treated and prevented. These "outcomes" review concepts covered in the chapter and illustrate the application of nutrition knowledge to clinical situations.

OUTCOME

Elliot did some reading about vegan diets that helped him understand the food combinations that would give him enough protein to meet his needs. Peanut butter and bread is inexpensive and provides complementary proteins. For more variety, he discovered that ethnic food provided some wonderful vegan options. The Chinese restaurant has a tasty tofu and vegetable stir-fry. The Indian restaurant offers a number of vegan options that contain lentils or chickpeas with rice or bread. He spent some extra time at the grocery store and discovered milks, including soy, almond, and rice, that do not contain animal products but give him something to put on his cereal. He also found a variety of vegan prepared foods such as veggie burgers and frozen enchiladas. His diet is now varied and provides plenty of protein, with less saturated fat and more fiber than his old diet. **Although he is enjoying vegan food, his reading also helped him realize that he does not necessarily need to eat vegan to improve his health and reduce his footprint on the planet.**

© Luca Manieri/iStockphoto

© Chris Schmidt/iStockphoto

Learning Objectives

Each chapter section begins with one or more learning objectives that indicate, in behavioral terms, what the student must be able to do to demonstrate mastery of the material in the chapter.

4.6 Meeting Recommendations for Carbohydrate Intake

LEARNING OBJECTIVES

- List recommendations for total carbohydrate, added sugars, and fiber intake.
- Choose unrefined sources of carbohydrate from each food group of MyPlate.
- Use food labels to choose foods that provide healthy carbohydrates.
- Discuss the role of alternative sweeteners in weight loss.

Critical-Thinking Exercises

These exercises, which appear in each chapter, use a critical-thinking approach to making decisions and solving problems regarding nutrition. They help students apply their nutrition knowledge to everyday situations by presenting a nutrition-related problem and then asking a series of provocative questions that lead the student through the logical progression of thought needed to collect information and solve the problem. Many of these exercises focus on modifying a diet to reduce disease risk or maintain health. For example, the exercise in Chapter 2 focuses on how the food choices of a 22-year-old man can fit on MyPlate and how to use this tool to plan a healthy diet. Chapter 11's Critical-Thinking exercise takes students through the process of assessing the risk of osteoporosis and modifying a diet to increase calcium intake.

CRITICAL THINKING — Fitting Foods, Meals, and Diets onto MyPlate

Joe Gough/Shutterstock

For the first time in his life, Lucas is shopping and cooking for himself. He wants to prepare healthy meals so he uses MyPlate as a guide for his food choices. He is a 22-year-old male who exercises less than 30 minutes a day.

For breakfast Lucas has cereal, milk, orange juice, and coffee with cream and sugar. When he puts the foods from this meal on MyPlate he notices that the sections for vegetables and protein foods are empty.

CRITICAL THINKING QUESTIONS

- Does he need to include foods from each group at every meal?
- Where do the cream and sugar he puts in his coffee fit onto MyPlate?
- Plan a lunch for Lucas that includes a food from each food group of MyPlate.

For dinner Lucas has lasagna. To see how it fits onto MyPlate, he breaks it down into individual ingredients.

INGREDIENTS IN LASAGNA	FOOD GROUP	AMOUNT ON MYPLATE
Noodles (1.5 oz dry)		
Tomato sauce (¾ cup)		
Ground beef (2 oz)		
Mozzarella cheese (3 T)	Dairy	¼ cup
Ricotta cheese (¼ cup)	Dairy	½ cup
Olive oil (1 tsp)	Oils	

CRITICAL THINKING QUESTIONS

- Complete the table to the left to see how his lasagna fits onto MyPlate. (Hint: Use Food-A-Pedia in Supertracker of MyPlate)
- If he used butter instead of olive oil, where would it fit on MyPlate?

iProfile Use iProfile to find nutrient-dense substitutions for foods that are high in empty calories

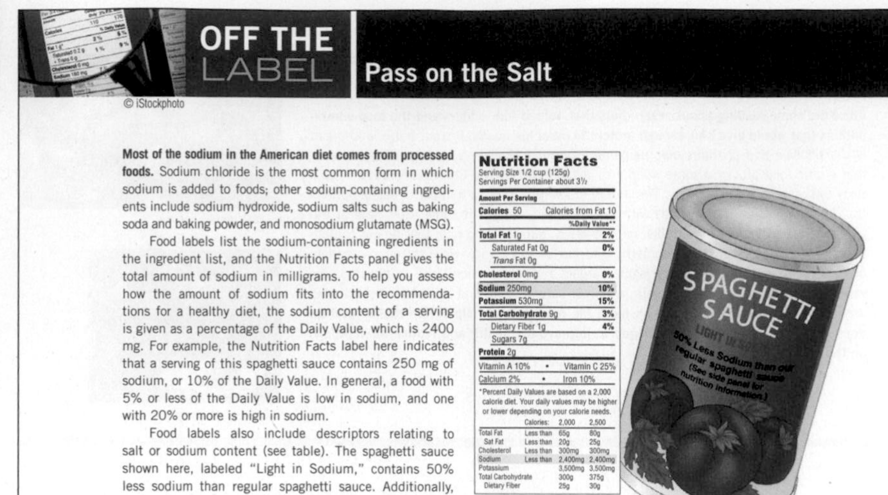

OFF THE LABEL Pass on the Salt

© iStockphoto

Most of the sodium in the American diet comes from processed foods. Sodium chloride is the most common form in which sodium is added to foods; other sodium-containing ingredients include sodium hydroxide, sodium salts such as baking soda and baking powder, and monosodium glutamate (MSG).

Food labels list the sodium-containing ingredients in the ingredient list, and the Nutrition Facts panel gives the total amount of sodium in milligrams. To help you assess how the amount of sodium fits into the recommendations for a healthy diet, the sodium content of a serving is given as a percentage of the Daily Value, which is 2400 mg. For example, the Nutrition Facts label here indicates that a serving of this spaghetti sauce contains 250 mg of sodium, or 10% of the Daily Value. In general, a food with 5% or less of the Daily Value is low in sodium, and one with 20% or more is high in sodium.

Food labels also include descriptors relating to salt or sodium content (see table). The spaghetti sauce shown here, labeled "Light in Sodium," contains 50% less sodium than regular spaghetti sauce. Additionally, a health claim that diets low in sodium may reduce the risk of high blood pressure can appear on foods that meet the definition of a low-sodium food and provide less than 20% of the Daily Value for fat, saturated fat, and cholesterol per serving. Some medications, such as pain relievers, antacids, and cough medications, can also contribute a significant amount of sodium. Drug facts labels on over-the-counter medications can help identify those that contain significant amounts of sodium.

Nutrition Facts
Serving Size 1/2 cup (125g)
Servings Per Container about 3½

Amount Per Serving	
Calories 50	Calories from Fat 10
	%Daily Value**
Total Fat 1g	2%
Saturated Fat 0g	0%
Trans Fat 0g	
Cholesterol 0mg	0%
Sodium 250mg	10%
Potassium 530mg	15%
Total Carbohydrate 9g	3%
Dietary Fiber 1g	4%
Sugars 7g	
Protein 2g	
Vitamin A 10% •	Vitamin C 25%
Calcium 2% •	Iron 10%

*Percent Daily Values are based on a 2,000 calorie diet. Your daily values may be higher or lower depending on your calorie needs.

	Calories:	2,000	2,500
Total Fat	Less than	65g	80g
Sat Fat	Less than	20g	25g
Cholesterol	Less than	300mg	300mg
Sodium	Less than	2,400mg	2,400mg
Potassium		3,500mg	3,500mg
Total Carbohydrate		300g	375g
Dietary Fiber		25g	30g

Light Spaghetti Sauce, 250 milligrams (mg) per serving
Regular Spaghetti Sauce, 500mg per serving

THINK CRITICALLY: If instead of a serving of this spaghetti sauce, which is light in sodium, you ate a serving of regular spaghetti sauce, how much more sodium would you be consuming?

Off the Label

"Off the Label" boxes present in-depth information on food, supplement, and even drug labels. Off the Label is designed to show students how to use labels to make wise choices. For example, the Off the Label in Chapter 4 shows how to use the ingredient list to choose whole-grain products and avoid foods that are high in added sugars. The one in Chapter 9 helps students use the information on Supplement Facts labels to make safe supplement decisions, and in Chapter 10, the Off the Label box discusses the sodium content of packaged foods and shows students how to use food labels to select low-sodium foods.

Science Applied

These boxed features included in every chapter focus on research studies that are key to our current knowledge of nutrition and our understanding of nutrition-related diseases. These help students appreciate how science is done and how the results apply to our understanding of nutrition and medicine. For example, the Science Applied in Chapter 8 discusses how studying the niacin-deficiency disease pellagra led to the enrichment of grains, and in Chapter 9 it describes how studying a bleeding disorder in cattle led scientists to a better understanding of vitamin K and coagulation. Other Science Applied topics include how weightlessness in space contributes to our grasp of osteoporosis, how the discovery of the leptin gene enhanced our understanding of the genetics of body weight regulation, and how the discovery of LDL receptors, which help remove LDL cholesterol from the blood, led to medications to treat high blood cholesterol.

SCIENCE APPLIED Saving Cows, Killing Rats, and Surviving Heart Attacks

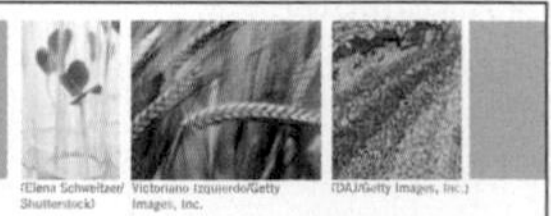

In the 1930s, hemorrhagic sweet clover disease was killing cows across the prairies of the midwestern United States and Canada. It occurred in cattle that were fed moldy, spoiled sweet clover hay. These animals died because their blood did not clot. Even a minor scratch from a barbed wire fence could be fatal; once bleeding began, it did not stop.

THE SCIENCE

On a snowy night in 1933, a disgruntled farmer delivered a bale of moldy clover hay, a pail of unclotted blood, and a dead cow to the laboratory of Dr. Carl Link at the University of Wisconsin. Link, the university's first professor of biochemistry, was already making a name for himself in the scientific community when he was presented with this challenge. What was in the clover that killed the man's cows? What could he do to stop the bleeding disease that was killing them? Link had just begun to research hemorrhagic sweet clover disease, but at the time the only advice he could offer the farmer was to find alternative feed and try blood transfusions to save his remaining animals. Link and his colleagues began a line of inquiry that ultimately led to the development of an anti-blood-clotting factor that today saves the lives of hundreds of thousands of people.

Six years after the farmer's challenge, Link and colleagues had isolated the anticoagulant *dicumarol* from moldy clover. Dicumarol is a derivative of coumarin, which gives clover its sweet scent; mold converts coumarin to dicumarol. Cows fed moldy clover consume dicumarol, which interferes with vitamin K activity and consequently prevents normal blood clotting (see figure).

THE APPLICATION

The discovery of dicumarol enhanced our understanding of the blood-clotting mechanism and led to the development of anticoagulant drugs. These drugs help eliminate blood clots and prevent their formation. In carefully regulated doses, they are used to treat heart attacks, which occur when blood clots block blood flow to the heart muscle. Dicumarol, first synthesized in 1940, was the first anticoagulant that could be administered orally to humans. Further work with dicumarol led Link to propose the use of a more potent derivative, called *warfarin*, as rat poison. The name warfarin comes from the initials of the Wisconsin Alumni Research Foundation (WARF), which patented the drug, and from coumarin, the compound from which it is derived. When rats consume the odorless, colorless warfarin, their blood fails to clot, and they bleed to death. Warfarin was used as a rodenticide for nearly a decade before it was introduced into clinical medicine in 1954. Sodium warfarin, also known by the brand name Coumadin, soon became the most widely prescribed anticoagulant drug in the nation. It was used to treat President Dwight D. Eisenhower when he suffered a heart attack in 1955.

Sodium warfarin and dicumarol have been administered to millions of patients to prevent blood clots, which can cause heart attacks and strokes. Thus a substance that killed hundreds of cattle across the Great Plains during the Great Depression and today efficiently kills rodents invading our homes also saves a multitude of human lives.

THINK CRITICALLY: Why should patients taking warfarin avoid supplements that contain vitamin K?

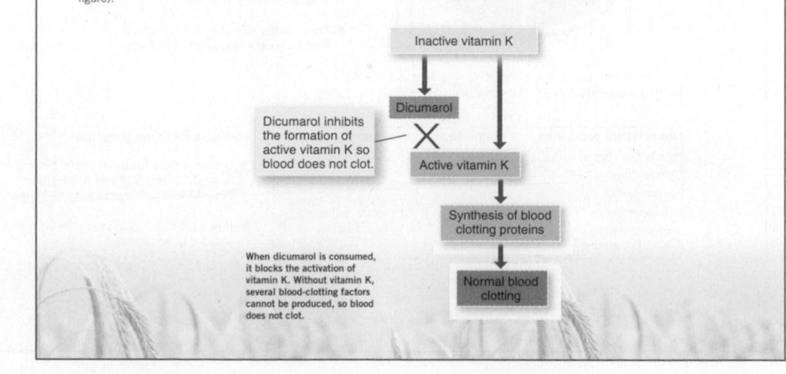

When dicumarol is consumed, it blocks the activation of vitamin K. Without vitamin K, several blood-clotting factors cannot be produced, so blood does not clot.

Debate

The Debate is an essay that explores a controversial nutrition issue. Both sides of the issue are presented and then Think Critically questions at the end ask students to integrate and evaluate the information presented. Debates address topics such as "How Involved Should the Government Be in Your Food Choices?," "Are Personalized Diets the Best Approach to Reducing Chronic Disease?," "Should You Be Gluten Free?," "Should You Avoid High-Fructose Corn Syrup?," and "Is Surgery a Good Solution to Obesity?".

Process Diagrams

Sometimes the hardest part of understanding an illustration is knowing where to start. To help students comprehend the more complex line art, each step in the processes is numbered and includes a narrative describing what happens. Process diagrams help students reduce intimidating topics such as digestion and absorption, metabolism, and lipid transport to a series of easy-to-follow steps.

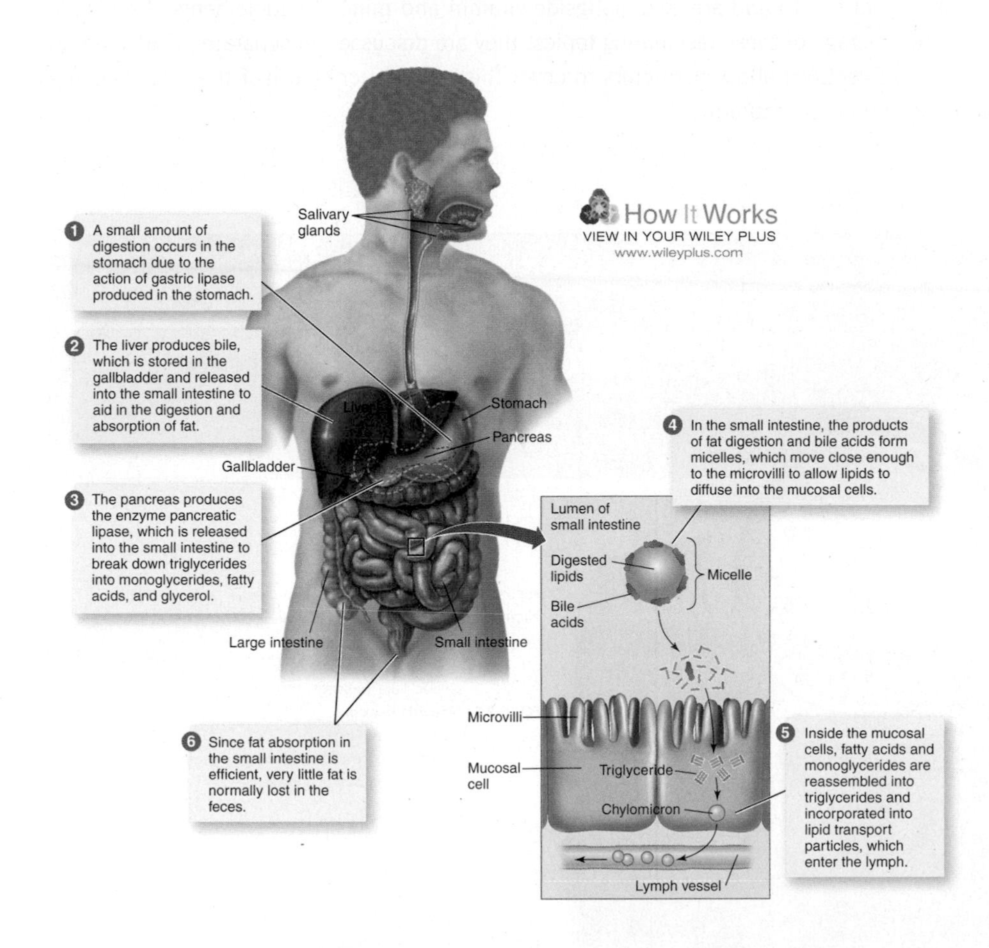

Applications

These exercises, which appear at the end of each chapter, are divided into three parts. "Assessing Your Diet" focuses on the student's personal diet and nutrition concerns, "Consumer Issues" addresses more general consumer nutrition issues such as reading food labels, and "Clinical Concerns" asks about more clinical nutrition and health issues as they relate to the chapter material. All require the student to think critically and apply key nutrition concepts, and all help reinforce the importance of nutrition in health promotion and disease prevention. Some of these can be done as collaborative learning exercises, which encourage students to work together and learn from one another to solve a problem.

Chapter Summary

Each chapter ends with a summary that highlights important concepts addressed in each section of the chapter.

Review Questions

A set of questions appears at the end of each chapter to test students' understanding of the key points covered. It can be used by students as a study tool to test their comprehension of the important information in the chapter.

Focus On

The text includes five Focus On "mini-chapters": Focus on Alcohol, Focus on Eating Disorders, Focus on Phytochemicals, Focus on Nonvitamin/Nonmineral Supplements, and Focus on Biotechnology. These discuss topics of great interest to students that are not necessarily part of the core curriculum in nutrition. For example, eating disorders have nutritional symptoms but are really psychological disorders; alcohol is consumed in the diet and affects nutrient metabolism, but is not itself a nutrient; and herbal supplements are not nutrients but may affect health and are sold alongside vitamin and mineral supplements. To provide adequate coverage of these fascinating topics, they are discussed in separate Focus On sections. These sections allow instructors to cover this material or skip it if they feel it is not pertinent to their curriculum.

Focus on Eating Disorders

© DNY59/iStockphoto

CHAPTER OUTLINE

Normal eating patterns are flexible. One day you may eat twice as much as on another. On some days your food choices are varied and nutritious, while on others you may survive on snacks and fast food. Lunch one day may include an appetizer, entree, and dessert and on the next it may be a carton of yogurt grabbed on the run. Normal eating patterns include eating more than we need at a party or other special occasion and less than we need when we are busy or stressed. Normal eating may also involve limiting intake in order to manage weight and meet recommendations for a healthy diet.

BBC, the New York Times or CBS video

Available in WileyPLUS, these videos help students explore issues like the Mediterranean diet, the local food movement, diet myths, and other relevant topics.

Marginal Features

The text includes a number of marginal features that aid comprehension, enhance interest, and point out where particular types of information can be found.

Definitions of New Terms

New terms are highlighted in bold in the text and defined in the margin, providing easy access to new terms as they appear. These and other terms are included in an extensive glossary at the back of the text.

> make wise nutrition decisions, whether they involve what to have for breakfast or whether a headline about vitamin E supplements is true.
>
> **The Scientific Method**
>
> Advances in nutrition are made using the **scientific method**. The scientific method offers a systematic, unbiased approach to evaluating the relationships among food, nutrients, and health. The first step of the scientific method is to make an observation and ask questions about the observation. The next step is to propose a **hypothesis**, or explanation for the observation. Once a hypothesis has been proposed, experiments can be designed to test it. The experiments must provide objective results that can be measured and repeated. If the experimental results do not prove the hypothesis to be wrong, a **theory**, or a scientific explanation based on experimentation, can be established (**Figure 1.13**). Scientific theories are accepted only as long as they cannot be disproved and continue to be supported by all new evidence that accumulates. Even a theory that has been accepted by the scientific community for years can be proved wrong.
>
> **scientific method** The general approach of science that is used to explain observations about the world around us.
>
> **hypothesis** An educated guess made to explain an observation or to answer a question.
>
> **theory** An explanation based on scientific study and reasoning.

How It Works Icons These icons appear by figures or concepts that students can explore further through animations in WileyPLUS.

iProfile Icons These show students where they can use the iProfile software to answer a nutrition-related question.

Video Icons These icons appear by topics that students can explore further by watching a BBC, the New York Times or CBS video on WileyPLUS.

Metabolism Icons These highlight discussions of how each nutrient fits into the metabolic processes of nutrition.

Life cycle Icons These highlight discussions of the differences in nutritional requirements, concerns, and effects at different life stages. This information helps students understand how nutrient requirements are affected by life stage as well as offering information relevant to students in all phases of life. Life cycle nutrition is also covered in depth in separate chapters (Chapters 14, 15, and 16).

New to this Edition

New Art

The art program has been completely revamped. To ensure that students "see" and assimilate the processes of nutrition, figures have been redesigned to present and explain complex processes in clear steps, organize information, and interconnect related pieces of information. The figures are integrated with the accompanying text to clarify and reinforce major concepts, while allowing students to understand the details. The photographs used are not merely decorative, but are purposefully used to focus attention on important concepts.

To make metabolism more intuitive, all metabolism illustrations have been redesigned. The new layout makes the steps involved in generating ATP more straightforward. Each step is distinguished by color and a numbered description. Repeated use of the same basic figure allows symbols such as the blue circle of the citric acid cycle, the purple spheres that represent electrons, and the icons used for glucose, fatty acids, and amino acids to become immediately recognizable.

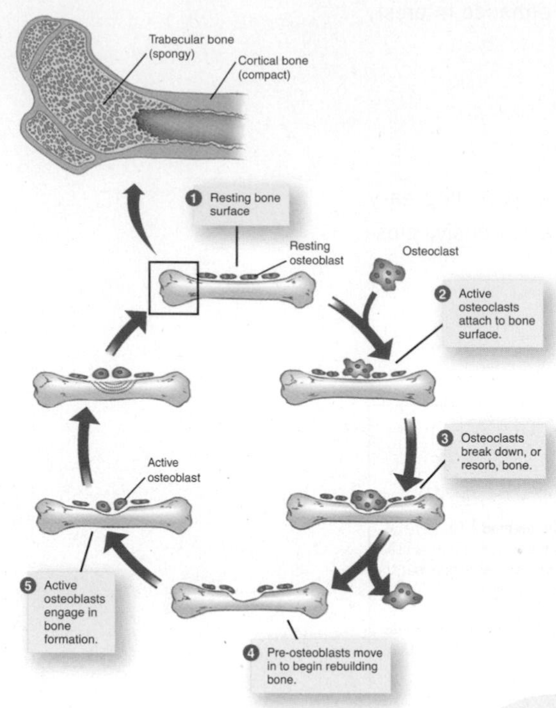

Peroxides

H_2O_2 H_2O_2 H_2O_2

Glutathione peroxidase (selenium)

H_2O

Neutralized peroxides

Free radicals

No free radical formation

Vitamin E

Neutralized free radicals

Cellular damage

No cellular damage

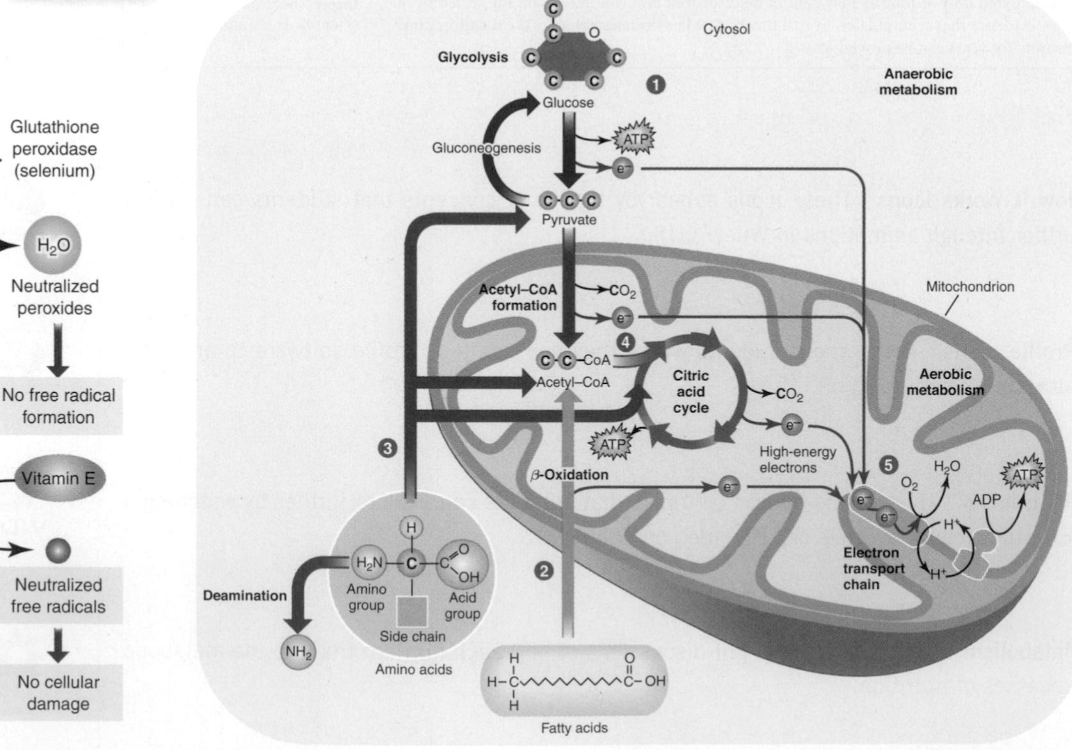

1 Glycolysis breaks glucose in half to yield two pyruvate molecules that are converted to acetyl-CoA.

2 β-Oxidation breaks fatty acids into two-carbon units that form acetyl-CoA.

3 After deamination, amino acids can break down to form acetyl-CoA, pyruvate, or other intermediates.

4 The acetyl-CoA from glucose, fatty acid, and amino acid breakdown can enter the citric acid cycle.

5 The electrons released are passed to the electron transport chain, and their energy is used to convert ADP to ATP.

To organize information visually and integrate related pieces of information, illustrations, photographs, graphs, and tables have been combined. For example, Nutrition Facts panels are presented along with tables that show how to convert IUs to micrograms and milligrams. Combinations of photographs and drawings are used to help students see how the structure of fatty acids affects their properties. In the example shown here, a graph and micrographs of the muscle and blood vessels provide an integrated picture of how exercise training affects physiology.

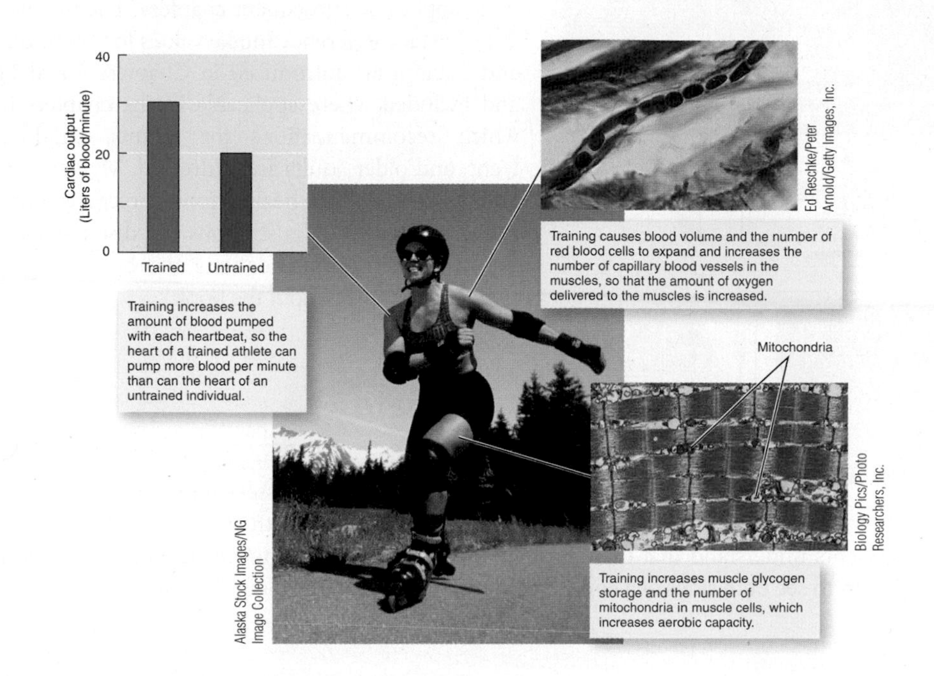

The new art program also includes new life cycle illustrations. *Nutrition Science and Applications* has always used life cycle icons to highlight the life cycle information integrated throughout. In addition to these, some life cycle information has been pulled out of the text of this third edition and presented in new visuals that highlight how nutrient needs and the ability to meet them change throughout life.

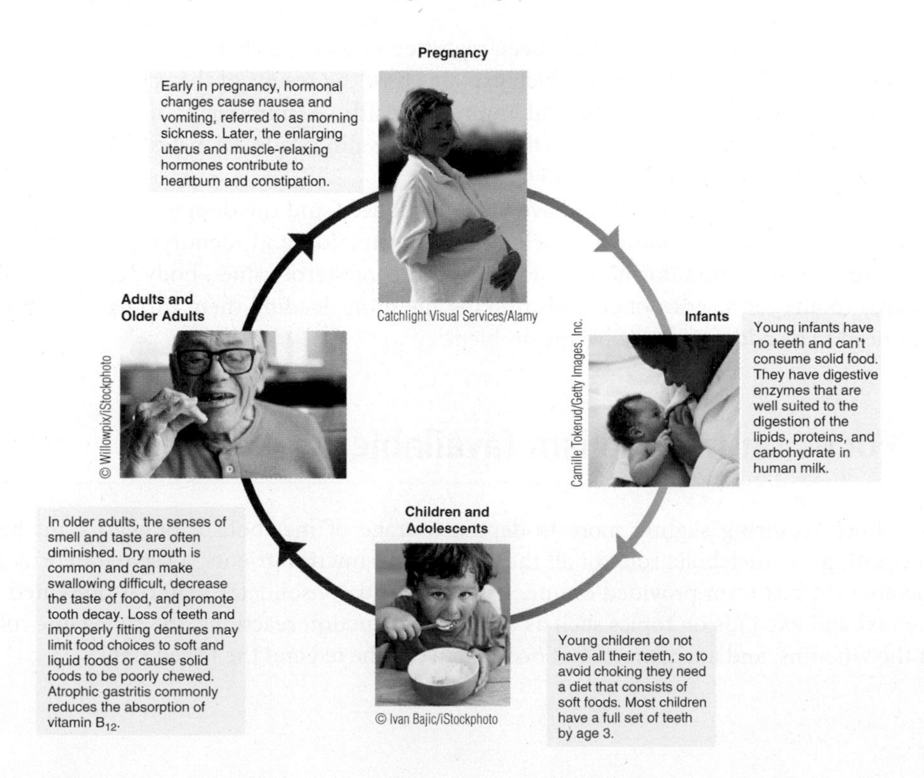

Up-to-Date Nutrition Guidelines and Recommendations

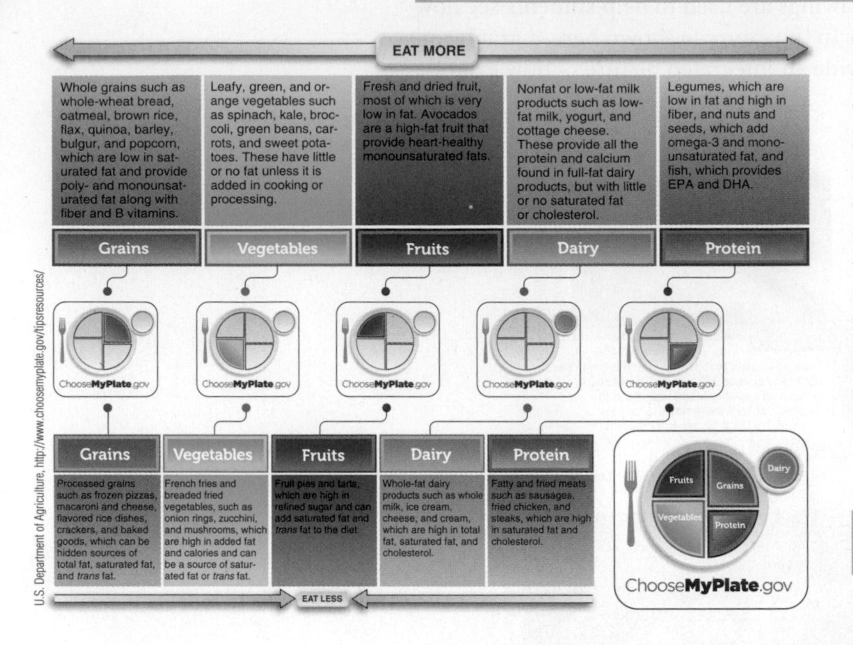

This up-to-date text includes the recommendations of the 2010 Dietary Guidelines for Americans, which are described in Chapter 2 and addressed in all applicable subsequent chapters. The updated 2011 Dietary Reference Intake values for vitamin D and calcium are introduced in Chapters 9 and 11 and included where applicable in the chapters in which recommendations for infants, children, teens, and older adults are addressed. MyPlate, the USDAs new guide to healthy food choices, is introduced in Chapter 2. Illustrations and discussions of the MyPlate recommendations and how to apply them appear throughout the book.

Debate

This new feature addresses controversial nutrition topics by presenting both sides of the issue. They help students understand that there are not always clear answers to nutrition questions and allow them to see both sides of a controversy and think critically about the issues.

New and Updated Feature Topics

A number of the features in *Nutrition Science and Applications*, Third Edition, have been improved and new topics have been introduced.

Over half of the chapter Case Study introductions and Outcome stories are new or have been updated.

The Science Applied boxes have been rewritten to distinguish **The Science** and how it was done from **The Application**, which explains how the results of the science have been used to enhance understanding and improve health. A number of new topics have been included, such as "How the Discovery of Insulin Made Diabetes Treatable" and "NHANES: A Snapshot of America's Health."

New Critical Thinking topics have been introduced and the design improved. These exercises start by giving students background information and identifying a nutritional problem. Students use information such as blood cholesterol values, body weight, or diet records to answer a series of critical-thinking questions leading them through the issues that need to be addressed to solve the problem.

Focus on Metabolism (available online)

For those requiring slightly more in-depth coverage of metabolism or who want help integrating the metabolic roles of all the macro- and micronutrients, a separate Focus on Metabolism has been provided online. It reviews and consolidates material presented in the text and expands on topics such as oxidation-reduction reactions, the coenzyme roles of the vitamins, and adaptations that occur between the fed and the fasted state.

Chapter-by-Chapter Overview

CHAPTER 1—"Nutrition: Food for Health," begins by discussing the American diet—how it has changed, and how healthy it is. It emphasizes that food choices affect current and future health. This chapter provides an overview of the nutrients and their roles in the body, and defines the basic principles of balance, variety, and moderation that are key to a healthy diet. It also introduces the scientific method and the steps students need to follow to sort accurate from inaccurate nutrition information.

CHAPTER 2—"Nutrition Guidelines: Applying the Science of Nutrition," takes the science out of the laboratory and shows how advances in nutrition knowledge have been used to develop the Dietary Reference Intakes (DRIs), the Dietary Guidelines for Americans, and tools for diet planning, including MyPlate, food labels, and exchange lists. The final section of this chapter discusses how these and other tools can be used to assess the nutritional health of populations and individuals.

CHAPTER 3—"Digestion, Absorption, and Metabolism," discusses how food is digested, how nutrients from foods are absorbed into the body and transported to the cells where they are broken down to provide energy or used to synthesize structural or regulatory molecules, and finally how wastes are removed. This chapter provides an overview of metabolism that serves as a launching pad for the more detailed metabolism information presented in subsequent chapters.

CHAPTERS 4, 5, AND 6—Feature the energy-yielding nutrients carbohydrates, lipids, and proteins. Each begins with a discussion of the sources and types of the respective macronutrient in our food. The body of each chapter discusses nutrient digestion, absorption, metabolism, functions, and the impact of different types and amounts of these nutrients on our health. Each chapter ends with a discussion of how to choose a diet that meets recommendations.

Chapter 4, "Carbohydrates: Sugars, Starches, and Fiber," discusses the health impact of refined grains and foods high in added sugar versus whole grains and unrefined sources of sugars. Chapter 5, "Lipids: Triglycerides, Phospholipids, and Cholesterol," points out that Americans are not eating too much fat, but are typically choosing the wrong types of fat for a healthy diet. Chapter 6, "Proteins and Amino Acids," discusses animal and plant sources of protein and points out that either plant or animal proteins can meet protein needs, but these protein sources bring with them different combinations of nutrients. In addition to discussing how to meet protein needs, this chapter includes information on how to plan healthy vegetarian diets.

FOCUS ON ALCOHOL discusses alcohol metabolism, the health risks associated with excessive alcohol consumption, and the health benefits associated with moderate intake.

CHAPTER 7—"Energy Balance and Weight Management," discusses the obesity epidemic and its causes, and the effects of excess body fat on health. The chapter explains energy balance and shows how small changes in diet and behavior can alter long-term energy balance. The most up-to-date information on how body weight is regulated and the role of

genetic versus environmental factors in determining body fatness is covered. The chapter includes recommendations for healthy body weight and composition and equations for determining energy needs. It also discusses weight-loss options that range from simple energy restriction to risky surgical approaches.

FOCUS ON EATING DISORDERS includes a comprehensive discussion of the different types of eating disorders, their causes, consequences, and treatment. This section also addresses the sociocultural factors that influence body ideal as well as what to do if you have a friend or relative you suspect has an eating disorder.

CHAPTER 8—"The Water-Soluble Vitamins," begins with a general overview of the vitamins—where they are found in the diet, factors affecting their bioavailability, and how they function. Each of the B vitamins and vitamin C is then discussed individually, providing information on sources in the diet, functions in the body, impact on health, recommended intakes, use as dietary supplements, and potential for toxicity. This chapter also discusses choline, a substance that is not currently classified as a vitamin but one for which DRIs have been established.

CHAPTER 9—"The Fat-Soluble Vitamins," introduces the fat-soluble vitamins within the context of the modern diet and then presents each one with a discussion of their sources in the diet, functions in the body, impact on health, recommended intakes, use as dietary supplements, and potential for toxicity.

FOCUS ON PHYTOCHEMICALS discusses the role of phytochemicals in nutrition and health. These substances are not dietary essentials but can positively impact health. Different categories of phytochemicals are presented, along with a discussion of how to maximize their intake.

CHAPTER 10—"Water and the Electrolytes," addresses water, a nutrient often overlooked, and sodium, potassium, and chloride, because they help regulate the distribution of body fluids. This chapter presents information on where these nutrients are found in the diet and describes their functions in the body and their relationship to health and disease. A discussion of hypertension illustrates the importance of sodium, potassium, and other minerals and dietary components in blood pressure regulation. Advances in our understanding of how dietary patterns affect hypertension are stressed in a discussion of the DASH diet, a dietary pattern recommended by the Dietary Guidelines that has been shown to lower blood pressure.

CHAPTER 11—"Major Minerals and Bone Health," provides an overview of the minerals and discusses the remaining major minerals, calcium, phosphorus, magnesium, and sulfur. Their functions in the body and availability in the diet as well as their relationship to health and disease are addressed. Because most of these play an important role in bone health, this chapter also includes a section on the relationship between nutrition and the development of osteoporosis.

CHAPTER 12—"The Trace Minerals," discusses the trace minerals in a format similar to that used for other micronutrients. Emphasis is placed on the unique roles of some minerals as well as the similarities that some have in their functions and the interactions among them. Discussions of the health issues related to these nutrients help create interest, as do discussions of the pros and cons of trace mineral supplements.

FOCUS ON NONVITAMIN/NONMINERAL SUPPLEMENTS targets dietary supplements that contain substances other than micronutrients. Micronutrient supplements are discussed in Chapters 8 through 12 with the appropriate nutrient, but many of the supplements Americans are taking include ingredients that are not vitamins or minerals. This

Focus will help students evaluate the benefits and risks associated with supplements containing substances such as coenzyme Q and glucosamine that are found naturally in the body as well as herbal ingredients such as echinacea.

CHAPTER 13—"Nutrition and Physical Activity," discusses the relationships among physical activity, nutrition, and health. It emphasizes the importance of exercise for health maintenance as well as the impact nutrition can have on exercise performance. Because nutrients fuel activity, this chapter serves as a review of energy metabolism. By this point in the text, students have studied all the essential nutrients, so a complete discussion of the macronutrients and micronutrients needed to generate ATP for various types of activity can be included. An expanded discussion of ergogenic aids for more competitive athletes directs students to use a risk-benefit analysis of these products before deciding whether or not to use them.

CHAPTER 14—"Nutrition During Pregnancy and Lactation," addresses the role of nutrition in human development by discussing the nutritional needs of women during pregnancy and lactation as well as the nutritional needs of infants. The benefits of breastfeeding are discussed.

CHAPTER 15—"Nutrition from Infancy to Adolescence," begins with a discussion of the rising rates of childhood obesity and other chronic diseases and the importance of learning healthy eating habits early in life. The chapter discusses nutrient needs and how they change from infancy through adolescence.

CHAPTER 16—"Nutrition and Aging: The Adult Years," addresses how nutrition affects health and how the physiological changes that occur with aging affect nutritional needs and the ability to meet them. The impact that chronic disease, medications, and socioeconomic changes have on the risk of malnutrition is discussed.

CHAPTER 17—"Food Safety," discusses the risks and benefits associated with the U.S. food supply and includes information on the impact of microbial hazards, chemical toxins, food additives, irradiation, and food packaging. This chapter discusses the use of HACCP (Hazard Analysis Critical Control Point) to ensure safe food and the role of the consumer in selecting, storing, and preparing food to reduce the risk of food-borne illness.

FOCUS ON BIOTECHNOLOGY explains how genetic engineering is used to modify the characteristics of organisms and create new products. It addresses the potential benefits and risks associated with this expanding technology.

CHAPTER 18—"World Hunger and Malnutrition," discusses the coexistence of hunger and malnutrition along with obesity in both developed and developing nations around the world. It examines the causes of world hunger and solutions that can impact the amounts and types of food and nutrients that are available.

Teaching and Learning Resources

Nutrition: Science and Applications is accompanied by a complete set of supplementary teaching and learning materials.

WileyPLUS

WileyPLUS is an innovative, research-based online environment for effective teaching and learning.

WileyPLUS builds students' confidence because it takes the guesswork out of studying by providing students with a clear roadmap: what to do, how to do it, if they did it right. Students will take more initiative so you'll have greater impact on their achievement in the classroom and beyond.

What do students receive with WileyPLUS?

- The complete digital textbook, saving students up to 60% off the cost of a printed text.
- Question assistance, including links to relevant sections in the online digital textbook.
- Immediate feedback and proof of progress, 24/7.
- Integrated multimedia resources—including MP3 downloads, videos, animations, and much more—that provide multiple study paths and encourage more active learning.

What do instructors receive with WileyPLUS?

- Reliable resources that reinforce course goals inside and outside of the classroom.
- The ability to identify easily those students who are falling behind.
- Media-rich course materials and assessment content, including—Instructor's Resources, Test Bank, PowerPoint® Slides, Learning Objectives, Computerized Test Bank, Pre- and Post-Lecture Quizzes, and much more.

Student Companion Website

[www.wiley.com/college/smolin] A dynamic website rich with many activities for review and exploration includes: Quizzes for each chapter, Flash Cards for learning concepts and terminology; Food For Thought articles; and Study Tools and Tips.

Student Resources in WileyPLUS

Videos

Videos produced by the New York Times, BBC, and CBS spark discussion and allow students to explore current and relevant topics in nutrition.

How It Works
VIEW IN YOUR WILEY PLUS
www.wileyplus.com

How It Works Animations

Wiley has developed a new set of animations for nutrition students and professors. We surveyed professors across North America to find out what topics were the most difficult to teach and learn and what processes were most essential to the introductory nutrition course. After much research and review, we developed animations on these topics, to make these difficult processes easier for students to learn and bring the process diagrams from the book to life.

These animations and accompanying quiz questions are available in WileyPLUS.

Absorption of Nutrients
Flow of Blood During Absorption
Digestion and Absorption of Carbohydrates
Glucose Metabolism
Regulation of Glucose Metabolism
Maintaining Normal Blood Glucose Levels
Blood Glucose Regulation
Digestion and Absorption of Lipids
Lipid Transport
Metabolism of Lipids
Digestion and Absorption of Proteins
Metabolism of Proteins
Role of B Vitamins in Metabolism
Action of Antioxidants Against Free Radicals
How Vitamin C Supports Immune Function
Acid-Base Balance

- **iProfile 3.0: Assessing Your Diet and Energy Balance**—New version of iProfile, featuring improved food searches. Available online, this dynamic, newly developed diet assessment software includes nutrient values for over 50,000 foods, inclusive of many of the most popular food choices of students today, as well as ethnic and cultural choices. It includes a feature that allows users to add foods to the database to keep pace with the ever-growing market of available products. Since exercise is such an important part of energy balance, the software also enables students to track and analyze their physical activity. In addition to the ability to track and analyze diets and exercise, some distinctive features include serving-size animations, a quick self-quiz, single nutrient reports, recipe builder, and an easy-to-use design.
- **Energy Acquisition: The Digestive System and Metabolism 2.0** [0-471-70754-6]—This CD-ROM, from the popular series Interactions: Exploring the Functions of the Human Body, uses animations, interactivity, and clinical correlations to enhance student understanding of the difficult concepts of metabolism and the structures and functions of the digestive system.
- **Your Learning Styles**—A questionnaire allows students to discover their preferred learning styles. This resource provides descriptions of all learning styles and study tips designed for each learning style.

The teaching materials available to ***instructors*** include the following:

- **Instructor Companion Website** [www.wiley.com/college/smolin]—A dedicated companion website for instructors provides many resources for preparing and presenting lectures. Also available are all of the illustrations and tables in the text already placed within PowerPoint Slides, as well as a set of Lecture PowerPoints that combine important images with major concepts from each chapter. Questions for use with Clicker Systems are also provided.
- **Test bank**—Available online, the Test Bank includes multiple-choice questions as well as short case studies with questions.
- **Computerized Test Bank**—This computerized version of the Test Bank makes preparing clear, concise tests quick and easy. It is available in both Windows and Macintosh formats.
- **Nutrition Visual Library 4.0**—This resource includes all of the illustrations from the textbook in labeled, unlabeled, and unlabeled with leader line formats. In addition, select illustrations and photographs from Wiley's Anatomy and Physiology textbooks are included. Search for images by chapter, or by using keywords. It is also available through WileyPLUS.

iProfile Assignments

Use the **iProfile: Case Study** Assignments provided in the Assignment section of WileyPLUS. Each chapter contains a case study that investigates a different dietary pattern. For example, case studies examine fast foods and nutrient density, the Mediterranean diet, and a vegetarian diet. Students are asked to enter the food choices of the case study subject and then analyze the impact of these choices. Each case study topic includes a question to generate discussion as well as gradable multiple-choice and true-false questions.

iProfile Essay Questions in WileyPLUS

Essay questions that ask about students' individual reports based on their food and activity journals are available in WileyPLUS. These essay questions ask students to reflect on their own diets and activities. For example, students are asked to look at their fiber intake, compare it to the recommendations for fiber and suggest foods that could increase the amount of fiber in their diets.

© Alejandro Rivera/iStockphoto

© Ben Blankenburg/iStockphoto

To the Student

Nutrition is a subject that all of you have a personal interest in, whether you are concerned about your own nutritional health, a parent with diabetes, or a friend with an eating disorder. You may enroll in a nutrition course to learn what to eat and how to choose healthy foods and then be surprised when the course talks about protein synthesis, lipid transport, and anaerobic metabolism. A good course and textbook should do both.

As authors, our goal is to provide you with tools that can be used throughout your life. We could tell you what to eat for breakfast, but if you didn't understand why the foods were healthy choices you would not be able to make your own healthy choices from a different set of breakfast foods, or use the same principles to choose a healthy dinner. On the other hand, for example, if you understand why saturated fat affects blood cholesterol or how sodium affects blood pressure, you will not forget how to choose a healthy diet.

The critical-thinking approach we have used in this text will help you understand the science of nutrition and give you the decision-making skills you need to navigate the scores of choices you face when deciding what to eat and which of the latest nutrition headlines to believe. By becoming a knowledgeable consumer, you will be able to make informed choices about diet and lifestyle, whether you use this information to improve your own health or to pursue a career in nutrition.

Acknowledgments

We are grateful to the editorial and production staff at John Wiley & Sons for their help and support. We thank our Executive Editor Bonnie Roesch for originally signing this book, our Senior Editor Kevin Witt for his enthusiasm and guidance in helping us navigate the nutrition market, our Project Editor Lorraina Raccuia for the endless hours she devoted to the book, our Senior Product Designer Linda Muriello for her help with creating the media, our Executive Marketing Manager Clay Stone for coordinating the advertising, marketing, and sales efforts for the book and its supplements, and our Assistant Content Editor Lauren Morris for all her help and support with creating the content of the WileyPLUS course. We also thank our Photo Editor Mary Ann Price for ensuring the outstanding quality of the photos in this text, our designer Wendy Lai for delivering an attractively designed text and cover, and our Production Editors Patricia McFadden and Barbara Russiello for their patience and efficiency in guiding this project through production. We would also like to add a special thank-you to our illustrator, Elizabeth Morales, whose creativity, talent, and ability to translate our sketches into understandable visual illustrations has elevated the visual impact of the text.

The authors wish to offer a special thanks to those who have worked on the preparation of the many resources found on the companion websites and within WileyPLUS. They are Katie Clark, The University of San Diego; James Collins, University of Florida, Gainesville; Timaree Hagenburger, Cosumnes River College; Laura Hutchinson, Holyoke Community College; Fran Lukacik, Community College of Philadelphia; Anna Page, Johnson County Community College; Terri Stilson, Edmonds Community College, Washington; Catherine Weinstein, Nassau Community College; and Shahla Wunderlich, Montclair State University.

We are grateful to the following reviewers, survey respondents and focus group attendees who offered their comments and suggestions.

Reviewers, Focus Group Participants, and Survey Respondents

Charalee Allen, *Cincinnati State Technical and Community College*
Valerie Amend, *Ball State University*
Patricia Andrews, *Allegheny College*
Karen Arbuckle, *Cossatot Community College of the University of Arkansas*
Cmeltem Arikan, *University of Massachusetts Dartmouth*
Malene Arnaud-Davis, *Delgado Community College*
Susan Asanovic, *Norwalk Community College*
Pat Bebo, *University of Massachusetts Boston*
Melinda Beck, *University of North Carolina at Chapel Hill*
Marian Benz, *Milwaukee Area Technical College*
R. Bibby, *Tyler Junior College*
Anne Black, *Austin Peay State University*
Debbie Bradney, *Lynchburg College*
Shelly Brandenburger, *South Dakota State University*
Kenneth Broughton, *University of Wyoming*
Carmen Bruni, *Texas A&M International University*
Sarah Brunnig, *Brevard Community College*
Audra Boehne, *Kishwaukee College*
Tracy Bonoffski, *University of North Carolina at Charlotte*
Angelina Boyce, *Hillsborough Community College*
Carmen Boyd, *Missouri State University*
Jim Burkard, *Nashville State Community College*
Barbara Canuel, *Bristol Community College*
Barbara Carlson, *Cedar Crest College*
Michael Carmel, *Trident Technical College*
Diane Carson, *California State University, Long Beach*
Elizabeth Casparro, *Foothill College*
Erin Caudill, *Southeast Community College*
J. Cerami, *University of New Mexico*
Melissa Chabot, *University at Buffalo*
Jo Carol Chezeem, *Ball State University*
Anne Cioffi, *Albany College of Pharmacy and Health Sciences*
Katie Clark, *Santa Rosa Junior College*
Nicole Clark, *Indiana University of Pennsylvania*
Candy Coffin, *Briar Cliff University*
Janet Colson, *Middle Tennessee State University*
Kathy Jo Cook, *Brigham Young University*
Jessica Coppola, *Sacramento City College*
Michael Crosier, *Framingham State University*
Joyce Curry, *Dallas County Community College District*
Amanda Dahl, *Mississippi University for Women*
Tammy Darke, *Mt. San Antonio College*
Ruth Davies, *Edison State College*
Karen Defries, *Erie Community College*
Sunnie Delano, *Pepperdine University*

J. Demarchi, *Saddleback College*
Phil Denette, *Delgado Community College*
Maggi Dorsett, *Butte College*
Kelly Eichmann, *Fresno City College*
Keith Erickson, *University of North Carolina at Greensboro*
Nastaran Faghihnia, *University of California, Berkeley*
Mary Flynn, *Brown University*
Laura Frank, *Immaculata University*
Trish Froehlich, *Palm Beach State College*
Karen Gabrielsen, *Everett Community College*
Andrea Geamanu, *Wayne State University*
Karen Geismar, *Texas A&M University*
Mary Gengler, *South Dakota State University*
Leonard Gerber, *University of Rhode Island*
Diana Goldberg, *Roger Williams University*
Jill Golden, *Orange Coast College*
Jana Gonsalves, *American River College*
Gloria Gonzalez, *Pensacola State College*
Aleida Gordon, *California State Polytechnic University, Pomona*
Virginia Gray, *California State University, Long Beach*
Kathy Guarino, *Orange County Community College, SUNY*
Ginger Hagen, *College of DuPage*
Sandra Haggard, *University of Maine*
Christina Hasemann, *Broome Community College*
Rosanne Heller, *Long Beach City College*
D. Hennager, *Kirkwood Community College*
Beckee Hobson, *College of the Sequoias*
Peter Horvath, *University at Buffalo*
Catherine Howard, *Texarkana College*
Donna Huisenga, *Illinois Central College*
Robert Humphrey, *Cayuga Community College*
Jasmin Hutchinson, *Springfield College*
Marlene Israelsen, *Utah State University*
Seema Jejurikar, *Bellevue College*
Steven Jenkins, *Saint Louis University*
Deirdra Johnson, *Hampton University*
Michelle Johnson, *East Tennessee State University*
Shanil Juma, *Texas Woman's University*
Judith Kasperek, *Pitt Community College*
Leslie Killeen, *Montclair State University*
Dee Kinney, *University of Cincinnati*
Nadine Kirkpatrick, *Sacramento City College*
Patty Lange-Otsuka, *Hawaii Pacific University*
Julie Leonard, *Ohio State University*
Sarah Leupen, *University of Maryland, Baltimore County*
Victoria Liu, *Columbia College Chicago*
Clara Lowden, *Riverside City College*
Nancy Ludwig, *Chemeketa Community College*
Fran Lukacik, *Community College of Philadelphia*
Marypat Maciolek, *Middlesex County College*
Anne Marietta, *Southeast Missouri State University*
Rose Martin, *Iowa State University*
Mary Martinez, *Central New Mexico Community College*
Kathleen Mcguigan, *Hudson Valley Community College*
Glen McNeil, *Fort Hays State University*
Paula Mendelsohn, *Florida Atlantic University*
Mark Meskin, *California State Polytechnic University, Pomona*
Gary Miller, *Wake Forest University*
Lisa Moran, *Kentucky Community & Technical College System*
Adrienne Morecraft, *Tulsa Community College*
Sarah Murray, *Missouri State University*
Judy Myhand, *Louisiana State University*
Susan O'Neill-Cook, *Ferris State University*
Nancy O'Sullivan, *College of DuPage*
Anna Page, *Johnson County Community College*
Anghela Paredes, *University of Florida*
Linda Parker, *University of Miami*
Janet Peterson, *Linfield College*
Kindra Peterson, *San Joaquin Delta College*
Linda Pezzolesi, *Hudson Valley Community College*
Kathryn Pinna, *City College San Francisco*
John Polagruto, *Sacramento City College*
Taylor Puryear, *Barstow Community College*
Elizabeth Quintana, *West Virginia University*
Rabia Rahman, *Saint Louis University*
Scott Rahschulte, *Ivy Tech State College Lawrenceburg*
Helen Reid, *Missouri State University*
Nuha Rice, *Clackamas Community College*
Kyle Richardson, *Joliet Junior College*
Leah Robinson, *Bucks County Community College*
Susan Sadler, *University of Denver*
Barbara Saulter, *Wayne County Community College District*
Jason Sawyer, *Central Connecticut State University*
Nancy Schmidt, *Miami University Middletown*
Margaret Schrama, *Massasoit Community College*
Janet Schwartz, *Framingham State University*
Bridget Scott, *Nicholls State University*
Joseph Scruggs, *Ozarks Technical Community College*
Judith Sharlin, *Palm Beach State College*
Vidya Sharma, *San Antonio College*
Denise Signorelli, *College of Southern Nevada*
Melissa Silva, *Bryant University*
Diana Simon, *Montclair State University*
Carole Sloan, *Henry Ford Community College*
Leslie Spencer, *Rowan University*
Deborah Stone, *Itawamba Community College in Northeast Mississippi*
Paul Strieleman, *University of Chicago*
Ronald Swisher, *Oregon Institute of Technology*
Linda Talley, *Texas A&M University*
Stephanie Taylor-Davis, *Indiana University of Pennsylvania*
Anna Tourkakis, *North Shore Community College*
Barbara Troy, *Marquette University*
Diane Wagoner, *Indiana University of Pennsylvania*
Kathy Wallace, *Community College of Allegheny County*
Daryle Wane, *Pasco-Hernando Community College*
Sue Ellen Warren, *El Camino College*
Christopher Watters, *Middlebury College*
Suzy Weems, *Baylor University*
Janet Westhoff, *Mott Community College*
John Wielichowski, *Milwaukee Area Technical College*
Frances Williams, *El Camino College*
Frederick Wolf, *Central Arizona College*
Cynthia Wright, *Southern Utah University*
John Young, *University of Nevada, Las Vegas*
Jennifer Zimmerman, *Tallahassee Community College*
Maureen Zimmerman, *Mesa Community College*

Nutrition Advisory Board Members

Jeanne Boone, *Palm Beach State College*
Linda Brown, *University of North Florida*
James F. Collins, *University of Florida, Gainesville*
Barbara Goldman, *Palm Beach State College*
Timaree Hagenburger, *Cosumnes River College*
Donna Handley, *University of Rhode Island*
Kirsten Hilpert, *State University of New York, Oneonta*
Laura Hutchinson, *Holyoke Community College*
Karen Israel, *Anne Arundel Community College*
Owen Murphy, *Central Oregon Community College*
Judith Myhand, *Louisiana State University*
Shahla Wunderlich, *Montclair State University*

Brief Contents

Contents

Pixtal/SuperStock

SPL/Photo Researchers, Inc.

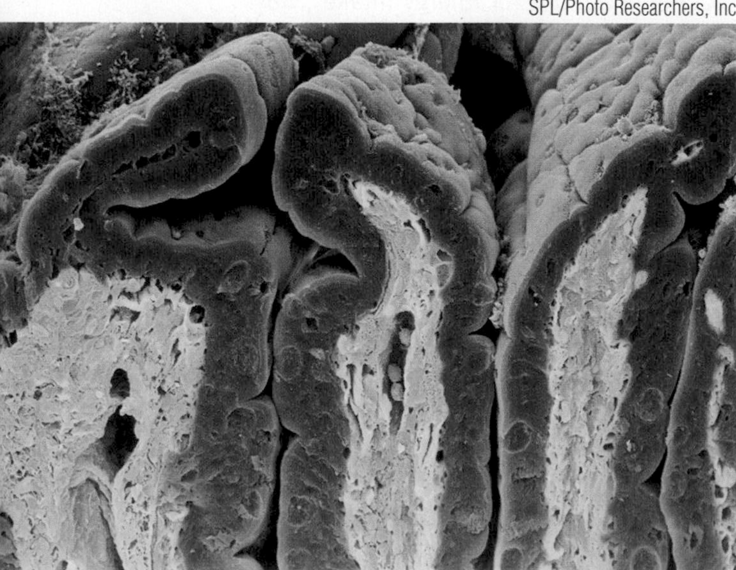

© Alexander Chernyakov/iStockphoto

4 Carbohydrates: Sugars, Starches, and Fiber 111

5 Lipids: Triglycerides, Phospholipids, and Cholesterol 157

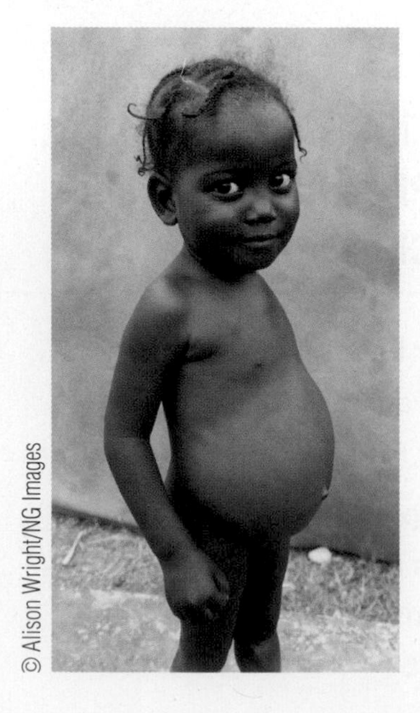
© Alison Wright/NG Images

Mike Blake/Reuters/NewsCom

From A. Mazzone and A. Dal Canton, The New England Journal of Medicine, 2002; 347-222-223

11 Major Minerals and Bone Health 439

Michael Klein/Peter Arnold/Getty Images, Inc.

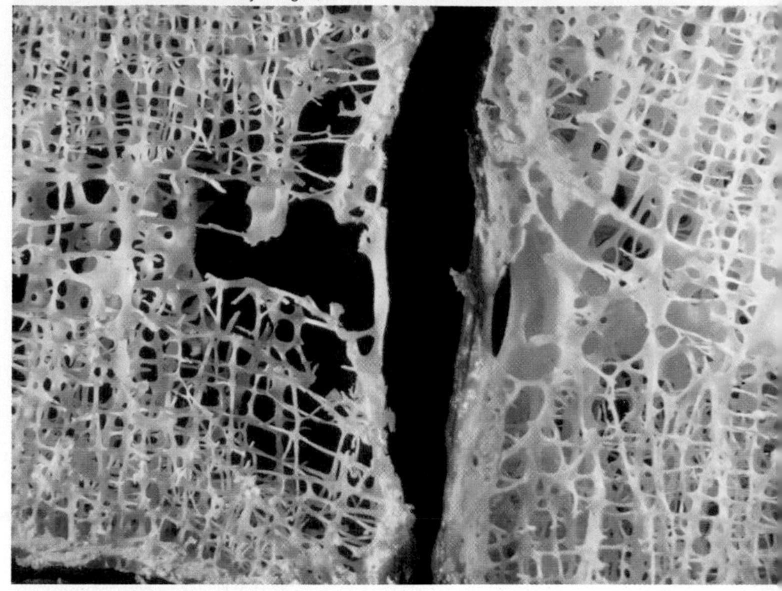

12 The Trace Minerals 473

© amana images Inc./Alamy Limited

courtesy Lori Smolin

Cusp and Flirt/Masterfile

Tony Freeman/PhotoEdit

Digital Vision/Getty Images, Inc.

18 World Hunger and Malnutrition 731

Alison Wright/NG Image Collection

CHAPTER OUTLINE

Mauritius Images/Age Fotostock America, Inc.

Nutrition: Food for Health

1

CASE STUDY

Kaitlyn peered into the tiny refrigerator in her dorm room looking for something to eat. The only thing she had was some leftover pizza and a couple of cookies. Once these were gone she headed down the hall to get a bag of potato chips and an energy drink from the vending machine. They weren't a good choice—but they were her only option. Kaitlyn had been in classes and at work until late in the evening. When she finally sat down to study, she realized that she had missed dinner. She needed to eat something to keep her going until she finished that chapter in her nutrition book, but her dorm offered few food choices late at night.

© Alejandro Rivera/iStockphoto

As a college freshman away from home for the first time, Kaitlyn has gained a few pounds and is beginning to be concerned about her weight. Her father recently had cardiac bypass surgery, and her mother takes medication for high blood pressure. Kaitlyn knows that because of this family history, her diet is a particularly important part of her future health. The dorm cafeteria, although not great, does offer a variety of choices. The problem is that Kaitlyn doesn't know how to choose a healthy diet. She planned to keep healthy choices in her refrigerator but never takes time to go to the store.

Several of her friends have started taking supplements like Mega B to give them more energy and ginkgo biloba to improve their memories. Kaitlyn is tempted to start taking them but she's not sure the claims about them are true. **To optimize her health Kaitlyn needs to learn the basics of nutrition science and perfect the art of making nutritionally sound decisions and healthy food choices—a goal that is a little overwhelming at first.**

1.1 Nutrition and Our Diet

LEARNING OBJECTIVES

- Define the terms *nutrition* and *nutrient*.
- Describe two changes in American eating patterns that have increased chronic disease.

nutrition A science that studies the interactions between living organisms and food.

nutrients Substances in foods that provide energy and structure and help regulate body processes.

Nutrition is a science that studies all the interactions that occur between living organisms and food. Food provides **nutrients** and energy, which are needed to keep us alive and healthy, to support growth, and to allow reproduction. A healthy diet includes the right combination of food to provide the correct amounts of energy and nutrients without deficiencies or excesses. However, we don't always choose the right combination of foods to optimize nutritional health. In America today, even when food is abundant, fast-paced lifestyles and food choices made available through modern technology contribute to a diet that contains too much of some nutrients and too little of others.

Food in 21st-Century America

For much of human history, in order to get enough to eat people needed to spend most of their day obtaining food and preparing meals. Even 100 years ago, the time spent for meal preparation was measured in hours—hours spent peeling, chopping, baking, roasting, stewing, and then cleaning. Today, a microwavable dinner that includes meat, rice, vegetables, and dessert can be ready in five minutes.

Our modern food supply includes an endless assortment of eating options. Many of these choices are foods that have been part of the human diet for centuries—fresh fruits and vegetables, meats, and grains. But others are newer additions—frozen vegetables complete with sauce, canned soups and stews, pre-cooked packaged meats, frozen prepared meals, and snack foods. Fifty years ago people ate most of their meals at home, with their families, at a leisurely pace. Today more single-parent households and families with both parents working mean the woman of the house is no longer at home in the afternoon preparing wholesome meals. Dinner is a rush because busy after-school schedules impinge on family mealtimes.

processed food A food that has been specially treated or changed from its natural state either at home or in a processing plant.

Today, few young adults even know how to prepare a full meal, and shoppers of all ages are choosing to buy more convenient, **processed foods** that can be boiled or heated in the microwave rather than raw ingredients that need to be chopped, seasoned, and cooked. The increase in the availability and variety of these convenience foods has made it easier and quicker to get something to eat. In addition, Americans today are replacing more and more home-cooked meals with meals from fast-food restaurants.[1]

These changes in the American food supply have made it easier and faster to obtain a meal or snack, but they have not necessarily improved our nutritional health (**Figure 1.1**). Over the past century the major nutrition concerns in the United States have shifted from providing enough nutrients to meet people's needs to limiting overconsumption and the chronic diseases related to excesses of energy and certain nutrients.

How Healthy Is the American Diet?

The American diet isn't as healthy as it could be. As it has become easier to obtain and prepare food, the amount of food consumed has increased.[2] American adults eat more calories than they did 30 years ago, primarily due to larger portion sizes, especially from fast foods, and an increase in the consumption of salty snacks, soft drinks, and pizza.[3] Americans today get 32% of their calories from meals eaten away from home; these meals tend to be higher in calories than foods prepared at home.[4]

In addition to eating more calories than we need, we are not eating enough of the foods that make up a healthy diet. Recommendations suggest a diet based on whole grains, vegetables, and fruits, with smaller amounts of low-fat dairy products and lean meats and

THINK CRITICALLY

How does a homemade turkey sandwich compare to a 12-inch sub from the shop on the corner in terms of convenience, cost, and calories?

© Juanmonino/iStockphoto

Breakfast at home
A cup of coffee with whole milk and sugar and an English muffin with butter would cost about 50 cents and provide about 200 kcalories.

P. Mittongtare/FoodPix/Getty Images

Breakfast out
A 16-ounce caramel mocha and a healthy-looking bran muffin at Dunkin' Donuts or the corner coffee bar would cost about $4 and provide about 770 kcalories.

Figure 1.1 **The costs of convenience**
If you stop for a muffin and coffee on your way to work or school, you can save a couple of minutes but you may pay a higher price than you think in terms of both dollars and calories. The impact of stopping for coffee and a muffin once in a while when you are running late is minimal, but making it an everyday habit can break your dollar and calorie budget.

limited amounts of sweets and certain types of fats, such as saturated fat. As a population, we don't eat enough whole grains, vegetables, fruits, seafood, or dairy products. We frequently choose potatoes, often fried, for a vegetable but consume few nutrient-rich dark green and deep yellow vegetables. Our diets are high in snack foods and desserts that supply us with more salt (sodium) and sugar than is recommended (**Figure 1.2a**).[1,5] Instead of milk, we are choosing sweetened beverages, especially carbonated soft drinks.[1,6] This dietary pattern along with a lack of physical activity increases the risk of developing chronic diseases such as diabetes, obesity, heart disease, and cancer, which are the major causes of illness and death in our population.[7] It has been estimated that over 15% of all deaths in the United States can be attributed to a poor diet and a sedentary lifestyle (**Figure 1.2b**).[8] Recommendations for reducing disease risk focus on changes in the foods we choose and the amount of exercise we get.[5,9]

A healthy diet does not need to exclude processed convenience foods, but it must involve wise choices. To choose a healthy diet that provides the right amounts of energy and each nutrient, we need to understand how our bodies obtain nutrients from food, which nutrients are essential, how much we need, and which foods provide healthy sources of nutrients. We also need to recognize which nutrition information to believe.

Figure 1.2 **Poor diets and nutrition-related deaths**

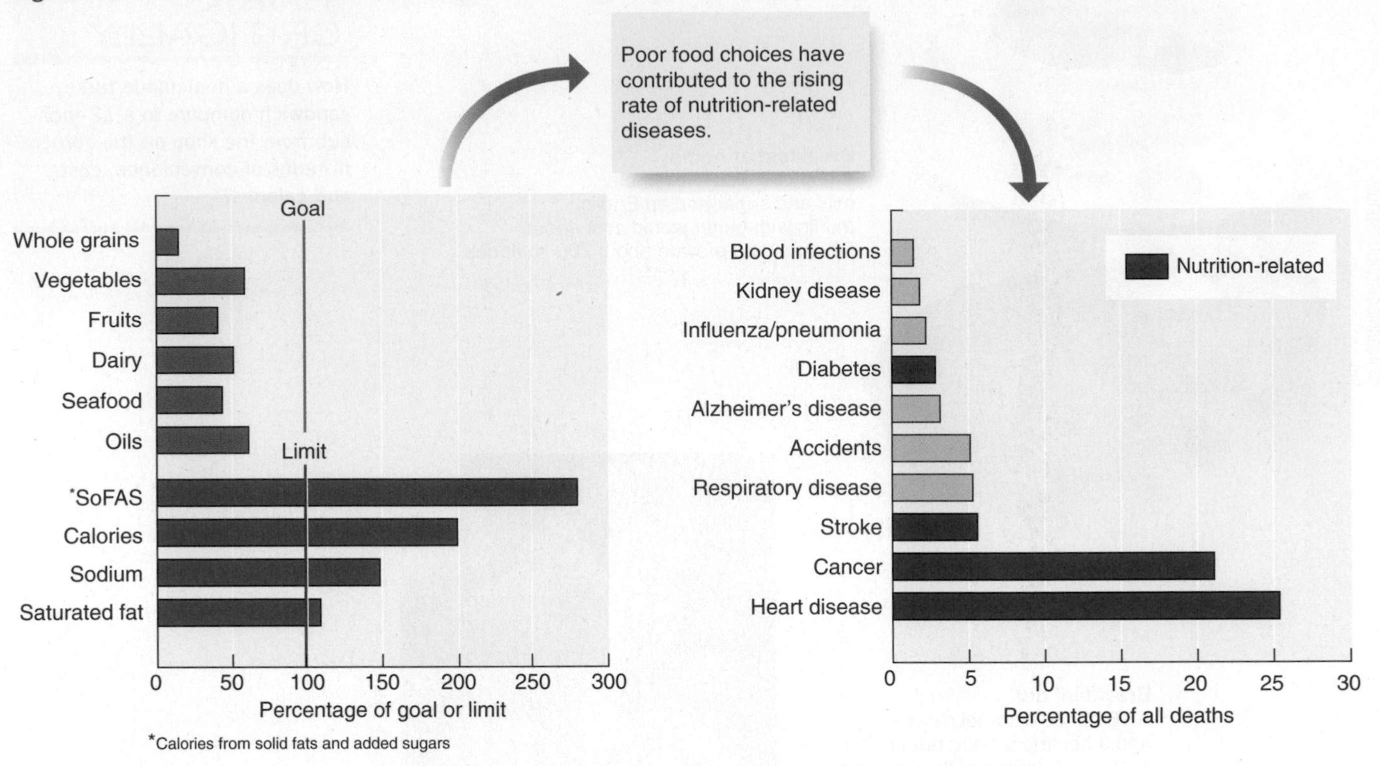

(a) The current U.S. dietary pattern is not as healthy as it could be. The graph shows the usual U.S. intake of selected foods and nutrients as a percentage of the recommended goal or limit.

(b) The graph shows the leading causes of death in the United States; those shown in purple are nutrition-related.

1.2 Food Provides Nutrients

LEARNING OBJECTIVES

- Define the term *essential nutrient* and list the six classes of nutrients.
- Describe the three general functions of nutrients.
- Discuss how nutrition can affect your health in the short term and in the long term.
- Define *malnutrition*.

essential nutrient A nutrient that must be provided in the diet because the body either cannot make it or cannot make it in sufficient quantities to satisfy its needs.

fortified food Food to which one or more nutrients have been added.

enriched grains Grain products to which specific amounts of thiamin, riboflavin, niacin, and iron have been added. Since 1998 folic acid has also been added to enriched grains.

dietary supplement A product intended for ingestion in the diet that contains one or more of the following: vitamins, minerals, plant-derived substances, amino acids, and concentrates or extracts.

To date, approximately 45 nutrients have been determined to be essential to human life. **Essential nutrients** must be supplied in the diet to support life; they either cannot be made by the body or cannot be made in large enough quantities to meet needs. For example, our bodies cannot synthesize vitamin C, but we need it to stay healthy. If we do not consume vitamin C in the foods we eat, we will begin to show signs of a vitamin C deficiency. If vitamin C is not added back into the diet, vitamin C deficiency will eventually be fatal.

Our intake of essential nutrients is determined by our food choices. Some foods are naturally high in nutrients, and some contain nutrients added during processing. Foods to which nutrients have been added are called **fortified foods**. Some fortified foods like milk with added vitamin A and **enriched grains** have been a part of our food supply for decades. The government mandated the fortification of these foods to eliminate nutrient deficiencies in the population, and the amounts and types of nutrients added are specified. Other foods, such as orange juice with added calcium and flavored water with added vitamins and minerals, are not part of mandated fortification programs. These foods are fortified with nutrients to increase sales by meeting consumer demand for nutrient-rich foods. The amounts and types of nutrients added to these foods are at the discretion of the manufacturer. **Dietary supplements** are another source of nutrients in the American food supply. National surveys indicate that over 50% of adults in the United States take some type of vitamin or mineral supplement to boost their nutrient intake.[10]

In addition to nutrients, food contains substances that are needed by the body but are not essential in the diet. Lecithin, for example, is a substance found in egg yolks that is needed for nerve function. It is not considered an essential nutrient because it can be manufactured in the body in adequate amounts. The diet also contains substances that are not made by the body and are not necessary for life, but that have health-promoting properties. Those that come from plants are called **phytochemicals**; those that come from animal foods are called **zoochemicals**. For example, a phytochemical found in broccoli called sulforaphane is not essential in the diet but has effects in the body that may help reduce the risk of cancer.

The Six Classes of Nutrients

Chemically, there are six classes of nutrients: carbohydrates, lipids, proteins, water, vitamins, and minerals. These classes can be grouped in a variety of ways—by whether they provide energy to the body, by how much is needed in the diet, and by their chemical structure. Carbohydrates, lipids, and proteins provide energy and thus are referred to as **energy-yielding nutrients**. Alcohol also provides energy but is not considered a nutrient because it is not needed to support life (see **Focus on Alcohol**). Along with water, the energy-yielding nutrients constitute the major portion of most foods and are required in relatively large amounts by the body. Therefore, they are referred to as **macronutrients** (*macro-* means large). Their requirements are measured in kilograms (kg) or grams (g). Vitamins and minerals are classified as **micronutrients** because they are needed in small amounts in the diet (*micro-* means small). The amounts required are expressed in milligrams (1 mg = 1/1000 g) or micrograms (1 μg = 1/1,000,000 g) (see Appendix G). Structurally, carbohydrates, proteins, lipids, and vitamins are **organic molecules,** so they are referred to as organic nutrients. Minerals and water are **inorganic molecules,** so they are referred to as inorganic nutrients.

The Energy-Yielding Nutrients The energy provided by carbohydrates, lipids (fat), and proteins is measured in **kilocalories** (abbreviated kcalories or kcals) or in **kilojoules** (abbreviated kjoules or kJs). The more common term, calorie (lowercase *c*), is technically 1/1000 of a kcalorie, but when spelled with a capital *C* Calorie means a kcalorie. In the popular press, calorie (small *c*) is often used to express the kcalorie content of a food or diet (**Figure 1.3**).

phytochemical A substance found in plant foods (*phyto-* means plant) that is not an essential nutrient but may have health-promoting properties.

zoochemical A substance found in animal foods (*zoo-* means animal) that is not an essential nutrient but may have health-promoting properties.

energy-yielding nutrient A nutrient that can be metabolized to provide energy in the body.

macronutrient A nutrient needed by the body in large amounts. These include water and the energy-yielding nutrients: carbohydrates, lipids, and proteins.

micronutrient A nutrient needed by the body in small amounts. These include vitamins and minerals.

organic molecule A molecule that contains carbon bonded to hydrogen.

inorganic molecule A molecule that contains no carbon-hydrogen bonds.

kilocalorie (kcalorie, kcal) The unit of heat used to express the amount of energy provided by foods. It is the amount of heat required to raise the temperature of 1 kg of water 1 degree Celsius (1 kcalorie = 4.18 kjoules).

kilojoule (kjoule, kJ) A unit of work that can be used to express energy intake and energy output. It is the amount of work required to move an object weighing 1 kg a distance of 1 m under the force of gravity (1 kjoule = 0.24 kcalorie).

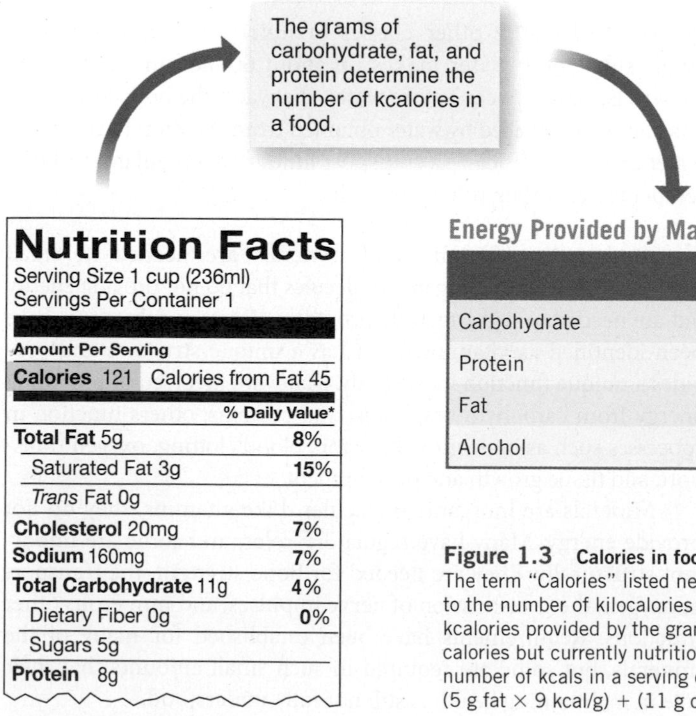

The grams of carbohydrate, fat, and protein determine the number of kcalories in a food.

Nutrition Facts
Serving Size 1 cup (236ml)
Servings Per Container 1

Amount Per Serving	
Calories 121 Calories from Fat 45	
	% Daily Value*
Total Fat 5g	8%
Saturated Fat 3g	15%
Trans Fat 0g	
Cholesterol 20mg	7%
Sodium 160mg	7%
Total Carbohydrate 11g	4%
Dietary Fiber 0g	0%
Sugars 5g	
Protein 8g	

Energy Provided by Macronutrients and Alcohol

	Kcalories/gram	Kjoules/gram
Carbohydrate	4	16.7
Protein	4	16.7
Fat	9	37.6
Alcohol	7	29.3

Figure 1.3 Calories in foods
The term "Calories" listed near the top of the Nutrition Facts panel of a food label technically refers to the number of kilocalories (kcalories) in a serving of the food. It is equal to the total number of kcalories provided by the grams of carbohydrate, fat, and protein in a serving. Alcohol also provides calories but currently nutrition information is not required on the labels of alcoholic beverages. The number of kcals in a serving of the food from the label shown here can be calculated as follows:
(5 g fat × 9 kcal/g) + (11 g carbohydrate × 4 kcal/g) + (8 g protein × 4 kcal/g) = 121 kcal per serving.

Figure 1.4 Carbohydrates, lipids, and proteins

(a) Starches are a type of carbohydrate made of sugars linked together. Pasta, rice, and bread contain mostly starch; whole-grain bread, oatmeal, and popcorn are high in fiber; and cookies, cakes, and carbonated beverages are high in added sugar.

(b) Most of the lipids we consume in our food are triglycerides; each contains three fatty acids. Plant sources of triglycerides in our diet include vegetable oils, avocados, olives, and nuts. Animal sources of triglycerides include cream, butter, meat, and whole milk; these animal foods also provide cholesterol.

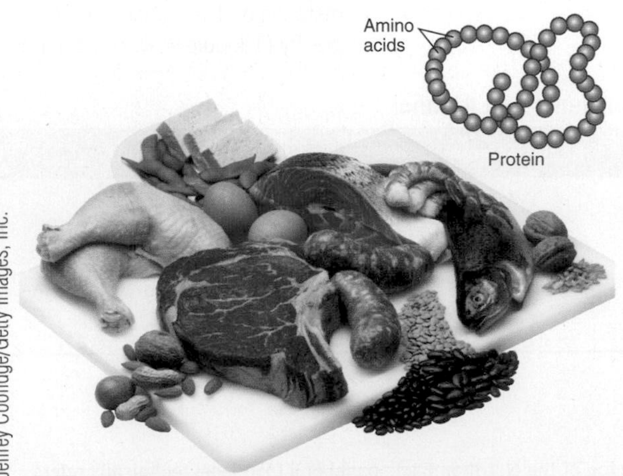

(c) Proteins are made of amino acids linked together. The proteins we obtain from animal foods, such as meat, fish, and eggs, better match our amino acid needs than do most individual plant proteins, such as those found in grains, nuts, and beans. However, when plant sources of protein are combined, they can provide all the amino acids we need.

Carbohydrates provide a readily available source of energy to the body. They contain 4 kcalories per gram (see Figure 1.3). Carbohydrates include sugars such as those in table sugar, fruit, and milk, and starches such as those in vegetables, legumes, and grains. Sugars are the simplest form of carbohydrate, and starches are more complex carbohydrates made of many sugars linked together (**Figure 1.4a**). Most fiber is also carbohydrate. Fiber cannot be digested and therefore provides very little energy. However, it is important for gastrointestinal health. Fiber is found in vegetables, fruits, legumes, and whole grains.

Lipids, commonly called fats and oils, provide 9 kcal per gram. They are a concentrated source of energy in food and a lightweight storage form of energy in the body. There are several types of lipids that are important in nutrition. Triglycerides are the type that is most abundant in foods and in the body. The fat on the outside of a steak, the butter and oil that is added to food during cooking, and the layer of fat under a person's skin are all composed almost entirely of triglycerides. Triglycerides are made up of fatty acids (**Figure 1.4b**). Different types of fatty acids have different health effects. Diets high in saturated fatty acids increase the risk of heart disease, whereas those high in unsaturated fatty acids may reduce these risks. Cholesterol is another type of lipid; high levels in the blood can increase heart disease risk (see Chapter 5).

Protein is needed for growth and maintenance of body structures and to regulate and facilitate body processes. It can also be used to provide energy—4 kcal per gram. Meat, fish, poultry, eggs, milk, grains, vegetables, and legumes all provide protein. Like carbohydrate and lipid, protein is not a single substance. There are thousands of different proteins in the human body and in the diet. All of these are made up of units called amino acids. Different combinations of amino acids are linked together to form different types of proteins (**Figure 1.4c**). Some amino acids can be made by the body, and others are essential in the diet. The proteins in animal products better match our need for amino acids than do plant proteins, but both plant and animal proteins can provide all the amino acids we need.

Water Unlike the other classes of nutrients, water is only a single substance. Water makes up about 60% of an adult's body weight. Because we can't store water, the water the body loses must constantly be replaced by water obtained from the diet. In the body, water acts as a lubricant, a transport fluid, and a regulator of body temperature, among other functions.

Micronutrients Vitamins and minerals are needed in small amounts. Vitamins are organic molecules that do not provide energy but are needed to regulate body processes. Thirteen substances have been identified as vitamins. Each has a unique structure and provides a unique function in the body. Many are involved in providing energy from carbohydrates, lipids, and proteins; others function in processes such as bone growth, vision, blood clotting, oxygen transport, and tissue growth and development.

Minerals are inorganic molecules. Like vitamins, they do not provide energy. Many have regulatory roles, and some are important structurally. They are needed for bone strength, the transport of oxygen, the transmission of nerve impulses, and numerous other functions. Requirements have been established for many of the minerals, but some are required in such small amounts that their role in maintaining health is still not fully understood.

Vitamins and minerals are found in most foods. Fresh foods are a good natural source, and many processed foods are fortified with vitamins and minerals. Food processing and preparation can also cause vitamin losses because some are destroyed by exposure to light, heat, and oxygen. Minerals are more stable but can still be lost along with vitamins when food components are separated during processing and in the water used in cooking and processing.

What Nutrients Do

Together, the macronutrients and micronutrients provide energy, structure, and regulation, which are needed for growth, maintenance and repair, and reproduction. Each nutrient provides one or more of these functions, but all nutrients together are needed to maintain health.

Provide Energy Inside the body, biochemical reactions release the energy contained in carbohydrates, lipids, and proteins. Some of this energy is used to synthesize new compounds and maintain basic body functions, some is used to fuel physical activity, and some is lost as heat. When the energy in the carbohydrates, lipids, and proteins consumed in the diet is not needed immediately, it can be stored, primarily as body fat. These stores can provide energy when dietary sources are unavailable. Over the long term, if more energy is consumed than is needed, body fat stores enlarge and body weight increases. If less energy is consumed than is needed, the body will burn its fat stores to meet its energy needs and body weight will decrease.

Form Structures With the exception of vitamins, all the classes of nutrients are involved in forming and maintaining the body's structure (**Figure 1.5**). Fat deposited under the skin contributes to our body shape, for instance, and proteins form the ligaments and tendons that hold our bones together and attach our muscles to our bones. Minerals harden bone. Protein and water make up the structure of the muscles, which help define our body contours, and protein and carbohydrates form the cartilage that cushions our joints. On a smaller scale, lipids, proteins, and water form the structure of individual cells. Lipids and proteins make up the membranes that surround each cell, and water and dissolved substances fill the cells and the spaces around them.

Regulate Body Processes All six classes of nutrients play important roles in regulating body processes (**Figure 1.6**). To maintain life, body temperature,

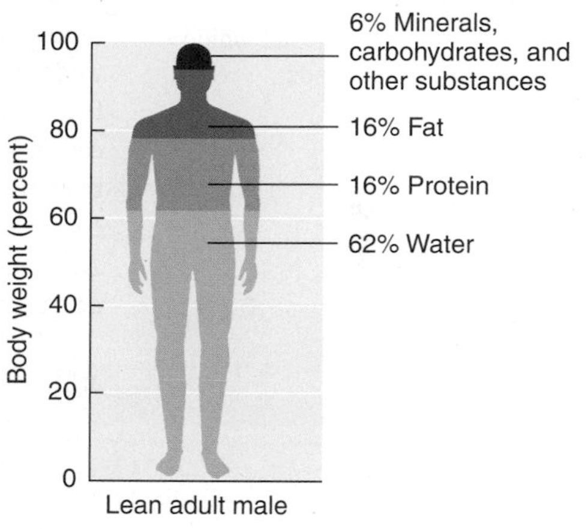

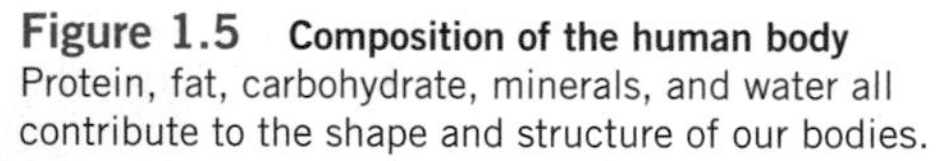
Figure 1.5 Composition of the human body
Protein, fat, carbohydrate, minerals, and water all contribute to the shape and structure of our bodies.

Figure 1.6 Examples of nutrient functions in the body

Function	Nutrient	Example
Energy	Carbohydrate	Glucose is a carbohydrate that provides energy to body cells.
	Lipid	Fat is the most plentiful source of stored energy in the body.
	Protein	Protein consumed in excess of protein needs will be used for energy.
Structure	Lipid	Lipids are the principal component of the membranes that surround each cell.
	Protein	Protein in connective tissue holds bones together and holds muscles to bones. Protein in muscles defines their shape.
	Minerals	Calcium and phosphorus are minerals that harden teeth and bones.
Regulation	Lipid	Estrogen is a lipid hormone that helps regulate the female reproductive cycle.
	Protein	Leptin is a protein that helps regulate the size of body fat stores.
	Carbohydrate	Sugar chains attached to proteins circulating in the blood signal whether the protein should remain in the blood or be removed by the liver.
	Water	Water in sweat helps cool the body to regulate body temperature.
	Vitamins	B vitamins regulate the use of macronutrients for energy.
	Minerals	Sodium is a mineral that helps regulate blood volume.

When body temperature increases, sweat is produced, cooling the body as the water evaporates from the skin.

homeostasis A physiological state in which a stable internal body environment is maintained.

metabolism The sum of all the chemical reactions that take place in a living organism.

blood pressure, blood sugar level, and hundreds of other parameters must be kept relatively constant, a condition referred to as **homeostasis**. Maintaining homeostasis involves thousands of chemical reactions and physiological processes. Together all of the chemical reactions that occur in the body are referred to as **metabolism**. Proteins, vitamins, and minerals are regulatory nutrients that help control how quickly metabolic reactions take place throughout the body. Lipids and proteins are needed to make regulatory molecules that stimulate or inhibit various body processes.

Nutrition and Health

malnutrition Any condition resulting from an energy or nutrient intake either above or below that which is optimal.

undernutrition Any condition resulting from an energy or nutrient intake below that which meets nutritional needs.

What we eat has an enormous impact on how healthy we are now and how likely we are to develop chronic diseases like heart disease, obesity, and diabetes in the future. Consuming either too little or too much of one or more nutrients or energy will cause **malnutrition**. Malnutrition can affect our health today and can impact our health 20, 30, or 40 years from now.

How Undernutrition Affects Health **Undernutrition** is a form of malnutrition caused by a deficiency of energy or nutrients. It may be caused by a deficient intake, increased requirements, or an inability to absorb or use nutrients. Starvation, the most severe form of undernutrition, is a deficiency of energy that causes weight loss, poor growth, the inability to reproduce, and if severe enough, death (**Figure 1.7a**). Deficiencies of individual nutrients can also cause serious health problems. The symptoms of nutrient deficiencies often reflect the functions of the deficient nutrient. For example, vitamin A is necessary for vision; a deficiency of vitamin A can result in blindness. Vitamin B_{12} is needed for normal nerve function. A deficiency of this vitamin, which is more common in older adults because absorption often decreases with age, causes changes in mental status.

Some nutrient deficiencies cause symptoms quickly. In only a matter of hours an athlete exercising in hot weather may become dehydrated due to a deficiency of water.

Figure 1.7 **Malnutrition**

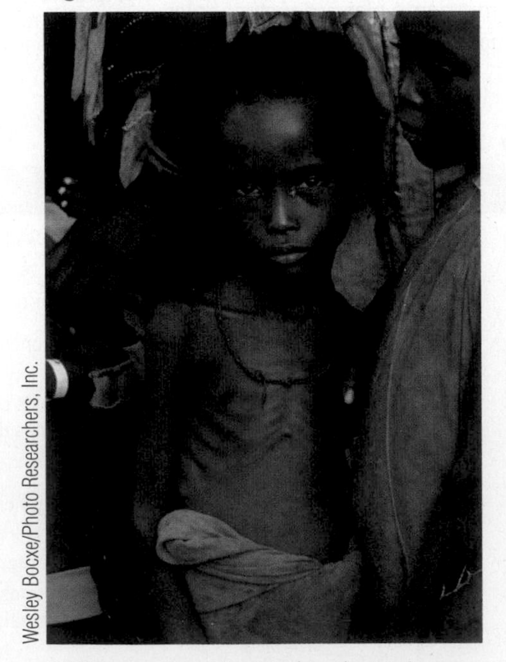

Wesley Bocxe/Photo Researchers, Inc.

(a) The symptoms of starvation, the most obvious form of undernutrition, occur gradually over time when the energy provided by the diet is too low to meet the body's needs. Body tissues are broken down to provide the energy to support vital functions, resulting in loss of body fat and wasting of muscles.

© Dennis MacDonald/Age Fotostock America, Inc.

(b) Obesity is a form of overnutrition that occurs when energy intake surpasses energy expenditure over a long period, causing the accumulation of an excessive amount of body fat. Today over a third of adults and an estimated 17% of U.S. children and adolescents ages 2 to 19 years have so much excess body fat that they are classified as obese.[11,12]

Drinking water relieves the headache, fatigue, and dizziness caused by dehydration almost as rapidly as these symptoms appeared. Other nutritional deficiencies may take much longer to become apparent. Symptoms of scurvy, a disease caused by a deficiency of vitamin C, appear after months of deficient intake; osteoporosis, a condition in which the bones become weak and break easily, occurs after years of consuming a calcium-deficient diet.

How Overnutrition Affects Health **Overnutrition**, an excess of energy or nutrients, is also a form of malnutrition. When excesses of specific nutrients are consumed, an adverse or toxic reaction may occur. For example, a single excessive dose of iron can cause liver failure and too much vitamin B_6 can cause nerve damage. These nutrient toxicities usually result from taking large doses of vitamin and mineral supplements. Foods generally do not contain high enough concentrations of nutrients to be toxic.

overnutrition Poor nutritional status resulting from an energy or nutrient intake in excess of that which is optimal for health.

The type of overnutrition that is most common in the United States today does not have immediate toxic effects but contributes to the development of chronic diseases in the long term. The typical U.S. diet, which provides more calories than are needed, has resulted in an epidemic of obesity in which more than 68% of adults are overweight or obese (**Figure 1.7b**).[12] Diets that are high in sodium contribute to high blood pressure; an excess intake of saturated fat contributes to heart disease; and a dietary pattern that is high in red meat and saturated fat and low in fruits, vegetables, and fiber may increase the risk of certain cancers.[5]

genes Units of a larger molecule called DNA that are responsible for inherited traits.

nutritional genomics or **nutrigenomics** The study of how diet affects our genes and how individual genetic variation can affect the impact of nutrients or other food components on health.

How Our Diets Interact with Our Genetic Makeup What we eat affects our health, but diet alone does not determine whether we will develop a particular disease. Each of us inherits a unique combination of **genes**. Some of these genes affect our risk of developing chronic diseases such as heart disease, cancer, high blood pressure, and diabetes, but their impact is affected by our diet and lifestyle (**Figure 1.8**). Our genetic makeup determines the impact a certain nutrient will have on us. For example, some people inherit a combination of genes that makes their blood pressure more sensitive to the amount of sodium in their diet. When these individuals consume even an average amount of sodium, their blood pressure increases (as discussed further in Chapter 10). Others inherit genes that allow them to consume more sodium without a rise in blood pressure. Those who inherit this "salt sensitivity" can reduce their blood pressure, and the complications associated with high blood pressure, by eating a diet that is low in sodium.

Our increasing understanding of human genetics has given rise to the discipline of **nutritional genomics** or **nutrigenomics**, which explores the interaction between genetic variation and nutrition.[13] This research has led to the development of the concept of "personalized nutrition," the idea that a diet based on the genes an individual has inherited can be used to prevent, moderate, or cure chronic disease. Although today we do not know enough to use a sample of your DNA to tell you what to eat to optimize your health, we do know that certain dietary patterns can reduce the risk of many chronic diseases (see **Debate: Are Personalized Diets the Best Approach to Reducing Chronic Disease?**).

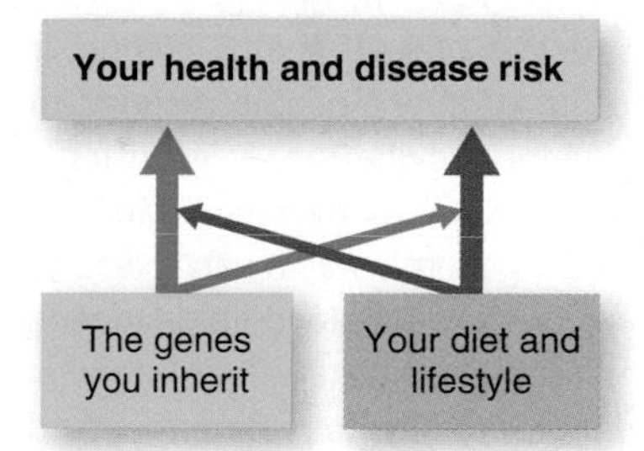

Figure 1.8 **Diet, genes, and health** Both the genes you inherit and your diet and lifestyle choices directly affect your health and disease risk. There is also interplay between these such that your diet and lifestyle choices can increase or decrease the effect genes have on your health and your genes can influence the impact that diet and lifestyle have on your health.

1.3 Food Choices for a Healthy Diet

LEARNING OBJECTIVES

- List factors other than nutrition that affect your food choices.
- Define *nutrient density.*
- Explain the importance of variety, balance, and moderation in selecting a healthy diet.

Each of the food choices we make contributes to our diet as a whole. This diet must provide enough energy to fuel the body and all the essential nutrients and other food components—in the right proportions—to prevent deficiencies, promote health, and protect against chronic disease. No single food choice is good or bad in and of itself, but all of our choices combined make up a dietary pattern that is either healthy or not so healthy.

DEBATE

Are Personalized Diets the Best Approach to Reducing Chronic Disease?

Alamy

We can't change our genetic background, but the impact that our genes have on our health can be altered by factors we can control, including our diet. The field of nutrigenomics suggests that we can reduce our risk of disease by tailoring our diets to our individual genetic makeup.[1] Someday we may be able to go to the doctor's office, have our genes analyzed, and then have specific foods and dietary supplements prescribed to optimize our health and prevent diseases to which we are susceptible. Are such genetically customized foods and diets the next big advance in public health, or are we better off sticking to general diet recommendations that can be applied to everyone?

Current nutrition guidelines are based on what is thought to be good for all healthy people in the population. Yet we know that different people respond differently to the same diet, so dietary advice that is good for the majority of people may not be optimal for everyone. Modern medicine is already practicing nutrigenomics at a very basic level. Dietitians instruct people with elevated blood lipids to reduce their intake of saturated fat and increase their fiber intake and show those with high blood pressure how to reduce their sodium intake.

These dietary interventions can be helpful in controlling current medical conditions, but many people believe that with nutrigenomics we can do better. They suggest that with more information on a person's genetic background, more individualized recommendations could be developed that would improve outcomes for heart disease, high blood pressure, diabetes, and other disorders.[2] These diets could be customized to take into consideration individual genetic variations in the disease as well as life stage, dietary preferences, and other aspects of health status.[3] Some propose that if followed, these personalized recommendations may supplement and even replace prescription drugs.

Nutrigenomics may also allow diet prescriptions to prevent, not just manage, disease. Finding out early in life that you are genetically predisposed to a particular condition would allow you to eat a diet specifically targeting your risk. But this information could also backfire. If people are told they are at very low risk for heart disease, they may interpret this to mean that they do not need to worry about overindulging in fatty animal products. This would then increase their lifestyle risk of developing heart disease as well as the risk of certain types of cancer and obesity.[4]

Some people believe that we should not spend time and money on nutrigenomics because it is unlikely to improve public health, or even individual health. Many people fail to follow current population-wide nutrition guidelines not because they lack the knowledge, money, or motivation to do so but simply because they choose not to. Therefore, it is unlikely that they will follow personalized guidelines any better or that genetic test results will motivate them to eat healthier.[5,6] The priority for public health should not be to fine-tune diet prescriptions but to find out what will make people change their diets and live healthier lives.

Other concerns with nutrigenomics are the ethics of widespread genetic testing and the possibility that commercial interests will drive nutrigenomics rather than any benefits to public health.[6] People who strictly follow their diet prescriptions will certainly benefit, but the big beneficiaries of personalized diet prescriptions will be biotech companies, which would profit from genetic testing needed to establish disease profiles, and the food industry, which would benefit from the creation and sale of foods designed to treat disease.[4] There is also concern that personalized nutrition will be very costly, perhaps reserved for those with money and education.[3]

In the future, will we select breakfast cereals and dietary supplements based on our genes? Will this result in better health than is achieved by following population-wide nutrition guidelines designed to improve public health? When considering individual diet prescription we must not lose sight of the fact that food is not just about health but also about pleasure, culture, sociability, identity, and beliefs.

THINK CRITICALLY: If genetic testing determines that you are at low risk for high blood pressure, does this mean that you can ignore the public health recommendation to limit salt intake? Why or why not?

---

[1]Trujillo, E., Davis, C., and Milner, J. Nutrigenomics, proteomics, metabolomics, and the practice of dietetics. *J Am Diet Assoc* 106:403–413, 2006.

[2]Simopoulos, A. P. Nutrigenetics/nutrigenomics. *Annu Rev Public Health* 31:53–68, 2010.

[3]Fenech, M., El-Sohemy, A., Cahill, L., et al. Nutrigenetics and nutrigenomics: Viewpoints on the current status and applications in nutrition research and practice. *J Nutrigenet Nutrigenomics* 4:69–89, 2011.

[4]Sherwood, D. Nutrigenomics: Public concerns and commercial interests. *Agro Food Ind Hi Tech* 17:56–57, 2006.

[5]Lampe, J. W. For debate: Investment in nutrigenomics will advance the role of nutrition in public health. *Cancer Epidemiol Biomarkers Prev* 15:2329–2330, 2006.

[6]Wallace, H. Your diet tailored to your genes: Preventing diseases or misleading marketing? GeneWatch UK, January 2006.

What Determines Food Choices?

There are hundreds of food choices to make and hundreds of reasons for making them. Even though the foods we eat provide the nutrients and energy necessary to maintain health, the foods we choose are not necessarily determined by the nutrients these foods contain. Our food choices and food intake are affected not only by nutrient needs but also by what is available to us, where we live, what is within our budget and compatible with our lifestyle, what we like, what is culturally acceptable, what our emotional and psychological needs are, and what we think we should eat.

Michael Hanson/NG Image Collection

Figure 1.9 **Modern transportation increases food availability**
In the United States and other developed countries, many nonnative and seasonal foods, such as these grapes, are available year-round because they can be stored and shipped from other parts of the world.

Availability The food available to an individual or a population is affected by location, socioeconomic status, and health. In developing parts of the world, dietary choices are often limited to foods produced locally. Nutrients that are lacking in local foods will be lacking in the population's diet. This is less of a factor in more developed countries because the ability to store, transport, and process food allows year-round access to seasonal foods and foods grown and produced at distant locations (**Figure 1.9**).

VIDEO

Even if foods are available in the store, it doesn't mean that they are available to all individuals. Socioeconomic factors such as income level, living conditions, and lifestyle as well as education affect the types and amounts of foods that are available. Individuals with limited incomes can choose only the types and amounts of foods that they can afford. Individuals who don't own cars can purchase only what they can carry home. Those without refrigerators or stoves are limited in what foods can be prepared at home. And those who can't or don't have time to cook are limited to raw foods, prepared foods, and restaurant meals.

Health status also affects the availability of food. People who cannot carry heavy packages are limited in what they can purchase. People with food allergies, digestive problems, and dental disease are limited in the foods that are safe and comfortable for them to eat. People consuming special diets to manage disease conditions are limited to foods that meet their dietary prescriptions.

Cultural and Family Background Food preferences and eating habits are learned as part of each individual's family, cultural, national, and social background. They are among the oldest and most entrenched features of every culture. In Japan rice is the focus of the meal, whereas in Italy pasta is included with every meal. Curries characterize Indian cuisine, and we expect refried beans and tortillas when we go out for Mexican food. The foods we are exposed to as children influence what foods we buy and cook as adults. If your mother never served artichokes or Swiss chard, you may not consider eating them as an adult (**Figure 1.10**).

What would a birthday be without a cake, or Thanksgiving without a turkey? Each of us associates holidays such as Christmas, Easter, Passover, New Year's Day, and Kwanzaa with specific foods that are traditional in our family, religion, and culture. Seventh-Day Adventists are vegetarians; Jews and Muslims do not eat pork; Sikhs and Hindus do not eat beef. Even for those who choose not to observe religious dietary rules, habit may dictate many mealtime decisions. Jewish kosher laws prohibit the consumption of meat and milk in the same meal. Often Jews who do not follow kosher law as adults may choose not to serve milk at dinner because they never had it as children.

AFP/Getty Images, Inc.

Figure 1.10 **Culture dictates food acceptability**
If you grew up in Asia or Africa, you might consider grasshoppers, termites, or the silkworms in this Vietnamese market an acceptable food choice. Most Americans would not be willing to include insects as part of a meal.

Social Acceptability In addition to being part of our cultural heritage, food is the centerpiece of our everyday social interactions. We get together with friends for a meal or for a cup of coffee and dessert. The dinner table is often the focal point for communication within the family—a place where the experiences of the day are shared. Social events dictate our food choices in a number of ways. When invited to a friend's house for dinner, we may eat foods we don't like out of politeness to our hosts. We sometimes alter our food choices because of peer pressure. For example, an adolescent may feel that stopping for a cheeseburger or taco after school is an important part of being accepted by his or her peers.

Personal Preference We eat what we like. Tradition, religion, and social values may dictate what foods we consider appropriate, but personal preferences for taste, smell, appearance, and texture affect which foods we actually consume. How would you feel about giving up your favorite foods? Probably not too good, and you are not alone. Even though most Americans understand that nutrition is important to their health, many do not choose a healthy diet because they don't want to give up their favorite foods and they don't want to eat foods they don't like.[14] Personal convictions also affect food choices; a vegetarian would not choose a meal that contains meat, and an environmentalist may not buy foods packaged in nonrecyclable containers.

Psychological and Emotional Factors Food represents comfort, love, and security. We learn to associate food with these feelings as infants suckling while cradled in our mothers' arms. As children and as adults, comfort foods such as hot tea and chicken soup help us to feel better when we are sick, sad, tired, or lonely. We use food as a reward when we are good—A's on a report card are celebrated with an ice cream cone. We sometimes take away food as punishment—a child who misbehaves is sent to bed without dessert. We consider ourselves good when we eat healthy foods and bad when we order a decadent dessert. We celebrate milestones and reward life's accomplishments with food. Food may also be an expression and a moderator of mood and emotional states. When we are upset, some of us turn to chocolate or overeat in general, while others eat less or stop eating altogether.

Health Concerns Individuals' perceptions of what makes a healthy diet affect their food and nutrition choices. For example, some people may choose low-carbohydrate foods if they believe that these choices will help them lose weight. They may limit red meat intake to reduce their risk of heart disease, or they may purchase organically produced foods if they believe that reducing pesticide exposure will prevent illness.

How to Choose a Healthy Diet

A healthy diet is one that provides the right amount of energy to keep weight in the desirable range; the proper types and balance of carbohydrates, proteins, and fats; plenty of water; and sufficient but not excessive amounts of essential vitamins and minerals. A healthy diet is rich in nutrient-dense foods such as whole grains, fruits, vegetables, lean meats, seafood, nuts and seeds, and low-fat dairy products. **Nutrient density** is a measure of the nutrients a food provides compared to its energy content. Nutrient-dense foods contain more nutrients per calorie than do foods with a lower nutrient density. For example, a glass of low-fat milk is a more nutrient-dense choice than a soft drink or a glass of sweetened iced tea (**Figure 1.11a**). If a large proportion of your diet is made up of foods that are low in nutrient density, such as candy, soft drinks, snack foods, and baked goods, you could have a hard time meeting your nutrient needs without exceeding your calorie needs. By choosing nutrient-dense foods, you can meet all your nutrient needs and have calories left over for occasional treats that are lower in nutrients and higher in calories (**Figure 1.11b**). A healthy diet is based on variety, balance, and moderation. Using these principles, you can develop a personal strategy for making better choices and maintain your health for the long term.

nutrient density An evaluation of the nutrient content of a food in comparison to the calories it provides.

Eat a Variety of Foods No one food can provide all the nutrients the body needs for optimal health. Eating a variety of foods, however, helps ensure an adequate nutrient intake. Variety means choosing foods from different food groups—vegetables, grains, fruits, dairy products, and high-protein foods. Some of these groups are good sources of carbohydrate, fiber, and vitamins; others are rich in minerals and phytochemicals. All are important.

Variety also means choosing different foods from within each food group. For instance, if you choose three servings of vegetables a day and they are all carrots, it is unlikely that you will meet your nutrient needs. Carrots provide fiber and vitamin A but are a poor source of vitamin C. If instead you have carrots, peppers, and broccoli, you will be getting vitamin C along with more vitamin A, vitamin K, fiber, and phytochemicals than carrots alone would provide. Likewise, if you always choose red meat as a protein source, you will be missing out on the fiber in beans and the healthy fats in nuts and fish. Variety comes from choosing different foods not only each day but also each week and each season. If you had apples and

Figure 1.11 Nutrient density

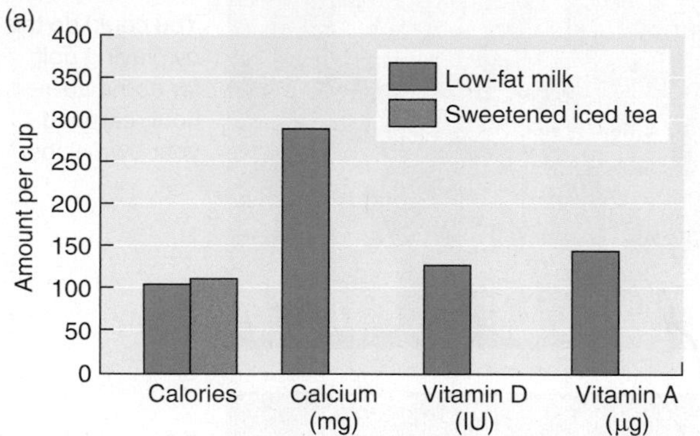

(a) Nutrient density is important in choosing a healthy diet. For example, low-fat milk is higher in nutrient density because it provides about the same number of calories per cup as sweetened iced tea, and also provides calcium, vitamin D, vitamin A, and other nutrients, including protein.

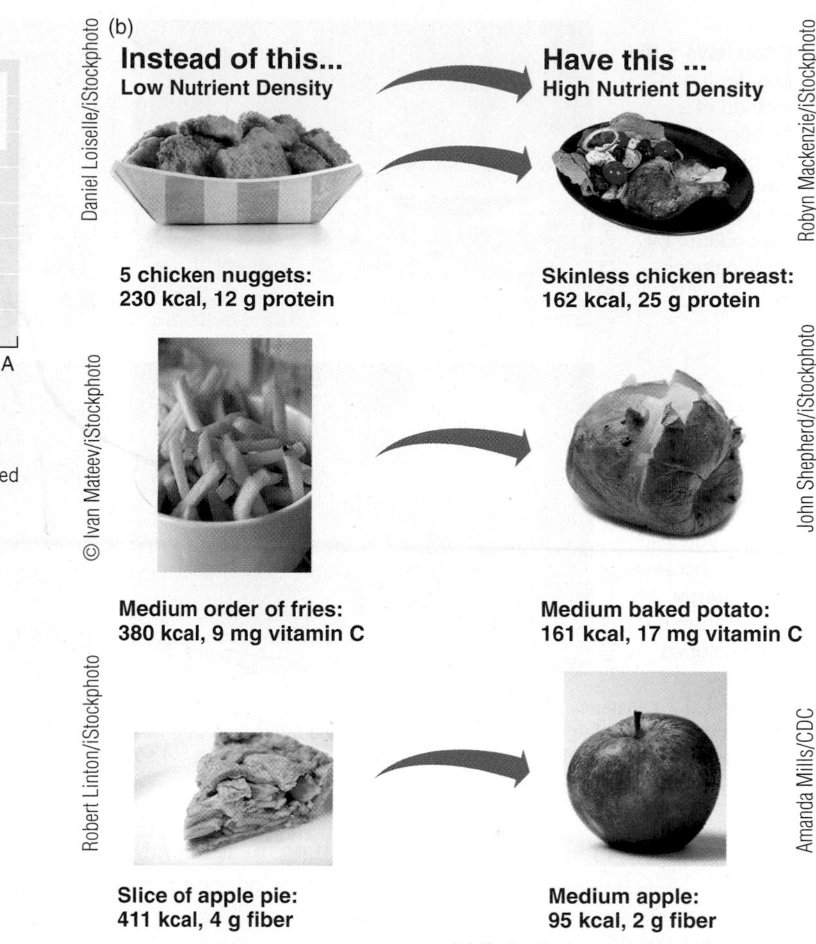

(b) Typically, less processed foods are higher in nutrient density. For example, a roasted chicken breast is more nutrient dense than chicken nuggets; a baked potato is more nutrient dense than French fries; and an apple is more nutrient dense than apple pie.

THINK CRITICALLY

The slice of apple pie has twice as much fiber as the apple, so why is the apple considered the more nutrient-dense choice?

grapes today, have blueberries and cantaloupe tomorrow. If you can't find tasty tomatoes in December, replace them with a winter vegetable like squash.

Choosing a varied diet is also important because there are interactions between different foods and nutrients. These interactions may be positive, enhancing nutrient utilization, or negative, inhibiting nutrient use. For example, consuming iron with orange juice enhances iron absorption, while consuming iron with milk may reduce its absorption. In a varied diet these interactions balance out. In addition, some foods may contain natural toxins or residues of pesticides, fertilizers, and other toxic substances (see Chapter 17). Choosing a variety of foods avoids an excess of any one of these substances. For example, tuna may contain traces of mercury, but as long as you don't eat tuna too often, you are unlikely to consume a toxic amount.

Balance Your Choices Choosing a healthy diet is a balancing act. Healthy eating doesn't mean giving up your favorite foods. There is no such thing as a good food or a bad food—only healthy diets and unhealthy diets. Any food can be part of a healthy diet as long as overall intake over the course of days, weeks, and months provides enough of all of the nutrients needed without excesses of any. When you choose a food, like white rice, that is lacking in fiber, balance this choice with one, like oatmeal, that provides lots of fiber. When you choose a food that is high in fat, like cheese, then balance that choice with a low-fat one, like a piece of fruit. Balancing your choices allows foods that would not usually be considered healthy choices to fit into an overall healthy diet. For example, baked goods, snack foods, and sodas should be balanced with nutrient-dense choices such as salads, fresh fruit, and low-fat dairy products. If your favorite meal is a burger, French fries, and a milkshake, enjoy it but balance it with asparagus, brown rice, and baked chicken at the next meal.

Balance involves mixing and matching foods in proportions that allow you to get enough of the nutrients you need and not too much of ones that might harm your health. A balanced diet provides plenty of whole grains, fruits, and vegetables. It contains enough but not too much of each of the vitamins and minerals, as well as protein, carbohydrate, fat, and water. It also balances the energy taken in with the energy used up in daily activities so body weight stays in the healthy range (**Figure 1.12**).

Figure 1.12 **Balance calories in with calories out**
To keep your weight stable, you need to burn the same number of calories as you consume. Extra calories you consume during the day can be balanced by increasing the calories you burn in physical activity.

Everything in Moderation Moderation means everything is okay, as long as you don't overdo it. Moderation means not consuming too much energy, too much fat, too much sugar, too much salt, or too much alcohol. It means watching your portion sizes. Have you ever sat down in front of the TV with a bag of chips, and before you knew it half the bag was gone? If you have, then you know how easy it is to let portion sizes get out of control. Choosing moderately will help you maintain a healthy weight and help prevent some of the chronic diseases like heart disease and cancer that are on the rise in the U.S. population. The fact that more Americans are obese than ever before demonstrates that we have not been practicing moderation when it comes to energy intake. Moderation will make it easier to balance your diet and will allow you to enjoy a greater variety of foods.

1.4 The Science Behind Nutrition

LEARNING OBJECTIVES

- List the steps of the scientific method.
- Compare the type of information obtained from epidemiology to that obtained from human intervention or laboratory studies.
- Discuss why animals and cells are used to study human nutrition.
- Discuss how the ethics of human and animal studies is monitored.

Nutrition, like all science, continues to develop as new discoveries provide clues to the right combination of nutrients needed for optimal health. As knowledge and technology advance, new nutrition principles are developed. Sometimes established beliefs and concepts must give way to new ideas, and recommendations change. Today more and more consumers are seeking information about nutrition and how to improve their diets.[14] But they may find this frustrating because the experts seem to change their minds so often. One day consumers are told margarine is better for them than butter; the next day a report says that it is just as bad. Developing an understanding of the process of science and how it is used to study the relationship between nutrition and health can help consumers make wise nutrition decisions, whether they involve what to have for breakfast or whether a headline about vitamin E supplements is true.

The Scientific Method

Advances in nutrition are made using the **scientific method**. The scientific method offers a systematic, unbiased approach to evaluating the relationships among food, nutrients, and health. The first step of the scientific method is to make an observation and ask questions about the observation. The next step is to propose a **hypothesis**, or explanation for the observation. Once a hypothesis has been proposed, experiments can be designed to test it. The experiments must provide objective results that can be measured and repeated. If the experimental results do not prove the hypothesis to be wrong, a **theory**, or a scientific explanation based on experimentation, can be established (**Figure 1.13**). Scientific theories are accepted only as long as they cannot be disproved and continue to be supported by all new evidence that accumulates. Even a theory that has been accepted by the scientific community for years can be proved wrong.

scientific method The general approach of science that is used to explain observations about the world around us.

hypothesis An educated guess made to explain an observation or to answer a question.

theory An explanation based on scientific study and reasoning.

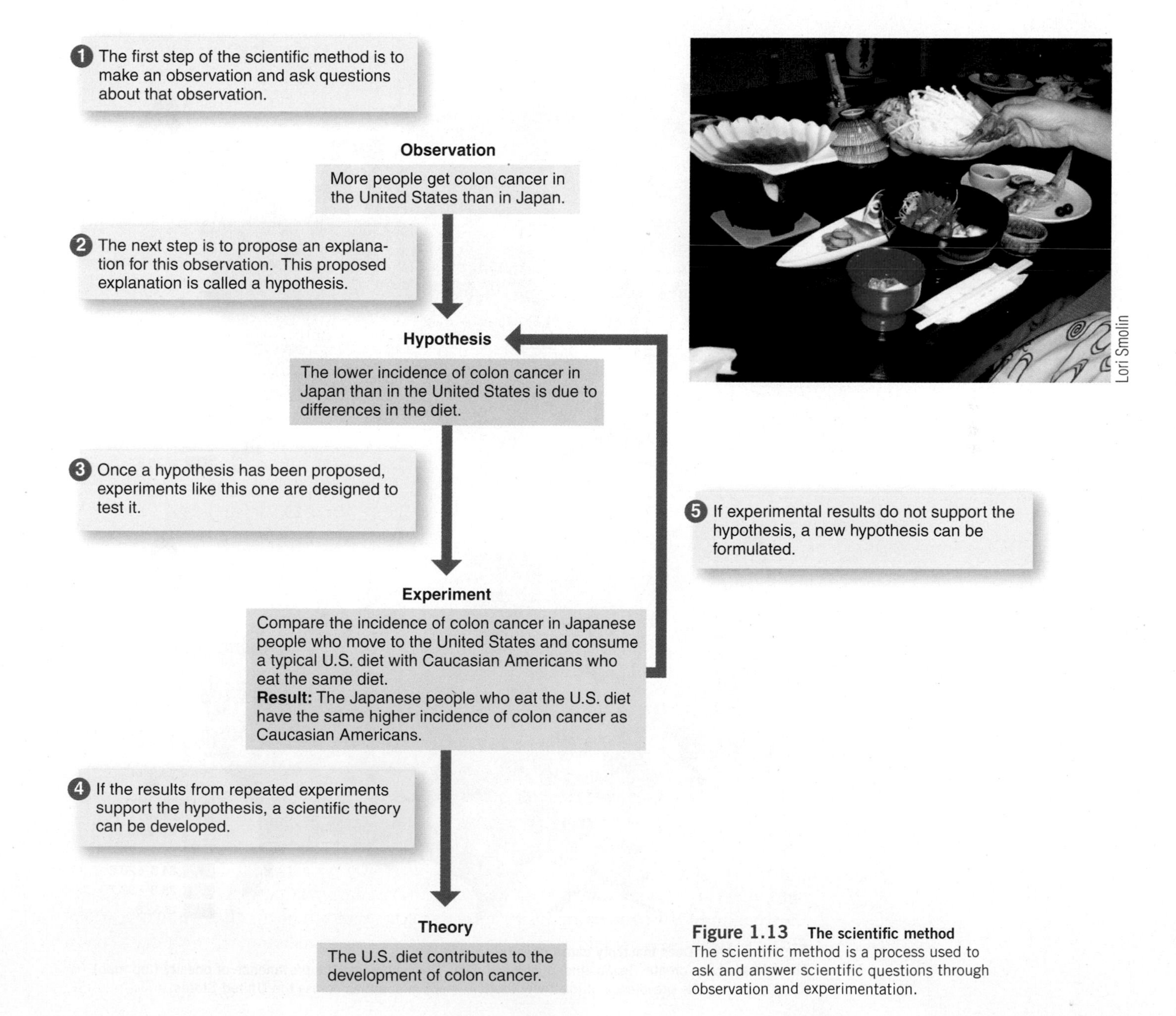

Figure 1.13 The scientific method
The scientific method is a process used to ask and answer scientific questions through observation and experimentation.

How Scientists Study Nutrition

Nutrition research studies are done to determine nutrient requirements, to learn more about the metabolism of nutrients, and to understand the role of nutrition in health and disease. Perfect tools do not exist for addressing all these questions. However, there are many types of research studies that can be useful in understanding the relationships between humans and their nutrient intake.

Observational Studies Some of our nutrition knowledge has been obtained by observing relationships between diet and health in different populations throughout the world. This study of diet, health, and disease patterns is called **epidemiology**. Epidemiology does not determine cause and effect relationships—it just identifies patterns (**Figure 1.14**). For instance, by comparing diets in different countries with the incidence of cancer, scientists were able to identify an association, or **correlation**, between diets high in fruits and vegetables and a lower incidence of cancer. Some epidemiological studies collect data from a cross section of the population at one point in time, whereas others collect data from the same individuals over a long period of time (see **Science Applied: How Epidemiology Led to Dietary Recommendations for Heart Disease**).

epidemiology The study of the interrelationships between health and disease and other factors in the environment or lifestyle of different populations.

correlation Two or more factors occurring together.

THINK CRITICALLY

Does the data from these maps mean that inactivity is the cause of obesity? Will everyone who is inactive be obese?

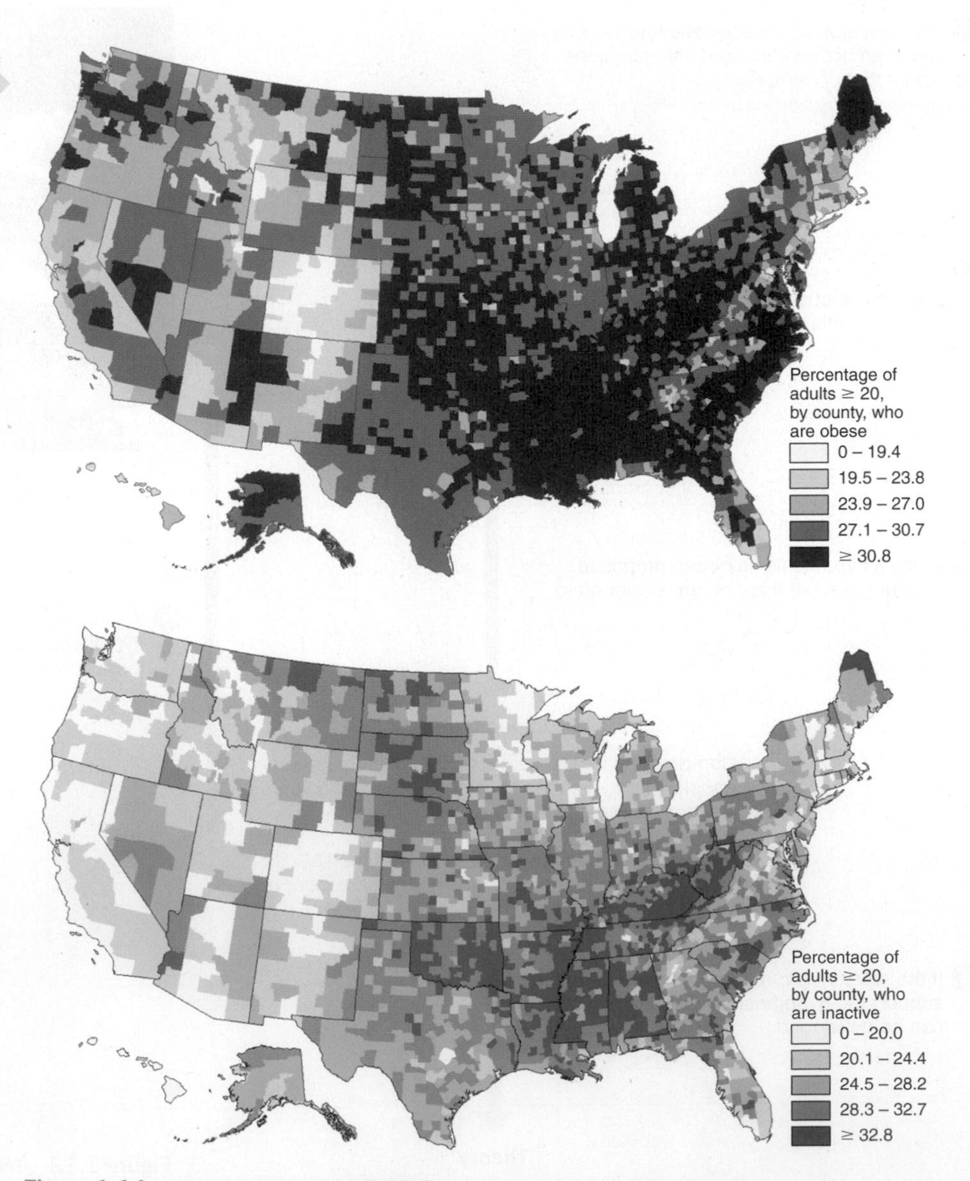

Figure 1.14 Does inactivity cause obesity?
These maps were created using epidemiological data. They show that the prevanence of obesity (top map) roughly mirrors the prevalence of inactivity (bottom map) in counties across the United States.

SCIENCE APPLIED

How Epidemiology Led to Dietary Recommendations for Heart Disease

(Elena Schweitzer/Shutterstock) Victoriano Izquierdo/Getty Images, Inc. (DAJ/Getty Images, Inc.)

As early as the 1930s, scientists and physicians noticed that the incidence of heart disease was different in different countries and varied with social class and occupation.[1] In the 1940s Professor Ancel Keys observed that well-fed American businessmen had high rates of heart disease, while the rate of heart disease had decreased in post-war Europe where food supplies were low. This led Keys to hypothesize that diet affected the risk of heart disease and to launch a series of epidemiological studies that changed the way we manage heart disease.

THE SCIENCE

To test the diet and heart disease hypothesis, Keys began a study that collected annual physiological and lifestyle data on 500 healthy businessmen in Minnesota. He knew that over time some would develop heart disease and others would not. He hoped that comparing the data from afflicted men with that from healthy men would identify which characteristics were related to the development of the disease.[2] Key's study was what we call a prospective study: one that follows people over time and relates outcomes, such as the development of heart disease, to other factors such as body weight, blood cholesterol, or diet. The aim was to identify how those who developed heart disease differed from those who escaped it.

The Minnesota businessmen study provided Keys with the groundwork for a much larger epidemiological study called the Seven Countries Study, which was designed to compare characteristics among populations across a wide spectrum of diet and lifestyle. Between 1958 and 1964 the Seven Countries Study enrolled 12,763 men, ages 40 to 59 years, from 16 different regions within 7 countries on 3 continents. The study evaluated health status, dietary intake, body weight, blood pressure, blood cholesterol level, and other health-related parameters at regular intervals. Patterns began to emerge. In northern European countries, the diet was high in dairy products; in the United States, it was high in meat; in southern Europe, it was high in vegetables, legumes, fish, and wine; and in Japan, it was high in cereals, soy products, and fish.[3] After 10 years, 1512 of the study participants were dead—413 of them from coronary heart disease[4]—but mortality differed strikingly with location. The island of Crete had only one coronary death out of 686 men studied, whereas eastern Finland had 78 coronary deaths among 817 participants.[5]

The coronary death rate was correlated with the average percentage of calories from saturated fat. In places such as Japan and the Greek islands, where the diet was low in saturated fat, blood cholesterol levels were lower, as was the risk of dying

Pixtal/SuperStock

The vegetables, fruit, fish, and olive oil shown here are plentiful in the diet of countries around the Mediterranean Sea—where the Seven Countries Study found the incidence of coronary heart disease to be low. Wine was routinely, but not excessively, consumed with meals. This dietary pattern, now known as the "Mediterranean diet," is one of the dietary patterns currently recommended for reducing the risk of heart disease.

from heart disease. In countries such as Finland and the United States, where the diet was higher in saturated fat, blood cholesterol levels were higher, as was the incidence of heart disease. The results showed that blood cholesterol was strongly correlated with coronary heart disease deaths both for populations and for individuals. Some of the difference in the incidence of heart disease between countries was not explained by risk factors, but within a population risk factors such as diet, blood cholesterol, blood pressure, and cigarette smoking did predict risk.

THE APPLICATION

The Seven Countries Study and other epidemiological investigations of heart disease have established the major risk factors for heart disease. These findings have helped develop individual dietary and lifestyle recommendations to treat people at high risk for heart disease as well as population-wide guidelines to promote public health. Heart disease is still the leading cause of death in the United States, but since 1950, age-adjusted death rates from cardiovascular disease have declined 60%. This is considered one of the most important public health achievements of the 20th century. The decline is due in part to fewer people smoking, better health care, and the use of antihypertensive and cholesterol-lowering drugs, but changes in diet based on the results of Seven Countries Study and other epidemiological studies have also had an impact.[6]

[1]Epstein, F. H., Cardiovascular disease epidemiology: A journey from the past into the future. *Circulation* 93:1755–1764, 1996.

[2]Blackburn, H. Cardiovascular disease epidemiology.

[3]Menotti, A., Kromhout, D., Blackburn, H., et al. Food intake patterns and 25-year mortality from coronary heart disease: Cross-cultural correlations in the Seven Countries Study. *Eur J Epidemiol* 15:507–515, 1999.

[4]The diet and all-causes death rate in the Seven Countries Study. *Lancet* 2:58–61, 1981.

[5]Keys, A. *Seven Countries: A Multivariate Analysis of Death and Coronary Heart Disease.* Commonwealth Fund. Cambridge, MA: Harvard University Press, 1980.

[6]Achievements in Public Health, 1900–1999: Decline in Deaths from Heart Disease and Stroke—United States, 1900–1999. *MMWR* 48:649–656, 1999.

Foodcollection/Getty Images

When explorers began making long sea voyages in the 1400s, it was observed that after about 12 weeks at sea the sailors began getting the symptoms of the disease scurvy. We know today that scurvy is due to a deficiency of vitamin C, but at the time Scottish physician James Lind was looking for a cure for the disease, vitamins had not even been discovered. Lind hypothesized that scurvy was caused by lack of acid in the body.

In 1747, Lind set up an experiment involving 12 sailors suffering from scurvy. He divided them into six groups. They were all fed the same diet that today we know was lacking in vitamin C, also known as ascorbic acid. He supplemented each group with a different substance. Most were acids, such as vinegar, dilute sulfuric acid, and cider. One group received lemons and oranges, which are high in ascorbic acid. After only a week this group was much improved, while sailors in the other groups remained ill and died.

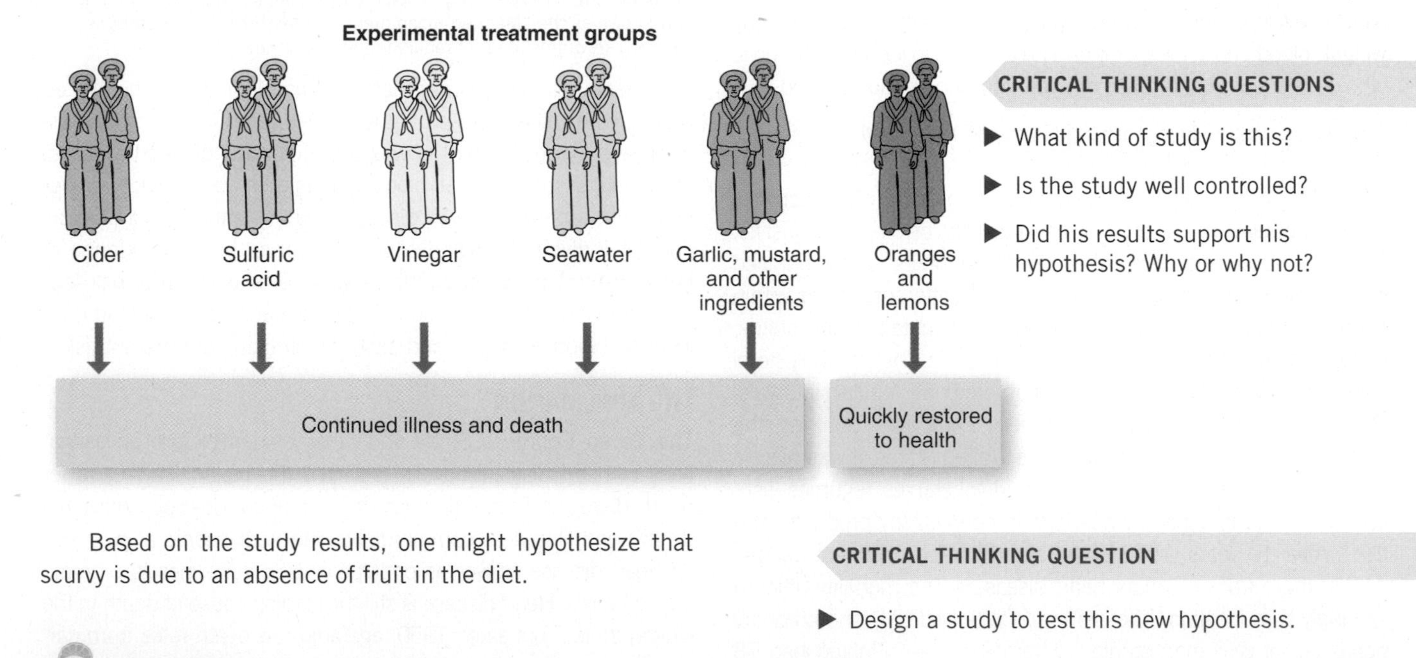

CRITICAL THINKING QUESTIONS

- What kind of study is this?
- Is the study well controlled?
- Did his results support his hypothesis? Why or why not?

Based on the study results, one might hypothesize that scurvy is due to an absence of fruit in the diet.

CRITICAL THINKING QUESTION

- Design a study to test this new hypothesis.

iProfile Use iProfile to find vegetables that are good sources of vitamin C.

case-control study A type of observational study that compares individuals with a particular condition under study with individuals of the same age, sex, and background who do not have the condition.

Case-control studies are a type of epidemiological study that compares individuals with a particular condition to similar individuals without the condition. For example, a case-control study of colon cancer might include a comparison of the dietary intake of a 45-year-old African American man with colon cancer to a man of the same age and ethnic background who is free of the disease. If a pattern, such as a higher fat intake among the cancer patients, is found in comparing cases to controls, a hypothesis can be proposed. Hypotheses based on case-control studies and other types of epidemiology must then be tested by controlled intervention and laboratory studies.

Human Intervention Studies The observations and hypotheses that arise from epidemiology can be tested using **human intervention studies**, often referred to as **clinical trials**. This type of experiment actively intervenes in the lives of a population and examines the effect of this intervention (see **Critical Thinking: Early Science**). Nutrition intervention studies generally explore the effects of altering people's diets. For example, if it is determined by epidemiology that populations eating a diet low in saturated fat have a lower incidence of heart disease, an intervention trial may be designed with an **experimental group** that consumes a diet lower in saturated fat than is typical in the population and a **control group** that consumes the typical diet. The groups can be monitored to see if the dietary intervention affects the incidence of heart disease over the long term.

human intervention study or **clinical trial** A study of a population in which there is an experimental manipulation of some members of the population; observations and measurements are made to determine the effects of this manipulation.

experimental group In a scientific experiment, the group of participants who undergo the treatment being tested.

control group In a scientific experiment, the group of participants used as a basis of comparison. They are similar to the participants in the experimental group but do not receive the treatment being tested.

Laboratory Studies Laboratory studies are conducted in research facilities such as hospitals and universities. They are used to learn more about how nutrients function and to evaluate the relationships among nutrient intake, levels of nutrients in the body, and health. They may study nutrient requirements and functions in whole organisms—either humans or animals—or they may focus on nutrient functions at the cellular, biochemical, or molecular level.

How Whole Organisms Are Used to Study Nutrition Many nutrition studies are done by feeding a specific diet to a person or animal and monitoring the physiological effects of that diet. **Depletion-repletion studies** are a classic method for studying the functions of nutrients and estimating the requirement for a particular nutrient. This type of study involves depleting a nutrient by feeding experimental subjects a diet devoid of that nutrient. After a period of time, if the nutrient is essential, symptoms of a deficiency will develop. The symptoms provide information on how the nutrient functions in the body. The nutrient is then added back to the diet, or repleted, until the symptoms are reversed. The requirement for that nutrient is the amount needed to reverse or prevent the deficiency symptoms. An example of a depletion-repletion study might be to feed animals a diet devoid of vitamin A for several weeks and examine the deficiency symptoms that appear. Then, if vitamin A is incrementally added back to the diet, the amount that prevents deficiency symptoms can be identified.

depletion-repletion study A study that feeds a diet devoid of a nutrient until signs of deficiency appear and then adds the nutrient back to the diet to a level at which symptoms disappear.

Another method for determining nutrient functions and requirements is to compare the intake of a nutrient with its excretion. This type of study is known as a **balance study**. If more of a nutrient is consumed than is excreted, balance is positive and it is assumed that the nutrient is being used or stored by the body. If more of the nutrient is excreted than is consumed, balance is negative, indicating that some is being lost from the body. When the amount consumed equals the amount lost, the body is neither gaining nor losing that nutrient and is said to be in a steady state or in balance (**Figure 1.15**). By varying the amount of a nutrient consumed and then measuring the amount excreted, it is possible to determine the minimum amount of that nutrient needed to replace body losses. This type of study can be used to determine protein requirements because protein is not stored in the body. It is not useful for determining the requirements for nutrients such as fat and iron that are stored when an excess is available.

balance study A study that compares the total amount of a nutrient that enters the body with the total amount that leaves the body.

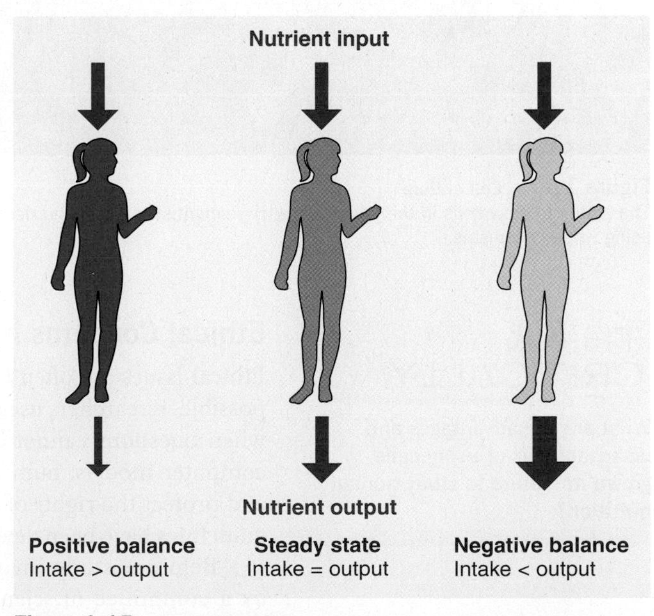

Figure 1.15 Balance studies
Nutrient balance studies compare the amount of a nutrient that is consumed with the amount excreted to determine whether the amount of the nutrient in the body is increasing (positive balance), remaining constant (balance, or steady state), or decreasing (negative balance).

Why Animals Are Used to Study Human Nutrition Ideally, studies of human nutrition would be done in humans. However, because studying humans is costly, time consuming, inconvenient for subjects, and in some cases impossible for ethical reasons, many studies are done using experimental animals.

An ideal animal model is one with metabolic and digestive processes similar to humans. For example, cows are rarely used in human nutrition research because they digest their food in four stomach-like chambers as opposed to a single stomach.

Lori Smolin

Figure 1.16 **Choosing animals to model human nutrition**
Guinea pigs are used to study vitamin C requirements and functions because the guinea pig is one of the few animals, other than humans, that cannot make vitamin C in its body.

Keith Weller/Courtesy USDA

Figure 1.17 **Cell culture**
The ability to grow cells in the laboratory allows scientists to study nutrients without using whole organisms.

Pigs, on the other hand, are a good model because they digest food in a manner similar to that of humans. In addition to digestion and metabolism, factors such as cost and time must be considered. Pigs and other large animals are expensive to use, and they take a long time to develop nutrient deficiencies. Smaller laboratory animals, such as rats and mice, are therefore the most common experimental animals. They are inexpensive, have short life spans, reproduce quickly, and show the effects of nutritional changes rapidly. Their food intake can be easily controlled, and their excretions can be measured accurately using special cages. Even when researchers use small animals, the species of animal must be carefully chosen (**Figure 1.16**). For example, rats are more resistant to heart disease than are humans, so they are not a good model for studying the effect of diet on heart disease. Rabbits, on the other hand, do develop heart disease and can be used to study diet–heart disease relationships. Even the best animal model is not the same as a human, and care must be taken when extrapolating the results to the human population. For example, a study that uses rats to show that a calcium supplement increases bone density can hypothesize, but not conclude, that the supplement will have the same effect in humans.

Studies Using Cells Another alternative to conducting studies in humans is to study cells either extracted from humans or animals or grown in the laboratory (**Figure 1.17**). Biochemistry can be used to study how nutrients are used to provide energy and how they regulate biochemical reactions in cells. Molecular biology can be used to study how genes regulate cell functions. The types and amounts of nutrients available to cells can affect the action of genes. For example, vitamin A can directly activate certain genes. Knowledge gained from biochemical and molecular biological research can be used to study nutrition-related conditions that affect the entire organism.

THINK CRITICALLY

What are the advantages and disadvantages of using cells grown in culture to study human nutrition?

Ethical Concerns in Scientific Study

Ethical issues are often raised in the process of conducting nutrition research. Whenever possible, researchers use alternatives to human subjects or experimental animals. However, when questions cannot be answered using alternatives such as cells grown in culture or computer models, human and animal experimentation is still necessary. To avoid harm and protect the rights of humans and animals used in experimental research, government guidelines have been developed.

Before a study involving human subjects can be conducted, it must be reviewed by a committee of scientists and nonscientists to ensure that the rights of the subjects are respected and that the risk of physical, social, and psychological injury is balanced against the potential benefit of the research. Before subjects participate in a study, an oral and written explanation of the purpose of the research, the procedures used, and the possible risks and benefits are provided. Those who choose to participate must then sign a consent stating exactly what they have agreed to do. Signing a consent form doesn't mean subjects must complete the study if it turns out to be more than they bargained for—subjects can leave a study at any time. This informed consent process is part of the strict safety and ethical regulations that must be followed when research involves human subjects. These regulations protect subjects but limit the type of study that can be done on humans. For example, much of what we know today about the effects of starvation in humans was determined during World War II by conducting depletion-repletion studies using conscientious objectors as experimental subjects. These subjects were monitored physically and psychologically while they were starved and then re-fed. These individuals experienced some level of suffering during the trials and risked longer-lasting physical and psychological harm. It is unlikely that this study would be approved if researchers wanted to repeat it today.

As with experiments involving humans, the federal government mandates that panels of scientists review experiments that propose to use animals. These panels consider whether the need for animals is justified and whether all precautions will be taken to avoid pain and suffering. Animal housing and handling are strictly regulated, and a violation of these guidelines can close a research facility.

The development of the techniques of molecular biology has given rise to ethical issues regarding the manipulation of genes. Guidelines for manipulating genes have been developed and are discussed in the **Focus on Biotechnology**.

1.5 Evaluating Nutrition Information

LEARNING OBJECTIVES

- Discuss why individual testimonies are not considered reliable sources of information.
- Name three points to consider when evaluating the reliability of nutritional information.
- Discuss the components of a well-designed study including experimental controls, control groups, and placebos.

We are bombarded with nutrition information. Some of what we hear is accurate and based on science and some of it is incorrect or exaggerated to sell products or make news headlines more enticing: oat bran lowers cholesterol, antioxidants prevent cancer, low-carb diets promote weight loss, vitamin C cures the common cold, vitamin E slows aging. Sifting through this information and distinguishing the useful from the useless can be overwhelming. Just as scientists use the scientific method to expand their understanding of the world around us, each of us can use an understanding of how science is done to evaluate nutrition claims by asking the questions discussed below and summarized in (**Table 1.1**).

Does It Make Sense?

The first question to ask yourself when evaluating nutrition information is: does the claim being made make sense? Some claims, we know, are too outrageous to be true or safe. For example, the hypothetical advertisement for StayWell illustrated on the next page states that this product will make illness a thing of the past (**Figure 1.18**). This is certainly appealing, but it is hard to believe, and common sense should tell you that it is too good to be true. The claim that StayWell will reduce cold symptoms, however, is not so outrageous. Common sense can also help determine whether health and nutrition information poses a risk. For example, a report that recommends you take supplements providing

TABLE 1.1 Sorting Out Nutrition Information

• **Does It Make Sense?**
○ **Is it too outrageous to believe?** If it is, disregard it.
○ **Does it pose a risk?** If it doesn't sound safe, don't do it.
• **What's the Source?**
○ **Is it selling something?** Many claims are made to sell products rather than present unbiased nutrition information.
○ **Is it someone's opinion or personal story?** Anecdotal information can be applied only to the person telling the story.
○ **Who can you believe?** Recommendations made by the government, universities, and nonprofit organizations are based on science and designed to provide information to improve health.
• **Is It Based on Good Science?**
○ **Is the study well controlled?** Have experimental controls been used so that each variable studied can be compared with a known situation?
○ **Were participants and researchers blinded to the treatment?** If the subjects or researchers know which group is receiving treatment, their beliefs and biases may alter the outcome of the study.
○ **Was the information interpreted accurately?** Compare news reports and advertisements to the information in peer-reviewed studies. Has the importance of the study been exaggerated to sell a product or make a headline more appealing? If the study was done in animals, can the results be applied to humans?
• **Has It Stood the Test of Time?**
○ **Is the study the first to support a particular finding?** If it is, wait before changing your diet based on the results.
○ **Has the finding been shown repeatedly in different studies over a period of years?** If it has, it will become the basis of reliable nutrition recommendations.

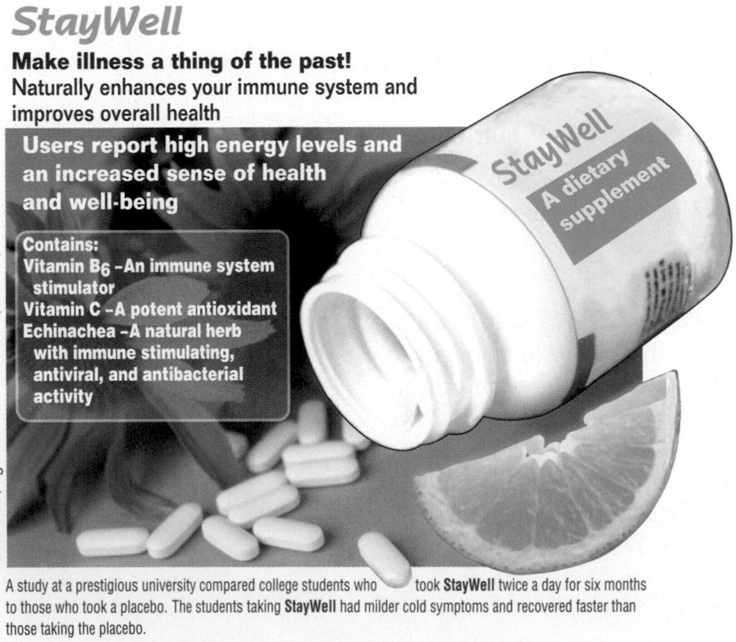

©Hernera/Age Fotostock America, Inc.

Figure 1.18 What's behind the claims? This hypothetical supplement advertisement illustrates the types of nutrition claims that consumers must be prepared to evaluate.

1000% of the recommended amount of a vitamin or mineral should be viewed skeptically because high levels can be toxic and are not found in nature.

What's the Source?

If the claim seems reasonable, look to see where it came from. Was it a personal testimony, a government recommendation, or advice from a health professional? Is it in a news story or an advertising promotion? Is it on television, in a magazine, or on a Web page?

Is It Selling Something? If a person or company will profit from the information presented, you should be wary. Information presented in newspapers and magazines and on television may be biased or exaggerated because it must help sell magazines or boost ratings. Consider whether the claim is making a magazine cover or newspaper headline more appealing. Claims that are part of an advertisement for a food product or dietary supplement should be viewed skeptically because advertisements are designed to increase product sales and the company stands to profit from your belief in that claim. For example, in an advertisement for a vitamin E supplement, a company may claim that vitamin E increases longevity even if the research supporting this claim was done in rats, not humans; used higher amounts of vitamin E than are provided by the supplement; and caused only a very small change in life span. Information on the Internet is also likely to be biased toward a product or service if it comes from a site where you can buy the product (.com sites). Understanding food labels can help evaluate nutrition claims made about packaged foods (see **Off the Label: Look beyond the Banner on the Label**).

OFF THE LABEL

Look Beyond the Banner on the Label

© iStockphoto

Food labels are designed to provide information that can help you select a healthy diet. Although they are standardized and follow federal guidelines, they can still be confusing and even misleading. You need to consider all the information on the label and remember that each food choice is only one part of your overall diet.

Many food labels highlight information about the amounts of individual nutrients. But, because no single nutrient makes a food good or bad for you, you must look beyond these descriptors to understand the overall contribution that the food makes to your diet. For example, chocolate cookies labeled "fat free" may not be your best choice if you are trying to reduce your sugar intake or increase your fiber consumption. Descriptors such as "fresh" and "healthy" imply that the product is better for you than a similar alternative. A food labeled "fresh" may sound appealing, but actually any raw food that has not been frozen, heat processed, or otherwise preserved can be labeled fresh. The term doesn't provide any information about the nutrient content of the food or how long it took the product to travel from the farm to the grocery store shelf. "Healthy" implies that a product is wholesome and nutritious. In fact, to use the term "healthy," a food must be low in fat and saturated fat, contain no more than 360 mg of sodium and 60 mg of cholesterol per serving, and be a good source of one or more important nutrients. Although all these qualities are indeed part of a healthy diet, foods that fit this definition are not necessarily the basis for a healthy diet. For instance, many fruit drinks fit the labeling definition of healthy. They are low in fat, saturated fat, cholesterol, and sodium, and they supply at least 10% of the recommended intake for vitamin C. But they are a good choice only in limited quantities because they are high in added refined sugars and contain few other nutrients. Likewise, a food that doesn't meet the labeling definition of healthy is not necessarily a poor choice. Vegetable soup, for example, contains more sodium than the definition of healthy will allow, but if the rest of your diet is not high in sodium, the soup can be a healthy choice.

Healthy-sounding product names can also be misleading. Product names must comply with legal definitions, but they don't have to make sense to consumers. Unless you have memorized the U.S. Department of Agriculture (USDA) and Food and Drug Administration (FDA) labeling regulations, you have no way of knowing how much beef is needed for a product to be called a beef enchilada, for example, or how much chicken must be in chicken soup and how much fruit is in a fruit roll-up. "Lasagna with meat sauce" must be 6% meat, but "lasagna with meat and sauce" must be 12% meat.

To get the whole picture, then, you need to look beyond the healthy-sounding descriptors and the product name. Read the whole label, which must include the food's nutrient content and information on how it fits into the diet as a whole (see Chapter 2 and the Off the Label boxes throughout this book).

Healthy diets are low in sugar, but foods such as these are not the basis for a healthy diet.

Jose Luis Pelaez/Getty Images, Inc.

Before

© Haveseen/iStockphoto

After

Figure 1.19 **Anecdotal claims** Weight-loss product advertisements often show before and after photos of people who have successfully lost weight using the product. These photos and personal stories of success are anecdotal evidence because they do not arise from controlled experiments that are evaluated scientifically. Therefore, it cannot be assumed that the product will produce the same results for you or anyone else.

Is It Someone's Opinion or Personal Story? Personal testimonies or opinions, referred to as anecdotal evidence, are not a source of reliable information that can be applied to the general public (**Figure 1.19**). Reliable information comes from scientific experiments providing measurable data that can be quantified and repeated. Measurements of body weight and blood pressure are examples of parameters that can be quantified reliably. In contrast, the claims in Figure 1.18 that StayWell improves energy levels and overall sense of health and well-being are anecdotes based on the individual experiences of supplement users rather than on measured parameters. In order to be useful in science, feelings and opinions must be quantified using standardized questionnaires.

Who Can You Believe? There are a number of sources you can rely on for credible nutrition information. Government recommendations regarding healthy dietary practices are developed by committees of scientists who interpret the latest well-conducted research studies and use their conclusions to develop recommendations for the population as a whole. The government provides information about food safety and recommendations on food choices and the amounts of specific nutrients needed to avoid nutrient deficiencies and excesses and to prevent chronic diseases. These recommendations are used to develop food-labeling standards and are the basis for public health policies and programs. They are published in pamphlets and brochures and are available on Web sites designed for consumers.

Nonprofit organizations such as the American Dietetic Association, the American Medical Association, and the American Institute for Cancer Research are also a good source of nutrition information. The purpose of the information they provide to the public is to improve health. Reports that come from universities are supported by research and are also a reliable place to look for information. Many universities provide information that targets the general public, and university research studies are usually published in respected journals and are well scrutinized.

To help you evaluate the credibility of information in an article in print or posted on a Web site, check the author's credentials. Where does the author work? Does this person have a degree in nutrition or medicine? Although "nutritionists" and "nutrition counselors" may provide accurate information, these terms are not legally defined and are used by a wide range of people, from college professors with doctoral degrees from reputable universities to health-food-store clerks with no formal training. One reliable source of nutrition information is registered dietitians. Registered dietitians (RDs) are nutrition professionals who have completed a four-year college degree in a nutrition-related field and have met established criteria to certify them in providing nutrition counseling.

Is It Based on Good Science?

Most of the information we get is based on research studies. For some nutrition claims, not enough information is given to evaluate the validity of the studies on which they are based. Others, however, do provide the details of how a study was done. For example, the StayWell ad describes a university study that supposedly supports the advertisers' claims. If this type of information is available, ask yourself if the study was well controlled and if the results were interpreted accurately.

variable A factor or condition that is changed in an experimental setting.

Is the Study Well Controlled? Experimental controls are used to ensure that each factor or **variable** studied can be compared with a known situation. A control group acts as a standard of comparison for the treatment being tested. A control group is treated in the same way as the experimental groups except no experimental treatment is implemented. For example, the university study described in the StayWell ad compares the severity of cold symptoms and the cold recovery time of college students who took the StayWell supplement (experimental group) with the same parameters in a group of students of similar age, gender, health status, and dietary and exercise habits who did not take the supplement (control group).

In the StayWell experiment, a **placebo** was used in order to make the control and experimental groups indistinguishable. A placebo is identical in appearance to the actual treatment but has no therapeutic value. Using a placebo prevents participants in the experiment from knowing whether or not they are receiving the actual supplement. When the subjects do not know which treatment they are receiving, the study is called a **single-blind study**. Using a placebo in a single-blind study helps to prevent the expectations of subjects from biasing the results. For example, if the students think they are taking a supplement that reduces cold symptoms, they may develop a positive outlook that makes their cold symptoms milder. Errors can also occur if investigator's expectations bias the results or the interpretation of the data. This type of error can be avoided by designing a **double-blind study** in which neither the subjects nor the investigators know who is in which group until after the results have been analyzed.

placebo A fake medicine or supplement that is indistinguishable in appearance from the real thing. It is used to disguise the control and experimental groups in an experiment.

single-blind study An experiment in which either the study participants or the researchers are unaware of which participants are in a control or an experimental group.

double-blind study An experiment in which neither the study participants nor the researchers know which participants are in a control or an experimental group.

Was the Information Interpreted Accurately? In science, the interpretation of results is as important as the way studies are done. Even well-designed, carefully executed experiments can be a source of misinformation if the experimental results are interpreted incorrectly or if the implications of the results are exaggerated. For example, the headline in Figure 1.18 states that StayWell improves overall health. However, the study cited investigates only cold symptoms; overall health is not addressed. Also, StayWell is tested using college students as experimental subjects, so an accurate interpretation of the results would not assume that StayWell would have the same effect in elderly people or young children.

One way to ensure that experiments are correctly interpreted is to use a peer-review system. Most scientific journals require that prior to publication, two or three experts (who did not take part in the research that is being evaluated) agree that the experiment under review was well conducted and that the results were interpreted fairly. For example, reviewers will assess whether or not the study included enough subjects to demonstrate the effect of the experimental treatment. To claim a treatment has a particular effect, an experiment must show that the treatment being tested causes a result to occur more frequently than it would occur by chance. More subjects are needed to demonstrate an effect that frequently occurs by chance. For example, if many people only have mild symptoms when they get a cold, then an experiment to see if StayWell reduces cold symptoms would need to include many subjects to demonstrate an effect. However, if everyone got very ill with a cold, only a few subjects would be needed to show that StayWell reduces cold symptoms.

Peer-reviewed journals that publish nutrition-related articles include the *American Journal of Clinical Nutrition*, the *Journal of Nutrition*, the *Journal of the American Dietetic Association*, the *New England Journal of Medicine*, and the *International Journal of Sport Nutrition*.

Has It Stood the Test of Time?

Often the results of a new scientific study are in the morning news the same day they are published in a peer-reviewed journal. Sometimes this information is correct, but a single study is never enough to develop a reliable theory. Results need to be reproducible before they can be used as the basis for nutrition decisions or used to make dietary recommendations. Headlines based on a single study should therefore be viewed skeptically. The information may be accurate, but there is no way to know because there has not been time to repeat the work and reaffirm the conclusions. If, for example, someone has found the secret to easy weight loss, the information will undoubtedly appear again if the finding is valid. If it is not, it will fade away with all the other weight-loss cures that have come and gone.

OUTCOME

Mauritius Images/Age Fotostock America, Inc.

© Alejandro Rivera/iStockphoto

Kaitlyn found it hard to eat a healthy diet during the first semester of her freshman year. At home, her favorite meal of burgers and fries was an occasional treat. Having these available every night made it hard for her to vary her diet. She knew these weren't a great everyday choice, but she didn't know how to choose a healthy diet. Now that she's read this chapter, she knows that a healthy diet includes a variety of foods. She aims for an assortment of choices from within each of the food groups. She recognizes that eating a healthy diet involves using moderation, so she doesn't consume too much of any one food or food group. She now heads for the salad bar and tries some different hot entrees. She makes a point of getting to the store where she can buy yogurt and fruit to keep in her dorm refrigerator. Kaitlyn now knows how to balance less healthy choices, like study snacks, with healthier foods at other times of the day. She is gaining confidence in her ability to balance her food intake with her activity. In addition, her understanding of the scientific method and how to evaluate nutrition information has given her the tools she needs to make decisions about following fads and using dietary supplements. **Now, in the middle of her second semester, Kaitlyn is maintaining her weight and using the principles of variety, balance, and moderation to choose a nutrient-dense diet that minimizes her risks of developing heart disease and high blood pressure.**

APPLICATIONS

ASSESSING YOUR DIET

1. **How healthy is your diet?**
   - **a.** How many different vegetables and fruits did you eat today? How about this week? If you average fewer than five a day, make some suggestions that would increase the amount and variety of fruits and vegetables in your diet.
   - **b.** If you had a treat such as a doughnut or an extravagant dessert, did you balance it with some healthier choices at other times during the day or the next day? Suggest a healthy choice you could have to balance two of your favorite treats.
   - **c.** Do you order large portions? Use iProfile to look up how many more calories are in a large burger, fries, and drink than in a medium-size order.
   - **d.** Do you ever eat foods right out of the package? If you do, it is hard to tell how much you really ate. Suggest some things you could do to control your portion size.
2. **What factors affect your food choices?**
   - **a.** List four food items you ate today or yesterday.
   - **b.** For each food listed, indicate the factor or factors that influenced your selection of that particular food. For example, if you ate a candy bar before your noon class, did you choose it because it was available in the vending machine outside the lecture hall, because you didn't have enough money for anything else, because you just like candy bars, because you were depressed, because all of your friends were eating them, because it is good for you, or for some other reason?
   - **c.** For each food, indicate what information you used in making the selection. For example, did you read the label on the product or consider something you had read or heard recently in the news media?
   - **d.** List three types of information you regularly use to make your food choices.

CONSUMER ISSUES

3. **For many college students, their freshman year is the first time they are making all their own food choices, and they don't always make the best ones. Learning to apply the principles of variety, balance, and moderation can help improve these choices. For each scenario, explain what is wrong with the dietary choices and what could be done to improve them.**
   - **a.** Amad loves fast food. He grabs a doughnut for breakfast, a burger and fries for lunch, and either tacos or pizza for dinner.

b. Every day Helen eats cereal for breakfast; a peanut butter sandwich for lunch; and chicken, broccoli, and rice for dinner.

c. Amy takes advantage of the variety of foods offered in the cafeteria. She often has two types of whole-grain cereals for breakfast, selects eight or more ingredients at the salad bar, and includes two types of meat and two or three vegetables with her dinner. She has gained a few pounds and is concerned that she will become a victim of the "freshman 15"—the 15 or so pounds often gained in the first year away from home.

CLINICAL CONCERNS

4. **The Good Heart Study is evaluating the relationship between a new miracle drug and heart disease. The study involves 200 participants who are divided into two equal groups. Pills identical in appearance are administered to all participants. Group 1 receives two tablets per day of the miracle drug and group 2 receives two placebo tablets per day. After one month, half of the subjects in group 1 have blood cholesterol levels at or below the recommended level. In group 2, 49 of the subjects have blood cholesterol levels at or below the recommended level.**
   a. Is the study blinded?
   b. What factors would you consider when dividing the subjects between the experimental and control groups?
   c. Based on the study results, does the miracle drug lower cholesterol?

5. **An ad for a supplement called PowerBoost claims that it will increase your muscle strength, decrease your body fat, and boost your drive and motivation. Does this nutritional supplement live up to its claims?**
   a. Could the claim that the supplement "boosts your drive and motivation" be supported by quantifiable data?
   b. The ad summarizes a university study that measured the amount of lean tissue and fat tissue in weightlifters before and four weeks after they began consuming the PowerBoost drink. Describe the control group that should be included in this study?
   c. The results report a gain of 5.2 lb of lean tissue and a loss of 4.5 lb of fat tissue in weightlifters taking PowerBoost. The graph below compares the results for the experimental group (taking PowerBoost) to those for the control group. Based on this result, do you think the supplement is effective?

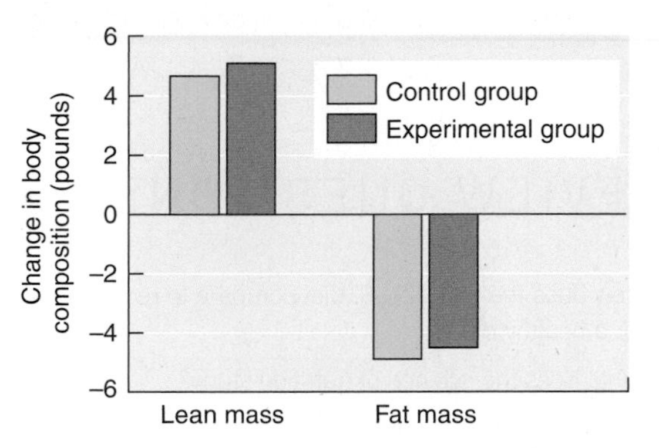

SUMMARY

1.1 Nutrition and Our Diet

- Nutrition is a science that encompasses all the interactions that occur between living organisms and food. Nutrients are substances in foods that provide energy and structure and help regulate body processes. In 21st- century America, we spend less time preparing food and eating at home and rely more on fast food, processed convience foods, and prepared foods than 50 years ago and at the same time consume more calories than we did in the past.
- The typical American diet does not meet the recommendations for a healthy diet and contributes to the development of chronic diseases such as diabetes, obesity, and heart disease.

1.2 Food Provides Nutrients

- About 45 nutrients are essential to human life. Nutrients consumed come from those naturally present in foods, those added to fortified foods, and those contained in dietary supplements. In addition to nutrients, food provides phytochemicals and zoochemicals, nonessential substances that may provide health benefits.
- There are six classes of nutrients: carbohydrates, lipids, proteins, water, vitamins, and minerals. Nutrients are needed by the body for growth, maintenance and repair, and reproduction. Carbohydrates, lipids, and proteins are energy-yielding nutrients. The energy they provide to the body is measured in kcalories or kjoules. Carbohydrates, lipids, protein, water, and minerals provide structure to the body, and all nutrient classes help regulate the biochemical reactions of metabolism to maintain homeostasis.
- When energy or one or more nutrients are deficient or excessive in the diet, malnutrition may result. Malnutrition includes both undernutrition and overnutrition. Undernutrition is caused by a deficiency of energy or nutrients. Overnutrition may be caused by a toxic dose of a nutrient or the chronic overconsumption of energy or of nutrients that increases the risk of chronic disease. Depending on the cause, the symptoms of malnutrition can occur in the short term or over the course of many weeks, months, or even years. The diet you consume can affect your genetic predisposition for developing a variety of chronic diseases.

1.3 Food Choices for a Healthy Diet

- Food choices are affected by food availability, sociocultural influences, personal tastes, emotional factors, and what we think we should eat to stay healthy. No one food choice is good or bad, and no one choice can make a diet healthy or unhealthy—each choice contributes to the diet as a whole.
- A healthy diet includes a variety of nutrient-dense foods from each food group as well as a variety of foods from within each group. It balances energy and nutrient intake

with needs and moderates choices to keep intakes of energy, fat, sugar, sodium, and alcohol within reason.

1.4 The Science behind Nutrition

- The science of nutrition uses the scientific method to determine the relationships between food and the nutrient needs and health of the body. The scientific method involves making observations of natural events, formulating hypotheses to explain these events, designing and performing experiments to test the hypotheses, and developing theories that explain the observed phenomenon based on the experimental results.
- The science of nutrition uses many different types of experimental approaches to determine nutrient functions and requirements. Observational studies identify relationships between diet and health. Intervention trials can test hypotheses developed from epidemiology. Laboratory studies use biochemical and molecular methods to study whole organisms or cells.
- Ethical guidelines protect humans and animals involved in research studies but limit the types of experiments that can be done.

1.5 Evaluating Nutrition Information

- When judging nutrition claims, first consider whether the information makes sense and is safe and whether it comes from a reliable source, such as an educational institution, the government, or a nonprofit organization. Information that promotes a product or in any other way benefits the person or organization providing it should be viewed with skepticism. Individual testimonies cannot be trusted because they have not been tested by experimentation.
- To be valid, nutrition information should be based on experiments that use quantifiable measurements, the right type and number of experimental subjects, appropriate controls, and a careful interpretation of experimental results. Reliable information will be supported by more than a single research study.

REVIEW QUESTIONS

1. How does the typical U.S. diet compare to recommendations for a healthy diet?
2. What does the science of nutrition study?
3. What is an essential nutrient?
4. List six classes of nutrients and indicate which provide energy.
5. List three functions provided by nutrients.
6. List three ways in which what you eat today can affect your health.
7. What is malnutrition?
8. List three factors other than biological need that influence what we eat.
9. Why is it important to choose a variety of foods?
10. How does moderation help maintain a healthy weight?
11. List the steps of the scientific method.
12. What type of information can be obtained using epidemiology?
13. Why are animals used to study human nutrition?
14. What factors should be considered when judging nutrition claims?
15. What is a control group?
16. What is a placebo?
17. What is a double-blind study?

REFERENCES

1. Briefel, R. R., and Johnson, C. L. Secular trends in dietary intake in the United States. *Annu Rev Nutr* 24:401–431, 2004.
2. Nielsen, S. J., Siega-Riz, A. M., and Popkin, B. M. Trends in energy intake in U.S. between 1977 and 1996: Similar shifts seen across age groups. *Obes Res* 10:370–378, 2002.
3. Nielsen, S. J., and Popkin, B. M. Patterns and trends in food portion sizes, 1977–1998. *JAMA* 289:450–453, 2003.
4. U.S. Department of Agriculture Office of Communications. Profiling food consumption in America. *Agriculture Fact Book, 2001–2002.*
5. U.S. Department of Agriculture and U.S. Department of Health and Human Services. *Dietary Guidelines for Americans, 2010*, 7th ed. Washington, DC: U.S. Government Printing Office, December 2010.
6. French, S. A., Lin, B. H., and Guthrie, J. F. National trends in soft drink consumption among children and adolescents age 6 to 17 years: Prevalence, amounts, and sources, 1977/1978 to 1994/1998. *J Am Diet Assoc* 103:1326–1331, 2003.
7. Frazao, E. America's eating habits: Changes and consequences. *Agriculture Information Bulletin*, no. 750. Beltsville, MD: ERS, USDA, 1999.
8. Mokdad, A. H., Marks, J. S., Stroup, D. F., and Gerberding, J. L. Correction: Actual causes of death in the United States, 2000. *JAMA* 291:1238–1245, 2004.
9. U.S. Department of Health and Human Services. Physical Activity Guidelines for Americans, 2008.
10. Picciano, M. F. *Who is Using Dietary Supplements? What Are They Using?* Office of Dietary Supplements, National Institutes of Health.
11. Ogden, C. L., Carroll, M. D., Kit, B. K., and Flegal, K. M. Prevalence of obesity and trends in body mass index among US children and adolescents, 1999–2010. *JAMA* 307:483–490, 2012.
12. Flegal, K. M., Carroll, M. D., Kit, B. K., and Ogden, C. L. Prevalence of obesity and trends in distribution of body mass index among U.S. adults, 1999–2010. *JAMA* 307:491–497, 2012.
13. Kaput, J. Nutrigenomics–2006 update. *Clin Chem Lab Med.* 45:279–287, 2007.
14. American Dietetic Association. *Nutrition and You: Trends 2008.*

***To access links to online sources, please go to www.wiley.com/college/smolin and select Nutrition: Science and Applications, 3rd edition. From this page, select either the student or instructor companion site. Once on the desired site, select References.**

CHAPTER OUTLINE

Nutrition Guidelines: Applying the Science of Nutrition

2

© Spaces Images/SuperStock

CASE STUDY

Joe was starting his freshman year at college on an athletic scholarship. Although it was only September, he wanted to make sure he ate well to stay in shape for track season in March. Now that he was away from home and had to make his own food choices at the dorm cafeteria, Joe was overwhelmed. Should breakfast be oatmeal, doughnuts, an omelet, or just fruit? He liked burgers and fries for lunch, but thought he should choose a salad and a cold sandwich. Dinner offered even more choices—there was always some type of meat dish, a pasta choice, fish, and a vegetarian entree. For dessert there were cakes, pies, and big vats of scoop-your-own ice cream to tempt him. Many of his friends piled their plates and served themselves two or three desserts, but he knew those choices were not the best. His track coach suggested he talk to the team nutrition advisor, who was also a professor in the university's Department of Nutritional Sciences.

Joe was surprised when the professor suggested that he go to the MyPlate Web site to find out how to choose a healthy diet. Joe knew that there was a lot of nutrition misinformation online, but this site provided reliable, user-friendly advice. By using this government site he could look up how much food from the different food groups he should eat to meet his needs based on his age and activity level. **Joe figured he could use the tools the site provided to help him make healthy choices in the cafeteria and select appropriate snacks to keep in his room so he would be ready for track in the spring.**

© Blend Images/SuperStock

2.1 The Development of Nutrition Recommendations

LEARNING OBJECTIVES

- List two reasons why population-wide nutrition recommendations are developed.
- Distinguish food guides from nutrient intake recommendations.
- Discuss why the focus of food guides has changed since the 1940s.

People need to eat to survive, but health-conscious individuals want to do more than survive. They want to choose diets that will optimize their health. An optimal diet would contain just the right amount of each nutrient to prevent deficiencies and maintain health and, for some people, to maintain a healthy pregnancy or to allow growth. The science of nutrition has determined which nutrients are necessary to keep humans alive and how much of each is needed at different stages of life. This information has been used to make general recommendations for the types and amounts of nutrients and foods that will maintain the health of individuals and populations. These recommendations are also used as a standard of comparison to assess whether populations and individuals are consuming diets that promote health.

Early U.S. Food Guidance

The U.S. government has been in the business of making nutritional recommendations for over 100 years. These recommendations have changed over time as our food intake patterns have changed and our knowledge of what constitutes a healthy diet has evolved.

The first dietary recommendations in the United States were published in 1894 by the U.S. Department of Agriculture (USDA). They suggested amounts of protein, carbohydrate,

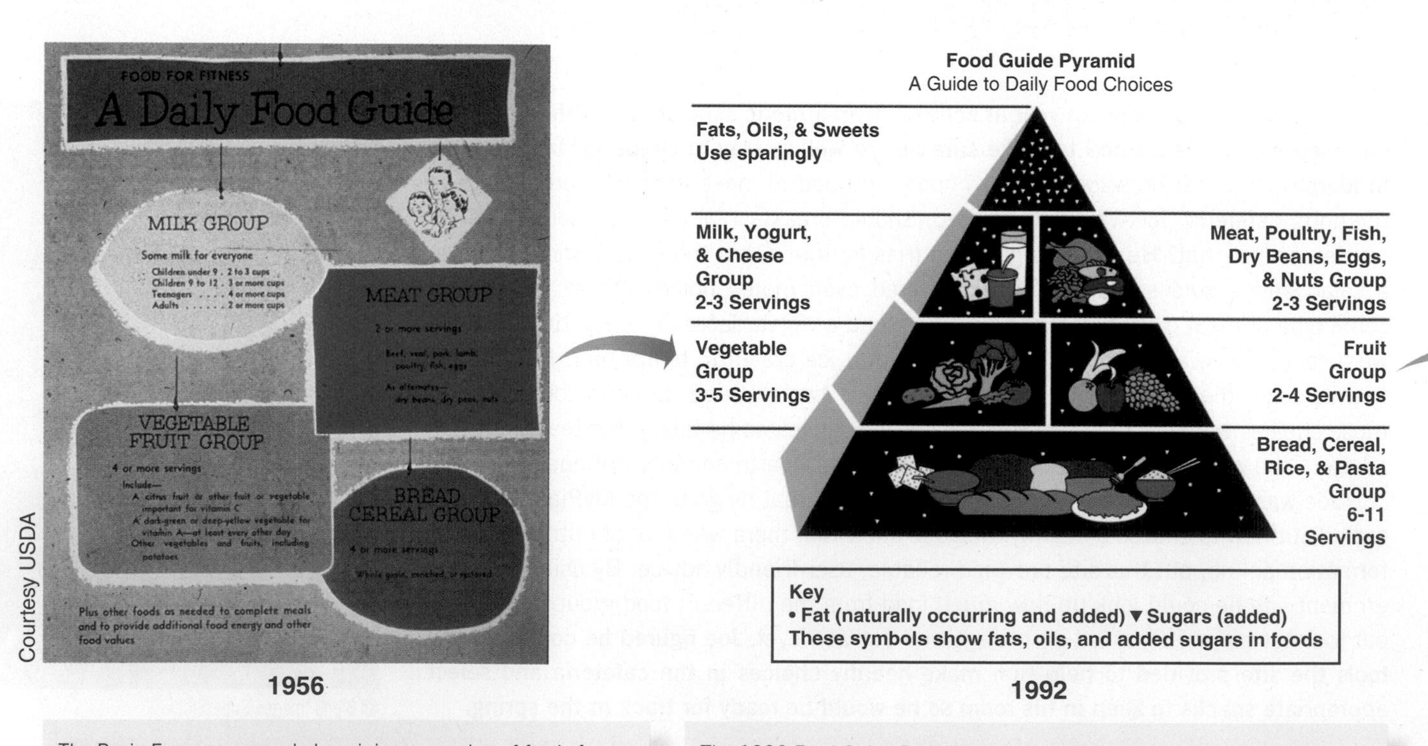

The Basic Four recommended a minimum number of foods from each of four food groups—milk, meat, vegetables and fruits, and bread and cereal—and focused on getting enough nutrients.

The 1992 Food Guide Pyramid organized food groups into a pyramid shape to emphasize the relative contribution of each group—more should be consumed from the larger food groups at the bottom.

Figure 2.1 A timeline of food guides
How food guides present recommendations has changed over the years, but the basic message has stayed the same: Choose the right combinations of foods to promote health.

fat, and "mineral matter" needed to keep Americans healthy.[1] At the time, specific vitamins and minerals essential for health had not been identified; nevertheless, this work set the stage for the development of the first food guides. Food guides help people choose foods that make up a healthy diet. The first food guide to target the general public, called *How to Select Foods*, was released in 1917. It made recommendations based on five food groups: meat and milk, cereals, vegetables and fruit, fats and fatty foods, and sugars and sugary foods. During the Great Depression of the 1930s a food guide was released that consisted of 12 food groups and focused on responding to food scarcities.

In the early 1940s, as the United States entered World War II, the Food and Nutrition Board was established to advise the army and other federal agencies regarding problems related to food and the nutritional health of the armed forces and the general population. The Food and Nutrition Board developed the first set of nutrient intake recommendations. This set of recommendations for specific amounts of nutrients came to be known as the Recommended Dietary Allowances (RDAs). The original RDAs made recommendations for amounts of energy and for nine essential nutrients that were most likely to be deficient in people's diets—protein, iron, calcium, vitamins A and D, thiamin, riboflavin, niacin, and vitamin C. Recommended intakes were based on amounts that would prevent nutrient deficiencies.[2] The same year, the USDA released a new food guide, the Basic Seven, which was revised in 1946 to become the National Food Guide. In 1956, the seven food groups were condensed to the "Basic Four." A guide with four food groups was used for the next two decades and eventually evolved into MyPlate, the current food guide (**Figure 2.1**).

How the Focus of Dietary Guidance Has Changed

Over the years since the first food guides and the original RDAs were developed, dietary habits and disease patterns have changed. Overt nutrient deficiencies are now rare in the United States, but the incidence of nutrition-related chronic diseases, such as heart disease, diabetes, osteoporosis, and obesity, has increased. The first response to the importance of nutrition in the changing disease patterns came in 1977 when the Senate Select Committee

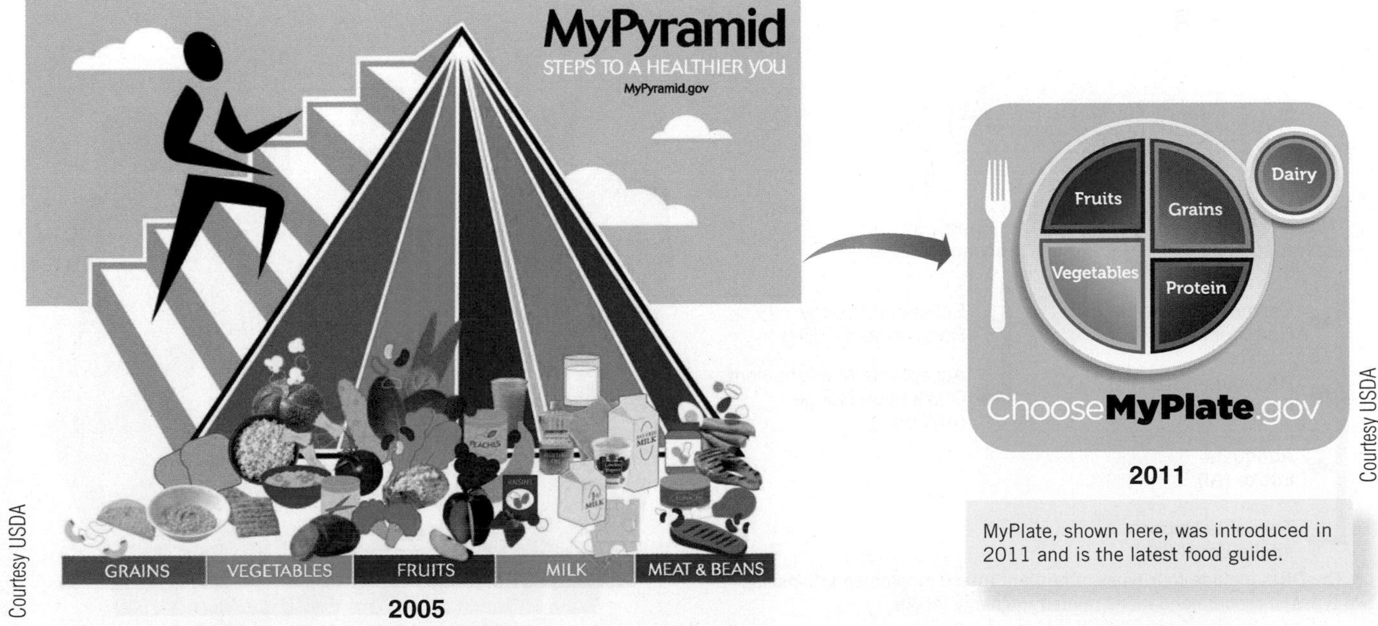

MyPyramid: Steps to a Healthier You kept the pyramid concept, but changed the shapes and arrangement of the food groups and added activity to the graphic.

MyPlate, shown here, was introduced in 2011 and is the latest food guide.

on Nutrition and Human Needs released the *Dietary Goals for the United States*.[3] This was the first set of recommendations that considered health promotion, rather than just deficiency prevention. The Dietary Goals were subsequently modified and published as the Dietary Guidelines for Americans in 1980 by the USDA and the U.S. Department of Health and Human Services. Since then, they have been revised approximately every five years to reflect advances in science and our understanding of what constitutes a diet that promotes health. The current (seventh) edition was released in 2010.[4]

Dietary Reference Intakes (DRIs) A set of reference values for the intake of energy, nutrients, and food components that can be used for planning and assessing the diets of healthy people in the United States and Canada.

Other dietary recommendations have also responded to these changes in diet and disease patterns. The original RDAs have been expanded into the **Dietary Reference Intakes (DRIs)**, which address concerns about excess as well as deficiency. Early food guides have evolved into MyPlate, which suggests amounts and types of food from five food groups to meet the recommendations of the 2010 Dietary Guidelines (see Figure 2.1). In addition, standardized food labels have been developed to help consumers choose foods that meet these recommendations.

2.2 Dietary Reference Intakes

LEARNING OBJECTIVES

- Describe the four types of nutrient intake recommendations included in the DRIs and explain the purpose of each.
- Name the five variables used to calculate the Estimated Energy Requirements.

The Dietary Reference Intakes (DRIs) are recommendations for the amounts of energy, nutrients, and other food components that healthy people should consume in order to stay healthy, reduce the risk of chronic disease, and prevent deficiencies. They are not required amounts that must be consumed each day but recommendations for what should be consumed on an average daily basis. They include several types of recommendations that address both nutrient intake and energy intake and include values that are appropriate for people of different genders and stages of life. These values take into account the physiological differences that affect the nutrient needs of men and women, infants, children, adolescents, adults, older adults, and pregnant and lactating women. The pregnancy and lactation recommendations are divided into age categories to distinguish the unique nutritional needs of pregnancy and lactation in teenagers and older mothers (**Figure 2.2**).

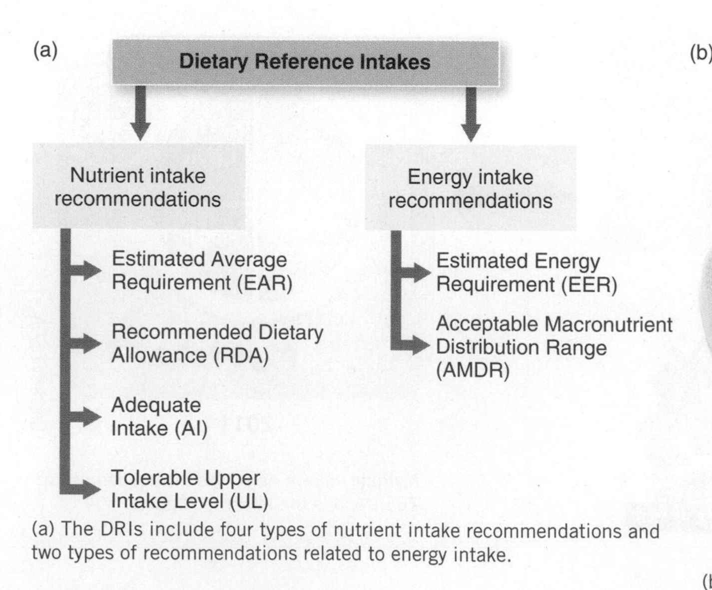

(a) The DRIs include four types of nutrient intake recommendations and two types of recommendations related to energy intake.

(b) Because gender and life stage affect nutrient needs, DRI values have been set for each gender and all life stages.

Figure 2.2 **Types of DRI recommendations**

The DRIs are standards developed for the United States and Canada. Different counties use different standards developed based on the nutritional issues in their country. The World Health Organization and the Food and Agriculture Organization of the United Nations, organizations concerned with international health, publish a set of dietary standards to apply worldwide.[5]

Recommendations for Nutrient Intake

The DRI recommendations for macronutrients and micronutrients include four different sets of reference values (see Figure 2.2a).[6] The **Estimated Average Requirements (EARs)** can be used to evaluate the nutrient intake of populations. The **Recommended Dietary Allowances (RDAs)** and the **Adequate Intakes (AIs)** can be used as goals for individual intake and to plan or evaluate individual diets, and the **Tolerable Upper Intake Levels (ULs)** recommend a limit above which nutrient toxicities are more likely.

Estimated Average Requirements (EARs) Intakes that meet the estimated nutrient needs of 50% of individuals in a gender and life-stage group.

Recommended Dietary Allowances (RDAs) Intakes that are sufficient to meet the nutrient needs of almost all healthy people in a specific life-stage and gender group.

Adequate Intakes (AIs) Intakes that should be used as a goal when no RDA exists. These values are an approximation of the average nutrient intake that appears to sustain a desired indicator of health.

Tolerable Upper Intake Levels (ULs) Maximum daily intakes that are unlikely to pose a risk of adverse health effects to almost all individuals in the specified life-stage and gender group.

Estimated Average Requirements (EARs)

An EAR is the amount of a nutrient that is estimated to meet the needs of 50% of people in the same gender and life-stage group (**Figure 2.3**). EAR values are useful for evaluating the adequacy of, and planning for, the nutrient intake of population groups. For example, the prevalence of low iron intake in a population can be estimated by looking at the proportion of the population with iron intakes below the EAR.

To set the EAR for a nutrient, scientists must establish a measurable indicator or criteria of adequacy for that nutrient. This may be the level of the nutrient or a metabolite of that nutrient in the blood, a body function that relies on the nutrient, or the appearance of a deficiency symptom. This functional indicator can then be evaluated to determine the biological effect of different levels of nutrient intake. Appropriate criteria of adequate intake must be established for each nutrient in each life-stage and gender group.

Recommended Dietary Allowances (RDAs)

The RDA values are higher than the EARs because they are calculated to meet the needs of nearly all healthy individuals in each gender and life-stage group. They are determined by starting with the EAR value and using the variability in the requirements among individuals to increase it to an amount that meets the needs of 97 to 98% of healthy individuals (see Figure 2.3). Because the RDA is set higher than the needs of most people, it serves as a target for individual intake.

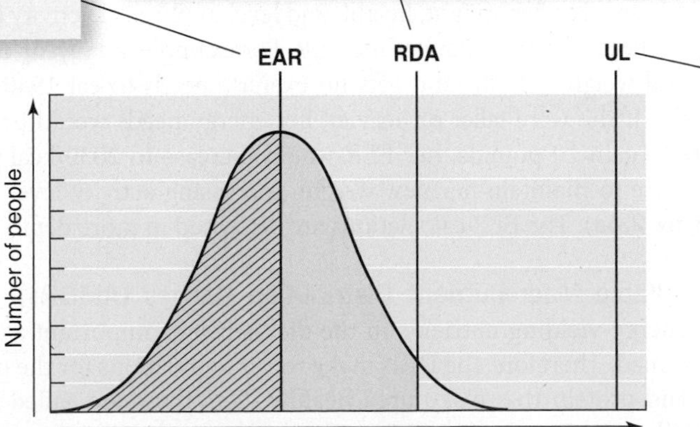

Figure 2.3 Understanding EARs, RDAs, and ULs
The EAR and RDA for a nutrient are determined by measuring the amount of the nutrient required by different individuals in a population group and plotting all the values. The resulting plot is a bell-shaped curve; a few individuals in the group need only a small amount of the nutrient, a few need a large amount, and the majority need an amount that falls between the extremes.

An intake less than the RDA does not necessarily indicate that the needs of that particular person have not been met; however, the risk of a deficiency is low if intake meets the RDA and increases as intake falls below the RDA.[7]

Adequate Intakes (AIs) Although the AIs are also used as a guide for individual intake, they are not based on EAR values. These are estimates used when there is insufficient scientific evidence to set an EAR and calculate an RDA. The AIs are based on observed or experimentally determined approximations of the average nutrient intake by a healthy population. When an AI value rather than an RDA is set, it targets the need for more research on the requirement for that nutrient. A healthy individual whose intake of a specific nutrient is at or above the AI is unlikely to be deficient in that nutrient.

Tolerable Upper Intake Levels (ULs) The fourth set of values, the Tolerable Upper Intake Levels, represent the maximum level of daily intake of a nutrient that is unlikely to pose a risk of adverse health effects to almost all individuals in the specified group. These are not recommended levels but rather levels of intake that can probably be tolerated (see Figure 2.3). ULs are used as a guide for limiting intake when people are planning diets and evaluating the possibility of overconsumption. The exact level of intake that will cause an adverse effect cannot be known with certainty for each individual, but if a person's intake is below the UL, there is good assurance that an adverse effect will not occur.

To establish a UL, a specific adverse effect or indicator of excess is considered. For instance, for niacin the ill effect is flushing, and for vitamin D it is elevated blood calcium levels. The lowest level of intake that causes the adverse effect is determined, and the UL is set far enough below this level that even the most sensitive people in the population are unlikely to be affected. If adverse effects have been associated only with intake from supplements, the UL is based only on this source. Therefore, for some nutrients these values represent intake from supplements alone; for some, intake from supplements and fortified foods; and for others, total intake from food, fortified food, water, nonfood sources, and supplements. For many nutrients, data are insufficient to establish a UL value.

Recommendations for Energy Intake

The DRIs make two types of recommendations regarding energy intake. One provides an estimate of how much energy is needed to maintain body weight, and the other provides information about the proportion of each of the energy-yielding nutrients from which this energy should come.

Estimated Energy Requirements (EERs) Average energy intakes predicted to maintain body weight in healthy individuals.

Estimated Energy Requirements (EERs) The recommendations for energy intake are called **Estimated Energy Requirements (EERs)**.[8] They provide an estimate of the number of calories needed to keep weight stable in a healthy person. Variables in the EER equations include age, gender, weight, height, and level of physical activity (see inside cover). Changing any of these variables changes the EER. For example, a 19-year-old girl who is 5 feet 4 inches tall and weighs 127 lbs and gets no exercise needs to eat 1940 kcal a day to maintain her weight. If she were taller or heavier, her energy needs would be greater. For example, if she were to gain 20 pounds, her EER would increase to 2030 kcal per day; she would need to eat more to maintain her new weight. Increasing activity level also increases energy needs (**Figure 2.4a**). The EER calculations are discussed in more depth in Chapter 7.

Acceptable Macronutrient Distribution Ranges (AMDRs) Ranges of intake for energy-yielding nutrients, expressed as a percentage of total energy intake, that are associated with reduced risk of chronic disease while providing adequate intakes of essential nutrients.

Acceptable Macronutrient Distribution Ranges (AMDRs) The proportion of each of the energy-yielding nutrients in the diet is just as important as the total amount of energy consumed. Therefore, the DRIs make recommendations for the proportions of carbohydrate, fat, and protein that make up a healthy diet. These are called **Acceptable Macronutrient Distribution Ranges (AMDRs)**. These recommendations are expressed as ranges because healthy diets can be made up of many different proportions of carbohydrate, protein, and fat.[8] According to the AMDRs, a healthy diet for an adult can contain from 45 to 65% of calories from carbohydrate, 20 to 35% from fat, and 10 to 35% from protein. When calorie intake stays the same, changing the proportion of one of these will change the proportion of the others as well. So, for example, in a diet that provides 2000 kcal with 50%

Figure 2.4 **Energy intake: EERs and AMDRs**

(a)

(a) This 20-year-old male is 6 feet tall and weighs 165 pounds. If he jogs for 30 minutes 5 days a week he needs to consume 3196 kcal per day to maintain his weight. If he joins the track team and practices for several hours a day, he will need to up his intake to 3698 kcal/day or more to maintain his weight.

(b)

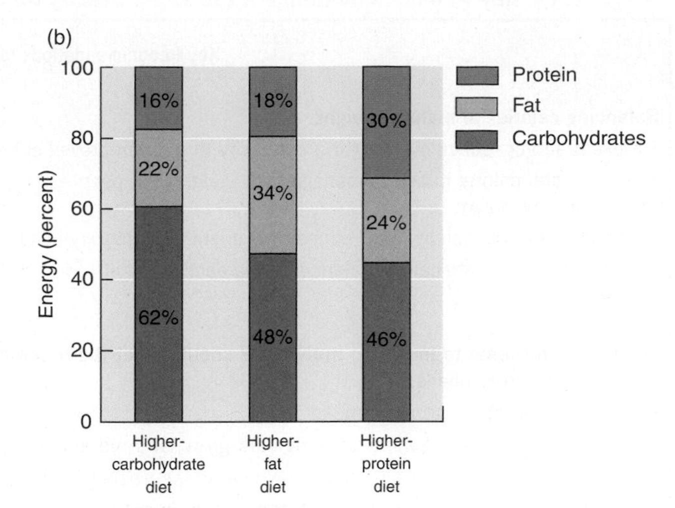

(b) If energy intake remains constant, changing the amount of one of the energy-yielding nutrients in the diet changes the proportions of the others.

from carbohydrate, the other 50% will come from protein and fat. If the calories are kept the same but carbohydrate intake is decreased, the percentage of fat and/or protein will increase (**Figure 2.4b**). The AMDRs allow flexibility in food choices based on individual preferences while still providing a diet that minimizes disease risk. AMDR values have also been set for specific amino acids and fatty acids (see Appendix A).

Applications of the Dietary Reference Intakes

The DRIs can be used to plan diets, to assess the adequacy of diets, and to make judgments about excessive intakes for individuals and populations.[6] For example, they can be used as a standard for meals prepared for schools, hospitals, and other health-care facilities; for government feeding programs for the elderly; and even for meals for astronauts. They can be used to determine standards for food labeling and to develop practical tools for diet planning, such as food guides. They can also be used to interpret information gathered about the food consumed by a population to help identify potential nutritional inadequacies that may be of public-health concern.

Despite their many uses, dietary standards cannot be used to identify with certainty whether a specific person has a nutritional deficiency or excess. To ascertain this, an evaluation of the person's nutritional status using dietary, clinical, biochemical, and body-size measurements is needed, as discussed later in this chapter.

2.3 The Dietary Guidelines for Americans

LEARNING OBJECTIVES

- Explain the purpose of the Dietary Guidelines.
- Discuss the recommendations of the Dietary Guidelines that address the rising rate of obesity.
- Name two foods or food components that the Dietary Guidelines recommend we increase and two that we should decrease in our diets.

What Are the Dietary Guidelines?

Unlike the DRIs, which provide recommendations for specific amounts of nutrients and food components, the Dietary Guidelines suggest overall diet and lifestyle choices that will promote health. The 2010 Dietary Guidelines for Americans provide evidence-based nutritional guidance to promote health and reduce the prevalence of overweight and

TABLE 2.1 Key Recommendations of the 2010 Dietary Guidelines

Key Recommendations for the General Population

Balancing calories to manage weight
- Prevent and/or reduce overweight and obesity through improved eating and physical activity behaviors.
- Control total calorie intake to manage body weight. For people who are overweight or obese, this means consuming fewer calories from foods and beverages.
- Increase physical activity and reduce time spent in sedentary behaviors.
- Maintain appropriate calorie balance during each stage of life—childhood, adolescence, adulthood, pregnancy and breastfeeding, and older age.

Foods and nutrients to increase. Individuals should meet the following recommendations as part of a healthy eating pattern while staying within their calorie needs:
- Increase vegetable and fruit intake.
- Eat a variety of vegetables, especially dark-green and red and orange vegetables and beans and peas.
- Consume at least half of all grains as whole grains. Increase whole-grain intake by replacing refined grains with whole grains.
- Increase intake of fat-free or low-fat milk and milk products, such as milk, yogurt, cheese, or fortified soy beverages.
- Choose a variety of protein foods, including seafood, lean meat and poultry, eggs, beans and peas, soy products, and unsalted nuts and seeds.
- Increase the amount and variety of seafood consumed by choosing seafood in place of some meat and poultry.
- Replace protein foods that are higher in solid fats with choices that are lower in solid fats and calories and/or are sources of oils.
- Use oils to replace solid fats where possible.
- Choose foods that provide more potassium, dietary fiber, calcium, and vitamin D, which are nutrients of concern in U.S. diets. These foods include vegetables, fruits, whole grains, and milk and milk products.

Foods and food components to reduce
- Reduce daily sodium intake to less than 2300 milligrams (mg), and further reduce intake to 1500 mg among persons who are 51 and older and those of any age who are African American or have hypertension, diabetes, or chronic kidney disease. The 1500-mg recommendation applies to about half of the U.S. population, including children, and the majority of adults.
- Consume less than 10% of calories from saturated fatty acids by replacing them with monounsaturated and polyunsaturated fatty acids.
- Consume less than 300 mg per day of dietary cholesterol.
- Keep *trans* fatty acid consumption as low as possible by limiting foods that contain synthetic sources of *trans* fats, such as partially hydrogenated oils, and by limiting other solid fats.
- Reduce intake of calories from solid fats and added sugars.
- Limit consumption of foods that contain refined grains, especially refined grain foods that contain solid fats, added sugars, and sodium.
- If alcohol is consumed, consume it in moderation—up to one drink per day for women and two drinks per day for men, and only by adults of legal drinking age.

Building healthy eating patterns
- Select an eating pattern that meets nutrient needs over time at an appropriate calorie level.
- Account for all foods and beverages consumed, and assess how they fit within a total healthy eating pattern.
- Follow food safety recommendations when preparing and eating foods to reduce the risk of food-borne illnesses.

Specific Lifecycle Recommendations

Women capable of becoming pregnant
- Choose foods that supply heme iron, which is more readily absorbed by the body, additional iron sources, and enhancers of iron absorption such as vitamin C-rich foods.
- Consume 400 micrograms per day of synthetic folic acid (from fortified foods and/or supplements) in addition to food forms of folate from a varied diet.

Women who are pregnant or breastfeeding
- Consume 8 to 12 ounces of seafood per week from a variety of seafood types.
- Due to their high methyl mercury content, limit white (albacore) tuna to 6 ounces per week and do not eat the following four types of fish: tilefish, shark, swordfish, and king mackerel.
- If pregnant, take an iron supplement, as recommended by an obstetrician or other health-care provider.

Individuals ages 50 years and older
- Consume foods fortified with vitamin B_{12}, such as fortified cereals, or dietary supplements.

obesity and the risk of chronic disease. The recommendations of this seventh edition of the Dietary Guidelines for Americans focus on balancing energy intake with physical activity and consuming nutrient-dense foods and beverages (**Table 2.1**). These recommendations are designed for Americans two years of age and older. Adopting the recommendations in the Dietary Guidelines will help Americans live healthier lives, which will lower health-care costs and help to strengthen America's long-term economic competitiveness and overall productivity.

Balancing Calories to Manage Weight

More Americans are overweight than ever before, and the numbers continue to grow. To address this problem, the 2010 Dietary Guidelines emphasize balancing the calories consumed in food and beverages with the calories expended through physical activity in order to achieve and maintain a healthy weight. Weight maintenance requires consuming the same number of calories as you burn; this means that if you eat more, you need to exercise more (see Chapter 7). Losing weight requires consuming fewer calories than you burn.

The Dietary Guidelines suggest that most Americans are in energy imbalance because they consume too many calories in their diet and expend too few in activity. To achieve calorie balance, adults should decrease energy intake by reducing portion sizes and limiting consumption of added sugars, solid fats, and alcohol, which provide calories but few essential nutrients, and increase their energy expenditure through exercise. To promote health and reduce disease risk, a minimum of 150 minutes of moderate exercise is recommended each week. Some adults will need a higher level of physical activity than others (the equivalent of more than 300 minutes of moderate-intensity activity) to achieve and maintain a healthy body weight (**Figure 2.5**).

Figure 2.5 **Healthy weight and exercise recommendations**

Jose Luis Pelaez/Getty Images, Inc.

(a) The 2010 Dietary Guidelines for Americans recommend choosing eating patterns that help maintain or achieve a healthy weight over time.

Alaska Stock Images/NG Image Collection

(b) The Dietary Guidelines suggest that most Americans increase their aerobic physical activity gradually over time to a minimum of 150 minutes per week.

© Matthew Chattle/Alamy

Figure 2.6 **Watch beverage choices**
It is important to consider beverages as part of the diet. Beverages such as these add calories and sugar to the diet, but few other nutrients. Currently, American adults consume about 400 kcal per day from beverages.[4] The Dietary Guidelines recommend replacing sugary drinks with water.

Foods and Nutrients to Increase

The Dietary Guidelines recommend that Americans increase their vegetable and fruit intake to at least 2 ½ cups per day and improve their choices by selecting more fruits than fruit juices and eating a variety of vegetables, especially dark-green and red and orange vegetables and beans and peas. The Dietary Guidelines suggest replacing refined grains with whole grains so that at least half of grain servings are whole grains. They also suggest that Americans increase their intake of fat-free or low-fat milk and milk products while limiting consumption of high-fat dairy products such as cheese. This pattern of fruit, vegetable, grain, and dairy consumption will increase our intake of potassium, dietary fiber, calcium, and vitamin D, which are nutrients of concern in American diets. Protein choices should include a variety of protein foods, such as lean meat, poultry, seafood, eggs, beans and peas, soy products, and unsalted nuts and seeds. The Dietary Guidelines also recommend using oils in place of solid fats when possible.

Foods and Food Components to Reduce

The Dietary Guidelines recommend reducing intake of saturated fat, *trans* fat, and cholesterol—the types of lipids that increase the risk of heart disease (see Chapter 5). In order to prevent high blood pressure, the Dietary Guidelines recommend limiting sodium intake to less than 2300 mg/day. Those who are 51 and older, and those of any age who are African American or have hypertension, diabetes, or chronic kidney disease, should limit sodium to less than 1500 mg/day. The Guidelines also recommend that Americans reduce their intake of refined grains by making half their grains whole and of added sugars by choosing and preparing foods and beverages with little added sugar (**Figure 2.6**). Added sugars provide calories but contribute few nutrients (see Chapter 4).

The Dietary Guidelines also advise those who drink alcohol to do so in moderation. Consuming one or two drinks per day is associated with lowest all-cause mortality as well as lowest coronary heart disease mortality. Alcohol consumption is not recommended for those who cannot restrict their intake, women of childbearing age who may become pregnant, pregnant and lactating women, children and adolescents, those taking medications that interact with alcohol, those with specific medical conditions, and those engaged in activities that require attention, skill, or coordination (see **Focus on Alcohol**).

Building Healthy Eating Patterns

There is no single diet that defines "healthy." Rather, there are a variety of healthy eating patterns that can accommodate differences in food preferences due to culture, ethnicity, tradition, personal tastes, and variations in food cost and availability. All these patterns are abundant in nutrient-dense foods such as vegetables, fruits, and whole grains; include moderate amounts of a variety of high-protein foods; and are low in full-fat dairy products. Healthy eating patterns include more oils than solid fats and limit added sugars and sodium. The Dietary Guidelines for Americans discuss three healthy eating patterns (**Figure 2.7**).

A fundamental premise of the Dietary Guidelines is that nutrients should come primarily from foods; supplements and fortified foods may be advantageous in specific situations to increase intake of a particular vitamin or mineral. Fortification can provide a food-based means for increasing the intake of individual nutrients.

© Gabor Izso/iStockphoto

© Jorgen Udvang/iStockphoto

Masterfile

The DASH Eating Plan

Food group	Servings
Grains	6–8/day
Vegetables	4–5/day
Fruits	4–5/day
Fat-free or low-fat milk and milk products	2–3/day
Lean meats, poultry, and fish	6 or less/day
Nuts, seeds, and legumes	4–5/week
Fat and oils	2–3/day
Sweets and added sugars	5 or less/week

The DASH Eating Plan is plentiful in fruits and vegetables, whole grains, low-fat dairy products, fish, poultry, seeds, and nuts. It was first developed for lowering blood pressure (see Chapter 10).

© mbirdy/iStockphoto

© Sabina Salihbasic/iStockphoto

USDA Food Patterns

Food group	Amount/day
Vegetables	2.5 cups
Fruit and juices	2.0 cups
Grains	6.0 ounces
Dairy products	3.0 cups
Protein foods	5.5 ounces
Oils	27 grams
Solid fats	16 grams
Added sugars	32 grams

The USDA Food Patterns suggest amounts of foods from different food groups for different calorie levels (2000 kcal shown here). The USDA Food Patterns and their vegetarian variations were developed to help individuals follow the Dietary Guidelines and are the basis for the MyPlate recommendations.

Mediterranean Eating Pattern

Foods	How often
Fruits, vegetables, grains (mostly whole), olive oil, nuts, legumes and seeds, herbs and spices	Every meal
Fish and seafood	At least twice a week
Cheese and yogurt	Moderate portions daily or weekly
Poultry and eggs	Moderate portions every 2 days or weekly
Meats and sweets	Less often

Traditional Mediterranean eating patterns are based on fruits, vegetables, grains, olive oil, fish, beans, nuts, legumes, and seeds. They include smaller amounts of red meat and sweets. The incidence of chronic diseases such as heart disease is low in populations consuming these diets (see Chapter 5).

Figure 2.7 Healthy eating patterns
The Dietary Guidelines for Americans suggest that there are many ways to choose a healthy diet, including the USDA Food Patterns, the DASH Eating Plan, and Mediterranean-style eating patterns. These patterns all focus on similar types of foods. (See Appendix D for more information.)

Paying attention to food safety is also part of healthful eating. Currently, food-borne illness affects about 48 million individuals in the United States every year and leads to 128,000 hospitalizations and 3000 deaths.[9] Consumers can prevent food-borne illness at home by washing hands, rinsing vegetables and fruits, preventing cross-contamination, cooking foods to safe internal temperatures, and storing foods safely (see Chapter 13).

2.4 MyPlate: Putting the Dietary Guidelines into Practice

LEARNING OBJECTIVES

- Discuss the MyPlate recommendations for proportionality, variety, moderation, and nutrient density.
- Plan a diet based on your MyPlate Daily Food Plan.

To help individuals apply the recommendations of the Dietary Guidelines to their own food choices, the USDA has developed MyPlate (**Figure 2.8**).[10] This educational tool translates the Dietary Guidelines recommendations into the choices that make up a healthy dietary and lifestyle pattern. Like previous food guides, MyPlate is a food group system; it divides foods into groups based on the nutrients they supply most abundantly. The plate icon illustrates the proportions of food recommended from each of five food groups: fruits, vegetables, grains, protein foods, and dairy. Half of your plate should be fruits and vegetables, about a quarter grains, and about a quarter protein foods. Dairy should accompany meals, as shown by the glass to the side (see Figure 2.8a). The amounts recommended from each food group for an individual depend on their energy needs.

Food guides developed in other countries have adopted different shapes to emphasize the proportions of choices that should come from different food groups. Korea and China use a pagoda shape; Mexico, Australia, and most European countries use a pie or plate shape; and Canada uses a rainbow.

Understanding the MyPlate Messages

MyPlate emphasizes the importance of proportionality, variety, moderation, and nutrient density in a healthy diet. Proportionality means eating more of some types of foods than others. The MyPlate icon shows how much of your plate should be filled with foods from various food groups (see Figure 2.8a).

Variety is important for a healthy diet because no one food or food group provides all the nutrients and food components the body needs. A variety of foods should be selected from within each food group. The vegetables food group includes choices from five subgroups: dark-green vegetables such as broccoli, collard greens, and kale; red and orange vegetables such as carrots, sweet potatoes, and red peppers; starchy vegetables such as corn, green peas, and potatoes; other vegetables such as cabbage, asparagus, and artichokes; and beans and peas such as lentils, chickpeas, and black beans. Beans and peas are good sources of the nutrients found in both vegetables and protein foods, so they can be counted in either food group. Protein foods include meat, poultry, seafood, beans and peas, eggs, processed soy products, nuts, and seeds. Grains include whole grains such as whole-wheat bread, oatmeal, and brown rice as well as refined grains such as white bread, white rice, and white pasta (see Chapter 4). Fruits include fresh, canned, or dried fruit and 100% fruit juice. Dairy includes all fluid milk products and many foods made from milk such as cheese, yogurt, and pudding, as well as calcium-fortified soy products.

Moderation involves limiting portion sizes and choosing nutrient-dense foods to balance calories consumed with calories expended. Tips such as make half your grains whole, choose whole or cut-up fruits more often than juice, select fat-free or low-fat dairy products, keep meat and poultry portions small and lean, and many more found on the MyPlate Web site are designed to help consumers make nutrient-dense choices.

How to Use Your Daily Food Plan

Your Daily Food Plan tells you how much food to eat from each food group (see Figure 2.8b). The amounts from the grains group are expressed in ounces. An ounce of grains is equivalent to 1 cup of cold cereal, ½ cup of cooked cereal or grains, or a slice of bread. So if you

Figure 2.8 **MyPlate recommendations**

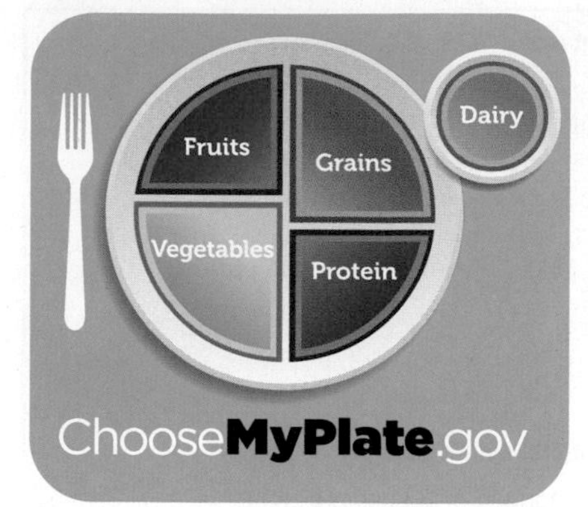

Balancing Calories

- Enjoy your food, but eat less.
- Avoid oversized portions.

Foods to Increase

- Make half your plate fruits and vegetables.
- Make at least half your grains whole grains.
- Switch to fat-free or low-fat (1%) milk.

Foods to Reduce

- Compare sodium in foods such as soup, breads, and frozen meals—and choose the foods with lower numbers.
- Drink water instead of sugary drinks.

(a) The MyPlate icon shows what a balanced meal should look like. Shown here next to the icon are some key messages from the Dietary Guidelines to help consumers balance calories and increase the nutrient density of their diet.

Grains 6 ounces	Vegetables 2 1/2cups	Fruits 2 cups	Dairy 3 cups	Protein 5 1/2 ounces
Make half your grains whole	**Vary your veggies**	**Focus on fruits**	**Get your calcium-rich foods**	**Go lean with protein**
Aim for at least **3 ounces** of whole grains a day	Aim for these amounts **each week:** **Dark-green veggies** = 1 1/2 cups **Red & orange veggies** = 5 1/2 cups **Beans & peas** = 1 1/2 cups **Starchy veggies** = 5 cups **Other veggies** = 4 cups	Eat a variety of fruit Choose whole or cut-up fruits more often than fruit juice	Drink fat-free or low-fat (1%) milk for the same amount of calcium and other nutrients as whole milk but less fat and Calories Select fat-free or low-fat yogurt and cheese, or try calcium-fortified soy products	Twice a week, make seafood the protein on your plate Vary your protein routine–choose beans, peas, nuts, and seeds more often Keep meat and poultry portions small and lean

Find your balance between food and physical activity
Be physically active for at least **150 minutes** each week.

Know your limits on fats, sugars, and sodium
Your allowance for oils is **6 teaspoons** a day. Limit calories from solid fats and added sugars to **260 Calories** a day. Reduce sodium intake to less than **2300 mg** a day.

(b) The MyPlate Daily Food Plan shown here is for a person who needs 2000 kcal per day You can find your own personalized food plan in Appendix D or at the MyPlate Web site **www.ChooseMyPlate.gov**.

have two cups of cereal and two slices of toast at breakfast, you have already consumed 4 ounces of grains for the day (two-thirds of the total for a 2000-kcal diet). The amounts recommended for protein foods are also expressed in ounces. One ounce is equivalent to an ounce of cooked meat, poultry, or fish; one egg; 1 tablespoon of peanut butter; ¼ cup of cooked dry beans; or ¼ cup of nuts or seeds. The amounts recommended for fruits, vegetables, and dairy are given in cups. **Table 2.2** provides examples of amounts of foods that are equivalent to a cup of vegetables, fruits, or dairy, or an ounce of grains or protein foods.

TABLE 2.2 What Counts as an Ounce or a Cup?

Grains	Food	Amount that counts as an ounce
One cup of rice is the size of a baseball	Bagel	¼ large or one mini bagel
	Bread	1 regular slice
	Crackers	5 whole wheat or 7 square or round
	Cooked cereal, rice, or pasta	½ cup (a typical serving size is 1 cup)
	Popcorn	3 cups, ¼ microwave bag
	Ready-to-eat breakfast cereal	1 cup flakes or rounds, 1 ¼ cup puffed

Vegetables	Food	Amount that counts as a cup
A medium potato is the size of computer mouse	Broccoli	1 cup chopped or 3 spears 5" long
	Carrots	1 cup strips, 2 medium carrots, or 12 baby carrots
	Greens (collards, mustard greens, turnip greens, kale, spinach)	1 cup cooked
	Raw leafy greens (spinach, romaine, dark-green leafy lettuce, endive)	2 cups raw
	Sweet potato	1 large baked (2 ¼" or more diameter)
	Corn on the cob	1 large ear (8 to 9" long)
	Dry beans	1 cup cooked whole or mashed
	White potatoes	1 medium boiled or baked (2 ½ to 3" diameter) french fried: 20 medium to long strips (2 ½ to 4" long)
	Tomatoes	1 large whole (3" diameter); 1 cup chopped raw, canned, or cooked; 1 cup juice

Fruits	Food	Amount that counts as a cup
A medium apple is the size of a baseball	Apple	½ large or 1 small
	Banana	1 large (8 to 9" long), 1 cup sliced
	Orange	1 large (3" diameter)
	Strawberries	8 large
	Grapes	32 seedless grapes
	Dried fruit	½ cup
	100% fruit juice	1 cup

Dairy	Food	Amount that counts as a cup
An ounce of cheese is the size of four dice	Milk and yogurt	1 cup
	Cheese	2 slices or 1 ½ ounces hard cheese, ⅓ cup shredded, ½ cup ricotta, 2 cups cottage cheese
	Pudding or frozen yogurt	1 cup
	Ice cream	1 ½ cups or 3 scoops

Protein	Food	Amount that counts as an ounce
Three ounces of meat or tofu is the size of a deck of cards	Meat, poultry, fish	1 ounce cooked, ⅓ small steak, ⅓ small hamburger, ⅓ chicken breast, ⅓ can tuna (a typical serving size is 3 oz)
	Eggs	1 egg
	Nuts and seeds	½ ounce, 12 almonds, 24 pistachios, 7 walnut halves
	Dry beans or peas	¼ cup cooked black, kidney, or pinto beans; ¼ cup cooked chick peas, lentils, or split peas; ½ cup lentil soup; ¼ cup tofu; 2 T hummus; 1 T peanut butter

Photos by Andy Washnik

TABLE 2.3 Estimated Energy Needs (kcalories)

	Males			Females		
Age (yrs)	Inactive	Moderately active	Active	Inactive	Moderately active	Active
16–18	2400	2800	3200	1800	2000	2400
19–20	2600	2800	3000	2000	2200	2400
21–25	2400	2800	3000	2000	2200	2400
26–30	2400	2600	3000	1800	2000	2400
31–35	2400	2600	3000	1800	2000	2200
36–40	2400	2600	2800	1800	2000	2200
41–45	2200	2600	2800	1800	2000	2200
46–50	2200	2400	2800	1800	2000	2200
51–55	2200	2400	2800	1600	1800	2200
56–60	2200	2400	2600	1600	1800	2200
61–65	2000	2400	2600	1600	1800	2000
66–70	2000	2200	2600	1600	1800	2000

Inactive = less than 30 minutes a day of moderate activity.
Moderately active = between 30 and 60 minutes of moderate activity a day.
Active = 60 minutes or more of moderate activity a day.
Source: www.ChooseMyPlate.gov

TABLE 2.4 Food Group Choices for Different Energy Levels

Energy level (kcal)	1800	2000	2200	2400	2600	2800	3000	3200
Grains (ounces)	6	6	7	8	9	10	10	10
Vegetables (cups)	2.5	2.5	3	3	3.5	3.5	4	4
Fruits (cups)	1.5	2	2	2	2	2.5	2.5	2.5
Dairy (cups)	3	3	3	3	3	3	3	3
Protein (ounces)	5	5.5	6	6.5	6.5	7	7	7
Oils (tsp)	5	6	6	7	8	8	10	11
Empty calorie allowance (kcal)	160	260	270	330	360	400	460	600

Source: www.ChooseMyPlate.gov

The calorie needs and recommended amounts from each food group for adults are summarized in **Tables 2.3** and **2.4**. To find your MyPlate Daily Food Plan, look up your calorie needs in Table 2.3 and then match them to your recommended food group amounts in Table 2.4. Once you know how much to choose from each group, you can plan your meals and snacks for the day. **Figure 2.9** shows a sample day's menu for a 2000-kcal diet.

It is easy to see where some foods in your diet fit on MyPlate. For example, a chicken breast is 3 ounces from the protein group; a scoop of rice is 2 ounces from the grains group. It is more difficult to see how much mixed foods such as pizza, stews, and casseroles contribute to each food group. To fit these on your plate, individual ingredients must be considered. For example, a slice of pizza provides 1 ounce of grains, ⅛ cup of vegetables, and ½ cup of dairy. Having meat on your pizza adds about ¼ ounce from the protein group (see Figure 2.9 and **Critical Thinking: Fitting Foods, Meals, and Diets onto MyPlate**).

A Daily Food Plan also includes recommendations about the amounts of oils (in teaspoons) that should be included in your diet (see Figures 2.8b and 2.9). Oils are fats that are liquid at room temperature; they come from plants and fish. They are rich in

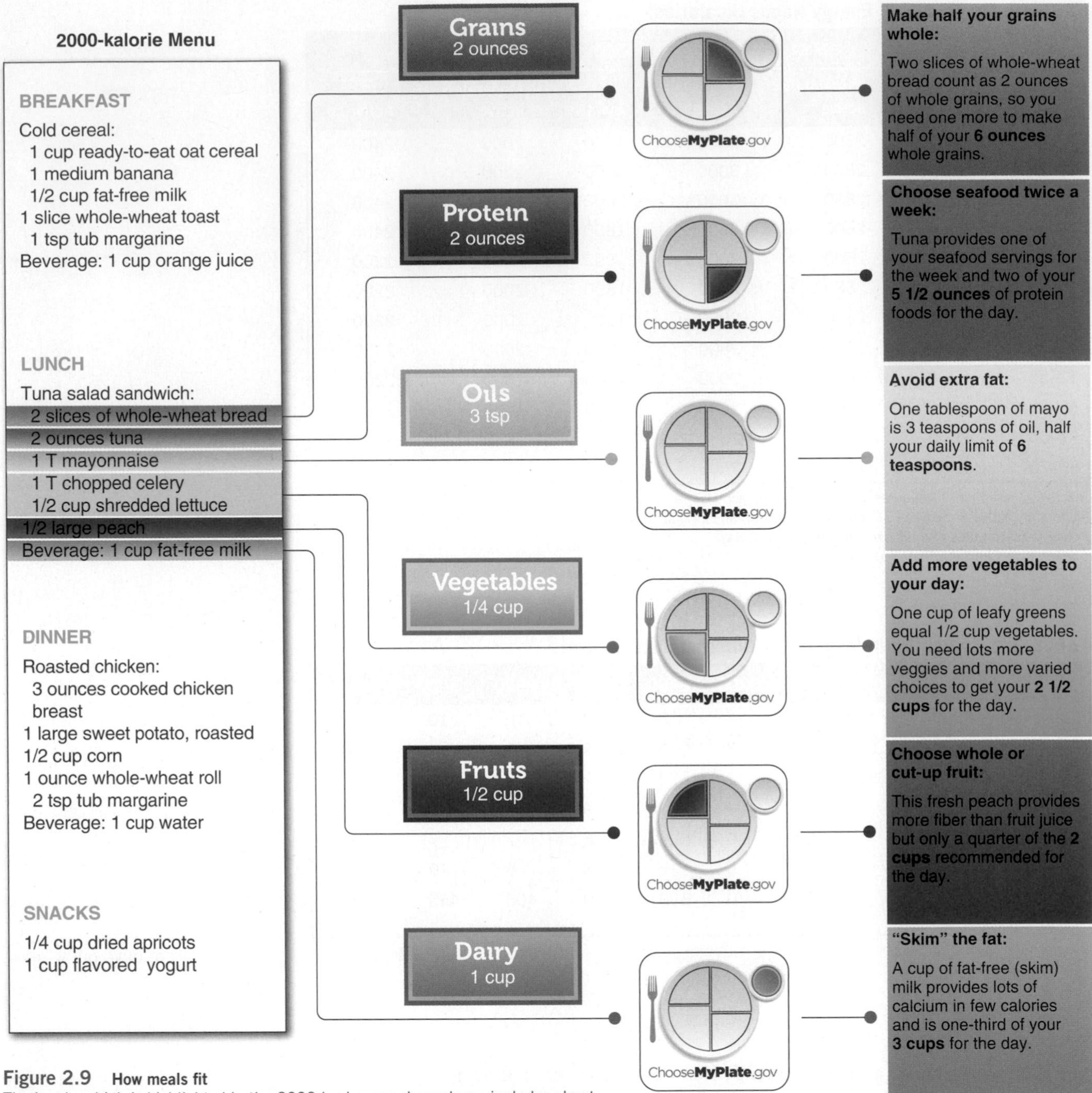

Figure 2.9 **How meals fit**
The lunch, which is highlighted in the 2000-kcal menu shown here, includes about a third of the amounts of grains, protein foods, and dairy recommended for the day, a quarter of the fruit, and half the oils but only a small proportion of the vegetables recommended. The lettuce and celery fit into to the "other vegetables" category, so many vegetable choices throughout the rest of the day and week should come from dark-green, red and orange, and starchy vegetables as well as beans and peas.

unsaturated fats, which help protect against heart disease. Solid fats are fats that are solid at room temperature, such as butter and shortening. They provide saturated and *trans* fat and should be limited in the diet.

A MyPlate Daily Food Plan recommends at least 150 minutes of activity each week to help balance food and physical activity and includes a calorie limit for **empty calories** from solid fats and added sugars. Your empty calorie allowance is the number of calories you can consume after you have satisfied all your nutrient needs from nutrient-dense food choices. These "extra" calories are needed to maintain your weight and can come from additional nutrient-dense choices or from foods that are high in added sugar or solid fats.

empty calories Calories from solid fats and/or added sugars, which add calories to the food but few nutrients.

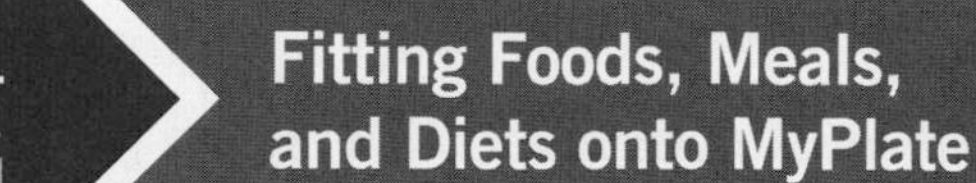

CRITICAL THINKING: Fitting Foods, Meals, and Diets onto MyPlate

Joe Gough/Shutterstock

For the first time in his life, Lucas is shopping and cooking for himself. He wants to prepare healthy meals so he uses MyPlate as a guide for his food choices. He is a 22-year-old male who exercises less than 30 minutes a day.

For breakfast Lucas has cereal, milk, orange juice, and coffee with cream and sugar. When he puts the foods from this meal on MyPlate he notices that the sections for vegetables and protein foods are empty.

CRITICAL THINKING QUESTIONS

- Does he need to include foods from each group at every meal?
- Where do the cream and sugar he puts in his coffee fit onto MyPlate?
- Plan a lunch for Lucas that includes a food from each food group of MyPlate.

For dinner Lucas has lasagna. To see how it fits onto MyPlate, he breaks it down into individual ingredients.

INGREDIENTS IN LASAGNA	FOOD GROUP	AMOUNT ON MYPLATE
Noodles (1.5 oz dry)		
Tomato sauce (¾ cup)		
Ground beef (2 oz)		
Mozzarella cheese (3 T)	Dairy	¼ cup
Ricotta cheese (¼ cup)	Dairy	½ cup
Olive oil (1 tsp)	Oils	

CRITICAL THINKING QUESTIONS

- Complete the table to the left to see how his lasagna fits onto MyPlate. (Hint: Use Food-A-Pedia in Supertracker of MyPlate)
- If he used butter instead of olive oil, where would it fit on MyPlate?

iProfile Use iProfile to find nutrient-dense substitutions for foods that are high in empty calories

Some empty calories come from foods, such as butter, table sugar, soft drinks, and candy, that don't belong in any food group because all their calories are empty. Some empty calories come from foods that belong to a food group but contain added sugars and solid fats (**Figure 2.10a**). Oils are healthy fats, so they are not considered empty calories. It is important to limit empty calories because consuming too many means you can't meet your nutrient needs without exceeding your calorie needs (**Figure 2.10b**).

One way to see how your daily intake matches the MyPlate recommendations is to use the interactive tools (SuperTracker) on the MyPlate Web site to help track your progress toward choosing a healthy diet.

Figure 2.10 Empty calories

Food group	Foods with few or no empty calories	Foods with about half their energy from empty calories
Grains	Whole-wheat bread	Blueberry muffin
	Brown rice	Croissant
	Plain bagel	Doughnuts
	Whole-wheat crackers	Chocolate chip cookies
Vegetables	Steamed broccoli	Broccoli and cream sauce
	Sweet potatoes	Candied sweet potatoes
	Green beans	Green bean casserole
Fruits	Raw apples	Apple pie
	Strawberries	Frozen sweetened strawberries
	Fresh peaches	Canned peaches in heavy syrup
Dairy	Fat-free or low-fat milk	Whole milk
	Nonfat plain yogurt	Fruit-flavored low-fat yogurt
	Nonfat mozzarella	Cheddar cheese
Protein	Extra lean ground beef	Sausage
	Roasted chicken breast without skin	Fried chicken with skin
	Kidney beans	Bologna

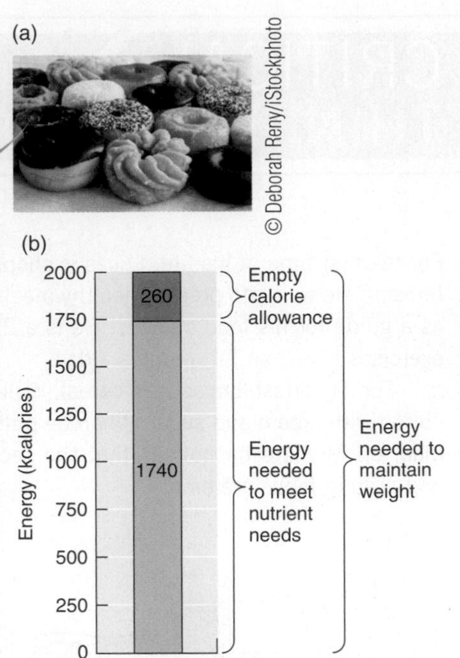

(a) Doughnuts are in the grains group, but about half of their calories are empty calories from solid fat and sugar. The foods included in the right column of the table all belong to food groups but have so much solid fat or added sugar that about half of their energy is from empty calories.

(b) If you are at a healthy weight and you choose nutrient-dense foods, you can satisfy all your nutrient needs with fewer calories than you need to maintain your weight. The extra calories needed to maintain your weight is your empty calorie allowance.

2.5 Food and Supplement Labels

LEARNING OBJECTIVES

- Assess the amount of saturated fat in a food by using the Nutrition Facts label.
- Review a food label and identify the ingredients that are present in the greatest and least amounts.
- Use a dietary supplement label to determine the types and amounts of ingredients the product provides.

Food labels are a tool that can help plan diets that meet the recommendations of the DRIs, Dietary Guidelines, and MyPlate. They are designed to help consumers make healthy food choices by providing information about the nutrient composition of foods and how they fit into the overall diet. Checking the food label can help you select a food that is a good source of fiber or one that is low in saturated fat or high in vitamin C. To make this information uniform and easy to use, food labeling standards are specified by the Nutrition Labeling and Education Act of 1990. Dietary supplements are required to carry a similar label to help consumers know what supplemental nutrients and other ingredients they are choosing.

Food Labels

More than 98% of all processed, packaged foods sold in the United States carry standard food labels.[11] The Food and Drug Administration (FDA) regulates the labeling of all foods except meat and poultry products, which are regulated by the USDA. All packaged foods except those produced by small businesses and those in packages too small to fit the information must be labeled. Raw fruits, vegetables, fish, meat, and poultry are not required to carry individual labels. The FDA has asked grocery stores to voluntarily provide nutrition information for the raw fruits, vegetables, and fish most frequently eaten in the United States, and the USDA encourages voluntary nutrition labeling of raw meat and poultry. The information can appear on large placards or in consumer pamphlets or brochures.

All labels are required to contain basic product information such as the name of the product; the net contents or weight; the date by which the product should be sold; and the name and place of business of the manufacturer, packager, or distributor. In addition, most food labels contain a Nutrition Facts panel and a list of the food's ingredients.

How to Understand the Nutrition Facts The nutrition information section of the label is entitled "Nutrition Facts" (**Figure 2.11**). It provides information about serving size; total Calories (on food labels "Calorie" with a capital "C" is used to represent kcal); Calories from fat; the amounts of total fat, saturated fat, *trans* fat, cholesterol, sodium, total carbohydrate, dietary fiber, sugars, and protein per serving; and how the food fits into the overall diet.

What Is the Serving Size? The serving size of a food product is given in common household and metric measures and is followed by the number of servings per container. Serving size is based on a standard list of serving sizes. The use of standard serving sizes allows comparisons to be made easily between products. For example, comparing the energy content of different types of cookies is simplified because all packages list energy values for a standard serving size of about an ounce and tell you the number of cookies per serving. However, these serving sizes are not always the same as the portions you put on your plate. If you eat two servings, you are consuming twice as many calories.

How to Use Daily Values Food labels list the amounts of most nutrients as a percentage of a standard called the **Daily Value**. Daily Values help consumers determine how a food fits into their overall diet. The % Daily Value is the amount of a nutrient in a food as a percentage of the recommendation for a 2000-kcal diet. For example, if a food provides 10% of the Daily Value for dietary fiber, then the food provides 10% of the recommended daily intake for dietary fiber in a 2000-kcal diet. As a general rule, a Daily Value of 5% or less indicates that the food is low in that nutrient and a Daily Value of 20% or more indicates that it is high. For most nutrients, the Daily Value is a target for intake, but for some, such as total fat, saturated fat, and cholesterol, it is a maximum recommended amount. For these you would want to select foods with a low % Daily Value. So, for example, if a food contains 5% of the Daily Value for saturated fat, then it contains only 5% of the maximum amount recommended for a 2000-kcal diet. Food labels must list the Daily Value for total fat, saturated fat, cholesterol, sodium, total carbohydrate, and dietary fiber, as well as for vitamin A, vitamin C, calcium, and iron.

Daily Value A nutrient reference value used on food labels to help consumers see how foods fit into their overall diets.

Standard serving sizes are required to allow consumers to compare products. For example, the number of calories in one serving of this macaroni and cheese can be compared to the number of calories in one serving of packaged rice because the values for both are for a standard 1-cup serving.

Food labels must list the "% Daily Value" for total fat, saturated fat, cholesterol, sodium, total carbohydrate, and dietary fiber, as well as for vitamins A and C, calcium, and iron. A % Daily Value of 5% or less is considered low, and a value of 20% or more is considered high.

The label provides information about the amounts of nutrients whose intake should be limited—total fat, saturated fat, *trans* fat, cholesterol, and sodium.

The label provides information about the amounts of nutrients that tend to be low in the American diet—fiber, vitamins A and C, calcium, and iron.

The footnote gives the Daily Values for 2,000- and 2,500-kcal diets to illustrate that for some nutrients the Daily Value increases with increasing caloric intake.

Nutrition Facts

Serving Size 1 cup (228 g)
Servings Per Container 2

Amount Per Serving	
Calories 250	**Calories from Fat** 110
	% Daily Value*
Total Fat 12g	18%
Saturated Fat 3g	15%
Trans Fat 1.5g	
Cholesterol 30mg	10%
Sodium 470mg	20%
Total Carbohydrate 31g	10%
Dietary Fiber 0g	0%
Sugars 5g	
Protein 5g	
Vitamin A	4%
Vitamin C	2%
Calcium	20%
Iron	4%

*Percent Daily Values are based on a 2,000 calorie diet. Your daily values may be higher or lower depending on your calorie needs:

	Calories:	2,000	2,500
Total Fat	Less than	65g	80g
Sat. Fat	Less than	20g	25g
Cholesterol	Less than	300mg	300mg
Sodium	Less than	2,400mg	2,400mg
Total Carbohydrate		300g	375g
Dietary Fiber		25g	30g

Figure 2.11 Nutrition Facts
This label from a macaroni-and-cheese package illustrates how the Nutrition Facts panel can help you evaluate the nutritional contribution this food will make to your diet.

THINK CRITICALLY

How much cholesterol would you consume if you ate the entire box of macaroni and cheese? What percentage of the Daily Value would that represent?

TABLE 2.5 Standards That Make Up the Daily Values

Nutrient	Daily Reference Value	Amount in 2000-kcalorie diet	Amount in 2500-kcalorie diet
Total fat	<30% of kcalories	<65 g	<80 g
Saturated fat	<10% of kcalories	<20 g	<25 g
Total carbohydrate	60% of kcalories	300 g	375 g
Dietary fiber	11.5 g/1000 kcalories	25 g	30 g
Protein	10% of kcalories	50 g	63 g
Cholesterol	<300 mg	<300 mg	<300 mg
Sodium	<2400 mg	<2400 mg	<2400 mg
Potassium	3500 mg	3500 mg	3500 mg

Nutrient	Reference Daily Intake*	Nutrient	Reference Daily Intake*
Vitamin A	5000 IU[†]	Vitamin E	30 IU[†]
Biotin	300 μg	Riboflavin	1.7 mg
Vitamin C	60 mg	Niacin	20 mg
Vitamin B_6	2.0 mg	Vitamin B_{12}	6 μg
Thiamin	1.5 mg	Chromium	120 μg
Folic acid	400 μg	Phosphorus	1000 mg
Pantothenic acid	10 mg	Selenium	70 μg
Vitamin K	80 μg	Calcium	1000 mg
Iodine	150 μg	Magnesium	400 mg
Molybdenum	75 μg	Manganese	2 mg
Iron	18 mg	Zinc	15 mg
Vitamin D	400 IU[†]	Chloride	3400 mg
Copper	2 mg		

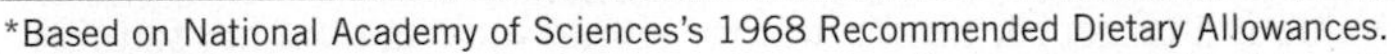

*Based on National Academy of Sciences's 1968 Recommended Dietary Allowances.

[†]The Reference Daily Intakes for some fat-soluble vitamins are expressed in International Units (IU). Equations for converting International Units (IU) to μg or mg are available in Appendix G.

Reference Daily Intakes (RDIs) Reference values established for vitamins and minerals that are based on the highest amount of each nutrient recommended for any adult age group by the 1968 RDAs.

Daily Reference Values (DRVs) Reference values established for protein and seven nutrients for which no original RDAs were established. The values are based on dietary recommendations for reducing the risk of chronic disease.

Because the Daily Values are a single set of standards for everyone, they may overestimate the amount of a nutrient needed for some groups, but they do not underestimate the requirement for any group (except pregnant and lactating women). Daily Values are based on two sets of standards, the **Reference Daily Intakes (RDIs)** and the **Daily Reference Values (DRVs)** (**Table 2.5**). To avoid confusion, only the term "Daily Value" appears on food labels.

What the Ingredient List Tells You The ingredients section of the label lists the contents of the product in order of their prominence by weight (**Figure 2.12**). An ingredient list is required on all products containing more than one ingredient. Food additives, including food colors and flavorings, must be listed among the ingredients (see **Off the Label: Using Food Labels to Choose Wisely**).

THINK CRITICALLY

If you drink a cup of this product, can you count it as a cup of fruit?

INGREDIENTS: Water, high-fructose corn syrup, pear and grape juice concentrates, citric acid, water-extracted orange and pineapple juice concentrates, natural flavor

Figure 2.12 Interpreting an ingredient list
The ingredient list shown here indicates that water and the sweetener high-fructose corn syrup are the first two ingredients and thus are the most abundant ingredients by weight.

© iStockphoto

OFF THE LABEL

Using Food Labels to Choose Wisely

Food labels can't help you include more vegetables and fruits in your diet each day or ensure that you select a varied diet, but they do provide you with a good source of nutrition information. For example, for breakfast you might choose granola with whole milk or oatmeal made with nonfat milk. Which is the better choice if you want your diet to be low in added sugar and saturated fat, and high in nutrient density? You can see on the label that a cup of whole milk provides 150 kcal and 5 g of saturated fat. These 5 g of saturated fat represents 25% of the Daily Value—that is, 25% of the maximum amount of saturated fat recommended per day for a 2000-kcal diet. A cup of nonfat milk, in contrast, contains no saturated fat and only 90 kcal. Both contain only the sugar that is naturally present in milk, are good sources of calcium and vitamins A and D, and are in the dairy group. But because nonfat milk contributes fewer calories and no saturated fat, choosing it makes it easier to meet the Dietary Guidelines recommendation to limit empty calories and consume less than 10% of calories from saturated fat.

Granola and oatmeal each provide a serving from the grains group—both contain whole grains—but the amounts of fat and added sugars differ. A serving of granola provides 230 kcal and 9 g of fat, 3.5 of which is saturated. A serving of oatmeal has only 150 kcal and 3 g of fat, 0.5 of which is saturated. The ingredient list reveals that oatmeal contains only rolled oats. No sugars are added. In contrast, the granola contains rolled oats plus brown sugar and honey in addition to other ingredients. The oatmeal is lower in calories from both saturated fat and added sugar, so it is higher in nutrient density.

Knowing how to interpret the information on food labels can help you choose a diet that meets the recommendations of the 2010 Dietary Guidelines and MyPlate. However, this doesn't mean you can never have a doughnut for breakfast because the label identifies it as high in fat, sugars, and calories. Even a sugary, high-fat food choice can be part of a healthy diet as long as it is balanced with healthy low-fat, low-sugar choices throughout the day. Remember, it is your total dietary pattern—not each choice—that counts.

THINK CRITICALLY: In the near future the FDA will be revising the current food label. Suggest some changes that you think would help consumers identify nutrient-dense foods.

Nonfat Milk

Nutrition Facts
Serving Size 1 cup (236 ml)
Servings per Container 16

Amount Per Serving	
Calories 90	Calories from Fat 0
	% Daily Value*
Total Fat 0g	0%
Saturated Fat 0g	0%
Trans Fat 0g	
Cholesterol less than 5mg	1%
Sodium 125mg	5%
Total Carbohydrate 13g	4%
Dietary Fiber 0g	0%
Sugars 12g	
Protein 9g	

INGREDIENTS: GRADE A FAT FREE SKIM MILK. VITAMIN A PALMITATE AND VITAMIN D3

Whole Milk

Nutrition Facts
Serving Size 1 cup (236 ml)
Servings per Container 16

Amount Per Serving	
Calories 150	Calories from Fat 70
	% Daily Value*
Total Fat 8g	12%
Saturated Fat 5g	25%
Trans Fat 0g	
Cholesterol 35mg	11%
Sodium 125mg	5%
Total Carbohydrate 12g	4%
Dietary Fiber 0g	0%
Sugars 12g	
Protein 8g	

INGREDIENTS: MILK

Old-Fashioned Oats

Nutrition Facts
Serving Size 1/2 cup (40 g) dry
Servings Per Container about 13

Amount Per Serving	Dry	Cereal with 1/2 cup Skim Milk
Calories	150	190
Calories from Fat	25	30
	% Daily Value*	
Total Fat 3g	5%	5%
Saturated Fat 0.5g	3%	3%
Trans Fat 0g		
Cholesterol 0mg	0%	0%
Sodium 0mg	0%	3%
Total Carbohydrate 27g	9%	11%
Dietary Fiber 4g	16%	16%
Sugars 0g		
Protein 5g		

INGREDIENTS: 100% ROLLED OATS

Natural Granola

Nutrition Facts
Serving Size 1/2 cup (51g)
Servings Per Container about 16

Amount Per Serving	
Calories 230	Calories from Fat 80
	% Daily Value*
Total Fat 9g	13%
Saturated Fat 3.5g	18%
Trans Fat 1g	
Polyunsaturated Fat 1g	
Monounsaturated Fat 4g	
Cholesterol 0mg	0%
Sodium 20mg	1%
Potassium 250mg	7%
Total Carbohydrate 34g	11%
Other Carbohudrate 15g	
Dietary Fiber 3g	13%
Sugars 16g	
Protein 5g	

INGREDIENTS: Whole grain rolled oats, whole grain rolled wheat, brown sugar, raisins, dried coconut, almonds, partially hydrogenated cottonseed and soybean oils, nonfat dry milk, glycerin, honey.

Dietary Supplement Labels

There are thousands of types of dietary supplements and about half of all American adults take some kind of supplement.[12] These products, which can include vitamins; minerals; herbs, botanicals, or other plant-derived substances; amino acids; enzymes; concentrates; and extracts, must carry a standard label that meets the specifications of the Dietary Supplement Health and Education Act of 1994.[13] Each product must include the words "dietary supplement" on the label and carry a Supplement Facts panel similar to the Nutrition Facts panel found on most processed foods (**Figure 2.13a**). This panel lists the recommended serving size and the name and quantity of each ingredient per serving. The source of the ingredient may be given with its name in the Supplement Facts panel or in the ingredient list below the panel. The nutrients for which Daily Values have been established are listed first, followed by other dietary ingredients for which Daily Values have not been established.[13]

Supplements are classified as foods and not drugs, so they are not bound by the strict laws that regulate drug manufacturing. To help ensure that these products contain the right ingredients and the right amount per dose, the FDA established dietary supplement "current Good Manufacturing Practice" regulations. These require manufacturers to test their products to ensure identity, purity, strength, and composition.[14] The U.S. Pharmacopeia (USP) Convention, which sets the standards for drug manufacturing, has also developed the USP Dietary Supplement Verification Program (DSVP).[15] This voluntary program evaluates and confirms the contents of dietary supplements, the manufacturing processes, and compliance with standards of purity. Products that have been reviewed and meet USP criteria can use the DSVP verification mark on the label or the statement "made to US Pharmacopeia (USP) quality, purity, and potency standards" (**Figure 2.13b**). Consumers can be assured that

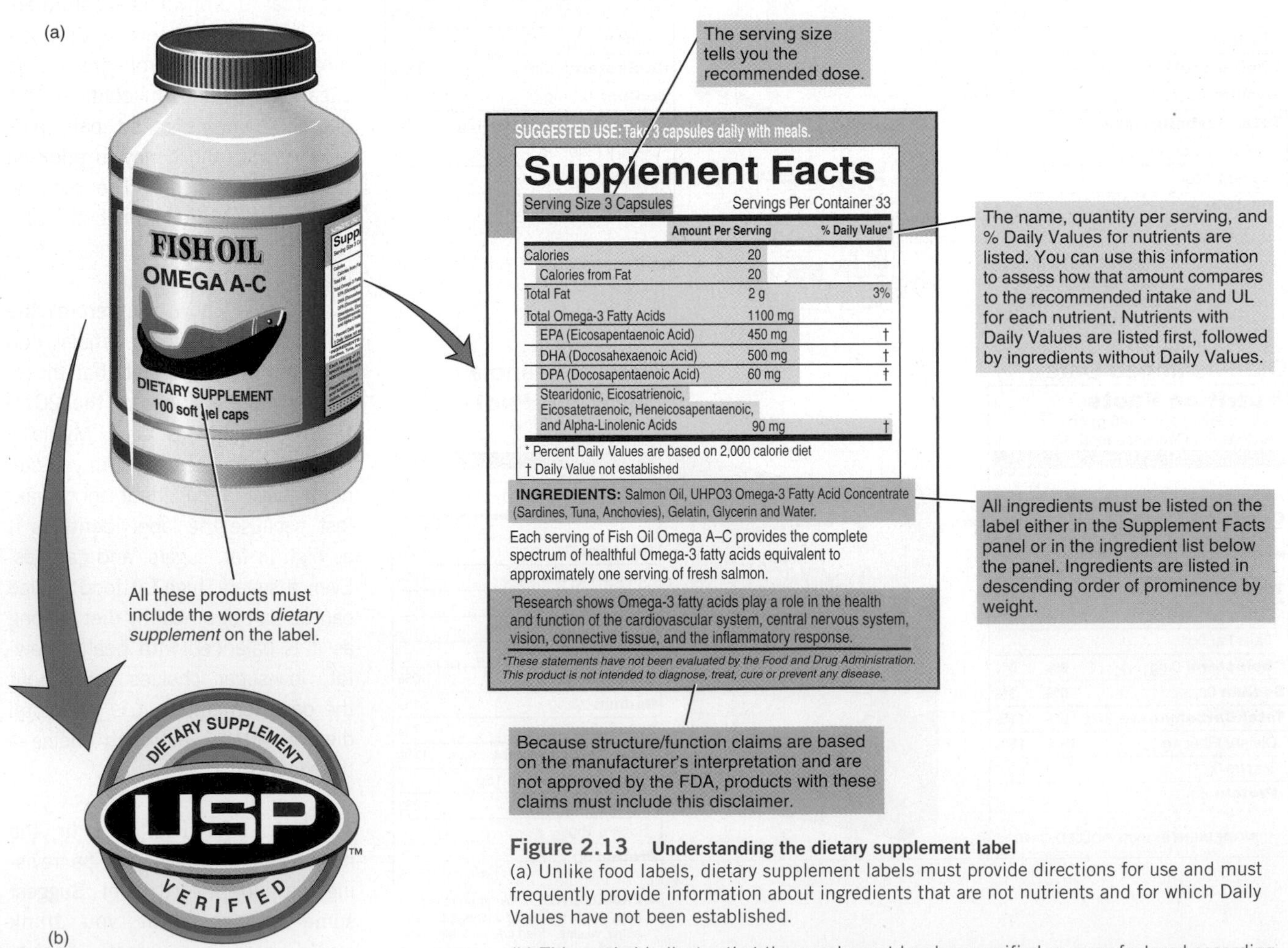

Figure 2.13 Understanding the dietary supplement label
(a) Unlike food labels, dietary supplement labels must provide directions for use and must frequently provide information about ingredients that are not nutrients and for which Daily Values have not been established.

(b) This symbol indicates that the supplement has been verified as manufactured according to the quality, purity, and potency standards set by the U.S. Pharmacopeia (USP) Convention.

products that have been USP-verified contain the ingredients in the amounts as listed on the label, will disintegrate or dissolve effectively to release nutrients for absorption, meet requirements for limits on contaminants, and comply with good manufacturing practices.

Products that carry the DSVP verification mark must meet certain manufacturing standards, but they are not necessarily safe or effective. Dietary supplements do not need to be approved by the FDA for safety and effectiveness before they are marketed. According to the Dietary Supplement Health and Education Act, the manufacturer is responsible for ensuring that a supplement is safe before it is sold. If a problem arises after the supplement is on the market, it is the responsibility of the FDA to prove that the supplement represents a risk and that it should be removed from the market (see **Focus on Dietary Supplements**).

Understanding Label Claims

Three different types of claims can be used on food and dietary supplement labels: nutrient content claims, health claims, and structure/function claims. The responsibility for ensuring the validity of these claims rests with the manufacturer, the FDA, or, in the case of advertising, with the Federal Trade Commission.[16]

Nutrient Content Claims Food and supplement labels often highlight the level of a nutrient or dietary substance in a product that might be of interest to the consumer, such as "low Calorie" or "high fiber." Definitions for nutrient content descriptors such as "free," "high," and "reduced" have been established by the FDA (**Table 2.6**). In selecting a product labeled

TABLE 2.6 Descriptors Commonly Used on Food Labels

Free	Product contains no amount of, or a trivial amount of, fat, saturated fat, cholesterol, sodium, sugars, or calories. For example, "sugar free" and "fat free" both mean less than 0.5 g per serving. Synonyms for "free" include "without," "no," and "zero."
Low	Used for foods that can be eaten frequently without exceeding the Daily Value for fat, saturated fat, cholesterol, sodium, or calories. Specific definitions have been established for each of these nutrients. For example, "low fat" means that the food contains 3 g or less per serving, and "low cholesterol" means that the food contains less than 20 mg of cholesterol per serving. Synonyms for "low" include "little," "few," and "low source of."
Lean and extra lean	Used to describe the fat content of meat, poultry, seafood, and game meats."Lean" means that the food contains less than 10 g fat, less than 4.5 g saturated fat, and less than 95 mg of cholesterol per serving and per 100 g. "Extra lean" means that the food contains less than 5 g fat, less than 2 g saturated fat, and less than 95 mg of cholesterol per serving and per 100 g.
High	Used for foods that contain 20% or more of the Daily Value for a particular nutrient. Synonyms for "high" include "rich in" and "excellent source of."
Good source	Food contains 10 to 19% of the Daily Value for a particular nutrient per serving.
Reduced	Nutritionally altered product contains 25% less of a nutrient or of energy than the regular or reference product.
Less	Food, whether altered or not, contains 25% less of a nutrient or of energy than the reference food. For example, pretzels may claim to have "less fat" than potato chips. "Fewer" may be used as a synonym for "less."
Light	Used in different ways. It can be used on a nutritionally altered product that contains one-third fewer calories or half the fat of a reference food. It can be used when the sodium content of a low-calorie, low-fat food has been reduced by 50%. The term "light" can also be used to describe properties such as texture and color as long as the label explains the intent—for example,"light and fluffy."
More	Serving of food, whether altered or not, contains a nutrient that is at least 10% of the Daily Value more than the reference food. This definition also applies to foods using the terms "fortified," "enriched," or "added."
Healthy	Used to describe foods that are low in fat and saturated fat and contain no more than 360 mg of sodium and no more than 60 mg of cholesterol per serving and provide at least 10% of the Daily Value for vitamins A or C, or iron, calcium, protein, or fiber.
Fresh	Used on foods that are raw and have never been frozen or heated and contain no preservatives.
High potency	Used on foods or supplements to describe individual vitamins or minerals that are present at 100% or more of the Daily Value or on multi-ingredient foods or supplement products that contain 100% or more of the Daily Value for at least two-thirds of the vitamins and minerals present in significant amounts (e.g.,"High-potency multivitamin, multimineral dietary supplement tablets").
Antioxidant	Used to describe foods or supplements that are "a good source of" or "high in" a nutrient for which there is an established Daily Value and for which there is scientific evidence of its function as an antioxidant.

Source: U.S. Food and Drug Administration.

with a descriptor such as "fat free," consumers can be assured that the food meets the defined criteria, in this case that the product contains less than 0.5 gram of fat per serving. Likewise, when selecting a supplement, the claim "excellent source of vitamin C" means that a serving of the product must contain at least 20% of the Daily Value for vitamin C. Terms like "high potency" or "antioxidant" used on supplement labels are also considered nutrient content claims. The specific definition of each of these descriptors is given in Table 2.6, and their use in relation to specific nutrients is discussed in Off the Label features throughout this textbook.

Health Claims Food and supplement labels are also permitted to include a number of health claims if they are relevant to the product. Health claims refer to a relationship between a food, food component, or dietary supplement ingredient and the risk of a disease or health-related condition. They can help consumers choose products that will meet their dietary needs or health goals. For example, low-fat milk, a good source of calcium, might include on the package label a statement indicating that a diet high in calcium will reduce the risk of developing osteoporosis.

The FDA reviews all health claims, but there are three different paths that can lead to the authorization of a health claim. The first is the most stringent. These health claims are authorized after an extensive review of the scientific evidence and are said to meet the *significant scientific agreement* standard (**Figure 2.14**). A second way to authorize a

Figure 2.14 Health claims on food labels
Oatmeal contains enough soluble fiber to be permitted to include on the label a health claim about the relationship between soluble fiber and the risk of heart disease. Other authorized and qualified health claims that can appear on food and supplement labels are listed in the table.

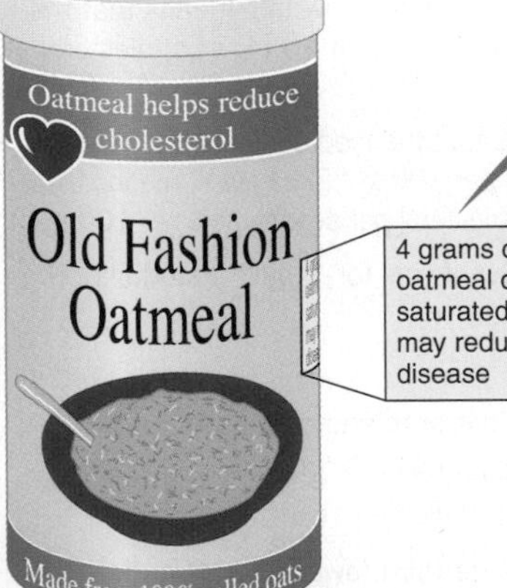

Authorized health claims: Based on significant scientific agreement
- Calcium intake and calcium and vitamin D intake and the risk of osteoporosis
- Sodium intake and the risk of high blood pressure
- Saturated fat and cholesterol intake and the risk of coronary heart disease
- Grain products, fruits, and vegetables that contain fiber and the risk of some types of cancer
- Grain products, fruits, and vegetables that contain fiber, particularly soluble fiber, and the risk of coronary heart disease
- Fruit and vegetable intake and the risk of certain types of cancer
- Soluble fiber from certain foods and the risk of coronary heart disease
- Soy protein and the risk of coronary heart disease
- Dietary fat and the risk of some cancers
- Folic acid and a woman's risk of having a child with a brain or spinal cord defect
- Sugar alcohols and other noncariogenic carbohydrate sweeteners and the risk of dental caries
- Plant sterol/stanol esters and risk of coronary heart disease

Authorized health claims: Based on an authoritative statement
- Whole-grain foods and the risk of heart disease and certain cancers
- Potassium and the risk of high blood pressure and stroke
- Fluoridated water and reduced risk of dental caries
- Saturated fat, cholesterol, and *trans* fat, and reduced risk of heart disease

Qualified health claims
- Qualified claims about cancer risk: Tomatoes and/or tomato sauce and prostate, ovarian, gastric, and pancreatic cancers; calcium and colon/rectal cancer and calcium and recurrent colon/rectal polyps; green tea and cancer; selenium and cancer; antioxidant vitamins and cancer
- Qualified claims about cardiovascular disease risk: Nuts and heart disease; walnuts and heart disease; omega-3 fatty acids and coronary heart disease; B vitamins and vascular disease; monounsaturated fatty acids from olive oil and coronary heart disease; unsaturated fatty acids from canola oil and coronary heart disease; corn oil and heart disease
- Qualified claims about cognitive function: Phosphatidylserine and cognitive dysfunction and dementia
- Qualified claims about diabetes: Chromium picolinate and diabetes
- Qualified claims about hypertension: Calcium and hypertension, pregnancy-induced hypertension, and preeclampsia
- Qualified claims about neural tube birth defects: 0.8 mg folic acid and neural tube birth defects

health claim on foods (but not supplements) is based on a statement of support, called an *authoritative statement*, from an appropriate scientific body of the U. S. government or the National Academy of Sciences. Finally, some health claims are approved when there is emerging but not well-established evidence for a relationship between a food, food component, or dietary supplement and reduced risk of a disease or health-related condition. These are called *qualified health claims* and must be accompanied by a statement explaining this so they do not mislead the consumer (see Figure 2.14 and Appendix E).

Structure/Function Claims Structure/function claims describe the role of a nutrient or dietary ingredient in maintaining normal structure or function in humans. For example, a structure/function claim about calcium may state that "calcium builds strong bones." These claims may also describe the general well-being that arises from consumption of a nutrient or dietary supplement, such as "fiber maintains bowel regularity." Some structure/function claims also describe a benefit in relation to a nutrient-deficiency disease, such as "vitamin C prevents scurvy." These statements must also tell how widespread the disease is in the United States. Structure/function claims are not approved by the FDA. They are based on the manufacturer's review and interpretation of the scientific literature and are not supposed to be untrue or misleading, but they can be confusing. For example, the health claim "lowers cholesterol" requires FDA approval, but the structure/function claim "helps maintain normal cholesterol levels" does not. It would not be unreasonable for consumers with high cholesterol to conclude that a product that "helps maintain normal cholesterol levels" would help lower their elevated blood cholesterol level to within the normal range. Structure/function claims must be accompanied by the disclaimer that "This statement has not been evaluated by the Food and Drug Administration. This product is not intended to diagnose, treat, cure, or prevent any disease" (see Figure 2.13).

Labeling of Food Served in Restaurants

Food served in restaurants, delicatessens, and bakeries is not required to carry labels unless the food is from a food establishment that has 20 or more locations (**Figure 2.15**). Vending machine operators who own or operate 20 or more vending machines must also disclose calorie content for certain items.[17] In any restaurant nutrition information must be available upon request when a claim is made about a menu item's nutritional content or

AFP. PHOTO/Emmanuel Dunand/Getty Images, Inc

Figure 2.15 Menu calorie labeling Food chains must list calorie content information for standard menu items on restaurant menus and menu boards, including drive-through menu boards. Other nutrient information, such as total fat, saturated fat, cholesterol, sodium, total carbohydrates, sugars, fiber, and total protein, must be made available in writing upon request.

THINK CRITICALLY

Do you think calorie labeling is an effective strategy for getting Americans to eat less? Why or why not?

TABLE 2.7 How to Choose Healthy Restaurant Meals

Fast food
• Skip the breaded and fried chicken or fish and choose broiled instead.
• Choose a plain burger rather than one with extra sauce and cheese.
• Split an order of fries with a friend or choose a salad instead.
• Top your pizza with vegetables rather than meat and skip the extra cheese.
Standard American fare
• Avoid buffets and all-you-can-eat specials.
• Order a baked potato instead of French fries.
• Avoid thick buttery, creamy, and cheesy sauces such as béarnaise, hollandaise, or Mornay sauce. If you are unsure what is in the sauce, ask the waitstaff.
• Trim the fat from meat and remove the skin from poultry.
• Order your protein grilled, broiled, flame-cooked, steamed, poached, roasted, or baked.
• Order your salad dressing on the side—then you can choose how much you add.
Ethnic choices
• Chinese: Choose items with large portions of vegetables and skip the fried wontons, dumplings, and egg rolls.
• Italian: Have a red sauce instead of high-fat alfredo or other cream sauce on your pasta.
• Mexican: Go easy on chips and order fajitas instead of the fried items such as chile rellenos, chorizo, chimichangas, or flautas.
• Indian: Choose dishes with lentils and vegetables and avoid ghee (clarified butter), coconut milk, and cream sauces such as korma. Choose tikka and tandoori chicken or fish and kabob preparations and avoid crispy dishes.
• Middle Eastern: Have hummus with pita, yogurt sauce (tzatziki sauce), eggplant, and other vegetable dishes, and avoid fatty gyros, fried falafel, and baklava.

health benefits, such as "low-fat" or "heart healthy."[18] The food labeling laws that regulate packaged foods also apply to menus, so these terms mean the same thing they do on food labels. For example, if you order a "low-fat" tostado, you can assume that low fat means that it contains 3 g or less of fat per serving, as it would on a food label (**Table 2.7**).

2.6 Other Nutrition Guidelines

LEARNING OBJECTIVES

- List the parameters that foods in the same Exchange List have in common.
- Describe the purpose of the Healthy People Initiative.

Exchange Lists A system of grouping foods based on their energy and macronutrient content.

In addition to the guidelines discussed in the previous section, there are a number of other types of recommendations made to promote a healthy diet and lifestyle. The **Exchange Lists** is a food group system that is useful in planning diets to meet specific energy and macronutrient goals. The Healthy People Initiative is a health promotion program that includes nutrition in its recommendations.

In addition to guidelines for a healthy diet for the general population, recommendations to populations at risk for certain diseases have been published by groups such as the American Heart Association, the American Diabetes Association, and the American Institute for Cancer Research. These groups base their recommendations on sound scientific literature, but because of their interest in preventing a specific disease, their recommendations may differ slightly from one another in emphasis and focus.

How the Exchange Lists Are Used

The Exchange Lists were first developed in 1950 by the American Dietetic Association and the American Diabetes Association as a meal-planning tool for individuals with diabetes. Since then, their use has been expanded to planning weight-loss diets and diets in general.

TABLE 2.8 Energy and Macronutrient Values of the Exchange Lists

Exchange group/lists	Serving size	Energy (kcal)	Carbohydrate (g)	Protein (g)	Fat (g)
Carbohydrate Group					
Starch	⅓ cup pasta, ½ cup potatoes; 1 slice bread	80	15	3	0–1
Fruit	1 small apple, peach, or pear; ½ banana; ½ cup canned fruit (in juice)	60	15	0	0
Milk	1 cup milk or yogurt				
Nonfat		90	12	8	0
Low-fat		110	12	8	3
Reduced fat		120	12	8	5
Whole		150	12	8	8
Other carbohydrates	Serving sizes vary	Varies	15	Varies	Varies
Vegetables	½ cup cooked vegetables, 1 cup raw	25	5	2	0
Meat/Meat Substitute Group	1 oz meat or cheese				
Very lean		35	0	7	0–1
Lean		55	0	7	3
Medium fat		75	0	7	5
High fat		100	0	7	8
Fat Group	1 tsp butter, margarine, or oil; 1 T salad dressing	45	0	0	5

The latest revision of the Exchange Lists divides foods into three main groups based on their macronutrient content: the carbohydrate group, the meat and meat-substitute group, and the fat group (**Table 2.8**).[19] The carbohydrate group includes exchange lists for foods that are sources of carbohydrates: starches, fruits, milk, and vegetables. It also defines a list of other high-carbohydrate foods and indicates how to fit these foods into a diet based on exchanges. The meat and meat-substitute group includes an exchange list with four subgroups: very lean, lean, medium-fat, and high-fat meat. The fat group includes an exchange list with subgroups of monounsaturated, polyunsaturated, and saturated fats (see Online Appendix: Exchange Lists).

The exchanges are designed so that each serving within a list contains approximately the same amount of energy, carbohydrate, protein, and fat. Therefore, any one of the foods on a list can be exchanged with any other food on the list without altering the calories or amounts of carbohydrate, protein, or fat in the diet (**Figure 2.16**). The food groupings of the Exchange Lists differ from the MyPlate groups because the lists are designed to meet energy and macronutrient criteria, whereas the MyPlate groups are designed to be good

Figure 2.16 Exchanging foods with the same macronutrient content Foods that are in the same exchange list, such as rice, bread, and potatoes, are similar in their energy and macronutrient content, so they can be exchanged for one another in a calorie-controlled diet.

sources of certain nutrients regardless of their energy content. For example, a potato is included in the starch exchange list because it contains about the same amount of energy, carbohydrate, protein, and fat as breads and grains, but in MyPlate a potato is in the vegetable group because it is a good source of vitamins, minerals, and fiber.

The Exchange Lists can be used to design diets to meet individual tastes and preferences at specific energy and macronutrient levels. For instance, a diet could be calculated to provide 1600 kcal with 80 grams of protein, 207 grams of carbohydrate, and 50 grams of fat. The consumer would meet these nutrient criteria by consuming a prescribed number of servings from each of the exchanges. For example, he or she would be instructed to choose six starch exchanges, two milk exchanges, three nonstarchy vegetable exchanges, and so on.

The Healthy People Initiative

The U.S. Public Health Service, along with hundreds of private and public organizations, has developed a set of public health objectives called Healthy People. The first set, Healthy People 2000, developed in 1990, was directed toward the year 2000. This set of health-promotion and disease-prevention objectives is revised every 10 years, with the goal of increasing the quality and length of healthy lives for the population as a whole and eliminating health disparities among different segments of the population. The latest version of these objectives has been released as Healthy People 2020. The long-term goal is to create a social climate in which everyone has a chance to live a long, healthy life. Healthy People 2020 includes objectives designed to improve the health of all people in the United States by promoting healthy behaviors, protecting health, assuring access to quality health care, and strengthening community prevention.[20] Many of the objectives of Healthy People 2020 address improving the nutritional status of the population (see Appendix D). For instance, Healthy People is working toward reducing the number of cancer and heart disease deaths and the prevalence of obesity in adults by promoting active lifestyles and diets that follow the MyPlate recommendations. It promotes a reduction in growth retardation in children by encouraging healthy feeding practices, including breast-feeding for infants. Other nutrition-related objectives are designed to improve the delivery of nutrition information and services (see **Debate: How Involved Should the Government Be in Your Food Choices?**).

2.7 Assessing Nutritional Health

LEARNING OBJECTIVES

- Describe three types of information used in assessing nutritional status.
- Compare three methods that are used to assess food intake.
- Explain the types of tools used to assess the nutritional health of populations.

nutritional status State of health as it is influenced by the intake and utilization of nutrients.

nutritional assessment An evaluation used to determine the nutritional status of individuals or groups for the purpose of identifying nutritional needs and planning personal health-care or community programs to meet those needs.

To be healthy, people need to consume combinations of foods that provide appropriate amounts of nutrients. Scientists and public health officials have developed standards for the amounts of nutrients needed and tools for planning diets to meet these needs. But how do we know if the nutritional requirements of an individual or the population are being met? Evaluating the **nutritional status** of individuals and populations can identify nutritional needs and be used to plan diets to meet these needs.

VIDEO

How to Assess an Individual's Nutritional Status

Are you losing weight? Gaining weight? Do you have a history of heart disease in your family? Are you at risk for a nutrient deficiency because you can't get to the store, can't afford to buy healthy foods, or don't know what to eat or how to cook? An individual **nutritional assessment** helps to determine if a person has a nutrient deficiency or excess or is at risk of one, or if that individual is at risk of chronic diseases that are affected by diet. It requires a review of past and present dietary intake, a clinical assessment that evaluates body size and includes a medical history and physical exam, and laboratory measurements. Even with

DEBATE

How Involved Should the Government Be in Your Food Choices?

The typical U.S. diet is not as healthy as it could be. Poor dietary habits in the United States have resulted in an unfit, unhealthy nation. Our overindulgence has contributed to our high rates of obesity, diabetes, high blood pressure, and heart disease.[1] This is the concern not only of the individuals whose lives it affects but also of the government. The dollar cost to our health-care system is huge; half of the $147 billion per year the United States spends on obesity comes from government-funded Medicare and Medicaid. Government concern is not just financial. The fact that almost one in four applicants to the military is rejected for being overweight is suggested as a threat to national security and military readiness.[2] So, who is responsible for our unhealthy diet, and who should be responsible for changing what we eat?

Paul J. Richards/AFP/Getty Images, Inc.

Because of the financial and societal costs of an unfit nation, some argue that the government should be more involved in directing our dietary choices. Proponents of more government involvement in our food choices suggest that our food environment is the cause of our unhealthy eating habits. Obesity expert Kelly Brownell believes that environment plays a more powerful role in determining food choices than does personal irresponsibility.[3] Brownell and other proponents of government intervention argue that the government should treat our noxious food environment like any other public health threat and develop programs to keep us safe and healthy. Just as government regulations help to ensure that our food is not contaminated with harmful bacteria, laws could ensure that what you order at a restaurant will not contribute to heart disease or cancer. Unfortunately, unlike bacteria, individual foods are difficult to classify as healthy or unhealthy. Almost all food has some nutritional benefits, and the arguments as to what is a "junk food" and what we should add or subtract from our diets are ongoing. However, many people believe there are things that could be done to ensure healthier choices.

One option to encourage healthier choices suggested by proponents of government intervention is to tax junk food, making it more expensive, and increase subsidies for fruits and vegetables, making them less expensive. Other suggestions include zoning restrictions to keep fast-food restaurants away from schools and child-care facilities and limitations on the types of foods that can be advertised on children's television. All these ideas have pros and cons, and none will absolve individuals of the responsibility for getting more exercise and making healthier food choices.

Opponents of government involvement believe it is an infringement on personal freedom and suggest that individuals need to take responsibility for their actions. They propose that the food industry work with the public to make healthier food more available and affordable. Many food companies have already responded to the need for a better diet; General Mills and Kellogg's offer whole-grain cereals. And the giant food retailer Wal-Mart has announced a major campaign to make healthy food more affordable.

Our current food environment makes unhealthy eating easy. Opportunities for fatty, salty, and sweet foods are available 24/7, and the portions offered are often massive. To preserve our public health, the United States needs to change the way it eats. This change could be driven by government regulations and taxes, it could come from changes in the food industry, or it could come from individuals taking more responsibility for their choices and their health. Should we as individuals take responsibility for our diet and health, or should the government intervene?

THINK CRITICALLY: Whose responsibility are your food choices? For example, is a restaurant that sells 32-ounce sodas resposible when a customer drinks this amount and gains weight?

---

[1]U.S. Department of Agriculture and U.S. Department of Health and Human Services. *Dietary Guidelines for Americans*, 2010, 7th ed. Washington, DC: U.S. Government Printing Office, December 2010.

[2]Cawley, J., and Maclean, J. C., "Report: Unfit for service: The implications of rising obesity for U.S. military recruitment," NBER Working Paper no. 16408, September 2010.

[3]Vaitheeswaran, V. Economist debates: Food policy. *The Economist*, December 8, 2009.

Figure 2.17 Methods of assessing food intake
There are a variety of ways to determine what someone eats. Which tool is best depends on the type of information needed, how many subjects need to be surveyed, and the time and cost constraints.

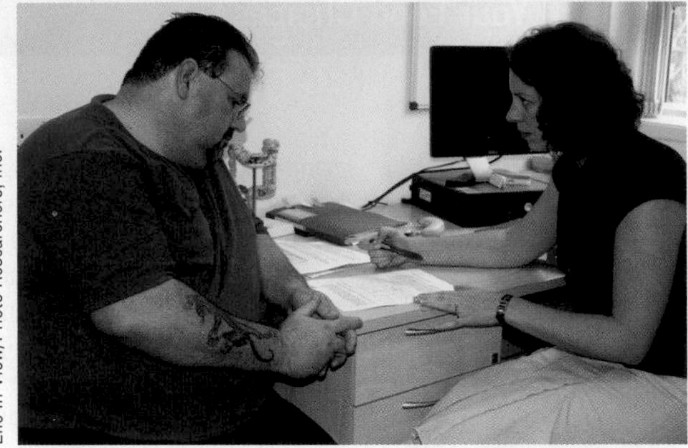

(a) 24-hour recall: The 24-hour recall method relies on a subject's memory, but because subjects don't record their intake, they are less likely than with a food record to change what they eat. It can survey large numbers of people in a short time and can be conducted by telephone.

FOOD DIARY

Record all the food and beverages you eat. Include the food, how it was prepared, the amount you ate and the brand name. Don't forget to list all fats used in cooking and all spreads and sauces added.

Time	Food	Kind and how prepared	Amount
7:00 A.M.	Eggs	scrambled	2
	Butter	in eggs	1 tsp.
	toast	whole wheat	2 slices
	Butter	on toast	2 tsp.
	Milk	non-fat	8 oz.
	Orange juice	from frozen concentrate	8 oz.
12:00 P.M.	Big Mac.	McDonald's	1

(b) Food intake record: Food records created as individuals consume their meals can be very reliable and provide detailed information, but they are time-consuming for participants and can be costly and time consuming to analyze.

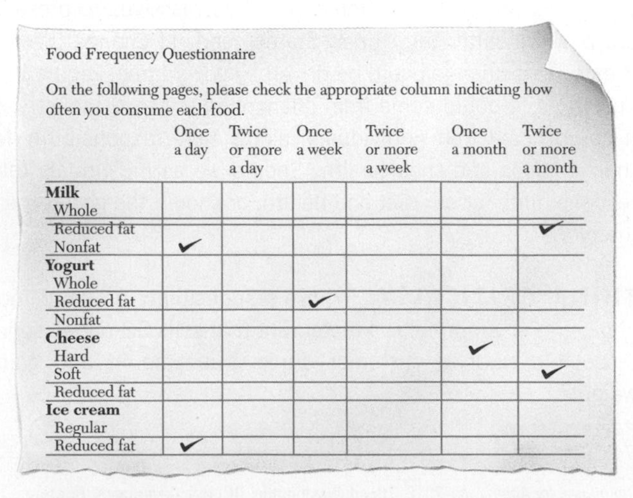

Food Frequency Questionnaire

On the following pages, please check the appropriate column indicating how often you consume each food.

	Once a day	Twice or more a day	Once a week	Twice or more a week	Once a month	Twice or more a month
Milk						
Whole						
Reduced fat						✓
Nonfat	✓					
Yogurt						
Whole						
Reduced fat			✓			
Nonfat						
Cheese						
Hard					✓	
Soft						✓
Reduced fat						
Ice cream						
Regular						
Reduced fat	✓					

(c) Food frequency questionnaire: Food frequencies can be used to study food intake patterns, such as typical fruit and vegetable intake. They are easy for participants and relatively inexpensive for researchers to analyze.

all these tools, diagnosing a nutritional deficiency or excess is not trivial. Estimates of dietary intake are not always accurate, and symptoms may be indistinguishable from other medical conditions.

How Dietary Intake Is Estimated Accurate information about food intake is necessary to study nutrient requirements and functions, to make nutrition recommendations for a population, and to evaluate an individual's nutritional status. Dietary intake information can be obtained by observing all the food and drink consumed by an individual for a specified period of time or by asking the person to record or recall his or her intake. Neither option is ideal. Being observed or asked to record intake can affect an individual's intake, and recording and recalling intake are imprecise measures because these methods rely on the memory and reliability of the consumer. For instance, a person who is attempting to lose weight may tend to report smaller portions than he or she actually ate.[21] Despite this problem, the commonly used methods described here are the best tools available for evaluating dietary intake to predict nutrient deficiencies or excesses.

24-Hour Recall The most common method of assessing dietary intake is a **24-hour recall,** in which a trained interviewer asks a person to recall exactly what he or she ate during the preceding 24-hour period. A detailed description of all food and drink, including descriptions of cooking methods and brand names of products, is recorded (**Figure 2.17a**). Since food intake varies from day to day, repeated 24-hour recalls by the same individual provide a more accurate estimate of typical intake.

Food Diary or Food Intake Record Food intake information can also be gathered by having a consumer keep a **food diary**, or record, of all the food and drink consumed for a set period of time. Typically, this is done for two to seven days, including at least one weekend day, since most people eat differently on weekends than during the school or work week. Foods may be weighed or measured or portion sizes just estimated (**Figure 2.17b**). To accurately assess intake, the record needs to be as complete as possible, including all beverages, condiments, and the brand names and preparation methods. Food records can provide detailed information, but the act of recording can affect intake. Someone may decide to skip the handful of chips they would have eaten rather than record that they ate it.

Food Frequency A **food frequency questionnaire** lists a variety of foods, and the consumer is asked to estimate the frequency with which he or she consumes each item or food group: For example, "How often do you drink milk?" or "How many times a week do you eat red meat?" This method cannot be used to itemize a specific day's intake, but it can give a general picture of a person's typical pattern of food intake (**Figure 2.17c**).

Diet History A **diet history** collects information about dietary patterns. It may review eating habits: Do you cook your own meals? Do you skip lunch? It may also include a combination of methods to assess food intake such as a 24-hour recall along with a food frequency questionnaire. The combination of two or more methods often provides more complete information than one method alone. For instance, if an individual's 24-hour recall does not include milk but a food frequency questionnaire suggests that the individual usually drinks milk five days a week, the two can be combined to provide a more accurate picture of this individual's typical intake.

24-hour recall A method of assessing dietary intake in which a trained interviewer helps an individual remember what he or she ate during the previous day.

food diary A method of assessing dietary intake that involves an individual keeping a written record of all food and drink consumed during a defined period.

food frequency questionnaire A method of assessing dietary intake that gathers information about how often certain categories of food are consumed.

diet history Information about dietary habits and patterns.

Analyzing Nutrient Intake Once information on food intake has been obtained, the nutrient content of the diet can be compared to recommended intakes. This can be done in a number of ways. To get a general picture of dietary intake, an individual's food record can be compared with a guide for diet planning such as MyPlate. For example, does the individual consume the recommended amount of milk each day? If an evaluation of the energy and macronutrient content of the diet is needed, it can be estimated using the Exchange Lists. A more precise and extensive analysis of dietary intake can be done by totaling the nutrients contributed by each food item.

Information on the nutrient composition of foods is available on food labels, in published food composition tables, and in computer databases. Food labels provide information for only some nutrients, and they are not available for all foods. Food composition tables generated by government and industry laboratories can provide more extensive information on food composition (see the *Nutrient Composition of Foods* supplement for an abbreviated list). The major source of food composition data in the United States is the USDA Nutrient Database for Standard Reference, which is available online.[22] Computer programs with food composition databases, such as **iProfile**, are available for professionals and for home use. An easy-to-use online nutrient analysis can be done by using MyPlate SuperTracker, which uses the USDA database to analyze foods and diets entered into the program.

To analyze nutrient intake correctly using a computer program, each food and the exact portion consumed must be entered into the program. If a food is not found in the computer database, an appropriate substitute can be used or the food can be broken down into its individual ingredients. For example, homemade vegetable soup could be entered as generic vegetable soup or as vegetable broth, carrots, green beans, rice, and so on. If a new product has come on the market, the information from the food label can be added to the database. The advantage of computer diet analysis is that it is fast and accurate. A program can calculate the nutrients for each day or average them over several days. It can also compare nutrient intake to recommended amounts. However, the information generated by computer diet analysis is useful only if it is entered correctly and interpreted appropriately. Also, a nutrient intake that is below the recommended amount does not always indicate a deficiency, and intake that meets recommendations does not ensure adequate nutritional status. **Figure 2.18** illustrates how iProfile diet analysis software compares the nutrients in a diet with recommendations.

Nutrient	My DRI	My Intake	Percent of My DRI (0% 50% 100%)
Vitamin A (RAE)	700 μg	525	75%
Vitamin C	75 mg	86	115%
Iron	18 mg	9.7	54%
Calcium	1000 mg	750	75%
Saturated fat	< 23.8 g	31.9	Above recommended range

Figure 2.18 Computerized diet analysis
In this example based on an iProfile report, which shows only a few nutrients, intake of vitamin C and saturated fat is above the recommended amounts and intake of vitamin A, calcium, and iron is below the recommended amounts.

anthropometric measurements External measurements of the body, such as height, weight, limb circumference, and skin-fold thickness.

Anthropometric Measurements Evaluating nutritional health also involves an assessment of an individual's height, weight, and body size. These **anthropometric measurements** can be compared with population standards (see Appendix B) or used to monitor changes in an individual over time (**Figure 2.19**). If an individual's measurements differ significantly from the standards, it could indicate a nutritional deficiency or excess; however, this information should be evaluated only within the context of that person's personal and family history. For example, children who are small for their age may have a nutritional deficiency or may simply have inherited their small body size. Individuals who weigh less than the standard may be adequately nourished if they have never weighed more than their current weight and are otherwise healthy.

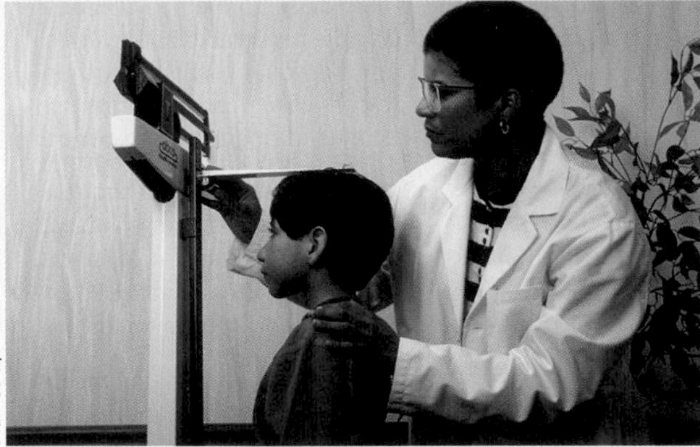

Blair Seitz/Photo Researchers

Figure 2.19 Body measurements
Nutritional assessment includes measures of body dimensions such as height and weight, which can be monitored over time or compared with standards for a given population.

Medical History A medical history is an important component of a nutritional assessment because dietary needs depend on genetic background, life stage, and health status. Family history is important because the risk of developing nutrition-related diseases is affected by an individual's genes (**Figure 2.20**). If your mother died of a heart attack at age 50, you have a higher than average risk of developing heart disease. If you have a family history of diabetes, you have an increased risk of developing this disease. If both of your parents are overweight, it increases the chances that you, too, will have a weight problem.

Information about life stage is important when assessing nutritional status because nutrient needs vary at different stages. Pregnant women need more of some nutrients and energy to support the development of a healthy newborn. Energy and protein needs per unit body weight are higher in infants than at any other time of life. The needs of older adults change as their body composition changes and the ability to digest, absorb, and metabolize certain nutrients declines.

Existing health conditions can impact nutritional health. Some conditions, such as arthritis, affect the ability to acquire and prepare food. Others affect the kinds of foods that should be consumed or the way nutrients are handled by the body. For example, gastrointestinal disorders may decrease the ability to digest foods and absorb nutrients. Kidney disease alters the ability to excrete nitrogen and so affects the amount of protein that should be consumed.

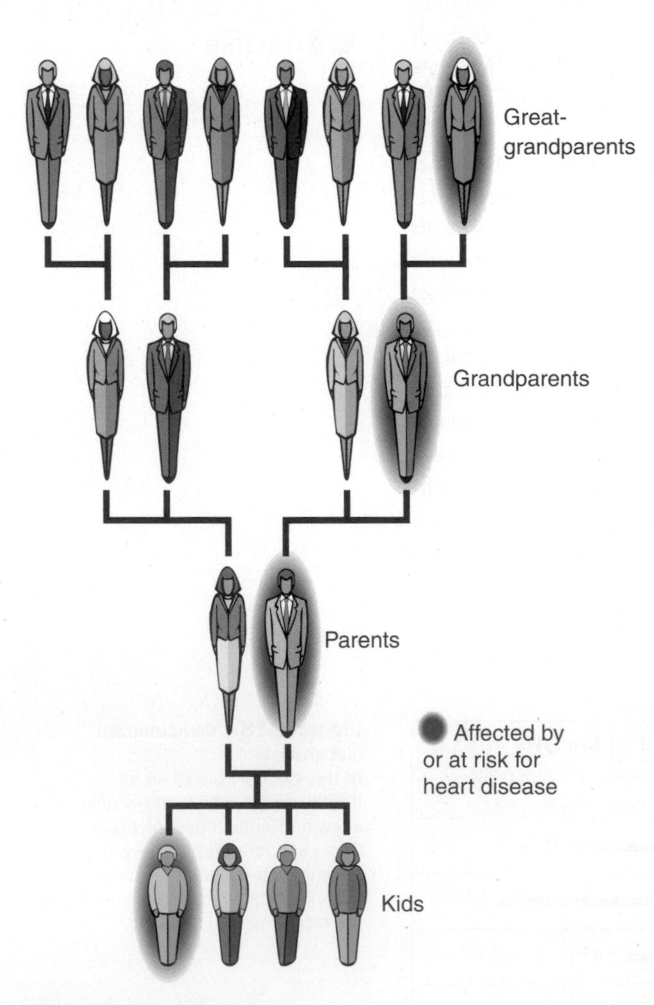

Figure 2.20 Family history
This family tree traces heart disease in one family. It illustrates how a person's family medical history can reveal his or her genetic predisposition for developing heart disease.

Physical Exam In conjunction with personal and family medical history, a careful physical exam can detect the symptoms of, and risk factors for, nutrition-related diseases. In a physical exam, all areas of the body including the mouth, skin, hair, eyes, and fingernails are examined for indications of poor nutritional status. Symptoms such as dry skin, cracked lips, or lethargy may indicate a nutritional deficiency, but these types of symptoms are nonspecific and may be due to factors unrelated to nutritional status. Determining whether the symptoms noted in a physical exam are due to malnutrition or another disease requires that they be evaluated not only within the context of each individual's medical history but in conjunction with the results of laboratory measurements.

Laboratory Measurements Measures of nutrients or their by-products in body cells or fluids such as blood and urine can be used to detect nutrient deficiencies and excesses (see Appendix C). For instance, levels of the blood protein albumin are often used to assess protein status. For some nutrients, blood levels reflect what has been absorbed from the most recent meal rather than the total body status of the nutrient. To assess the status of these nutrients, it may be necessary to measure nutrient functions rather than just blood levels. For example, vitamin B_6 is needed for chemical reactions involved in amino acid metabolism. Measuring the rates of chemical reactions that require vitamin B_6 can be used to assess vitamin B_6 status.

Laboratory data can also be used to evaluate risk for nutrition-related chronic diseases. For instance, heart disease risk can be assessed by measuring blood cholesterol. Measuring the amount of glucose in the blood can be used to diagnose diabetes. More sophisticated medical tests can be used to obtain additional information about the risk and progression of nutrition-related diseases. For example, procedures are available to determine the extent of coronary artery blockage in an individual with heart disease or to assess bone density in someone at risk for osteoporosis.

How a Nutrient Deficiency Progresses A nutritional deficiency usually takes time to develop. For example, an individual who is not meeting the requirement for protein may not suffer any physical signs of protein deficiency for months. Deficiencies generally progress through a number of stages. Appropriate nutritional assessment tools can identify deficiencies at any of these stages and allow intervention to restore nutritional health (**Figure 2.21**).

The earliest stage of nutrient deficiency is intake that is inadequate to meet needs. This may occur due to a deficient diet, poor absorption, increased need, or increased losses from the body. Assessment of dietary intake can help identify low levels in the diet, and a physical exam, medical history, and laboratory tests can identify conditions that might reduce absorption or increase need or losses. The next stage of deficiency is declining nutrient stores in the body. Anthropometric measures that assess body weight and body fat can be used to monitor energy stores. Laboratory tests that measure levels of nutrients and their by-products in blood and tissues can be used to detect decreases in nutrient stores. For instance, measuring levels of the iron-containing protein ferritin in the blood can be used to detect low iron stores. The next stage of deficiency is altered biochemical or physiological functions, such as low enzyme activities or reduced amounts of regulatory or structural molecules. Finally, if a deficiency persists, function is disrupted enough

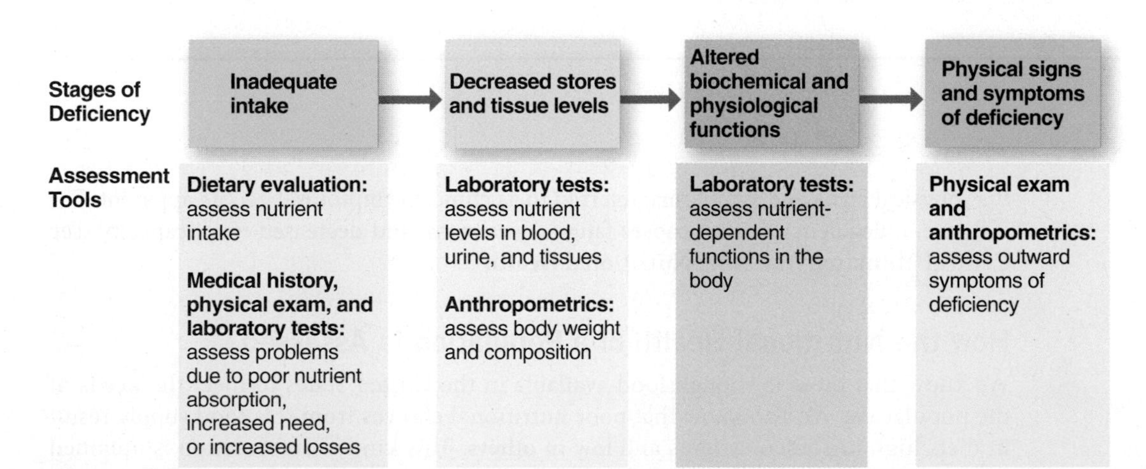

Figure 2.21 **Assessing different stages of nutrient deficiency**
Different assessment tools yield different types of information that can be used to evaluate the severity of a nutrient deficiency.

Assessing Nutritional Health

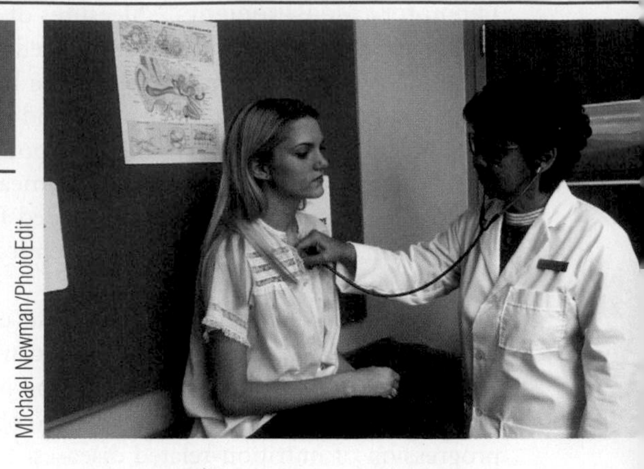
Michael Newman/PhotoEdit

Darra is a 23 year-old freshman in college who has been feeling tired and has had difficulty concentrating in class. She goes to the health clinic looking thin and pale. Anthropometric measurements show that she is 5'4" tall and weighs 114 lbs; a year ago she weighed 120 lbs. She tells the physician that she hasn't been trying to lose weight and that she stopped eating red meat a year ago. The physician suspects iron deficiency anemia, so she checks Darra's blood and reviews her diet.

LABORATORY ASSESSMENT

The results of her blood test indicate that her blood hemoglobin level is 11.2 g per 100 mL and her hematocrit is 35 mL per 100 mL.

CRITICAL THINKING QUESTION

- Compare Darra's hemoglobin and hematocrit with normal values (Appendix C).

DIET ANALYSIS

The following table shows the nutrients in Darra's diet that are above or below recommendations.

NUTRIENT	AMOUNT
Protein	46 g
Iron	6 mg
Calcium	1300 mg

CRITICAL THINKING QUESTIONS

- Compare Darra's iron intake with her RDA.
- Does Darra have iron deficiency anemia? Why or why not?
- Her calcium intake exceeds the RDA. Is this a concern? Why or why not?

iProfile Use iProfile to find foods that are good sources of iron.

that physical signs and symptoms, referred to as clinical symptoms, become apparent. For instance, a deficiency of iron causes fatigue, weakness, and decreased work capacity (see **Critical Thinking: Assessing Nutritional Health**).

How the Nutritional Health of a Population Is Assessed

We know that there is enough food available in the United States to meet the needs of the population. We also know that poor nutritional choices from this food supply result in diets high in some nutrients and low in others. This kind of information is obtained by monitoring what foods are available and what is consumed. In the United States, the National Nutrition Monitoring and Related Research Program is responsible for providing an ongoing description of nutrition conditions in the population by collecting information about food availability and consumption; food composition; and the eating behaviors, health, and nutritional status of the population.[23] These epidemiological data are gathered from population surveys. These surveys are key in establishing relationships among diet, nutritional status, and the health of the U.S. population. The information is used for the purpose of planning nutrition-related policies and programs and predicting future trends of public health importance. For example, they may help identify the need for nutrition education, food assistance programs, or the addition of a specific nutrient to the food supply.

Monitoring the Food Supply The food available to a population is estimated using **food disappearance surveys**. The food supply includes all that is grown, manufactured, or imported for sale in the country. Food use or "disappearance" is estimated by measuring what food is sold. These types of surveys are used to estimate what is available to the population, provide year-to-year comparisons, and identify trends in the diet; but they tend to overestimate actual intake because they do not consider losses that occur during processing, marketing, and home use. Also, the surveys do not consider how food is distributed throughout the population. For example, **Figure 2.22** illustrates the food disappearance data on milk consumption between 1910 and 2005. It shows that the consumption of whole milk, which is high in fat, has declined since the 1950s and the consumption of lower-fat milks has increased. From this it can be concluded that fat intake from milk has declined. But the graph also indicates that total milk consumption has been declining. This may alert the government that calcium intake from milk has decreased and that there may be a risk for low calcium intake in the population. The numbers in this graph do not give any information about how much milk each person is drinking or who is at risk for inadequate calcium intake.

food disappearance surveys
A survey that estimates the food use of a population by monitoring the amount of food that leaves the marketplace.

Monitoring Nutritional Status The nutritional status of the population is monitored by examining and comparing trends in food intake and health. This is done by interviewing individuals within the population to determine what food is actually consumed and by collecting information on health and nutritional status. The Department of Health and Human Services conducts the **National Health and Nutrition Examination Survey (NHANES)**, which combines information on food consumption with medical histories, physical examinations, and laboratory measurements to monitor both nutritional and health information. Data on food, energy, and nutrient intake can be assessed by comparing population intake with reference values such as the DRIs or with other guidelines such as the recommendations of the Dietary Guidelines. For example, the NHANES data indicate that few Americans eat the recommended amounts of fruits and vegetables and that the number of people who are overweight has increased in all adult age groups in the past decade (see **Science Applied: NHANES: A Snapshot of America's Nutritional Health**).

National Health and Nutrition Examination Survey (NHANES)
A survey that collects information about the health and nutritional status of individuals in the population.

A system that has been developed to evaluate the adequacy of American diets is the Healthy Eating Index (HEI).[24] It provides a measure that summarizes overall diet quality by scoring 12 components of the diet. An individual who carefully follows the Dietary Guidelines would have an HEI score of 100. An HEI score over 80 implies that a person has a good diet; a score of 51 to 80, a diet that needs improvement; and a score less than 51, a poor diet. The typical American diet has not changed much since it was evaluated in 1994.[24] The average score was 58.2 for 2001–2002, suggesting a diet that needs improvement.

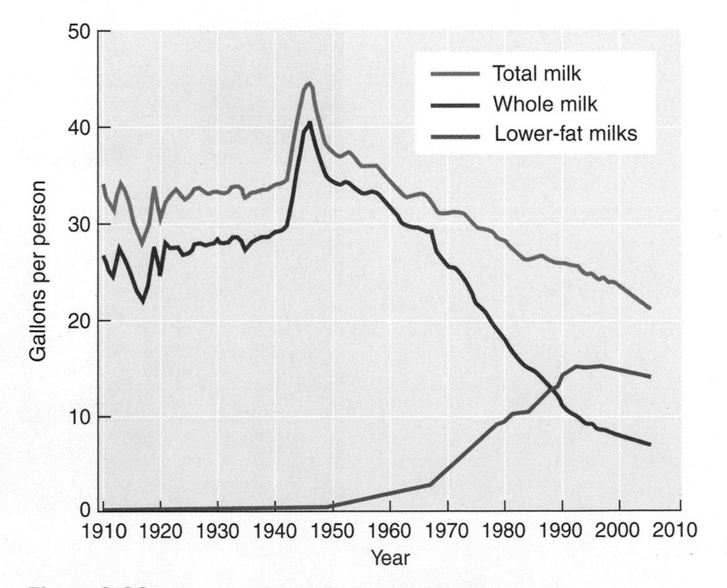

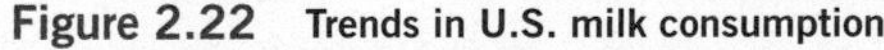

Figure 2.22 Trends in U.S. milk consumption
Food disappearance data in the United States from 1910 to 2005 illustrates that there was an increase in the consumption of lower-fat milks and a decrease in whole-milk consumption during the second half of the 20th century.

SCIENCE APPLIED

NHANES: A Snapshot of America's Nutritional Health

How do we know that 68% of adults are overweight or obese, that obesity among children is on the rise, that almost half of women over the age of 50 have low bone density, or that American adults consume only about half the recommended amount of fruit?[1,2,3] Statistics such as these, along with a multitude of other health and nutrition data, are derived from the National Health and Nutrition Examination Survey, or NHANES. NHANES is the largest and longest-running survey of health *and* nutrition data for the U.S. population.

NHANES evolved from the National Health Survey Act of 1956, which established a continuing National Health Survey to obtain information about the health status of individuals in the United States. Beginning in 1970, a nutrition emphasis was introduced because researchers began to discover links between dietary habits and disease. Since then, five NHANES have been conducted, and in 1999 it became a continuous survey that is conducted on a yearly basis.

THE SCIENCE

To provide a "snapshot" of the health and nutritional status of the U.S. population, NHANES collects data using individual interviews and physical examinations. Since it is not possible to collect data from every person residing in the United States, NHANES tries to obtain a nationally representative yet manageable sample of people. To achieve this goal, a complex sampling design is used.[4] Each survey participant theoretically represents about 50,000 other U.S. residents.

Once sampling has identified a potential survey participant, a trained interviewer conducts an initial in-home interview. Selected participants must agree to participate and sign a consent form.

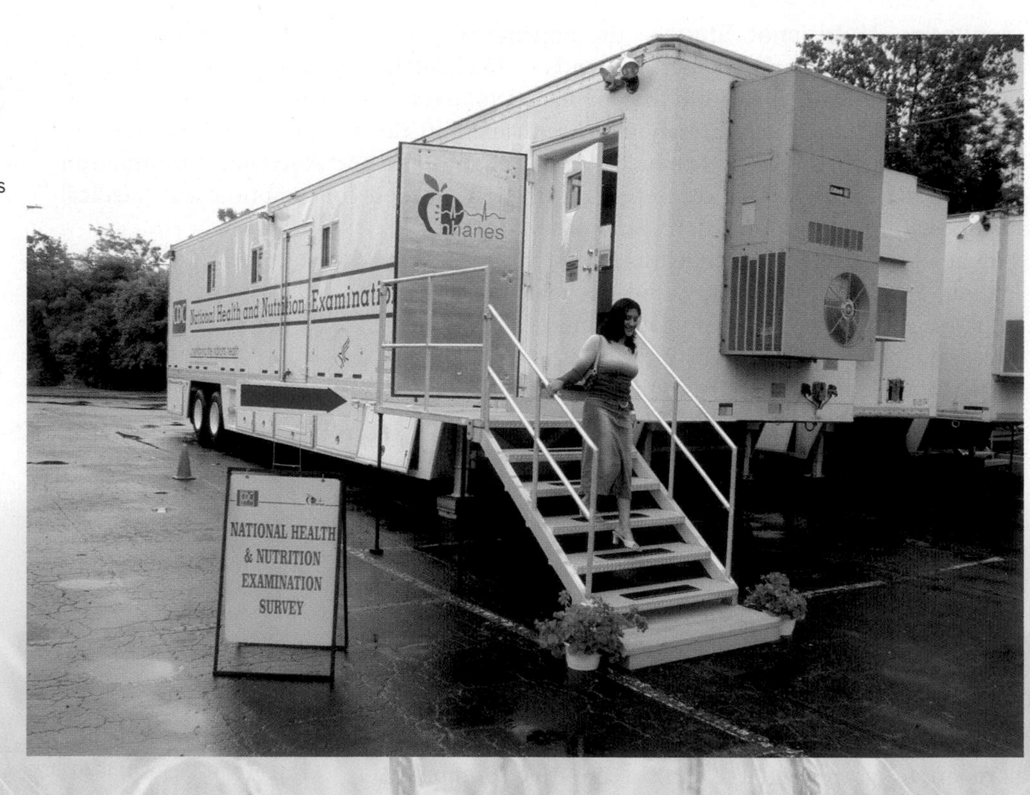

The NHANES mobile examination centers consist of four large trailers that travel around the country. They contain all the diagnostic equipment necessary to conduct a wide range of physical and biochemical evaluations. The survey team consists of a physician, medical and health technicians, and dietary and health interviewers.

National Center for Health Statistics/CDC

(Elena Schweitzer/Shutterstock) Victoriano Izquierdo/Getty Images, Inc. (DAJ/Getty Images, Inc.)

Then each participant visits a mobile examination center for the complete battery of medical and nutritional tests and measurements. Data are collected from physical exams, laboratory tests, and additional survey questionnaires.[5] The nutritional assessment component of NHANES includes several types of data, as shown in the table.

Nutritional Assessment Components in NHANES

Method of assessment	What type of data is obtained?
24-hour recall and food frequency questionnaire	Food intake information
Nutrition-related interview	Information about dietary supplements, water intake, meal and snack patterns, infant feeding practices, alcohol intake, and food sufficiency
Anthropometric measurements	Height, body weight, body composition, and body circumference measures to determine body weight-fat distribution
Laboratory measurements	Blood lipid levels, blood glucose levels, and measures of protein, vitamin, and mineral status
Clinical assessments	Chronic disease risk assessment for conditions such as heart disease, diabetes, osteoporosis, and gallbladder disease

THE APPLICATION

NHANES provides health professionals and policymakers with the data needed to determine the incidence of major diseases and health conditions such as cardiovascular disease, diabetes, obesity, osteoporosis, and cancer, as well as to identify and monitor trends in medical conditions and risk factors. It helps identify emerging public health issues so that the appropriate public health policies and preventative interventions can be developed. NHANES also provides information on the prevalence and trends of risk factors and other key health behaviors, including alcohol use, tobacco use and exposure, drug use, sexual experience, immunization histories, and physical activity.

Data from NHANES have been instrumental in the development and implementation of a number of health-related guidelines and reforms and public policy initiatives. NHANES data were used to develop growth standards for children and were instrumental in reducing the manufacture and sale of lead-containing products. Nutrient intake data from the first two NHANESs indicated that certain segments of the U.S. population were not consuming adequate amounts of iron. This information led to the fortification of grain products with iron. Similarly, NHANES data that supported the connection between the B vitamin folic acid and the incidence of neural tube defects, a type of birth defect, led to the fortification of grain products with folic acid. Ongoing collection and critical analysis of NHANES data will continue to provide information on the nutritional health of the U.S. population that can be used to guide nutrient recommendations and direct nutrition policies.

THINK CRITICALLY: If NHANES data found that the incidence of type 2 diabetes has increased at about the same rate as the consumption of sugar-sweetened beverages, does this mean that sugar causes diabetes? Why or why not?"

[1]Flegal, K. M., Carroll, M. D., Ogden, C. L., and Curtin, L. R. Prevalence and trends in obesity among U.S. adults, 1999–2008. *JAMA* 303:235–241. 2010.
[2]Ogden, C. L., and Carroll, M. D. Prevalence of obesity among children and adolescents: United States, trends 1963–1965 through 2007–2008, June 2010.
[3]Centers for Disease Control and Prevention, Key Statistics from NHANES.
[4]Centers for Disease Control and Prevention, National Center for Health Statistics, Continuous NHANES Web tutorial, key concepts about NHANES survey design.
[5]Centers for Disease Control and Prevention, National Center for Health Statistics, Continuous NHANES Web tutorial, key concepts about survey methodology.

OUTCOME

© Spaces Images/SuperStock

© Blend Images/SuperStock

Joe wanted to make sure that he was not only making nutritious food choices, but also that he was choosing a diet to optimize his fitness for the track season. To construct his training diet, he learned to use MyPlate, which promotes a diet that meets the recommendations of the DRIs and incorporates the Dietary Guidelines for Americans. By comparing his diet to his MyPlate recommendations, he figured out what he was consuming in excess and what types of foods he needed more of. This prepared him to better navigate the seemingly endless variety of selections offered by the dorm cafeteria. Although he was tempted by the unlimited portions and all those great desserts, he knew that he would regret the extra pounds when it came time to run track. MyPlate helped him decide how much food to put on his plate and how many treats he could indulge in without putting on weight. When he bought snacks, he used the information on food labels to determine nutritional benefits and pitfalls. **By the time the track season started, Joe was in excellent physical as well as nutritional health. He had kept his weight where he wanted it and felt fit to compete. His diligence paid off, and he qualified for the state championships.**

APPLICATIONS

ASSESSING YOUR DIET

1. **What do you eat? To find out, make a form similar to the sample food record shown below or use one provided by your instructor to keep a food diary of everything you eat for three consecutive days. Since you may eat differently on weekends, record for two weekdays and one weekend day. To make sure you don't forget anything, carry your record with you and record food as it is consumed. This record will be used in Applications throughout this book to focus on particular nutrients. Make the record as complete as possible by using the following tips:**
   - **a.** Include all food and drink, and be as specific as possible in describing the food and drink you consume. For example, did you eat a chicken breast or thigh? Was there skin on it?
   - **b.** Measure or estimate as carefully as possible the portion size that you ate, for example, ½ cup of rice, 10 potato chips, 2 ounces of tofu, and 6 ounces of milk.
   - **c.** Record the preparation or cooking method. For example, was your potato peeled? Was it baked or fried?
   - **d.** Include anything added to your food, for instance, butter, ketchup, or salad dressing.
   - **e.** Don't forget snacks, beverages, and desserts.
   - **f.** If the food is from a fast-food chain, list the name.
   - **g.** You may have to break down mixed dishes into their ingredients. For example, a tuna sandwich can be listed as 2 slices of whole-wheat bread, 1 tablespoon of mayonnaise, and 3 ounces of tuna packed in water.
   - **h.** Use iProfile to analyze your food record.

Sample Food Record

Food or beverage	Kind/how prepared	Amount
Chicken salad sandwich:		
Wheat bread		2 slices
Chicken	Skinless breast	2 oz
Mayonnaise	Low-fat	1 T
Diet cola		1 12-oz can

2. **Does your diet meet the recommendations of your MyPlate Daily Food Plan? To find out, go to www.ChooseMyPlate.gov and select "Daily Food Plans" in the "SuperTracker and other tools" menu. Print out your Meal Tracking Worksheet.**
   - **a.** Select one day of your food intake record. On the left side of the Meal Tracking Worksheet, list the foods and amounts you consumed that day.
   - **b.** On the right side of the form, record each food next to the MyPlate food group to which it belongs. In the far right column, record the total amount of food you consumed from each group in cups or ounces. (Hint: Use the information for each food group under the MyPlate tab, such as the food gallery, to help determine your serving sizes.)
   - **c.** How do the amounts you consumed from each food group compare to the recommendations? Are there any food groups you need to eat more or less of?
   - **d.** Are your food choices consistent with the selection tips listed on the worksheet for each group? How might you

modify your food choices to more closely follow these suggestions?

e. List foods you consumed that are mostly solid fat or added sugar.

f. List the foods you consumed from each group that contribute empty calories.

CONSUMER ISSUES

3. What's in packaged foods? Select three packaged foods and check the label.

a. What is the percentage of calories from fat in each of these foods?

b. How much total carbohydrate, total fat, and fiber are in a serving of each?

c. How does each of these foods fit into an overall daily diet with regard to sodium? saturated fat? dietary fiber?

d. If you consumed a serving of each of these three foods, how much more sodium and saturated fat could you consume during the day without exceeding the Daily Value? How much more fiber should you consume that day to meet the Daily Value for a 2000-kcal diet?

4. Food intake data from population surveys provided the following information about the intake of milk and soda in 1977–1978 and 1994–1996.

a. How might these trends have affected the number of calories from added sugars in the teen diet?

b. How might these trends affect calcium intake in teenagers?

Population	Year	Milk consumption (ounces per day)	Soda consumption (ounces per day)
Teenage boys (ages 12–19)	1977–1978	16	7
	1994–1996	10	20
Teenage girls (ages 12–19)	1977–1978	10	7
	1994–1996	6	14

5. Are "health foods" different from standard products?

a. Compare the label from a product such as cereal, crackers, or cookies purchased at the grocery store to the label from a comparable product from a "health" or "natural" food store.

b. Which is higher in total fat? saturated fat? *trans* fat? cholesterol?

c. Which has more calories per serving?

d. Which has more sugars?

e. How do the ingredients differ?

f. What other differences or similarities do you notice?

6. Nutrition scientists at the Harvard School of Public Health and Harvard Medical School have developed the Healthy Eating Plate shown here. They believe it offers more specific and more accurate recommendations for following a healthy diet than MyPlate.

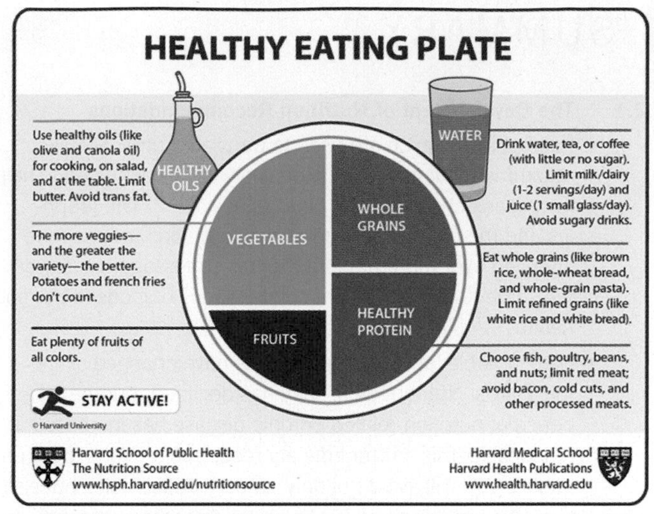

a. Compare the food groups of the Healthy Eating Plate with the food groups on MyPlate.

b. Compare the proportions of food from each food group on the Healthy Eating Plate and MyPlate.

c. The Healthy Eating Plate icon includes basic information about what to choose from each group for a healthy nutrient-dense diet. The MyPlate icon is simpler. The additional food-selection recommendations can be found on the MyPlate Web site. Which approach do you think will better help Americans improve their diets?

CLINICAL CONCERNS

7. To encourage healthy eating in her family, Sylvia hangs a MyPlate poster on the refrigerator. Her family is enthusiastic but don't know how much they should eat from each group. Use the ChooseMyPlate Web site to find out the calorie level and the amounts of food recommended from each of the food groups for the following family members:

a. Her 45-year-old inactive mother.

b. Her 12-year-old sister who plays outside for about 45 minutes a day.

c. Her 16-year-old brother who spends 2 hours a day practicing with the track team.

d. Her 70-year-old grandmother who is inactive.

8. Now that spring has arrived and the outdoor temperature is increasing, Liza begins training for a cycling event that will be held in early September. The event will involve riding 60 to 70 miles a day for 7 days in a row.

a. Liza is 5' 5" tall and weighs 130 pounds. Over the winter she was getting only about 20 minutes of exercise a day. Now she is riding her bike for two to three hours a day. How does this change the amounts recommended from each MyPlate food group?

b. If Lisa lost 10 lbs during her training, how would it change her MyPlate recommendations?

SUMMARY

2.1 The Development of Nutrition Recommendations

- Food guides, which are used to translate nutrient-intake recommendations into food choices, have been used to help Americans choose healthy diets since 1917. The graphics and the number and composition of food groups have changed over the years, but the basic message has stayed the same: Choose the right combinations of foods to promote health.
- Dietary habits and disease patterns have changed since the 1940s. Nutrient deficiency has declined, but the incidence of nutrition-related chronic disease has increased. In response to this, current dietary recommendations focus on intakes that will avoid not only nutrient deficiencies but also avoid excesses and prevent chronic diseases in the majority of healthy persons.

2.2 Dietary Reference Intakes

- The DRIs include four sets of nutrient intake recommendations. The Estimated Average Requirements (EARs) are amounts of nutrients estimated to meet the needs of half of the people in a particular gender and life-stage group. The Recommended Dietary Allowances (RDAs), which are based on the EARs, are recommendations calculated to meet the needs of nearly all healthy individuals (97–98%) in a specific group. Adequate Intakes (AIs) estimate nutrient needs based on average intakes by healthy populations when there is insufficient scientific evidence to calculate an EAR and RDA. Tolerable Upper Intake Levels (ULs) provide a guide for a safe upper limit of intake.
- Energy recommendations of the DRIs include Estimated Energy Requirements (EERs), which provide a recommendation for energy intakes that will maintain body weight, and Acceptable Macronutrient Distribution Ranges (AMDRs), which recommend the proportions of energy intake that should come from carbohydrate, protein, and fat.
- The Dietary Reference Intakes can be used to evaluate and plan nutrient intakes for populations, as a guide for individual intake, and to make judgments about excessive intakes for individuals and populations.

2.3 The Dietary Guidelines for Americans

- The Dietary Guidelines for Americans are a set of diet and lifestyle recommendations designed to promote health and reduce the risk of overweight and obesity and chronic disease in the U.S. population.
- The 2010 Dietary Guidelines emphasize balancing the calories consumed in food and beverages with the calories expended through physical activity in order to achieve and maintain a healthy weight. To accomplish this, they recommend that Americans increase their activity level and choose a healthy eating pattern. Healthy eating patterns are higher in fruits, vegetables, whole grains, low-fat dairy products, and seafood than current American diets, and they are lower in saturated fat, *trans* fat, cholesterol, salt, and added sugar.

2.4 MyPlate: Putting the Dietary Guidelines into Practice

- MyPlate is the USDA's current food guide. It shows the proportions of foods from five food groups that make up a healthy diet. MyPlate stresses using variety, proportionality, and moderation in choosing a healthy diet and promotes the physical activity recommendations included in the Dietary Guidelines.
- Individual Daily Food Plans available at www.ChooseMyPlate.gov recommend amounts of food from each food group and oils based on individual energy needs. Recommendations are also made for the number of empty calories that can be included in an individual's diet. Exceeding the empty calorie limit makes it difficult to meet nutrient needs without exceeding energy needs.

2.5 Food and Supplement Labels

- Standardized food labels are designed to help consumers make healthy food choices by providing information about the nutrient composition of foods and about how a food fits into the overall diet. The Nutrition Facts panel presents information about the amounts of various nutrients in a standard serving. For most nutrients, the amount is also given as a percentage of the Daily Value. A food label's ingredient list states the contents of the product in order of prominence by weight.
- Labels on dietary supplements are designed to help consumers make educated decisions about these products. The safety of dietary supplements is not evaluated by the FDA before they are marketed.
- Food and supplement labels often include FDA-defined nutrient content descriptors, such as "low fat" or "high fiber," and health claims, which refer to a relationship between a nutrient, food, food component, or dietary supplement and the risk of a particular disease or health-related condition. All health claims are reviewed by the FDA and permitted only when they are supported by scientific evidence, but the level of scientific support for such claims varies.
- Nutrition information about food served in restaurants must be available upon request when a claim is made about a menu item's nutritional content or health benefits. Food establishments with 20 or more locations must list calorie-content information on menus and menu boards and other nutrient information must be made available in writing upon request.

2.6 Other Nutrition Guidelines

- The Exchange Lists are used to plan individual diets that provide specific amounts of energy, carbohydrate, protein, and fat.
- The Healthy People Initiative establishes a set of health promotion and disease prevention objectives for the U.S. population.

2.7 Assessing Nutritional Health

- Individual nutritional status is assessed by evaluating dietary intake and by examining and interpreting anthropometric measures, physical exam results, and laboratory values within the context of an individual's medical history.
- The nutritional status of populations is monitored by measuring what foods are available, what foods are consumed, and how nutrient intake is related to overall health.

REVIEW QUESTIONS

1. Which types of DRI standards can be used as a goal for individual intake?
2. What type of DRI standard can be used to evaluate the adequacy of nutrient intake in a population?
3. Which DRI standard can help you determine if a supplement contains a toxic level of a nutrient?
4. What is the purpose of the Dietary Guidelines?
5. How does the MyPlate graphic promote a varied diet?
6. Why is variety important to a healthy diet?
7. Explain how the MyPlate graphic illustrates proportionality.
8. Name two foods from each food group that are low in empty calories and two foods that are high in empty calories.
9. Why do the Dietary Guidelines and MyPlate recommend limiting solid fats and added sugar?
10. Why are serving sizes standardized on food labels?
11. How do the Daily Values help consumers determine how foods fit into their overall diets?
12. What determines the order in which food ingredients are listed on a label?
13. How are the Exchange Lists used in planning diets?
14. What is nutritional status?
15. List the components of individual nutritional assessment.
16. How does food disappearance data and NHANES help monitor the nutritional health of populations?

REFERENCES

1. Davis, C., and Saltos, E. Dietary recommendations and how they have changed over time. In *America's Eating Habits: Changes and Consequences*. U.S. Department of Agriculture, Economic Research Service, Food and Rural Economics Division. *Agriculture Information Bulletin* no. 750.
2. National Research Council, Food and Nutrition Board. *Recommended Dietary Allowances*, 10th ed. Washington, DC: National Academies Press, 1989.
3. U.S. Senate., *Eating in America: Dietary Goals for the United States*. Report of the Select Committee on Nutrition and Human Needs. Cambridge, MA: MIT Press, 1977.
4. U.S. Department of Agriculture and U.S. Department of Health and Human Services. *Dietary Guidelines for Americans, 2010*, 7th ed. Washington, DC: U.S. Government Printing Office, December 2010.
5. World Health Organization/Food and Agriculture Organization. Diet, nutrition and the prevention of chronic diseases. Report of a joint WHO/FAO expert consultation.
6. Institute of Medicine, Food and Nutrition Board. *Dietary Reference Intakes for Calcium, Phosphorus, Magnesium, Vitamin D, and Fluoride*. Washington, DC: National Academies Press, 1997.
7. Institute of Medicine, Food and Nutrition Board. *Dietary Reference Intakes for Thiamin, Riboflavin, Niacin, Vitamin B-6, Folate, Vitamin B-12, Pantothenic Acid, Biotin, and Choline*. Washington, DC: National Academies Press, 1998.
8. Institute of Medicine, Food and Nutrition Board. *Dietary Reference Intakes for Energy, Carbohydrates, Fiber, Fat, Protein, and Amino Acids*. Washington, DC: National Academies Press, 2002, 2005.
9. Centers for Disease Control and Prevention. CDC estimates of foodborne illness in the United States.
10. U.S. Department of Agriculture. MyPlate. Available online at http://www.choosemyplate.gov/. Accessed December 20, 2011.
11. Legault, L., Brandt, M. B., McCabe, N., et al. 2000–2001 food label and package survey: An update on prevalence of nutrition labeling and claims on processed, packaged foods. *J Am Diet Assoc* 104:952–958, 2004.
12. Picciano, M. F. Office of Dietary Supplements, National Institutes of Health. Who is using dietary supplements and what are they using?
13. U.S. Food and Drug Administration, Center for Food Safety and Applied Nutrition /Office of Nutrition, Labeling, and Dietary Supplements. Guidance for food industry, a dietary supplement labeling guide, April 2005.
14. U.S. Food and Drug Administration. Current good manufacturing practice in manufacturing, packaging, labeling, or holding operations for dietary supplements; final rule. *Fed Regist* 72:34751–34958, June 25, 2007.
15. U.S. Pharmacopeia. USP Dietary Supplement Verification Program.
16. U.S. Food and Drug Administration, Center for Food Safety and Applied Nutrition. Label claims.
17. U.S. Food and Drug Administration. New menu and vending machine labeling requirements.
18. U.S. Food and Drug Administration, Center for Food Safety and Applied Nutrition. Guidance for industry: A labeling guide for restaurants and other retail establishments selling away-from-home foods, April 2008.
19. American Diabetes Association. *Choose Your Foods: Exchange Lists for Diabetes*. Alexandria, VA: American Dietetic Association 2007.
20. U.S. Department of Health and Human Services, Office of Disease Prevention and Health Promotion. Topics and objectives index.
21. Johansson, L., Solvoll, K., Bjørneboe, G-E. A., and Drevon, C. A. Under- and overreporting of energy intake related to weight status and lifestyle in a nationwide sample. *Am J Clin Nutr* 68:226–274, 1998.
22. U.S. Department of Agriculture. USDA nutrient database for standard reference.
23. Interagency Board of Nutrition Monitoring and Related Research. Bialostosky, K., ed. *Nutrition Monitoring in the United States: The Directory of Federal and State Nutrition Monitoring and Related Research Activities*. Hyattsville, MD: National Center for Health Statistics, 2000.
24. U.S. Department of Agriculture, Center for Nutrition Policy and Promotion. Diet quality of Americans in 1994–96 and 2001–02 as measured by the Healthy Eating Index, 2005.

***To access links to online sources, please go to www.wiley.com/college/Smolin and Nutrition: Science and Applications, 3rd edition. From this page, select the student or instructor companion site. Once on the desired site, select References.**

CHAPTER OUTLINE

© Bojan Kontrec/iStockphoto

Digestion, Absorption, and Metabolism

3

CASE STUDY

Gillian popped two more antacids into her mouth as she hurried to the next table with a piping hot pizza balanced on her shoulder. The burning sensation in the center of her chest accompanied by burping and a sour taste in her throat and mouth was becoming a regular evening occurrence, and she often woke up with a sore throat and a bad taste in her mouth. Sometimes the chest pain was so severe that it hurt right through to her back, and she was afraid she was having a heart attack. When she talked to her friends, they all said it was just heartburn and they also experienced it occasionally.

Gillian is a 25-year-old college student who waitresses evenings to help pay her expenses. She usually skips breakfast and just has coffee in the morning so she can grab a few more minutes of sleep before her first class. She has a small lunch and consumes most of her food in the evening at work, where she gets meals for free. She typically gulps down three or four slices of pizza during her breaks. After she gets off work at 11 PM, she goes right to bed.

David Woolley/Taxi/Getty Images, Inc.

The heartburn is interfering with her sleep and her class work, so she makes an appointment with her doctor. He schedules a diagnostic test to monitor the acidity in her esophagus. The results indicate that she has gastroesophageal reflux disease, or GERD. GERD is a condition in which the stomach contents leak backward from the stomach into the esophagus, causing damage to the esophageal lining. Treatment is important because, in addition to being uncomfortable, acid in her esophagus can cause esophageal ulcers, dental problems, hoarseness, and cough and can also lead to changes that increase her risk of esophageal cancer. **Her doctor instructs her about some lifestyle modifications that he hopes will allow her to manage the condition without medication.**

3.1 Food Becomes Us

LEARNING OBJECTIVES

- Describe the organization of life from atoms to organisms.
- Name three organ systems that are involved in obtaining and using the nutrients in food.

The old adage that you are what you eat is not literally true, but biochemically it is a fact. The food we eat provides all the energy we need to stay alive and active and all the raw materials we need to build and maintain our body structures and synthesize regulatory molecules.

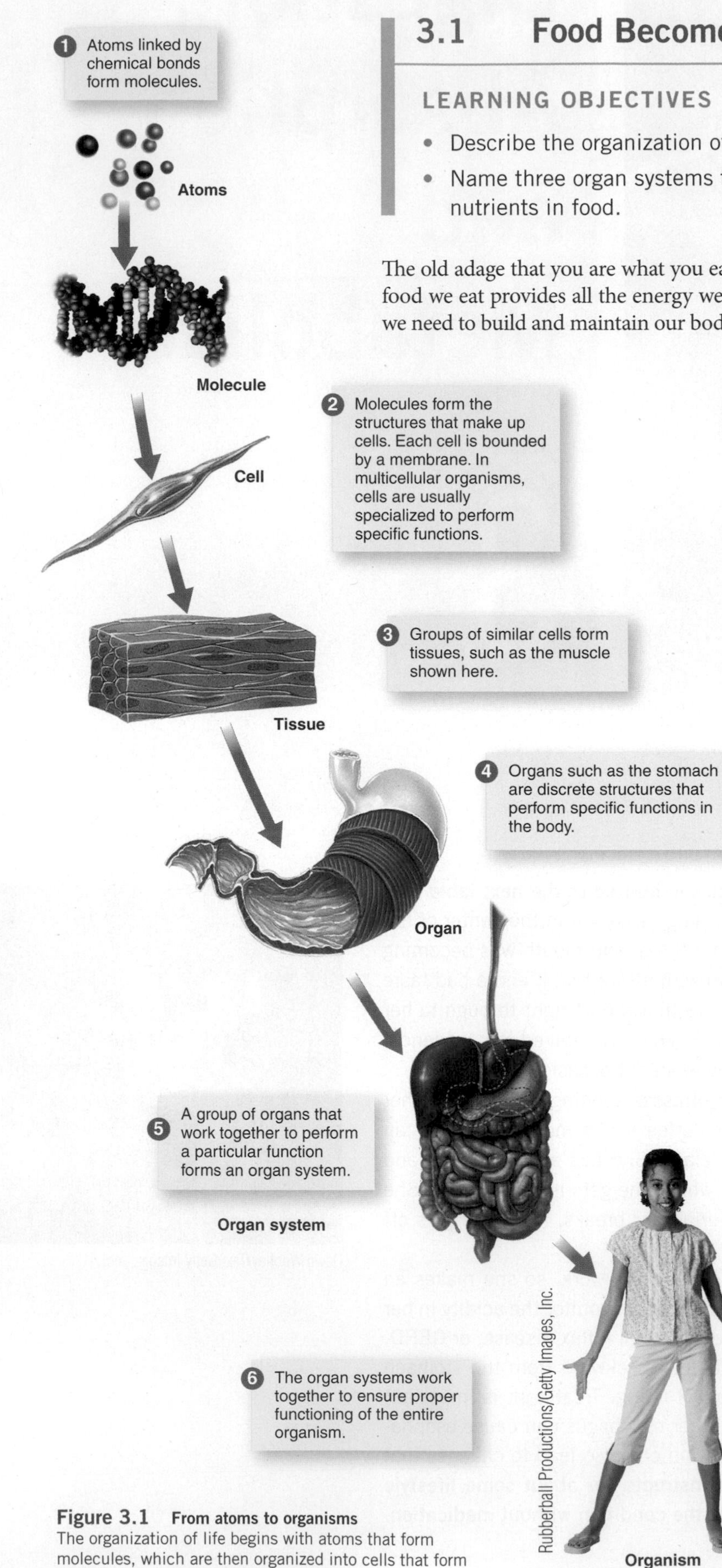

Figure 3.1 From atoms to organisms The organization of life begins with atoms that form molecules, which are then organized into cells that form tissues, organs, and whole organisms.

Atoms Form Molecules

To be useful to us, the food we eat must be broken into smaller components, absorbed into the body, and then converted into forms that can be used. Our bodies and the food we eat, like all matter on Earth, are made of units called **atoms**. Atoms cannot be further broken down by chemical means. Atoms of different **elements** have different characteristics. Carbon, hydrogen, oxygen, and nitrogen are the most abundant elements in our bodies and in the foods we eat. Atoms can be linked by forces called **chemical bonds** to form **molecules** (**Figure 3.1** and Appendix H). The chemistry of all life on Earth is based on organic molecules, which are those that contain carbon bonded to hydrogen. As discussed in Chapter 1, carbohydrates, lipids, proteins, and vitamins are nutrient classes that are made up of organic molecules, whereas water and minerals are inorganic nutrients because they do not contain carbon-hydrogen bonds.

Cells Form Tissues and Organs

In any living system, molecules are organized into structures that form **cells**, the smallest unit of life. Cells of similar structure and function are organized into tissues. The human body contains four types of tissue: muscle, nerve, epithelial, and connective. These tissues are organized in varying combinations into **organs**, which are discrete structures that perform specialized functions in the body (see Figure 3.1). The stomach is an example of an organ; it contains all four types of tissue.

Organs Cooperate as Organ Systems

Most organs do not function alone but are part of a group of cooperative organs called an *organ system*. The organ systems in humans are described in **Table 3.1**. An organ

TABLE 3.1 Organ Systems and Their Functions

Organ system	What it includes	What it does
Nervous	Nerves, sense organs, brain, and spinal cord	Responds to stimuli from the external and internal environments; conducts impulses to activate muscles and glands; integrates activities of other systems.
Respiratory	Lungs, trachea, and air passageways	Supplies the blood with oxygen and removes carbon dioxide.
Urinary	Kidneys and their associated structures	Eliminates wastes and regulates the balance of water, electrolytes, and acid in the blood.
Reproductive	Testes, ovaries, and associated structures	Produces offspring.
Cardiovascular/circulatory	Heart and blood vessels	Transports blood, which carries oxygen, nutrients, and wastes.
Lymphatic/Immune	Lymph and lymph structures, white blood cells	Defends against foreign invaders; picks up fluid leaked from blood vessels; transports fat-soluble nutrients.
Muscular	Skeletal muscles	Provides movement and structure.
Skeletal	Bones and joints	Protects and supports the body; provides a framework for the muscles to use for movement.
Endocrine	Pituitary, adrenal, thyroid, pancreas, and other ductless glands	Secretes hormones that regulate processes such as growth, reproduction, and nutrient use.
Integumentary	Skin, hair, nails, and sweat glands	Covers and protects the body; helps control body temperature.
Digestive	Mouth, pharynx, esophagus, stomach, intestines, pancreas, liver, and gallbladder	Ingests and digests food, absorbs nutrients, eliminates unabsorbed food residues and other wastes.

Source: Adapted from Marieb, E. N., and Hoehn, K. *Human Anatomy and Physiology*, 7th ed. Menlo Park, CA: Benjamin Cummings, 2007.

may be part of more than one organ system. For example, the pancreas is part of the endocrine system as well as the digestive system. Organ systems work together to support the entire organism.

The digestive system is the organ system primarily responsible for the movement of nutrients into the body; however, several other organ systems are also important in using nutrients. The endocrine system secretes **hormones** that help regulate food intake and the function of digestive organs. The nervous system aids in digestion by sending nerve signals that help control the passage of food through the digestive tract. Once absorbed, nutrients are transported to individual cells by the cardiovascular system. The body's urinary, respiratory, and integumentary systems allow the elimination of metabolic waste products.

3.2 The Digestive System

LEARNING OBJECTIVES

- Define *digestion* and *absorption*.
- Explain the roles of mucus, enzymes, nerves, and hormones in digestion.

The digestive system provides two major functions: **digestion** and **absorption**. Most food must be digested in order for the nutrients it contains to be absorbed into the body. When you eat a taco, for example, the tortilla, meat, cheese, lettuce, and tomato are broken apart, releasing the nutrients and other food components they contain. Water, vitamins, and minerals are absorbed without being broken into smaller units, but proteins, carbohydrates, and fats must be digested further. Proteins are broken down into amino acids, most of the carbohydrate is broken down into sugars, and most fats are digested to produce fatty acids. The amino acids, sugars, and fatty acids can then be absorbed into the body. The fiber in whole grains, fruits, and vegetables cannot be digested and therefore is not absorbed from the digestive tract into the body. It and other unabsorbed substances pass through the digestive tract and are eliminated in **feces**.

atom The smallest unit of an element that still retains the properties of that element.

element A substance that cannot be broken down into products with different properties.

chemical bond The force that holds atoms together.

molecule A group of two or more atoms of the same or different elements bonded together.

cells The basic structural and functional units of plant and animal life.

organs Discrete structures composed of more than one tissue that perform a specialized function.

hormones Chemical messengers that are produced in one location, released into the blood, and elicit responses at other locations in the body.

digestion The process of breaking food into components small enough to be absorbed into the body.

absorption The process of taking substances into the interior of the body.

feces Body waste, including unabsorbed food residue, bacteria, mucus, and dead cells, which is excreted from the gastrointestinal tract by passing through the anus.

Organs of the Digestive System

gastrointestinal tract A hollow tube consisting of the mouth, pharynx, esophagus, stomach, small intestine, large intestine, and anus, in which digestion and absorption of nutrients occurs.

The digestive system consists of the **gastrointestinal tract** and accessory organs (**Figure 3.2a**). The gastrointestinal tract is a hollow tube, about 30 feet in length, that runs from the mouth to the anus. It is also called the GI tract, gut, digestive tract, intestinal tract, and alimentary canal. The organs of the gastrointestinal tract include the mouth, pharynx, esophagus, stomach, small intestine, large intestine, and anus. The inside of the tube that these organs form is called the *lumen* (**Figure 3.2b**). Food within the lumen of the gastrointestinal tract has not been absorbed and is therefore technically still outside the body. When you swallow something that cannot be digested, such as an apple seed, it passes through your digestive tract and exits in the feces without ever being broken down or entering your blood or cells. Only after substances have been absorbed into the cells that line the intestine can they be said to be "inside" the body.

transit time The time between the ingestion of food and the elimination of the solid waste from that food.

How Long Is Food in the GI Tract? The amount of time it takes for food and its waste products to pass the length of the GI tract from mouth to anus is referred to as **transit time**. In a healthy adult, transit time is about 24 to 72 hours. It is affected by the composition of the diet, physical activity, emotions, medications, and illnesses. To measure transit time, researchers add a nonabsorbable dye to a meal and measure the time between consumption of the dye and its appearance in the feces. The shorter the transit time, the more rapidly material is passing through the digestive tract.

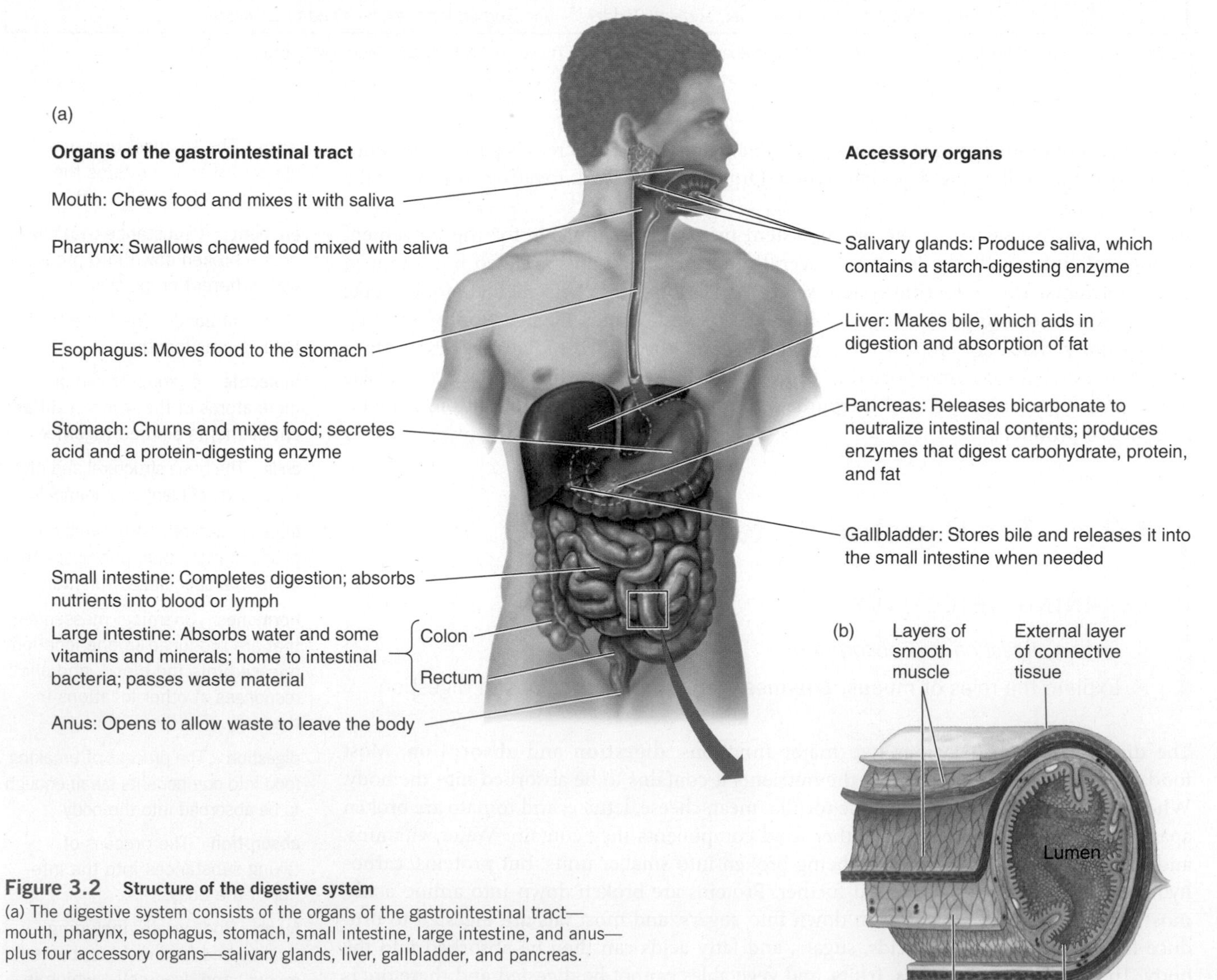

Figure 3.2 Structure of the digestive system
(a) The digestive system consists of the organs of the gastrointestinal tract—mouth, pharynx, esophagus, stomach, small intestine, large intestine, and anus—plus four accessory organs—salivary glands, liver, gallbladder, and pancreas.

(b) A cross section through the wall of the small intestine shows the four tissue layers: mucosa, connective tissue, smooth muscle layers, and outer connective tissue layer.

Structure of the Gut Wall The wall of the GI tract contains four layers of tissue (see Figure 3.2b). Lining the lumen is the **mucosa**, a layer of mucosal cells that serves as a protective layer and is responsible for the absorption of the end products of digestion. The cells of the mucosa are in direct contact with churning food and harsh digestive secretions. Therefore, these cells have a short life span—only about two to five days. When these cells die, they are sloughed off into the lumen, where some components are digested and absorbed and the remainder are excreted in the feces. Because mucosal cells reproduce rapidly, the mucosa has high nutrient requirements and is therefore one of the first parts of the body to be affected by nutrient deficiencies. Surrounding the mucosa is a layer of connective tissue containing nerves and blood vessels. This layer provides support, delivers nutrients to the mucosa, and provides the nerve signals that control secretions and muscle contractions. Layers of smooth muscle—the type over which we do not have voluntary control—surround the connective tissue. The contraction of smooth muscles mixes food, breaks it into smaller particles, and propels it through the digestive tract. The final, external layer is also made up of connective tissue and provides support and protection.

mucosa The layer of tissue lining the GI tract and other body cavities.

mucus A viscous fluid secreted by glands in the GI tract and other parts of the body. It acts to lubricate, moisten, and protect cells from harsh environments.

enzyme A protein molecule that accelerates the rate of a chemical reaction without itself being changed.

Digestive Secretions

Digestion inside the lumen of the GI tract is aided by digestive secretions. One of these substances is **mucus**, a viscous material produced by glands in the mucosal lining of the gut. Mucus moistens, lubricates, and protects the digestive tract. **Enzymes**, another component of digestive system secretions, are protein molecules that speed up chemical reactions without themselves being consumed or changed by the reactions (**Figure 3.3**). In digestion, enzymes accelerate the breakdown of nutrients. Different enzymes are needed to breakdown different nutrients. For example, an enzyme that digests carbohydrate has no effect on fat, and one that digests fat has no effect on carbohydrate. Digestive enzymes and their actions are summarized in **Table 3.2**.

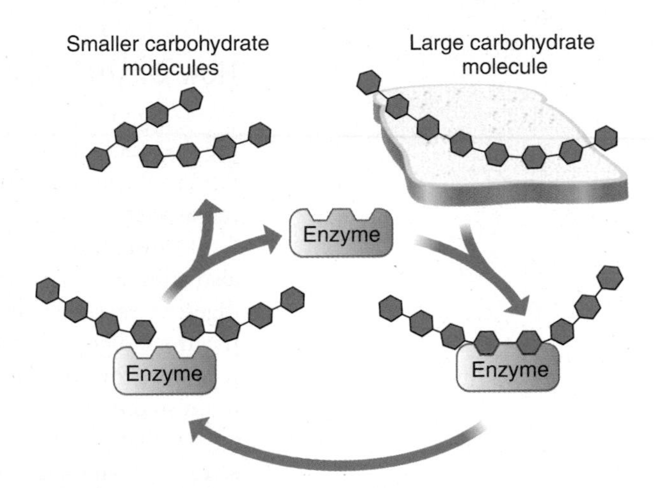

Figure 3.3 **Enzyme activity**
Enzymes speed up chemical reactions without themselves being altered by the reaction. In this example, the enzyme amylase helps break a large carbohydrate molecule (a starch molecule from the bread) into two smaller ones.

TABLE 3.2 Digestive Enzyme Functions

Enzyme	Where it comes from	Where it works	What it does
Salivary amylase	Salivary glands	Mouth	Breaks starch into shorter chains of glucose.
Rennin	Stomach	Stomach	Causes the milk protein casein to curdle.
Pepsin			Breaks proteins into polypeptides and amino acids.
Trypsin	Pancreas	Small intestine	Breaks proteins and polypeptides into shorter polypeptides.
Chymotrypsin			Breaks proteins and polypeptides into shorter polypeptides.
Carboxypeptidase			Breaks polypeptides into amino acids.
Lipase			Breaks triglycerides into monoglycerides, fatty acids, and glycerol.
Pancreatic amylase			Breaks starch into shorter glucose chains and maltose.
Carboxypeptidase, aminopeptidase, and dipeptidase	Small intestine	Small intestine	Break polypeptides and dipeptides into amino acids.
Sucrase			Breaks sucrose into glucose and fructose.
Lactase			Breaks lactose into glucose and galactose.
Maltase			Breaks maltose into glucose.
Dextrinase			Breaks short chains of glucose into individual glucose molecules.

TABLE 3.3 Digestive Hormone Functions

Hormone	Where it comes from	What it does
Gastrin	Stomach mucosa	Stimulates secretion of hydrochloric acid and pepsinogen by gastric glands in the stomach and increases gastric motility and emptying.
Somatostatin	Stomach and duodenal mucosa	Inhibits the following: stomach secretion, motility, and emptying; pancreatic secretion; absorption in the small intestine; gallbladder contraction; and bile release.
Secretin	Duodenal mucosa	Inhibits gastric secretion and motility; increases output of water and bicarbonate from the pancreas; increases bile output from the liver.
Cholecystokinin (CCK)		Stimulates contraction of the gallbladder to expel bile; increases output of enzyme-rich pancreatic juice.
Gastric inhibitory peptide		Inhibits gastric secretion and motility.

How Gastrointestinal Activity Is Regulated

Nerve signals help regulate activity in the GI tract. The sight and smell of food, as well as the presence of food in the gut, stimulate nerves throughout the GI tract. For example, food in the mouth can trigger a nerve impulse that signals the stomach to prepare itself for the arrival of food. Nerve signals cause muscle contractions that churn, mix, and propel food through the gut at a rate that allows optimal absorption of nutrients. Nerve signals also stimulate or inhibit digestive secretions. For example, food in the mouth stimulates digestive secretions in the stomach. After food has passed through a section of the digestive tract, digestive secretions decrease and muscular activity slows to conserve energy and resources for other body processes. The nerves in the GI tract also communicate with the brain so digestive activity can be coordinated with other body needs.

Activity in the digestive tract is also regulated by hormones released into the bloodstream. Hormones that affect gastrointestinal function are produced both by cells lining the digestive tract and by a number of accessory organs. Hormonal signals help prepare different parts of the gut for the arrival of food and thus regulate the digestion of nutrients and the rate that food moves through the system. Some of the hormones released by the GI tract and their functions are summarized in **Table 3.3**.

3.3 Digestion and Absorption

LEARNING OBJECTIVES

- Describe what happens in each of the organs of the gastrointestinal tract.
- Discuss factors that influence how quickly food moves through the gastrointestinal tract.
- Explain how the structure of the small intestine enhances its function.
- Distinguish simple diffusion, osmosis, facilitated diffusion, and active transport.

Imagine the smell of freshly baked cookies. Is your mouth watering? You don't even need to put food in your mouth for activity to begin in the digestive tract. Sensory input such as the sight and smell of the cookies, and even just the thought of food, may make your mouth water and your stomach rumble and begin to secrete digestive substances. This cephalic (pertaining to the head) response occurs when the nervous system signals the digestive system to ready itself for a meal. It can occur even when the body is not in need of food. In order for food to be used by the body, however, you need to do more than smell your meal. The food must be consumed and digested, and the nutrients must be absorbed and transported to the body's cells. This involves the combined functions of all the organs of the digestive system, as well as the help of some other organ systems.

The Mouth

The mouth is the entry point for food into the digestive tract. Here, food is tasted and the mechanical breakdown and chemical digestion of food begin. The presence of food in the mouth stimulates the flow of **saliva** from the salivary glands located internally at the sides of the face and immediately below and in front of the ears (see Figure 3.2a). Saliva plays many roles: It moistens the food so that it can easily be tasted and swallowed, it begins the enzymatic digestion of starch, it cleanses the mouth and protects teeth from decay, and it lubricates the upper GI tract. By moistening food, saliva helps us taste because molecules dissolve in the saliva and are carried to the taste buds, most of which are located on the tongue. Signals from the taste buds, along with the aroma of food, allow us to enjoy the taste of the food we eat. Saliva begins the chemical digestion of carbohydrate because it contains the enzyme **salivary amylase**. Salivary amylase can break the long sugar chains of starch in foods like bread and cereal into shorter chains of sugars (see Figure 3.3). Saliva also protects against tooth decay because it helps wash away food particles and it contains **lysozyme**—an enzyme that kills bacteria that may cause tooth decay (see Chapter 4).

saliva A watery fluid produced and secreted into the mouth by the salivary glands. It contains lubricants, enzymes, and other substances.

salivary amylase An enzyme secreted by the salivary glands that breaks down starch.

lysozyme An enzyme in saliva, tears, and sweat that is capable of destroying certain types of bacteria.

Chewing food begins the mechanical aspect of digestion. Adult humans have 32 teeth, specialized for biting, tearing, grinding, and crushing foods. Chewing breaks food into small pieces. This makes the food easier to swallow and increases the surface area in contact with digestive juices. The tongue helps mix food with saliva and aids chewing by constantly repositioning food between the teeth. Chewing also breaks apart fiber, which traps nutrients in some foods. If the fiber is not broken apart, some nutrients cannot be absorbed. For example, if the fibrous skin of a raisin is not broken open by the teeth, the nutrients within the raisin will remain inaccessible and the raisin will travel undigested through the intestines and be eliminated in the feces.

The Pharynx

The tongue initiates swallowing by moving the rounded mass, or bolus, of chewed food mixed with saliva back toward the **pharynx**. The pharynx is shared by the digestive tract and the respiratory tract: Food and liquid pass through the pharynx on their way to the stomach, and air passes here on its way to and from the lungs. We are able to start the muscular contractions of swallowing by choice, but once initiated it becomes involuntary and proceeds under the control of nerves.

pharynx A funnel-shaped opening that connects the nasal passages and mouth to the respiratory passages and esophagus. It is a common passageway for food and air and is responsible for swallowing.

During swallowing, the air passages are blocked by a valve-like flap of tissue called the **epiglottis**, which ensures that the bolus of food passes to the stomach, not the lungs (**Figure 3.4a**). Sometimes food can pass into an upper air passageway. It is usually dislodged with a cough, but if it becomes stuck it can block the flow of air and cause choking.

epiglottis A piece of elastic connective tissue at the back of the throat that covers the opening of the passageway to the lungs during swallowing.

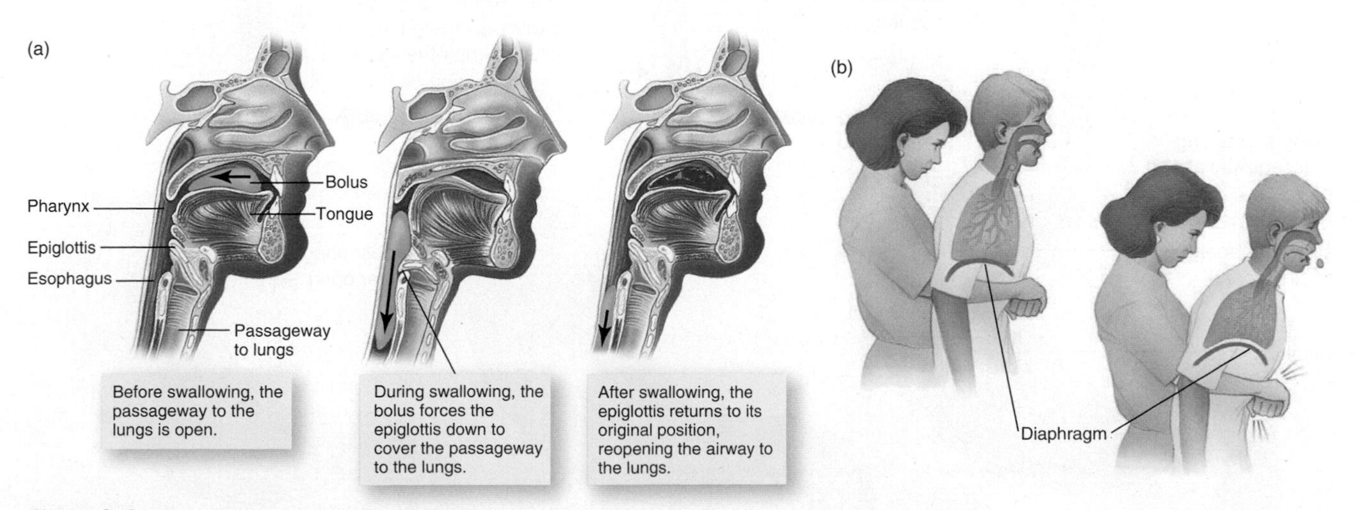

Figure 3.4 **The role of the epiglottis**
(a) When a bolus of food is swallowed, it pushes the epiglottis down over the opening to the air passageways.

(b) If food becomes lodged in the passageway leading to the lungs, it can block the flow of air. The Heimlich maneuver, which involves a series of thrusts directed upward from under the diaphragm (the muscle separating the chest and abdominal cavities), forces air out of the lungs, blowing the lodged food out of the air passageway.

A quick response is required to save the life of a person whose airway is completely blocked. The Heimlich maneuver, which forces air out of the lungs by using a sudden application of pressure just below the diaphragm, can blow an object out of the blocked air passage (**Figure 3.4b**).

The Esophagus

esophagus A portion of the GI tract that extends from the pharynx to the stomach.

peristalsis Coordinated muscular contractions that move food through the GI tract.

The **esophagus** is a tube that passes through the diaphragm, a muscular wall separating the abdominal cavity from the chest cavity where the lungs are located, to connect the pharynx and stomach. In the esophagus the bolus of food is moved along by rhythmic contractions of the smooth muscles, called **peristalsis** (**Figure 3.5**). Peristalsis is like an ocean wave that moves through the muscle, producing a narrowing in the lumen that pushes food and fluid in front of it. It takes only about four to eight seconds for solid food to move from the pharynx down the esophagus to the stomach, and even less time for liquids to make the trip. The peristaltic waves in the esophagus are so powerful that food and fluids will reach your stomach even if you are upside down. This contractile movement, which is controlled automatically by the nervous system, occurs throughout the GI tract, pushing the food bolus along from the pharynx through the large intestine.

sphincter A muscular valve that helps control the flow of materials in the GI tract.

To move from the esophagus into the stomach, food must pass through a **sphincter**, a muscle that encircles the tube of the digestive tract and acts as a valve. When the muscle contracts, the valve is closed. The gastroesophageal sphincter, also called the cardiac or lower esophageal sphincter, relaxes reflexively just before a peristaltic wave reaches it, allowing food to enter the stomach (see Figure 3.5). This valve normally prevents foods from moving back out of the stomach. Occasionally, however, materials do pass out of the stomach through this valve. Heartburn occurs when some of the acidic stomach contents leaks up and out of the stomach into the esophagus, causing a burning sensation. Vomiting is the result of a reverse peristaltic wave that causes the sphincter to relax and allow the food to pass upward out of the stomach toward the mouth.

The Stomach

chyme A mixture of partially digested food and stomach secretions.

The stomach is an expanded portion of the GI tract that serves as a temporary storage place for food. While held in the stomach, the bolus is mashed and mixed with highly acidic stomach secretions to form a semiliquid food mass called **chyme**. Some digestion

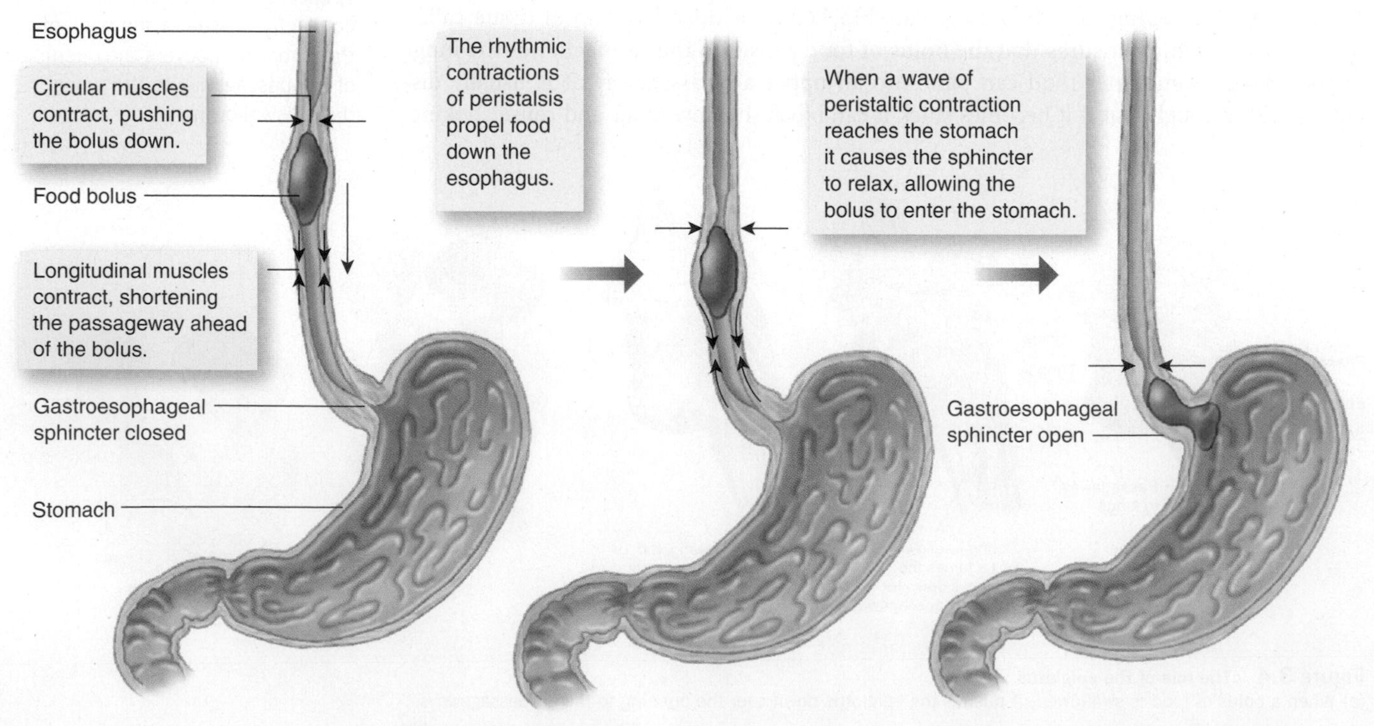

Figure 3.5 Peristalsis
The food we swallow is pushed along by the rhythmic muscular contractions of peristalsis and enters the stomach in response to the opening and closing of the gastroesophageal sphincter.

takes place in the stomach but, with the exception of some water, alcohol, and a few drugs, including aspirin and acetaminophen (Tylenol), very little absorption occurs here.

How the Structure of the Stomach Supports Its Function The stomach walls are thicker and have stronger muscles than other segments of the GI tract. Two layers of muscle, one running longitudinally down the tract and one running around it, surround most of the GI tract. The stomach has an additional layer that circles it diagonally, allowing for powerful contractions that thoroughly churn and mix the stomach contents (**Figure 3.6a**).

The lining of the stomach is interrupted by millions of tiny openings called gastric pits. At the bottom of these pits are the gastric glands. A number of different types of cells cover the surface of the stomach and extend into the gastric pits and glands (**Figure 3.6b**). Some secrete substances into the stomach. These stomach secretions are collectively referred to as *gastric juice*. Other cells in the gastric glands secrete a variety of hormones and hormone-like compounds into the blood.

What Is in Gastric Juice? Gastric juice is a mixture of water, mucus, hydrochloric acid, and an inactive enzyme called **pepsinogen**. Hydrochloric acid, produced by parietal cells, acidifies the stomach contents and as a result kills most bacteria present in food. Parietal cells also produce intrinsic factor, which is needed for the absorption of vitamin B_{12} (see Chapter 8). Pepsinogen is produced by chief cells. It is an inactive form of the protein-digesting enzyme **pepsin**, which breaks proteins into shorter chains of amino acids called *polypeptides*. Pepsin is secreted in an inactive form so that it will not damage the gastric glands that produce it. It is activated by hydrochloric acid when it enters the stomach. Pepsin functions best in the acidic environment of the stomach. This acidic environment stops the function of salivary amylase. Therefore, the digestion of starch from foods such as bread and potatoes stops in the stomach, and digestion of the protein from foods such as meat, milk, and legumes begins. The protein of the stomach wall is protected from the acid and pepsin by a thick layer of mucus. In infants, the stomach glands also produce the enzyme rennin. Rennin acts on the milk protein casein to convert it into a curdy substance resembling sour milk.

pepsinogen An inactive protein-digesting enzyme produced by gastric glands and activated to pepsin by acid in the stomach.

pepsin A protein-digesting enzyme produced by the gastric glands. It is secreted in the gastric juice in an inactive form and activated by acid in the stomach.

THINK CRITICALLY

Why is the protein-digesting enzyme pepsin produced in an inactive form?

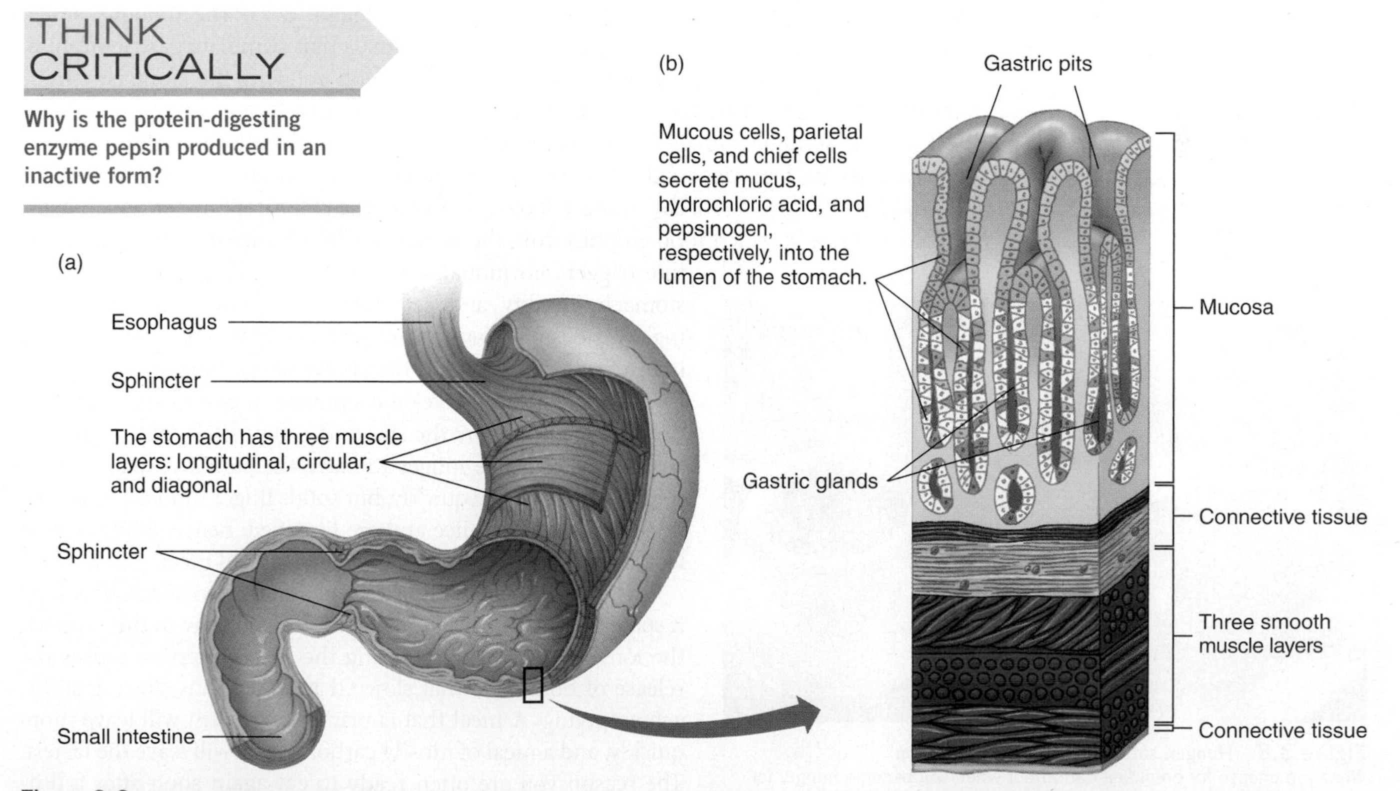

Figure 3.6 Structure of the stomach
(a) The stomach wall contains three layers of smooth muscle, which contract powerfully to mix food.
(b) The lining of the stomach is covered with gastric pits. Inside these pits are the gastric glands, made up of cells that produce the components of gastric juice.

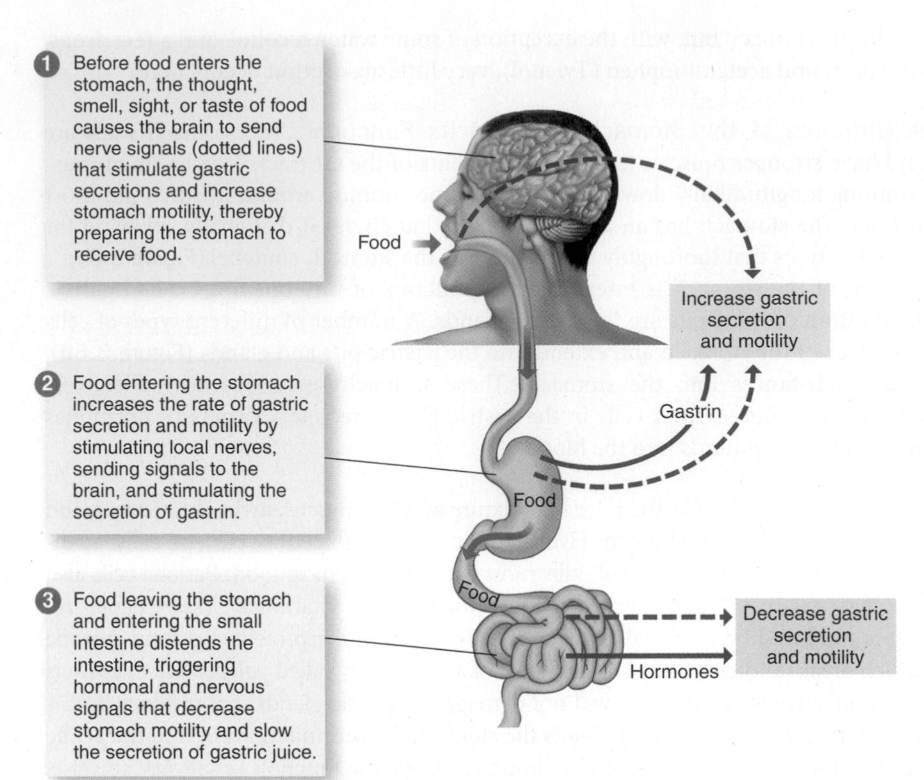

Figure 3.7 Regulation of stomach motility and secretion
Stomach activity is affected by food that has not yet reached the stomach, by food that is in the stomach, and by food that has left the stomach.

How Stomach Activity Is Regulated How much your stomach churns, how much gastric juice is released, and how fast material empties out of the stomach are regulated by signals from both nerves and hormones. These signals originate from three different sites—the brain, the stomach, and the small intestine (**Figure 3.7**). The thought, smell, sight, or taste of food causes the brain to send nerve signals that stimulate gastric motility and secretion, preparing the stomach to receive food. Food entering the stomach stimulates the release of gastric secretions and an increase in motility. It does this by stretching local nerves, sending signals to the brain, and promoting the secretion of the hormone **gastrin**. Gastrin then triggers the release of gastric juice and increases stomach motility.

gastrin A hormone secreted by the stomach mucosa that stimulates the secretion of gastric juice.

As chyme moves out of the stomach it passes through the pyloric sphincter, which helps regulate the rate at which food empties from the stomach. Chyme entering the small intestine triggers hormonal and nervous signals that can decrease stomach motility and secretions and slow stomach emptying. This ensures that the amount of chyme entering the small intestine does not exceed the ability of the intestine to process it. Chyme normally leaves the stomach in two to six hours, but this rate is affected by the size and composition of the meal. A large meal will take longer to leave the stomach than a small meal. Liquids empty quickly, but solids linger until they are well mixed with gastric juice and are liquefied; hence, a liquid meal leaves the stomach more quickly than a solid meal.

The nutritional composition of a meal also affects how long it stays in the stomach. A high-fat meal will stay in the stomach the longest because fat entering the small intestine causes the release of hormones that slow GI motility, thus slowing stomach emptying. A meal that is primarily protein will leave more quickly, and a meal of mostly carbohydrate will leave the fastest. The reason you are often ready to eat again soon after a dinner of vegetables and rice is that this high-carbohydrate, low-fat meal leaves the stomach rapidly (**Figure 3.8**).

Figure 3.8 Hunger, satiety, and meal composition
What you choose for breakfast can affect when you become hungry for lunch. Toast and coffee will leave your stomach far more quickly than a larger meal with more protein and fat. The bacon, eggs, and buttered toast, shown here leave the stomach more slowly, keeping you full longer.

Other factors such as exercise and emotions also affect gastric emptying. Sadness and fear tend to slow emptying, while aggression tends to increase gastric motility and speed emptying. Exercise will delay stomach emptying because the body directs its resources to the exercising muscles.

The Small Intestine

The small intestine is the main site of digestion of food and absorption of nutrients. It is a narrow tube about 20 feet in length. It is divided into three segments. The first 12 inches are the duodenum, the next 8 feet are the jejunum, and the last 11 feet are the ileum.

How the Structure of the Small Intestine Maximizes Function The structure of the small intestine is specialized to allow maximal absorption of the nutrients. In addition to its length, the small intestine has three other features that increase the area of its absorptive surface (**Figure 3.9**). First, the intestinal walls are arranged in large circular folds, which increase the surface area in contact with nutrients. Second, its entire inner surface is covered with finger-like projections called **villi** (singular *villus*). And last, the mucosal cells on the surface of each of these villi are covered with tiny **microvilli**, often referred to as the **brush border**. Within each villus are a blood vessel and a lymph vessel, called a **lacteal**,

villi (villus) Finger-like protrusions of the lining of the small intestine that participate in the digestion and absorption of nutrients.

microvilli or **brush border** Minute brush-like projections on the mucosal cell membrane that increase the absorptive surface area in the small intestine.

lacteal A small lymph vessel in the intestine that absorbs and transports the products of fat digestion.

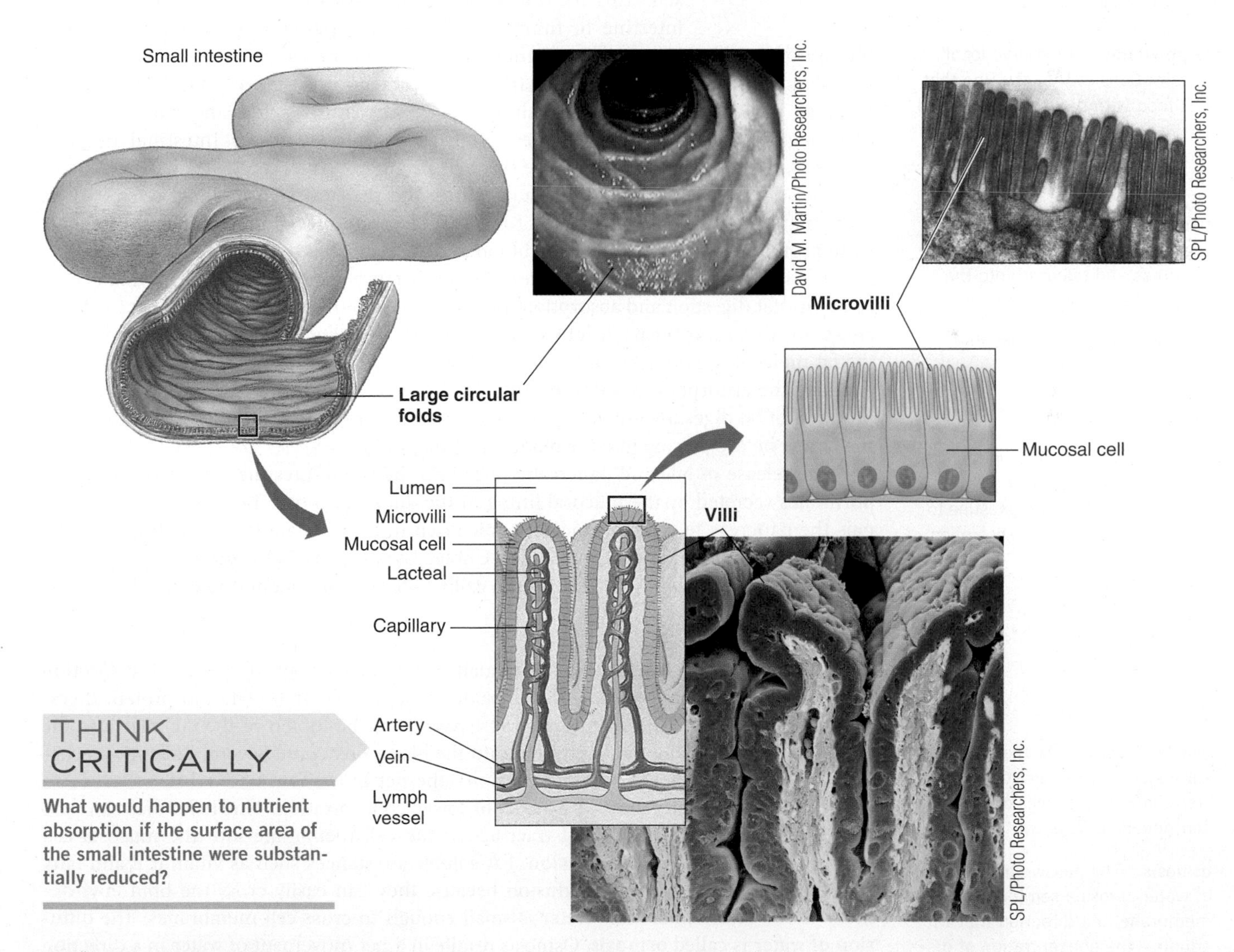

THINK CRITICALLY

What would happen to nutrient absorption if the surface area of the small intestine were substantially reduced?

Figure 3.9 Structure of the small intestine
The small intestine contains large circular folds, called *villi*, and smaller folds, called *microvilli*, both of which increase the absorptive surface area. Together these features provide a surface area that is about the size of a tennis court (250 m², or 2700 ft²).

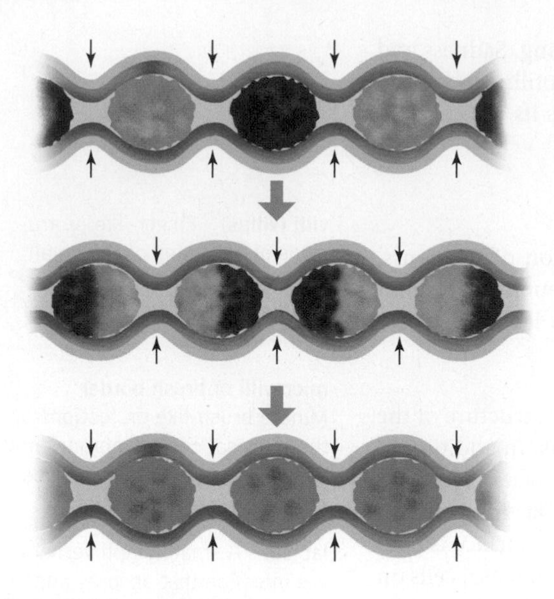

Figure 3.10 Segmentation
The alternating contraction and relaxation of segments of the small intestine move food forward and backward, mixing it rather than propelling it forward.

which are located only one cell layer away from the nutrients in the intestinal lumen. Nutrients must cross the mucosal cell layer to reach the bloodstream or lymphatic system for delivery to the tissues of the body.

Motility and Secretions in the Small Intestine Chyme is propelled through the small intestine by peristalsis, and the mixing of chyme with digestive secretions is aided in the small intestine by rhythmic local constrictions called **segmentation** (**Figure 3.10**). Segmentation also enhances absorption by repeatedly moving chyme over the surface of the intestinal mucosa.

The cells of the small intestine produce some digestive enzymes as well as a watery mucus-containing fluid called *intestinal juice* that aids in absorption. However, normal digestion and absorption in the small intestine also require secretions from the pancreas and gallbladder. The pancreas secretes pancreatic juice, which contains both bicarbonate ions and digestive enzymes. The bicarbonate ions neutralize the acid in chyme, making the environment in the small intestine neutral rather than acidic, as it is in the stomach. This neutrality allows enzymes from the pancreas and small intestine to function. The enzyme **pancreatic amylase** continues the job of breaking starch into sugars that was started in the mouth by salivary amylase. Pancreatic **proteases**, including trypsin and chymotrypsin (see Table 3.2), continue to break protein into shorter and shorter chains of amino acids, and pancreatic fat-digesting enzymes, called **lipases**, break triglycerides into fatty acids. Intestinal digestive enzymes, found attached to or inside the cells lining the small intestine, are involved in the digestion of sugars into single sugar units and the digestion of small polypeptides into single amino acids. The sugars from carbohydrate digestion and the amino acids from protein digestion pass into the blood and are delivered to the liver (**Figure 3.11**).

segmentation Rhythmic local constrictions of the intestine that mix food with digestive juices and speed absorption by repeatedly moving the food mass over the intestinal wall.

pancreatic amylase Starch-digesting enzyme produced in the pancreas and released into the small intestine.

protease A protein-digesting enzyme

lipase A fat-digesting enzyme.

bile A substance made in the liver and stored in the gallbladder. It is released into the small intestine to aid in fat digestion and absorption.

The gallbladder stores and secretes **bile**, a substance produced in the liver that is necessary for fat digestion and absorption. Bile secreted into the small intestine mixes with fat and emulsifies it, or breaks it into smaller globules, allowing lipases to access and digest the fat molecules more efficiently. The bile and digested fats then form small droplets that facilitate the absorption of fat into the mucosal cells. Once inside the mucosal cells, the products of fat digestion are incorporated into transport particles. These are absorbed into the lymph before passing into the blood (see Chapter 5).

The release of bile and pancreatic juice into the small intestine is controlled by two hormones secreted by the mucosal lining of the duodenum (see Table 3.3). Secretin signals the pancreas to secrete pancreatic juice rich in bicarbonate ions and stimulates the liver to secrete bile into the gallbladder. Cholecystokinin (CCK) signals the pancreas to secrete digestive enzymes and causes the gallbladder to contract and release bile into the duodenum.

How Nutrients Are Absorbed The small intestine is the primary site of absorption for water, vitamins, minerals, and the products of carbohydrate, fat, and protein digestion. To be absorbed, these nutrients must pass from the lumen of the GI tract into the mucosal cells lining the tract and then into the blood or lymph. Several different mechanisms are involved. Some rely on diffusion—the net movement of a substance from an area of higher concentration to an area of lower concentration. Nutrients that can pass freely from the lumen of the GI tract across the cell membrane into the mucosal cell can be absorbed by **simple diffusion**. Fat-soluble substances such as vitamin E and fatty acids are absorbed by simple diffusion because they can easily cross the lipid environment of the cell membranes. Water is small enough to cross cell membranes. The diffusion of water is called **osmosis**. Osmosis results in a net movement of water in a direction that will balance the concentration of dissolved substances on either side of a membrane.

simple diffusion The movement of substances from an area of higher concentration to an area of lower concentration. No energy is required.

osmosis The passive movement of water across a semipermeable membrane in a direction that will equalize the concentration of dissolved substances on both sides.

Figure 3.11 **Digestion and absorption in the small intestine**
Starch digestion begins in the mouth, and proteins are broken into smaller polypeptides in the stomach, but most digestion and absorption occurs in the small intestine.

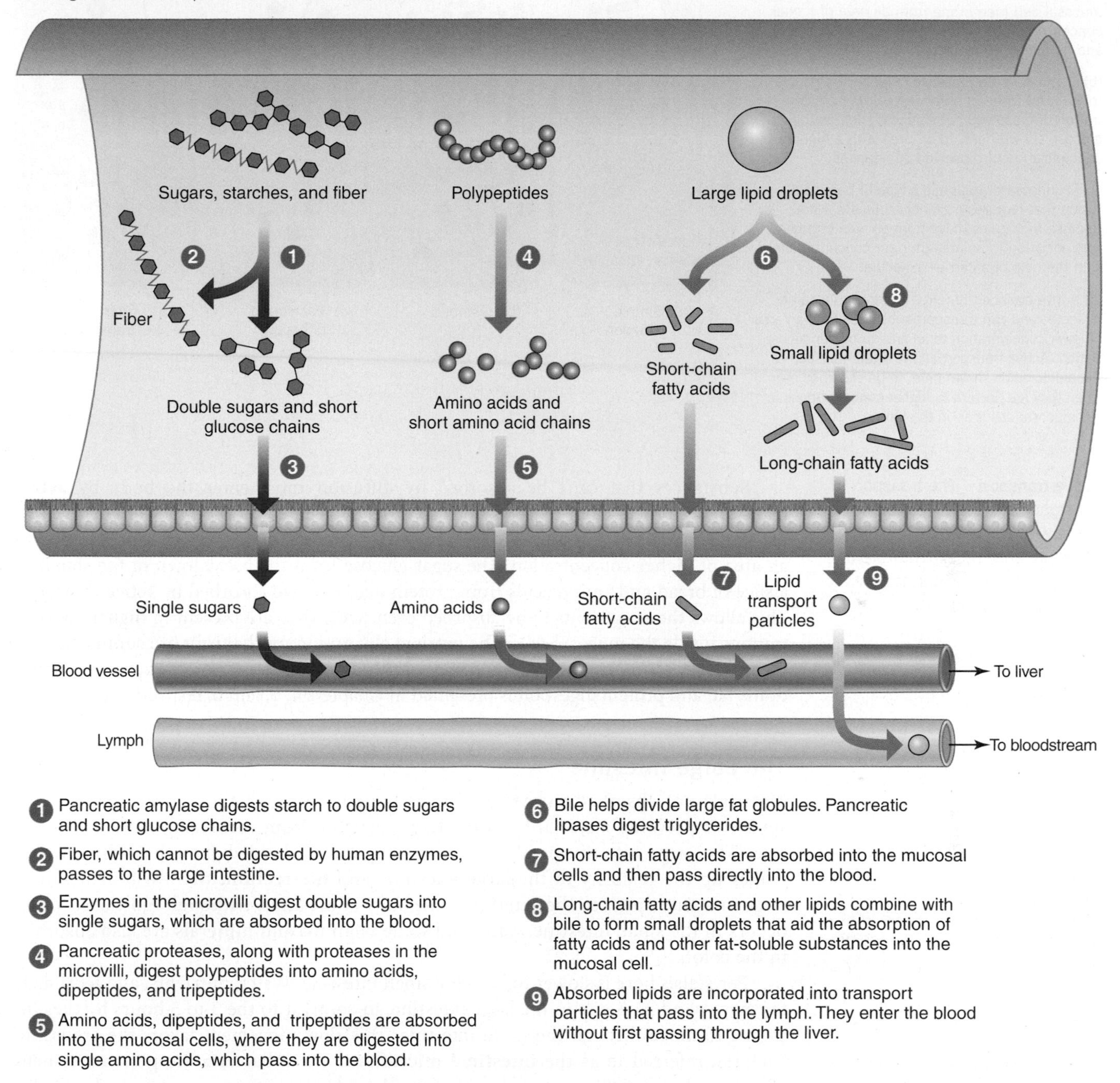

1. Pancreatic amylase digests starch to double sugars and short glucose chains.
2. Fiber, which cannot be digested by human enzymes, passes to the large intestine.
3. Enzymes in the microvilli digest double sugars into single sugars, which are absorbed into the blood.
4. Pancreatic proteases, along with proteases in the microvilli, digest polypeptides into amino acids, dipeptides, and tripeptides.
5. Amino acids, dipeptides, and tripeptides are absorbed into the mucosal cells, where they are digested into single amino acids, which pass into the blood.
6. Bile helps divide large fat globules. Pancreatic lipases digest triglycerides.
7. Short-chain fatty acids are absorbed into the mucosal cells and then pass directly into the blood.
8. Long-chain fatty acids and other lipids combine with bile to form small droplets that aid the absorption of fatty acids and other fat-soluble substances into the mucosal cell.
9. Absorbed lipids are incorporated into transport particles that pass into the lymph. They enter the blood without first passing through the liver.

For example, if there is a high concentration of sugar in the lumen of the intestine, water will actually move from the mucosal cells into the lumen. As the sugar is absorbed and the concentration of sugar in the lumen decreases, water will move back into the mucosal cells by osmosis. Some nutrients that cannot pass freely across cell membranes are absorbed by **facilitated diffusion**, which uses carrier molecules to help substances cross the cell membrane. Even though these nutrients are helped across the cell membrane, they still move from an area of higher concentration to one of lower concentration; the sugar fructose found in fruit is absorbed by facilitated diffusion.

facilitated diffusion The movement of substances across a cell membrane from an area of higher concentration to an area of lower concentration with the aid of a carrier molecule. No energy is required.

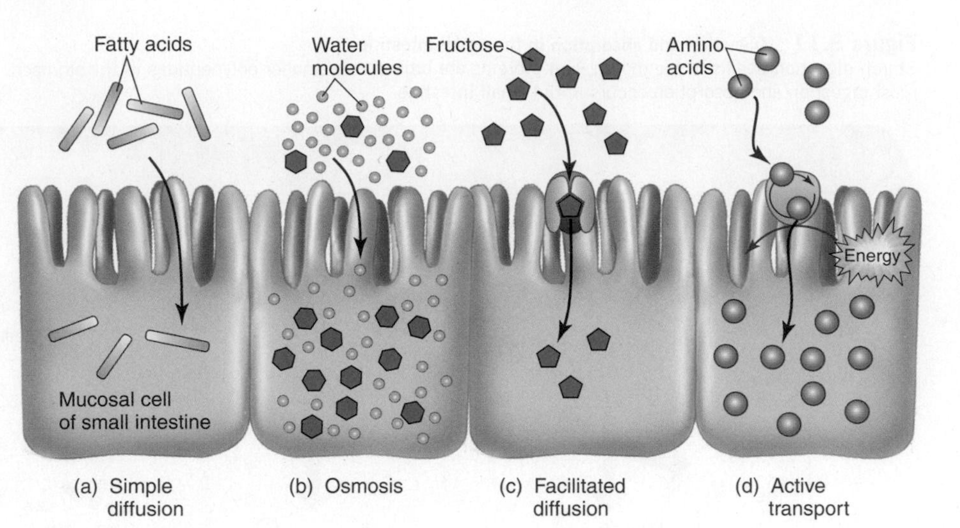

Figure 3.12 Absorption mechanisms
(a) In simple diffusion, substances such as the fatty acids shown here pass freely across the mucosal cell membrane from an area of higher concentration to an area of lower concentration, and no energy is required.

(b) Osmosis is the passage of water molecules (blue dots) from an area with a lower concentration of dissolved substances, such as the glucose shown here, to an area with a higher concentration of dissolved substances.

(c) Facilitated diffusion is a type of passive diffusion that requires a carrier molecule. Here, fructose molecules move from an area of higher concentration to an area of lower concentration, with the help of a carrier molecule.

(d) Active transport requires energy and a carrier molecule and can transport substance from an area of lower concentration to an area of higher concentration. Active transport allows nutrients, such as the amino acids shown here, to be absorbed even when they are present in higher concentrations in the mucosal cell than in the lumen.

active transport The transport of substances across a cell membrane with the aid of a carrier molecule and the expenditure of energy. This may occur against a concentration gradient.

Substances that can't be absorbed by diffusion must enter the body by **active transport**, a process that requires both a carrier molecule and the input of energy. This use of energy allows substances to be transported from an area of lower concentration to an area of higher concentration. The sugar glucose from the breakdown of the starch in a slice of bread and amino acids from protein digestion are absorbed by active transport. This allows these nutrients to be absorbed even when they are present in higher concentrations inside the mucosal cells. The nutrient absorption mechanisms are summarized in **Figure 3.12**. More specific information about the absorption of the products of carbohydrate, fat, and protein digestion is presented in Chapters 4, 5, and 6, respectively.

The Large Intestine

colon The largest portion of the large intestine.

rectum The portion of the large intestine that connects the colon and anus.

Materials not absorbed in the small intestine pass through the ileocecal valve. This sphincter prevents material from the large intestine from reentering the small intestine. The large intestine is about 5 feet long and is divided into the **colon**, which makes up the majority of the large intestine, and the **rectum**, the last 8 inches. The large intestine opens to the exterior of the body at the anus. Although most absorption occurs in the small intestine, water and some vitamins and minerals are also absorbed in the colon.

intestinal microflora Microorganisms that inhabit the large intestine.

Peristalsis here is slower than in the small intestine. Water, nutrients, and fecal matter may spend 24 hours in the large intestine, in contrast to the 3 to 5 hours it takes for chyme to move through the small intestine. This slow movement favors the growth of bacteria, referred to as the **intestinal microflora**. These bacteria are permanent beneficial residents of this part of the GI tract. The microflora act on unabsorbed portions of food, such as the fiber, producing nutrients that the bacteria themselves can use or, in some cases, that can be absorbed into the body. For example, the microflora synthesize small amounts of fatty acids, some B vitamins, and vitamin K, some of which can be absorbed. One additional by-product of bacterial metabolism is gas, which causes flatulence. In normal adult humans, between 200 and 2000 mL of intestinal gas are produced per day.

There are 300 to 500 species of bacteria that reside in the large intestine. The right mix of these intestinal bacteria is important for immune function, proper growth and development of colon cells, and optimal intestinal motility and transit time.[1] Having healthy microflora can inhibit the growth of harmful bacteria and has been shown to prevent the diarrhea associated with antibiotic use and to reduce the duration of diarrhea resulting from intestinal infections and other causes.[2,3] There is also evidence that having healthy microflora may relieve constipation, reduce allergy symptoms, and modify the risk of

Figure 3.13 **Probiotics**

Andy Washnik

(a) Ads claim that eating these specialized probiotic yogurts will help regulate the digestive system. These products as well as most other yogurts contain active cultures of beneficial bacteria, including *Lactobacillus* and *Bifidobacterium*.

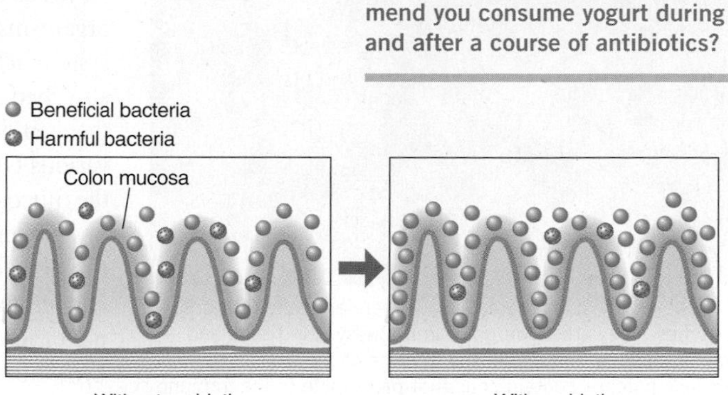

(b) When living beneficial bacteria are consumed in adequate amounts, they live temporarily in the colon, where they inhibit the growth of harmful bacteria and confer other health benefits on the host. However, the bacteria must be consumed frequently because they are flushed out in the feces.

THINK CRITICALLY

Why might your doctor recommend you consume yogurt during and after a course of antibiotics?

inflammatory bowel disease and colon cancer.[4,5,6] One way of promoting healthy microflora is to actually consume the beneficial bacteria, called **probiotics**, in foods such as yogurt or in tablets (**Figure 3.13**). Another way is to consume **prebiotics**, substances that serve as a food supply for beneficial bacteria. Prebiotics are found naturally in onions, bananas, garlic, and artichokes and are sold as dietary supplements.

probiotic Live bacteria that when consumed live temporarily in the colon and confer health benefits on the host.

prebiotic A substance that passes undigested into the colon and stimulates the growth and/or activity of certain types of bacteria.

Material that is not absorbed in the colon passes into the rectum, where it is stored temporarily and then evacuated through the anus as feces. The feces are a mixture of undigested unabsorbed matter, dead cells, secretions from the GI tract, water, and bacteria. The amount of bacteria in the feces varies but can make up more than half the weight of the feces. The amount of water in the feces is affected by fiber and fluid intake. Fiber retains water, so when adequate fiber and fluid are consumed, feces have a high water content and are easily passed. When inadequate fiber or fluid is consumed, feces are hard and dry, and difficult to eliminate.

3.4 Digestion and Health

LEARNING OBJECTIVES

- Explain the role of the GI tract in preventing infection.
- Describe what happens when someone consumes food to which they are allergic.
- Name three common digestive complaints, and explain their causes and consequences.
- Explain how changes in the digestive system throughout life affect digestion and absorption.

The health of the GI tract is essential to overall health. The gut not only takes nutrients into the body but also acts as a defense against invasion by potentially dangerous contaminants. Food allergies, which can be life threatening, have their origins in the GI tract, but most common gastrointestinal problems are minor and do not affect long-term health.

Figure 3.14 Immune system tissue in the small intestine

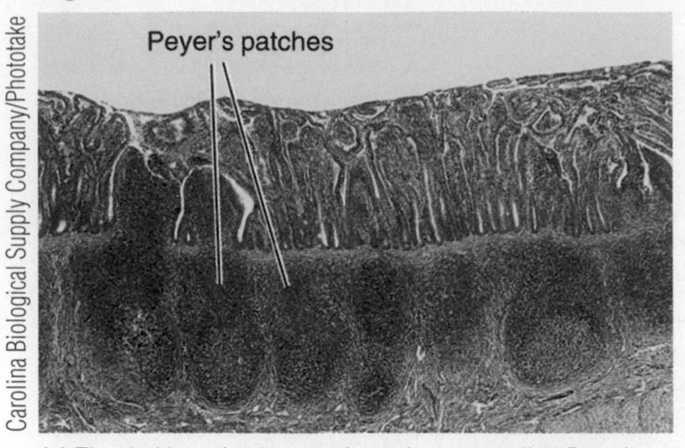

(a) The darkly stained areas shown here are called Peyer's patches. They are made up of immune-system tissue and are embedded throughout the mucosa of the small intestine. Peyer's patches contain cells that participate in the immune system's efforts to prevent harmful organisms or materials present in the GI tract from making us ill.

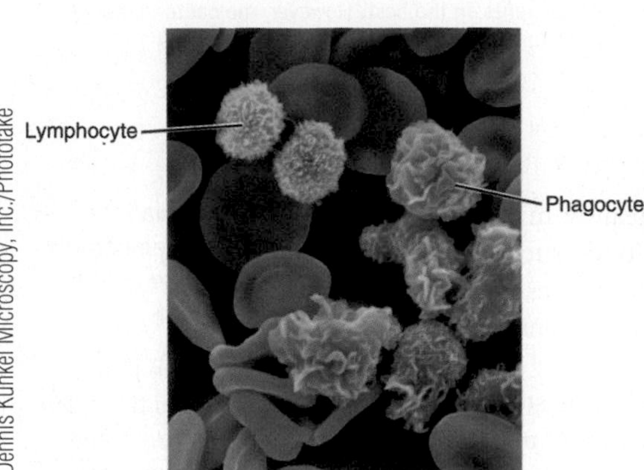

(b) The cells shown here in pink are phagocytes, which can engulf and destroy invading substances. The cells shown in green are lymphocytes, which are specific with regard to which invaders they can attack. Some lymphocytes directly kill invaders, while others secrete antibodies that help destroy antigens.

Immune Functions of the Gastrointestinal Tract

Food almost always contains bacteria and other contaminants, but it rarely makes us sick. Acid in the stomach kills most bacteria, and the mucosa serves as a barrier, preventing the absorption of toxins and disease-causing organisms. The gastrointestinal tract is also an important site of immune system activity. If an invading substance, or **antigen**, enters the lumen or is absorbed into the mucosa, the immune system has a number of weapons that can destroy it before it spreads to other parts of the body. These include various types of white blood cells, which circulate in the blood and reside in the mucosa of the gastrointestinal tract (**Figure 3.14a**).

When an antigen is present, phagocytes are the first types of cells to come to the body's defense (**Figure 3.14b**). They target any invader and engulf and destroy it. These cells also release chemicals that signal other immune system cells called *lymphocytes* to come join the fight. Lymphocytes are very specific about what they attack. Some lymphocytes bind directly to a specific invader or antigen and destroy it. This type of lymphocyte helps eliminate cancer cells, foreign cells, and cells that have been infected by viruses and bacteria. Other lymphocytes produce and secrete protein molecules called **antibodies**. Antibodies bind to invading antigens and help to destroy them. Each antibody is designed to fight off only one specific antigen. Once the body has made antibodies to a specific antigen, it remembers and can rapidly produce these antibodies to fight that antigen any time it enters the body.

The Immune Response and Food Allergies The immune system protects us from many invaders without our even being aware that the battle is going on. Unfortunately, the response of the immune system to a foreign substance is also responsible for allergic reactions. Food allergies occur when the body sees a food protein, called an **allergen**, as a foreign substance. The first time the protein is consumed and a piece of it is absorbed intact, the immune system is stimulated. When the protein is consumed again, the immune system recognizes it and mounts an attack, causing an allergic reaction. Allergic reactions to food can cause symptoms that range from hives to life-threatening breathing difficulties (see Chapter 6). Food allergies affect 3 to 4% of adults and about 6% of children under three years of age. They are responsible for 150 to 200 deaths each year.[7] The most common sources of food allergens are seafood, peanuts, tree nuts, fish, soy, wheat, milk, and eggs. Food labels must indicate if the product contains any of these major food allergens (see Chapter 6, Off the Label: Is It Safe for You?).

antigen A foreign substance (almost always a protein) that, when introduced into the body, stimulates an immune response.

antibody A protein produced by cells of the immune system that destroys or inactivates foreign substances in the body.

allergen A substance that causes an allergic reaction.

The Immune Response and Celiac Disease Celiac disease is a condition in which the protein gluten, found in wheat, barley, and rye, triggers an immune system response that damages or destroys the villi of the small intestine. For people with this disease, also called gluten intolerance, gluten sensitivity, celiac sprue, nontropical sprue, and gluten-sensitive enteropathy, consuming even a tiny amount of gluten can cause abdominal pain, diarrhea, and fatigue. If not managed properly, the intestinal damage caused by celiac disease reduces nutrient absorption and can lead to malnutrition, weight loss, anemia, osteoporosis, intestinal cancer, and other chronic illnesses. Celiac disease is an inherited condition that affects an estimated 1 in 141 people in the population.[8] It can be diagnosed by a blood test or an intestinal biopsy. Although gluten-free diets are currently popular, they are necessary only for people with diagnosed celiac disease (see **Debate: Should You Be Gluten Free?**).

For people with celiac disease, consuming a diet that eliminates gluten provides relief from symptoms and long-term complications. This means eliminating all products made from wheat, barley, or rye, including most breads, crackers, pastas, cereals, cakes, and cookies. It also requires eliminating foods ranging from packaged gravies to soy sauce that contain trace amounts of these grains.

DEBATE

Should You Be Gluten Free?

You see the term "gluten free" on breakfast cereals, cake mixes, pastas, soups, and a host of other products. Celebrities are touting the benefits of going gluten free. Chelsea Clinton even had her wedding cake baked without gluten. Gluten is a protein found in wheat, barley, and rye. Gluten-free diets are essential for people who have celiac disease, a condition in which consuming gluten triggers an immune response that damages the lining of the small intestine. But a switch to gluten-free products has also been promoted for weight loss and to treat a host of other ailments. Is gluten free a healthy alternative for everyone?

Advocates claim that a gluten-free diet will help those suffering with joint pain, rheumatoid arthritis, osteoporosis, anemia, and diabetes. They contend that individuals with these symptoms have undiagnosed celiac disease. Although a small number of people may benefit from a gluten-free diet because they have celiac disease but no obvious symptoms, eating a gluten-free diet is unlikely to cure these conditions in people who do not have celiac disease.

What about going gluten free for weight loss? Gluten-free foods are not any lower in calories than other foods, but eliminating everything that contains gluten from your diet—most types of cereal, bread, pasta, cakes, and cookies—may help cut calories. For those who do not have celiac disease, there is no nutritional benefit to gluten-free foods, but carefully choosing everything you put in your mouth (in the process of avoiding gluten) will force you to plan your diet carefully and may help improve the nutrient density of your diet and promote weight loss.

Is a gluten-free diet harmful? Eliminating gluten involves avoiding carefully checking each ingredient in the foods you eat to eliminate products made not only from wheat, which is the major grain in the American diet, but also from barley and rye. It also requires carefully checking each ingredient in foods to exclude the myriad of food products to which wheat has been added (see figure). The major problem with a gluten-free diet is that it eliminates most flours, breads, pasta, and breakfast cereals, which are important sources of B vitamins and iron. This creates a risk of nutrient deficiencies. A gluten-free diet is not harmful as long as it provides enough of all the nutrients typically consumed in gluten-containing foods. People diagnosed with celiac disease generally work with a dietitian to make sure they have a well-balanced, varied diet that meets all their nutrient needs. People trying a gluten-free diet for other reasons generally do not.

Although there is no research to support eliminating gluten if you do not have celiac disease, anything that makes you consider your diet carefully is a good thing. Individuals with gluten sensitivity are benefitting from the gluten-free craze because it has increased the availability and quality of gluten-free foods, improved the labeling of gluten-free products, and heightened awareness of celiac disease. Proponents think a gluten-free diet will improve everyone's health. Skeptics consider gluten-free another trend like the low-carb fad of a few years back.

THINK CRITICALLY: Neither potatoes nor onions contain gluten. If you had celiac disease, which would be a better choice: the French fries or the onion rings shown in the photo?

Masterfile

Causes and Consequences of Digestive Problems

Minor digestive problems, such as heartburn, constipation, and diarrhea, are common and, in most cases, have little effect on nutritional status. However, more long-term or severe problems can have serious consequences for nutrition and overall health. Some digestive problems and their causes, consequences, and solutions are given in **Table 3.4**.

gastroesophageal reflux disease (GERD) A chronic condition in which acidic stomach contents leak back up into the esophagus, causing pain and damaging the esophagus.

Heartburn and GERD Heartburn is one of the most common digestive complaints. It occurs when the acidic stomach contents leak back into the esophagus, causing a burning sensation in the chest or throat (**Figure 3.15**). The more technical term for the leakage of stomach contents back into the esophagus is *gastroesophageal reflux*. Occasional heartburn from reflux is common, but if it occurs more than twice a week, it may indicate a condition called **gastroesophageal reflux disease (GERD)**. If left untreated,

TABLE 3.4 Digestive Problems and Nutritional Consequences

Problem	Causes	Consequences	Treatment/Management
Dry mouth	Disease, medications	Decreased food intake due to changes in taste, difficulty chewing and swallowing, increased tooth decay, and gum disease.	Change medications, use artificial saliva.
Dental pain and loss of teeth	Tooth decay, gum disease	Reduced food intake due to impaired ability to chew, reduced nutrient absorption due to incomplete digestion.	Change consistency of foods consumed.
Heartburn, gastroesophageal reflux disease (GERD)	Stomach acid leaking into esophagus due to overeating, anxiety, stress, pregnancy, hiatal hernia, or disease processes	Pain and discomfort after eating, ulcers, increased cancer risk.	Reduce meal size, avoid high-fat foods, consume liquids between rather than with meals, remain upright after eating, take antacids and other medications.
Hiatal hernia	Pressure on the abdomen from persistent or severe coughing or vomiting, pregnancy, straining while defecating, or lifting heavy objects	Heartburn, belching, GERD, and chest pain	Reduce meal size, avoid high-fat foods, consume liquids between rather than with meals, remain upright after eating, take antacids and other medications, weight loss.
Ulcers	Infection of stomach by *H. pylori*, which damages the epithelial lining; chronic use of drugs such as aspirin and ibuprofen that erode the mucosa; GERD	Pain, bleeding, and possible abdominal infection	Antibiotics to treat infection, antacids to reduce acid, change in medications.
Vomiting	Bacterial and viral infections, medications, other illnesses, eating disorders, pregnancy, food allergies	Dehydration and electrolyte imbalance; if chronic, can damage the mouth, gums, esophagus, and teeth.	Medications to treat infection, fluid and electrolyte replacement.
Diarrhea	Bacterial and viral infections, medications, food intolerance	Dehydration and electrolyte imbalance	Medications to treat infection, fluid and electrolyte replacement.
Constipation	Low fiber intake, low fluid intake, high fiber in combination with low fluid intake, weak intestinal muscles	Discomfort, intestinal blockage, formation of outpouchings in the intestinal wall called diverticula (see Chapter 4)	High-fiber, high-fluid diet, exercise, medications.
Irritable bowel syndrome	Unusual pattern of muscle contractions in the intestines, may be stronger and last longer or slower and weaker than normal.	Abdominal pain or cramping and changes in bowel function—including bloating, gas, diarrhea, and constipation.	Manage stress and make changes in diet and lifestyle, fiber supplements, antidiarrheal medications, other medications.
Pancreatic disease	Cystic fibrosis or pancreatitis	Malabsorption of fat, fat-soluble vitamins, and vitamin B_{12} due to reduced availability of pancreatic enzymes and bicarbonate.	Oral supplements of digestive enzymes.
Gallstones	Deposits of cholesterol, bile pigments, and calcium in the gallbladder or bile duct	Pain and poor fat digestion and absorption.	Low-fat diet, surgical removal of the gallbladder.

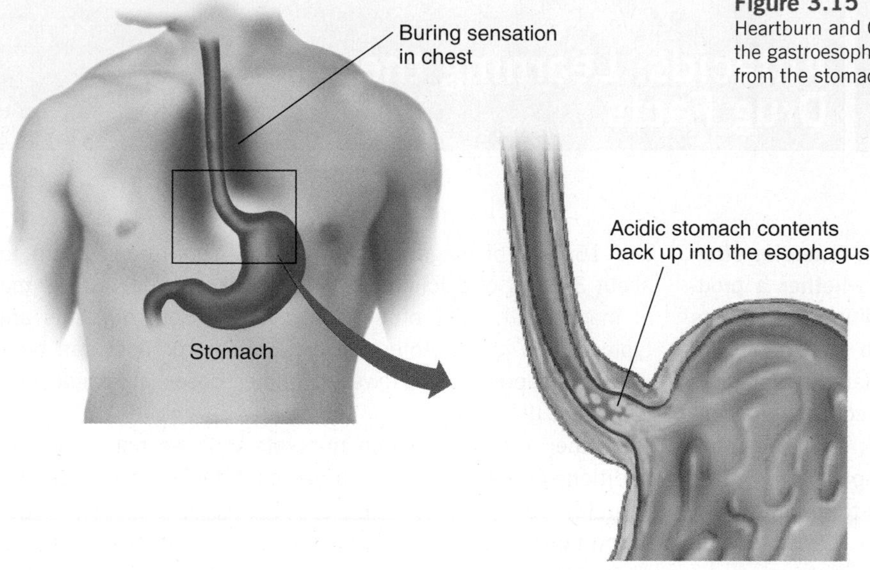

Figure 3.15 **Gastroesophageal reflux**
Heartburn and GERD occur when stomach acid leaks through the gastroesophageal sphincter, which separates the esophagus from the stomach, and irritates the lining of the esophagus.

GERD can eventually lead to more serious health problems such as **peptic ulcers** and cancer. Whether you have occasional heartburn or GERD, it can cause discomfort after eating a large meal and limit the types and amounts of food you can comfortably consume. Symptoms can be reduced by eating small meals and avoiding spicy foods, fatty and fried foods, citrus fruits, chocolate, caffeinated beverages, tomato-based foods, garlic, onions, and mint. Remaining upright after eating, wearing loose clothing, avoiding smoking and alcohol, and losing weight may also help relieve symptoms. A number of medications are also available to reduce reflux symptoms. These decrease the acidity of stomach contents by either neutralizing stomach acid or blocking its production. Some of these medications can add calcium and other nutrients to the diet, but by decreasing acidity they may also reduce the absorption of nutrients that require acid to aid their absorption, particularly vitamin B_{12}, iron, and calcium (see **Off the Label: Antacids: Learning the Drug Facts**).

peptic ulcer An open sore in the lining of the stomach, esophagus, or small intestine.

Heartburn and GERD are sometimes caused by a *hiatal hernia*. The opening in the diaphragm that the esophagus passes through is called the hiatus. A hiatal hernia occurs when the upper part of the stomach bulges through this opening into the chest cavity. Hiatal hernias are common, occurring in about one-quarter of people older than 50. They're more common in women, in smokers, and in people who are overweight. Most hiatal hernias cause no signs or symptoms, but larger ones may allow food and acid to back up into the esophagus, causing heartburn and chest pain. Hiatal hernias can usually be treated by heartburn medications, but in severe cases surgery may be needed.

Peptic Ulcers Peptic ulcers are open sores that develop in the lining of the esophagus, stomach, or upper portion of the small intestine. They occur when the mucosa is eroded away, exposing the underlying tissues to pepsin and acid in the gastric juices. If the damage reaches the nerve layer, it causes pain, and if capillaries are damaged, gastrointestinal bleeding can occur. If the ulcer perforates the wall of the GI tract, a serious abdominal infection can occur. The leading cause of ulcers is acid-resistant bacteria called *Helicobacter pylori (H. pylori)* that infect the lining of the stomach, destroying the protective mucosal layer and causing damage to the stomach wall (see **Science Applied: Discovering What Causes Ulcers**).[9]

Ulcers can also result from GERD or the chronic use of drugs such as aspirin and ibuprofen that erode the mucosa of the GI tract. Mild ulcers can affect nutrient intake by limiting food choices to those that do not cause discomfort, but more severe ulcers can cause bleeding that can be life threatening.

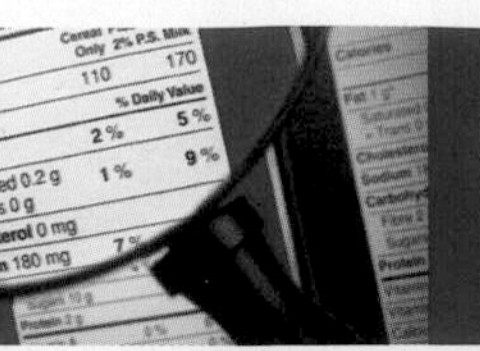

© iStockphoto

OFF THE LABEL

Antacids: Learning the Drug Facts

Did you know that over-the-counter heartburn remedies often contain nutrients? How can you determine whether a product will affect your nutritional health in addition to relieving your heartburn? When choosing a product to treat heartburn (or any other ailment), be sure to check the label. Drug Facts labels on nonprescription drugs are designed to help consumers take medications correctly, as well as understand the drug's benefits, risks, and nutritional impact. They must present information in a standardized, easy-to-follow format that is as readable and as consistent as the Nutrition Facts labels on food products. The active ingredients must be listed first, along with the purpose for each, followed by uses, warnings, directions, and then inactive ingredients.[1]

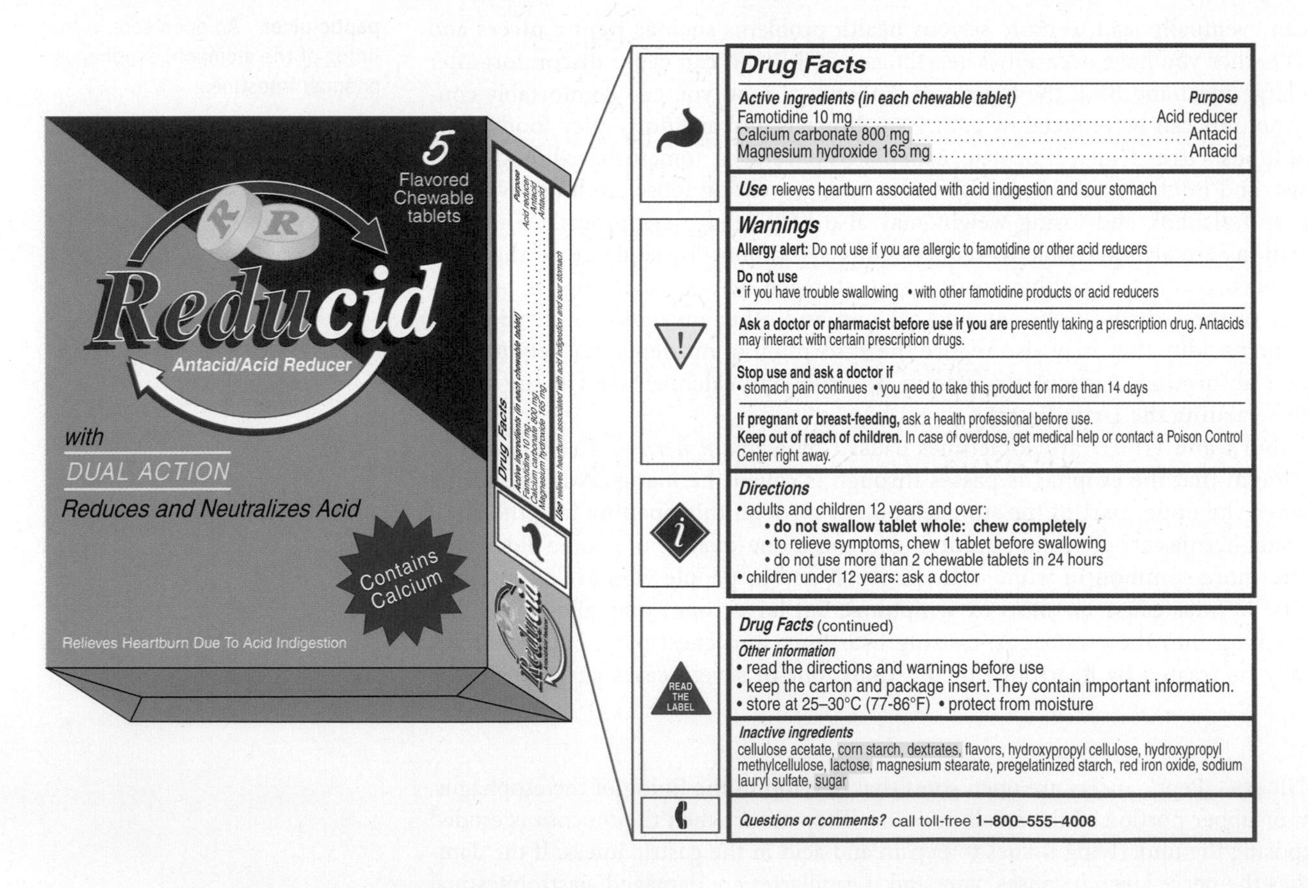

The label shown here indicates that Reducid contains famotidine, a drug that reduces the amount of acid secreted by the stomach. It also contains calcium carbonate and magnesium hydroxide, antacids that contribute minerals as well as neutralizing the acidity of the stomach contents. The label tells you that a single tablet contains 800 mg of calcium carbonate and 165 mg of magnesium hydroxide. This is equivalent to about 320 mg of calcium (32% of the Daily Value) and 69 mg of magnesium (17% of the Daily Value). Both nutrients are typically low in the American diet, so this product can be a welcome supplement. However, higher doses of magnesium can cause diarrhea.

Other antacids contain minerals that are not beneficial additions to the diet. For example, each tablet of Alka-Seltzer contains 567 mg of sodium, about 24% of the Daily Value. Aluminum is also found in a number of antacids; it binds phosphorus and limits its absorption and may cause constipation. The list of inactive ingredients shows that Reducid contains sugar, dextrates (a form of sugar), lactose, and starch, important information for individuals who have diabetes, lactose intolerance, or are consuming low-calorie diets.

If you are looking for an over-the-counter medication to treat occasional heartburn, understanding how to read the label will help you to use it correctly and to know how the product will affect your nutrient intake.

[1]U.S. Food and Drug Administration. OTC drug facts label.

Pancreatic and Gallbladder Problems Abnormalities in the pancreas and gallbladder can affect digestion and cause discomfort. If the pancreas is not functioning normally, the availability of enzymes needed to digest carbohydrate, fat, and protein may be reduced, limiting the ability to digest and absorb these essential nutrients. If the gallbladder is not releasing bile, fat absorption can be impaired. A common condition that affects the gallbladder is gallstones. These are clumps of solid material that form in the gallbladder or bile duct and can cause pain when the gallbladder contracts in response to fat in the intestine. Gallstones can interfere with bile secretion and reduce fat absorption. They are usually treated by removing the gallbladder. Once the gallbladder has been removed, bile simply drips into the intestine as it is produced rather than being stored and squeezed out in larger amounts in response to fat in the intestine.

Diarrhea and Constipation Common discomforts that are related to problems in the intestines include diarrhea and constipation. Diarrhea refers to frequent watery stools. It occurs when material moves through the colon too quickly for sufficient water to be absorbed or when water is drawn into the lumen from cells lining the intestinal tract. Diarrhea may occur when the lining of the small intestine is inflamed so nutrients and water are not absorbed. The inflammation may be caused by infection with a microorganism or by other irritants, including foods. When harmful microorganisms infect the GI tract, diarrhea (as well as vomiting) helps to flush them out.

Constipation refers to hard, dry stools that are difficult to pass. It can occur when the diet is low in fiber or high in fiber and low in fluids, both of which cause the stool to become hard and dry and difficult to evacuate. Certain medications, lack of exercise, and a weakening of the muscles of the large intestine can also cause constipation. Constipation increases pressure in the colon and can lead to outpouches in the colon wall called diverticula (see Chapter 4). Irritable bowel syndrome (IBS) is a condition that can cause either diarrhea or constipation. In IBS abnormal contractions of the intestinal wall cause bloating, gas, abdominal pain, and cramping, in addition to diarrhea and constipation.

How Nutrients Are Provided to People Who Can't Eat For individuals who are unable to consume food or digest and absorb the nutrients needed to meet their requirements, several alternative feeding methods have been developed. People who are unable to swallow can be fed a liquid diet through a tube inserted into the stomach or intestine. This **enteral** or **tube-feeding** can provide all the essential nutrients for patients who are unconscious or have suffered an injury to the upper GI tract (**Figure 3.16a**). For individuals whose GI tract is not functional, nutrients can be provided directly into the bloodstream. This is referred to as **total parenteral nutrition (TPN)** (**Figure 3.16b**). Carefully planned TPN can provide

enteral or **tube-feeding** A method of feeding by providing a liquid diet directly to the stomach or intestine through a tube placed down the throat or through the wall of the GI tract.

total parenteral nutrition (TPN) A technique for nourishing an individual by providing all needed nutrients directly into the circulatory system.

THINK CRITICALLY

What type of carbohydrate must be used in TPN solutions? Why?

Figure 3.16 **Alternate feeding methods**

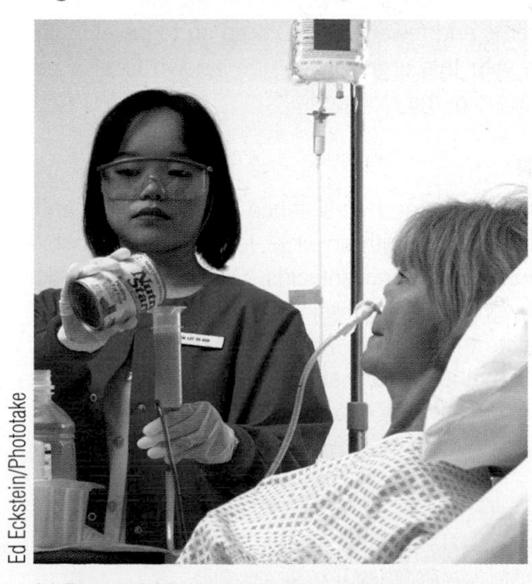

Ed Eckstein/Phototake

(a) Enteral feeding can be delivered by a narrow tube passed through the nasal passages and into the stomach or intestine so that a liquid diet can be delivered for digestion and absorption.

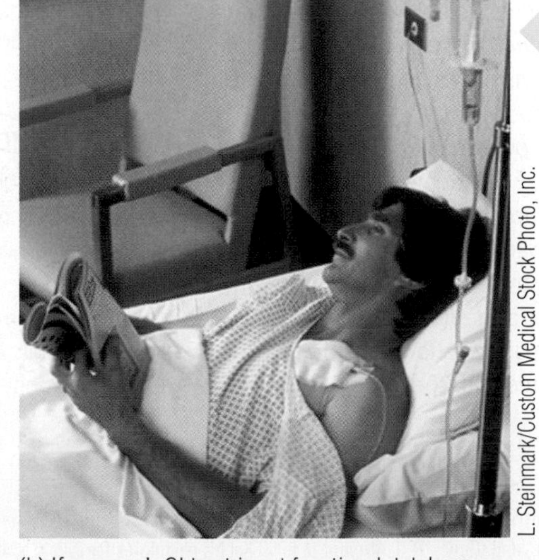

L. Steinmark/Custom Medical Stock Photo, Inc.

(b) If a person's GI tract is not functional, total parenteral nutrition can deliver a solution providing all of the essential nutrients directly into the bloodstream through a large vein in the upper arm or chest.

SCIENCE APPLIED

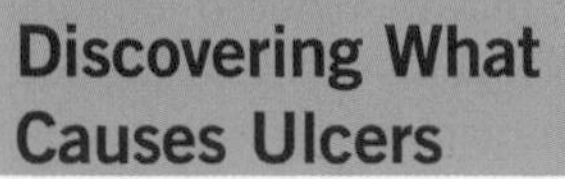

Discovering What Causes Ulcers

(Elena Schweitzer/Shutterstock)

Victoriano Izquierdo/Getty Images, Inc.

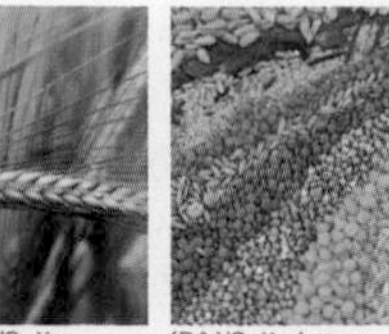

(DAJ/Getty Images, Inc.)

Peptic ulcers are lesions in the wall of the stomach or other part of the GI tract. They cause pain and, sometimes, bleeding. For many years they were thought to be caused by excess stomach acid. Treatments included a bland diet to avoid irritation, milk to coat the GI lining, stress reduction to decrease acid secretion, and drugs to neutralize acid or reduce its secretion. These treatments usually reduced irritation and decreased symptoms, but ulcers were a chronic condition even with treatment. Today, due to the observations and perseverance of Australian Nobel laureates B. J. Marshall and J. R. Warren, many ulcers can be cured.

Helicobacter pylori

Science Photo Library/Photo Researchers, Inc.

CNRI/SPL/Photo Researchers, Inc.

This electron micrograph shows the helical-shaped bacteria *Helicobacter pylori* attached to the gastric mucosa.

Infection with *Helicobacter pylori* is the most common cause of ulcers, such as this one in the lining of the stomach.

THE SCIENCE

While examining stomach biopsies under a microscope, Dr. Warren observed curvy-shaped bacteria. He noticed that the bacteria were always present in tissue that was inflamed and that the number of organisms correlated with the degree of inflammation.[1] In 1982 he and Dr. Marshall isolated and grew this bacteria, later named *Helicobacter pylori (H. pylori)*, in the laboratory.[2] They believed that the bacteria were the cause of peptic ulcers.

This hypothesis, however was not immediately embraced by the scientific community. At the time, it was assumed that ulcers were caused by too much stomach acid and it was widely accepted that bacteria could not survive in the strong acid of the stomach. To persuade other scientists that the bacteria hypothesis was correct, Marshall decided to gather data using himself as an experimental subject.

Marshall's experiment involved first having a small sample of his gastric mucosa examined to confirm that it was not infected with bacteria. He then drank a vial containing *H. pylori* that had been isolated from a patient with chronic gastric inflammation. Ten days later he developed symptoms of gastric inflammation. A follow-up sample of his gastric mucosa confirmed that his stomach lining was now inflamed and *H. pylori* bacteria could be seen attached to the mucosa (see figure).[3] Fortunately, his symptoms resolved quickly with antibiotic therapy. This confirmed the connection between *H. pylori* and stomach inflammation, but since Marshall did not develop an ulcer, that link was still unproven. The connection between *H. pylori* and ulcers was eventually supported by epidemiological studies that showed an increased incidence of ulcers in persons infected with the bacteria.[1]

As our understanding of how *H. pylori* survive in the acid environment of the stomach grew, the idea that bacteria caused ulcers became more universally accepted. In the stomach, *H. pylori* use a tail-like structure, called a flagella, to swim through the protective mucus layer and adhere to the mucosal cells. Once attached to the mucosal cells, they use an enzyme to produce substances that neutralize the acid immediately around them.[4] The bacteria damage mucosal cells and increase the release of gastrin, a hormone that increases stomach acid secretion.[5] The presence of bacteria causes inflammation and other immune responses that lead to additional tissue damage. About 30 to 50% of the world's population is infected with *H. pylori*.[1] In the United States about 20% of people under 40 years of age and half of those over 60 are infected.[6] Most people who are infected with *H. pylori* do not have symptoms and fewer than 20% go on to develop an ulcer, but infection with this organism is now known to also be associated with cancers of the stomach.[1,6]

THE APPLICATION

Because of the observations and persistence of Drs. Warren and Marshall, a patient diagnosed with an ulcer today is given antibiotics rather than a bland diet and antacids. Successful antibiotic therapy combined with acid-suppression therapy can allow the ulcer to heal, eliminate the bacteria that caused the disease, and cure the patient. A hypothesis that was at first rejected by the scientific community is now a theory supported by scientific evidence.

[1]Lynch, N. A. *Helicobacter.pylori* and ulcers: A paradigm revised.

[2]Marshall, B. J., and Warren, J. R. Unidentified curved bacilli on gastric epithelium in active chronic gastritis. *Lancet* 1:1311–1315, 1984.

[3]Marshall, B. J., Armstrong, J. A., McGechie, D. B., and Glancy, R. J. Attempt to fulfill Koch's postulates for pyloric. *Campylobacter Med J Aust* 142:436–439, 1985.

[4]McGee, D. J., and Mobley, H. L. T. Mechanisms of *Helicobacter pylori* infection: Bacterial factors. *Curr Top Microbiol Immunol* 241:156–180, 1999.

[5]Joseph, I. M., and Kirschner, D. A. Model for the study of *Helicobacter pylori* interaction with human gastric acid secretion. *J Theor Biol* 228:55–80 2004.

[6]National Digestive Disease Information Clearinghouse, National Institutes of Diabetes and Digestive and Kidney Diseases. *H. pylori* and peptic ulcer.

all the nutrients essential to life. When all nutrients are not provided in a TPN solution, nutrient deficiencies develop quickly. Inadvertently feeding patients incomplete TPN solutions has helped demonstrate the essentiality of several trace minerals.

The Digestive System Throughout Life

There are some differences in the way the digestive system functions during pregnancy and infancy and with advancing age. These changes affect the ability to ingest and digest food and absorb nutrients (**Figure 3.17**). A well-planned diet is needed to maintain nutritional status at all stages of life.

Why Pregnancy Causes Digestive Discomforts Physiological changes that occur during pregnancy may cause gastrointestinal problems. During the first three months, many women experience nausea, referred to as morning sickness. This term is a misnomer because the nausea can occur at any time of the day. Morning sickness is believed to be due to pregnancy-related hormonal changes. In most cases, it can be managed by consuming frequent small meals, eating dry crackers or cereal, and avoiding foods and smells that cause nausea. In severe cases where uncontrollable vomiting occurs, nutrients may need to be given intravenously to maintain nutritional health.

Figure 3.17 Gastrointestinal changes throughout the life cycle

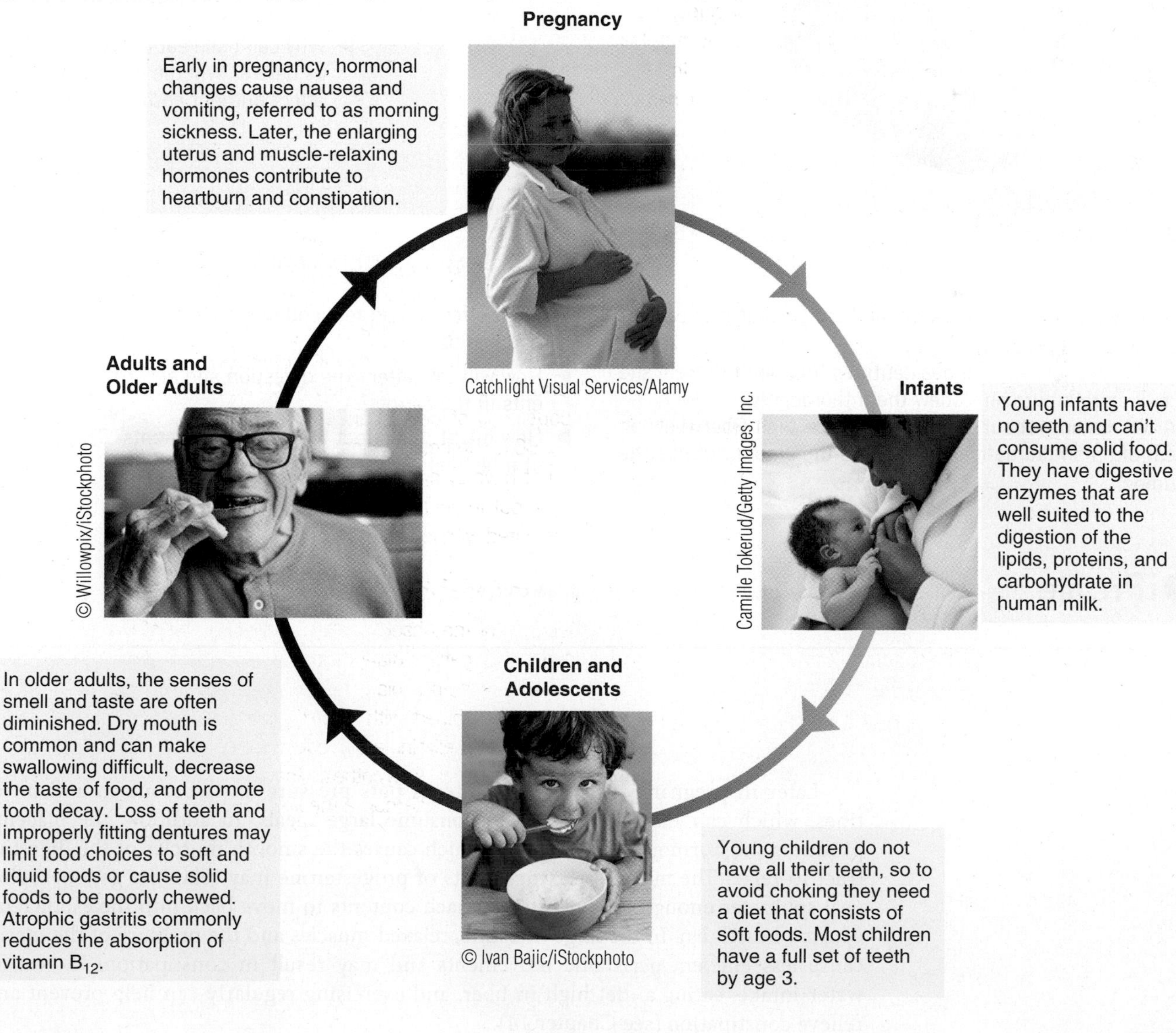

How Gastrointestinal Problems Can Affect Nutrition

Many factors affect how well we digest food and absorb nutrients. For each situation described below, consider how digestion and absorption are affected and what the consequences are for nutritional status.

A 40-year-old woman weighing 350 lbs has undergone a surgical procedure called gastric banding to help her lose weight. The diagram shows how her stomach was altered.

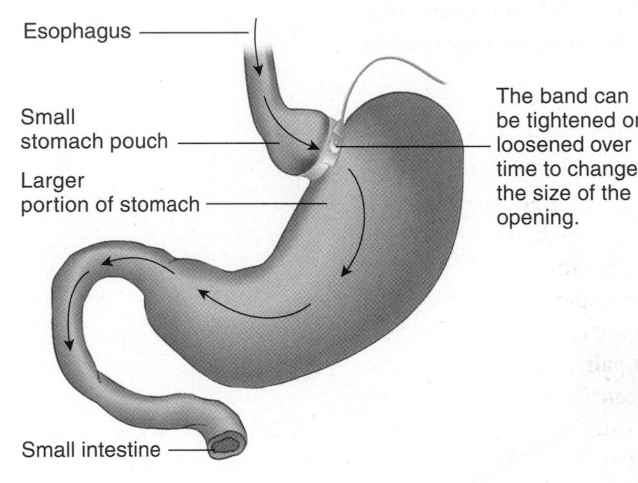

CRITICAL THINKING QUESTIONS

- Why can't she eat as much food as before? Will the procedure affect nutrient absorption?

CRITICAL THINKING QUESTION

A 50-year old man is taking medication that reduces the amount of saliva produced.

- What effect would reduced saliva have on his nutrition and health?

An 80-year-old woman has dentures that don't fit well; she likes raw carrots but can't chew them thoroughly.

- How will this affect the digestion and absorption of nutrients in the carrots?

After reading about the benefits of a high-fiber diet, an 18-year-old man dramatically increases the amount of fiber he consumes.

- How might this affect his bowel movements? The amount of intestinal gas?

iProfile Use iProfile to find out how much fiber is in a cup of raw carrots.

Later in pregnancy, the enlarged uterus puts pressure on the stomach and intestines, which can make it difficult to consume large meals. In addition, the placenta produces the hormone progesterone, which causes the smooth muscles of the digestive tract to relax. The muscle-relaxing effects of progesterone may relax the gastroesophageal sphincter enough to allow the stomach contents to move back into the esophagus, causing heartburn. In the large intestine, relaxed muscles and the pressure of the uterus cause less efficient peristaltic movements and may result in constipation. Increasing water intake, eating a diet high in fiber, and exercising regularly can help prevent and relieve constipation (see Chapter 14).

How the Infant Digestive Tract Differs The digestive system is one of the last to fully mature in developing humans. At birth, the digestive tract is functional, but a newborn is not ready to consume an adult diet. The most obvious difference between the infant and adult digestive tracts is that newborns are not able to chew and swallow solid food. Levels of digestive enzymes also differ between infants and adults. In infants, the digestion of milk protein is aided by rennin, an enzyme produced in the infant stomach that is not found in adults.[10] The stomachs of newborns also produce the enzyme gastric lipase. This enzyme is present in adults but plays a more important role in infants, where it begins the digestion of the fats in human milk. Low levels of pancreatic enzymes in infants limit starch digestion; however, enzymes at the brush border of the small intestine allow the milk sugar lactose to be digested and absorbed.

The ability to absorb intact proteins is greater in infants than that in adults. The absorption of whole proteins can cause food allergies (see Chapter 15), but it also allows infants to absorb immune factors from their mothers' milk. These proteins provide temporary immunity to certain diseases.

The bacteria in the large intestine of infants are also different from those in adults. At birth the infant gut has no bacteria, but the gut is quickly colonized. Because of their all-milk diet, infants' gut microflora differ from adults until the second year of life.

How the Digestive System Changes with Aging Although there are few dramatic changes in nutrient requirements with aging, changes in the digestive tract and other systems may affect the palatability of food and the ability to obtain proper nutrition. The senses of smell and taste are often diminished or even lost with age, reducing the appeal of food. A reduction in the amount of saliva may make swallowing difficult, decrease the taste of food, and also promote tooth decay. Loss of teeth and improperly fitting dentures may limit food choices to soft and liquid foods or cause solid foods to be poorly chewed. Gastrointestinal secretions may also be reduced, but this rarely impairs absorption because the levels secreted in healthy elderly adults are still sufficient to break down food into forms that can be absorbed. A condition called **atrophic gastritis** that causes a reduction in the secretion of stomach acid is also common in the elderly. This may decrease the absorption of several vitamins and minerals and may allow bacterial growth to increase (see Chapters 8 and 16). Constipation is a common complaint among the elderly. It may be caused by decreased motility and elasticity in the colon, weakened abdominal and pelvic muscles, and a decrease in sensory perception (see **Critical Thinking: How Gastrointestinal Problems Can Affect Nutrition**).

atrophic gastritis An inflammation of the stomach lining that causes a reduction in stomach acid and allows bacterial overgrowth.

3.5 Delivering Nutrients to Body Cells

LEARNING OBJECTIVES

- Explain why the cardiovascular system is important in nutrition.
- Discuss the role of the lymphatic system in nutrition.
- Compare the path of an amino acid to that of a large fatty acid from absorption to delivery to a cell.

Nutrients absorbed into the mucosal cells of the intestine enter the blood circulation by either the **hepatic portal circulation** or the **lymphatic system**. The hepatic portal circulation is part of the cardiovascular system, which consists of the heart and blood vessels. Amino acids from protein digestion, simple sugars from carbohydrate digestion, and the water-soluble products of fat digestion are absorbed into **capillaries** that are part of the hepatic portal circulation. The products of fat digestion that are not water-soluble are taken into lacteals, which are small vessels of the lymphatic system, before entering the blood.

hepatic portal circulation The system of blood vessels that collects nutrient-laden blood from the digestive organs and delivers it to the liver.

lymphatic system The system of vessels, organs, and tissues that drains excess fluid from the spaces between cells, transports fat-soluble substances from the digestive tract, and contributes to immune function.

capillary A small, thin-walled blood vessel where the exchange of gases and nutrients between blood and body cells occurs.

Why the Cardiovascular System Is Important in Nutrition

The cardiovascular system is a closed network of tubules through which blood is pumped. Blood carries nutrients and oxygen to the cells of all the organs and tissues of the body and removes waste products from these same cells. Blood also carries other substances, such as hormones, from one part of the body to another.

vein A vessel that carries blood toward the heart.

artery A vessel that carries blood away from the heart.

The Heart and Blood Vessels The heart is the workhorse of the cardiovascular system. It is a muscular pump with two circulatory loops—one that delivers blood to the lungs and one that delivers blood to the rest of the body (**Figure 3.18**). The blood vessels that transport blood and dissolved substances toward the heart are called **veins**, and those that transport blood and dissolved substances away from the heart are called **arteries**. As arteries carry blood away from the heart, they branch many times to form smaller and smaller blood vessels. The smallest arteries are called arterioles. Arterioles then branch to form capillaries. Blood flows from capillaries into the smallest veins, the venules, which converge to form larger and larger veins for return to the heart.

The exchange of nutrients and gases occurs across the thin walls of the capillaries. In most body tissues, oxygen and nutrients carried by the blood pass from the

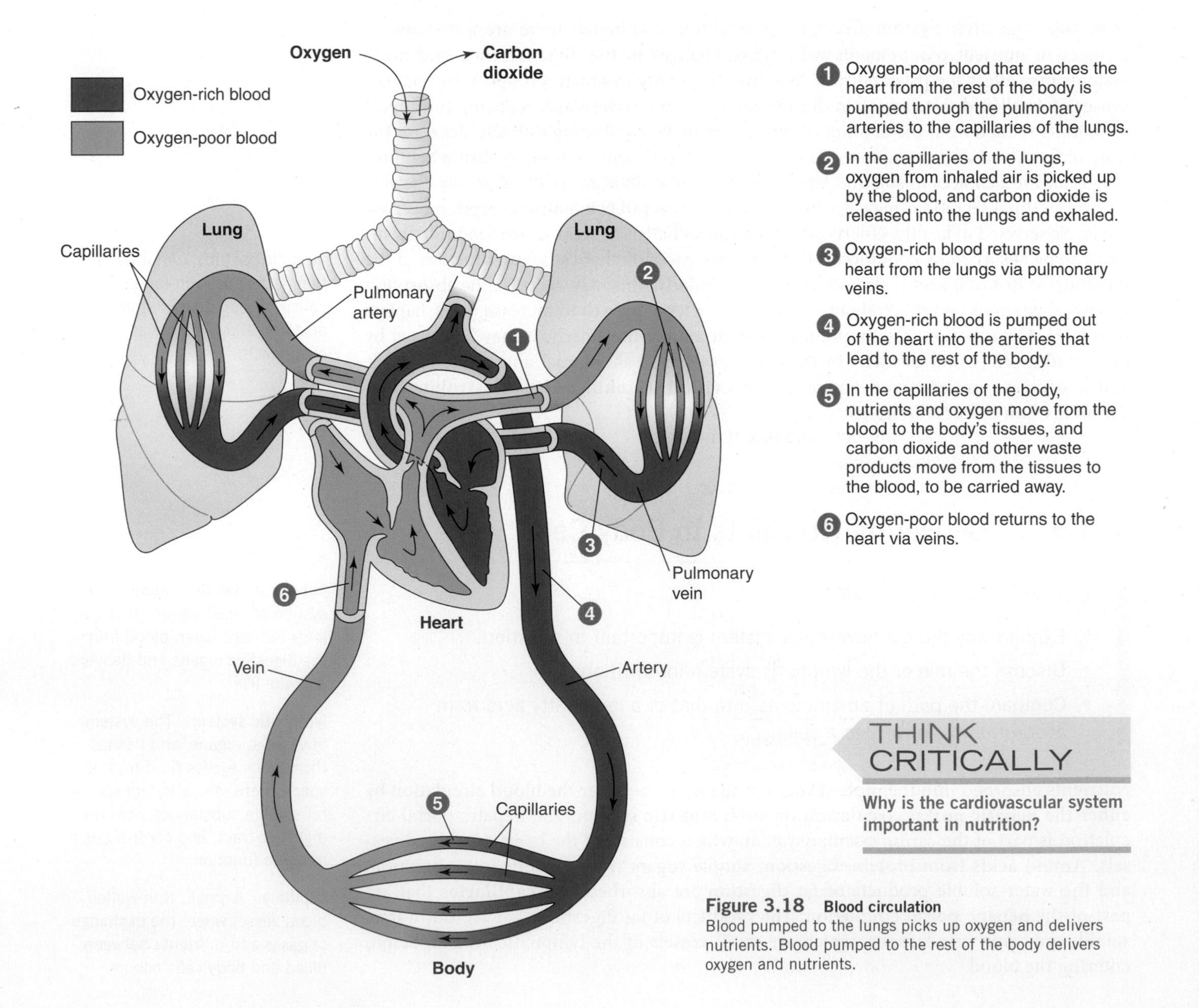

Figure 3.18 Blood circulation
Blood pumped to the lungs picks up oxygen and delivers nutrients. Blood pumped to the rest of the body delivers oxygen and nutrients.

THINK CRITICALLY

Why is the cardiovascular system important in nutrition?

capillaries into the cells, and carbon dioxide and other waste products pass from the cells into the capillaries. In the capillaries of the lungs, blood releases carbon dioxide to be exhaled and picks up oxygen to be delivered to the cells. In capillaries in the villi of the small intestine, blood delivers oxygen and picks up water-soluble nutrients absorbed from the diet.

The volume of blood flow, and hence the amounts of nutrients and oxygen that are delivered to an organ or tissue, depends on the need. When you are resting, about 25% of your blood goes to the digestive system, about 20% to the skeletal muscles, and the rest to the heart, kidneys, brain, skin, and other organs.[10] This distribution changes when you eat or exercise. When you have eaten a large meal, a greater proportion of your blood goes to your digestive system to provide the oxygen and nutrients needed by the GI muscles and glands for digestion of the meal and absorption of nutrients. When you are exercising strenuously, about 70% of your blood is directed to your skeletal muscles to deliver nutrients and oxygen and remove carbon dioxide and other waste products. Attempting to exercise after a large meal creates a conflict. The body cannot supply both the intestines and the muscles with enough blood to support their respective activities. The muscles win, and food remains in the intestines, often resulting in cramps.

Most Nutrients Go Directly to the Liver In the small intestine, water-soluble molecules, including amino acids, sugars, water-soluble vitamins, and water-soluble products of fat digestion, cross the mucosal cells of the villi and enter capillaries. These capillaries merge to form venules at the base of the villi. The venules then merge to form larger and larger veins, which eventually form the **hepatic portal vein**. The hepatic portal vein transports blood directly to the liver, where absorbed nutrients are processed before they enter the general circulation (**Figure 3.19**).

hepatic portal vein The vein that transports blood from the GI tract to the liver.

The liver acts as a gatekeeper between substances absorbed from the intestine and the rest of the body. Depending on the immediate needs of the body, some nutrients are stored in the liver, some are broken down or changed into different forms, and others are allowed to pass through unchanged for delivery to other body cells. For example, the liver, with the help of hormones from the pancreas, keeps the concentration of glucose in the blood constant. The liver modulates blood glucose by removing absorbed glucose from the blood and storing it, by sending absorbed glucose on to the tissues of the body, or by releasing glucose (from liver stores or synthesis) into the blood. The liver also plays an important role in the synthesis and breakdown of amino acids, proteins, and fats. It modifies the products of protein breakdown to form molecules that can be safely transported to the kidneys for excretion. The liver also contains enzyme systems that protect the body from toxins that are absorbed by the GI tract.

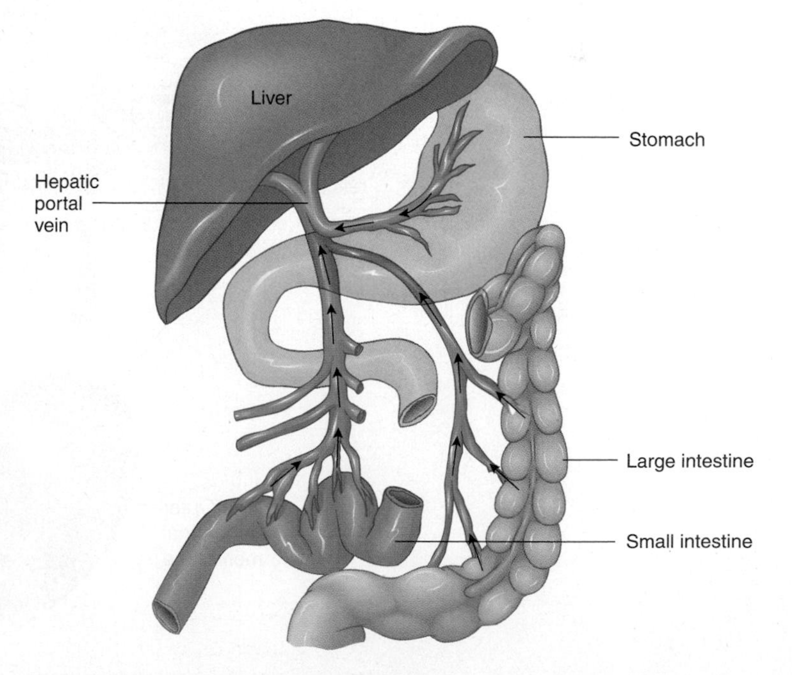

Figure 3.19 Hepatic portal circulation
The hepatic portal circulation carries blood from the stomach and intestines to the liver. Nutrients absorbed directly into the blood reach the liver via the hepatic portal vein.

The Roles of the Lymphatic System

The lymphatic system is important for fluid balance and immune function as well as for the absorption of lipids. It includes a network of tubules, called lymph vessels, as well as lymphatic organs and tissues, which contain infection-fighting cells. Lymph vessels transport a clear fluid, called lymph. Lymph originates from the fluid that squeezes out of blood vessels and accumulates in tissues. This excess fluid drains into the lymph vessels. Unlike blood, which circulates, lymph flows only one way, draining fluid from the tissues and returning it to the blood. Before entering the blood steam, lymph flows through lymph nodes, where infection-fighting lymphocytes and phagocytes congregate. Here foreign invaders, such as viruses and bacteria or chemicals that are harmful to the body, are removed and can trigger an immune response.

The lymphatic system is important for absorption because fat-soluble materials such as triglycerides, cholesterol, and fat-soluble vitamins are incorporated into particles that are too large to enter the intestinal capillaries (see Chapter 5). These particles pass from the intestinal mucosa into the lacteals, small lymph vessels in the villi, which drain into larger lymph vessels. Lymph vessels from the intestines and other parts of the body eventually empty into the bloodstream. Therefore, substances that are absorbed via the lymphatic system do not pass through the liver before entering the general blood circulation.

cell membrane The membrane that surrounds the cell contents.

selectively permeable Describes a membrane or barrier that will allow some substances to pass freely but will restrict the passage of others.

cytosol The liquid found within cells.

organelles Cellular organs that carry out specific metabolic functions.

mitochondrion (mitochondria) Cellular organelle that is responsible for providing energy in the form of adenosine triphosphate (ATP) for cellular activities.

How Nutrients Enter Body Cells

In order for nutrients to be used by the body, they must enter the cells. To enter a cell, substances must first cross the **cell membrane**. The cell membrane surrounds the cell and maintains homeostasis in the cell by controlling what enters and what exits. It is **selectively permeable**, allowing some substances, such as water, to pass freely back and forth, while limiting the passage of others. Nutrients and other substances are transported from the blood into cells by simple and facilitated diffusion and active transport. Inside the cell membrane are the **cytosol**, or cell fluid, and **organelles** that perform functions necessary for cell survival. The largest organelle is the nucleus, which contains the cell's genetic material. The **mitochondrion** (plural **mitochondria**) is the organelle where the metabolic reactions that provide energy occur. Its structure, which contains an outer and an inner membrane, facilitates its role in energy metabolism (**Figure 3.20**).

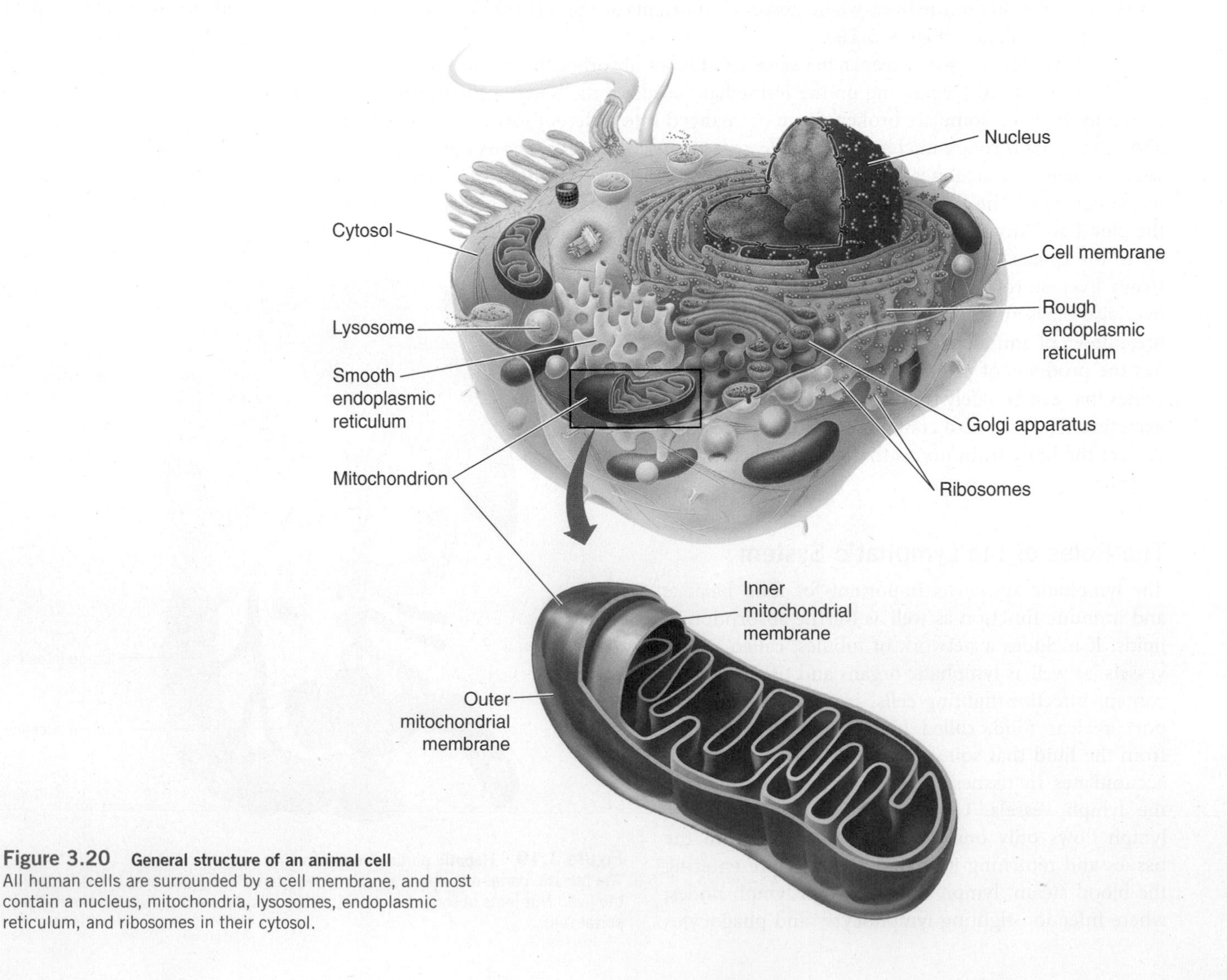

Figure 3.20 General structure of an animal cell
All human cells are surrounded by a cell membrane, and most contain a nucleus, mitochondria, lysosomes, endoplasmic reticulum, and ribosomes in their cytosol.

3.6 Metabolism of Nutrients: An Overview

LEARNING OBJECTIVES

- Compare anabolic and catabolic pathways.
- Name the dietary fuel sources used to produce ATP.
- Describe the types of molecules that are synthesized from glucose, amino acids, and fatty acids.

Foods are digested and the products of digestion absorbed, transported, and delivered to body cells by the mechanisms described thus far. Each nutrient plays a unique role in metabolism. If the proper amounts and types of nutrients are not delivered to cells, the reactions of metabolism cannot proceed optimally, resulting in poor health. The following discussion provides a brief overview of how glucose, fatty acids, and amino acids are used to provide energy. Details about the metabolism of these energy-yielding nutrients and the role of other nutrients in metabolism will be discussed in appropriate chapters throughout this book and are reviewed in the Online Appendix: Focus on Metabolism.

metabolic pathway A series of chemical reactions inside a living organism that results in the transformation of one molecule into another.

coenzyme A small organic molecule (not a protein but sometimes a vitamin) that is necessary for the proper functioning of many enzymes.

catabolic pathways The biochemical reactions by which substances are broken down into simpler molecules releasing energy.

ATP (adenosine triphosphate) The high-energy molecule used by the body to perform energy-requiring activities.

Anabolic and Catabolic Pathways

Depending on the body's needs, the glucose, fatty acids, and amino acids absorbed from the diet are broken down to provide energy, used to synthesize essential structural or regulatory molecules, or transformed into energy-storage molecules. The conversion of one molecule into another often involves a series of reactions. The series of biochemical reactions needed to go from a raw material to the final product is called a **metabolic pathway**. For each of the reactions of a metabolic pathway to proceed at an appropriate rate, an enzyme is required. These enzymes often need help from **coenzymes**. The B vitamins are important coenzymes in energy metabolism.

Some metabolic pathways break large molecules into smaller ones. These **catabolic pathways** release energy trapped in the chemical bonds that hold molecules together. Some of this energy is lost as heat, but some is used to synthesize a molecule called **adenosine triphosphate (ATP)** that can be used as an energy source by the body (**Figure 3.21**). ATP can be thought of as the energy currency of the cell. The chemical bonds of ATP are very

Figure 3.21 **ATP links and catabolic and anabolic reactions**

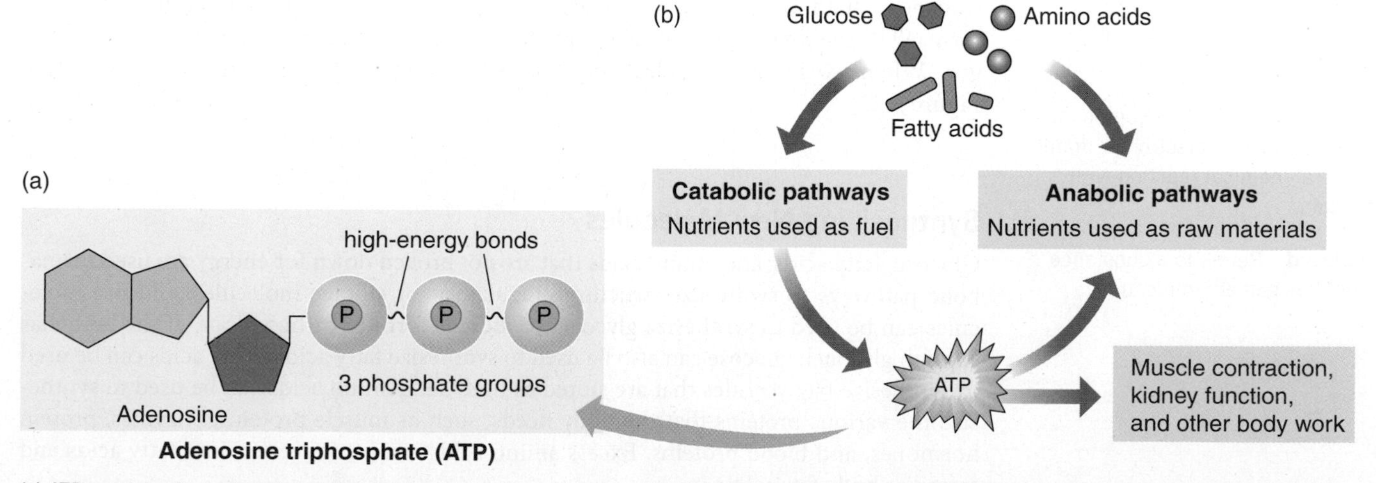

(a) ATP consists of an adenosine molecule attached to three phosphate groups. The bonds between the phosphate groups are very high in energy, which is released when the bonds are broken.

(b) Nutrients delivered to body cells can be used either in catabolic reactions to produce ATP or as raw materials in anabolic reactions that use ATP to synthesize molecules needed by the body.

anabolic pathways Energy-requiring biochemical reactions in which simpler molecules are combined to form more complex substances.

high in energy, and when they break, the energy is released and can be used to power body processes, such as muscle contraction or nerve conduction—or it can be used to synthesize new molecules needed to maintain and repair body tissues. Metabolic pathways that use energy from ATP to build body compounds are referred to as **anabolic pathways** . The anabolic and catabolic pathways of metabolism occur in the body continually and simultaneously (see Figure 3.21).

Producing ATP

Inside cells, glucose, fatty acids, and amino acids derived from carbohydrates, fats, and proteins, respectively, can be broken down in the presence of oxygen to produce carbon dioxide and water. These reactions release energy that is used to add a phosphate group to adenosine diphosphate (ADP) to form ATP. This catabolic pathway is called **cellular respiration**. In cellular respiration, oxygen brought into the body by the respiratory system and delivered to cells by the circulatory system is used and carbon dioxide is released. This carbon dioxide is then transported to the lungs, where it is eliminated in exhaled air.

cellular respiration The reactions that break down glucose, fatty acids, and amino acids in the presence of oxygen to produce carbon dioxide, water, and energy in the form of ATP.

acetyl-CoA A metabolic intermediate formed during the breakdown of glucose, fatty acids, and amino acids. It is a two-carbon compound attached to a molecule of CoA.

citric acid cycle Also known as the Krebs cycle or the tricarboxylic acid cycle, this is the stage of cellular respiration in which two carbons from acetyl-CoA are oxidized, producing two molecules of carbon dioxide.

electron High-energy particle carrying a negative charge that orbits the nucleus of an atom.

electron transport chain The final stage of cellular respiration in which electrons are passed down a chain of molecules to oxygen to form water and produce ATP.

oxidized Refers to a compound that has lost an electron or undergone a chemical reaction with oxygen.

reduced Refers to a substance that has gained an electron.

The reactions of cellular respiration are key to providing energy for all body processes. Without available oxygen, only glucose can be used to produce ATP (see Chapters 4, 13, and the Online Appendix: Focus on Metabolism). When oxygen is available, glucose, fatty acids, and amino acids can all be broken down to yield two-carbon units that form a molecule called **acetyl-CoA** (**Figure 3.22**). To form acetyl-CoA from glucose, the glucose must first be split in half by a pathway in the cell cytosol called *glycolysis* (see Chapter 4). To form acetyl-CoA from fatty acids, the carbon chains that make up fatty acids are broken into two-carbon units by a pathway in the mitochondria called *beta-oxidation (ß-oxidation)* (see Chapter 5). Amino acids vary in structure, but after the amino group is removed by *deamination*, they can be broken down into units that can form acetyl-CoA (see Chapter 6). Acetyl-CoA from all of these sources is broken down inside the mitochondria via the metabolic pathway known as the **citric acid cycle**. In this pathway, the two carbons of acetyl-CoA are removed one at a time, forming carbon dioxide molecules, releasing **electrons**, and generating a small amount of ATP. The electrons, which are high in energy, are passed to shuttling molecules for transport to the **electron transport chain**.

The electron transport chain consists of a series of molecules that accept electrons from the shuttling molecules and pass them from one to another down the chain. When one substance loses an electron, another must pick up that electron. A substance that loses an electron is said to be **oxidized** and one that gains an electron is said to be **reduced**. Reactions that transfer electrons are called *oxidation-reduction* reactions and are very important in energy metabolism. As electrons are passed along the electron transport chain, their energy is released and used to add a phosphate group to ADP to form ATP. The final molecule to accept electrons in the electron transport chain is oxygen. When oxygen accepts electrons it is reduced and forms a molecule of water (see Figure 3.22).

Synthesizing New Molecules

Glucose, fatty acids, and amino acids that are not broken down for energy are used in anabolic pathways to synthesize structural, regulatory, or storage molecules. Glucose molecules can be used to synthesize glycogen, a storage form of carbohydrate. If the body has enough glycogen, glucose can also be used to synthesize fatty acids. Fatty acids can be used to synthesize triglycerides that are stored as body fat. Amino acids can be used to synthesize the various proteins that the body needs, such as muscle proteins, enzymes, protein hormones, and blood proteins. Excess amino acids can be converted into fatty acids and stored as body fat.

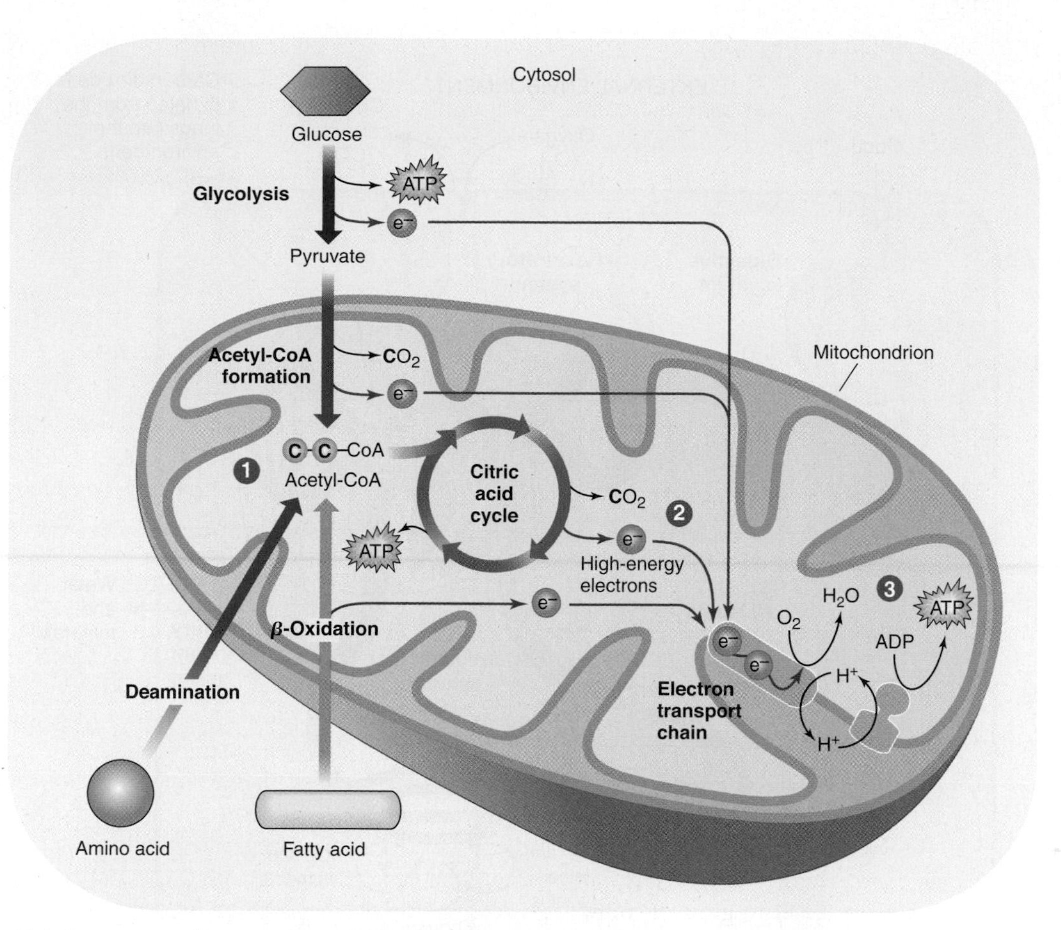

Figure 3.22 Cellular respiration Cellular respiration uses oxygen to convert glucose, fatty acids, and amino acids into carbon dioxide, water, and energy, in the form of ATP.

3.7 Elimination of Metabolic Wastes

LEARNING OBJECTIVES

- Distinguish the substances that are eliminated in the feces from those lost in the urine.
- List four routes for eliminating waste products from the body.

Substances that cannot be absorbed by the body, such as fiber, are excreted from the GI tract in feces. Metabolic wastes, such as carbon dioxide, nitrogen, and water, which are generated by nutrient metabolism, must also be removed from the body. These are eliminated by the lungs, the skin, and the kidneys.

The Respiratory System and Skin

Carbon dioxide produced by cellular respiration leaves the cells and is transported to the lungs by red blood cells. At the lungs, red blood cells release their load of carbon dioxide, which is then exhaled. In addition to carbon dioxide, a significant amount of water is lost

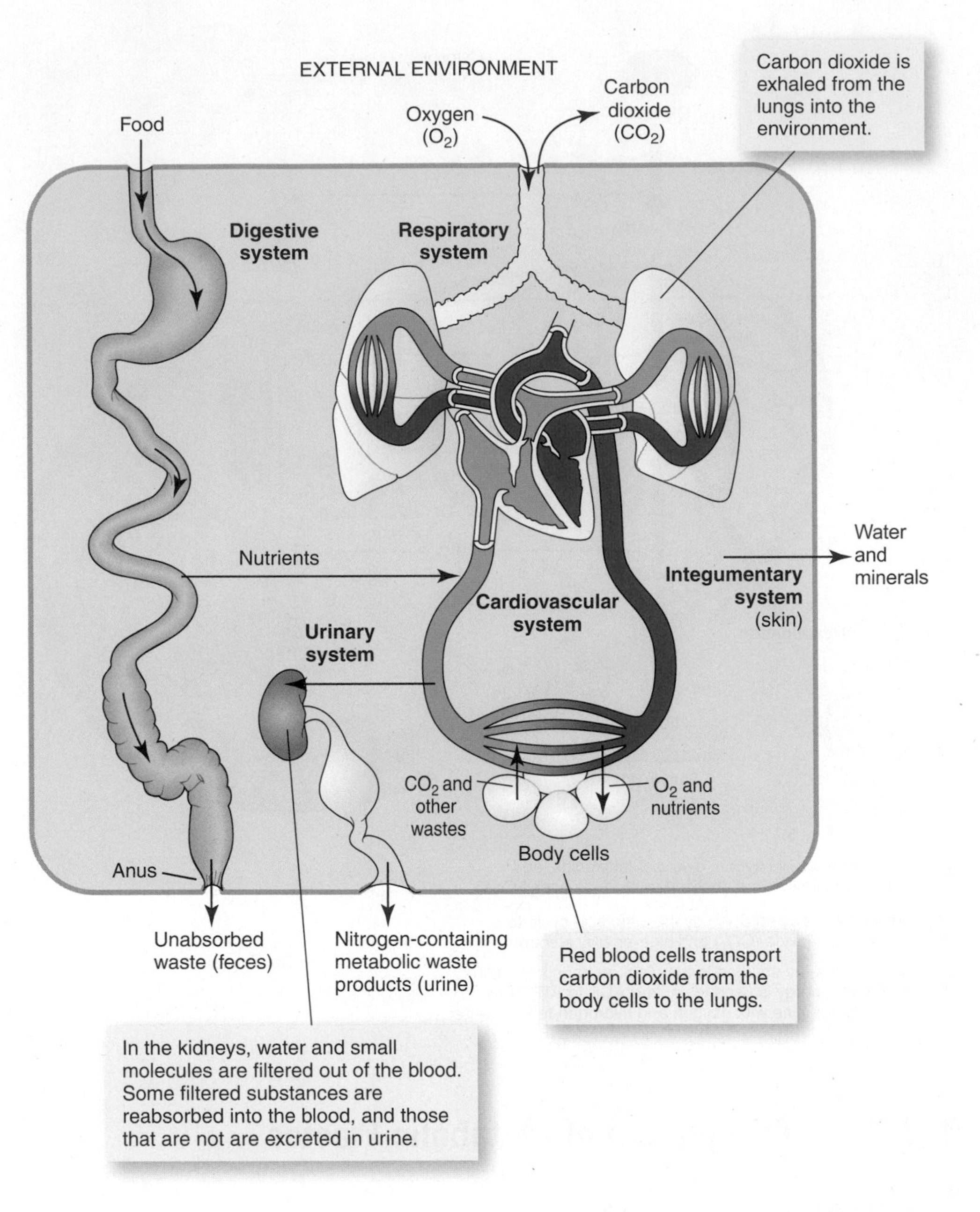

Figure 3.23 Organ systems involved in the elimination of wastes Substances in food that cannot be absorbed are eliminated in the feces. Nutrients that are absorbed from the digestive system and oxygen taken in by the respiratory system are distributed to all the cells in the body by the cardiovascular system. Wastes generated from nutrient metabolism, called metabolic wastes, are eliminated from the body by the integumentary, the urinary, and respiratory systems.

from the lungs by evaporation. Water, along with protein-breakdown products and minerals, is also lost through the skin in perspiration or sweat (**Figure 3.23**).

The Kidneys

nephron The functional unit of the kidney that performs the job of filtering the blood and maintaining fluid balance.

glomerulus A ball of capillaries in the nephron that filters blood during urine formation.

The kidneys, which are part of the urinary system, are the primary means for the excretion of water, metabolic waste products, and excess minerals. Each kidney consists of about 1 million **nephrons**. The nephrons consist of a **glomerulus** where the blood is filtered and a series of tubules where molecules that have been filtered out of the blood can be reabsorbed. As blood flows through the glomerulus, most of the small dissolved molecules are filtered out. Protein molecules and blood cells are too large to be removed by the glomerulus. Filtered substances that are needed are then reabsorbed back into the blood. Components that are not needed are not reabsorbed and are passed down the ureters to the bladder and excreted in the urine. The amounts of water and other substances excreted in the urine are regulated so that homeostasis is maintained (see Chapter 10).

OUTCOME

© Bojan Kontrec/iStockphoto

David Woolley/Taxi/Getty Images, Inc.

Gillian's pattern of eating large amounts of food in the evening combined with the bending, lifting, and rushing around necessary for her job contributed to her heartburn. Her choice of foods and eating right before going to bed made the symptoms worse. To manage her GERD, Gillian began eating evenly spaced meals and snacks. She woke earlier to make time for breakfast and packed a sandwich so she had a healthy lunch. At work she ate a smaller meal and consumed it early in her shift to make sure her stomach was no longer full when she lay down for bed. She was advised that acidic foods like orange juice and tomatoes, spicy foods like peppers, and high fat-foods like cheese and fatty meats, as well as caffeine and alcohol, increased the risk of reflux. So, instead of pizza for her evening meal, she tried to eat some of the lighter chicken and pasta dishes served at the restaurant. By not eating for a few hours before going to bed and elevating the head of her bed, Gillian was able to prevent reflux while sleeping. She was advised to use antacids to manage occasional symptoms and that other medications were available if symptoms persisted. After a few weeks, her symptoms have all but disappeared. **Gillian has had to change her lifestyle to manage this condition. If she reverts to her previous habits symptoms will likely return.**

APPLICATIONS

ASSESSING YOUR DIET

1. **Look at the iProfile report from the food record you kept in Chapter 2, and review the protein, carbohydrate, and fat content of each food.**
   - **a.** List foods that do not begin chemical digestion until they have left the mouth.
   - **b.** List foods that might require bile for digestion and explain why.
   - **c.** List foods that might begin their digestion in your stomach and explain why.
2. **Imagine you wake up on a Sunday morning and join some friends for a large breakfast consisting of a cheese omelet and sausage (foods high in fat and protein), a croissant with butter (which contains carbohydrate but is also very high in fat), and a small glass of orange juice. After the meal, you remember that you have plans to play basketball with a friend in just an hour.**
   - **a.** If you keep your plans and play basketball, what problems might you experience while exercising?
   - **b.** Had you remembered your plans for strenuous exercise before you had breakfast, what type of meal might you have selected to ensure that your stomach would empty more quickly?

CONSUMER ISSUES

3. **There are hundreds of products available to aid digestion. Go to the drugstore or search the Internet, and select a product claiming to aid digestion.**
   - **a.** List the claims made for the product.
   - **b.** Use the information in Chapter 1 on judging nutritional claims to analyze the information given.
   - **c.** What nutrients, if any, does the product provide?
   - **d.** What risks, if any, does it carry?
   - **e.** Would you take it if you were experiencing digestive problems? Why or why not?
4. **Plan a lunch menu for a senior center.**
   - **a.** Include foods from each MyPlate food group.
   - **b.** Choose foods that you think most seniors would enjoy.
   - **c.** How would you modify your menu if you are told that many of the diners have difficulty chewing?
   - **d.** How would you change this menu if it were for a group of teens rather than seniors?

CLINICAL CONCERNS

5. **Mr. Jones has gallstones.**
   - **a.** Why does this cause pain when he eats?
   - **b.** What type of foods should he avoid and why?
6. **A 47-year-old woman undergoes treatment for colon cancer that requires that most of her large intestine be surgically removed.**
   - **a.** How would this affect her fluid needs?
   - **b.** Will the absorption of any nutrients other than water be affected? Why or why not?

SUMMARY

3.1 Food Becomes Us

- Our bodies and the foods we eat are all made from the same building blocks—atoms. Atoms are linked by chemical bonds to form molecules.
- Molecules can be organized to form cells, which are the smallest unit of life. Cells of similar structure and function are organized into tissues, and tissues into organs, and organs cooperate to form organ systems.

3.2 The Digestive System

- The digestive system is the organ system primarily responsible for the movement of nutrients into the body. The digestive system provides two major functions: digestion and absorption. Digestion is the process by which food is broken down into units that are small enough to be absorbed. Absorption is the process by which nutrients are transported into the body.
- The digestive system consists of the gastrointestinal tract and accessory organs. The gastrointestinal (GI) tract consists of a hollow tube that begins at the mouth and continues through the pharynx, esophagus, stomach, small intestine, and large intestine.
- The digestion of food and the absorption of nutrients in the lumen of the GI tract are aided by the secretion of mucus and enzymes.
- The passage of food and the secretion of digestive substances are regulated by nervous and hormonal signals.

3.3 Digestion and Absorption

- The processes involved in digestion begin in response to the smell or sight of food and continue as food enters the digestive tract at the mouth.
- In the mouth, food is broken into smaller pieces by the teeth and mixed with saliva. The chemical digestion of carbohydrate is begun in the mouth by the enzyme salivary amylase.
- From the mouth, food passes through the pharynx and into the esophagus. The rhythmic contractions of peristalsis propel it down the esophagus to the stomach.
- The stomach acts as a temporary storage tank for food. The muscles of the stomach mix the food into a semiliquid mass called chyme, and gastric juice containing hydrochloric acid and pepsin begins protein digestion. Stomach emptying is affected by the amount and composition of food consumed and regulated by nervous and hormonal signals from the stomach and small intestine.
- The small intestine is the primary site of nutrient digestion and absorption. The small intestine contains large circular folds, villi, and microvilli, all of which increase the absorptive surface area. Chyme is propelled through the small intestine by peristalsis and mixed by the contractions of segmentation. In the small intestine, bicarbonate from the pancreas neutralizes stomach acid, and pancreatic and intestinal enzymes digest carbohydrate, fat, and protein. The digestion of fat in the small intestine is aided by bile from the gallbladder. Bile helps make fat available to fat-digesting enzymes by breaking it into small droplets and also facilitates fat absorption.
- The absorption of food across the intestinal mucosa occurs by several different processes. Simple diffusion, osmosis, and facilitated diffusion do not require energy but can only move substances from an area of higher concentration to an area of lower concentration. Active transport requires energy but can transport substances against a concentration gradient.
- Components of chyme that are not absorbed in the small intestine pass on to the large intestine, where some water and nutrients are absorbed. The large intestine is populated by bacteria that digest some of these unabsorbed materials, such as fiber, producing small amounts of nutrients and gas. The remaining unabsorbed materials are excreted in feces.

3.4 Digestion and Health

- Immune system cells and tissues located in the gastrointestinal tract help eliminate disease-causing organisms or toxins. Abnormalities in immune function are the cause of food allergies and celiac disease.
- Heartburn and GERD are common digestive problems that are caused by the leakage of stomach contents into the esophagus. Peptic ulcers are caused by infection with *Helicobacter pylori*, GERD, or medications that damage the mucosa. Gallstones can cause pain and interfere with fat digestion and absorption. Diarrhea may occur when the lining of the small intestine is inflamed so nutrients and water are not absorbed. Constipation can be caused by a diet that is low in fiber or high in fiber and low in fluids.
- Tube-feeding can nourish patients who are unable to swallow food on their own. Total parenteral nutrition is necessary when the gut is not able to digest and or absorb nutrients.
- Digestive system function is affected by life stage. During pregnancy, physiological changes cause morning sickness, heartburn, and constipation. During infancy, the immaturity of the GI tract limits what foods can be ingested and digested. In older adults, changes in the digestive tract may decrease the appeal of food and the ability to digest and absorb nutrients.

3.5 Delivering Nutrients to Body Cells

- Absorbed nutrients are delivered to the cells of the body by the cardiovascular system. The heart pumps blood to the lungs to pick up oxygen and eliminate carbon dioxide. From the lungs, blood returns to the heart and is pumped to the rest of the body to deliver oxygen and nutrients and remove carbon dioxide and other wastes before returning to the heart. Blood is pumped away from the heart in arteries and returned to the heart in veins. Exchange of nutrients and gases occurs at the smallest blood vessels, the capillaries.
- The products of carbohydrate and protein digestion and the water-soluble products of fat digestion enter capillaries in the intestinal villi and are transported to the liver via the hepatic portal circulation. The liver serves as a processing center, removing the absorbed substances for storage, converting them into other forms, or allowing them to pass unaltered. The liver also protects the body from toxic substances that may have been absorbed.
- The lymphatic system is important for fluid balance, immune function, and the absorption of the fat-soluble products of digestion. Lipid transport particles are absorbed into lacteals in the intestinal villi. Lacteals join larger lymph vessels. The

nutrients absorbed via the lymphatic system enter the blood circulation without first passing through the liver.

- Cells are the final destination of absorbed nutrients. To enter the cells, nutrients must be transported across cell membranes.

3.6 Metabolism of Nutrients: An Overview

- Within the cells, glucose, fatty acids, and amino acids absorbed from the diet can be broken down by catabolic pathways, which release energy in the form of ATP, or used in anabolic pathways, which use ATP to build body compounds. The sum of all the chemical reactions of the body is called metabolism.
- The reactions that completely break down macronutrients in the presence of oxygen to produce water, carbon dioxide, and ATP are referred to as cellular respiration. Glucose, fatty acids, and amino acids can all be broken down into two-carbon molecules that form acetyl-CoA. The reactions of the citric acid cycle and the electron transport chain complete the breakdown of acetyl-CoA to form carbon dioxide and water and generate ATP.
- Dietary glucose, fatty acids, and amino acids are used to synthesize structural, regulatory, or storage molecules.

3.7 Elimination of Metabolic Wastes

- Unabsorbed materials are excreted in the feces. Carbon dioxide is eliminated in exhaled air. Water is lost via the lungs and skin.
- Water, metabolic waste products, and excess minerals are excreted by the kidneys.

REVIEW QUESTIONS

1. What is the smallest unit of plant and animal life?
2. List three organ systems involved in the digestion and absorption of food.
3. How do teeth function in digestion?
4. What is peristalsis? What is segmentation?
5. List two functions of the stomach.
6. How is the movement of material through the digestive tract regulated?
7. List three mechanisms by which nutrients are absorbed.
8. Where do most digestion and absorption occur?
9. How does the structure of the small intestine aid absorption?
10. What products of digestion are transported by the lymphatic system?
11. How does the digestive tract protect us from harmful microorganisms?
12. What causes food allergies?
13. Explain what causes heartburn and GERD.
14. What path does an amino acid follow from absorption to delivery to the cell? Compare this to the path a large fatty acid would follow from absorption to delivery to the cell.
15. What is the form of energy used by cells?
16. Explain what occurs during the citric acid cycle and the electron transport chain.
17. What happens to material that is not absorbed in the small intestine?
18. How do the lungs and kidneys help eliminate metabolic waste products?

REFERENCES

1. Heselmans, M., Reid, G., Akkermans, L. M., et al. Gut flora in health and disease: Potential role of probiotics. *Curr Issues Intest Microbiol* 6:1–7, 2005.
2. Johnston, B. C., Supina, A. L., Ospina, M., and Vohra, S. Probiotics for the prevention of pediatric antibiotic-associated diarrhea. *Cochrane Database Syst Rev* 18: CD004827, 2007.
3. Gill, H., and Prasad, J. Probiotics, immunomodulation, and health benefits. *Adv Exp Med Biol* 606:423–454, 2008.
4. Isolauri, E., and Salminen, S. Probiotics: Use in allergic disorders: A Nutrition, Allergy, Mucosal Immunology, and Intestinal Microbiota (NAMI) Research Group report. *J Clin Gastroenterol* 42(suppl 2):S91–S96, 2008.
5. Rescigno, M. The pathogenic role of intestinal flora in IBD and colon cancer. *Curr Drug Targets* 9:395–403, 2008.
6. Bosscher, D., Breynaert, A., Pieters, L., and Hermans, N. Food-based strategies to modulate the composition of the intestinal microbiota and their associated health effects. *J Physiol Pharmacol* 60(suppl 6):5–11, 2009.
7. American Academy of Allergy Asthma and Immunology. Allergy statistics.
8. Rubio-Tapia, A., Ludvigsson, J.F., Brantner, T.L., et al. The prevalence of celiac disease in the United States. *Am J Gastroenterol* Jul 31, 2012. [Epub ahead of print]
9. Konturek, S. J., Konturek, P. C., Konturek, J. W., et al. *Helicobacter pylori* and its involvement in gastritis and peptic ulcer formation. *J Physiol Pharmacol* 57(suppl 3): 29–50, 2006.
10. Marieb, E. N., and Hoehn, K. *Human Anatomy and Physiology*, 7th ed. Menlo Park, CA: Benjamin Cummings, 2007.

***To access links to online sources, please go to www.wiley.com/college/smolin and select Nutrition: Science and Applications, 3rd edition. From this page, select either the student or instructor companion site. Once on the desired site, select References.**

CHAPTER OUTLINE

Cole Group/Photodisc/Getty Images, Inc.

Carbohydrates: Sugars, Starches, and Fiber

4

CASE STUDY

Shamara was excited when she lost 15 pounds in only a few weeks on a low-carbohydrate diet. She had cut out all the chips, cookies, and soda that she frequently snacked on while studying and in between classes. She had also cut out the bagel she ate for breakfast, her sandwich at lunch, and the potatoes that she enjoyed with dinner. But, after a month, she was so tired of the diet that she went back to her old ways of eating. She quickly regained half of the weight she had lost. Now Shamara has decided to forget the low-carb diet and focus just on eating healthy foods. She looked up her MyPlate recommendations and was surprised to find that not all carbohydrates are bad—MyPlate recommended that she eat six servings of grain products and at least three of them should be whole grains. She also saw that she needed to increase her intake of fruits and vegetables. Most fruits and vegetables had been forbidden on her low-carb diet because they contain carbohydrate, but MyPlate recommends fruits and vegetables fill half your plate because they are good sources of vitamins, minerals, and fiber.

The first step Shamara took to improve her diet was to keep a bag of oranges and apples in her dorm room. With those healthy snacks available, she would be less tempted to hit the vending machine if she wanted to snack at night. She went back to eating a bagel or cereal for breakfast and made sure she chose whole-grain varieties. When she had her diet analyzed at a health fair a few weeks later, she was dismayed to see that even after all the changes she had made, her diet was still high in added sugars. She started looking at food labels and found that the bran muffin she had after her first class had 15 grams of sugar. Her breakfast cereal box had a big banner saying it had 5 g of fiber, but the food label showed that it also contained 18 g of sugar, about half the amount in a can of soda. **Choosing healthy carbohydrates was turning out to be almost as difficult as eliminating them.**

© Elena Elisseeva/Alamy

4.1 Carbohydrates in Our Food

LEARNING OBJECTIVES

- Identify the sources of refined and unrefined carbohydrates in the typical American diet.
- Discuss why the recommendations for a healthy diet promote unrefined carbohydrates.
- Explain why added sugars are considered empty calories.

Foods that are high in carbohydrates are the basis of our diet; carbohydrates provide more than half of calories in the American diet. Foods as diverse as oatmeal, whole-wheat bread, fresh fruit, milk, chocolate cake, potato chips, and carbonated soft drinks are high in carbohydrates (**Figure 4.1**). The carbohydrates in these foods provide a readily available source of energy; they supply 4 kcal per gram. However, the additional nutritional impact they deliver varies depending on whether the carbohydrate is **refined** or in its natural state.

refined Refers to foods that have undergone processing that changes or removes various components of the original food.

Oatmeal, whole-wheat bread, fresh fruit, and milk are considered unrefined or whole food sources of carbohydrates because they have not been altered from their natural state. These foods contain vitamins, minerals, and other health-promoting substances as well as carbohydrates. Potato chips, cake, and soda provide carbohydrates that have been refined. Refining separates carbohydrates from many of the other essential nutrients and food components present in the whole food.

Over the last century the amounts and sources of carbohydrates in the American diet have changed. Our total carbohydrate intake decreased between 1909 and 1963. Most of

Figure 4.1 **Sources of carbohydrate**

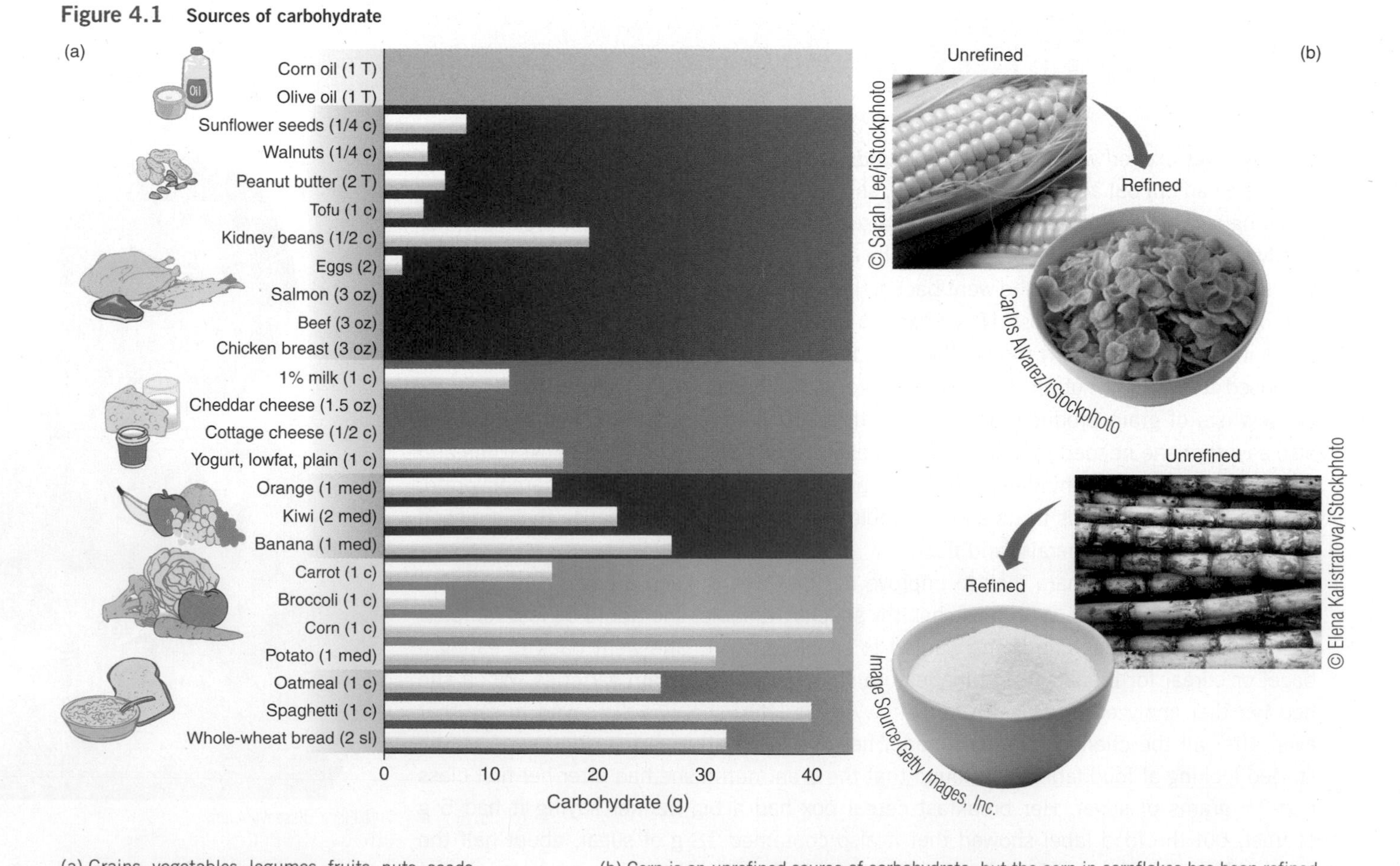

(a) Grains, vegetables, legumes, fruits, nuts, seeds, and milk are all sources of naturally occurring dietary carbohydrates.

(b) Corn is an unrefined source of carbohydrate, but the corn in cornflakes has been refined through grinding, cooking, extruding, and drying. The sugar you sprinkle on cornflakes is also a refined carbohydrate; it has been processed from sugarcane or sugar beets.

this drop was due to a decrease in the consumption of whole grains, and with it came a 40% drop in the amount of fiber consumed.[1] Since the 1960s, our total carbohydrate intake has increased, but our fiber intake did not rise with it, suggesting an increase in the intake of refined carbohydrates. Much of the carbohydrate added back to our diet between 1960 and 2000 came from sugars; over this time period per-capita sugar consumption rose by 33%.[2] Whole-grain breads and cereals were replaced by white bread, snack foods, and sugared soft drinks. The types of sweeteners also changed. In the 1960s we sweetened food with cane and beet sugar, but today most of the foods we buy are sweetened with corn sweeteners (see **Debate: Should You Avoid High-Fructose Corn Syrup**).

Today, recommendations for a healthy diet tell us to choose more of the unrefined carbohydrates we used to eat, whole grains, vegetables, **legumes**, and fruits, and fewer foods high in refined carbohydrates and **added sugars** such as baked goods and soft drinks.

legumes Plants in the pea or bean family, which produce an elongated pod containing large starchy seeds. Examples include green peas, kidney beans, and peanuts.

added sugars Sugars and syrups that have been added to foods during processing or preparation.

How Refined and Unrefined Carbohydrates Differ

Unrefined sources of carbohydrate such as whole grains, legumes, vegetables, fruits, and milk contain a variety of nutrients in addition to carbohydrates. Whole grains, legumes, and vegetables provide B vitamins, some minerals, fiber, and phytochemicals. Fruits provide vitamins A and C along with fiber. Milk is a good source of the B vitamin riboflavin and the mineral calcium. In contrast, refined sources of carbohydrate, such as the corn flakes you may have had for breakfast, are made from corn kernels that have been processed into crunchy flakes (see Figure 4.1b). During these refining steps, many of the nutrients and other healthful components of the corn kernel are lost. When we eat the entire kernel or seed of a grain, such as corn, rice, or wheat, we are eating an unrefined or **whole-grain** product. The whole-grain kernel includes the bran, the germ, and the endosperm (**Figure 4.2**).

The outermost **bran** layers of a kernel of grain contain most of the fiber and are a good source of vitamins. The **germ**, which lies at the base of the kernel, is the plant embryo where sprouting occurs. It is the source of some vegetable oils such as corn oil, and is rich in vitamin E. It also contains protein, fiber, and the B vitamins riboflavin, thiamin, and vitamin B_6. The remainder of the kernel is the **endosperm**, which is primarily starch but also contains most of the protein and some vitamins and minerals. During the milling of grain into flour, the grinding detaches the germ and bran from the endosperm. Whole-grain flours such as whole-wheat flour include most of the bran, germ, and endosperm.

whole grain The entire kernel of grain including the bran layers, the germ, and the endosperm.

bran The protective outer layers of whole grains. It is a concentrated source of dietary fiber.

germ The embryo or sprouting portion of a kernel of grain. It contains oil, protein, fiber, and vitamins.

endosperm The largest portion of a kernel of grain. It is primarily starch and serves as a food supply for the sprouting seed.

Figure 4.2 **Whole grains**

(a) A kernel of grain, such as the wheat shown here, is made up of three major components: the bran, the germ, and the endosperm.

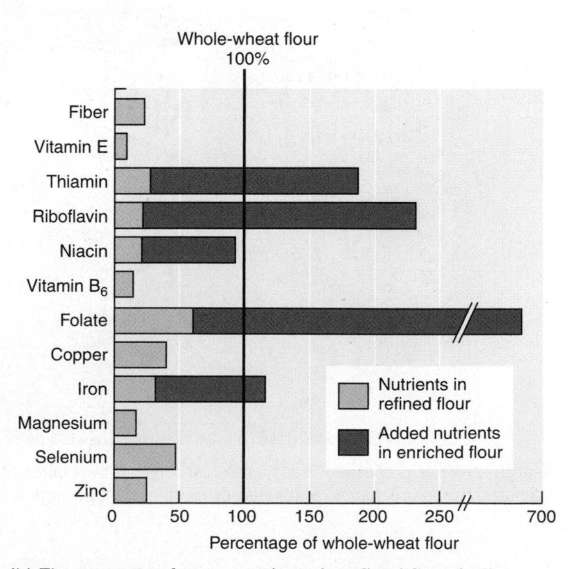

(b) The amounts of many nutrients in refined flour (yellow bars) are much lower than the amounts originally present in the whole grain (100% line). In enriched flour, thiamin, riboflavin, niacin, iron, and folate have been added back in amounts that equal or exceed the original levels (brown bars).

Should You Avoid High-Fructose Corn Syrup?

High-fructose corn syrup (HFCS) is one of the most successful food ingredients in modern history.[1] Developed in the 1960s as a convenient inexpensive way to sweeten food, it has now become the most common added sweetener in the American diet. Between 1970 and 1990, HFCS consumption increased more than 1000%.[2] The ubiquitous use of this sweetener has created concern about the effect it has on our health. Increased consumption of HFCS has been related to the development of obesity, heart disease, and diabetes, among other disorders.[3] Is HFCS just a convenient way to sweeten our food or is it a threat to our health?

HFCS is a syrup made by extracting starch from corn and treating it to break the bonds between the glucose molecules. The resulting corn syrup is then treated to convert about half the glucose to fructose (hence "high-fructose" corn syrup). Manufacturers prefer HFCS as an added sweetener because it is cheaper and more stable during storage than other sweeteners.[1] In 1970, the most common sweetener in

Since 1970, HFCS intake has increased dramatically, while sucrose use has declined. Over this same time period, the incidence of obesity has more than doubled.

A large range of processed foods, from carbonated beverages and fruit drinks to cereals, crackers, barbecue sauce, and salad dressings, contain high-fructose corn syrup.

Andy Washnik photo and composition. Background corn photo from Bill Grove/ istockphoto.com

the American diet was sucrose (see graph). Today HFCS has almost completely replaced sucrose in soft drinks and is found in many other foods, ranging from breakfast cereals to canned soups and salad dressings.

HFCS has been implicated as a cause of obesity because the increase in its use parallels the increase in obesity (see graph). Obesity in turn increases the risk of diabetes and heart disease. There is some physiological basis for the relationship between HFCS and obesity. When excess energy is consumed, fructose is converted to fat more readily than glucose. In addition, fructose is not as effective as glucose at stimulating the release of hormones that suppress appetite or at inhibiting the release of hormones that stimulate appetite.[4] So, when compared to glucose, fructose consumption contributes more to fat synthesis and less to appetite suppression, potentially leading to overeating and weight gain.

The counterargument to the contention that HFCS contributes to obesity more than other sweeteners is that many foods contain fructose and fructose has been part of the human diet for centuries. Fructose, or fruit sugar, is the primary sugar in most fruits and vegetables and is a component of the table sugar (sucrose) we add to food. Sucrose is glucose bound to fructose in equal proportions, so it is half fructose, about the same proportion of fructose to glucose as in HFCS. So, why single out the fructose in HFCS? One reason is that the HFCS used in soft drinks has slightly more fructose than glucose, so increasing HFCS intake increases fructose consumption more than increasing sucrose intake. Another is that the fructose molecules in HFCS are unbound, ready for absorption and utilization. In contrast, the fructose molecules in sucrose are bound to glucose and must be released by digestion before they can be absorbed and utilized. It is unclear whether the small differences in the proportion of glucose to fructose in HFCS versus sucrose or the need for digestion lead to differences in the metabolic effects of HFCS and sucrose.

Studies that have been done to examine the impact of HFCS consumption on weight gain have neither condemned nor vindicated HFCS. Studies in humans have shown an increase in weight gain with diets high in fructose compared to glucose, but a comparison of weight gain in people consuming HFCS versus those consuming sucrose has not been done.[5] Animal studies have made this comparison. In one study, rats were fed a diet supplemented with water sweetened with either HFCS or sucrose. The rats that received the HFCS over an eight-week period gained more weight, more fat, and had higher triglyceride levels than the rats fed sucrose, despite actually consuming fewer calories.[6] Can this animal data be extrapolated to humans?

Is HFCS worse than other sweeteners? We gain excess body weight by consuming more energy than we expend. Modern lifestyles have provided us with virtually unlimited access to food and limited reasons to burn calories. Blaming the obesity crisis on a single food or additive is an oversimplification that draws attention away from the goal of preventing obesity by encouraging increased activity and moderate energy intake. When consumed in large amounts, HFCS has the potential to both increase energy intake and promote the deposition of body fat. But will eliminating HFCS from our food supply necessarily make our diets healthier? Will replacing HFCS with sucrose help reduce obesity?

THINK CRITICALLY: Compare the relationship between the percentage of adults who are obese and the intake of HFCS from 1970 to 2000 and from 2000 to 2005. Do they correlate with each other over both these time periods? What do these relationships tell you about the role of HFCS in obesity?

[1]White, J. S. Straight talk about high-fructose corn syrup: What it is and what it ain't. *Am J Clin Nutr* 88:1716S–1721S, 2008.

[2]Bray, G. A., Nielsen, S. J., Popkin, B.M. Consumption of high-fructose corn syrup in beverages may play a role in the epidemic of obesity. *Am J Clin Nutr* 79:537–543, 2004.

[3]Gaby, A. R. Adverse effects of dietary fructose. *Altern Med Rev* 10:294–306, 2005.

[4]Tappy, L., and Lê, K. A. Metabolic effects of fructose and the worldwide increase in obesity. *Physiol Rev* 90:23–46, 2010.

[5]Stanhope, K. L., Schwarz, J. M., Keim, N. L., et al. Consuming fructose-sweetened, not glucose-sweetened, beverages increases visceral adiposity and lipids and decreases insulin sensitivity in overweight/obese humans. *J Clin Invest* 119:1322–1334, 2009.

[6]Bocarsly, M. E., Powell, E. S., Avena, N. M., and Hoebel, B. G. High-fructose corn syrup causes characteristics of obesity in rats: Increased body weight, body fat and triglyceride levels. *Pharmacol Biochem Behav* 97:101–106, 2010.

THINK CRITICALLY

If you choose the soda, what food or foods could you add to provide the same amount of vitamin C and folate you would have gotten from the kiwis?

Figure 4.3 Added sugar and nutrient density
A 12-ounce can of soda contains about 140 kcal from sugar but almost no other nutrients. Three medium kiwis also provide about 140 kcal, along with plenty of other nutrients, including vitamin C, folate, and calcium, making the kiwis more nutrient dense than the soda.

White flour, however, is produced from just the endosperm. Fiber and some vitamins, minerals, and phytochemicals naturally found in the whole grain are therefore lost. In order to restore some of the lost nutrients, refined grains sold in the United States are fortified with some, but not all, of the nutrients lost in processing. **Enriched grains** contain added thiamin, riboflavin, niacin, and iron and are fortified with folate. However, they do not contain added vitamin E, magnesium, vitamin B_6, or a number of other nutrients, whose levels are also reduced by milling (see Figure 4.2b).

enriched grain Grain products to which specific amounts of thiamin, riboflavin, niacin, and iron have been added. Since 1998, folic acid has also been added to enriched grains.

Why We Should Limit Added Sugars

The sugar you sprinkle on your cereal adds calories without adding any nutrients other than carbohydrate. This reduces the nutrient density of your breakfast. But the sugar you add to food isn't the only source of added sugars in the diet—much of the added sugar we consume comes from desserts, beverages, and snacks that we purchase already prepared. Refined added sugars make up about 16% of the calories in the American diet.[3] Added sugars are not nutritionally or chemically different from sugars occurring naturally in foods. The only difference is that they have been separated from their plant sources and therefore are not consumed with all of the fiber, vitamins, minerals, and other substances found in the original plant. Because added sugars provide few nutrients for the number of calories they contain, they have a low nutrient density and are considered empty calories. Unrefined or whole food sources of sugar such as fruit provide vitamins, minerals, and phytochemicals as well as calories. For example, three kiwis provide the same number of calories as a 12-ounce soda, but the kiwis are more nutrient dense because they provide a variety of other nutrients, in addition to sugar (**Figure 4.3**).

4.2 Types of Carbohydrates

LEARNING OBJECTIVES

- Compare the structures of simple and complex carbohydrates.
- List food sources of simple carbohydrates and complex carbohydrates.
- Distinguish between soluble and insoluble fiber, and name food sources of each.

Chemically, carbohydrates are compounds made up of one or more sugar units that contain carbon (*carbo-*), as well as hydrogen and oxygen in the same two-to-one proportion found in water (hydrate, H_2O). Carbohydrates made up of only one sugar unit are called

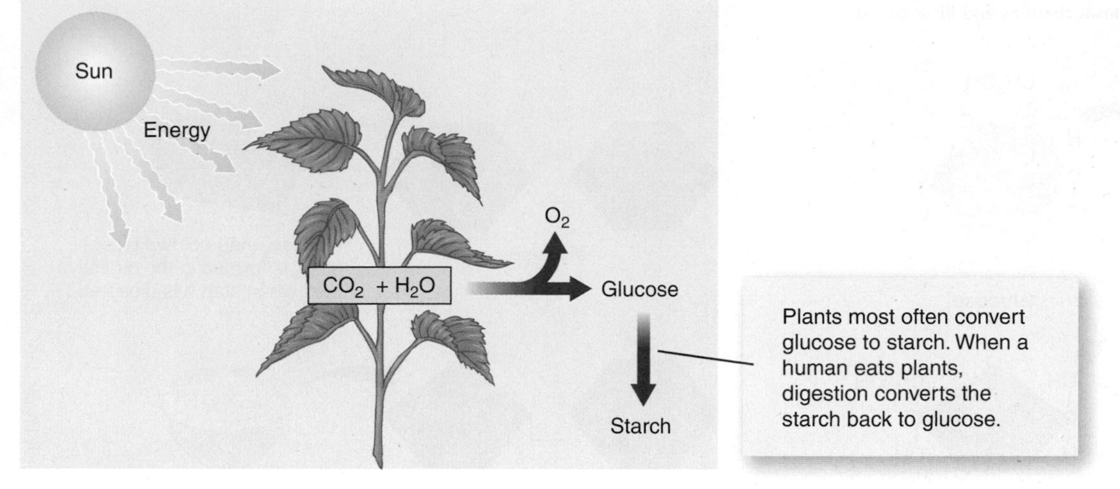

Figure 4.4 **Photosynthesis**
Glucose is produced in plants through the process of photosynthesis, which uses energy from the sun to convert carbon dioxide and water to glucose.

monosaccharides (*mono-* means one), those made up of two sugar units are called **disaccharides** (*di-* means two), and those made up of more than two sugar units are called **polysaccharides** (*poly-* means many).

monosaccharide A carbohydrate made up of a single sugar unit.

disaccharide A carbohydrate made up of two sugar units.

polysaccharide A carbohydrate made up of many sugar units.

Simple Carbohydrates

Monosaccharides and disaccharides are classified as **simple carbohydrates** or sugars. Fruits, vegetables, and milk are sources of simple carbohydrates. The sugars we add to food such as white table sugar, brown sugar, molasses, and confectioner's sugar, as well as the high-fructose corn syrup added in processing are simple carbohydrates. These are produced by refining and processing the carbohydrates from plants such as sugarcane, sugar beets, and corn.

simple carbohydrates A class of carbohydrates, known as sugars, that includes monosaccharides and disaccharides.

Monosaccharides The monosaccharide **glucose**, commonly referred to as blood sugar, is the most important carbohydrate fuel for the body. Glucose is produced in plants by the process of photosynthesis, which uses energy from the sun to combine carbon dioxide and water (**Figure 4.4**). Glucose rarely occurs as a monosaccharide in food. It is most often found as part of a disaccharide or starch.

glucose A monosaccharide that is the primary form of carbohydrate used to provide energy in the body. It is the sugar referred to as blood sugar.

Two other common monosaccharides in the diet are fructose and galactose. Like glucose, each contains 6 carbon, 12 hydrogen, and 6 oxygen atoms, but they differ in their arrangement (**Figure 4.5a**). **Fructose** is a monosaccharide that tastes sweeter than glucose. It is found in fruits and vegetables and makes up more than half the sugar in honey. Because fructose does not cause as great a rise in blood glucose as other sugars, it is sometimes used in products for people with diabetes. However, because fructose causes an increase in blood lipids, its use should be limited. Fructose consumed in fruits or juices can also cause diarrhea in children. Most of the fructose in our diet comes from high-fructose corn syrup, which is used in most soft drinks in the United States (see Debate: Should You Avoid High-Fructose Corn Syrup?). **Galactose** is rarely present as a monosaccharide in the food supply but is a part of lactose, the disaccharide in milk.

fructose A monosaccharide that is the primary form of carbohydrate found in fruit.

galactose A monosaccharide that combines with glucose to form lactose, or milk sugar.

Disaccharides The most common disaccharides in our diet are maltose, sucrose, and lactose (**Figure 4.5b**). **Maltose** consists of two molecules of glucose. This sugar is made whenever starch is broken down. For example, it is responsible for the slightly sweet taste experienced when bread is held in the mouth for a few minutes. As salivary amylase begins digesting the starch, some sweeter-tasting maltose is formed. **Sucrose**, or common white table sugar, is the disaccharide formed by linking glucose to fructose. It is found in sugarcane, sugar beets, honey, and maple syrup. Sucrose is the only sweetener that can be called "sugar" in the ingredient list on food labels in the United States.

maltose A disaccharide made up of two molecules of glucose. It is formed in the intestines during starch digestion.

sucrose A disaccharide that is formed by linking fructose and glucose. It is commonly known as table sugar or white sugar.

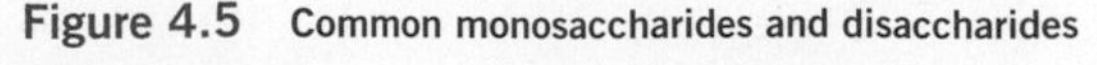

Figure 4.5 Common monosaccharides and disaccharides

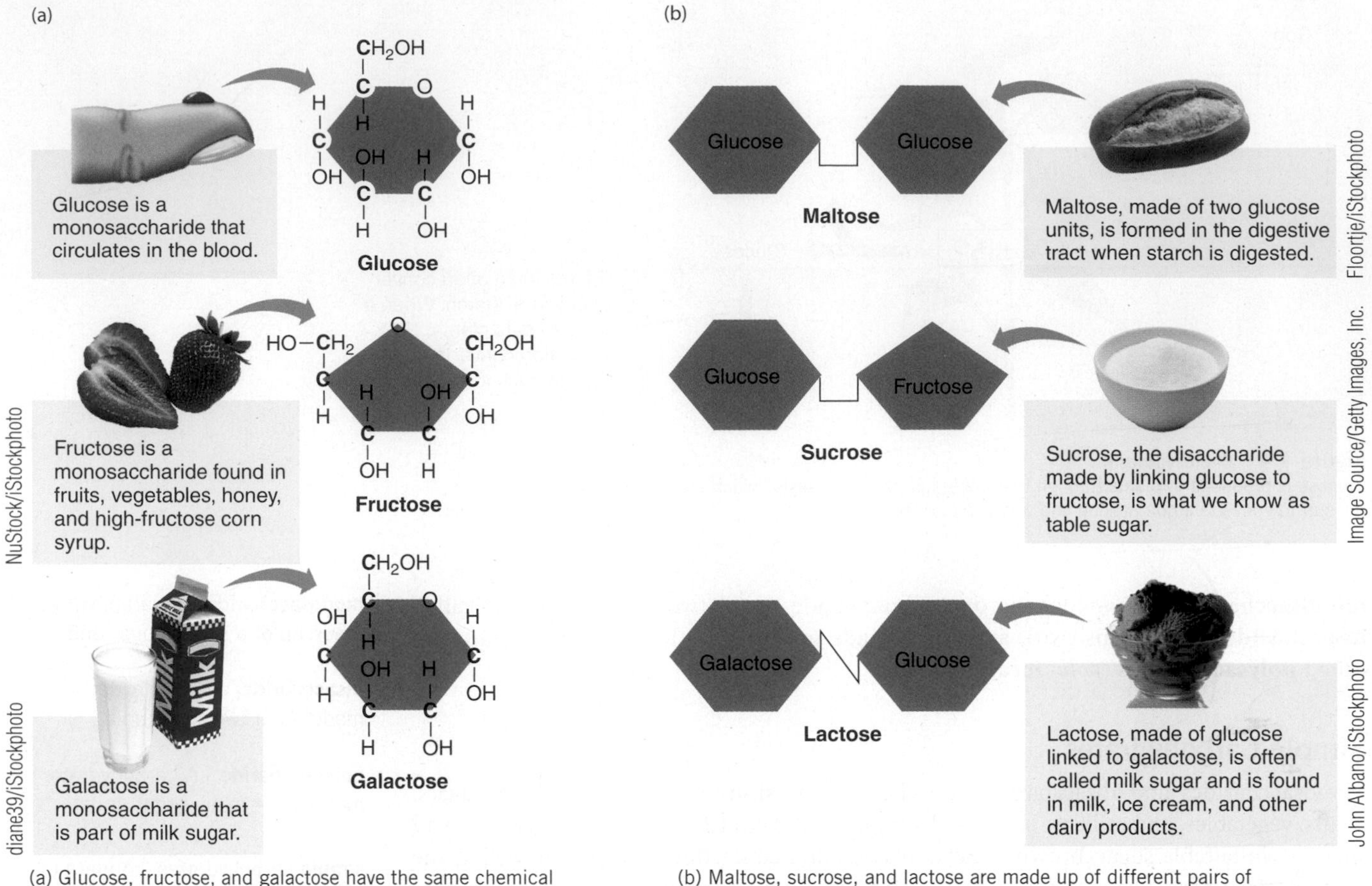

(a) Glucose, fructose, and galactose have the same chemical formulas, but the atoms are arranged differently.

(b) Maltose, sucrose, and lactose are made up of different pairs of monosaccharides.

lactose A disaccharide that is formed by linking galactose and glucose. It is commonly known as milk sugar.

Lactose, or milk sugar, is glucose linked to galactose (see Figure 4.5b). Lactose is the only sugar found naturally in animal foods. It contributes about 30% of the energy in whole cow's milk and about 40% of the energy in human milk.

Making and Breaking Sugar Chains The type of chemical reaction that breaks the bonds between sugar molecules is called a hydrolysis reaction (**Figure 4.6**). Hydrolysis reactions use water to add a hydroxyl group (OH) to one sugar and a hydrogen atom (H) to the other. The type of reaction that links two sugars together is called a condensation or dehydration reaction. Condensation reactions release a molecule of water by taking a hydroxyl group from one sugar and a hydrogen atom from the other.

Complex Carbohydrates

complex carbohydrates Carbohydrates composed of sugar molecules linked together in straight or branching chains. They include oligosaccharides, glycogen, starches, and fibers.

oligosaccharides Short-chain carbohydrates containing 3 to 10 sugar units.

Complex carbohydrates are generally not sweet to the taste like simple carbohydrates. They include short chains of 3 to 10 monosaccharides called **oligosaccharides** and longer-chain polysaccharides. The longer polysaccharides include glycogen in animals and starch and fiber in plants.

Oligosaccharides Some oligosaccharides are formed in the gut during the breakdown of polysaccharides. These are then further digested to simple sugars. Other oligosaccharides are found naturally in foods such as beans and other legumes, onions, bananas, garlic, and artichokes. Many of these are not digested by human enzymes in the digestive tract and pass into the colon where they are broken down by the intestinal microflora. They can affect the types of bacteria that grow in the colon and have beneficial effects on gastrointestinal (GI) health. Oligosaccharides present in human milk make the infant stool easier

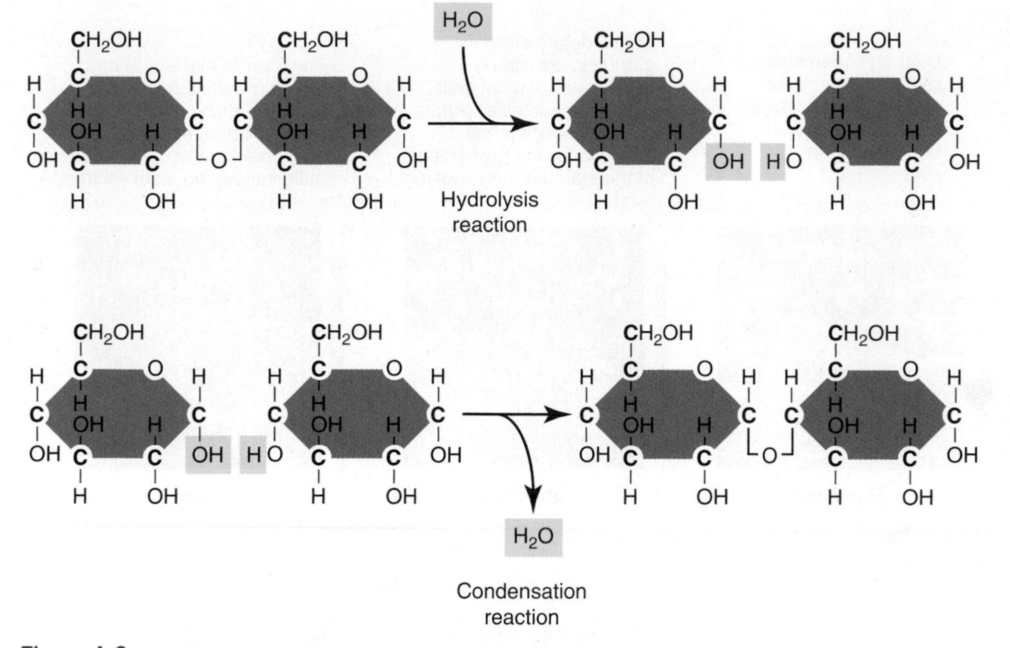

Figure 4.6 Hydrolysis and condensation reactions
In the hydrolysis reaction shown on top, the addition of a molecule of water breaks maltose into its component glucose molecules. In the condensation reaction below it, two glucose molecules are joined to form the disaccharide maltose and a molecule of water.

to pass, help promote the growth of a healthy intestinal microflora, and may protect the infant from infections that cause diarrhea.[4]

Glycogen **Glycogen** is the storage form of carbohydrate in animals. It is a polysaccharide made up of highly branched chains of glucose molecules (**Figure 4.7**). The branched structure allows it to be broken down quickly when glucose is needed. In humans, glycogen is stored in the muscles and in the liver. Muscle glycogen provides glucose to the muscle as a source of energy during activity; liver glycogen releases glucose into the bloodstream for delivery to cells throughout the body. We don't consume glycogen in our food because glycogen present in animal muscles is broken down soon after slaughter and thus is not present when the meat is consumed.

glycogen A carbohydrate made of many glucose molecules linked together in a highly branched structure. It is the storage form of carbohydrate in animals.

The amount of glycogen that can be stored in the body is relatively small—about 200 to 500 g. The amount of glycogen stored in muscle can be temporarily increased by a regimen called *carbohydrate loading* or *glycogen supercompensation*. This regimen is often used by endurance athletes to build up glycogen stores before an event. Extra glycogen can mean the difference between running only 20 miles or finishing a 26-mile marathon before exhaustion takes over. Glycogen supercompensation is discussed in more detail in Chapter 13.

Starches **Starch** is the storage form of carbohydrate in plants. Starch is made up of two types of molecules: amylose, which consists of long straight chains of glucose molecules, and amylopectin, which consists of branched chains of glucose molecules (see Figure 4.7). Starch accumulates in roots and tubers (the underground energy-storage organ of some plants), where it provides energy for the growth and reproduction of the plant. Therefore we consume starch in roots and tubers such as potatoes, sweet potatoes, beets, turnips, and cassava. Starch accumulates in seeds as an energy source for the developing plant embryo. We consume the starch in grain seeds such as wheat, barley, and rye. We also consume starch in legumes, such as lentils, soybeans, and pinto and kidney beans. In addition to the starch naturally present in foods, the diet also contains refined starch such as cornstarch, which is added to thicken to get rid of the second "such as" sauces, puddings, and gravies. Starch can be used to thicken foods because starch granules absorb water and swell when they are heated in water (see Figure 4.7). As a starch-thickened mixture cools, high-amylose starches form bonds

starch A carbohydrate made of many glucose molecules linked in straight or branching chains. The bonds that hold the glucose molecules together can be broken by human digestive enzymes.

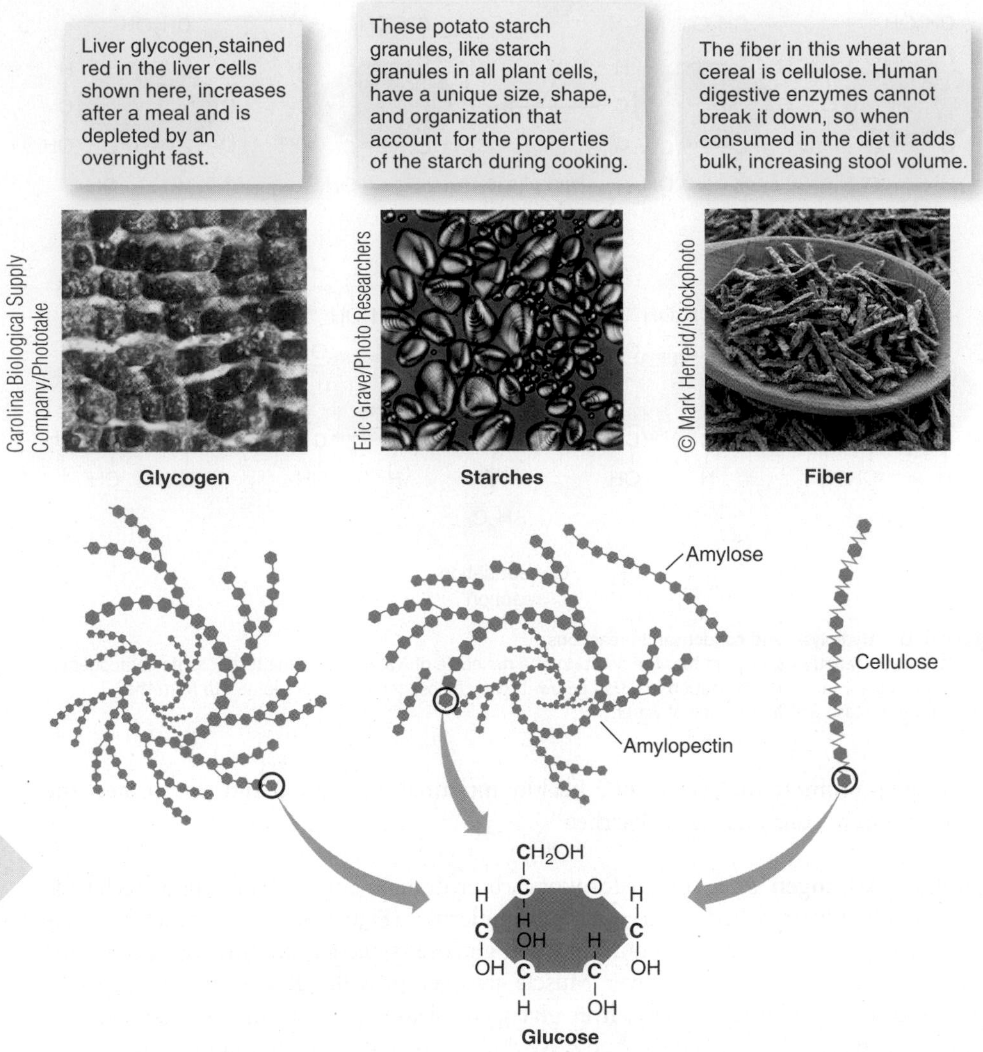

Liver glycogen,stained red in the liver cells shown here, increases after a meal and is depleted by an overnight fast.

These potato starch granules, like starch granules in all plant cells, have a unique size, shape, and organization that account for the properties of the starch during cooking.

The fiber in this wheat bran cereal is cellulose. Human digestive enzymes cannot break it down, so when consumed in the diet it adds bulk, increasing stool volume.

Carolina Biological Supply Company/Phototake

Glycogen

Eric Grave/Photo Researchers

Starches

© Mark Herreid/iStockphoto

Fiber

Amylose

Amylopectin

Cellulose

Glucose

Figure 4.7 **Complex carbohydrates** Glycogen, starches, and the fiber cellulose are made up of straight or branching chains of glucose.

THINK CRITICALLY

How do the bonds that link the glucose units in a molecule of starch differ from those in a molecule of cellulose fiber?

between the molecules, forming a gel. Some starches are treated to enhance their ability to form a gel. These modified food starches are added to processed foods as thickeners.

fiber A mixture of indigestible carbohydrates and lignin that is found in plants.

soluble fiber Fiber that dissolves in water or absorbs water to form viscous solutions and can be broken down by the intestinal microflora. It includes pectins, gums, and some hemicelluloses.

insoluble fiber Fiber that, for the most part, does not dissolve in water and cannot be broken down by bacteria in the large intestine. It includes cellulose, some hemicelluloses, and lignin.

Fiber **Fiber** includes certain complex carbohydrates and lignins (substances in plants that are not carbohydrates but are classified as fiber) that cannot be digested by human enzymes (see Figure 4.7). Since they cannot be digested, they cannot be absorbed into the body. However, fiber consumed in the diet can have beneficial health effects, from reducing constipation to lowering blood cholesterol. Fiber is consumed intact in plant foods and purified or isolated fibers can be added to foods or supplements. For example, oat bran is often added to bread.[3]

Fiber includes a number of different chemical substances that have different physical and physiological properties. **Soluble fibers** form viscous solutions when mixed with water. Although human enzymes can't digest soluble fiber, bacteria in the large intestine can break it down, producing gas and short-chain fatty acids, small quantities of which can be absorbed. Soluble fibers are found around and inside plant cells. They include pectins, gums, and some hemicelluloses. Some of the foods that contain soluble fibers include oats, apples, beans, and seaweed.

Fibers that cannot be broken down by bacteria in the large intestine and do not dissolve in water are called **insoluble fibers**. They are primarily derived from the structural parts of plants, such as the cell walls, and include cellulose, some hemicelluloses,

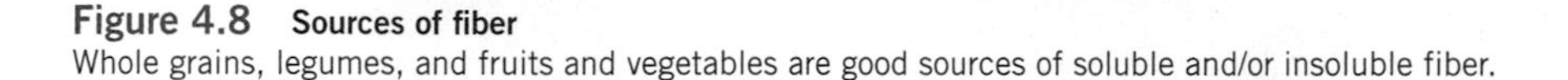

Figure 4.8 **Sources of fiber**
Whole grains, legumes, and fruits and vegetables are good sources of soluble and/or insoluble fiber.

and lignins. Food sources of insoluble fibers include wheat bran and rye bran, which are mostly hemicellulose and cellulose, and vegetables such as broccoli, which contain woody fibers composed partly of lignins. Most foods of plant origin contain mixtures of soluble and insoluble fibers (**Figure 4.8**).

In addition to the soluble and insoluble fibers found in whole grains, fruits, and vegetables, our diet contains fibers that are added to foods during processing. Pectin is a soluble fiber found in fruits and vegetables that forms a gel when sugar and acid are added. It is used to thicken yogurt and to form jams and jellies. Carbohydrate gums such as xanthan gum and locust-bean gum are also soluble fibers. They combine with water and are used to keep solutions from separating. Commercially prepared gravies, puddings, reduced-fat salad dressings, and frozen desserts are examples of products that contain carbohydrate gums. Pectins and gums are also used to mimic the texture of fat in the production of reduced-fat products (see Chapter 5). Insoluble fibers such as wheat bran are added to foods like breads and muffins to reduce calorie content and meet consumer demand for high-fiber foods.

4.3 Carbohydrates in the Digestive Tract

LEARNING OBJECTIVES

- Describe the steps of carbohydrate digestion.
- Explain why lactose intolerance causes gas and bloating when milk is consumed.
- Discuss the effects of dietary fiber and other indigestible carbohydrates on gastrointestinal function and health.

Disaccharides and complex carbohydrates must be digested to monosaccharides to be absorbed. Some people are unable to digest the disaccharide lactose: It spills into the colon, causing uncomfortable side effects. All humans lack the digestive enzymes needed to completely break down a variety of oligosaccharides, certain forms of starch, and fiber. These indigestible carbohydrates have important effects on the health and function of the digestive system and the body as a whole.

Carbohydrate Digestion and Absorption

Digestion of starch begins in the mouth, where the enzyme salivary amylase starts breaking it into shorter polysaccharides. The majority of starch and disaccharide digestion occurs in the small intestine. Here, pancreatic amylases complete the job of breaking down starch into disaccharides and oligosaccharides. The digestion of disaccharides and oligosaccharides is completed by enzymes attached to the microvilli in the small intestine (**Figure 4.9**). Here maltose is broken down into two glucose molecules by the enzyme maltase, sucrose is broken down by sucrase to yield glucose and fructose, and lactose is broken down by **lactase** to form glucose and galactose. The resulting monosaccharides—glucose, galactose, and fructose—are then absorbed and transported to the liver via the hepatic portal vein.

lactase An enzyme located in the microvilli of the small intestine that breaks the disaccharide lactose into glucose and galactose.

lactose intolerance The inability to digest lactose because of a reduction in the levels of the enzyme lactase. It causes symptoms including intestinal gas and bloating after dairy products are consumed.

Lactose Intolerance

Lactose intolerance is a condition in which there is not enough of the enzyme lactase in the small intestine to digest the milk sugar lactose. Most infants produce enough of the enzyme lactase to digest the lactose in milk, but in many people the enzyme's activity decreases with

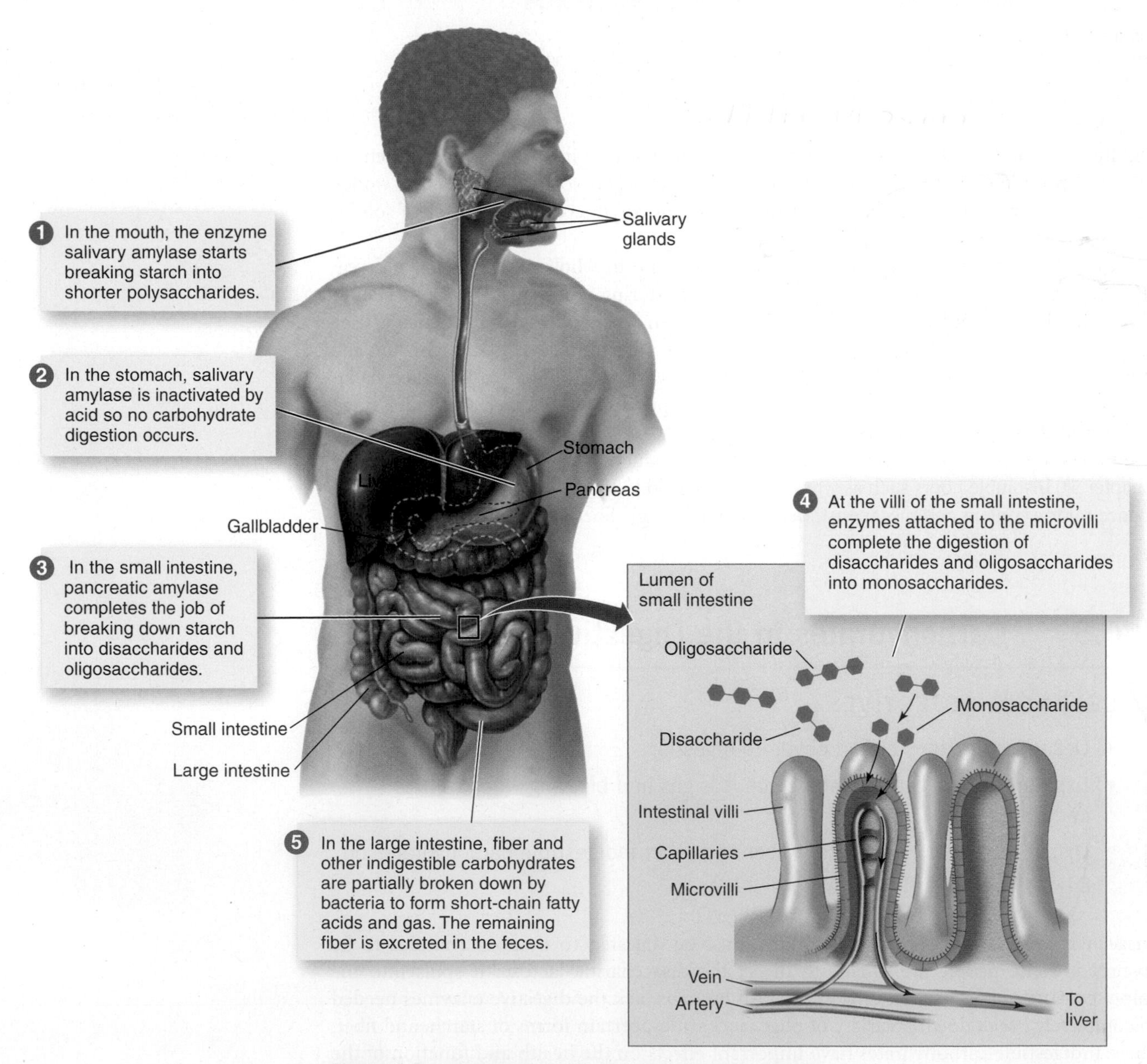

Figure 4.9 **Overview of carbohydrate digestion and absorption**
During digestion, enzymes break starches and sugars into monosaccharides, which are absorbed. Most of the fiber and other indigestible carbohydrates are excreted in the feces.

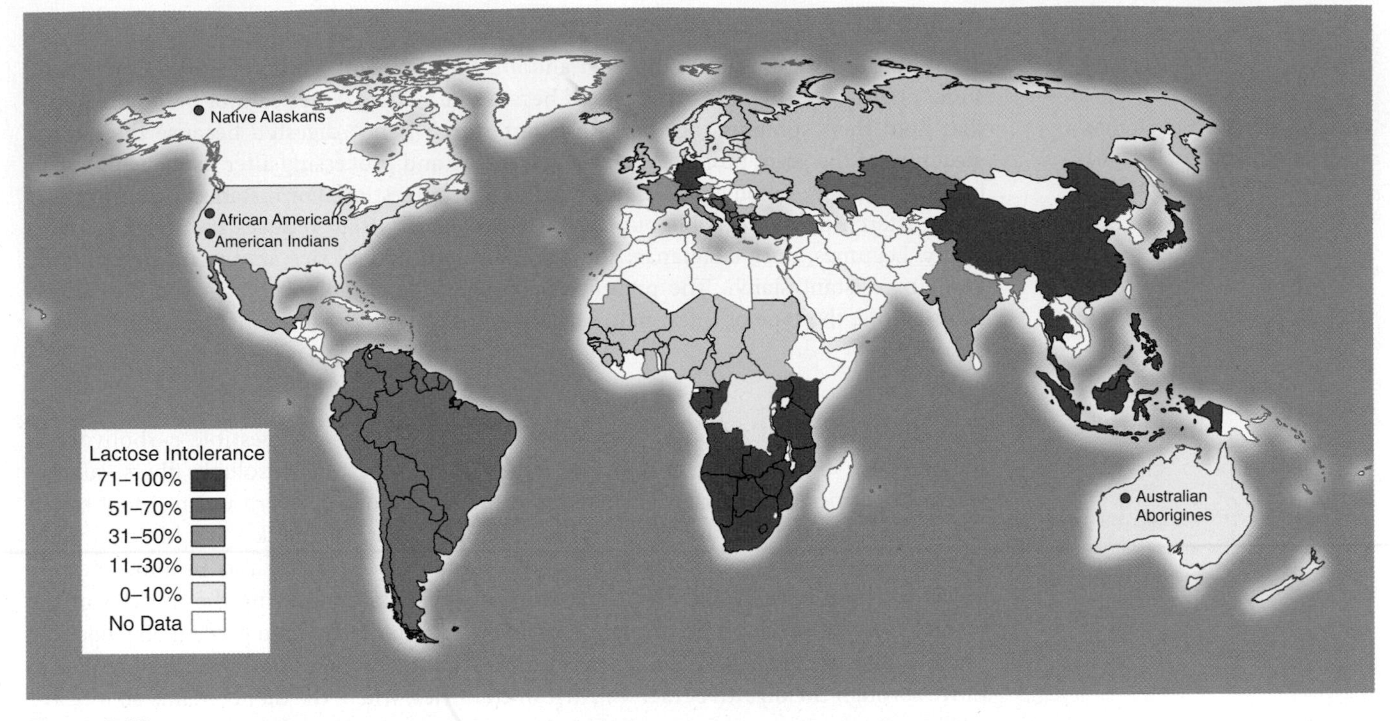

Figure 4.10 **Incidence of lactose intolerance**
The incidence of lactose intolerance around the world varies dramatically. Whether or not individuals retain the ability to digest lactose into adulthood depends on their racial and ethnic background.

age. In the United States, between 30 and 50 million people are lactose intolerant; it is more common in some ethnic and racial groups than in others. Up to 80% of African Americans, 80 to 100% of Native Americans, and 90 to 100% of Asian Americans, but only about 15% of Caucasian Americans, are lactose intolerant (**Figure 4.10**).[5] Lactose intolerance may also occur as a result of an intestinal infection or other disease. It is then referred to as secondary lactose intolerance and may disappear when the other condition is resolved.

Why Lactose Intolerance Causes GI Distress When individuals with lactose intolerance consume milk or other dairy products, the lactose is not digested in the small intestine. Instead, the undigested lactose passes into the large intestine where it increases the number of small molecules, which draws in water by osmosis. The lactose is rapidly metabolized by intestinal bacteria, producing acids and gas. Together the increase in fluid in the intestine along with the acid and gas causes symptoms that include abdominal distention, flatulence, cramping, and diarrhea. In some people, this occurs only when they consume a large amount of lactose such as a glass of milk. In others, symptoms occur even with consumption of foods containing small amounts of this sugar.

How People with Lactose Intolerance Can Meet Calcium Needs In the United States, dairy products are the main source of calcium; the Dietary Guidelines recommend 3 cups of low-fat milk or milk products each day.[6] Without dairy products, it can be difficult to get enough calcium. Some people with lactose intolerance can consume small amounts of dairy products without symptoms and so can meet their calcium needs by dividing the 3 cups into many smaller portions throughout the day. Those who cannot tolerate any lactose can meet their calcium needs with foods like tofu, fish, vegetables, and fermented dairy products. These foods provide dietary calcium in cultures where lactose intolerance is common. For example, in Asia, tofu and fish consumed with bones supply calcium, and in the Near East, cheese and yogurt provide much of the calcium. These fermented products are more easily tolerated than milk because some of the lactose originally present is digested by bacteria or lost in processing. Calcium-fortified foods, calcium supplements, milk treated with the enzyme lactase, and lactase tablets, which can be consumed with or before milk products, are also available for those with lactose intolerance.

Indigestible Carbohydrates

resistant starch Starch that escapes digestion in the small intestine of healthy people.

Some carbohydrates are not digested and are therefore not readily absorbed. Fiber and some oligosaccharides are not digested because human enzymes cannot break the bonds that hold their subunits together. **Resistant starch** is not digested because the natural structure of the grain protects it, because cooking and processing alter its digestibility, or because it has intentionally been modified to resist digestion. For instance, heating makes potato starch more digestible but cooling the cooked potato reduces the starch's digestibility. Legumes, unripe bananas, and cold cooked potatoes, rice, and pasta are naturally high in resistant starch. The presence of indigestible carbohydrates in the diet affects transit time, the type of intestinal microflora, the amount of intestinal gas, and nutrient absorption.

How Indigestible Carbohydrates Affect Transit Time Indigestible carbohydrates increase the volume of material in the lumen of the intestine. Insoluble fibers, such as wheat bran, increase the bulk of material in the feces. Soluble fibers and resistant starch draw water into the intestine. The combination of the increased bulk and additional water allows easier evacuation of the stool. Indigestible carbohydrates also promote healthy bowel function because the extra bulk stimulates peristalsis, causing the muscles of the colon to work more, become stronger, and function better, helping to prevent constipation. The increase in peristalsis reduces transit time—the time it takes food and fecal matter to move through the digestive tract. In African countries, where the diet contains 40 to 150 g of fiber per day, the transit time is 36 hours or less. In the United States, where the usual fiber intake is only about 15 g per day,[7] it is not uncommon for transit time to be as long as 96 hours (**Figure 4.11**).

Why Indigestible Carbohydrates Promote a Healthy Microflora When soluble fibers, resistant starch, and oligosaccharides reach the colon, they serve as a food source for the bacteria that reside there. As these carbohydrates are broken down by the bacteria, short-chain fatty acids are produced and the colonic contents become more acidic. The short-chain fatty acids produced serve as a fuel source for cells in the colon as well as for other body tissues and may play a role in regulating cellular processes. The acid inhibits the growth of undesirable bacteria and favors the growth of beneficial species of bacteria, such as *Lactobacilli* and *Bifidobacteria*, which are well adapted to acid conditions. In addition to inhibiting the growth of disease-causing bacteria, these beneficial bacteria and their metabolic by-products may help prevent inflammation in the bowel and potentially protect against colon cancer.[9,10]

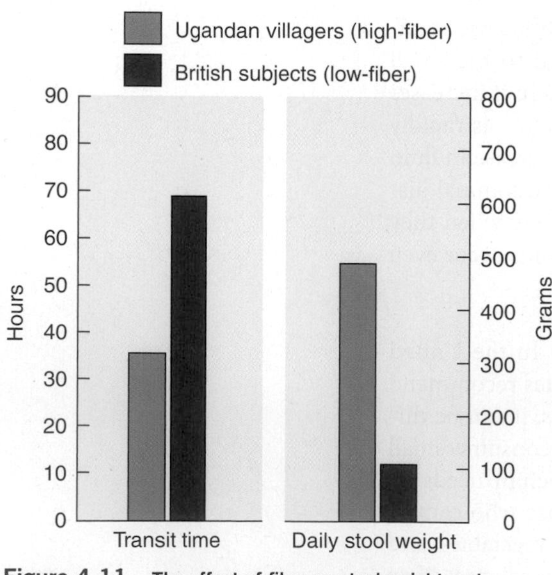

Figure 4.11 The effect of fiber on stool weight and transit time
In a study done in Africa, Ugandan villagers who consumed a diet high in fiber were compared to British subjects living in Uganda who consumed a more refined, low-fiber diet. Stool weights were greater and transit times shorter for Ugandan villagers than for British subjects.[8]

Why Indigestible Carbohydrates Increase Intestinal Gas Anyone who has ever eaten beans knows of their potentially embarrassing side effect of flatulence. The reason beans cause gas is that they are particularly high in the oligosaccharides raffinose and stachyose, which cannot be digested by enzymes in the human stomach and small intestine. They pass into the large intestine, where the bacteria that live there digest them, producing gas and other by-products. This gas can cause abdominal discomfort and flatulence. To alleviate the problem, over-the-counter enzyme tablets and solutions (such as Beano®) can be consumed to break down oligosaccharides before they reach the intestinal bacteria, thereby reducing the amount of gas produced.

As with oligosaccharides, intestinal gas is a by-product of the bacterial breakdown of fiber and resistant starch. A sudden increase in the fiber content of the diet can cause abdominal discomfort, gas, and diarrhea. Constipation can also be a problem if fiber intake is increased without an increase in fluid intake. To avoid these problems, the fiber content of the diet should be increased gradually and fluid intake should be increased along with it.

How Indigestible Carbohydrates Affect Nutrient Absorption Fiber binds certain minerals, preventing their absorption. For instance, wheat bran fiber binds the minerals zinc, calcium, magnesium, and iron. Too much fiber can reduce the absorption of these essential minerals. However, when mineral intake meets recommendations, a reasonable intake of high-fiber foods does not compromise mineral status. Soluble fiber also binds cholesterol and bile, which is made from cholesterol, reducing their absorption. This is beneficial because it can lower blood cholesterol and help reduce the risk of heart disease.

In the stomach, fiber causes distention and slows emptying. This is beneficial for people who are trying to lose weight because they feel satiated after fewer calories are consumed. A high-fiber diet may be a disadvantage for people with a small stomach capacity because they may satisfy their hunger before their nutrient requirements are met. Generally, this is a problem only when the diet is low in protein or micronutrients or when high-fiber diets are consumed by young children, whose small stomachs limit the amount of food they can eat.

Fiber and other indigestible carbohydrates also moderate nutrient absorption. They increase the volume of the intestinal contents and absorb water, forming viscous solutions. This slows nutrient absorption by speeding passage through the GI tract, by decreasing the amount of contact between nutrients and the absorptive surface of the small intestine, and by reducing contact between digestive enzymes and food. In the small intestine, the added volume and viscosity slows the absorption of glucose and other nutrients (**Figure 4.12**). This can be beneficial because slowing the absorption of glucose reduces fluctuations in blood glucose levels.

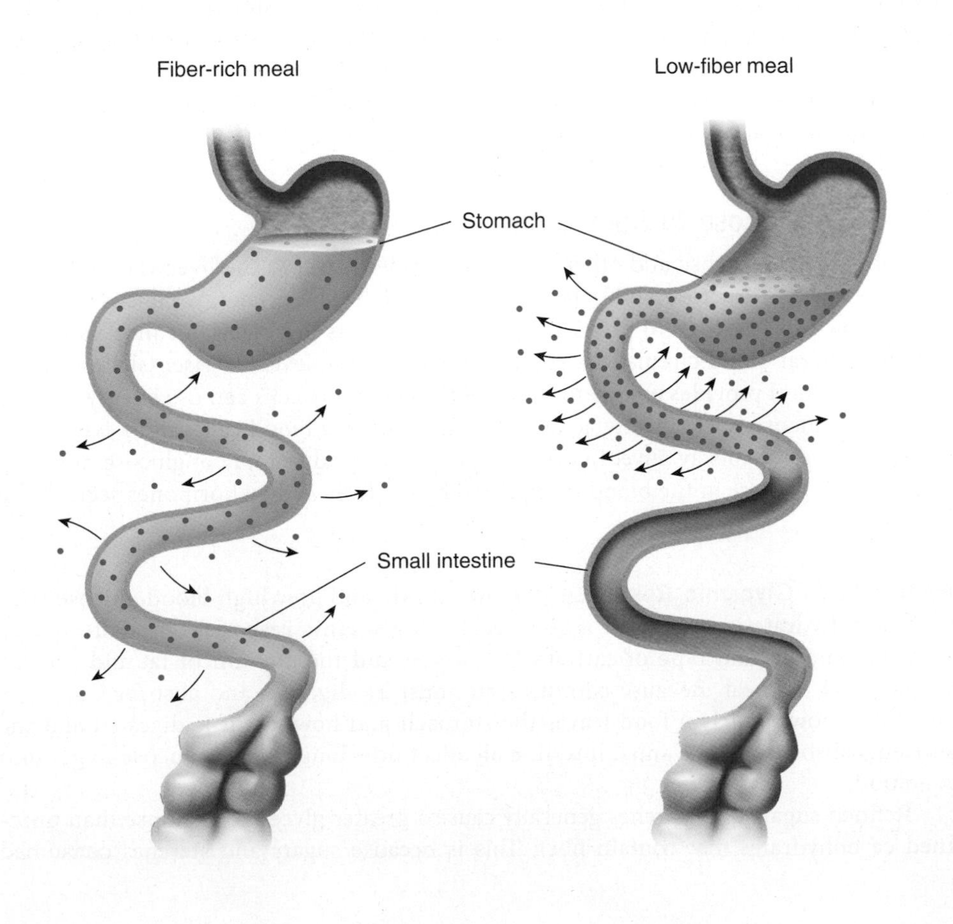

Figure 4.12 The effect of fiber on digestion and absorption
As shown on the left, the bulk and volume of a fiber-rich meal dilute the gastrointestinal contents. This dilution slows the digestion of food and absorption of nutrients (shown as green dots moving slowly out of the intestine). With a low-fiber meal, as shown on the right, nutrients are more concentrated; digestion and absorption occur more rapidly (shown green dots moving quickly out of the intestine).

4.4 Carbohydrates in the Body

LEARNING OBJECTIVES

- Name the main function of glucose in the body.
- Discuss factors that affect blood glucose levels after eating and when fasting.
- Describe the steps involved in metabolizing glucose to produce ATP.
- Explain how metabolism changes when carbohydrate is limited.

Carbohydrate is an important source of energy for body cells. Some cells, such as those in the brain, rely almost exclusively on glucose for energy. It is therefore important to regulate levels of glucose in the blood to ensure adequate glucose delivery to cells. Once inside body cells, glucose can be completely oxidized and the energy it contains used to synthesize ATP.

Carbohydrate Functions

The main function of glucose is to provide energy to body cells, but carbohydrates also provide other essential functions. The monosaccharide galactose is an important molecule in nervous tissue. It also combines with glucose to make lactose in women who are producing breast milk. Deoxyribose and ribose are monosaccharides that are components of DNA and RNA (ribonucleic acid), respectively, which contain the genetic information for the synthesis of proteins. Deoxyribose and ribose can be synthesized by the body and are not found in significant amounts in the diet. Ribose is also a component of the vitamin riboflavin. Oligosaccharides are also important in our bodies. They are found attached to proteins or lipids on the surface of cells, where they help to signal information about the cells. Glycogen provides a storage form of glucose. Another type of carbohydrate that is important in the body is mucopolysaccharides. They are a type of polysaccharide that functions with proteins in body secretions and structures. Mucopolysaccharides give mucus its viscous consistency and provide cushioning and lubrication in connective tissue.

Delivering Glucose to Body Cells

After absorption, glucose and other monosaccharides travel to the liver via the hepatic portal vein. Here fructose and galactose are metabolized for energy. Glucose may be broken down by the liver to provide energy or stored as glycogen, but most is passed into the general blood circulation, causing blood glucose levels to rise. Glucose delivered in the blood provides energy to body cells. Many body cells can use energy sources other than glucose, such as fatty acids, but brain cells, red blood cells, and a few others must have glucose to stay alive. In order to provide a steady supply of glucose, the concentration of glucose in the blood is regulated by the liver and by hormones secreted by the pancreas.

glycemic response The rate, magnitude, and duration of the rise in blood glucose that occurs after a particular food or meal is consumed.

What Affects Glycemic Response? How quickly and how high blood glucose rises after carbohydrate is consumed is referred to as **glycemic response**. It is affected by both the amount and type of carbohydrate eaten and the amount of fat and protein in that food or meal. Because carbohydrate must be digested and absorbed to enter the blood, how quickly a food leaves the stomach and how fast it is digested and the nutrients absorbed in the small intestine all affect how long it takes glucose to get into the blood.

Refined sugars and starches generally cause a greater glycemic response than unrefined carbohydrates that contain fiber. This is because sugars and starches consumed

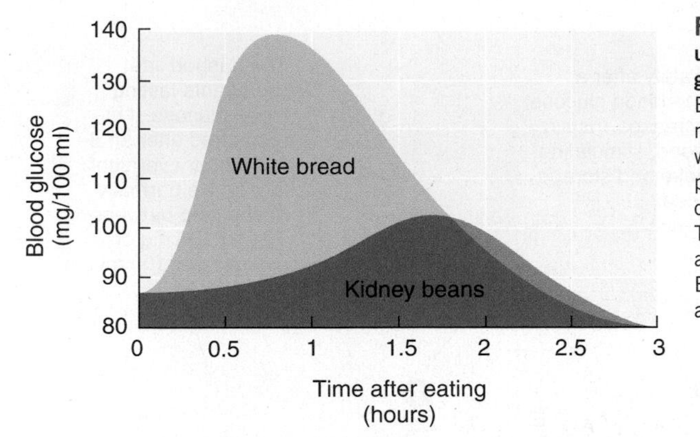

Figure 4.13 Effect of refined and unrefined carbohydrates on blood glucose levels
Blood glucose rises higher and more rapidly after a person eats white bread than it does after the person eats an equivalent amount of carbohydrate from kidney beans. The kidney beans contain fiber and are higher in protein than the bread. Both slow digestion and absorption and reduce glycemic response.

THINK CRITICALLY

How would the graph of blood glucose levels after eating a peanut butter sandwich differ from the graph shown here for just the bread?

alone leave the stomach quickly and are rapidly digested and absorbed, causing a sharp, swift rise in blood sugar. For example, when you drink a can of soda or eat a slice of white bread on an empty stomach, your blood sugar increases within minutes. Eating a high-fiber food causes a slower, lower increase in blood sugar (**Figure 4.13**). The presence of fat and protein also slows stomach emptying, and therefore foods high in these macronutrients generally cause a smaller glycemic response than foods containing sugar or starch alone. For example, ice cream is high in sugar but also contains fat and some protein, so it causes a smaller rise in blood glucose than sorbet, which contains sugar but no fat or protein.

The glycemic response of a specific food can be quantified by its **glycemic index**. Glycemic index is a ranking of how a food affects blood glucose compared to the response of an equivalent amount of carbohydrate from a reference food such as white bread or pure glucose. The reference food is assigned a value of 100 and the values for other foods are expressed relative to this. Foods that have a glycemic index of 70 or more compared to glucose are considered high-glycemic-index foods; those with an index of less than 55 are considered low-glycemic-index foods.

glycemic index A ranking of the effect on blood glucose of a food of a certain carbohydrate content relative to an equal amount of carbohydrate from a reference food such as white bread or glucose.

The glycemic index can be used to evaluate the effect of a specific food on blood glucose, but it is not based on amounts of carbohydrate in a typical portion of food. For example, watermelon has a high glycemic index, but this is based on a larger piece of watermelon than is typically consumed. The actual rise in blood glucose after eating a slice of watermelon is not large. **Glycemic load** is a method of assessing glycemic response that takes into account both the glycemic index of the food and the amount of carbohydrate in a typical portion. To calculate glycemic load, the grams of carbohydrate in a serving of food are multiplied by that food's glycemic index expressed as a percentage. A glycemic load of 20 or more is considered high, whereas a value of less than 11 is considered low.

glycemic load An index of the glycemic response that occurs after eating specific foods. It is calculated by multiplying a food's glycemic index by the amount of available carbohydrate in a serving of the food.

A shortcoming of both glycemic index and glycemic load is that they are determined for individual foods, but we typically eat meals containing mixtures of foods. Knowing the glycemic index or glycemic load of a specific food does not tell us much about what blood glucose levels will be after eating this food as part of a mixed meal. For example, bowl of white rice has a high glycemic index and glycemic load, but if rice is part of a meal that contains chicken and broccoli, the rise in blood glucose is much less.

Supplying Glucose to Body Cells In order to provide energy, glucose in the bloodstream must enter cells. A rise in blood glucose triggers the pancreas to secrete the hormone **insulin**, which allows glucose to be taken into muscle and adipose tissue cells. In the liver, insulin promotes the storage of glucose as glycogen and, to a lesser extent, fat. In muscle, it stimulates not only the uptake of glucose for ATP production but also the synthesis of muscle glycogen for energy storage (**Figure 4.14**). Insulin also stimulates protein synthesis, and in fat-storing cells, it stimulates lipid synthesis. These actions remove glucose from the blood, decreasing its level.

insulin A hormone secreted by the pancreas that allows the uptake of glucose by body cells and has other metabolic effects such as stimulating protein and fat synthesis and the synthesis of glycogen in liver and muscle.

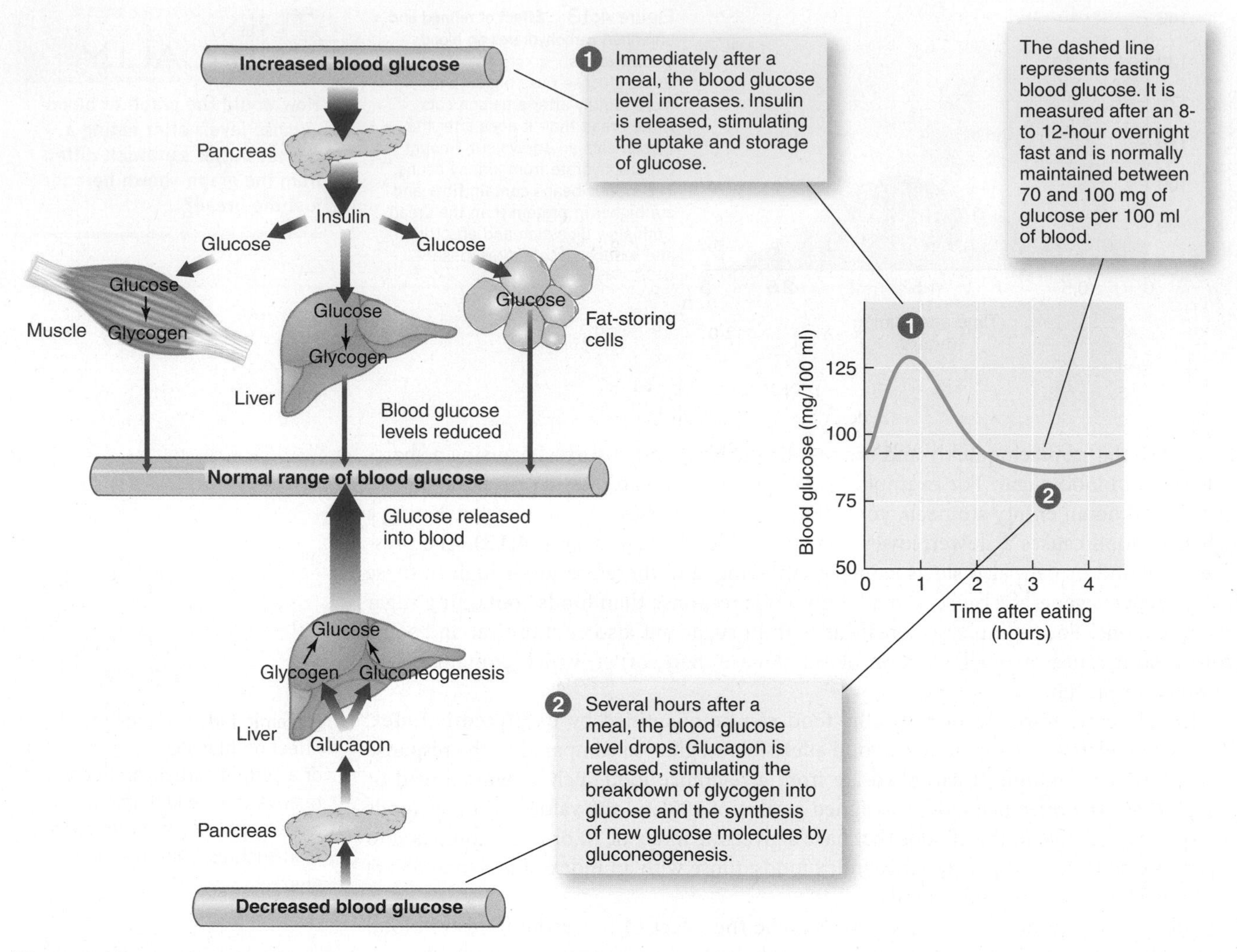

Figure 4.14 Blood glucose regulation
Blood glucose levels are regulated by the hormones insulin and glucagon, secreted by the pancreas.

glucagon A hormone secreted by the pancreas that stimulates the breakdown of liver glycogen and the synthesis of glucose to increase blood sugar.

gluconeogenesis The synthesis of glucose from simple noncarbohydrate molecules. Amino acids from protein are the primary source of carbons for glucose synthesis.

cellular respiration The reactions that break down carbohydrates, fats, and proteins in the presence of oxygen to produce carbon dioxide, water, and ATP.

When no carbohydrate has been consumed for a few hours, the glucose level in the blood—and consequently the glucose available to the cells—begins to decrease. This triggers the pancreas to secrete the hormone **glucagon**. Glucagon signals liver cells to break down glycogen into glucose, which is released into the bloodstream. Glucagon also stimulates the liver to synthesize new glucose molecules by **gluconeogenesis** (see Figure 4.14). Newly synthesized glucose is released into the blood to prevent blood glucose from dropping below the normal range. Gluconeogenesis can also be stimulated by the hormone epinephrine, also known as adrenaline. This hormone, which is released in response to dangerous or stressful situations, enables the body to respond to emergencies. It causes a rapid release of glucose into the blood to supply the energy needed for action.

How Glucose Provides Energy

Once glucose reaches body cells it can be metabolized through **cellular respiration** to generate ATP. Cellular respiration uses 6 molecules of oxygen to oxidize 1 molecule of glucose, forming 6 molecules of carbon dioxide, 6 molecules of water, and about 38 molecules of ATP:

$$\underset{\text{glucose}}{C_6H_{12}O_6} + \underset{\text{oxygen}}{6\,O_2} \rightarrow \underset{\text{carbon dioxide}}{6\,CO_2} + \underset{\text{water}}{6\,H_2O} + \text{ATP}$$

The carbon dioxide produced by cellular respiration is transported to the lungs, where it is eliminated in exhaled air. Providing energy through cellular respiration involves four interconnected stages.

Glycolysis The first stage of cellular respiration takes place in the cytosol of the cell and is called **glycolysis**, meaning glucose breakdown. Because oxygen isn't needed for this reaction, glycolysis is sometimes called **anaerobic metabolism**. In glycolysis, the six-carbon sugar glucose is broken into 2 three-carbon molecules called pyruvate (**Figure 4.15**). The reactions generate two molecules of ATP for each molecule of glucose and release high-energy electrons that are passed to shuttling molecules, which can transport them to the last stage of cellular respiration. When oxygen is limited, no further metabolism of glucose and production of ATP occur (see Online Appendix: Focus on Metabolism).

Acetyl-CoA Formation When oxygen is present, **aerobic metabolism** can proceed. In the mitochondria, one carbon is removed from pyruvate and released as CO_2. The remaining two-carbon compound combines with a molecule of coenzyme A (CoA) to form

glycolysis (also called **anaerobic metabolism**). Metabolic reactions in the cytosol of the cell that split glucose into 2 three-carbon pyruvate molecules, yielding two ATP molecules.

aerobic metabolism Metabolism in the presence of oxygen, which can completely break down glucose to yield carbon dioxide, water, and as many as 38 ATP molecules.

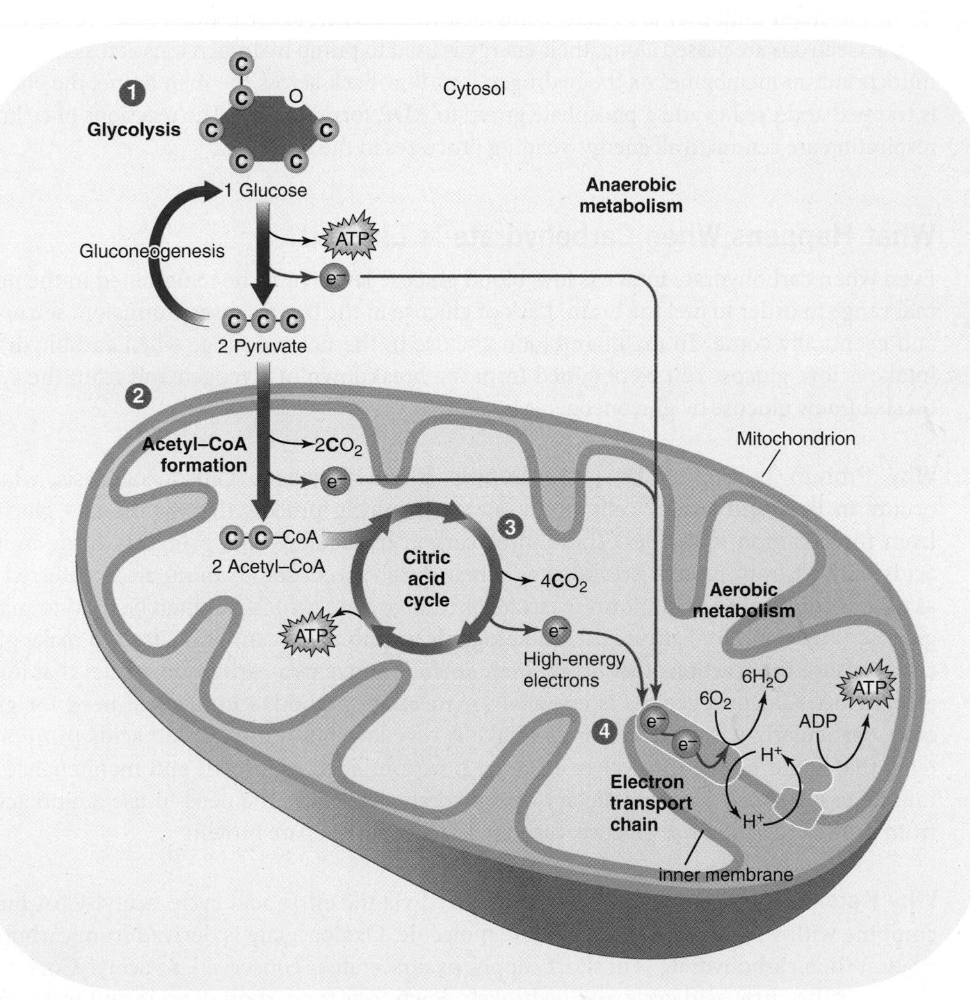

Figure 4.15 **Cellular respiration** The reactions of cellular respiration split the bonds between carbon atoms in glucose, releasing energy that is used to add a phosphate to ADP, to form ATP. ATP is used to power the energy-requiring processes in the body.

1. In the cytosol of the cell, glycolysis splits glucose, a six-carbon molecule, into two molecules of pyruvate, a three-carbon molecule. This step produces high-energy electrons (e^-) and two molecules of ATP per glucose. Each pyruvate is then either broken down to produce more ATP or used to make glucose via gluconeogenesis.
2. When oxygen is available, pyruvate can be used to produce more ATP. The first step is to remove one carbon as carbon dioxide from each pyruvate. This produces a two-carbon molecule that combines with coenzyme A to form acetyl-CoA and releases high-energy electrons.
3. Acetyl-CoA enters the citric acid cycle, where two carbons are lost as carbon dioxide, high-energy electrons are released, and a small amount of ATP is produced.
4. Most ATP is produced in the final step of cellular respiration, the electron transport chain. Here, the energy from the high-energy electrons released in previous steps pumps hydrogen ions across the inner mitochondrial membrane. As the hydrogen ions flow back, the energy is used to convert ADP to ATP. The electrons are combined with oxygen and hydrogen to form water.

acetyl-CoA (see Figure 4.15). High-energy electrons are released and passed to shuttling molecules for transport to the last stage of cellular respiration. Acetyl-CoA then enters the third stage of cellular respiration, the citric acid cycle.

Citric Acid Cycle In the third stage, acetyl-CoA combines with oxaloacetate, a four-carbon molecule derived from carbohydrate, to form a six-carbon molecule called citric acid and begin the citric acid cycle (see Figure 4.15). The reactions of the citric acid cycle then remove one carbon at a time, to produce carbon dioxide. After two carbons have been removed in this manner, a four-carbon oxaloacetate molecule is reformed and the cycle can begin again. These chemical reactions produce two ATP molecules per glucose molecule and also remove electrons, which are passed to shuttling molecules for transport to the fourth and last stage of cellular respiration, the electron transport chain.

Electron Transport Chain The electron transport chain consists of a series of molecules, most of which are proteins, associated with the inner membrane of the mitochondria. These molecules accept electrons from the shuttling molecules and pass them from one to another down the chain until they are finally combined with oxygen to form water (see Figure 4.15). As the electrons are passed along, their energy is used to pump hydrogen ions across the inner mitochondrial membrane. As the hydrogen ions flow back across the membrane, the energy is trapped and used to add a phosphate group to ADP, forming ATP. The reactions of cellular respiration are central to all energy-yielding processes in the body.

What Happens When Carbohydrate Is Limited?

Even when carbohydrate intake is low, blood glucose levels must be maintained in the normal range in order to fuel the brain. Lack of glucose in the brain causes confusion, seizures, and eventually coma. To maintain blood glucose in the normal range when carbohydrate intake is low, glucose can be obtained from the breakdown of glycogen and from the synthesis of new glucose by gluconeogenesis.

Why Protein Is Broken Down to Supply Blood Glucose Gluconeogenesis, which occurs in liver and kidney cells, is an energy-requiring process that synthesizes glucose from three-carbon molecules. These three-carbon molecules come primarily from amino acids derived from protein breakdown. When metabolized, some amino acids, referred to as **glucogenic amino acids**, form pyruvate or oxaloacetate, which can then be used to make glucose (**Figure 4.16**). Fatty acids and **ketogenic amino acids** cannot be used to make glucose because the reactions that break them down produce two-carbon molecules that form acetyl-CoA. Gluconeogenesis is essential for meeting the body's immediate need for glucose, particularly when carbohydrate intake is very low, but it uses amino acids from proteins that could be used for other essential functions such as growth and maintenance of muscle tissue. Since adequate dietary carbohydrate eliminates the need to use amino acids from protein to synthesize glucose, carbohydrate is said to spare protein.

glucogenic amino acid An amino acid that can be used to synthesize glucose through gluconeogenesis.

ketogenic amino acid An amino acid that breaks down to form acetyl-CoA and thus contributes to ketone synthesis.

Why Ketones Are Formed To be metabolized via the citric acid cycle, acetyl-CoA must combine with a four-carbon oxaloacetate molecule. Oxaloacetate is derived from carbohydrate. When carbohydrate is in short supply, oxaloacetate is conserved, so acetyl-CoA cannot enter the citric acid cycle and be broken down to form carbon dioxide and water and produce ATP. Instead, the liver converts it into compounds known as **ketones** or **ketone bodies**, which are released into the blood (see Figure 4.16). When fatty acids (or ketogenic amino acids) are broken down for energy, they yield two-carbon molecules that form acetyl-CoA. Without carbohydrate, fatty acids and ketogenic amino acids cannot be completely oxidized and ketones are formed.

ketones or **ketone bodies** Molecules formed in the liver when there is not sufficient carbohydrate to completely metabolize the two-carbon units produced from fat breakdown.

Ketone production is a normal response to starvation or to a diet very low in carbohydrate. Ketones can be used as an energy source by tissues, such as those in the heart, muscle, and kidney. After about three days of fasting, even the brain adapts and can obtain about half of its energy from ketones. The use of ketones for energy helps spare glucose and decreases the amount of protein that must be broken down to synthesize glucose.

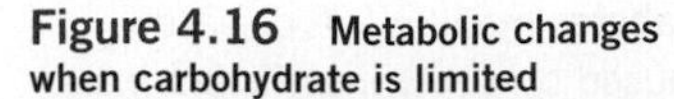

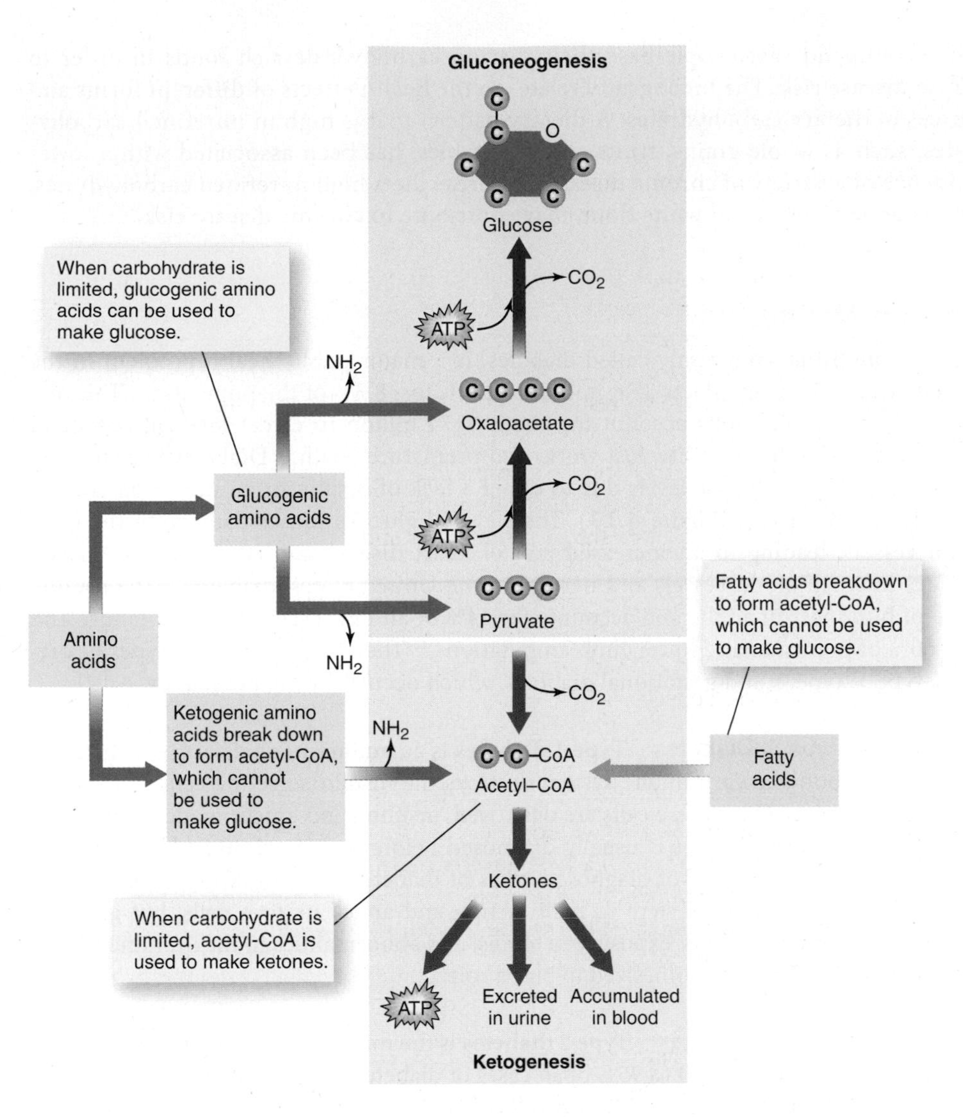

Figure 4.16 Metabolic changes when carbohydrate is limited Amino acids can be used to make glucose or to make ketones, depending on the type of amino acids and the availability of glucose. Fatty acids cannot be used to make glucose.

Ketones not used for energy can be excreted by the kidney in urine. However, if ketone production is high, ketones can build up in the blood, causing **ketosis**. Mild ketosis can occur during starvation or when consuming a low-carbohydrate weight-loss diet and can cause symptoms such as reduced appetite, headaches, dry mouth, and odd-smelling breath. Severe ketosis can occur with untreated diabetes and, as discussed below, it can increase the blood's acidity so much that normal body processes cannot proceed, eventually causing coma and even death.

ketosis High levels of ketones in the blood.

4.5 Carbohydrates and Health

LEARNING OBJECTIVES

- Discuss the causes and health consequences of diabetes.
- Describe how carbohydrate intake contributes to the development of dental caries.
- Discuss the role of carbohydrates in weight management.
- Explain how fiber intake affects the risk of heart disease, diverticulitis, and colon cancer.

Are carbohydrates good for you or bad for you? They have been blamed for a host of chronic health problems, from diabetes to obesity, yet carbohydrate-rich foods are the basis of healthy diets around the world and U.S. guidelines for a healthy

diabetes mellitus A disease caused by either insufficient insulin production or decreased sensitivity of cells to insulin. It results in elevated blood glucose levels.

type 1 diabetes A form of diabetes that is caused by the autoimmune destruction of insulin-producing cells in the pancreas, usually leading to absolute insulin deficiency; previously known as insulin-dependent diabetes or juvenile-onset diabetes.

type 2 diabetes A form of diabetes characterized by insulin resistance and relative insulin deficiency; previously known as noninsulin-dependent diabetes or adult-onset diabetes.

diet recommend that people base their diet on carbohydrate-rich foods in order to reduce disease risk. The incongruity relates to the health effects of different forms and sources of dietary carbohydrates: A dietary pattern that is high in unrefined carbohydrates, such as whole grains, fruits, and vegetables, has been associated with a lower incidence of a variety of chronic diseases, whereas diets high in refined carbohydrates, such as added sugars and white flour, may contribute to chronic disease risk.[11]

Diabetes Mellitus

Diabetes mellitus, commonly called diabetes, is a major public health problem in the United States. Diabetes affects 25.8 million Americans, 8.3% of the population. This disease and its complications account for about $174 billion in direct medical costs and indirect costs due to disability, lost work, and premature death.[12] Diabetes is characterized by high blood glucose levels due to either a lack of insulin or an unresponsiveness or resistance to insulin (**Figure 4.17**). The elevated glucose causes damage to the large blood vessels, leading to an increased risk of heart disease and stroke. It also causes changes in small blood vessels and nerves. In the United States, diabetes is the leading cause of blindness in adults and accounts for 44% of all new cases of kidney failure and over 60% of nontraumatic lower-limb amputations.[12] There are three main types of diabetes: type 1; type 2; and gestational diabetes, which occurs during pregnancy.

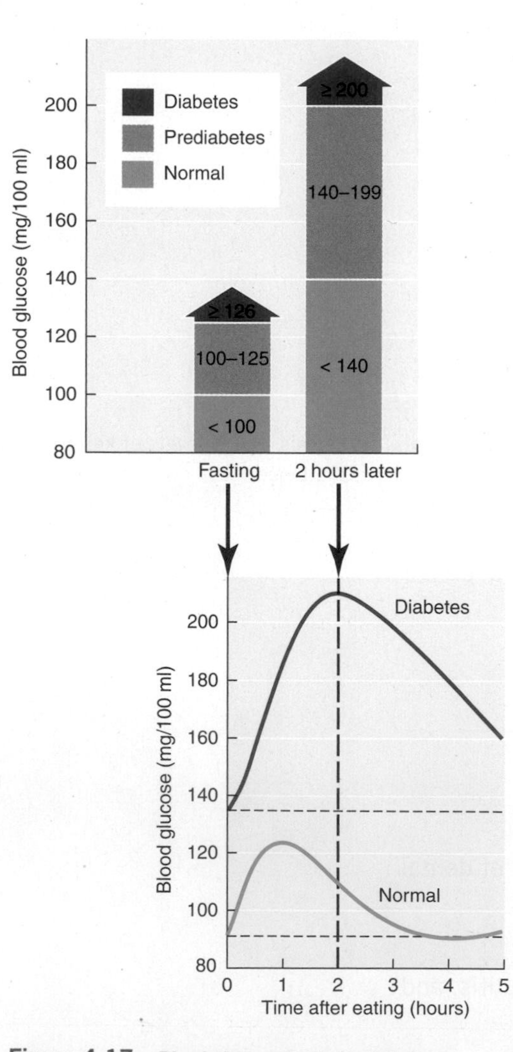

Figure 4.17 Blood glucose levels in diabetes Blood glucose levels measured after an eight-hour fast and two hours after consuming 75 g of glucose determine whether an individual has normal blood glucose levels, prediabetes, or diabetes.

Type 1 Diabetes **Type 1 diabetes** is an autoimmune disease in which the body's own immune system destroys the insulin-secreting cells of the pancreas. Once these cells are destroyed, insulin is no longer made in the body. Type 1 diabetes is usually diagnosed before the age of 30 and accounts for only 5 to 10% of diagnosed cases of diabetes. It is not known what causes the immune system to malfunction and attack its own cells, but genetics, viral infections, exposure to toxins, and abnormalities in the immune system have been hypothesized to play a role.

Type 2 Diabetes **Type 2 diabetes** is the more common form of diabetes. It accounts for 90 to 95% of all cases of diabetes in the United States. It occurs when the body does not produce enough insulin to keep blood glucose in the normal range. This can occur because body cells lose their sensitivity to insulin, a condition called *insulin resistance*, or when the amount of insulin secreted is reduced. When tissues are resistant to insulin, large amounts are required for cells to take up enough glucose to meet their energy needs. Type 2 diabetes is believed to be due to a combination of genetic and lifestyle factors. Risk of developing this disease is increased in people with a family history of diabetes; in those who are overweight, particularly if they carry their extra body fat in the abdominal region; and in those who have a sedentary lifestyle. Evidence is accumulating that the risk of developing type 2 diabetes is increased by diets high in refined carbohydrates.[13] The incidence of type 2 diabetes is higher among minority groups. Compared with non-Hispanic white adults, the risk of diabetes is 18% higher among Asian Americans, 66% higher among Hispanics/Latinos, and 77% higher among non-Hispanic blacks.[12] Type 2 diabetes may occur as part of a combination of conditions called **metabolic syndrome** that includes obesity, elevated blood pressure, altered blood lipid levels, and insulin resistance.

Type 2 diabetes is often preceded by a condition called **prediabetes** or **impaired glucose tolerance** in which blood glucose levels are above normal but not high enough to be diagnosed as diabetes (see Figure 4.17). An estimated 79 million adults ages 20 years and older have prediabetes and are therefore at increased risk for developing diabetes, as well as heart disease and stroke.[12] Progression to diabetes among those with prediabetes is not inevitable. Weight loss and increased physical activity among people with prediabetes can prevent or delay diabetes and may return blood glucose levels to normal.

Type 2 diabetes has typically been diagnosed in persons over the age of 40, but its incidence is increasing among younger individuals (**Figure 4.18**). This change is thought to be due to the increasing incidence of obesity and overweight in younger age groups. Before 1994, less than 5% of children with newly diagnosed diabetes were classified as type 2, but recently this has increased to 30 to 50%.[14]

Gestational Diabetes **Gestational diabetes** is a form of diabetes that occurs in women during pregnancy. It may be caused by the hormonal changes of pregnancy. The high levels of glucose in the mother's blood increase the risk of complications for the unborn child (see Chapter 14). Gestational diabetes usually disappears once the pregnancy is complete and hormones return to nonpregnant levels. However, individuals who have had gestational diabetes have an increased risk for developing type 2 diabetes later in life.

Why Diabetes Causes Symptoms and Complications The symptoms and complications of diabetes result from the fact that, without sufficient insulin, glucose cannot be used normally. Cells that require insulin for glucose uptake are starved for glucose, and cells that can use glucose without insulin are exposed to damaging high levels.

Immediate Symptoms The immediate symptoms of diabetes may include excessive thirst, frequent urination, blurred vision, and weight loss. Excessive thirst and frequent urination occur because blood glucose levels rise so high that the kidneys excrete glucose, which draws water with it, increasing the volume of urine. Blurred vision occurs when excess glucose enters the lens of the eye, drawing in water and causing the lens to swell. Weight loss, and impaired growth in children, occur because glucose cannot enter muscle and adipose tissue cells to be used for energy, so the body responds as it does in starvation,

metabolic syndrome A collection of health risks, including high blood pressure, altered blood lipids, high blood glucose, and a large waist circumference, that increases the chance of developing heart disease, stroke, and diabetes. The condition is also known by other names, including Syndrome X, insulin resistance syndrome, and dysmetabolic syndrome.

prediabetes or **impaired glucose tolerance** A fasting blood glucose level above the normal range but not high enough to be classified as diabetes.

gestational diabetes A form of diabetes that occurs during pregnancy and resolves after the baby is born.

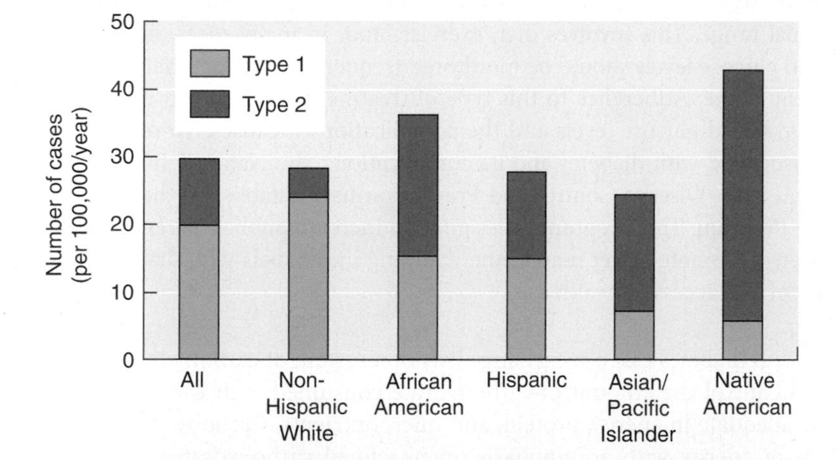

© Jeff Greenberg/Alamy

Figure 4.18 Type 1 and type 2 diabetes in adolescents This graph compares the number of cases of type 1 and type 2 diabetes diagnosed per year in adolescents (ages 10 to 19) by race/ethnicity. Twenty years ago, type 2 diabetes was rare in this age group, but as in adults, the incidence is rising, especially in certain minority groups. Reducing type 2 diabetes in minority groups will require culturally sensitive strategies to modify diet and lifestyle.

THINK CRITICALLY

Based on your knowledge of the connection between type 2 diabetes and obesity, which population group would you predict has the highest incidence of obesity?

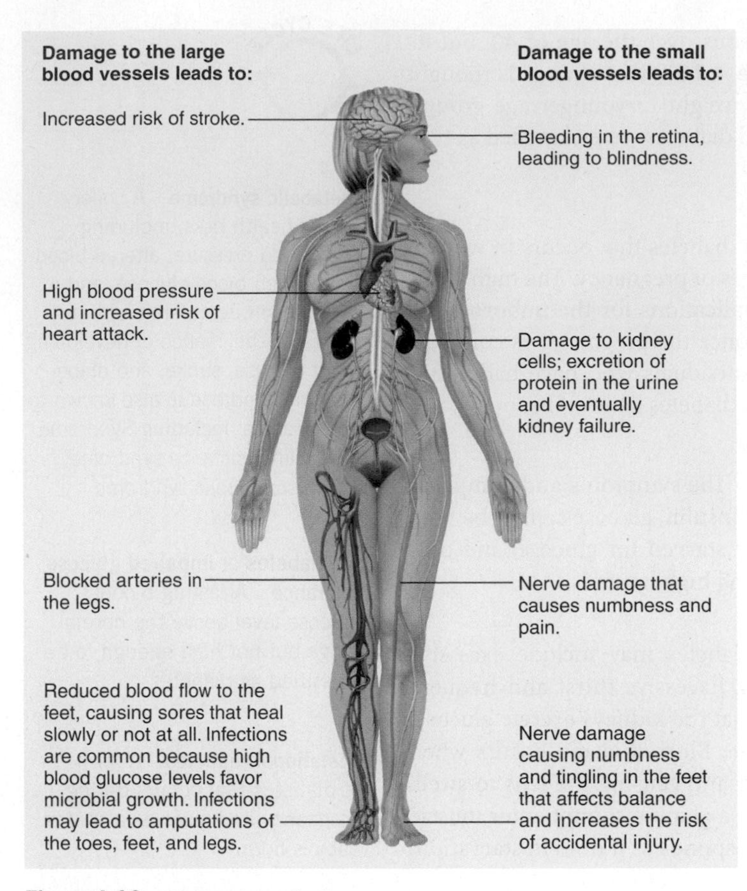

Figure 4.19 **Diabetes complications** The long-term complications of diabetes result from damage to both the large blood vessels, which leads to an increased risk of heart attack and stroke, and changes in small blood vessels, which can cause blindness, kidney failure, and nerve dysfunction.

breaking down fat and protein to supply fuel. With limited carbohydrate available for fatty acid metabolism, ketones are formed and released into the blood. Some ketones are used as fuel by muscle and adipose tissue, but in type 1 diabetes, they are produced more rapidly than they can be used and thus accumulate in the blood. This elevation of ketones causes an increase in the acidity of the blood called *ketoacidosis*. In type 2 diabetes, ketoacidosis usually does not develop because there is enough insulin to allow some glucose to be used, so fewer ketones are produced.

Long-Term Complications The long-term complications of diabetes include damage to the heart, blood vessels, kidneys, eyes, and nerves. This damage is thought to be a result of prolonged exposure to high levels of blood glucose. When glucose is high it can bind to proteins, contributing to blood vessel damage and abnormalities in blood cell function. Damage to the large blood vessels leads to increased risk of heart disease and stroke. The risk for heart disease and stroke is two to four times higher among people with diabetes.[12] Heart disease is the leading cause of premature death among people with diabetes. High blood glucose causes changes in small blood vessels and nerves that lead to kidney failure, blindness, and nerve dysfunction (**Figure 4.19**).

How Diabetes Is Treated The goal of diabetes treatment is to keep blood glucose levels within the normal range. This involves diet, exercise, and, in many cases, medication (**Figure 4.20**). Blood glucose levels should be monitored frequently to assure that levels are staying in the healthy range. Adherence to this type of treatment regimen can reduce the incidence of elevated blood glucose levels and the complications it causes. To reduce disability and death associated with diabetes and its complications, the National Institutes of Health and the Centers for Disease Control and Prevention have established the National Diabetes Education Program. This program is designed to increase public awareness of the seriousness of diabetes, promote better management among individuals with diabetes, and improve the quality of and access to health care.[15]

Diet Individuals with diabetes can use exchange lists or a system of carbohydrate counting to estimate and control the amount of carbohydrate consumed with each meal. The diet should also be adequate in energy, protein, and micronutrients. Carbohydrate should provide 40 to 50% of energy with an emphasis on unrefined carbohydrates, which do not cause a rapid rise in blood glucose. To help prevent heart disease, fat intake should be limited to no more than 30% of energy, with no more than 10% from saturated fat. Overweight individuals may need to restrict energy intake to promote weight loss, which can be beneficial for maintaining blood glucose levels in the normal range.

Exercise Exercise is an important component of diabetes management. Exercise increases the sensitivity of body cells to insulin so that glucose is available to fuel exercising muscles. Exercise also promotes weight loss, which further reduces insulin resistance. A single bout of exercise increases glucose uptake for up to 48 hours. Regular exercise results in a continuous increase in insulin sensitivity in the skeletal muscle.[17] Exercise patterns must be coordinated with meal and medication schedules to ensure that glucose is available during exercise.

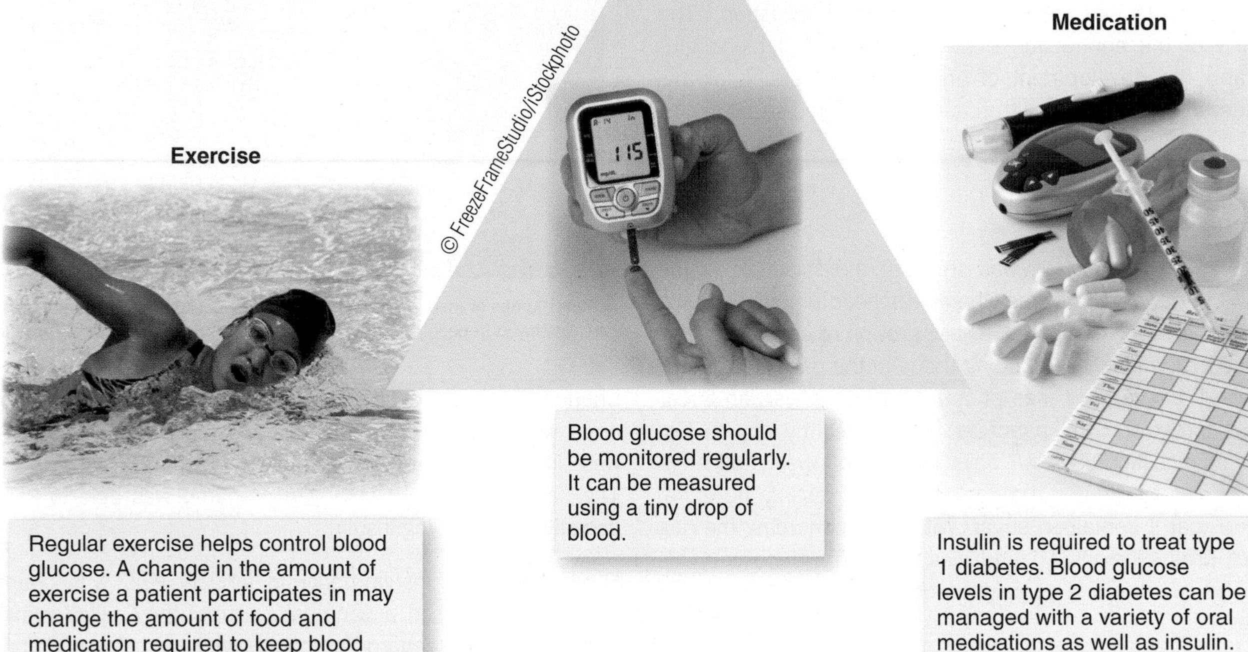
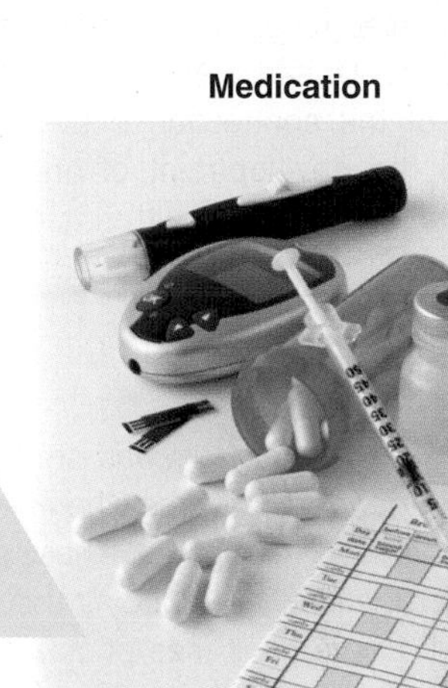

Figure 4.20 **Diabetes treatment**
The treatment of diabetes involves monitoring blood glucose, controlling diet and exercise patterns, and in some cases using medication. Carbohydrate intake must be coordinated with medication and exercise schedules so that glucose and insulin are available in the proper proportions at the same time to maintain normal blood glucose levels.[16]

Medication When diet and exercise cannot keep blood glucose in the normal range, drug treatments are needed. In type 1 diabetes, insulin production is absent, so insulin must be injected (see **Science Applied: How the Discovery of Insulin Made Diabetes Treatable**). Insulin cannot be taken orally because it is a protein that would be broken down in the GI tract, losing its ability to function. Type 2 diabetes can often be treated with medications that increase pancreatic insulin production, decrease glucose production by the liver, enhance insulin action, or slow carbohydrate digestion to keep blood glucose in the normal range. In some cases of type 2 diabetes, injected insulin is needed to achieve normal blood glucose levels.

Hypoglycemia

Hypoglycemia is a condition in which blood sugar drops low enough to cause symptoms, including irritability, nervousness, sweating, shakiness, anxiety, rapid heartbeat, headache, hunger, weakness, and sometimes seizure and coma. It can occur in people with diabetes

hypoglycemia A low blood glucose level, usually below 40 to 50 mg of glucose per 100 mL of blood.

SCIENCE APPLIED

How the Discovery of Insulin Made Diabetes Treatable

Diabetes was first described by Aratacus of Cappadocia (in what is now Turkey) in the first century AD, but its cause was a mystery. Before the discovery of insulin, diabetes was a death sentence. Today diabetes affects 25.8 million people of all ages in the United States, or 8.3% of the population.[1] The discovery of the connection between insulin and blood glucose, and the development of an injectable version of this hormone, allowed people with this once-deadly disease to live normal lives.

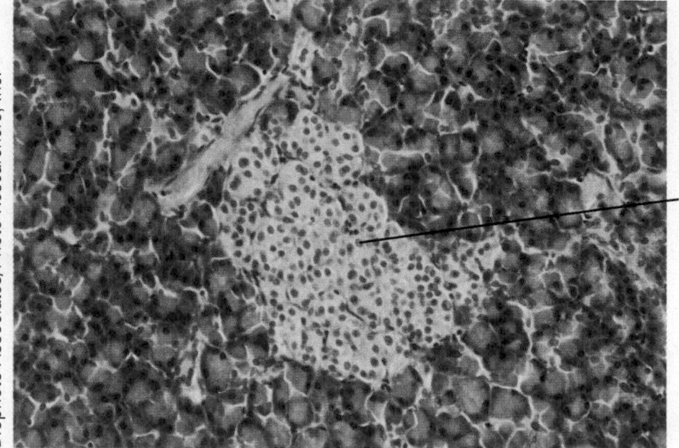

These clumps of insulin-producing cells, called the islets of Langerhans, are destroyed in people with type 1 diabetes so no insulin can be produced.

THE SCIENCE

In 1869, Paul Langerhans was studying the structure of the pancreas under a microscope when he noticed unusual clumps of tissue scattered throughout. These groups of cells were later identified as the insulin-producing tissue in the pancreas, called the islets of Langerhans (see photo). In 1889, physician Oscar Minkowski removed the pancreas from a healthy dog to test its role in digestion. Several days later, Minkowski noticed a swarm of flies feeding on the dog's urine. On testing the urine, he found that it contained sugar. Without a pancreas, the dog was not producing insulin and was excreting glucose in his urine. This was the first connection between the pancreas and diabetes.

Over the next two decades, research into diabetes and the pancreas continued. In 1921, Frederick Banting, intrigued by reading about the association between the pancreas and diabetes, became convinced that he could find the antidiabetic substance. Banting and two other Canadian researchers, John Macleod and Charles Best, first conducted experiments on dogs. The pancreas was removed from 10 dogs and these now-diabetic dogs were injected with an extract made from pancreatic islet cells. The pancreatic extract they had prepared was able to keep a dog alive by providing insulin.

Human studies of insulin began in January 1922 when Banting and colleagues used a pancreatic extract they called isletin to treat 14-year-old Leonard Thompson, who was dying of diabetes; his blood sugar decreased. After further purifying the extract, Banting went from bed to bed treating children dying from diabetes.[2] One of the first American patients to receive insulin treatment was Elizabeth Gossett, the daughter of the governor of New York; with insulin treatment, she lived to the age of 73.[3] In 1923

as a result of overmedication or an imbalance between insulin level and carbohydrate intake. People with diabetes must learn to recognize the symptoms of hypoglycemia and immediately treat them by consuming a source of quickly absorbed carbohydrate, such as juice or hard candy. Following this, a meal should be consumed within about 30 minutes to keep glucose in the healthy range.

In individuals without diabetes, hypoglycemia can result from abnormalities in the production of or response to insulin or other hormones involved in blood sugar regulation. There are two forms of hypoglycemia. The first, reactive hypoglycemia, occurs in response to the consumption of high-carbohydrate foods. The rise in blood glucose from the carbohydrate stimulates insulin release. However, too much insulin is secreted, resulting in a rapid fall in blood glucose to an abnormally low level. The treatment for reactive

(Elena Schweitzer/Shutterstock) Victoriano Izquierdo/Getty Images, Inc. (DAJ/Getty Images, Inc.)

Banting and Macleod were awarded the Nobel Prize for the discovery of insulin.[4]

THE APPLICATION

The first insulins used to treat humans were isolated and purified from the pancreatic islets of slaughtered animals, first dogs, then horses and pigs. They were used to treat diabetes for 50 years. For most patients they worked very well, but some developed allergies to the foreign proteins. In 1978, the biotechnology company Genentech produced the first synthetically manufactured insulin that could be mass-produced. It was produced by inserting the gene for human insulin into bacterial DNA. This recombinant DNA was then transferred into either bacteria or yeast cells, which served as miniature "factories" to produce the human insulin. The genetically engineered human insulin did not cause the problems that animal insulin sometimes did.

Today in the United States, the insulin used by tens of thousands of Americans is made by recombinant DNA technology. It is available in a variety of forms, some of which release quickly and some slowly. It can be delivered with a syringe, an insulin pen, or an insulin pump (see photo). Without this drug, thousands would die; with it, they can live healthy, productive lives.

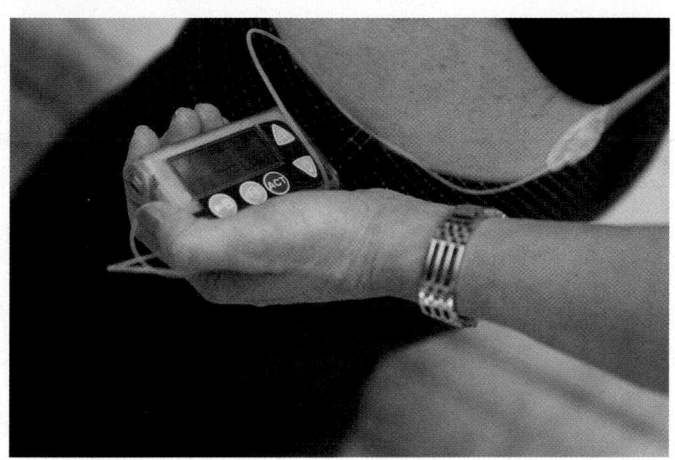

© Mark Hatfield/iStockphoto

With the press of a button, people with diabetes today can pump in a dose of genetically engineered human insulin that matches the amount of carbohydrate consumed.

[1]Centers for Disease Control and Prevention. National Diabetes Fact Sheet: National estimates and general information on diabetes and prediabetes in the United States, 2011. Atlanta, GA: U.S. Department of Health and Human Services, Centers for Disease Control and Prevention, 2011.

[2]Nordqvist, C. All about diabetes: Discovery of insulin. *Med News Today*, April 2010.

[3]Zuger, A. Rediscovering the first miracle drug. *New York Times*, October 4, 2010.

[4]The Nobel Prize in Physiology or Medicine 1923. Nobelprize.org. February 19, 2012.

hypoglycemia is a diet that prevents rapid changes in blood glucose. Small, frequent meals low in simple carbohydrates and high in protein and fiber are recommended. A second form of hypoglycemia, fasting hypoglycemia, is not related to food intake. In this disorder, abnormal insulin secretion results in episodes of low blood glucose levels. This condition is often caused by pancreatic tumors.

Dental Caries

dental caries The decay and deterioration of teeth caused by acid produced when bacteria on the teeth metabolize carbohydrate.

The most well-documented health problem associated with a diet high in carbohydrates is **dental caries**, or tooth cavities. It is one of the most common childhood diseases in the United States; 85% of people 18 years of age and older have had caries.[18] Cavities are

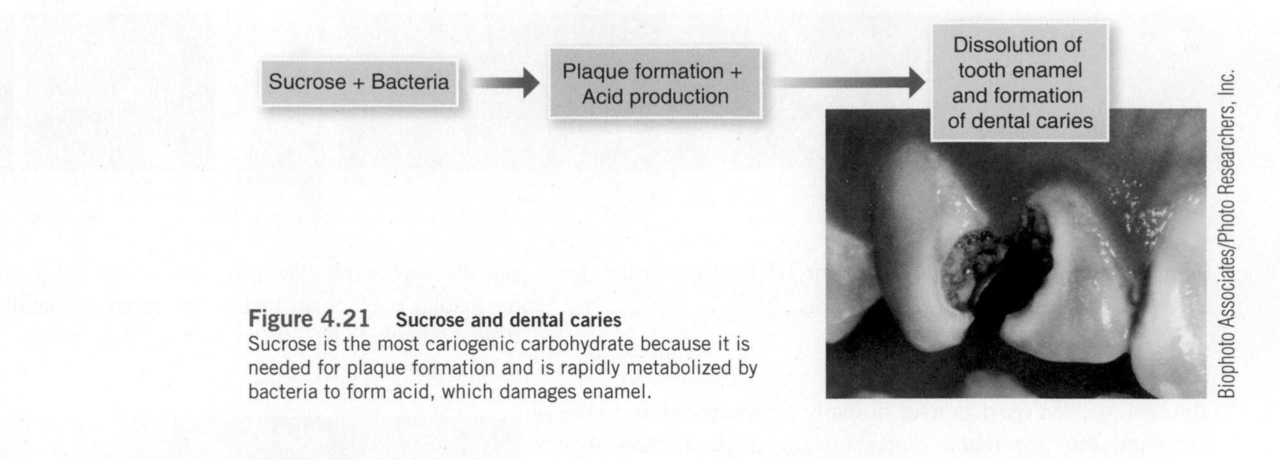

Figure 4.21 **Sucrose and dental caries**
Sucrose is the most cariogenic carbohydrate because it is needed for plaque formation and is rapidly metabolized by bacteria to form acid, which damages enamel.

caused when bacteria that live in the mouth form colonies, known as plaque, on the tooth surface. If the plaque is not brushed, flossed, or scraped away, the bacteria metabolize carbohydrate from the food we eat, producing acid. The acid can then dissolve the enamel and underlying structure of the teeth, forming cavities. Sucrose is considered the most cariogenic carbohydrate because, in addition to being metabolized to acid by oral bacteria, it is needed for the synthesis of materials that help bacteria stick to the teeth and form plaque (**Figure 4.21**).[19] Once plaque forms, simple carbohydrates are the most rapidly used food source for bacteria and therefore easily produce tooth-damaging acids, but starchy foods that stick to the teeth can also promote tooth decay. Foods such as gummy candies, cereals, crackers, cookies, and raisins and other sticky dried fruits tend to remain on the teeth longer, providing a continuous supply of nutrients to decay-causing bacteria. Other foods, such as chocolate, ice cream, and bananas, are rapidly washed away from the teeth and therefore are less likely to promote cavities. Frequent snacking, sucking on hard candy, or slowly sipping soda can also increase the risk of cavities by providing a continuous food supply for the bacteria. Limiting sugar intake can help prevent dental caries, but other dietary factors and proper dental hygiene are important even if the diet is low in sugar. Dairy products, sugarless gums (sweetened with sugar alcohols), and fluoride reduce caries formation. Brushing teeth after eating reduces cavity risk no matter what food is consumed.

Carbohydrates and Weight Management

The popularity of low-carbohydrate diets for weight loss has given carbohydrates a reputation for being fattening. Carbohydrates in and of themselves are not "fattening." They provide 4 kcal per gram compared with 9 kcal per gram provided by fat (**Figure 4.22**). This is not to say that carbohydrate consumed in excess of energy needs will not add pounds. Any energy source consumed in excess of requirements can cause weight gain. But carbohydrates are no more fattening than any other energy source. In fact, excess carbohydrate in the diet is less efficient at producing body fat than excess fat in the diet (see Chapter 7). But, even though carbohydrates are not high in calories, the amount and type of carbohydrate can affect how easily body weight is managed.

Figure 4.22 **Carbohydrates and calorie intake**
It is the fat we add to high-carbohydrate foods that increases their calorie tally. A medium-size baked potato provides about 160 kcal. Adding a tablespoon of butter brings the total to 260 kcal. A plate of plain pasta has about 200 kcal, but with a high-fat sauce, the kcalorie rise to 300; add sausage and the meal is now 450 kcal.

Why Low-Carbohydrate Diets Promote Weight Loss The rationale behind consuming a low-carbohydrate diet for weight loss is that foods high in carbohydrate stimulate the release of insulin, which is a hormone that promotes energy storage. It is suggested that the more insulin you release, the more fat you will store. Refined carbohydrates cause the greatest glycemic response and subsequent rise in insulin. They are therefore hypothesized to shift metabolism toward fat storage. Low-carbohydrate diets allow only a small amount of carbohydrate and therefore cause less of a glycemic response and less insulin release, which is suggested to promote fat loss. Weight loss while consuming a very low-carbohydrate diet may also be affected by ketone levels and the amount of protein in the diet. Ketones help suppress appetite and the high-protein content of a low-carbohydrate diet can be satiating, so both help the dieter eat less. Low-carbohydrate weight-loss diets have been shown to produce weight loss as effectively as other types of weight-loss diets (see Chapter 6 Debate: Are High-Protein Diets a Safe Way to Lose Weight?).[20] The weight loss on these diets, as with any weight-loss diet, is caused by reducing energy intake.[21,22] Weight loss can be maintained only by continuing the lifestyle changes that led to weight loss.

How Unrefined Carbohydrates Help Manage Weight Diets high in unrefined carbohydrates can help make weight maintenance and weight loss easier. A diet high in unrefined carbohydrates is high in fiber, which increases the sense of fullness by adding bulk and slowing digestion and absorption. This allows you to feel satisfied with less food and can help promote weight loss.[23] Unrefined foods that are good sources of fiber have a lower glycemic response and therefore cause a smaller rise in insulin than refined carbohydrates. Diets high in unrefined foods therefore have glycemic effects similar to diets low in carbohydrate and high in protein.

Carbohydrates and Heart Disease

The impact of carbohydrates on heart disease risk depends on the type of carbohydrate. There is evidence that diets high in sugar can raise blood lipid levels and thereby increase the risk of heart disease,[24] whereas diets high in whole grains and fiber have been found to reduce the risk of heart disease.[23,25,26] Whole grains provide fiber, resistant starch, oligosaccharides, omega-3 fatty acids, vitamins, minerals, antioxidants, and other phytochemicals that may be protective against heart disease (see Chapter 5).

The soluble fiber in whole grains and other unrefined sources of carbohydrates may reduce the risk of heart disease by lowering blood cholesterol levels. Soluble fiber binds cholesterol and bile acids, which are made from cholesterol, in the digestive tract. Normally, bile acids secreted into the GI tract are absorbed and reused. When bound to fiber, they are excreted in the feces rather than being absorbed (**Figure 4.23**). The liver must then use cholesterol from the blood to synthesize new bile acids. This provides a mechanism for eliminating cholesterol from the body and reducing blood cholesterol levels. Soluble fibers from legumes, oats, guar gum, pectin, flax seed, and psyllium (a grain used in bulk-forming laxatives such as Metamucil) are effective at reducing cholesterol, but insoluble fibers such as wheat bran or cellulose are not.[27] Soluble fiber may also help lower blood cholesterol because the by-products of the bacterial breakdown of the fiber may inhibit cholesterol synthesis in the liver or increase its removal from the blood.[26]

Because of the beneficial effects of fiber on heart disease risk, the FDA permits health claims on food products containing soluble fiber and fiber in general, which state that

Figure 4.23 **Effect of soluble fiber on cholesterol absorption**

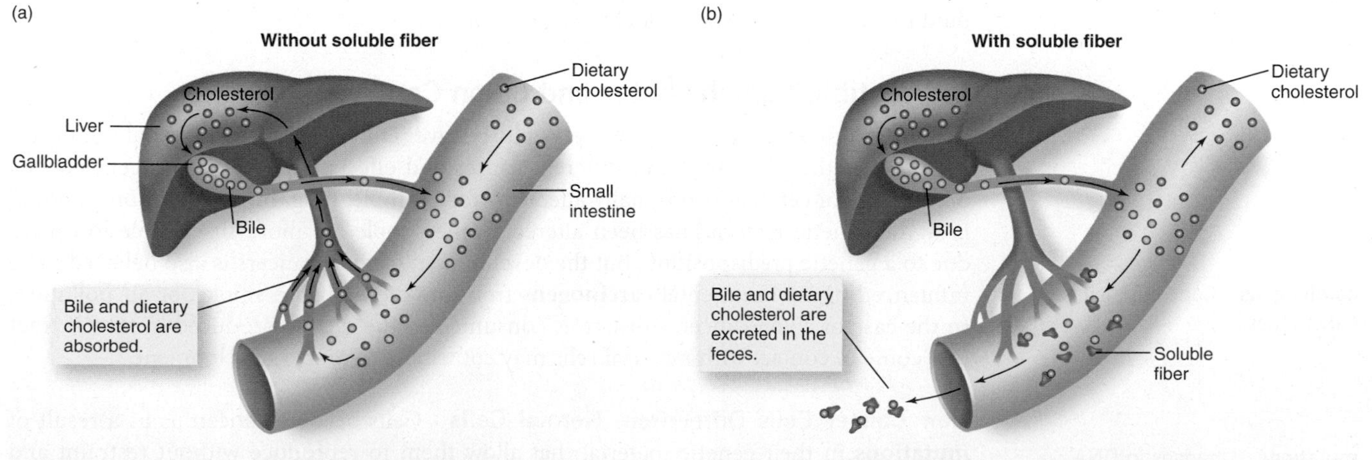

(a) When the diet is low in soluble fiber, dietary cholesterol and bile, which contains cholesterol and bile acids made from cholesterol, are absorbed into the blood and transported to the liver, where they are reused.

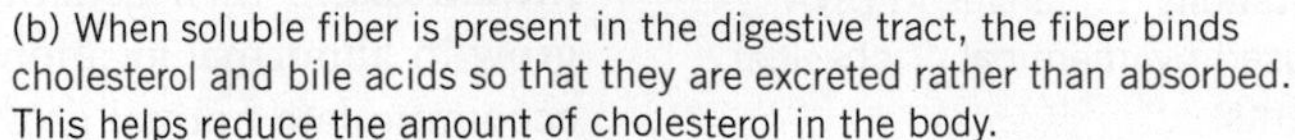

(b) When soluble fiber is present in the digestive tract, the fiber binds cholesterol and bile acids so that they are excreted rather than absorbed. This helps reduce the amount of cholesterol in the body.

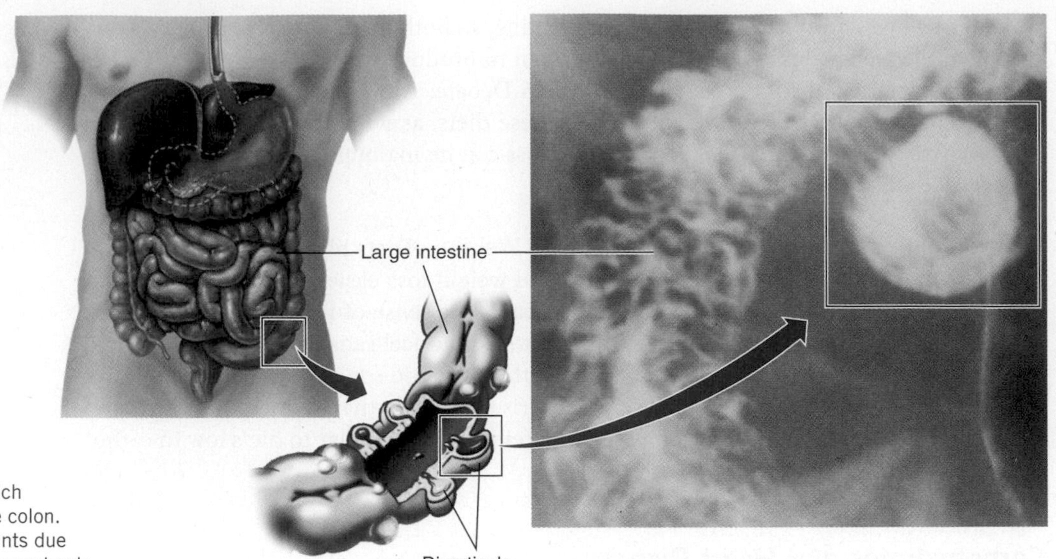

Figure 4.24 Diverticulosis
Diverticulosis is a condition in which outpouches form in the wall of the colon. These diverticula form at weak points due to pressure exerted when the colon contracts.

these products may reduce the risk of coronary heart disease. In addition to lowering blood cholesterol levels, diets high in whole grains and other unrefined carbohydrates help lower blood pressure, normalize blood glucose levels, prevent obesity, and affect a number of other parameters, all of which help reduce the risk of heart disease.[27]

Indigestible Carbohydrates and Bowel Health

hemorrhoids Swollen veins in the anal or rectal area.

diverticulosis A condition in which pouches called **diverticula** protrude from the wall of the large intestine. When these become inflamed, the condition is called **diverticulitis.**

Fiber and other indigestible carbohydrates add bulk and absorb water in the gastrointestinal tract, making the feces larger and softer and reducing the pressure needed for defecation. This helps reduce the incidence of constipation and **hemorrhoids**, the swelling of veins in the rectal or anal area. It also lessens symptoms associated with **diverticulosis**, a condition in which outpouches called **diverticula** (the singular is *diverticulum*) form in the wall of the colon (**Figure 4.24**).[28] Fecal matter may occasionally accumulate in these outpouchings, causing irritation, pain, inflammation, and infection, a condition known as **diverticulitis**. Treatment of diverticulitis usually includes antibiotics to reduce bacterial growth and a temporary decrease in fiber intake to prevent irritation of the inflamed tissues. Once the inflammation is resolved, however, a high-fiber intake is recommended to increase fecal bulk, decrease transit time, ease stool elimination, and reduce future attacks of diverticulitis.[29]

Although fiber helps soften stools and prevent constipation, if fiber is consumed without sufficient fluid, it can also cause constipation. The more fiber there is in the diet, the more water is needed to keep the stool soft. When too little fluid is consumed, the stool becomes hard and difficult to eliminate. In severe cases when fiber intake is excessive and fluid intake is low, intestinal blockage can occur.

Indigestible Carbohydrates and Colon Cancer

Cancer is a disease that affects the way cells behave. Different cancers originate in different parts of the body and have different causes and effects. The type of cancer depends on the type of cell that is originally affected—for example, lung, breast, or colon—and on how the genetic material has been altered. Some people are more susceptible to cancer due to a genetic predisposition, but the development of most cancers is also believed to be influenced by environmental **carcinogens** from the diet, tobacco smoke, or air pollution. In the case of colon cancer, substances consumed in the diet or produced in the GI tract that come in contact with mucosal cells may contribute to cancer development.

carcinogens Cancer-causing substances.

How Cancer Cells Differ from Normal Cells Cells become cancerous as a result of **mutations** in their genetic material that allow them to reproduce without restraint and grow in abnormal locations. Normal body cells reproduce only to replace lost cells or to accommodate normal growth, but cancer cells divide continuously, forming enlarged cell

mutations Changes in DNA caused by chemical or physical agents.

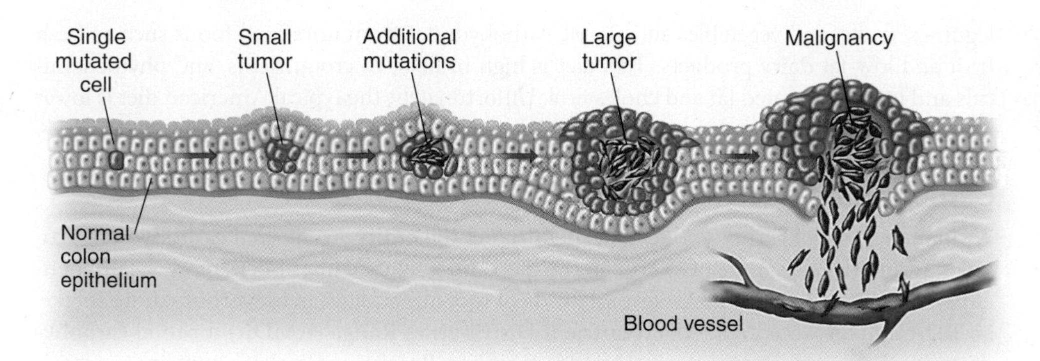

Figure 4.25 Development of colon cancer
Colon cancer, like other cancers, progresses from a single mutated cell to a tumor to a malignancy.

masses known as tumors. Further mutations form a mass of cells that can invade and colonize areas reserved for other cells, referred to as a **malignancy** (**Figure 4.25**). The cancer cells eventually crowd out the normal cells, robbing them of nourishment and preventing them from functioning properly. Some carcinogens act by damaging DNA and inducing mutations. These are usually referred to as *tumor initiators*, since the induction of mutations in key genes is thought to be the initiating event in cancer development. Other carcinogens contribute to cancer development by stimulating cells to divide. Such compounds are referred to as *tumor promoters*, since the increased cell division they induce enlarges the population of mutated cells, which is necessary for cancer to progress.

malignancy A mass of cells showing uncontrolled growth, a tendency to invade and damage surrounding tissues, and an ability to seed daughter growths to sites remote from the original growth.

Why Fiber May Protect against Colon Cancer Epidemiological studies have found that the incidence of colon cancer is lower in populations consuming diets high in fiber and whole grains.[30] Several hypotheses have been suggested to explain how fiber might affect the development of colon cancer. One is related to its ability to decrease contact between the mucosal cells of the large intestine and the fecal contents, which may contain tumor initiators or tumor promoters. Fiber decreases contact by increasing fecal bulk, diluting the colon contents, and speeding transit. Another hypothesis relates to the effect of fiber on the intestinal microflora and the by-products of microbial metabolism, such as fatty acids, that accumulate there. These by-products may directly support the health of colon cells or may cause changes in the environment of the colon that can inhibit the development of colon cancer. It has also been hypothesized that high-fiber diets protect against colon cancer because of the antioxidant vitamins and phytochemicals that are present along with fiber in plant foods.

Intervention studies have not supported the epidemiological observations of a connection between fiber intake and colon cancer.[23] A number of reasons have been suggested for this discrepancy—the interventions were not long enough, the fiber dose was not high enough, the type of cancer monitored was not appropriate, or the fiber itself is not really protective but rather some other component in the diet of the low-cancer populations may have a protective effect. Although the evidence that diets high in fiber protect against colon cancer is mixed, high fiber intake is still recommended to reduce the risk of chronic disease.[6]

4.6 Meeting Recommendations for Carbohydrate Intake

LEARNING OBJECTIVES

- List recommendations for total carbohydrate, added sugars, and fiber intake.
- Choose unrefined sources of carbohydrate from each food group of MyPlate.
- Use food labels to choose foods that provide healthy carbohydrates.
- Discuss the role of alternative sweeteners in weight loss.

The average American diet provides plenty of carbohydrate, but whether or not this carbohydrate promotes or harms our health depends on the food sources and types of carbohydrates we choose. A healthy diet is high in complex carbohydrates from whole grains, legumes, fruits, and vegetables and simple carbohydrates from unrefined foods such as fresh fruit and low-fat dairy products. This diet is high in fiber, micronutrients, and phytochemicals and low in saturated fat and cholesterol. Unfortunately, the typical American diet is lower in whole grains, fruits, and vegetables and higher in added sugars than is recommended.

Carbohydrate Recommendations

A small amount of carbohydrate is needed to fuel the brain. Additional carbohydrate provides an important source of energy in the diet, and adequate fiber offers many health benefits. Therefore the DRIs make several kinds of recommendations for carbohydrate intake: an RDA and an Acceptable Macronutrient Distribution Range (AMDR) for total carbohydrate intake and an Adequate Intake (AI) for fiber. Because no specific toxicity is associated with high intakes of carbohydrate in general or of different types of carbohydrates, no UL has been established for total carbohydrate, added sugars, or fiber.

How Much Total Carbohydrate? The RDA for carbohydrate for adults and children has been set at 130 g per day based on the average minimum amount of glucose used by the brain.[3] In a diet that meets energy needs, this amount will provide adequate glucose and prevent ketosis. This amount of carbohydrate provides only 520 kcal; that's about 25% of the energy in a 2000-kcal diet. It is equivalent to the amount in a breakfast of a cup of juice, two slices of toast with jam, and a bowl of cereal with half a banana and milk. Most people consume well in excess of this amount over the course of a day. Additional carbohydrate provides an important source of energy in the diet, and carbohydrate-containing foods can add vitamins, minerals, fiber, and phytochemicals. A diet that includes only 130 g of carbohydrate and meets calorie needs would be very high in protein and fat.

The AMDR for carbohydrate intake for a healthy diet has been set at 45 to 65% of energy (225 to 325 g in a 2000-kcal diet). Choosing a diet in this range will allow you to meet your energy needs without consuming excessive amounts of protein or fat. The percent of energy from carbohydrate in your diet can be calculated by using the equations shown in **Table 4.1**. This same calculation can be used to determine the percent of energy as carbohydrate in individual foods. The amount of carbohydrate in a food or in a diet can be estimated from the Exchange Lists (**Table 4.2**) or determined using values from food labels, iProfile or other diet-analysis programs, or food composition tables and databases (see Nutrient Composition of Foods Supplement or iProfile).

What Types of Carbohydrates Are Recommended? The sources of carbohydrate are more important than the absolute amount. In order to promote a diet that meets needs for all nutrients, most carbohydrate should come from unrefined food sources with limited amounts from added sugars.

TABLE 4.1 Calculating Percent of Energy from Carbohydrate

Determine

- The total energy (kcal) intake for the day
- The g of carbohydrate in the day's diet

Calculate energy from carbohydrate

- Carbohydrate provides 4 kcal per g
- Multiply g of carbohydrate by 4 kcal per g

$$\text{Energy (kcal) from carbohydrate} = \text{g carbohydrate} \times 4 \text{ kcal/g carbohydrate}$$

Calculate percent of energy from carbohydrate

Divide energy from carbohydrate by total energy and multiply by 100 to express as a percentage

$$\text{Percent of energy from carbohydrate} = \frac{\text{kcal from carbohydrate}}{\text{Total kcal}} \times 100$$

For example:

A diet contains 2500 kcal and 350 g of carbohydrate

$$350 \text{ g of carbohydrate} \times 4 \text{ kcal/g} = 1400 \text{ kcal of carbohydrate}$$

$$\frac{1400 \text{ kcal of carbohydrate}}{2500 \text{ kcal}} \times 100 = 56\% \text{ of energy (kcal) from carbohydrate}$$

TABLE 4.2 Using Exchange Lists to Estimate Carbohydrate Content

Exchange groups/lists	Serving size	Carbohydrates (g)
Carbohydrate group		
Starch	⅓ cup pasta; ½ cup potatoes; 1 slice bread	15
Fruit	1 small apple, peach, pear; ½ banana; ½ cup canned fruit (in juice)	15
Milk	1 cup milk or yogurt	
Nonfat		12
Lowfat		12
Reduced fat		12
Whole		12
Other carbohydrates	Serving sizes vary	15
Vegetables	½ cup cooked vegetables, 1 cup raw	5
Meat/meat substitute group	1 oz meat or cheese, ½ cup legumes	
Very lean		0
Lean		0
Medium-fat		0
High-fat		0
Fat group	1 tsp butter, margarine, or oil; 1 T salad dressing	0

Unrefined carbohydrates are good sources of fiber. The AI for fiber is based on the amount of fiber needed to reduce heart disease risk.[3] For young adult men, it is 38 g/day, and for young adult women, it is 25 g/day. The AI for fiber depends on energy intake so it is lower for children and older adults, who require fewer calories (see inside cover and Appendix A). Eating a bowl of Raisin Bran with a half-cup of strawberries for breakfast, a sandwich on whole-wheat bread with lettuce and tomatoes and an apple for lunch, eggplant parmesan for dinner, and popcorn for a snack will provide about 25 g of fiber. **Table 4.3** offers an exchange system for estimating fiber in foods.

There is no RDA or Daily Value for added sugars, but the 2010 Dietary Guidelines recommend reducing the consumption of added sugars, which add calories without contributing to the overall nutrient adequacy of the diet. The amount that can be included in

TABLE 4.3 Estimating the Fiber Content of Foods

Food group/serving	High fiber	Medium fiber	Low fiber
Fiber per serving	*4–5 g*	*2–3 g*	*0.5–1 g*
Grain group			
Breads (1 slice)	—	Whole wheat, rye	White bread, bagel (½), tortilla, roll (½), English muffin (½), graham crackers (2)
Cereals (½ cup)	All Bran, Bran Buds 100% Bran Flakes	40% Bran, Shredded Wheat	Cheerios, Rice Krispies
Rice and pasta (½ cup)	—	Whole-wheat pasta, brown rice	Macaroni, pasta, white rice
Fruit group			
Fruits (1 medium or ½ cup)	Berries, prunes	Apple, apricot, banana, orange, raisins	Melon, canned fruit, Juice
Vegetable group			
Vegetables (½ cup)	Peas, broccoli, spinach	Green beans, carrots, eggplant, cabbage, potatoes with skin, corn	Asparagus, cauliflower, celery, lettuce, tomatoes, zucchini, peppers, potatoes without skin, onions
Dry bean group			
Beans (½ cup)	Pinto beans, red kidney beans, black-eyed peas	—	—

Sources: Adapted from Bright-See, E., Benda, C.,Vartouhi, J., et al. Development and testing of a dietary fibre exchange system. *Can Diet Assoc J* 47:199–205, 1986; Marlett, J. A. Content and composition of dietary fiber in 117 frequently consumed foods. *J Am Diet Assoc* 92:175–186, 1992.

a healthy diet depends on the total calories that can be consumed without weight gain.[6] A diet that includes no more than the recommended number of empty calories will include only about 10% of calories from added sugars. This coincides with the recommendation of the World Health Organization that no more than 10% of energy come from added sugars.[31] Reducing added sugars reduces calorie intake without reducing essential nutrients.[6]

Translating Recommendations into Healthy Diets

The typical North American diet meets some but not all of these carbohydrate recommendations. We consume about 50% of energy from carbohydrate, but most of this comes from refined sources. We consume too little fiber, too few whole grains and other unrefined carbohydrate sources, and too much added sugar. Our average fiber intake is only 15 g per day, which is well below the AI.

Using MyPlate to Make Healthy Choices To promote a healthy, balanced diet, MyPlate and the 2010 Dietary Guidelines recommend increasing consumption of whole grains, fruits and vegetables, and low-fat dairy products while limiting foods high in refined grains and added sugars, such as soft drinks and other sweetened beverages, sweet bakery products, and candy (**Figure 4.26**).

Eat More Whole Grains, Fruits, and Vegetables To encourage the consumption of healthy sources of carbohydrates, MyPlate recommends that you choose half of your grains from

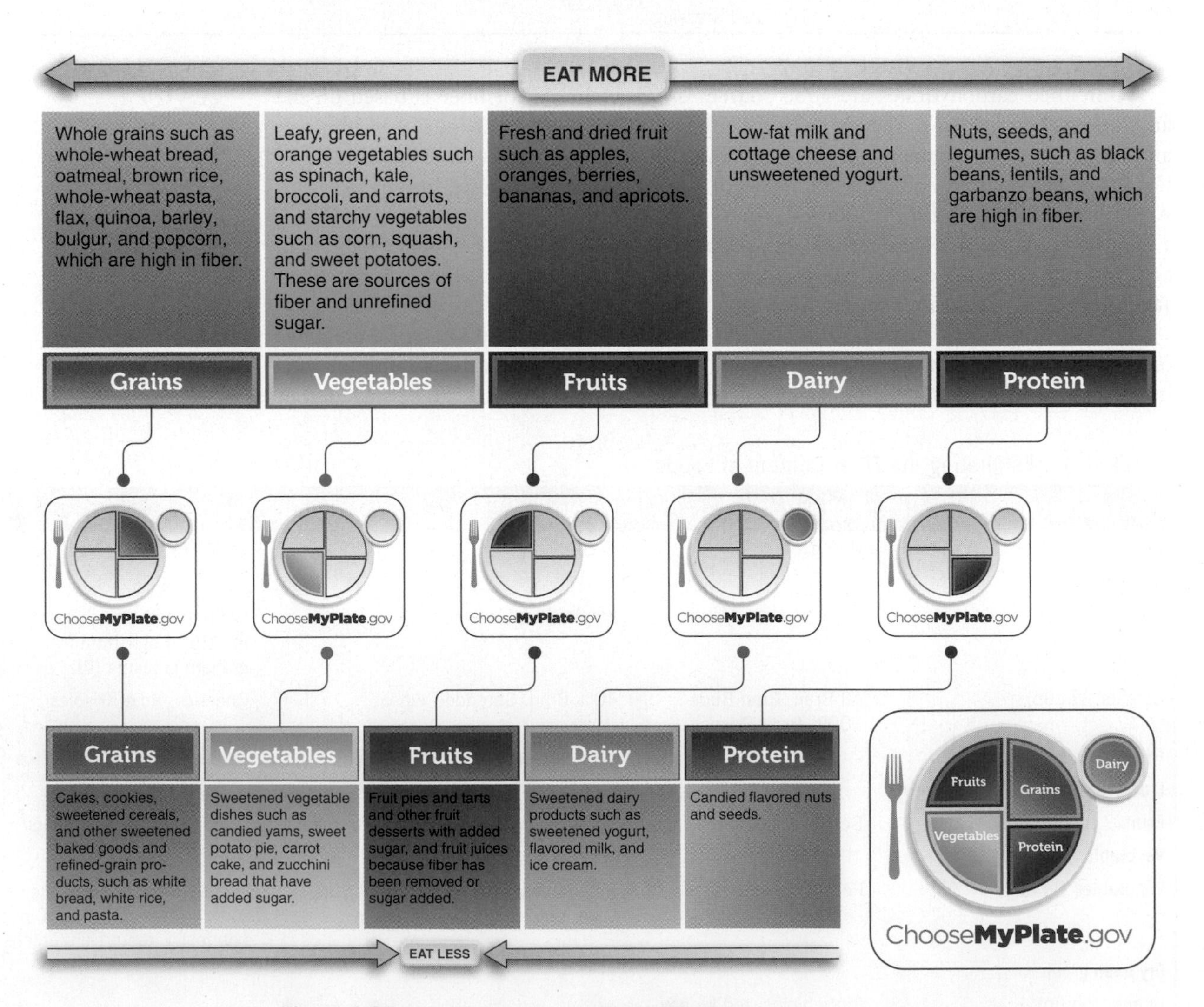

Figure 4.26 Healthy MyPlate carbohydrate choices
The healthiest carbohydrate choices are whole grains, legumes, and fresh fruits and vegetables, which are low in added sugar and often are good sources of fiber. Foods containing refined carbohydrates should be limited because they are typically low in fiber and contain added sugars, which add empty calories.

Ken Karp

Food	Amount of added sugar (tsp)	Energy from added sugar (kcal)
Cookies, 2 medium chocolate chip	3	48
Frosted corn flakes, 1 oz	3	48
Cake, frosted, 2 oz	6	96
Pie, fruit, 2-crust, ⅛ of 9"	6	96
Fruit, canned in heavy syrup, 1 cup	4	64
Low-fat fruit yogurt, 1 cup	7	112
Ice cream, vanilla, 1 cup	3	48
Chocolate bar, 1 oz	5	80
Fruit drink, 12 oz	12	192
Cola, not diet, 20-oz bottle	15	240

Figure 4.27 How much added sugar do you eat?
These foods all contribute to the added sugar content of your diet. Drinking a 20-oz bottle of sweetened iced tea uses up about 240 of the 260 empty calories allowed in a 2000-kcal diet.

whole sources or eat at least three servings of whole grains per day. Whole grains are good sources of fiber as well as micronutrients and phytochemicals. Fruits and vegetables are also excellent unrefined food sources of both complex and simple carbohydrates. For a 2000-kcal diet, MyPlate recommends 2 cups of fruit and 2½ cups of vegetables. To maximize fiber intake, most fruit choices should be whole fruits rather than juices. An apple provides about 80 to 90 kcal and 2.7 g of fiber, whereas a cup of apple juice provides the same amount of energy but almost no fiber (0.2 g). Choosing legumes as part of the protein or vegetables servings will increase fiber intake—one-half cup of cooked black beans has about 7 g of fiber.

Limit Added Sugars Most of the added sugars in the American diet come from sugar-sweetened beverages such as soda, energy drinks, sports drinks, and sugar-sweetened fruit drinks. Candy, cakes, cookies, pies, dairy desserts, and thousands of other processed products also contain added sugars. Sugars are also added at home when we sprinkle sugar on our cereal or spoon honey into our tea. The greater your consumption of foods high in added sugars, the harder it is to meet your nutrient needs without gaining weight. People who eat foods and beverages high in added sugars tend to consume more calories and fewer micronutrients. To avoid exceeding your empty calorie limit, added sugar intake must be kept to a minimum. For example, a 2000-kcal diet can include only about 260 empty calories (**Figure 4.27**).

Putting It All Together To meet the recommendations for a healthy diet, refined carbohydrates should be replaced with unrefined sources of simple and complex carbohydrates (**Table 4.4**). For

TABLE 4.4 How to Choose Healthy Carbohydrates

Make half your grains whole
• Have your sandwich on whole-wheat, oat bran, rye, or pumpernickel bread. • Switch to whole-wheat pasta and brown rice. • Fill your cereal bowl with plain oatmeal and add a few raisins for sweetness. • Check the ingredient list for the words "whole" or "whole grain" before the grain ingredient's name.
Increase your fruits and veggies
• Don't forget beans. Kidney beans, chickpeas, black beans, and others have more fiber and resistant starch than any other vegetables. • Add berries, bananas, or dried fruit to cereal and desserts. • Pile veggies, such as lettuce, tomatoes, cucumbers, avocados, and peppers, on your sandwich. • Have more than one vegetable at dinner, and try a new fruit.
Limit added sugars
• Switch to a 12-oz can instead of a 20-oz bottle when you grab a soft drink or, better yet, have a glass of water or low-fat milk. • Use one-quarter less sugar in your recipe next time you bake. • Snack on a piece of fruit instead of a candy bar. • Check the ingredient list on breakfast cereals to find ones that have no added sugar.

example, choosing a stir-fry meal of a few ounces of beef and plenty of vegetables on brown rice can provide the same calories but more fiber and less fat than a dinner of steak, white rice, and a small salad with dressing. To limit added sugars, foods high in added sugar should be replaced with natural sources of sugar such as fruits and unsweetened dairy products. If instead of a 20-oz bottle of soda you have an 8-ounce glass of low-fat milk, you will consume 140 fewer kcal and no added sugar, as well as getting plenty of high-quality protein, calcium, and other micronutrients. Using fresh instead of canned fruit can also help increase fiber and decrease added refined sugars. For example, one-half cup of pear halves canned in heavy syrup provides 90 kcal, 1 g of fiber, and almost 20 g of sugar, most of which is added in the syrup. One large fresh pear provides 90 kcal, 4 g of fiber, and no refined sugar (see **Critical Thinking: Becoming Less Refined**).

Using Food Labels to Make Healthy Choices Food labels can help in choosing the right mix of carbohydrates. The Nutrition Facts panel helps consumers find foods that are good sources of fiber and low in total sugars (**Figure 4.28**). The ingredient list helps

Figure 4.28 Carbohydrates on food labels
The Nutrition Facts Panel lists the grams of total carbohydrate, fiber, and sugars in foods. This label from Raisin Bran tells you that it is high in fiber but also high in sugars. The table defines descriptors used to highlight the sugar and fiber content of foods.

Raisin Bran®

Nutrition Facts
Serving Size 1 Cup (59g)

Amount Per Serving	Cereal	with 1/2 cup skim milk
Calories	190	230
Calories from Fat	10	10
	% Daily Value**	
Total Fat 1g*	2%	2%
Saturated Fat 0g	0%	0%
Trans Fat 0g		
Polyunsaturated Fat 0g		
Monounsaturated Fat 0g		
Cholesterol 0mg	0%	0%
Sodium 210mg	9%	12%
Potassium 310mg	9%	15%
Total Carbohydrate 46g	15%	17%
Dietary Fiber 7g	28%	28%
Sugars 18g		
Protein 5g		
Vitamin A	10%	15%
Vitamin C	0%	0%
Calcium	2%	15%
Iron	25%	25%
Vitamin D	10%	25%
Thiamin	25%	30%
Riboflavin	25%	35%
Niacin	25%	25%
Vitamin B_6	25%	25%
Folic Acid	25%	25%
Vitamin B_{12}	25%	35%
Phosphorus	20%	30%
Magnesium	20%	25%
Zinc	10%	15%

* Amount in cereal. One-half cup of skim milk contributes an addtional 40 calories, 65mg sodium, 6g total carbohydrates (6g sugars), and 4g protein.
** Percent Daily Values are based on a 2,000 calorie diet. Your daily values may be higher or lower depending on your calorie needs:

		2,000	2,500
	Calories	2,000	2,500
Total Fat	Less than	65g	80g
Sat. Fat	Less than	20g	25g
Cholesterol	Less than	300mg	300mg
Sodium	Less than	2,400mg	2,400mg
Potassium		3,500mg	3,500mg
Total Carbohydrate		300g	375g
Dietary Fiber		25g	30g

Total carbohydrate content is listed in grams and as a percent of the Daily Value.

The total amount of fiber in the product is listed in grams and as a percent of the Daily Value. The amounts of soluble and insoluble fiber may be listed if manufacturers choose to include them but are not required.

The number of grams of sugars listed in the Nutrition Facts is the total grams of monosaccharides plus disaccharides in a serving. This number does not distinguish between added sugar and the sugar occurring naturally in the food. No Daily Value has been established for sugars.

The Daily Value for total carbohydrate is calculated as 60% of the energy. For a 2000-kcal diet, this represents 300 g of carbohydrate ([2000 kcal x 0.6]/4 kcal/g of carbohydrate = 300 g)

The Daily Value for fiber is 25 g in a 2000-kcal diet.

Sugar and Fiber Descriptors on Food Labels

Descriptor	Definition
Sugar-free	Product contains no amount, or a trivial amount, of sugars (less than 0.5 g per serving). Synonyms for "free" include "without," "no," and "zero."
Reduced sugar	Nutritionally altered product contains 25% less sugar than the regular or reference product.
Less sugar	Whether altered or not, a food contains 25% less sugar than the reference food. "Fewer" may be used as a synonym for "less."
No added sugars or without added sugars	No sugar or sugar-containing ingredient is added during processing.
High fiber	Food contains 20% or more of the Daily Value for fiber per serving. Synonyms for "high" include "rich in" and "excellent source of."
Good source of fiber	Food contains 10 to 19% of the Daily Value for fiber per serving. Synonyms for "good source of" include "contains" and "provides."
More fiber	Food contains 10% or more of the Daily Value for fiber per serving than an appropriate reference food. Synonyms for "more" include "added" (or "fortified" and "enriched"), "extra," or "plus."

CRITICAL THINKING

Becoming Less Refined

Emma thinks that a good diet is important. She is concerned that she eats too much added sugar and not enough fiber. She records her food intake for a day and calculates its nutrient content.

DIET ANALYSIS

Emma's typical diet provides 2340 kcal and 350 g of carbohydrate. The amounts of fiber and sugars in each food are shown in the table.

FOOD	SERVING	FIBER (g)	SUGARS (g)
Breakfast			
White bread	2 slices	1.2	2
with jelly	1 T	0.1	6
and margarine	1 tsp	0	0
Fruit punch	8 fl oz	0	22
Lunch			
Macaroni and cheese	1 cup	1.4	8
Milk	1 cup	0	12
Apple	1 medium	3.7	18
Snack			
Soda	20 oz	0	63
3 Musketeers bar	1 regular	1	34
Dinner			
Roast beef	3 oz	0	0
Flour tortillas	1, 8-inch size	1.4	1
Pinto beans	1 cup	10	1
Snack			
Ice cream	⅔ cup	0.8	17
with cherry syrup	2 T	0	16
Total		**19.6**	**200**

CRITICAL THINKING QUESTIONS

- Calculate the percent of calories from carbohydrate in Emma's diet. Does it meet the AMDR?
- Does her diet meet fiber recommendations?
- Find three foods in her diet that are low in fiber and high in sugar.
- Suggest an alternative for each of these that will provide less added refined sugar and/or more fiber.

iProfile Use iProfile to see how much fiber you would add to your diet if you made all your bread whole grain.

identify whole-grain products and the sources of added sugars (see **Off the Label: What Does the Ingredient List Tell You about Carbohydrates?**).

Many food labels indicate that the product is made from whole grains or include nutrient content claims such as "high fiber" or "excellent source of fiber," which can help you find high-fiber products (see Figure 4.28). If you are looking for low-carbohydrate products, be aware that the terms "low carbohydrate" and "low carb" have not been defined by the FDA, so at the moment the definition is up to the manufacturer. Some products may also advertise the number of "net carbs" they provide. Again, there is no official definition of net carbs,

© iStockphoto

OFF THE LABEL What Does the Ingredient List Tell You about Carbohydrates?

How can you tell whether your breakfast cereal is made with whole grains or whether the sugar in your canned peaches is added or natural? It turns out that much of the information you need to make healthy carbohydrate choices must be found in the ingredient list, rather than on the Nutrition Facts panel. The key to understanding the ingredient list is that ingredients are listed in order of their prominence by weight, so the closer to the front of the list an ingredient appears, the more of it there is in the food.

To see if your choice is a whole-grain product, look to see if the first ingredient is a whole grain. Whole wheat, whole oats, oatmeal, rolled oats, whole-grain corn, popcorn, brown rice, whole rye, whole-grain barley, wild rice, buckwheat, triticale, bulgur, cracked wheat, millet, quinoa, and sorghum are whole grains. Wheat flour is not. "Wheat" refers to the type of grain, not how refined it is. Products whose first ingredient is wheat flour, enriched flour, or degerminated cornmeal are not made predominantly from whole grains. Don't be fooled by healthy-sounding descriptors like "multi-grain" or "seven-grain." These simply mean the product contains more than one type of grain, not that these grains are necessarily whole grains. "Stone ground" refers to how the grain was processed, not whether the bran and germ are included; and terms like "bran" and "oat" may refer to things added to the product, not whether the product is made predominantly from a whole grain. Don't forget to look at the rest of the ingredient list, too. The presence of whole grain is not the only criterion for a healthy choice. For example, Lucky Charms cereal is a whole grain, but marshmallows are the second ingredient, making it high in sugar as well.

Checking the ingredient list is key to identifying foods high in added sugar because the single value listed for "sugars" on the Nutrition Facts panel doesn't differentiate between added and natural sugar. The sweeteners in the ingredient list are those added as ingredients, not the sugar that is naturally present in the food. Foods that are high in added sugar will have a sweetener listed early in the ingredient list. If the first ingredient is sugar, then that product contains more sugar by weight than anything else. Remember, though, that the sum of the added sweeteners hiding farther down in the ingredient list may be considerable. Recognizing added sugars can be a challenge. The only sweetener that can be called "sugar" on the ingredient list is sucrose; sucrose is found in brown, powdered, granulated, and raw sugar, and each form may be listed separately. There also are many other sugars added to foods. Dry sugars that are added include invert sugar, dextrose, glucose, maltose, lactose, and fructose. Corn syrup, honey, molasses, malt syrup, sugar syrup, fruit juice concentrates, and high-fructose corn syrup are added as syrups. Increasing your sugar vocabulary will help you keep track of the high-sugar foods in your diet.

Nutrition Facts
Serving Size 1 Container

Amount Per Serving	
Calories 190	**Calories from Fat** 30
Amount/Serving	% **DV***
Total Fat 3.5g	**5%**
Saturated Fat 2g	**10%**
Trans Fat 0g	
Cholesterol 15mg	**4%**
Sodium 100mg	**4%**
Potassium 310g	**9%**
Total Carbohydrate 32g	**11%**
Dietary Fiber 0g	**0%**
Sugars 28g	
Protein 7g	**14%**
Vitamin A 15% •	Calcium 30%

*Percent Daily Values (DV) are based on a 2,000 calorie diet.

INGREDIENTS: CULTURED PASTEURIZED GRADE A REDUCED FAT MILK, SUGAR, NONFAT MILK, HIGH FRUCTOSE CORN SYRUP, STRAWBERRY PUREE, MODIFIED CORN STARCH, KOSHER GELATIN, TRICALCIUM PHOSPHATE, NATURAL FLAVOR, COLORED WITH CARMINE, VITAMIN A ACETATE, VITAMIN D_3

The yogurt label shown here lists sugar as the second ingredient and high-fructose corn syrup as the fourth. Together these two contribute much of the sugar in this food.

The Nutrition Facts on this strawberry yogurt indicate that is contains 28 grams of sugars (monosaccharides and disaccharides). This amount includes both the sugar found naturally in the strawberries and milk (about 12 grams), and plus the sugar added for more sweetness (about 16 grams).

but usually this number is calculated by subtracting the grams of fiber, and sometimes other poorly absorbed carbohydrates, from the grams of total carbohydrate in the food.

Health claims can also help identify foods that are good sources of whole grains and fiber. Foods that are low in fat and contain at least 51% of their weight as whole grains can include the health claim that states, "Diets rich in whole-grain foods and other plant foods and low in total fat, saturated fat, and cholesterol may help reduce the risk of heart disease and certain cancers."[32] Fiber-containing grain products, fruits, and vegetables that contain at least 2.5 g of fiber per serving and are low in fat may claim to reduce the risk of cancer. Fruits, vegetables, and grain products that are low in total fat, saturated fat, and cholesterol and that contain at least 0.6 g of soluble fiber per serving, and foods that contain at least 0.75 g of soluble fiber per serving from whole oats or psyllium husks, may claim to reduce the risk of heart disease (see Appendix E).

The Role of Alternative Sweeteners

America's love of sweets and the bad press surrounding sugar have driven the technological development of an increasing number of alternative or artificial sweeteners. These sugar substitutes, which provide little or no energy, are added to a host of low-calorie and "light" foods such as yogurts, ice creams, and soft drinks. Although many sugar substitutes are technically not carbohydrates, they were developed to replace simple sugars in food products or as an alternative for table sugar at home.

When alternative sweeteners are used to replace added sugars in the diet, they can help reduce the incidence of dental caries and manage blood sugar levels. Their usefulness for weight loss is more controversial. The average American eats about 32 teaspoons of added sweeteners per day.[33] The rising consumption of sugared beverages has been blamed for the increasing prevalence of obesity in America.

If individuals trying to lose weight replace sugar and high-sugar foods, such as soft drinks, with artificially sweetened products, they will lower their calorie intake. Whether use of these products promotes weight loss, however, depends on whether the calories they spare are added back from other food sources. When the effect of nonnutritive sweeteners on body weight was examined, their use was actually associated with weight gain, not weight loss. One hypothesis to explain this is that the sweet taste increases appetite, causing consumers actually to increase their food intake.[34]

Even if replacing foods high in added sugars with foods sweetened with sugar substitutes will cut down on calories and decrease sugar intake, switching to alternative sweeteners will not necessarily make your diet healthier. Using alternative sweeteners will not increase the intake of whole grains or fresh fruits and vegetables—key components of a healthy diet. Foods that are high in added sugar tend to be nutrient poor. Replacing them with artificially sweetened alternatives does not necessarily increase the nutrient density of the diet or improve overall diet quality. However, these products can be part of a healthy diet when used in moderation as part of a diet that is based on whole grains, vegetables, and fruits.

Types of Nonnutritive Sweeteners Alternative sweeteners that provide no calories are often referred to as nonnutritive sweeteners. The main competitors in the nonnutritive sweetener market in the United States today are saccharin, aspartame, sucralose, acesulfame K (acesulfame potassium), and rebiana (**Table 4.5**). These are used alone or in combination to sweeten a variety of foods. Cyclamate, an alternative sweetener that was popular in the 1960s, was banned by the FDA in 1969 so is no longer available in the United States. It is still sold in Canada and some 50 other countries.

Alternative sweeteners consumed in reasonable amounts are generally safe for healthy people.[35] Aspartame can be a special concern for individuals who have a genetic disorder called phenylketonuria (PKU). These individuals must restrict their intake of phenylalanine, one of the amino acids that make up aspartame, to prevent brain damage (see Chapter 6). The FDA has defined acceptable daily intakes (ADIs)—levels that should not be exceeded when using these products (see Table 4.5). The ADI is an estimate of the amount of the sweetener per kilogram of body weight that an individual can safely consume every day over a lifetime with minimal risk.

TABLE 4.5 Nonnutritive Sweeteners

Sweetener	Brand names	What is it?	ADI
Saccharin	Sweet'N Low, SugarTwin Andy Washnik	The oldest of the nonnutritive sweeteners, developed in 1879. It was once considered a carcinogen but was taken off the government's list of cancer-causing substances in 2000. It is 300 times sweeter than sucrose and has a bitter aftertaste.	5 mg/kg of body weight/day. One packet contains 36 mg of saccharin. A 154-lb (70-kg) person would exceed the ADI by consuming 10 packets or about three 12-oz saccharin-sweetened beverages.
Aspartame	Equal, NutraSweet Andy Washnik	Made of two amino acids (phenylalanine and aspartic acid). Because it breaks down when heated, it is typically used in cold products or added after cooking. It is 200 times sweeter than sucrose.	50 mg/kg of body weight/day. One packet contains 37 mg of aspartame. To exceed the ADI, a 154-lb (70-kg) person would have to consume 95 packets or sixteen 12-oz aspartame-sweetened beverages. It must be limited in the diets of people with phenylketonuria (see Chapter 6).
Acesulfame K	Sunett, Sweet One © Sara Wight	A heat-stable sweetener that is often used in combination with other sweeteners. It is 200 times sweeter than sucrose.	15 mg/kg of body weight/day. A 154-lb (70-kg) person could consume 2 gallons of beverages containing acesulfame K without exceeding the ADI.
Neotame	No brand name. Neotame is not sold as a table sweetener.	Made from the same two amino acids as aspartame, but because the bond between them is harder to break than the bond in aspartame, it is heat stable and can be used in baking. It is used in soft drinks, dairy products, and gum. It is 8000 times sweeter than sucrose.	18 mg/kg of body weight/day.
Sucralose	Splenda Andy Washnik	Made from sucrose molecules that have been modified so that they cannot be digested or absorbed. It is heat stable, so it can be used in cooking. It is 600 times sweeter than sucrose.	5 mg/kg of body weight/day. One packet contains about 12 mg of sucralose. A 154-lb (70-kg) person could consume 29 packets without exceeding the ADI.
Rebiana	Truvia, Pure Via © Sara Wight	A natural sweetener made from the leaf of the stevia plant.[36] It is the newest sweetener on the market and is about 300 times sweeter than sucrose.	12 mg/kg of body weight/day. To exceed the ADI, a 154-lb (70-kg) person would have to consume more than 30 packets of a rebiana sweetener or drink about six 12-oz cans of a rebiana-sweetened soda.

Sugar Alcohols **Sugar alcohols** are another type of alternative sweetener. Also called polyols, sugar alcohols such as sorbitol, mannitol, lactitol, and xylitol are chemical derivatives of sugar that are not digested, absorbed, or metabolized to the same extent as monosaccharides and disaccharides and thus generally provide less energy than sucrose. Maltitol provides 3 kcal per g, lactitol 2 kcal per g, and erythritol only 0.2 kcal per g. Because they are not monosaccharides or disaccharides, sugar alcohols can be used in products labeled "sugar free" or "no sugar added." If a product uses these descriptors, the grams of sugar alcohols in a serving must be listed in the Nutrition Facts portion of the food label under carbohydrates.

sugar alcohols Sweeteners that are structurally related to sugars but provide less energy than monosaccharides and disaccharides because they are not well absorbed.

Products sweetened with sugar alcohols, such as chewing gums, candies, ice creams, and baked goods, may carry the health claim statement that they do not promote tooth decay. These products are less likely to promote tooth decay because the bacteria in the mouth cannot metabolize sugar alcohols as rapidly as sucrose. Consumption of large amounts of sugar alcohols (more than 50 g of sorbitol or 20 g of mannitol per day) can cause diarrhea.

OUTCOME

Cole Group/Photodisc/Getty Images, Inc.

When Shamara abandoned her low-carb diet in favor of the dietary pattern recommended by MyPlate, she was pleased to be able to eat many of her favorite foods again, but she still struggled to limit her added sugars. She now understands that the meats, vegetables, and fresh fruits she eats at dinner don't contain added sugars. The added sugar in her diet comes from the packaged cereals, snacks, and baked goods she eats for breakfast and between meals.

She has learned to identify added sugar on food labels and found a few snack choices that are high in whole grains and not in added sugar. Now she grabs one of these instead of a muffin or pastry when she stops at the coffee shop. To keep her sugar intake down she has learned to avoid the desserts in the cafeteria and to drink water or milk rather than sweetened beverages. She has stocked her room with healthy snacks, such as fruit, baby carrots, and whole-grain crackers. She is surprised that by cutting out the sugar and watching her portion sizes she has been able to slowly lose some weight without the drastic changes her low-carb diet had required. **Shamara's new diet is flavorful and varied, so she can more easily stick with it in the long term.**

© Elena Elisseeva/Alamy

APPLICATIONS

ASSESSING YOUR DIET

1. **How much carbohydrate is in your diet?**
   a. Use iProfile and the three-day diet record you kept in Chapter 2 to calculate your average carbohydrate and energy intake.
   b. What is the percent of energy from carbohydrate in your diet?
   c. How does your percent of energy from carbohydrate compare with the recommended 45 to 65%? If your intake is above or below this range, suggest some changes that would move your carbohydrate intake into the recommended range.

2. **Look at the sources of carbohydrate in your diet.**
   a. Indicate which sources are refined and which are unrefined.
   b. Suggest an unrefined choice to replace each refined choice.
   c. Note which foods in your diet contain added refined sugars.
   c. Suggest an alternative that is low in added sugar for each food that is high in added sugar.

3. **How much fiber is in your diet?**
   a. Use iProfile to calculate the grams of fiber in your original diet.

b. Does your intake meet the AI for fiber for someone of your age and gender?

c. How do the changes you suggested in question 2b affect the fiber content of your diet? If your modified diet still does not meet fiber recommendations, what foods could you add to further increase your intake?

CONSUMER ISSUES

4. Cheryl is trying to increase her fiber intake. For lunch she typically orders a sandwich from a local deli. Her favorite is turkey on a Kaiser roll with lettuce and mayo. Below, column A lists the types of bread available, column B lists the fillings, and column C lists the vegetables that can be added to the sandwiches. Can you suggest three different sandwiches for Cheryl that would provide more fiber than her typical sandwich?

Column A	Column B	Column C
Kaiser roll	Sliced turkey	Lettuce
Sourdough bread	Soyburger	Tomatoes
Whole-wheat bread	Ham	Green and red peppers
Rye bread	Hummus	Onions
Sesame bagel	Falafel	Cucumber
Pumpernickel bread	Tuna salad	Olives
Pita bread	Peanut butter	Pickles
Oat bread	Grilled eggplant	Alfalfa sprouts

5. Breakfast cereal manufacturers are making more of their products with whole grains, but this does not mean they are low in added sugar. Select three breakfast cereal labels to answer the following questions.

a. Which ingredients in the ingredient list are forms of added sugar?

b. Where do they appear in the list (i.e., second, fifth, etc.)? If they achieved the same sweetness by adding only one type of sweetener, where might it appear in the list?

c. How much total sugar is in a serving of the cereal? What percent of the calories does this represent?

CLINICAL CONCERNS

6. Bob weighs about 30 pounds more than he wants to weigh, so he decides to try to shed pounds quickly with a low-carbohydrate weight-loss diet. The diet allows an unlimited amount of beef, chicken, and fish as well as limited fruits and vegetables; breads, grains, and cereals are not allowed. Bob is overjoyed with his initial rapid weight loss, but after about a week his weight loss slows down and he begins to feel tired and light-headed. He is having headaches and notices a funny smell on his breath. A nutritional assessment suggests that Bob needs about 2500 kcal a day to maintain his weight. His weight-loss diet provides about 1000 kcal, 25 g of carbohydrate, 125 g of protein, and 44 g of fat per day. He consumes about 3 cups of fluid daily.

a. Explain why Bob is tired, is light-headed, and has headaches and an unusual odor on his breath.

b. What dietary changes could you suggest to reduce these symptoms?

7. When Jordan was 12 years old he began losing weight, despite the fact that he seemed to be hungry all the time. He was also thirsty and usually had to get up at least twice during the night to urinate. He was soon diagnosed with type 1 diabetes.

a. Why was Jordan losing weight?

b. What caused his thirst and excessive urination?

c. Can Jordan be treated with just diet and exercise? Why or why not?

d. Plan a meal for Jordan that a 12-year-old would enjoy. Use the iProfile diet analysis computer program to include about 60 g of carbohydrate in the meal.

SUMMARY

4.1 Carbohydrates in Our Food

- Some carbohydrates in our diets are from unrefined whole foods such as whole grains, fruits, and vegetables. Others are from refined grain products like white breads and baked goods and added sugars such as those found in candies and soft drinks.
- Unrefined sources of carbohydrate such as whole grains, legumes, vegetables, fruit, and milk contain a variety of nutrients and phytochemicals. Much of the fiber and vitamins and minerals is lost when the food is refined.
- Added refined sugars provide energy but few nutrients, so they reduce the nutrient density of the diet.

4.2 Types of Carbohydrates

- Carbohydrates are chemical compounds that contain carbon, hydrogen, and oxygen. Simple carbohydrates, also called sugars, include monosaccharides and disaccharides. They are found in foods such as table sugar, honey, milk, and fruit.
- Complex carbohydrates include oligosaccharides and polysaccharides. Glycogen is a polysaccharide found in animals, and starch and fiber are polysaccharides found in plants. Neither soluble nor insoluble fiber can be digested by human enzymes, but soluble fiber, which forms viscous solutions in water, can be broken down by bacteria in the large intestine.

4.3 Carbohydrates in the Digestive Tract

- Sugars and starches consumed in food are broken down in the digestive tract to monosaccharides, which can be absorbed into the bloodstream.
- Lactose intolerance occurs when the enzyme lactase is not available in sufficient quantities to digest lactose. Undigested lactose passes into the colon where it draws in water and is metabolized by bacteria, producing gas and acids and causing abdominal distension, flatulence, cramping, and diarrhea. Lactose intolerance is more common in some populations than others.
- Fiber, some oligosaccharides, and resistant starch are carbohydrates that are not broken down by human digestive enzymes in the stomach and small intestine and therefore pass into the colon. These indigestible carbohydrates benefit health by increasing the amount of water and bulk in the intestine, which stimulates gastrointestinal motility; promoting the growth of a healthy microflora; and slowing nutrient absorption.

4.4 Carbohydrates in the Body

- In the body, glucose provides a source of energy—4 kcal per gram. Carbohydrates are also needed for nerve function, for cell communication, and to synthesize DNA and RNA and molecules that provide cushioning and lubrication.
- Blood glucose rises after eating. How quickly and how high blood glucose rises is referred to as glycemic response. Glycemic response can be quantified using glycemic index or glycemic load.
- To ensure a steady supply of glucose to body cells, blood glucose levels are maintained within normal limits by the liver and the hormones insulin and glucagon. When blood glucose rises, insulin is released from the pancreas, allowing muscle and adipose tissue cells to take up the glucose. When blood glucose falls, glucagon is released to increase blood glucose.
- Glucose is metabolized through cellular respiration, which begins with glycolysis or anaerobic metabolism. Glycolysis breaks each six-carbon glucose molecule into 2 three-carbon pyruvate molecules, producing ATP even when oxygen in unavailable. When oxygen is available, aerobic metabolism can proceed. Pyruvate loses a carbon as carbon dioxide to form acetyl-CoA, which is then broken down by the citric acid cycle to form two carbon dioxide molecules. Electrons released at each step pass to the electron transport chain, where their energy is used to generate ATP and water is formed.
- When carbohydrate is limited, metabolism must shift to make sure that glucose is available to the brain and other cells that require glucose as an energy source. Glucose can be obtained from the breakdown of glycogen or from gluconeogenesis, which synthesizes glucose from three-carbon molecules most often derived from amino acid breakdown. When carbohydrate is limited, acetyl-CoA cannot to enter the citric acid cycle and instead is used to make ketones.

4.5 Carbohydrates and Health

- Diabetes is an abnormality in blood sugar regulation resulting in high blood glucose levels, which damage tissues and increase the risk of heart disease, kidney failure, blindness, and amputations. Diabetes occurs either because insufficient insulin is produced or because there is a decrease in the sensitivity of body cells to insulin. Treatment to maintain glucose in the normal range includes diet, exercise, and medication.
- Hypoglycemia is a condition in which blood glucose falls to abnormally low levels, causing symptoms such as sweating, headaches, and rapid heartbeat. It is a common complication of diabetes treatment.
- Bacteria in the mouth can form colonies that stick to the teeth. When carbohydrates—particularly simple carbohydrates—are consumed, they are metabolized by the bacteria, producing acids that dissolve the tooth enamel, forming dental caries.
- Carbohydrates contribute to weight gain when total energy intake exceeds needs. Low-carbohydrate diets cause less insulin release, which is believed to favor fat loss. They also cause ketosis, which suppresses appetite and therefore food intake, leading to weight loss. Unrefined carbohydrates can support weight management by reducing glycemic response and causing a feeling of fullness.
- High-sugar diets can increase heart disease risk by raising blood lipids. Unrefined carbohydrate sources help lower blood lipids and provide other dietary components that help protect against heart disease.
- Indigestible carbohydrates make the stool larger and softer. This reduces the pressure needed to move material through the colon, lowering the risk of constipation, hemorrhoids, and the symptoms of diverticular disease.
- Cancer cells differ from normal cells because they divide without restraint and are able to grow in areas reserved for other cells. Cells in the colon may be exposed to carcinogens in the colon contents. Fiber may help reduce the risk of colon cancer by decreasing the amount of contact between the cells lining the colon and these carcinogenic substances.

4.6 Meeting Recommendations for Carbohydrate Intake

- Guidelines for healthy diets recommend 45 to 65% of energy from carbohydrates and a fiber intake of 38 g/day for men and 25 g/day for women. Most dietary carbohydrate should come from unrefined food sources, with limited amounts from added sugars.
- Sources of carbohydrate can be found in all the MyPlate food groups. Unrefined carbohydrates such as whole grains, legumes, fresh fruits and vegetables, and low-fat milk should be selected, and foods high in added sugars, such as baked goods, candy, and soft drinks should be limited. Food labels can help identify foods that are high in fiber, low in sugar, and sources of whole grains.
- Alternative sweeteners can be used to reduce the amount of added sugar in the diet. They do not contribute to tooth decay and can help keep blood sugar in the normal range. They reduce the energy content of the diet if the calories they eliminate are not added back in other foods.

REVIEW QUESTIONS

1. What foods are good sources of unrefined complex carbohydrates? Unrefined simple carbohydrates?
2. What is the basic unit of carbohydrate?
3. List three common simple carbohydrates. Where are they found in the diet? Where are they found in the body?
4. Describe three types of complex carbohydrates.
5. Why is added sugar considered a source of empty calories?
6. How much energy is provided by a gram of carbohydrate?
7. Explain how fiber affects gastrointestinal health.
8. What is the main function of glucose in the body?
9. Explain why drinking a sugar-sweetened beverage causes a person's blood glucose to rise faster than it does after eating a bean burrito.
10. Compare the roles of insulin and glucagon in regulating blood glucose.
11. Describe what happens during the process of glycolysis.
12. What are the end products of cellular respiration? During which step is each produced?
13. Explain why carbohydrate is said to spare protein.
14. What is diabetes, and what are the long-term complications of this disease?
15. Why is ketosis a problem only in type 1 diabetes?
16. Why does a diet high in sucrose promote tooth decay?
17. What health benefits are associated with a diet high in unrefined carbohydrates?
18. How can you use the information on food labels to help you identify foods that are high in added sugars? In fiber?
19. What are the advantages and disadvantages of alternative sweeteners?

REFERENCES

1. Gross, L. S., Li, L., Ford, E. S., and Liu, S. Increased consumption of refined carbohydrates and the epidemic of type 2 diabetes in the United States: An ecologic assessment. *Am J Clin Nutr* 79:774–779, 2004.
2. Jensen, H. H., and Beghin, J. C. U.S. sweetener consumption trends and dietary guidelines. *Iowa Ag Rev* 11:10–11, 2005.
3. Institute of Medicine. Food and Nutrition Board. *Dietary Reference Intakes for Energy, Carbohydrates, Fiber, Fat, Protein and Amino Acids.* Washington, DC: National Academies Press, 2002, 2005.
4. Newburg, D. S., Ruiz-Palacios, G. M., and Morrow, A. L. Human milk glycans protect infants against enteric pathogens. *Annu Rev Nutr* 25:37–58, 2005.
5. National Digestive Diseases Information Clearinghouse (NDDIC), National Institute of Diabetes and Digestive and Kidney Diseases (NIDDK), National Institutes of Health of the U.S. Department of Health and Human Services. Lactose intolerance.
6. U.S. Department of Agriculture and U.S. Department of Health and Human Services. *Dietary Guidelines for Americans, 2010,* 7th ed. Washington, DC: U.S. Government Printing Office, December 2010.
7. U.S. Department of Agriculture. ARS data tables: Results from CSFII, 1996 ARS Food Surveys Research Group.
8. Burkitt, D. P., Walker, A. R. P., and Painter, N. S. Dietary fiber and disease. *JAMA* 229:1068–1074, 1974.
9. Saulnier, D. M., Kolida, S., and Gibson, G. R. Microbiology of the human intestinal tract and approaches for its dietary modulation. *Curr Pharm Des* 15:1403–1414, 2009.
10. Rose, D. J., DeMeo, M. T., Keshavarzian, A., and Hamaker, B. R. Influence of dietary fiber on inflammatory bowel disease and colon cancer: Importance of fermentation pattern. *Nutr Rev* 65: 51–62, 2007.
11. Walker, C., and Reamy, B. V. Diets for cardiovascular disease prevention: What is the evidence? *Am Fam Physician* 79: 571–578, 2009.
12. National Diabetes Information Clearinghouse. National diabetes statistics, 2011.
13. Barclay, A. W., Petocz, P., McMillan-Price, J., et al. Glycemic index, glycemic load, and chronic disease risk—a meta-analysis of observational studies. *Am J Clin Nutr* 87:627–637, 2008.
14. National Diabetes Education Program. Overview of diabetes in children and adolescents.
15. National Diabetes Education Program.
16. American Diabetes Association. Standards of medical care in diabetes—2009. *Diabetes Care* 32:S13–S60, 2009.
17. Hawley, J. A., and Lessard, S. J. Exercise training-induced improvements in insulin action. *Acta Physiol* (Oxf) 192:127–135, 2008.
18. U.S. Department of Health and Human Services, U.S. Public Health Service. *Oral Health in America: A Report of the Surgeon General.* Rockville, MD: National Institutes of Health, 2000.
19. Paes Leme, A. F., Koo, H., Bellato, C. M., et al. The role of sucrose in cariogenic dental biofilm formation—new insight. *J Dent Res* 85:878–887, 2006.
20. Shai, I., Schwarzfuchs, D., Henkin, Y., et al. Weight loss with a low-carbohydrate, Mediterranean, or low-fat diet. *N Engl J Med* 359:229–241, 2008.
21. Sacks, F. M., Bray, G. A., Carey, V. J., et al. Comparison of weight-loss diets with different compositions of fat, protein, and carbohydrates. *N Engl J Med* 360:859–873, 2009.

22. Buchholz, A. C., and Schoeller, D. A. Is a calorie a calorie? *Am J Clin Nutr* 79:899S–906S, 2004.

23. Slavin, J. L. Position of the American Dietetic Association: Health implications of dietary fiber. *J Am Diet Assoc* 108: 1716–1731, 2008. Erratum in *J Am Diet Assoc* 109:350, 2009.

24. Chong, M. F., Fielding, B. A., and Frayn, K. N. Metabolic interaction of dietary sugars and plasma lipids with a focus on mechanisms and de novo lipogenesis. *Proc Nutr Soc* 66:52–59, 2007.

25. Anderson, J. W., Baird, P., Davis, R. H., Jr., et al. Health benefits of dietary fiber. *Nutr Rev* 67:188–205, 2009.

26. Anderson, J. W. Whole grains protect against atherosclerotic cardiovascular disease. *Proc Nutr Soc* 62:135–142, 2003.

27. Institute of Medicine Food and Nutrition Board. *Dietary Reference Intakes: Proposed Definition of Dietary Fiber.* Washington, DC: National Academies Press, 2001.

28. Strate, L. L. Lifestyle factors and the course of diverticular disease. *Dig Dis* 30:35–45, 2012.

29. National Digestive Disease Information Clearinghouse. Diverticulosis and diverticulitis.

30. Aune, D., Chan, D. S., Lau, R., et al. Dietary fibre, whole grains, and risk of colorectal cancer: Systematic review and dose-response meta-analysis of prospective studies. *BMJ* 343:d6617, 2011.

31. World Health Organization. Diet and chronic disease prevention report of the WHO FAO Expert Commission Geneva 2002.

32. U.S. Department of Health and Human Services. Food and Drug Administration qualified health claims subject to enforcement discretion, September 2003; updated August 2005, November 2005, and April 2007, Washington, DC, 2005.

33. U.S. Department of Agriculture. Profiling food consumption in America. *The Agriculture Fact Book 2001–2002.*

34. Yang, Q. Gain weight by "going diet"? Artificial sweeteners and the neurobiology of sugar cravings: Neuroscience 2010. *Yale J Biol Med* 83:101–108, 2010.

35. American Dietetic Association. Position of the American Dietetic Association: Use of nutritive and non-nutritive sweeteners. *J Am Diet Assoc* 104:255–275, 2004.

36. Stevia sweetener gets US FDA go-ahead. Food Navigator-USA.com, December 18, 2008.

***To access links to online sources, please go to www.wiley.com/college/smolin and select Nutrition: Science and Applications, 3rd edition. From this page, select either the student or instructor companion site. Once on the desired site, select References.**

CHAPTER OUTLINE

Studio R Schmitz/Age Fotostock America, Inc.

Lipids: Triglycerides, Phospholipids, and Cholesterol

5

CASE STUDY

Kyle is a 20-year-old college student who is worried about his risk of heart disease. Because his grandmother died of a heart attack at age 50, Kyle knows he is at an increased risk. He recently completed a heart disease risk assessment at a health fair and found that he was about 25 lb overweight, his percent body fat was higher than recommended, and his blood cholesterol was slightly elevated at 220 mg/100 mL. He was told that he should see a physician to evaluate his cardiovascular health. How could a 20-year-old be at risk for heart disease, and how can he lower that risk?

Kyle has begun to think about changes he could make in his diet and lifestyle. He eats a lot of red meat and has only one or two servings of fruits and vegetables a day. He drinks whole milk and snacks on ice cream every night. His only exercise is a Tuesday afternoon game of Frisbee with friends and lifting weights on Friday nights.

When he tells his friends and family about his health concerns, everyone is quick to offer advice. His girlfriend, who is a vegetarian, recommends that he eliminate meat from his diet. His biology lab partner tells him to cut out all fat. His sister tells him about the Mediterranean diet and recommends he eat pasta with plenty of olive oil every night. His roommate tells him to eat more fish. His mother says he should stop using margarine because of all the *trans* fat it contains. **All the advice Kyle has received has made him even more confused about how to change his diet.**

© Ben Blankenburg/iStockphoto

5.1 Fats in Our Food

LEARNING OBJECTIVES

- Name two qualities that fat adds to foods.
- Discuss the sources of fat in the American diet.
- Describe how dietary fat affects the risk of chronic disease.

lipids A group of organic molecules, most of which do not dissolve in water. They include fatty acids, triglycerides, phospholipids, and sterols.

Lipids, the chemical term for what we commonly call fats, make foods appealing by contributing to their texture, flavor, and aroma. Butterfat gives ice cream, cheesecake, and cream cheese a smooth texture and rich taste. Olive oil imparts a unique flavor to salads and many traditional Italian and Greek dishes. Sesame oil gives egg rolls and other Chinese foods their distinctive aroma. But while the fats in our foods contribute to their appeal, they also add calories; fat has 9 kcalories per gram, compared with 4 kcalories per gram for carbohydrate or protein. In addition, the types of fats we eat can affect our health; too much of the wrong type of fat can increase the risk of chronic disease.

Sources of Fat in Our Diet

The typical American diet contains about 34% of calories from fat.[1] Some of this fat is naturally present in foods such as meats, dairy products, nuts, seeds, and vegetable oils, but much of it is added to our food in processing or at the table (**Figure 5.1a**).

Some of the fats in our food are obvious—the stripes of fat in a slice of bacon sizzling in a frying pan, for example, or the layer of fat around the outside of a sirloin steak. Other visible sources of fat in our diets are the fats we add to foods at the table—the pat of butter melting on a slice of warm toast and the dressing poured over a salad.

Not all sources of dietary fat are obvious. Whole milk, cheese, avocados, and nuts are naturally abundant in fat, and foods such as croissants, doughnuts, cookies, and muffins, which we think of as sources of carbohydrate, may also be quite high in fat. We also add invisible fat when we fry foods (**Figure 5.1b**).

How Dietary Fat Affects Health

Fat has a bad reputation. Beginning in the 1950s, Americans were told that too much fat made them fat, increased their risk of heart disease, and maybe even increased their risk of cancer. In response to these messages, over the next few decades many Americans switched from whole milk to low-fat, chose chicken in place of beef, consumed fewer eggs, and used less butter and high-fat salad dressing. Nonetheless, the incidence of heart disease and cancer did not decrease and the number of obese people doubled.

Part of the reason for this is that when Americans reduced obvious fats in their diets, they consumed more hidden fats from foods such as pizza, pasta dishes, snack foods, and fried potatoes.[2] Our portion sizes also got bigger, we snacked more, and instead of preparing meals at home we began to eat out and bring home prepared foods that were higher in calories than what we would cook at home.[3,4] Changing the kinds of foods we chose without paying attention to calories promoted weight gain. Changing the sources of fat without paying attention to the type of fat did not protect us from chronic disease. Over the past 40 years the types of fat and the number of grams of fat Americans consume daily have changed little (**Figure 5.2**). Efforts to reduce chronic disease risk by cutting fat from our diets have failed not only because we haven't really cut our fat intake but also because fat does not deserve its bad reputation. Today, we understand that the types of fat in the diet as well as the overall dietary pattern are more important than total fat intake in terms of chronic disease risk. High intakes of fat from meats, dairy products, and processed fats used in shortening and margarine are associated with a higher incidence of heart disease and certain types of cancer. Diets high in fats from fish, nuts, and vegetable oils seem to protect against chronic disease. A healthy diet includes the right kinds of fats along with plenty of whole grains, fruits, and vegetables.

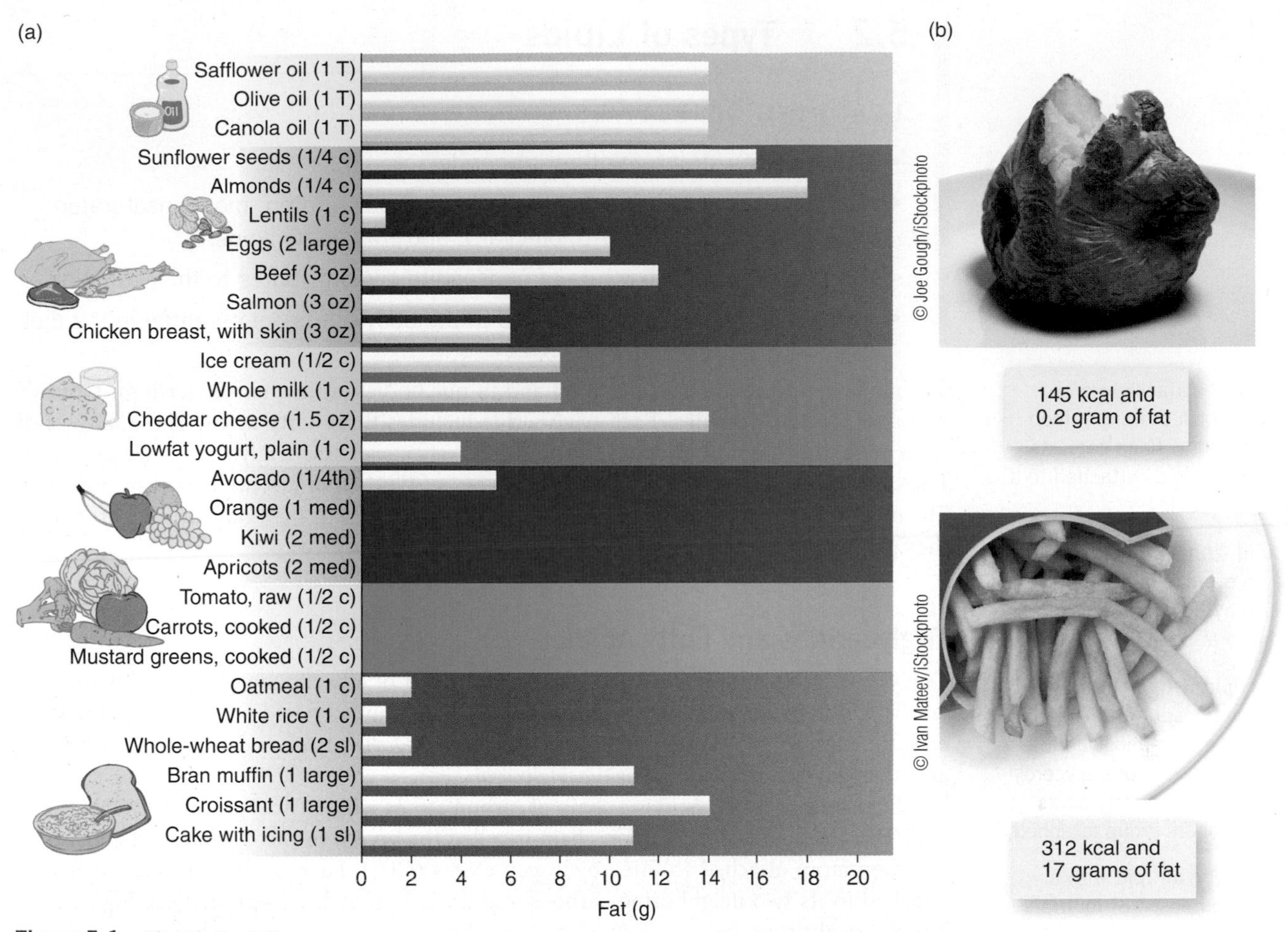

Figure 5.1 Where's the fat?

(a) Most of the fat that is naturally present in foods is found in meats, dairy products, nuts, seeds, and vegetable oils. Processed foods and baked goods are sources of less visible fats. In general fruits, vegetables, and grains are low in fat unless we add fat to them.

(b) French fries start as potatoes, which are low in fat, but when they are immersed in hot oil for frying, they soak up fat. You would have to add four pats of butter to your baked potato for it to have as much fat as a medium order of French fries.

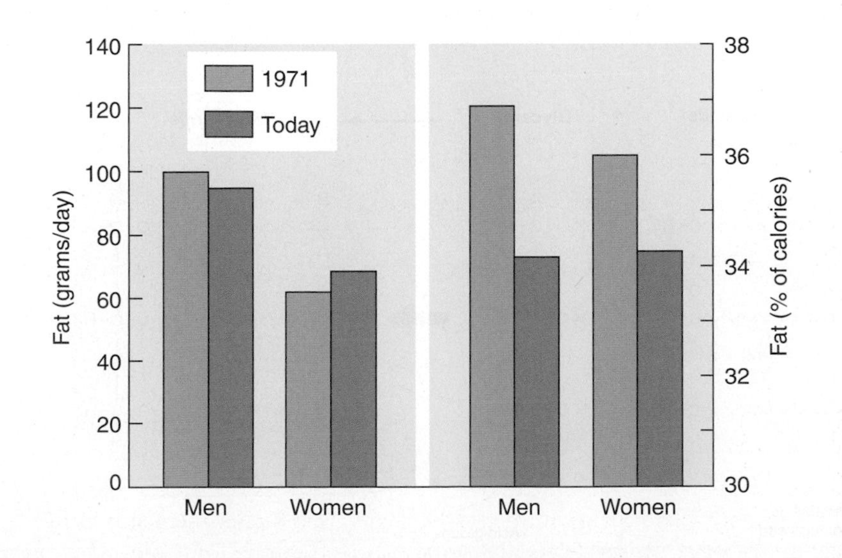

Figure 5.2 Changes in fat and energy intake
While the number of calories we eat has increased, the number of grams of fat we eat has remained constant. This caused the percentage of calories from fat to drop from about 37% in 1971 to about 34% today.[1,5,6]

5.2 Types of Lipids

LEARNING OBJECTIVES

- Describe the structure of a triglyceride.
- Compare the structures and food sources of saturated, monounsaturated, polyunsaturated, omega-6, omega-3, and *trans* fatty acids.
- Describe how the structure of phospholipids contributes to their function.
- Explain why cholesterol is needed in the body but is not essential in the diet.

triglyceride (triacylglycerol) The major form of lipid in food and in the body. Each consists of three fatty acids attached to a glycerol molecule.

fatty acid An organic molecule made up of a chain of carbons linked to hydrogen atoms with an acid group at one end.

phospholipid A type of lipid containing phosphorus. The most common is a phosphoglyceride, which is composed of a glycerol backbone with two fatty acids and a phosphate group attached.

sterol A type of lipid with a structure composed of multiple chemical rings.

The majority of the lipids in our food and in our bodies, and what the term *fat* typically refers to, are **triglycerides**. Each triglyceride includes three **fatty acids**. These fatty acids determine the physical properties and health effects of the triglycerides we consume. **Phospholipids** and **sterols** are two other types of lipids that are important in nutrition. The different structures of these lipids affect their function in the body and the properties they give to food.

Triglycerides and Fatty Acids

Triglycerides, also known as triacylglycerols, consist of a backbone of glycerol with three fatty acids attached, as shown in **Figure 5.3**. If only one fatty acid is attached to the glycerol, the molecule is called a *monoglyceride* or *monoacylglycerol*, and when two fatty acids are attached, it is a *diglyceride* or *diacylglycerol*.

Structurally, a fatty acid is a chain of carbon atoms with an acid group (COOH) at one end. The other end of the carbon chain is called the *omega* or *methyl end*, and consists of a carbon atom attached to three hydrogen atoms (CH_3). Each of the carbons between is attached to its two neighboring carbons and up to two hydrogen atoms (see Figure 5.3). The physical properties of a fatty acid depend on the length of the carbon chain and the type and location of the bonds between the carbon atoms.

Fatty Acid Chain Length Fatty acids vary in length from a few to 20 or more carbons. Most fatty acids in plants and animals, including humans, contain between 14 and 22 carbons. Short-chain fatty acids range from 4 to 7 carbons in length. They remain liquid at colder temperatures. For example, the short-chain fatty acids in whole milk remain liquid even in the refrigerator. Medium-chain fatty acids, such as those in coconut oil, range from

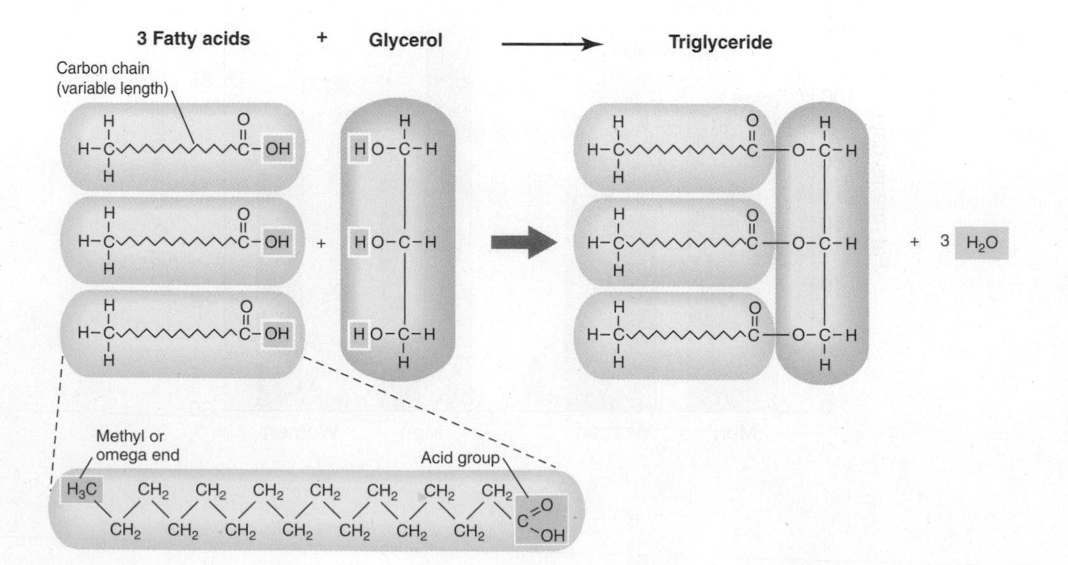

Figure 5.3 Formation and structure of triglycerides
A triglyceride is formed when three fatty acids bind to a molecule of glycerol. As each bond is formed, a hydrogen atom (H) from the glycerol and a hydroxyl group (OH) from the fatty acid combine to form a molecule of water. The three fatty acids that make up a triglyceride molecule may vary in structure and physical properties.

8 to 12 carbons. They solidify in the refrigerator but remain liquid at room temperature. Long-chain fatty acids (greater than 12 carbons), such as those in beef fat, usually remain solid at room temperature.

Saturated Fatty Acids Each carbon atom forms four bonds to link it to four other atoms. If a carbon is not bound to four other atoms, double bonds are formed. A fatty acid in which each carbon in the chain is bound to two hydrogen atoms is saturated with hydrogens and is therefore called a **saturated fatty acid** (**Figure 5.4**). Diets high in saturated fatty acids have been shown to increase the risk of heart disease. The most common saturated fatty acids are palmitic acid, which has 16 carbons, and stearic acid, which has 18 carbons. These are found most often in animal foods such as meat and dairy products. Plant sources of saturated fatty acids include palm oil, palm kernel oil, and coconut oil.

saturated fatty acid A fatty acid in which the carbon atoms are bound to as many hydrogen atoms as possible and that therefore contains no carbon-carbon double bonds.

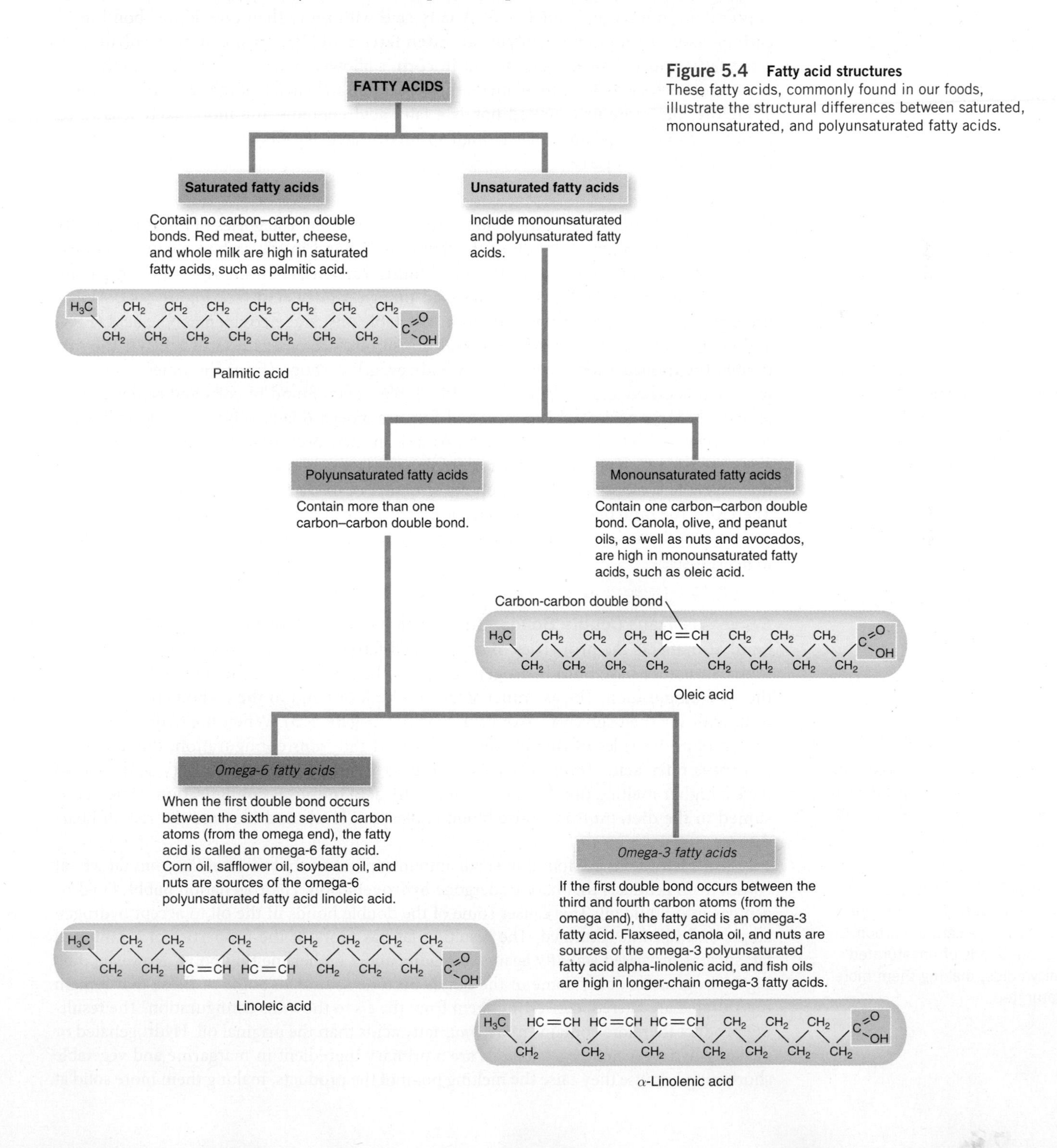

Figure 5.4 **Fatty acid structures**
These fatty acids, commonly found in our foods, illustrate the structural differences between saturated, monounsaturated, and polyunsaturated fatty acids.

tropical oils A term used in the popular press to refer to the saturated oils—coconut, palm, and palm kernel oil—that are derived from plants grown in tropical regions.

These are often called **tropical oils** because they are found in plants common in tropical climates. Tropical oils can be useful in processed foods such as breakfast cereals, crackers, salad dressings, and cookies because they are less susceptible to spoilage than are more unsaturated oils. Spoilage of fats and oils, referred to as rancidity, occurs when the unsaturated bonds in fatty acids are damaged by oxygen. When fats go rancid, they give food an "off" flavor.

monounsaturated fatty acid A fatty acid that contains one carbon-carbon double bond.

polyunsaturated fatty acid A fatty acid that contains two or more carbon-carbon double bonds.

Unsaturated Fatty Acids Unsaturated fatty acids contain one or more pairs of carbons that are not saturated with hydrogen atoms. The carbon pairs within the chain that are bound to only one hydrogen form carbon-carbon double bonds (see Figure 5.4). A fatty acid containing one double bond in its carbon chain is called a **monounsaturated fatty acid**. In our diets, the most common monounsaturated fatty acid is oleic acid, which is prevalent in olive and canola oils. A fatty acid with more than one double bond in its carbon chain is said to be a **polyunsaturated fatty acid**. The most common polyunsaturated fatty acid is linoleic acid, found in corn, safflower, and soybean oils. Unsaturated fatty acids melt at cooler temperatures than saturated fatty acids of the same chain length. Therefore, the more unsaturated bonds a fatty acid contains, the more likely it is to be liquid at room temperature. Diets high in unsaturated fatty acids are associated with a lower risk of heart disease.

omega-3 fatty acid A fatty acid containing a carbon-carbon double bond between the third and fourth carbons from the omega end.

omega-6 fatty acid A fatty acid containing a carbon-carbon double bond between the sixth and seventh carbons from the omega end.

Omega-3 and Omega-6 Fatty Acids There are different categories of unsaturated fatty acids depending on the location of the first double bond in the carbon chain. If the first double bond occurs between the third and fourth carbons, counting from the omega end (CH_3) of the chain, the fat is said to be an **omega-3 fatty acid** (see Figure 5.4). Alpha-linolenic acid (α-linolenic acid), found in vegetable oils, and eicosapentaenoic acid (EPA) and docosahexaenoic acid (DHA), found in fish oils, are omega-3 fatty acids. If the first double bond occurs between the sixth and seventh carbons (from the omega end), the fatty acid is called an **omega-6 fatty acid**. Linoleic acid, found in corn and safflower oils, and arachidonic acid, found in meat and fish, are omega-6 fatty acids. Linoleic acid is the major omega-6 fatty acid in the North American diet. Both omega-3 and omega-6 fatty acids are used to synthesize regulatory molecules in the body, and the biological effect of the molecule synthesized depends on the structure of the fatty acid from which it is made. Therefore, the ratio of omega-3 to omega-6 fatty acids in the diet is important in processes such as blood pressure regulation, blood clotting, and immune function that affect the risk of heart disease.

***trans* fatty acid** An unsaturated fatty acid in which the hydrogen atoms are on opposite sides of the double bond.

***Cis* versus *Trans* Double Bonds** The position of the hydrogen atoms around a double bond also affects the properties of unsaturated fatty acids. Most unsaturated fatty acids found in nature have both hydrogen atoms on the same side of the double bond, called the *cis* configuration. The asymmetry forces a kink or bend in the carbon chain, making it difficult for the triglycerides to pack together (**Figure 5.5**). When the hydrogen atoms are on opposite sides of the double bond, called the *trans* configuration, the fatty acid is a ***trans* fatty acid**. *Trans* fatty acids can pack tightly like saturated fatty acids and so have a higher melting point than the same fatty acid in the *cis* configuration. When consumed in the diet, *trans* fats raise blood cholesterol levels and increase the risk of heart disease.

hydrogenation The process whereby hydrogen atoms are added to the carbon-carbon double bonds of unsaturated fatty acids, making them more saturated.

Trans fatty acids are found in small amounts in nature, but most of the *trans* fat we eat comes from products that have undergone **hydrogenation**. Hydrogenation bubbles hydrogen gas into liquid oil. This causes some of the double bonds in the oil to accept hydrogen atoms and become saturated. The resulting fat has more of the properties of a saturated fat, such as increased stability against rancidity and a higher melting point. However, during hydrogenation, only some of the bonds become saturated. Some of those that remain unsaturated are altered, converting them from the *cis* to the *trans* configuration. The resulting product therefore contains more *trans* fatty acids than the original oil. Hydrogenated or partially hydrogenated vegetable oils are a primary ingredient in margarine and vegetable shortening because they raise the melting point of the products, making them more solid at

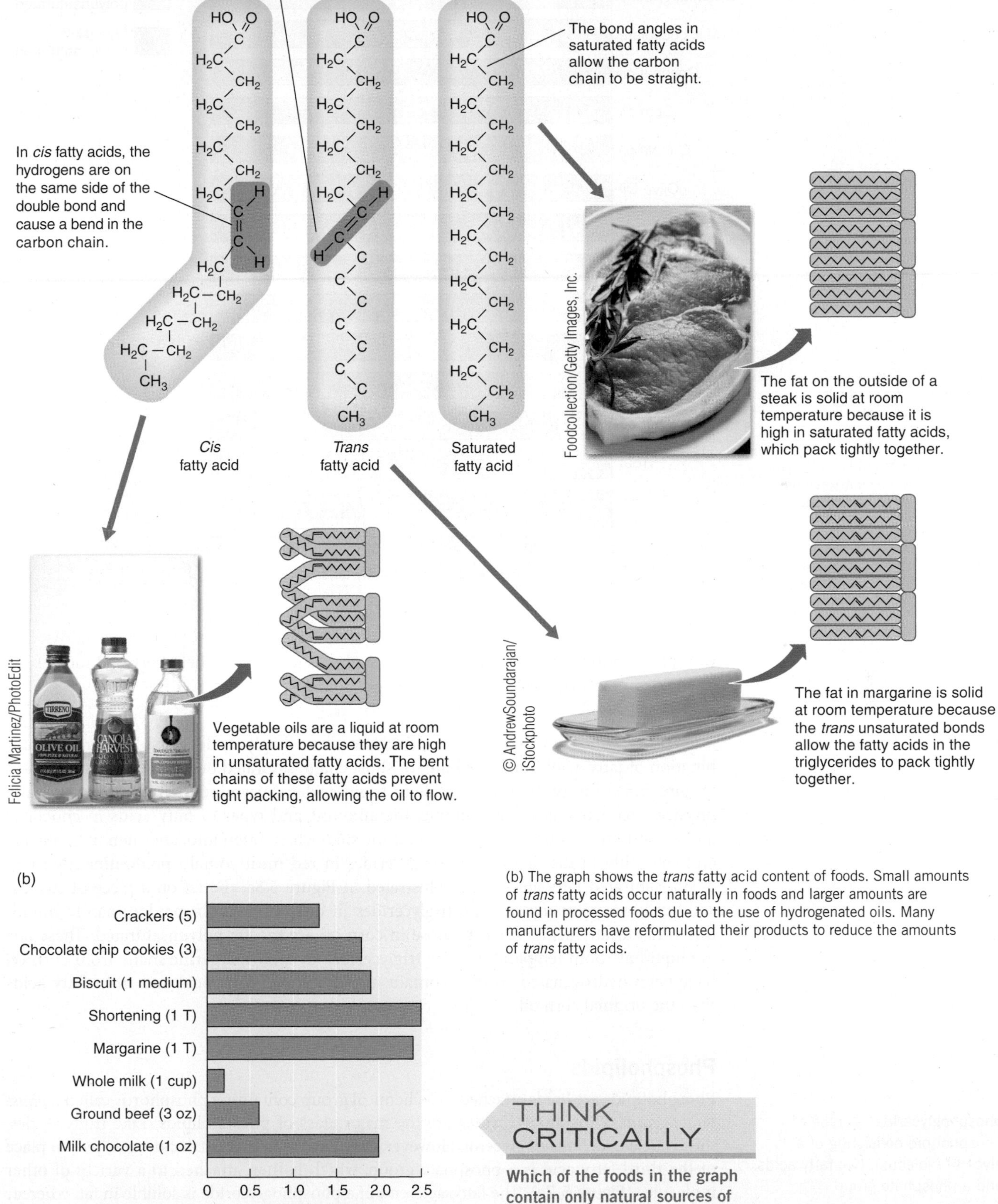

Figure 5.5 ***Cis* versus *trans* fatty acids**
(a) The orientation of the hydrogen atoms around the carbon-carbon double bond in fatty acids determines how "bent" or "straight" the carbon chain is and how tightly the triglycerides can pack together. This affects the melting point of the fat.

(b) The graph shows the *trans* fatty acid content of foods. Small amounts of *trans* fatty acids occur naturally in foods and larger amounts are found in processed foods due to the use of hydrogenated oils. Many manufacturers have reformulated their products to reduce the amounts of *trans* fatty acids.

THINK CRITICALLY

Which of the foods in the graph contain only natural sources of *trans* fatty acids?

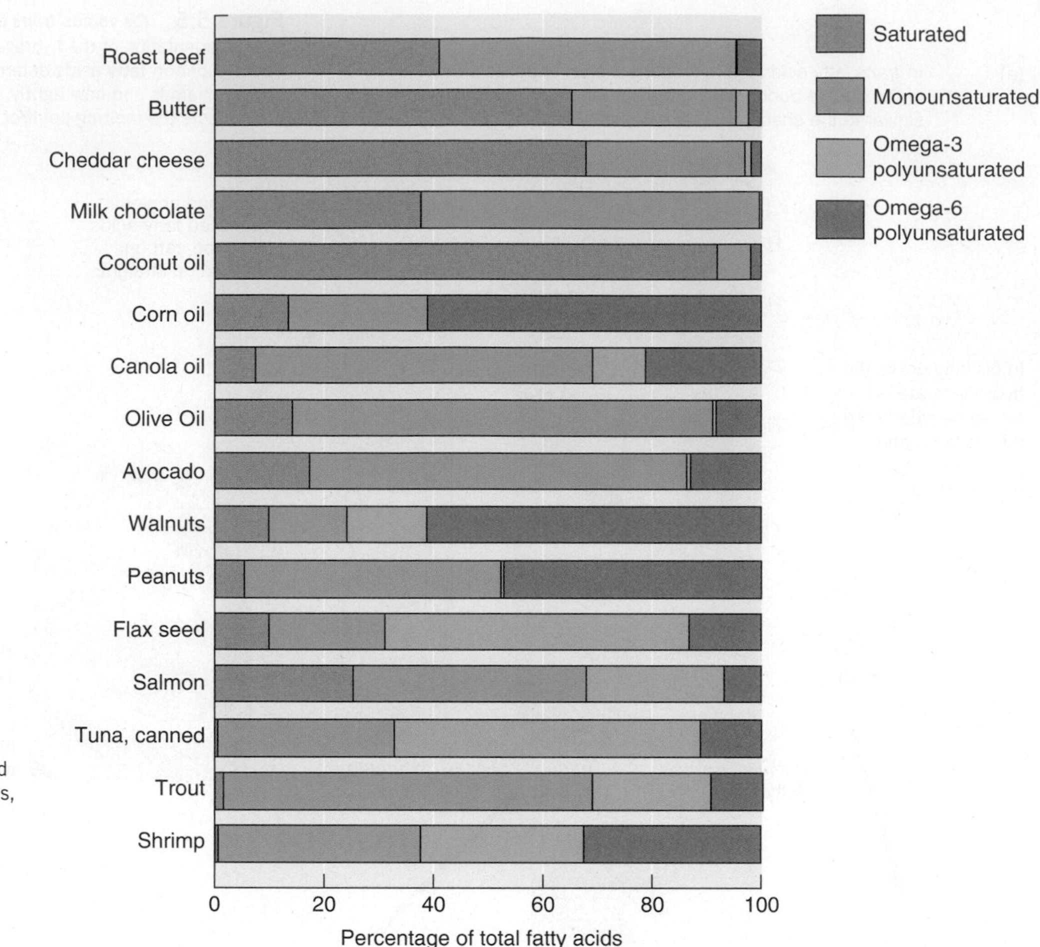

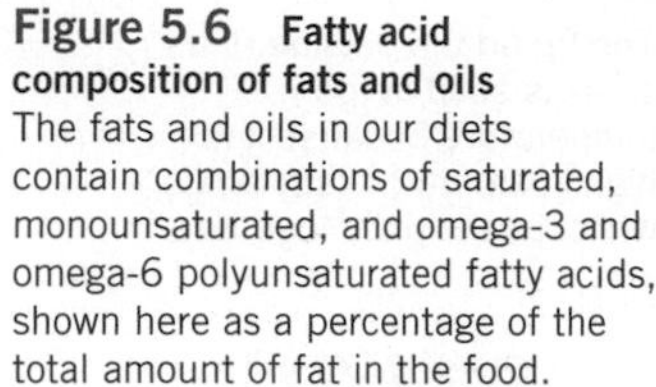

Figure 5.6 Fatty acid composition of fats and oils The fats and oils in our diets contain combinations of saturated, monounsaturated, and omega-3 and omega-6 polyunsaturated fatty acids, shown here as a percentage of the total amount of fat in the food.

room temperature. They are also used to lengthen shelf life in other processed foods such as cookies, crackers, breakfast cereals, and potato chips.

Fatty Acids and the Properties of Triglycerides Triglycerides may contain any combination of fatty acids: long, medium, or short chain; saturated or unsaturated; *cis* or *trans* (**Figure 5.6**). The types of fatty acids in triglycerides determine their texture, taste, and physical characteristics. For example, the amounts and types of fatty acids in chocolate allow it to remain solid at room temperature, snap when bitten into, and then melt quickly and smoothly in the mouth. The triglycerides in red meat contain predominantly long-chain, saturated fatty acids so, as illustrated in Figure 5.5a; the fat on a piece of steak is solid at room temperature. The triglycerides in olive oil contain predominantly monounsaturated fatty acids, whereas those in corn oil are mostly polyunsaturated. These fats are liquid at room temperature. The triglycerides in solid margarine made from corn oil have been hydrogenated so they contain more saturated fatty acid and *trans* fatty acids than the original corn oil.

Phospholipids

Phospholipids are lipids attached to a chemical group containing phosphorus called a *phosphate group*. **Phosphoglycerides** are the major class of phospholipids. Like triglycerides, they have a backbone of glycerol. However, they have only two fatty acids attached. In place of the third fatty acid is a phosphate group, which is then attached to a variety of other molecules (**Figure 5.7a**). The fatty-acid end of a phosphoglyceride is soluble in fat, whereas the phosphate end is water-soluble. This allows phosphoglycerides to mix in both water and fat—a property that makes them important for many functions in foods and in the body.

phosphoglyceride A type of phospholipid consisting of a glycerol molecule, two fatty acids, and a phosphate group.

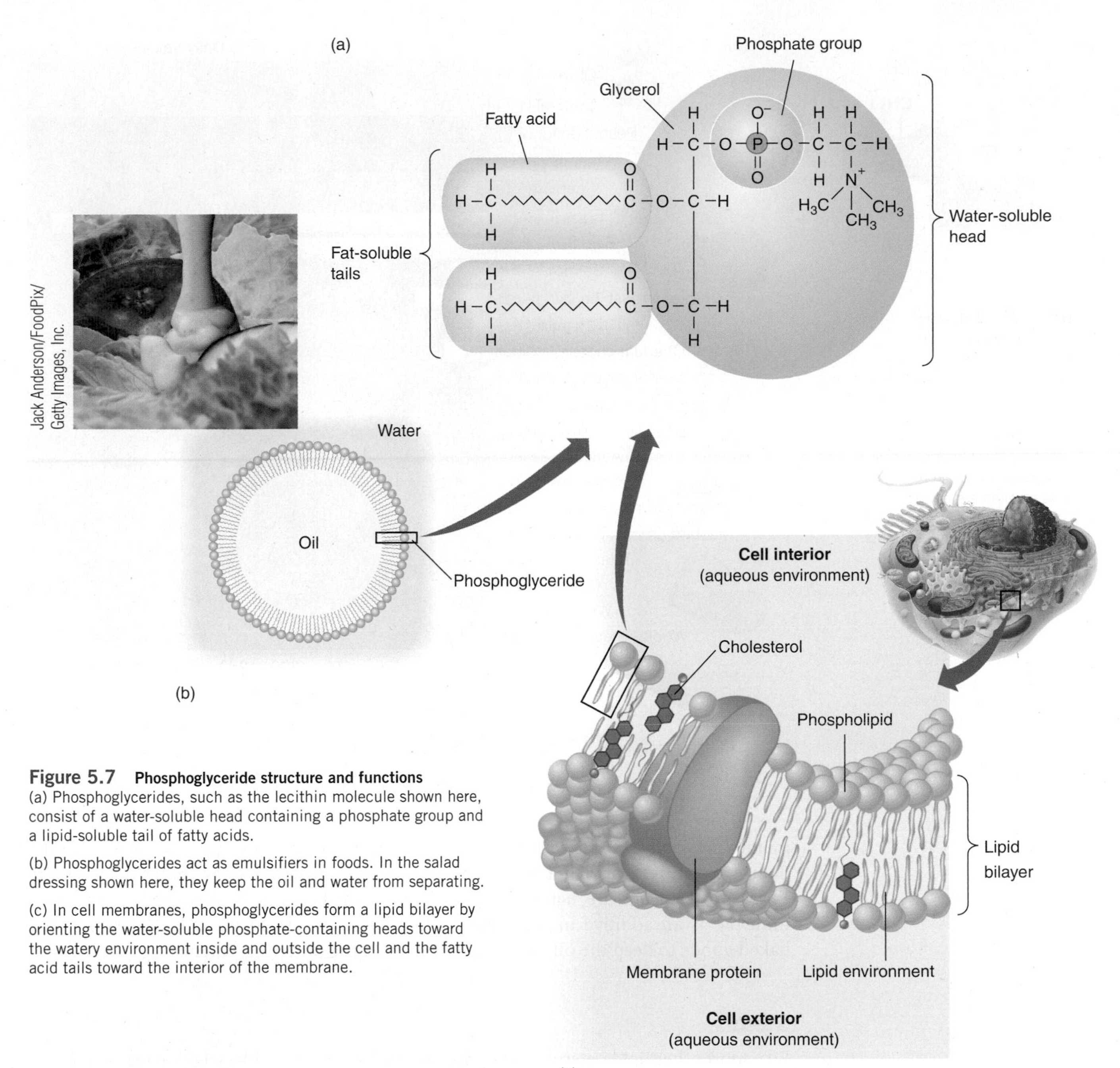

Figure 5.7 Phosphoglyceride structure and functions
(a) Phosphoglycerides, such as the lecithin molecule shown here, consist of a water-soluble head containing a phosphate group and a lipid-soluble tail of fatty acids.

(b) Phosphoglycerides act as emulsifiers in foods. In the salad dressing shown here, they keep the oil and water from separating.

(c) In cell membranes, phosphoglycerides form a lipid bilayer by orienting the water-soluble phosphate-containing heads toward the watery environment inside and outside the cell and the fatty acid tails toward the interior of the membrane.

In foods, the ability of phosphoglycerides to mix with both water and fat allows them to act as **emulsifiers** (**Figure 5.7b**). For example, egg yolks, which contain phosphoglycerides, function as an emulsifier in cake batter, where they allow the oil and water to mix. In the body, phosphoglycerides are an important component of cell membranes. The phosphoglycerides in membranes form a **lipid bilayer**, in which the water-soluble phosphate groups orient toward the aqueous environment inside and outside the cell, while the water-insoluble fatty acids stay in the lipid environment sandwiched in between (**Figure 5.7c**). This forms the barrier that helps regulate which substances can pass into and out of the cell.

The specific function of a phosphoglyceride depends on the molecule that is attached to the phosphate group. If a molecule of choline is attached, the phospholipid is called **lecithin** (see Figure 5.7a). In the body, lecithin is a major constituent of cell

emulsifier A substance that allows water and fat to mix by breaking large fat globules into smaller ones.

lipid bilayer Two layers of phosphoglyceride molecules oriented so that the fat-soluble fatty acid tails are sandwiched between the water-soluble phosphate-containing heads.

lecithin A phosphoglyceride composed of a glycerol backbone, two fatty acids, a phosphate group, and a molecule of choline.

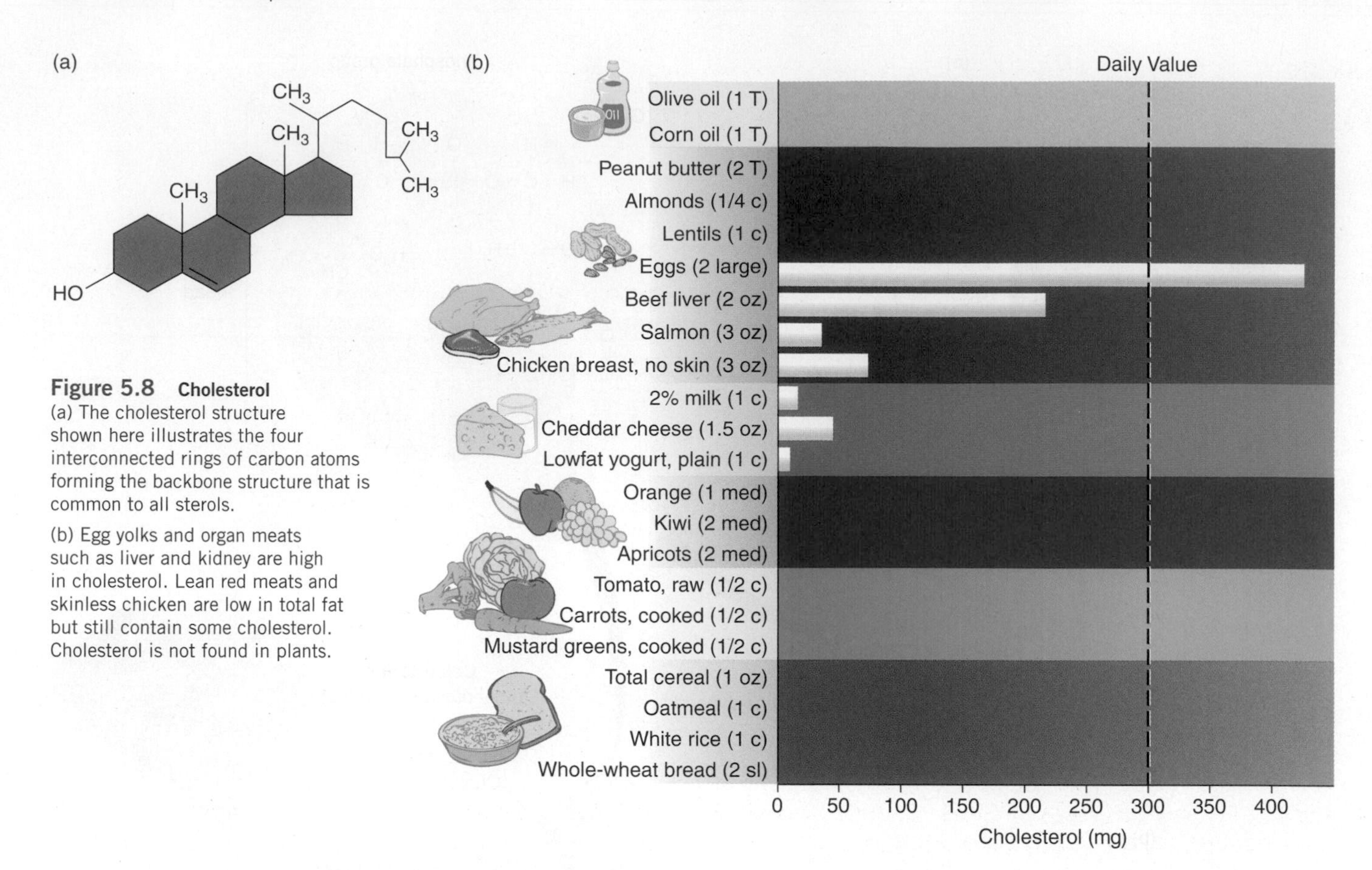

Figure 5.8 Cholesterol
(a) The cholesterol structure shown here illustrates the four interconnected rings of carbon atoms forming the backbone structure that is common to all sterols.

(b) Egg yolks and organ meats such as liver and kidney are high in cholesterol. Lean red meats and skinless chicken are low in total fat but still contain some cholesterol. Cholesterol is not found in plants.

membranes and is required for their optimal function. It is also used to synthesize the neurotransmitter acetylcholine, which is important in the memory center of the brain. Eggs and soybeans are natural sources of lecithin. Lecithin is also used by the food industry as an additive in margarine, salad dressings, chocolate, frozen desserts, and baked goods to keep the oil from separating from the other ingredients.

Sterols

cholesterol A lipid that consists of multiple chemical rings and is made only by animal cells.

Like most other lipids, sterols do not dissolve well in water. Unlike triglycerides and phosphoglycerides, their structure consists of multiple chemical rings (**Figure 5.8a**). Sterols are found in both plants and animals. **Cholesterol**, probably the best-known sterol, is only found in animals (**Figure 5.8b**). Cholesterol is necessary in the body, but because the liver manufactures it, it is not essential in the diet. More than 90% of the cholesterol in the body is found in cell membranes (see Figure 5.7c). It is also part of myelin, the coating on many nerve cells (see Figure 8.19). Cholesterol is needed to synthesize vitamin D in the skin; cholic acid, a component of bile; and steroid hormones. The steroid hormones include testosterone and estrogen, which promote growth and the development of sex characteristics, and cortisol, which is released in response to stress and promotes glucose synthesis in the liver.

In the diet, cholesterol is found only in foods from animal sources (see Figure 5.8b). Plant foods do not contain cholesterol unless animal products are combined with them in cooking or processing. Plants do contain other sterols, however, and these plant sterols have a role similar to that of cholesterol in animals: They help form plant cell membranes. Plant sterols are found in small quantities in most plant foods; when consumed in the diet, they can help reduce cholesterol levels in the body by decreasing cholesterol absorption from the diet (see **Debate: Good Egg, Bad Egg?**).

DEBATE

Good Egg, Bad Egg?

Dietary recommendations in the United States have been telling us to limit egg consumption since the 1960s. The reason is that one egg has over 200 mg of cholesterol. An ounce of lean meat has only about 30 mg. The Dietary Guidelines and the American Heart Association recommend limiting cholesterol intake to less than 300 mg per day. So, is it okay to eat eggs for breakfast?

The cholesterol in the body comes from the diet and from cholesterol synthesized in the liver. Even if you don't eat any cholesterol, your liver will make all you need. For many people, when they eat cholesterol, their liver production slows, so blood levels don't rise; others don't regulate cholesterol as well. For them, an increase in cholesterol intake results in an increase in blood cholesterol. However, the increase is typically due to increases in both "good" HDL cholesterol and "bad" LDL cholesterol, so the risk of atherosclerosis does not change.[1] Furthermore, the LDL particles that form when dietary cholesterol increases are large. These larger LDL particles are thought to be less of a cardiovascular risk than smaller, denser ones.[1] But these changes in blood lipoproteins were measured in the fasting state. There is evidence that dietary cholesterol may cause harm just after a meal by affecting blood lipid levels, increasing the susceptibility of LDL cholesterol to oxidation, and by potentiating the adverse effects of dietary saturated fat.[2]

Currently, the vast majority of epidemiological studies do not find a relationship between dietary cholesterol or egg consumption and cardiovascular disease.[1,3] For example, an evaluation of more than 20,000 male physicians participating in the Physicians' Health Study found that eating up to six eggs per week did not affect the risk or incidence of cardiovascular disease.[4] However, eating seven or more eggs per week caused an increased risk of death from cardiovascular disease, and eating any eggs was found to increase the risk of cardiovascular disease in people with type-2 diabetes.[4,5]

Nutrition professionals recognize that it is the overall dietary pattern, not the avoidance of particular foods, that is most important for health and wellness. Eggs are part of the Mediterranean dietary pattern, which is associated with good cardiovascular health. One large egg contains 6 g of high-quality protein, and unlike many other sources of cholesterol, eggs are low in cholesterol-raising saturated fat (see table). Eggs are also a good source of zinc, B vitamins, vitamin A, and iron. The yolk is rich in lutein and zeaxanthin, two phytochemicals that help protect against macular degeneration and cataracts. Eggs may also help you maintain your weight. A recent study found that people who eat an egg-based breakfast ate fewer overall calories during the day than people who have a bagel-based breakfast.[6]

Masterfile

The 2010 Dietary Guidelines has concluded that eating one egg per day is not harmful and does not result in increased risk of cardiovascular disease in healthy individuals. Despite this, they continue to recommend limiting dietary cholesterol to less than 300 mg per day, with further reductions to less than 200 mg per day for persons with or at high risk for cardiovascular disease.[7] If you eat an egg every day, is your diet likely to exceed 300 mg of cholesterol per day?

THINK CRITICALLY: What food or foods in this table other than eggs are high in cholesterol but low in saturated fat? How might they impact the risk of cardiovascular disease?

Cholesterol and Saturated Fat Content of Animal Foods

Food	Cholesterol (mg)	Saturated fat (g)[a]
Egg, one, hard boiled	212	1.6
Shrimp, 3 oz, raw	129	0.3
Salmon, 3 oz, cooked	57	0.6
Hamburger patty, 3 oz, broiled	71	7.5
Skinless chicken breast, 3 oz, roasted	72	0.9
Bacon, 3 oz, pan fried	94	11.7
Pork sausage, 3 oz	71	7.8
Butter, 2 tablespoons	61	14.6
Milk, whole, 8 fluid oz	24	4.6
Cheese, cheddar, 1 oz	30	6
Ice cream, vanilla, ½ cup	32	4.9

[a]Values are from iProfile 2.0.

[1]Fernandez, M. L., and Calle, M. Revisiting dietary cholesterol recommendations: Does the evidence support a limit of 300 mg/d? *Curr Atheroscler Rep* 12:377–383, 2010.

[2]Spence, J. D., Jenkins, D. J., and Davignon, J. Dietary cholesterol and egg yolks: Not for patients at risk of vascular disease. *Can J Cardiol* 26:e336-e339, 2010.

[3]Kritchevsky, S. B. A review of scientific research and recommendations regarding eggs. *J Am Coll Nutr* 23(6 suppl):596S–600S, 2004.

[4]Djoussé, L., and Gaziano, J. M. Egg consumption and cardiovascular disease and mortality in the Physicians' Health Study. *Am J Clin Nutr* 87:964–969, 2008.

[5]Djoussé, L., and Gaziano, J. M. Egg consumption and risk of heart failure in the Physicians' Health Study. *Circulation* 117:512–516, 2008.

[6]Ratliff, J., Leite, J. O., de Ogburn, R., et al. Consuming eggs for breakfast influences plasma glucose and ghrelin, while reducing energy intake during the next 24 hours in adult men. *Nutr Res* 30:96–103, 2010.

[7]U.S. Department of Agriculture and U.S. Department of Health and Human Services. *Dietary Guidelines for Americans, 2010*, 7th ed. Washington, DC: U.S. Government Printing Office, December 2010.

5.3 Lipids in the Digestive Tract

LEARNING OBJECTIVES

- Describe the steps involved in the digestion of triglycerides.
- Explain how micelles facilitate lipid absorption.

In healthy adults, most of the digestion of dietary fat takes place in the small intestine due to the action of lipid-digesting enzymes called *lipases*. A small amount of triglyceride digestion also occurs in the stomach due to the action of gastric lipase, an enzyme that can act in the acidic environment of the stomach. Gastric lipase is particularly important in infants because it helps digest the fat in milk.[7]

In the small intestine, bile from the gallbladder helps break fat into small globules. The triglycerides in these globules are digested by lipases from the pancreas, which break them down into fatty acids and monoglycerides (**Figure 5.9**). These products of triglyceride digestion,

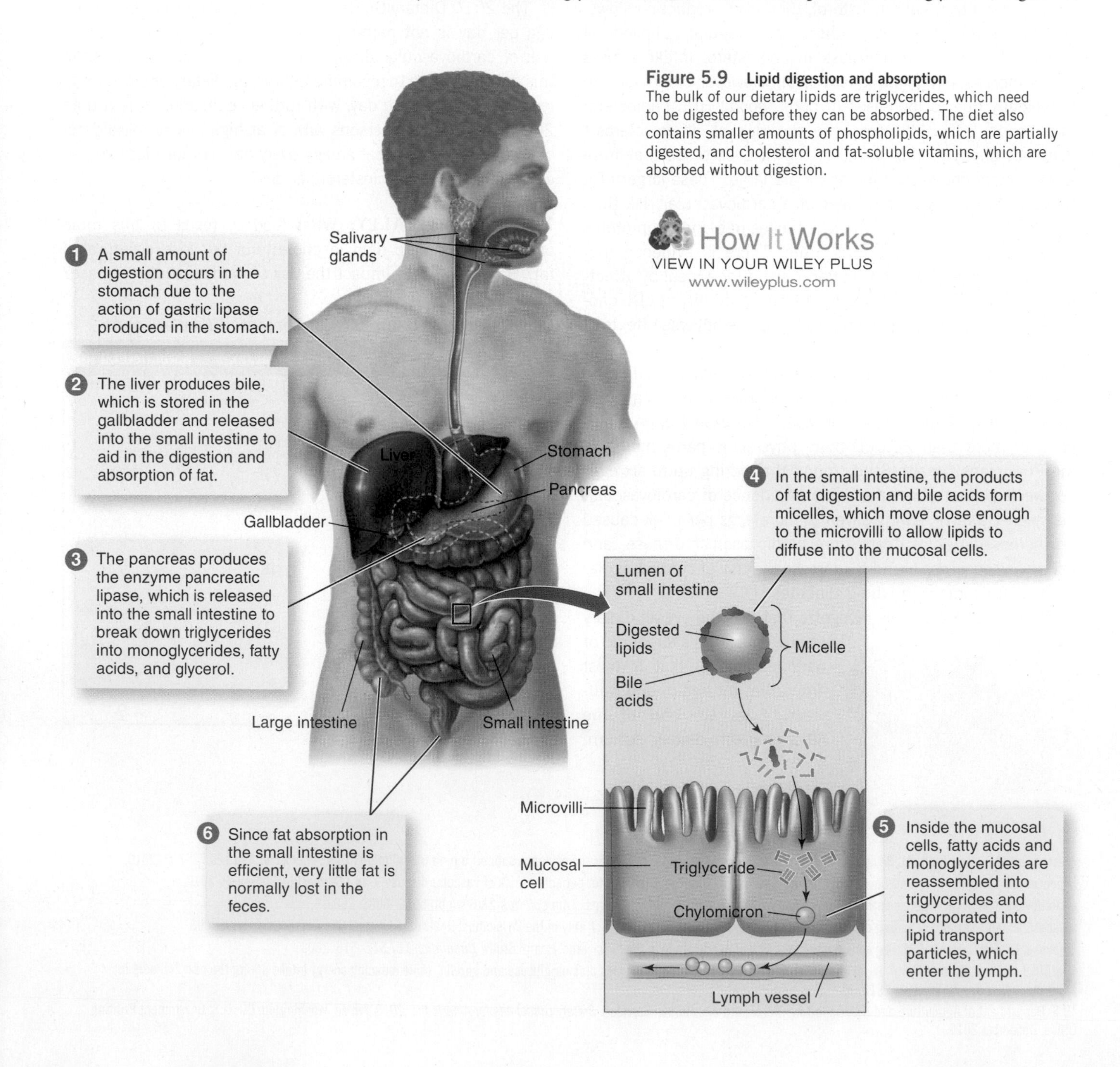

Figure 5.9 **Lipid digestion and absorption**
The bulk of our dietary lipids are triglycerides, which need to be digested before they can be absorbed. The diet also contains smaller amounts of phospholipids, which are partially digested, and cholesterol and fat-soluble vitamins, which are absorbed without digestion.

cholesterol, and other fat-soluble substances, including fat-soluble vitamins, mix with bile to form smaller droplets called **micelles**. Micelles have a fat-soluble center surrounded by a coating of bile acids. They facilitate the absorption of lipids into the mucosal cells of the small intestine by allowing these substances to get close enough to the microvilli to diffuse across into the mucosal cells. Most of the bile acids in micelles are also absorbed and returned to the liver to be reused. Once inside the mucosal cell, long-chain fatty acids, cholesterol, and other fat-soluble substances require further processing before they can be transported in the blood.

micelles Particles formed in the small intestine when the products of fat digestion are surrounded by bile acids. They facilitate the absorption of fat.

The fat-soluble vitamins (A, D, E, and K) are absorbed through the same process as other lipids. These vitamins are not digested but must be incorporated into micelles to be absorbed. The amounts absorbed can be reduced if dietary fat is very low or if disease, other dietary components, or medications such as the diet drug Alli, interfere with fat absorption.

5.4 Lipid Transport in the Body

LEARNING OBJECTIVES

- Describe how lipids are transported in the blood and enter cells.
- Compare and contrast the functions of LDLs and HDLs.

How lipids are absorbed and transported throughout the body depends on their solubility in water. Short- and medium-chain triglycerides are water soluble. They are easily digested and the resulting short- and medium-chain length fatty acids do not require micelles to be absorbed. Once absorbed, they pass directly into the blood and travel to the liver and other cells throughout the body. Lipids that are not soluble in water, such as long-chain fatty acids, cholesterol, and fat-soluble vitamins, cannot enter the bloodstream directly so they must be packaged for transport. They are covered in a water-soluble envelope of protein, phospholipids, and cholesterol to form **lipoproteins** (**Figure 5.10**).

lipoprotein A particle containing a core of triglycerides and cholesterol surrounded by a shell of protein, phospholipids, and cholesterol that transports lipids in blood and lymph.

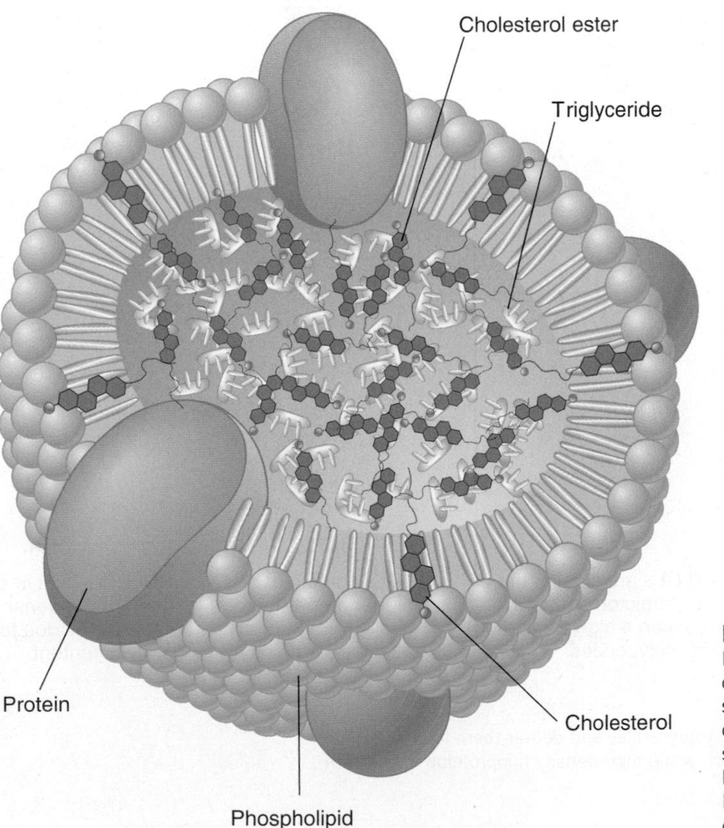

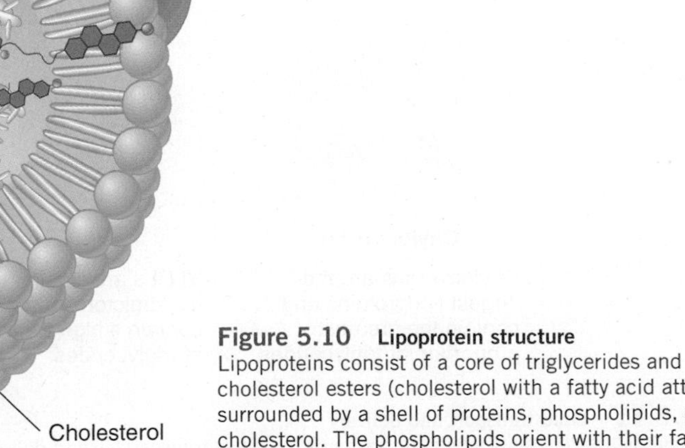

Figure 5.10 Lipoprotein structure
Lipoproteins consist of a core of triglycerides and cholesterol esters (cholesterol with a fatty acid attached) surrounded by a shell of proteins, phospholipids, and cholesterol. The phospholipids orient with their fat-soluble tails toward the interior and their water-soluble heads toward the outside, allowing water-insoluble lipids to be transported in the aqueous environment of the blood.

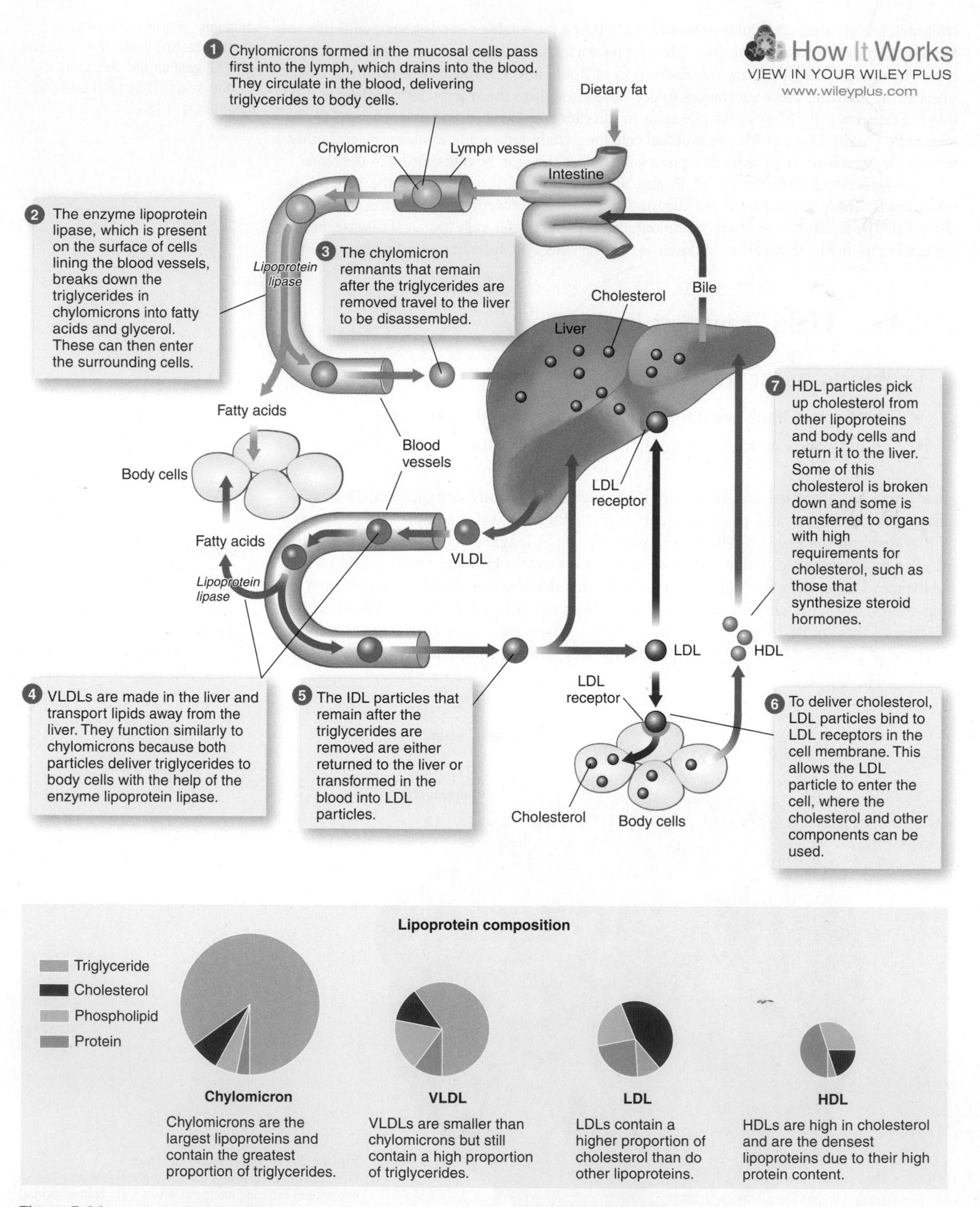

Figure 5.11 Lipid transport and delivery
Chylomicrons and very-low-density lipoproteins transport triglycerides and deliver them to body cells. Low-density lipoproteins transport and deliver cholesterol, while high-density lipoproteins help return cholesterol to the liver for reuse or elimination.

Different types of lipoproteins transport dietary lipids from the small intestine to body cells, from the liver to body cells, and from body cells back to the liver for disposal.

How Lipids Are Transported from the Small Intestine

To be transported from the small intestine, long-chain fatty acids and monoglycerides must first be assembled into triglycerides by the mucosal cell. These triglycerides are then combined with cholesterol, phospholipids, and a small amount of protein to form lipoproteins called **chylomicrons** (**Figure 5.11**). Chylomicrons are too large to enter the capillaries in the villi of the small intestine, so they pass from the intestinal mucosa into the lymphatic system, which then delivers them to the blood without first passing through the liver.

As chylomicrons circulate in the blood, the enzyme **lipoprotein lipase**, present on the surface of the cells lining the blood vessels, breaks the triglycerides down into fatty acids and glycerol, which enter the surrounding cells. The fatty acids can be either used as fuel or resynthesized into triglycerides for storage. What remains of the chylomicron is a chylomicron remnant composed mostly of cholesterol and protein. This goes to the liver and is disassembled (see Figure 5.11).

chylomicron A lipoprotein that transports lipid from the mucosal cells of the small intestine and delivers triglycerides to other body cells.

lipoprotein lipase An enzyme that breaks down triglycerides into fatty acids and glycerol; attached to the cell membranes of cells that line the blood vessels.

How Lipids Are Transported from the Liver

The liver is the major lipid-producing organ in the body. Here excess protein and carbohydrate, as well as alcohol, can be broken down and used to make triglycerides or cholesterol. Triglycerides made in the liver are incorporated into lipoprotein particles called **very-low-density lipoproteins (VLDLs)**. Cholesterol synthesized in the liver or delivered in chylomicron remnants can be incorporated into VLDLs or can be used to make bile. VLDLs transport lipids out of the liver and deliver triglycerides to body cells. As with chylomicrons, the enzyme lipoprotein lipase breaks down the triglycerides in VLDLs so that the fatty acids can be taken up by surrounding cells. Once the triglycerides are removed from the VLDLs, a denser, smaller, intermediate-density lipoprotein (IDL) remains. About two-thirds of the IDLs are returned to the liver, and the rest are transformed in the blood into **low-density lipoproteins (LDLs)**. LDLs contain less triglyceride and therefore proportionally more cholesterol than VLDLs. LDLs are the primary cholesterol delivery system for cells (see Figure 5.11).

For cholesterol to enter cells, the LDLs must be taken up by cells. For this to occur, a protein on the surface of the LDL particle called *apoprotein B* (*apo B*) must bind to a receptor protein on the cell membrane, called an **LDL receptor**. This binding allows LDLs to be removed from the blood circulation and enter cells where they are broken apart and their cholesterol and other components can be used (see Figure 5.11). If the amount of LDL cholesterol in the blood exceeds the amount that can be taken up by cells—due to either too much LDL cholesterol or too few LDL receptors—the result is a high level of LDL cholesterol.[8] High levels of LDL particles in the blood have been associated with an increased risk for heart disease (see **Science Applied: How a Genetic Disease Led to Treatment for High Blood Cholesterol**).

very-low-density lipoprotein (VLDL) A lipoprotein assembled by the liver that carries lipids from the liver and delivers triglycerides to body cells.

low-density lipoprotein (LDL) A lipoprotein that transports cholesterol to cells. Elevated LDL cholesterol increases the risk of cardiovascular disease.

LDL receptor A protein on the surface of cells that binds to LDL particles and allows their contents to be taken up for use by the cell.

How Cholesterol Is Eliminated

Because most body cells have no system for breaking down cholesterol, it must be returned to the liver to be eliminated from the body. This reverse cholesterol transport is accomplished by the densest of the lipoprotein particles, called **high-density lipoproteins (HDLs)** (see Figure 5.11). These particles originate from the intestinal tract and liver and circulate in the blood, picking up cholesterol from other lipoproteins and body cells. They function as a temporary storage site for lipids. Some of the cholesterol in HDLs is taken directly to the liver for disposal, and some is transferred to organs that have a high requirement for cholesterol, such as those involved in steroid hormone synthesis. High levels of HDL in the blood help prevent cholesterol from depositing in the artery walls and are associated with a reduction in heart disease risk.

high-density lipoprotein (HDL) A lipoprotein that picks up cholesterol from cells and transports it to the liver so that it can be eliminated from the body. A high level of HDL decreases the risk of cardiovascular disease.

SCIENCE APPLIED

How a Genetic Disease Led to Treatment for High Blood Cholesterol

(Elena Schweitzer/Shutterstock) Victoriano Izquierdo/Getty Images, Inc. (DAJ/Getty Images, Inc.)

Children with a rare form of the inherited disease familial hypercholesterolemia have blood cholesterol levels that range from 650 to 1000 mg/100 mL—six times the normal level. Their blood cholesterol levels are so high that it deposits in the tissues, forming soft, raised bumps on the skin called *xanthomas* (see photo). The elevated cholesterol damages blood vessels, leading to premature atherosclerosis. Chest pain, heart attacks, and death are common before the age of 15, and it is rare for individuals with this disease to survive past age 30.[1,2] Research into the causes of this disease led Drs. Michael Brown and Joseph Goldstein to a discovery that is key to our understanding of how blood cholesterol is regulated.

THE SCIENCE

To study familial hypercholesterolemia, Brown and Goldstein grew cells in culture. They observed that when LDL cholesterol was added to the culture medium, the rate of cholesterol synthesis in cells from normal subjects decreased. But, in cells from individuals with familial hypercholesterolemia, the presence of LDL in the medium did not cause a decrease in cholesterol synthesis.[3] By using radioactively labeled LDL particles, Brown and Goldstein were able to demonstrate that in cells from normal individuals, LDL particles bind to the cells and are removed from the surrounding media. In cells from individuals with familial hypercholesterolemia, the LDL particles are unable to bind to the cells and therefore cholesterol cannot be removed from the surrounding media.[4] The binding of LDL to the cell was found to be due to a protein on the surface of the cell membrane that was named the *LDL receptor*.

THE APPLICATION

The discovery of the LDL receptor led to an understanding of how the body controls the concentration of LDL cholesterol in the blood. The LDL receptor is a cell surface protein; it is synthesized by the cell and inserted into the cell membrane. The binding of an LDL particle to the receptor results in the movement of the LDL particle into the cell (see figure). When cells are deprived of cholesterol, the number of receptors increases. Conversely, when cholesterol accumulates in cells, the synthesis of receptors decreases.

Goldstein and Brown recognized that the number of receptors in the liver could be increased by lowering the cholesterol content of liver cells. This can be done by consuming a diet that is low in cholesterol and saturated fats and by drugs that block liver synthesis of cholesterol. Such a drug was isolated from mold in 1976.[5] When taken by people, this class of drug, collectively called statins, was shown to lower plasma LDL levels, which in turn led to protection from heart attacks. By 2003, the statin drug Lipitor was the best-selling drug in history.[6] Statins work by increasing the number of LDL receptors, a mechanism of action that we understand because of the work of Brown and Goldstein.

THINK CRITICALLY: Does someone who is taking statin drugs to lower their blood cholesterol need to limit the types of fat in their diet?

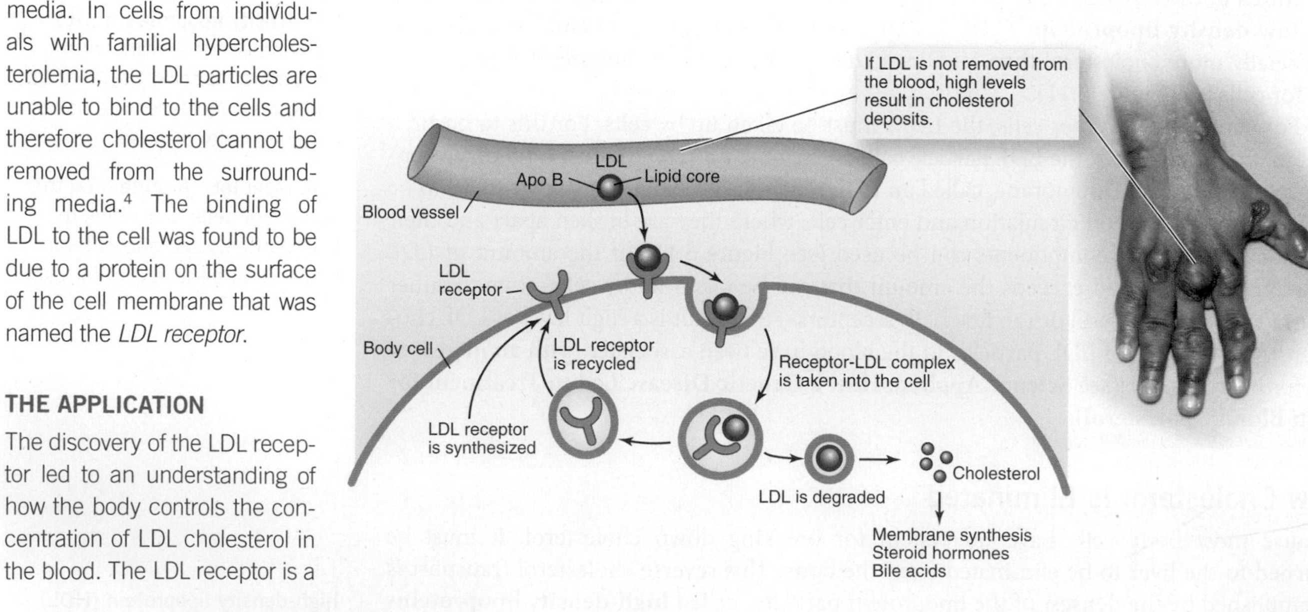

Kumar et al. Cases Journal 2008 1:71 doi:10.1186/1757-1626-1-71/

[1]Fredrickson, D. S., Goldstein, J. L., and Brown, M. S. The familial hypercholesterolemias. In *The Metabolic Basis of Inherited Disease*, 4th ed. J. B. Stanbury, J. B. Wyngaarden, and D. S. Fredrickson, eds. New York: McGraw-Hill, 1974, pp. 604–655.
[2]Goldstein, J. L., and Brown, M. S. The LDL receptor locus and the genetics of familial hypercholesterolemia. *Ann Rev Genet* 13:259–289, 1979.
[3]Goldstein, J. L., and Brown, M. S. The low-density lipoprotein pathway and its relation to atherosclerosis. *Ann Rev Biochem* 46:897–930, 1977.
[4]Brown, M. S., and Goldstein, J. L. Familial hypercholesterolemia: Defective binding of lipoproteins to cultured fibroblasts associated with impaired regulation of 3-hydroxy-3-methylglutaryl coenzyme A reductase activity. *Proc Nat Acad Sci* 71:788–792, 1974.
[5]Endo, A. The discovery and development of HMG-CoA reductase inhibitors. *J Lipid Res* 33:1569–1582, 1992.
[6]Doing things differently. *Pfizer 2008 Annual Review*, April 23, 2009, p. 15.

5.5 Lipid Functions in the Body

LEARNING OBJECTIVES

- List four functions of lipids in the body.
- Explain why enough of, and the right balance of, omega-3 and omega-6 fatty acids are needed in the diet.
- Describe how fatty acids are used to generate ATP.
- Discuss how fat is stored after a meal and how stored fat is retrieved between meals and during fasting.

Lipids are necessary to maintain health. In our diet, fat is needed to absorb fat-soluble vitamins and is a source of essential fatty acids. In our bodies, lipids form structural and regulatory molecules, are stored as an energy reserve, and are broken down via cellular respiration to provide energy in the form of ATP.

Lipids Provide Structure and Lubrication

Most of the lipids in the human body are triglycerides stored in **adipose tissue**, which is composed of **adipocytes**. Adipose tissue lies under the skin and around internal organs. Triglycerides in adipose tissue provide a lightweight energy storage molecule, cushion our internal organs, insulate the body from changes in temperature, and contribute to our body shape and contours (**Figure 5.12**). Triglycerides also make up the oils that lubricate body surfaces. For example, oil glands in the skin help to waterproof it and keep it soft and supple.

adipose tissue Tissue found under the skin and around body organs that is composed of fat-storing cells.

adipocyte A fat-storing cell.

Cholesterol and phospholipids are also important structural molecules. Cholesterol is particularly important in the brain and nervous system, where it is used to form myelin, an insulating coating around nerves. Cholesterol and phospholipids form the structure of all cell membranes and therefore define the boundaries of cells and partition off their organelles.

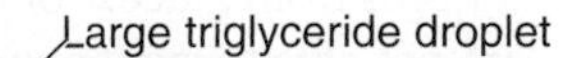
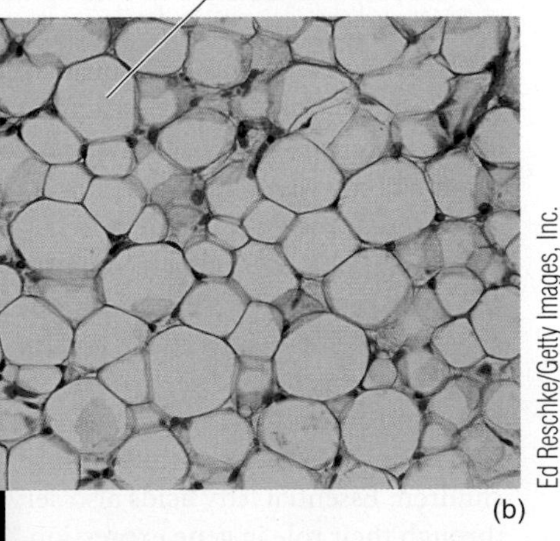

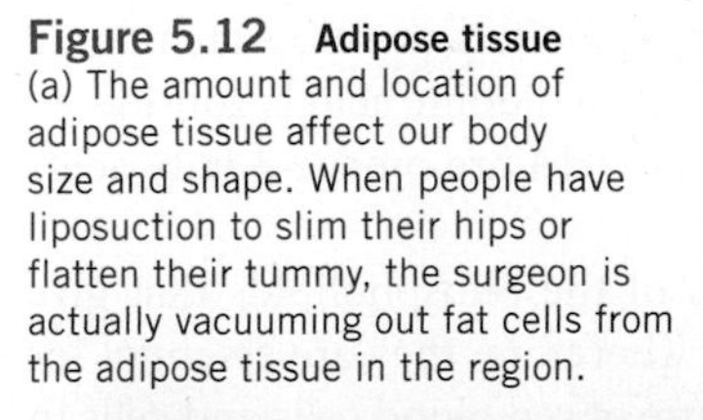
Figure 5.12 Adipose tissue
(a) The amount and location of adipose tissue affect our body size and shape. When people have liposuction to slim their hips or flatten their tummy, the surgeon is actually vacuuming out fat cells from the adipose tissue in the region.

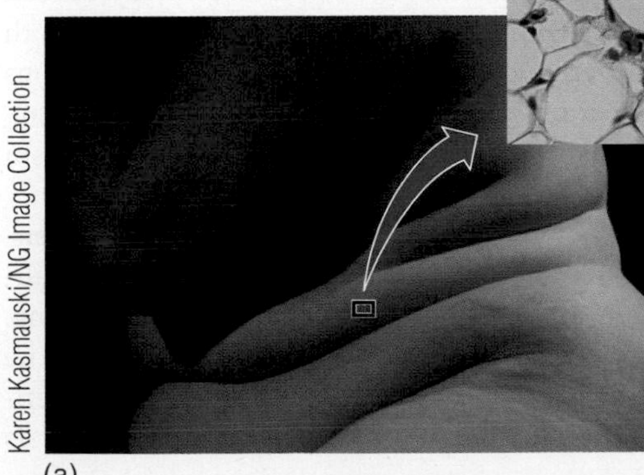

(b) Adipose tissue cells contain large droplets of triglyceride that push the other cell components to the perimeter of the cell. As weight is gained, the triglyceride droplets enlarge.

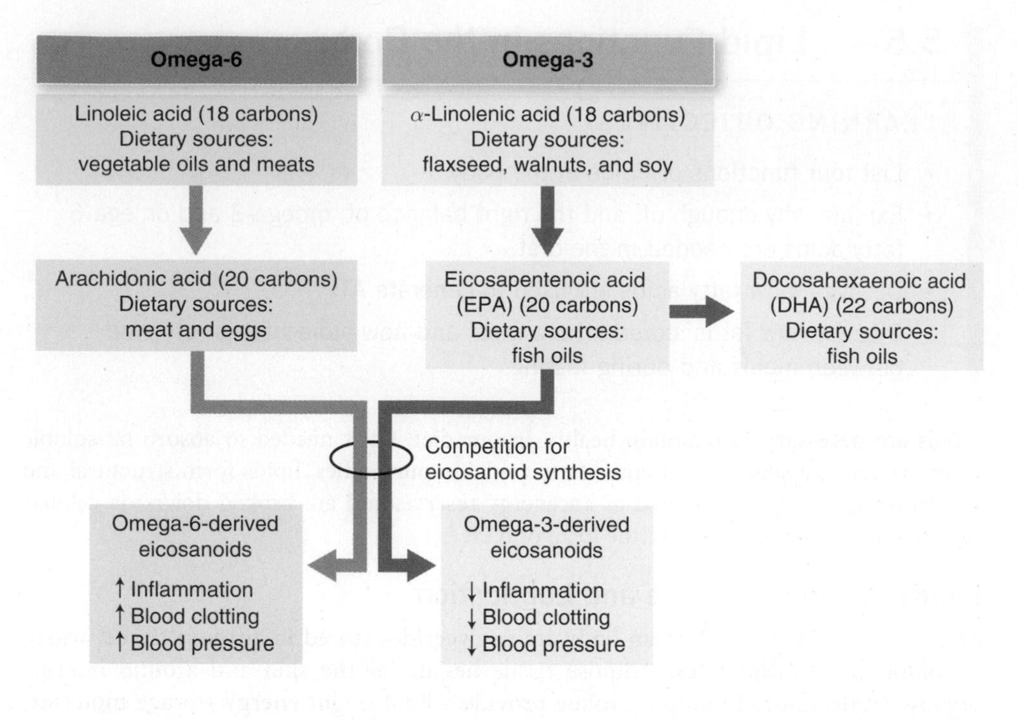

Figure 5.13 Essential fatty acids and eicosanoid synthesis
If the diet contains enough of the essential fatty acids linoleic acid and α-linolenic acid, enough of the longer-chain omega-6 and omega-3 fatty acids can usually be synthesized. Arachidonic acid and EPA compete for the enzymes that synthesize eicosanoids, so more EPA-derived eicosinoids will be made when more EPA is consumed or synthesized.

How Lipids Regulate Body Processes

Cholesterol and fatty acids are both used to synthesize regulatory molecules in the body. Cholesterol, either consumed in the diet or made in the liver, is used to make steroid hormones, which include the sex hormones estrogen and testosterone and the stress hormone cortisol. Polyunsaturated fatty acids are used to make hormone-like molecules that help regulate blood pressure, blood clotting, and other body parameters. However, unlike cholesterol, which is synthesized by the liver, the body cannot make all the fatty acids it needs, so certain ones must be consumed in the diet.

Essential Fatty Acids The human body is capable of synthesizing most of the fatty acids it needs from glucose or other sources of carbon, hydrogen, and oxygen. Humans, however, are not able to synthesize double bonds in the omega-6 and omega-3 positions. Therefore, the fatty acids linoleic acid (omega-6) and α-linolenic acid (omega-3) are **essential fatty acids**. If the diet is low in linoleic acid and/or α-linolenic acid, other fatty acids that the body would normally synthesize from them become dietary essentials as well. Arachidonic acid is an omega-6 fatty acid synthesized from linoleic acid. Arachidonic acid is considered essential only when the diet is low in linoleic acid. EPA and DHA are omega-3 fatty acids synthesized from α-linolenic acid (**Figure 5.13**).

essential fatty acid A fatty acid that must be consumed in the diet because it cannot be made by the body or cannot be made in sufficient quantities to meet needs.

Essential fatty acids are important for the formation of the phospholipids that give cell membranes their structure and functional properties. Therefore, they are essential for growth, development, fertility, and maintaining the structure of red blood cells and cells in the skin and nervous system. DHA is particularly important in the retina of the eye. Both DHA and arachidonic acid are needed to synthesize cell membranes in the central nervous system and are therefore important for normal brain development in infants and young children. Essential fatty acids also serve as regulators of glucose and fatty acid metabolism through their role in gene expression.[9]

If adequate amounts of linoleic and α-linolenic acid are not consumed, an **essential fatty acid deficiency** will result. Symptoms include scaly, dry skin, liver abnormalities, poor healing of wounds, growth failure in infants, and impaired vision and hearing. Because the requirement for essential fatty acids is well below the typical intake in the United States, essential fatty acid deficiencies are rare in this country. However, deficiencies have been seen in infants and young children fed very low-fat diets and in individuals who are unable to absorb lipids.

essential fatty acid deficiency A condition characterized by dry scaly skin and poor growth that results when the diet does not supply sufficient amounts of the essential fatty acids.

Eicosanoid Synthesis and Function Getting enough essential fatty acids in your diet will prevent deficiency, but the ratio of dietary omega-6 to omega-3 fatty acids also affects your health. This is because the omega-6 and omega-3 polyunsaturated fatty acids made from them are used to make hormone-like molecules called **eicosanoids**. Eicosanoids help regulate blood clotting, blood pressure, immune function, and other body processes. The effect an eicosanoid has on these functions depends on the fatty acid from which it is made. For example, when the omega-6 fatty acid arachidonic acid is the starting material, the eicosanoid synthesized increases blood clotting; whereas when the eicosanoid is made from the omega-3 fatty acid EPA, it decreases blood clotting (see Figure 5.13). Omega-3 fatty acids, particularly EPA and DHA, have anti-inflammatory properties. Some of this effect is due to the amounts and types of eicosanoids made, and some is due to the effects these fatty acids have on other aspects of cell function. Since **inflammation** plays a role in the progression of heart disease, a nutrient with anti-inflammatory properties would protect against heart disease. Epidemiological studies and feeding trials both provide evidence for a beneficial effect of omega-3 fatty acids on the manifestations of heart disease and stroke.[10] The anti-inflammatory properties of omega-3 fatty acids have also been shown to be of benefit in other chronic diseases that involve inflammatory processes such as rheumatoid arthritis, inflammatory bowel disorders, and cancer.[11]

eicosanoids Regulatory molecules, including prostaglandins and related compounds, that can be synthesized from omega-3 and omega-6 fatty acids.

inflammation The response of a part of the body to injury that increases blood flow with an influx of white blood cells and other chemical substances to facilitate healing. It is characterized by swelling, heat, redness, pain, and loss of function.

The ratio of dietary omega-6 to omega-3 essential fatty acids affects the balance of these fatty acids in the tissues and therefore the ratio of the types of eicosanoids made from them (see Figure 5.13). In order to maintain a healthy balance in the body, a dietary ratio of linoleic to α-linolenic acid of 5:1 to 10:1 is recommended. To provide this ratio, a diet that contains 20 grams of linoleic acid would need to include 2 to 4 grams of α-linolenic acid. However, if the diet contains plenty of arachidonic acid, EPA, and DHA, which are made from linoleic and α-linolenic acid, the actual ratio of linoleic to α-linolenic is of less concern.[5] The American diet contains plenty of omega-6 fatty acids, so to get a healthier mix Americans should increase their intake of omega-3s from foods such as fish, walnuts, flaxseed, leafy green vegetables, and soybean and canola oils (see Figure 5.6).[12]

How Triglycerides Provide Energy

Triglycerides consumed in the diet can be used as an immediate source of energy or stored in adipose tissue for future use. The majority of ATP produced from triglyceride metabolism is from the breakdown of fatty acids. Only a small amount is derived from glycerol. In the first step of fatty acid breakdown, called **beta-oxidation (β-oxidation)**, the carbon chain of fatty acids is split into two-carbon units that form acetyl-CoA and release high-energy electrons (**Figure 5.14**).

beta-oxidation (β-oxidation) The first step in the production of ATP from fatty acids. This pathway breaks the carbon chain of fatty acids into two-carbon units that form acetyl-CoA and releases high-energy electrons that are passed to the electron transport chain.

To enter the citric acid cycle, acetyl-CoA from β-oxidation combines with oxaloacetate, a four-carbon molecule derived from carbohydrate, to form a six-carbon molecule called *citric acid*. The reactions of the citric acid cycle then remove one carbon at a time to produce carbon dioxide (see Figure 5.14). After two carbons have been removed in this manner, a four-carbon oxaloacetate molecule is re-formed and the cycle can begin again. The high-energy electrons released in β-oxidation and the citric acid cycle are shuttled to the last stage of cellular respiration, the electron transport chain. Molecules here accept electrons and pass them down the chain until they are finally combined with oxygen and hydrogen to form water. The energy in the electrons is used to convert ADP to ATP.

The glycerol from triglyceride breakdown can also be used to produce ATP or to make glucose via gluconeogenesis. Glycerol makes up only a small proportion of the carbon in a triglyceride molecule, so the amount of glucose that can result from triglyceride breakdown is small.

How Fat Is Stored and Retrieved

Throughout the day, triglycerides are continuously stored and then broken down, depending on the immediate energy needs of the body. For example, after a meal some triglycerides will be stored; then, between meals some of the stored triglycerides will be broken down to provide energy. When the energy in the diet equals the body's energy requirements, the net amount of stored triglyceride in the body does not change.

Figure 5.14 Triglyceride metabolism
The breakdown of triglycerides yields fatty acids and a small amount of glycerol. Fatty acids provide most of the energy stored in a triglyceride molecule.

How It Works
VIEW IN YOUR WILEY PLUS
www.wileyplus.com

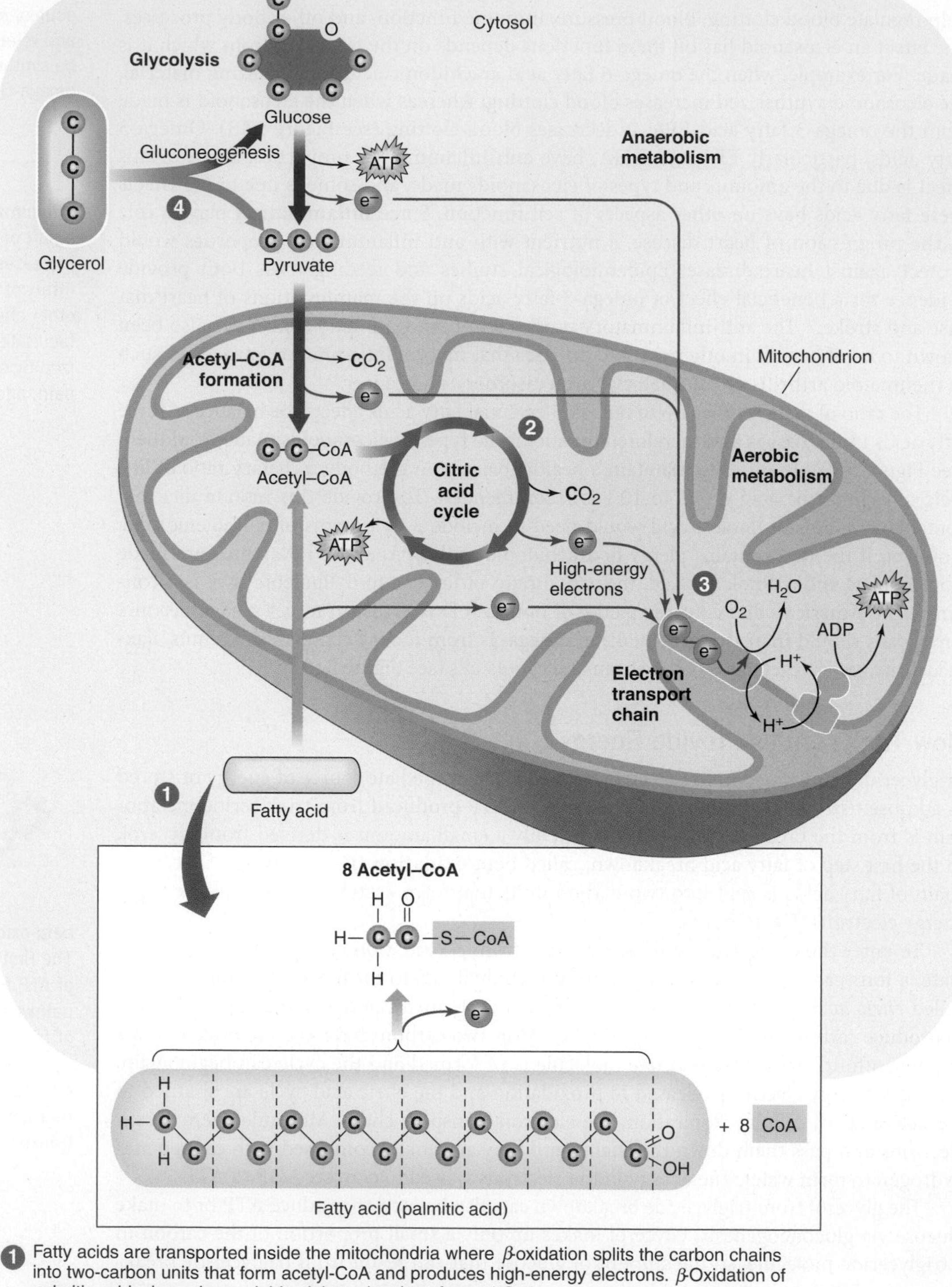

1. Fatty acids are transported inside the mitochondria where β-oxidation splits the carbon chains into two-carbon units that form acetyl-CoA and produces high-energy electrons. β-Oxidation of palmitic acid, shown here, yields eight molecules of acetyl-CoA.
2. If oxygen and enough carbohydrate are available, acetyl-CoA combines with oxaloacetate to enter the citric acid cycle, producing two molecules of carbon dioxide and releasing high-energy electrons that are shuttled to the electron transport chain.
3. In the final step of aerobic metabolism, the energy from the high-energy electrons released from β-oxidation and the citric acid cycle pumps hydrogen ions across the inner mitochondrial membrane. As the hydrogen ions flow back, the energy is used to convert ADP to ATP. The electrons are combined with oxygen and hydrogen to form water.
4. Glycerol molecules, from triglyceride breakdown, contain three carbon atoms. They can be used to produce small amounts of ATP or form glucose.

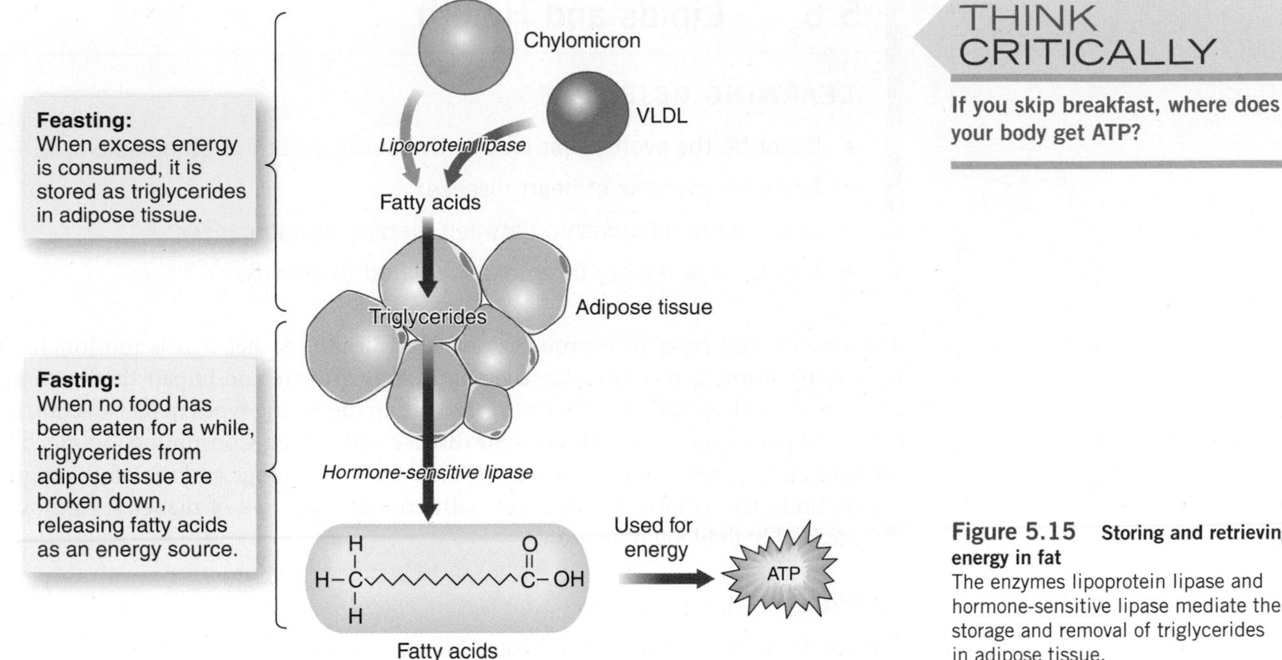

Figure 5.15 **Storing and retrieving energy in fat**
The enzymes lipoprotein lipase and hormone-sensitive lipase mediate the storage and removal of triglycerides in adipose tissue.

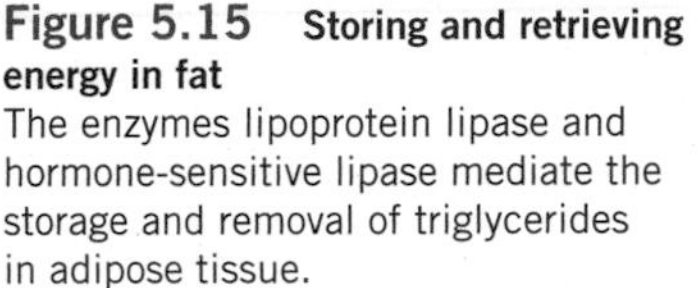

If you skip breakfast, where does your body get ATP?

Feasting When energy is ingested in excess of needs, the excess can be stored as fat in adipose tissue. Excess energy consumed as fat is packaged in chylomicrons and transported directly from the intestines to the adipose tissue. Because the fatty acids in our body fat come from the fatty acids we eat, what we eat affects the fatty acid composition of our adipose tissue; therefore, if you eat more saturated fat, there will be more saturated fat in your adipose tissue.

Excess energy consumed as carbohydrate or protein must first go to the liver, where the carbohydrate and protein can be used, although inefficiently, to synthesize fatty acids; these fatty acids are then assembled into triglycerides, which are transported to the adipose tissue in VLDLs. Lipoprotein lipase at the membrane of cells lining the blood vessels breaks down the triglycerides from both chylomicrons and VLDLs so that the fatty acids can enter the cells, where they are reassembled into triglycerides for storage (**Figure 5.15**).

The ability of the body to store fat is theoretically limitless. Adipocytes can increase in weight by about 50 times, and new adipocytes can be synthesized when existing cells reach their maximum size (see Chapter 7). Because each gram of fat provides 9 kcalories, compared with only 4 kcalories per gram from carbohydrate or protein, a large amount of energy can be stored in the body as fat without a great increase in size or weight. Even a lean man, whose body fat is only about 10% of his weight, stores over 50,000 kcalories of energy as fat.

Fasting When less energy is consumed than is needed, the body uses energy from fat stores. In this situation, the enzyme **hormone-sensitive lipase** inside the adipocytes receives a hormonal signal that turns on enzyme activity so it begins breaking down stored triglycerides (see Figure 5.15). The fatty acids and glycerol are released directly into the blood, where they can be taken up by cells throughout the body to produce ATP.

hormone-sensitive lipase An enzyme present in adipocytes that responds to chemical signals by breaking down triglycerides into fatty acids and glycerol for release into the bloodstream.

If there is not enough carbohydrate to allow acetyl-CoA from fat breakdown to enter the citric acid cycle, it will be used to make ketones (see Figure 4.16). Ketones can be used as an energy source by muscle and adipose tissue. During prolonged starvation, the brain can adapt to use ketones to meet about half of its energy needs. For the other half, it must use glucose. Fatty acids cannot be used to make glucose, and, as seen in Figure 5.14, only a small amount of glucose can be made from glycerol.

5.6 Lipids and Health

LEARNING OBJECTIVES

- Describe the events that lead to the development of atherosclerosis.
- Evaluate your risk of heart disease.
- Discuss the relationship between dietary fat and cancer.
- Explain how dietary fat intake is related to obesity.

The amount and types of fat you eat can affect health. A diet that is too low in fat can reduce the absorption of fat-soluble vitamins, slow growth, and impair the functioning of the skin, eyes, liver, and other body organs. Eating the wrong types of fat can contribute to chronic diseases such as **cardiovascular disease** and cancer. Consuming too much fat can increase energy intake and contribute to extra body fat storage and therefore weight gain. Excess body fat, in turn, is associated with an increased risk of diabetes, cardiovascular disease, and high blood pressure.

cardiovascular disease Any disease affecting the heart and blood vessels.

Cardiovascular Disease

Almost 83 million adults in the United States, more than one-third, suffer from one or more types of cardiovascular disease.[13] It is the number-one cause of death of both men and women in the United States. **Atherosclerosis** is a type of cardiovascular disease in which lipids and fibrous material are deposited in the artery walls, reducing their elasticity and eventually blocking the flow of blood. The development of atherosclerosis has been linked to diets that are high in cholesterol, saturated fat, and *trans* fat.[5]

atherosclerosis A type of cardiovascular disease that involves the buildup of fatty material in the artery walls.

How Does Atherosclerosis Develop? Inflammation, which is the process whereby the body responds to injury, is what drives the formation of **atherosclerotic plaque**. For example, cutting yourself triggers an inflammatory response. White blood cells, which are part of the immune system, rush to the injured area, blood clots form, and new tissue grows to heal the wound. Similar inflammatory responses occur when an artery is injured, but instead of resulting in healing, they lead to the development of atherosclerotic plaque (**Figure 5.16**).

atherosclerotic plaque The cholesterol-rich material that is deposited in the arteries of individuals with atherosclerosis. It consists of cholesterol, smooth muscle cells, fibrous tissue, and eventually calcium.

The atherosclerotic process begins as a response to an injury that causes changes in the lining of the artery wall. The exact cause of the injury is not known but may be related to elevated blood levels of LDL cholesterol, glucose, or the amino acid homocysteine; high blood pressure; free radicals caused by cigarette smoking; diabetes; genetic alterations; or infectious microorganisms.[14] The specific cause may be different in different people.

Once the initial injury has occurred, the lining of the artery becomes more permeable to LDL particles, which migrate into the artery wall (see Figure 5.16). Once inside, the LDL particles are modified chemically, often by oxidation, to form **oxidized LDL cholesterol**. Oxidized LDL cholesterol is harmful, and its presence promotes inflammation in a number of ways. It triggers the production and release of substances that cause immune-system cells to stick to the lining of the artery and then to migrate into the artery wall. Once inside, these cells become large white blood cells called *macrophages*, which have **scavenger receptors** on their surface. Just as LDL receptors bind to LDL cholesterol, scavenger receptors bind to and transport oxidized LDL cholesterol into the interior of the cell. As macrophages fill with more and more oxidized LDL cholesterol, they are transformed into cholesterol-filled foam cells (named because of their foamy appearance under a microscope). Foam cells accumulate in the artery wall and then burst, depositing cholesterol to form a fatty streak (see Figure 5.16).[14,15]

oxidized LDL cholesterol A substance formed when the cholesterol in LDL particles is oxidized by reactive oxygen molecules. It is key in the development of atherosclerosis because it contributes to the inflammatory process.

scavenger receptor A protein on the surface of macrophages that binds to oxidized LDL cholesterol and allows it to be taken up by the cell.

Macrophages and foam cells secrete growth factors and other chemicals that continue the inflammatory response and promote growth of the plaque. The release of growth factors signals smooth muscle cells from the wall of the artery to migrate into the fatty streak and secrete fibrous proteins. Platelets, which are cell fragments involved in blood clotting, become sticky and clump together around the lesion. As the lesion enlarges it causes the

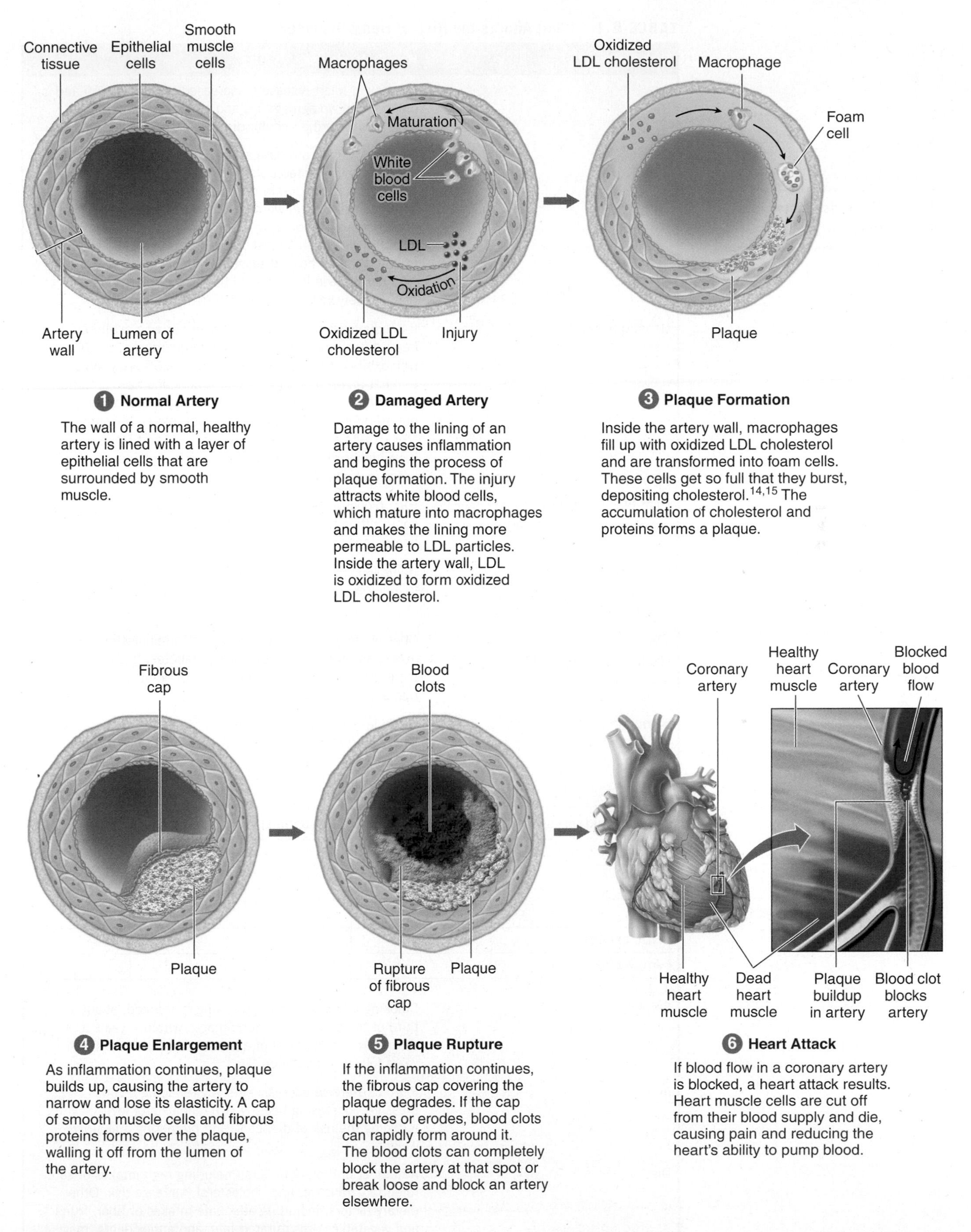

Figure 5.16 **Development of atherosclerosis**
The inflammation that occurs in response to an injury to the artery wall precipitates the development of atherosclerotic plaque. The buildup of plaque can eventually lead to a heart attack or stroke.

THINK CRITICALLY

Which of these risk factors are modifiable?

TABLE 5.1 What Affects the Risk of Heart Disease?

Risk factor	How it affects risk
Age	The risk of heart disease is increased in men age 45 and older and in women age 55 and older. Almost half of people with cardiovascular disease are 60 or older.[13]
Gender	Men and women are both at risk for heart disease, but men are generally affected a decade earlier than are women. This difference is due in part to the protective effect of the hormone estrogen in women. As women age, the effects of menopause—including a decline in estrogen level and a gain in weight—increase heart disease risk. Although we tend to think of heart disease as a man's disease, cardiovascular disease has claimed the lives of more women than men in every year since 1984.[13]
Genetic background	Individuals with a male family member who exhibited heart disease before age 55 or a female family member who exhibited heart disease before age 65 are considered to be at increased risk. African Americans have a higher risk of heart disease than the general population, in part due to their higher incidence of high blood pressure. Mexican Americans, Native Americans, and Hawaiians have higher risk, due in part to a higher incidence of diabetes and obesity in these groups.[13]
High blood pressure (Blood pressure ≥140/90 mm Hg)	High blood pressure can damage blood vessel walls, contributing to atherosclerosis. It forces the heart to work harder, causing it to enlarge and weaken over time.
Diabetes (Fasting blood glucose ≥126 mg/100 mL)	High blood glucose damages blood vessel walls, contributing to atherosclerosis.
Obesity (Body mass index[a] ≥30)	Obesity increases blood pressure, blood cholesterol levels, and the risk of developing diabetes. It also increases the amount of work the heart must do to pump blood throughout the body.
Blood lipid levels	Blood levels of total cholesterol, LDL cholesterol, HDL cholesterol, and triglycerides all affect risk.[17] Currently, about 16.2% of American adults have blood cholesterol levels of 240 mg per 100 mL or greater.[13]

Type of lipid (mg/100 mL)	Low risk/optimal	Near optimal	Borderline high	High risk
Total cholesterol	<200		200–239	≥240
LDL cholesterol	<100 <70 in high-risk individuals	100–129	130–159	≥160
HDL cholesterol	≥60			<40
Triglycerides	<150			

Risk factor	How it affects risk
Smoking	Smoking increases risk. If smoking is stopped, about a third of the excess risk is eliminated within 2 years and risk returns to the level of a nonsmoker within 10 to 14 years following cessation.[18]
Activity	Regular exercise decreases risk by reducing blood pressure, increasing healthy HDL cholesterol levels, reducing the risk of diabetes, and promoting a healthy weight.
Diet	A number of dietary factors, including high intakes of saturated fat, *trans* fat, and cholesterol, increase risk. Other dietary factors, including adequate intakes of fiber, fruits and vegetables, unsaturated fats, and antioxidants, may offer a protective effect.

[a]See Chapter 7 for information on body mass index (BMI) and how it can be calculated.

artery to narrow and lose its elasticity, hampering blood flow. As the process progresses, a fibrous cap of smooth muscle cells and fibrous proteins forms over the mixture of white blood cells, lipids, and debris, walling it off from the lumen of the artery. The formation of the cap is a way of healing the injury, but if the inflammation continues substances secreted by immune system cells can degrade this cap. If the cap becomes too thin and ruptures, the material leaks out and causes blood clots to form.[16] The clots can completely block the artery at that spot, or break loose and block an artery elsewhere. If this occurs in blood vessels that supply the heart muscle, blood flow to the heart muscle is interrupted and heart cells die, resulting in a heart attack or myocardial infarction (see Figure 5.16). If the blood flow to the brain is interrupted, a stroke results.

Risk Factors for Heart Disease High blood pressure, diabetes, obesity, and abnormal blood lipid levels are considered primary risk factors because they directly increase the risk of developing heart disease. Other factors that affect risk include age, gender, genetic background, and lifestyle factors such as smoking, exercise, and diet. These may directly affect risk or act indirectly by altering blood cholesterol levels, blood pressure, body weight, or the risk of diabetes (**Table 5.1**). While you can't change your gender, age, or genetic makeup, many lifestyle risk factors for heart disease are modifiable.

How Diet Affects the Risk of Heart Disease The risk of heart disease is affected by individual nutrients and particular whole foods. For example, diets high in sodium and saturated fat increase heart disease risk. Diets high in fiber and certain vitamins can reduce heart disease risk. Consuming fish, nuts, and whole grains may decrease risk, while diets that are high in red meat may increase risk. Evidence now suggests that a complicated set of many nutrients and dietary components interact to influence the risk of heart disease (**Figure 5.17**).[19] Therefore, it is important to focus on whole foods and dietary patterns to reduce the risk of heart disease. For example, diets that are plentiful in fruits, vegetables, and whole grains and low in high-fat meats and dairy products reduce the risk of heart disease.[20] The heart-protective effect of dietary patterns has prompted nutrition experts to promote a Mediterranean or DASH dietary pattern (see Chapter 10) to reduce the risk of heart disease in the United States.[12]

How Dietary Cholesterol Affects Risk The extent to which cholesterol intake affects blood levels depends on an individual's genes. Cholesterol in the blood comes from cholesterol both consumed in the diet and made by the liver. Generally, the body makes about three to four times more cholesterol than is consumed in the diet. In some individuals, as dietary cholesterol increases, liver cholesterol synthesis decreases so that blood levels do not change.[21] In others, however, liver synthesis does not decrease in response to an increase in dietary cholesterol, so blood cholesterol levels rise.

Why Saturated Fat Increases Risk When the diet is high in saturated fatty acids, liver production of cholesterol-carrying lipoproteins increases and the activity of LDL receptors in the liver is reduced, so that LDL cholesterol cannot be removed from the blood.[22] Therefore diets high in saturated fat increase LDL cholesterol in the blood. Increased LDL then increases the risk of atherosclerosis. When the diet is low in saturated fats, lipoprotein production decreases and the number of LDL receptors increases, allowing more cholesterol to be removed from the bloodstream. This lowers LDL cholesterol levels and the risk of heart disease. The saturated fatty acid stearic acid, found in chocolate and beef, is an exception because it does not increase blood cholesterol levels. However, diets high in stearic acid may contribute to heart disease by affecting other factors involved in the development of atherosclerosis, such as blood platelets and blood clotting.[23]

How *Trans* Fat Increases Risk Both clinical and epidemiological studies provide evidence that a high *trans* fatty acid intake increases the risk of heart disease. Many studies have found that diets high in *trans* fatty acids cause a greater increase in heart disease risk than those high in saturated fatty acids.[24] A 2% increase in energy intake from *trans* fat, the

Individual nutrients and food components	
Factors That Increase Risk	**Factors That Reduce Risk**
Cholesterol	Polyunsaturated fat
Saturated fat	Monosaturated fat
Trans fat	Fiber
Sodium	B vitamins
Excess sugar	Antioxidants
Excess energy	Moderate alcohol

Nutrients are found in foods.

Whole foods

© hlphoto/iStockphoto

Fish is high in omega-3 fatty acids, which reduce the risk of heart disease. In addition to lowering LDL cholesterol and triglyceride levels, omega-3 fatty acids protect against heart disease by decreasing blood clotting, lowering blood pressure, improving the function of the cells lining blood vessels, reducing inflammation, and modulating heartbeats.[34]

© Donald Erickson/iStockphoto

Nuts are a good source of monounsaturated fat, which lowers LDL cholesterol and makes it less susceptible to oxidation. Nuts are also high in omega-3 fatty acids, fiber, vegetable protein, antioxidants, and plant sterols. Diets containing nuts may improve blood lipids and the functioning of cells lining the artery wall.[35]

© Svetl/iStockphoto

Oatmeal and brown rice are good sources of soluble fiber, which has been shown to reduce blood cholesterol levels. Whole grains also provide omega-3 fatty acids, B vitamins, and antioxidants, as well as other phytochemicals that may protect against heart disease.[36]

Photo Researchers, Inc.

Plant sterols resemble cholesterol chemically, making it difficult for the digestive tract to distinguish them from cholesterol. Consuming plant sterols reduces cholesterol absorption, lowering total and LDL cholesterol levels.[39] Plant sterols are found in vegetable oils, nuts, seeds, cereals, legumes, and many fruits and vegetables and are added to special margarines, salad dressings, and orange juice.

© Pierre-Luc Bernier/iStockphoto

Modest consumption of dark chocolate is associated with reduced risk of heart disease. This is attributed to the phytochemicals in dark chocolate.[37] In addition, most of the fat in chocolate is from stearic acid, a saturated fatty that does not cause an increase in blood levels of LDL cholesterol.

© Florea Marius Catalin/iStockphoto

Moderate alcohol consumption—that is, one drink a day for women and two a day for men (one drink is equivalent to 5 ounces wine, 12 ounces beer, or 1.5 ounces distilled spirits)—reduces blood clotting and increases HDL cholesterol but also raises blood triglyceride levels. Higher alcohol intake increases the risk of heart disease and causes other health and societal problems.

© Igor Dutina/iStockphoto

Soy can lower the risk of heart disease. It has a small LDL cholesterol-lowering effect and is high in healthy polyunsaturated fat, fiber, vitamins, and minerals and low in unhealthy saturated fat.[38]

Foods make up dietary patterns.

Dietary patterns

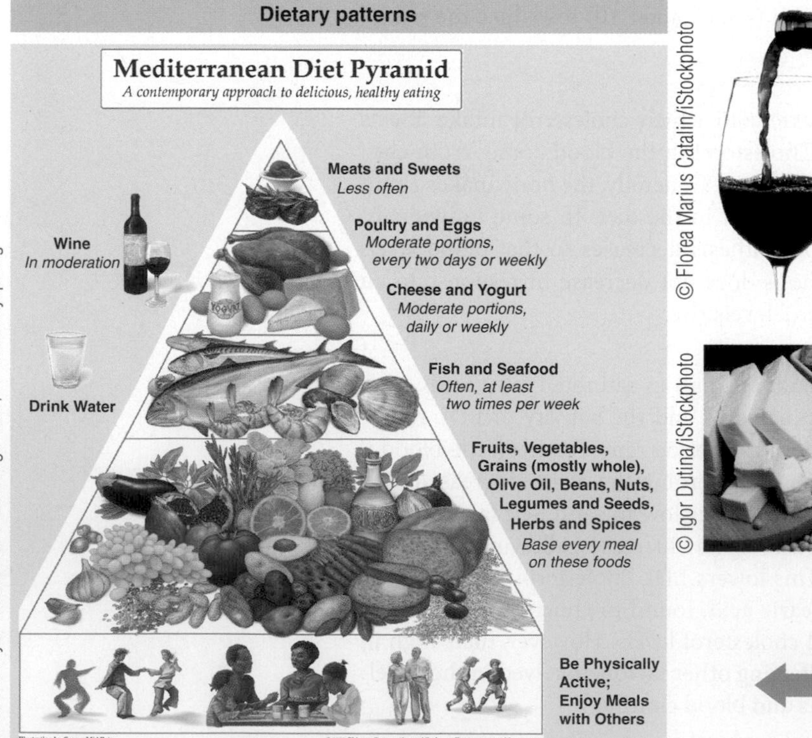

©2009 Oldways Preservation & Exchange Trust, www.oldwayspt.org

In the Mediterranean region, the main source of dietary fat is olive oil, and the typical diet is high in nuts, vegetables, and fruits. Fish is consumed routinely and red meat rarely. Despite a fat intake that is similar to that of the U.S. diet, the incidence of heart disease is much lower. This diet pyramid is based on the dietary patterns of Crete, Greece, and southern Italy around 1960, when the rates of chronic disease in this region were among the lowest in the world. This is one of the dietary patterns recommended in the 2010 Dietary Guidelines.

Figure 5.17 Nutrients, foods, and dietary patterns affect heart disease risk

Much of the impact that a dietary pattern has on heart disease risk depends on the combinations of nutrients and whole foods that make up the dietary pattern.

equivalent of adding 4 g of *trans* fat to a 2,000-kcal diet, was associated with a 23% increase in cardiovascular risk.[25] Some of the increase in risk is due to the effect of *trans* fatty acids on blood cholesterol levels; they raise LDL cholesterol levels and, in some studies, have been shown to lower HDL cholesterol. *Trans* fat further increases risk because it increases inflammation and interferes with fat metabolism.[26]

How Polyunsaturated Fat Reduces Risk When saturated fat in the diet is replaced by any type of polyunsaturated fat (except *trans*), blood levels of LDL cholesterol decrease.[17,19] However, a high intake of omega-6 polyunsaturated fatty acids may also cause a slight decrease in HDL cholesterol, which is undesirable in terms of heart disease risk. Increasing omega-3 fatty acid intake lowers levels of LDL cholesterol but not HDL cholesterol.[17] In addition to their effects on blood lipids, omega-3 fatty acids may also reduce heart disease risk by preventing the growth of atherosclerotic plaque and by promoting the synthesis of eicosanoids that protect against heart disease. Populations that consume a diet high in omega-3 fatty acids from seafood, such as the Inuits in Greenland, have a low incidence of heart disease, despite a high intake of total fat.[27,28]

How Monounsaturated Fat Reduces Risk As with polyunsaturated fatty acids, substituting monounsaturated fat for saturated fat reduces unhealthy LDL cholesterol. However, because fewer double bonds are present in monounsaturated fatty acids, they are less susceptible to oxidation than polyunsaturated fatty acids. Therefore dietary monounsaturated fatty acids decrease the oxidative susceptibility of LDL compared to polyunsaturated fatty acids. So monounsaturated fatty acids will reduce the formation of oxidized LDL, which is known to induce an inflammatory response and stimulate production of other reactive oxygen species, processes that contribute to the progression of atherosclerosis.[17]

Populations with diets high in monounsaturated fats, such as those in Mediterranean countries where olive oil is commonly used, have a mortality rate from heart disease that is half of that in the United States. This is true even when total fat intake provides 40% or more of energy intake.[29] However, the high intake of monounsaturated fat from olive oil is unlikely to be the only factor responsible for the difference in the incidence of heart disease between the Mediterranean countries and the United States. The typical Mediterranean diet is higher in fruits and vegetables and lower in animal products than the current U.S. diet, and is consumed in countries where the lifestyle includes more day-to-day activity and has fewer of the stresses of modern life (see Figure 5.17).

VIDEO

Why Plant Foods Reduce Risk Higher intakes of fruits and vegetables are associated with a lower incidence of cardiovascular disease.[30] The reason may be that many of the dietary components that protect you from heart disease and few of those that increase risk are found in fruits and vegetables and other plant foods. Fruits, vegetables, whole grains, and legumes are good sources of fiber, omega-3 fatty acids (α-linolenic acid), antioxidant vitamins, and phytochemicals. Soluble fibers, such as those in oat bran, legumes, psyllium, and flaxseed, have been shown to reduce blood cholesterol levels and therefore reduce heart disease risk (see Chapter 4). The antioxidant vitamins and phytochemicals in plant foods protect against heart disease because they decrease the oxidation of LDL cholesterol and therefore are hypothesized to prevent development of plaque in artery walls.[19]

Why B Vitamins Reduce Risk Adequate intakes of vitamin B_6, vitamin B_{12}, and folic acid can help protect against heart disease because they keep blood levels of the amino acid homocysteine low (see Chapter 8). Elevated homocysteine levels are associated with a higher incidence of heart disease.[31] Higher levels of these B vitamins in the blood are related to lower levels of homocysteine. The fortification of enriched grains with folic acid in 1998 increased folic acid intake in the United States and Canada and has resulted in a reduction in blood homocysteine levels and the risk of stroke in this population.[32] There is currently insufficient evidence to recommend supplements of B vitamins to reduce the risk of heart disease.

Niacin is another B vitamin that may affect heart disease risk. When consumed in extremely high doses, the nicotinic acid form of niacin can be used to lower blood cholesterol. Nicotinic acid is inexpensive and widely available without a prescription, but at the high doses needed to lower cholesterol it should be considered a drug, not a nutrient. Because of the potential side effects, an individual using nicotinic acid as a drug to lower cholesterol should be monitored by a physician.

Moderate Alcohol Consumption Alcohol can be a dangerous drug, but moderate alcohol consumption has been shown to reduce stress, to raise levels of HDL cholesterol, and to reduce blood clotting and thus reduce the risk of cardiovascular disease.[33] These effects are greater when red wine is consumed, as it commonly is in Mediterranean countries (see Figure 5.17). Red wine is high in phytochemicals called *polyphenols*, which are antioxidants that protect against LDL oxidation, and may have other effects that protect against the development of atherosclerosis[33] (see **Focus on Alcohol**).

Dietary Excess In addition to the lipids discussed above that increase heart disease risk, too much added sugar and salt can increase risk by contributing to the development of high blood triglycerides or high blood pressure, respectively (see Chapters 4 and 10). Excess energy can also increase the risk of heart disease by contributing to obesity (see Chapter 7).

Dietary Fat and Cancer

Cancer is the second leading cause of death in the United States. As with cardiovascular disease, there is evidence that the risk of cancer can be reduced with changes in diet and activity patterns.[40] Populations consuming diets high in fruits and vegetables tend to have a lower cancer risk. These foods are high in fiber and provide antioxidant vitamins and phytochemicals. In contrast, populations that consume diets high in fat, particularly animal fats, have a higher cancer incidence.

The mechanism whereby a high intake of dietary fat increases the incidence of various cancers is less well understood than the relationship between dietary fat and cardiovascular disease; however, dietary fat has been suggested to be both a tumor promoter and tumor initiator, depending on the type of cancer. The good news is that for the most part the same type of diet that protects you from cardiovascular disease will also reduce the risk of certain forms of cancer. For example, the Mediterranean diet, which is high in monounsaturated fat from olive oil and omega-3 fatty acids from fish, is associated with a low risk of cancers of the breast, ovary, colon, and upper digestive and respiratory tracts.[41,42,43] *Trans* fatty acids, on the other hand, not only raise LDL cholesterol levels but are also believed to increase the risk of breast cancer.[44]

How Dietary Fat Affects Breast Cancer Breast cancer is the leading form of cancer in women worldwide. In the United States, about 230,480 new cases of invasive breast cancer occur among women annually, and approximately 39,520 women will die from the disease.[45]

The mechanism by which diet affects breast cancer has been studied in laboratory animals. Since most laboratory animals do not get breast cancer tumors, studies are conducted by implanting breast tumors and examining how diet affects their growth. These experiments demonstrate that dietary fat is a tumor promoter: The tumors are more likely to grow in mice fed a high-fat diet than in those fed a low-fat diet. The type of fat also affects tumor growth: In animals fed diets high in linoleic acid, which is found in polyunsaturated vegetable oils, the tumors grow faster than in rats fed diets high in saturated fatty acids or omega-3 fatty acids. However, support of these animal studies with trials in humans has been equivocal.[46,47,48]

How Dietary Fat Affects Colon Cancer Epidemiology has correlated the incidence of colon cancer with high-fat, low-fiber diets. Diets high in red meat, which is high in saturated fat, are associated with a higher incidence of colon cancer.[49] Diets that are high in omega-3 fatty acids from fish and monounsaturated fatty acids from olive oil are associated with a lower incidence of colon cancer.[50] The connection between dietary fat and colon cancer may be related to the breakdown of fat in the large intestine. Here, bacteria metabolize dietary fat and bile, producing substances that may act as tumor initiators. A high intake of fiber tends to dilute these carcinogens by increasing the volume of feces. High-fiber diets also decrease transit time. Both of these effects reduce the exposure of the intestinal mucosa to hazardous substances in the colon contents (see Chapter 4).[51]

Dietary Fat and Obesity

Dietary fat has been postulated to contribute to weight gain and obesity. One reason is that fat has 9 kcalories per gram, more than twice as much as either carbohydrate or protein. Therefore, a high-fat meal contains more calories in the same volume as a lower-fat meal. Because people have a tendency to eat a certain weight or volume of food, a high-fat meal will add more calories than a lower-fat meal.[52] High-fat diets also tend to promote overconsumption because energy from fat is less satiating than energy from carbohydrate, so when eating a high-fat meal you will eat more calories before you feel full.[53] Dietary fat may also contribute to weight gain because it is stored very efficiently as body fat (see Chapter 7).

Despite the fact that fat is fattening, the fat content of the U.S. diet is unlikely to be the sole reason for the high rate of obesity in the United States.[54] Weight gain occurs when energy intake exceeds energy expenditure, regardless of whether the extra energy comes from fat, carbohydrate, or protein. The increasing prevalence of overweight and obesity in the United States and worldwide is likely due to a general increase in calorie intake combined with a decrease in energy expenditure.[12]

5.7 Meeting Recommendations for Fat Intake

LEARNING OBJECTIVES

- List the recommendations for total fat, saturated fat, *trans* fat, and cholesterol intake.
- Discuss why the Dietary Guidelines recommend less solid fat and more liquid oils.
- Choose heart-healthy foods from each section of MyPlate.
- Use food labels to choose foods that provide healthy fats.

Recommendations for amounts and types of lipids that promote health and prevent disease are made by the DRIs and the 2010 Dietary Guidelines as well as organizations and programs that target specific diet-related chronic diseases such as heart disease and cancer. Following the guidelines of MyPlate and using the information on food labels can help translate these recommendations into appropriate food choices and overall healthy diets.

Fat and Cholesterol Recommendations

Only a small amount of fat is required for life, but a diet that provides only the minimum amount would be very high in carbohydrate, would not be very palatable, and would not necessarily be any healthier than diets with more fat. Therefore, the recommendations for fat intake focus on getting enough to meet the need for essential fatty acids and choosing a diet that provides the amounts and types of fat that will promote health and prevent disease.

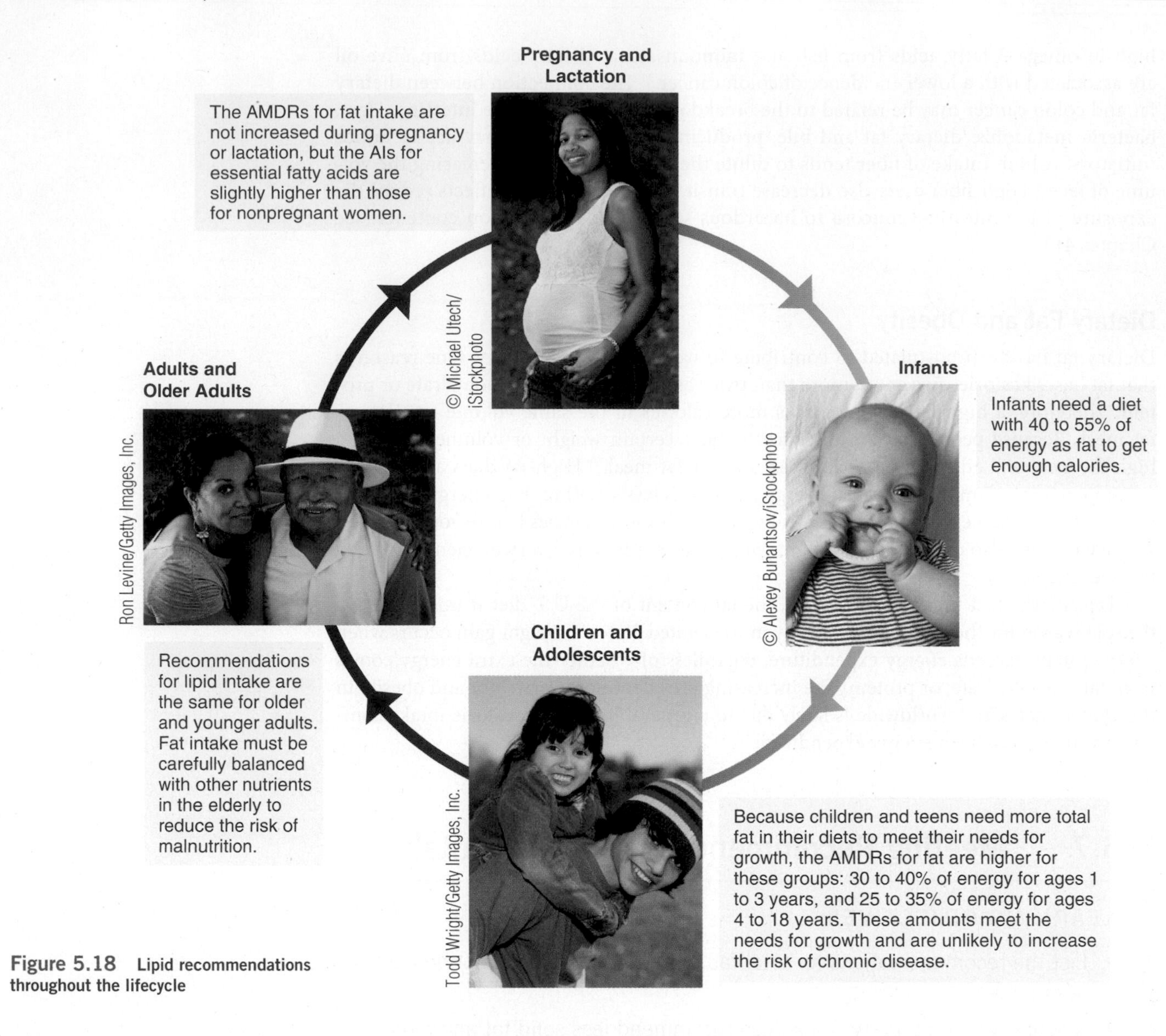

Figure 5.18 **Lipid recommendations throughout the lifecycle**

How Much Total Fat? The DRIs set an AMDR for total fat intake of 20 to 35% of calories for adults, allowing a range of fat intakes to meet individual food preferences while still promoting health. Recommendations differ somewhat for different life stages (**Figure 5.18**). Fat intakes above the AMDR generally result in a higher intake of saturated fat and make it more difficult to avoid consuming excess calories. Intakes below the 20% lower level of the AMDRs increase the probability that vitamin E and essential fatty acid intakes will be low and may contribute to unfavorable changes in HDL and triglyceride levels.

The percent of energy from of fat in your diet can be calculated by using the calculations shown in **Table 5.2**. This same calculation can be used to determine the percent of energy as saturated fat in the diet, or of total fat or saturated fat in individual foods. The amount of fat in a food or in a diet can be estimated from the Exchange Lists (**Table 5.3**) or determined using values from food labels, food composition tables, or iProfile or other diet analysis programs (see **Nutrient Composition of Foods Supplement** or **iProfile**).

What Types of Lipids Are Recommended? The 2010 Dietary Guidelines recommends that the fats in our diet come primarily from liquid oils and foods containing liquid fats such as nuts and fish. These sources are rich in essential fatty acids. The guidelines also recommend we limit our intake of solid fats to help reduce the risk of heart disease.

TABLE 5.2 Calculating Percent of Energy from Fat

Determine
• The total energy (kcalorie) intake for the day
• The grams of fat in the day's diet
Calculate energy from fat
• Fat provides 9 kcal/g
• Multiply grams of fat by 9 kcal/g
Energy (kcal) from fat = grams fat × 9 kcal/g fat
Calculate % energy from fat
• Divide energy from fat by total energy and multiply by 100 to express as a percent
$\text{Percent of energy from fat} = \frac{\text{kcal from fat}}{\text{Total kcal}} \times 100$
For example:
A diet contains 2000 kcal and 75 g of fat
75 g of fat × 9 kcal/g = 675 kcal from fat
$\frac{\text{675 kcal from fat}}{\text{2000 kcal}} \times 100 = 34\%$ of energy (kcals) from fat

TABLE 5.3 Using Exchange Lists to Estimate Fat Content

Exchange groups/lists	Serving size	Fat (g)
Carbohydrate Group		
Starch	⅓ cup pasta; ½ cup potatoes; 1 slice bread	0–1
Fruit	1 small apple, peach, pear; ½ banana; ½ cup canned fruit (in juice)	0
Milk	1 cup milk or yogurt	
Nonfat		0
Low-fat		2–3
Reduced-fat		5
Whole		8
Other carbohydrates	Serving sizes vary	Varies
Vegetables	½ cup cooked vegetables, 1 cup raw	0
Meat/Meat Substitute Group	1 oz meat or cheese, ½ cup legumes	
Very lean		0–1
Lean		3
Medium fat		5
High fat		8
Fat Group	1 tsp butter, margarine, or oil; 1 T salad dressing	5

How Much of the Essential Fatty Acids? The amounts of the essential fatty acids recommended by the DRIs are based on the amounts consumed by the healthy U.S. population. The AI for linoleic acid is 12 grams per day for women and 17 grams per day for men. You can meet your requirement by consuming a half-cup of almonds or 2 tablespoons of corn oil. For α-linolenic acid, the AI is 1.1 grams per day for women and 1.6 grams per day for men. Your requirement can be met by eating a quarter-cup of walnuts or a tablespoon of canola oil or ground flaxseeds. Consuming these amounts provides the recommended ratio of linoleic to α-linolenic acid of between 5:1 and 10:1.[5] AMDRs of 5 to 10% of energy for linoleic acid and 0.6 to 1.2% of energy for α-linolenic acid (with 10% or less of this as EPA and DHA) have been set.[5]

Choose Liquid Fats To meet recommendations for essential fatty acids the Dietary Guidelines recommend that most of the fats in the diet come from oils. Oils such as olive, canola, peanut, safflower, corn, and soybean are high in polyunsaturated and monounsaturated fats. Foods containing healthy oils such as nuts, avocados, and fish also provide plenty of monounsaturated fat and omega-3 fatty acids to help meet the recommended ratio of omega-6 to omega-3. Omega-3 fatty acid supplements can help improve this ratio, but because the beneficial effects are greater when the omega-3 fatty acids are consumed in fish as opposed to supplements, the American Heart Association recommends consuming fish twice a week to increase omega-3 intakes and the Dietary Guidelines recommend that Americans increase their intake of seafood.[12,28]

Limit Solid Fats The 2010 Dietary Guidelines recommend limiting solid fats, which are generally sources of saturated fat, *trans* fat, and cholesterol. To reduce the risk of heart disease it is recommended that everyone over age 2 limit saturated fatty acid intake to less than 10% of total calories by replacing foods high in saturated fat with sources of mono or polyunsaturated fat. Reducing saturated fat intake to 7% of calories and keeping *trans* fat intake as low as possible can lower risk even more. Limiting sources of solid fats in the diet will help to reduce saturated fat, *trans* fat, and excess calories.[12] The guidelines recommend limiting cholesterol to less than 300 mg per day for the general population and suggest that those at high risk for heart disease reduce their cholesterol intake to less than 200 mg per day.

Guidelines for Prevention of Specific Diseases

In addition to these general recommendations for fat intake, some dietary recommendations target populations at risk for specific diseases. For example, to reduce the risk of cancer the American Cancer Society provides diet and lifestyle recommendations similar to the Dietary Guidelines, but also recommends limiting the amount of processed meat and red meat consumed.[55] The American Heart Association and the National Cholesterol Education Program (NCEP) have developed dietary and lifestyle recommendations to lower heart disease risk (see Appendix D and **Figure 5.19**).[17] For individuals with extremely high cholesterol levels or for those for whom lifestyle changes are not effective, the NCEP recommends drug therapy. The most common drugs used to treat elevated blood cholesterol are the statins; these work by blocking cholesterol synthesis in the liver and by increasing the capacity of the liver to remove cholesterol from the blood. Other cholesterol-lowering drugs act in the gastrointestinal tract by preventing cholesterol and bile absorption.

Dietary Recommendations to Reduce the Risk of Heart Disease

Dietary factor	Recommendation
Saturated fat	<7% of calories
Polyunsaturated fat	Up to 10% of calories
Monounsaturated fat	Up to 20% of calories
Cholesterol	<200 mg/day
Total fat	25–35% of calories
Protein	Approximately 15% of calories
Carbohydrate	50–60% of calories
Soluble fiber	10–25 g/day
Plant stanols	2 g/day
Sodium	≤2300 mg/day
Total calories	Balance to maintain a desirable body weight

Figure 5.19 Recommendations of the National Cholesterol Education Program
The National Cholesterol Education Program recommends that all adults have their blood cholesterol levels checked at least every five years. Those with elevated blood cholesterol or at risk for abnormal lipid levels should follow these dietary recommendations.

TABLE 5.4 How to Choose: Fat and Cholesterol

Limit your intake of cholesterol, *trans* fat, and saturated fat
• Choose low-fat milk and yogurt. • Trim the fat from your meat, and serve chicken and fish but don't eat the skin. • Cut in half your usual amount of butter and use soft rather than stick margarine. • Watch your fast-food choices—choose grilled chicken over burgers and skip the special sauce.
Increase the proportion of polyunsaturated and monounsaturated fats
• Snack on nuts and seeds. • Add olives and avocados to your salads. • Use olive, peanut, or canola oil for cooking and salad dressing. • Use corn, sunflower, or safflower oil for baking.
Up your omega-3 intake
• Sprinkle ground flax seeds on your cereal or yogurt. • Have a serving mackerel, lake trout, sardines, tuna, or salmon. • Pick a leafy green vegetable with dinner. • Add walnuts to your salad or cereal.

Translating Recommendations into Healthy Diets

The typical U.S. diet meets the recommendations of 20 to 35% of calories from fat. *Trans* fat intake has declined, for the most part because food manufacturers have reduced the *trans* fat content of fats and oils used in processing. However, cholesterol and saturated fat intake often exceeds recommendations, and most people don't get enough omega-3 polyunsaturated fatty acids.[1,12] Shifting the sources of dietary fat can improve the proportion of healthy fats in your diet. **Table 5.4** gives some suggestions on how to make healthy fat choices.

Using MyPlate to Make Healthy Choices

Your choices from each food group can have a significant impact on the amounts and types of fats in your diet (**Figure 5.20**). Generally, grains, fruits, and vegetables are low in total fat and saturated fat, and they contain no cholesterol. However, choices from these groups need to be made with care to avoid fats that are added in processing or preparation. For example, flavored rice dishes and baked goods such as doughnuts, cookies, and muffins are grain products that are sources of dietary fat. In fact, most baked goods are so high in added fats and sugars that about half of their calories are considered empty calories. Fresh vegetables are very low in fat, but French fries and breaded fried vegetables, such as onion rings and mushrooms, are high in fat and energy, and depending on the fat used for frying, can be a source of saturated fat or *trans* fat.

Smart choices from the protein and dairy groups can reduce your intake of unhealthy fats and boost healthy ones. For example, limiting fatty cuts of meat and high-fat processed meats, trimming the fat from meat, removing the skin from poultry, and choosing low-fat dairy products can reduce the amount of total fat, saturated fat, and cholesterol in the diet. Choosing fish and shellfish will provide a meal with less saturated fat and more heart-protective long-chain omega-3 fatty acids. Levels of omega-3 fatty acids are higher in oilier fish such as salmon, trout, and herring. Choosing legumes as a protein source will add

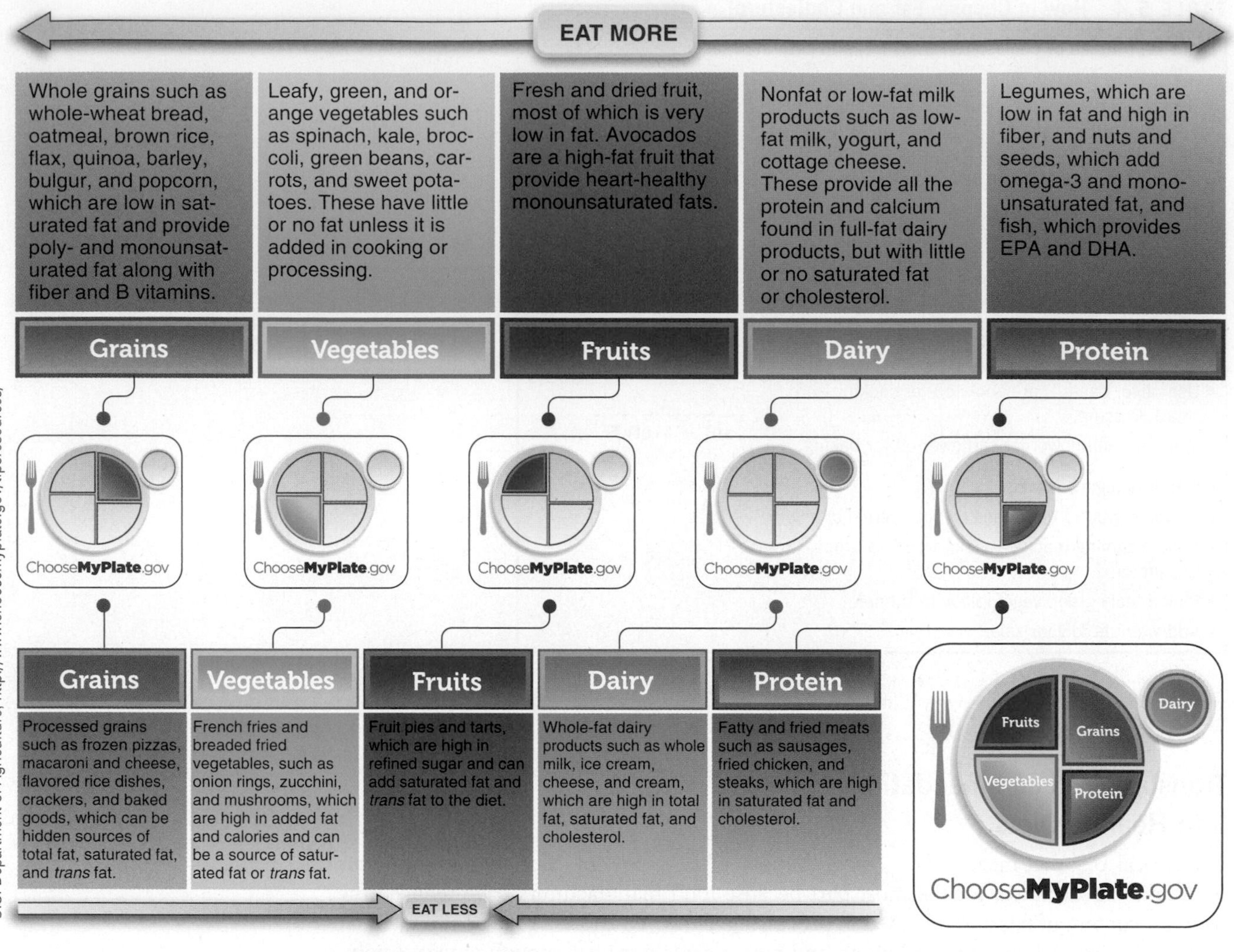

Figure 5.20 Healthy MyPlate choices
MyPlate recommends limiting intake of solid fats, which includes fats that are high in saturated or *trans* fat, and choosing liquid oils, which are high in mono- and polyunsaturated fats.

fiber without adding total fat, saturated fat, or cholesterol, and choosing nuts and seeds will add a source of omega-3 and monounsaturated fat.

Oils, butter, margarine, fatty sauces, and salad dressings used in cooking or added at the table are the most concentrated sources of fat in the diet. Limiting these can reduce your total fat and calorie intake, and choosing liquid oils rather than solid fats can increase the proportion of unsaturated fats in your diet. Solid fats such as butter, shortening, beef fat, and lard are high in saturated or *trans* fats. These solid fats provide the same number of calories per gram as oils but few essential nutrients. Therefore MyPlate considers them to be empty calories. Consuming too many empty calories makes it difficult to meet your nutrient needs without gaining weight. If you consume a 2000-kcalorie diet, you can include only about 260 empty calories. Spreading 1 tablespoon of butter on your morning bagel uses up 100 kcalories, over one-third of your empty calorie allowance. Using vegetable oils such as canola and olive oil that are high in monounsaturated fat, or corn and soybean oils that are high in polyunsaturated fat instead of butter will increase the proportion of unsaturated fats. Soybean oil and canola oil are also good plant sources of omega-3 unsaturated fatty acids. Choosing margarines and other spreads that are low in *trans* fat can reduce *trans* fat intake (see **Critical Thinking: Choosing Healthy Fats**).

CRITICAL THINKING

Choosing Healthy Fats

© Paul Johnson/iStockphoto

Isabel has a busy schedule—working full time and going to college. She picks up breakfast and lunch on the run and makes a quick dinner when she gets home at night. She would like to improve her diet by choosing healthier fats. An analysis of Isabel's fat intake is shown in the table.

FOOD	SERVING	FAT (g)	SATURATED FAT (g)	*TRANS* FAT (g)
Breakfast				
Bran muffin	1 large (3 oz)	6	2.6	0.5
Margarine	2 tsp	18	1.3	2
Coffee	1 cup	0	0	0
Whole milk	1 cup	8	5	0.2
Lunch				
Big burger	1 (2 oz bun, 5 oz beef, 1 oz cheese, 2 T dressing)	30	12.5	1
French fries	1.5 cup	22	5	2
Diet Coke	16 oz	0	0	0
Snack				
Apple	1 med	0	0	0
Dinner				
Fish sticks	6 (3 oz)	10	2.5	0
Potato puffs	¾ cup	8	4	4
Hot tea	8 oz	0	0	0
Apple pie	1 slice	13	3.4	3.3
TOTAL		115	36.3	13

CRITICAL THINKING QUESTIONS

- Isabel's diet provides 2200 kcals. What is her percent of calories from fat? Saturated fat? *Trans* Fat? Are these within the recommended ranges?
- Find the solid fats. Which two foods contain the most saturated fat? *Trans* fat?
- Suggest a whole-grain breakfast item that has less saturated fat than the bran muffin.

Fat is not the only factor in a healthy diet. Isabel compares her current diet to her MyPlate Daily Food Plan.

FOOD GROUP	DAILY FOOD PLAN	ISABEL'S DIET
Grains	7 oz (3.5 whole grains)	6 oz (3 whole)
Vegetables	3 cups	1¾ cup
Fruit	2 cups	1 cup
Dairy	3 cups	2 cups
Protein	6 oz	6 oz
Oils	6 tsp	6 tsp

CRITICAL THINKING QUESTIONS

- Isabel's weight is stable on this diet. How can she be short on fruits, vegetables, grains, and dairy, but still meet her calorie needs?
- To increase the variety of her vegetables, suggest a more nutrient-dense vegetable to replace the potato puffs.
- Suggest a substitution that would add a fruit to her diet without increasing the calories.

iProfile Use iProfile to find a fast-food sandwich that has less saturated fat than a Big Mac.

© iStockphoto

OFF THE LABEL Choosing Lean Meat

Looking for lean meat? Although fresh meats are not required to carry a Nutrition Facts label, they often do carry information about fat content. Understanding the terminology can help you choose meats that will fit into your diet plan.

The terms "lean" and "extra lean" describe the fat content of packaged meats such as hot dogs and lunch meat, as well as fresh meats such as pork chops and steaks. "Lean" means the meat contains less than 10% fat by weight, and "extra lean" means it contains no more than 5% fat by weight. Ground beef is an exception to these labeling rules. The USDA allows ground beef to be labeled "lean" even if as much as 22% of its weight is fat. To further complicate your shopping, the amount of fat in ground beef labeled "lean" and "extra lean" can vary from store to store.

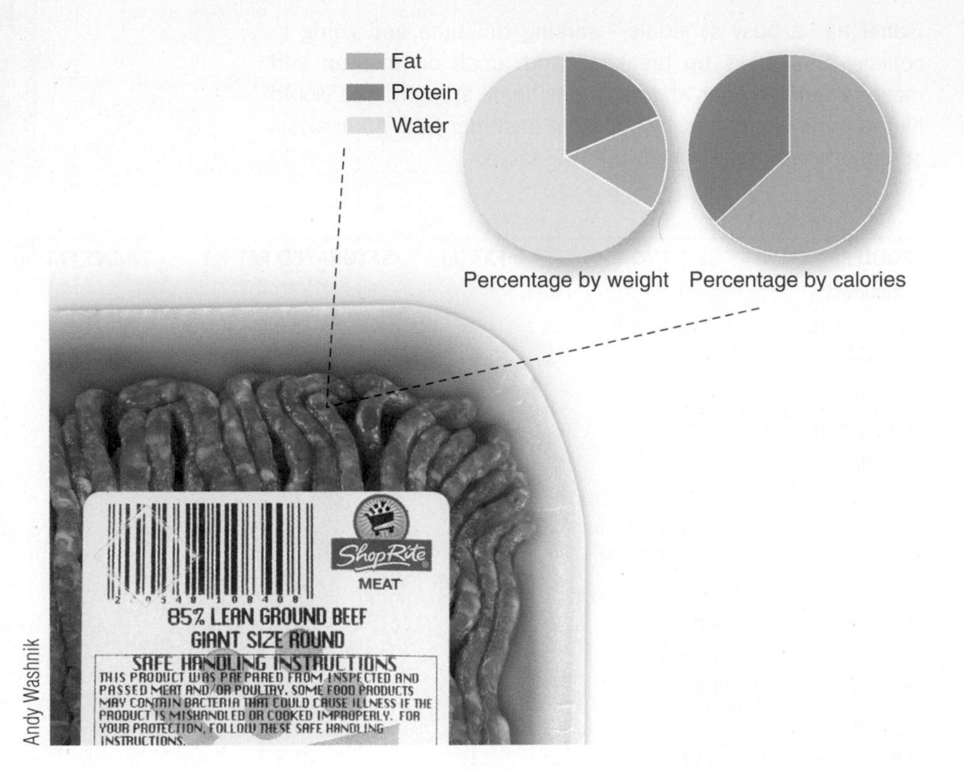

Andy Washnik

You can still figure out how much fat is in lean ground beef because ground meats labeled "lean" or "extra lean" must indicate the actual percentage of fat versus lean by stating that it is a certain "% lean" (see figure). But "% lean" claims can be misleading. A food that is 85% lean has only 15% fat, which might not sound like much. However, "% lean" refers to the weight of the meat that is lean. So when the label says the meat is 85% lean, it means that 15% of the weight of the meat is fat, or that there are 15 grams of fat in 100 grams (3.5 ounces) of raw hamburger. This is a relatively small percentage by weight, but because fat contains 9 kcalories per gram, the fat contributes 63% of the calories in the meat (see charts).

Should you pass on the ground beef and select ground turkey instead? Check the label. If the package is labeled "ground turkey," it may contain skin and leg meat and actually have more fat (about 15 to 20%) than lean ground beef. Only poultry labeled "ground turkey (or chicken) breast" is made with just the lean breast meat and contains only about 3% fat by weight.

THINK CRITICALLY: If you want to purchase ground beef that fits the definition of "lean" meat, what "% lean" should you look for on the label?

Using Food Labels to Make Healthy Choices

Food labels provide an accessible source of information on the fat content of packaged foods. Understanding how to use this information can help you make more informed choices about the foods you include in your diet. Unfortunately, food labels are not always available on fresh meats, which are one of the main contributors of saturated fat and cholesterol in our diets (see **Off the Label: Choosing Lean Meat**).

The Nutrition Facts section provides information about the amounts of total fat, saturated fat, *trans* fat, and cholesterol in a food (**Figure 5.21**). The % Daily Values for total fat, saturated fat, and cholesterol help consumers to tell at a glance how much of

Figure 5.21 **Fat and cholesterol on food labels** Understanding how to use food labels can help you make more informed choices. This label, from chocolate chip cookies, tells us that they are low in cholesterol, moderately high in saturated fat, and contain less than 0.5 g of *trans* fat in a serving.

Chocolate Chip Cookies

Nutrition Facts
Serving Size 3 Cookies (33g)
Servings Per Container About 13

Amount Per Serving	
Calories 160	Calories from Fat 70
	% Daily Value*
Total Fat 8g	12%
Saturated Fat 3g	15%
Trans Fat 0g	
Polyunsat Fat 2.5g	
Monounsat Fat 2g	
Cholesterol 0mg	0%
Sodium 110mg	5%
Potassium 45mg	1%
Total Carbohydrate 22g	7%
Dietary Fiber Less than 1gram	3%
Sugars 11g	
Protein 2g	
Vitamin A 0% • Vitamin C 0%	
Calcium 0% • Iron 4%	

*Percent Daily Values are based on a 2,000 calorie diet. Your daily values may be higher or lower depending on your calorie needs:

	Calories:	2,000	2,500
Total Fat	Less than	65g	80g
Sat Fat	Less than	20g	25g
Cholesterol	Less than	300mg	300mg
Sodium	Less than	2,400mg	2,400mg
Potassium	Less than	3,500mg	3,500mg
Total Carbohydrate		300g	375g
Dietary Fiber		25g	30g

INGREDIENTS: ENRICHED FLOUR (WHEAT FLOUR, NIACIN, REDUCED IRON, THIAMINE MONONITRATE (VITAMIN B1), FOLIC ACID, SEMISWEET CHOCOLATE CHIPS (SUGAR, CHOCOLATE, COCOA BUTTER, DEXTROSE, SOY LECITHIN AN EMULSIFIER), SUGAR, SOY BEAN OIL AND/OR PARTIALLY HYDROGENATED COTTONSEED OIL, HIGH FRUCTOSE CORN SYRUP, LEAVENING (BAKING SODA AND/OR AMMONIUM PHOSPHATE), SALT, WHEY (FROM MILK), NATURAL AND ARTIFICIAL FLAVOR, CARAMEL COLOR.
CONTAINS: WHEAT, SOY, MILK

A % Daily Value is listed for total fat, saturated fat, and cholesterol. This allows consumers to tell how a food fits the recommendations. Generally, ≤5% of the Daily Value is low, and ≥20% is high. There are no Daily Values for *trans*, polyunsaturated, and monounsaturated fats.

The Nutrition Facts panel lists calories from fat; grams of total fat, saturated fat, and *trans* fat; and milligrams of cholesterol in a serving. The amount of monounsaturated and polyunsaturated fat is voluntarily included on the labels of some products.

If the product has less than 0.5 gram of *trans* fat per serving, the Nutrition Facts panel will list the amount of *trans* fat as 0, even if partially hydrogenated oil is an ingredient.

The Daily Values recommend consuming less than 30% of calories as fat, no more than 300 mg of cholesterol per day, and no more than 10% of calories as saturated fat. It is recommended that *trans* fat intake be limited to the amounts present naturally in meats and dairy products (≤0.5% of calories).

The sources of fat in a product are listed in the ingredient list with the other ingredients, in order of prominence by weight.

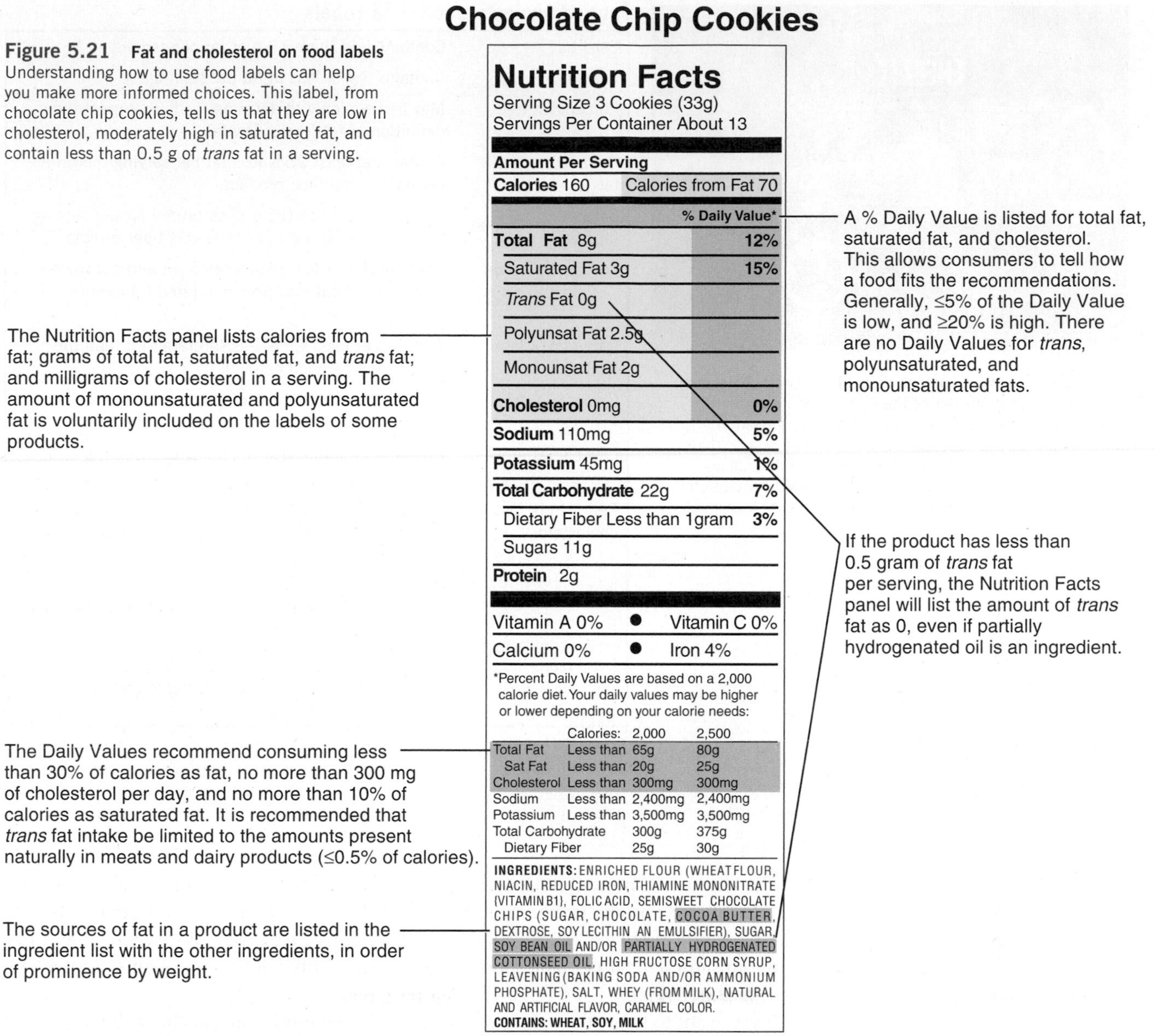

the recommended daily limit a serving provides. For example, if a serving provides 50% of the Daily Value for saturated fat—that is, half the recommended maximum daily intake for a 2000-kcalorie diet—the rest of the day's foods will have to be carefully selected to not exceed the recommended maximum. To choose foods low in saturated fat and cholesterol, look for foods with 5% of the Daily Value or less of these in a serving and avoid those with 20% or more.

Although the amount of *trans* fat in a serving has been included on food labels since 2006, it may mislead many consumers because products with less than 0.5 grams per serving may be labeled as having 0 g of *trans* fat. As a result, individuals may ingest significant quantities of *trans* fats while believing they have consumed none.[56] The American Heart Association recommends limiting *trans* fat to 1% energy (or 2 g based on a 2000-kcal diet).[57] So, if throughout the day someone consumes two servings of products containing 0.49 g of *trans* fat per serving, they will have consumed almost half the recommended maximum while thinking they have not consumed any.

Andy Washnik

Figure 5.22 **Nutrient content claims for fat and cholesterol** Food labeling regulations define the nutrient content claims used on these labels and prevent them from being used in ways that might confuse consumers. For instance, a food that contains no cholesterol but is high in saturated fat, such as crackers containing coconut oil, cannot be labeled "cholesterol free" because saturated fat in the diet raises blood cholesterol.

Fat and Cholesterol on Food Labels

Fat-free	Contains less than 0.5 g of fat per serving.
Low-fat	Contains 3 g or less of fat per serving.
Percent fat free	May be used only to describe foods that meet the definition of fat-free or low-fat.
Reduced or less fat	Contains at least 25% less fat per serving than the regular or reference product.
Saturated fat-free	Contains less than 0.5 g of saturated fat per serving and less than 0.5 g *trans* fatty acids per serving.
Low saturated fat	Contains 1 g or less of saturated fat and not more than 15% of calories from saturated fat per serving.
Reduced or less saturated fat	Contains at least 25% less saturated fat than the regular or reference product.
Cholesterol-free	Contains less than 2 mg of cholesterol and 2 g or less of saturated fat per serving.
Low cholesterol	Contains 20 mg or less of cholesterol and 2 g or less of saturated fat per serving.
Reduced or less cholesterol	Contains at least 25% less cholesterol than the regular or reference product and 2 g or less of saturated fat per serving.
Lean	Contains less than 10 g of fat, 4.5 g or less of saturated fat, and less than 95 mg of cholesterol per serving and per 100 g.
Extra lean	Contains less than 5 g of fat, less than 2 g of saturated fat, and less than 95 mg of cholesterol per serving and per 100 g.

Source: FDA, Center for Food Safety and Applied Nutrition.

The amount of monounsaturated and polyunsaturated fat is voluntarily included on the labels of some products. For example, in addition to listing the 2 g of saturated fat, the label on a bottle of olive oil may indicate that it contains 2 g of polyunsaturated fat and 10 g of monounsaturated fat per tablespoon. There are no Daily Values for polyunsaturated and monounsaturated fat.

If you want to know the source of fat in a packaged food, you can check the ingredient list. This will show you, for example, if a food contains corn oil, soybean oil, coconut oil, or partially hydrogenated vegetable oil (see Figure 5.21). Labels may also include nutrient content claims such as "low-fat," "fat-free," and "low cholesterol" that describe their fat content (**Figure 5.22**). Food labeling regulations have developed standard definitions for these terms. Food labels may also include health claims related to their fat content. For example, foods low in saturated fat and cholesterol may state that they help to reduce the risk of coronary heart disease (see Appendix E).

The Role of Reduced-Fat Foods and Fat Replacers

People often choose low-fat and reduced-fat products in order to reduce the total amount of fat in their diets. Some of these foods, such as low-fat and nonfat milk and yogurt, are made by simply removing the fat, but in other products, the fat is replaced with ingredients that mimic the taste and texture of the fat. Some reduced-fat foods contain added sugars to improve the taste and texture. Some contain soluble fiber or modified proteins that simulate fat, and others contain fats that have been altered to reduce or prevent absorption (**Figure 5.23**).[58] A problem with nonabsorbable fats is that they reduce the absorption of the fat-soluble substances in the diet, including the fat-soluble vitamins, A, D, E, and K. To avoid depleting these vitamins, products made with the nonabsorbable fat substitute Olestra have been fortified with them. However,

The artificial fat Olestra (sucrose polyester) is made by attaching fatty acids to sucrose molecules. Olestra cannot be digested by either human enzymes or bacterial enzymes in the gastrointestinal tract. Therefore, it is excreted in the feces without being absorbed and so provides no calories.

Polysaccharides such as pectins and gums are often used in baked goods, as well as salad dressings, sauces, and ice cream. These thicken foods and mimic the slippery texture that fat provides. They reduce the amount of fat in a product and at the same time add soluble fiber. They typically add only about 1 to 2 kcalories per gram.

The sugar sucrose is usually added to low-fat and nonfat baked goods to improve flavor and add volume. Sucrose adds 4 kcalories per gram.

Protein-based fat replacers are made from milk and egg proteins processed to form millions of microscopic balls that slide over each other, mimicking the creamy texture of fat. They provide less than 2 kcalories per gram and are used in frozen desserts, cheese foods, and other products but cannot be used for frying because they break down at high temperatures.

Figure 5.23 Fat replacers
Carbohydrates and proteins added to replace fat add calories to foods. In most cases, however, the low-fat food is lower in calories than the original product. Some fat replacers are made from fats that have been modified to reduce how well they can be digested and absorbed. The calories they provide depend on how much is absorbed.

these products are not fortified with β-carotene and other fat-soluble substances that may be important for health. Another potential problem with Olestra is that it can cause abdominal cramping and loose stools in some individuals because it passes into the colon without being digested.

In addition to modifying fats to reduce absorption, food manufacturers are modifying them to improve their healthfulness. Only about 10% of the glycerides naturally present in plant oils are diglycerides. However, food manufacturers are able to modify these oils to create oils that contain 70% diglycerides. These high-diglycerides oils (marketed as ENOVA oil) are not lower in calories, but when they replace similar oils containing triglycerides, blood lipid levels are lower after eating and body weight and fat accumulation in the abdominal region are reduced.[59]

Will using low-fat and reduced-fat products improve your diet? Some low-fat foods make an important contribution to a healthy diet. Low-fat dairy products are recommended because they provide all the essential nutrients contained in the full-fat versions but have fewer calories and less saturated fat and cholesterol. Using these products increases the nutrient density of the diet as a whole. However, not all reduced-fat foods are nutrient dense. Low-fat baked goods often have more sugar than the full-fat versions because extra sugar is needed to add volume and make up for the flavor that is lost when the fat is removed. Some are just lower-fat versions of nutrient-poor choices, such as cookies and chips. If these reduced-fat desserts and snack foods replace whole grains, fruits, and vegetables, the resulting diet could be low in fat but also low in fiber, vitamins, minerals, and phytochemicals.

Using low-fat foods does not necessarily transform a poor diet into a healthy one or improve overall diet quality, but if used appropriately, fat-modified foods can be part of a healthy diet.[58] For example, if a low-fat salad dressing replaces a full-fat version, it allows you to enhance the appeal of a nutrient-rich salad without as much added fat and calories from the dressing. Low-fat products can also be used in conjunction with weight-loss diets because they are often lower in calories. But check the label. Although most are lower in calories, they are by no means calorie free and cannot be consumed liberally without adding calories to the diet and possibly contributing to weight gain.

OUTCOME

Studio R Schmitz/Age Fotostock America, Inc.

© Ben Blankenburg/iStockphoto

Kyle's friends and relatives had offered lots of advice on how to reduce his risk of heart disease. He tried a few of their suggestions. He consumed a vegetarian diet for only three days before he decided that wasn't the approach for him. He got tired of fish after eating tuna three days a week. He tried going to the gym every day for a week but got behind in his schoolwork. Finally he decided to make more sensible changes. He plans to have one vegetarian meal each week, eat a tuna sandwich once a week, and choose fish when he eats out to increase his intake of heart-healthy omega-3 fatty acids. He knows it will be easy for him to eat more whole grains and at least six servings of fruits and vegetables each day if he makes a list before he goes to the grocery store so he has the food in his kitchen. He now carefully reads food labels, avoids *trans* fat, and uses canola oil when he cooks. The biggest change Kyle made was to increase his activity. He walks everywhere and plays racquetball or swims twice a week. **Kyle can't do anything about his family history but has changed his lifestyle to reduce the risks he can control.**

APPLICATIONS

ASSESSING YOUR DIET

1. **How do the fats in your diet compare to the recommendations?**

   a. Use iProfile to calculate your average fat, saturated fat, and cholesterol intake using the three-day food record you kept in Chapter 2.
   b. How does your fat intake compare with the recommendation of 20 to 35% of energy from total fat?
   c. How does your percent of energy from saturated fat compare to the Dietary Guidelines recommendation of less than 10% of calories?
   d. What foods do you typically consume that are high in saturated fat? Suggest food substitutions that will decrease the amount of saturated fat in your diet.
   e. How does your cholesterol intake compare to the Dietary Guidelines recommendation of less than 300 mg per day? If it is greater than 300 mg, suggest some substitutions that will decrease your cholesterol intake.
   f. List some packaged foods in your diet that contain *trans* fat. What substitutions could you make to decrease your *trans* fat intake?
2. **Do you make healthy choices from all the MyPlate food groups?**
   a. List the dairy products in your diet and indicate if they are full fat or reduced fat.
   b. List the grain products you typically consume. How many of them are baked goods with added fats? How many of them are eaten with added solid fats such as butter or margarine? Suggest changes you could make to reduce the solid fats and empty calories added to your carbohydrates.

c. List the vegetables in your diet. Underline those that are cooked or prepared in a way that increases their fat intake. For example, are they fried or topped with butter or salad dressing? Are any of these solid fats?

d. List the high-protein foods in your diet. Which ones are naturally high in fat? Which, if any, are high in fat because fat is added in cooking or as sauces or gravies? Which are low in fat?

3. Are you getting your omega-3's?

a. Review all three days of the food record you kept in Chapter 2 and list foods you eat that are good sources of omega-3 fatty acids.

b. If your diet does not contain good sources of omega-3 fatty acids, suggest some substitutions that would boost your omega-3 intake.

CONSUMER ISSUES

4. How much fat is in packaged foods? Examine the food labels on four packaged foods. If you consumed only this food for an entire day, how many servings could you eat before exceeding the Daily Value for:

a. Total fat?

b. Saturated fat?

c. Cholesterol?

5. The table shown here illustrates the percentage of calories from fat in the diets of American men and women and the incidence of obesity among adults between 1971 and 2000. Use these values to construct a graph. Based on your graph, discuss what happened to each over time and how these two factors might be related.

Year[6,60,61]	1971–1974	1976–1980	1988–1994	1999–2000
% Calories from fat (men)	36.9	36.8	33.9	32.8
% Calories from fat (women)	36.1	36	33.4	32.8
Percentage of adults who are obese	14.5	15	23.3	30.5

CLINICAL CONCERNS

6. Sam is supposed to lower his intake of saturated fat. Use iProfile to help him find substitutions for the following foods that are lower in saturated fat.

a. Scrambled eggs

b. Ham and cheese sub

c. Meatball sandwich

d. Fried chicken leg

e. Pepperoni pizza

f. Cheeseburger and fries

7. Jose is on his high-school track team. Practices are held in the afternoon; an hour after school is out. He eats a sports bar and a soda right after school so he is not hungry during practice. The first meet is held on a Saturday morning. He has a hearty breakfast of bacon, sausage, and eggs. During the meet he feels bloated and sluggish and his times are much slower than during practice. What should Jose have done differently? Suggest a better pre-meet breakfast.

SUMMARY

5.1 Fats in Our Food

- Lipids add calories, texture, and flavor to our foods.
- The typical American diet contains about 34% of calories from fat. Some of the fats we eat are visible, but others are less obvious. The sources of fat in the American diet have changed over the past 40 years. Americans have reduced their intake of eggs, red meat, whole milk, and butter but have increased the amount of fat they consume from pizza, pasta dishes, snack foods, and fast food, so the amount and type of fat in the diet have changed little.
- Too much of the wrong types of fat increases the risk of chronic disease. A healthy diet includes the right kinds of fats along with plenty of whole grains, fruits, and vegetables.

5.2 Types of Lipids

- Lipids are a diverse group of organic compounds, most of which do not dissolve in water. Triglycerides, commonly referred to as fat, are the most abundant lipid in our diet and our bodies. They consist of a backbone of glycerol with three fatty acids attached. The physical properties and health effects of triglycerides depend on the fatty acids they contain.
- Fatty acids consist of a carbon chain with an acid group at one end. Fatty acids that are saturated with hydrogen atoms are saturated fatty acids and those that contain carbon-carbon double bonds are unsaturated. The length of the carbon chain and the number and position of the carbon-carbon double bonds as well as the configuration of the double bonds determine the physical properties and health effects of fatty acids.
- Phosphoglycerides are a type of phospholipid that consist of a backbone of glycerol, two fatty acids, and a phosphate group. Phosphoglycerides allow water and oil to mix. They are used as emulsifiers in the food industry and are an important component of cell membranes and lipoproteins.
- Sterols, of which cholesterol is the best known, are made up of multiple chemical rings. Cholesterol is made by the body and consumed in animal foods in the diet. In the body, it is a component of cell membranes and is used to synthesize vitamin D, bile acids, and a number of hormones.

5.3 Lipids in the Digestive Tract

- Some triglyceride digestion begins in the stomach due to the action of gastric lipase. In the small intestine, muscular churning mixes chyme with bile from the gallbladder to break fat into small globules. This allows pancreatic lipase to access these fats for digestion.
- The products of fat digestion (primarily fatty acids and monoglycerides), cholesterol, phospholipids, and other fat-soluble substances combine with bile to form micelles, which facilitate the absorption of these materials into the cells of the small intestine.

5.4 Lipid Transport in the Body

- In body fluids, water-insoluble lipids are transported as lipoproteins. Lipids absorbed from the intestine are incorporated into chylomicrons, which enter the lymphatic system before entering the blood. The triglycerides in chylomicrons are broken down by lipoprotein lipase on the surface of cells lining the blood vessels. The fatty acids released are taken up by surrounding cells and the chylomicron remnants that remain are taken up by the liver.
- Very-low-density lipoproteins (VLDLs) are synthesized by the liver. With the help of lipoprotein lipase, they deliver triglycerides to body cells. Once the triglycerides have been removed, intermediate-density lipoproteins (IDLs) remain. These are transformed in the blood into low-density lipoproteins (LDLs). LDLs deliver cholesterol to tissues by binding to LDL receptors on the cell surface.
- High-density lipoproteins (HDLs) are made by the liver and small intestine. They help remove cholesterol from cells and transport it to the liver for disposal.

5.5 Lipid Functions in the Body

- Dietary fat is needed for the absorption of fat-soluble vitamins and to provide essential fatty acids. In the body, triglycerides in adipose tissue provide a concentrated source of energy and insulate the body against shock and temperature changes. Oils lubricate body surfaces.
- Linoleic acid (omega-6) and α-linolenic acid (omega-3) are considered essential fatty acids because they cannot be synthesized by the body. Essential fatty acids are needed for normal structure and function of cell membranes, particularly those in the retina and central nervous system. Hormones synthesized from cholesterol and eicosanoids synthesized from omega-6 and omega-3 polyunsaturated fatty acids have important regulatory roles. Eicosanoids help regulate blood clotting, blood pressure, immune function, and other body processes. The ratio of dietary omega-6 to omega-3 fatty acids affects the balance of omega-6 and omega-3 eicosanoids made and hence their overall physiological effects.
- Triglycerides provide a concentrated source of energy. Throughout the day triglycerides are continuously stored in adipose tissue and then broken down to release fatty acids, depending on the immediate energy needs of the body. After we eat, chylomicrons and VLDLs transport triglycerides to cells. Lipoprotein lipase breaks down the triglycerides so the fatty acids can be taken up by cells and used for energy or storage. When we fast, triglycerides stored in adipocytes are broken down by hormone-sensitive lipase, and the fatty acids and glycerol are released into the blood. To generate ATP from fatty acids, β-oxidation first breaks fatty acids into two carbon units that form acetyl-CoA. In the presence of oxygen, acetyl-CoA can be broken down by the citric acid cycle and ATP produced by the electron transport chain.

5.6 Lipids and Health

Go to WileyPLUS to view a video clip on the Mediterranean diet.

- Atherosclerosis is a disease characterized by deposits of lipids and fibrous material in the artery wall. It is begun by an injury to the artery wall that triggers an inflammatory response that leads to plaque formation. A key event in plaque formation is the oxidation of LDL cholesterol in the artery wall. Oxidized LDL cholesterol is taken up by macrophages, forming foam cells. These burst and deposit cholesterol in the artery wall. Oxidized LDL cholesterol also contributes to plaque formation by promoting inflammation and sending signals that lead to fibrous deposits and blood clot formation.
- High blood levels of total and LDL cholesterol are a risk factor for heart disease. High blood HDL cholesterol protects against heart disease. The risk of heart disease is also increased by diabetes, high blood pressure, and obesity.
- Individual nutrients and dietary components, whole foods, and dietary patterns affect the risk of heart disease. Diets high in saturated fat, *trans* fatty acids, and cholesterol increase the risk of heart disease. Diets high in omega-6 and omega-3 polyunsaturated fatty acids, monounsaturated fatty acids, certain B vitamins, and plant foods containing fiber, antioxidants, and phytochemicals reduce the risk of heart disease. The total dietary and lifestyle pattern is more important than any individual dietary factor in reducing heart disease risk.
- Diets high in fat are associated with an increased incidence of certain types of cancer. In some types of cancer, such as breast cancer, fat may act as a tumor promoter, increasing the rate of tumor growth. In the case of colon cancer, dietary fat in the colon may act as a tumor initiator by forming compounds that cause mutations.
- Fat contains 9 kcalories per gram. A high-fat diet therefore increases the likelihood of weight gain, but it is not the primary cause of obesity. Consuming more energy than expended leads to weight gain regardless of whether the energy is from fat, carbohydrate, or protein.

5.7 Meeting Recommendations for Fat Intake

- The DRI recommendations regarding fat include AIs and AMDRs for essential fatty acids and an AMDR for total fat intake of 20 to 35% of energy for adults. The DRIs also advise keeping *trans* fats, saturated fats, and cholesterol to a minimum to reduce the risk of heart disease. The Dietary Guidelines recommend that total fat account for no more than 30% of energy, that saturated fat account for no more than 10% of energy, and that dietary cholesterol be no more than 300 mg per day.
- To keep the amount and type of fat in the diet healthy added fats, protein sources, and processed foods must be chosen carefully. Following the MyPlate recommendations to choose liquid oils rather than solid fats will help provide a healthy

amount and ratio of essential fatty acids. Limiting animal fats from the protein and dairy groups reduces saturated fat intake. Choosing fish increases intake of omega-3 fatty acids. Eating nuts and seeds increase both monounsaturated fats and omega-3 fatty acids. Processed foods can be high in saturated and *trans* fat. A diet based on whole grains, fruits, vegetables, and lean meats and low-fat dairy products will meet the recommendations for fat intake.

- Low-fat foods are made by removing the fat or by including fat replacers, which simulate the taste and texture of fat. Some fat replacers are made by using mixtures of carbohydrates or proteins to simulate the properties of fat, and some use lipids that are modified to reduce absorption. Low-fat foods and products containing fat replacers can help reduce fat and energy intake when used in moderation as part of a balanced diet.

REVIEW QUESTIONS

1. How have the sources of fat in the American diet changed over the past 40 years?
2. Name two functions of fat in foods.
3. What is a lipid?
4. Name four types of lipids found in the body.
5. What distinguishes a saturated fat from a monounsaturated fat? From a polyunsaturated fat?
6. What type of processing increases the amounts of *trans* fatty acids?
7. List four functions of fat in the body.
8. Is essential-fatty-acid deficiency common in developed countries? Why or why not?
9. Why is the ratio of omega-6 to omega-3 fatty acids in the diet important?
10. What is the advantage of storing energy as body fat rather than as carbohydrate?
11. What is the function of bile in fat digestion and absorption?
12. How do HDLs differ from LDLs?
13. How are blood levels of LDLs and HDLs related to the risk of cardiovascular disease?
14. What types of foods contain cholesterol?
15. How does an atherosclerotic plaque form?
16. What is the AMDR for total dietary fat intake for adults?
17. What information about dietary fat is included on food labels?
18. List two foods that are sources of monounsaturated fatty acids, two that are sources of omega-3 fatty acids, and two that are sources of cholesterol.

REFERENCES

1. U.S. Department of Agriculture, Agricultural Research Service. 2010. Energy intakes: Percentages of energy from protein, carbohydrate, fat, and alcohol, by gender and age, *What We Eat in America*, NHANES 2007–2008.
2. Popkin, B. M., Siega-Riz, A. M., Haines, P. S., and Jahns, L. Where's the fat? Trends in U.S. diets 1965–1996. *Prev Med* 32:245–254, 2001.
3. Poti, J. M., and Popkin, B. M. Trends in energy intake among US children by eating location and food source, 1977–2006. *J Am Diet Assoc* 111:1156–1164, 2011.
4. Guthrie, J. F., Lin, B. H., and Frazao, E. Role of food prepared away from home in the American diet, 1977–78 versus 1994–96: Changes and consequences. *J Nutr Educ Behav* 34:140–150, 2002.
5. Institute of Medicine, Food and Nutrition Board. *Dietary Reference Intakes for Energy, Carbohydrates, Fiber, Fat, Protein and Amino Acids.* Washington, DC: National Academies Press, 2002, 2005.
6. Centers for Disease Control and Prevention. Trends in intake of energy and macronutrients—United States, 1971–2000. *MMWR Morb Mortal Wkly Rep* 53:80–82, 2004.
7. Lindquist, S., and Hernell, O. Lipid digestion and absorption in early life: An update. *Curr Opin Clin Nutr Metab Care* 13:314–320, 2010.
8. Brown, M. S., and Goldstein, J. L. How LDL receptors influence cholesterol and atherosclerosis. *Sci Am* 251:58–66, 1984.
9. Varga, T., Czimmerer, Z., and Nagy, L. PPARs are a unique set of fatty acid regulated transcription factors controlling both lipid metabolism and inflammation. *Biochim Biophys Acta* 1812:1007–1022, 2011.
10. Cottin, S. C., Sanders, T. A., and Hall, W. L. The differential effects of EPA and DHA on cardiovascular risk factors. *Proc Nutr Soc* 70:215–231, 2011.
11. Wall, R., Ross, R. P., Fitzgerald, G. F., and Stanton, C. Fatty acids from fish: The anti-inflammatory potential of long-chain omega-3 fatty acids. *Nutr Rev* 68:280–289, 2010.
12. U.S. Department of Agriculture and U.S. Department of Health and Human Services. *Dietary Guidelines for Americans, 2010*, 7th ed. Washington, DC: U.S. Government Printing Office, December 2010.
13. Roger, V. L., Go, A. S., Lloyd-Jones, D. M., et al. American Heart Association Statistics Committee and Stroke Statistics

Subcommittee. Heart disease and stroke statistics—2012 update: A report from the American Heart Association. *Circulation.* 125:e2–e220, 2012.

14. Libby, P., Okamoto, Y., Rocha, V. Z., and Folco, E. Inflammation in atherosclerosis: Transition from theory to practice. *Circ J* 74:213–220, 2010.

15. Miller, Y. I., Choi, S. H., Fang, L., and Tsimikas, S. Lipoprotein modification and macrophage uptake: Role of pathologic cholesterol transport in atherogenesis. *Subcell Biochem* 51:229–251, 2010.

16. Choi, S. Y., and Mintz, G. S. What have we learned about plaque rupture in acute coronary syndromes? *Curr Cardiol Rep* 12:338–343, 2010.

17. National Cholesterol Education Program. Third report of the Expert Panel on Detection, Evaluation, and Treatment of High Blood Cholesterol in Adults (Adult Treatment Panel III), 2001, 2004.

18. Kawachi, I., Colditz, G. A., Stampfer, M. J., et al. Smoking cessation and time course of decreased risks of coronary heart disease in middle-aged women. *Arch Intern Med* 154:169–175. 1994.

19. Bhupathiraju, S. N., and Tucker, K. L. Coronary heart disease prevention: Nutrients, foods, and dietary patterns. *Clin Chim Acta* 412:1493–1514. 2011.

20. Fung, T. T., Chiuve, S. E., McCullough, M. L., et al. Adherence to a DASH-style diet and risk of coronary heart disease and stroke in women. *Arch Intern Med* 168:713–720, 2008.

21. Denke, M. A. Review of human studies evaluating individual dietary responsiveness in patients with hypercholesterolemia. *Am J Clin Nutr* 62(suppl): 471S–477S, 1995.

22. Fernandez, M. L., and West, K. L. Mechanisms by which dietary fatty acids modulate plasma lipids. *J Nutr* 135: 2075–2078, 2005.

23. Hunter, J. E., Zhang, J., and Kris-Etherton, P. M. Cardiovascular disease risk of dietary stearic acid compared with trans, other saturated, and unsaturated fatty acids: A systematic review. *Am J Clin Nutr* 91:46–63, 2010.

24. Van Horn, L., McCoin, M., Kris-Etherton, P. M., et al. The evidence for dietary prevention and treatment of cardiovascular disease. *J Am Diet Assoc* 108:287–331, 2008.

25. Mozaffarian, D., Katan, M. B., Ascherio, A., et al. Trans fatty acids and cardiovascular disease. *N Engl J Med* 354:1601–1613, 2006.

26. Remig, V., Franklin, B., Margolis, S., et al. Trans fats in America: A review of their use, consumption, health implications, and regulation. *J Am Diet Assoc* 110:585–592, 2010.

27. Chateau-Degat, M. L., Dewailly, E., Louchini, R., et al. Cardiovascular burden and related risk factors among Nunavik (Quebec) Inuit: Insights from baseline findings in the circumpolar Inuit health in transition cohort study. *Can J Cardiol* 26:190–196, 2010.

28. American Heart Association. Fish and omega-3 fatty acids.

29. Willett, W. C., Sacks, F., Trichopouluo, A., et al. Mediterranean diet pyramid: A cultural model for healthy eating. *Am J Clin Nutr* 61(suppl):1402S–1406S, 1995.

30. He, F. J., Nowson, C. A., Lucas, M., and MacGregor, G. A. Increased consumption of fruit and vegetables is related to a reduced risk of coronary heart disease: Meta-analysis of cohort studies. *J Hum Hypertens* 21:717–728, 2007.

31. Humphrey, L. L., Fu, R., Rogers, K., et al. Homocysteine level and coronary heart disease incidence: A systematic review and meta-analysis. *Mayo Clin Proc* 83:1203–1212, 2008.

32. Yang, Q., Botto, L. D., Erickson, J. D., et al. Improvement in stroke mortality in Canada and the United States, 1990 to 2002. *Circulation* 113:1335–1343, 2006.

33. Klatsky, A. L. Alcohol and cardiovascular health. *Physiol Behav* 100:76–81, 2010.

34. Massaro, M., Scoditti, E., Carluccio, M. A., et al. Omega-3 fatty acids, inflammation and angiogenesis: Basic mechanisms behind the cardioprotective effects of fish and fish oils. *Cell Mol Biol* 56:59–82, 2010.

35. Sabaté, J., and Wien, M. Nuts, blood lipids and cardiovascular disease. *Asia Pac J Clin Nutr* 19:131–136, 2010.

36. Mellen, P. B., Walsh, T. F., and Herrington, D. M. Whole grain intake and cardiovascular disease: A meta-analysis. *Nutr Metab Cardiovasc Dis* 18:283–290, 2008.

37. Corti, R., Flammer, A. J., Hollenberg, N. K., and Lüscher, T. F. Cocoa and cardiovascular health. *Circulation* 119:1433–1441, 2009.

38. Sacks, F. M., Lichtenstein, A., Van Horn L., et al. Soy protein, isoflavones, and cardiovascular health: An American Heart Association Science Advisory for professionals from the Nutrition Committee. *Circulation* 113:1034–1044, 2006.

39. Gupta, A. K., Savopoulos, C. G., Ahuja, J., and Hatzitolios, A. I. Role of phytosterols in lipid-lowering: current perspectives. *QJM* 104:301–308, 2011.

40. World Cancer Research Fund/American Institute for Cancer Research (AICR). *Food, Nutrition, Physical Activity, and the Prevention of Cancer: A Global Perspective.* Washington, DC: AICR, 2007.

41. Bosetti, C., Pelucchi, C., and La Vecchia, C. Diet and cancer in Mediterranean countries: Carbohydrates and fats. *Public Health Nutr* 12:1595–1600, 2009.

42. Hall, M. N., Chavarro, J. E., Lee, I. M., et al. A 22-year prospective study of fish, n-3 fatty acid intake, and colorectal cancer risk in men. *Cancer Epidemiol Biomarkers Prev* 17:1136–1143, 2008.

43. Gerber, M. Background review paper on total fat, fatty acid intake and cancers. *Ann Nutr Metab* 55:140–161, 2009.

44. Chajès, V., Thiébaut, A. C. M., Rotival, M., et al. Association between serum trans-monounsaturated fatty acids and breast cancer risk in the E3N-EPIC Study. *Am J Epidemiol* 167:1312–1320, 2008.

45. American Cancer Society. Breast cancer facts and figures 2011–2012.

46. Zock, P. L., and Katan, M. B. Linoleic acid intake and cancer risk: A review and meta-analysis. *Am J Clin Nutr* 68:142–153, 1998.

47. Zhang, C. X., Ho, S. C., and Lin, F. Y. Dietary fat intake and risk of breast cancer: A case–control study in China. *Eur J Cancer Prev* 20:199–206, 2011.

48. American Cancer Society. What are the risk factors for breast cancer?

49. Kushi, L., and Giovannucci, E. Dietary fat and cancer. *Am J Med* 113(suppl 9B):63S–70S, 2002.

50. Hall, M. N., Chavarro, J. E., Lee, I. M., et al. A 22-year prospective study of fish, n-3 fatty acid intake, and colorectal cancer risk in men. *Cancer Epidemiol Biomarkers Prev* 17:1136–1143, 2008.

51. Physicians Committee for Responsible Medicine. Cancer facts: Meat consumption and cancer risk.

52. Devitt, A. A., and Mattes, R. D. Effects of food unit size and energy density on intake in humans. *Appetite* 42:213–220, 2004.

53. Ledikwe, J. H., Blanck, H. M., Kettel Khan, L., et al. Dietary energy density is associated with energy intake and weight status in US adults. *Am J Clin Nutr* 83:1362–1368, 2006.

54. Melanson, E. L., Astrup, A., and Donahoo, W. T. The relationship between dietary fat and fatty acid intake and body weight, diabetes, and the metabolic syndrome. *Ann Nutr Metab* 55:229–243, 2009.

55. American Cancer Society. ACS guidelines on nutrition and physical activity for cancer prevention.

56. Remig, V., Franklin, B., Margolis, S., et al. Trans fats in America: A review of their use, consumption, health implications, and regulation. *J Am Diet Assoc* 110:585–592, 2010.

57. American Heart Association Nutrition Committee, Lichtenstein, A. H., Appel, L. J., et al. Diet and lifestyle recommendations revision 2006: A scientific statement from the American Heart Association Nutrition Committee. Circulation 114:82–96, 2006.

58. American Dietetic Association. Position of the American Dietetic Association: Fat replacers. *J Am Diet Assoc* 105:266–275, 2005.

59. Tada, N., and Yoshida, H. Diacyglycerol on lipid metabolism. *Curr Opin Lipidol* 14:29–33, 2003.

60. Flegal, K. M., Carroll, M. D., Ogden, C. L., and Johnson, C. L. Prevalence and trends in obesity among US adults, 1999–2000. *JAMA* 288:1723–1727, 2002.

61. Ogden, C. L., Carroll, M. D., Curtin, L. R., et al. Prevalence of overweight and obesity in the United States, 1999–2004. *JAMA* 295:1549–1555, 2006.

***To access links to online sources, please go to www.wiley.com/college/smolin and select Nutrition: Science and Applications, 3rd edition. From this page, select either the student or instructor companion site. Once on the desired site, select References.**

Focus on Alcohol

© Bill Grove/iStockphoto

CHAPTER OUTLINE

Almost every human culture since the dawn of civilization has produced and consumed some type of alcoholic beverage. These intoxicating beverages are part of religious ceremonies, social traditions, and even medical prescriptions. Sumerian clay tablets dating back to 2100 BCE record physicians' prescriptions for beer. In ancient Egypt, beer and wine were prescribed as part of medical treatment.

Depending on the times and the culture, alcohol use has been touted, casually accepted, denounced, and even outlawed. Today, some people refrain from its use for religious, cultural, personal, and medical reasons, while others enjoy the relaxing effects afforded by drinking these beverages. Whether alcohol consumption represents a risk or provides some benefits depends on who is drinking and how much is consumed. In some groups, moderate alcohol consumption provides health advantages, but excessive alcohol consumption always has medical and social consequences that negatively impact drinkers and their families. Alcohol can reduce nutrient intake and affect the absorption, storage, mobilization, activation, and metabolism of nutrients. The breakdown of alcohol produces toxic compounds that damage tissues, particularly the liver.

F1.1 What's in Alcoholic Beverages?

LEARNING OBJECTIVES

- Explain the difference between alcohol and ethanol.
- Name the sources of the calories provided by alcoholic beverages.

Chemically, any molecule that contains a hydroxyl group (—OH) is an alcohol. Although there are many molecules present in our diet and our bodies that can be classified as alcohols, the term *alcohol* refers almost always to **ethanol**, also known as grain alcohol, and often to any beverage that contains ethanol (**Figure F1.1**). Alcoholic beverages, whether beer, wine, or distilled liquor, consist primarily of water, ethanol, and varying amounts of sugars. The amounts of other nutrients such as protein, vitamins, and minerals are almost negligible. Carbohydrate and alcohol provide the calories in these beverages. An average drink, defined as about 5 fl oz of wine, 12 fl oz of beer, or 1.5 fl oz of distilled spirits, contains about 12 to 14 grams (about 0.5 oz) of alcohol, which contributes about 90 kcalories (7 kcals per gram). The amount of energy contributed by carbohydrate depends on the type of beverage (**Table F1.1**).

ethanol The type of alcohol in alcoholic beverages. It is produced by yeast fermentation of sugar.

Figure F1.1 Sources and structure of ethanol
Ethanol, the type of alcohol in all alcoholic beverages, is a small water-soluble molecule. The photo illustrates the amounts of beer, wine, or distilled spirits in an average drink.

TABLE F1.1 Energy, Carbohydrate, and Alcohol Content of Alcoholic Beverages

Beverage	Typical serving (fluid oz)	Energy (kcals)	Carbohydrate (g)	Alcohol (g)
Log Island iced tea	7	170	24.3	17.8
Gin and tonic	10	114	10.5	10.7
Wine cooler	12	170	20.2	13.2
Beer	12	146	13.2	12.8
Light beer	12	100	4.6	11.3
White wine	5	100	1.2	13.7
Red wine	5	106	2.5	13.7
Bourbon	1.5	96	0	13.9
Whiskey	1.5	96	0	13.9
Vodka	1.5	96	0	13.9

F1.2 Alcohol Absorption and Excretion

LEARNING OBJECTIVES

- Explain how food in the stomach affects alcohol absorption.
- Discuss how alcohol is eliminated from the body.

When alcohol is consumed, it is absorbed along the entire gastrointestinal tract by simple diffusion. Only a small amount is absorbed in the mouth and esophagus. Larger amounts are absorbed in the stomach, and the majority of absorption occurs in the duodenum and jejunum of the small intestine. Because alcohol is absorbed rapidly and significant amounts can be absorbed directly from the stomach, the effects of alcohol consumption are almost immediate, especially if it is consumed on an empty stomach. If there is food in the stomach, absorption is slowed because the stomach contents dilute the alcohol, reducing the amount in direct contact with the stomach wall. Food in the stomach also slows absorption because it slows stomach emptying and therefore decreases the rate at which alcohol enters the small intestine, where absorption is the most rapid.

Once absorbed, alcohol enters the bloodstream. It is a small water-soluble molecule and therefore is rapidly distributed throughout all body water compartments. It crosses cell membranes by simple diffusion, so the amount that enters a cell depends on the concentration gradient across the cell membrane. Blood alcohol level therefore reflects the total amount of alcohol in the body and is dependent on the difference between the rates of alcohol absorption and elimination. Peak blood concentrations are attained approximately 1 hour after ingestion (**Figure F1.2a**). Many variables, including the kind and quantity of alcoholic beverage consumed, the speed at which the beverage

Figure F1.2 **Blood alcohol concentration**

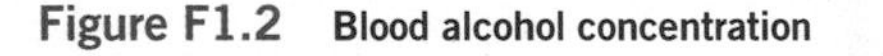

(a) Alcohol is absorbed quickly and enters the bloodstream, where levels can be measured. This graph shows blood alcohol levels in men and women after consumption of 0.5 grams of alcohol per kg of body weight. Equivalent amounts of alcohol cause higher blood alcohol concentrations in women.

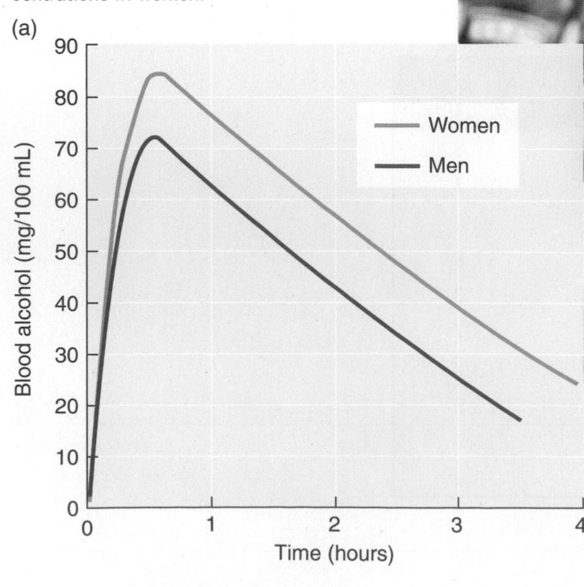

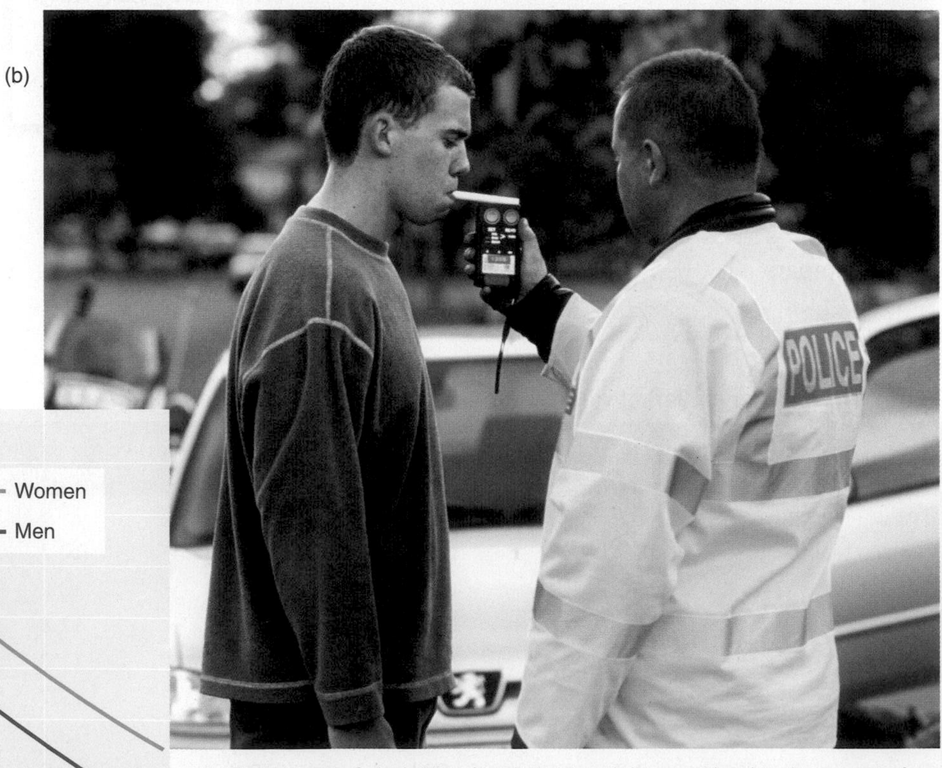

(b) The amount of alcohol lost in exhaled breath is proportional to the amount of alcohol in the blood. Therefore, a measurement of breath alcohol can be used to estimate blood alcohol level and determine if an individual is driving under the influence of alcohol.

TABLE F1.2 Factors Affecting Blood Alcohol Level

Factor	Effect
Weight	The more people weigh, the more body water they have, so the more dilute the alcohol in their blood is after consuming a given amount.
Gender	Men have more body water and more stomach alcohol dehydrogenase (ADH) activity and thus have a lower blood alcohol level after consuming a standard amount of alcohol than women of the same size.
Food	Food in the stomach slows alcohol absorption, so the more food people eat before drinking, the lower their blood alcohol level will be.
Drinking rate	The body metabolizes alcohol slowly. As the number of drinks per hour increases, blood alcohol level steadily rises.
The type of drink	The amount of alcohol in the drink affects how fast the blood alcohol level rises. When carbonated mixers (such as tonic water or club soda) are used, the body absorbs alcohol more quickly.

is drunk, the food consumed with it, the weight and gender of the consumer, and the activity of alcohol-metabolizing enzymes in the body, determine blood alcohol level (**Table F1.2**).

Alcohol can be eliminated by the liver, kidney, and lungs. About 90% of the alcohol consumed is metabolized by the liver. Because alcohol is a toxin and cannot be stored in the body, it must be metabolized rapidly. When alcohol reaches the liver, it is given metabolic priority and is therefore broken down before carbohydrate, protein, or fat. About 5% of consumed alcohol is excreted into the urine. The alcohol that reaches the kidney acts as a diuretic, increasing water excretion. Therefore, excessive alcohol intake can cause dehydration. Alcohol is also eliminated via the lungs during exhalation. The amount lost through the lungs is predictable and reliable enough to be used to estimate blood alcohol level from a measure of breath alcohol (**Figure F1.2b**).

F1.3 Alcohol Metabolism

LEARNING OBJECTIVES

- Compare the two metabolic pathways that metabolize alcohol.
- Explain why alcohol intake increases fatty acid synthesis.

There are two primary metabolic pathways for alcohol metabolism: the **alcohol dehydrogenase (ADH)** pathway located in the cytosol of the cell and the **microsomal ethanol-oxidizing system (MEOS)** located in small vesicles called *microsomes* that form from a membranous organelle called the *smooth endoplasmic reticulum*.

Alcohol Dehydrogenase

In people who consume moderate amounts of alcohol and/or consume alcohol only occasionally, most of the alcohol is broken down via the ADH pathway. Although liver cells have the highest levels of ADH activity, this enzyme has also been found in all parts of the gastrointestinal tract, with the greatest amounts in the stomach.[1] The amount of alcohol broken down in the stomach may be significant when small amounts of alcohol are consumed. One hypothesis on why women become intoxicated after consuming less alcohol than men is that women have lower activities of this stomach enzyme.[2] Another explanation for why women have higher blood-alcohol concentrations than men following the consumption of a standard amount of alcohol is the fact that they have a higher proportion of body fat and thus less body water than men.[3] The alcohol they do consume therefore is distributed in a smaller amount of body water (see Figure F1.2a).

alcohol dehydrogenase (ADH) An enzyme found primarily in the liver and stomach that helps break down alcohol into acetaldehyde, which is then converted to acetyl-CoA.

microsomal ethanol-oxidizing system (MEOS) A liver enzyme system located in microsomes that converts alcohol to acetaldehyde. Activity increases with increases in alcohol consumption.

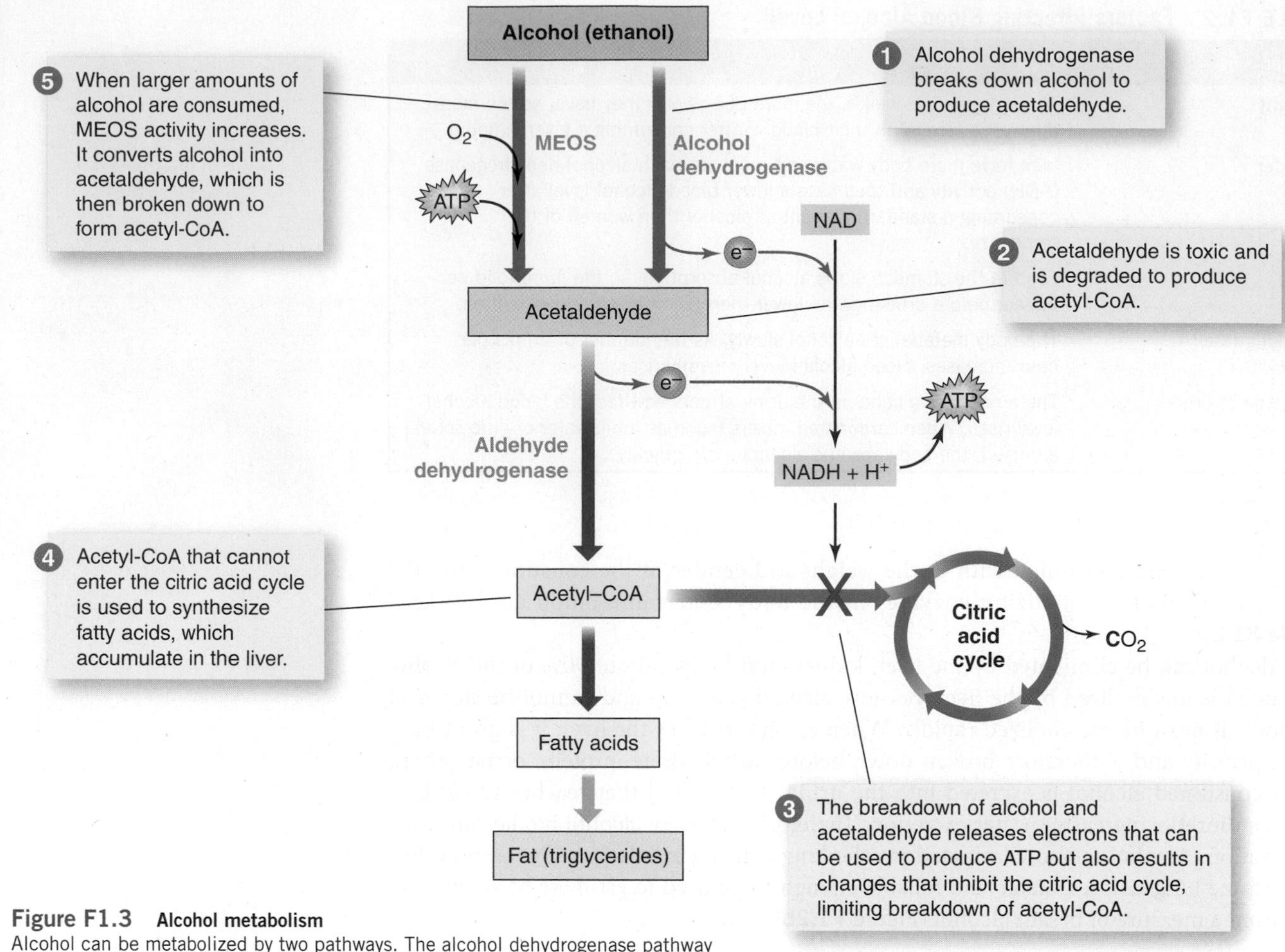

Figure F1.3 **Alcohol metabolism**
Alcohol can be metabolized by two pathways. The alcohol dehydrogenase pathway predominates when small amounts of alcohol are consumed, and the MEOS pathway becomes important when larger amounts are consumed.

ADH converts alcohol to acetaldehyde. Acetaldehyde is a toxic compound that is further degraded by the mitochondrial enzyme aldehyde dehydrogenase to a two-carbon molecule called *acetate* that forms acetyl-CoA (**Figure F1.3**). These reactions release high-energy electrons and hydrogen ions that are picked up by NAD to form NADH. NAD (nicotinamide adenine dinucleotide) is a form of the vitamin niacin (see Chapter 8 and Online Appendix: Focus on Metabolism). Although these processes produce ATP, high levels of NADH slow the citric acid cycle, preventing acetyl-CoA from being further broken down. Instead, the acetyl-CoA generated by alcohol breakdown, as well as acetyl-CoA from carbohydrate or fat metabolism, is used to synthesize fatty acids that accumulate in the liver. Fat accumulation can be seen in the liver after only a single bout of heavy drinking.

Microsomal Ethanol-Oxidizing System

Alcohol can also be metabolized in the liver by the MEOS pathway. This pathway is particularly important when greater amounts of alcohol are consumed. As with alcohol dehydrogenase, MEOS converts alcohol to acetaldehyde, which is then broken down by aldehyde dehydrogenase in the mitochondria (see Figure F1.3). The MEOS pathway requires oxygen and the input of energy to break down alcohol. In addition to forming acetaldehyde and water, reactive oxygen molecules are generated. These reactive molecules can contribute to liver disease. The rate that ADH breaks down alcohol is fairly constant, but MEOS activity increases when more alcohol is consumed. The MEOS pathway also metabolizes other drugs, so as activity increases in response to high alcohol intake the metabolism of other drugs may be altered.

F1.4 Adverse Effects of Alcohol Consumption

LEARNING OBJECTIVES

- Describe the short-term symptoms of alcohol intoxication.
- Explain why chronic excessive alcohol consumption can lead to malnutrition.
- Describe the long-term effects of alcoholism on the liver.

The consumption of alcohol has short-term effects that interfere with organ function for several hours after ingestion. It also has long-term effects that result from chronic alcohol consumption. Chronic alcohol consumption can cause disease both because it interferes with nutritional status and because it produces toxic compounds during its breakdown.

The effects of alcohol vary with life stage. When consumed during pregnancy, alcohol can cause abnormal brain development and other birth defects in the fetus (see Chapter 14). When consumed during childhood and adolescence, when the brain is still developing and changing, alcohol can cause permanent reductions in learning and memory.[4] In everyone, alcohol either directly or indirectly affects every organ in the body and increases the risk of malnutrition and many chronic diseases (**Table F1.3**). Almost 40,000 deaths per year are due to accute alcohol toxicity or alcohol-related liver disease.[5] Thousands more result from accidents and homicides associated with alcohol.

What Are the Acute Effects of Alcohol Consumption?

Depending on body size, amount of previous drinking, food intake, and general health, the liver can break down about 0.5 oz of alcohol per hour. This is the amount of

TABLE F1.3 Health Effects of Chronic Alcohol Use

Health Effect	Role of Alcohol
Birth defects	Increases the risk of fetal alcohol spectrum disorders, including fetal alcohol syndrome, when consumed during pregnancy.
Gastrointestinal problems	Damages the lining of the stomach and small intestine, and contributes to the development of pancreatitis.
Liver disease	Causes fatty liver, alcoholic hepatitis, and cirrhosis.
Malnutrition	Decreases nutrient absorption and alters the storage, metabolism, and excretion of some vitamins and minerals. Associated with a poor diet because alcohol replaces more nutrient-dense energy sources in the diet.
Neurological disorders	Contributes to impaired memory, dementia, and peripheral neuropathy.
Cardiovascular disorders	Associated with cardiovascular diseases such as cardiomyopathy, hypertension, arrhythmias, and stroke.
Blood disorders	Increases the risk of anemia and infection.
Immune function	Depresses the immune system and results in a predisposition to infectious diseases, including respiratory infections, pneumonia, and tuberculosis.
Cancer	Increases the risk for cancer, particularly of the upper digestive tract—including the esophagus, mouth, pharynx, and larynx—and of the liver, pancreas, breast, and colon.
Sexual dysfunction	Can lead to inadequate functioning of the testes and ovaries, resulting in hormonal deficiencies, sexual dysfunction, and infertility, It is also related to a higher rate of early menopause and a higher frequency of menstrual irregularities (duration, flow, or both).
Psychological disturbances	Causes depression, anxiety, and insomnia, and is associated with a higher incidence of suicide.

TABLE F1.4 Effects of Alcohol on the Central Nervous System

Number of drinks[a]	Blood alcohol[b] (%)	Effect on central nervous system
2	0.05	Impaired judgment, altered mood, relaxed inhibitions and tensions, increased heart rate
4	0.10	Impaired coordination, delayed reaction time, impaired peripheral vision
6	0.15	Unrestrained behavior, slurred speech, blurred vision, staggered gait
8	0.20	Double vision, inability to walk, lethargy
12	0.30	Stupor, confusion, coma
≥14	0.35–0.60	Unconsciousness, shock, coma, death

[a]Each drink contains 0.5 oz of ethanol and is equivalent to 5 fl oz of wine, 12 fl oz of beer, or 1.5 fl oz of distilled liquor.

[b]Values represent blood alcohol approximately 1 hour after consumption for a 150-lb individual. Actual blood alcohol values depend on the amount of alcohol in the beverage, the rate of consumption, foods consumed with the alcohol, gender, and body weight.

alcohol in one drink (5 fl oz of wine, 12 fl oz of beer, or 1.5 fl oz of distilled liquor). When alcohol intake exceeds the ability of the liver to break it down, the excess accumulates in the bloodstream until the liver enzymes can metabolize it. The circulating alcohol affects the brain, resulting in impaired mental and physical abilities. In the brain, alcohol acts as a depressant, slowing the rate at which neurological signals are received. First, it affects reasoning; if drinking continues, the vision and speech centers of the brain are affected. Next, large-muscle control becomes impaired, causing lack of coordination. Finally, if alcohol consumption continues, it can result in **alcohol poisoning**, a serious condition that can slow breathing, heart rate, and the gag reflex, leading to loss of consciousness, choking, coma, and even death (**Table F1.4**). This occurs most frequently with **binge drinking**. Binge drinking is a problem on college campuses that causes about 50 deaths and hundreds of cases of alcohol poisoning annually. It is estimated that about 40% of college students "binge" on alcohol at least once during a two-week period.[6]

alcohol poisoning When the quantity of alcohol consumed exceeds an individual's tolerance for alcohol and impairs mental and physical abilities.

binge drinking The consumption, within two hours, of five or more drinks for men or four or more drinks for women.

The effects of alcohol on the central nervous system are what make driving while under the influence of alcohol so dangerous. Alcohol affects reaction time, eye-hand coordination, accuracy, and balance. Not only does alcohol impair one's ability to operate a motor vehicle, but it also impairs one's judgment in the decision to drive. The abuse of alcohol is involved in 40% of all traffic fatalities and also contributes to domestic violence.[7]

Even if the individual does not lose consciousness, excess drinking may still result in memory loss. Drinking enough alcohol to cause amnesia is called **blackout drinking**. Blackout drinking puts people at risk because they have no memory of events that occurred during the blackout. During blackouts, individuals may engage in risky behaviors such as unprotected sexual intercourse, property vandalism, or driving a car—and have no memory of it afterward.

blackout drinking Amnesia following a period of excess alcohol consumption.

What Are the Chronic Effects of Alcohol Use?

One risk associated with regular alcohol consumption is the possibility of addiction. Alcohol addiction is referred to as **alcoholism**. The risk of addiction is increased in individuals who begin drinking at a younger age.[6] Alcohol addiction, like any other drug addiction, is a physiological problem that needs treatment. Alcoholism is believed to have a genetic component that makes some people more likely to become addicted, but environment also plays a significant role.[8] Thus, someone with a genetic

alcoholism A chronic disorder characterized by dependence on alcohol and development of withdrawal symptoms when alcohol intake is reduced.

predisposition toward alcoholism whose family and peers do not consume alcohol is much less likely to become addicted than someone with the same genes who drinks regularly with friends.

Why Alcohol Abuse Contributes to Malnutrition One of the complications of long-term excessive alcohol consumption is malnutrition. Alcoholic beverages contribute energy but few nutrients; they may replace more nutrient-dense energy sources in the diet, thereby reducing overall nutrient intake. As the percentage of calories from alcohol increases, the risk of nutrient deficiencies rises. With more moderate alcohol intakes, the drinker typically substitutes alcohol for carbohydrate in the diet and total energy intake increases slightly. When intake of alcohol exceeds 30% of calories, protein and fat intake as well as carbohydrate intake decrease and consumption of essential micronutrients such as thiamin and vitamins A and C may fall below recommended amounts.[9] Therefore, a diet that is high in alcohol causes primary malnutrition because nutrient-dense energy sources in the diet are replaced by alcoholic beverages, decreasing overall nutrient intake.

In addition to decreasing nutrient intake, alcohol can contribute to a secondary malnutrition by interfering with nutrient absorption, even when adequate amounts of nutrients are consumed. Alcohol causes inflammation of the stomach, pancreas, and intestine, which impairs the digestion of food and absorption of nutrients into the blood. Alcohol damage to the lining of the small intestine decreases the absorption of several B vitamins and vitamin C.[1] Deficiency of the B vitamin thiamin is a particular concern with chronic alcohol consumption (see Chapter 8). The mucosal damage caused by alcohol also increases the ability of large molecules to cross the mucosa. This allows toxins from the gut lumen to enter the portal blood, increasing the liver's exposure to these toxins and, consequently, the risk of liver injury. Alcohol also contributes to malnutrition by altering the storage, metabolism, and excretion of other vitamins and some minerals.

In addition to contributing to undernutrition, alcohol consumption may be related to obesity. Alcohol may contribute to weight gain in some situations because liquids are less satiating than solid food and drinking may stimulate appetite, promoting consumption of additional energy sources. Weight gain is more likely with higher intakes of alcohol and may be affected by the type of alcoholic beverage.[10] Calories consumed as alcohol are more likely to be deposited as fat in the abdominal region; excess abdominal fat increases the risk of high blood pressure, heart disease, and diabetes.

How Alcohol Affects the Liver Chronic alcohol consumption is associated with hypertension, heart disease, and stroke, but the most significant physiological effects occur in the liver. In addition to causing liver damage from malnutrition, alcohol causes liver damage through its toxic effects. Metabolism via ADH produces excess amounts of the electron carrier NADH. NADH inhibits the citric acid cycle and affects the metabolism of lipids, carbohydrates, and proteins (see Figure F1.3). High levels of NADH favor fat synthesis and inhibits fatty acid breakdown, leading to fat accumulation in the liver. Metabolism by MEOS generates free radicals, resulting in oxidative stress, which causes oxidation of lipids, membrane damage, and altered enzyme activities. Whether broken down by ADH or MEOS, toxic acetaldehyde is formed. Acetaldehyde exerts its toxic effects by binding to proteins and by inhibiting reactions and functions of the mitochondria. This decreases the metabolism of acetaldehyde to acetyl-CoA, allowing more acetaldehyde to accumulate causing further liver damage.

Chronic overconsumption of alcohol leads to alcoholic liver disease. The first phase of this disease is **fatty liver**, a condition that occurs when alcohol consumption increases the synthesis and deposition of fat in the liver. It occurs in almost all people who drink heavily due to increased synthesis and deposition of fat. If drinking continues, this condition may progress to **alcoholic hepatitis**, which is an inflammation of the liver. Both of these conditions are reversible if alcohol consumption is stopped and good nutritional and health

fatty liver The accumulation of fat in the liver.

alcoholic hepatitis Inflammation of the liver caused by alcohol consumption.

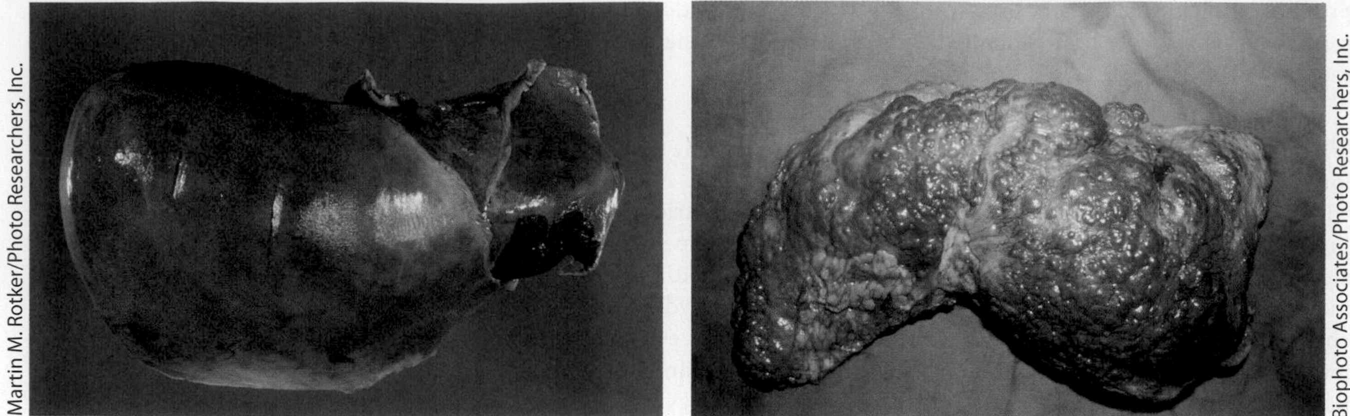

Figure F1.4 **Alcoholic cirrhosis**
The liver on the left is normal. The one on the right is from a patient with cirrhosis. Cirrhosis is an irreversible condition in which fibrous scar tissue replaces normal liver tissue and interferes with liver function.

cirrhosis Chronic liver disease characterized by the loss of functioning liver cells and the accumulation of fibrous connective tissue.

practices are followed. If alcohol consumption continues, **cirrhosis** may develop. This is an irreversible condition in which fibrous deposits scar the liver and interfere with its function (**Figure F1.4**). Because the liver is the primary site of many metabolic reactions, cirrhosis is often fatal.

Alcohol and Cancer Alcohol consumption is associated with an increased risk of cancer of the oral cavity, pharynx, esophagus, larynx, breast, liver, colon, rectum, and stomach.[11] Ingestion of all types of alcoholic beverages is associated with an increased cancer risk, supporting the hypothesis that it is alcohol consumption and not some other factor that is responsible.[12] Cancer risk is increased even with modest alcohol intake. For example, epidemiology shows that breast cancer risk is increased with consumption of as few as three to six drinks per week.[13]

The production of acetaldehyde, a known carcinogen, is one mechanism proposed for alcohol's carcinogenic effects. In the case of colon cancer, the acetaldehyde is produced by bacterial metabolism. Alcohol consumed in the diet is absorbed before it reaches the colon, but alcohol can pass from the blood into the lumen of the colon. The alcohol that enters the large intestine is then metabolized by bacteria to yield acetaldehyde. Toxic acetaldehyde accumulates in the colon, where it may contribute to mucosal injury and colon cancer. Acetaldehyde absorbed back into the blood contributes to liver damage. The American Institute for Cancer Research suggests that based on current evidence there is no truly "safe" level of alcohol consumption that does not promote cancer.[14]

F1.5 Benefits of Alcohol Consumption and Safe Drinking

LEARNING OBJECTIVES

- Describe how moderate alcohol intake reduces the risk of cardiovascular disease.
- List three things you could do to reduce the chances of becoming intoxicated while drinking alcoholic beverages.

For some people, moderate alcohol consumption (defined by the Dietary Guidelines as no more than one drink per day for women and two drinks per day for men) may be beneficial. Consuming alcoholic beverages before or with meals can stimulate appetite and improve mood. It can be relaxing, producing a euphoria that can enhance social interactions. It can also reduce the risk of heart disease.

Epidemiological and clinical studies have shown that light to moderate drinking reduces the risk for heart disease and stroke when compared to not consuming alcohol at all.[15] This reduction in heart disease risk results in a reduced mortality in middle-age and older adults who consume moderate amounts of alcohol.[16] However, at high levels of consumption, the mortality from heart disease and other alcohol-related problems is increased[17] (**Figure F1.5**). Wine consumption has been suggested as a reason for the lower incidence of heart disease in certain cultures. The Mediterranean diet, which has been associated with a reduced risk of heart disease, includes daily consumption of wine in moderation (see Chapter 1, Science Applied: How Epidemiology Led to Dietary Recommendations for Heart Disease). And one explanation for the French paradox—the fact that the French eat a diet that is as high or higher in saturated fat than the American diet but suffer from far less heart disease—is the glass of wine they drink with meals. The particular benefit of red wine is likely due not only to the alcohol but also to the phytochemicals (phenols) it contains[18,19] (see Figure F1.5).

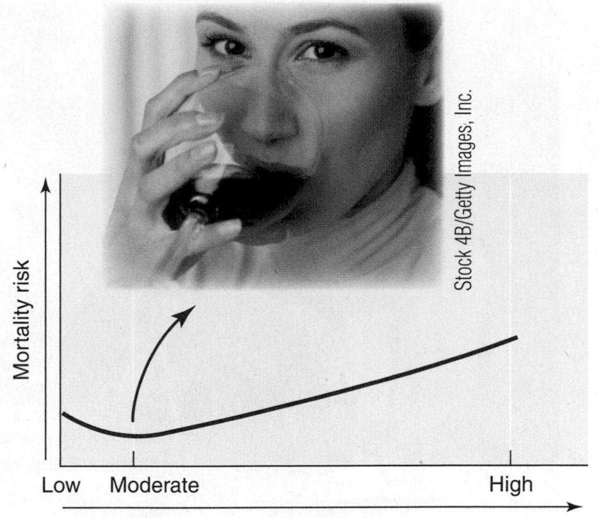

Figure F1.5 **Alcohol consumption and mortality**
The risk of mortality plotted against the amount of alcohol consumed generally results in a J-shaped curve, with lowest mortality at the level that corresponds to moderate alcohol consumption. The actual shape of the curve varies with age and gender.

There are a number of ways that moderate alcohol consumption reduces the risk of heart disease[17,20] (**Figure F1.6**). The most significant is its effect on HDL cholesterol. Moderate alcohol consumption can increase HDL cholesterol by 30% and is believed to account for about half of alcohol's protective effect. Alcohol also lowers

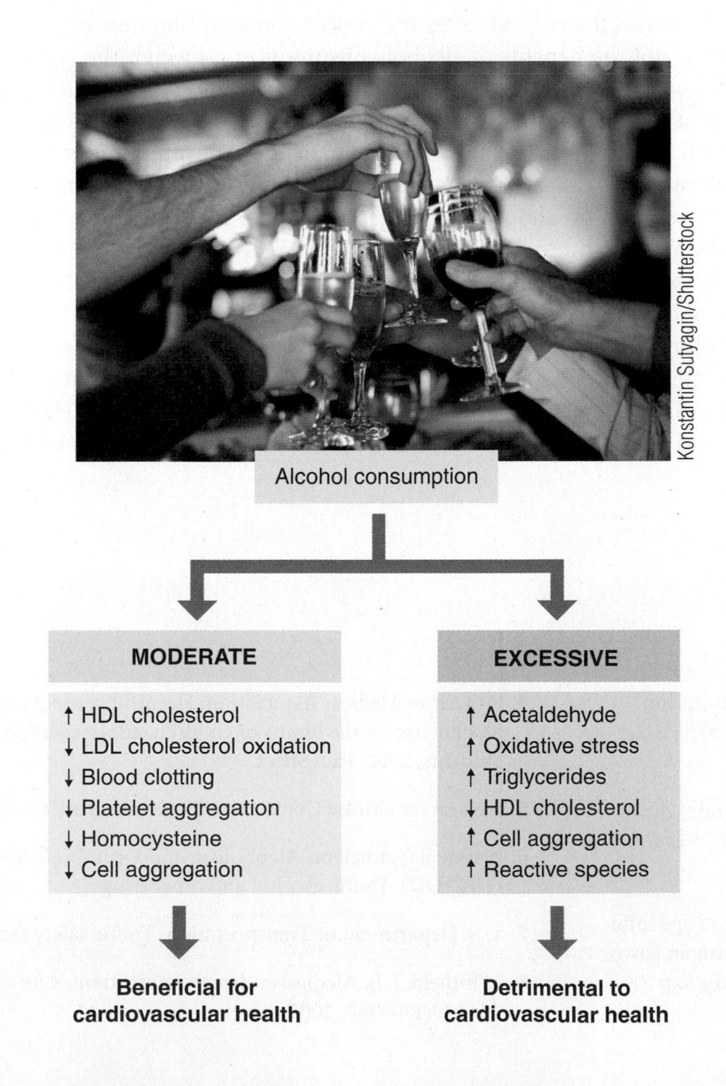

Figure F1.6 **Cardiovascular benefits and risks of alcohol consumption**
Moderate alcohol consumption has effects that reduce the risk of heart disease, but excessive intake contributes to cardiovascular disease risk.

Thomas Tolstrup/Getty Images, Inc.

Figure F1.7 **Moderate alcohol consumption**
When a moderate amount of alcohol is consumed with food, the risk of intoxication is reduced and the potential benefits are enhanced.

the risk of heart disease by reducing platelet stickiness and levels of the blood-clotting protein fibrinogen. Both of these effects reduce the formation of blood clots that can block blood flow to the heart resulting in a heart attack. Smaller benefits have been attributed to alcohol's role in increasing insulin sensitivity and reducing inflammation.[20,21] Although all of these effects have been documented, the benefit any one individual will gain from moderate alcohol consumption depends on his or her genetic background, health status, and overall lifestyle.

The risks posed by alcohol depend on the consumer and the amount consumed. Some people should not consume any alcohol. For instance, women who are pregnant or trying to conceive should not consume alcohol because it can damage the fetus (see Chapter 14). Children and adolescents should not consume alcohol because they are more likely to suffer its toxic effects—drunkenness and poisoning leading to seizures, coma, and death.[6] Individuals who plan to drive or operate machinery should not consume alcohol because it can impair coordination and reflexes. Alcoholics should avoid alcohol because they cannot restrict their drinking to moderate levels. Finally, individuals taking medications that can interact with alcohol should avoid alcohol. For everyone, the risks of excessive alcohol consumption outweigh the benefits.

Whether or not the benefits of alcohol consumption outweigh the risks, drinking is a personal decision that must take into account medical and social issues. It is not recommended that anyone begin drinking or drink more frequently to obtain alcohol's potential health benefits because even moderate alcohol consumption is associated with increased risk of breast cancer, violence, drowning, motor vehicle accidents, and injuries from falls.[22] Those who do choose to drink should do so in moderation. Alcohol should be consumed slowly, no more than one drink every 1.5 hours. Sipping, not gulping, allows the liver time to break down the alcohol that has already been consumed. Alternating nonalcoholic and alcoholic drinks will also slow down the rate of alcohol intake and prevent dehydration. Alcohol absorption is most rapid on an empty stomach. Consuming alcohol with meals slows its absorption and may also enhance its protective effects on the cardiovascular system (**Figure F1.7**). Also, the effect of alcohol on HDL cholesterol levels is believed to be greater when the liver is processing nutrients from a meal.

REFERENCES

1. Bode, C., and Bode, J. C. Effect of alcohol consumption on the gut. *Best Pract Res Clin Gastroenterol* 17:575–592, 2003.
2. Baraona, E., Abittan, C. S., Dohmen, K., et al. Gender differences in pharmacokinetics of alcohol. *Alcohol Clin Exp Res* 25:502–507, 2001.
3. Lai, C. L., Chao, Y. C., Chen, Y. C., et al. No sex and age influence on the expression pattern and activities of human gastric alcohol and aldehyde dehydrogenases. *Alcohol Clin Exp Res* 24:1625–1632, 2000.
4. American Medical Association. Harmful consequences of alcohol use on the brains of children, adolescents and college students, 2002. Fact Sheet.
5. Centers for Disease Control and Prevention. FastStats, Alcohol use.
6. National Council on Alcoholism and Drug Dependence (NCADD). Youth, alcohol and other drugs.
7. U.S. Department of Transportation. Traffic safety facts.
8. Whitfield, J. B. Alcohol and gene interactions. *Clin Chem Lab Med* 43:480–487, 2005.

9. Lieber, C. S. Relationships between nutrition, alcohol use, and liver disease. *Alcohol Res Health* 27:220–231, 2003.

10. Sayon-Orea, C., Martinez-Gonzalez, M. A., and Bes-Rastrollo, M. Alcohol consumption and body weight: A systematic review. *Nutr Rev* 69:419–431, 2011.

11. Bagnardi, V., Blangiardo, M., La Vecchia, C., and Corrao, G. A meta-analysis of alcohol drinking and cancer risk. *Br J Cancer* 85:1700–1705, 2001.

12. Testino, G., and Borro, P. Alcohol and gastrointestinal oncology. *World J Gastrointest Oncol* 2:322–325, 2010.

13. Chen, W. Y., Rosner, B., Hankinson, S. E., et al. Moderate alcohol consumption during adult life, drinking patterns, and breast cancer risk. *JAMA* 306:1884–1890, 2011.

14. World Cancer Research Fund. Food, nutrition, physical activity, and prevention of cancer: A global perspective. AICR, Washington, DC, 2010.

15. Djousse, L., Ellison, R. C., Beiser, A., et al. Alcohol consumption and the risk of ischemic stroke: The Framingham Study. *Stroke* 4:907–912, 2002.

16. Tolstrup, J., and Grønbaek. M. Alcohol and atherosclerosis: Recent insights. *Curr Atheroscler Rep* 9:116–124, 2007.

17. Lucas, D. L., Brown, R. A., Wassef, M., and Giles, T. D. Alcohol and the cardiovascular system: Research challenges and opportunities. *J Am Coll Cardiol* 45:1916–1924, 2005.

18. Saremi, A., and Arora, R. The cardiovascular implications of alcohol and red wine. *Am J Ther* 15:265–277, 2008.

19. Goldfinger, T. M. Beyond the French paradox: The impact of moderate beverage alcohol and wine consumption in the prevention of cardiovascular disease. *Cardiol Clin* 21:449–457, 2003.

20. Klatsky, A. L. Alcohol and cardiovascular diseases. *Expert Rev Cardiovasc Ther* 7:499–506, 2009.

21. Nova, E., Baccan, G. C., Veses, A., et al. Potential health benefits of moderate alcohol consumption: current perspectives in research. *Proc Nutr Soc* 71:307–15, 2012.

22. U.S. Department of Health and Human Services and U.S. Department of Agriculture. *Dietary Guidelines for Americans, 2010*, 7th ed. Washington, DC: US Government Printing Office, December 2010.

***To access links to online sources, please go to** www.wiley.com/college/smolin **and select Nutrition: Science and Applications, 3rd edition. From this page, select either the student or instructor companion site. Once on the desired site, select References.**

CHAPTER OUTLINE

© Luca Manieri/iStockphoto

Proteins and Amino Acids

6

CASE STUDY

Elliot thinks that it will be better for his health and the health of the planet if he stops eating animal products. He is a college student who eats some of his meals at the student union and does some of his own cooking. He doesn't know much about nutrition, but he knows that the union offers vegetarian choices at every meal. What he doesn't realize is that, although the vegetarian meals served at the student union do not contain meat, they may contain eggs and dairy products. He had planned to eliminate these from his diet as well as meat.

© Chris Schmidt/iStockphoto

In starting his new eating plan, Elliot thought breakfast would be easy, but as he poured himself a bowl of cereal, he realized that milk was an animal product. He switched to toast, but the butter he had in the refrigerator was an animal product so he had to settle for jam alone. He likes cream in his coffee, but when he read the label on his nondairy coffee creamer it said it contained milk protein. For lunch he usually goes out with friends for a burger or a sandwich. The burger restaurant didn't have any vegetarian options, so he tried the sandwich shop. His only choice there was to order the veggie and cheese sub, without the cheese. This meal wasn't very filling, so on the way home he bought a big bag of chips—at least these didn't contain any animal products.

For dinner that night the cafeteria offered vegetarian lasagna, but it was full of cheese. To stick with his animal-product-free diet he just had some pasta with marinara sauce. Ice cream for dessert was out so he opted for a slice of apple pie once he was assured it didn't contain any animal products. **By the end of the day he was frustrated, hungry, and felt his diet had no variety and was actually less healthy than what he ate before he tried to become a vegetarian.**

6.1 Protein in Our Food

LEARNING OBJECTIVES

- Identify the foods groups that provide the most concentrated sources of protein.
- Compare the nutrients provided by animal and plant sources of protein.

Protein is plentiful in the American food supply; our typical protein intake is about 90 g/day and has changed little over the past 30 years.[1,2] Most people in the United States and other developed countries have access to a variety of protein sources: animal sources including meats, eggs, and dairy products; and plant sources from legumes, grains, and vegetables. With the variety of foods available, even those who choose a vegetarian diet can easily meet their protein needs. Yet many people still worry about getting enough protein. Protein drinks, pills, and powders fill the shelves of health-food stores. Consumers often choose high-protein foods and supplements because protein is associated with good health. Protein is promoted to keep minds sharp and hair shiny and to aid in weight loss. Unlike carbohydrate and fat, the popular press associates protein only with positive effects. It has not been accused of being fattening, causing tooth decay, or increasing the risk of heart disease.

Sources of Protein in Our Diet

Most people around the world rely on plant proteins from grains and vegetables to meet their needs. These foods tend to be more available and less expensive than animal sources of protein. For example, in rural Mexico, most of the protein in the diet comes from beans, rice, and tortillas; in India protein comes from lentils and rice; in China rice with small amounts of meat and soy provides the protein. As the economic prosperity of a population grows, the proportion of animal foods in its diet typically increases.

In the United States, about two-thirds of the dietary protein comes from meat, poultry, fish, eggs, dairy products, and other animal sources.[3] These animal products are the most concentrated sources of protein. One egg contains about 7 grams of protein, a cup of milk contains 8 g, and a 3-ounce serving of meat provides over 20 g. But plant foods such as grains and legumes are also important sources of protein. Legumes, such as lentils, soybeans, peanuts, black-eyed peas, chickpeas, and dried beans, provide about 6 to 10 g of protein per half-cup serving. Nuts and seeds provide about 5 to 10 g per quarter cup. Even bread, rice, and pasta provide protein—about 2 to 3 g per slice or half cup (**Figure 6.1**). Although most plant sources of protein are not used as efficiently by the human body as animal sources, they can still easily meet most people's needs.

Nutrients That Accompany Protein Sources

The source of the protein in the diet determines what other nutrients are consumed along with it. Animal products provide an excellent source of B vitamins and minerals such as iron, zinc, and calcium. But animal products are low in fiber and are often high in saturated fat and cholesterol—a nutrient mix that increases the risk of heart disease. Plant sources of protein are a good source of most, but not all, B vitamins, and they also supply iron, zinc, and calcium but in less absorbable forms. But foods that provide good sources of plant proteins also contain fiber, phytochemicals, and unsaturated fats—dietary components that should be increased to promote health. Recommendations for a healthy diet suggest that our diets be based on whole-grains, vegetables, and fruits and provide smaller amounts of lean meats and low-fat dairy products. Following these guidelines will provide plenty of protein but without an over-reliance on animal sources of protein.

Figure 6.1 **Protein sources**

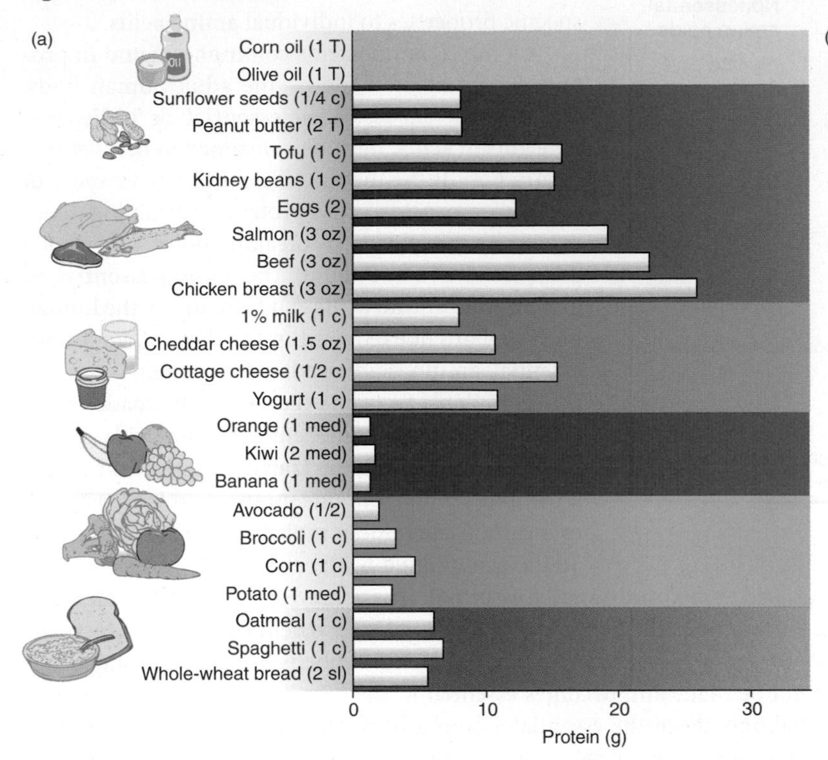

(b)

(a) Animal products such as meat, fish, poultry, eggs, and dairy products are good sources of protein, but some plant sources, such as legumes, nuts, and seeds, provide just as much. Grains are lower in protein but make an important protein contribution because we consume multiple servings daily.

(b) Animal sources of protein are high in iron, zinc, and calcium but also add saturated fat and cholesterol to the diet. Plant sources of protein are rich in fiber, phytochemicals, and unsaturated oils.

6.2 Protein Molecules

LEARNING OBJECTIVES

- Describe the general structure of an amino acid, a polypeptide, and a protein.
- Distinguish between essential and nonessential amino acids.
- Discuss how the order of amino acids in a polypeptide chain affects protein structure and function.

A protein molecule, whether found in a steak, a kidney bean, or a part of the human body, is constructed of one or more folded, chainlike strands of **amino acids**. The amino acid chains of each type of protein molecule contain a characteristic number and proportion of amino acids that are bound together in a precise order. The chains fold into specific configurations, giving each protein a unique three-dimensional shape that is essential to its particular function. Variations in the number, proportion, and order of amino acids in the chains allow for an infinite number of different protein structures.

amino acids The building blocks of proteins. Each contains a central carbon atom bound to a hydrogen atom, an amino group, an acid group, and a side chain.

Like carbohydrates and lipids, protein molecules contain the elements carbon, hydrogen, and oxygen, but proteins are distinguished from carbohydrates and triglycerides by the presence of the element nitrogen in their structure.

Amino Acid Structure

Approximately 20 amino acids are commonly found in proteins. Each amino acid consists of a carbon atom bound to four chemical groups: a hydrogen atom; an amino group, which contains nitrogen; an acid group; and a fourth group or side chain that varies in length

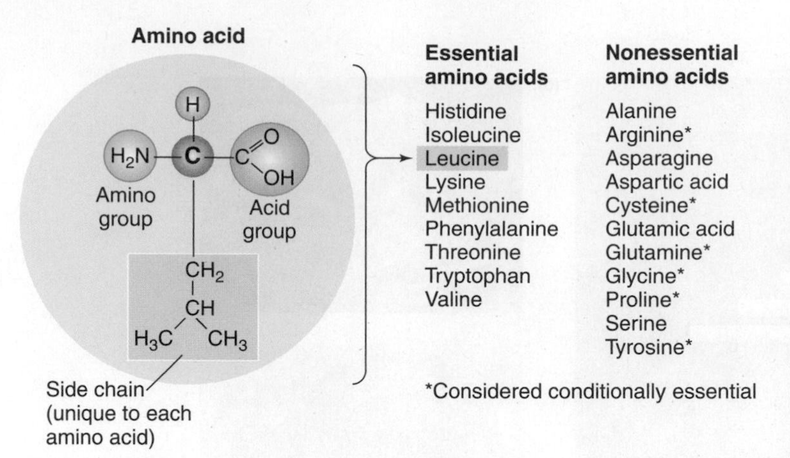

Figure 6.2 Amino acid structure
All amino acids have a similar structure, but each has a different side chain. Those that cannot be made in the body, such as leucine shown here, are considered essential.

and structure (**Figure 6.2**). Different side chains give specific properties to individual amino acids.

Of the 20 amino acids commonly found in protein, 9 cannot be made by the adult human body. These amino acids, called **essential** or **indispensable amino acids**, must be consumed in the diet (see Figure 6.2). If the diet is deficient in one or more of these amino acids, new proteins containing them cannot be made without breaking down other body proteins to provide them. The 11 **nonessential** or **dispensable amino acids** can be made by the human body and are not required in the diet. When a nonessential amino acid needed for protein synthesis is not available from the diet, it can be made in the body. Most of the nonessential amino acids can be made by the process of **transamination**, in which an amino group from one amino acid is transferred to a carbon-containing molecule to form a different amino acid (**Figure 6.3**).

Some amino acids are **conditionally essential**. These are essential only under certain conditions. For example, the conditionally essential amino acid tyrosine can be made in the body from the essential amino acid phenylalanine. If phenylalanine is in short supply, tyrosine cannot be made and becomes essential in the diet. Likewise, the amino acid cysteine is essential only when the essential amino acid methionine is in short supply or cannot be converted to cysteine. Other amino acids may be essential at certain times of life, such as premature infancy, or due to certain conditions, such as metabolic abnormalities or physical stress.

essential or **indispensable amino acid** An amino acid that cannot be synthesized by the human body in sufficient amounts to meet needs and therefore must be included in the diet.

nonessential or **dispensable amino acid** An amino acid that can be synthesized by the human body in sufficient amounts to meet needs.

transamination The process by which an amino group from one amino acid is transferred to a carbon compound to form a new amino acid.

conditionally essential amino acid An amino acid that is essential in the diet only under certain conditions or at certain times of life.

dipeptide Two amino acids linked by a peptide bond. A **tripeptide** is three amino acids linked by peptide bonds, and a **polypeptide** refers to any chain of three or more amino acids linked by peptide bonds.

Protein Structure

To form proteins, amino acids are linked together by peptide bonds, which are the linkages that form between the acid group of one amino acid and the nitrogen atom of the next amino acid (**Figure 6.4**). When two amino acids are linked with a peptide bond, the molecule formed is called a **dipeptide**; when three amino acids are linked, they form a **tripeptide**. Many amino acids bonded together constitute a **polypeptide**.

Why Proteins Have Three-Dimensional Shapes A protein is made of one or more polypeptide chains folded into a complex three-dimensional shape (see Figure 6.4). The order and chemical properties of the amino acids in a polypeptide chain determine the three-dimensional shape of the protein. Folds and bends in the chain occur when some of the amino acids attract each other and others repel each other. For example, amino

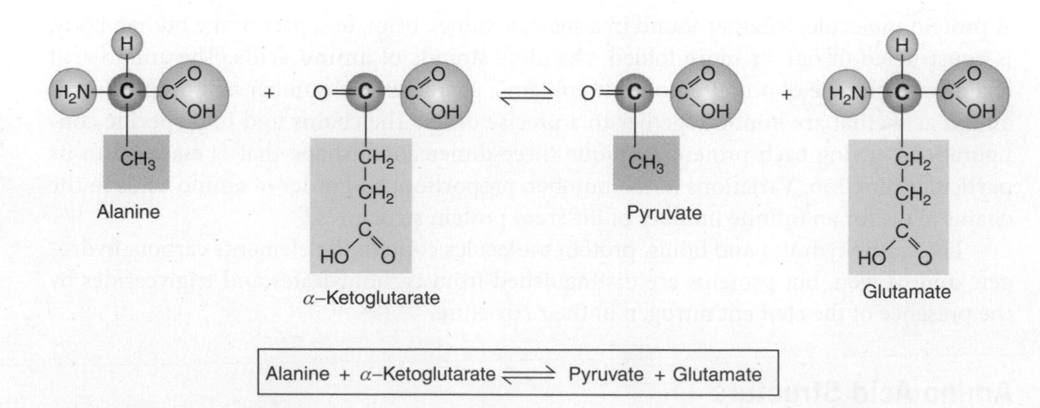

Figure 6.3 Transamination
In this example, transamination transfers the amino group from the nonessential amino acid alanine to the carbon compound alpha-ketoglutarate (α-ketoglutarate) to form the three-carbon compound pyruvate and the essential amino acid glutamate.

acids at various places along the chain that are attracted to each other may cause segments of the chain to coil like a telephone cord. Amino acids that are attracted to water will orient to the outside of the structure to be in contact with body fluids, whereas amino acids that repel water will fold to the inside to be away from body fluids. After polypeptide chains have folded, several chains may bind together to form the final protein.

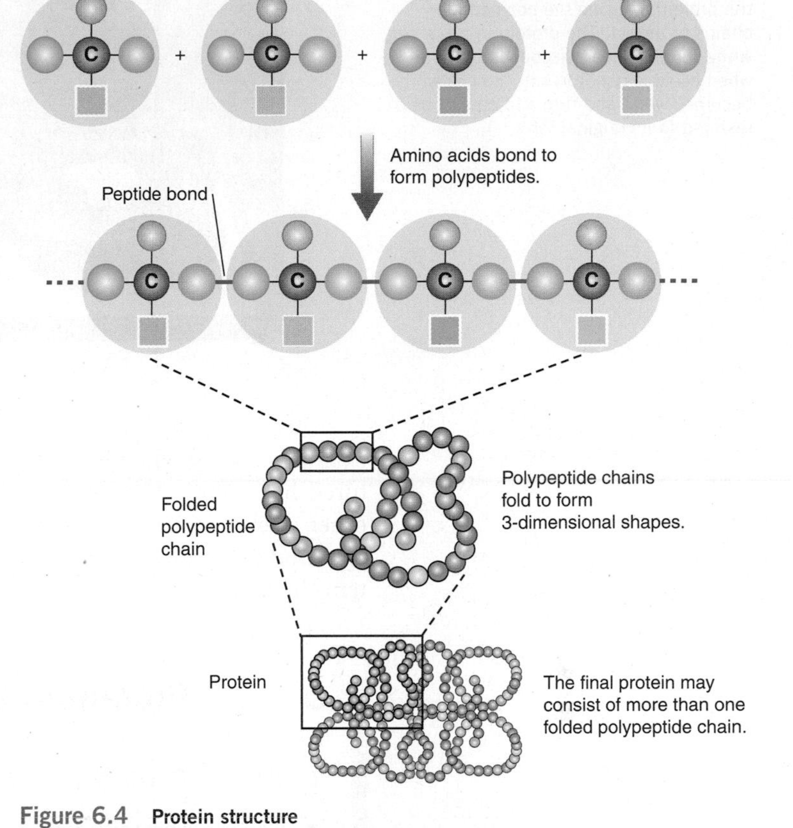

Figure 6.4 Protein structure
When many amino acids are joined by peptide bonds, which link the acid group (shown in purple) of one amino acid to the amino group (shown in turquoise) of another, they form a polypeptide. The chemical properties of the amino acids in the polypeptide chain cause it to fold, contributing to the three-dimensional structure of the protein. The final protein may include one or more folded polypeptide chains.

How Protein Structure Contributes to Function

It is the shape of the final protein that determines its function. For example, the elongated shape of the connective tissue proteins, collagen and α-keratin, allows them to give strength to fingernails and ligaments. The oxygen-carrying protein hemoglobin has a spherical shape, which allows proper functioning of the red blood cells. If the shape of a protein is altered, its function may be disrupted. For example, in the genetic disease sickle cell anemia, an abnormality in DNA causes a single amino acid in the hemoglobin molecule to be altered. As a result, the hemoglobin molecules bind together in long chains. Thus, rather than having the disc shape characteristic of normal red blood cells, red blood cells containing these rope-like strands of sickle cell hemoglobin have a distorted shape that resembles a crescent or sickle (**Figure 6.5**). Sickle-shaped red blood cells can block capillaries, causing inflammation and pain. They also rupture easily, leading to anemia from a shortage of red blood cells.

Changes in protein structure can be caused by changes in the physical environment of the protein, such as an increase in temperature or a change in acidity. Such changes cause protein **denaturation**. In food, the heat of cooking denatures protein, thereby changing its shape and physical properties. For

denaturation The alteration of a protein's three-dimensional structure.

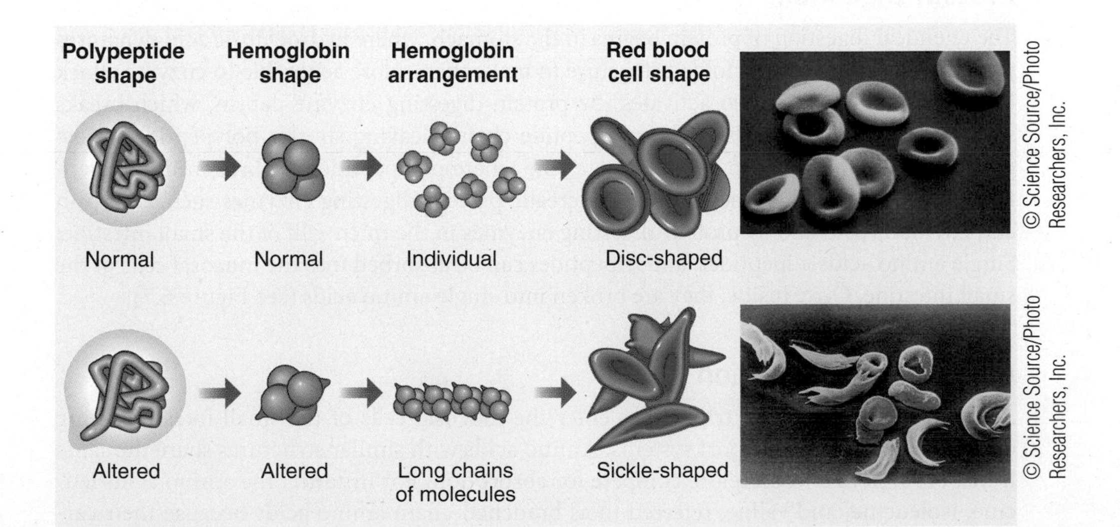

Figure 6.5 Sickle cell anemia
In sickle cell anemia, a change in the sequence of amino acids in hemoglobin alters the shape and function of the protein molecule. Sickle-cell hemoglobin forms long chains that distort the shape of red blood cells.

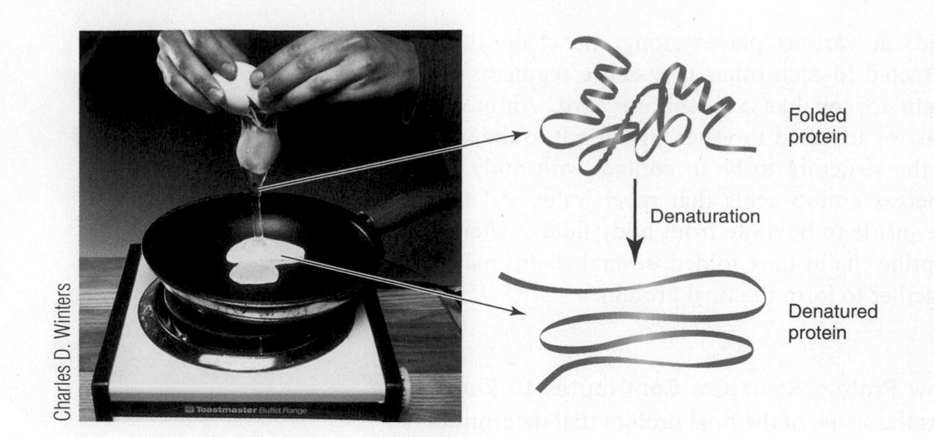

Figure 6.6 **Protein denaturation** When an egg is cooked, the heat denatures the protein, causing the polypeptide chains to unfold. The protein in a raw egg white forms a clear, viscous liquid, but when cooking denatures it, the egg white becomes white and firm and cannot be restored to its original form.

example, an egg white becomes hard and opaque after the protein it contains is denatured by cooking (**Figure 6.6**). The denaturation of proteins in our food also creates other characteristics we desire. For example, whipped cream is made when mechanical agitation denatures the protein in cream and yogurt forms when acid denatures the proteins in milk.

6.3 Protein in the Digestive Tract

LEARNING OBJECTIVES

- Describe protein digestion and amino acid absorption.
- Explain why consuming an excess of one amino acid can cause a deficiency of another amino acid.
- Discuss how protein digestion is related to food allergies.

Protein enters the digestive tract from food, from digestive secretions, and from sloughed gastrointestinal cells. No matter what the source, the protein must be broken down to amino acids before entering the bloodstream.

Protein Digestion

The chemical digestion of protein begins in the stomach, where hydrochloric acid denatures proteins, opening up their folded structure to make them more accessible to enzyme attack (**Figure 6.7**). The acid also activates the protein-digesting enzyme pepsin, which breaks some of the peptide bonds in the polypeptide chains, leaving shorter polypeptides. Most protein digestion occurs in the small intestine where polypeptides are broken into tripeptides, dipeptides, and amino acids by pancreatic protein-digesting enzymes such as trypsin and chymotrypsin and by protein-digesting enzymes in the microvilli of the small intestine. Single amino acids, dipeptides, and tripeptides can be absorbed into the mucosal cells of the small intestine. Once inside, they are broken into single amino acids (see Figure 6.7).

Amino Acid Absorption

Amino acids and di- and tripeptides enter the mucosal cells of the small intestine using one of several active transport systems. Amino acids with similar structures share the same transport system and therefore compete for absorption. For instance, the amino acids leucine, isoleucine, and valine, referred to as branched-chain amino acids because their carbon side chains have a branching structure (see Appendix H), share the same transport system. If there is an excess of any one of the amino acids sharing a transport system, more of it will be absorbed, slowing the absorption of the other competing amino acids

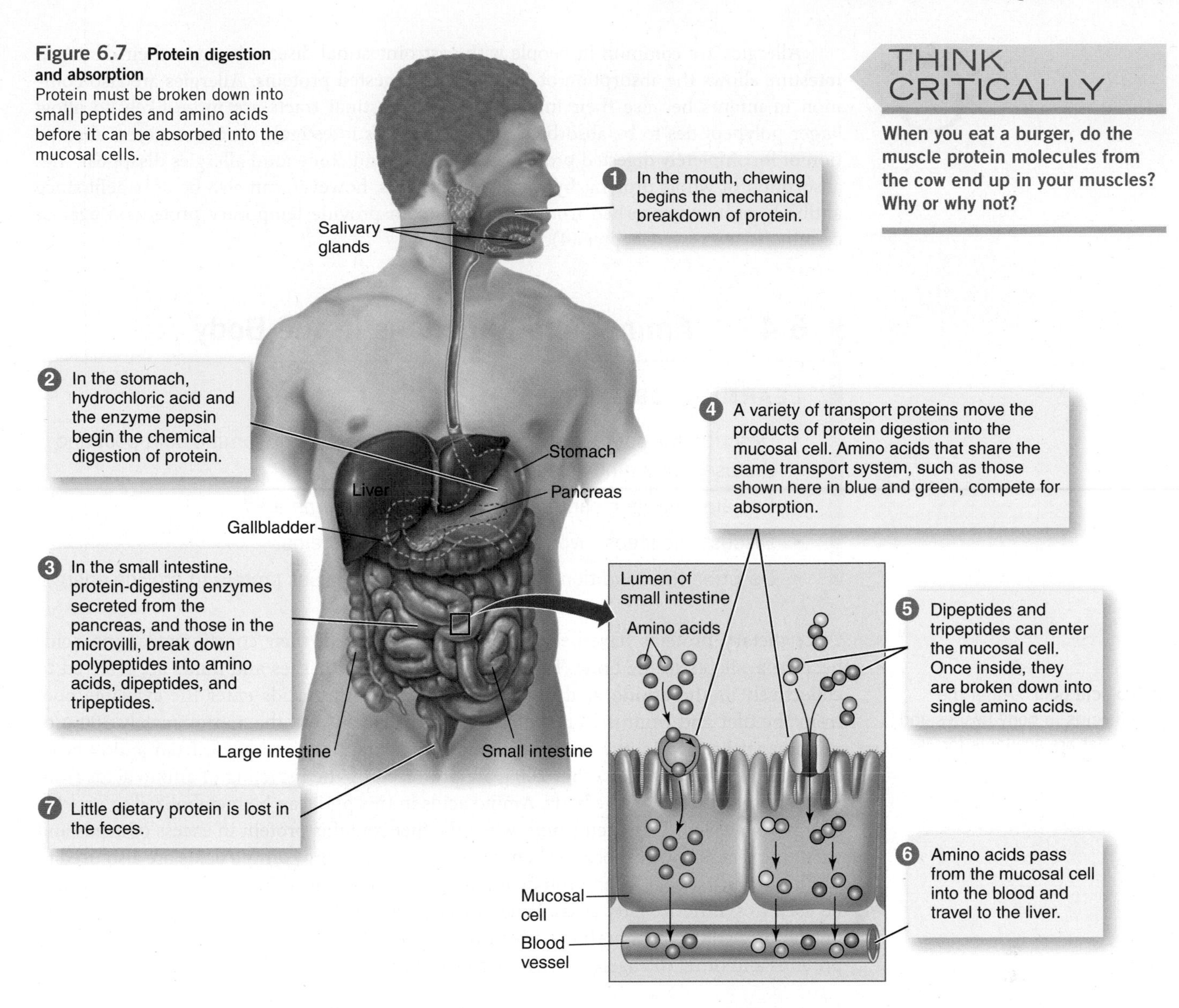

Figure 6.7 Protein digestion and absorption
Protein must be broken down into small peptides and amino acids before it can be absorbed into the mucosal cells.

THINK CRITICALLY

When you eat a burger, do the muscle protein molecules from the cow end up in your muscles? Why or why not?

(see Figure 6.7). This is generally not a problem when amino acids are consumed in foods because they contain a variety of amino acids without disproportionately large amounts of any one. However, taking a supplement of one amino acid can provide enough of it to impair absorption of other amino acids that share the same transport system. For example, weight lifters often take supplements of the amino acid arginine, which shares the same transport system as lysine. When large doses of arginine are ingested, the absorption of lysine will be reduced.

Why Undigested Protein Can Cause Food Allergies

Food allergies are triggered when a protein from the diet is absorbed without being completely digested. The proteins from milk, eggs, peanuts, tree nuts, wheat, soy, fish, and shellfish are common causes of food allergies. The first time the protein is consumed and a piece of it is absorbed intact, the immune system is stimulated (see Chapter 3). When the protein is consumed again, the immune system sees it as a foreign substance and mounts an attack, causing an allergic reaction. Allergic reactions to food can cause symptoms throughout the body. These can involve the digestive system, causing vomiting or diarrhea; the skin, causing a rash or hives; the respiratory tract, causing difficulty in breathing; or the cardiovascular system, causing a drop in blood pressure. A rapid severe allergic reaction that involves more than one part of the body is called **anaphylaxis**. A severe anaphylactic reaction can cause breathing difficulty or a dangerous drop in blood pressure and can be fatal.

anaphylaxis An immediate and severe allergic reaction to a substance (e.g., food or drugs). Symptoms include breathing difficulty, loss of consciousness, and a drop in blood pressure and can be fatal.

Allergies are common in people with gastrointestinal disease because their damaged intestine allows the absorption of incompletely digested proteins. Allergies are also common in infants because their immature gastrointestinal tracts are more likely to allow larger polypeptides to be absorbed. Once an infant's intestinal mucosa matures, absorption of incompletely digested proteins is less likely and some food allergies disappear. The absorption of whole proteins by very young infants, however, can also be of benefit since antibody proteins absorbed from breast milk can provide temporary protection against certain diseases (see Chapter 14).

6.4 Amino Acid Functions in the Body

LEARNING OBJECTIVES

- Describe the sources of the amino acids entering the amino acid pool and the uses for amino acids that leave the pool.
- Explain what is meant by the term *limiting amino acid.*
- Discuss the steps involved in synthesizing proteins.
- Describe the conditions under which the body uses protein to provide energy.

amino acid pool All of the amino acids in body tissues and fluids that are available for use by the body.

Once dietary proteins have been digested and absorbed, their constituent amino acids become available to the body. The amino acids in body tissues and fluids are referred to collectively as the **amino acid pool** (**Figure 6.8**). Amino acids enter the available pool from the diet and from the breakdown of body proteins. Of the approximately 300 g of protein synthesized by the body each day, only about 100 g are made from amino acids consumed in the diet. The other 200 g are produced by the recycling of amino acids from protein broken down in the body. Amino acids in this pool can be metabolized to provide energy—4 kcals/g. This occurs both when the diet contains protein in excess of needs and when the diet is low in energy. When the diet is low in energy amino acids are also used to synthesis glucose and when protein and energy are consumed in excess, amino acids can be used to synthesize fatty acids. When dietary intake of protein and energy are adequate but not excessive, most amino acids in the amino acid pool are used to synthesize body proteins and other nitrogen-containing compounds.

Proteins Are Continually Broken Down and Resynthesized

protein turnover The continuous synthesis and breakdown of body proteins.

Body proteins are not static but rather are continually broken down and resynthesized. This process, referred to as **protein turnover**, is necessary for normal growth and maintenance of body tissues and for adaptation to changing situations. The rate at which proteins are

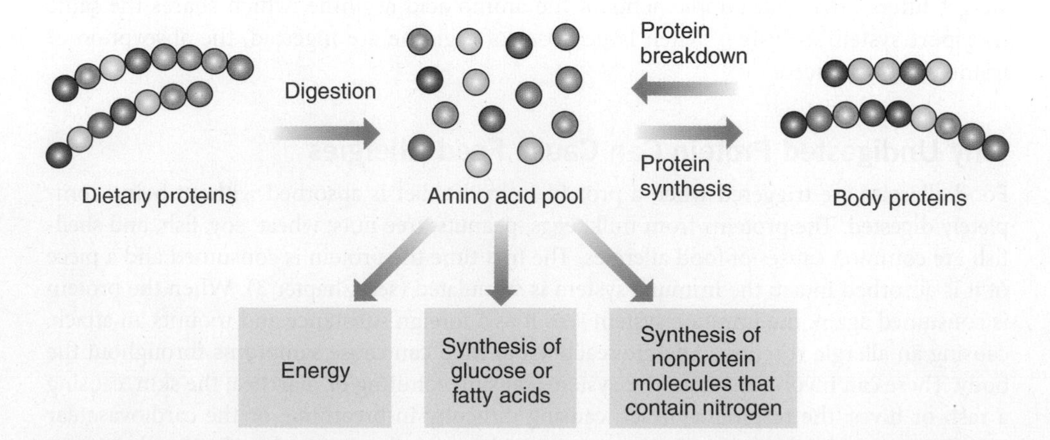

Figure 6.8 **Amino acid pool**
Amino acids in the available pool come from the diet and from the breakdown of body proteins. They are used to synthesize body proteins and nonprotein molecules, and to generate ATP or to synthesize glucose or fatty acids.

made and degraded varies with the protein and is related to its function. Proteins whose concentration must be regulated and proteins that act as chemical signals in the body tend to have high rates of turnover—that is, synthesis and degradation. For example, the level of insulin in the blood can be rapidly increased by increasing its synthesis and decreasing its degradation. To decrease insulin levels, breakdown can be increased and synthesis slowed. Varying levels of synthesis and degradation allow the amounts of specific proteins to adjust quickly to maintain homeostasis as body conditions change. Structural proteins, such as collagen in connective tissue, have slower rates of turnover. The total amount of body protein broken down each day is large; twice as many of the amino acids in the amino acid pool come from recycled proteins as from dietary proteins.

How Amino Acids Are Used to Synthesize Proteins

The amino acids used to synthesize body proteins come from the amino acid pool. The instructions that dictate which amino acids are needed, and in what order they should be combined, are contained in stretches of DNA called **genes**. When a protein is needed, the process of protein synthesis begins.

The first step in protein synthesis involves copying, or transcribing, the DNA code from the gene into a molecule of *messenger RNA (mRNA)*. This process is called **transcription** (**Figure 6.9**). The mRNA then takes this information from the nucleus of the cell to ribosomes in the cytosol, where proteins are made. Here the information in mRNA is translated through another type of RNA, called *transfer RNA (tRNA)*. Transfer RNA reads the code and delivers the needed amino acids to form a polypeptide chain. This process is called **translation**. After translation, polypeptides typically undergo further chemical modifications before achieving their final protein structure and function (see **Science Applied: How Bacteria Created the Field of Biotechnology**).

gene A length of DNA containing the information needed to synthesize RNA, which may translate into a polypeptide sequence.

transcription The process of copying the information in DNA to a molecule of mRNA.

translation The process of translating the mRNA code into the amino acid sequence of a polypeptide chain.

Why Genes Are Regulated **Gene expression** is the process whereby the information coded in a gene is used to produce a protein or other gene product. When a gene is expressed, the product for which the gene codes is synthesized. Sometimes the final product is a molecule of RNA, and sometimes it is a protein. Which gene products are made and when they are made are important to the health of the cell and the organism, and therefore gene expression is carefully regulated. Not all genes are expressed in all cells or at all times. For example, the hormone glucagon is a protein that is made by cells in the pancreas. Glucagon is not made by other body cells because the gene is not expressed in cells other than those in the pancreas. The expression of some genes changes depending on

gene expression The process by which the information coded in a gene is used to synthesize a product, either a protein or a molecule of RNA.

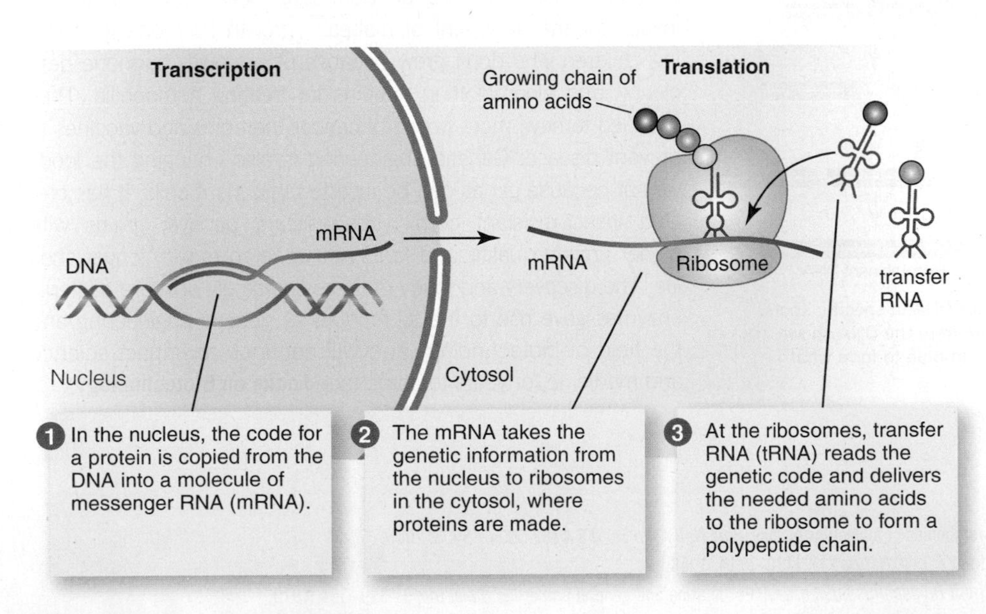

Figure 6.9 Transcription and translation
DNA in cell nuclei provides the information needed to assemble proteins (transcription and translation).

SCIENCE APPLIED

How Bacteria Created the Field of Biotechnology

(Elena Schweitzer/Shutterstock) Victoriano Izquierdo/Getty Images, Inc. (DAJ/Getty Images, Inc.)

By using the techniques of genetic engineering, scientists today can create bacteria that make human insulin and plants that synthesize their own insecticide. These techniques can modify the genes in an organism, thus changing the proteins made by that organism. This creates unlimited possibilities for the production of hormones, drugs, and foods that can improve human lives. This technology did not exist before 1970 because we didn't have the "tools" needed to cut DNA. The ability to cut DNA and paste the pieces together emerged from basic laboratory research that studied bacteria.[1]

THE SCIENCE

In the 1950s, it was observed that some strains of bacteria were able to slice viral DNA into pieces.[1] This ability was found to be due to bacterial enzymes, called *restriction enzymes.* Restriction enzymes are present in bacteria as a form of protection. If a virus invades the bacterium, it can defend itself by cutting the viral DNA into little pieces before the virus can cause harm. By the late 1960s and early 1970s, restriction enzymes were being isolated and characterized.[2,3] At the same time, bacterial enzymes that repair breaks in DNA, called *DNA ligases,* were also being studied. DNA ligases had the ability to paste together two strands of DNA.

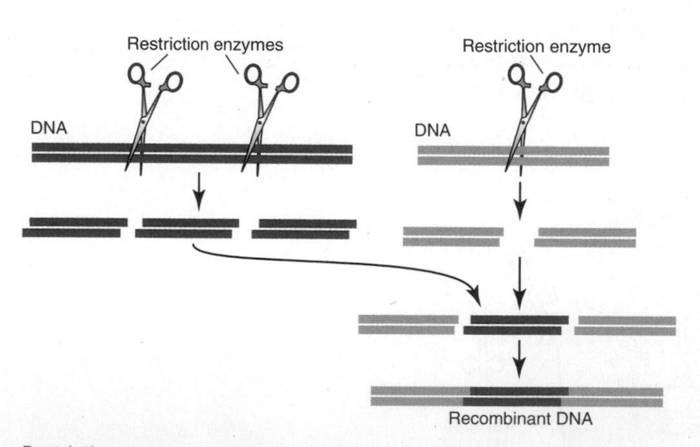

Restriction enzymes act like "scissors" that cut DNA in specific spots. The restriction enzymes shown here cut a gene from the DNA shown in red, which can be pasted into the DNA shown in blue to form what is referred to as recombinant DNA.

It was hypothesized and confirmed by experimentation that restriction enzymes cut at specific sites on DNA molecules.[3,4] The practical uses for restriction enzymes for the analysis of DNA became apparent in 1971 when Katherine Danna and Daniel Nathans took a small amount of purified viral DNA, incubated it with a restriction enzyme, and were able to separate the resulting DNA fragments based on their size.[5] Researchers were quick to recognize that restriction enzymes provided them with a remarkable new tool for investigating gene organization, function, and expression. These enzymes could cleave DNA at specific sites, generating gene-size fragments that could then be re-joined in the laboratory. Today restriction enzymes that cut DNA at more than 250 different specific sequences have been purified and scientists can purchase them from scientific suppliers and use them to locate, isolate, prepare, and study small segments of DNA. At the time restriction enzymes were discovered, their impact on the fields of biology and science was impossible to predict.

THE APPLICATION

In the laboratory, restriction enzymes act like precision scissors that can clip out a gene that codes for a specific protein (see figure). Because DNA in all forms of life is made of the same building blocks, restriction enzymes from bacteria can cut the DNA from a virus, a cow, a soybean plant, or a human cell with equal efficiency. When the gene for a protein produced by a human cell is cut out and pasted into the DNA inside a bacterial cell, the resulting bacterial cells can produce the human protein. These techniques have created an abundant, safe supply of human insulin for the treatment of diabetes, growth hormone for treating children who don't grow because of a growth hormone deficiency, and blood-clotting proteins for treating hemophilia. They have led to new, more powerful cancer therapies and vaccines to prevent disease. Genetic engineering is also changing the foods we eat because genes can be inserted into plant cells. It has created insect-resistant corn, virus-resistant papaya, grains with higher protein quality, and fruits and vegetables with longer shelf life. The discovery and study of these seemingly obscure bacterial enzymes gave rise to the techniques of genetic engineering and the field of biotechnology and will continue to impact science and medicine for years to come (see **Focus on Biotechnology**).

[1] Roberts, R. J. How restriction enzymes became the workhorses of molecular biology. *Proc Natl Acad Sci U S A* 102:5905–5908, 2005.
[2] Meselson, M., and Yuan, R. DNA restriction enzyme from *E. coli. Nature* 217:1110–1114, 1968.
[3] Smith, H. O., and Wilcox, K. W. A restriction enzyme from *Hemophilus influenzae*. I. Purification and general properties. *J Mol Biol* 51:379–391, 1970.
[4] Arber, W. Host-controlled modification of bacteriophage. *Annu Rev Microbiol* 19:365–378, 1965.
[5] Danna, K., and Nathans, D. Specific cleavage of simian virus 40 DNA by restriction endonuclease of *Hemophilus influenzae. Proc Natl Acad Sci U S A* 68:2913–2917, 1971.

the need for the proteins for which they code. For example, when zinc intake is high, the expression of a gene that codes for metallothioneine, a metal-binding protein, is turned on. This allows more of this protein to be synthesized, increasing the capacity to bind zinc and other metal ions. The levels of nutrients in the body also affect the expression of other genes. For example, vitamin A levels affect the expression of genes involved in the maturation and specialization of cells and vitamin D affects the expression of genes that code for calcium transport proteins (see Chapter 9).

What Are Limiting Amino Acids? During the synthesis of a protein, a shortage of one needed amino acid can stop the process. Just as on an assembly line, if one part is missing, the line stops—a different part cannot be substituted. If the missing amino acid is a nonessential amino acid, it can be synthesized, most often by transamination, and protein synthesis can continue. If the missing amino acid is an essential amino acid, the body cannot make the amino acid, but it can break down some of its own proteins to obtain it. If an amino acid cannot be supplied, protein synthesis stops.

The essential amino acid present in shortest supply relative to need is called the **limiting amino acid** because lack of this amino acid limits the ability to synthesize the needed protein (**Figure 6.10**). Proteins from different foods provide different combinations of amino acids, so the limiting amino acids differ in different foods. Animal proteins generally provide enough of all of the amino acids and thus do not limit protein synthesis. Plant proteins are generally low in one or more of the essential amino acids. For example, lysine is the limiting amino acid in rice, whereas methionine is the limiting amino acid in beans. An exception is soy protein, which is as good as animal protein when it comes to supplying essential amino acids. When the combination of foods in the diet provides adequate amounts of all the essential amino acids needed to synthesize a specific protein, synthesis of the polypeptide chains that make up the protein can be completed and the chains can be released for further processing by the cell.

limiting amino acid The essential amino acid that is available in the lowest concentration in relation to the body's needs.

Synthesis of Nonprotein Molecules

Amino acids are needed for the synthesis of a variety of nonprotein molecules that contain nitrogen. For example, the amino acid tryptophan is used to synthesize the neurotransmitter serotonin, which acts in the relaxation center of the brain. Amino acids are also needed for the synthesis of the nitrogen-containing compounds that are the building blocks of DNA and RNA. Other molecules synthesized from amino acids include the skin pigment melanin; the vitamin niacin; creatine, needed for muscle contraction; and histamine, which causes blood vessels to dilate.

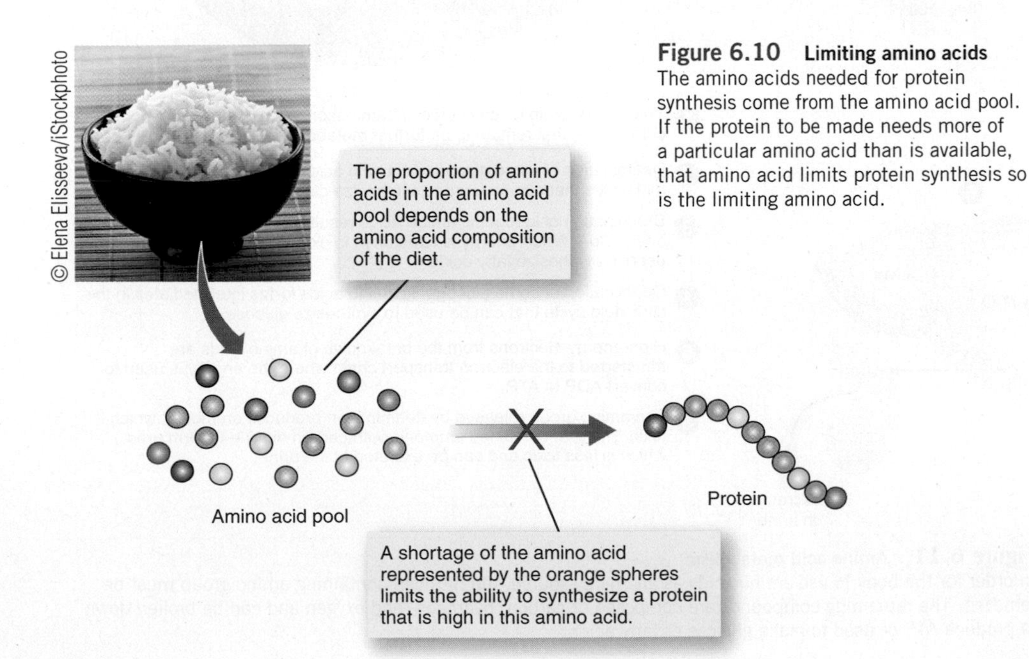

Figure 6.10 **Limiting amino acids** The amino acids needed for protein synthesis come from the amino acid pool. If the protein to be made needs more of a particular amino acid than is available, that amino acid limits protein synthesis so is the limiting amino acid.

How Amino Acids Provide Energy

deamination The removal of the amino group from an amino acid.

urea A nitrogen-containing waste product formed from the breakdown of amino acids that is excreted in the urine.

Although carbohydrate and fat are more efficient energy sources, amino acids from the diet and from body proteins are also used to provide energy (**Figure 6.11**). Before this can occur, the nitrogen-containing amino group must be removed from the amino acids in a process called **deamination**. The amino group that is released is converted into **urea** and excreted in the urine. The carbon compounds remaining after deamination can be used in a number of ways, depending on the needs of the body. If energy is needed, the carbon structure of the amino acids can be converted into acetyl-CoA or compounds that directly enter the citric acid cycle to produce ATP (see Figure 6.11). Amino acids that break down to form acetyl-CoA are called ketogenic amino acids

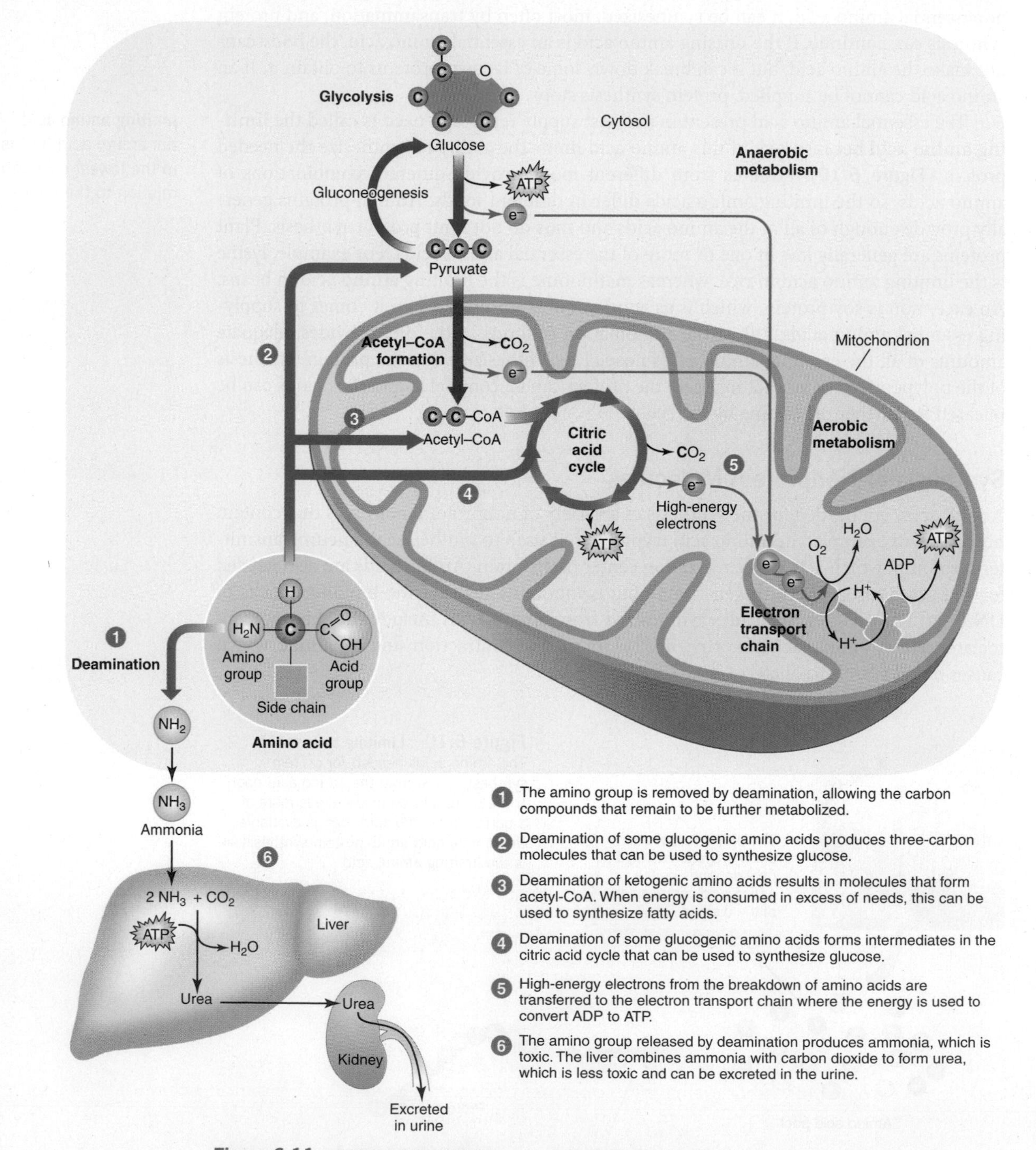

Figure 6.11 Amino acid metabolism
In order for the body to use amino acids as an energy source, the nitrogen-containing amino group must be removed. The remaining compounds are composed of carbon, hydrogen, and oxygen and can be broken down to produce ATP or used to make glucose or fatty acids.

(see Chapter 4). Amino acids that break down to form three-carbon compounds or intermediates in the citric acid cycle are called glucogenic amino acids. These can be used by the liver to synthesize glucose via gluconeogenesis. The use of amino acids as an energy source increases both when the diet does not provide enough total energy to meet needs, as in starvation, and when protein is consumed in excess of needs.

When Energy Intake Is Low When energy is deficient, body proteins, such as enzymes and muscle proteins, are broken down into amino acids that can then be used to generate ATP or synthesize glucose. This provides energy in times of need, but it also robs the body of functional proteins. The most dispensable proteins are broken down first, conserving others for the numerous critical roles they play, but if the energy deficit continues, more critical proteins, such as those that make up the heart and other internal organs, will also be degraded. A loss of more than 30% of the body's protein reduces the strength of the muscles required for breathing and heart function, depresses immune function, and causes a general loss of organ function that is great enough to cause death.

When Protein Intake Exceeds Needs Amino acids are used for energy when protein intake exceeds protein needs. If the diet is adequate in energy and high in protein, the amino acids from the excess protein are deaminated and used to produce ATP. If both energy and protein intake exceed needs, the extra amino acids can be converted into acetyl-CoA and used to synthesize fatty acids that can be stored as triglycerides in adipose tissue and can contribute to weight gain.

6.5 Functions of Body Proteins

LEARNING OBJECTIVES

- Give an example of how proteins contribute to body structure.
- Explain why proteins consumed in the diet do not function in the body.
- Describe four types of proteins that help regulate and facilitate body processes.

When you think of the protein in your body, you probably think of muscle, but muscle contains only a few of the more than 500,000 proteins in the human body, each with a specific function. Some perform important structural roles and others help facilitate and regulate specific body processes (**Figure 6.12**). All the proteins that provide structure and facilitate body activities are synthesized in the body from the amino acids in the amino acid pool.

How Proteins Provide Structure

Proteins provide structure to individual cells and to the body as a whole. In cells, proteins are an integral part of the cell membrane, the cytosol, and the organelles. Skin, hair, and muscle are body structures that are composed largely of protein. The most abundant protein in the body is collagen; it holds cells together and forms the protein framework of bones and teeth. It also forms tendons and ligaments, strengthens artery walls, and is a major constituent of scar tissue. When the diet is deficient in protein, these structures break down. The muscles shrink, the skin loses its elasticity, and the hair becomes thin and can easily be pulled out by the roots.

How Proteins Facilitate and Regulate Body Processes

Proteins are the molecules that do most of the body's work. They are necessary for virtually every activity in the body, from promoting chemical reactions and transporting substance into and out of cells to protecting us from disease and allowing us to move (see Figure 6.12).

Enzymes and Protein Hormones Enzymes are protein molecules that speed up metabolic reactions but are not used up or destroyed in the process. Without the help of enzymes, the metabolic reactions that break down molecules to provide energy and build molecules needed

Figure 6.12 What proteins do

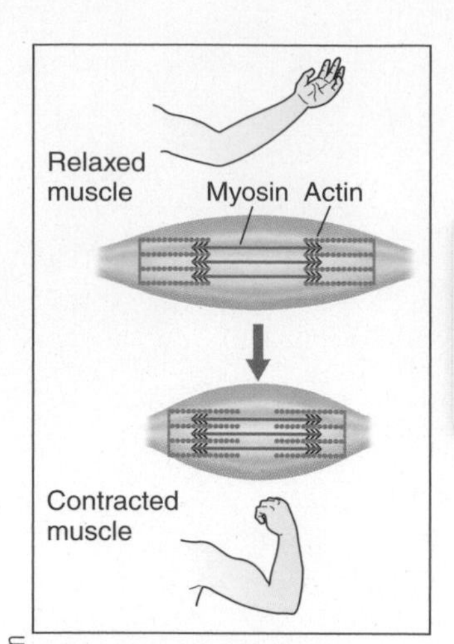

Proteins help us move. During muscle contraction the proteins actin and myosin slide past each other, causing the muscle fibers to shorten.

The protein keratin gives strength to hair. It can be made only inside the body, so a healthy diet will do more for hair quality than expensive protein shampoos.

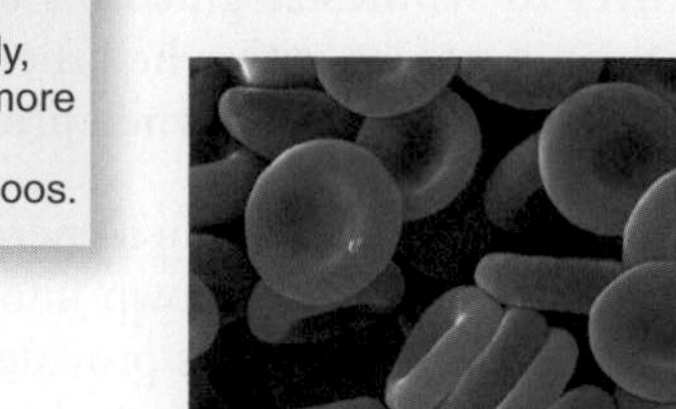

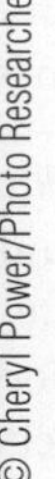

© Cheryl Power/Photo Researchers

Proteins help transport substances in the body. The protein hemoglobin, which gives red blood cells their color, shuttles oxygen to body cells and carries away carbon dioxide.

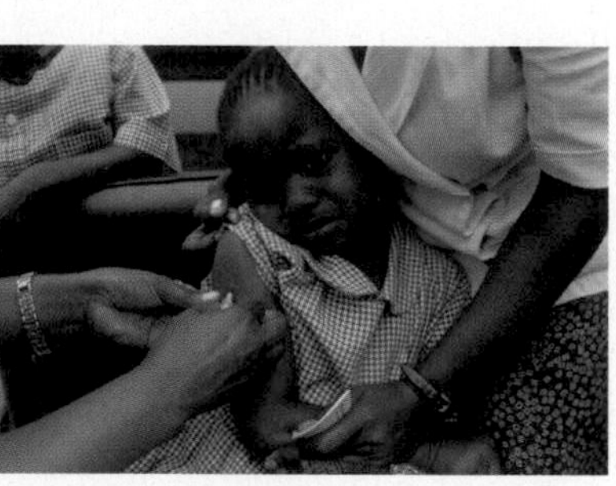

© Karen Kasmauski/NG Image Collection

Some hormones are proteins; insulin and glucagon are protein hormones that help regulate blood sugar.

Antibody proteins help protect us from disease. The vaccination this child is receiving contains dead or inactivated measles virus. It will stimulate the immune system to make antibodies that help destroy the measles virus.

Collagen is the major protein in skin and ligaments. It also gives strength to other tissues and organs and forms the protein framework of bones and teeth.

Photodisc/Getty Images, Inc.

Proteins help regulate fluid balance. If protein levels in the blood fall too low, water leaks out of the blood vessels and accumulates in the tissues, causing swelling known as edema (below).

Almost all the chemical reactions in the body require the help of enzymes, which are proteins. Each enzyme has a unique structure that allows it to interact with the specific molecules in the reaction it accelerates.

Membrane proteins are transporters and receptors. Transport proteins help move amino acids and other molecules into and out of cells and receptor proteins bind to specific molecules to relay messages.

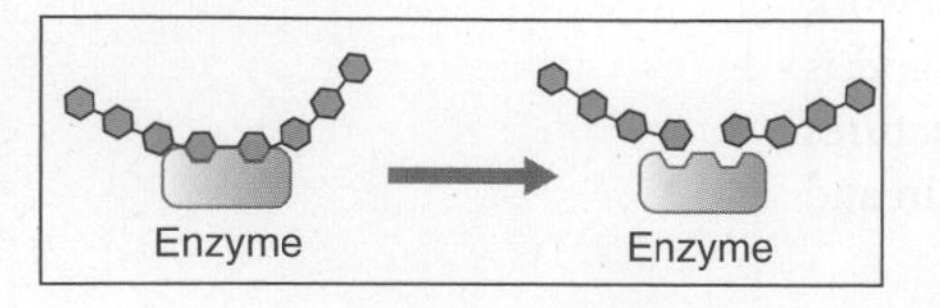

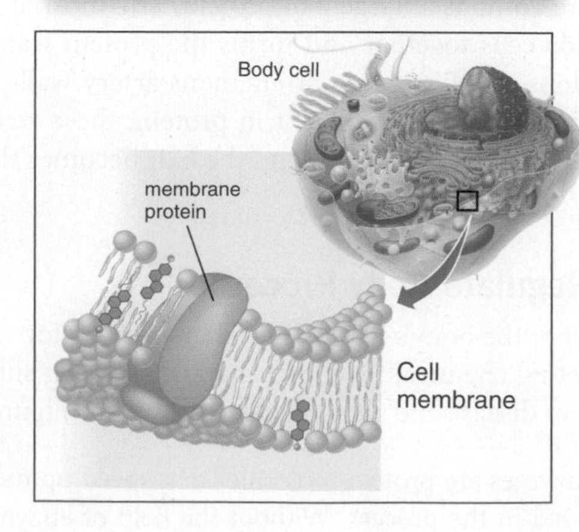

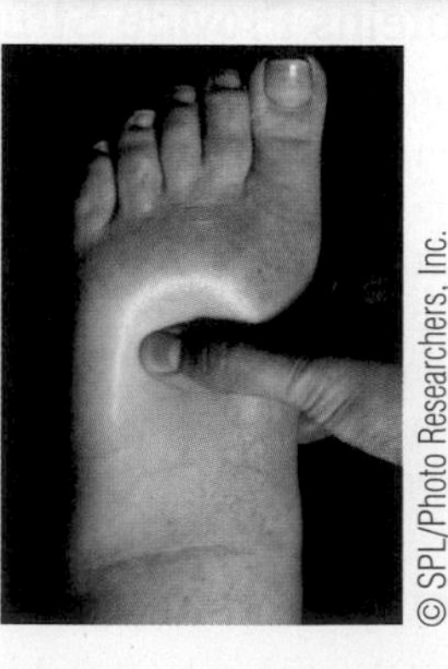

© SPL/Photo Researchers, Inc.

by the body would occur too slowly to support life. Each of the reactions involved in the production of ATP and the synthesis and breakdown of carbohydrates, lipids, and proteins requires a specific enzyme with a unique structure (see Figure 6.12). If the structure of the enzyme molecule is changed, it can no longer function in the reaction it is designed to accelerate.

Hormones are chemical messengers that are secreted into the blood by one tissue or organ and act on target cells in other parts of the body. Hormones made from amino acids are classified as protein or peptide hormones. For instance, insulin and glucagon are peptide hormones involved in maintaining a steady level of blood glucose. Unlike steroid hormones, which are made from cholesterol and can diffuse across the cell membrane and enter the cell, peptide hormones act by binding to protein receptors on the surface of the cell membrane.

The enzymes and hormones that function in the body are made by the body. When these substances are consumed in food or supplements, they are broken down during digestion and are absorbed from the gastrointestinal tract as amino acids. This is the reason that people with diabetes must inject insulin; taken orally, it would not end up in the bloodstream as insulin. The same is true of enzyme supplements; they are broken down in the gut and do not enter the bloodstream intact. Enzymes intended to function while in the gut can provide function. For example, lactase, taken by individuals with lactose intolerance, remains functional long enough to break down lactose consumed at the same time. Eventually, even digestive enzymes are also digested and absorbed as amino acids.

Transport Proteins Proteins transport substances into and out of individual cells and throughout the body. At the cellular level, transport proteins present in cell membranes help move substances such as glucose and amino acids across the cell membrane into and out of cells. Transport proteins in the blood carry substances from one organ to another (see Figure 6.12). For example, the proteins that form lipoproteins are needed to transport lipids from the intestines and liver to body cells. Some nutrients must be bound to a specific protein to be transported in the blood. When protein is deficient, the nutrients that require proteins for transport cannot travel to the cells. For this reason, a protein deficiency can cause a vitamin A deficiency: even if vitamin A is consumed in the diet, without protein it cannot be transported to the cells where it is needed.

How Proteins Protect Us Proteins play an important role in protecting the body from injury and invasion by foreign substances. Skin, which is made up primarily of protein, is the first barrier against infection and injury. Foreign particles such as dirt or bacteria that are on the skin cannot enter the body and can be washed away. If the skin is broken and blood vessels are injured, blood-clotting proteins, including fibrin and thrombin, help prevent too much blood from being lost. If a foreign particle such as a virus or bacterium enters the body, the immune system fights it off by synthesizing proteins called **antibodies**. Each antibody has a unique structure that allows it to attach to a specific invader. When an antibody binds to an invading substance, the production of more antibodies is stimulated, and other parts of the immune system are signaled to help destroy the invading substance. The next time the same type of invading bacterium or virus enters the body, the immune system is already primed to produce specific antibodies to fight off that particular invader. Immunizations against diseases, such as measles, stimulate the immune system to produce antibodies to the virus (see Figure 6.12). When the body comes in contact with the live virus, the immune system is already primed, so a large-scale immune system attack is mounted and the infection is prevented. When the immune system malfunctions as a result of protein deficiency, human immunodeficiency virus (HIV) infection, or other causes, the ability to protect the body from infection is compromised.

antibody A protein produced by the body's immune system that recognizes foreign substances in the body and helps destroy them.

Contractile Proteins The proteins in muscles allow us to move. When you climb a flight of stairs, walk across the room, or run around the block, you are relying on the muscle proteins actin and myosin, which interact to cause muscles to contract. For example, when you do a biceps curl, the muscles in your biceps shorten as the alternating actin and myosin protein fibers slide past one another (see Figure 6.12). Actin and myosin also cause contraction of the heart muscle and the muscles that cause contraction in the digestive tract, blood vessels, and glands. Actin and myosin can also cause contraction in nonmuscle cells. For example, this contraction helps white blood cells change shape and move so they can reach infected tissues in the body. The energy for contraction comes from ATP, which is derived primarily from the metabolism of carbohydrate and fat.

Proteins That Regulate Fluid Balance The distribution of fluid among body cells, the bloodstream, and the spaces between cells is important for homeostasis. Osmosis causes water to move back and forth across membranes to maintain appropriate concentrations of dissolved substances inside and outside cells and tissues (see Chapter 10). Proteins help regulate this fluid balance in two ways. First, protein pumps located in cell membranes transport substances from one side of a membrane to the other. Second, large protein molecules present in the blood hold fluid in the blood by contributing to the concentration of dissolved particles in the bloodstream. In cases of protein malnutrition, the concentration of these large proteins in the blood decreases, so fluid is no longer held in the blood by osmosis and it accumulates in tissues and in the abdomen (see Figure 6.12).

pH A measure of acidity.

Proteins That Regulate Acid–Base Balance The chemical reactions of metabolism require a specific level of acidity, or **pH**, to function properly. In the gastrointestinal tract, pH varies widely. The digestive enzyme pepsin works best in the acid environment of the stomach, whereas the pancreatic enzymes operate best in the more neutral environment of the small intestine. Inside the body, the range of optimal pH is much tighter. The acids and bases produced by metabolic reactions must be neutralized in order to prevent changes in pH, which could prevent life-sustaining metabolic reactions from proceeding normally. The lungs and kidneys help maintain a normal pH by eliminating some of these waste products. Proteins both in the blood and within the cells act as buffers to prevent changes in pH. They function by attracting or releasing hydrogen ions. For instance, the protein hemoglobin in red blood cells helps neutralize acid produced when carbon dioxide reacts with water. Untreated type 1 diabetes is an example of what happens when the amount of acid produced exceeds the ability of the body's proteins and other systems to neutralize it. In type 1 diabetes, the inability to get glucose into cells results in the breakdown of fats and the buildup of ketones, which are acidic. As ketones accumulate, they cause a drop in pH called ketoacidosis (see Chapter 4). The acidic pH denatures proteins, and they are unable to perform their functions, resulting in coma and eventually, if not treated, death.

6.6 Protein, Amino Acids, and Health

LEARNING OBJECTIVES

- Compare kwashiorkor with marasmus.
- Explain why protein-energy malnutrition develops more rapidly in young children than in adults.
- Discuss the potential risks associated with a high-protein diet.

protein-energy malnutrition (PEM) A condition characterized by wasting and an increased susceptibility to infection that results from the long-term consumption of insufficient amounts of energy and protein to meet needs.

kwashiorkor A form of protein-energy malnutrition in which only protein is deficient.

marasmus A form of protein-energy malnutrition in which a deficiency of energy in the diet causes severe body wasting.

A diet adequate in protein is essential to health. Dietary protein is needed for growth and to replace body protein that is broken down and lost each day. If too little protein is consumed, the consequences can be dramatic and devastating. Too much protein, particularly if it is derived primarily from animal sources, may also have negative health effects. In addition, some people are sensitive to specific proteins and amino acids.

Protein Deficiency

Because of the availability and variety of foods in developed countries, protein deficiency is uncommon. However, in developing nations, concerns about inadequate protein are very real. Diets deficient in protein are most often deficient in energy as well, but a pure protein deficiency can occur when food choices are extremely limited and the staple food of a population is very low in protein. The term **protein-energy malnutrition (PEM)** is used to refer to the continuum of protein deficiency conditions ranging from pure protein deficiency, called **kwashiorkor**, to overall energy deficiency, called **marasmus**. Most protein-energy malnutrition is a combination of the two.

LIFE CYCLE

Kwashiorkor Kwashiorkor is typically a disease of children. The word "kwashiorkor" comes from the Ga language of coastal Ghana. It means the disease that the first child gets when a second child is born.[4] When the new baby is born, the older child is no longer breastfed. Rather than receiving protein-rich breast milk, the young child is fed a watered-down version of the diet eaten by the rest of the family. This diet is low in protein and is often high in fiber and difficult to digest. The child, even if able to get adequate energy, is not able to eat a large enough quantity to get adequate protein. Because children are growing, their protein needs per unit of body weight are higher than those of adults, and the effects of a deficiency become evident much more quickly. Although kwashiorkor only occurs in those whose diets are deficient in protein, it is associated with, or even triggered by, infectious diseases.

The symptoms of kwashiorkor can be explained by examining the roles that proteins play in the body. Because protein is needed for the synthesis of new tissue, kwashiorkor hampers growth in height and weight in children. Because proteins are important in immune function, there is an increased susceptibility to infection. There are changes in hair color because less of the skin pigment melanin is made; the skin flakes because structural proteins are not available to provide elasticity and support. Cells lining the digestive tract die and cannot be replaced, so nutrient absorption is impaired. The bloated belly typical of this condition is a result of both fat accumulating in the liver, because there is not enough protein to transport it to other tissues, and fluid accumulating in the abdomen, because there is not enough protein in the blood to keep water from diffusing out of the blood vessels (**Figure 6.13**).

Figure 6.13 **Protein-energy malnutrition**

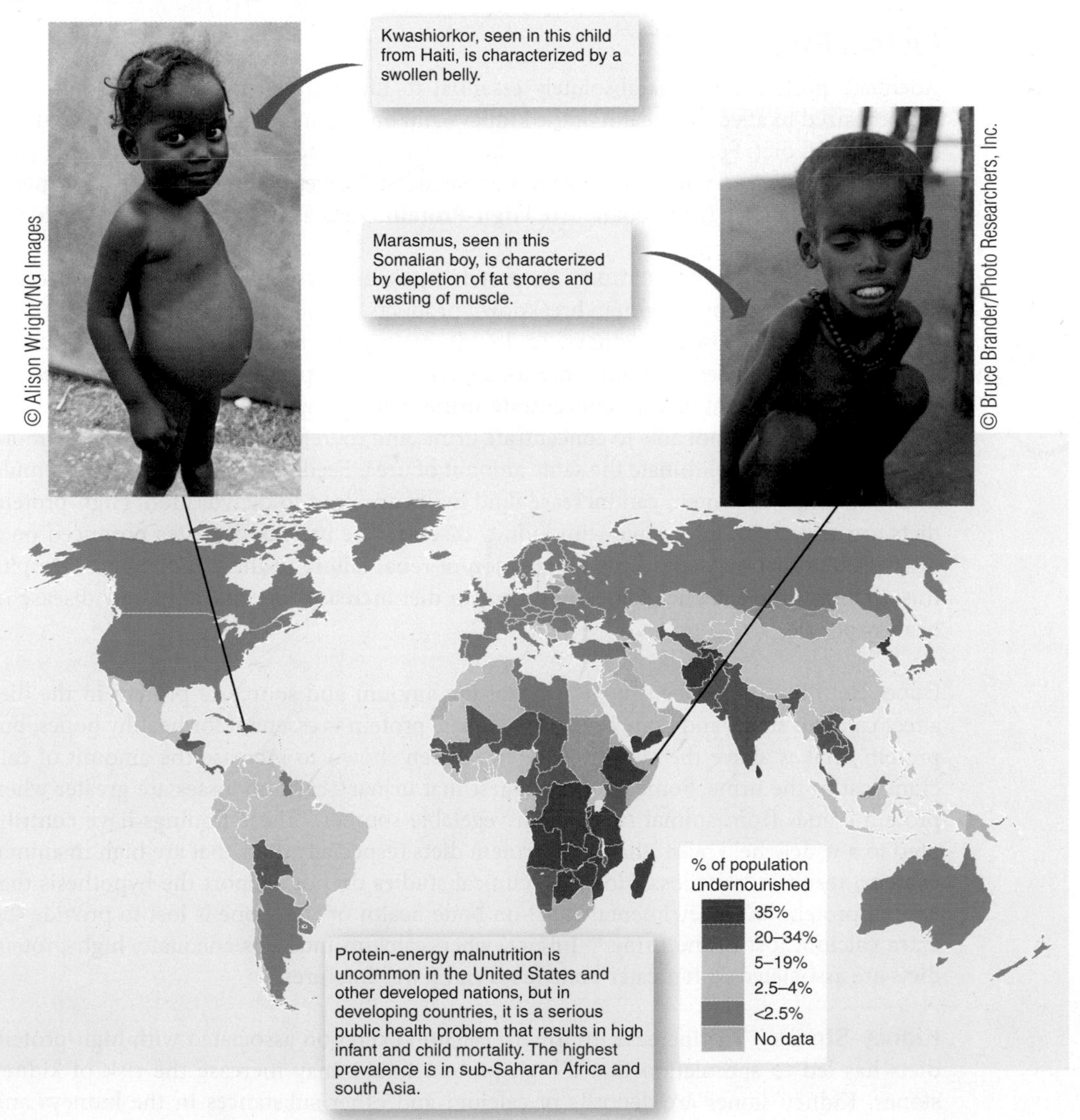

Kwashiorkor occurs most commonly in Africa, South and Central America, and South Asia. It has also been reported in poverty-stricken areas in the United States. Although kwashiorkor is thought of as a disease of children, it is seen in hospitalized adults who have high-protein needs due to infection or trauma and low-protein intake because they are unable to eat.

Marasmus At the other end of the continuum of protein-energy malnutrition is marasmus, meaning to waste away. Marasmus is due to a deficiency of energy, but protein and other nutrients are usually also insufficient to meet needs. Marasmus may have some of the same symptoms as kwashiorkor, but there are also differences. In kwashiorkor, some fat stores are retained, since energy intake is adequate. In contrast, marasmic individuals appear emaciated because their body fat stores have been used to provide energy (see Figure 6.13). Since fat is a major energy source and carbohydrate is limited, ketosis may occur in marasmus. This is not so in kwashiorkor because carbohydrate intake is adequate—only protein is deficient.

Marasmus occurs in individuals of all ages but is most devastating in infants and children because adequate energy is essential for growth. Most brain growth takes place in the first year of life, so malnutrition early in life causes a decrease in intelligence and learning ability that persists throughout life. Marasmus often occurs in children who are fed diluted infant formula prepared by caregivers trying to stretch limited supplies. It occurs less often in breast-fed infants. Marasmus is the form of malnutrition that occurs with eating disorders.

Protein Excess

Adequate protein intake is absolutely essential to life, but too much protein has been hypothesized to affect the health of the kidneys and bones, and to impact the healthfulness of the overall diet. For a healthy person, there are no short-term problems associated with consuming a diet very high in protein, but we are still investigating whether the same is true in the long term (see **Debate: Are High-Protein Diets a Safe Way to Lose Weight?**).

Hydration and Kidney Function As protein intake increases above the amount needed, so does the production of protein breakdown products, such as urea, which must be eliminated from the body by the kidneys. To do this, more water must be excreted in the urine, increasing water losses. Although not a concern for most people, this can be a problem if the kidneys are not able to concentrate urine. For example, the immature kidneys of newborn infants are not able to concentrate urine, and therefore they need to excrete more water than adults to eliminate the same amount of urea. Feeding a newborn infant formula that is too high in protein can increase fluid losses and lead to dehydration. High-protein diets are also a risk for people with kidney disease. The increased wastes produced on a high-protein diet may speed the progression of renal failure in these individuals. Despite this, there is little evidence that a high-protein diet increases the risk of kidney disease in healthy people.[5,6]

Bone Health It has been suggested that the amount and source of protein in the diet affect calcium status and bone health.[7] Adequate protein is essential for healthy bones, but protein intakes above the current RDA have been shown to increase the amount of calcium lost in the urine. Some studies suggest that urinary calcium losses are greater when protein comes from animal rather than vegetable sources.[7] These findings have contributed to a widely held belief that high-protein diets (especially diets that are high in animal protein) result in bone loss. However, clinical studies do not support the hypothesis that animal protein has a detrimental effect on bone health or that bone is lost to provide the extra calcium lost in the urine.[8,9] In fact, when calcium intake is adequate, high-protein diets are associated with greater bone mass and fewer fractures.[9]

Kidney Stones The increase in urinary calcium excretion associated with high-protein diets has led to speculation that a high protein intake may increase the risk of kidney stones. Kidney stones are deposits of calcium and other substances in the kidneys and

DEBATE

Are High-Protein Diets a Safe Way to Lose Weight?

© Danny Hooks/iStockphoto

Eat all the bacon, eggs, and steak you want and still lose weight! High-protein, low-carbohydrate diets limit food choices to avoid carbohydrates but don't count calories, so you can have all those high-fat, high-calorie foods that other diets restrict. For decades, low-carb diets, such as the Atkins diet, have been popular for weight loss, but do you really lose weight and are they healthy?

To lose weight, you need to eat fewer calories than you expend. So, any diet that limits calorie intake will promote weight loss. High-protein, low-carbohydrate diets seem to defy the laws of thermodynamics because they don't limit calories but they do result in rapid weight loss. The real reason high-protein diets promote weight loss is that protein in the diet promotes satiety and a lack of carbohydrate causes ketone production, which suppresses appetite. So, even though dieters aren't counting calories, they still eat less. When compared to people on a low-fat diet, people on a high-protein, low-carbohydrate diet are less bothered by hunger.[1] A two-year comparison of three diets, a low-carbohydrate diet that did not specifically limit calories, a Mediterranean-style diet with a calorie restriction, and a low-fat plan with a calorie restriction, demonstrated that all three caused a significant drop in body weight, blood pressure, and waist circumference. The drops were greater on the low-carbohydrate and Mediterranean diets than on the low-fat diet.[2]

One criticism of high-protein diets is that they are high in animal products, which make the diet high in total fat, saturated fat, and cholesterol, dietary components that increase the risk of cardiovascular disease. Despite this, improvements in blood lipid levels have been demonstrated in people consuming these diets for up to two years.[2] One study concluded that high-protein, low-carbohydrate diets were as effective as low-fat diets for weight loss and caused a reduction in blood triglyceride levels and a rise in healthy high-density lipoprotein (HDL) cholesterol levels.[3] High-protein diets can cause an increase in low-density lipoprotein (LDL) cholesterol, but the increase is due to an increase LDL particles that are larger in size, which pose less of a risk for heart disease than smaller LDLs.[4]

Another concern with high-protein diets is that restrictions on the consumption of many fruits, vegetables, and whole grains may lead to micronutrient deficiencies. After eight weeks, people consuming a high-protein, low-carbohydrate diet were at an increased risk for deficiencies of vitamin C, the B vitamins thiamin and folic acid, and the minerals iron and magnesium.[5] This dietary pattern, which is high in animal products and low in grains, fruits, and vegetables, may also be detrimental to colon health. One study showed that after four weeks, high-protein, low-carbohydrate weight-loss diets resulted in a significant decrease in substances in the feces that protect against cancer and a rise in the concentrations of hazardous substances in the colon. Therefore, long-term adherence to such diets may increase the risk of colon cancer.[6] This is backed up by a body of literature that shows an increased risk of cancer, especially colorectal cancer, with a high intake of meat.[7]

The most significant concern with high-protein diets is that there is still no real long-term data on their effectiveness or safety. High-protein diets promote weight loss in the short term. Over time, more vegetables and limited whole grains are typically introduced into these diets. But as carbohydrate is reintroduced, ketone levels fall, and a behavior change is needed to maintain a reduced calorie intake and long-term weight loss. Maintaining a strict low-carbohydrate diet means eliminating whole food groups from the diet. Is this healthy for life? Will blood lipids remain healthy after 5 or 10 years? Will the risk of cancer be affected? There is no argument that reducing weight into the healthy range is important, but can the risks of a weight-loss diet outweigh the benefits of weight loss?

THINK CRITICALLY: If you were following a low-carbohydrate diet, what foods would you need to eliminate from the breakfast shown in the photo above?

---

[1]Martin, C. K., Rosenbaum, D., Han, H., et al. Change in food cravings, food preferences, and appetite during a low-carbohydrate and low-fat diet. *Obesity* 19:1963–1970, 2011.

[2]Shai, I., Schwarzfuchs, D., Henkin, Y., et al. Weight loss with a low-carbohydrate, Mediterranean, or low-fat diet. *N Engl J Med* 359:229–241, 2008.

[3]Nordmann, A. J., Nordmann, A., Briel M., et al. Effects of low-carbohydrate vs low-fat diets on weight loss and cardiovascular risk factors: A meta-analysis of randomized controlled trials. *Arch Intern Med* 166:285–293, 2006.

[4]Samaha, F. F., Foster, G. D., and Makris, A. P. Low-carbohydrate diets, obesity, and metabolic risk factors for cardiovascular disease. *Curr Atheroscler Rep* 9:441–447, 2007.

[5]Gardner, C. D., Kim, S., Bersamin, A., et al. Micronutrient quality of weight-loss diets that focus on macronutrients: Results from the A TO Z study. *Am J Clin Nutr* 92:304–312, 2010.

[6]Russell, W. R., Gratz, S. W., Duncan, S. H., et al. High-protein, reduced-carbohydrate weight-loss diets promote metabolite profiles likely to be detrimental to colonic health. *Am J Clin Nutr* 93:1062–1072, 2011.

[7]Ferguson, L. R. Meat and cancer. *Meat Sci* 84:308–313, 2010.

© iStockphoto

OFF THE LABEL Is It Safe for You?

"What is food to one, is to others bitter poison," said the Roman philosopher and scientist Lucretius. Today, fortunately, the information in ingredient lists can help those with food allergies tell the difference. Food manufacturers are required to state clearly if a product contains any of the eight major food allergens: milk, eggs, peanuts, tree nuts, fish, shellfish, soy, and wheat.[1] Products that contain tree nuts also must list the type of nut on the label, and foods containing fish or shellfish must give the species. This information helps quickly identify foods that contain these allergens so they can be avoided.

The information about allergens can be presented on the label in one of three ways. The first is to simply list it clearly in the ingredient list. For example, MILK can be listed with other ingredients. A second way is to use a parenthetical statement. Sometimes this is helpful when listing the actual ingredient is not clear to the consumer. For example, people who are allergic to milk might be allergic to the milk protein casein. They must not only avoid milk and other dairy products, but also foods to which casein has been added. Casein, often listed as sodium caseinate, is used in coffee whiteners and frozen dessert toppings. Foods containing casein can identify it as a milk-derived allergen by using a parenthetical statement, such as CASEIN (MILK).

A third way food labels may indicate the presence of one or more of these eight common allergens is to use the word "contains" followed by the name of the major food allergen printed at the end of the ingredient list or next to it, for example: CONTAINS WHEAT AND SOY INGREDIENTS. This is helpful because some foods are unexpected sources of allergens. For example, isolated soy protein may be added to canned soups or gravies to aid in emulsification or add texture. A person trying to avoid soy would not necessarily think of soup as a soy product, but the statement at the end of the ingredients warns them that this product CONTAINS SOY INGREDIENTS (see figure).

Keep in mind that even a food that lists zero grams of protein in the Nutrition Facts panel may include protein sources in the ingredient list. The amounts may be too small to add a significant amount of protein but can be enough to cause an allergic reaction. For example, small amounts of *protein hydrolysates* or *hydrolyzed proteins*—proteins that have been treated with acid or enzymes to break them down into amino acids and small peptides—are often added as flavorings, flavor enhancers, stabilizers, or thickening agents in foods such as soups and packaged rice and potato products. To help individuals avoid specific proteins, foods containing protein hydrolysates are required to list the source of the hydrolysate—HYDROLYZED CORN PROTEIN in the soup example.

Finally, products that do not intentionally contain proteins that cause allergies may provide a special warning if there is a potential for cross-contamination from equipment or foods processed in the same facility. For example, you may see products declaring that they MAY CONTAIN ALMONDS or that they were PROCESSED IN A FACILITY THAT ALSO PROCESSES PEANUTS—a potentially life-saving warning for individuals with peanut allergy, who can have a severe reaction from minute amounts of peanut protein.

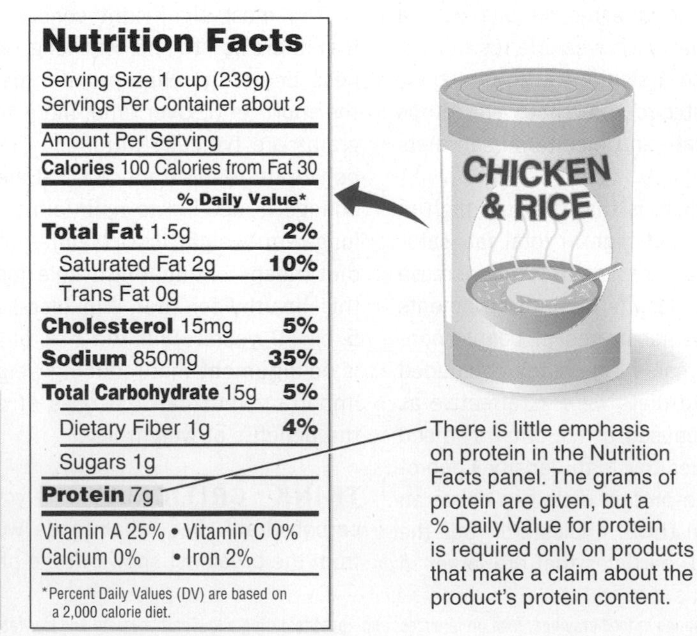

The Nutrition Facts panel lists the grams of protein in a serving, but the ingredient list tells you which amino acids or protein-containing ingredients have been added to the product.

[1]U.S. Food and Drug Adminsitration, Center for Food Safety and Applied Nutrition. Food Allergen Labeling and Consumer Protection Act of 2004.

urinary tract. Higher concentrations of calcium and acid in the urine increase the likelihood that the calcium will be deposited, forming these stones. Epidemiological studies suggest that diets that are rich in animal protein and low in fluid contribute to the formation of kidney stones.[10]

Heart Disease and Cancer Risk Another concern with high-protein diets is related more to the dietary components that accompany animal versus plant proteins. Typically, high-protein diets are also high in animal products, which are high in saturated fat and cholesterol. Although this dietary pattern is associated with an increased risk of heart disease, when consumed for up to two years, high-protein, low-carbohydrate weight-loss diets do not appear to cause blood lipid patterns that increase the risk of heart disease.[11] In addition to being high in meat, these diets are also typically low in vegetables and fruits, a pattern associated with a greater risk of colon cancer.[12]

Protein and Amino Acid Intolerances

Unlike lipids and carbohydrates, people don't think of protein as contributing to health problems, but for some people, the wrong protein can be harmful. In some cases, this is because a protein in food is recognized and targeted by the immune system, causing an allergic reaction. There are no cures for food allergies so, to avoid symptoms, allergic individuals need to avoid eating foods that contain proteins that cause an allergic reaction (see **Off the Label: Is It Safe for You?**). Not all adverse reactions to proteins and amino acids are due to allergies; some are due to food intolerances, also called food sensitivities. These reactions do not involve the immune system. The symptoms of a food intolerance can range from minor discomfort, such as the abdominal distress some people feel after eating raw onions, to more severe reactions.

Why People with Celiac Disease Must Avoid Gluten Gluten intolerance, also called **celiac disease**, celiac sprue, or gluten-sensitive enteropathy, is a form of food intolerance. Individuals with celiac disease cannot tolerate gluten, a protein found in wheat, rye, and barley. Celiac disease is an autoimmune disease in which gluten triggers the immune system to attack the villi in the small intestine, causing symptoms such as diarrhea, abdominal bloating and cramps, weight loss, and anemia (see Chapter 3, Debate: Shoud You Be Gluten Free?). Once thought to be a rare childhood disease, it is now known to affect more than 2 million people in the United States.[13] The only treatment is to avoid gluten by eliminating all products containing gluten from the diet.

celiac disease A disorder that causes damage to the intestines when the protein gluten is eaten.

Why Aspartame Is Dangerous for People with Phenylketonuria Aspartame is a sugar substitute composed of two amino acids, aspartic acid and phenylalanine. Aspartame is used in a wide variety of foods, including carbonated beverages, gelatin desserts, and chewing gum. Digestion breaks aspartame into aspartic acid, phenylalanine, and an alcohol called methanol. Because the phenylalanine released from aspartame digestion is absorbed into the blood, food products containing this alternative sweetener must be avoided by individuals with a genetic disorder called **phenylketonuria (PKU)**.

phenylketonuria (PKU) An inherited disease in which the body cannot metabolize the amino acid phenylalanine. If the disease is untreated, toxic by-products called *phenylketones* accumulate in the blood and interfere with brain development.

Individuals with PKU inherit a defective gene for an enzyme called *phenylalanine hydroxylase* that is needed to metabolize phenylalanine. In those with this faulty gene, the enzyme does not function properly, and they are unable to convert the essential amino acid phenylalanine to the semiessential amino acid tyrosine. Instead, phenylalanine is converted to compounds called *phenylketones*, which build up in the blood (**Figure 6.14a**). In infants and children, high phenylketone levels can interfere with brain development, causing intellectual disability.

PKU afflicts about 1 in 12,000 newborn infants.[14] Infants are tested for this disorder at birth because brain damage can be avoided by a special low-phenylalanine diet. Pregnant women with PKU must be especially careful to consume a low-phenylalanine diet in order to protect their unborn children from high phenylketone levels, which can cause brain abnormalities and other birth defects.[15] The diet for people with PKU must provide just enough phenylalanine to meet the body's need for protein synthesis but not so much that

THINK CRITICALLY

Why do you think this warning appears on diet soda labels but not on labels for high-protein foods such as meat and milk?

Figure 6.14 Phenylketonuria

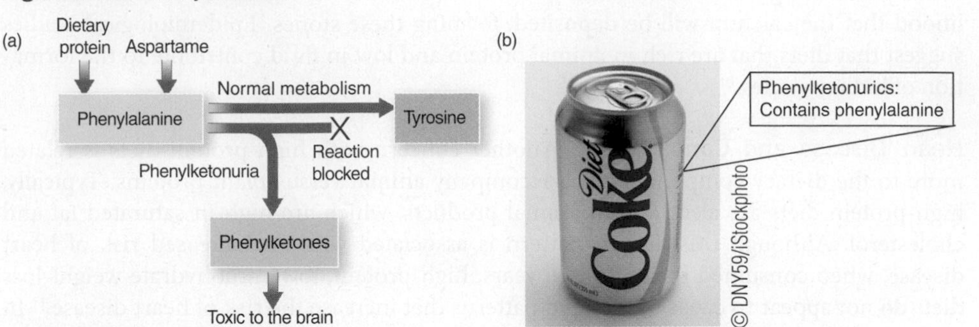

(a) In individuals with phenylketonuria, phenylalanine cannot be converted to tyrosine. Instead it forms phenylketones, which accumulate and can interfere with brain development.

(b) Warnings for individuals with PKU are included on the labels of all products sweetened with aspartame, though few of them specify the actual amount of the sweetener contained in the product.

the buildup of phenylketones occurs. The diet must also provide sufficient tyrosine because PKU prevents conversion of phenylalanine to tyrosine, causing it to become an essential amino acid. All proteins naturally contain phenylalanine, so a low-phenylalanine diet must carefully regulate overall protein intake. Special low-phenylalanine or phenylalanine-free formulas are manufactured for infants with this disease. Because foods such as diet soda and other aspartame-sweetened beverages do not contain protein, individuals with this disease might not expect them to contain phenylalanine (**Figure 6.14b**).

Monosodium Glutamate Monosodium glutamate (MSG) is a flavor enhancer best known for its use in Chinese cooking. MSG consists of the amino acid glutamic acid (or glutamate) bound to sodium. In addition to being added to Chinese food, it is used in meat tenderizers, sold as a seasoning, and added to a variety of packaged foods such as potato chips, canned soups, cured meats, and packaged entrees. Some people report adverse reactions including a flushed face, tingling or burning sensations, headache, rapid heartbeat, chest pain, and general weakness after consuming MSG. These symptoms are referred to as *MSG symptom complex*, commonly termed *Chinese restaurant syndrome*. Despite anecdotal reports, research has been unable to confirm that MSG ingestion causes any adverse reactions.[16] Individuals who wish to avoid it can check the ingredient list on packaged foods for monosodium glutamate or potassium glutamate. When eating out, you can avoid this additive by asking the restaurant to prepare your food without added MSG.

6.7 Meeting Recommendations for Protein Intake

LEARNING OBJECTIVES

- Discuss how protein needs are determined.
- Explain what is meant by protein quality.
- Review a diet and substitute complementary plant proteins for the animal proteins it contains.
- Discuss the health benefits and nutritional risks of vegetarian diets.

Protein consumed in the diet must supply amino acids to replace losses that occur during protein turnover, to repair damaged tissues, and to synthesize new body proteins for growth. Current recommendations do not suggest a change in the amount of protein Americans consume but do advise increasing the amount of plant protein and decreasing the amount of animal protein. The recommended dietary pattern is based on whole grains, legumes, fruits, and vegetables, with smaller amounts of lean meats and low-fat milk products.

How Protein Requirements Are Determined

Historically, recommendations for protein intake were estimated from the amount of protein consumed by healthy workingmen in the general population. These protein levels were often as high as 150 grams per day. Current recommendations are generally lower than this and are based on **nitrogen balance** studies (**Figure 6.15**). Since protein is the only macronutrient that contains a significant amount of nitrogen, the amount of protein used by the body can be estimated by comparing nitrogen intake with nitrogen loss. Nitrogen intake is calculated from dietary protein intake. Nitrogen loss or output is measured by totaling the amounts of nitrogen excreted in urine and feces and that lost from skin, sweat, hair, and nails. The majority of the nitrogen lost is excreted in the urine as urea (see Figure 6.11). Comparing the amount of nitrogen consumed with the amount lost provides information about the amount of protein being synthesized and broken down within the body. An individual who is consuming enough protein to meet body needs is in protein or nitrogen balance. This means the individual is consuming enough protein to replace the amount that is lost from the body. The protein requirement is the smallest amount of dietary protein that will maintain balance when energy needs are met by carbohydrate and fat.

nitrogen balance The amount of nitrogen consumed in the diet compared with the amount excreted by the body over a given period.

If the body breaks down more protein than it synthesizes, then nitrogen balance is negative; this means more nitrogen is lost than ingested (see Figure 6.15). This indicates that body protein is being lost. Negative nitrogen balance can occur when intake is too low or when the amount of protein breakdown has been increased by a stress such as injury, illness, or surgery.

If the body is synthesizing more protein than it breaks down, nitrogen balance is positive; this indicates that the body is using dietary protein for the synthesis of new body proteins (see Figure 6.15). Positive nitrogen balance occurs when new tissue is synthesized, such as during growth, pregnancy, wound healing, or muscle building.

The protein requirement of a specific individual can be determined by doing a nitrogen balance study for that individual. Because this procedure cannot be done for everyone, the protein needs of populations must be estimated from balance study data.

Figure 6.15 **Nitrogen balance**

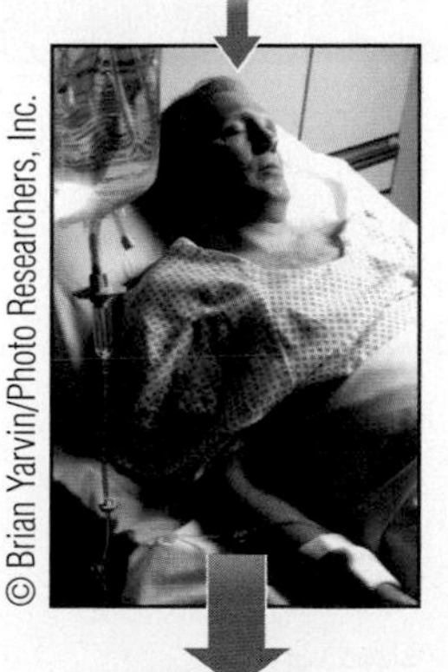

Recommendations for protein intake for the general public are actually higher than the requirements determined by nitrogen balance studies for individuals. This is to allow a margin of safety that will ensure that the needs of the majority of the population are met.

Protein Recommendations

There are several ways to consider protein recommendations. They can be given as the absolute amount of protein in the diet, the percentage of total calories consumed as protein, and the amount of protein ingested per kilogram of body weight. Regardless of how it is presented, most Americans get plenty of protein.

The RDA for protein is expressed per unit of body weight because protein is needed to maintain and repair the body. The more a person weighs, the more protein he or she needs for those purposes. The RDA for protein for adults is 0.8 g of protein per kg of body weight per day.[6] For an adult weighing 70 kg (154 lb), the recommended intake would be 56 g of protein per day (70 kg × 0.8 g/kg/d = 56 g); for an adult weighing 59 kg (130 lb) it would be 47 g/day (59 kg × 0.8 g/kg/d = 47 g). Typical protein intake in the United States is about 90 g/day.[2] Protein needs are increased if protein is being deposited in the body, as it is during growth, or when protein losses are increased, such as during lactation or when the body is injured. RDAs have also been established for each of the essential amino acids (see Appendix A); these are not a concern in a typical diet but are important when developing intravenous feeding solutions.

The Acceptable Macronutrient Distribution Range (AMDR) for protein is 10 to 35% of energy for adults.[6, 17] This range allows for different food preferences and eating patterns. A protein intake in this range will meet needs, balance with carbohydrate and fat calories, and not increase health risks. A diet that provides only 10% of calories from protein meets the RDA, but is a relatively low-protein diet compared with the 16% of energy from protein in the typical U.S. diet.[17] Very few people consume more protein than the upper end of the healthy range—35% of calories. If the proportion of protein goes higher than this, the diet will likely be higher in fat and lower in carbohydrate than is recommended.

The amount of protein in a food or in a diet can be estimated from the Exchange Lists (**Table 6.1**) or determined using values from food labels, food composition tables, iProfile or other diet analysis programs or databases (see *Nutrient Composition of Foods* supplement or iProfile).

TABLE 6.1 Using Exchange Lists to Estimate Protein Content

Exchange group/list	Serving size	Protein (g)
Carbohydrate Group		
Starch	⅓ cup pasta, ½ cup potatoes; 1 slice bread	3
Fruit	1 small apple, peach, pear; ½ banana; ½ cup canned fruit (in juice)	0
Milk	1 cup milk or yogurt	
Nonfat		8
Low-fat		8
Reduced-fat		8
Whole		8
Other carbohydrates	Serving sizes vary	Varies
Vegetables	½ cup cooked vegetables, 1 cup raw	2
Meat/Meat Substitute group	1 oz meat or cheese, ½ cup legumes	
Very lean		7
Lean		7
Medium-fat		7
High-fat		7
Fat Group	1 tsp butter, margarine, or oil; 1 T salad dressing	0

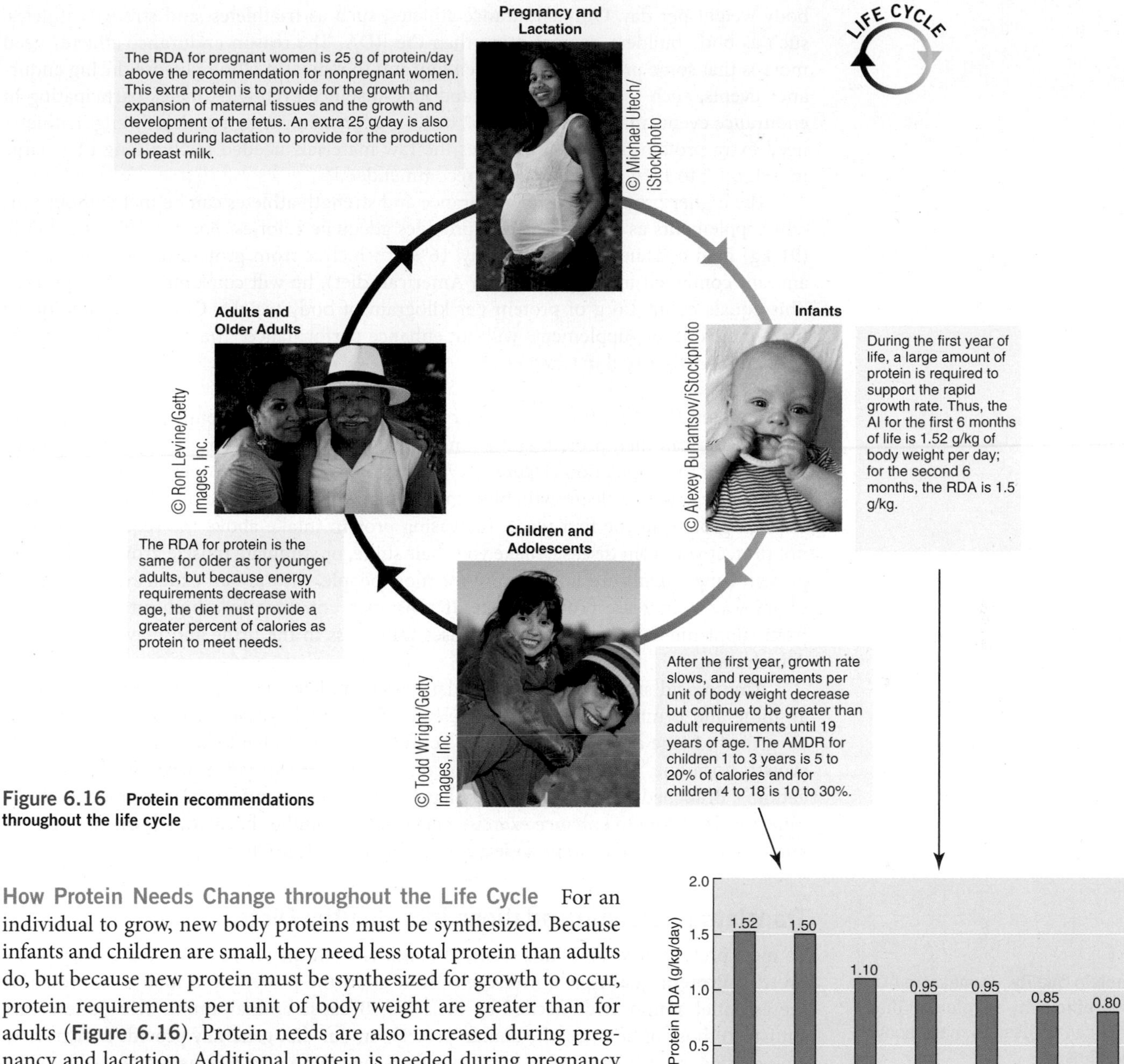

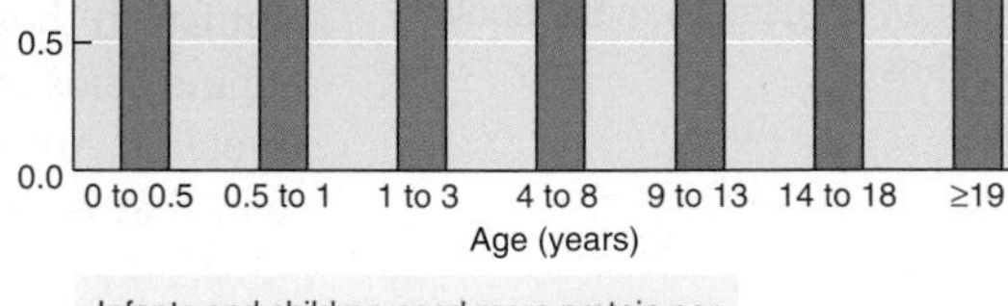

Figure 6.16 **Protein recommendations throughout the life cycle**

How Protein Needs Change throughout the Life Cycle For an individual to grow, new body proteins must be synthesized. Because infants and children are small, they need less total protein than adults do, but because new protein must be synthesized for growth to occur, protein requirements per unit of body weight are greater than for adults (**Figure 6.16**). Protein needs are also increased during pregnancy and lactation. Additional protein is needed during pregnancy to support the expansion of maternal blood volume, the growth of the uterus and breasts, the formation of the placenta, and the growth and development of the fetus. Additional protein is needed during lactation to support the production and secretion of milk, which is high in protein. Most women in North America already consume enough protein in their typical diets to meet the increased needs of pregnancy and lactation.

Why Protein Needs Are Increased by Illness and Injury Extreme stresses on the body such as infections, fevers, burns, or surgery increase the amount of protein that is broken down. For the body to heal and rebuild, these losses must be replaced by dietary protein. The extra amount needed depends on the injury. A severe infection may increase protein needs by about 30%; a serious burn can increase protein requirements by 200 to 400%.

How Exercise Affects Protein Needs The marketing of protein powders and amino acid supplements to athletes might lead people to believe that protein is in short supply in an athlete's diet. In fact, athletes can obtain plenty of protein in their diets without supplements. Most athletes can meet their protein needs by consuming the RDA of 0.8 g/kg of body weight per day. Only endurance athletes, such as triathletes, and strength athletes, such as body builders, require more than the RDA. The reason endurance athletes need more is that some protein is used for energy and to maintain blood glucose during endurance events, such as ultramarathons and long-distance cycling. Athletes participating in endurance events may benefit from 1.2 to 1.4 g of protein per kg per day. Strength athletes need extra protein because it provides the raw materials needed for building their large muscles; 1.2 to 1.7 g per kg per day is recommended.[18]

The higher protein needs of endurance and strength athletes can be met without protein supplements as long as the diet provides adequate calories. For example, if a 200-lb (91-kg) man consumes 3600 kcals/day, 16% of which is from protein (approximately the amount contained in a typical North American diet), he will consume 144 g of protein. This equals about 1.6 g of protein per kilogram of body weight. Consuming additional protein as food or supplements will not enhance performance. The protein needs of athletes are also discussed in Chapter 13.

Do We Need Protein and Amino Acid Supplements? Even though protein is plentiful in the American diet, protein and amino acid supplements remain popular among some segments of the population (**Figure 6.17**). Protein is needed for proper immune function, healthy hair, and muscle growth, but supplements will impact these only if the diet is deficient in protein in the first place. Increasing protein intake above the requirement does not protect you from disease, make your hair shine, or stimulate muscle growth. Although protein supplements are not harmful for most people, they are an expensive and unnecessary way to increase protein intake. If consumed consistently, a high intake of protein from supplements or from foods increases water loss in the urine and may contribute to dehydration.

Amino acid supplements are popular among athletes. For instance, the amino acids arginine and ornithine stimulate the release of growth hormone, which promotes muscle growth. Large doses of these amino acids have been shown to stimulate the release of growth hormone, but the effect this has on exercise performance needs further investigation.[19] Branched-chain amino acids (leucine, isoleucine, and valine), arginine, and alanine are also taken to enhance exercise performance. Studies have not shown any of these amino acid supplements to provide a consistent benefit (see Chapter 13).

Translating Recommendations into Healthy Diets

protein quality A measure of how efficiently a protein in the diet can be used to make body proteins.

To meet protein needs, it is important to consider both the amount and the quality of the protein. **Protein quality** is a measure of how good the protein in a food is at providing the essential amino acids needed by the body. Because animal amino acid patterns are similar to those of humans, the animal proteins in our diet generally provide a mixture of amino acids that better matches our needs than the amino acid mixtures provided by plant proteins. Animal proteins also tend to be digested more easily than plant proteins; only protein that is digested can contribute amino acids to meet requirements.[20] Because they are

A typical protein drink provides 10 to 35 g of protein per serving, or 20 to 70% of the Daily Value. It can add about 100 to 250 kcals to the diet and thus can contribute to weight gain.

Because some amino acids are absorbed using the same transport systems, supplementing one amino acid can cause a deficiency of others that share the same transport system (see Figure 6.7).

© Charles D. Winters/Photo Researchers, Inc.

Figure 6.17 Protein and amino acid supplements
Supplements marketed to increase the intake of protein or specific amino acids are expensive and usually unnecessary.

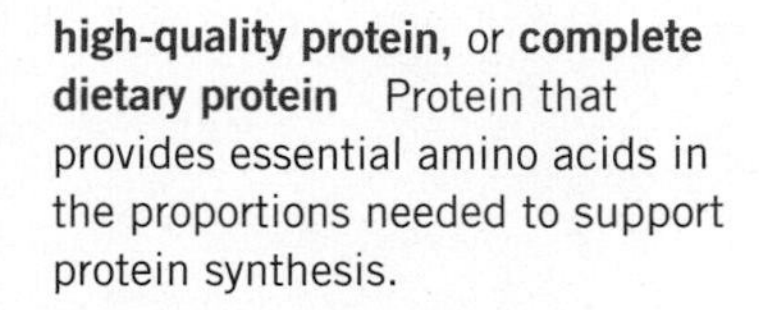

easily digested and supply essential amino acids in the proper proportions for human use, foods of animal origin are generally sources of **high-quality protein,** or **complete dietary protein**. When your diet contains high-quality protein, you don't have to eat as much total protein to meet your needs.

Compared to animal proteins, plant proteins are usually more difficult to digest and are lower in one or more of the essential amino acids. They are therefore generally referred to as **incomplete dietary protein**. Exceptions include quinoa and soy protein, which are both high-quality plant proteins. The RDA for protein is calculated assuming that the diet contains a mixture of plant and animal proteins and therefore is of mixed quality.

high-quality protein, or **complete dietary protein** Protein that provides essential amino acids in the proportions needed to support protein synthesis.

incomplete dietary protein Protein that is deficient in one or more essential amino acids relative to body needs.

protein digestibility-corrected amino acid score (PDCAAS) A measure of protein quality that reflects a protein's digestibility as well as the proportions of amino acids it provides.

How Protein Quality Is Measured Protein quality is measured experimentally using a number of methods. Some evaluate the amino acid composition of the protein, while others assess how well the protein supports growth (**Figure 6.18**). A chemical or amino acid score compares the amino acid pattern of the food protein of interest with that found in a reference protein known to be of high quality, such as egg protein. It is calculated by comparing the amount of the limiting amino acid in the test protein with the amount of that amino acid in egg protein. In this analysis, proteins with the most desirable proportions of amino acids will have the highest scores.

Amino acid score is a useful measure, but it does not consider digestibility. A measure that considers both amino acid composition and digestibility is the **protein digestibility-corrected amino acid score (PDCAAS)** (see Figure 6.18). PDCAAS measures the quality of a protein by comparing its amino acid composition to the

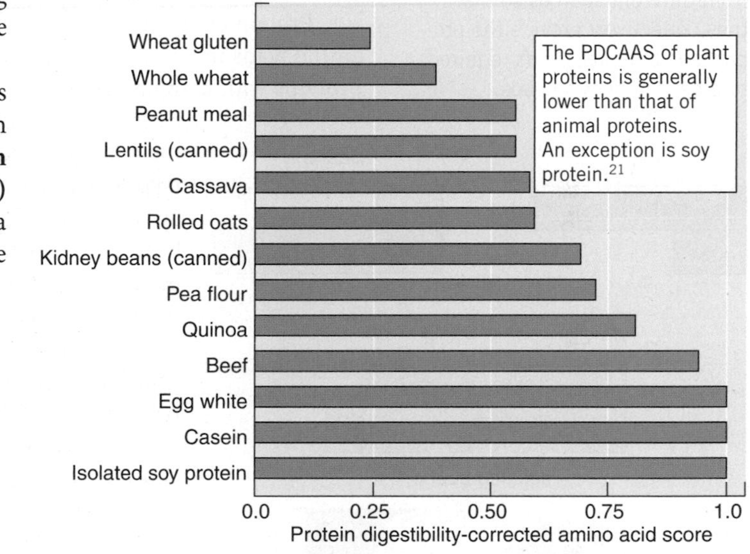

Figure 6.18 Measures of protein quality
Being able to evaluate the protein quality of a food or a diet is valuable when assessing the adequacy of human diets throughout the world. For example, knowing the quality of the protein in cassava is extremely important in determining the adequacy of the diet in countries where it is a staple and both food and protein are scarce.

Measures of Protein Quality

Type of measure	What it is	How it is calculated
Chemical or amino acid score	A measure of protein quality determined by comparing the essential amino acid content of the protein in a food with that in a reference protein. The lowest amino acid ratio calculated is the chemical score.	$= \frac{\text{mg of limiting amino acid/g of test protein}}{\text{mg of limiting amino acid/g of reference protein}} \times 100$
Protein digestibility-corrected amino acid score (PCDAAS)	A measure of protein quality that reflects a protein's digestibility as well as the proportions of amino acids it provides.	= amino score × digestibility factor
Protein efficiency ratio (PER)	A measure of protein quality determined by comparing the weight gain of a laboratory animal fed a test protein with the weight gain of an animal fed a reference protein.	$= \frac{\text{wt gain when fed test protein}}{\text{wt gain when fed reference protein}}$
Net protein utilization (NPU)	A measure of protein quality determined by comparing the amount of nitrogen retained in the body with the amount eaten in the diet.	$= \frac{\text{nitrogen retained}}{\text{nitrogen consumed}} \times 100$
Biological value (BV)	A measure of protein quality determined by comparing the amount of nitrogen retained in the body with the amount absorbed from the diet.	$= \frac{\text{nitrogen retained}}{\text{nitrogen absorbed}} \times 100$

amino acid requirements of a 2- to 5-year-old child (the age group with the highest needs relative to size) and then adjusting this for digestibility. A higher PDCAAS means that less of the protein is needed to provide all the needed amino acids. This method is currently used to assess the protein quality of foods for humans and is the standard used by the FDA to determine the % Daily Value for protein for food labels on products intended for people over 1 year of age.

Complementary Proteins Assessing protein quality is important in planning diets for regions of the world where protein is scarce. It is less important in industrialized countries, where high-quality protein is readily available to most people. A more appropriate way of evaluating protein quality in an individual diet is to look at the sources of the protein. As discussed previously, foods of animal origin are sources of complete dietary protein, whereas most plant foods contain proteins that are incomplete. If the protein in a diet comes from both animal and plant sources, it most likely contains adequate amounts of all the essential amino acids needed for protein synthesis. If the protein in a diet comes only from incomplete plant sources, a technique called **protein complementation** can be used to meet protein needs.

protein complementation The process of combining proteins from different sources so that they collectively provide the proportions of amino acids required to meet the body's needs.

Protein complementation combines foods containing proteins with different limiting amino acids in order to improve the protein quality of the diet as a whole. By eating plant proteins with complementary amino acid patterns, essential amino acid requirements can be met without consuming any animal proteins. The amino acids that are most often limited in plant proteins are lysine, methionine, cysteine, and tryptophan. As a general rule, legumes are deficient in methionine and cysteine but high in lysine. Grains, nuts, and seeds are deficient in lysine but high in methionine and cysteine. Corn is deficient in tryptophan as well as lysine but is a good source of methionine. Combining plant foods with complementary proteins provides all of the essential amino acids. For example, consuming rice, which is limited in the amino acid lysine but high in methionine and cysteine, with beans, which are high in lysine but limited in methionine and cysteine, provides enough of all the amino acids needed by the body (**Figure 6.19**). Plant proteins can also be complemented with animal proteins in order to meet the need for essential amino acids. For example, in Asia rice is often flavored with a small amount of spiced beef, chicken, or fish. It is not necessary to consume complementary proteins at each meal, but the entire day's diet should include proteins from complementary sources in order to satisfy the daily need for amino acids.[22]

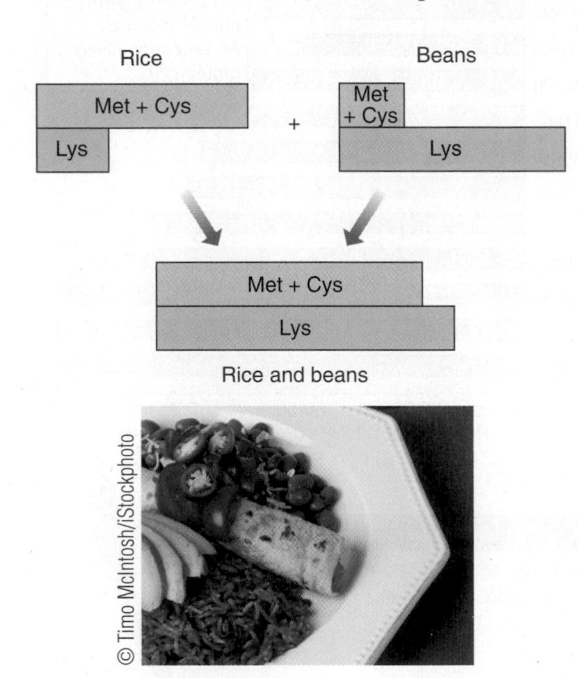

Figure 6.19 Protein complementation
When rice, which is limited in lysine (Lys) but high in methionine (Met) and cysteine (Cys), is eaten with beans, which are high in Lys but limited in Met and Cys, the combination provides all the amino acids needed by the body. Beans and rice or beans and tortillas are common complementary protein combinations in Central and South America.

Using MyPlate to Make Healthy Choices MyPlate and the Dietary Guidelines include recommendations regarding both animal and plant sources of protein to meet people's need for protein and essential amino acids (**Figure 6.20**). The MyPlate food groups that provide the most protein per serving are the dairy and protein groups; 1 cup of milk provides about 8 g of protein, 1 ounce of meat about 7 g, and ½ cup of beans 7 to 10 g. Each serving from the grains group and the vegetables group provides 2 to 4 g.

Animal foods are the major source of saturated fat and the only source of cholesterol in the diet. To assure an overall healthy diet, choose a variety of protein foods including seafood, lean meat and poultry, eggs, beans, soy products, and unsalted nuts and seeds. Choosing lean meats and skinless chicken provides iron, zinc, and B vitamins without adding too much saturated fat. Increasing consumption of seafood adds iron and zinc and also contributes heart-healthy omega-3 fatty acids. Choosing fat-free or low-fat dairy products provides high-quality protein and calcium without much saturated fat.

Replacing animal sources of protein that are high in solid fats with plant sources will reduce saturated fat, cholesterol, and calories. Plant sources of protein bring with them poly- and monounsaturated fats and dietary fiber. Legumes, for example, provide about 15 g of fiber per cup. Much of this is soluble fiber, which helps lower blood cholesterol. Choosing nuts and seeds from the protein group increases intake of heart-healthy monounsaturated fats and omega-3 fatty acids as well as fiber. Whole grains and vegetables add fiber, phytochemicals,

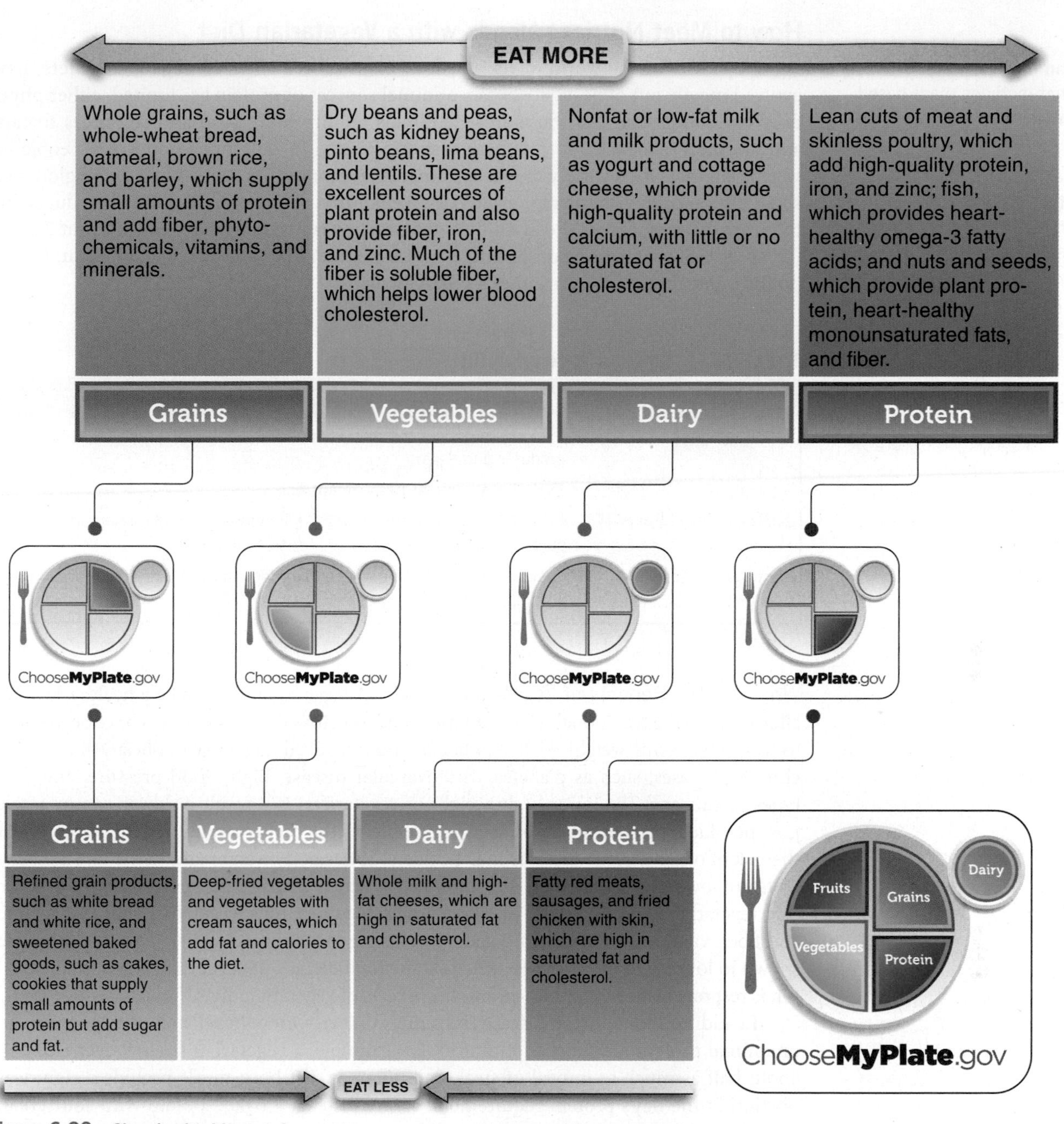

Figure 6.20 **Choosing healthy protein sources**
The protein group and the dairy group provide the most concentrated sources of protein. Nuts, dry beans, and peas are the most concentrated sources of plant protein. When you eat dry beans or peas, you can count them in either the vegetables group or the protein group. The fruit group is not shown here because there is very little protein in fruit.

vitamins, and minerals. They are lower in protein, but because we eat a larger number of servings from these groups, they make an important contribution to our protein intake.

Using Food Labels to Make Healthy Choices Food labels provide a ready source of information about the protein content of packaged foods; however, since the labeling of raw meats and fish is voluntary, many of the greatest sources of protein in the diet do not carry food labels. The Nutrition Facts section lists the number of grams of protein per serving, but the % Daily Value is generally not included. It is required only on labels that carry a claim related to the food's protein content, such as "high protein." The ingredient list provides information on the protein and amino acid–containing ingredients in the food. This information can be important for people with allergies and those trying to avoid certain additives (see Off the Label: Is It Safe for You?).

How to Meet Nutrient Needs with a Vegetarian Diet

vegetarian diet A pattern of food intake that includes plant-based foods and eliminates some or all foods of animal origin.

vegan diet A pattern of food intake that eliminates all animal products.

In many parts of the world, diets based on plant proteins, called **vegetarian diets**, have evolved mostly out of necessity because animal sources of protein are limited, either physically or economically, in those areas. Animals require more land and resources to raise and are more expensive to purchase than are plants. In developed countries, people eat vegetarian diets for a variety of reasons other than economics, such as health, religion, personal ethics, or environmental awareness. **Vegan diets** eliminate all animal products, but there are other types of vegetarian diets that are less restrictive (**Table 6.2**). About 2.3% of the adult U.S. population consistently consumes a diet that does not include meat, fish, or poultry, and about 1.4% consumes a vegan diet.[22]

TABLE 6.2 Types of Vegetarian Diets

Diet	What it excludes and includes
Semivegetarian	Excludes red meat but may include fish and poultry, as well as dairy products and eggs
Pescetarian	Excludes all animal flesh except fish
Lacto-ovo vegetarian	Excludes all animal flesh but does include eggs and dairy products such as milk and cheese
Lacto-vegetarian	Excludes animal flesh and eggs but does include dairy products
Vegan	Excludes all food of animal origin

What Are the Benefits of Vegetarian Diets? A vegetarian diet can be a healthy, low-cost alternative to the traditional American meat-and-potatoes diet. Vegetarians have been shown to have lower body weight relative to height and a reduced incidence of obesity and of other chronic diseases, such as diabetes, cardiovascular disease, high blood pressure, and some types of cancer.[22] The lower body weight of vegetarians is a result of lower energy intake, possibly due to higher intake of fiber, which makes the diet more filling. The reductions in the risk of other chronic diseases may be due to lower body weight and to the fact that these diets are lower in saturated fat and cholesterol, which increase disease risk. Or it could be that vegetarian diets are higher in whole grains, legumes, nuts, vegetables, and fruits, which add fiber, vitamins, minerals, antioxidants, and phytochemicals—substances that have been shown to lower disease risk. It is likely that the total dietary pattern, rather than a single factor, is responsible for the health-promoting effects of vegetarian diets.

In addition to reducing disease risks, diets that rely more heavily on plant protein than on animal protein are more economical. For example, a vegetarian stir-fry over rice costs about half as much as a meal of steak and potatoes. Yet both meals provide a significant portion of the day's protein requirement. A small steak, a baked potato with sour cream, and a tossed salad provides about 50 g of protein, whereas a dish of rice with tofu and vegetables provides about 30 g.

Why Some Nutrients Are at Risk of Deficiency in Vegetarian Diets Despite the health and economic benefits of vegetarian diets, a poorly planned vegetarian diet can cause nutrient deficiencies. Protein deficiency is a risk when poorly planned vegan diets that contain little high-quality protein are consumed by small children or by adults with increased protein needs, such as pregnant women and those recovering from illness or injury. Most people can easily meet their protein needs with a well-planned vegan diet (**Figure 6.21**) or with lacto- or lacto-ovo vegetarian diets, which contain high-quality animal proteins from eggs or milk. The egg and milk proteins complement the limiting amino acids in the plant proteins.

Vitamin and mineral deficiencies are a greater concern for vegetarians than is protein deficiency.[22] Of primary concern to vegans is vitamin B_{12}. Because this B vitamin is found almost exclusively in animal products, vegans must take vitamin B_{12} supplements or consume foods fortified with vitamin B_{12} to meet their needs for this nutrient. Another nutrient of concern is calcium. Dairy products are the major source of calcium in the North American diet, so diets that eliminate these foods must rely on plant sources of calcium. Likewise, because

Figure 6.21 Combining plant proteins
Combining complementary sources of plant proteins can provide enough of all the essential amino acids, without animal proteins. Grain and legume combinations that have become cultural staples include rice or tortillas and beans in Central and South America, rice and tofu in China and Japan, rice and lentils in India, rice and black-eyed peas in the southern United States, and bread and peanut butter (peanuts are legumes) throughout the United States.

much of the dietary vitamin D comes from fortified milk, this vitamin must be made in the body from exposure to sunlight or consumed in other sources. Iron and zinc may be deficient in vegetarian diets because they exclude red meat, which is an excellent source of these minerals, and iron and zinc are poorly absorbed from plant sources. Because dairy products are low in iron and zinc, lacto-ovo and lacto-vegetarians as well as vegans are at risk for deficiencies of these minerals. Vegan diets may also be low in iodine and the omega-3 fatty acids EPA and DHA (see Chapter 5).[22] Diets that do not include fish, eggs, or large amounts of sea vegetables may need to include higher levels of α-linolenic acid to ensure adequate amounts of EPA and DHA. By including flaxseed or flax and canola oil, which are good sources of α-linolenic acid, vegetarians can have a healthy ratio of omega-6 to omega-3 fatty acids and be able to synthesize enough of the longer-chain-length omega-3 fatty acids, EPA and DHA, without consuming animal products. **Table 6.3** provides suggestions for how vegans can meet the need for the nutrients just discussed.

TABLE 6.3 Meeting Nutrient Needs with a Vegan Diet[22]

Nutrient at risk	Foods to include in a vegan diet
Protein	Soy-based products, legumes, seeds, nuts, grains, and vegetables
Vitamin B_{12}	Products fortified with vitamin B_{12}, such as soy beverages, rice milk, and breakfast cereals; fortified nutritional yeast; dietary supplements
Calcium	Tofu processed with calcium; calcium-rich vegetables such as broccoli, kale, bok choy, and legumes; products fortified with calcium, such as soy beverages, rice milk, grain products, and orange juice
Vitamin D	Sunshine; products fortified with vitamin D, such as soy beverages, rice milk, breakfast cereals, and margarine
Iron	Legumes, tofu, dark green leafy vegetables, dried fruit, whole grains, iron-fortified grain products (absorption is improved when iron-containing foods are consumed with vitamin C found in citrus fruit, tomatoes, strawberries, and dark green vegetables)
Zinc	Whole grains, wheat germ, legumes, nuts, tofu, and fortified breakfast cereals
Iodine	Iodized salt, sea vegetables (seaweed), and foods grown near the sea
Omega-3 fatty acids	Canola oil, flaxseed and flaxseed oil, soybean oil, walnuts, and sea vegetables (seaweed), which contain EPA or fatty acids that can be used to synthesize EPA and DHA; DHA-rich microalgae

Choosing a Healthy Vegetarian Diet

© Robyn Mackenzie/iStockphoto

BACKGROUND

Ajay decided to stop eating meat a year ago. Now that he is taking a nutrition class, he wonders if his vegetarian diet is as healthy as he thought. He is 22 years old, is 5'9" tall, weighs 154 pounds, and gets little exercise. His diet analysis is shown here.

FOOD	SERVING	ENERGY (kcal)	PROTEIN (g)	FIBER (g)	SAT FAT (g)
Breakfast					
Frosted flakes	2 cup	320	3	2.7	0
Low-fat milk	1 cup	103	8	0	2
Orange juice	1 cup	112	2	0.5	0
Whole-wheat toast	2 slices	144	6	2.2	0
Butter	2 tsp	68	0	0	5
Coffee	1 cup	0	0	0	0
Powdered coffee creamer	1 T	30	0	0	2
Lunch					
Cheese pizza (regular crust)	2 slices	542	24	4	10
Potato chips	1 oz	155	2	1.2	3
Cola	12 oz	136	0	0	0
Ice cream sandwich	1	143	3	1	3
Dinner					
Vegetable lasagna	10 oz	400	18	4.8	7
Garlic bread	2 slices	189	4	1.5	1
Green salad	1 cup	22	2	3	0
Salad dressing	2 T	118	0	0.3	2
Ice cream	¾ cup	345	6	0	8
Iced tea	8 oz	90	0	0	0
Total		2917	78	21.2	43

CRITICAL THINKING QUESTIONS

- Does Ajay's diet provide enough protein to meet his RDA of 0.8 grams of protein per kilogram of body weight?
- How does the percentage of calories from saturated fat in Ajay's diet compare with recommendations?
- How does Ajay's fiber intake compare to recommendations?
- How does Ajay's intake of fruits and vegetables compare to the recommendations of MyPlate? (Hint: Use Food-A-Pedia on the MyPlate website.)

Ajay is shocked by how much saturated fat and how little fiber there is in his vegetarian diet. He decides to replace some of the high-fat dairy products in his diet with complementary plant proteins and to eat more fruits, vegetables, and whole grains.

- Suggest a lunch to replace the pizza that will include complementary plant proteins and provide at least a serving of whole grains and a cup of vegetables.

iProfile Use iProfile to find the protein content of your favorite vegetarian entree.

How to Plan Healthy Vegetarian Diets Well-planned vegetarian diets, including vegan diets, can meet nutrient needs at all stages of the life cycle, from infancy, childhood, and adolescence to early, middle, and late adulthood, and during pregnancy and lactation (see **Critical Thinking: Choosing a Healthy Vegetarian Diet**).[22]

One way to plan a healthy vegetarian diet is to modify the selections from MyPlate. The food choices and recommended amounts from the grains, vegetables, and fruits groups should stay the same for vegetarians. Including a cup of dark green and colorful vegetables daily will help meet iron and calcium needs. The dairy group and the protein group include foods of animal origin. Vegetarians who consume eggs and milk can still choose these foods. Those who avoid all animal foods can choose dry beans, nuts and seeds, and soy products from the protein group. Fortified soymilk can be chosen from the dairy group. To obtain adequate vitamin B_{12}, vegans must take supplements or use products fortified with vitamin B_{12}. Obtaining plenty of omega-3 fatty acids from foods such as canola oil, nuts, sea vegetables, and flaxseed ensures adequate synthesis of the long-chain omega-3 fatty acids DHA and EPA.

OUTCOME

Elliot did some reading about vegan diets that helped him understand the food combinations that would give him enough protein to meet his needs. Peanut butter and bread is inexpensive and provides complementary proteins. For more variety, he discovered that ethnic food provided some wonderful vegan options. The Chinese restaurant has a tasty tofu and vegetable stir-fry. The Indian restaurant offers a number of vegan options that contain lentils or chickpeas with rice or bread. He spent some extra time at the grocery store and discovered milks, including soy, almond, and rice, that do not contain animal products but give him something to put on his cereal. He also found a variety of vegan prepared foods such as veggie burgers and frozen enchiladas. His diet is now varied and provides plenty of protein, with less saturated fat and more fiber than his old diet. **Although he is enjoying vegan food, his reading also helped him realize that he does not necessarily need to eat vegan to improve his health and reduce his footprint on the planet.**

© Luca Manieri/iStockphoto

© Chris Schmidt/iStockphoto

APPLICATIONS

ASSESSING YOUR DIET

1. **How much protein do you eat?**
   - **a.** Use iProfile to calculate your average daily protein intake using the three-day food record you kept in Chapter 2.
   - **b.** How does your protein intake compare to the RDA for protein for someone of your weight, age, and life stage? How does your protein intake compare to the recommended percentage of calories from protein of between 10 and 35%? If you consumed just the RDA for protein, what percentage of calories would this represent?
   - **c.** If your protein intake is greater than the RDA, do you think you should decrease it? Why or why not?
   - **d.** If your protein intake is less than the RDA, modify one day of your diet to meet your protein needs.
2. **Do your protein sources add saturated fat to your diet?**
   - **a.** Using the three-day food record you kept in Chapter 2, record your total protein and saturated fat intake for each day in the table on the next page.

	Protein (g)	Saturated fat (g)
Day 1		
Day 2		
Day 2		

b. What is the relationship between the amount of saturated fat and protein in your diet?

c. For each day, list the three foods that contribute the most protein to your diet. Are they animal or plant foods?

d. What percentage of your total saturated fat for that day do these three foods provide?

3. **What changes would make your diet vegetarian?**

a. Make a list of each of the meats or other nondairy animal foods in your diet. Then list a food you could substitute for each and still have a meal you would eat.

b. Once you have made these substitutions, use iProfile to determine how much protein your new diet provides. Does it meet your RDA for protein?

c. Now take a look at your dairy sources of protein. Convert this lacto-vegetarian diet into a vegan diet by substituting plant sources of protein for dairy products. Use protein complementation (see Figure 6.21) to be sure that you meet your need for essential amino acids.

CONSUMER ISSUES

4. **Anna consumes a vegan diet. Her weekly grocery list includes the foods below. Using the foods from her list, suggest a breakfast, lunch, and dinner menu that uses protein complementation. You may want to add some other vegetables and fruit to make your meals more complete.**

Rice
Whole-wheat bread
Tofu
Oatmeal
Corn tortillas
Peanut butter
Black-eyed peas
Lentils
Cashews
Hummus (chickpeas and sesame seeds)
Soymilk

5. **Mark is trying to improve his protein choices. MyPlate recommends he choose lean meats and low-fat dairy products, and eat more legumes and nuts and seeds. Listed below are five of Mark's favorite meals. Suggest how they could be modified to follow the recommendations of MyPlate and still include some of Mark's favorite protein sources.**

Fried chicken and potatoes
Fish and chips
Kung pao chicken
Beef and cheese burrito
Pepperoni and sausage pizza

CLINICAL CONCERNS

6. **A friend of yours is a weight lifter. He has read that if he eats a high-protein diet he will build muscle more quickly. He is 5'8" tall and weighs 160 lbs. He drinks two or three protein shakes daily and always has two eggs for breakfast, a 4-ounce hamburger for lunch, and a 6-ounce steak for dinner.**

a. Use iProfile to determine how much protein the eggs, hamburger, and steak contribute to his diet.

b. Use the Internet to determine how much protein a typical "protein shake" contains.

c. How does the protein he consumes from these sources compare to his requirement? (Remember that protein needs may be higher for weight lifters.)

d. Does he need the protein shakes to meet his protein needs?

7. **The Amecht Company wants to include nitrogen balance studies in the assays it performs in its clinical laboratory. To test their methodology, company technicians analyze nitrogen balance in three individuals. The technicians are given information about the daily nitrogen intake of these subjects and analyze samples of urine and feces to determine daily nitrogen losses. Nitrogen balance is equal to nitrogen intake minus nitrogen output.**

a. Subject A consumed 6.4 g of nitrogen. The laboratory determines that she lost 8.0 g of nitrogen in her urine and feces. What is subject A's nitrogen balance? List some possible reasons that would explain this result.

b. Subject B is a 29-year-old man who weighs 82 kg and consumes an adequate diet providing 2700 kcals and 11.2 g of nitrogen a day. The laboratory determines that he lost 11.2 g of nitrogen in his urine and feces. What can you tell about the subject based on his nitrogen balance?

c. Subject C is a 31-year-old pregnant woman of average pre-pregnancy weight who is consuming 2500 kcals and 12.8 g of nitrogen a day. Considering that she is pregnant, would you expect her nitrogen excretion to be greater than or less than 12.8?

8. **One of the most common examples of protein complementation is a peanut butter sandwich. The table here gives the grams of each essential amino acid in 100 g of peanut protein and 100 g of wheat protein and in 100 g of a reference amino acid pattern.**

Amino acid	Peanut protein	Wheat protein	Reference amino acid pattern
Isoleucine	4.0	3.4	5.9
Leucine	7.7	6.2	9.0
Lysine	3.9	1.7	7.2
Methionine + cysteine	2.4	3.6	6.3
Phenylalanine + tyrosine	10.8	6.4	10.3
Threonine	3.0	2.4	5.0
Trptophan	1.2	1.0	1.3
Valine	4.6	3.8	6.7

a. For both peanut protein and wheat protein, calculate the percentage of each amino acid supplied relative to the amino acid reference pattern.
b. Which is the limiting amino acid in peanut protein (the amino acid present in the smallest amount relative to the reference pattern)? Which is the limiting amino acid in wheat protein? Which of the two food proteins has the higher amino acid score?

SUMMARY

6.1 Protein in Our Food

- Dietary protein comes from both animal and plant sources. Most of the protein in the American diet comes from animal foods, but in developing countries most dietary protein comes from plant sources.
- Animal sources of protein are high in iron, zinc, and calcium but also add saturated fat and cholesterol to the diet. Plant sources of protein are rich in fiber, phytochemicals, and monounsaturated and polyunsaturated fats.

6.2 Protein Molecules

- Amino acids consist of a carbon atom with a hydrogen atom, a nitrogen-containing group, an acid group, and a unique side chain attached. The amino acids that the body is unable to make in sufficient amounts are essential amino acids and must be consumed in the diet.
- Proteins are made of amino acid chains that fold over on themselves to create unique three-dimensional structures. The shape of a protein determines its function.

6.3 Protein in the Digestive Tract

- Protein digestion begins in the stomach where acid denatures protein and pepsin breaks polypeptide chains into shorter chains. In the small intestine the remaining polypeptide chains are digested to tripeptides, dipeptides, and amino acids, which are absorbed into the mucosal cell, where dipeptide and tripeptides are broken into single amino acids.
- Amino acids are absorbed into the mucosal cell using one of several active transport systems. Amino acids that share the same transport system compete for absorption.
- Undigested protein fragments that are absorbed can trigger a food allergy.

6.4 Protein in the Body

- The amino acids in body tissues and fluids that are available for the synthesis of protein and other nitrogen-containing molecules or for ATP production are known as the amino acid pool; they come from both dietary protein and the degradation of body proteins. The continuous breakdown and resynthesis of body proteins are referred to as protein turnover and are necessary for growth, maintenance, and regulation.
- DNA in the nucleus of cells contains the information needed to make body proteins. In transcription, this information is copied into a molecule of mRNA, which carries it to the cytosol. In translation, tRNA translates the mRNA code into a sequence of amino acids. Which genes are expressed determines which proteins are made. If an amino acid needed for protein synthesis is not available, this limiting amino acid can stop protein synthesis.
- Amino acids are used to make nonprotein molecules that contain nitrogen, such as DNA, RNA, and neurotransmitters.
- Amino acids can be used to provide energy when the diet doesn't meet energy needs or when protein intake exceeds needs. Amino acids that are used for energy are first deaminated. The amino groups removed by deamination are converted into urea, which can safely be excreted. The carbon compounds that remain can be broken down to generate ATP or be used to synthesize glucose or fatty acids, depending on the needs of the body.

6.5 Functions of Body Proteins

- Proteins provide structure at the cellular level as an integral component of cell membranes. The proteins in skin, hair, muscle, and connective tissue provide a structural framework for the whole body.
- In the body, proteins facilitate and regulate body functions. Regulatory proteins include enzymes, hormones, transport proteins, antibodies, contractile proteins, and proteins that affect fluid balance and acid balance.

6.6 Protein, Amino Acids, and Health

- Protein-energy malnutrition (PEM) is a concern, primarily in developing countries. Kwashiorkor is a form of PEM that occurs when the protein content of the diet is deficient but energy intake is adequate. It is most common in children. Marasmus is a form of PEM that occurs when total energy intake is deficient.
- High-protein diets increase the production of urea and other waste products that must be excreted in the urine and therefore can increase water losses. High protein intakes increase urinary calcium losses, but when calcium intake is adequate, high-protein diets are associated with greater bone mass and fewer fractures. Diets high in animal proteins and low in fluid are associated with an increased risk of kidney stones. High-protein diets can be high in saturated fat and cholesterol.
- People with food allergies must avoid certain protein sources. Some proteins and amino acids trigger food intolerances. Those with the genetic disease phenylketonuria and sensitivities to gluten must avoid specific foods.

6.7 Meeting Recommendations for Protein Intake

- Protein requirements can be determined using nitrogen balance, which compares the amount of nitrogen consumed in the diet with the amount excreted.

- The RDA for protein for healthy adults is 0.8 g/kg of body weight per day. Healthy diets can include 10 to 35% of energy from protein. Growth, pregnancy, lactation, illness, and injury can increase requirements. Certain types of physical activity may also increase protein needs. Protein supplements are not needed to meet the protein needs of healthy people.
- Animal proteins contain a pattern of amino acids that matches the needs of the human body more closely than the pattern of amino acids in plant proteins. Animal proteins are therefore said to be of higher quality than plant proteins. Diets that include little or no animal protein can provide adequate protein if the sources of protein are complemented to supply enough of all the essential amino acids.
- MyPlate recommends choosing a variety of protein foods including seafood, lean meat and poultry, eggs, beans, soy products, and unsalted nuts and seeds. Choosing fat-free or low-fat dairy products provides high-quality protein and calcium without much saturated fat.
- Many vegetarian diets include some animal products. Vegan diets exclude all foods of animal origin. Vegetarian diets are associated with a lower risk for obesity, diabetes, cardiovascular disease, high blood pressure, and some types of cancer. Vegetarian diets can easily meet protein needs, but care must be taken to include enough iron and zinc in lacto-ovo vegetarian diets. Well-planned vegan diets can provide adequate amounts of calcium, vitamin D, iron, zinc, and omega-3 fatty acids but must include supplements or fortified foods to meet the need for vitamin B_{12}.

REVIEW QUESTIONS

1. List some plant sources of protein.
2. What are amino acids?
3. Describe the general structure of a protein.
4. What is an essential amino acid?
5. What is the amino acid pool, and where do these amino acids come from?
6. List three structural and/or regulatory functions of proteins in the body.
7. Explain how proteins are synthesized.
8. Why is protein deficiency most common in infants and children?
9. Compare and contrast the causes and symptoms of kwashiorkor and marasmus.
10. How does the typical protein intake in the United States compare to recommendations?
11. What health problems are associated with a diet high in animal proteins?
12. What effect does moderate exercise have on protein needs?
13. What does nitrogen balance suggest about the balance between protein synthesis and protein breakdown in the body?
14. What is protein quality?
15. What is protein complementation?
16. What nutrients are at risk of deficiency in vegan diets?

REFERENCES

1. Wright, J. D., Kennedy-Stephenson, J., Wang, C. Y., et al. Trends in intake of energy and macronutrients—United States, 1971–2000. *MMWR Morb Mortal Wkly Rep* 6:80–82, 2004.
2. Fulgoni, V. L. Current protein intake in America: Analysis of the National Health and Nutrition Examination Survey, 2003–2004. *Am J Clin Nutr* 87:1554S–1557S, 2008.
3. Smit, E., Nieto, F. J., Crespo, C. J., and Mitchell, P. Estimates of animal and plant protein intake in U.S. adults: Results from the Third National Health and Nutrition Examination Survey, 1988–1991. *J Am Diet Assoc* 99:813–820, 1999.
4. Williams, C. D. Kwashiorkor: Nutritional disease of children associated with maize diet. *Lancet* 2:1151–1154, 1935.
5. Friedman, A. N. High-protein diets: Potential effects on the kidney in renal health and disease. *Am J Kidney Dis* 44:950–962, 2008.
6. Institute of Medicine, Food and Nutrition Board. *Dietary Reference Intakes for Energy, Carbohydrates, Fiber, Fat, Protein and Amino Acids.* Washington, DC: National Academies Press, 2002, 2005.
7. Heaney, R. P., and Layman, D. K. Amount and type of protein influences bone health. *Am J Clin Nutr* 87:1567S–1570S, 2008.
8. Calvez, J., Poupin, N., Chesneau, C., et al. Protein intake, calcium balance and health consequences. *Eur J Clin Nutr* 66:281–295, 2012.
9. Cao, J. J., and Nielson, F. H. Acid diet (high-meat protein) effects on calcium metabolism and bone health. *Curr Opin Clin Nutr Metab Care* 13:698–702, 2010.
10. Meschi, T., Nouvenne, A., and Borghi, L. Lifestyle recommendations to reduce the risk of kidney stones. *Urol Clin North Am* 38:313–320, 2011.

11. Shai, I., Schwarzfuchs, D., Henkin, Y., et al. Weight loss with a low-carbohydrate, Mediterranean, or low-fat diet. *N Engl J Med* 359:229–241, 2008.

12. Randi, G., Edefonti, V., Ferraroni, M., et al. Dietary patterns and the risk of colorectal cancer and adenomas. *Nutr Rev* 68:389–408, 2010.

13. National Institute of Diabetes, Digestive and Kidney Diseases. Celiac disease.

14. Seymour, C. A., Cockburn, F., Thomason, M. J., et al. Newborn screening for inborn errors of metabolism: A systematic review. *Health Technol Assess* 1:1–95, 1997.

15. Brown, A. S., Fernhoff, P. M., Waisbren, S. E., et al. Barriers to successful dietary control among pregnant women with phenylketonuria. *Genet Med* 4:84–89, 2002.

16. Williams, A. N., and Woessner, K. M. Monosodium glutamate "allergy": Menace or myth? *Clin Exp Allergy* 39:640–646, 2009.

17. U.S. Department of Agriculture, Agricultural Research Service. Table 5. Energy intakes: Percentages of energy from protein, carbohydrate, fat, and alcohol, by gender and age in the United States, 2009–2010. *What We Eat in America*, NHANES 2009–2010.

18. American Dietetic Association. Position of the American Dietetic Association, Dietitians of Canada, and the American College of Sports Medicine: Nutrition and athletic performance. *J Am Diet Assoc* 109:509–527, 2009.

19. Zajac, A., Poprzecki, S., Zebrowska, A., et al. Arginine and ornithine supplementation increases growth hormone and insulin-like growth factor-1 serum levels after heavy-resistance exercise in strength-trained athletes *J Strength Cond Res* 24:1082–1090, 2010.

20. Stipanuk, M. H. Protein and amino requirements. In *Biochemical and Physiological Aspects of Human Nutrition*. M. H. Stipanik, ed. St. Louis: Saunders Elsevier, 2006, pp. 419–448.

21. Food and Agriculture Organization/World Health Organization. Protein quality evaluation. Report of the joint FAO/WHO expert consultation. FAO/WHO. 1989.

22. Craig, W. J., Mangels, A. R., and American Dietetic Association. Position of the American Dietetic Association: Vegetarian diets. *J Am Diet Assoc* 109:1266–1282, 2009.

***To access links to online sources, please go to** www.wiley.com/college/smolin **and select Nutrition: Science and Applications, 3rd edition. From this page, select either the student or instructor companion site. Once on the desired site, select References.**

CHAPTER OUTLINE

Mike Blake/Reuters/NewsCom

Energy Balance and Weight Management

7

CASE STUDY

Spring break was only a week away, and Bethany wanted to take off a few pounds to look good in a bikini on the beach in Bermuda. She looked on the Internet and found the One-Week All-You-Can Eat Diet. It promised a loss of five pounds in seven days. She couldn't believe she could really eat all she wanted and still lose weight.

The diet did allow her to eat all she wanted, but only of certain foods. On day 1, Bethany was allowed to eat all the fruit she wanted, except bananas. Day 2 was all the vegetables she could eat, as long as they didn't have anything on them but vinegar or soy sauce. On day 3, she could eat all the fruits and vegetables she wanted, but all she wanted was a burger and fries. By day 4, she got to have milk—five glasses—and five bananas. On the last three days she was allowed up to 12 ounces of meat and unlimited fresh vegetables. Bethany followed the diet carefully. It was kind of like eating a different section of MyPlate every day. She got bored with the food choices some days, but there was always something she could eat if she was hungry. She lost 5 pounds before she left for Bermuda and felt great her first day in a bikini. To her disappointment, however, by the time the week was over she had regained all the weight she had lost and more. **Bethany had gained 10 pounds since the beginning of her freshman year. She was hoping that the weight loss she achieved on her one-week crash diet would be permanent.**

Exactostock/SuperStock

7.1 The Obesity Epidemic

LEARNING OBJECTIVES

- Discuss the increasing incidence of obesity in the United States and throughout the world.
- Describe contemporary lifestyle factors that have contributed to our high incidence of obesity.
- Compare the influence of hunger versus appetite on food intake.

overweight Being too heavy for one's height. It is defined as having a body mass index (a ratio of weight to height squared) of 25 to 29.9 kg/m^2.

Obese Having excess body fat. Obesity is defined as a body mass index (a ratio of weight to height squared) of 30 kg/m^2 or greater.

In the United States today, a staggering 68.8% of adults are **overweight** or **obese**.[1] The numbers have increased dramatically over the past five decades (**Figure 7.1**). In 1960, only 13.4% of American adults were obese. By 1990, about 23% were obese, and today, only two decades later, almost 36% are obese.[1] Obesity affects both men and women and all racial and ethnic groups. Obesity rates for minorities often exceed those in the general population: Among African Americans, more than 58% of women and 38% of men are obese, and among Hispanic Americans, 37% of men and about 41% of women are obese.[1] The problem is not limited to adults. Almost 17% of U.S. children and adolescents ages 2 through 19 are obese.[2]

The repercussions of this rise in obesity have led public health officials to call it an epidemic. Carrying excess body fat is not just a cosmetic concern. It increases the risk of a host of chronic health problems such as diabetes, heart disease, and cancer. It also burdens us financially: The medical-care costs of obesity in the United States are estimated to total about \$147 billion.[3]

THINK CRITICALLY

What regions of the United States have the highest obesity rates in 2010? Suggest why this might be the case.

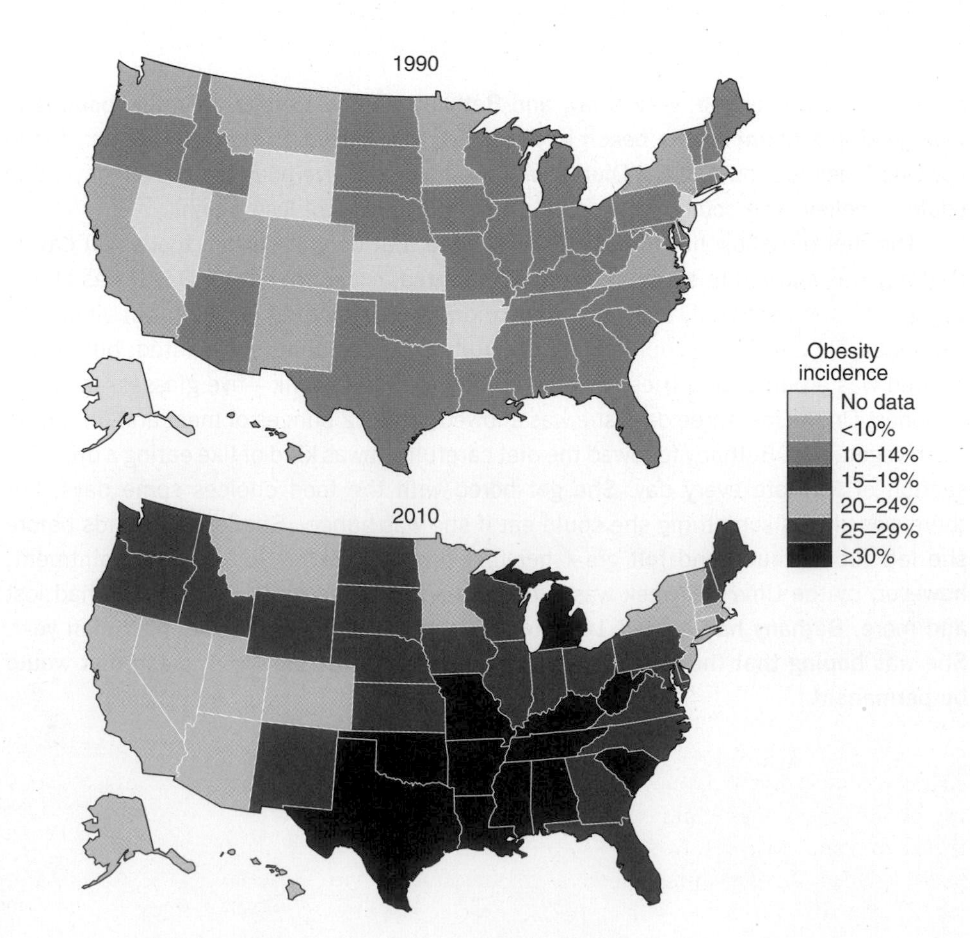

Figure 7.1 Incidence of obesity by state
These maps show the percentage of the adult population classified as obese in each state in 1990 and 2010. In 1990, the percentage of the adult population that was obese was less than 20% in all 50 states. By 2010, no states had an adult obesity rate less than 20%. (Source: Behavior risk factor surveillance system, CDC.)

Obesity around the World

The increasing rates of obesity are not only a concern in the United States. Around the world, approximately 1.5 billion adults are overweight, and of these, 500 million are obese. The World Health Organization projects that by 2015, approximately 2.3 billion adults will be overweight and more than 700 million will be obese.[4] It is such an important trend that the term *globesity* has been coined to reflect the escalation of global obesity and overweight. Once considered problems only in high-income countries, overweight and obesity are now on the rise in low- and middle-income countries, particularly in urban settings.

Why Are We Getting Fatter?

The reason Americans are getting fatter is that changes in our food supply and lifestyle over the past 40 years have affected what we eat, how much we eat, and how much exercise we get. Simply put, more Americans are overweight than ever before because we are eating more and burning fewer calories than we did 50 years ago.[5] Food is plentiful and continuously available, and little activity is required in our daily lives.

Americans Are Eating More In America today, supermarkets, fast-food restaurants, and convenience marts make palatable, affordable food readily available to the majority of the population 24 hours a day. We are constantly bombarded with cues to eat: Advertisements entice us with tasty, inexpensive foods, and convenience stores, food courts, and vending machines tempt us with the sights and smells of fatty, sweet, high-calorie snacks. As a result, since 1970 the amount of energy available to us has increased by about 600 kcals/day, with the greatest increases in added fats, grains, dairy products, and sweeteners.[5] The accessibility of tasty treats stimulates **appetite**. Because appetite is triggered by external cues such as the sight or smell of food, it is usually appetite, and not **hunger**, that makes us stop for an ice cream cone on a summer afternoon or give in to the smell of freshly baked chocolate-chip cookies while strolling through the mall. Studies examining the relationship between the food environment and weight have found that people in communities with more fast-food or quick-service restaurants tend to have a higher risk of weight gain, overweight, and obesity.[5]

appetite A desire to consume specific foods that is independent of hunger.

hunger A desire to consume food that is triggered by internal physiological signals.

In addition to having more enticing choices available to us, we consume more calories today because portion sizes have increased (**Figure 7.2**). The more food that is put in front of people, the more they eat.[5] So if you increase the amount of pasta on your plate or cereal in your bowl, you will most likely consume more calories. People tend to eat in units, such as one cookie, one sandwich, or one bag of chips, regardless of the size of the unit. So, when presented with larger units, such as bigger muffins, burgers, or bottles of soda, people still eat or drink the whole thing. As the portions offered have increased, so has the amount Americans consume. You probably wouldn't order an extra hamburger, but you will finish the one you order even if it is the size of two burgers. Portion size is associated with body weight; being served and consuming larger portions are associated with weight gain, whereas small portions are associated with weight loss.[5]

Figure 7.2 Portion distortion
The burgers and French fries served in fast-food restaurants today are over twice the size they were when fast food first appeared about 40 years ago. Soft-drink serving sizes have also escalated. A large fast-food soft drink today contains 32 ounces, providing about 300 kcals, and 20-oz bottles have replaced 12-oz cans in many vending machines.

Social changes over the past few decades have also contributed to the increase in the number of calories Americans consume. Busy schedules and an increase in the number of single-parent households and households with two working parents mean that families are often too rushed to cook meals at home. As a result, prepackaged, convenience, and fast-food meals have become mainstays. These foods are typically higher in fat and energy than foods prepared at home.

Figure 7.3 Activity reduces the risk of obesity
A typical office worker today walks only about 3000 to 5000 steps per day (2000 steps = approximately 1 mile). In contrast, in the Amish community—where automobiles and other modern conveniences are not allowed—a typical adult takes 14,000 to 18,000 steps a day. The overall incidence of obesity among the Amish is only 4%.[6]

Americans Are Moving Less Along with America's rising energy intake, there has been a decline in the amount of energy Americans expend, both at work and at play. Fewer American adults today work in jobs that require physical labor. People drive to work rather than walk or bike, take elevators instead of stairs, use dryers rather than hang clothes outside, and cut the lawn with riding mowers rather than with push mowers. All these modern conveniences reduce the amount of energy expended daily (**Figure 7.3**). Americans are also less active during their leisure time because busy schedules and long days at work and commuting leave little time for active recreation. Instead, at the end of the day, people tend to sit in front of television sets, video games, and computers.

Inactivity is also contributing to excess body weight among children. In the 1960s, schools provided daily physical education classes, and children spent their after-school hours playing outdoors; today, they are more likely to spend their afternoons indoors with televisions, video games, and computers. As a result, they burn fewer calories, snack more, and consequently gain weight.

7.2 Exploring Energy Balance

LEARNING OBJECTIVES

- Explain the principle of energy balance.
- Describe the processes involved in generating ATP from food.
- Describe the components of energy expenditure.
- Explain how excess energy consumed in the diet is stored in the body.

energy balance The amount of energy consumed in the diet compared with the amount expended by the body over a given period.

The principle of **energy balance** states that, when energy consumption equals energy expenditure, body weight remains constant. Energy balance can be achieved at any weight—fat, thin, or in between (**Figure 7.4**); it simply means that body weight is not changing. If, however, the amount of energy taken in exceeds the amount expended, energy balance is positive and the extra energy will be stored in the body, causing weight to increase. On the other hand, if less energy is taken in than expended, energy balance is negative and weight will be lost.

If you increase the amount of exercise you get, what happens to your energy needs?

Figure 7.4 Energy balance
Maintaining your weight requires a balance between how much energy you consume and how much energy you expend.

Energy In: Calories Consumed in Food

The energy taken into the body comes from the energy-yielding nutrients (carbohydrates, fats, and proteins) and alcohol consumed in food and beverages. Individuals who struggle with weight loss often think of the calories in food as an enemy—something to be avoided. However, food and the energy it provides are essential for life. Just as gasoline is necessary to run an engine, the energy in food is necessary to run the body. The amount of energy (number of calories) taken in depends on the total amount of food consumed and the nutrient composition of that food. The energy content of food can be measured precisely in the laboratory or estimated from its nutrient composition.

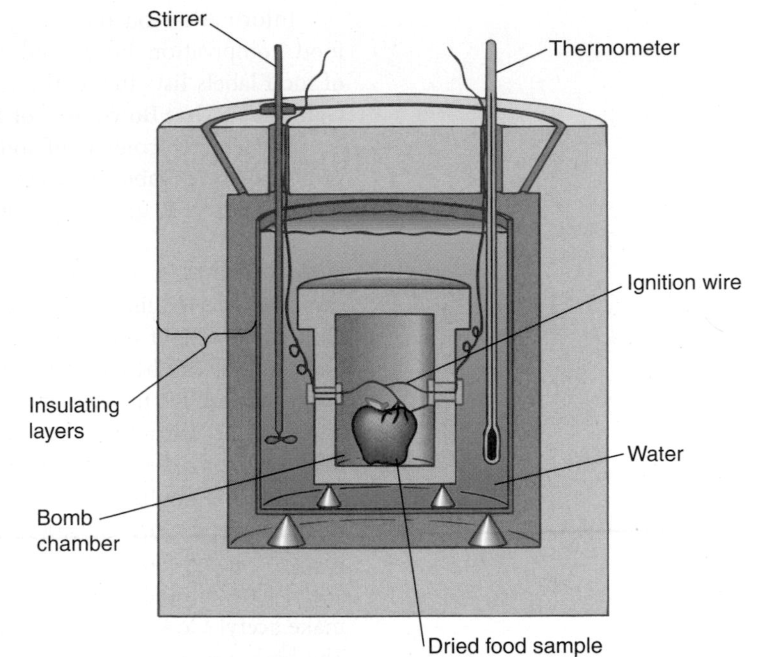

Figure 7.5 Bomb calorimeter When dried food is combusted inside the chamber of a bomb calorimeter, the rise in temperature of the surrounding water can be used to determine the energy content of the food.

How the Amount of Energy in Food Is Determined

The amount of energy in a food or a mixture of foods can be determined in the laboratory using a **bomb calorimeter**. A bomb calorimeter consists of a chamber surrounded by a jacket of water (**Figure 7.5**). Food is dried, placed in the chamber, and burned. As the food combusts, heat is released, raising the temperature of the water. The increase in water temperature can be used to calculate the amount of energy in the food based on the fact that 1 kcal is the amount of heat needed to increase the temperature of 1 kg of water by 1 degree Celsius.

Combusting a food in a bomb calorimeter determines the total amount of energy contained in that food. However, because the body cannot completely digest, absorb, and utilize all of the substances in a food, bomb calorimeter values are slightly higher than the amount of energy the body can obtain from that food. To correct for this difference, feeding experiments have been done to measure the amount of energy that is not available to the body, such as that lost in urine and feces. Subtracting this unavailable energy from the values determined in the bomb calorimeter gives a more accurate estimate of the energy obtained from food. These types of experiments were used to determine the amount of energy provided by the carbohydrate (4 kcals/g), fat (9 kcals/g), protein (4 kcals/g), and alcohol (7 kcals/g) in a mixed diet.

bomb calorimeter An instrument used to determine the energy content of food. It measures the heat energy released when a dried food is combusted.

When the nutrient composition of a food is known, the energy content of the food can be calculated by totaling the energy from the grams of carbohydrate, fat, and protein in the food (**Figure 7.6**). Vitamins, minerals, and water, though essential nutrients, do not provide energy to the body.

Figure 7.6 Estimating the energy content of food Most foods are of mixed composition; for instance, this slice of pizza contains about 15 g of protein, 50 g of carbohydrate, and 10 g of fat. Its energy content can be calculated as shown in the table.

Calculate the energy provided by each macronutrient:
15 g protein × 4 kcal/g = 60 kcals
50 g carbohydrate × 4 kcal/g = 200 kcals
10 g fat × 9 kcal/g = 90 kcals
Calculate the total amount of energy:
Total energy = 60 kcals + 200 kcals + 90 kcals = 350 kcals per slice

If you took off the three pieces of pepperoni on this slice of pizza, which provide 1 gram of protein, no carbohydrate, and 3 grams of fat, how many kcals would this eliminate?

Information on the energy content of foods can be found on food labels and in food composition tables and databases such as iProfile. The Nutrition Facts portion of food labels lists the total calories in a serving of food (see **Off the Label: How Many Calories in That Bowl, Box, or Bottle?**).

The energy content of foods in a diet can also be estimated from the Exchange Lists shown in **Table 7.1**. For example, one starch exchange, whether a slice of bread, one-half cup (100 g) of cereal, or six saltines, provides about 80 kcals.

How the Energy in Food Is Converted into ATP Just as the energy in flowing water can be converted into electrical energy, which can then be converted into the light energy emitted by a light bulb, the energy stored in the chemical bonds of carbohydrates, fats, and proteins can be converted into ATP, which can be used to keep you alive and moving. To generate ATP, the carbohydrates, proteins, and triglycerides consumed in the diet are digested and absorbed. The resulting glucose, amino acids, and fatty acids are then broken down, or oxidized, to generate ATP (**Figure 7.7**, see Chapters 4 through 6 and Online Focus on Metabolism). Each of these nutrients can be converted into the common intermediate acetyl-CoA. Glycolysis converts glucose into pyruvate, which then loses a carbon to form acetyl-CoA. β-Oxidation breaks fatty acids into two-carbon units that form acetyl-CoA, and after deamination the carbon skeletons from amino acid can be used to make acetyl-CoA or other intermediates. Acetyl-CoA can then enter the citric acid cycle. The high-energy electrons released at various metabolic steps are passed to the electron transport chain where their energy is trapped and used to generate ATP. The ATP is then used to fuel metabolic reactions that build and maintain body components and to power other cellular and body activities. Much of the energy consumed in food is also converted to and lost from the body as heat.

total energy expenditure (TEE) The sum of the energy used for basal metabolism, activity, processing food, deposition of new tissue, and production of milk.

Energy Out: Calories Used by the Body

Energy is defined as the ability to do work. In the body, energy is used to do body work. The total amount of energy used by the body each day, or **total energy expenditure (TEE)**, includes the energy needed to maintain basic bodily functions such as the beating of your heart, as well as that needed to fuel activity and process food. In individuals who are growing or pregnant, total energy expenditure also includes the energy used to deposit new

TABLE 7.1 Using Exchange Lists to Estimate Energy Content

Exchange groups/lists	Serving size	Energy (kcals)
Carbohydrate Group		
Starch	⅓ cup pasta; ½ cup potatoes; 1 slice bread	80
Fruit	1 small apple, peach, pear; ½ banana; ½ cup canned fruit (in juice)	60
Milk	1 cup milk or yogurt	
Nonfat		90
Low-fat		110
Reduced-fat		120
Whole		150
Other carbohydrates	Serving sizes vary	Varies
Vegetables	½ cup cooked vegetables, 1 cup raw	25
Meat/Meat Substitute Group	1 oz meat or cheese; ½ cup legumes	
Very lean		35
Lean		55
Medium-fat		75
High-fat		100
Fat Group	1 tsp butter, margarine, or oil; 1 T salad dressing	45

Figure 7.7 Producing ATP from glucose, fatty acids, and amino acids
Glucose, fatty acids, and amino acids can be broken down by the reactions of cellular respiration to yield carbon dioxide, water, and energy in the form of ATP.

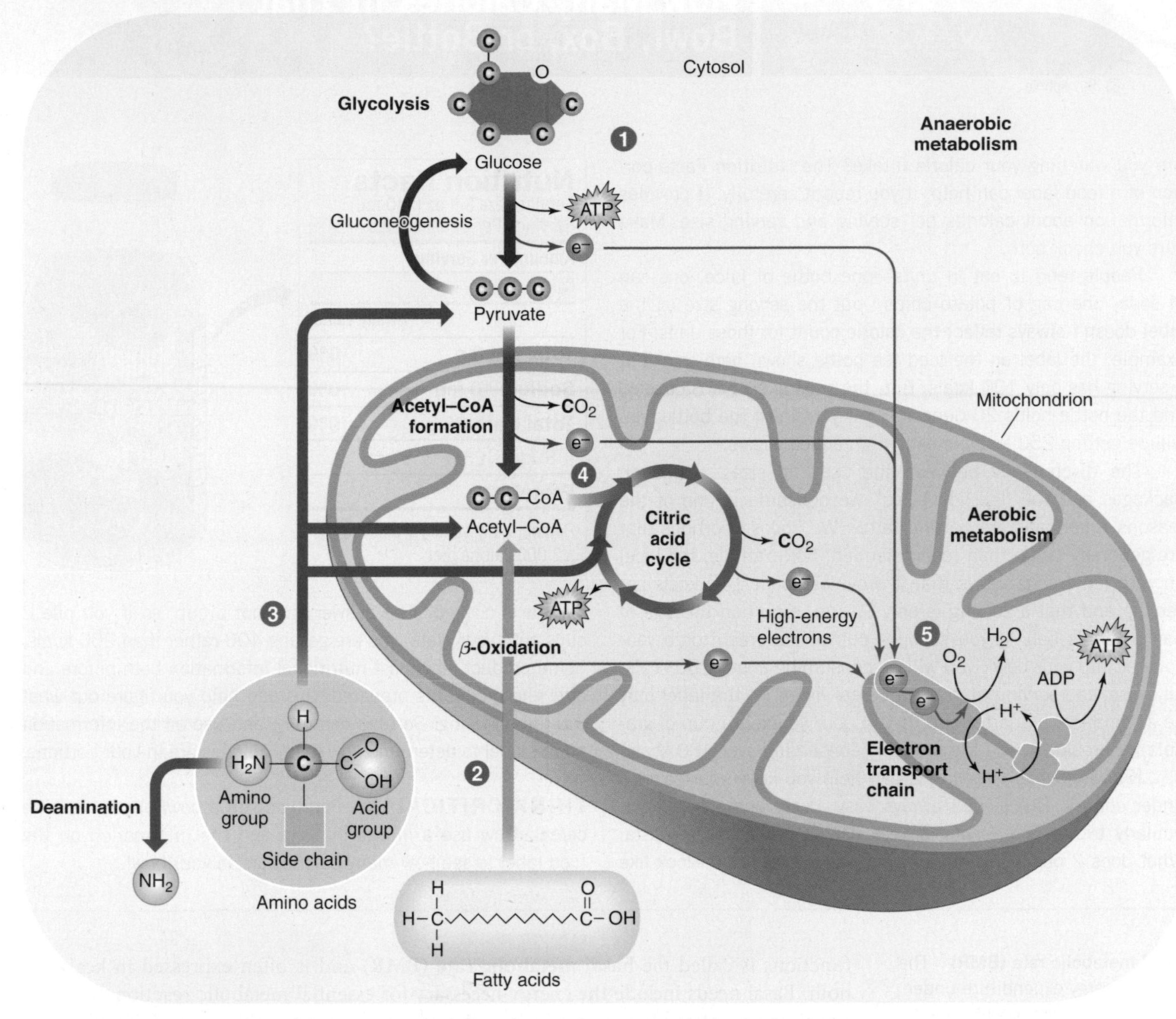

1. Glycolysis breaks glucose in half to yield two pyruvate molecules that are converted to acetyl-CoA.
2. β-Oxidation breaks fatty acids into two-carbon units that form acetyl-CoA.
3. After deamination, amino acids can break down to form acetyl-CoA, pyruvate, or other intermediates.
4. The acetyl-CoA from glucose, fatty acid, and amino acid breakdown can enter the citric acid cycle.
5. The electrons released are passed to the electron transport chain, and their energy is used to convert ADP to ATP.

tissues. In women who are lactating, it includes the energy used to produce milk. A small amount of energy is also used to maintain body temperature in a cold environment.

Energy for Basal Metabolism For most people, about 60 to 75% of the body's total energy expenditure is used for basal metabolism. Basal metabolism, or **basal energy expenditure (BEE)**, includes all the involuntary things your body does to stay alive such as breathing, circulating blood, regulating body temperature, synthesizing tissues, removing waste products, and sending nerve signals. The rate at which energy is used for these basic

basal energy expenditure (BEE) The energy expended to maintain an awake resting body that is not digesting food.

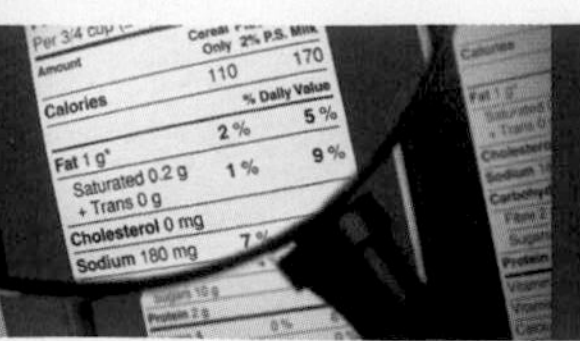

© iStockphoto

OFF THE LABEL

How Many Calories in That Bowl, Box, or Bottle?

Are you watching your calorie intake? The Nutrition Facts portion of a food label can help, if you read it carefully. It provides information about calories per serving and serving size. Make sure you check both.

People tend to eat in units—one bottle of juice, one can of soda, one bag of potato chips—but the serving size on the label doesn't always reflect the calorie count for those units. For example, the label on the iced tea bottle shown here says that a serving has only 100 kcals. But, the serving size is 8 ounces, and the bottle holds 20 ounces. So if you finish the bottle, you will be getting 250 kcals, mostly from added sugars.

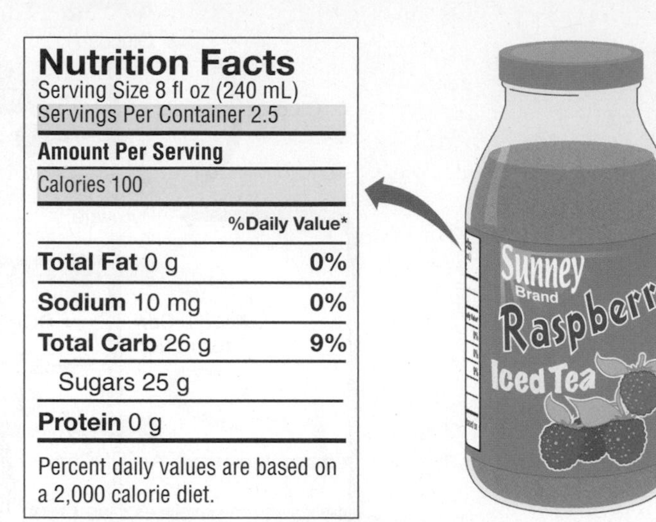

Nutrition Facts
Serving Size 8 fl oz (240 mL)
Servings Per Container 2.5

Amount Per Serving	
Calories 100	
	%Daily Value*
Total Fat 0 g	**0%**
Sodium 10 mg	**0%**
Total Carb 26 g	**9%**
Sugars 25 g	
Protein 0 g	

Percent daily values are based on a 2,000 calorie diet.

The discrepancy between the "serving sizes" listed on packages and the "portion sizes" we consume is one of the reasons Americans are getting fatter. We choose portions that are generally larger than recommended. For example, the label on your ice cream shows that it provides about 140 kcals per serving and that a serving is only ½ cup, a portion the size of half a tennis ball. If you scoop a cup of ice cream onto your cone or into your bowl, you will be consuming about 280 kcals. Likewise, the serving of granola cereal listed on the label may be as small as a quarter cup. If you pour yourself a cup of granola for breakfast, you are probably consuming over 400 kcals.

Knowing what a serving is can help you keep your calories under control. But it isn't always easy. Pasta portions are particularly tricky. The serving size is usually given as dry pasta. What does 2 ounces of dry spaghetti—for 200 kcals—look like once it is cooked? The answer is about a cup, so if you pile 2 cups onto your plate, you are getting 400 rather than 200 kcals. Some product labels list nutritional information both before and after the product is prepared; this can help you figure out what you are choosing. So read carefully, and use all the information on the label to determine how many calories are in your portions.

THINK CRITICALLY: Pour yourself a bowl of your favorite cereal. Now use a measuring cup and the information on the food label to see how many calories are in your bowl.

---

functions is called the **basal metabolic rate (BMR)** and is often expressed in kcals per hour. Basal needs include the energy necessary for essential metabolic reactions and life-sustaining functions but do not include the energy needed for physical activity or for the digestion of food and the absorption of nutrients. Therefore, to minimize residual energy expenditure for activity or processing food, BMR is measured in the morning, in a warm room before the subject rises, and at least 12 hours after food intake or activity (**Figure 7.8**). Because of the difficulty of achieving these conditions, measures are often made after about five to six hours without food or exercise. When done under these conditions it is reported as **resting energy expenditure (REE)** or **resting metabolic rate (RMR)**. RMR values are about 10 to 20% higher than BMR values.[7]

Basal needs are affected by factors such as body weight, gender, growth rate, and age. BMR increases with increasing body weight, so it is higher in heavier individuals. It also rises with increasing **lean body mass**; thus, BMR is generally higher in men than in women because men have more lean tissue. BMR increases during periods of rapid growth because energy is required to produce new body tissue. It decreases with age, partly due to the decrease in lean body mass that usually occurs in older adults.

Basal needs can be altered by certain abnormal conditions. An elevation in body temperature increases BMR. It is estimated that for every 1 degree Fahrenheit above normal body temperature, there is a 7% increase in BMR. This extra energy use explains why a fever can cause weight loss. Abnormal levels of thyroid hormones can also affect BMR. Individuals who overproduce these hormones burn more energy; in fact, a symptom used to diagnose thyroid hormone excess is unexplained weight loss. Individuals with an underproduction

basal metabolic rate (BMR) The rate of energy expenditure under resting conditions. BMR measurements are performed in a warm room in the morning before the subject rises and at least 12 hours after the last food or activity.

resting energy expenditure (REE) or **resting metabolic rate (RMR)** Terms used when an estimate of basal metabolism is determined by measuring energy utilization after five to six hours without food or exercise.

lean body mass Body mass attributed to nonfat body components such as bone, muscle, and internal organs. It is also called fat-free mass.

of thyroid hormones require less energy. The fact that hormones produced by the thyroid gland affect energy expenditure is the reason that obesity was once explained as a glandular problem. It is now known that obesity due to a thyroid hormone deficiency is rare.

Metabolic rate may also be affected by low-energy diets. Energy intake below needs may depress resting metabolic rate by 10 to 20%, or the equivalent of 100 to 400 kcals/day.[8] This drop in basal needs decreases the amount of energy needed to maintain weight. It is a beneficial adaptation in starvation, but it makes intentional weight loss more difficult.

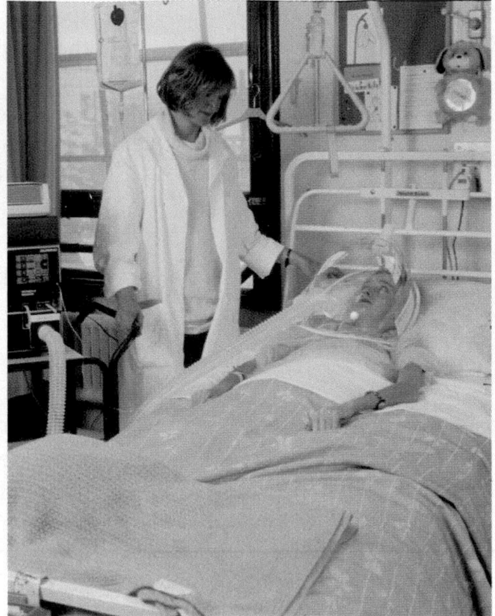

Figure 7.8 Measuring BMR
To assess BMR, expired gases can be collected and measured by having the subject breathe into a hood. Because aerobic metabolism uses oxygen and produces carbon dioxide, the amounts of these gases in expired air can be used to estimate the amount of energy that is being used.

Physical Activity Physical activity is the second major component of energy expenditure. It represents the metabolic cost of external work, which includes both the energy needed for planned exercise and that needed for daily activities such as walking to work, typing, performing yard work, house cleaning, and even fidgeting.[9] The energy expended for daily activities is called **nonexercise activity thermogenesis (NEAT)** and includes the energy expended for everything that is not sleeping, eating, or sports-like exercise. NEAT accounts for the majority of the energy expended for activity and varies enormously, depending on an individual's occupation and daily movements.[10] For most people, physical activity accounts for 15 to 30% of energy requirements, but this varies greatly (**Figure 7.9**).

nonexercise activity thermogenesis (NEAT) The energy expended for everything we do other than sleeping, eating, or sports-like exercise.

The energy required for an activity depends on how strenuous the activity is and the length of time it is performed. For example, walking at a speed of 3 to 4 miles per hour requires a moderate degree of exertion and uses about 300 kcal/hr for a 70-kg man. The energy required increases progressively as the walking

Figure 7.9 Activity as a portion of total energy requirement

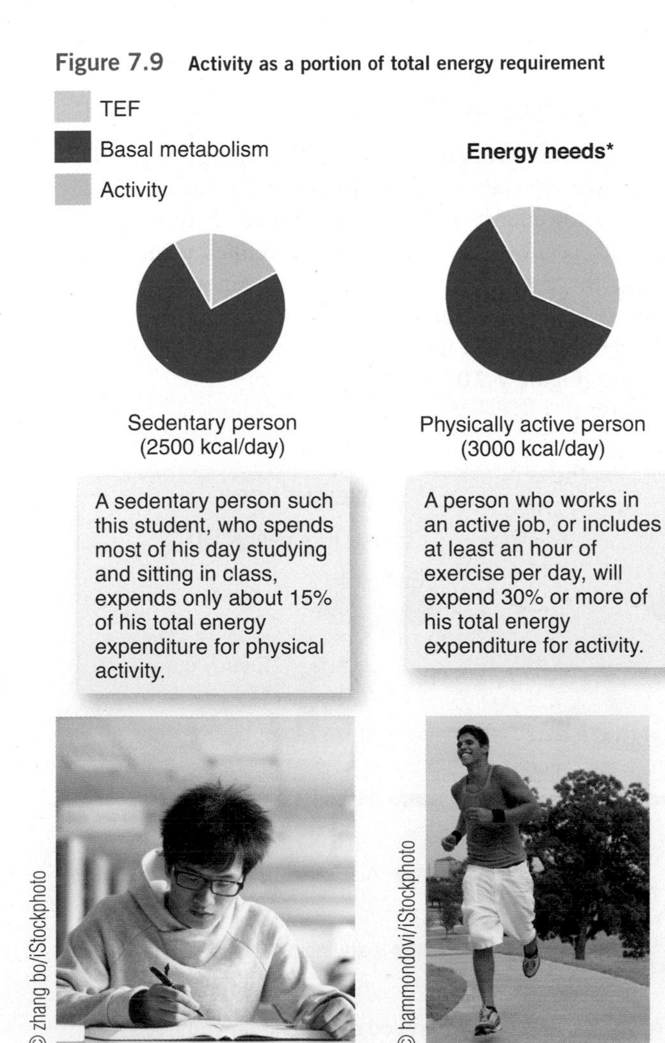

* Approximate energy expenditure for a 19-year-old male in various activity categories.

speed and length of time of the activity continue to increase and as the body weight of the exerciser rises. In addition to the energy expended during exercise, there is a small increase in energy expenditure for a period of time after exercise has been completed. The energy expended for activity is the one component of our total energy needs over which we have control. To maintain health, people today need to consciously increase their physical activity. This does not mean they have to run marathons. Choosing to take the stairs rather than the elevator, walking rather than taking the bus, and riding a bike rather than driving to the store all increase activity. The energy costs of specific activities are listed in Appendix F.

thermic effect of food (TEF) or **diet-induced thermogenesis** The energy required for the digestion of food and the absorption, metabolism, and storage of nutrients. It is equal to approximately 10% of daily energy intake.

Thermic Effect of Food Our energy comes from food, but we also need energy to digest food and to absorb, metabolize, and store the nutrients from this food. The energy used for these processes is called the **thermic effect of food (TEF)** or **diet-induced thermogenesis**. This energy expenditure causes body temperature to rise slightly for several hours after eating. The energy required for TEF is estimated to be about 10% of energy intake but can vary depending on the amounts and types of nutrients consumed. Because it takes energy to store nutrients, TEF increases with the size of the meal. A meal that is high in fat has a lower TEF than a meal high in carbohydrate or protein because dietary fat can be used or stored more efficiently than either protein or carbohydrate. The metabolic cost of either oxidizing or storing dietary fat is only 2 to 3% of the energy consumed, whereas the cost of using amino acids by either oxidizing them or incorporating them into proteins is 15 to 30% of the energy consumed, and the cost of breaking down carbohydrate or storing it as glycogen is 6 to 8%.[11] The difference in the cost of storing different nutrients as fat means that a diet high in fat may produce more body fat than a diet high in carbohydrate.[12]

Storing Energy and Using Energy Stores

To function normally the body needs a steady supply of energy. This energy is provided by our food and our body stores. People typically eat three to six times during the day. When we eat, some nutrients are used for energy and some are stored. Between meals, stored energy is used to meet needs. Typically, these stores are then refilled using energy consumed in the next meal so that there is no net change in the amount of stored energy. However, when we eat in excess, more energy is stored as fat than is needed to supply energy for the day and we gain weight. When our intake is too low, we use our fat stores to meet our daily needs and lose weight (**Figure 7.10**).

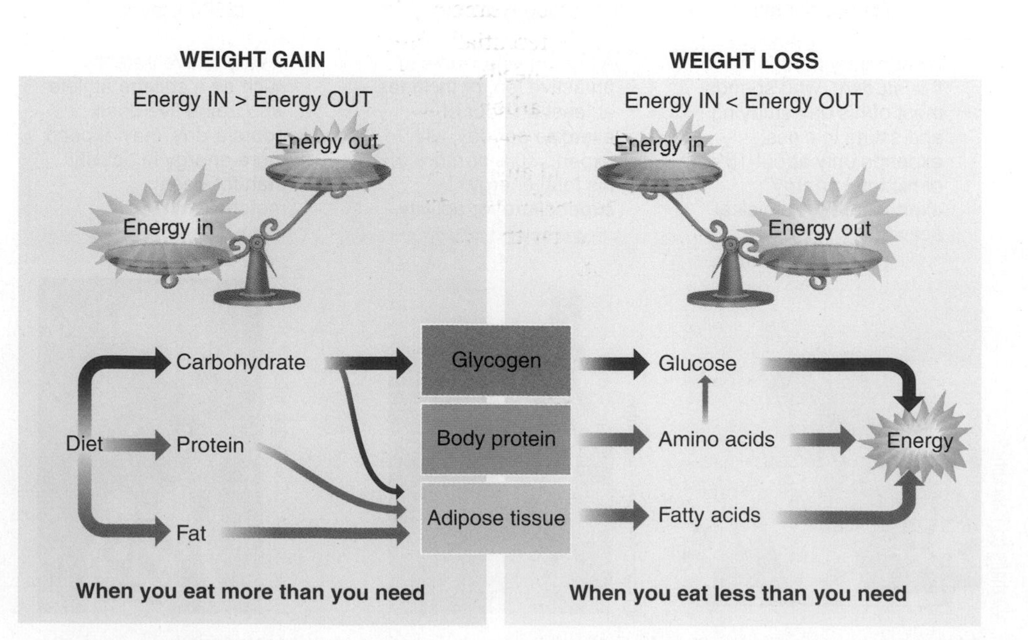

Figure 7.10 Feasting and fasting When you eat more than you need at that time, the extra energy will be stored as glycogen in your liver and muscle and as fat in your adipose tissue. When you haven't eaten in a while, you retrieve energy from these stores. During fasting, body protein is also broken down to amino acids to make glucose and provide energy.

TABLE 7.2 Sources of Energy in the Body[13,14]

Energy source	Primary location	Energy (kcals)[a]
Glycogen	Liver and muscle	1,400
Glucose or lipid	Body fluids	100
Triglyceride	Adipose tissue	115,000
Protein	Muscle	25,000

[a]Values represent the approximate amounts in a 70-kg male.

How Feasting Results in Weight Gain When excess energy is consumed (feasting), we generally say this excess is stored as fat. This is an oversimplification of a complex situation. After eating, the body prioritizes how nutrients are used based on body needs, which nutrients can be stored, and how efficiently they can be stored.

Energy is stored in the body as glycogen and triglycerides. Glycogen stores are located in the liver and muscle, and fill when dietary carbohydrate is adequate. The body generally stores only about 200 to 500 grams of glycogen—enough to provide glucose for about 24 hours (**Table 7.2**). Triglycerides are stored in adipose tissue, which is made up of adipocytes. Adipocytes grow in size as they accumulate more triglycerides and shrink as triglycerides are removed from them. The greater the number of adipocytes an individual has, the greater the ability to store fat. Although most adipocytes are formed between infancy and adolescence, excessive weight gain can cause the production of new adipocytes in adults.

Why Most Body Fat Comes from Dietary Fat There is a metabolic hierarchy of how fuels are used by the body. Alcohol, although not a nutrient, does supply energy. Because it is toxic and cannot be stored in the body, it is rapidly oxidized. Amino acids from dietary protein are next in the hierarchy. They are first used to synthesize needed body proteins and nitrogen-containing nonprotein molecules; any excess is then broken down to provide energy because there is no mechanism for storing them as amino acids or proteins. Carbohydrate is used to maintain blood glucose and to build glycogen stores. Once glycogen stores are full, the remaining carbohydrate is oxidized for energy. Fat, unlike the other energy-yielding nutrients, is not needed as a fuel for a particular tissue or to build tissues and can be stored in the body in virtually unlimited amounts. Therefore, if the energy consumed is in excess of immediate needs, dietary fat is preferentially stored. For example, after a meal the body's energy needs are met by first breaking down dietary protein and carbohydrate that are not needed for essential functions. To meet any remaining energy needs, dietary fat is oxidized, and any dietary fat that is left is stored as triglycerides, primarily in adipose tissue (**Figure 7.11**). Therefore, most of the fat that is stored in the body comes directly from dietary fat.

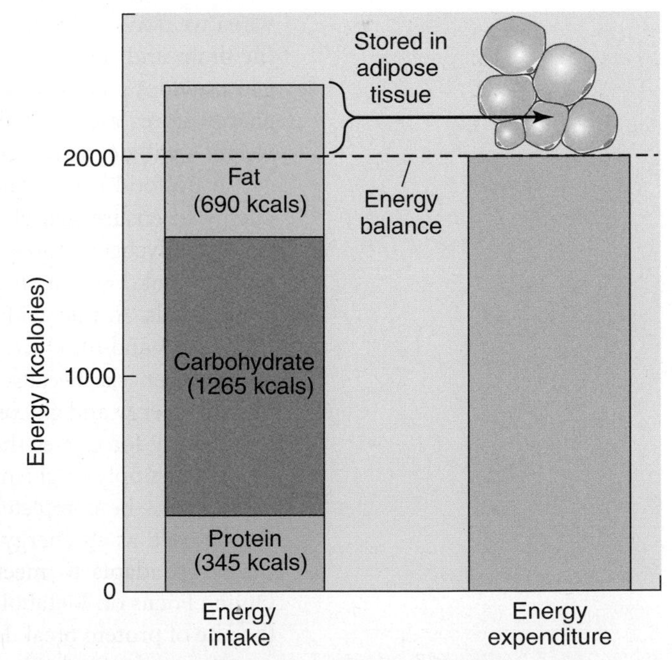

Figure 7.11 How dietary energy-yielding nutrients are used If energy intake exceeds expenditure, dietary fat is stored in adipose tissue. In this example, in which 2000 kcals are needed for energy balance, all the calories from carbohydrate and protein are used to meet the body's need for synthesis and energy, but only some of the calories from fat are used to meet energy needs. The remaining fat calories are stored in adipose tissue.

How Carbohydrate and Protein Can Be Used to Synthesize Fat The body is capable of converting glucose and amino acids into triglycerides for storage. However, under normal dietary circumstances, this rarely occurs because these conversions are energetically costly.[15] Making fat from glucose involves converting glucose to acetyl-CoA and then assembling fatty acids from the two-carbon acetyl-CoA units. The fatty acids must then be joined to a molecule of glycerol to make triglycerides for storage. To convert amino acids to fat, the amino group must first be removed and then the carbon skeleton must be

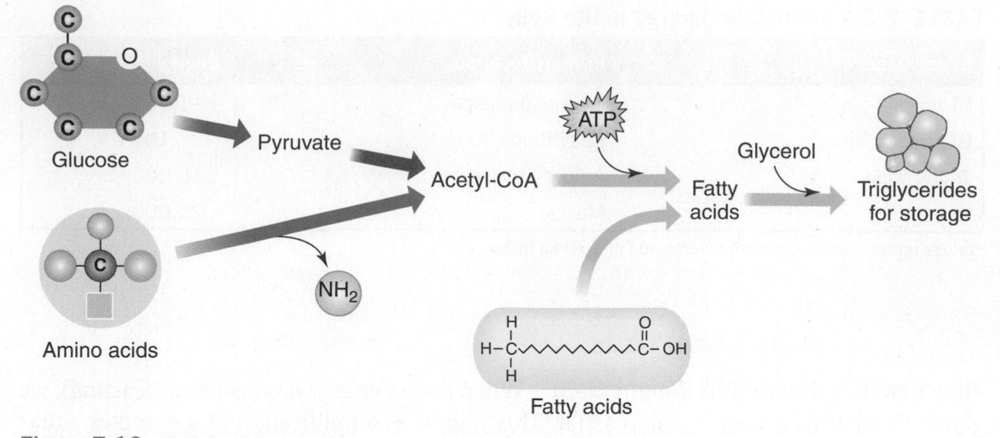

Figure 7.12 Storing energy as fat
It is metabolically much simpler and less costly to convert dietary fatty acids to triglycerides for storage than to convert glucose or amino acids to triglycerides.

broken down to yield acetyl-CoA that can be used for fatty acid synthesis. In contrast, the conversion of dietary fat to body fat requires only the removal and reattachment of fatty acids from the glycerol backbone (see Chapter 5); it takes only a small percentage of the energy in the fat to store the fat in adipose tissue (**Figure 7.12**). The low metabolic cost of converting dietary fat to stored body fat makes it more efficient for the body to use dietary fat to make body fat and to oxidize carbohydrate and protein to meet immediate energy needs rather than convert them to fat. The conversion of carbohydrate to fat becomes important only when the diet is composed primarily of carbohydrate and energy intake exceeds expenditure.[13] Regardless of the composition of the diet, however, when excess energy is consumed over the long term, fat stores will enlarge and weight will increase.

How Fasting Results in Weight Loss The body needs a steady supply of energy even when we don't eat. Some of this energy must come from glucose, which is needed to fuel the brain and several other types of body cells. Between meals, the breakdown of glycogen provides glucose and the breakdown of stored fat meets other energy needs. If these energy stores are not fully replenished, the amount of stored energy—and hence, body weight—will decrease (see Figure 7.10).

If no food is eaten for more than several hours, the body must shift the way it uses energy to ensure that glucose continues to be available to the brain and other cells that need it. Glycogen stores can provide glucose but are limited, so glucose is also supplied by the breakdown of small amounts of body protein, primarily muscle protein, to yield amino acids. Amino acids can then be used to synthesize glucose via gluconeogenesis (see Chapters 4 and 6). Once glycogen stores are depleted, all of the glucose must come from gluconeogenesis. Because protein is not stored in the body, the breakdown of protein to provide energy and glucose results in the loss of functional body proteins.

Energy for tissues that don't require glucose is provided by the breakdown of stored fat. If the supply of glucose is limited, such as during fasting, fatty acids delivered to the liver cannot be completely oxidized, so ketones are produced (see Chapter 4). Ketones can be used as an energy source by many tissues. After about three days of fasting, even the brain adapts to meet some of its energy needs from ketones (see Chapter 5 and Online Focus on Metabolism). This reduces the amount of glucose needed and thus slows the rate of protein breakdown.

If energy intake is less than energy output for a prolonged period, substantial amounts of fat are used to provide energy and some protein is degraded to provide glucose. This results in weight loss (see Figure 7.10). The magnitude of the weight loss depends on the degree of energy deficit and the length of time over which it occurs. It is estimated that an energy deficit of about 3500 kcals will result in the loss of a pound of adipose tissue.

7.3 Estimating Energy Requirements

LEARNING OBJECTIVES

- Describe three ways in which energy expenditure is measured.
- Determine your physical activity level and PA value by tallying your daily activity.
- Calculate your EER using your PA value and the appropriate EER equation.

The amount of energy expended by the body, and hence the energy needed to maintain body weight, can be measured using a variety of techniques. Data from these measurements can then be used to estimate energy needs for a variety of people under a variety of circumstances. Calculations of energy expenditure have been used to generate the **Estimated Energy Requirements (EERs)** (see Chapter 2), which are the current recommendations for energy intake in the United States An EER is the number of calories needed for a healthy individual to maintain his or her weight.

Estimated Energy Requirements (EERs) The amount of energy recommended by the DRIs to maintain body weight in a healthy person based on age, gender, size, and activity level.

direct calorimetry A method of determining energy use that measures the amount of heat produced.

indirect calorimetry A method of estimating energy use that compares the amount of oxygen consumed to the amount of carbon dioxide expired.

doubly labeled water technique A method for measuring energy expenditure based on measuring the disappearance of isotopes of hydrogen and oxygen in body fluids after consumption of a defined amount of water labeled with both isotopes.

isotopes Alternative forms of an element that have different atomic masses and may or may not be radioactive.

How Energy Expenditure Is Measured

Energy expenditure can be measured by calorimetry, which is the science of measuring heat flow. Calorimetry can determine energy expenditure directly or indirectly. Measuring doubly labeled water is another technique for assessing energy expenditure; this method can be used over prolonged periods.

Direct Calorimetry Measuring the amount of heat produced when a food is combusted in a bomb calorimeter is a type of **direct calorimetry** (see Figure 7.5). In humans, direct calorimetry measures the amount of heat given off by the body; the heat produced is proportional to the amount of energy used. This heat is generated by metabolic reactions that both convert food energy into ATP and use ATP to power body processes. Direct calorimetry is an accurate method for measuring energy expenditure, but it is expensive and impractical because it requires that the individual being assessed remain in an insulated chamber throughout the procedure in order to measure the heat produced.

Indirect Calorimetry **Indirect calorimetry**, which estimates energy use by assessing oxygen utilization, is somewhat less cumbersome than direct calorimetry. To obtain a measurement, the researcher has the subject breathe into a mouthpiece, mask, or ventilated hood (**Figure 7.13**). Oxygen use and carbon dioxide production are measured by analyzing the difference between the composition of inhaled and exhaled air. The body's energy use can be calculated from these values because the burning of fuels by the body in cellular respiration uses oxygen and produces carbon dioxide. This method can measure the energy used for individual components of expenditure, such as physical activity or BMR. It can also be used to estimate total energy needs, but it is not practical in free-living individuals because the equipment is too cumbersome for long-term use.

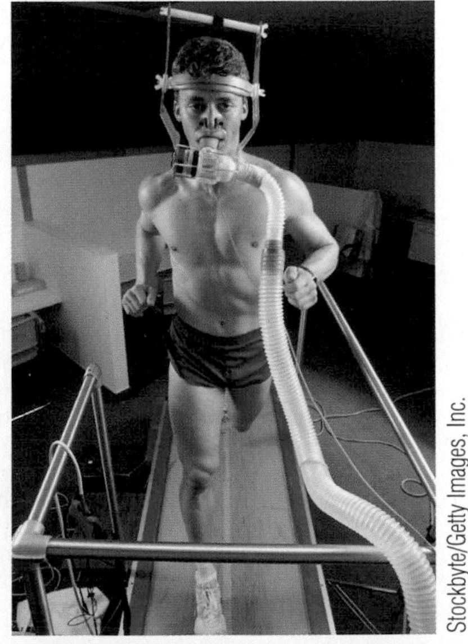

Stockbyte/Getty Images, Inc.

Figure 7.13 Collecting expired gases to measure energy use
Indirect calorimetry can be used to assess the energy expended for specific activities. Subjects breathe into a mouthpiece or through a mask that is placed over their nose and mouth. The amounts of O_2 and CO_2 in the inhaled and exhaled air are measured.

Doubly Labeled Water A more practical method for measuring energy needs is the **doubly labeled water technique**. This involves having the subject ingest or be injected with water labeled with **isotopes** of oxygen and hydrogen. The labeled oxygen and hydrogen are used by the body in metabolism. The labeled hydrogen leaves the body as part of water molecules and the labeled oxygen leaves the body as part of both water and carbon dioxide molecules. The difference in the rates of disappearance of these two isotopes can be used to calculate the amount of carbon dioxide produced by metabolic reactions in the body.

The doubly labeled water technique does not require the individual to carry any equipment and can be used to measure energy expenditure in free-living subjects for periods up to two weeks. It is now the preferred method for determining the total daily energy expenditures of both healthy and clinical populations.[7, 16] However, it is not helpful in determining the proportion of energy used for BMR, physical activity, or TEF.

How to Calculate EER

Measurements of energy expenditure using doubly labeled water have been used to develop equations for estimating individuals' energy needs. These EER equations are used to estimate the amount of energy needed to maintain energy balance in a healthy person of a given age, gender, weight, height, and level of physical activity.[7] No specific RDAs or ULs for energy have been established.

Determine Physical Activity Level In order to calculate an individual's energy needs using the EER equations, his or her physical activity level must be estimated. This can be done by keeping a daily log of activities and the amount of time spent at each. Physical activity level can then be categorized as sedentary, low active, active, or very active. **Figure 7.14**

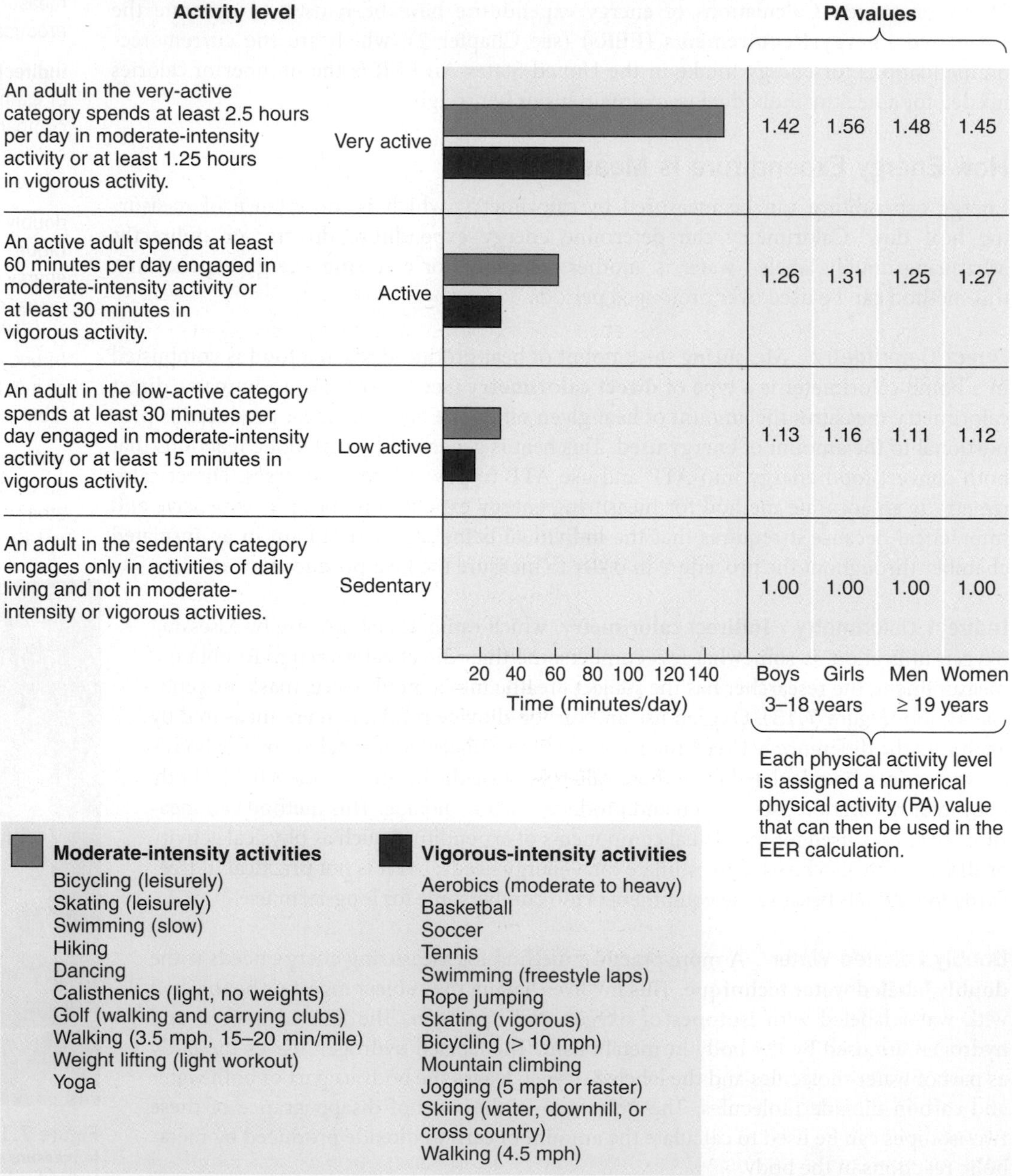

Activity level		PA values: Boys (3–18 years)	Girls (3–18 years)	Men (≥ 19 years)	Women (≥ 19 years)
An adult in the very-active category spends at least 2.5 hours per day in moderate-intensity activity or at least 1.25 hours in vigorous activity.	Very active	1.42	1.56	1.48	1.45
An active adult spends at least 60 minutes per day engaged in moderate-intensity activity or at least 30 minutes in vigorous activity.	Active	1.26	1.31	1.25	1.27
An adult in the low-active category spends at least 30 minutes per day engaged in moderate-intensity activity or at least 15 minutes in vigorous activity.	Low active	1.13	1.16	1.11	1.12
An adult in the sedentary category engages only in activities of daily living and not in moderate-intensity or vigorous activities.	Sedentary	1.00	1.00	1.00	1.00

Time (minutes/day): 20 40 60 80 100 120 140

Each physical activity level is assigned a numerical physical activity (PA) value that can then be used in the EER calculation.

Moderate-intensity activities
- Bicycling (leisurely)
- Skating (leisurely)
- Swimming (slow)
- Hiking
- Dancing
- Calisthenics (light, no weights)
- Golf (walking and carrying clubs)
- Walking (3.5 mph, 15–20 min/mile)
- Weight lifting (light workout)
- Yoga

Vigorous-intensity activities
- Aerobics (moderate to heavy)
- Basketball
- Soccer
- Tennis
- Swimming (freestyle laps)
- Rope jumping
- Skating (vigorous)
- Bicycling (> 10 mph)
- Mountain climbing
- Jogging (5 mph or faster)
- Skiing (water, downhill, or cross country)
- Walking (4.5 mph)

Figure 7.14 Physical activity level and PA values
Physical activity level, which is used to calculate EER, is categorized as sedentary, low active, active, or very active. A sedentary person spends about 2.5 hours per day engaged in the activities of daily living, such as housework, homework, and yard work. Adding activity moves the person into the low-active, active, or very-active category. Activity can be moderate or vigorous or a combination of the two; compared to moderate-intensity activity, vigorous activity will burn the same number of calories in less time.

TABLE 7.3 Calculating Your EER

- **Find your weight in kilograms (kg) and your height in meters (m):**

  Weight in kilograms = weight in pounds ÷ 2.2 lbs/kg
  Height in meters = height in inches × 0.0254 m/in
  For example: 160 pounds = 160 lbs ÷ 2.2 lbs/kg = 72.7 kg
  5' 9" = 69 in × 0.0254 m/in = 1.75 m

- **Estimate the amount of physical activity you get per day and use Figure 7.14 to find the PA value for someone your age, gender, and activity level:**

  For example, if you are a 19-year-old male who performs 40 minutes of vigorous activity a day, you are in the active category and have a PA value of 1.25.

- **Choose the appropriate EER prediction equation below, and calculate your EER:**

  For example, if you are an active 19-year-old male:
  EER = 662 − (9.53 × Age in yrs) + PA [(15.91 × Weight in kg) + (539.6 × Height in m)]
  where Age = 19 yrs, Weight = 72.7 kg, Height = 1.75 m, Active PA value = 1.25
  EER = 662 − (9.53 × 19) × 1.25 ([15.91 × 72.7] + [539.6 × 1.75]) = 3107 kcal/day

Life stage	EER prediction equation[a]
Boys 9–18 yrs	EER = 88.5 − (61.9 × Age in yrs) + PA[(26.7 × Weight in kg) + (903 × Height in m)] + 25
Girls 9–18 yrs	EER = 135.3 − (30.8 × Age in yrs) + PA[(10.0 × Weight in kg) + (934 × Height in m)] + 25
Men ≥ 19 yrs	EER = 662 − (9.53 × Age in yrs) + PA[(15.91 × Weight in kg) + (539.6 × Height in m)]
Women ≥ 19 yrs	EER = 354 − (6.91 × Age in yrs) + PA[(9.36 × Weight in kg) + (726 × Height in m)]

[a]These equations are appropriate for determining EERs in normal-weight individuals. Equations that predict the amount of energy needed for weight maintenance in overweight and obese individuals are also available (see Appendix A).

can be used to translate the amount of time spent engaged in moderate-intensity or vigorous activities into one of these four physical activity levels. Each activity level corresponds to a numerical **PA (physical activity) value** that can be used in the EER equations. For example, if you spend about an hour a day walking (a moderate-intensity activity) or about 30 minutes jogging (a vigorous activity), you are in the active category and should use the active PA value corresponding to your age and gender when calculating your EER.

PA (physical activity) value A numeric value associated with activity level that is a variable in the EER equations used to calculate energy needs.

Activity level has a significant effect on energy needs. For example, a 30-year-old woman who is 5′ 5″ tall and weighs 130 lbs needs about 1900 kcals/day if she is at the sedentary activity level. If she increases her activity to "active," the level recommended by the DRIs, her energy needs increase to 2370 kcals/day.

Choose the Appropriate EER Equation Energy needs are affected by gender, height, weight, life stage, and age as well as level of physical activity. These factors are all taken into consideration in the EER equations (**Table 7.3** or inside cover). Separate equations for men and women reflect gender differences in energy requirements. As age increases in adults, energy needs decrease. Height and weight are variables in the equations, and when larger numbers are entered, calculated results reflect the higher energy needs of taller, heavier individuals (**Figure 7.15**). EERs predict energy expenditure in normal-weight individuals. Equations that predict the amount of energy needed for weight maintenance for overweight and obese individuals are given in Appendix A.

The EER values for infants, children, and adolescents include the energy used to deposit tissues associated with growth. Beginning at age 3, there are separate EER equations for boys and girls because of differences in growth and physical activity. The EER for pregnancy is determined as the sum of the total energy expenditure of a nonpregnant woman plus the energy needed to maintain pregnancy and deposit maternal and fetal tissue. During lactation, EER is the sum of the total energy expenditure of nonlactating women and the energy in the milk produced, minus the energy mobilized from maternal tissue stores.

Figure 7.15 **Variables that affect energy requirements**
A sedentary 25-year-old man who is 5' 11" tall and weighs 170 lbs requires 2624 kcals to maintain his weight. If he begins exercising an hour a day he will need to eat 3174 kcals to prevent weight loss. As this man ages, his EER declines. If he gains weight his energy needs will increase. If he decreases his level of activity, the number of calories he can eat without gaining weight will drop.

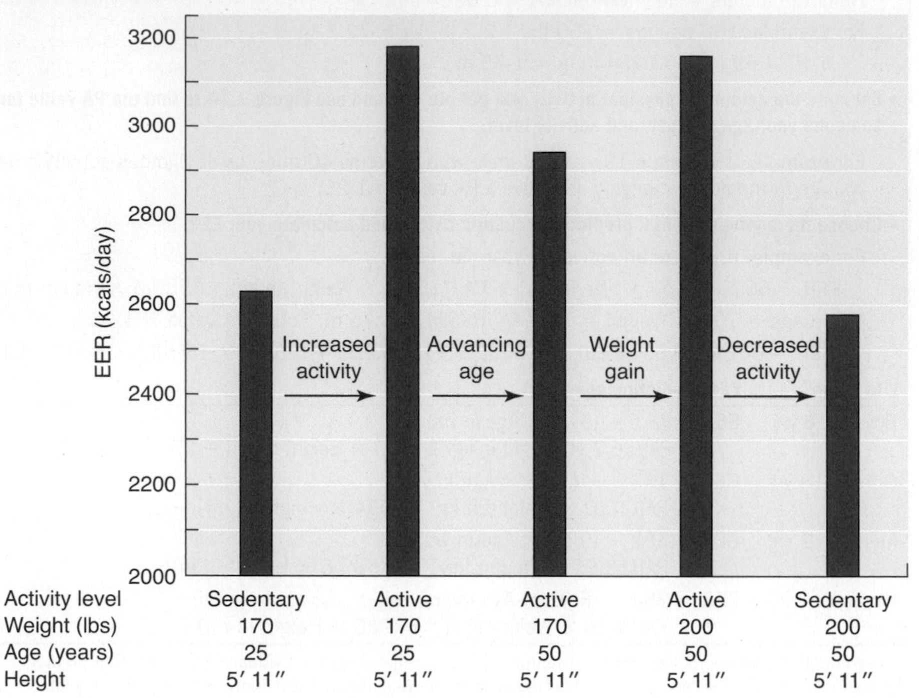

7.4 Body Weight and Health

LEARNING OBJECTIVES

- Name some health problems that are common in overweight individuals.
- Explain the relationship among obesity, heart disease, and diabetes.
- Discuss some circumstances when being lean increases health risks.

Some body fat is essential for health, since it provides an energy store, cushions internal organs, and insulates against changes in temperature, but too much body fat can increase the risk of disease and create psychological and social problems. Individuals who have little stored fat also have a greater risk for early death than individuals whose body fat is within the normal range.[17]

How Excess Body Fat Affects Health

Having too much body fat increases a person's risk of developing a host of chronic health problems, including heart disease, high blood pressure, stroke, high blood cholesterol, diabetes, gallbladder disease, arthritis, sleep disorders, respiratory problems, menstrual irregularities, and certain cancers, including those of the breast, uterus, and colon (**Figure 7.16**). The presence of some of these conditions then increases the risk of premature death. Obesity also increases the incidence and severity of infectious disease and has been linked to poor wound healing and surgical complications. The more excess body fat people have, the greater their health risks. The longer they carry excess fat, the greater the risks; individuals who gain excess weight at a young age and remain overweight throughout life face the greatest health risks. The sedentary lifestyle

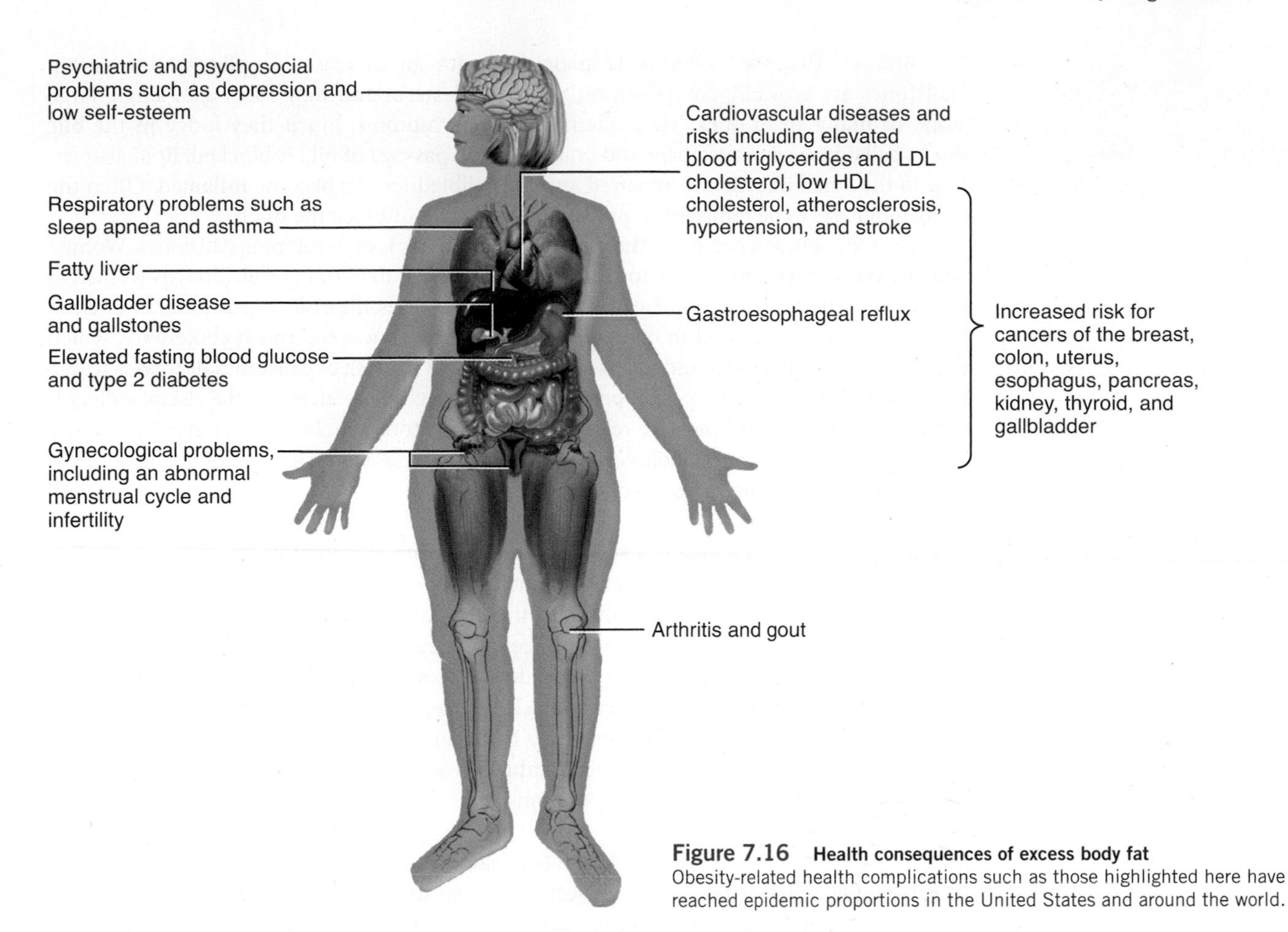

Figure 7.16 **Health consequences of excess body fat**
Obesity-related health complications such as those highlighted here have reached epidemic proportions in the United States and around the world.

associated with obesity further increases the risk of diabetes and heart disease and their complications.

Being overweight causes problems throughout life. During pregnancy, carrying excess body fat increases risks both for the mother and her baby (see Chapter 14). In children and adolescents, being overweight contributes to the development of high blood cholesterol levels, high blood pressure, and elevated blood glucose (see Chapter 15). The high rate of childhood obesity is a major threat to this generation because the longer a person is overweight the greater the risks; those who gain excess weight at a young age and remain overweight throughout life have the greatest health risks.

Heart Disease, Stroke, and Diabetes Obesity is considered a primary risk factor for cardiovascular disease and type 2 diabetes. Carrying excess body fat increases the amount of work required by the heart and the risk of developing high blood pressure and abnormal blood lipid levels. When excess body weight and fat decrease blood cholesterol levels improve, blood pressure decreases, and the risk of heart disease is reduced.[18] Being overweight also contributes to angina (chest pain caused by decreased oxygen to the heart) and sudden death from heart disease or stroke without any signs or symptoms.

More than 85% of people with type 2 diabetes are overweight or obese.[19] Excess body fat increases insulin resistance, which increases the amount of insulin needed to keep blood glucose in the normal range. Over time, the body is no longer able to regulate blood sugar levels, and type 2 diabetes results. Having diabetes, in turn, increases the risks of heart disease and stroke. Atherosclerosis is the most common, long-term complication of diabetes, and cardiovascular risk factors including hypertension, which predisposes to stroke, are more frequent in diabetics. In overweight individuals who have diabetes, weight loss can help maintain blood glucose levels in the normal range.[19]

Gallbladder Disease Obesity is associated with an increase in gallstone formation. Gallstones are typically composed mostly of cholesterol and may form as a single large stone or many small ones. They often cause no symptoms, but if they lodge in the bile ducts, gallstones can cause pain and cramps. If the passage of bile is blocked, lipid absorption in the small intestine is impaired and the gallbladder can become inflamed. Often the gallbladder has to be removed to prevent the pain and unblock the duct.

The more obese a person is, the greater his or her risk is of developing gallstones. Women who are obese have about three to seven times the risk of developing gallstones as women at a healthy body weight.[20] The reason that obesity increases the risk of gallstones is unclear, but researchers believe that in obese people the liver produces too much cholesterol, which deposits in the gallbladder and forms stones. Although the risk of gallstones decreases with a lower body weight, weight loss, in particular rapid weight loss, increases the risk of gallbladder disease because, as lipids are released from body stores, cholesterol synthesis increases, increasing the tendency for cholesterol stone formation. Gallstones are one of the most medically important complications of voluntary weight loss.[20]

Sleep Apnea Sleep apnea is a serious, potentially life-threatening condition characterized by brief interruptions of breathing during sleep. These short stops in breathing can happen up to 400 times every night, and therefore patients with sleep apnea sleep very poorly and wake up in the morning still feeling tired. Sufferers remain tired throughout the day. Sleep apnea has also been associated with other conditions such as high blood pressure, abnormal blood lipids, heart attack, stroke, and type 2 diabetes.[21, 22] It is a common complication of obesity because fatty tissue in the pharynx and neck compress the airway and block airflow. In obese individuals, weight loss may decrease both the frequency and severity of sleep apnea symptoms.[19]

Cancer Excess body weight and fat are associated with an increased risk of certain forms of cancer. Several mechanisms have been proposed to explain these associations.[23] Many involve hormones. Adipose tissue produces the hormone estrogen. It has been hypothesized that the greater risk of postmenopausal breast cancer seen in obese women is due to elevated estrogen levels caused by their excess body fat.[24] It is unclear why being overweight seems to protect against the form of breast cancer that occurs in women in the childbearing years.[25] Fat cells also produce protein hormones called adipokines. Some of these may contribute to the development of cancer because they promote cell proliferation. Levels of the hormones insulin and insulin-like growth factor, which may promote tumor development, are higher in obese individuals. Adipocytes may also directly or indirectly affect other molecules that regulate tumor growth. Other mechanisms proposed to play a role in the increased cancer risk seen with obesity include inflammation, increased oxidative stress, and altered immune response.

Joint Disorders Excess weight and fat can increase the risk of developing osteoarthritis. This is a type of arthritis that occurs when the cartilage cushioning the joints breaks down and gradually becomes rougher and thinner. As the process continues, a substantial amount of cartilage wears away so the bones in the joints rub against each other, causing pain and reducing movement. Being overweight is the most common cause of excess pressure on the joints and can speed the rate at which the cartilage wears down. Losing weight reduces the pressure and strain on the joints and slows the wear and tear of cartilage.[19] In individuals suffering from osteoarthritis, weight loss can help reduce pain and stiffness in the affected joints, especially those in the hips, knees, back, and feet.[26]

Gout is a joint disease caused by high levels of uric acid in the blood. Uric acid is a nitrogenous waste product resulting from the breakdown of DNA and similar types of molecules. It sometimes forms into solid stones or crystal masses that deposit in the joints, causing pain. Gout is more common in overweight people, and the risk of developing the disorder increases with higher body weight.

Psychological and Social Problems Carrying excess body fat has psychological and social consequences. Our society puts a high value on physical appearance. Being thin is considered attractive, and being fat is not. Those who do not conform to standards may pay a high psychological and social price. For example, overweight children are often teased and ostracized.[27] Their psychological profile is characterized by low body satisfaction, low self-esteem, depression, and social isolation.[28] If obese children grow into obese adolescents and adults, and most of them do, they may be discriminated against in college admissions, in the job market, in the workplace, and even on public transportation. Obese individuals of every age are more likely to experience depression, a negative self-image, and feelings of inadequacy.[29] The physical health consequences of obesity may not manifest themselves as disease for years, but the psychological and social problems experienced by the obese are felt every day.

Health Implications of Being Underweight

Being underweight is associated with increased risk of early death,[15] but this does not mean that all thin people are at risk. People who are naturally lean have a lower incidence of certain chronic diseases and do not face increased health risks due to their low body weight. However, low body fat due to starvation, eating disorders, or disease reduces body fat and muscle mass, affects electrolyte balance, and decreases the ability of the immune system to fight disease.

Too little body fat can cause problems at all stages of life. During adolescence, it can delay sexual development. During pregnancy, too little weight gain increases the risk that the baby will have health complications, and in the elderly, too little body fat increases the risk of malnutrition. In developed countries, socioeconomic conditions may create isolated pockets of undernutrition, but severe cases of wasting are usually a result either of self-starvation due to an eating disorder, such as anorexia nervosa (see **Focus on Eating Disorders**), or of a disease process, such as AIDS or cancer.

7.5 Guidelines for a Healthy Body Weight

LEARNING OBJECTIVES

- Describe three methods used to estimate percent body fat.
- Calculate your BMI, and determine if it is in the healthy range.
- Contrast visceral and subcutaneous fat.

Guidelines for a healthy body weight are based on the weight at which the risks of illness and death are lowest. These risks are associated not only with body weight but also with the amount and location of body fat; therefore assessment of a healthy body weight must consider body composition.

Body Mass Index

The current standard for evaluating body weight is **body mass index (BMI)**, which is calculated from body weight and height. Although BMI does not directly assess body composition, BMI values correlate well with the amount of body fat in most people.[30] BMI is calculated from a ratio of weight to height according to either of the following equations:

body mass index (BMI) A measure of body weight in relation to height that is used to compare body size with a standard.

$$\text{BMI} = \text{weight in kg}/(\text{height in m})^2$$

or

$$\text{BMI} = \text{weight in lbs}/(\text{height in inches})^2 \times 703^*$$

For example, someone who is 6 feet (72 in or 1.83 m) tall and weighs 180 lb (81.8 kg) has a BMI of 24.5 kg/m^2.

*The multiplier 703 is used by the Center for Disease Control and Prevention in developing growth charts, 704.5 is used by the National Institutes of Health, and 700 is used by the American Dietetic Association. The variation in outcome is insignificant.

TABLE 7.4 Is Your BMI in the Healthy Range?*

	Under-weight		Normal						Overweight					Obese										Extremely obese		
BMI	17	18	19	20	21	22	23	24	25	26	27	28	29	30	31	32	33	34	35	36	37	38	39	40	41	42
Height (feet/inches)	Body weight (pounds)																									
4' 10"	81	86	91	96	100	105	110	115	119	124	129	134	138	143	148	153	158	162	167	172	177	181	186	191	198	201
4' 11"	84	89	94	99	104	109	114	119	124	128	133	138	143	148	153	158	163	168	173	178	183	188	193	198	203	208
5' 0"	87	92	97	102	107	112	118	123	128	133	138	143	148	153	158	163	168	174	179	184	189	194	199	204	209	215
5' 1"	90	95	100	106	111	116	122	127	132	137	143	148	153	158	164	169	174	180	185	190	195	201	206	211	217	222
5' 2"	93	98	104	109	115	120	126	131	136	142	147	153	158	164	169	175	180	186	191	196	202	207	213	218	224	229
5' 3"	96	102	107	113	118	124	130	135	141	146	152	158	163	169	175	180	186	191	197	203	208	214	220	225	231	237
5' 4"	99	105	110	116	122	128	134	140	145	151	157	163	169	174	180	186	192	197	204	209	215	221	227	232	238	244
5' 5"	102	108	114	120	126	132	138	144	150	156	162	168	174	180	186	192	198	204	210	216	222	228	234	240	246	252
5' 6"	105	112	118	124	130	136	142	148	155	161	167	173	179	186	192	198	204	210	216	223	229	235	241	247	253	260
5' 7"	108	115	121	127	134	140	146	153	159	166	172	178	185	191	198	204	211	217	223	230	236	242	249	255	261	268
5' 8"	112	119	125	131	138	144	151	158	164	171	177	184	190	197	203	210	216	223	230	236	243	249	256	262	269	276
5' 9"	115	122	128	135	142	149	155	162	169	176	182	189	196	203	209	216	223	230	236	243	250	257	263	270	277	284
5' 10"	119	126	132	139	146	153	160	167	174	181	188	195	202	209	216	222	229	236	243	250	257	264	271	278	285	292
5' 11"	122	129	136	143	150	157	165	172	179	186	193	200	208	215	222	229	236	243	250	257	265	272	279	286	293	301
6' 0"	125	133	140	147	154	162	169	177	184	191	199	206	213	221	228	235	242	250	258	265	272	279	287	294	302	309
6' 1"	129	137	144	151	159	166	174	182	189	197	204	212	219	227	235	242	250	257	265	272	280	288	295	302	310	318
6' 2"	132	140	148	155	163	171	179	186	194	202	210	218	225	233	241	249	256	264	272	280	287	295	303	311	319	326
6' 3"	136	144	152	160	168	176	184	192	200	208	216	224	232	240	248	256	264	272	279	287	295	303	311	319	327	335
6' 4"	140	148	156	164	172	180	189	197	205	213	221	230	238	246	254	263	271	279	287	295	304	312	320	328	336	344

*Locate your height in the leftmost column, and read across to your weight. Follow the column containing your weight up to the top line to find your BMI.

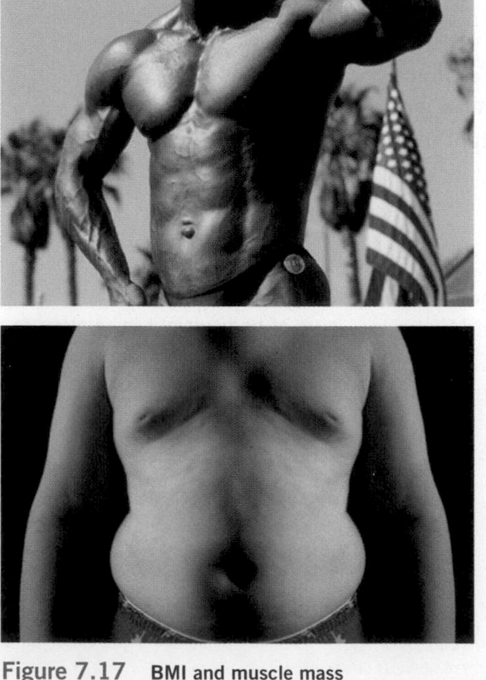

Jodi Cobb/NG Image Collection

Joel Sartore/NG Image Collection

Figure 7.17 BMI and muscle mass
Both of these men have a BMI of 33, but only the man in the bottom photo has excess body fat. The high body weight of the man in the top photo is due to his large muscle mass. The amount of body fat he has, and hence his disease risk, are low.

What Is a Healthy BMI? A healthy body weight is defined as a BMI of 18.5 to 24.9 kg/m^2. In general, people with a BMI within this range have the lowest health risks. Underweight is defined as a BMI of less than 18.5 kg/m^2, overweight is identified as a BMI of 25 to 29.9 kg/m^2, and obese as a BMI of 30 kg/m^2 or greater. A BMI of 40 or over is classified as extreme or morbid obesity. **Table 7.4** can be used to find your BMI and determine whether it is in the healthy range.

Limitations of BMI Even though BMI correlates well with the amount of body fat; it is not a perfect tool for evaluating the health risks associated with obesity. This is particularly true in athletes who have highly developed muscles; their BMI may be above the healthy body weight range because they have an unusually large amount of lean body mass. In these individuals, BMI is high but body fat and hence disease risk are low (**Figure 7.17**).

BMI is also not suitable for evaluating weight in pregnant and lactating women because of their rapidly changing weight and body composition (see Chapter 14). It is also less accurate in individuals who have lost muscle, such as many older adults. Because of these limitations, BMI should not be the only measure used to determine nutritional health and fitness. For example, someone who is in the overweight category based on BMI but consumes a healthy diet and exercises regularly may be more fit and have a lower risk of chronic disease than someone with a BMI in the healthy range who is sedentary and eats a poor diet.

Assessing Body Composition

The human body is composed of lean tissue and body fat. Lean tissue, referred to as lean body mass or fat-free mass, includes bones, muscles, and all tissue except adipose tissue. Adipose tissue lies under the skin and around internal organs. The amount of adipose tissue an individual carries and where that fat is deposited are affected by age, gender, and genetics as well as by energy balance (**Figure 7.18**). Adult women have more stored body fat than men, so the level that is healthy for women is somewhat higher. A healthy level of body fat for young adult women is between 21 and 32% of total weight; for young adult men, it is between 8 and 19%.[31]

A number of techniques are available for assessing body composition. Many require expensive equipment and must be performed in a research setting by trained technicians. Others are more portable, so they are appropriate for use in a clinic, office, or health club.

bioelectric impedance analysis A technique for estimating body composition that measures body fat by directing a low-energy electric current through the body and calculating resistance to flow.

Bioelectric Impedance Analysis **Bioelectric impedance analysis** is the most popular way to measure body composition. It estimates body fat by directing a painless, low-energy electrical current through the body. The difference between the current applied to

Figure 7.18 Changes in body composition throughout the life cycle

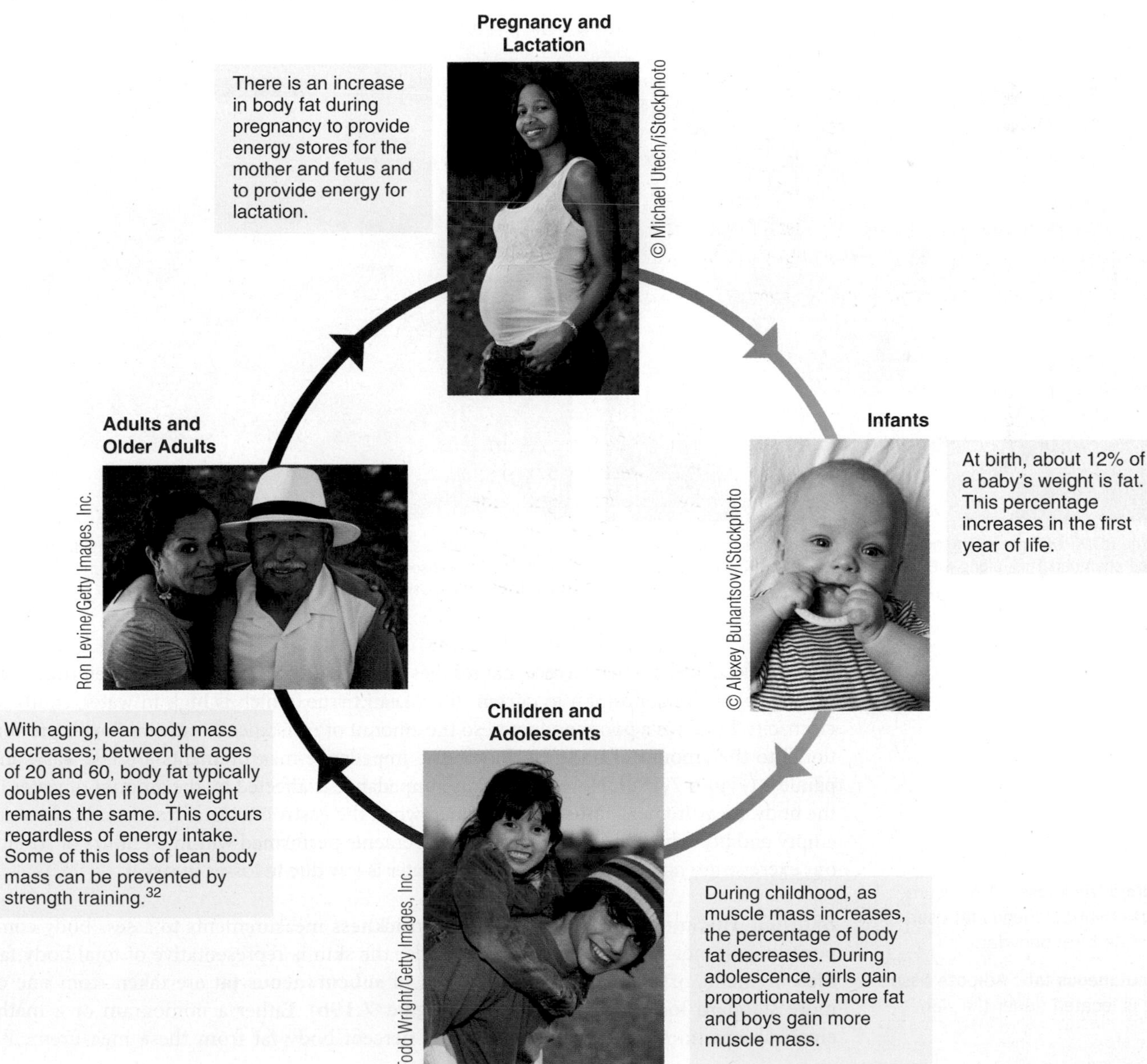

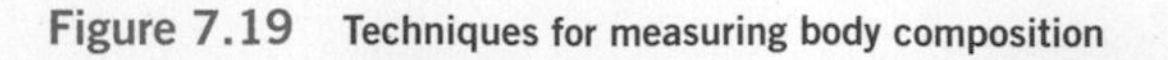

Figure 7.19 **Techniques for measuring body composition**

Yoav Levy/Phototake/Alamy

(a) Bioelectric impedance devices such as this one measure the resistance of body tissues to the flow of electric current. They can be used to assess body fat because current moves more quickly through lean tissue, which is high in water, than it does through fat, which resists current flow.

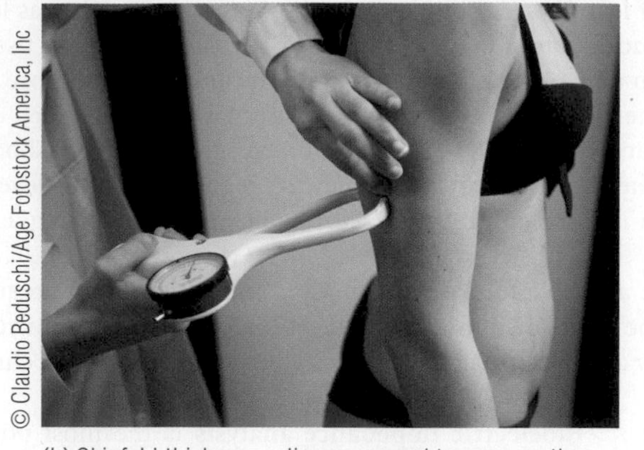

© Claudio Beduschi/Age Fotostock America, Inc

(b) Skinfold thickness calipers are used to measure the thickness of the fat layer under the skin at several locations. The most common sites are the triceps, shown here, and the subscapular area (just below the shoulder blade).

© Yoav Levy/Phototake

(c) To measure underwater weight, subjects must sit on a scale, expel the air from their lungs, and be lowered into a tank of water. A comparison of body weight on land and underwater is used to determine body density.

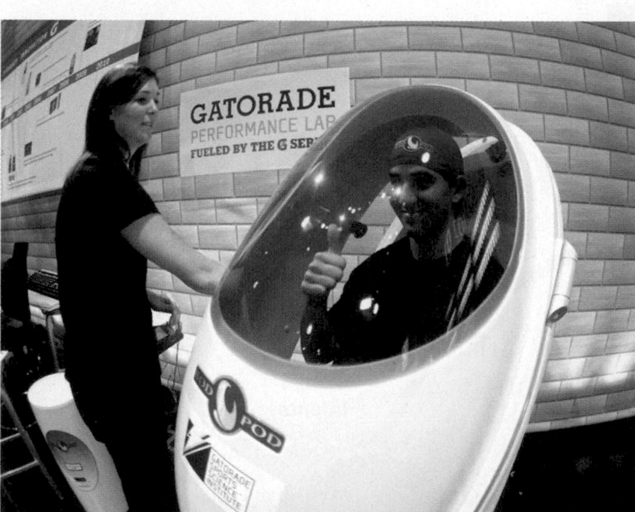

Jonathan Kirn/Feature Photo Service/NewsCom

(d) The BOD POD® measures the amount of air displaced by the body in a closed chamber. This, along with body weight, is used to determine body density.

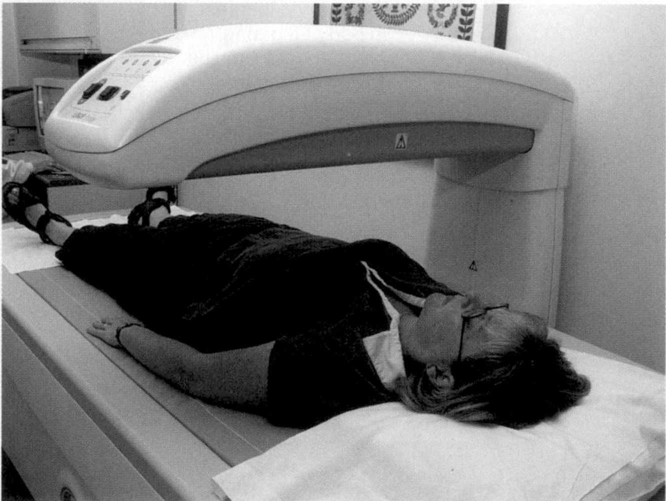

NewsCom

(e) DXA distinguishes among various body tissues by measuring differences in levels of X-ray absorption. When having the scan done, one must lay still in the supine position on what looks like an X-ray table.

the first electrode and the current that reaches the second electrode is used to determine the resistance, or the opposition to current flow. Lean tissue, which is high in water, conducts electricity, but fat is a poor conductor, so the amount of resistance to current flow is proportional to the amount of body fat. Bioelectric impedance measurements are fast, easy, and painless (**Figure 7.19a**). However, because impedance is affected by the amount of water in the body, measurements must be performed when the gastrointestinal tract and bladder are empty and body hydration is normal. Measurements performed within 24 hours of strenuous exercise are not accurate because body water is low due to losses in sweat.

skinfold thickness A measurement of subcutaneous fat used to estimate total body fat.

subcutaneous fat Adipose tissue that is located under the skin.

Skinfold Thickness The use of **skinfold thickness** measurements to assess body composition assumes that the amount of fat under the skin is representative of total body fat. Measurements of the thickness of the layer of **subcutaneous fat** are taken from one or more standard locations using calipers (**Figure 7.19b**). Either a nomogram or a mathematical equation is then used to estimate percent body fat from these measurements.

Skinfold measurements are noninvasive and, when performed by a trained individual, can accurately predict body fat in normal-weight individuals. These measures are more difficult to perform and less accurate in obese and elderly subjects.

Water and Air Displacement **Underwater weighing** can be used to assess body composition because lean tissue is denser than fat tissue. The difference between a person's weight on land and underwater is used to calculate body volume and body density. Body density is proportional to fat-free mass, so it can be used to determine the amount of body fat. Although this method is accurate and noninvasive, it requires special equipment and cannot be used for some groups, such as small children or frail adults (**Figure 7.19c**). A newer method for estimating body composition measures air displacement rather than water displacement to determine body volume. The individual is assessed in an air-filled chamber (known as the BOD POD®) rather than in water. It is accurate and more convenient than underwater weighing but expensive and not readily available (**Figure 7.19d**).[33]

underwater weighing A technique that uses the difference between body weight underwater and body weight on land to estimate body density and calculate body composition.

Dual-Energy X-Ray Absorptiometry Dual-energy X-ray absorptiometry (DXA) is a method that uses two X-ray energies for assessing body composition (**Figure 7.19e**). A single investigation can accurately determine total body mass, bone mineral mass, and the amount and percentage of body fat.[34] Results can also assess body composition in specific regions of the body. It can be used to accurately estimate the amount of **visceral fat**, which is associated with the risk of heart disease and other chronic diseases.[35]

visceral fat Adipose tissue that is located in the abdomen around the body's internal organs.

Dilution Methods Body fat can also be assessed by using the principle of dilution. Because water is present primarily in lean tissue and not in fat, a water-soluble isotope can be ingested or injected into the bloodstream and allowed to mix with the water throughout the body. The concentration of the isotope in a sample of body fluid, such as blood, can then be measured. The extent to which the isotope has been diluted can be used to calculate the amount of lean tissue in the body, and body fat can then be calculated by subtracting lean weight from total body weight. Another technique measures a naturally occurring isotope of potassium. Because potassium is found primarily in lean tissue, a measure of the amount of this isotope in the body can be used to determine the total amount of body potassium, which can then be used to estimate the amount of lean tissue. Dilution techniques are expensive and invasive, usually requiring injections. They are used primarily for research purposes.

How Location of Body Fat Affects Health Risk

The location of body fat stores affects the risks associated with having too much fat (**Figure 7.20**). Excess subcutaneous fat, which is adipose tissue located under the skin, does not increase health risk as much as does excess visceral fat, which is adipose tissue located around the organs in the abdomen. Generally, fat in the hips and lower body is subcutaneous, whereas fat in the abdominal region is primarily visceral. Visceral fat is more metabolically active than subcutaneous fat, releasing dozens of biologically active substances that can contribute to disease.[36,37] An increase in visceral fat is associated with a higher incidence of heart disease, high blood cholesterol, high blood pressure, stroke, diabetes, and breast cancer.

Where your extra fat is deposited is determined primarily by your genes.[38] Visceral fat storage is more common in men than women. African American men store less visceral fat than white men of the same age and level of total body fat.[39] After menopause, visceral fat increases in women. Other factors that affect the amount of visceral fat include stress, tobacco use, and alcohol consumption, all of which predispose people to visceral fat deposition, and physical activity, which reduces it.

Distinguishing the relative amounts of visceral and subcutaneous fat requires sophisticated imaging techniques. However, the risk associated with visceral fat deposition can be estimated by measuring waist circumference. For men whose BMI is greater than 25 kg/m^2, a waist circumference greater than 40 inches is associated with an increased risk. For women in this BMI range, a waist circumference of greater than 35 inches

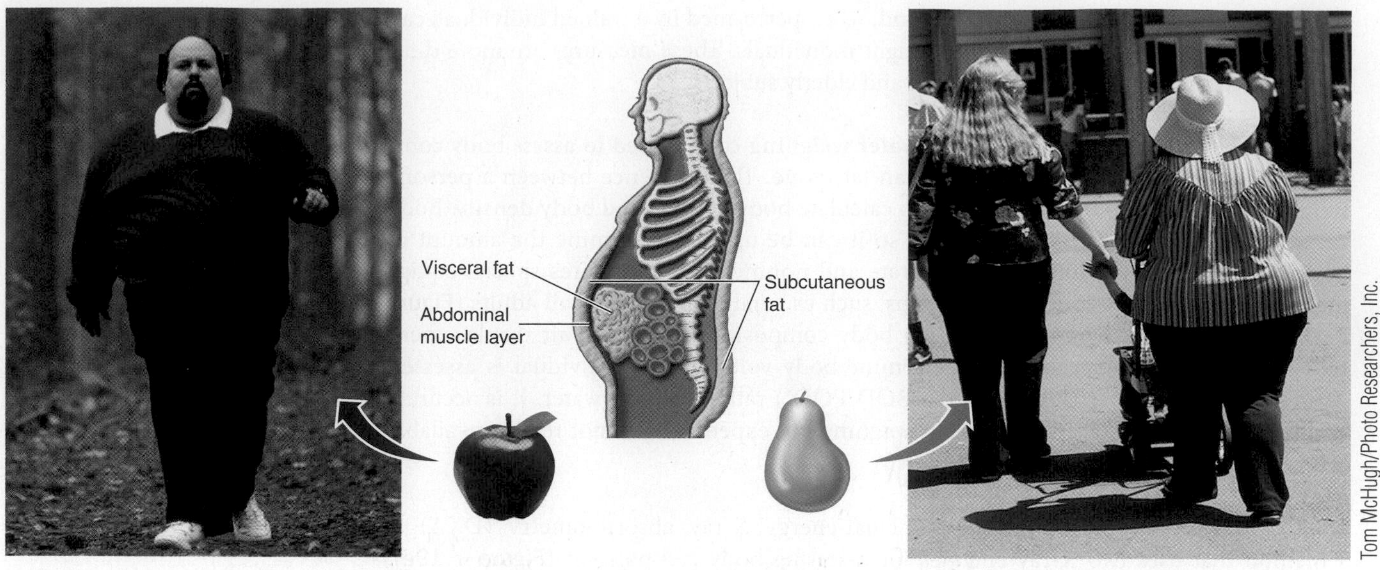

Figure 7.20 Visceral and subcutaneous body fat
People who carry their excess fat around and above the waist have more visceral fat (left). Those who carry their extra fat below the waist, in the hips and thighs, have more subcutaneous fat (right). In the popular literature, these body types have been dubbed "apples" and "pears," respectively.

TABLE 7.5 BMI, Waist Circumference, and Disease Risk[30]

		Disease risk[b]	
	BMI (kg/m²)[a]	Men, waist ≤ 40 inches, and women, waist ≤ 35 inches	Men, waist > 40 inches, and women, waist > 35 inches
Underweight	<18.5		
Normal weight	18.5–24.9		
Overweight	25.0–29.9	Increased	High
Obesity (class I)	30.0–34.9	High	Very high
Obesity (class II)	35.0–39.9	Very high	Very high
Extreme or morbid obesity (class III)	≥40	Extremely high	Extremely high

[a]BMI = body weight (kg)/height squared (m^2)

[b]Disease risk for type 2 diabetes, hypertension, and cardiovascular disease relative to individuals with a normal weight and normal waist circumference.

increases risks[30] (**Table 7.5**). In individuals under 5 feet in height or with a BMI greater than or equal to 35 kg/m^2, these cutoff points are not helpful in predicting health risks.

7.6 What Determines Body Size and Shape?

LEARNING OBJECTIVES

- Discuss genetic and environmental factors that affect body weight.
- Describe what is meant by a set point for body weight.
- List four physiological signals that determine whether you feel hungry or full.
- Explain why leptin levels might be higher in an obese than in a lean individual.

Courtesy Lori Smolin

Bruce Ayres/Stone/Getty Images, Inc.

Figure 7.21 Genes and body shape
The genes we inherit from our parents are important determinants of our body size and shape. The boy on the left inherited his father's long, lean body, whereas the boy on the right has his father's huskier build and will likely have a tendency to be overweight throughout his life.

Most people are shaped like their mother or father. This is because the information that determines body size and shape is contained in the genes people inherit from their parents. Some of us inherit long, lean bodies, and others inherit huskier builds and the tendency to put on pounds (**Figure 7.21**). Genes involved in regulating body fatness have been called **obesity genes**. More than 100 genes that are associated with body-weight regulation have been identified; it is estimated that 20 to 30 of these may contribute to obesity in humans.[40] Obesity genes are responsible for the production of proteins that affect how much food people eat, how much energy they expend, and how efficiently their body fat is stored. The combined effects of all these genes help to determine and regulate what people weigh and how much fat they carry. But genes are not the only factor; regardless of your genetic background, the lifestyle choices you make play an important role in determining what you weigh. The drastic rise in obesity that has occurred over the past several decades demonstrates the impact of lifestyle factors on weight. The frequency of genes in a population takes many generations to change, but environmental conditions that affect lifestyle can change quickly.

obesity genes Genes that code for proteins involved in the regulation of food intake, energy expenditure, or the deposition of body fat. When they are abnormal, the result is abnormal amounts of body fat.

Genes vs. Environment

The genes you inherit are an important determinant of what you weigh. If one or both of your parents is obese, your risk of becoming obese is increased. Individuals with a family history of obesity are two to three times more likely to be obese, and the risk increases with the magnitude of the obesity.[41] The influence of genes on body weight was demonstrated dramatically by a study done with identical twins. During the study, the pairs of identical twins were overfed to the same extent. Each set of twins tended to gain the same amount of weight and to deposit fat in the same parts of their bodies. In contrast, large differences were seen between sets of twins—some of the twin sets gained only 9 lbs, whereas others gained as much as 29 lbs.[38] By studying identical twins, researchers have been able to determine that about 75% of the variation in BMI between individuals can be attributed to genes.[42] This means that the remaining 25% is determined by the environment in which you live and the lifestyle choices you make. So even if you inherit genes that predispose you to being overweight, your actual weight is determined by the balance between the genes you inherit and your lifestyle choices.

When genetically susceptible individuals find themselves in an environment where food is appealing and plentiful and physical activity is easily avoided, obesity is a likely outcome but not the only possible one. If people who inherit genes that predispose them

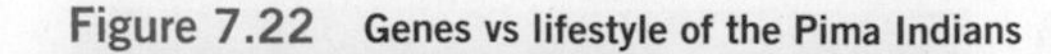
Figure 7.22 Genes vs lifestyle of the Pima Indians

(a)

Matt York/AP/Wide World Photos

(a) Genetic analysis of the Pima Indian population living in Arizona has identified a number of genes that may be responsible for this group's tendency to store excess body fat.[44] When this genetic susceptibility is combined with an environment that fosters a sedentary lifestyle and consumption of high-calorie, high-fat processed foods, the outcome is the strikingly high incidence of obesity seen in this population.

(b)

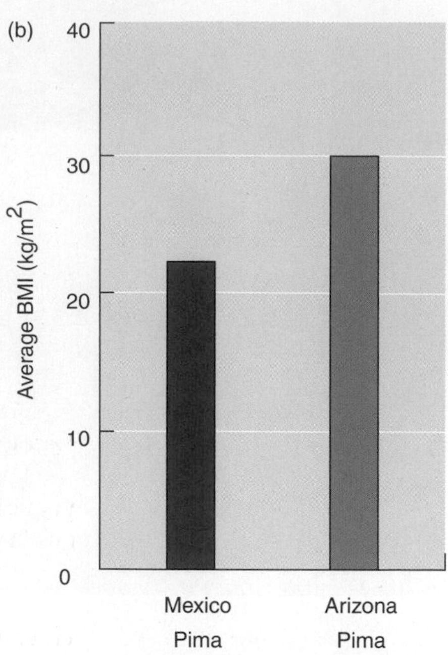

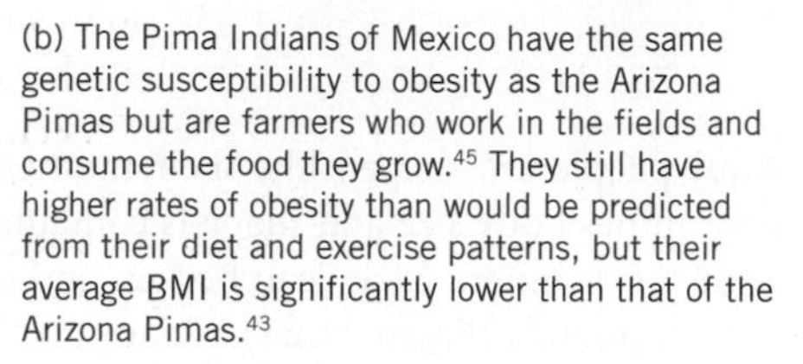
(b) The Pima Indians of Mexico have the same genetic susceptibility to obesity as the Arizona Pimas but are farmers who work in the fields and consume the food they grow.[45] They still have higher rates of obesity than would be predicted from their diet and exercise patterns, but their average BMI is significantly lower than that of the Arizona Pimas.[43]

to being overweight carefully plan their food intake and exercise regularly, they can maintain a healthy weight. It is also possible for individuals with no genetic tendency toward obesity to end up overweight if they consume a high-calorie diet and get little exercise. The interplay between genetics and lifestyle is illustrated by the higher incidence of obesity in Pima Indians living in Arizona than in a genetically similar group of Pima Indians living in Mexico (**Figure 7.22**).[43]

How Food Intake and Body Weight Are Regulated

What we eat and how much we exercise vary from day to day, but body weight tends to remain relatively constant for long periods. The body compensates for variations in diet and exercise by adjusting energy intake and expenditure to keep weight at a particular level, or **set point**. This set point, which is believed to be determined in part by genes, explains why your weight remains fairly constant, despite the added activity of a weekend hiking trip, or why most people gain back the weight they lose when they follow a weight-loss diet.[46]

set point A level at which body fat or body weight seems to resist change despite changes in energy intake or output.

To regulate weight and fatness at a constant level, the body must be able to respond both to changes in food intake that occur over a short time frame and to changes in the amount of stored body fat that occur in the long term. Signals related to food intake affect hunger and **satiety** over a short period of time—from meal to meal—whereas signals from the adipose tissue trigger the brain to adjust both food intake and energy expenditure for long-term regulation.

satiety The feeling of fullness and satisfaction, caused by food consumption, that eliminates the desire to eat.

Short-Term: How Food Intake Is Regulated from Meal to Meal How do you know how much to eat for breakfast or when it is time to eat lunch? To some extent, your level of hunger or satiety determines how much you eat at each meal. These physical sensations that tell people to eat or stop eating are triggered by signals from the GI tract, levels of circulating nutrients, and messages from the brain.[47] Some signals are sent before food is eaten, some are sent while food is in the GI tract, and some occur once nutrients are circulating in the bloodstream.

The simplest type of signal about how much food has been eaten comes from local nerves in the walls of the stomach and small intestine that sense the volume or pressure of food and send a message to the brain to either start or stop eating. Once food is consumed, the presence of nutrients in the GI tract triggers the release of gastrointestinal peptide hormones that interact with the nervous system to affect hunger or satiety. Once nutrients have been absorbed, circulating levels of nutrients, including glucose, amino acids, ketones, and fatty acids, are monitored by the brain and may trigger signals to eat or not to eat.[48] Nutrients that are taken up by the brain may affect neurotransmitter concentrations, which then affect the amounts and types of nutrients consumed. For example, some studies suggest that when brain levels of the neurotransmitter serotonin are low, carbohydrate is craved, but when it is high, protein is preferred.[48] Absorbed nutrients also affect metabolism in the liver because absorbed water-soluble nutrients go there directly. Changes in liver metabolism, in particular the amount of ATP, are believed to be involved in regulating food intake.

There are many different peptide hormones that regulate food intake. The hormone **ghrelin** may be the reason people typically feel hungry around lunchtime regardless of when and how much they had for breakfast. It is produced by the stomach and is believed to stimulate the desire to eat meals at usual times. Blood levels of ghrelin rise an hour or two before a meal and drop very low after a meal. Levels have been found to rise in people who have lost weight, increasing their desire to eat more.[49] Cholecystokinin, released when chyme enters the small intestine, slows stomach emptying, which induces satiety, causing us to stop eating. Another hormone that causes a reduction in appetite is peptide YY. It is released from the GI tract after a meal, and the amount released is proportional to the number of calories in the meal.[50]

ghrelin A peptide hormone produced by the stomach that stimulates food intake.

Psychological factors can also affect hunger and satiety. For example, some people eat for comfort and to relieve stress. Others may lose their appetite when these same emotions are felt. Psychological distress can alter the mechanisms that regulate food intake.

Long-Term: How the Amount of Body Fat Is Regulated Short-term regulators of energy balance affect the size and timing of individual meals, but if a change in food intake is sustained over a long period it can affect long-term energy balance and, hence, body weight and fatness. To regulate the amount of fat at a set level, the body must be able to monitor how much fat is present. This information is believed to come from hormones, such as insulin and **leptin**, which are secreted in proportion to the amount of body fat.[51,52] Insulin is secreted from the pancreas when blood glucose levels rise; its circulating concentration is proportional to the amount of body fat. Insulin interacts with a portion of the brain called the hypothalamus to reduce food intake and body weight, and insulin levels are believed to affect the amount of leptin produced and secreted. Leptin is a hormone that is produced by the adipocytes and acts in the hypothalamus. The amount of leptin produced is proportional to the size of adipocytes—more leptin is released as fat stores increase. Leptin exerts its effect on food intake and energy expenditure by binding to leptin receptors present in the hypothalamus. This triggers mechanisms that affect energy intake and expenditure. When leptin levels are high, mechanisms that increase energy expenditure and decrease food intake are stimulated, and pathways that promote food intake and hence weight gain are inhibited. When fat stores shrink, less leptin is released. Low leptin levels in the brain allow pathways that decrease energy expenditure and increase food intake to become active.[51] Unfortunately, leptin regulation, like other regulatory mechanisms, is much better at preventing weight loss than at defending against weight gain. Obese individuals generally have high levels of leptin, but these levels are not effective at reducing calorie intake and increasing energy expenditure[53] (**Figure 7.23**; see **Science Applied: Leptin: Discovery of an Obesity Gene**).

leptin A peptide hormone produced by adipocytes that signals information about the amount of body fat.

Hormonal signals involved in the long-term regulation of body weight act in the brain not only to favor shifts in energy balance but also to affect the sensitivity of the brain to short-term signals of energy balance. For example, during weight loss, low levels of these hormones are hypothesized to decrease the efficacy of satiety signals, suppress pathways that cause weight loss, and activate pathways that contribute to weight gain.[54]

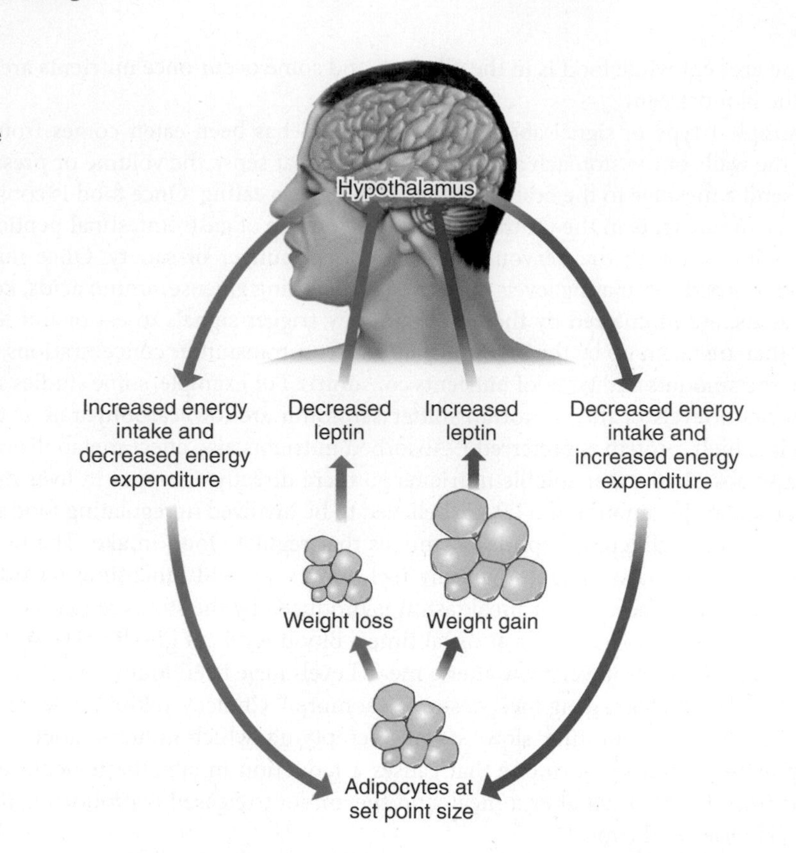

Figure 7.23 Leptin regulation of body fat Changes in the size of the adipocytes affect the amount of leptin released. The amount of leptin reaching the hypothalamus determines the response and helps return body fat stores to a set level.

THINK CRITICALLY

What might happen to body weight in a person who makes leptin normally but whose leptin receptor in the brain is defective so it always acts as if large amounts of leptin are bound to it? to a person who makes more leptin than normal?

Despite regulatory mechanisms that act to keep our weight at a stable set point, changes in physiological, psychological, and environmental circumstances cause the level at which body weight is maintained to change, usually increasing it over time. For example, body weight increases in most adults between the ages of 30 and 60 years, and after childbearing, most women return to a weight that is 1 to 2 lbs higher than their prepregnancy weight. This supports the hypothesis that the mechanisms that defend against weight loss are stronger than those that prevent weight gain.[54]

Why Some People Gain Weight Easily

When a gene is defective, the protein it codes for is not made or is made incorrectly. When an obesity gene, such as the gene for leptin, is defective, the signals to decrease food intake and/or increase energy expenditure are not received, and weight gain results. A few cases of human obesity have been linked directly to defects in the genes for leptin and leptin receptors,[55] but mutations in single genes such as these are not responsible for most human obesity. Rather, variations in many genes interact with one another and affect metabolic rate, food intake, fat storage, and activity level. These in turn affect body weight, determining why some of us stay lean and others put on pounds easily.

Thrifty Metabolism Throughout human history, starvation has threatened survival. Over time, the human body has evolved ways to conserve body fat stores and prevent weight loss. Individuals who are more efficient at using energy and storing fat, what has been called a "thrifty" metabolism, would have been more likely to survive. In the United States today, however, food is abundant, so people who inherited these "thrifty genes" are more likely to be obese. Some people, such as the Pima Indians discussed earlier, gain weight more easily because they inherited genes that give them a thrifty metabolism.

How much weight people gain in response to overeating also depends on the ability of their metabolism to respond to the increase in energy intake relative to expenditure. When people overeat occasionally, their metabolism speeds up to burn the extra energy and prevent weight gain.[56, 57] Conversely, when body weight is reduced by restricting food intake, energy expenditure decreases to conserve energy and promote weight gain.[58] These changes in the amount of energy expended in response to changes in circumstance, such as over- or undereating, changes in environmental temperature, or trauma, are referred

SCIENCE APPLIED

Leptin: Discovery of an Obesity Gene

(Elena Schweitzer/Shutterstock) Victoriano Izquierdo/Getty Images, Inc. (DAJ/Getty Images, Inc.)

A discovery made by Rockefeller University physician Jeffrey Friedman and his colleagues in 1994 brought hope to millions of people. Perhaps the cause of obesity had been found, and a cure might be close at hand. Was relief in sight for those who suffer from the physical and social consequences of obesity?

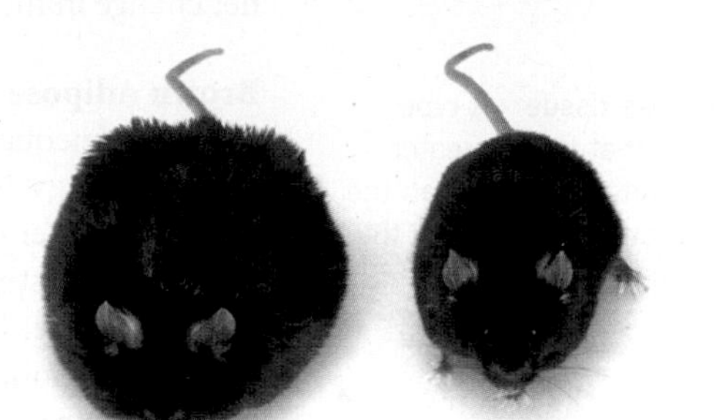

A mouse with a defect in the leptin gene (ob/ob) may weigh three times as much as a normal mouse. Both of these mice have defective ob genes, but the one on the right was treated with leptin injections.

THE SCIENCE

Dr. Friedman's work began with a strain of mice called ob/ob, because they have 2 copies of a gene that predisposes them to obesity. Ob/ob mice become grossly obese, gaining up to three times the normal body weight. The ob/ob strain arose spontaneously in 1950 in the mouse colony at the Jackson Laboratory in Bar Harbor, Maine. Friedman and his colleagues unraveled the cause for the obesity in this strain of mice when they identified and cloned the gene that was responsible.[1]

Researchers used a series of breeding experiments to localize the gene to a particular stretch of DNA. They then looked to see if any of the genes in this stretch of DNA were expressed in adipose tissue. The search yielded a single gene. Evidence that this gene was involved in the regulation of body weight was obtained by examining the gene and the protein it codes for in the ob/ob mice. Researchers found that this protein, which they named leptin, was either not produced or produced in an inactive form in the obese mice. Soon afterward a similar gene was identified in humans.

THE APPLICATION

Optimism about the role of the peptide hormone leptin in human obesity was so great that a biotechnology firm (Amgen) paid $25 million for the commercial rights to leptin in the hope that it could be used to treat human obesity. Those hopes grew even higher when Friedman and his colleagues were able to demonstrate that injections of the hormone could restore the genetically obese mice to normal weight (see figure).[2,3] Unfortunately, the role of leptin in human obesity has not been as dramatic. Mutations in this gene are not responsible for most human obesity.[4] In fact, obese humans generally have high blood leptin levels.[5] High doses of leptin administered to obese humans produce only modest weight loss.[6]

The leptin receptor—a protein in the brain to which leptin must bind to produce weight reduction—was also identified.[7] The fact that obese humans have high levels of leptin suggested that the cause of human obesity might involve an abnormality in leptin receptors. If leptin receptors were defective, the leptin produced would have no place to bind and would not be able to signal mechanisms to promote weight reduction. Thus far, however, defective leptin receptors have not been found to be an important cause of human obesity.[8]

Continued study of the role of leptin in obesity has confirmed that it is an important signal involved in the long-term regulation of body fat, but it does not act alone. There are many steps, involving many genes, that occur between the production of leptin in adipose tissue and alterations in food intake and energy expenditure that are mediated by the brain. Researchers have discovered about a dozen peptides that interact with leptin in the brain to control appetite.[9]

Despite the fact that the identification of leptin has not produced a cure for human obesity, its discovery energized the field of obesity research. This discovery was an important advance in our understanding of the genetics of body-weight regulation. Continued work will someday answer the questions that remain about why some of us are obese and some of us are lean.

[1]Zhang, Y., Proenca, R., Maffei, M., et al. Positional cloning of the mouse obese gene and its human homologue. *Nature* 372:425–432, 1994.
[2]Halaas, J. L., Gajiwala, K. S., Maffei, M., et al. Weight-reducing effects of the plasma protein encoded by the obese gene. *Science* 269:543–546, 1995.
[3]Pelleymounter, M. A., Cullen, M. J., Baker, M. B., et al. Effects of the obese gene product on body weight regulation in ob/ob mice. *Science* 269:540–543, 1995.
[4]Montague, C. T., Farooqi, I. S., Whitehead, J. P., et al. Congenital leptin deficiency is associated with severe early onset obesity in children. *Nature* 387:903–908, 1997.
[5]Considine, R. V., Sinha, M. K., Heiman, M. L., et al. Serum immunoreactive-leptin concentrations in normal weight and obese humans. *N Engl J Med* 334:292–295, 1996.
[6]Gura, T. Obesity research: Leptin not impressive in clinical trial. *Science* 286:881–882, 1999.
[7]Tartaglia, L. A., Dembski, M., Weng X., et al. Identification and expression cloning of a leptin receptor, *OB-R. Cell* 83:1263–1271, 1995.
[8]Tsigos, C., Kyrou, I., and Raptis, S. A. Monogenic forms of obesity and diabetes mellitus. *J Pediatr Endocrinol Metab* 15:241–253, 2002.
[9]Gura, T. Tracing leptin's partners in regulating body weight. *Science* 287:1738–1741, 2000.

adaptive thermogenesis The change in energy expenditure induced by factors such as changes in ambient temperature and food intake.

to as **adaptive thermogenesis**. Genetic variations in the adaptive response to overeating may help explain why some people gain weight more easily than others.

Futile Cycling Several biochemical mechanisms have been proposed to explain adaptive thermogenesis. One is substrate cycling or futile cycling, which wastes energy by allowing opposing biochemical reactions to occur simultaneously. For example, a molecule is formed, consuming ATP, and then is quickly broken down again. Energy is consumed, but there is no net change in the number of molecules in the body and therefore no storage of energy as fat.

brown adipose tissue A type of fat tissue that has a greater number of mitochondria than the more common white adipose tissue. It can waste energy as heat.

Brown Adipose Tissue A second way that excess energy might be dissipated is by separating or uncoupling the electron transport chain from the production of ATP. When this occurs, energy is lost as heat rather than being used to produce ATP. For example, the increase in energy expenditure that occurs when mice are injected with leptin is believed to be due to the stimulation of receptors on a specialized type of adipose tissue called **brown adipose tissue**. Brown adipose tissue can waste energy as heat. This tissue contains many more mitochondria than regular white adipose tissue, and these mitochondria can be uncoupled from the electron transport chain, allowing the energy in food to be released as heat. In rats, brown adipose tissue generates heat to prevent weight gain during overfeeding and to provide warmth when the ambient temperature is low. Significant amounts of brown adipose tissue are present in human infants, but until recently it was believed that adults did not have enough brown adipose tissue for it to be relevant physiologically. Newer technology now allows researchers to measure the amount of brown adipose tissue in adults. In a study of young men, brown adipose tissue activity was found to be lower in overweight and obese men than in lean men, suggesting that it may play a role in human obesity.[59]

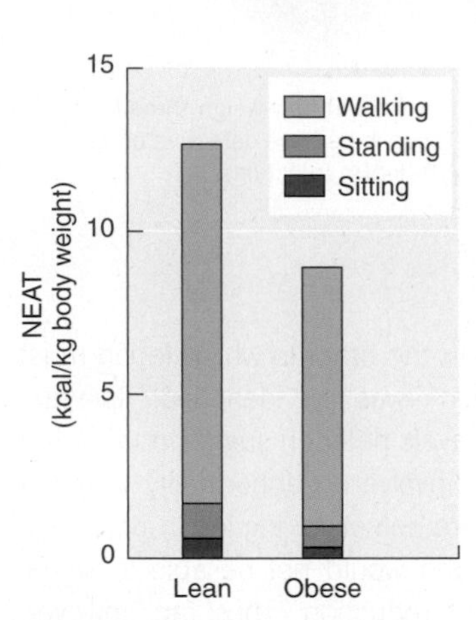

Figure 7.24 NEAT in obese vs. lean individuals
Obese individuals use fewer calories for NEAT than their lean counterparts. When obese and lean individuals who do not engage in any planned exercise were compared, the lean people walked more and sat less, by about two hours per day, than the obese study subjects. If obese individuals could adopt the same patterns as the lean subjects, they would expend an extra 350 kcals/day.[60]

Low Levels of Activity Activity burns calories; this is true whether the activity is planned exercise or NEAT (nonexercise activity thermogenesis) activities such as housework, walking between classes, fidgeting, and moving to maintain posture. How active you are is affected by your genes and your personal choices. Some people may gain weight more easily because they inherit a tendency to expend less energy on activity. Even if they spend the same amount of time as a lean person in planned exercise, their total energy expenditure may be lower because they expend less energy for NEAT activities[60] (**Figure 7.24**). The impact of NEAT on weight gain was demonstrated by a study that overfed normal-weight individuals. There was a 10-fold variation in the amount of fat they gained. Some subjects were able to increase energy expenditure to a greater extent and so gained less fat. About two-thirds of the increase in energy expenditure that occurred with overfeeding was found to be due to an increase in unplanned exercise.[61] Those who gained the least weight had the greatest levels of involuntary exercise. The mechanisms that cause some people to respond to excess energy intake by becoming restless and increasing NEAT activity while others remain lethargic are still not understood.

7.7 Recommendations for Managing Body Weight

LEARNING OBJECTIVES

- Evaluate an individual's weight and medical history to determine if weight loss is recommended.
- Discuss the recommendations for the rate and amount of weight loss.
- Name the three components of an ideal weight-management plan.
- Describe methods for addressing the population-wide obesity crisis.

Managing body weight to keep it within the healthy range involves a series of lifestyle choices. It requires maintaining a balance between calorie intake and exercise. For some people, this means making healthy food choices, controlling portion sizes, and maintaining an active lifestyle to avoid weight gain as they age. For many others, it may mean developing a meal and exercise plan that will allow their weight to decrease. And for some, it may mean working to increase weight and then keep it in the healthy range. The goal for everyone is to achieve a healthy weight and stay there.

Who Should Lose Weight?

Not everyone who is a few pounds overweight will benefit from weight loss. The risks associated with overweight and obesity are related to the degree of excess weight or fat, the distribution of the excess body fat, the presence of other diseases or risk factors that often accompany obesity, and the age and life stage of the individual. To determine if someone should lose weight, the first step is to evaluate the person's current weight and weight history and review his or her medical conditions (**Figure 7.25**). A BMI above the healthy range generally means that weight loss would improve long-term health, but this is not always the case. Some people with a high BMI, such as weightlifters, may have a large amount of muscle mass but not more body fat than is recommended. If, however, the high BMI is accompanied by an elevation in the proportion of body fat or an increase in waist circumference, weight loss is usually recommended.

Consider Medical Risk Factors A key factor in whether or not weight loss is recommended is the presence of diseases and abnormalities associated with excess body fat (see Figure 7.16). For example, elevated blood pressure, blood glucose, and blood cholesterol levels all increase the risk associated with excess body weight. A person whose BMI is above the healthy range and has two or more of these risk factors will probably benefit from weight loss. A family history of these conditions is also a consideration in determining whether or not weight loss is recommended. On the other hand, a person who has a BMI in the overweight range (25 to 29.9 kg/m^2) but has none of the health conditions associated with excess body fat and has a healthy lifestyle may not benefit from the weight loss. For example, a person with a BMI of 28 kg/m^2 and a waist circumference of 33 inches, whose blood pressure and cholesterol levels are normal and who exercises regularly would not reduce his or her health risks by losing weight. For this individual, weight management may mean simply preventing further weight gain. For others, this risk assessment may indicate that the excess body weight is a health risk and a weight-loss plan should be developed (see Figure 7.25).

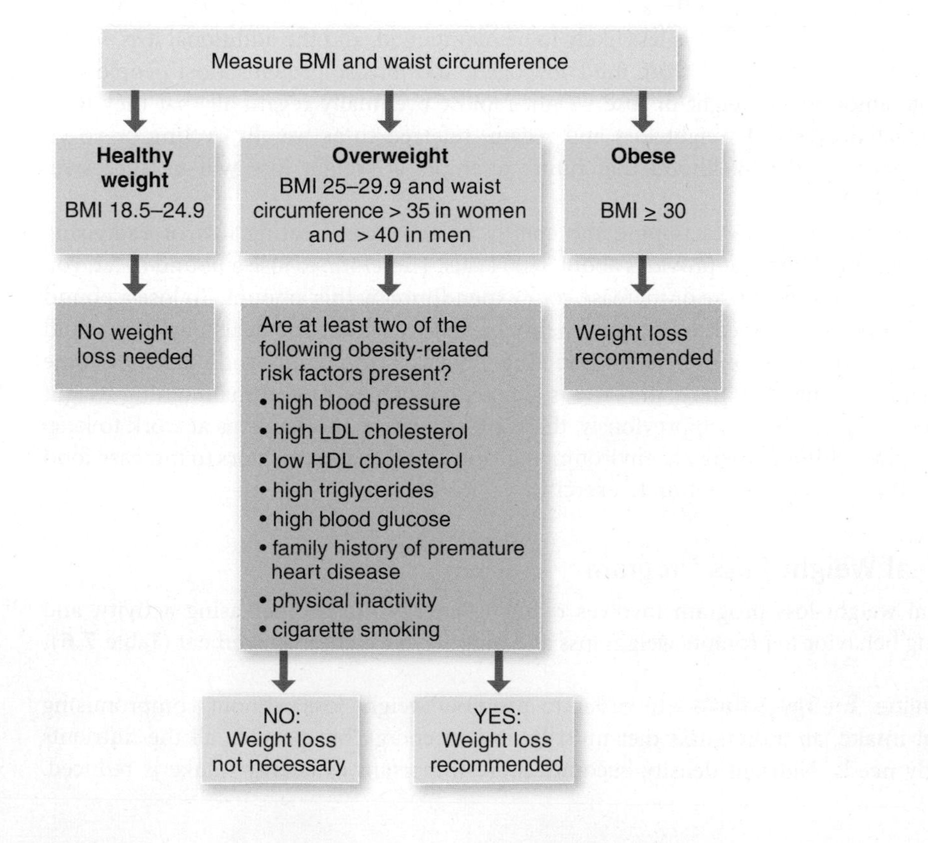

Figure 7.25 **Determining who would benefit from weight loss**
This flow chart illustrates some of the factors to consider when deciding if an individual will benefit from weight loss. People who have a BMI of 25 to 29.9 kg/m^2, do not have a high waist measurement, and have fewer than two risk factors may not benefit from weight loss but should avoid further weight gain.

THINK CRITICALLY

Would weight loss be recommended for a woman with a BMI of ≥ 27 kg/m^2 and a waist circumference of 42, who is a smoker and has high blood pressure?

Consider Life Stage Recommendations about weight loss are affected by life stage; at certain times of life weight loss is not recommended. For example, obesity and overweight is a growing problem among children. However, strict weight-loss diets are generally not recommended for children or adolescents because a reduction in intake can interfere with growth. A better approach is to encourage an increase in physical activity, along with a moderate restriction in energy intake. This will allow the child to grow in height with little additional weight gain (see Chapter 15).

Weight-loss diets are also not recommended during pregnancy. Even women who are overweight or obese at the start of pregnancy should gain at a slow, steady rate over the course of pregnancy. A weight-loss program can be initiated after the baby is born and the mother has recovered (see Chapter 14). Slow weight loss is appropriate during lactation, but rapid weight loss can decrease milk production.

In older adults a few extra pounds may provide a reserve that is beneficial in the event of a long-term illness. In addition, the risks associated with excess body fat are lower for older adults than they are for younger adults.[62] However, the decision to treat obesity should not be based on age alone. Weight loss can enhance day-to-day functioning and improve cardiovascular disease risk factors at all ages.[37] Older people tend to lose muscle and replace it with fat; therefore, weight-training activities are an important part of a weight-loss program in the elderly (see Chapter 16).

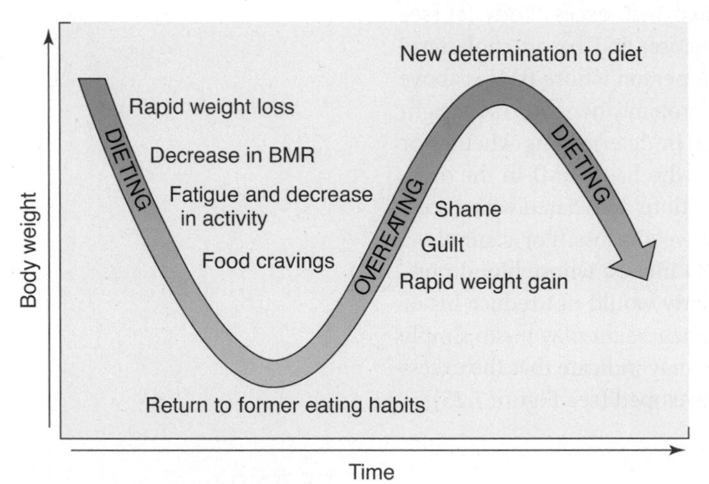

Figure 7.26 Weight cycling Drastic reductions in food intake lead to rapid weight loss but also cause a drop in BMR and may result reduced activity and increased food cravings. These make it difficult to maintain the weight loss and contribute to weight cycling.

Weight-Loss Goals and Guidelines

The medical goal for weight loss in an overweight person is to reduce the health risks associated with being overweight. For most people, a loss of 5 to 15% of body weight will significantly reduce disease risk. The initial goal of weight loss should therefore be to reduce body weight by about 10% over a period of about six months.[30] A slow loss of 10% of body weight is considered achievable for most people. After this initial weight loss, the person's health risks can be reassessed to determine if additional weight loss would be beneficial. To ensure that most of what is lost is fat and not lean tissue, the weight should be lost slowly at a rate of between 0.5 and 2 lbs per week. If weight is lost more rapidly, the loss is less likely to be maintained, and the additional loss will be from fluid, glycogen, and muscle protein. Most people who lose large amounts of weight or lose weight rapidly eventually regain all that they have lost. Repeated cycles of weight loss and regain, referred to as **weight cycling** or **yo-yo dieting**, decrease the likelihood that future attempts at weight loss will be successful (**Figure 7.26**).[63]

weight cycling or **yo-yo dieting** The repeated loss and regain of body weight.

Losing weight requires tipping the energy balance scale: eating less or exercising more. A pound of body fat provides about 3500 kcals. Therefore, to lose a pound of fat, you need to decrease your intake or increase your expenditure by this amount. To lose a pound in a week, you would need to tip your energy balance by about 500 kcals/day. This could mean adding 500 kcals of exercise, subtracting 500 kcals from your food intake, or some combination of the two. The arithmetic is simple, but achieving and maintaining weight loss is not easy. As discussed previously, there are regulatory mechanisms at work to keep body weight stable and there are environmental and emotional motivators to increase food intake and reduce the inclination to exercise.

An Ideal Weight-Loss Program

An ideal weight-loss program involves reducing energy intake, increasing activity, and changing behavior to promote weight loss and long-term weight management (**Table 7.6**).

Decreasing Energy Intake In order to promote weight loss without compromising nutrient intake, an individual's diet must be low in energy but provide all the nutrients the body needs. Nutrient density becomes more important as energy intake is reduced.

TABLE 7.6 How to Choose: Weight Management

Balance your intake and output
• Know your calorie needs and monitor what you eat.
• Weigh yourself once a week; if the number goes up, cut down your calories.
• When you add dessert, add extra exercise.
• Watch your alcohol consumption, and count the calories in alcoholic beverages.
Cut down on calories
• Replace your sugar-sweetened soft drink with a glass of water with lemon.
• Have a plain burger, not one with a special sauce or an extra-large patty.
• Portion your chips rather than eating right from the bag.
• Bring your own lunch rather than eating out.
• Don't supersize—choose a small drink and a small order of fries.
• When you eat out, share an entree with a friend or take some home for lunch the next day.
Don't get too hungry
• Eat breakfast—you'll eat less later in the day.
• Fill up on high-fiber foods.
• Increase your servings of low-calorie vegetables.
• Choose nutrient-dense snacks, such as cut-up veggies and fruit.
Increase activity
• Go for a bike ride.
• Try bowling or miniature golf instead of watching TV on Friday nights.
• Take a walk during your lunch break or after dinner.
• Play tennis; you don't have to be good to get plenty of exercise.
• Get off the bus one stop early to increase the distance you walk.

The Dietary Guidelines suggest reducing empty calories from added sugars, solid fats, and alcohol.[5] Even when choosing nutrient-dense foods, it is difficult to meet nutrient needs with intakes of less than 1200 kcals/day; therefore dieters consuming less than this should take a multivitamin and mineral supplement. Medical supervision is recommended if intake is 800 kcals per day or less.

Increasing Physical Activity Physical activity is an important component of any weight-management program. Exercise promotes fat loss and weight maintenance. It increases energy expenditure, so if intake remains the same, energy stored as fat is used for fuel. An increase in activity of 200 kcals five times a week will result in the loss of 1 lb in about 3½ weeks. In addition to increasing energy expenditure, exercise also promotes muscle development. This is important for promoting weight loss because muscle is metabolically active tissue. Increasing muscle mass helps to prevent the drop in metabolic rate that occurs as body weight decreases. In addition to promoting and maintaining weight loss, physical activity improves overall fitness and relieves boredom and stress.

To promote health and prevent gradual weight gain, adults should engage in 150 minutes of moderate activity per week.[5,64] Some individuals may need the equivalent of 60 minutes of moderate exercise per day to achieve and maintain a healthy weight. Data that track people who have lost weight and maintained the loss for over five years indicate that they exercise, on average, about an hour a day.[65] Exercise guidelines and the benefits of exercise are discussed further in Chapter 13.

Modifying Behavior People tend to think of weight loss as something they can accomplish by going on a diet. When the weight is lost, they go off the diet. The problem is that when their eating patterns return to what they were previously, they regain the weight. This "on a diet, off a diet" pattern may allow you to look good for spring break, but it's not what is needed for long-term weight management. To manage your weight at a healthy level, you need to establish a pattern of food intake and exercise that allows you to enjoy foods and activities you like without your weight climbing. It should be a pattern that you can comfortably adopt for life.

behavior modification A process used to gradually and permanently change habitual behaviors.

Changing food consumption and exercise patterns requires identifying the patterns that led to weight gain and replacing them with new ones to promote and maintain weight loss. This can be accomplished through a process called **behavior modification**, which is based on the theory that behaviors involve (1) antecedents or cues that lead to the behavior, (2) the behavior itself, and (3) consequences of the behavior. These are referred to as the ABCs of behavior modification.

The first step in a behavior modification program is to identify cues that lead to eating. You can do this by keeping a log of everything you eat or drink, where you were when you ate, what else you were doing at the time, and what motivated you to eat at that time. Then by analyzing this log, you can see what prompted you to eat excessive amounts or high-calorie foods. For instance, sitting in front of the television and mindlessly demolishing a bag of potato chips may cause you to overeat and then feel bad because you consumed the extra calories (**Figure 7.27**). In this case, the antecedent is watching TV, the behavior is mindlessly eating the chips, and the consequence is feeling remorse and gaining weight. The key to modifying this behavior is to recognize the antecedent, change the behavior, and replace the negative consequence with a positive one. In this example, not taking food with you to the television, or taking only the portion of food you want to consume, eliminates the antecedent and the behavior. The consequence is that you have consumed only the food you planned, you do not gain weight, and you feel a sense of accomplishment. Applying behavior modification techniques to change eating behaviors has been shown to improve long-term weight maintenance.

Managing America's Weight

Although successful weight management ultimately depends on an individual's choices, changes in what has been termed our obesogenic environment can reduce obesity by helping all Americans improve their food choices, reduce serving sizes, and increase their physical activity.[66] Food manufacturers and restaurants can help us cut calories by offering healthier foods and packaging or serving foods in smaller portions. Communities can help increase activity by providing parks, bike paths, and other recreational facilities for people of all ages. Businesses and schools can contribute by offering more opportunities for physical activity at the workplace and during the school day.

Even small changes, if they are consistent, can arrest the increase in obesity in the population. It has been estimated that a population-wide shift in energy balance of only 100 kcals a day, the equivalent of walking a mile or cutting out a scoop of ice cream, would prevent further weight gain in the majority of the population[67] (see **Critical Thinking: Balancing Energy: Genetics and Lifestyle**).

Figure 7.27 Behavior modification

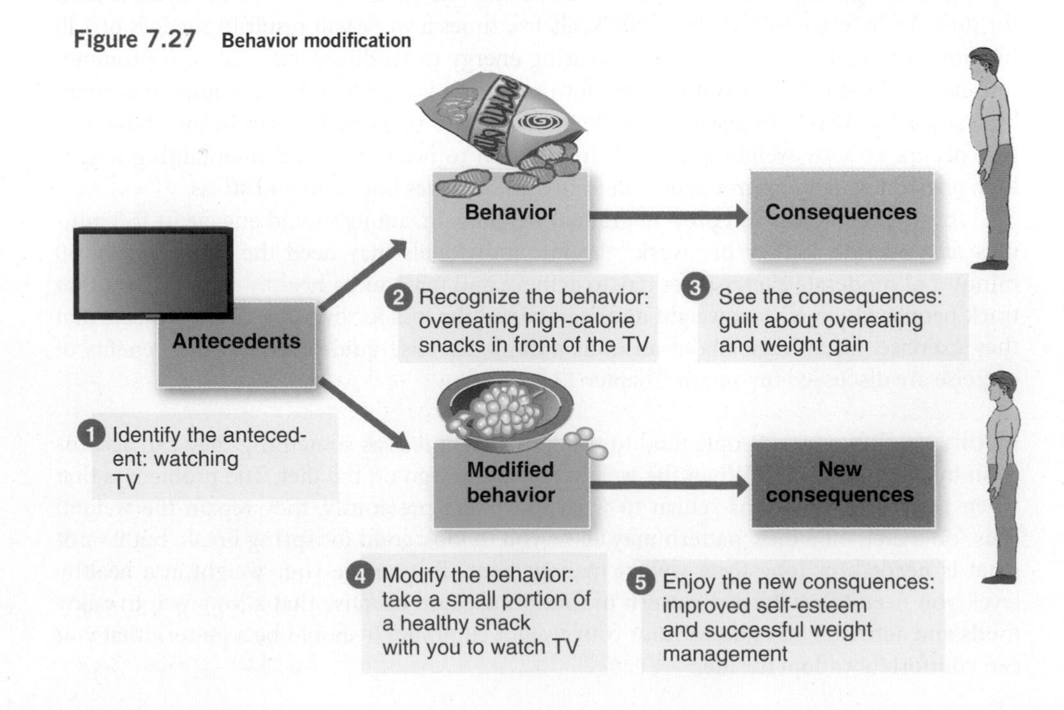

CRITICAL THINKING Balancing Energy: Genetics and Lifestyle

Tim Mantoani/Masterfile

Aysha was always slightly overweight. Because her parents are both obese, no one was surprised. During her freshman year at college, she noticed how her thinner friends ate and spent their free time. She decided to make some changes.

Aysha, now 23 years old, is 5 feet 4 inches tall and weighs 155 lbs.

By recording and analyzing her food intake for 3 days, Aysha determines that her typical intake is 2450 kcals/day. By keeping an activity log, she estimates that she gets 30 minutes of moderate-intensity activity per day. This puts her activity level in the "low-active" category.

Aysha has heard that changing her energy balance by even 100 kcals a day can help manage weight. She lists some of her high-calorie choices.

FOOD	ENERGY (KCALS)
Glazed doughnut twist, 5 inches	490
Potato chips, 2 oz	300
KFC chicken thigh, extra crispy	380
Tombstone 4-Meat pizza, 1 slice	400
Peanut butter chocolate granola bar	190
Breyers Reese's Ice Cream, 1 cup	340
Snickers, 4 oz	543

To increase her activity to about an hour a day, she begins walking with her roommate three days a week and enrolls in an aerobic class two nights a week.

As Aysha begins to lose weight, she wonders if her weight problem is due to the genes she inherited or her lifestyle.

CRITICAL THINKING QUESTIONS

- What is her BMI? Is it in the healthy range?
- How does Aysha's EER compare with her intake? Is she in energy balance?
- Suggest an alternative for each food from the same food group that would provide 100 kcals less.
- If she maintains the higher activity level and reduced intake, how long will it take her to lose 10 pounds (assume 3500 kcals per pound)?
- What would you tell her?

iProfile Use iProfile to find a snack for her that would provide less than 100 kcals.

Suggestions for Weight Gain

As difficult as weight loss is for some people, weight gain can be equally elusive for underweight individuals. The first step toward weight gain is a clinical evaluation to rule out medical reasons for low body weight. This is particularly important when weight loss occurs unexpectedly. If the low body weight is due to low intake or high expenditure, gradually increasing consumption of energy-dense foods is suggested. More frequent meals and high-calorie snacks such as nuts, peanut butter, or milkshakes between meals can help increase energy intake. Replacing low-calorie fluids like water and diet beverages with fruit juices and milk may also help. To encourage a gain in muscle rather than fat, strength-training exercise should be a component of any weight-gain program. This approach requires extra calories to fuel the activity needed to build muscles. These recommendations apply to individuals who are naturally thin and have trouble gaining weight on the recommended energy intake. This dietary approach may not promote weight gain for those who limit intake because of an eating disorder (see **Focus on Eating Disorders**).

7.8 Approaches to Weight Loss

LEARNING OBJECTIVES

- Compare the benefits of an exchange-type diet with one that relies on prepared meals and drinks.
- Describe how the proportions of carbohydrate and fat relate to the effectiveness of a weight-loss diet.
- Explain how weight-loss drugs and surgery can promote weight loss.
- Discuss when weight-loss drugs or surgery might be considered appropriate.

Anyone who has battled a weight problem has probably dreamed of a solution that would melt away the pounds without requiring him or her to measure every morsel that passes his or her lips. People desperate to lose weight are prey to all sorts of diets and products that promise quick fixes. They willingly eat a single food for days at a time, select foods based on special fat-burning qualities, and consume odd combinations at specific times of the day. Most diets, no matter how outlandish, do promote weight loss, at least in the short term, because they reduce energy intake. When weight-loss diets fail, medications and surgery can be acceptable weight-management tools for high-risk individuals. The true test of the effectiveness of a weight-loss diet, drug, or surgical procedure is whether those who use it can safely maintain their weight loss over the long term.

Methods to Reduce Energy Intake

An ideal diet is one that is part of a weight-management program that promotes weight loss and then maintenance of that loss over the long term. To be successful, the program needs to encourage changes in the lifestyle patterns that led to weight gain. When selecting a program, look for one that is based on sound nutrition and exercise principles; suits your individual food preferences; promotes long-term lifestyle changes; and meets your needs in terms of cost, convenience, and time commitment (**Table 7.7**). While quick fixes are tempting, if the program's approach is not one that can be followed for a lifetime, it is unlikely to promote successful weight management.

The following sections discuss some of the more common methods for reducing calorie intake. The advantages and disadvantages of a number of popular diets and commercial weight-management programs are given in **Table 7.8**.

Food Exchanges Diets that are based on exchanges instruct the dieter to select a certain number of servings from specific food groups in order to limit calorie intake while providing an adequate balance of nutrients. For example, for breakfast the dieter might be instructed to choose one food from group A, one from group B, and two from group C. The foods in these groups or exchanges are similar in their energy and nutrient content, so they can be exchanged for one another. Some diet plans use the Exchange Lists established by the American Diabetes Association and American Dietetic Association (see Table 7.1). Others, such as Weight Watchers, have developed their own "systems" by which to limit calorie intake. These programs offer variety from meal to meal and from day to day and if carefully planned are likely to meet nutrient needs. In addition, they teach meal-planning skills that are easy to apply away from home and can be used over the long term.

Prepared Meals and Drinks It is easier to eat less when someone else decides what you're having and puts appropriate portions of that food in front of you. This is the idea behind diet plans that sell prepackaged meals designed to replace some or all of your usual meals. These diets are easy to follow as long as you are not traveling or eating out, but they can be expensive and are not practical over the long term. Because all meals are provided, they do not teach the food-selection skills needed to make a long-term lifestyle change.

Rather than a prepackaged meal, many diet plans replace some or all meals with special beverages. They can make reducing intake easy because they eliminate the problem of

TABLE 7.7 Distinguishing between Healthy Diets and Fad Diets

A healthy diet . . .	A fad diet . . .
Promotes a healthy dietary pattern that meets nutrient needs, includes a variety of foods, suits food preferences, and can be maintained throughout life.	Limits food selections to a few food groups or promotes rituals such as eating only specific food combinations. As a result, it may be limited in certain nutrients and in variety.
Promotes a reasonable weight loss of 0.5 to 2 pounds per week and does not restrict energy intake to under 1200 kcal/day.	Promotes rapid weight loss of much more than 2 pounds per week.
Promotes or includes physical activity.	Advertises weight loss without the need to exercise.
Is flexible enough to be followed when eating out and includes foods that are easily obtained.	May require a rigid menu or avoidance of certain foods or may include "magic" foods that promise to burn fat or speed up metabolism.
Does not require costly supplements.	May require the purchase of special foods, weight-loss patches, expensive supplements, creams, or other products.
Promotes a change in behavior. Teaches new eating habits. Provides social support.	Does not recommend changes in activity and eating habits, recommends an eating pattern that is difficult to follow for life, or provides no support other than a book that must be purchased.
Is based on sound scientific principles and may include monitoring by qualified health professionals.	Makes outlandish and unscientific claims, does not support claims that it is clinically tested or scientifically proven, claims that it is new and improved or based on some new scientific discovery, or relies on testimonials from celebrities or connects the diet to trendy places such as Beverly Hills.

TABLE 7.8 Pros and Cons of Some Commercial Weight-Loss Diets

Diet	Approach	Pros	Cons
Weight Watchers	Low calorie, social support	Safe, inexpensive, flexible	Requires group participation for optimal results
Jenny Craig	Low calorie	Safe, convenient	Expensive; relies on purchase of special foods
SlimFast	Low calorie	Safe	Does not promote long-term behavior change
The New Beverly Hills Diet	Specific timing and combinations of foods	Inexpensive	Based on unusual principles; does not promote long-term behavior change; nutritionally unsound
Optifast	Very-low-calorie formula	Rapid weight loss	Expensive; dangerous if does not include medical supervision
Fit or Fat	Increased exercise	Safe, inexpensive	No social support
The Zone (and Mastering the Zone) Diet	Low carbohydrate (40% of energy)	Inexpensive, flexible	Based on questionable principles; no social support
Eating Thin for Life	Moderation—written as weight-loss success stories, recipes, and menu ideas	Inexpensive	No social support
Dieting with the Duchess	Simple nutrition and exercise tips	Inexpensive, flexible	No social support
Cabbage soup diet	Unlimited amounts of cabbage soup, fruit, coffee, and tea	Rapid weight loss	No social support; does not promote long-term behavior change; lacks variety
Grapefruit diet	Some foods have special qualities that burn fat	Inexpensive	Based on unsound principles
Sugar Busters	Eliminates sugar; low calorie—1200 kcals/day	Inexpensive	No social support; based on unsound principles
Volumetrics Weight Control Plan	Emphasizes foods high in water, fiber, and air to promote fullness with few calories	Safe, inexpensive	No social support or exercise component
Atkins' Diet	Very low carbohydrate	Inexpensive; rapid initial weight loss	Difficult to follow in the long term; no social support
South Beach Diet	Initially very low carbohydrate; then more healthy carbohydrates are allowed	Safe, inexpensive, heart-healthy	Initial weight loss is mostly water; no social support

Figure 7.28 **Preportioned meals and beverages**
Frozen prepared meals, meal-replacement bars, and drinks are popular with dieters because the food is preportioned and no decisions or measuring is required.

choosing appropriate portions of low-calorie foods. Many of the liquid weight-loss diets that are available over the counter recommend a combination of food and the liquid formula to provide about 800 to 1200 kcals/day. These formula plans promote weight loss as long as the foods eaten with them are low in calories. They are easy to use and relatively inexpensive, but they do little to change eating habits for life (**Figure 7.28**). Most diet programs that rely exclusively on liquid formulas have high dropout rates and poor long-term weight-maintenance results. These are not recommended without medical supervision.

very-low-calorie diet A weight-loss diet that provides fewer than 800 kcals/day.

protein-sparing modified fast A very-low-calorie diet with a high proportion of protein, designed to maximize the loss of fat and minimize the loss of protein from the body.

Very-Low-Calorie Diets A **very-low-calorie diet** is defined as one that provides fewer than 800 kcals/day, usually by replacing all food with liquid shakes or meal-replacement bars. These are usually medically supervised programs administered through hospitals or clinics. One type of very-low-calorie diet is the **protein-sparing modified fast**, which provides a high proportion of protein but little energy. The concept behind this is that the protein in the diet will be used to meet the body's protein needs and will, therefore, prevent excessive loss of body protein.

Very-low-calorie diets will cause rapid weight loss; initial weight loss is 3 to 5 lbs per week. This can provide a psychological boost and motivate the dieter to continue losing weight; however, in most cases, much of this initial weight loss is from water loss. Once the initial water loss ends, weight loss slows. The dieter's basal metabolism slows to conserve energy, and physical activity decreases because the dieter often does not have the energy to continue his or her typical level of activity.

Very-low-calorie diets are no more effective than other methods of weight loss in the long term and carry more risks. At these low-energy intakes, body protein is broken down and potassium is excreted. Depletion of potassium can result in an irregular heartbeat and is potentially deadly. Other side effects include gallstones, fatigue, nausea, cold intolerance, light-headedness, nervousness, constipation or diarrhea, anemia, hair loss, dry skin, and menstrual irregularities. These diets are not recommended for people with a BMI less than 30 kg/m^2, children, adolescents, pregnant or breast-feeding women, or those with severe medical problems.[68] Since 1984, the FDA has required that all very-low-calorie diet formulas carry a warning that they can cause serious illness and should be used only under medical supervision.

Low-Fat Diets Low-fat weight-loss diets have been popular for decades. Fat is high in energy: 9 kcals/g—almost twice as much as either carbohydrate or protein. Low-fat

Figure 7.29 **Low-fat vs low-carbohydrate meals**
Low-fat diets can include high-carbohydrate foods like pasta and potatoes, but quantities must be limited to avoid exceeding calorie limits. Low-carbohydrate diets exclude all grains and starchy vegetables, but do not mandate limiting calories. Which diet is more effective in the long term depends on an individual's preferences and ability to comply long term.

THINK CRITICALLY

The meal on the left contains 500 kcals, while the one of the right contains only 350 kcals. Why might the dieter eating the meal on the right be satisfied with fewer calories?

diets therefore tend to provide a greater volume of food for less energy than a diet with more fat. Because people have a tendency to eat a certain weight or volume of food, if that food is low in fat it will contribute fewer calories. Low-fat diets that are high in unrefined carbohydrates may satisfy hunger after less energy is consumed. Differences in the way dietary fat and dietary carbohydrate are used by the body also explain why low-fat diets are effective for weight loss. Excess calories from dietary fat are stored more efficiently than excess calories from carbohydrate, so consuming excess energy from fat leads to a greater accumulation of body fat than consuming excess energy as carbohydrate. As with any diet, effectiveness is related more to adherence to the diet than to the type of diet (**Figure 7.29**).

Problems with low-fat diets occur when people eat large quantities of low-fat foods without considering that these foods are not necessarily low in calories. Even a diet low in fat will result in weight gain if energy intake exceeds energy output. In the 1990s, the food industry flooded the market with reduced-fat cookies and cakes. These foods were low in fat but not in energy, so when consumed in large amounts they caused weight gain, not weight loss.

Low-Carbohydrate Diets If you cut out the pasta and potatoes you will lose weight, right? The Atkins Diet, South Beach Diet, Sugar-Busters, Calories Don't Count, Scarsdale Diet, and Zone are just a few of the low-carbohydrate weight-loss diets that have been promoted over the past 50 years. In addition to promising weight loss, these diets claim to improve athletic performance and promote overall health. Low-carbohydrate diets are all based on the premise that a high-carbohydrate intake causes an increase in insulin levels, which promotes the storage of body fat. Restricting carbohydrate intake is hypothesized to reduce insulin, thereby reducing fat storage and promoting fat loss. Unfortunately, the relationship between carbohydrate intake, insulin levels, and body fat is not that simple. Some of these diets severely limit carbohydrate intake, while others are less restrictive and are more concerned with the type of carbohydrate.

Very-Low-Carbohydrate Diets Very restrictive low-carbohydrate diets prohibit foods such as breads, grains, and fruits and limit vegetable intake while allowing unlimited quantities of meat and high-fat foods that are low in carbohydrate (see Figure 7.29). These diets cause a rapid initial weight loss, most of which is water. This occurs because when carbohydrate intake is low, glycogen stores, along with the water they hold, are lost quickly. Ketones are produced because fat is not completely broken down in the absence of carbohydrate. Excretion of these ketones in the urine causes further water loss.

Weight loss continues on low-carbohydrate diets because total energy intake is reduced. High-protein foods are satiating and elevated blood ketone levels suppress appetite, making it easier to reduce food intake. In addition, these diets limit food choices to such an extent that the monotony results in a spontaneous reduction in energy intake.[69] The availability of carbohydrate-modified products such as low-carbohydrate bread and pasta makes these diets more palatable but may also make them less effective. As with special low-fat products, low-carbohydrate foods are not necessarily low in calories and cannot be eaten liberally without affecting energy balance.

Although these diets do promote weight loss, more research is needed to determine the health consequences of consuming very-low-carbohydrate diets for long periods. They are higher in fat and protein and lower in fruits, vegetables, whole grains, and milk than is recommended. Although diets high in animal fats generally increase the risk of heart disease, improvements in blood lipids levels have been observed in people following low-carbohydrate diets for up to two years.[70] There is, however, concern that the high intake of meat may contribute to an increased risk of colon cancer.[71] Low intakes of fruits, vegetables, and whole grains could affect health by reducing intakes of essential nutrients, phytochemicals, and fiber. (See Chapter 6, Debate: Are High-Protein Diets a Safe Way to Lose Weight?)

Including Unrefined Carbohydrate Not all low-carbohydrate diets severely restrict carbohydrate, and most allow the dieter to increase carbohydrate intake over time. The carbohydrates that are included come from unrefined sources such as vegetables and whole grains. These foods are high in fiber and do not increase blood glucose and insulin levels nearly as much as refined carbohydrates such as white flour and sugar. Including whole grains and vegetables in the diet increases the intake of fiber, phytochemicals, and micronutrients and has only a modest effect on insulin levels.

Weight-Loss Drugs and Supplements

An ideal drug for the treatment of obesity would permit an individual to lose weight and maintain the loss, be safe when used for long periods of time, have no side effects, and not be addictive. Many attempts have been made to develop such a drug, but weight-loss drugs still carry risks. For example, in 1996, approximately 18 million prescriptions were written for a drug combination called fen-phen (fenfluramine and phentermine) that was being used to reduce food intake. The following year, it was linked to serious heart-valve damage. As a result, fenfluramine and the related drug dexfenfluramine were withdrawn from the market.[72]

Prescription and Nonprescription Drugs Prescription drugs available for the treatment of obesity include those that decrease food intake by affecting the activity of brain neurotransmitters (for example, phentermine, trade name Adipex, and a combination of amphetamine and phentermine, trade name Qnexa) and those that decrease energy intake by reducing fat absorption (for example, orlistat, brand name Xenical). Prescription drugs such as these are recommended only for those whose health is seriously compromised by their body weight: those with a BMI greater than 30 kg/m^2 and those with a BMI greater than or equal to 27 kg/m^2 who have obesity-related risk factors or diseases.[30] One of the major disadvantages of drug treatment is that, even if the drug promotes weight loss in the short term, the weight is regained when the drug is discontinued.

Like prescription drugs, nonprescription medications sold over the counter to promote weight loss are regulated by the FDA and must adhere to strict guidelines regarding the dose per pill and the effectiveness of the ingredients. Only a limited number of substances have been approved by the FDA for sale as nonprescription weight-loss medications. One of these is a nonprescription version of orlistat, called Alli (**Figure 7.30**). It acts by disabling the enzyme lipase, preventing triglycerides from being digested into monoglycerides, fatty acids, and glycerol. The undigested triglycerides continue through the intestines and are eliminated in the feces. This cuts the number of calories available to the body, but the fat in the stool may cause gas, diarrhea, and more frequent and hard-to-control bowel movements. As with prescription medications, any weight loss that occurs with over-the-counter weight-loss medications is usually regained when the product is no longer consumed.

Dietary Supplements There are many dietary supplements commonly used for weight loss.[72] Dieters are attracted to these supplements for a number of reasons, including

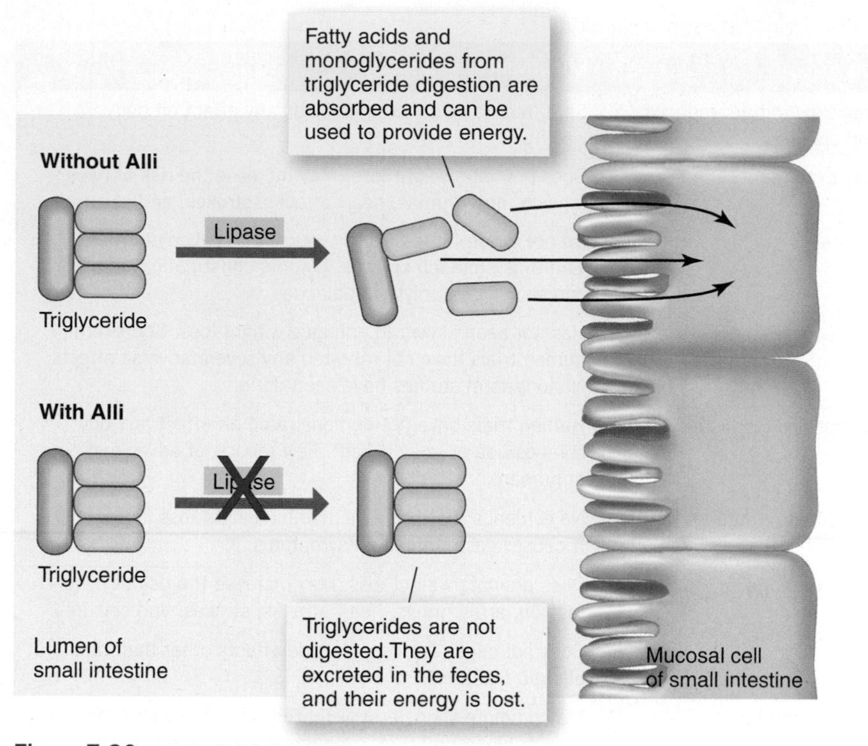

Figure 7.30 Alli and fat digestion
Alli is an over-the-counter weight-loss aid that acts by preventing triglyceride digestion.

THINK CRITICALLY

Will Alli be an effective weight-loss aid for someone who eats a low-fat diet?

cost and availability. Some see them as an alternative to conventional approaches that have failed them in the past. Many believe that since they are natural they are safer than prescription drugs. Unfortunately, it cannot be assumed that a product is safe simply because it is labeled "herbal" or "all natural." Some of these are powerful drugs with dangerous side effects. Like other dietary supplements, weight-loss supplements are not strictly regulated by the FDA, so their safety and effectiveness may not have been carefully tested. Manufacturers are not required to provide proof of their products' safety and efficacy before they are marketed. Some products claiming to be weight-loss supplements have been found to contain hidden prescription drugs or compounds that have not been adequately studied in humans.[74]

Weight-loss supplements use a variety of approaches to either reduce food intake or increase energy expenditure. Most are ineffective and a few are dangerous (**Table 7.9**). Weight-loss supplements that contain soluble fiber promise to reduce the amount you eat by absorbing water and filling up your stomach. Although they are safe, there is little evidence that they promote weight loss.[74] Hydroxycitric acid, conjugated linoleic acid, and chromium picolinate are weight-loss supplements that promise to enhance fat loss by altering metabolism so as to prevent the synthesis and deposition of fat. None of these has been shown to be effective for promoting weight loss in humans.[73, 76, 77]

Supplements that boost energy expenditure, often called "fat burners," can be effective for weight loss but have serious and potentially life-threatening side effects. One of the most popular and controversial herbal fat burners is ephedra, a stimulant that increases blood pressure and heart rate and constricts blood vessels. Use of ephedra-containing products has been associated with an increased risk of psychiatric, nervous, and gastrointestinal symptoms, as well as other serious health problems such as heart palpitations, hypertension, arrhythmias, heart attacks, strokes, and seizures.[78] Due to safety concerns, the FDA banned it in 2004. After the ban was instituted, supplement manufacturers began substituting other herbal products, such as bitter orange, that contain similar stimulants and therefore may have similar side effects.[79] Fat burners also typically contain guarana, an herbal source of caffeine. Green tea extract is another popular supplement used to boost metabolism and aid weight loss. It appears to be safe if used in appropriate amounts, but studies have not shown it to enhance weight loss.[80]

TABLE 7.9 Common Weight-Loss Supplements

Supplement	Proposed action	Effectiveness and safety
Apple cider vinegar	Increases energy expenditure, reduces hunger and food cravings	Safe, but no evidence that it has any effect on body weight.
Bitter orange	Increases energy expenditure	Does promote weight loss. May increase the risk of hypertension, arrhythmias, heart attacks, strokes, and seizure.
Cascara, senna, aloe, buckthorn, rhubarb root, and castor oil	Increase water loss	Do not cause fat loss, and overuse can cause diarrhea, vomiting, stomach cramps, chronic constipation, fainting, and severe electrolyte imbalances.
Chitosan	Blocks fat absorption	Has not been shown to enhance weight loss. Short-term human trials have not reported any severe adverse effects. No long-term studies have been done.
Chromium	Affects carbohydrate metabolism	Human trials have not demonstrated an effect on body composition or body weight. Few reports of adverse effects in humans.
Conjugated linoleic acid (CLA)	Increases fat oxidation or reduces fat synthesis	No evidence that it is effective for weight loss in humans. It causes gastrointestinal symptoms.
Country mallow	Increases energy expenditure	Does promote weight loss. May increase the risk of hypertension, arrhythmias, heart attacks, strokes, and seizure.
Dandelion	Increases water loss	Does not cause fat loss. No side effects other than rare allergic reactions.
Ephedra (ma huang, Chinese ephedra)	Increases energy expenditure	May promote short-term weight loss but increases the risk of high blood pressure, irregular heartbeat, heart attack, stroke, and death.
Ginseng	Affects carbohydrate metabolism	No evidence that it enhances weight loss. Side effects include diarrhea, headache, insomnia, changes in blood pressure, and altered bleeding time.
Glucomannan	Increases satiety	Safe, but has not been shown to promote weight loss.
Green tea	Increases fat oxidation or reduces fat synthesis	Safe, but no evidence from controlled clinical trials that tea or tea extracts promote weight loss.
Guar gum	Increases satiety	Safe, but has not been shown to promote weight loss.
Guggul	Boosts metabolism by stimulating thyroid activity	No human studies support the claim that it boosts thyroid activity. Side effects include gastrointestinal upset, headache, nausea, and hiccups.
Guarana	Suppresses appetite	Contains caffeine (about twice as much as coffee beans) and has been shown to be effective for short-term weight loss when used in combination with ephedra. Side effects include anxiety, nervousness, and difficulty sleeping.
Hoodia (kalaharii cactus, xhoba)	Suppresses appetite	Safety unknown. No reliable scientific evidence supports its effectiveness.
Hydroxycitric acid (extract from *Garcinia cambogia*)	Increases fat oxidation or reduces fat synthesis	Not found to be effective. Side effects include a laxative effect, abdominal pain, and vomiting.
L-Carnitine	Increases fat oxidation	Not found to significantly affect total body mass or fat mass. Side effects include nausea, vomiting, abdominal cramps, diarrhea, and body odor.
Licorice	Increases fat oxidation or reduces fat synthesis	Has been shown to reduce body fat in normal-weight subjects, but causes high blood pressure and low blood potassium.
Psyllium	Increases satiety	Safe, but has not been shown to promote weight loss.
Pyruvate	Increases fat oxidation or reduces fat synthesis	Some studies have shown a weight-loss benefit in individuals on weight-loss diets, but the dosage used in the studies was very high. No known side effects
Spirulina (blue-green algae)	Suppresses appetite	Safe, but no evidence that it aids weight loss.
St. John's wort	Enhances mood	No evidence it promotes weight loss. Contains similar ingredients to the antidepressant drug fluoxetine (Prozac) and should not be used by people taking antidepressants.

Several supplements cause weight loss by increasing the amount of water lost from the body—either because these supplements are diuretics or because they cause diarrhea. Water loss decreases body weight but does not cause a decrease in body fat. Herbal laxatives found in weight-loss teas and supplements include senna, aloe, buckthorn, rhubarb root, cascara, and castor oil. Overuse of these substances can have serious side effects, including diarrhea, electrolyte imbalances, and potential liver and kidney toxicity.[81]

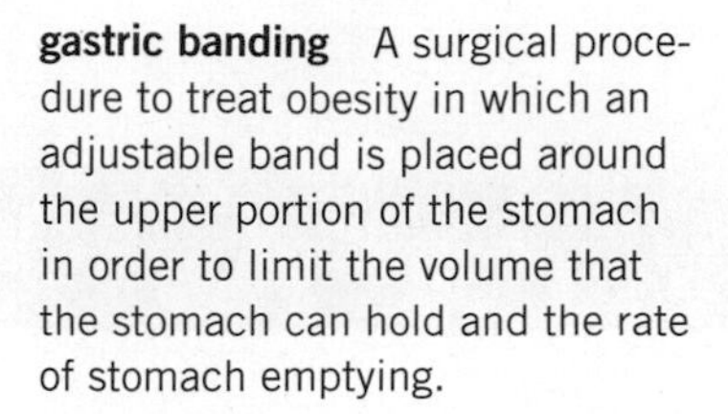

gastric banding A surgical procedure to treat obesity in which an adjustable band is placed around the upper portion of the stomach in order to limit the volume that the stomach can hold and the rate of stomach emptying.

gastric sleeve A surgical procedure to treat obesity that involves removing a large part of the stomach so the remainder resembles a tube or sleeve in order to limit the volume that the stomach can hold.

Bariatric Surgery

There are a number of surgical procedures used to promote weight loss. These are referred to as bariatric surgeries. They cause weight loss because they alter the GI tract to restrict the amount of food that can be consumed, limit the amount that can be absorbed, or both. Such surgical approaches are recommended only for individuals in whom the risk of dying from obesity and its complications is great. Generally this includes those with a BMI greater than or equal to 40 kg/m^2 (extreme obesity) and those with a BMI between 35 and 40 kg/m^2 who have other life-threatening conditions that could be remedied by weight loss. Each case must be evaluated individually to assess the potential risks and benefits of the surgery, but it is usually recommended only for those who have tried other methods and failed and whose weight severely limits their quality of life and ability to perform daily activities. To be successful, the individual must understand the procedure and its risks and be aware of how his or her life may change after the operation. Even after surgery, success requires a lifelong behavioral commitment that includes well-balanced eating and physical activity (see **Debate: Is Surgery a Good Solution to Obesity?**).

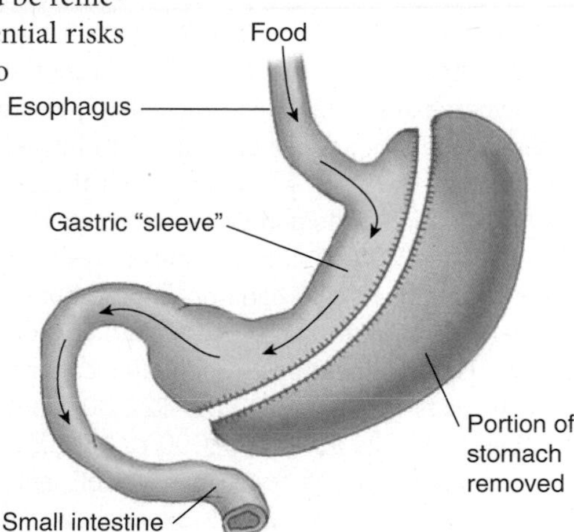

(b) Gastric sleeve surgery involves surgically removing a large part of the stomach. The procedure is not reversible and since it is relatively new the long-term risks and effectiveness are not known.

How Gastric Banding and Gastric Sleeve Surgery Promote Weight Loss **Gastric banding** and **gastric sleeve** surgery are procedures that just restrict food intake. Gastric banding uses an adjustable band to create a small pouch at the upper end of the stomach (**Figure 7.31a**). This new smaller stomach, which typically holds about 1 ounce, limits the amount of food that can be comfortably consumed at one time and slows the rate at which food leaves the stomach. Gastric sleeve surgery removes about 70 to 80% of the stomach so that it takes the shape of a tube or sleeve (**Figure 7.31b**). The size of the stomach

Figure 7.31 Types of bariatric surgery

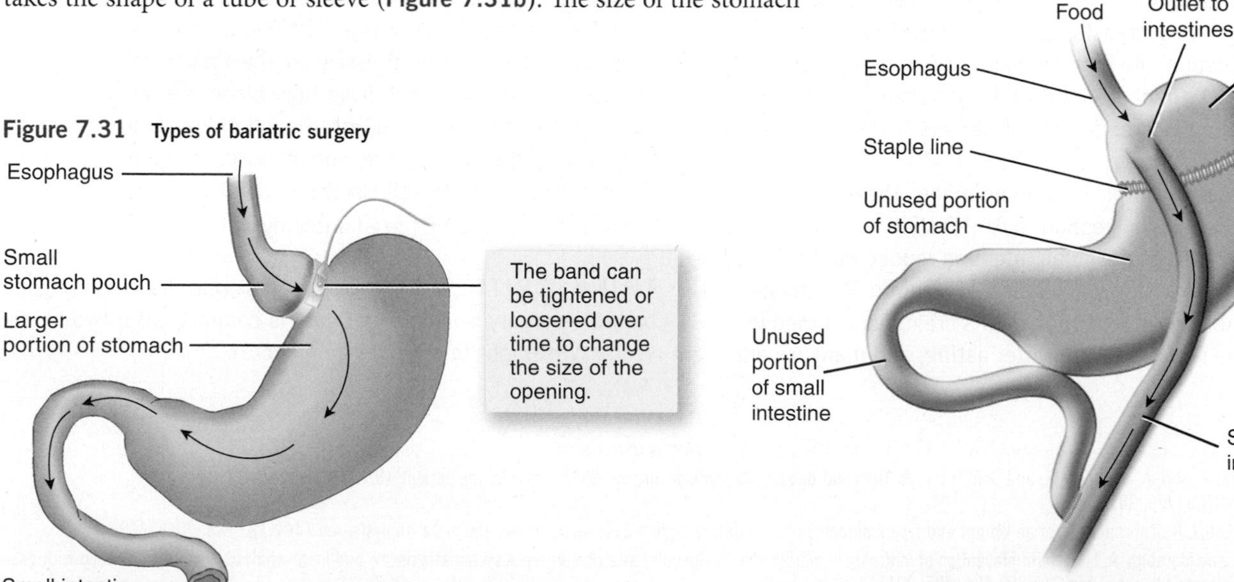

(a) Gastric banding involves surgically placing an adjustable band around the upper part of the stomach, creating a small pouch. Gastric banding entails less surgical risk and is more easily reversible than other types of weight-loss surgery.

(c) Gastric bypass surgery bypasses most of the stomach. The procedure illustrated here is the Roux-en-Y gastric bypass, which creates a stomach pouch about the size of an egg and connects it to the jejunum.

Is Surgery a Good Solution to Obesity?

The medical, social, and financial costs of obesity are well known. So is the fact that eating less and moving more will promote weight loss. However, these conventional methods do not always work. When they fail, is weight-loss surgery, known as bariatric surgery, a good weight-loss option?

Bariatric surgery is becoming increasingly common; over the past decade, there has been a 15-fold increase in the frequency of these procedures.[1] They are effective, and for some they can be lifesaving. When compared with conventional treatment, weight loss was greater at 2 years and at 10 years in those who had bariatric surgery. The incidence and control of type 2 diabetes, hypertension, high blood cholesterol, gastrointestinal reflux disease, and sleep apnea all improved with weight-loss surgery. In addition, quality of life improved in those who had surgery, and medication costs and overall mortality decreased.[2,3]

About 1 in 20 Americans is severely obese and meets the criteria for treatment with bariatric surgery, but only 0.6% of eligible patients have the surgery each year. This is because bariatric surgery is not without costs and complications.[4] Gastric bypass surgery costs about $18,000 to $22,000, and adjustable gastric banding costs about $17,000 to $30,000.[5] Complications are common; although fewer than 1% of patients die during or immediately after surgery, 20% experience adverse events.[6] The major complications, some of which require follow-up surgery, include ulcers, blockage of the opening from the stomach, leakage at the connection between the stomach and intestine, and pneumonia and blood clots. The incidence of gallstones and subsequent gallbladder removal is high because of the rapid weight loss that occurs following surgery. The reasons for the high complication rate include both the fact that anesthesia and surgery are risky in obese individuals and that there are a limited number of surgeons and surgical centers that have experience with these procedures.

In addition to surgical complications, there are long-term issues that affect digestion and lifestyle. Procedures that reduce the length of the small intestine reduce nutrient absorption and can lead to deficiencies of vitamin B_{12}, folate, calcium, and iron if dietary supplements are not consumed for life. Some people feel very sleepy after eating, and many experience chronic diarrhea, gas, foul-smelling stools, and other changes in bowel habits. People who have this surgery must be willing to commit to changes in the types and amounts of food they consume. Eating too much or eating the wrong foods may induce vomiting or cause food to dump rapidly into the small intestine, leading to symptoms such as nausea, rapid pulse, and diarrhea. There are also emotional consequences to the changes in body size and eating habits that contribute to a post–bariatric surgery divorce rate that is higher than the national average.[7]

Nick Elgar/Getty Images News and Sport Services

Gregory Pace/©Corbis

Significant weight loss is usually achieved 18 to 24 months after weight-loss surgery. NBC's *Today Show* weather anchor Al Roker lost 100 lb after undergoing gastric bypass surgery.

So is bariatric surgery a good option? It is expensive and risky, but results are impressive in the first decade after the procedures. For those who have high blood pressure, diabetes, high blood cholesterol, and arthritis, all of which are helped by weight loss, the surgery can enhance quality of life and even be lifesaving. But we still do not understand what the consequences of rearranging the GI anatomy will be in 20 or 30 years.

THINK CRITICALLY: In patients who have undergone bariatric surgery, some weight gain is common after two to five years. Why might this weight gain occur?

[1] Lautz, D., Goebel-Fabbri, A., Halperin, F., and Goldfien, A. B. The great debate: Medicine or surgery. What is best for the patient with type 2 diabetes? *Diabetes Care* 34:763–770, 2011.

[2] Buchwald, H., Estok, R., Fahrbach, K., et al. Weight and type 2 diabetes after bariatric surgery: Systematic review and meta-analysis. *Am J Med* 122:248–256, 2009.

[3] Pontiroli, A. E., and Morabito, A. Long-term prevention of mortality in morbid obesity through bariatric surgery: A systematic review and meta-analysis of trials performed with gastric banding and gastric bypass. *Ann Surg* 253:484–487, 2011.

[4] McEwen, L. N., Coelho, R. B., Baumann, L. M., et al. The cost, quality of life impact, and cost-utility of bariatric surgery in a managed care population. *Obes Surg* 20:919–928, 2010.

[5] Mann, D. Weight loss surgery insurance coverage. Consumer guide to bariatric surgery.

[6] Maggard, M. A., Shugarman, L. R., Suttorp, M., et al. Meta-analysis: Surgical treatment of obesity. *Ann Intern Med* 142:547–559, 2005.

[7] Online Surgery. Advantages and disadvantages of bariatric surgery.

is reduced and thus restricts the amount of food that can be consumed. Both these procedures can be done laparoscopically, through a series of small incisions, so that recovery time is kept to a minimum. They also both leave intact the pyloric sphincter, which regulates stomach emptying, so there are fewer complications.

How Gastric Bypass Promotes Weight Loss **Gastric bypass** causes a reduction in both food intake and nutrient absorption by bypassing part of the stomach and small intestine. The procedure bypasses the lower portion of the stomach and connects the small stomach pouch to a portion of the small intestine. The smaller stomach limits intake, and the shorter intestine reduces absorption (**Figure 7.31c**). Individuals who have this procedure, or any bariatric surgery, must make permanent changes in their eating habits and experience permanent changes in their bowel habits. Some weight regain is common after two to five years.

gastric bypass A surgical procedure to treat morbid obesity that both reduces the size of the stomach and bypasses a portion of the small intestine.

liposuction A procedure that suctions out adipose tissue from under the skin; used to decrease the size of local fat deposits such as on the abdomen or hips.

Does Liposuction Cause Weight Loss? **Liposuction** is another surgical procedure, but it is primarily a cosmetic one. It involves inserting a large hollow needle under the skin into a localized fat deposit and literally vacuuming out the fat. It is often advertised as a way to remove cellulite, which is just fat that has a lumpy appearance because of the presence of connections to the tissue layers below. The risks of liposuction include those associated with general anesthesia and the possibility of infection. The procedure can reduce the amount of fat in a specific location but will not significantly reduce overall body weight. Liposuction has not been found to affect obesity-related metabolic abnormalities, and therefore, unlike overall weight loss, it does not reduce the risk of heart disease and diabetes that is associated with excess body fat.[82]

OUTCOME

When Bethany got back to school, she enrolled in a nutrition class for the following semester. The class taught her that weight loss is not about going on a diet. She found out that her Bermuda diet worked because limiting her food choices limited her calorie intake but it was not something she could do for life. Weight management is about developing a lifestyle that shifts energy balance to maintain or lose weight. Now instead of having fruit one day and vegetables the next, Bethany makes half her plate fruits and vegetables and fills the other half with whole grains and protein foods, and drinks water or low-fat milk instead of soda. She can't eat all she wants, but because she is making nutrient-dense choices by avoiding foods with lots of fat or added sugar, she can eat enough volume to feel full. She is slowly losing weight at a rate of about a half a pound per week. **When she reaches her goal weight, she does not plan to make any changes in her diet other than to add enough calories to prevent further weight loss.**

Mike Blake/Reuters/NewsCom

Exactostock/SuperStock

APPLICATIONS

ASSESSING YOUR DIET

1. **Is your weight in the healthy range or should you lose weight?**
   a. Use Table 7.4 to determine your BMI and measure your waist circumference.
   b. If your BMI is 25 or more, use Figure 7.25 to determine whether you should lose weight. If you should lose weight, how much weight would you need to lose to reduce your weight by 10%?
   c. If your BMI is less than 18.5 kg/m^2, how much should you gain to move your BMI into the healthy range?
2. **Are you in energy balance?**
   a. Use iProfile to calculate your average daily energy intake from the three-day food record you kept in Chapter 2.
   b. Keep an activity log for several days. Use Figure 7.14 to determine your physical activity level and PA value. Calculate your EER (see Table 7.3).
   c. Compare your EER with your energy intake.
   d. If you consumed and expended this amount of energy every day, would your weight increase, decrease, or stay the same?
   e. If your intake does not equal your EER, how much would you gain or lose in a month? (Assume that 1 lb of fat is equal to 3500 kcals.)
   f. If your energy intake does not equal your EER, list some specific changes you could make in your diet or the amount of activity you get to make the two balance.
3. **Are genetic or environmental factors a larger influence on your energy balance? Answer the following questions to help you decide.**
   a. How has your weight changed over the past year?
   b. How much have your parents' weights changed since they were 21?
   c. Do you eat more servings of snack foods, such as chips and candy bars, or of fruits and vegetables daily? Has this changed over the past year?
   d. How has your activity changed over the past year?
   e. What patterns do you see emerging that can predict if your weight will change over the next year? the next 10 years? the next 20 years? What are your predictions?

CONSUMER ISSUES

4. **How useful are low-calorie products?**
   a. Go to the grocery store and select 5 to 10 products labeled "light," "reduced calorie," or "low calorie." Record the number of calories per serving for each.
   b. Use iProfile or food labels to compare the number of calories in a serving of each of the reduced-calorie foods to the number of calories in a serving of the original product.
   c. Can these foods be consumed in unlimited amounts without significantly increasing energy intake?
   d. Explain why you think each of these products would or would not be useful for someone on a weight-loss diet.
5. **Rose has gained 40 lbs over the past 8 years. Her BMI is now 31 kg/m^2. She wants to lose weight and is considering the three weight-loss programs described below.**

   Low-carbohydrate plan: This plan limits carbohydrate intake to 30 g/day—this means she cannot eat any grains, milk, or fruit. Many of her friends have used this diet to shed pounds and say they were never hungry.

   Liquid formula plan: This plan uses cans of formula to replace two meals a day. A one-week supply costs $10. It is easy because she doesn't need to do much meal planning and she can still eat dinner with her family.

   Exchange plan: This plan is run through her community center. It includes a walking group that meets five days a week and weekly meetings for weigh-ins and nutrition lectures. The cost is $25 per week.

   a. Are the diets she is considering nutritionally sound? Do they recommend activity and include social support? What about cost?
   b. Which plan do you think will be the most successful in the long term? Explain your choice.

CLINICAL CONCERNS

6. **Roger just celebrated his 40th birthday. He is about 40 lbs heavier than he was on his 30th birthday. At his current rate of weight gain, he will be 80 lbs overweight when he reaches age 50. Roger is 5' 8" tall and weighs 210 lbs. He has three young children at home and works full-time as a salesperson. His day starts at about 6 AM when he gets up and has breakfast with his wife and children. He spends most of the morning in his car traveling to visit his clients. He usually has at least one doughnut-and-coffee break. Lunch is fast food if he is alone or a restaurant meal if he is with clients. Most evenings he is home for dinner. By the time his children are in bed it is about 8 PM. He sits down with his wife for a glass of wine or a bowl of ice cream.**
   a. What is Roger's BMI? How many pounds would he need to lose to get his BMI into the healthy range?
   b. If Roger loses weight at a healthy rate, how long should it take him to get his BMI back into the healthy range?
   c. Recommend three changes Roger could make to reduce his calorie intake.
   d. Recommend an exercise program for Roger that fits his schedule.
   e. Should Roger consider surgery or drugs? Why or why not?
7. **What weight patterns and influences are present in your class?**
   a. Do a class survey by collecting everyone's answers to question 3 in the Assessing Your Diet section above. Tabulate the patterns that you see.
   b. Is weight generally increasing or decreasing?
   c. Is exercise increasing or decreasing?
   d. Which are the more popular snacks—prepackaged snack foods or fruits and vegetables?
   e. What percentage of your classmates has one parent whose weight has increased by 20 lbs or more since they were 21?
   f. What percentage of your classmates has two parents whose weight has increased by 20 lbs or more since they were 21?

SUMMARY

7.1 The Obesity Epidemic

- The rapid rise in the incidence of overweight and obesity in the United States over the past 50 years has been called an epidemic. It is also a growing problem around the world.
- More Americans are obese today because, as a population, we are taking in more calories and expending fewer. We eat more today because we are exposed to large portions of a wide variety of tasty, convenient foods that stimulate our appetite. We move less due to modern conveniences, busy lifestyles, and the availability of sedentary ways to spend our leisure time.

7.2 Exploring Energy Balance

- The principle of energy balance states that if energy intake equals energy needs, body weight will remain constant. Energy is provided to the body by the carbohydrate (4 kcals/g), fat (9 kcals/g), protein (4 kcals/g), and alcohol (7 kcals/g) in food and beverages. To power the body, these need to be broken down and their energy converted into ATP. The ATP then fuels metabolic reactions that build and maintain body components and provides energy for other cellular and body activities.
- Total energy expenditure (TEE) includes the energy required for basal metabolism, physical activity, the thermic effect of food, growth, milk production in lactating women, and maintenance of body temperature in a cold environment. The largest component of energy expenditure in most people is basal energy expenditure (BEE), which varies depending on body size, body composition, age, and gender. The energy needed for activity, which includes planned exercise and nonexercise activity thermogenesis (NEAT), typically accounts for 15 to 30% of energy expenditure but varies greatly depending on the individual. The thermic effect of food (TEF) is the energy required for the digestion of food and the absorption, metabolism, and storage of nutrients. It is equal to about 10% of energy consumed.
- When the diet contains more calories than are needed, the extra energy is stored for later use, primarily as fat. When the diet does not meet needs, body stores are used. Glucose is provided by the breakdown of glycogen stores or synthesized from amino acids by gluconeogenesis. Energy for tissues that don't require glucose is provided by the breakdown of stored fat.

7.3 Estimating Energy Requirements

- The amount of energy expended by the body can be determined by direct calorimetry, which measures the heat produced by the body, and indirect calorimetry, which measures oxygen utilization. The doubly labeled water method estimates energy expenditure by using water labeled with isotopes of hydrogen and oxygen to calculate carbon dioxide production. Current recommendations for energy needs were determined using the doubly labeled water method.
- Energy needs can be predicted by calculating Estimated Energy Requirements (EER). EER calculations take into account age, gender, life stage, height, weight, and activity level. As physical activity increases, more calories need to be consumed to maintain body weight.

7.4 Body Weight and Health

- Some body fat is essential for health, but too much increases the risk of developing chronic diseases such as diabetes, heart disease, high blood pressure, gallbladder disease, sleep apnea, arthritis, and certain types of cancer. Excess body fat also creates psychological and social problems.
- Being naturally lean decreases health risks, but if low body weight is due to starvation or eating disorders, it can affect electrolyte balance and immune function and increase the risk of early death.

7.5 Guidelines for a Healthy Body Weight

- A healthy body weight is a weight at which the risks of illness and death are lowest. These risks are associated not only with body weight but also with the amount and location of body fat. The most common way to evaluate the healthfulness of body weight is body mass index (BMI), which is calculated from a ratio of weight to height squared. In general, people with a BMI of 18.5 to 24.9 kg/m^2 have the lowest health risks.
- The amount of body fat can be assessed using techniques such as bioelectric impedance, skinfold thickness, underwater weighing, air displacement, dual-energy X-ray absorptiometry, and isotope dilution techniques.
- Assessment of a healthy body weight must consider body composition and the location of body fat. Health risks increase when too little or too much body fat is stored.
- An increase in visceral fat is associated with a greater incidence of heart disease, high blood pressure, stroke, and diabetes than an increase in subcutaneous fat. Waist measurement can be used to assess the presence of too much visceral fat.

7.6 What Determines Body Size and Shape?

- The genes people inherit determine their body shape and characteristics. Genes involved in regulating body fatness are referred to as obesity genes. Body weight is determined by the balance between the genes a person inherits and the lifestyle choices he or she makes.
- Signals from the GI tract, hormones, and levels of circulating nutrients regulate body weight in the short term by affecting hunger and satiety. Signals that relay information about the size of body fat stores, such as the release of leptin from adipocytes, regulate long-term energy intake and expenditure. Although body weight appears to be regulated around a set point, these regulatory mechanisms are much better at preventing weight loss than at defending against weight gain.
- Variations in the genes people inherit interact with one another and affect metabolic rate, food intake, fat storage, and activity level. Inheriting a thrifty metabolism and less ability to respond to an increase in calorie intake by dissipating the excess energy in metabolism or NEAT expenditure have been hypothesized to contribute to obesity.

7.7 Recommendations for Managing Body Weight

- Whether individuals need to lose weight depends on how much body fat they have, where the fat is located, and what their health risks are.
- A slow, steady weight loss of 0.5 to 2 lb per week is more likely to be maintained than rapid weight loss.
- Weight management involves adjusting energy intake and expenditure and modifying long-term behaviors. To lose a pound of fat, expenditure must exceed intake by approximately 3500 kcals.
- To reduce the incidence of obesity, Americans need to move more and eat less. This will require the involvement of food manufacturers, restaurants, schools, businesses, and communities. Even a shift in energy balance of 100 kcals a day would prevent further weight gain in most people.
- If underweight is not due to a medical condition, weight gain can be accomplished by increasing energy intake and lifting weights to increase muscle mass.

7.8 Approaches to Weight Loss

- There are thousands of diets that promise weight loss. All cause short-term weight loss because they decrease energy intake. Common methods for reducing calorie intake include the use of food exchanges, prepared meals or meal-replacement drinks, and decreasing the fat or carbohydrate content of the diet. A good weight-loss diet is one that allows a wide range of food choices, does not require the purchase and consumption of special foods or combinations of foods, and can be followed for life.
- Go to WileyPLUS to view a video clip on evaluating diets.

VIDEO

- Prescription weight-loss drugs are recommended only for individuals who are significantly overweight or have accompanying health risks. Those currently available act by suppressing appetite or blocking fat absorption. Nonprescription weight-loss medications and dietary supplements are also available. Some herbal weight-loss supplements may cause serious side effects.
- Weight-loss surgery is a drastic measure that is considered only for those whose health is seriously at risk because of their obesity. Common procedures include gastric banding, gastric sleeve, and gastric bypass. These surgeries cause changes in the GI tract that affect the amount of food that can be consumed and the absorption of nutrients. Even after surgery, weight loss requires changes in eating patterns and behavior.

REVIEW QUESTIONS

1. What changes in the U.S. environment and lifestyle have contributed to the obesity epidemic?
2. Explain how energy balance is related to body weight.
3. Which nutrients provide energy? How much does each provide?
4. What is basal metabolic rate?
5. What is NEAT? How does it affect energy balance?
6. What is the thermic effect of food?
7. Explain why the energy in dietary fat is stored in body fat more efficiently than the energy in dietary carbohydrate.
8. Describe three methods for measuring energy expenditure.
9. What is EER, and what variables are used in its calculation?
10. List five health problems that are associated with excess body fat.
11. Explain what is meant by a healthy body weight.
12. How is BMI calculated and why is it commonly used to assess body weight?
13. List five methods for determining the amount of body fat a person has.
14. How does the distribution of body fat affect the risks of excess body fat?
15. List some social and environmental factors that affect energy balance, and discuss how these might interact with an individual's genetic predisposition to a particular body weight.
16. Explain three mechanisms that make you stop eating when you have eaten enough at a meal.
17. Discuss the role of leptin in regulating body weight.
18. Explain why weight loss might be recommended for one overweight individual but not for another of the same BMI.
19. How many calories must be expended to lose a pound of fat?
20. What is the best approach to weight management? Why?
21. List some risks and benefits of weight-loss drugs.
22. How does weight-loss surgery cause negative energy balance?

REFERENCES

1. Flegal, K. M., Carroll, M. D., Kit, B. K., and Ogden, C. L., Prevalence of obesity and trends in the distribution of body mass index among US adults, 1999–2010. *JAMA* 307:491–497, 2012.
2. Ogden, C. L., Carroll, M. D., Kit, B. K., and Flegal, K. M., Prevalence of obesity and trends in body mass index among US children and adolescents, 1999–2010. *JAMA* 307:483–490, 2012.
3. Finkelstein, E. A., Trogdon, J. G., Cohen, J. W., and Dietz, W. Annual medical spending attributable to obesity: Payer- and service-specific estimates. *Health Affairs* 28:w822–w831, 2009.

4. World Health Organization. Obesity and overweight, September 2006. Fact sheet no. 311.

5. U.S. Department of Agriculture and U.S. Department of Health and Human Services. *Dietary Guidelines for Americans, 2010*, 7th ed. Washington, DC: U.S. Government Printing Office, December 2010.

6. Bassett, D. R., Schneider, P. L., and Huntington, G. E. Physical activity in an Old Order Amish community. *Med Sci Sports Exerc* 36:79–85, 2004.

7. Institute of Medicine, Food and Nutrition Board. *Dietary Reference Intakes for Energy, Carbohydrate, Fiber, Fat, Protein and Amino Acids.* Washington, DC: National Academies Press, 2002/2005.

8. Weinsier, R. L., Nagy, T. R., Hunter, G. R., et al. Do adaptive changes in metabolic rate favor weight regain in weight-reduced individuals? An examination of the set-point theory. *Am J Clin Nutr* 72:1088–1094, 2000.

9. Levine, J. A. Non-exercise activity thermogenesis (NEAT). *Best Pract Res Clin Endocrinol Metab* 16:679–702, 2002.

10. Levine, J. A., and Kotz, C. M. NEAT—non-exercise activity thermogenesis—egocentric and geocentric environmental factors vs. biological regulation. *Acta Physiol Scand* 184:309–318, 2005.

11. Kriketos, A. D., Peters, J. C., and Hill, J. O. Cellular and whole-animal energetics. In *Biochemical and Physiological Aspects of Human Nutrition*, M. H. Stipanuk, ed. Philadelphia: W. B. Saunders Company, 2000, 411–424.

12. Galgani, J., and Ravussin, E. Energy metabolism, fuel selection and body weight regulation. *Int J Obes (Lond)* 32 (suppl 7):S109-S119, 2008.

13. Cahill, G. F., Starvation in man. *N Eng J Med* 282:668–675, 1970.

14. Frayn, K. Metabolic Regulation: *A Human Perspective.* London: Portland Press, 1996, pp. 78–102.

15. Schutz, Y. Dietary fat, lipogenesis and energy balance. *Physiol Behav* 83:557–564, 2004.

16. Schoeller, D.A. Insights into energy balance from doubly labeled water. *Int J Obes* (Lond) 32(suppl 7):S72–S75, 2008.

17. Flegal, K. M., Graubard, B. I., Williamson, D. F., and Gail, M. H. Excess deaths associated with underweight, overweight, and obesity. *JAMA* 293:1861–1867, 2005.

18. NIDDK Weight Control Information Network. Understanding adult obesity.

19. NIDDK Weight Control Information Network. Do you know the risks of being overweight?

20. NIDDK Weight Control Information Network. Dieting and gallstones.

21. Lurie, A. Metabolic disorders associated with obstructive sleep apnea in adults. *Adv Cardiol* 46:67–138, 2011.

22. Fava, C., Montagnana, M., Favaloro, E. J., et al. Obstructive sleep apnea syndrome and cardiovascular diseases. *Semin Thromb Hemost* 37:280–297, 2011.

23. National Cancer Institute, Fact sheet, obesity and cancer risk.

24. Endogenous Hormones and Breast Cancer Collaborative Group, Key, T. J., Appleby, P. N., et al. Circulating sex hormones and breast cancer risk factors in postmenopausal women: Reanalysis of 13 studies. *Br J Cancer* 105:709–722, 2011.

25. La Vecchia, C., Giordano, S. H., Hortobagyi, G. N., and Chabner, B. Overweight, obesity, diabetes, and risk of breast cancer: Interlocking pieces of the puzzle. *Oncologist* 16:726–729, 2011.

26. Lementowski, P. W., and Zelicof, S. B. Obesity and osteoarthritis. *Am J Orthop* 37:148–151, 2008.

27. Gunnarsdottir, T., Njardvik, U., Olafsdottir, A. S., et al. Teasing and social rejection among obese children enrolling in family-based behavioural treatment: Effects on psychological adjustment and academic competencies. *Int J Obes* (Lond) 36:35–44, 2012.

28. Vander Wal, J. S., and Mitchell, E. R. Psychological complications of pediatric obesity. *Pediatr Clin North Am* 58:1393–1401, 2011.

29. Faith, M. S., Butryn, M., Wadden, T. A., et al. Evidence for prospective associations among depression and obesity in population-based studies. *Obes Rev* 12:e438–e453, 2011.

30. National Institutes of Health. NHLBI. *The Practical Guide: Identification, Evaluation and Treatment of Overweight and Obesity in Adults.* NIH Publication no. 02-4084. Bethesda, MD: National Institutes of Health, 2002.

31. Gallagher, D., Heymsfield, S., Heo, M., et al. Healthy percentage body fat ranges: An approach for developing guidelines based on body mass index. *Am J Clin Nut* 72: 694–701, 2000.

32. Evans, W. J. Protein nutrition, exercise and aging. *J Am Coll Nut* 23:601S–609S, 2004.

33. Fields, D. A., Hunter, G. R., and Goran. M. I. Validation of the BOD POD with hydrostatic weighing: Influence of body clothing. *Int J Obes Relat Metab Disord* 24:200–205, 2000.

34. Andreoli, A., Scalzo, G., Masala, S., et al. Body composition assessment by dual-energy X-ray absorptiometry (DXA). *Radiol Med* 114:286–300, 2009.

35. Kaul, S., Rothney, M. P., Peters, D. M., et al. Dual-energy X-ray absorptiometry for quantification of visceral fat. *Obesity* (Silver Spring) 20:1313–1318, 2012.

36. Redinger, R. N. The physiology of adiposity. *J Ky Med Assoc* 106:53–62, 2008.

37. Hamdy, O., Porramatikul, S., and Al-Ozairi E. Metabolic obesity: The paradox between visceral and subcutaneous fat. *Curr Diabetes Rev* 2:367–373, 2006.

38. Bouchard, C., Tremblay, A., Després, J.-P., et al. The response to long term feeding in identical twins. *N Eng J Med* 322:1477–1482, 1990.

39. Hoffman, D. J., Wang, Z., Gallagher, D., and Heymsfield, S. B. Comparison of visceral adipose tissue mass in adult African Americans and whites. *Obes Res* 13:66–74, 2005.

40. Rankinen, T., Zuberi, A., Chagnon, Y. C., et al. The human obesity gene map: The 2005 update. *Obesity* (Silver Spring) 14:529–644, 2006.

41. Bouchard, C. Genetics of human obesity: Recent results from linkage studies. *J Nutr* 127:1887S–1890S, 1997.

42. Wardle, J., Carnell, S., Haworth, C. M., and Plomin, R. Evidence for a strong genetic influence on childhood adiposity despite the force of the obesogenic environment. *Am J Clin Nutr* 87:398–404, 2008.

43. Ravussin, E., Valencia, M. E., Esparza, J., et al. Effects of a traditional lifestyle on obesity in Pima Indians. *Diabetes Care* 17:1067–1074, 1994.

44. Norman, R. A., Thompson, D. B., Foroud, T., et al. Genomewide search for genes influencing percent body fat in Pima Indians: Suggestive linkage at chromosome 11q21–q22. *Am J Hum Genet* 60:166–173, 1997.

45. Esparza, J., Fox, C., Harper, I. T., et al. Daily energy expenditure in Mexican and USA Pima Indians: Low physical activity as a possible cause of obesity. *Int J Obes Relat Metab Disord* 24:55–59, 2000.

46. Major, G. C., Doucet, E., Trayhurn, P., et al. Clinical significance of adaptive thermogenesis. *Int J Obes* (Lond.) 31:204–212, 2007.

47. Delzenne, N., Blundell, J., Brouns, F., et al. Gastrointestinal targets of appetite regulation in humans. *Obes Rev* 11:234–250, 2010.

48. Smith, G. P. Controls of food intake. In *Modern Nutrition in Health and Disease*, 10th ed. M. E. Shils, M. Shike, A. C. Ross, et al., eds. Philadelphia: Lippincott Williams & Wilkins, 2006, pp. 751–770.

49. Adams, C. E., Greenway, F. L., and Brantley, P. J. Lifestyle factors and ghrelin: critical review and implications for weight loss maintenance. *Obes Rev* 12:e211–e218, 2011.

50. Moran, T. H. Gut peptides in the control of food intake. *Int J Obes* (Lond) 33(suppl 1):S7–S10, 2009.

51. Shan, X., and Yeo, G. S. Central leptin and ghrelin signalling: Comparing and contrasting their mechanisms of action in the brain. *Rev Endocr Metab Disord* 12:197–209, 2011.

52. Belgardt, B. F., and Brüning, J. C. CNS leptin and insulin action in the control of energy homeostasis. *Ann N Y Acad Sci* 1212:97–113, 2010.

53. Myers, M. G. Jr., Leibel, R. L., Seeley, R. J., and Schwartz, M. W. Obesity and leptin resistance: Distinguishing cause from effect. *Trends Endocrinol Metab* 21:643–651, 2010.

54. Peters, J. C. Control of energy balance. In *Biochemical and Physiological Aspects of Human Nutrition*, 2nd ed. M. H. Stipanuk, ed. Philadelphia: W. B. Saunders, 2006, pp. 618–639.

55. Tsigos, C., Kyrou, I., and Raptis, S. A. Monogenic forms of obesity and diabetes mellitus. *J Pediatr Endocrinol Metab* 15:241–253, 2002.

56. Tremblay, A., Després, J.-P., Thriault, G., et al. Overfeeding and energy expenditure in humans. *Am J Clin Nutr* 56:857–862, 1992.

57. Diaz, E. O., Prentice, A. M., Goldberg, G. R., et al. Metabolic response to experimental overfeeding in lean and overweight healthy volunteers. *Am J Clin Nutr* 56:641–655, 1992.

58. Rosenbaum, M., and Leibel, R. L. Adaptive thermogenesis in humans. *Int J Obes* (Lond) 34:S47–S55, 2010.

59. van Marken Lichtenbelt, W. D., Vanhommerig, J. W., Smulders, N. M., et al. Cold-activated brown adipose tissue in healthy men. *N Engl J Med* 360:1500–1508, 2009.

60. Levine, J. A., Lanningham-Foster L. M, McCrady S. K., et al. Interindividual variation in posture allocation: Possible role in human obesity. *Science* 307:584–586, 2005.

61. Levine, J. A., Eberhardt, N. L., and Jensen, M. D. Role of non-exercise activity thermogenesis in resistance to fat gain in humans. *Science* 283:212–214, 1999.

62. Shea, M. K., Nicklas, B. J., Houston, D. K., et al. The effect of intentional weight loss on all-cause mortality in older adults: Results of a randomized controlled weight-loss trial. *Am J Clin Nutr* 94:839–846, 2011.

63. Kroke, A., Liese, A. D., Schulz, M., et al. Recent weight changes and weight cycling as predictors of subsequent two-year weight change in a middle-aged cohort. *Int J Obes Relat Metab Disord* 26:403–409, 2002.

64. United States Department of Health and Human Services. 2008 physical activity guidelines for Americans.

65. The National Weight Control Registry. NWCR facts.

66. Office of the Surgeon General. The surgeon general's call to action to prevent and decrease overweight and obesity. U.S. Department of Health and Human Services. Rockville, MD, 2001.

67. Hill, J. O. Can a small-changes approach help address the obesity epidemic? A report of the Joint Task Force of the American Society for Nutrition, Institute of Food Technologists, and International Food Information Council. *Am J Clin Nutr* 89:477–484, 2009.

68. American Dietetic Association. Position of the American Dietetic Association: Weight management. *J Am Diet Assoc* 109:330–346, 2009.

69. Westman, E. C., Feinman, R. D., Mavropoulos, J. C., et al. Low-carbohydrate nutrition and metabolism. *Am J Clin Nutr* 86:276–284, 2007.

70. Shai, I., Schwarzfuchs, D., Henkin, Y., et al. Weight loss with a low-carbohydrate, Mediterranean, or low-fat diet. *N Engl J Med* 359:229–241, 2008.

71. Russell, W. R., Gratz, S. W., Duncan, S. H., et al. High-protein, reduced-carbohydrate weight-loss diets promote metabolite profiles likely to be detrimental to colonic health. *Am J Clin Nutr* 93:1062–1072, 2011.

72. Frackelmann, K. Diet drug debacle: How two federally approved weight-loss drugs crashed. *Science News* 152:252–253, 1997.

73. Saper, R. B., Eisenberg, D. M., and Phillips, R. S. Common dietary supplements for weight loss. *Am Fam Physician* 70:1731–1738, 2004.

74. U.S. Food and Drug Administration. Beware of fraudulent weight-loss "dietary supplements."

75. Beck, E. J., Tapsell, L. C., Batterham, M. J., et al. Oat beta-glucan supplementation does not enhance the effectiveness of an energy-restricted diet in overweight women. *Br J Nutr* 103:1212–1222, 2010.

76. Li, J. J., Huang, C. J., and Xie, D. Anti-obesity effects of conjugated linoleic acid, docosahexaenoic acid, and eicosapentaenoic acid. *Mol Nutr Food Res* 52:631–645, 2008.

77. Yazaki, Y., Faridi, Z., Ma, Y., et al. A pilot study of chromium picolinate for weight loss. *J Altern Complement Med* 16:291–299, 2010.

78. Shekelle, P. G., Hardy, M. L., Morton, S. C., et al. Efficacy and safety of ephedra and ephedrine for weight loss and athletic performance: A meta-analysis. *JAMA* 289:1537–1545, 2003.

79. Hess, A. M., and Sullivan, D. L. Potential for toxicity with use of bitter orange extract and guarana for weight loss. *Ann Pharmacother* 39:574–575, 2005.

80. Sarma, D. N., Barrett, M. L., Chavez, M. L., et al. Safety of green tea extracts: a systematic review by the U.S. Pharmacopeia. *Drug Saf* 31:469–484, 2008.

81. Chan, T. Y. Potential risks associated with the use of herbal anti-obesity products. *Drug Saf* 32:453–456, 2009.

82. Klein, S., Fontana, L., Young, V. L., et al. Absence of an effect of liposuction on insulin action and risk factors for coronary heart disease. *N Engl J Med* 350:2549–2557, 2004.

***To access links to online sources, please go to** www.wiley.com/college/smolin **and select Nutrition: Science and Applications, 3rd edition. From this page, select either the student or instructor companion site. Once on the desired site, select References.**

Focus on Eating Disorders

© DNY59/iStockphoto

CHAPTER OUTLINE

Normal eating patterns are flexible. One day you may eat twice as much as on another. On some days your food choices are varied and nutritious, while on others you may survive on snacks and fast food. Lunch one day may include an appetizer, entree, and dessert, and on the next it may be a carton of yogurt grabbed on the run. Normal eating patterns include eating more than we need at a party or other special occasion and less than we need when we are busy or stressed. Normal eating may also involve limiting intake in order to manage weight and meet recommendations for a healthy diet.

What and how much people eat vary in response to social occasions, emotions, time limitations, hunger, and the availability of food, but generally people eat when they are hungry, choose foods they enjoy, and stop eating when they are satisfied. Abnormal eating occurs when a person is overly concerned with food, eating, and body size and shape. **When the emotional aspects of food and eating overpower the role of food as nourishment an eating disorder may develop.**

eating disorder A persistent disturbance in eating behavior or other behaviors intended to control weight that affects physical health and psychosocial functioning.

anorexia nervosa An eating disorder characterized by self-starvation, a distorted body image, and below-normal body weight.

bulimia nervosa An eating disorder characterized by the consumption of large amounts of food at one time (binge eating), followed by purging behaviors such as vomiting or the use of laxatives to eliminate calories from the body.

bingeing or **binge eating** The rapid consumption of a large amount of food in a discrete period of time associated with a feeling that eating is out of control.

F2.1 What Are Eating Disorders?

LEARNING OBJECTIVES

- Define eating disorder.
- Distinguish among anorexia, bulimia, and binge-eating disorder.
- Explain what is meant by the binge/purge cycle.

Eating disorders are psychological disorders that involve a persistent disturbance in eating patterns or other behaviors intended to control weight. They affect physical and nutritional health and psychosocial functioning. If untreated, eating disorders can be fatal.

According to mental health guidelines, there are three categories of eating disorders (**Table F2.1**). The first, **anorexia nervosa**, is characterized by self-starvation to reduce weight or prevent weight gain. **Bulimia nervosa**, the second category, involves frequent episodes of **bingeing** or **binge eating**, during which extremely large amounts of high-calorie foods

TABLE F2.1 Diagnostic Criteria for Eating Disorders[1]

Anorexia nervosa

- Refusal to maintain body weight at or above 85% of normal weight for age and height.
- Intense fear of gaining weight or becoming fat, even though underweight.
- Disturbance in the way body weight or shape is experienced, or denial of the seriousness of the current low body weight.
- Absence of at least three consecutive menstrual cycles without other known cause.

Restricting Type: During the current episode of anorexia nervosa, the person does not regularly engage in binge-eating or purging behavior (i.e., self-induced vomiting or the misuse of laxatives, diuretics, or enemas).

Binge-Eating Type or **Purging Type:** During the current episode of anorexia nervosa, the person regularly engages in binge-eating or purging behavior (i.e., self-induced vomiting or the misuse of laxatives, diuretics, or enemas).

Bulimia nervosa

- Recurrent episodes of binge eating.
- Recurrent inappropriate compensatory behavior to prevent weight gain, such as self-induced vomiting; misuse of laxatives, diuretics, enemas, or other medications; fasting; or excessive exercise.
- Occurrence, on average, of binge eating and inappropriate compensatory behaviors at least twice a week for three months.
- Undue influence by body shape and weight on self-evaluation.
- Disturbance does not occur exclusively during episodes of anorexia nervosa.

Purging Type: During the current episode of bulimia nervosa, the person regularly engages in self-induced vomiting or the misuse of laxatives, diuretics, or enemas.

Nonpurging Type: During the current episode of bulimia nervosa, the person uses other inappropriate compensatory behaviors, such as fasting or excessive exercise, but does not regularly engage in self-induced vomiting or the misuse of laxatives, diuretics, or enemas.

Eating disorders not otherwise specified (EDNOS)

- Criteria for anorexia nervosa are met except the individual menstruates regularly.
- Criteria for anorexia nervosa are met except that, despite substantial weight loss, the individual's current weight is in the normal range.
- Criteria for bulimia nervosa are met except binges occur at a frequency of less than twice a week and for a duration of less than three months.
- Inappropriate compensatory behavior after eating small amounts of food in individuals of normal body weight.
- Regularly chewing and spitting out, without swallowing, large amounts of food.
- Recurrent episodes of binge eating in the absence of regular use of inappropriate compensatory behaviors characteristic of bulimia (binge eating disorder).

purging Behaviors such as self-induced vomiting and misuse of laxatives, diuretics, or enemas to rid the body of calories.

eating disorders not otherwise specified (EDNOS) A category of eating disorders that includes abnormal eating behaviors that don't fit into the anorexia or bulimia nervosa categories.

binge-eating disorder An eating disorder characterized by recurrent episodes of binge eating in the absence of purging behavior.

are consumed. These episodes are almost always followed by depression, guilt, and **purging** behaviors, such as self-induced vomiting, to rid the body of the extra energy. The final category is **eating disorders not otherwise specified (EDNOS)**, which includes abnormal eating behaviors that don't fit into the other two categories. Over 50% of all people who seek treatment for an eating disorder are categorized as EDNOS. For example, someone who diets constantly and has a body weight that is very low but not low enough to be classified as anorexic would fit into the EDNOS category. **Binge-eating disorder**, which involves bingeing without purging, is also included in the EDNOS category.[1]

F2.2 What Causes Eating Disorders?

LEARNING OBJECTIVES

- Describe the genetic, psychological, and sociocultural factors that influence the development of eating disorders.
- Discuss how body ideal and the media affect the incidence of eating disorders.

We do not completely understand what causes eating disorders, but we do know that genetic, psychological, and sociocultural factors contribute to their development (**Figure F2.1**). Eating disorders can be triggered by traumatic events such as sexual abuse or by day-to-day occurrences such as teasing or judgmental comments by a friend or a coach. Eating disorders typically begin in adolescence when physical, psychological, and social development is occurring rapidly, but they occur in people of all ages, races, and socioeconomic backgrounds. They are more common in women than men. Although typically associated with white middle-class females, eating disorders are also a growing problem among African American and Hispanic women. They also occur in other minority groups, but the data concerning eating disorders in minority populations remain very limited. In the United States, the lifetime prevalence of anorexia, bulimia, and binge-eating disorder among girls and women is estimated to be 0.9, 1.5, and 3.5%, respectively.[2]

Eating disorders occur most frequently in groups that are concerned about maintaining a low body weight, such as professional dancers and models.[3] They are on the rise among athletes, especially those involved in sports that require the athlete to be thin, such as gymnastics and figure skating, or to fit into a particular weight class, such as wrestling.

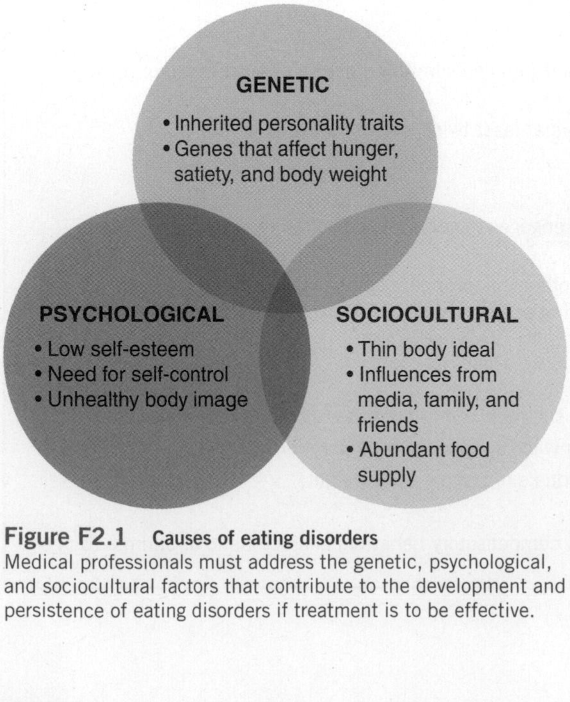

Figure F2.1 **Causes of eating disorders**
Medical professionals must address the genetic, psychological, and sociocultural factors that contribute to the development and persistence of eating disorders if treatment is to be effective.

The Role of Genetics

Eating disorders are not necessarily passed from parent to child, but the genes you inherit contribute to personality traits and other biological characteristics that might predispose you to developing an eating disorder. For example, inherited abnormalities in the levels of neurotransmitters such as serotonin, which affects food intake, have been hypothesized to contribute to the behaviors typical of anorexia and bulimia.[4] Binge-eating disorder may be linked to a defect in a gene called the melanocortin 4 receptor gene. The protein made by this gene helps control hunger and satiety. If this gene is abnormal and makes too little protein, the body feels too much hunger. In one study, all carriers of the mutant gene were binge eaters and mutations were found in 5% of obese subjects.[5] Genes such as this contribute to eating disorders, but a single gene is not likely to be the sole cause. These are complex diseases that are the result of the interaction of multiple genes with the environment. Each gene may have a small effect, but when taken together, they can increase risk severalfold. When placed in an environment conducive to eating disorders, individuals who carry such genes will be more likely to develop one.

Psychological Characteristics

Certain personality characteristics and psychological problems are common among individuals with eating disorders.[6] For example, people with eating disorders often have low **self-esteem**. Self-esteem refers to the judgments people make and maintain about themselves—a general attitude of approval or disapproval that indicates if the people think they are worthy and capable. Eating disorders are also rooted in the need for self-control. Those with eating disorders are often perfectionists who set very high standards for themselves and others. In order to be perfect they strive to be in control of their bodies and their lives. They view everything as either a success or a failure. Being fat is seen as failure, thin as success, and thinner as even more successful. In spite of their many achievements, those with eating disorders feel inadequate, defective, and worthless.

self-esteem The general attitude of approval or disapproval that people make and maintain about themselves.

body image The way a person perceives and imagines his or her body.

People with eating disorders often try to use their relationship with food to gain control over their lives and boost their self-esteem. They believe that controlling their food intake and weight demonstrates their ability to control other aspects of their lives and solve other problems. Their fixation with food and weight loss and their ability to control their intake and weight help them to feel better about themselves. Even if they feel insecure, helpless, or dissatisfied in other aspects of life, if they are in control of their food intake, weight, and body size they can associate this control with success. This feeling of being in control can become addictive.

SCPhotos/Alamy

Figure F2.2 Different body ideals A fuller figure is still desirable in many cultures. Young women in these cultures, such as the Zulu of South Africa, may try to gain weight in order to achieve what is viewed as the ideal female body. As television images of very thin Western women become more accessible, the Zulu cultural view of plumpness as desirable may be changing.[8]

Sociocultural Factors

While genetic and psychological issues may predispose individuals to eating disorders, sociocultural factors are important triggers for the onset of these disorders. An important sociocultural factor is body ideal. What is viewed as an ideal body differs across cultures and has changed throughout history. Ancient drawings and figurines show women with large breasts and swollen abdomens. This plump body ideal is still prevalent today in cultures where food is not readily available. Young women in these cultures may struggle to gain weight to achieve what is viewed as the ideal female body (**Figure F2.2**). In contrast, women in the United States strive to achieve a thin, lean body. The sociocultural ideals about body size are linked to **body image** and the incidence of eating disorders.[7] Eating disorders occur in societies where food is abundant and the body ideal is thin. They do not occur where food is scarce and people must worry about where their next meal is coming from.

Body Ideal in Modern America

From television and movies to magazines and advertisements and even toys, the culture in America today is a culture of thinness (**Figure F2.3**).

Figure F2.3 America's body ideal

Michael Newman/PhotoEdit

(a) The toys that children play with set a cultural standard for body ideal. Little girls playing with Barbie dolls want to be like Barbie when they grow up, and boys playing with Superman, Batman, or GI Joe action figures want to be like them. This includes looking like them. Unfortunately, Barbie's measurements would be virtually unachievable if Barbie were life-sized. The same is true of the big chest, muscular arms and legs, and flat stomach with "six-pack" abs seen on male action figures.

Andy Washnik

(b) Magazine covers and advertisements emphasize thinness as a standard for female beauty. These "ideal" bodies are frequently atypical of normal, healthy women.

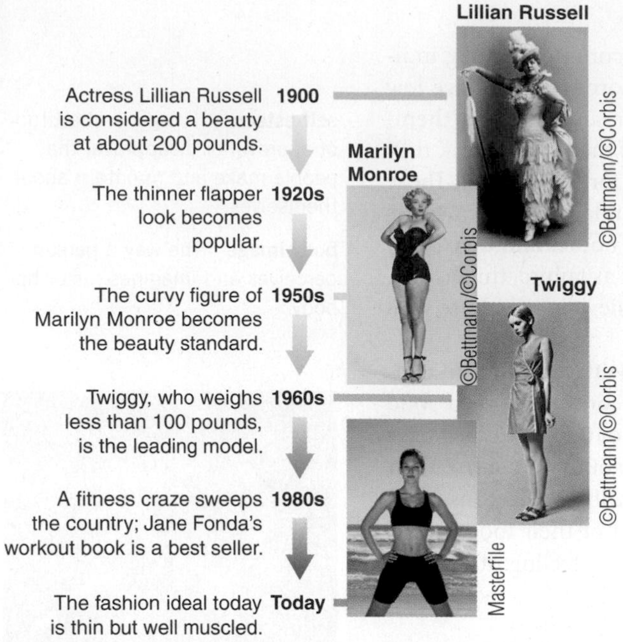

Lillian Russell

Actress Lillian Russell is considered a beauty at about 200 pounds. 1900

Marilyn Monroe

The thinner flapper look becomes popular. 1920s

Twiggy

The curvy figure of Marilyn Monroe becomes the beauty standard. 1950s

Twiggy, who weighs less than 100 pounds, is the leading model. 1960s

A fitness craze sweeps the country; Jane Fonda's workout book is a best seller. 1980s

The fashion ideal today is thin but well muscled. Today

©Bettmann/©Corbis ©Bettmann/©Corbis ©Bettmann/©Corbis Masterfile

Figure F2.4 **The changing female body ideal**
Thinness has not always been the beauty standard in the United States. This time line shows how the female body ideal has changed over the years.

Messages about what society views as a perfect body—the ideal we should strive for—are constantly delivered by the mass media. The perfect body is long, lean, and muscled. The tall dark muscular man gets the girl; the thin athletic woman gets her man. Thin fashion models adorn billboards and magazine covers to show off the latest clothes.

Being thin is associated with beauty, success, intelligence, and vitality. Being plump, on the other hand, is associated with failure, stupidity, and clumsiness. What young woman would want to be plump when exposed to these negative associations? Although men are also affected by these messages, American culture still places much more emphasis on the appearance of women's bodies. A young woman facing a future where she must be independent, have a prestigious job, maintain a successful love relationship, bear and nurture children, manage a household, and stay in fashion can become overwhelmed. Unable to master all these roles, she may look for some aspect of her life that she can control. Food intake and body weight are natural choices, since being thin brings the societal associations of success. These messages about how we should look are difficult to ignore and can create pressure to achieve this ideal body. But it is a standard that is very hard to meet—a standard that is contributing to disturbances in body image and eating behavior. This is illustrated by the fact that, as the body dimensions of female models, actresses, and other cultural icons have become thinner over the past several decades, the incidence of eating disorders has increased (**Figure F2.4**).

© Ted Fox/Alamy Limited

Figure F2.5 **Distorted body image**
When people with a distorted body image look in the mirror they see themselves as fat even when they are normal or underweight.

Body Image and Eating Disorders What individuals think they should look like or wish they looked like is affected by the ideals of their culture and society. Many women and girls, particularly teenage girls, are dissatisfied with their bodies because they look different from what they and their culture see as ideal. Fashion models today weigh 23% less than the average female. Although many women strive for this thin ideal, only 1% of young women have a chance of being as thin as a supermodel.[9] Almost everyone has some degree of body dissatisfaction, something that they would like to change about their bodies. For some, however, this becomes a pathological concern with body weight and shape, and as a result, body image may become distorted. A distorted body image means that an individuals are unable to judge the size of their own body and do not see themselves as they really are (**Figure F2.5**). Body image distortion is common with eating disorders, so even if these individuals acheive a body weight comparable to that of a fashion model, they may continue to see themselves as fat and strive to lose more weight (**Table F2.2**).

TABLE F2.2 To Maintain a Healthy Body Image

Try to . . .

- Accept that healthy bodies come in many shapes and sizes.
- Recognize your positive qualities.
- Remember that you can be your worst critic.
- Explore your internal self, as well as your external appearance.
- Spend your time and energy enjoying the positive things in your life.
- Be aware of your own weight prejudice. Explore how your feelings may affect your self-esteem.

And try not to . . .

- Let your body define who or what you are.
- Judge others on the basis of appearance, body size, or shape.
- Forget that society changes its ideals of beauty over the years.
- Forget that you are not alone in your pursuit of self-acceptance.
- Be afraid to enjoy life.

F2.3 Anorexia Nervosa

LEARNING OBJECTIVES

- Describe the physical and psychological features of anorexia nervosa.
- Discuss the health consequences of anorexia.
- Explain how anorexia is treated.

Anorexia means lack of appetite, but in the case of the eating disorder anorexia nervosa it is a desire to be thin, rather than a lack of appetite, that causes individuals to decrease their food intake. Anorexia nervosa was first recognized by physicians in the second half of the nineteenth century, and the characteristics they described are still true of the syndrome today: severe weight loss, **amenorrhea**, constipation, and restlessness. The overall prevalence of anorexia nervosa is estimated to be about 1% of the population.[10] The average age of onset is 17 years. There is a 5% death rate in the first two years, and this can reach 20% in untreated individuals.[3]

amenorrhea Delayed onset of menstruation or absence of three or more consecutive menstrual cycles.

Psychological Issues

The psychological component of anorexia nervosa revolves around an overwhelming fear of gaining weight, even in those who are already underweight. It is not uncommon for individuals with anorexia to feel that they would rather be dead than fat. Anorexia is also characterized by disturbances in body image or perception of body size that prevent those affected from seeing themselves as underweight even when they are dangerously thin. Those with this disorder may use body weight and shape as a means of self-evaluation: "If I weren't so fat then everyone would like and respect me and I wouldn't have other problems." However, no matter how much weight they lose, individuals with anorexia nervosa do not gain self-respect, inner assurance, or the happiness they seek. Therefore, they continue to restrict their food intake to promote weight loss.

Behaviors Associated with Anorexia Nervosa

The most obvious behaviors associated with anorexia are those that contribute to the maintenance of a body weight that is 15% or more below normal body weight (**Figure F2.6**).

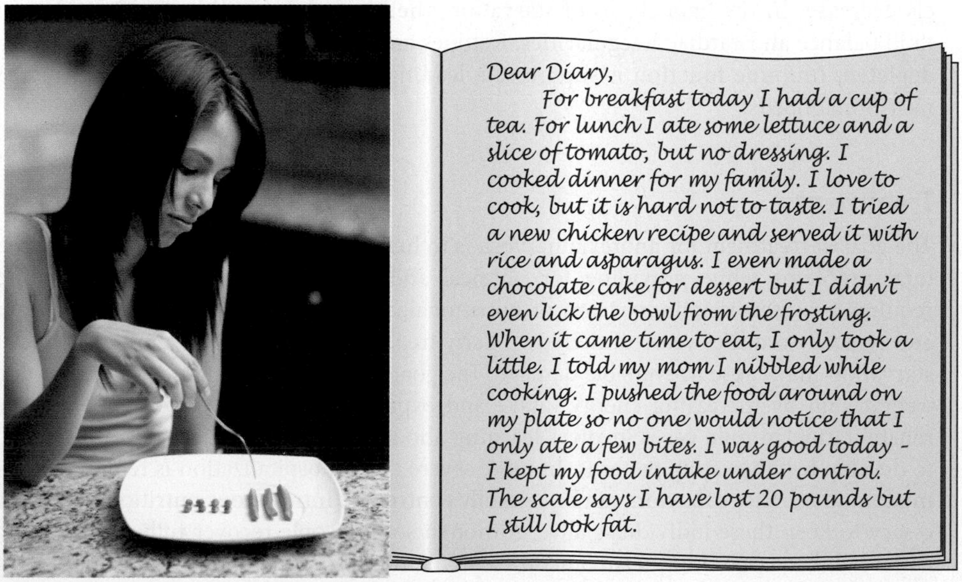

© Aldo Murillo/iStockphoto

Dear Diary,
For breakfast today I had a cup of tea. For lunch I ate some lettuce and a slice of tomato, but no dressing. I cooked dinner for my family. I love to cook, but it is hard not to taste. I tried a new chicken recipe and served it with rice and asparagus. I even made a chocolate cake for dessert but I didn't even lick the bowl from the frosting. When it came time to eat, I only took a little. I told my mom I nibbled while cooking. I pushed the food around on my plate so no one would notice that I only ate a few bites. I was good today - I kept my food intake under control. The scale says I have lost 20 pounds but I still look fat.

Figure F2.6 A day in the life of an anorexic
People with anorexia nervosa carefully regulate what they eat to maintain a very low body weight.

These behaviors include restriction of food intake, binge-eating and purging episodes, strange eating rituals, and excessive exercise. Based on these behaviors, anorexia is subdivided into two subtypes. Those with the *Restricting Type* maintain their low body weight solely by restricting their food intake and increasing their activity. Those with the *Binge-Eating/Purging Type* also typically restrict their food intake but, in addition, regularly engage in binge-eating and/or purging behaviors (see Table F2.1). It is estimated that about half of people with anorexia use purging as a means of weight control.[11]

For individuals with anorexia, food and eating become an obsession. In addition to restricting the total amount of food consumed, people with anorexia develop personal diet rituals, limiting certain foods and eating them in specific ways. Although they do not consume very much food, they are preoccupied with food and spend an enormous amount of time thinking about food, talking about food, and preparing food for others. Instead of eating, they move the food around the plate and cut it into tiny pieces (see Figure F2.6).

Both hyperactivity and overactivity are behaviors that are also typical of anorexia. This is in contrast to the decrease in activity and fatigue characteristic of other starvation states associated with weight loss. Many people with anorexia exercise excessively to burn calories. For some, the activity is surreptitious, such as going up and down stairs repeatedly or getting off the bus a few stops early. For others, the activity takes the form of strenuous physical exercise. They may become fanatical athletes and feel guilty if they cannot exercise. The exercise is typically done alone and is performed as a regular rigid routine. They may link exercise and eating, so a certain amount of exercise earns them the right to eat and if they eat too much they must pay the price by adding extra exercise. Those who use exercise to increase energy expenditure do not stop when they are tired; instead, they train compulsively beyond reasonable endurance (**Figure F2.7**).

Stockbyte/Getty Images, Inc.

Figure F2.7 Anorexia and exercise People with anorexia nervosa often use exercise as well as food restriction to achieve and maintain a very low body weight.

Physical Symptoms of Anorexia Nervosa

The first obvious physical manifestation of anorexia is weight loss. As weight loss becomes severe, symptoms of starvation begin to appear. Starvation affects mental function, causing those with anorexia to become apathetic, dull, exhausted, and depressed. Physical symptoms include depletion of fat stores; wasting of muscles; inflammation and swelling of the lips; flaking and peeling of skin; and growth of fine hair, called lanugo hair, on the body, and dry, thin, brittle hair on the head that may fall out. In females, estrogen levels drop and menstruation becomes irregular or stops. This can delay sexual maturation and can have long-term effects on bone density. In males, testosterone levels decrease. In the final stages of starvation, there are abnormalities in electrolyte and fluid balance and cardiac irregularities. Ketones are typically absent because fat stores are depleted. Immune function is suppressed, leading to infections, which further increase nutritional needs.

Treatment

The goal of treatment for anorexia nervosa is to help resolve the psychological and behavioral problems while providing for physical and nutritional rehabilitation. Treatment requires an interdisciplinary team of nutritional, mental health, and medical specialists and typically requires years of therapy.[12] Early treatment of anorexia is important because starvation may cause irreversible damage. The goal of nutrition intervention is to promote weight gain by increasing energy intake and expanding dietary choices.[12, 13] Nutritional rehabilitation in mild cases involves learning about nutrition and meal planning in order to develop healthy eating patterns. In more severe cases, hospitalization is required so food intake and exercise behaviors can be carefully controlled. Intravenous nutrition may be necessary to keep these individuals alive. Although some people recover fully from anorexia, about half have poor long-term outcomes—remaining irrationally concerned about weight gain and never achieving normal body weight. Some patients with anorexia also transition to bulimia nervosa.[14]

F2.4 Bulimia Nervosa

LEARNING OBJECTIVES

- Describe the physical and psychological features of bulimia nervosa.
- Discuss the health consequences of bulimia
- Explain how bulimia is treated.

Bulimia is from the Greek *bous* (ox) and *limos* (hunger), denoting hunger of such intensity that a person could eat an entire ox. The modern concept of bulimia nervosa as an eating disorder arose in the early 1970s, when a set of symptoms was identified and distinguished from anorexia and obesity. Many different names were used for this disorder, including dysorexia, bulimarexia, thin-fat syndrome, binge/purge syndrome, and dietary chaos syndrome. The term *bulimia nervosa* was coined in 1979 by a British psychiatrist who suggested that bulimia consisted of powerful urges to overeat in combination with a morbid fear of becoming fat and the avoidance of the fattening effects of food by inducing vomiting or abusing purgatives or both.[15] Today, the estimated overall prevalence for bulimia nervosa is 4%.[10] A diagnosis of bulimia is based on the frequency with which episodes of binge eating and purging or other types of inappropriate compensatory behaviors occur (see Table F2.1).

Psychological Issues

As with anorexia, people with bulimia have an intense fear of becoming fat. They have a negative body image accompanied by a distorted perception of their body size. Since their self-esteem is highly tied to their impressions of their body shape and weight, they blame all of their problems on their appearance; this allows them not to face the real problems in their life. People with bulimia are preoccupied with the fear that once they start eating they will not be able to stop. They may engage in continuous dieting, which leads to a preoccupation with food. They also think they are the only person in the world with this problem. As a result they are often socially isolated. In addition, they may avoid situations that will expose them to food, such as going to parties or out to dinner, further isolating themselves.

Behaviors Associated with Bulimia Nervosa

Bulimia typically begins with food restriction motivated by the desire to be thin. Overwhelming hunger may finally cause the dieting to be interrupted by a period of overeating. Eventually a pattern develops involving semi-starvation interrupted by periods of gorging. During a food binge, a person with bulimia experiences a sense of lack of control. The amount of food consumed during a binge varies, but is typically on the order of 3400 kcalories, while a normal teenager may consume 2000 to 3000 kcalories in an entire day. One study found that bulimics consumed an average of about 7000 kcalories in a 24-hour period.[16] Binges usually last less than two hours and occur in secrecy. They stop when the food runs out or when pain, fatigue, or an interruption intervenes. The amount of food consumed in a binge may not always be enormous, but it is perceived by the bulimic individual as a binge episode (**Figure F2.8**).

After binge episodes, individuals with bulimia use various inappropriate compensatory behaviors to eliminate the extra calories and prevent weight gain. Bulimia is subdivided into two types based on the type of compensatory behavior used. *Nonpurging bulimia* involves behaviors such as fasting or excessive exercise to prevent weight gain, whereas *purging bulimia* involves regularly engaging in behaviors that may include self-induced vomiting and misuse of enemas, laxatives and diuretics, or other medications (see Table F2.1). Self-induced vomiting is the most common purging behavior. It is used at the end of a binge but also after normal eating to eliminate food before it is absorbed and the energy it provides

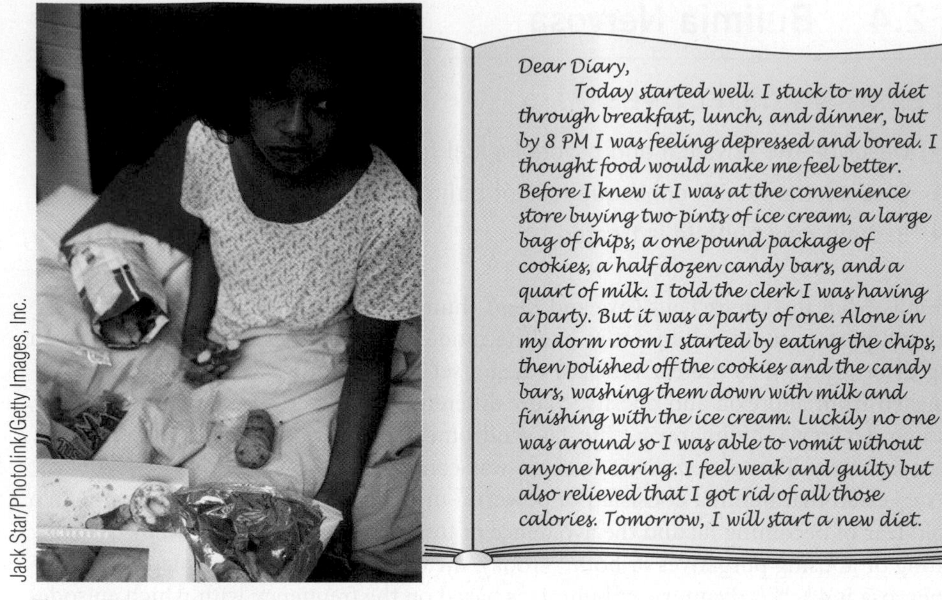

Jack Star/Photolink/Getty Images, Inc.

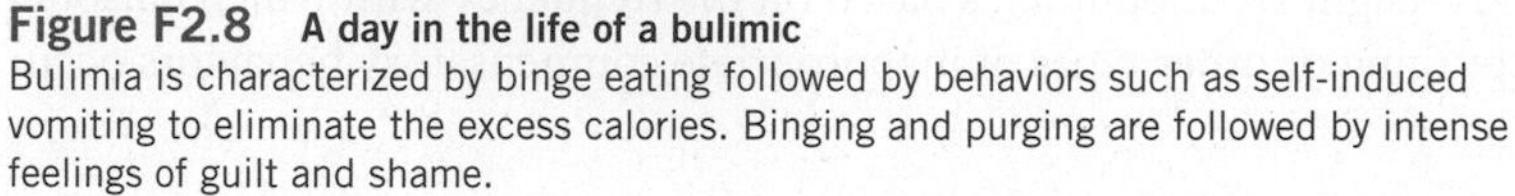

Figure F2.8 A day in the life of a bulimic
Bulimia is characterized by binge eating followed by behaviors such as self-induced vomiting to eliminate the excess calories. Binging and purging are followed by intense feelings of guilt and shame.

can cause weight gain (**Figure F2.8**). At first a physical maneuver such as sticking a finger down the throat is needed to induce vomiting, but patients eventually learn to vomit at will. Vomiting does not purge all calories consumed in a binge. After a binge containing 3530 kcalories, on average 1209 kcalories were retained. Interestingly, after a smaller binge of only 1549 kcalories, almost the same amount of energy remained in the stomach, 1128 kcalories.[17] Some bulimic individuals take laxatives to induce diarrhea. Although the patients believe the diarrhea prevents calories from being absorbed, in fact, nutrient absorption is almost complete before food enters the colon, where laxatives have their effect. The weight loss associated with laxative abuse is due to dehydration. Diuretics also cause water loss but via the kidney rather than the GI tract. They do not cause fat loss. Some bulimia sufferers use a combination of purging and nonpurging methods to eliminate excess calories.

Physical Complications of Bulimia Nervosa

It is the purging portion of the binge-purge cycle that is most hazardous to health in bulimia nervosa. Purging by vomiting brings stomach acid into the mouth. Frequent vomiting affects the GI tract by causing tooth decay, sores in the mouth and on the lips, swelling of the jaw and salivary glands, irritation of the throat, inflammation of the esophagus, and changes in stomach capacity and stomach emptying.[3] It also causes broken blood vessels in the face from the force of vomiting, electrolyte imbalance, dehydration, muscle weakness, and menstrual irregularities. Laxative and diuretic abuse can also cause dehydration and electrolyte imbalance. Rectal bleeding may occur from laxative overuse.

Treatment

The overall goal of therapy for people with bulimia nervosa is to separate eating from their emotions and from their perceptions of success and to promote eating in response to hunger and satiety. Psychological counseling is needed to address issues related to body image and a sense of lack of control over eating. Nutritional therapy addresses physiological imbalances caused by purging episodes as well as providing education on nutrient needs and how to meet them. Antidepressant medications may be beneficial in reducing the frequency of binge episodes. Treatment has been found to speed recovery, especially if it is provided soon after symptoms begin, but for some individuals this disorder may remain a chronic problem throughout life.[18]

F2.5 Binge-Eating Disorder

LEARNING OBJECTIVES

- Distinguish eating behavior in binge-eating disorder from that in anorexia and bulimia.
- Describe the health consequences of binge-eating disorder.

Binge-eating disorder, which is in the EDNOS category, is probably the most prevalent eating disorder. It affects about 3.5% of women and 2.0% of men during their lifetimes.[2] Unlike anorexia and bulimia, binge-eating disorder is not uncommon in men, who account for about 40% of cases.[19] Individuals with binge-eating disorder engage in recurrent episodes of binge eating but do not regularly engage in inappropriate compensatory behaviors such as vomiting, fasting, or excessive exercise (**Table F2.3**). As a result, overweight and obesity are common among people with binge-eating disorder (**Figure F2.9**).

TABLE F2.3 Diagnostic Criteria for Binge Eating Disorder[1]

Recurrent episodes of binge eating. An episode is characterized by:
• Eating a larger amount of food than normal during a short period of time (within any two-hour period). • Lack of control over eating during the binge episode (i.e., the feeling that one cannot stop eating).
Binge eating episodes are associated with three or more of the following:
• Eating until feeling uncomfortably full. • Eating large amounts of food when not physically hungry. • Eating much more rapidly than normal. • Eating alone because you are embarrassed by how much you're eating. • Feeling disgusted, depressed, or guilty after overeating.
Marked distress regarding binge eating is present.
Binge eating occurs, on average, at least two days a week for six months.
The binge eating is not associated with the regular use of inappropriate compensatory behavior (i.e., purging, excessive exercise, etc.) and does not occur exclusively during the course of bulimia nervosa or anorexia nervosa.

Ken Ross/Getty Images, Inc.

Dear Diary,

I got on the scale today. What a mistake! My weight is up to 250 pounds. I hate myself for being so fat. Just seeing that I gained more weight made me feel ashamed - all I wanted to do was bury my feelings in a box of cookies or a carton of ice cream. Why do I always think the food will help? Once I started eating I couldn't stop. When I finally did I felt even more disgusted, depressed, and guilty. I am always on a diet but it is never long before I lose control and pig out. I know my eating and my weight are not healthy but I just can't seem to stop.

Figure F2.9 **A day in the life of a binge eater**
People with binge-eating disorder often seek help for their weight rather than for their disordered eating pattern. It is estimated that 10 to 15% of people enrolled in commercial weight-loss programs suffer from this disorder.[20]

The major complications of binge-eating disorder are the conditions that accompany obesity, which include diabetes, high blood pressure, high blood cholesterol levels, gallbladder disease, heart disease, and certain types of cancer.[19] The primary emphasis of behavioral therapy for binge-eating disorder is on reducing binge eating, with a secondary focus on weight loss.[12] Treatment also involves counseling to improve body image and self-acceptance, a healthy nutritious diet, and increased exercise to promote weight loss.

F2.6 Eating Disorders in Special Groups

LEARNING OBJECTIVES

- Discuss how eating disorders differ in men and women.
- Describe some eating disorders that occur in children, pregnant women, and athletes.

Although anorexia and bulimia are most common in women in their teens and 20s, eating disorders occur in both genders and all age groups. They can be a complication in pregnant women and a problem for athletes and young children. They also occur in individuals with diseases that have a nutrition component such as diabetes. In addition, there are a number of less common eating disorders that appear in special groups and in the general population. These are listed in **Table F2.4**.

Eating Disorders in Men

Jim Cummins/Getty Images, Inc.

Figure F2.10 The male body ideal The ideal male body is as difficult for most men to achieve as the thin athletic ideal is for women.

The incidence of anorexia and bulimia is much lower among men than women. One reason for the lower incidence is that the cultural pressure for males to be thin is less intense. Women are encouraged to be thin to attract friends and romantic partners and to be successful at school and work, whereas men are encouraged to be strong and powerful. The male ideal is a V-shaped upper body that is muscular, moderate in weight, and low in body fat. This difference in societal expectations is reflected in the BMI of men and women when they first "feel fat" and begin dieting. Women who develop eating disorders generally feel fat and begin dieting when their BMI is in the healthy range, whereas men who develop eating disorders usually do not start dieting until their BMI is in the overweight range.[21] Men also often develop eating disorders at a greater age than women.

Although men currently represent a small percentage of those with eating disorders, the numbers seem to be on the rise.[22] This is likely due to increasing pressure to achieve an ideal male body. Advertisements directed at men today are showing more and more exposed skin with a focus on well-defined abdominal and chest muscles (**Figure F2.10**). Just as the Barbie doll set an unrealistic standard for young women, male action figures, superhero cartoons, and media ideals set a standard that is impossible for young men to achieve.

The physical consequences of eating disorders are similar in men and women. Both lose bone, but men are more severely affected by disorders related to bone loss and tend to have lower bone mineral density than women with the same disorder.[23] Rather than amenorrhea, men experience a gradual drop in testosterone levels. This causes a loss of sexual desire. Men with eating disorders have psychiatric conditions that are similar to those affecting women, including mood and personality disorders. Like women, men with eating disorders require professional help in order to recover, and the outcome of treatment is similar in men and women. Men, however, are less likely to seek treatment because they do not want to be perceived as having a "woman's disease."

Eating Disorders during Pregnancy

Eating disorders are common in women in their 20s, an age when many people choose to start a family. If the eating disorder interrupts the menstrual cycle, it will cause infertility.

TABLE F2.4 Other Eating Disorders

Eating Disorder	Characteristics	Who Is Affected	Consequences
Anorexia athletica	Engaging in compulsive exercise to lose weight or maintain a very low body weight.	Athletes	Can lead to more serious eating disorders and serious health problems including kidney failure, heart attack, and death.
Avoidance emotional disorder	Similar to anorexia nervosa in that the child avoids eating and experiences weight loss and the physical symptoms of anorexia. However, there is no distorted body image or fear of weight gain.	Children	Weight loss, reduced body fat, malnutrition.
Bigorexia (muscle dysmorphia or reverse anorexia)	Obsession with being small and underdeveloped. Individuals believe their muscles are inadequate even when they have a good muscle mass.	Bodybuilders and avid gym-goers, more common in men than women	Sufferers are at risk if they take steroids or other muscle-enhancing drugs.
Body dysmorphic disorder	An obsession with a perceived defect in the sufferer's body or appearance.	Affects males and females equally	Increased risk for depression and suicide.
Chewing and spitting	The person puts food in his/her mouth, tastes it, chews it, and then spits it out.	Those with other eating disorders	Since the food is not swallowed it can result in the same symptoms as starvation dieting.
Diabulimia (insulin misuse)	Withholding insulin to cause weight loss or prevent weight gain.	People with type I diabetes	Uncontrolled blood sugar, which can lead to blindness, kidney disease, heart disease, nerve damage, and amputations.
Female athlete triad	A triad of disordered eating, amenorrhea, and osteoporosis.	Female athletes in weight-dependent sports	Low estrogen levels, which interfere with calcium balance, eventually causing reductions in bone mass and an increased risk of bone fractures.
Night-eating syndrome	Most of the day's calories are eaten late in the day or at night. A similar disorder, in which a person may eat while asleep and have no memory of the events, is called nocturnal sleep-related eating disorder. It is considered a sleep disorder, not an eating disorder.	Obese adults and those experiencing stress	Obesity
Orthorexia nervosa	Obsession with eating food considered to be healthy or beneficial. Focus on the quality of the food, not the quantity.	No particular group	Harmful to interpersonal relationships.
Pica	Craving and eating nonfood items such as dirt, clay, paint chips, plaster, chalk, laundry starch, coffee grounds, and ashes.	Pregnant women, children, people whose family or ethnic customs include eating certain nonfood substances.	Mineral deficiencies, perforated intestines, intestinal infections.
Rumination syndrome	Eating, swallowing, and then regurgitating food back into the mouth where it is chewed and swallowed again.	Infants and adults with mental and emotional impairment	Bad breath, indigestion, chapped lips, damage to dental enamel and tissues in the mouth, aspiration of food leading to pneumonia, weight loss and failure to grow (children), electrolyte imbalance, and dehydration.
Selective eating disorder	Eating only a few foods, mostly carbohydrate.	Children	Malnutrition

Some women with eating disorders are able to conceive. Pregnancy is a stressful and anxious time for any woman. In someone with an eating disorder, the weight gain and change in body shape can worsen or cause a recurrence of the disorder.[24] In other women, symptoms may improve because of their concern for the welfare of the baby. Pregnancy can also make other medical problems related to the eating disorder worse, such as liver, heart, and kidney damage. Pregnant women with eating disorders are at increased risk of caesarean delivery and have a higher rate of miscarriages, premature birth, babies who are small for their age, and congenital malformations.[24]

What and how much a woman eats during pregnancy influences her health and the health of the baby before and after birth. Babies born to women with eating disorders are more likely to be slower to grow and develop, and they may lag behind intellectually and emotionally and remain dependent. They may also have difficulty developing social skills and relationships with other people.

pica An abnormal craving for and ingestion of nonfood items.

An eating disorder that is more common in pregnancy is **pica** (see Chapters 12 and 14). This is an abnormal craving for and ingestion of nonfood substances having little or no nutritional value. Commonly consumed substances include dirt, clay, chalk, paint chips, laundry starch, ice and freezer frost, baking soda, cornstarch, coffee grounds, and ashes. Pica during pregnancy is potentially dangerous. The consumption of large amounts of nonfood substances may cause micronutrient deficiencies by reducing the intake of nutrient-dense foods and by interfering with the absorption of certain minerals from food. Substances consumed could also cause intestinal obstruction or perforation and could contain toxins or harmful bacteria. The cause of pica is unknown, but there is some evidence that it results from cultural beliefs related to pregnancy, along with changes in food preferences that occur during pregnancy.

Eating Disorders in Children

Although the incidence is much lower than it is in late adolescence and early adulthood, anorexia and bulimia also occur in children under the age of 13.[10] There is concern that the prevalence of eating disorders is increasing in younger children in the United States as they are exposed to our cultural values about food and body weight. More girls than boys are affected, but the proportion of boys is greater than the proportion of men in the older age groups.[10] The implications for growth and development are even greater when eating disorders begin at a younger age.

Eating disorders can be difficult to diagnose in children because most children are finicky eaters at some point in their development. But when being finicky becomes extreme and growth is impaired, an eating disorder may be the cause. Disorders may involve consumption of a very limited number of foods or a general restriction of food intake (see Table F2.4). Diagnosing anorexia in girls under 13 is more challenging than in older girls because many have not started menstruating, so amenorrhea cannot be used as a diagnostic criterion. It is also difficult to determine the degree of underweight in children suspected of having anorexia because their growth may have slowed. Despite these problems, there is little doubt that childhood-onset anorexia does occur and is a serious illness.

Children with anorexia, like older patients, are generally perfectionists, conscientious, and hardworking, but eating disorders have also been recognized in children who are average or poor students. Eating disorders occur in children of all races and ethnicities and are even now seen in children living in poverty.[10]

Children with anorexia exhibit similar symptoms to adolescents with this disorder, including weight loss, food avoidance, preoccupation with food and calories, fear of fatness, excessive exercise, self-induced vomiting, and laxative abuse. Other physical changes that may accompany the weight loss in children with anorexia include growth of lanugo hair, low blood pressure, slow heart rate, poor peripheral circulation, cold peripheries, and delayed or arrested growth. Bone density may be reduced and bone age delayed. Vitamin and mineral deficiencies are common.

As with older patients, the prognosis for children with anorexia is variable.[25] If treatment is not effective, the resulting undernutrition is even more likely to cause physical complications such as heart and circulatory failure in children than in older patients. Growth is affected, but the long-term consequences depend on the outcome of treatment. If treatment restores normal eating, these children will catch up in growth. Complications that may persist include amenorrhea, delayed growth, impaired fertility, and osteoporosis.

Pornchai Kittiwongsakul/AFP/Getty Images, Inc.

Figure F2.11 **Gymnastics and eating disorders**
In sports such as gymnastics, the advantages offered by a small, light body can motivate athletes to diet to stay thin and can potentially contribute to the development of an eating disorder.

Eating Disorders in Athletes

The relationship between body weight and performance in certain sports contributes to the higher prevalence of eating disorders in athletes than in the general population.[26] It is higher in female athletes than in male athletes, and more common among those competing in leanness-dependent and weight-dependent sports such as ballet and other dance, figure skating, gymnastics, track and field, swimming, cycling, crew, wrestling, and horse racing (**Figure F2.11**). The prevalence of female college athletes with disordered eating has been estimated to range from 15 to 62%; significantly higher than the 5 to 10% of the general adult population that is estimated to be affected by eating disorders.[26] The most problematic women's sports are cross-country, gymnastics, swimming, and track and field. The male sports with the highest number of participants with eating disorders are wrestling and cross-country.[27]

Both anorexia nervosa and bulimia occur in athletes. The regimented schedule of an athlete makes it easy to use training diets and schedules, travel, or competition as an excuse to not eat normally and hide the eating disorder. Over time the continued starvation characteristic of anorexia leads to serious health problems as well as a decline in athletic performance. Starvation can lead to abnormal heart rhythms, low blood pressure, and atrophy of the heart muscle. The lack of food means that there is not adequate energy and nutrients to support activity and growth. Bulimia nervosa is more common in athletes than anorexia. It may begin because an athlete binges as a result of hunger caused by trying to stick with a very-low-calorie diet. As in nonathletes, most of the health complications associated with bulimia are the result of purging. It causes fluid loss and low potassium levels, which can result in extreme weakness as well as dangerous and sometimes lethal heart rhythms.

Anorexia Athletica Compulsive exercise, which has been termed *anorexia athletica*, is a type of eating disorder that is a particular problem in athletes. People with this disorder focus on exercise rather than food, but anorexia athletica is considered an eating disorder because the goal of the behavior is to expend calories to control weight. Extreme training is easy to justify because it is a common belief that serious athletes can never work too hard or too long and pain is accepted as an indicator of achievement. Compulsive exercisers will force themselves to exercise even when they don't feel well and may miss social events in order to fulfill their exercise quota. They often calculate exercise goals based on how much they eat. They believe that any break in the training schedule will cause them to gain weight and performance will suffer. Compulsive exercise can lead to more severe eating disorders such as anorexia and bulimia as well as serious health problems including kidney failure, heart attack, and death.

Female Athlete Triad Female athletes with eating disorders are at risk for a syndrome of interrelated disorders referred to as the **female athlete triad** (see Chapter 13). This syndrome includes disordered eating, amenorrhea, and **osteoporosis**. The three are linked because the extreme energy restriction that occurs in eating disorders creates a physiological condition similar to starvation, which leads to menstrual irregularities. High levels of exercise can also affect the menstrual cycle by increasing energy demands and causing hormonal changes.[28] When combined, energy restriction and excessive exercise contribute to amenorrhea. The low estrogen levels associated with amenorrhea then interfere with calcium balance, leading to reductions in bone mass and bone-mineral density.[29] Low estrogen levels also reduce calcium absorption and, when combined with poor calcium intake (common in female athletes and females in general), lead to

female athlete triad The combination of disordered eating, amenorrhea, and osteoporosis that occurs in some female athletes, particularly those involved in sports in which low body weight and appearance are important.

osteoporosis A bone disorder characterized by a decrease in bone mass, an increase in bone fragility, and an increased risk of fractures.

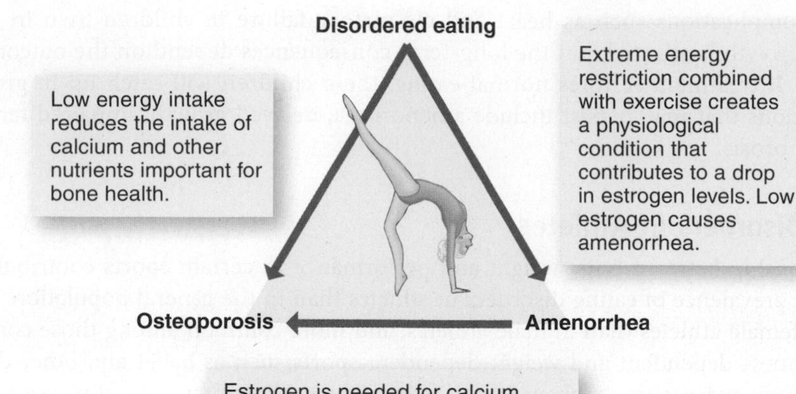

Figure F2.12 **Female athlete triad**
Women with female athlete triad typically have low body fat, do not menstruate regularly, and may experience multiple or recurrent stress fractures. Neither adequate dietary calcium nor the increase in bone mass caused by weight-bearing exercise can compensate for the bone loss caused by low estrogen levels.

premature bone loss, failure to reach maximal peak bone mass, and an increased risk of stress fractures (**Figure F2.12**).

Eating Disorders and Diabetes

Diabetes does not cause eating disorders, but it may set the stage for them both physically and emotionally and can be used to hide them. Diabetes is a disease characterized by a chronic elevation in blood glucose that is due to abnormalities in the production or effectiveness of the hormone insulin (see Chapter 4). Treatment involves paying careful attention to diet, exercise, body weight, and blood glucose levels. The timing of exercise and timing and composition of meals are crucial to good glucose control. This regimentation, which is part of routine diabetes management, may contribute to the development of eating disorders because it places attention on food portions and body weight; a focus similar to that seen in women with eating disorders who do not have diabetes.[30]

Control is a central issue in diabetes, as it is in eating disorders. People with diabetes may feel guilty or out of control if their blood sugar is too high. People with anorexia feel the same way if their weight increases. People with diabetes become consumed with strategies to control blood sugar, and those with an eating disorder become consumed with ways to control weight. Both are preoccupied with weight, food, and diet. Because this is expected in diabetes, people with diabetes can use it to hide anorexia or bulimia. They are supposed to watch what they eat, and the diabetes can be blamed for weight loss.

Those who take insulin to control their diabetes are at particular risk because they can misuse it to control their weight, a condition that has been termed *diabulimia*. Insulin is responsible for allowing glucose to enter muscle and adipose tissue cells. If patients cut back on the amount of insulin they take, the sugar in their blood cannot enter these cells, blood levels rise, and glucose is excreted in the urine. This causes weight loss, but at a very high cost. The long-term complications of high levels of blood glucose include blindness, kidney disease, cardiovascular disease, impaired circulation, and nerve death that can lead to limb amputations. Once people with diabetes start to control their weight by withholding insulin, they are reluctant to stop and may also begin using other inappropriate behaviors to control weight. Sometimes the weight loss seems to improve the diabetes, at least temporarily, by reducing the need for insulin, but if the weight loss continues, it can lead to organ failure and death.

F2.7 Preventing and Getting Treatment for Eating Disorders

LEARNING OBJECTIVES

- Describe factors that predispose people to eating disorders.
- List the steps you could take if you had a friend with an eating disorder.

Reducing the incidence of and morbidity from eating disorders involves action on a number of levels. The first step is to recognize individuals who are at risk. Early intervention can help prevent those who are at risk from developing serious eating disorders, and the actions of family and friends can help those who are affected get help before their health is impaired. To reduce the overall incidence of eating disorders, changes in social attitudes that contribute to their development need to occur.

How to Recognize the Risks

To prevent eating disorders, it is important first to recognize factors that increase risk. Excessive concerns about body weight, having friends who are preoccupied with weight, being teased by peers about weight, and problems with one's family all predispose people to eating disorders. There is an association between parental criticism and children's weight preoccupation. Dieting also increases risk. Girls and women who diet are more likely to develop an eating disorder than those who don't diet.[10] Those who have a mother, sister, or friend who diets are also at increased risk. Exposure to media pressure to be thin is also associated with the development of eating disorders.

How to Get Help for a Friend or Family Member

Those who are at risk for eating disorders can be targeted for intervention. For teens, parents play an important role. Arranging an evaluation with a physician and a mental health specialist when the first symptoms are discovered may help prevent the disorder.

Once an eating disorder has developed, people usually do not get better by themselves. Helping them to get medical and psychological treatment can avoid severe physical consequences. But getting a friend or relative with an eating disorder to agree to seek help is not always easy. People with eating disorders are good at hiding their behaviors and denying the problem and often do not want help (**Table F2.5**).

If you suspect a friend has an eating disorder you can alert a parent, teacher, coach, religious leader, school nurse, or other trusted adult about your concerns or confront your friend or relative yourself and express your concern. If you approach the person yourself, you need to be firm but supportive and caring. The goal in discussing a person's eating disorder is to encourage them to seek help. But, help is effective only if it is desired. People with eating disorders are likely to refuse help initially. When you approach someone about an eating disorder it is therefore important to make it clear that you are not trying to force them to do anything they don't want to do. Continued encouragement can help some people to seek professional help. The first reaction of someone confronted about an eating disorder is often to deny that they have a problem. Support your suspicions with examples of things you have seen that make you believe your friend has a problem. You should be

TABLE F2.5 Eating Disorders: How to Help

• Get the person to a doctor; the sooner the illness is treated, the more likely there will be a successful outcome.
• Talk to the parents, spouse, or other family members.
• Explain your concerns and the potential hazards of the disease.
• Do not expect the person to cooperate; denial is common.
• If you work with the person, contact your employee assistance program.

prepared for all possible reactions. People with eating disorders usually try to hide their behaviors, so it is traumatic for them when someone discovers their secret. One person may be relieved that you are concerned and willing to help, whereas another may be angry and defensive.

Reducing the Prevalence of Eating Disorders

Eating disorders are easier to prevent than to cure. The biggest impact on prevention can be made by social interventions that target the elimination of weight-related teasing and criticism from peers and family members. Another important target for reducing the incidence of eating disorders is the media. If the unrealistically thin body ideal presented by the media could be altered, the incidence of eating disorders would likely decrease. Even with these interventions, however, eating disorders are unlikely to go away, but education through schools and communities about the symptoms and complications of eating disorders can help people identify friends and family members at risk and persuade those with early symptoms to seek help.

REFERENCES

1. American Psychiatric Association. *Diagnostic and Statistical Manual of Mental Disorders*, 4th ed., Washington, DC: American Psychiatric Association, 2000.
2. Hudson, J. I., Hiripi, E., Pope, H. G. Jr, and Kessler, R. C. The prevalence and correlates of eating disorders in the National Comorbidity Survey Replication. *Biol Psychiatry* 61:348–358, 2007.
3. American Dietetic Association. Position of the American Dietetic Association: Nutrition intervention in the treatment of anorexia nervosa, bulimia nervosa, and other eating disorders. *J Am Diet Assoc* 106:2073–2082, 2006.
4. Bailer, U. F., and Kaye, W. H. Serotonin: Imaging findings in eating disorders. *Curr Top Behav Neurosci* 6:59–79, 2011.
5. Branson, R., Potoczna, N., Kral, J. G., et al. Binge eating as a major phenotype of melanocortin 4 receptor gene mutations. *N Engl J Med* 348:1096–1103, 2003.
6. Stice, E., Ng, J., and Shaw, H. Risk factors and prodromal eating pathology. *J Child Psychol Psychiatry* 51:518–525, 2010.
7. Stice, E. Sociocultural influences on body weight and eating disturbance. In *Eating Disorders and Obesity: A Comprehensive Handbook*, 2nd ed., C. G. Fairburn and K. D. Brownell, eds. New York: Guilford Press 2002, pp. 103–107.
8. Body image worries hit Zulu women. *BBC News*, April 16, 2004.
9. HealthyPlace. Eating Disorders: Body image and Advertising, December 11, 2008.
10. Rome, E. S. Eating disorders in children and adolescents. *Curr Probl Pediatr Adolesc Health Care* 42:28–44, 2012.
11. Peat, C., Mitchell, J. E., Hoek, H. W., and Wonderlich, S. Validity and utility of subtyping anorexia nervosa. *Int J Eat Disord* 42:1-7, 2009.
12. American Dietetic Association. Position of the American Dietetic Association: Nutrition intervention in the treatment of eating disorders. *J Am Diet Assoc* 111:1236–1241, 2011.
13. Fitzgibbon, M., and Stolley, M. Minority women: The untold story. NOVA Online: Dying to be thin.
14. Tozzi, F., Thornton, L. M., Klump, K. L., et al. Symptom fluctuation in eating disorders: Correlates of diagnostic crossover. *Am J Psychiatry* 162:732–740, 2005.
15. Vandereycken, W. History of anorexia nervosa and bulimia nervosa. In *Eating Disorders and Obesity A Comprehensive Handbook*, 2nd ed., C. G. Fairburnand K. D. Brownell, eds. New York: Guilford Press, 2002, pp. 151–154.
16. Kaye, W. H., Weltzin, T. E., McKee, M., et al. Laboratory assessment of feeding behavior in bulimia nervosa and healthy women. Methods of developing a human feeding laboratory. *Am J Clin Nutr* 55:372–380, 1992.
17. Kaye, W. H., Weltzin, T. E., Hsu, L. K., et al. Amount of calories retained after binge eating and vomiting. *Am J Psychiatry* 150:969–971, 1993.
18. Ebeling, H., Tapanainen, P., and Joutsenoja, A. A practice guideline for treatment of eating disorders in children and adolescents. *Ann Med* 35:488–501, 2003.
19. National Eating Disorders Association. Binge eating disorder.
20. Healthier You. Binge Eating Disorder.
21. Andersen, A. E., and Holman, J. E. Males with eating disorders: Challenges for treatment and research. *Psychopharmacol Bull* 33:391-397, 1997.
22. Spann, N., and Pritchard, M. Disordered eating in men: a look at perceived stress and excessive exercise. *Eat Weight Disord* 13:e25–e27, 2008.

23. Mehler, P. S., Sabel, A. L., Watson, T., and Andersen, A. E. High risk of osteoporosis in male patients with eating disorders. *Int J Eat Disord* 41:666-672, 2008.

24. Ward, V. B. Eating disorders in pregnancy. *BMJ* 336:93–96, 2008.

25. Casper, R. C. Eating disturbances and eating disorders in childhood. *Psychopharmacology—The Fourth Generation of Progress.*

26. Sundgot-Borgen, J., and Torstveit, M. K. Prevalence of eating disorders in elite athletes is higher than in the general population. *Clin J Sports Med* 14:25–32, 2004.

27. National Eating Disorders Information Center.

28. Warren, M. P., and Goodman, L. R. Exercise-induced endocrine pathologies. *J Endocrinol Invest* 26:873–878, 2003.

29. Lambrinoudaki, I., and Papadimitriou, D. Pathophysiology of bone loss in the female athlete. *Ann N Y Acad Sci* 1205: 45–50, 2010.

30. Larrañaga, A., Docet, M. F., and García-Mayor, R. V. Disordered eating behaviors in type 1 diabetic patients. *World J Diabetes* 2:189–195, 2011.

CHAPTER OUTLINE

Lew Robertson/FoodPix/Getty Images, Inc.

The Water-Soluble Vitamins

8

CASE STUDY

Gary is a graduate student in the history department. His program pays him a small stipend, but it barely covers his rent. To afford school without too many loans, he tries to keep his food costs low. For breakfast, he has oatmeal with milk. For lunch, he brings a tuna sandwich or occasionally goes out with his friends for a burger, fries, and soda. For dinner, he has ramen noodles, macaroni and cheese, or hot dogs. He snacks on soda, candy bars, and chips and has ice cream for dessert.

This eating plan worked for Gary because it was inexpensive and filling, but some of his fellow grad students started to question how healthy it was. Gary had to admit he wasn't feeling his best, and he started to suspect that his friends might be right about his diet. His legs had been aching, and he was tired a lot of the time. He decides that a visit to the health center for a medical evaluation would be a good idea.

© Cat London/iStockphoto

The doctor takes a blood sample to check for anemia and finds that Gary has low levels of the iron-containing protein hemoglobin. She schedules an appointment for him to meet with the dietitian. A quick diet evaluation with a food frequency questionnaire reveals that Gary's limited diet includes little meat and does not include any fruits and vegetables. Because he doesn't eat any fruits and vegetables, his vitamin C intake is very low. Vitamin C promotes iron absorption and is needed for the integrity of small blood vessels, so the dietitian thinks that Gary's low intake of vitamin C may have contributed to his anemia and possibly his leg aches. **As a history major, Gary knows that in centuries past the vitamin C–deficiency disease scurvy afflicted sailors and explorers who had limited access to fresh fruits and vegetables.**

8.1 What Are Vitamins?

LEARNING OBJECTIVES

- Define vitamin.
- Name the sources of vitamins in the U.S. diet.
- Describe how bioavailability affects vitamin requirements.
- Explain the role of coenzymes.

vitamins Organic compounds needed in the diet in small amounts to promote and regulate the chemical reactions and processes needed for growth, reproduction, and maintenance of health.

water-soluble vitamins Vitamins that dissolve in water.

fat-soluble vitamins Vitamins that dissolve in fat.

Vitamins are organic compounds that are essential in the diet in small amounts to promote and regulate the processes necessary for growth, reproduction, and the maintenance of health. When a vitamin is lacking in the diet, deficiency symptoms occur. When the vitamin is restored to the diet, the symptoms resolve. Vitamins have traditionally been grouped based on their solubility in water or fat. This chemical characteristic allows generalizations to be made about how they are absorbed, transported, excreted, and stored in the body. The **water-soluble vitamins** include the B vitamins and vitamin C. The **fat-soluble vitamins** include vitamins A, D, E, and K (**Figure 8.1**).

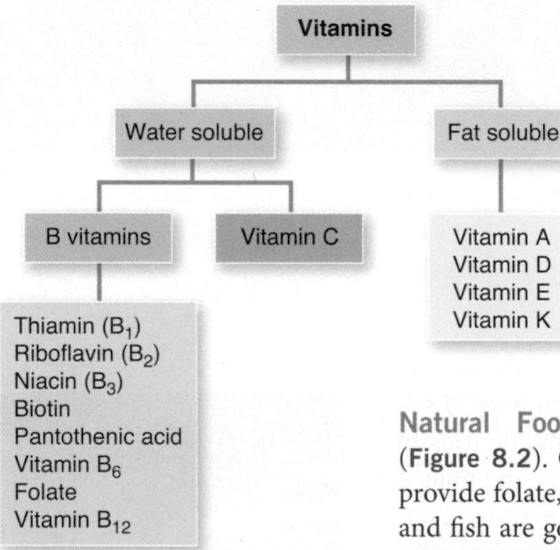

Figure 8.1 The vitamins The vitamins were initially named alphabetically in approximately the order in which they were identified: A, B, C, D, and E. The B vitamins were first thought to be one chemical substance but were later found to be many different substances, so the alphabetical name was broken down by numbers. For example, thiamin is B_1, riboflavin is B_2, and niacin is B_3. Vitamins B_6 and B_{12} are the only ones that are still commonly referred to by their numbers.

Vitamins in Our Diet

The last of the 13 compounds recognized as vitamins today was characterized in 1948. The ability to isolate and purify vitamins has allowed them to be added to the food supply and incorporated into pills. As a result, our diet includes not only vitamins that are naturally present in food but also those that have been added to foods and those consumed as dietary supplements. Despite the variety of options for obtaining vitamins, it is still possible to consume too little of some vitamins, and the popularity of supplements has increased the likelihood that certain vitamins will be consumed in excess.

Natural Food Sources of Vitamins Almost all foods contain some vitamins (**Figure 8.2**). Grains are good sources of most of the B vitamins. Leafy green vegetables provide folate, vitamin A, vitamin E, and vitamin K; citrus fruit provides vitamin C. Meat and fish are good sources of all of the B vitamins, and milk provides riboflavin and vitamins A and D. Even oils provide vitamins; vegetable oils are high in vitamin E. How much of each of these vitamins remains in a food when it reaches the table depends on how the food is handled. Cooking and storage methods can cause vitamin losses. Processing can cause vitamin losses but can also add vitamins and other nutrients to food.

Fortified Foods Fortified foods contain added nutrients. Consuming these foods can be beneficial if the added nutrients are deficient in the diet, but it can also increase the risk of toxicity.

fortification A general term used to describe the process of adding nutrients to foods.

enriched Refers to a food that has had nutrients added to restore those lost in processing to a level equal to or higher than originally present.

Sometimes nutrients are added to foods to comply with government fortification programs that mandate such additions in order to prevent vitamin or mineral deficiencies and promote health in the population. **Fortification** is an effective way to supplement nutrients that are deficient in the population's diet without having to rely on consumers to alter their food choices or to take nutrient supplements. Which foods are fortified, which nutrients are added, and how much of a nutrient is added depend on the food supply, the needs of the population, and public health policies. In the United States, iodine was first added to salt in the 1920s and vitamin D has been added to cow's milk since the early 1930s. By 1943, most refined grain products were **enriched** with thiamin, riboflavin, niacin, and iron. Today, fortification programs are used throughout the world to increase the intake of nutrients likely to be deficient (see Chapter 18). The levels of nutrients added are based on an amount that is high enough to benefit those who need to increase their intake but not so high as to increase the risk of excessive intakes in others.

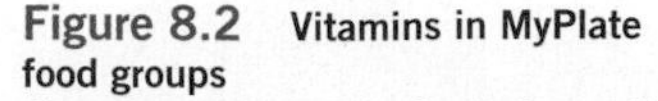

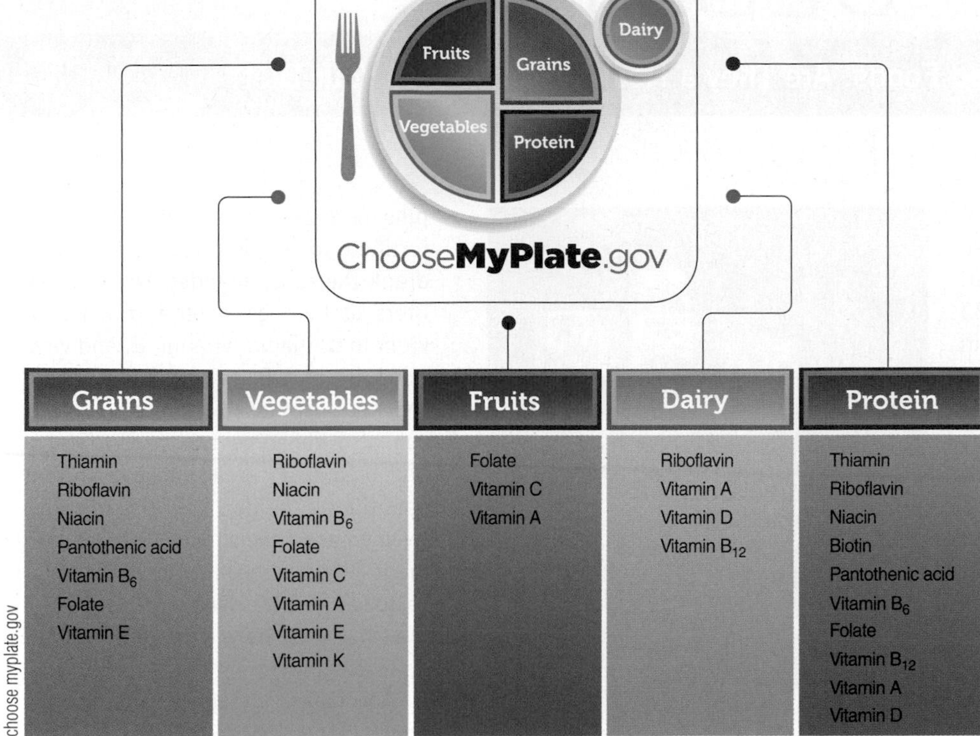

Figure 8.2 **Vitamins in MyPlate food groups**
Vitamins are found in foods from all groups, but some groups are lacking in specific vitamins. For example, grains, fruits, and vegetables lack vitamin B_{12}, and grains and protein foods are low in vitamin C.

Fortified Breakfast Cereal

Nutrition Facts
Serving Size 1 Cup (50g/1.8 oz.)
Servings per Container About 10

Amount Per Serving	Cereal	Cereal with 1/2 Cup Vitamins A&D Fat Free Milk
Calories	180	220
Calories from Fat	5	5
	% Daily Value**	
Total Fat 0.5g*	1%	1%
Saturated Fat 0g	0%	0%
Trans Fat 0g		
Cholesterol 0mg	0%	0%
Sodium 280mg	12%	14%
Potassium 100mg	3%	9%
Total Carbohydrate 35g	12%	14%
Dietary Fiber 2g	9%	9%
Sugars 7g		
Other Carbohydrate 26g		
Protein 3g		
Vitamin A	15%	20%
Vitamin C	25%	25%
Calcium	0%	15%
Iron	100%	100%
Vitamin D	10%	25%
Vitamin E	100%	100%
Thiamin	100%	100%
Riboflavin	100%	110%
Niacin	100%	100%
Vitamin B_6	100%	100%
Folic Acid	100%	100%
Vitamin B_{12}	100%	110%
Pantothenate	100%	100%
Phosphorus	10%	20%
Magnesium	8%	10%
Zinc	100%	100%
Copper	4%	6%

Figure 8.3 **Nutrients in fortified breakfast cereal**
Almost all breakfast cereals are fortified with iron and calcium and a complement of vitamins, many of which are not deficient in the U.S. diet. The amounts and variety of nutrients added to some fortified breakfast cereals are so great that they resemble multivitamin/mineral supplements.

Fortification today extends beyond government-mandated programs. Manufacturers are now fortifying their products with a variety of nutrients. Much of this discretionary fortification involves nutrients that are of public health concern. For example, because many Americans do not consume enough calcium, the food industry has supplied us with calcium-fortified orange juice, breakfast cereal, and cheese. Other nutrients that are commonly added are those that are easy and inexpensive to add to a food, and may promote sales of the product. For example, fortifying a snack food such as a Fruit Roll-Up with vitamin C may boost sales because consumers believe it is a healthy choice. But this food is still high in added sugars and much lower in nutrient density than a piece of fresh fruit. The amounts of nutrients added may be arbitrary. This indiscriminate fortification of foods, such as breakfast cereals, can increase the risk of nutrient toxicities (see **Debate: Super-Fortified Foods: Are They a Healthy Addition to Your Diet?**).

A recent analysis of nutrient intakes in toddlers and preschoolers showed that a significant percentage had intakes of preformed vitamin A, folate, and zinc that exceeded the ULs. Much of the excess zinc is likely to be from fortified breakfast cereals (**Figure 8.3**).[1] Extensive fortification in the breakfast cereal industry has also made it difficult for those who must limit iron intake to find breakfast cereals that are not fortified with iron.

Dietary Supplements Dietary supplements are another source of vitamins in our diet. People take them to increase nutrient intake, as well as to enhance athletic performance, promote weight loss, alleviate existing symptoms and conditions, extend life, and prevent chronic disease. Supplements come as pills, tablets, liquids, and powders. While they provide specific nutrients and can help some people meet their nutrient needs, they do not provide all the benefits of a diet containing a wide variety of foods (see Chapter 9, Off the Label: Read the Label before You Supplement). A varied diet provides phytochemicals and other substances that are not nutrients but that have health-promoting properties. Epidemiological studies show that people who eat more fruits and vegetables have a lower incidence of a host of chronic diseases. These benefits are not duplicated by taking supplements of nutrients found in these foods. Scientists have not yet identified all the substances contained in foods, nor have they determined all of their effects on human health. What is

VIDEO

DEBATE

Super-Fortified Foods: Are They a Healthy Addition to Your Diet?

An orange, a tomato, a slice of bread, and a piece of grilled salmon—these are foods that are part of a healthy diet. What about an energy drink with 23 added vitamins and minerals, a protein bar with 100% of your daily vitamin requirements, soft drinks with echinacea and green tea extract, fruit juice with added phytochemicals, and bottled water fortified with vitamin C? Are products such as these that are fortified with large amounts of nutrients foods, or are they supplements (see photo)? Are they a safe, healthy addition to your diet, or do they pose a risk of toxicity?

Foods like these are fortified with large amounts of nutrients. Do they enhance our health, or create a risk for toxicity?

One could argue that fortified protein bars and juices are foods, not supplements, because they provide calories like traditional foods and the substances added to them, such as vitamin C, fish oil, or phytochemicals, are also naturally found in food. On the other hand, by definition, a supplement is a product intended to add nutrients or other substances to the diet, which these products certainly do. Does it matter if these supplemental substances come in a food or in a pill? Opponents of these foods would argue that it does because our decisions about eating foods are different from our decisions about supplements. Typically, we consider the dose when taking a supplement pill. But we eat to satisfy our sensory desires, fill our stomachs, and quench our thirsts. We don't think about whether the food or beverage might provide toxic amounts of nutrients.

In traditional foods, the amounts of nutrients are small, and the way they are combined limits absorption, making the risk of consuming a toxic amount of a nutrient almost nonexistent. In contrast, it is not difficult to swallow a very high dose of one or more nutrients from an excess of supplement pills or excessive servings of super-fortified foods. For example, if you drank the recommended two to three liters of fluid as water fortified with vitamin C, niacin, vitamin E, and vitamins B_6 and B_{12}, you could exceed the UL for these vitamins. Then if you also consume two cups of fortified breakfast cereal and two protein bars during the day, your risk of toxicity increases even more. The government labels these fortified products as foods and we eat them like foods, but they may have the same toxicity risks as supplements.

Advocates of super-fortified foods point out that they add health-promoting substances to the diet. But do super-fortified foods provide the benefits that the original food would have? In some cases they do. For example, if you are getting your calcium from orange juice, studies show that you are getting just about as much calcium as you would from milk.[1] On the other hand, fish oil consumed in capsules does not have all the heart-health benefits of fish oil consumed in a piece of fish.[2]

So, are these products foods, or are they supplements? It is a fine line. Whether they are helpful or harmful depends on what is in them and how much you consume. Should the government get involved in regulating the amounts of nutrients that can be added to all foods? These answers depend on your view of the government's role in food regulation. Should we be gobbling down super-fortified foods without a thought? Probably not.

THINK CRITICALLY: Should foods fortified with more than the RDA for one or more nutrients carry a warning to avoid overconsumption?

[1] Schroder, B. G., Griffin, I., Specker, B. L., and Abrams, S. A. Absorption of calcium from the carbonated dairy soft drink is greater than that from fat-free milk and calcium fortified orange juice in women. *Nutr Res* 25:737–742, 2005.

[2] He, K. Fish, long-chain omega-3 polyunsaturated fatty acids and prevention of cardiovascular disease—eat fish or take fish oil supplement? *Prog Cardiovasc Dis* 52: 95–114, 2009.

clear is that a wholesome, varied diet is important for optimal health. If chosen with care, supplements are unlikely to be harmful, but consumers should not rely heavily on them to meet their needs (**Figure 8.4**).

Despite their healthful intentions, many people taking supplements may still not be getting the nutrients they need the most. When the types of products people take are compared to the nutrients at risk for deficiency in the American diet, the two don't match. An analysis of the current U.S. diet has shown that average intakes of thiamin, riboflavin, niacin, and vitamin C from food meet or exceed recommendations. However, the average calcium intake in the American diet is below the recommended intake.[2] The multivitamin/mineral pills that most people take provide plenty of B vitamins and vitamin C but little calcium. Additional information on micronutrient supplements is included in the discussion of each vitamin and mineral in this chapter and in Chapters 9 through 12.

Figure 8.4 Supplements cannot replace food
Vitamin supplements cannot take the place of a balanced diet. Even a pill that meets all vitamin needs does not provide the energy, water, protein, minerals, fiber, or phytochemicals that would have been supplied by food sources of these vitamins.

Absorption, Storage, and Excretion of Vitamins

Whether vitamins come from foods, fortified foods, or supplements, they must be absorbed into the body to perform their functions. About 40 to 90% of the vitamins in food are absorbed, primarily in the small intestine (**Figure 8.5**).

THINK CRITICALLY

Why would a very-low-fat diet decrease the absorption of fat-soluble vitamins?

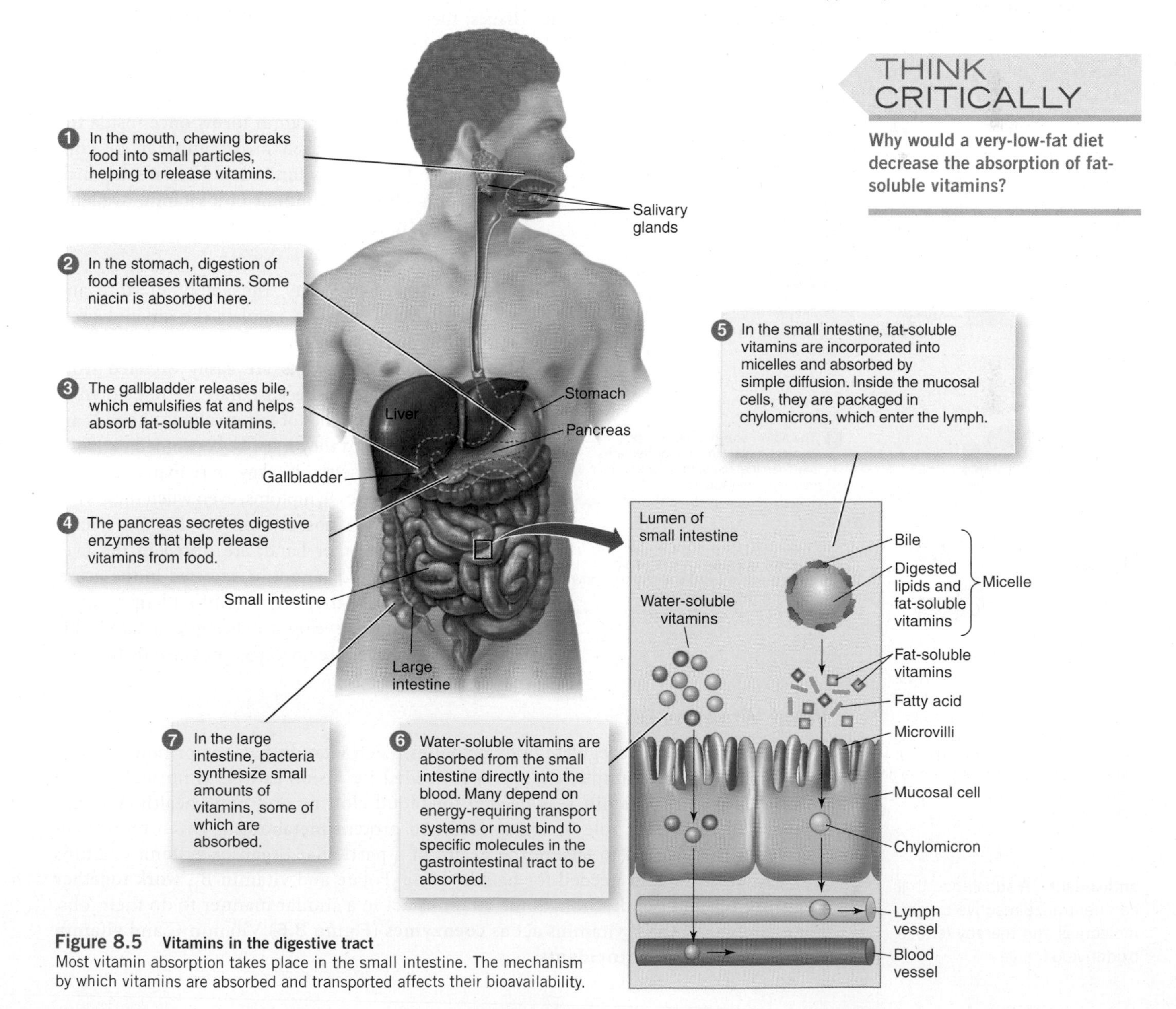

Figure 8.5 Vitamins in the digestive tract
Most vitamin absorption takes place in the small intestine. The mechanism by which vitamins are absorbed and transported affects their bioavailability.

bioavailability A general term describing how well a nutrient can be absorbed and used by the body.

provitamin or **vitamin precursor** A compound that can be converted into the active form of a vitamin in the body.

coenzyme A small nonprotein organic molecule that acts as a carrier of electrons or atoms in metabolic reactions and is necessary for the proper functioning of many enzymes.

The composition of the diet and conditions in the body, however, may influence how much of a vitamin is available in the body. **Bioavailability** refers to how readily a nutrient can be absorbed and utilized by the body.

Bioavailability One of the key factors affecting bioavailability is whether the vitamin is soluble in fat or water. Fat-soluble vitamins are absorbed by simple diffusion. They require fat in the diet for absorption and are poorly absorbed when the diet is very low in fat. The water-soluble vitamins do not require fat for absorption, but many depend on energy-requiring transport systems or must be bound to specific molecules in the gastrointestinal tract in order to be absorbed. For example, thiamin and vitamin C are absorbed by energy-requiring transport systems, riboflavin and niacin require carrier proteins for absorption, and vitamin B_{12} must be bound to a protein produced in the stomach before it can be absorbed in the small intestine.

Once absorbed into the blood, vitamins must be transported to the cells. Most of the water-soluble vitamins are bound to blood proteins for transport. Fat-soluble vitamins must be incorporated into lipoproteins or bound to transport proteins in order to be transported in the aqueous environment of the blood. For example, vitamins A, D, E, and K are all incorporated into chylomicrons for transport from the intestine (see Chapter 5). Vitamin A is stored in the liver, but it must be bound to a specific transport protein to be transported in the blood to other tissues; therefore, the amount delivered to the tissues depends on the availability of the transport protein.

Some vitamins are absorbed in inactive **provitamin** or **vitamin precursor** forms that must be converted into active vitamin forms once inside the body. How much of each provitamin can be converted into the active vitamin and the rate at which this occurs affect the amount of a vitamin available to function in the body.

How It Works
VIEW IN YOUR WILEY PLUS
www.wileyplus.com

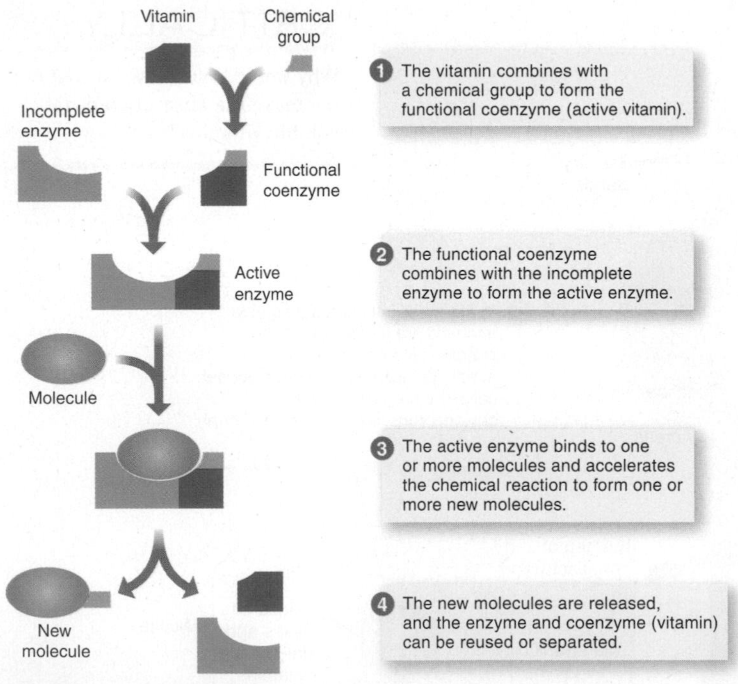

1. The vitamin combines with a chemical group to form the functional coenzyme (active vitamin).
2. The functional coenzyme combines with the incomplete enzyme to form the active enzyme.
3. The active enzyme binds to one or more molecules and accelerates the chemical reaction to form one or more new molecules.
4. The new molecules are released, and the enzyme and coenzyme (vitamin) can be reused or separated.

Figure 8.6 Coenzymes
Coenzymes are needed for enzyme activity. They act as carriers of electrons, atoms, or chemical groups that participate in the reaction. All the B vitamins are coenzymes, but there are also coenzymes that are not dietary essentials and therefore are not vitamins.

Storage and Excretion The ability to store and excrete vitamins helps to regulate the amount present in the body. With the exception of vitamin B_{12}, the water-soluble vitamins are easily excreted from the body in the urine. Because they are not stored to any great extent, supplies of water-soluble vitamins are rapidly depleted and they must be consumed regularly in the diet. Nevertheless, it takes more than a few days to develop deficiency symptoms, even when these vitamins are completely absent from the diet. Fat-soluble vitamins, on the other hand, are stored in the liver and fatty tissues and cannot be excreted in the urine. In general, because they are stored to a larger extent, it takes longer to develop a deficiency of fat-soluble vitamins when they are no longer provided by the diet.

What Vitamins Do

Vitamins promote and regulate body activities. Each vitamin has one or more important functions. For example, vitamin A is needed for vision as well as normal growth and development. Vitamin K is needed for blood clotting and bone health. Vitamin B_6 plays an important role in amino acid and protein metabolism. Often more than one vitamin is needed to ensure the health of a particular organ or system; vitamins A, D, K, and C are all needed for healthy bone. Folate and vitamin B_{12} work together to ensure normal cell division. Some vitamins act in a similar manner to do their jobs. For example, all the B vitamins act as **coenzymes** (**Figure 8.6**).Vitamin C and vitamin E both function as **antioxidants**.

antioxidant A substance that can neutralize reactive oxygen molecules and thereby reduce oxidative damage.

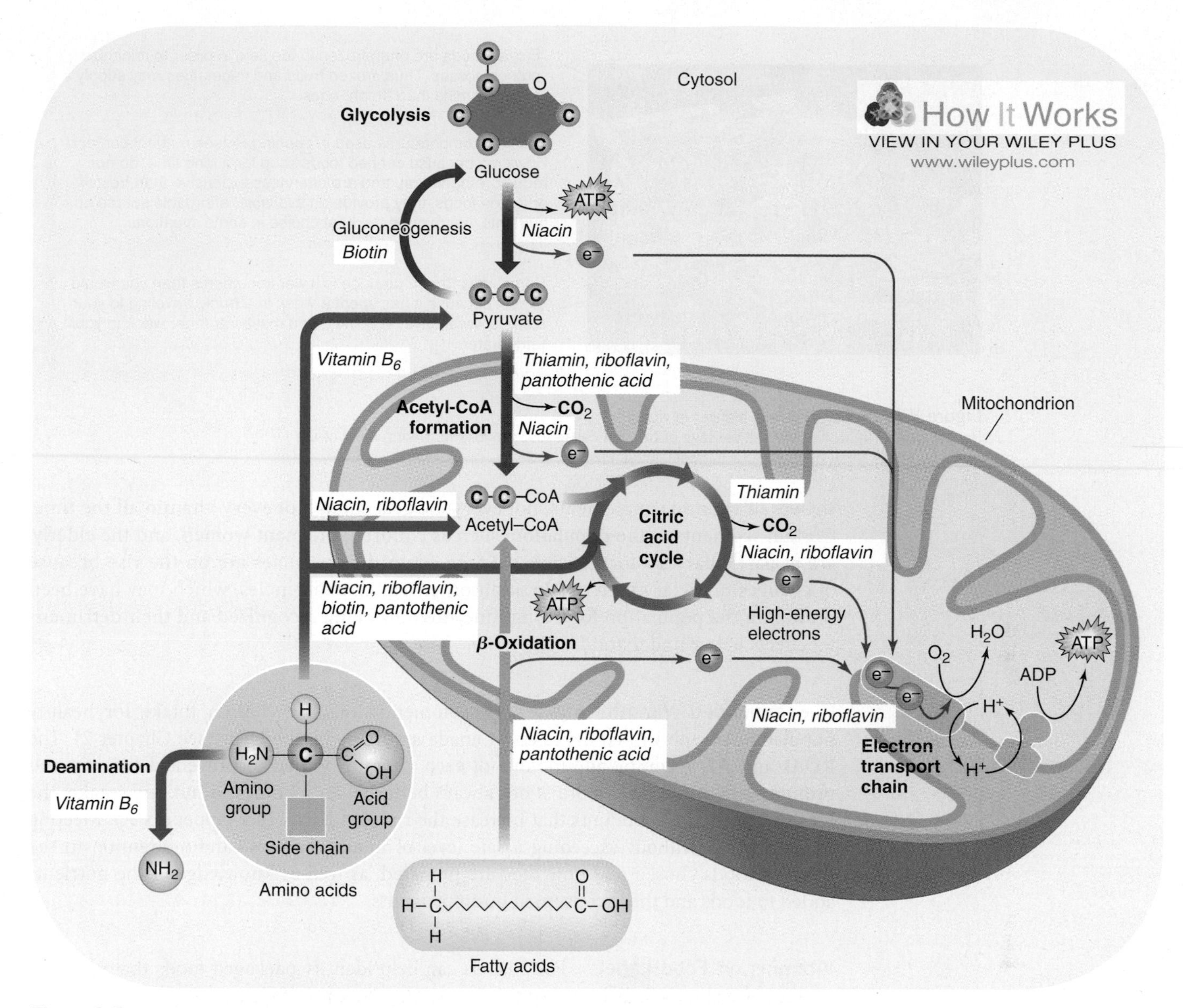

Figure 8.7 B vitamins and energy metabolism
Reactions that require thiamin, riboflavin, niacin, biotin, pantothenic acid, or vitamin B_6 as coenzymes are important in the production of ATP from glucose, fatty acids, and amino acids. The enzymes that catalyze the reactions would not function without the vitamin coenzymes shown.

Vitamins do not provide energy. However many of the B vitamins are coenzymes essential for the proper functioning of enzymes involved in the metabolism of the energy-yielding nutrients. Without these coenzymes, the reactions that produce ATP cannot proceed (**Figure 8.7** and Online **Focus on Metabolism**).

Metabolism

Meeting Vitamin Needs

Today, in the United States and other industrialized countries, an understanding of the sources and functions of the vitamins, a varied food supply, and the ability to fortify foods and supplement nutrients have helped to eliminate severe vitamin deficiencies as a public health problem. For example, niacin deficiency, which was common in the southern United States in the early 1900s, is now almost unheard of; vitamin C deficiency, which has killed countless sailors and soldiers throughout history, is now a rarity; and vitamin A deficiency, which remains a major public health concern worldwide, rarely occurs in developed countries. However, despite all our knowledge, our varied diet, and

George Semple

Frozen foods are often frozen in the field in order to minimize nutrient losses. Thus, frozen fruits and vegetables may supply more vitamins than "fresh" ones.

The high temperatures used in canning reduce nutrient content. However, because canned foods keep for a long time, do not require refrigeration, and are often less expensive than fresh or frozen foods, they provide an available, affordable source of nutrients that may be the best choice in some situations.

Sometimes "fresh" produce is lower in nutrients than you would expect because it has spent a week in a truck, traveling to your store, several days on a shelf, and maybe another week in your refrigerator.

Figure 8.8 **Which choice is highest in vitamins?**
Because heat, light, air, and the passage of time all cause foods to lose nutrients, most of us try to purchase fresh produce, but is fresh always best?

shelves of vitamin supplements, not everyone gets enough of every vitamin all the time. Certain segments of the population, such as children, pregnant women, and the elderly, are at particular risk for deficiency. Some vitamin deficiencies are on the rise because of changes in dietary patterns. In addition, marginal deficiencies, which may have been present in the population for a long time, are now being recognized and their detrimental effects better understood.

Recommended Vitamin Intakes Recommendations for vitamin intake for healthy populations in the United States and Canada are made by the DRIs (see Chapter 2). The RDAs and AIs recommend amounts of each vitamin needed to prevent deficiency and promote health. Because more is not always better when it comes to nutrient intake, the ULs caution against amounts that increase the risk of toxicity (see Appendix A). Meeting vitamin needs without exceeding a safe level of intake requires careful attention to the kinds of foods chosen and how they are prepared, as well as knowledge of the nutrients added to foods and those consumed in supplements.

Vitamins on Food Labels Food labels can help identify packaged foods that are good sources of vitamins. Labels are required to list the amounts of vitamin A and vitamin C in foods as a percentage of the Daily Values (see **Off the Label: How Much Vitamin C Is in Your Orange Juice?**). The % Daily Values of other vitamins are often provided voluntarily. These values include vitamins naturally found in food as well as those added in processing. Fresh fruits, vegetables, fish, meat, and poultry, which are excellent sources of many vitamins, do not carry food labels, but most stores voluntarily provide nutrition information for the raw fruits, vegetables, fish, meat, and poultry that are most frequently purchased.

The label doesn't tell everything about the vitamin content of foods because vitamins can be lost by exposure to light or oxygen, washed away during preparation, or destroyed by cooking (**Figure 8.8**). Vitamin losses can be minimized through food preparation methods that reduce exposure to heat and light (**Table 8.1**).

TABLE 8.1 Tips for Preserving the Vitamins in Your Food

- Store food away from heat and light, and eat it soon after purchasing it.
- Cut fruits and vegetables as close as possible to the time when they will be cooked or served.
- Don't soak vegetables.
- Cook vegetables with as little water as possible by microwaving, pressure-cooking, roasting, grilling, stir-frying, or baking rather than boiling them.
- If foods are cooked in water, use the cooking water to make soups and sauces so that you can retrieve some of the nutrients.
- Don't rinse rice before cooking in order to avoid washing away water-soluble vitamins.

© iStockphoto

OFF THE LABEL

How Much Vitamin C Is in Your Orange Juice?

How much vitamin C is in your orange juice? How much folate is in your breakfast cereal? And how much iron is in a box of raisins? It can be difficult to tell from the Nutrition Facts section of a food label exactly how much of a micronutrient (except sodium) is in a food. Food labels are required to provide the % Daily Values for vitamin A, vitamin C, iron, and calcium but not the actual amount. Daily Values for other vitamins and minerals are provided voluntarily on some foods. Since orange juice is not a significant source of either iron or vitamin A, this information is provided as a footnote in the label shown. To determine the amount of a nutrient in a serving of food from the % Daily Value, you need to know its Daily Value (see Table 2.5 and Appendix D). Once you know the Daily Value, you can multiply it by the % Daily Value on the label to determine the amount in a serving of the food. So, follow these steps to find out how much vitamin C is in a cup of orange juice:

1. Look up the Daily Value:

Vitamin	Daily Value
Vitamin A	5000 IU[a]
Vitamin D	400 IU[a]
Vitamin E	30 IU[a]
Vitamin K	80 μg
Biotin	300 μg
Pantothenic acid	10 mg
Vitamin C	**60 mg**
Thiamin	1.5 mg
Riboflavin	1.7 mg
Niacin	20 mg
Vitamin B_6	2.0 mg
Folic acid	400 μg
Vitamin B_{12}	6 μg

[a]The Daily Values for some fat-soluble vitamins are expressed in International Units (IU).

2. Find the % Daily Value (%DV) on the food label (see figure): %DV for vitamin C in orange juice = 120%.
3. Multiply the % Daily Value by the Daily Value to find out how much is in a serving: 60 mg × 120%= 60 × 1.2 = 72 mg vitamin C.

Orange Juice

Nutrition Facts
Serving Size 8 fl oz (250 mL)
Servings Per Container 8

Amount Per Serving
Calories 110 Calories from Fat 0

	%Daily Value**
Total Fat 0g	**2%**
Sodium 0mg	**0%**
Potassium 450mg	**13%**
Total Carbohydrate 26g	**9%**
Sugars 7g	
Protein 2g	

Vitamin C	120%	Calcium	2%
Thiamin	10%	Riboflavin	4%
Niacin	4%	Vitamin B_6	6%
Folate	15%	Magnesium	6%

Not a significant source or saturated fat, cholesterol, dietary fiber, vitamin A and iron.

*Percent Daily Values are based on a 2,000 calorie diet.

Foodcollection/Getty Images, Inc.

Even if you don't look up the Daily Value and calculate the exact amount of vitamin C or some other vitamin or mineral in a food, the % Daily Value on the food label helps you judge how much the food contains. As a general guideline, if the % Daily Value of a nutrient is 5% or less, the food is a poor source; if it is 10 to 19%, the food is a good source; and if it is 20% or more, the food is an excellent source. Whether you are converting a % Daily Value into the amount of a vitamin or are just looking at the % Daily Value, be sure to consider how many servings you plan to eat. Remember that doubling the serving doubles the nutrients and calories.

THINK CRITICALLY: How many micrograms of folate are provided by 4 fluid ounces of orange juice?

8.2 Thiamin

LEARNING OBJECTIVES

- Name two food groups that provide good sources of thiamin.
- Discuss the role of thiamin in providing energy.
- Explain why a thiamin deficiency causes neurological symptoms.

beriberi The disease resulting from a deficiency of thiamin.

Thiamin was the first of the B vitamins to be identified and is therefore sometimes called *vitamin* B_1. **Beriberi**, the disease that results from a deficiency of this vitamin, came to the attention of Western medicine in colonial Asia in the 19th century. Beriberi became such a problem that the Dutch East India Company sent a team of scientists to find its cause. A young physician named Christian Eijkman worked on the problem for over 10 years. His success came as a twist of fate. He ran out of food for his experimental chickens, and instead of the usual brown rice, he fed them white rice. Shortly thereafter, the chickens came down with beriberi-like symptoms. When he fed them brown rice again, they got well. The events provided evidence that the cause of beriberi was not a poison or a microorganism, as had previously been thought, but rather something missing from the diet.

Knowledge gained from Eijkman's studies made it possible to prevent and cure beriberi by feeding people a diet adequate in thiamin; however, the vitamin itself was not isolated until 1926. We now know that polishing the bran layer off rice kernels to make white rice removes the thiamin-rich portion of the grain. Therefore, in populations where white rice was the staple of the diet, beriberi became a common health problem. The incidence of beriberi in eastern Asia increased dramatically in the late 1800s due to the rising popularity of polished rice.

Thiamin in the Diet

Thiamin is widely distributed in foods (**Figure 8.9**). A large proportion of the thiamin consumed in the United States comes from enriched grain products such as pasta, rice

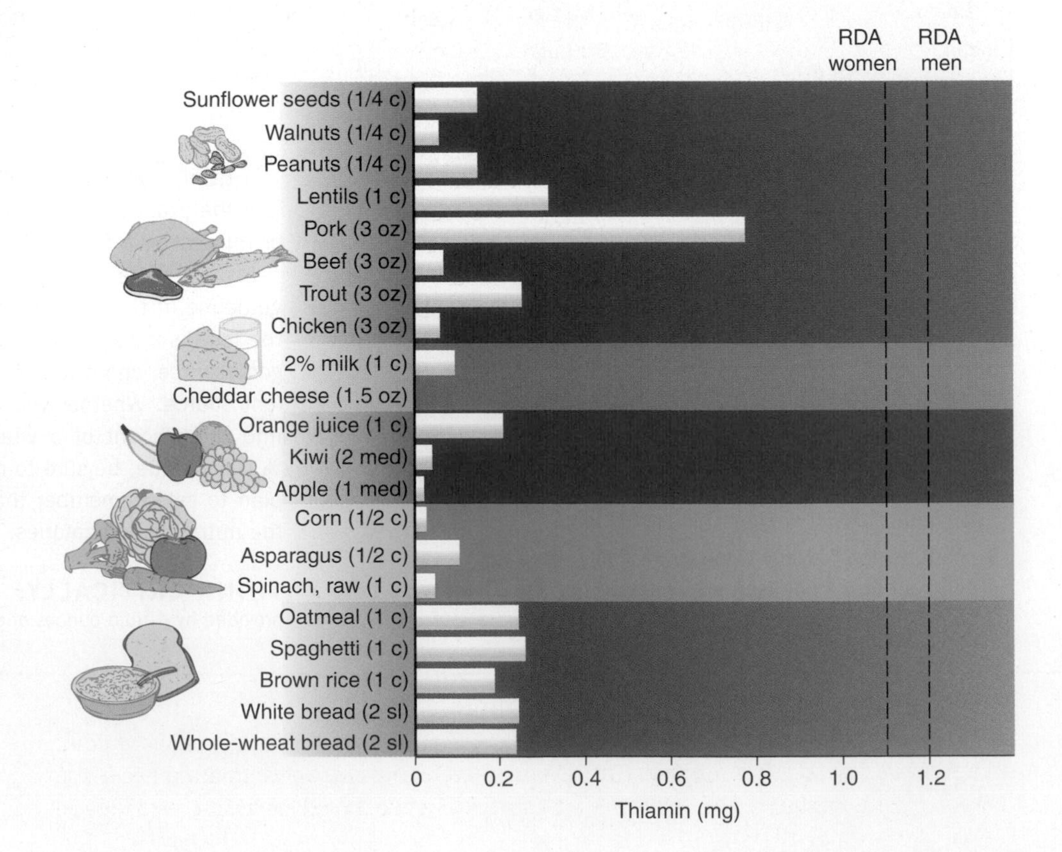

Figure 8.9 Thiamin content of MyPlate food groups
A combination of foods from the grains and protein food groups can easily supply the RDA for thiamin (dashed lines).

dishes, baked goods, and breakfast cereals. Pork, whole grains, legumes, nuts, seeds, and organ meats (liver, kidney, heart) are also good sources.

The thiamin in foods may be destroyed during cooking or storage because it is sensitive to heat, oxygen, and low-acid conditions. Thiamin bioavailability is also affected by the presence of anti-thiamin factors that destroy the vitamin. For instance, there are enzymes in raw shellfish and freshwater fish that degrade thiamin during food storage and preparation and during passage through the gastrointestinal tract. These enzymes are destroyed by cooking, so they are a concern only in foods consumed raw. Other anti-thiamin factors that are not inactivated by cooking are found in tea, coffee, betel nuts, blueberries, and red cabbage. Because these make thiamin unavailable to the body, habitual consumption of foods containing anti-thiamin factors increases the risk of thiamin deficiency.[3]

thiamin pyrophosphate The active coenzyme form of thiamin. It is the predominant form found inside cells, where it aids reactions in which a carbon-containing group is lost as CO_2.

How Thiamin Functions in the Body

Thiamin is a vitamin, so it does not provide energy, but it is important in energy metabolism. The active form, **thiamin pyrophosphate**, is a coenzyme for reactions in which a carbon is lost from larger molecules as carbon dioxide. For instance, the reaction that forms acetyl-CoA from pyruvate and one of the reactions of the citric acid cycle require thiamin pyrophosphate (see Figure 8.7). Thiamin is therefore essential to the production of ATP from glucose.

Thiamin is also needed for the metabolism of other sugars and certain amino acids; the synthesis of the neurotransmitter acetylcholine; and the production of the sugar ribose, which is needed to synthesize RNA (ribonucleic acid).

Wernicke–Korsakoff syndrome A form of thiamin deficiency associated with alcohol abuse that is characterized by mental confusion, disorientation, loss of memory, and a staggering gait.

Thiamin Deficiency

Thiamin deficiency results in the disease beriberi. Beriberi causes lethargy, fatigue, and other neurological symptoms. It can also cause cardiovascular problems such as rapid heartbeat, enlargement of the heart, and congestive heart failure.

Some, but not all, of the symptoms of beriberi can be explained by the roles of thiamin in glucose metabolism and in the synthesis of the neurotransmitter acetylcholine (**Figure 8.10**). The earliest symptoms, depression and weakness, which occur after only about 10 days on a thiamin-free diet, are probably related to the inability to use glucose completely. Since brain and nerve tissue rely on glucose for energy, the inability to rapidly form acetyl-CoA from pyruvate affects nervous system activity. Poor coordination, tingling in the arms and legs, and paralysis may also be caused by the lack of acetylcholine. The reason thiamin deficiency causes cardiovascular symptoms is not well understood.

Overt beriberi is rare in North America today, but thiamin deficiency does occur in alcoholics. They are particularly vulnerable because thiamin absorption is decreased due to the effect of alcohol on the GI tract. In addition, the liver damage that occurs with chronic alcohol consumption reduces conversion of thiamin to the active coenzyme form; thiamin intake also may be low due to a diet high in alcohol and low in nutrient-dense foods. Thiamin-deficient alcoholics may develop a neurological condition known as **Wernicke–Korsakoff syndrome** and characterized by mental confusion, psychosis, memory disturbances, and coma.

Figure 8.10 Thiamin deficiency causes beriberi
(a) The active thiamin coenzyme (thiamin pyrophosphate) is needed to convert pyruvate into acetyl-CoA. Acetyl-CoA can be used to produce ATP or to synthesize the neurotransmitter acetylcholine. Without thiamin, the body cannot synthesize acetylcholine or properly use glucose, which is the primary fuel for the brain and nerve cells.

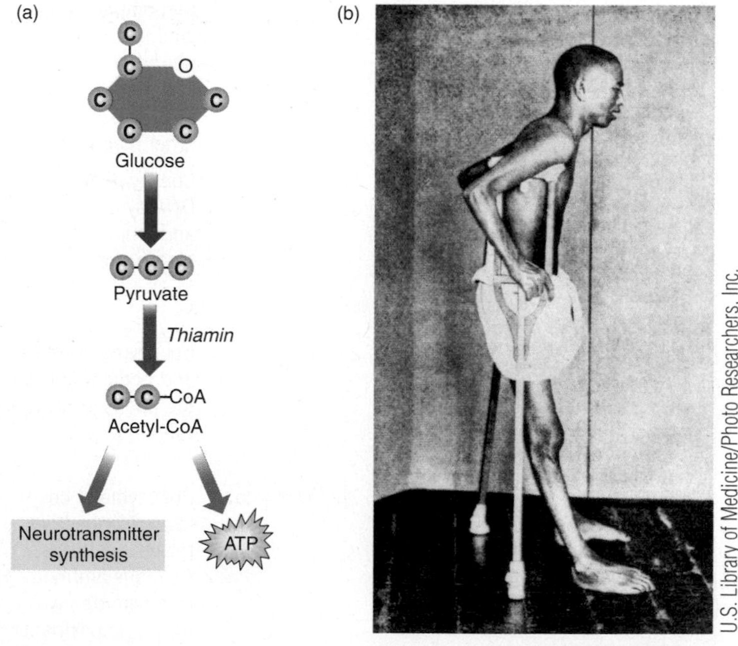

(b) For over 1000 years, beriberi flourished in East Asian countries. In Sri Lanka, the word *beriberi* means "I cannot," referring to the extreme weakness and depression that are the earliest symptoms of the disease.

Recommended Thiamin Intake

The RDA for thiamin for adult men ages 19 and older is set at 1.2 mg/day and for adult women 19 and older at 1.1 mg/day. The RDA is based on the amount of thiamin needed to achieve and maintain normal activity of a thiamin-dependent enzyme found in red blood

TABLE 8.2 A Summary of the Water-Soluble Vitamins and Choline (All Values Are RDAs Unless Otherwise Noted)

Vitamin	Sources	Recommended intake for adults	Major functions	Deficiency diseases and symptoms	Groups at risk of deficiency	Toxicity	UL
Thiamin (vitamin B_1, thiamin mononitrate)	Pork, whole and enriched grains, seeds, nuts, legumes	1.1–1.2 mg/day	Coenzyme in glucose and energy metabolism; needed for neurotransmitter synthesis and normal nerve function	Beriberi: weakness, apathy, irritability, nerve tingling, poor coordination, paralysis, heart changes	Alcoholics, those living in poverty	None reported	ND
Riboflavin (vitamin B_2)	Dairy products, whole and enriched grains, dark green vegetables, meats	1.1–1.3 mg/day	Coenzyme in energy and lipid metabolism	Inflammation of the mouth and tongue, cracks at corners of the mouth	None	None reported	ND
Niacin (nicotinamide, nicotinic acid, vitamin B_3)	Beef, chicken, fish, peanuts, legumes, whole and enriched grains; can be made from tryptophan	14–16 mg NE/day	Coenzyme in energy metabolism and lipid synthesis and breakdown	Pellagra: diarrhea, dermatitis on areas exposed to sun, dementia	Those consuming a limited diet based on corn; alcoholics	Flushing nausea, rash, tingling extremities	35 mg/day from fortified foods and supplements
Biotin	Liver, egg yolks; synthesized in the gut	30 μg/day[a]	Coenzyme in glucose synthesis and energy and fatty acid metabolism	Dermatitis, nausea, depression, hallucinations	Those consuming large amounts of raw egg whites; alcoholics	None reported	ND
Pantothenic acid (calcium pantothenate)	Meat, legumes, whole grains; widespread in foods	5 mg/day[a]	Coenzyme in energy metabolism and lipid synthesis and breakdown	Fatigue, rash	Alcoholics	None reported	ND
Vitamin B_6 (pyridoxine, pyridoxal phosphate, pyridoxamine)	Meat, fish, poultry, legumes, whole grains, nuts and seeds	1.3–1.7 mg/day	Coenzyme in protein and amino acid metabolism, neurotransmitter and hemoglobin synthesis, many other reactions	Headache, convulsions, other neurological symptoms, nausea, poor growth, anemia	Alcoholics	Numbness, nerve damage	100 mg/day
Folate (folic acid, folacin, pteroylglutamic acid)	Leafy green vegetables, legumes, seeds, enriched grains, orange juice	400 μg DFE/day	Coenzyme in DNA synthesis and amino acid metabolism	Macrocytic anemia, inflammation of tongue, diarrhea, poor growth, neural tube defects	Pregnant women, alcoholics	Masks B_{12} deficiency	1000 μg/day from fortified food and supplements
Vitamin B_{12} (cobalamin, cyanocobalamin)	Animal products	2.4 μg/day	Coenzyme in folate and homocysteine metabolism; nerve function	Pernicious anemia, macrocytic anemia, nerve damage	Vegans, elderly, people with stomach or intestinal disease	None reported	ND
Vitamin C (ascorbic acid, ascorbate)	Citrus fruit, broccoli, strawberries, leafy green vegetables, peppers	75–90 mg/day	Coenzyme in collagen (connective tissue) synthesis; hormone and neurotransmitter synthesis; antioxidant	Scurvy: poor wound healing, bleeding gums, loose teeth, bone fragility, joint pain, pinpoint hemorrhages	Alcoholics, elderly people	GI distress, diarrhea	2000 mg/day
Choline[b]	Egg yolks, organ meats, wheat germ, meat, fish, nuts, synthesis in the body	425–550 mg/day[a]	Synthesis of cell membranes and neurotransmitters	Fatty liver, muscle damage, abnormal prenatal development	None	Sweating, low blood pressure, liver damage	3500 mg/day

[a]Adequate intake (AI).

[b]Choline is technically not a vitamin, but recommendations have been made for its intake.

Note: UL, Tolerable Upper Intake Level; NE, niacin equivalent; DFE, dietary folate equivalent; ND, not determined due to insufficient data.

cells and normal urinary thiamin excretion.[4] For an average adult, half of the RDA can be obtained from 4 ounces of pork or one-quarter cup of shelled sunflower seeds.

The requirement for thiamin is increased during pregnancy to accommodate the needs of growth and energy utilization, and during lactation to meet the need for increased energy for milk production and to replace the thiamin secreted in milk. There is not enough information to establish an RDA for infants, so an AI has been set based on the thiamin intake of infants fed human milk. Recommended intakes for all groups are provided on the inside cover and a summary of the sources, recommended intakes, functions, deficiencies, and toxicities of thiamin and other water-soluble vitamins is provided in **Table 8.2**.

Thiamin Toxicity and Supplements

Since no toxicity has been reported when excess thiamin is consumed from either food or supplements, a UL for thiamin intake has not been established.[4] This does not mean that high intakes are necessarily safe. Intakes of thiamin above the RDA have not been shown to provide health benefits.

Thiamin supplements containing up to 50 mg/day are widely available and are marketed with the promise that they will provide "more energy." Thiamin is needed to produce ATP, but unless thiamin is deficient, increasing thiamin intake does not increase the ability to produce ATP. Because thiamin deficiency causes mental confusion and damages the heart, supplements often promise to improve mental function and prevent heart disease. However, in the absence of a deficiency, supplements do not have these effects. Thiamin is also included in supplements referred to as B-complex supplements (**Table 8.3**).

TABLE 8.3 Benefits and Risks of Water-Soluble Vitamin Supplements

Supplement	Claim	Actual benefits or risks
B complex (thiamin, riboflavin, niacin, pantothenic acid, biotin,vitamin B_6, vitamin B_{12})	Increases energy, needed during stress	Needed for energy metabolism but does not provide energy. Low risk of toxicity except for vitamin B_6.
Niacin (nicotinic acid form)	Lowers cholesterol	Medicinal doses may reduce cholesterol levels. Causes flushing, tingling and potentially liver damage. Doses above 35 mg should be taken only under medical supervision.
Vitamin B_6	Prevents heart disease; relieves carpal tunnel syndrome (CTS), autism, and premenstrual syndrome (PMS); enhances immune function	Adequate amounts needed to maintain immune function and normal homocysteine levels, which reduces heart disease risk—excess provides no additional benefit. Supplements are not beneficial for CTS but may help reduce the symtoms of PMS. Levels above the UL may cause tingling, numbness, and muscle weakness.
Folate (folic acid)	Prevents birth defects, protects against heart disease and cancer	Adequate amounts needed to keep homocysteine levels normal, which reduces heart disease risk. Low folate may increase cancer risk. Supplemental sources reduce the risk of birth defects and are recommended for women of childbearing age. May mask a vitamin B_{12} deficiency at high intakes.
Vitamin B_{12}	Prevents heart disease, prevents dementia, reduces fatigue	Adequate amounts needed for nerve function, for red blood cell synthesis, and to keep homocysteine levels low, which reduces heart disease risk. Supplemental sources recommended for older adults and vegans. No benefit of excess. Low risk of toxicity.
Vitamin C	Prevents colds, reduces cold symptoms, enhances immunity, protects against heart disease and cancer, enhances antioxidant protection	May reduce duration of colds. Important antioxidant but extra as supplements has not been shown to provide additional benefits. Too much can cause GI distress, damage teeth, promote kidney stone formation, and interfere with anticoagulant medications.

8.3 Riboflavin

LEARNING OBJECTIVES

- Describe the role of riboflavin in energy metabolism.
- Explain why milk sold in clear bottles might be low in riboflavin.

While searching for a cure for beriberi, scientists isolated riboflavin and several other B vitamins in addition to thiamin. This occurred because the extracts they made from vegetables and grains could be separated into two components: One contained thiamin, the antiberiberi factor they sought, and cured beriberi; the other was a mix of B vitamins that was later determined to contain riboflavin along with vitamin B_6, niacin, and pantothenic acid.

Riboflavin in the Diet

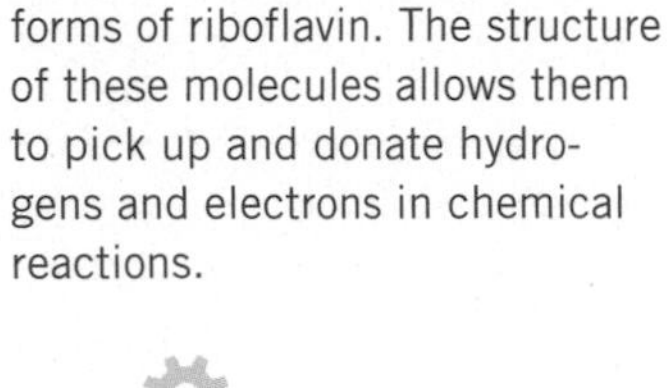

flavin adenine dinucleotide (FAD) and **flavin mononucleotide (FMN)** The active coenzyme forms of riboflavin. The structure of these molecules allows them to pick up and donate hydrogens and electrons in chemical reactions.

Milk is the best source of riboflavin in the North American diet. Other important sources of this B vitamin include liver, red meat, poultry, fish, and whole- and enriched grain products. Vegetable sources include asparagus, broccoli, mushrooms, and leafy green vegetables such as spinach (**Figure 8.11a**). Because riboflavin is destroyed by exposure to light, poor handling decreases a food's riboflavin content. This is a problem when milk is stored in clear containers and exposed to light. Cloudy plastic milk bottles block some light, partially protecting the riboflavin, but cardboard or opaque plastic milk containers are even better at preventing losses[5] (**Figure 8.11b**).

How Riboflavin Functions in the Body

Riboflavin forms the active coenzymes **flavin adenine dinucleotide (FAD)** and **flavin mononucleotide (FMN)**. FAD functions in the citric acid cycle and is important for the breakdown of fatty acids. Both FAD and FMN function as electron carriers in the electron transport chain (see Figure 8.7). Therefore, adequate riboflavin is crucial in providing energy from carbohydrate,

Figure 8.11 **Riboflavin in foods**

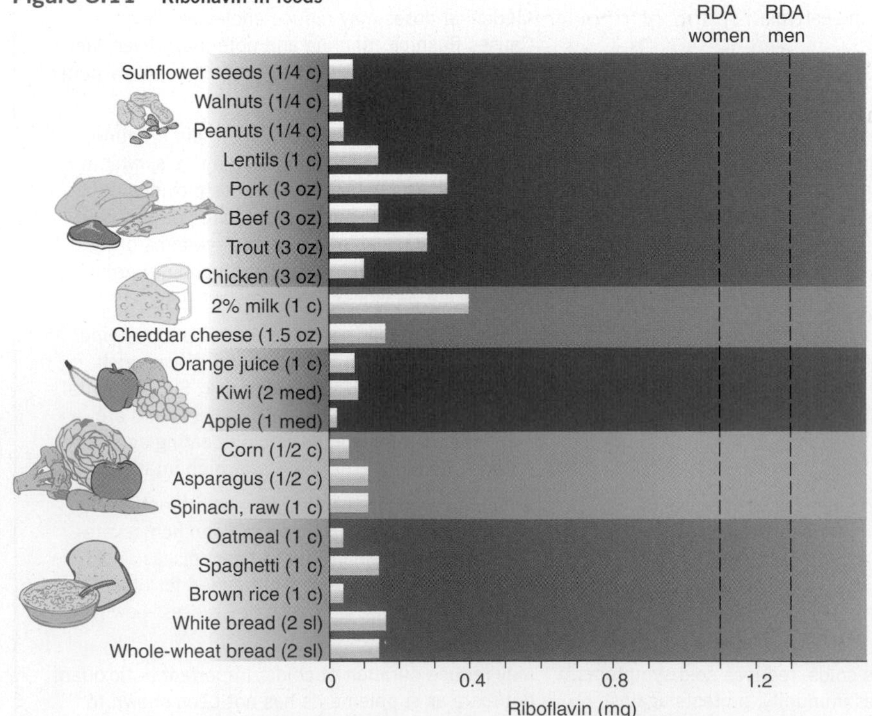

(a) This graph of the riboflavin content of foods from different MyPlate food groups shows that milk, meats, and fortified grains are good sources of riboflavin but a combination of foods are needed to supply the RDA (dashed lines).

Jupiter Images/FoodPix/Getty Images, Inc.

(b) Ever wonder why milk doesn't come in glass bottles anymore? One reason is that riboflavin is destroyed by light. Exposure to light can also cause an "off" flavor in the milk and losses of vitamins A and D.[5]

fat, and protein. Riboflavin is also involved directly or indirectly in converting a number of other vitamins, including folate, niacin, vitamin B_6, and vitamin K, into their active forms.

Riboflavin Deficiency

When riboflavin is deficient, injuries heal poorly because new cells cannot grow to replace the damaged ones. Tissues that grow most rapidly, such as the skin and the linings of the eyes, mouth, and tongue, are the first to be affected by a deficiency. Symptoms of riboflavin deficiency, called **ariboflavinosis**, include inflammation of the eyes, lips, mouth, and tongue; scaly, greasy skin eruptions; cracking of the tissue at the corners of the mouth; and confusion. Deficiency symptoms may develop after approximately two months on a riboflavin-poor diet.

ariboflavinosis The condition resulting from a deficiency of riboflavin.

A deficiency of riboflavin is rarely seen alone. It usually occurs in conjunction with deficiencies of other B vitamins. One reason is that the food sources of B vitamins are similar (see Table 8.2). Therefore, a deficiency of riboflavin due to poor diet will likely lead to multiple vitamin deficiencies. Because riboflavin is needed to convert other vitamins into their active forms, some of the symptoms seen with riboflavin deficiency are actually due to deficiencies in these other nutrients.

Recommended Riboflavin Intake

The RDA for riboflavin for adult men ages 19 and older is 1.3 mg/day and for adult women 19 and older, 1.1 mg/day. This recommendation is based on the amount of riboflavin needed to maintain normal activity of a riboflavin-dependent enzyme in red blood cells and normal riboflavin excretion in the urine.[4] Two cups of milk provide about half the daily amount of riboflavin recommended for a typical adult. The recommended intake can be met without milk if the daily diet includes 2 to 3 servings of meat and 4 to 5 servings of enriched grain products and high-riboflavin vegetables, such as spinach.

Additional riboflavin is recommended during pregnancy to support growth and increased energy utilization, and during lactation to allow for the riboflavin secreted in milk. There is not enough information to establish an RDA for infants, so an AI has been set based on the amount of riboflavin consumed by infants fed human milk (see inside cover).

Riboflavin Toxicity and Supplements

No adverse effects have been reported from overconsumption of riboflavin from foods or supplements, and there are not sufficient data to establish a UL for this vitamin. Large doses of riboflavin are not well absorbed, and it is readily excreted in the urine. A harmless side effect of high riboflavin intake, such as may be obtained from over-the-counter supplements, is bright yellow urine.

As with thiamin, the role of riboflavin in energy metabolism has led to claims that supplements containing riboflavin, such as B-complex supplements, will provide an energy boost (see Table 8.3). Although riboflavin is needed for energy metabolism, it does not provide energy. Since a deficiency causes skin and eye symptoms, riboflavin has also been suggested as a cure for eye diseases and skin disorders. However, in the absence of a deficiency, supplementation does not affect the eyes or skin.

8.4 Niacin

LEARNING OBJECTIVES

- Discuss why a niacin deficiency is more likely in someone consuming a limited diet that is based on corn.
- Explain the role of niacin in energy metabolism.
- List the three Ds of pellagra.

A deficiency of the B vitamin niacin results in a disease called **pellagra**, which causes progressive physical and mental deterioration. It was first observed in Europe in the

pellagra The disease resulting from a deficiency of niacin.

18th century, and in the early 20th century it became endemic in the southeastern United States. At the time, many believed that it was caused by an infection rather than a deficiency in the diet (see **Science Applied: Pellagra: Infectious Disease or Dietary Deficiency?**). The emergence of pellagra can be traced to the cultivation of corn, which is a poor source of niacin, as a dietary staple.[6] It primarily affects the poor who cannot afford a varied diet.

Niacin in the Diet

Meat and fish are good sources of niacin (**Figure 8.12a**). Other sources include legumes, mushrooms, wheat bran, asparagus, and peanuts. Niacin added to enriched flours used in baked goods provides much of the usable niacin in the North American diet. Niacin can also be synthesized in the body from the essential amino acid tryptophan (**Figure 8.12b**).

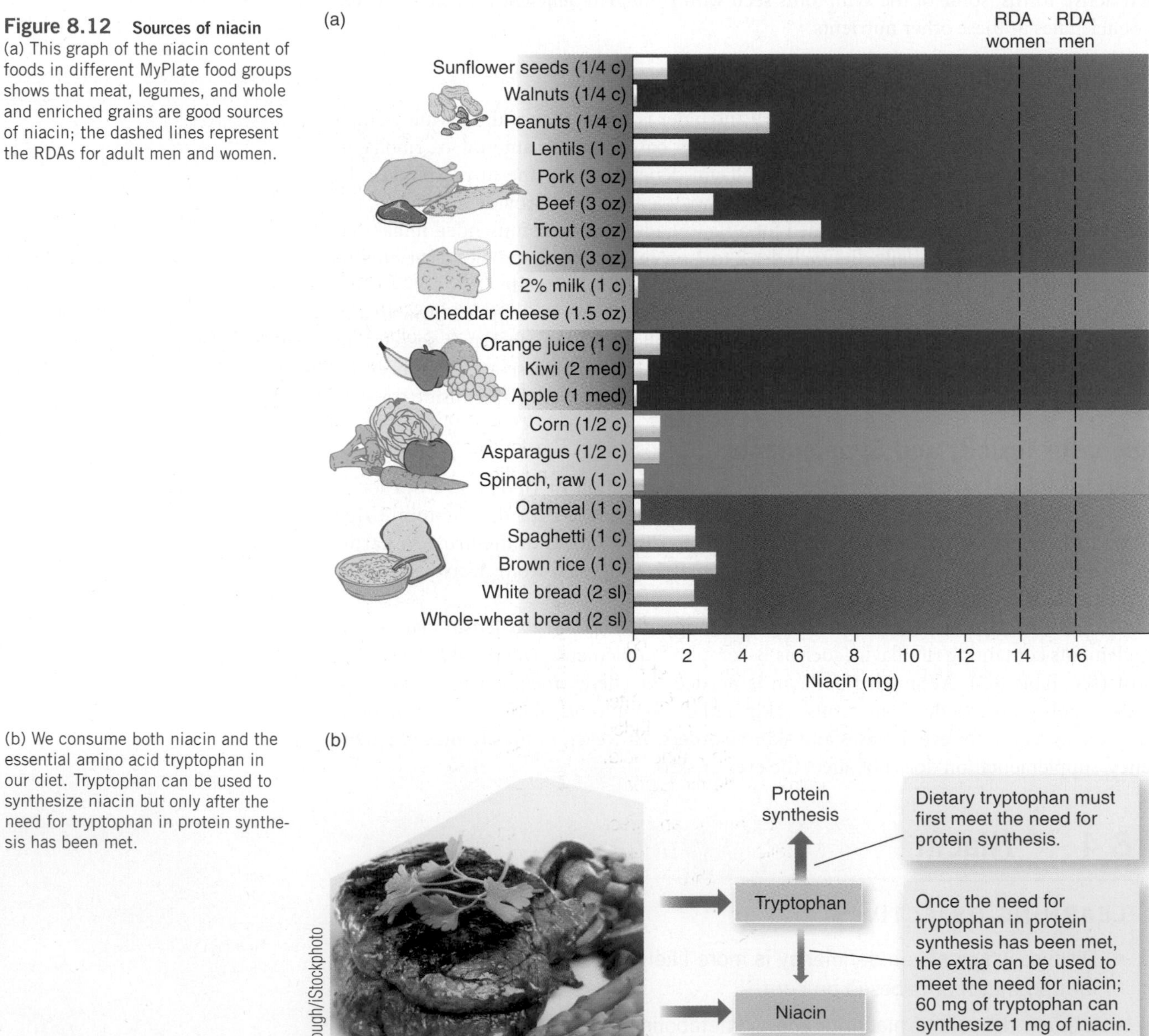

Figure 8.12 **Sources of niacin**
(a) This graph of the niacin content of foods in different MyPlate food groups shows that meat, legumes, and whole and enriched grains are good sources of niacin; the dashed lines represent the RDAs for adult men and women.

(b) We consume both niacin and the essential amino acid tryptophan in our diet. Tryptophan can be used to synthesize niacin but only after the need for tryptophan in protein synthesis has been met.

SCIENCE APPLIED

Pellagra: Infectious Disease or Dietary Deficiency?

(Elena Schweitzer/Shutterstock) Victoriano Izquierdo/Getty Images, Inc. (DAJ/Getty Images, Inc.)

In the early 1900s, psychiatric hospitals in the southeastern United States were filled with patients with dementia due to a disease called *pellagra*. Although we now know pellagra is due to niacin deficiency, at the time it was thought to be caused by an infectious agent or toxin. Annually 100,000 people were affected by pellagra, 3000 of them fatally. The federal government established the Thompson–McFadden Pellagra Commission and the U.S. Public Health Service selected Dr. Joseph Goldberger to investigate the cause of the disease and hopefully find a cure.

THE SCIENCE

Investigators with the Pellagra Commission believed that pellagra was caused by an infection. They injected monkeys and baboons with blood, urine, and other extracts from patients with pellagra, but none developed the disease. These results supported the view that pellagra was not due to an infectious agent, yet they still believed that pellagra could be transmitted in some way from a pellagrous to a nonpellagrous person: Perhaps monkeys and baboons were not susceptible to pellagra.[1]

Goldberger studied the communities and institutions where pellagra was common: Pellagra was prevalent among patients in institutions but not among attendants and nurses—a fact that did not support an infectious nature. Goldberger observed that the patients were fed a diet typical of the Southern poor: corn meal, molasses, and fatback or salt pork. The staff ate a more varied diet. By adding milk, eggs, and more meat to the patients' diets, Goldberger was able to cure pellagra. The next step in supporting his hypothesis that pellagra was caused by a dietary deficiency was to bring about pellagra in healthy people by feeding them the deficient institutional diet. In 1915, at a Mississippi State Penitentiary work farm, 12 convicts volunteered to be fed a diet of corn meal, grits, cornstarch, white wheat flour, white rice, cane syrup, sugar, sweet potatoes, a few green vegetables, and a liberal amount of pork fat. After six months, six men developed pellagra. Goldberger concluded that the cause of the disease was a deficiency of an amino acid, a mineral, a fat-soluble vitamin, or some unknown vitamin factor.

To provide the final proof that pellagra was not due to an infectious agent, Goldberger and 15 of his colleagues voluntarily injected themselves with blood, swabbed their throats with nasal secretions, and swallowed urine, feces, and skin cells from patients who were severely ill with pellagra (later in the experiment, they put the feces and other materials in capsules). After six months, neither he nor his colleagues had become ill. Goldberger proved that pellagra was not an infectious disease and continued searching for the missing dietary factor.

Despite Goldberger's efforts, the epidemic raged on. Experiments with dogs had demonstrated that yeast and liver could cure pellagra, but nothing had been done to change the Southern diet that produced the disease. Joseph Goldberger died in 1929, the year the epidemic reached its peak. In 1937, another research team identified the pellagra-preventative factor as nicotinic acid.

THE APPLICATION

The work of Goldberger and his successors showed us the cause of pellagra and that it could be prevented and cured by adding meat, eggs, milk, yeast, and other niacin-rich foods to the diet. Despite this understanding, poor dietary habits, poverty, and chronic malnutrition allowed this curable disease to continue until the 1940s when the enrichment of grains helped to eliminate it.[2]

In 1936, the Council on Foods and Nutrition of the American Medical Association recommended the fortification of food. In 1938, bakers began using high-vitamin yeast, which added protein, niacin, and other B vitamins to baked goods (see graph).[3] In the 1940s, the recommendations of the Committee on Food and Nutrition, now the Food and Nutrition Board, led to the enrichment of flour with thiamin, riboflavin, niacin, and iron.[4] The fortification of foods with nutrients that are deficient in a population's diet is now used throughout the world to improve nutrient intake and prevent deficiency diseases.

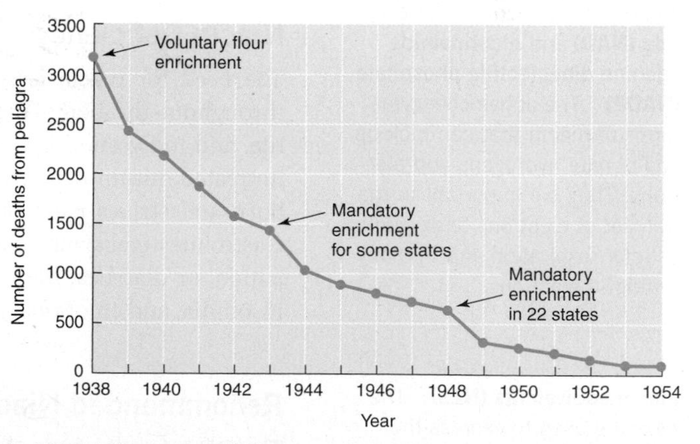

In 1938, bakers voluntarily began enriching flour with B vitamins, a move that led to a decline in mortality from pellagra. By the early 1940s, enrichment of flour was becoming mandatory. This measure, along with the economic boom created by World War II, led to a rapid and dramatic decline in pellagra mortality.

[1]Roe, D. A. *A Plague of Corn: The Social History of Pellagra.* Ithaca, NY: Cornell University Press, 1973.

[2]Syndenstricker, V. P. The history of pellagra, its recognition as a disorder of nutrition and its conquest. *Am J Clin Nutr* 6:409–441, 1958.

[3]Park, Y. K., Sempos, C. T., Barton, C. N., et al. Effectiveness of food fortification in the United States: The case of pellagra. *Am J Public Health* 90:727–738, 2000.

[4]Institute of Medicine, Food and Nutrition Board. *Dietary Reference Intakes: Guiding Principles for Nutrition Labeling and Fortification.* Washington, DC: National Academies Press, 2003.

Figure 8.13 Beans and tortillas provide niacin The treatment of corn with lime water during the preparation of tortillas improves niacin bioavailability. Serving tortillas alongside beans, which are a good source of niacin, has helped prevent pellagra in Mexico and other Latin American countries with a corn-based diet.

In a diet that contains high-protein foods such as milk and eggs, which are poor sources of niacin but good sources of tryptophan, much of the need for niacin can be met by tryptophan. Tryptophan, however, is used to make niacin only if enough is available first to meet the needs of protein synthesis. When the diet is low in tryptophan, it is not used to synthesize niacin. Food composition tables and databases list only preformed niacin in a food, not the amount of niacin that can be made from tryptophan contained within the food.

The association between niacin deficiency and a limited diet based on corn has been attributed to the low-tryptophan content of corn and the fact that the niacin found naturally in corn (and to a lesser extent in other cereal grains) is bound to other molecules and therefore not well absorbed. The treatment of corn with lime water (water and calcium hydroxide), as is done in Mexico and Central America during the making of tortillas, enhances the availability of niacin. The diet in these regions also contains legumes, which provide both niacin and a source of tryptophan for the synthesis of niacin (**Figure 8.13**). As a result, despite their corn-based diet, populations in these regions have not suffered from pellagra. Today, pellagra remains common in India and parts of China and Africa.[7] Efforts to eradicate this deficiency include the development of new varieties of corn that provide more available niacin and more tryptophan than traditional varieties.

How Niacin Functions in the Body

nicotinamide adenine dinucleotide (NAD) and **nicotinamide adenine dinucleotide phosphate (NADP)** The active coenzyme forms of niacin that can pick up and donate hydrogens and electrons. They are important in the transfer of electrons to oxygen in cellular respiration and in many synthetic reactions.

Niacin is important in the production of ATP from the energy-yielding nutrients as well as in reactions that synthesize other molecules. There are two forms of niacin: nicotinic acid and nicotinamide. Either form can be used by the body to make the two active coenzymes **nicotinamide adenine dinucleotide (NAD)** and **nicotinamide adenine dinucleotide phosphate (NADP)**. NAD functions in glycolysis and the citric acid cycle, accepting released electrons and passing them on to the electron transport chain where their energy is trapped and used to convert ADP to ATP (see Figure 8.7). NADP acts as an electron carrier in reactions that synthesize fatty acids and cholesterol.

Niacin Deficiency

The need for niacin is so widespread in metabolism that a deficiency causes damage throughout the body. The early symptoms of pellagra include fatigue, decreased appetite, and indigestion, followed by the three Ds: dermatitis, diarrhea, and dementia. If left untreated, niacin deficiency results in a fourth D—death. The dermatitis resembles sunburn and strikes parts of the body exposed to sunlight, heat, or injury (**Figure 8.14**). Gastrointestinal symptoms include a bright-red tongue and may include vomiting, constipation, or diarrhea. Mental symptoms begin with irritability, headaches, loss of memory, insomnia, and emotional instability and progress to psychosis and acute delirium.

Recommended Niacin Intake

niacin equivalents (NEs) The measure used to express the amount of niacin present in food, including that which can be made from its precursor, tryptophan. One NE is equal to 1 mg of niacin or 60 mg of tryptophan.

The RDA for niacin is expressed as **niacin equivalents (NEs)**. One NE is equal to 1 mg of niacin or 60 mg of tryptophan. This allows for the fact that some of the requirement for niacin can be met by the synthesis of niacin from tryptophan. To estimate the niacin contributed by high-protein foods, protein is assumed to be about 1% tryptophan. The criterion used to estimate the average niacin requirement is urinary excretion of niacin metabolites. The RDA for adult men and women of all ages is 16 and 14 mg NE per day, respectively.[4] A meal containing a medium chicken breast and a cup of steamed asparagus provides this amount.

Niacin needs are increased during pregnancy to account for the increase in energy expenditure and during lactation to account for both the increase in energy expenditure and the niacin secreted in milk. There is not enough information to establish an RDA for infants, so an AI has been set based on the amount of niacin found in human milk (see inside cover).

Niacin Toxicity and Supplements

There is no evidence of any adverse effects from consumption of niacin naturally occurring in foods, but supplements can be toxic. The adverse effects of high intakes of niacin include flushing of the skin, a tingling sensation in the hands and feet, a red skin rash, nausea, vomiting, diarrhea, high blood sugar levels, abnormalities in liver function, and blurred vision. Since flushing is the first toxicity symptom to appear as the dose is increased, the UL for adults has been set at 35 mg/day, the highest level that is unlikely to cause flushing in the majority of healthy people. This value applies to the forms of niacin contained in supplements and fortified foods but does not include niacin naturally occurring in foods.

Niacin is a commonly used vitamin supplement that is included in multivitamins as well as in B-complex vitamin supplements marketed to give you more energy. High doses of this vitamin are also used as a medication to treat elevated blood cholesterol (see Table 8.3). Doses of 50 mg/day or greater of the nicotinic acid form of niacin have been found to decrease blood levels of LDL cholesterol and triglycerides and increase HDL cholesterol.[8] Since the UL for niacin is only 35 mg/day, many people are unable to take niacin as a cholesterol-lowering drug because they experience side effects due to niacin toxicity. Because niacin supplements are available over the counter, people may try to treat themselves for high blood cholesterol, but high doses of vitamins are as dangerous as drugs and should be used only with medical supervision.

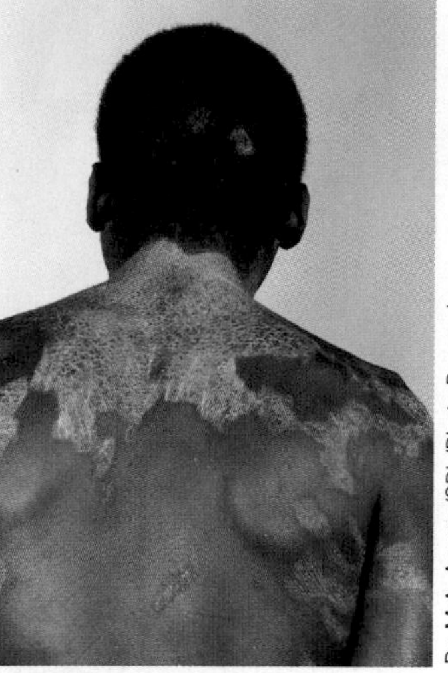

Dr. M.A. Ansary/SPL/Photo Researchers

Figure 8.14 Pellagra
The cracked, inflamed skin characteristic of pellagra most commonly appears on areas exposed to sunlight or other stresses.

8.5 Biotin

LEARNING OBJECTIVES

- Explain why consuming raw eggs might cause a biotin deficiency.
- Discuss why an AI rather than an RDA has been established for biotin.

Biotin was discovered when rats fed protein from raw egg white developed a syndrome of hair loss, dermatitis, and neuromuscular dysfunction. The symptoms were due to a deficiency of biotin. The deficiency was caused by a protein in raw egg white, called *avidin*, which tightly binds biotin and prevents its absorption.

Figure 8.15 Raw egg shakes
Drinking raw eggs or shakes containing raw eggs is a common practice among body builders. Cooked eggs provide just as much protein and do not reduce biotin absorption or pose a risk for *Salmonella* infection.

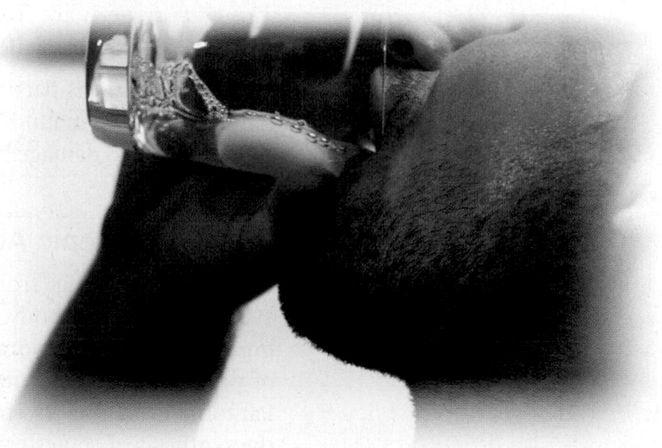

A Carmichael/Getty Images, Inc.

Biotin in the Diet

Good dietary sources of biotin include liver, egg yolks, yogurt, and nuts. Fruit and meat are poor sources. Foods containing raw egg whites should be avoided not only because avidin binds biotin and prevents its absorption but because raw eggs also may be contaminated with *Salmonella* bacteria, which can cause food-borne illness (**Figure 8.15**). Thoroughly cooking eggs destroys bacteria and denatures avidin so that it cannot bind biotin.

How Biotin Functions in the Body

Biotin is a coenzyme for a group of enzymes that add the acid group COOH to molecules. It functions in energy metabolism because it is needed to make a four-carbon molecule necessary in the citric acid cycle and in gluconeogenesis. This B vitamin is also important in the synthesis of fatty acids and some amino acids (see Figure 8.7).

Biotin Deficiency

Although biotin deficiency is uncommon, it has been observed in people with malabsorption or protein-energy malnutrition, those receiving tube feedings or total parenteral nutrition without biotin, those taking anticonvulsant drugs for long periods, and those frequently consuming raw egg whites.[4] When biotin intake is deficient, symptoms including nausea, thinning hair, loss of hair color, a red skin rash, depression, lethargy, hallucinations, and tingling of the extremities gradually appear.

Recommended Biotin Intake

A dietary requirement for biotin has been difficult to estimate because some biotin is produced by bacteria in the gastrointestinal tract and absorbed into the body. Therefore no RDA could be determined, but an AI of 30 μg/day has been established for adult men and women based on the amount of biotin found in a typical North American diet.[4]

LIFE CYCLE

No additional biotin is recommended for pregnancy, but the AI is increased during lactation to account for the amount secreted in milk. The AI for infants is based on the amount of biotin consumed by infants fed human milk.

Biotin Toxicity

No toxicity has been reported in patients given 200 mg/day of biotin to treat various disease states, and sufficient data are not available to establish a UL.[4]

8.6 Pantothenic Acid

LEARNING OBJECTIVES

- Discuss why pantothenic acid deficiency is rare.
- Explain why pantothenic acid is said to be needed "everywhere" in the body.

Pantothenic acid, which gets its name from the Greek word *pantothen* (meaning "from everywhere"), is a B vitamin that is widely distributed in foods.

Pantothenic Acid in the Diet

Pantothenic acid is particularly abundant in meat, eggs, whole grains, and legumes. It is found in lesser amounts in milk, vegetables, and fruits (**Figure 8.16**). Pantothenic acid is susceptible to damage by exposure to heat and low- or high-acid conditions.

How Pantothenic Acid Functions in the Body

In addition to being "from everywhere" in the diet, pantothenic acid seems to be needed everywhere in the body. It is part of coenzyme A (CoA), a coenzyme needed for the breakdown of carbohydrates, fatty acids, and amino acids, as well as for the modification of proteins and the synthesis of neurotransmitters, steroid hormones, and hemoglobin. Pantothenic acid is also needed for the activity of acyl carrier protein, which is needed for the synthesis of cholesterol and fatty acids (see Figure 8.7).

Pantothenic Acid Deficiency

The wide distribution of pantothenic acid in foods makes deficiency rare in humans. A deficiency of this vitamin alone has not been reported, but it may occur as part of a multiple–B vitamin deficiency resulting from malnutrition or chronic alcoholism.

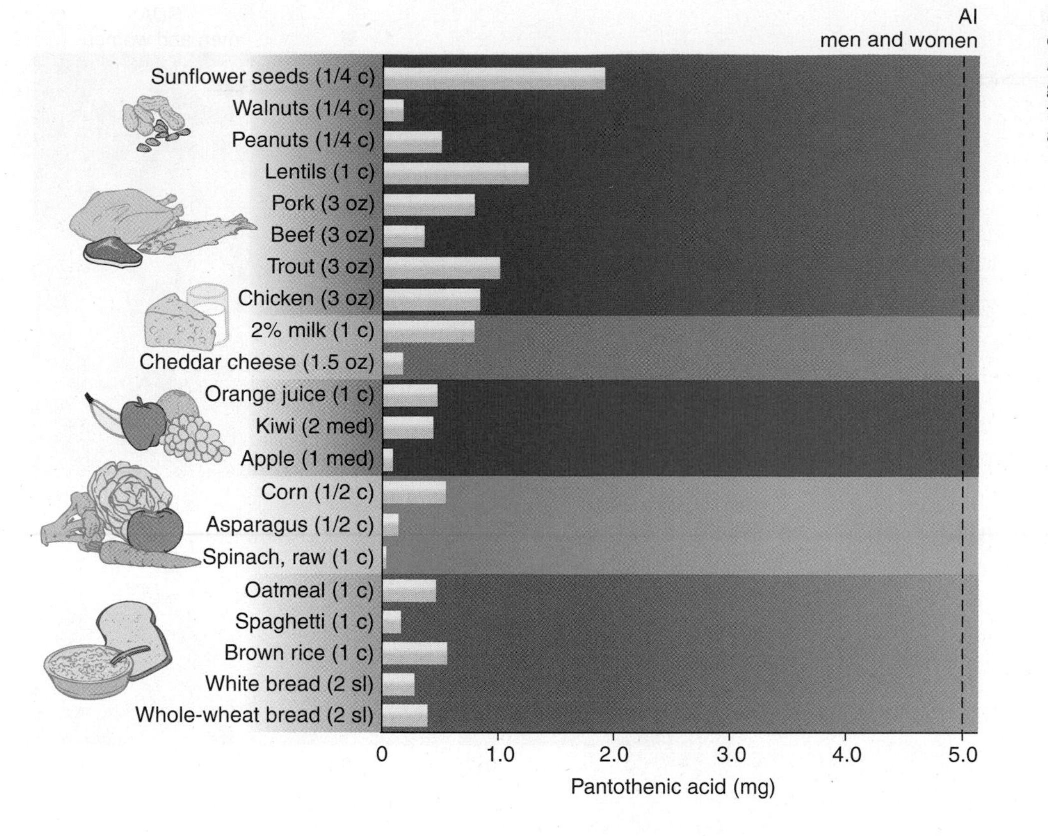

Figure 8.16 Pantothenic acid content of MyPlate food groups All food groups in MyPlate contain good sources of pantothenic acid; the dashed line represents the AI for adult men and women.

Recommended Pantothenic Acid Intake

There is no RDA for pantothenic acid, but an AI of 5 mg/day has been recommended for adult men and women.[4] This value is based on the intake of pantothenic acid sufficient to replace urinary losses. The AI is increased to 6 and 7 mg/day to meet the needs of pregnancy and lactation, respectively.

Pantothenic Acid Toxicity

Pantothenic acid is relatively nontoxic. No toxicity symptoms were reported in a study that fed young men 10 grams of pantothenic acid per day for six weeks. Another study found that doses of 10 to 20 grams per day may result in diarrhea and water retention.[4] Data are not sufficient to establish a UL for pantothenic acid.

8.7 Vitamin B_6

LEARNING OBJECTIVES

- Name two animal products and two plant products that are good sources of vitamin B_6.
- Explain the role of vitamin B_6 in amino acid metabolism.
- Describe the relationship between vitamin B_6 and heart disease.

Vitamin B_6 is particularly important for amino acid and protein metabolism. This B vitamin was identified only when a deficiency syndrome was discovered that did not respond to thiamin or riboflavin supplementation.

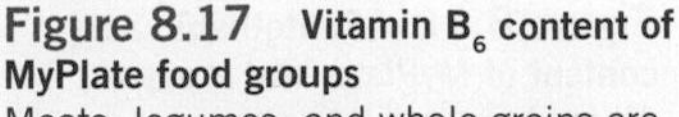

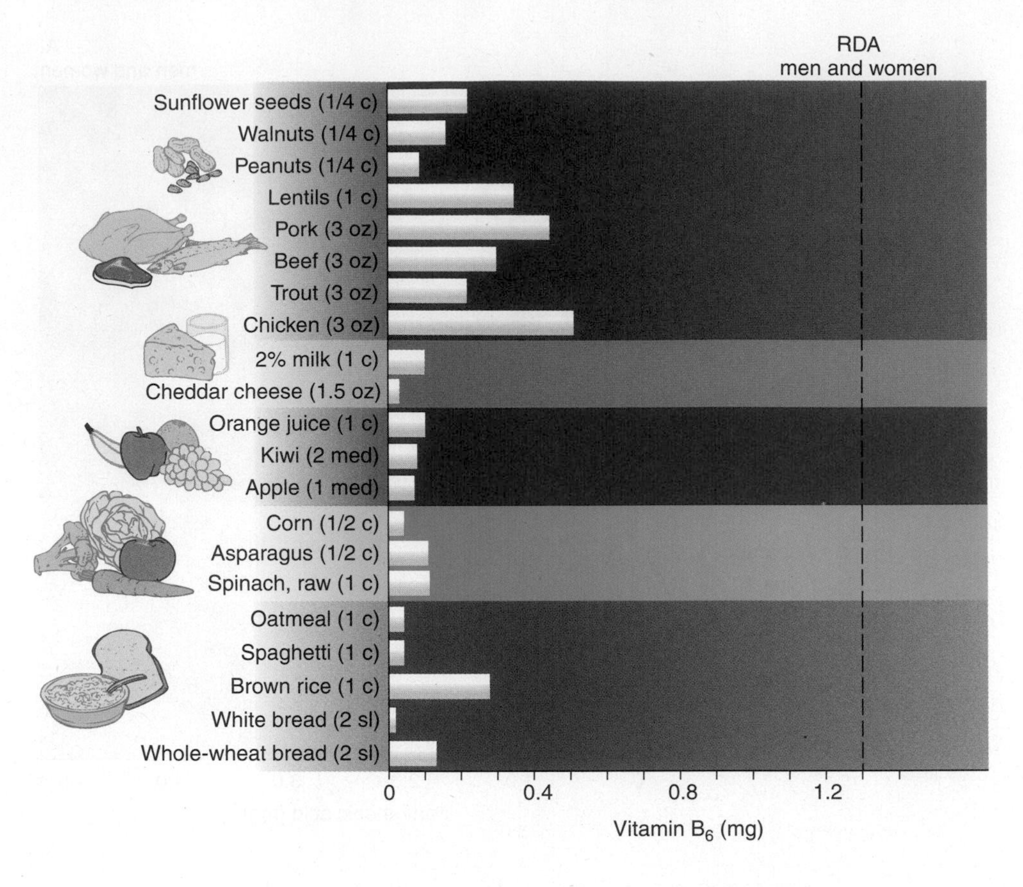

Figure 8.17 Vitamin B_6 content of MyPlate food groups
Meats, legumes, and whole grains are the best sources of vitamin B_6; the dashed line represents the RDA for adults age 50 and younger.

Vitamin B_6 in the Diet

Vitamin B_6 is found in both animal and plant foods. Animal sources include chicken, fish, pork, and organ meats. Good plant sources include whole-wheat products, brown rice, soybeans, sunflower seeds, and some fruits and vegetables such as bananas, broccoli, and spinach (**Figure 8.17**). Vitamin B_6 is easily destroyed by exposure to heat and light and is easily lost in processing. It is not added back in the enrichment of grain products, but fortified breakfast cereals make an important contribution to vitamin B_6 intake.[9]

How Vitamin B_6 Functions in the Body

Vitamin B_6, also known as **pyridoxine**, comprises a group of compounds including pyridoxal, pyridoxine, and pyridoxamine. All three forms can be converted into the active coenzyme form, **pyridoxal phosphate** (see Appendix H). Pyridoxal phosphate is needed for the activity of more than 100 enzymes involved in the metabolism of carbohydrate, fat, and protein. It is a coenzyme for transamination reactions that synthesize nonessential amino acids, for deamination reactions needed for amino acid breakdown, and for the synthesis of neurotransmitters from amino acids (**Figure 8.18**). Pyridoxal phosphate is also a coenzyme needed to synthesize the conditionally essential amino acid cysteine from methionine and to synthesize hemoglobin, the oxygen-carrying protein in red blood cells. Pyridoxal phosphate is important for the immune system because it is needed to form white blood cells. It is also needed for the conversion of tryptophan to niacin, the metabolism of glycogen, and the synthesis of the lipids that are part of the myelin coating on nerves (**Figure 8.19**).

pyridoxine The chemical term for vitamin B_6.

pyridoxal phosphate The major coenzyme form of vitamin B_6 that functions in more than 100 enzymatic reactions, many of which involve amino acid metabolism.

Vitamin B_6 Deficiency

A vitamin B_6 deficiency syndrome was defined in 1954 when an infant formula was overheated in the manufacturing process, destroying the vitamin B_6. The infants who consumed only this formula developed abdominal distress, convulsions, and other neurological symptoms.[10] The neurological symptoms associated with vitamin B_6 deficiency include depression, headaches, confusion, numbness and tingling in the extremities, and

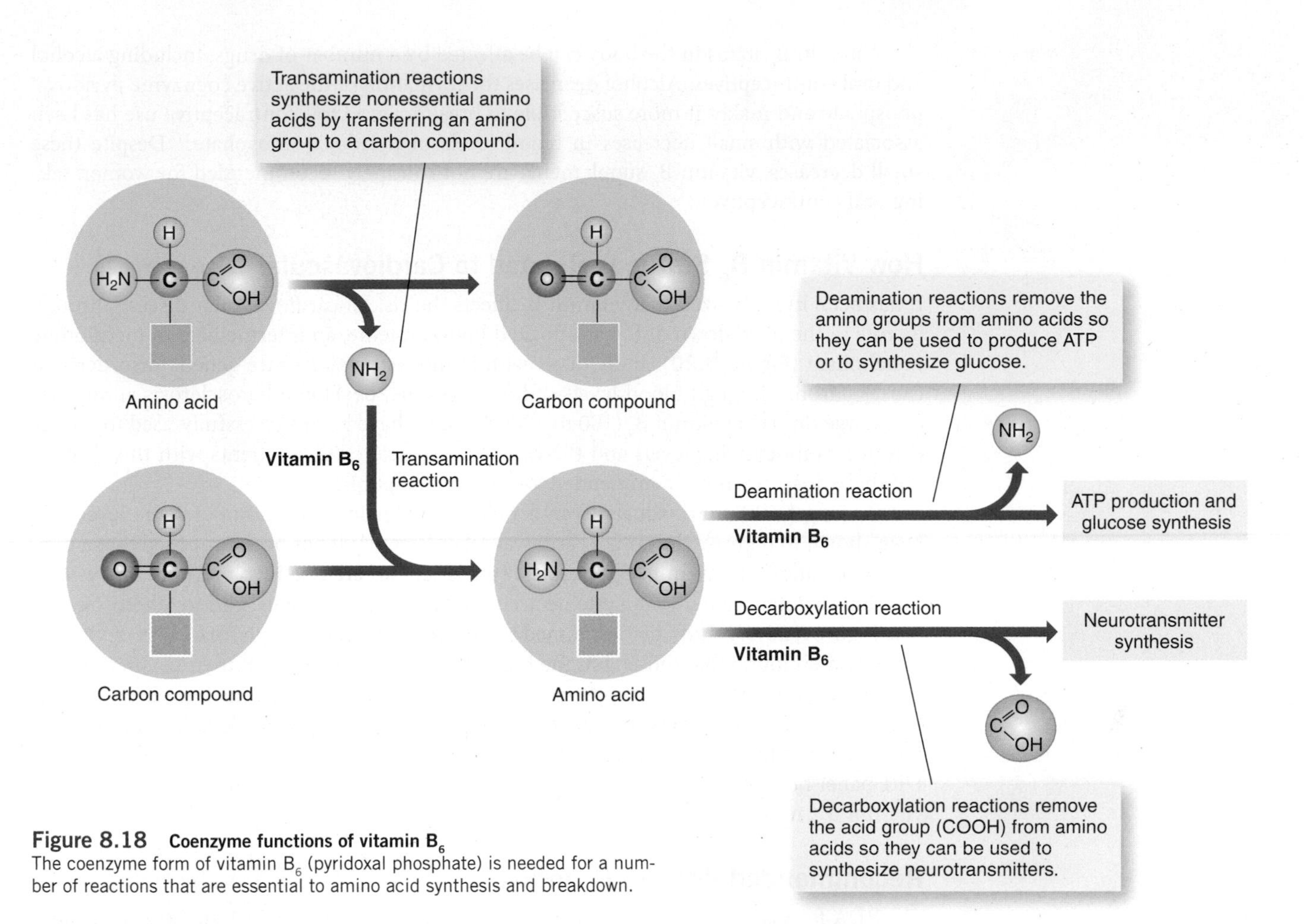

Figure 8.18 **Coenzyme functions of vitamin B_6**
The coenzyme form of vitamin B_6 (pyridoxal phosphate) is needed for a number of reactions that are essential to amino acid synthesis and breakdown.

seizures. These may be related to the role of vitamin B_6 in neurotransmitter synthesis and myelin formation. Anemia also occurs when vitamin B_6 is deficient due to impaired hemoglobin synthesis; red blood cells are small (microcytic) and pale due to the lack of hemoglobin. Other deficiency symptoms such as poor growth, skin lesions, and decreased antibody formation may occur because vitamin B_6 is important in protein and energy metabolism. Since vitamin B_6 is needed for amino acid metabolism, the onset of a deficiency can be hastened by a diet that is low in vitamin B_6 but high in protein.

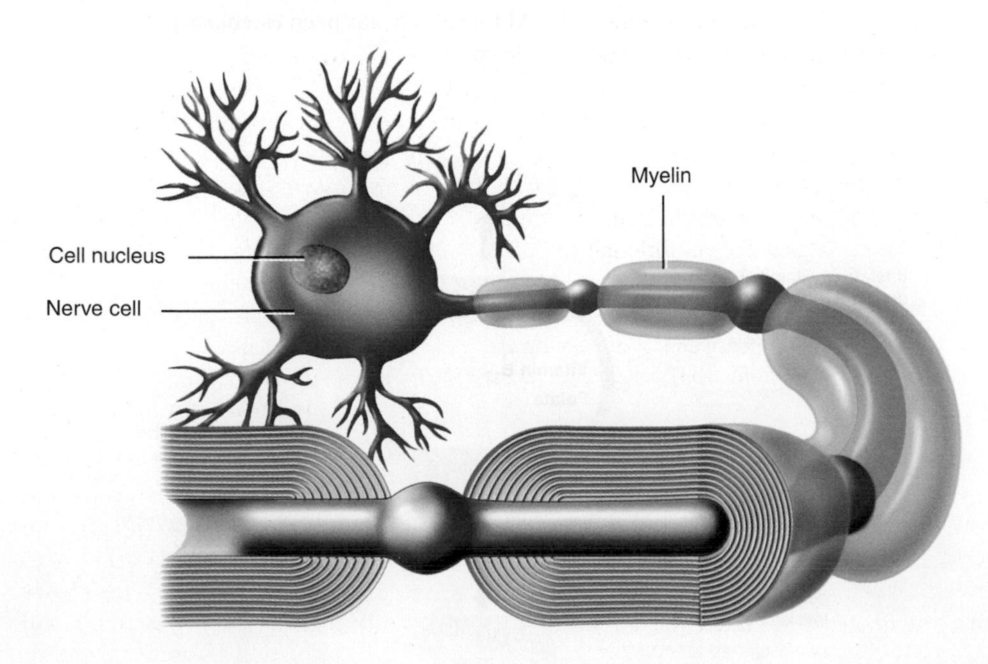

Figure 8.19 **Myelin**
Both vitamin B_6 and vitamin B_{12} (discussed later) are needed to synthesize and maintain the mylein coating on nerve cells. Myelin is essential for normal nerve transmission.

Vitamin B_6 status in the body can be affected by a number of drugs, including alcohol and oral contraceptives. Alcohol decreases the formation of the active coenzyme pyridoxal phosphate and makes it more susceptible to breakdown.[4] Oral contraceptive use has been associated with small decreases in blood levels of pyridoxal phosphate.[11] Despite these small decreases, vitamin B_6 supplements are not routinely recommended for women taking oral contraceptives.

How Vitamin B_6 Status Is Related to Cardiovascular Disease

It has been hypothesized that vitamin B_6 affects the risk of cardiovascular disease through its role in the breakdown of the amino acid homocysteine, an intermediate in methionine metabolism (**Figure 8.20**). Individuals with homocystinuria, a rare genetic disorder that causes chronically high blood levels of homocysteine, develop atherosclerosis at an early age. Large doses of vitamin B_6 (100 to 1000 mg/day) have been successfully used to reduce elevated homocysteine levels and the risk of atherosclerosis in patients with this disease, but these doses are not recommended for the general public.

Among healthy individuals, even a mild elevation in blood homocysteine levels has been shown to increase the risk of cardiovascular disease.[12] It has been proposed that a deficiency of vitamin B_6, vitamin B_{12}, or folate (the latter two are also involved in homocysteine metabolism) may cause homocysteine accumulation and eventually lead to atherosclerosis (see Figure 8.20). A study that examined the relationship between plasma levels of vitamin B_6 in women found that those with the highest levels of vitamin B_6 in their plasma had the lowest risk of heart attack.[13] Supplements of all three of these vitamins have been used to reduce homocysteine levels in individuals with mild homocysteine elevation, but supplementation has failed to significantly reduce overall cardiovascular risk.[12,14] At this time, the DRI panel believes that it is premature to conclude that increased intakes of vitamin B_6, vitamin B_{12}, or folate could decrease the risk of cardiovascular disease.[4]

Recommended Vitamin B_6 Intake

The RDA for vitamin B_6 is 1.3 mg/day for both adult men and women 19 to 50 years of age.[4] This is the amount needed to maintain adequate blood concentrations of the active coenzyme pyridoxal phosphate. In adults 51 years and older, the RDA is increased to 1.7 mg/day in men and 1.5 mg/day in women to maintain normal blood levels of pyridoxal phosphate. A 3-ounce (85-g) serving of chicken, fish, or pork, or half a baked potato, provides about a quarter of the RDA for an average adult; a banana provides about a third.

The RDA for vitamin B_6 is increased during pregnancy to provide for metabolic needs and growth of the mother and fetus. Because the vitamin B_6 concentration in breast milk is dependent on the mother's intake, the RDA is increased during lactation to assure adequate levels are supplied to the infant. An AI for infants has been established based on the vitamin B_6 content of human milk (see inside cover).

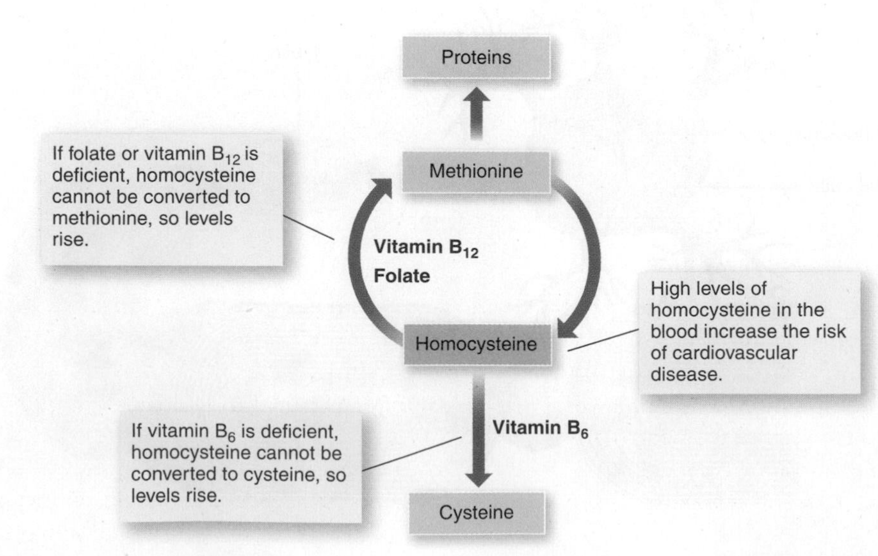

Figure 8.20 B vitamins and the risk of cardiovascular disease
Vitamin B_6, folate, and vitamin B_{12} are required in the metabolism of homocysteine. Without adequate amounts of these B vitamins, homocysteine accumulates and increases the risk of cardiovascular disease.

Vitamin B_6 Toxicity and Supplements

No adverse effects have been associated with high intakes of vitamin B_6 from foods, but large doses found in supplements can cause serious toxicity symptoms. This was first recognized in the 1980s when severe nerve impairment was reported in individuals taking 2 to 6 g of supplemental pyridoxine per day.[14] Symptoms were serious enough in some subjects that they were unable to walk. These symptoms improved when the pyridoxine supplements were stopped. To avoid toxicity when taking supplements containing vitamin B_6, it is important that intake not exceed the UL of 100 mg/day from food and supplements.[4] The UL is based on an amount that will not cause nerve damage in the majority of healthy people. Since high-dose supplements of vitamin B_6 containing 100 mg per dose (5000% of the Daily Value) are available over the counter, it is easy to obtain a dose that exceeds the UL.

Despite the potential for toxicity, people take vitamin B_6 supplements to treat carpal tunnel syndrome, reduce the symptoms of premenstrual syndrome (PMS), and strengthen immune function. Some of the marketing claims are founded in science but are often exaggerated to sell products (see Table 8.3).

Can Vitamin B_6 Alleviate Carpal Tunnel Syndrome? Vitamin B_6 supplements have been suggested to be useful in treating carpal tunnel syndrome, in which pressure on the nerves in the hand causes pain and weakness. Vitamin B_6 is essential for nerve function, but supplements have not been found to be any more effective than a placebo for the treatment of carpal tunnel syndrome.[15]

Does Vitamin B_6 Prevent Premenstrual Syndrome? PMS is a collection of physical and emotional symptoms that some women experience prior to menstruation. It causes mood swings, food cravings, bloating, tension, depression, headaches, acne, breast tenderness, anxiety, temper outbursts, and over 100 other symptoms. The proposed connection between these symptoms and vitamin B_6 is the fact that the vitamin is needed for the synthesis of the neurotransmitters serotonin and dopamine. Insufficient vitamin B_6 has been suggested to reduce levels of these neurotransmitters and cause the anxiety, irritability, and depression associated with PMS. Daily vitamin B_6 supplementation has been found in one study to significantly reduce PMS symptoms such as moodiness, irritability, forgetfulness, bloating, and, anxiety.[16] More research is needed to confirm that vitamin B_6 supplements are safe and effective for treating PMS.

Will Vitamin B_6 Boost Immunity? Immune function can be impaired by a deficiency of any nutrient that hinders cell growth and division. Therefore, one of the most common claims for vitamin supplements in general is that they improve immune function. Vitamin B_6 is no exception, and there are data to support the claim. Vitamin B_6 supplements have been found to improve immune function in older adults.[4] However, since elderly individuals frequently have low intakes of vitamin B_6, it is unclear whether the beneficial effects of supplements are due to an improvement in vitamin B_6 status or immune system stimulation.

8.8 Folate (Folic Acid)

LEARNING OBJECTIVES

- Compare the bioavailability of the folate occurring naturally in food to that found in fortified foods and supplements.
- Explain how folate deficiency causes anemia.
- Discuss why supplemental folic acid is recommended for women of child-bearing age.

It has been known for over 100 years that anemia often occurs during pregnancy. In 1937, a pregnant woman with anemia was successfully treated with a yeast preparation named *Wills Factor* (after Dr. Lucy Wills, who treated this patient). The Wills Factor was later isolated from spinach and named *folate*, after the Latin word for foliage. **Folate** and **folacin** are general terms for compounds that have chemical structures and nutritional properties

folate and **folacin** General terms for the many forms of this vitamin.

Figure 8.21 Structure of folate

(a)

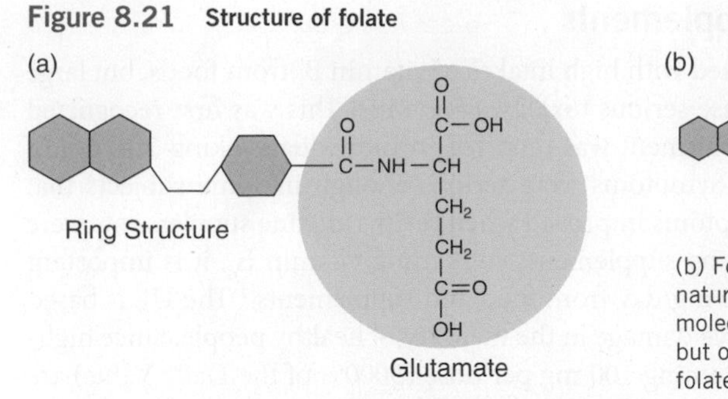

(a) Folic acid, or folate monoglutamate, is the form of folate that has only one molecule of glutamate in its structure. It is easily absorbed and is the form found in fortified foods and supplements.

(b)

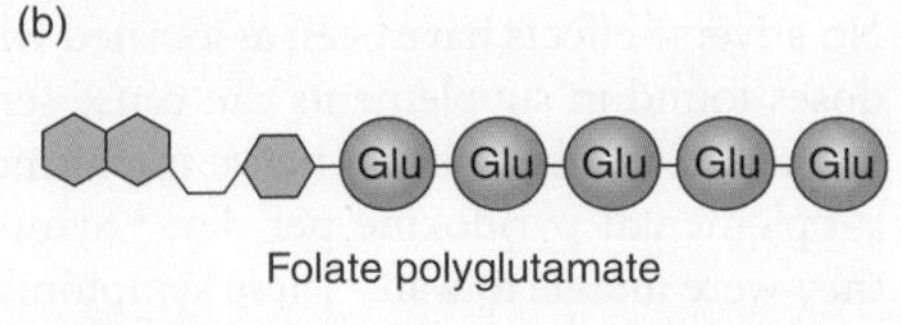

(b) Folate polyglutamate is the form found naturally in foods; it includes many glutamate molecules (Glu) attached to form a chain. All but one glutamate must be removed before the folate can be absorbed.

folic acid The monoglutamate form of folate, which is used in fortified foods and supplements.

similar to those of **folic acid** (**Figure 8.21**). The chemical name for folate is pteroylglutamic acid.

Folate in the Diet and the Digestive Tract

Excellent dietary sources of folate include liver, yeast, asparagus, oranges, legumes, and fortified grain products. Fair sources include vegetables such as corn, snap beans, mustard greens, and broccoli, as well as some nuts and seeds. Small amounts are found in meats, cheese, and milk (**Figure 8.22**). Most folate found naturally in food contains a string of glutamate molecules (see Figure 8.21). Glutamate is an amino acid, and folate bound to many glutamates is referred to as folate polyglutamate. Before this form can be absorbed, all but one of the glutamate molecules must be removed by enzymes in the microvilli of the small intestine to yield folic acid, the monoglutamate form. It is estimated that about 50% of the folate in food is absorbed.[4] Folic acid rarely occurs naturally in food but is used in supplements and fortified foods. It is more easily absorbed because it does not require

THINK CRITICALLY

Which of the food sources of folate shown in the graph are fortified?

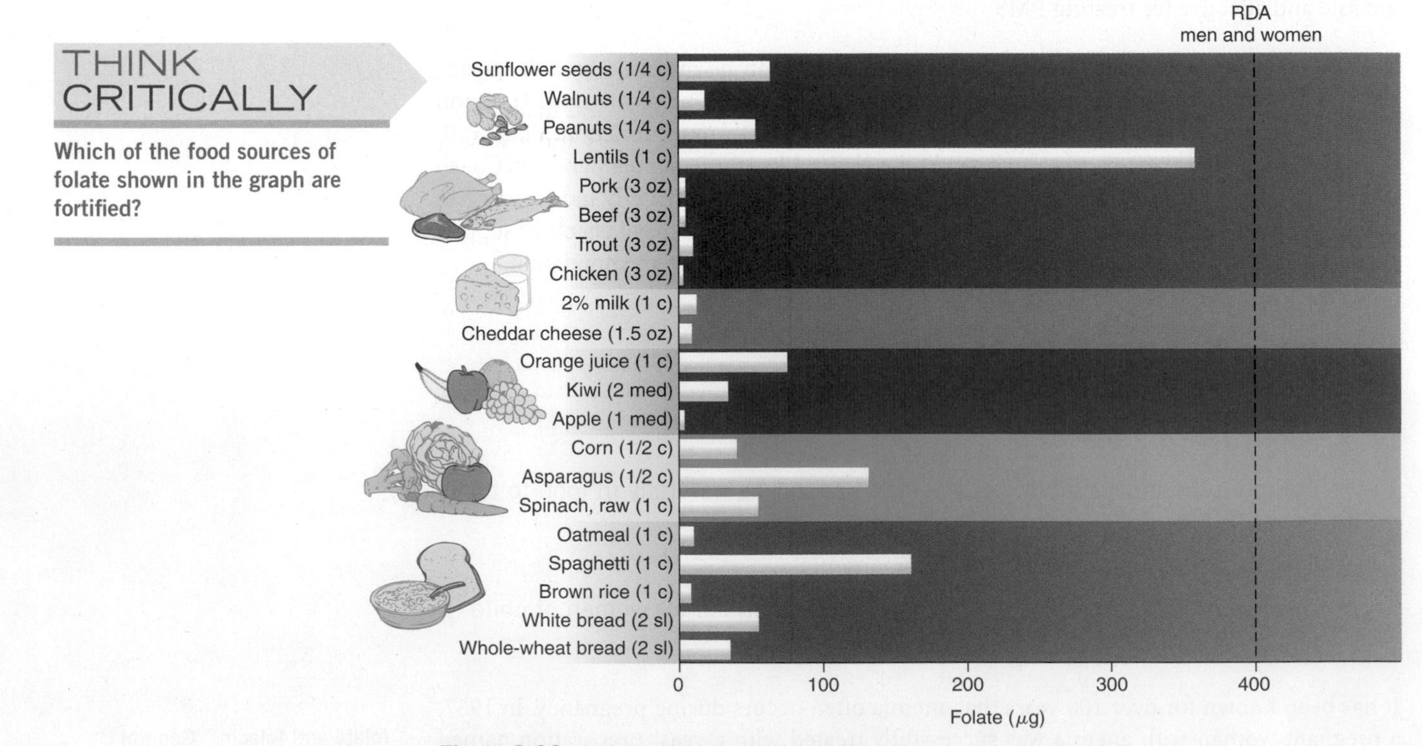

Figure 8.22 Folate content of MyPlate food groups
Adults can obtain the RDA for folate (dashed line) by eating legumes, fortified foods, and fruits and vegetables.

the enzymatic removal of a string of glutamate molecules. The bioavailability of the synthetic folic acid added to grain products and used in supplements is about twice that of the folate naturally found in food. Since 1998, enriched grain products, including breads, flours, corn meal, pasta, grits, and rice, have been fortified with folic acid.

neural tube defects Birth defects caused by abnormal development of the neural tube, the portion of the embryo that gives rise to the central nervous system.

How Folate Functions in the Body

A number of different active coenzyme forms of folate are involved in reactions that transfer chemical groups containing a single carbon atom. Folate coenzymes are needed for the synthesis of DNA and the metabolism of some amino acids. Before a cell divides, its DNA must replicate. Therefore, the role of folate in DNA synthesis makes it particularly important in tissues where cells are rapidly dividing such as bone marrow, where red blood cells are made, intestines, and skin and during periods of rapid growth, such as early in embryonic life.

Folate Deficiency

A deficiency of folate leads to a drop in blood folate levels and a rise in blood homocysteine followed by changes that affect rapidly dividing cells. Deficiency symptoms include anemia, poor growth, problems in nerve function, diarrhea, and inflammation of the tongue. Low folate intake in early pregnancy is associated with an increased risk of **neural tube defects.** Low folate status may also increase the risk of developing heart disease and certain types of cancer. Groups at risk of folate deficiency include pregnant women and premature infants because of their rapid rate of cell division and growth, the elderly because of their limited intake of foods high in folate, alcoholics because alcohol inhibits folate absorption, and tobacco smokers because smoke inactivates folate in the cells lining the lungs.[4]

megaloblasts Large, immature red blood cells that are formed when developing red blood cells are unable to divide normally.

macrocytes Larger-than-normal mature red blood cells that have a shortened life span.

megaloblastic or **macrocytic anemia** A condition in which there are abnormally large immature and mature red blood cells in the bloodstream and a reduction in the total number of red blood cells and hence in the oxygen-carrying capacity of the blood.

Why Folate Deficiency Causes Anemia Folate is needed for cells to divide. When folate is deficient, cells in the bone marrow that develop into red blood cells cannot replicate their DNA and so cannot divide. Instead, they just grow bigger. These large immature cells are known as **megaloblasts** and can be converted into large red blood cells called **macrocytes.** The result is that fewer mature red blood cells are produced and the oxygen-carrying capacity of the blood is reduced. This condition is called **megaloblastic** or **macrocytic anemia** (Figure 8.23).

anencephaly A birth defect due to failure of the neural tube to close that results in the absence of a major portion of the brain, skull, and scalp.

How Folate Is Related to the Risk of Neural Tube Defects Neural tube defects are not true folate-deficiency symptoms because not every pregnant woman with inadequate folate levels will bear a child with a neural tube defect. Instead, neural tube defects are probably due to a combination of factors that include low folate levels and a genetic predisposition. The exact role of folate in neural tube development is not known, but it is necessary for a critical step called *neural tube closure*. When neural tube closure does not occur normally, portions of the brain or spinal cord are not adequately protected. **Anencephaly**, is a neural tube defect that affects the brain. Babies with anencephaly are usually blind, deaf, and die soon after birth. **Spina bifida** is a neural tube defect that affects

spina bifida A type of neural tube defect resulting from the incorrect development of the spinal cord that can leave the spinal cord exposed.

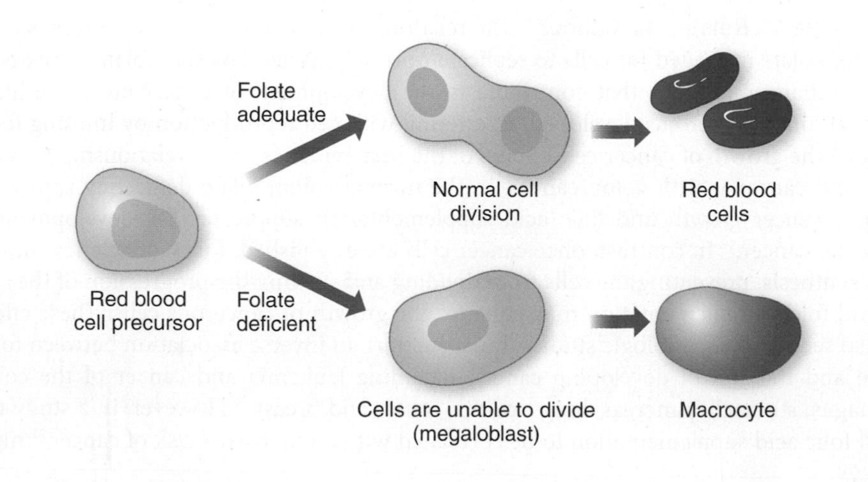

THINK CRITICALLY

Why does the inability to replicate DNA prevent cells from dividing?

Figure 8.23 Role of folate in macrocytic anemia
Folate is needed for DNA replication. Without folate, red blood cell precursors are unable to divide, resulting in the formation of abnormally large red blood cells called *macrocytes*.

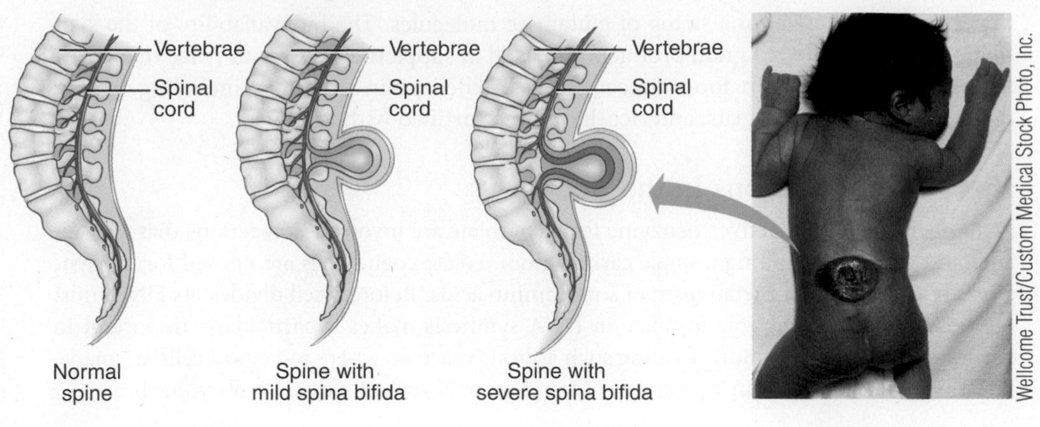

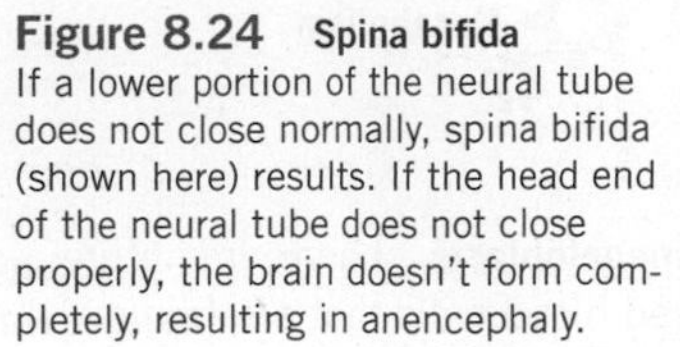

Figure 8.24 Spina bifida
If a lower portion of the neural tube does not close normally, spina bifida (shown here) results. If the head end of the neural tube does not close properly, the brain doesn't form completely, resulting in anencephaly.

the spine. It can cause nerve damage that leads to varying degrees of paralysis of the lower limbs and in some cases learning disabilities (**Figure 8.24**).

Neural tube closure is complete by 28 days after conception; therefore, folate status should be adequate even before a pregnancy begins to assure an adequate supply during early development (see Chapter 14). Studies in which supplemental folic acid was given to women before and during early pregnancy showed that 360 to 800 μg/day of synthetic folic acid in addition to food folate was associated with a reduced incidence of neural tube defects.[4] Because it is not known whether a diet naturally rich in folate offers the same protection as folic acid, supplemental folic acid from supplements or fortified foods is recommended. To be effective, folate must be adequate before most women are aware that they are pregnant, and, therefore, supplemental folic acid is recommended for all women of childbearing age. To help ensure adequate folate intake in women of childbearing age, in 1998 the FDA mandated that folic acid be added to all enriched grains and cereal products. Since then, the incidence of neural tube defects in the United States has decreased by almost 50%, and a similar reduction has been observed in other countries where folic acid fortification has been introduced (see Chapter 14: Science Applied: Folate: From Epidemiology to Health Policy).[17,18]

How Folate Is Related to the Risk of Cardiovascular Disease Low intakes of folate have been associated with an increased risk of cardiovascular disease. Its effect on cardiovascular disease is related to its role in the metabolism of the amino acid homocysteine (see Figure 8.20). Folate is needed to convert homocysteine to methionine. When folate is deficient, homocysteine levels rise because less is converted to methionine. The risk of cardiovascular disease increases with elevated homocysteine (see discussion above of vitamin B_6 and cardiovascular disease). Although supplementation with folic acid and vitamins B_6 and B_{12} lowers homocysteine levels, it has not been shown to reduce the risk of cardiovascular disease.[12,14]

How Folate Is Related to Cancer The relationship between folate and cancer is complicated. Folate is needed for cells to replicate normally. When levels of folate in the body are low, changes in DNA that contribute to the development of cancer are more likely. However, once cancer has developed, interfering with cell reproduction by limiting folate can slow the growth of cancer cells. Some of the best evidence of the relationship between folate and cancer is with colon cancer. In the normal colon, folate deficiency appears to enhance cancer growth and folic acid supplementation suppresses the development of colorectal cancer.[19] In contrast, once cancer cells are established, folate deficiency impairs DNA synthesis, preventing the cells from dividing and slowing the progression of the cancer, and folate supplementation may enhance the growth of cancerous cells. These effects are also seen in epidemiologic studies. Most support an inverse association between folate intake and the risk of developing cancer, including leukemia and cancer of the colon, esophagus, stomach, pancreas, lungs, cervix, ovary, and breast.[20] However, in a study that found folic acid supplementation to be correlated with an increased risk of cancer,[21] much of the increase was due to an increase in lung cancer in smokers—a population likely to have precancerous cells in the lungs.

Recommended Folate Intake

The RDA for folate is set at 400 μg **dietary folate equivalents (DFEs)** per day for adult men and women. One DFE is equal to 1 μg of food folate, 0.6 μg of synthetic folic acid from fortified food or supplements consumed with food, or 0.5 μg of synthetic folic acid consumed on an empty stomach. In order to reduce the risk of neural tube defects, a special recommendation is made for women capable of becoming pregnant.[4] A daily intake of 400 μg of synthetic folic acid from fortified foods and/or supplements is recommended in addition to the food folate consumed in a varied diet. Therefore, the total folate intake of this group should exceed the RDA (see **Critical Thinking: Meeting Folate Recommendations**).

dietary folate equivalents (DFEs) The unit used to express folate recommendations. One DFE is equivalent to 1 μg of folate naturally occurring in food, 0.6 μg of synthetic folic acid from fortified food or supplements consumed with food, or 0.5 μg of synthetic folic acid consumed on an empty stomach.

The RDA for folate during pregnancy is increased to 600 μg/day to provide for the increase in cell division. Although this level can be met by a carefully selected diet, folate is typically supplemented during pregnancy. The RDA is increased during lactation to account for folate secretion in milk. Needs per unit of body weight are higher for infants and children than for adults because of their rapid growth (see inside cover). Human and cow's milk provide enough folate to meet infant needs, but goat's milk does not. Infants and children given goat milk may not receive adequate folate unless it is provided from other sources.

Meeting Folate Needs

Women of childbearing age can meet folate recommendations by taking a multivitamin containing folic acid or by including fortified foods in their diet. To get the recommended 400 μg DFE of synthetic folic acid from fortified foods would require eating about one to four servings of fortified breakfast cereal or four to six servings of other fortified grain products each day. If it is not possible to meet recommendations from fortified foods, supplements should be used.

The Daily Values on food labels and values in some food composition tables are listed as μg total folate, not as μg DFE. DFEs correct for differences in the bioavailability of different forms of folate. For foods that are natural sources of folate, the folate content can be determined by multiplying the Daily Value (400 μg) by the % Daily Value listed on the label (see Off the Label: How Much Vitamin C Is in Your Orange Juice?). So, to calculate the folate in a package of frozen spinach that provides 25% of the Daily Value, multiply 400 μg folate $\times$ 0.25 = 100 μg DFE. In foods fortified with folic acid, the amount of folate indicated by the % Daily Value must be multiplied by 1.7 to account for the greater bioavailability of folic acid (**Table 8.4**).

TABLE 8.4 Calculating Dietary Folate Equivalents in Fortified Foods

The folate listed on labels of fortified foods is primarily folic acid, which is more available than natural forms of folate. In order to compare the folate content of these foods to recommendations, the amount of folic acid must be converted to dietary folate equivalents, expressed as μg DEF. This calculation assumes that all of the folate in these foods is from added folic acid.

Determine the amount of folic acid in the fortified food:

- Multiply the Daily Value for folate by the % Daily Values listed on the label.
  Daily Value is 400 μg

Convert the μg folic acid into μg DFE:

- Multiply the μg folic acid by 1.7.
  Folic acid added to a food in fortification provides 1.7 times more available folate per μg than folate naturally present in foods.

For example:

- A serving of English muffins provides 6% of the Daily Value for folate:
  To find the μg folic acid: 400 μg $\times$ 0.06 = 24 μg folic acid
  To convert to μg DFE: 24 μg folic acid $\times$ 1.7 = 40 μg DFE

Meeting Folate Recommendations

© Lynn Bendickson/iStockphoto

Mercedes would like to have a baby. At her annual physical, her physician advises her that women who are capable of becoming pregnant should consume 400 μg of folic acid from fortified foods or supplements each day in addition to the folate found in a varied diet. Mercedes records her food intake for a day to estimate her folate intake.

CURRENT DIET	FOOD AMOUNT	TOTAL FOLATE (μg)
Breakfast		
Oatmeal, regular	1 cup	2
Milk	1 cup	12
Banana	1 medium	22
Orange juice	1 cup	75
Coffee	1 cup	0
Lunch		
Hamburger	3 oz	11
Hamburger bun	1	32
French fries	20 pieces	24
Coke	12 oz	0
Apple	1 medium	4
Dinner		
Chicken	3 oz	4
Refried beans	½ cup	106
White rice	1 cup	80
Flour tortilla	1	60
Salad	1 cup	64
Salad dressing	1 T	1
Milk	1 cup	12
White cake	1 piece	32
Total		**541**

CRITICAL THINKING QUESTIONS

- Why is folate a concern for women planning a pregnancy?
- Which five foods in her diet are highest in folate? Of these, which are fortified with folic acid?
- Does Mercedes consume the recommended amount of folic acid from fortified foods?

Mercedes eats five servings of grains, but only one is a whole grain.

CRITICAL THINKING QUESTIONS

- How would replacing all her refined grain products with whole grains affect her folate intake?
- Would you recommend that Mercedes take a folic acid supplement? Why or why not?

iProfile Use iProfile to find out how much folate is in the folate-fortified grain products you consume each day.

Folate Toxicity

Although there is no known folate toxicity, a high intake may mask the early symptoms of vitamin B_{12} deficiency, allowing it to go untreated and resulting in irreversible nerve damage. This is discussed below. The UL for folic acid for adults is set at 1000 μg/day from supplements and/or fortified foods. This value was determined based on the progression of neurological symptoms seen in patients who are deficient in vitamin B_{12} and taking folic acid supplements.

8.9 Vitamin B_{12}

LEARNING OBJECTIVES

- Name foods that are good sources and poor sources of vitamin B_{12}.
- List the steps involved in vitamin B_{12} absorption.
- Compare the functions of folate and vitamin B_{12}.

Pernicious means deadly or fatal. A type of anemia that did not respond to iron supplementation was first described in 1820. This pernicious anemia could not be treated and was fatal. In the 1920s, researchers George Minot and William Murphy pursued their belief that pernicious anemia could be cured by something in the diet. They discovered that they could restore patients' health by feeding them about 4 to 8 ounces of slightly cooked liver at every meal. Today we know that pernicious anemia is caused by an inability to absorb sufficient vitamin B_{12} and that eating liver cured the disease because liver is a concentrated source of vitamin B_{12}. Pernicious anemia is now treated with injections or megadoses of vitamin B_{12} rather than with plates full of liver.

Vitamin B_{12} in the Diet

Vitamin B_{12} is found almost exclusively in animal products (**Figure 8.25**). It can be made by bacteria, fungi, and algae but not by plants and animals. It accumulates in animal tissue from the diet or from synthesis by bacterial microflora. Bacteria in the human colon produce vitamin B_{12}, but it cannot be absorbed. Vitamin B_{12} is not found in plant products unless they have been contaminated with bacteria, soil, insects, or other sources of vitamin B_{12} or have been fortified with it. Diets that do not include animal products must include supplements or foods fortified with vitamin B_{12} in order to meet needs.[22]

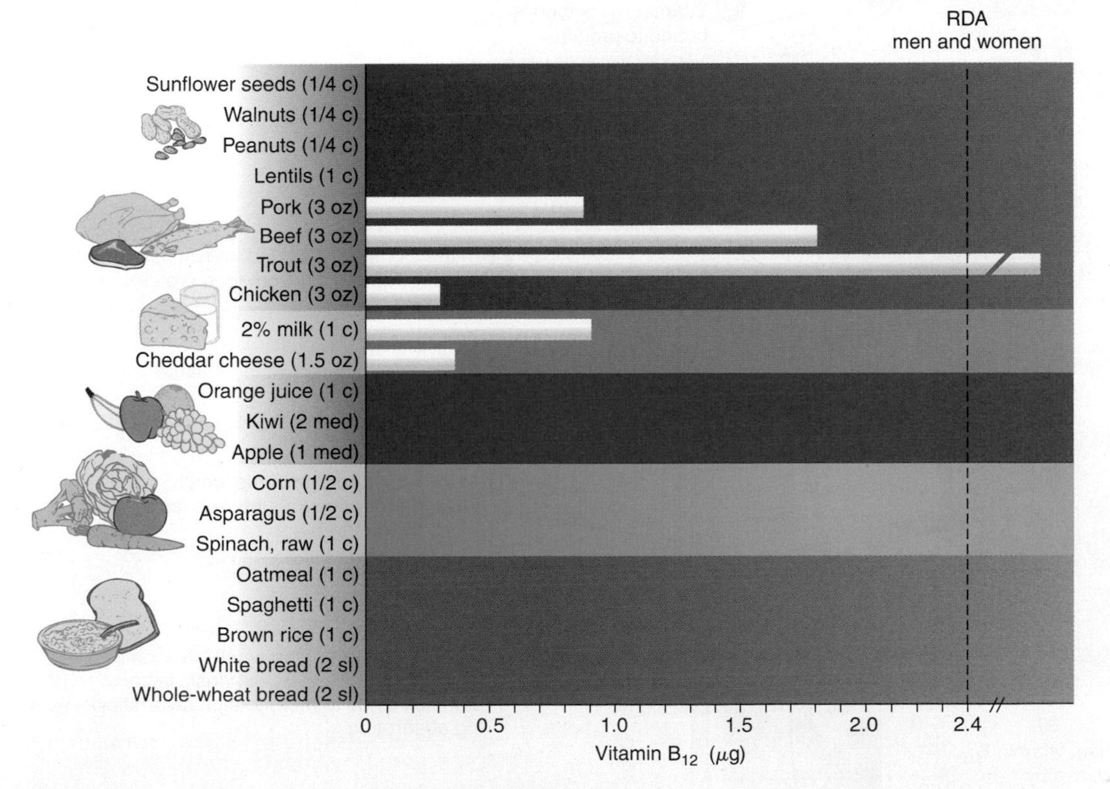

Figure 8.25 Vitamin B_{12} content of MyPlate food groups
Animal products provide vitamin B_{12}, but plant products do not unless they have been fortified with it or contaminated by bacteria, soil, insects, or other sources of the vitamin; the dashed line represents the RDA for adult men and women.

Vitamin B_{12} in the Digestive Tract

The vitamin B_{12} naturally present in food is bound to proteins and must be released before it can be absorbed. In the stomach, acid and the protein-digesting enzyme pepsin unfold and begin to break down proteins, releasing the vitamin B_{12}. If stomach acid is low, the vitamin cannot be released from food proteins and absorption is reduced. In the small intestine, vitamin B_{12} binds to **intrinsic factor**. Intrinsic factor is a protein secreted by the **parietal cells** in the gastric glands lining the stomach. The intrinsic factor–vitamin B_{12} complex binds to receptor proteins in the ileum—the last 11 feet of the small intestine—allowing the vitamin to be absorbed (**Figure 8.26**). Only a small amount of vitamin B_{12} can be absorbed when intrinsic factor is absent. Vitamin B_{12} absorption is also reduced when pancreatic secretions are insufficient.

intrinsic factor A protein produced in the stomach that is needed for the absorption of adequate amounts of vitamin B_{12}.

parietal cells Large cells in the stomach lining that produce and secrete intrinsic factor and hydrochloric acid.

Vitamin B_{12} is secreted in bile, but most of this is reabsorbed rather than being lost in the feces. Because of this efficient recycling, it can take many years of a deficient diet before the symptoms of vitamin B_{12} deficiency appear.

How Vitamin B_{12} Functions in the Body

Vitamin B_{12}, also known as **cobalamin**, can be converted into either of two active cobalamin coenzyme forms, methylcobalamin and adenosylcobalamin (see Appendix H). These coenzymes function in two important metabolic reactions. One rearranges carbon atoms

cobalamin The chemical term for a cobalt-containing compound commonly known as vitamin B_{12}.

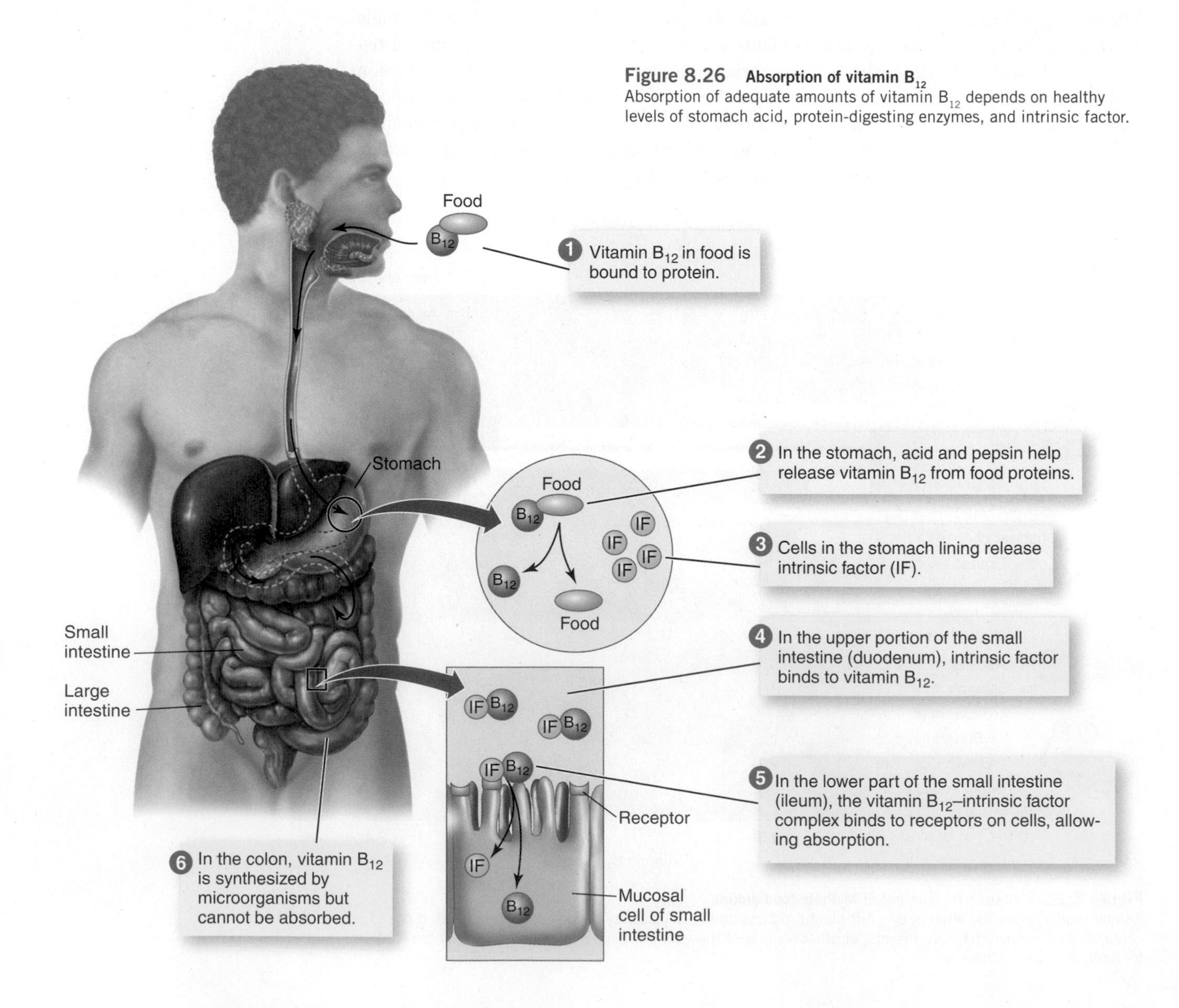

Figure 8.26 **Absorption of vitamin B_{12}**
Absorption of adequate amounts of vitamin B_{12} depends on healthy levels of stomach acid, protein-digesting enzymes, and intrinsic factor.

so that the breakdown products of some fatty acids and amino acids can enter the citric acid cycle and be used to provide energy. This reaction is necessary for the maintenance of healthy myelin, which insulates nerves and is essential for normal nerve transmission (see Figure 8.19). A second vitamin B_{12}-dependent reaction synthesizes the amino acid methionine from homocysteine. This reaction also requires folate (see Figure 8.20).

Vitamin B_{12} Deficiency

The body stores and reuses vitamin B_{12} more efficiently than it does most other water-soluble vitamins, so deficiency is typically caused by poor absorption rather than by low intake alone. Symptoms of vitamin B_{12} deficiency include an increase in blood homocysteine levels and macrocytic, megaloblastic anemia that is indistinguishable from that seen in folate deficiency. This anemia occurs because vitamin B_{12} is needed to convert folate into the form that is active for DNA synthesis (**Figure 8.27**). Lack of vitamin B_{12} causes a secondary folate deficiency and consequently megaloblastic anemia. Vitamin B_{12} deficiency also interferes with the maintenance of the myelin that coats the nerves, spinal cord, and brain. Myelin degeneration causes neurological symptoms, which include numbness and tingling, abnormalities in gait, memory loss, and disorientation. If not treated, this eventually causes paralysis and death.

Blatant deficiencies of vitamin B_{12} are rare because the body stores and reuses it efficiently. However, marginal vitamin B_{12} status is of public health concern, particularly for older adults and vegetarians who consume no animal products. Because of the efficient recycling of vitamin B_{12}, it can take many years of a deficient diet before the symptoms of deficiency appear. However, when absorption is impaired, neither dietary vitamin B_{12} nor the vitamin B_{12} secreted in the bile is absorbed, so deficiency symptoms appear more rapidly. This occurs both in individuals with **pernicious anemia** and in those with **atrophic gastritis**.

pernicious anemia A macrocytic anemia resulting from vitamin B_{12} deficiency that occurs when dietary vitamin B_{12} cannot be absorbed due to a lack of intrinsic factor.

atrophic gastritis An inflammation of the stomach lining that results in reduced secretion of stomach acid and bacterial overgrowth.

Why Pernicious Anemia Causes B_{12} Deficiency Pernicious anemia is the major cause of severe vitamin B_{12} deficiency. It is an autoimmune disease in which the parietal cells in the stomach that produce intrinsic factor are destroyed. Without intrinsic factor, vitamin B_{12} cannot be absorbed normally. This anemia can be treated with injections of the

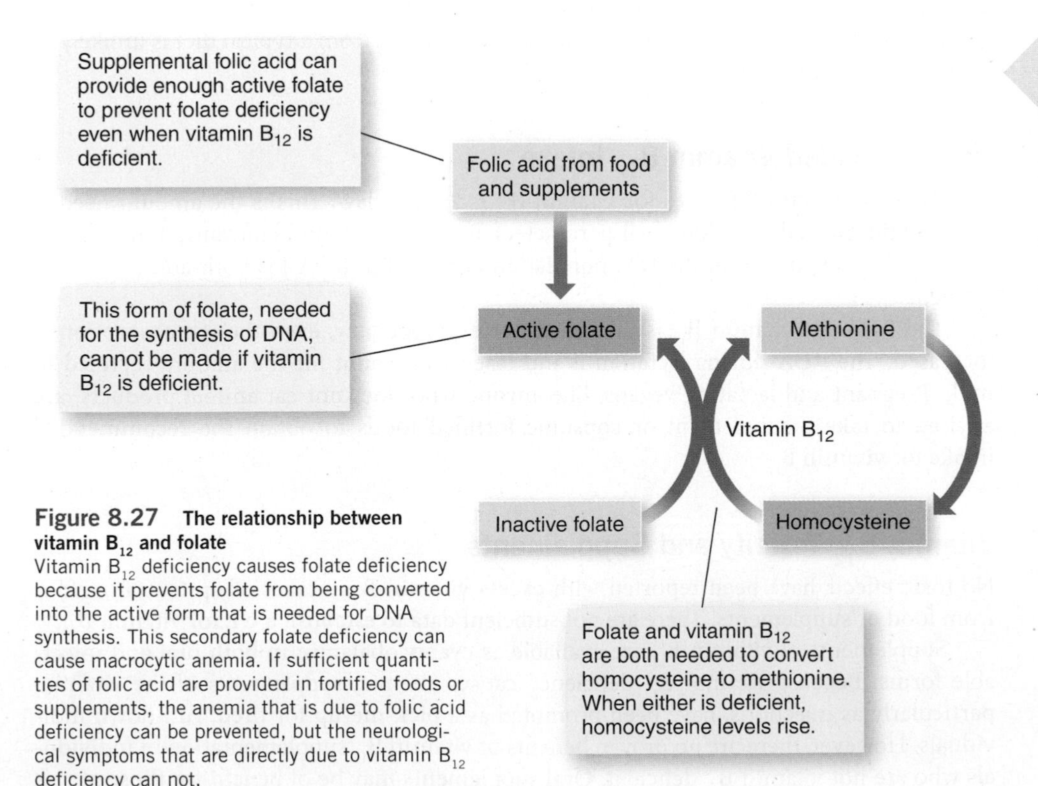

Figure 8.27 The relationship between vitamin B_{12} and folate
Vitamin B_{12} deficiency causes folate deficiency because it prevents folate from being converted into the active form that is needed for DNA synthesis. This secondary folate deficiency can cause macrocytic anemia. If sufficient quantities of folic acid are provided in fortified foods or supplements, the anemia that is due to folic acid deficiency can be prevented, but the neurological symptoms that are directly due to vitamin B_{12} deficiency can not.

THINK CRITICALLY

If you were deficient in vitamin B_{12} but took large amounts of folic acid from supplements, would you develop macrocytic anemia? Why or why not?

vitamin, with a B_{12}-containing nasal gel, or with oral megadoses. The injections, which deliver vitamin B_{12} to the subcutaneous fat or muscle, and the nasal gel, which allows the vitamin to enter the bloodstream through the nasal mucosa, bypass the gastrointestinal tract and thus the need for intrinsic factor. Megadoses can treat pernicious anemia because they allow adequate amounts of vitamin B_{12} to be absorbed by passive diffusion, which does not require intrinsic factor.

How Atrophic Gastritis Causes B_{12} Deficiency About 10 to 30% of individuals over 50 years of age are unable to absorb food-bound vitamin B_{12} normally because they have atrophic gastritis. Atrophic gastritis is an inflammation of the stomach lining that results in a reduction in stomach acid and bacterial overgrowth. When stomach acid is reduced, the enzymes that release vitamin B_{12} that is bound to food protein cannot function properly and the bound vitamin B_{12} cannot be released and absorbed. Atrophic gastritis also causes microbial overgrowth in the intestine. These microbes reduce the amount of vitamin B_{12} that is absorbed by competing for available vitamin B_{12}. In severe cases of atrophic gastritis, the production of intrinsic factor is also reduced, further impairing vitamin B_{12} absorption. It is recommended that individuals over the age of 50 meet their RDA by consuming foods fortified with vitamin B_{12} such as breakfast cereals or soy-based products or by taking a vitamin B_{12}-containing supplement. Because the vitamin B_{12} in these products is not bound to proteins, it is absorbed even when stomach acid is low.

Vegan Diets Vitamin B_{12} deficiency is also a concern among vegans since vitamin B_{12} is found only in foods of animal origin. Severe deficiency has been observed in breast-fed infants of vegan women, but marginal deficiency is a concern for all vegans if supplements or fortified foods are not included in the diet.

Why Supplemental Folic Acid Can Mask Vitamin B_{12} Deficiency If individuals with vitamin B_{12} deficiency consume enough folate, they will not develop anemia, which is an easily identified and reversible symptom of vitamin B_{12} deficiency (see Figure 8.27). Without this symptom, diagnosis can be delayed, allowing more serious and irreversible symptoms, such as nerve damage, to progress. Although the fortification of grain products with folic acid has raised concern that additional folate in the food supply could delay diagnosis of vitamin B_{12} deficiency, the amount consumed from a typical diet is unlikely to be high enough to cause problems.

Recommended Vitamin B_{12} Intake

The RDA for adults of all ages for vitamin B_{12} is 2.4 μg/day.[4] This is the amount needed to maintain normal red blood cell parameters and normal blood concentrations of vitamin B_{12}. Average intake in the U.S. population exceeds the RDA for both adult men and women.

The RDA for vitamin B_{12} is increased during pregnancy, even though absorption is increased. The RDA during lactation is increased to account for the amount secreted in milk. Pregnant and lactating vegans, like anyone who does not eat animal products, are advised to take a supplement or consume fortified foods to obtain the recommended intake for vitamin B_{12}.[22]

Vitamin B_{12} Toxicity and Supplements

No toxic effects have been reported with excess vitamin B_{12} intakes of up to 100 μg/day from food or supplements. There are not sufficient data to establish a UL for vitamin B_{12}.

Supplements of vitamin B_{12} are available as cyanocobalamin in both oral and injectable forms. Because vitamin B_{12} deficiency causes anemia, supplements of the vitamin, particularly as injections, have been promoted as a pick-me-up for tired, run-down individuals. However, there are no proven benefits of vitamin B_{12} supplementation in individuals who are not vitamin B_{12} deficient. Oral supplements may be of benefit for those at risk for vitamin B_{12} deficiency, such as vegans and individuals over 50 years of age.[4]

8.10 Vitamin C

LEARNING OBJECTIVES

- List the food groups that include good sources of vitamin C and ones that are lacking in foods that provide vitamin C.
- Relate the role of vitamin C in the body to the symptoms of scurvy.
- Explain how vitamin C prevents oxidative damage.

The vitamin C deficiency disease **scurvy** was the scourge of armies, navies, and explorers throughout history. It was described by ancient Greeks, Egyptians, and Romans. The reason obtaining enough vitamin C, also known as **ascorbic acid** or **ascorbate**, has been a particular problem for armies and explorers is that fresh fruits and vegetables are its main sources; these foods spoil quickly and don't transport well on long voyages. In the mid-1500s, the Indians of eastern Canada knew that an extract from white cedar needles would cure the disease. In 1594, after a voyage to the South Seas, Sir Richard Hawkins recommended that this sickness be treated by including citrus fruit in the diet. Despite this recommendation, 10,000 British sailors died of scurvy that same year. Over 100 years later, James Lind, a Scottish physician serving in the British navy, tested various agents for their effectiveness at curing scurvy and reported that two patients given citrus fruits recovered within six days. However, it was another 48 years before lime or lemon juice became a requirement in the rations of the mercantile service, earning British sailors the name *limeys*. Unfortunately, the rest of the world did not heed the lesson of the limeys. In the mid-19th century, during the U.S. Civil War, scurvy was rampant.

scurvy The vitamin C deficiency disease.

ascorbic acid or **ascorbate** The chemical term for vitamin C.

Vitamin C in the Diet

Citrus fruits, such as oranges, lemons, and limes, are excellent sources of vitamin C. Other fruits high in vitamin C include strawberries and cantaloupe. Vegetables in the cabbage family, such as broccoli, cauliflower, bok choy, and Brussels sprouts, as well as green leafy vegetables, green and red peppers, okra, tomatoes, and potatoes, are also good sources (**Figure 8.28**). Meat, fish, poultry, eggs, dairy products, and grains are poor sources.

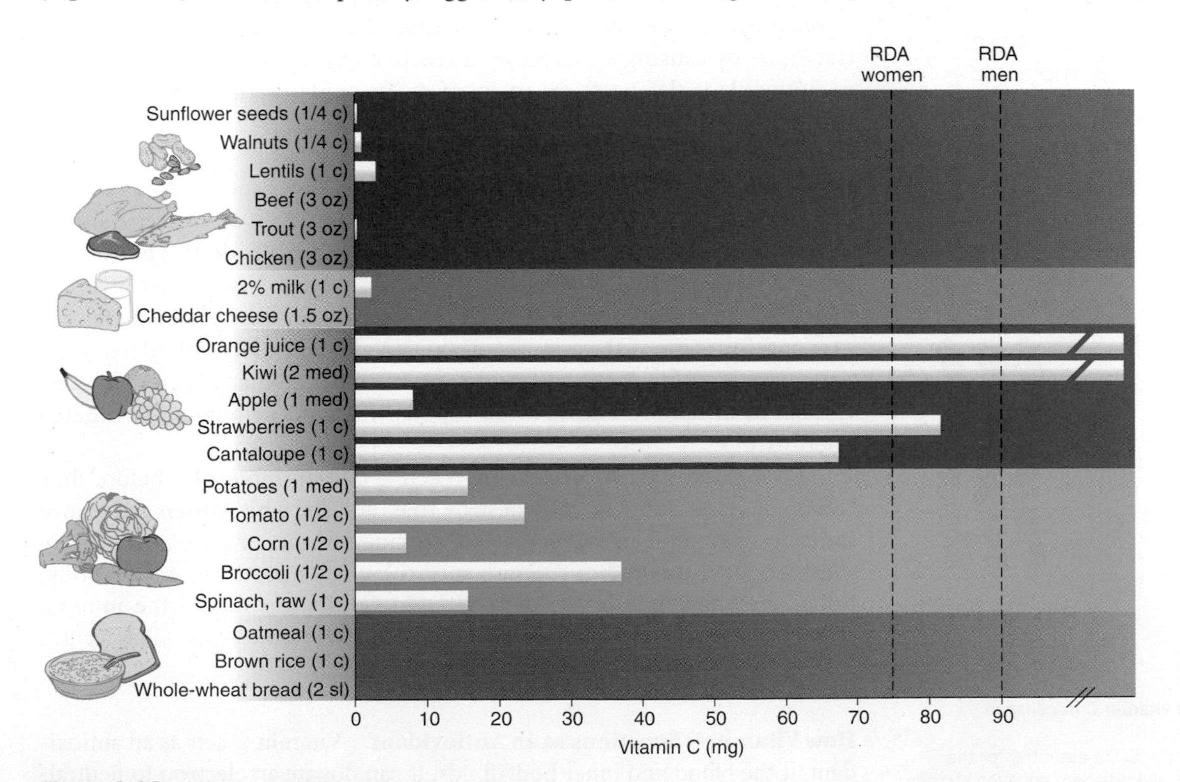

Figure 8.28 Vitamin C content of MyPlate food groups
Fruits and vegetables are the best sources of vitamin C; the dashed lines represent the RDAs for adult men and women.

collagen The major protein in connective tissue.

oxidative damage Damage caused by highly reactive oxygen molecules that steal electrons from other compounds, causing changes in structure and function.

The amount of vitamin C in packaged foods must be listed on food labels as a percentage of the Daily Value. This information can be used to identify packaged foods, such as frozen strawberries and orange juice, that are good sources of vitamin C. Fresh fruits and vegetables, which are the best sources of vitamin C, do not carry food labels.

Vitamin C is unstable and is destroyed by oxygen, light, and heat, so it is readily lost in cooking. This loss is accelerated by contact with copper or iron cooking utensils and by low-acid conditions.

How Vitamin C Functions in the Body

Vitamin C is a water-soluble vitamin that donates electrons in biochemical reactions, including those needed for the synthesis and maintenance of connective tissue. Because it can donate electrons, vitamin C also has a more general role as an antioxidant that, along with other antioxidants, protects the body from reactive oxygen molecules. Vitamin C also serves as a coenzyme in reactions needed for the synthesis of neurotransmitters, hormones such as the thyroid and steroid hormones, bile acids, and carnitine needed for fatty acid breakdown. It also helps maintain the immune system and aids in the absorption of iron.

oxidative stress A condition that occurs when there are more reactive oxygen molecules than can be neutralized by available antioxidant defenses.

pro-oxidant A substance that promotes oxidative damage.

free radical An atom or group of atoms that has at least one unpaired electron and is therefore unstable and highly reactive and can cause cellular damage.

dietary antioxidant A substance in food that significantly decreases the adverse effects of reactive species on normal physiological function in humans.

Vitamin C and Collagen Synthesis Many of the biochemical reactions requiring vitamin C add a hydroxyl group (OH) to other molecules. Two such reactions are essential for the formation of **collagen**, the protein that forms the base of all connective tissue in the body. Vitamin C is a coenzyme needed for activity of the enzymes that add the hydroxyl groups to the amino acids proline and lysine to form hydroxyproline and hydroxylysine, respectively. These hydroxyl groups are necessary for the formation of chemical bonds that cross-link the polypeptide strands of collagen to give it strength (**Figure 8.29**).

Vitamin C as an Antioxidant Vitamin C also functions as an antioxidant. Antioxidants protect against **oxidative damage**, which is damage caused by reactive oxygen molecules. **Oxidative stress** refers to a serious imbalance between the amounts of reactive oxygen molecules and the availability of antioxidant defenses. Oxidative stress has been related to the aging process as well as to the development of cancer and heart disease. Agents that can induce oxidative stress by causing an increase in reactive oxygen molecules, a decrease in antioxidant defenses, or an increase in oxidative damage are called **pro-oxidants**.

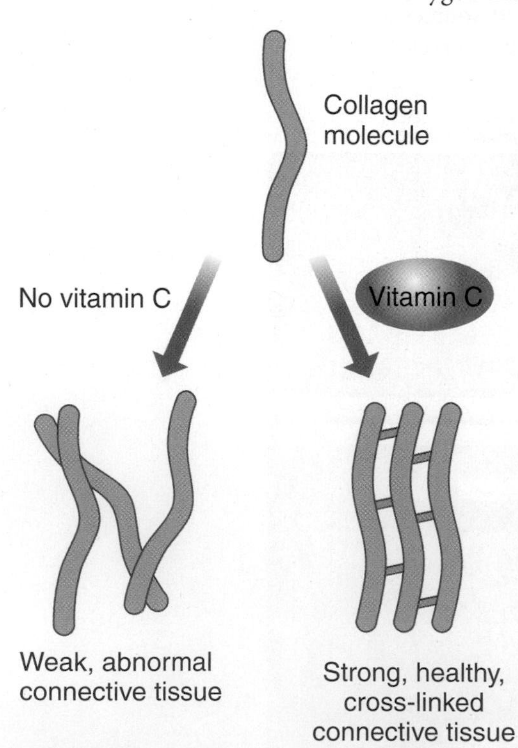

Figure 8.29 Role of vitamin C in collagen synthesis
Reactions requiring vitamin C are essential for the formation of bonds that hold adjacent collagen strands together and give the protein strength and stability.

How Antioxidants Work Reactive oxygen molecules such as **free radicals** come from environmental sources such as air pollution or cigarette smoke as well as from normal oxygen-requiring reactions inside the body. Free radicals cause damage by snatching electrons from DNA, proteins, carbohydrates, or unsaturated fatty acids. This results in changes in the structure and function of these molecules. DNA damage is hypothesized to be a major reason for the increase in cancer incidence that occurs with age. Oxidation of lipoproteins is a critical step in the development of atherosclerosis.

Antioxidants act by destroying reactive oxygen molecules before they can do damage. Some directly destroy free radicals, while others neutralize other reactive molecules, such as superoxide radicals or peroxides, before they can form free radicals. Some antioxidants are produced in the body; others are consumed in the diet. Vitamin C, vitamin E, and the mineral selenium have been classified as **dietary antioxidants** (see Chapter 12, Debate: Are Antioxidant Supplements Beneficial?).[23]

How Vitamin C Functions as an Antioxidant Vitamin C acts as an antioxidant in the blood and other body fluids. It can donate an electron to neutralize superoxide radicals and free radicals before they can damage lipids and

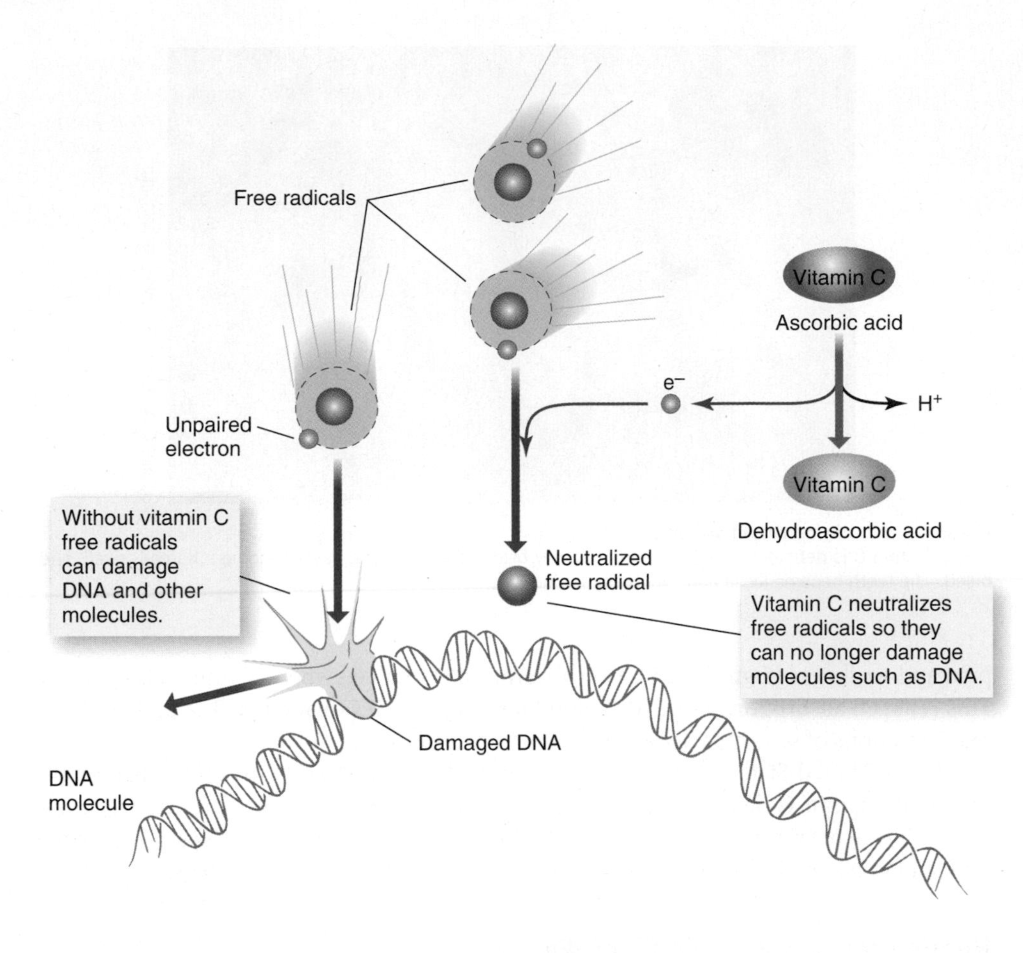

Figure 8.30 **Antioxidant role of vitamin C**
Vitamin C functions as an antioxidant by donating electrons to neutralize free radicals so that they are no longer damaging. When ascorbic acid donates an electron, it also loses a hydrogen ion to form a molecule of dehydroascorbic acid. This reaction is reversible and vitamin C can be restored when electrons and hydrogens are provided by other antioxidants such as glutathione.

DNA (**Figure 8.30**). Vitamin C has also been shown to scavenge reactive oxygen molecules in white blood cells, the lungs, and the stomach mucosa.[23] The antioxidant properties of vitamin C are also important for the functioning of other nutrients. Vitamin C donates an electron to regenerate the active antioxidant form of vitamin E and enhances iron absorption by keeping iron in its more readily absorbed reduced form (Fe^{+2}). When about 50 mg of vitamin C—the amount in a small glass of orange juice—is consumed in a meal containing iron, iron absorption is enhanced sixfold (see Chapter 12).

Vitamin C may also act as a pro-oxidant by converting iron and copper to reduced forms that can then generate free radicals. There is some evidence that vitamin C supplements can lead to oxidative damage to DNA. However, studies of this pro-oxidant effect of vitamin C are inconsistent, most showing that vitamin C either reduces or has no effect on oxidative DNA damage.[24] More research is needed to determine what factors influence whether the antioxidant or pro-oxidant properties of vitamin C predominate.

Vitamin C Deficiency

When vitamin C intake is below 10 mg/day, the symptoms of scurvy may appear. These symptoms reflect the role of vitamin C in the maintenance of collagen. Like all other body proteins, collagen is continuously being broken down and reformed. Without vitamin C, the bonds holding adjacent collagen molecules together cannot be formed, so the collagen that is broken down is replaced with abnormal collagen. The inability to form healthy collagen prevents wounds from healing normally and causes bone and joint aches, bone fractures, and improperly formed and loose teeth. Connective tissue is also important for blood vessel integrity. A vitamin C deficiency therefore causes weakened blood vessels and ruptured capillaries. This leads to easy bruising, tiny bleeds under the skin around the hair follicles, bleeding gums, and bleeding into the joints that causes joint

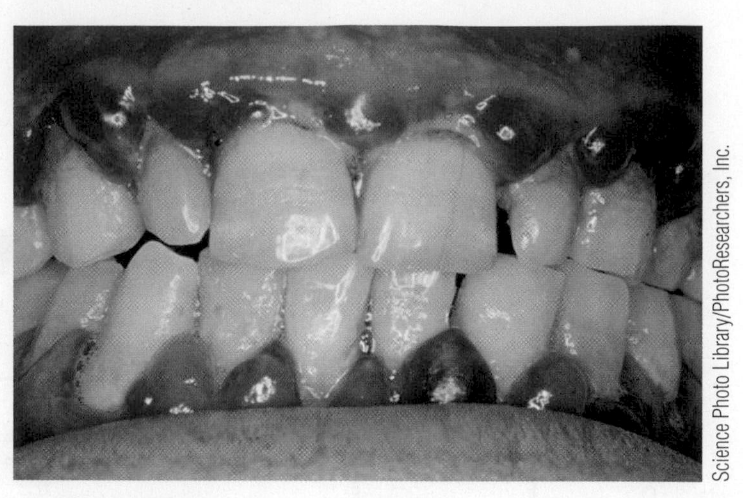

Figure 8.31 **Symptoms of scurvy**
When vitamin C is deficient, the symptoms of scurvy begin to appear. The gums become inflamed, swell, and bleed. The teeth become loose and eventually fall out.

pain and weakness (**Figure 8.31**). Iron absorption is reduced when vitamin C is deficient. This reduced absorption, along with blood loss, contributes to anemia. The psychological manifestations of scurvy include depression and hysteria.

In the United States today, severe vitamin C deficiency leading to scurvy is rare, but marginal vitamin C deficiency is a concern for individuals who consume few fruits and vegetables. Scurvy has been reported in infants fed diets consisting exclusively of cow's milk and in alcoholics and elderly individuals consuming nutrient-poor diets.

Recommended Vitamin C Intake

Humans are one of only a few animal species that require vitamin C in the diet. Most animals can synthesize vitamin C in their bodies. For example, a pig makes 8 g of the vitamin each day. The recommendations for vitamin C intake are based on the amount needed to maximize concentrations in neutrophils, a type of white blood cell, with minimal excretion of the vitamin in the urine. The RDA is 90 mg/day for men and 75 mg/day for women. This amount is easily obtained by drinking an 8-ounce glass of orange juice.

The RDA for vitamin C is increased during pregnancy and lactation. The recommendation for infants is based on the vitamin C content of human milk (see inside cover). Cigarette smoking increases the requirement for vitamin C because vitamin C is used to break down compounds in cigarette smoke. It is recommended that cigarette smokers consume an extra 35 mg of vitamin C daily[23]—the amount in a half-cup of broccoli. Exercise and mental and emotional stress have not been found to affect the need for vitamin C.

Vitamin C Toxicity and Supplements

Vitamin C is generally considered nontoxic. Large increases in intake do not cause large increases in the amount of vitamin C in body fluids. This is because the percentage of the dose absorbed decreases as the size of the dose increases and vitamin C absorbed in excess of need is excreted by the kidney. The most common symptoms that occur with consumption of vitamin C doses of 3 g or more are diarrhea, nausea, and abdominal cramps.[23] These are caused when unabsorbed vitamin C draws water into the intestine. The UL for vitamin C is based on an amount that is unlikely to cause gastrointestinal symptoms in healthy individuals and has been set at 2000 mg/day from food and supplements.

Vitamin C supplements should be avoided by some individuals. In people prone to kidney stones, excess vitamin C can increase stone formation. In individuals who are unable to regulate iron absorption, taking vitamin C supplements, which increase iron absorption, increases the risk that toxic amounts of iron will be stored. For those with

© Will Selarep/iStockphoto

Figure 8.32 **Vitamin C and the common cold**
Vitamin C supplements won't prevent you from catching a cold, but they may help you recover faster.

sickle cell anemia, excess vitamin C can worsen symptoms. In those taking medication to reduce blood clotting, taking more than 3000 mg/day of vitamin C can interfere with the effectiveness of the medication. Because the structure of vitamin C is similar to that of glucose, high doses can interfere with urine tests used to monitor glucose levels. A concern with high doses of chewable vitamin C supplements is that they can dissolve tooth enamel.

Do Vitamin C Supplements Cure the Common Cold? One-third of the population of the United States take supplements of vitamin C in hope of preventing or reducing symptoms of the common cold. Studies examining the relationship between vitamin C and the common cold date back to the 1930s. A review of placebo-controlled trials that supplemented diets with vitamin C failed to find that routine vitamin C supplementation reduced the incidence of colds in the healthy population.[25] Regular supplementation with vitamin C did, however, reduce the duration of cold symptoms (**Figure 8.32**). To have this effect, the supplements needed to be introduced before the onset of the cold symptoms. The effect of vitamin C on cold symptoms may be due to its direct antiviral effect, its antioxidant effect, its role in stimulating various aspects of immune function, its ability to increase the breakdown of histamine (a molecule that causes inflammation), or a combination of these.

Can Vitamin C Supplements Protect against Cardiovascular Disease? Vitamin C supplements have been suggested to reduce the risk of cardiovascular disease by promoting the health and growth of the cells lining the blood vessels, preventing the oxidation of LDL cholesterol, and reducing blood cholesterol levels. Vitamin C is suggested to prevent LDL oxidation because of its antioxidant function. Vitamin C may reduce blood cholesterol because it is involved in the synthesis of bile acids from cholesterol in the liver. Adequate vitamin C allows cholesterol to be used for bile synthesis and therefore may reduce the amount of cholesterol in the blood. Despite these roles of vitamin C in protecting LDL cholesterol from oxidation and modulating blood cholesterol levels, data thus far from epidemiology and human intervention trials have provided little evidence to support the use of vitamin C supplements in preventing atherosclerosis in humans.[26]

Do Vitamin C Supplements Protect against Cancer? It has been suggested that high doses of vitamin C both treat and prevent cancer. Although controlled trials have not found any benefits of vitamin C in the treatment of patients with advanced cancer,[27] there is evidence supporting a role for vitamin C in cancer prevention. Higher plasma vitamin C levels are associated with a lower cancer mortality in men.[28] As an antioxidant, vitamin C may protect against cancers caused by oxidative damage. In the case of gastrointestinal cancers, vitamin C may prevent cancer by inhibiting the formation of carcinogenic nitrosamines in the gut (see Chapter 17). Despite the association between higher intakes of vitamin C and a lower incidence of various cancers, clinical trials have not provided evidence to justify the use of supplemental vitamin C or other antioxidant supplements for cancer prevention.[29]

8.11 Choline

LEARNING OBJECTIVES

- Explain why choline is considered an essential nutrient.
- Name two foods that are good sources of choline.

Choline is a water-soluble substance that you may see included in supplements called *vitamin B complex*. Although it is not currently classified as a vitamin, it is recognized as an essential nutrient. It is needed for the synthesis of the neurotransmitter acetylcholine, the structure and function of cell membranes, lipid transport, and homocysteine metabolism. Because of these important roles, it is believed that deficiency could play a role in liver disease, atherosclerosis, and neurological disorders. Deficiency during pregnancy can interfere with brain development in the fetus, and deficiency in adults causes fatty liver and muscle damage.[30]

Choline can be synthesized in the human body, but the amounts are generally not enough to meet needs. Choline is found in many foods. The best dietary sources of choline are egg yolks, liver, meat and fish, whole grains, and nuts (**Figure 8.33**).[31] The DRIs have set AIs for this compound: 550 mg/day for men and 425 mg/day for women.[4] There is concern that current choline intakes are below recommendations for certain segments of the population, but it is not clear how this is affecting public health.

Excess choline intake can cause a fishy body odor, sweating, reduced growth rate, low blood pressure, and liver damage. The amounts needed to cause these symptoms are much higher than can be obtained from foods. The UL for choline for adults is 3.5 g/day.

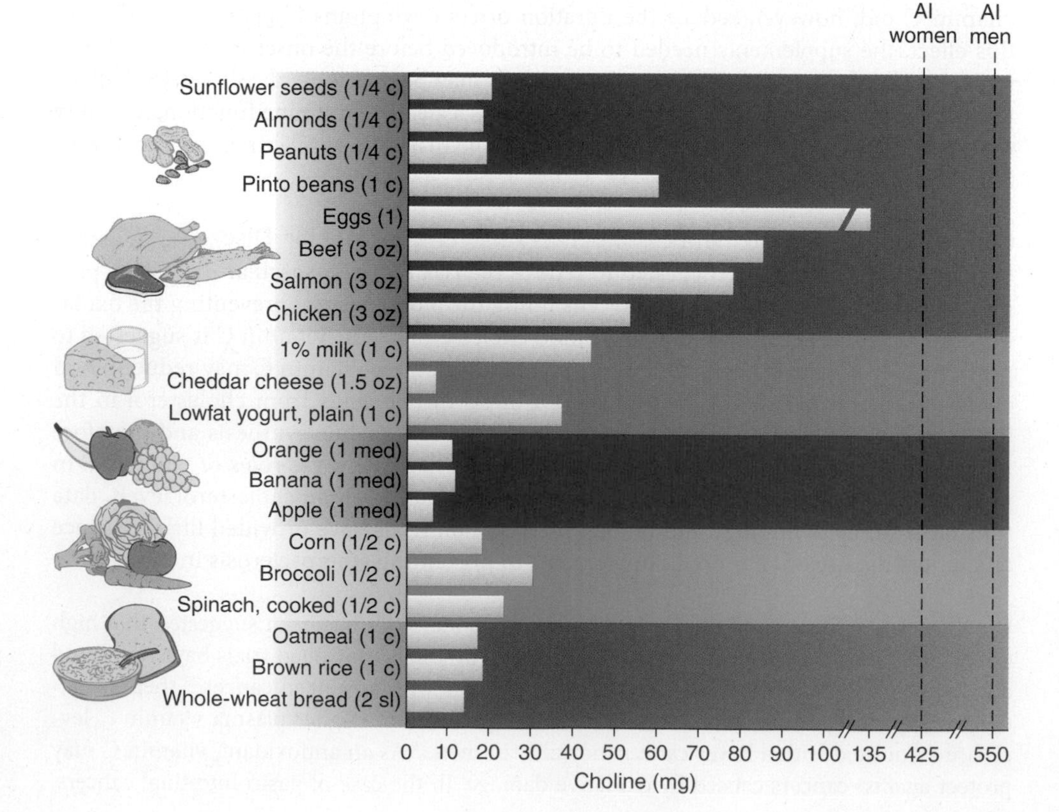

Figure 8.33 Choline content of MyPlate food groups[31]
Most foods provide some choline, but eggs and meats are the best sources. The dashed lines represent the AIs for men and women.

OUTCOME

Lew Robertson/FoodPix/Getty Images, Inc.

Gary's anemia and vitamin C deficiency were treated with a multivitamin and an iron supplement. He started feeling better almost immediately. He worked with the dietitian to plan a diet that fit his budget and met his need for iron and vitamin C as well as other essential nutrients. She taught him that red meat and fortified cereals are good sources of iron. She explained that without fruits and vegetables, his diet provided very little vitamin C. Without vitamin C, he was not efficiently absorbing the iron in the fortified grain products he did consume and his iron losses were increased. To address his fruit and vegetable intake, she suggested that he add frozen vegetables to his noodles, watch for seasonal produce on sale, and buy canned fruits and vegetables, which are generally less expensive. She also suggested baked potatoes topped with salsa and cheese, and pasta with ground beef and tomato sauce as low-cost meals that provide iron and vitamin C. By following her suggestions he was able to eat a few servings of fruits and vegetables every day and still keep to a budget. Gary's diet still doesn't meet recommendations for fruits and vegetables, but he is getting plenty of iron and vitamin C. He expects to complete his degree in the spring. He is lucky enough to have a job lined up and plans to use some of his paycheck to further improve his diet.

© Cat London/iStockphoto

APPLICATIONS

ASSESSING YOUR DIET

1. **Do you get enough of all the B vitamins?** Use iProfile and your food intake record from Chapter 2 to determine how much of each of the B vitamins your diet contains.
   - **a.** Make a list of the B vitamins, and compare your intake to the recommended amount.
   - **b.** For any of the B vitamins for which your intake is below the recommended amount, suggest food substitutions that would increase your intake to meet the recommendation.
   - **c.** Which foods in your diet provide the highest amounts of thiamin, riboflavin, and niacin. Were any of these sources fortified?
   - **d.** Where do you get your folate? List several natural sources of folate in your diet and several foods that are fortified with folic acid.
   - **e.** Do you get enough vitamin B_{12}? If you eliminated meat, fish, poultry, and eggs from your diet, would you be getting enough vitamin B_{12}? List the foods remaining in your diet that contribute vitamin B_{12}.
2. **Do you get enough vitamin C?** Use iProfile and your food intake record from Chapter 2 to determine how much vitamin C your diet contains.
   - **a.** How does your intake compare with the RDA?
   - **b.** Which foods are the best sources of vitamin C?
   - **c.** If your diet doesn't meet recommendations, suggest some dietary modifications that will help you meet the RDA for vitamin C.

CONSUMER ISSUES

3. **How much vitamin C is in your fast-food favorites?**
   - **a.** Use iProfile or information from company Web sites to determine the amount of vitamin C in five items you commonly order in fast-food restaurants.
   - **b.** What percentage of your RDA for vitamin C does each provide? How does this compare with the percentage of your daily energy needs that each provides?
   - **c.** Which of these fast-food items are highest in vitamin C? What component of the food provides the vitamin C?
4. **Take a look at the Nutrition Facts label on a box of breakfast cereal.**
   - **a.** What percentage of the Daily Value is included in a serving for each of the water-soluble vitamins?
   - **b.** What percentage of the Daily Value would you consume for each water-soluble vitamin if you ate the portion of cereal you typically have for breakfast?
   - **c.** Does the amount in your portion exceed the UL for any of these vitamins? Which ones?

CLINICAL CONCERNS

5. **Hazel is suffering through her third cold of the winter. When she complains at a local health food store, the clerk recommends several supplements to keep her healthy. These include a vitamin C supplement, a stress formula B vitamin supplement called B_{50}, a supplement called *Prevention Plus*, and the herbal supplement echinacea. Hazel checks the labels of the supplements she has purchased. The information is summarized in the table.**

Supplement/ Ingredients	Dose %	DV/Dose	Frequency
Vitamin C			3/day
Vitamin C	500 mg	833%	
B_{50}			3/day
Thiamin	50 mg	3333%	
Niacin	50 mg	250%	
Vitamin B_6	60 mg	2500%	
Riboflavin	50 mg	2941%	
Biotin	50 mg	16.66%	
Pantothenic acid	50 mg	500%	
Folic acid	50 mg	12.5%	
Vitamin B_{12}	50 mg	833%	
Prevention Plus			1/day
Vitamin C	1000 mg	1667%	
Zinc	15 mg	100%	
Echinacea			6/day
Echinacea	125 mg	—	

a. Review the ingredients in Hazel's supplements. If she takes all these products at the frequency recommended in the table, will she be exceeding the UL for any nutrients? List them.

b. Did you find a UL for echinacea? Why not?

c. Would you recommend that Hazel take these supplements? Why or why not?

6. **Harold is 72 years old. Recently he has been feeling tired and forgetful. Then he began having tingling in his hands and feet, difficulty walking, and diarrhea. Laboratory tests indicated that he had low levels of vitamin B_{12}. A diet history revealed that he eats lots of grains, fruits, and vegetables but very little meat or dairy products. The doctor explained that Harold's symptoms were due to a vitamin B_{12} deficiency.**

a. Why might Harold's diet increase his risk for vitamin B_{12} deficiency?

b. Harold's wife eats the same diet but has normal vitamin B_{12} levels. Why might Harold be at greater risk?

7. **Simone is taking a chemotherapeutic drug called methotrexate to treat her lung cancer. Methotrexate acts by inhibiting the metabolism of folic acid so the cancer cells can't divide.**

a. Why might she be instructed not to take a multivitamin or any supplement containing folate?

b. Why does methotrexate cause side effects such as low numbers of white blood cells and hair loss?

SUMMARY

8.1 What Are Vitamins?

- Vitamins are essential organic nutrients that do not provide energy and are required in small quantities in the diet. Americans consume vitamins that are naturally present in foods, added to foods by fortification, and supplied by supplements. Some foods are fortified with nutrients according to government guidelines to promote public health. Others are fortified according to the manufacturer's perceptions of what will sell in the marketplace.
- The bioavailability of a vitamin depends on how much can be absorbed and used by the body. The ability to store and excrete vitamins helps to regulate the amount present in the body. Water-soluble vitamins are generally excreted more easily than and not stored as well as fat-soluble vitamins.
- Vitamins are needed to promote and regulate body processes that are essential for growth, reproduction, and tissue maintenance. Each vitamin has one or more functions: many are coenzymes, some are antioxidants. More than one vitamin is typically needed to ensure the health of a particular organ or system.
- Vitamin deficiencies remain a major health problem worldwide, but severe deficiencies are rare in developed countries. Recommended intakes for vitamins are expressed as RDAs or AIs. UL values estimate the highest dose that is unlikely to cause toxicity. Food labels can help identify packaged foods that are good sources of vitamins.
- *Go to WileyPLUS to view a video clip on vitamin supplements.*

8.2 Thiamin

- The best food sources of thiamin are lean pork, legumes, and whole or enriched grains.
- Thiamin is required for the formation of acetyl-CoA from pyruvate and for a reaction in the citric acid cycle and is therefore particularly important for the production of ATP from glucose. It is also needed for the synthesis of the neurotransmitter acetylcholine.
- Thiamin deficiency, or beriberi, causes nervous system abnormalities. Deficiencies are common in alcoholics. No toxicity has been identified.

8.3 Riboflavin

- Milk, meat, and enriched grain products are the best food sources of riboflavin.
- Riboflavin coenzymes are needed for the generation of ATP from carbohydrate, fat, and protein.
- Riboflavin deficiency is rarely seen alone because food sources of riboflavin are also sources of other B vitamins and because riboflavin is needed for the utilization of several other vitamins. No toxicity has been identified.

8.4 Niacin

- Beef, chicken, turkey, fish, and enriched grain products are the best food sources of niacin. The amino acid tryptophan can be converted into niacin, so tryptophan from dietary protein can meet some of the niacin requirement.
- Niacin coenzymes are important in the breakdown of carbohydrate, fat, and protein to provide energy and in the synthesis of fatty acids and sterols.
- Niacin deficiency results in pellagra, which is characterized by dermatitis, diarrhea, dementia, and finally, if untreated, death.
- Supplements of the nicotinic acid form of niacin can lower elevated blood cholesterol but frequently cause toxicity symptoms such as flushing, tingling sensations, nausea, and a red skin rash.

8.5 Biotin

- Liver and egg yolks are good sources of biotin.
- Biotin is needed for the synthesis of glucose and fatty acids and for the metabolism of certain amino acids.
- An RDA has not been established for biotin because some of our requirement for this vitamin is met by bacterial synthesis in the GI tract. However, an AI has been set. Toxicity has not been reported.

8.6 Pantothenic Acid

- Pantothenic acid is abundant in the food supply and deficiency is rare.
- Pantothenic acid is part of coenzyme A (CoA), which is required for the production of ATP from carbohydrate, fat, and protein and the synthesis of cholesterol and fatty acids. There is no RDA, but an AI has been established.

8.7 Vitamin B_6

- Food sources of vitamin B_6 include chicken, fish, liver, and whole grains.
- Pyridoxal phosphate, the coenzyme form of vitamin B_6, is needed for the activity of more than 100 enzymes involved in the metabolism of carbohydrate, fat, and protein. Vitamin B_6 is a coenzyme for transamination, deamination, and decarboxylation reactions and is therefore particularly important for amino acid metabolism.
- Vitamin B_6 deficiency causes neurological symptoms, anemia due to impaired hemoglobin synthesis, poor immune function, and elevated levels of homocysteine, which can increase the risk of cardiovascular disease. High intakes from supplements can cause nervous system abnormalities.

8.8 Folate or Folic Acid

- Food sources of folate include liver, legumes, oranges, leafy green vegetables, and fortified grains.
- Folate is necessary for the synthesis of DNA, so it is especially important for rapidly dividing cells.
- Folate deficiency results in macrocytic anemia and can cause an increase in homocysteine levels. Low levels of folate before and during early pregnancy are associated with an increased incidence of neural tube defects in the offspring. A high intake of folate can mask the anemia caused by vitamin B_{12} deficiency. Low intakes of folate have been associated with an increased risk of cardiovascular disease and certain types of cancer.
- It is recommended that women of childbearing age consume 400 μg of folic acid from fortified foods and supplements in addition to the folate found in a varied diet.

8.9 Vitamin B_{12}

- Vitamin B_{12} is found almost exclusively in animal products.
- The absorption of vitamin B_{12} from food requires adequate levels of stomach acid, intrinsic factor, and pancreatic secretions.
- Vitamin B_{12} is needed for the metabolism of folate and fatty acids and to maintain the insulating layer of myelin surrounding nerves.
- Vitamin B_{12} deficiency increases homocysteine levels and can result in anemia and permanent nerve damage. Pernicious anemia is caused by severe vitamin B_{12} deficiency due to an absence of intrinsic factor. Vitamin B_{12} deficiency may also occur in vegans, who consume no animal products, and in older individuals with low stomach acid due to atrophic gastritis.

8.10 Vitamin C

- The best food sources of vitamin C are citrus fruits.
- Vitamin C is necessary for the synthesis and maintenance of collagen and for the synthesis of hormones and neurotransmitters. Vitamin C is also a water-soluble antioxidant. Antioxidants protect the body from reactive oxygen molecules such as free radicals. These molecules are generated from normal body reactions and come from the environment. They cause damage by stealing electrons from DNA, proteins, carbohydrates, and unsaturated fatty acids.
- Vitamin C deficiency, called scurvy, is characterized by poor wound healing, bleeding, and other symptoms related to the improper formation and maintenance of collagen.
- Vitamin C supplements are the most commonly taken vitamin supplements and are usually used to reduce the symptoms of the common cold.

8.11 Choline

- Choline is considered an essential nutrient but is not currently classified as a vitamin. Choline can be synthesized in the human body, but the amounts are generally not enough to meet needs. The best dietary sources of choline are egg yolks and meats.

REVIEW QUESTIONS

1. What is a vitamin?
2. List four factors that affect how much of a vitamin is available to the body.
3. Define *coenzyme*, and describe the coenzyme functions of five vitamins.
4. Why is thiamin deficiency common in alcoholics?
5. Why should milk be packaged in opaque containers?
6. What is pellagra?
7. How is vitamin B_6 involved in amino acid and protein metabolism?
8. Why is low folate intake of particular concern for women of childbearing age?
9. Why would someone who has had his stomach removed (or had gastric bypass surgery) need to receive injections of vitamin B_{12} to meet his needs?
10. Why are vegans at risk for vitamin B_{12} deficiency? The elderly?
11. Explain why deficiencies of vitamin B_6, folate, and vitamin B_{12} can all cause an increase in homocysteine levels.
12. Why does vitamin C deficiency cause poor wound healing?
13. What are reactive oxygen molecules and how do they cause damage?
14. What is the role of antioxidants and pro-oxidants in oxidative stress?
15. Does choline fit the definition of a vitamin? Why or why not?

REFERENCES

1. Butte, N. F., Fox, M. K., Briefel, R. R., et al. Nutrient intakes of U.S. infants, toddlers, and preschoolers meet or exceed dietary reference intakes. *J Am Diet Assoc* 110 (12 suppl):S27–S37, 2010.
2. Ervin, R. B., Wang, C. Y., Wright, J. D., and Kennedy-Stephenson, J. Dietary intake of selected minerals for the United States population: 1999–2000. Advance data from *Vital and Health Statistics*, no. 341. Hyattsville, Maryland: National Center for Health Statistics, 2004.
3. Butterworth, R. F. Thiamin. In *Modern Nutrition in Health and Disease*, 10th ed. M. E. Shils, M. Shike, A. C. Ross, et al., eds. Philadelphia: Lippincott Williams & Wilkins, 2006, pp. 426–433.
4. Institute of Medicine, Food and Nutrition Board. *Dietary Reference Intakes for Thiamin, Riboflavin, Niacin, Vitamin B-6, Folate, Vitamin B-12, Pantothenic Acid, Biotin, and Choline.* Washington, DC: National Academies Press, 1998.
5. Saffert, A., Pieper, G., and Jetten, J. Effect of package light transmittance on vitamin content of milk. *Technol Sci* 21:47–55, 2008.
6. Roe, D. A. *A Plague of Corn: The Social History of Pellagra.* Ithaca, NY: Cornell University Press, 1973.
7. Seal, A. J., Creeke, P. I., Dibari, F., et al. Low and deficient niacin status and pellagra are endemic in postwar Angola. *Am J Clin Nutr* 85:218–242, 2007.
8. Hochholzer, W., Berg, D. D., and Giugliano, R. P. The facts behind niacin. *Ther Adv Cardiovasc Dis* 5:227–240, 2011.
9. Galvin, M. A., Kiely, M., and Flynn, A. Impact of ready-to-eat breakfast cereal (RTEBC) consumption on adequacy of micronutrient intakes and compliance with dietary recommendations in Irish adults. *Public Health Nutr* 6:351–363, 2003.
10. Bessey, O. A., Adam, D. J., and Hansen, A. E. Intake of vitamin B_6 and infantile convulsions: A first approximation of requirements of pyridoxine in infants. *Pediatrics* 20:33–44, 1957.
11. Wilson, S. M., Bivins, B. N., Russell, K. A., and Bailey, L. B. Oral contraceptive use: Impact on folate, vitamin B_6, and vitamin B_{12} status. *Nutr Rev* 69:572-583, 2011.
12. Di Minno, M. N., Tremoli, E., Coppola, A., et al. Homocysteine and arterial thrombosis: Challenge and opportunity. *Thromb Haemost* 103:942–961, 2010.
13. Page, J. H., Ma, J., Chiuve, S. E., et al. Plasma vitamin B(6) and risk of myocardial infarction in women. *Circulation* 120:649–655, 2009.
14. McNulty, H., Pentieva, K., Hoey, L., and Ward, M. Homocysteine, B-vitamins and CVD. *Proc Nutr Soc* 67:232–237, 2008.
15. LeBlanc, K. E., and Cestia, W. Carpal tunnel syndrome. *Am Fam Physician* 83:952–958, 2011.
16. Kashanian, M., Mazinani, R., and Jalalmanesh, S. Pyridoxine (vitamin B_6) therapy for premenstrual syndrome. *Int J Gynaecol Obstet* 96:43–44, 2007.
17. Berry, R. J., Bailey, L., Mulinare, J., et al. Fortification of flour with folic acid. *Food Nutr Bull* 31(1 suppl):S22–S35, 2010.
18. Blencowe, H., Cousens, S., Modell, B., and Lawn, J. Folic acid to reduce neonatal mortality from neural tube disorders. *Int J Epidemiol* 39(suppl 1):i110–i121, 2010.
19. Kim, Y. I. Folate and colorectal cancer: An evidence-based critical review. *Mol Nutr Food Res* 51:267–292, 2007.
20. Kim, Y. I. Folic acid fortification and supplementation—good for some but not so good for others. *Nutr Rev* 65:504–511, 2007.
21. Ebbing, M., Bønaa, K. H., Nygård, O., et al. Cancer incidence and mortality after treatment with folic acid and vitamin B_{12}. *JAMA* 302:2119–2126, 2009.
22. American Dietetic Association. Position of the American Dietetic Association and Dietitians of Canada: Vegetarian diets. *J Am Diet Assoc* 103:748–765, 2003.

23. Food and Nutrition Board, Institute of Medicine. *Dietary Reference Intakes for Vitamin C, Vitamin E, Selenium, and Carotenoids*. Washington, DC: National Academies Press, 2000.

24. Duarte, T. L., and Lunec, J. Review: When is an antioxidant not an antioxidant? A view of novel actions and reactions of vitamin C. *Free Radic Res 39*:671–686, 2005.

25. Douglas, R. M., Hemila, H., Chalker, E., and Treacy, B. Vitamin C for preventing and treating the common cold. *Cochrane Database Syst Rev* 18(3): CD000980, 2007.

26. Farbstein, D., Kozak-Blickstein, A., and Levy, A. P. Antioxidant vitamins and their use in preventing cardiovascular disease. *Molecules* 15:8090–8110, 2010.

27. Cabanillas, F. Vitamin C and cancer: What can we conclude—1,609 patients and 33 years later? *P R Health Sci J* 29:215–217, 2010.

28. Khaw, K. T., Bingham, S., Welch, A., et al. Relation between plasma ascorbic acid and mortality in men and women in EPIC-Norfolk prospective study: A prospective population study, European Prospective Investigation into Cancer and Nutrition. *Lancet* 357:657–663, 2001.

29. Goodman, M., Bostick, R. M., Kucuk, O., and Jones, D. P. Clinical trials of antioxidants as cancer prevention agents: Past, present, and future. *Free Radic Biol Med* 51:1068–1084, 2011.

30. Zeisel, S. H., and da Costa, K. A. Choline: An essential nutrient for public health. *Nutr Rev* 67:615–623, 2009.

31. Patterson, K. Y., Bhagwat, S. A., Williams, J. R., et al. USDA database for the choline content of common foods, release 2, January, 2008.

***To access links to online sources, please go to** www.wiley.com/college/smolin **and select Nutrition: Science and Applications, 3rd edition. From this page, select either the student or instructor companion site. Once on the desired site, select References.**

CHAPTER OUTLINE

© DNY59/iStockphoto

The Fat-Soluble Vitamins

9

CASE STUDY

Ayana's pediatrician is concerned because at age 1 she has just one tooth. Her mother has breast-fed her since birth, and she has eaten some solid food in addition to the breast milk since she was about six months old. After noticing that she has only one tooth and examining her ribs and legs, the doctor tells Ayana's parents that she wants to draw a blood sample to help confirm her suspicion that Ayana has a vitamin D deficiency disease called rickets. The pediatrician explains that this deficiency is relatively rare in the United States, but risks are higher in children who are dark skinned, live in cold climates, and/or are breast-fed without vitamin D supplementation. Ayana's family lives in Idaho where the winters are long and the skies cloudy, reducing time spent outdoors in the sun. Although breast milk is best for babies, it is low in vitamin D. Vitamin D can be synthesized in the skin, but because Ayana has dark skin pigmentation, she requires greater sun exposure than lighter-skinned babies to synthesize adequate amounts.

Vitamin D is needed for proper formation and maintenance of bones and teeth. Without sufficient vitamin D, a child's legs bow under the weight of standing and bony bumps appear on each of the ribs. The poorly formed bones can break easily. The teeth erupt late and are very prone to decay. Ayana is beginning to show these symptoms. Rickets is diagnosed with bone x-rays and blood tests that measure levels of calcium and phosphorus, as well as levels of an enzyme produced by cells that break down bone.

Like Ayana's parents, most people in the United States are not familiar with rickets. **Recent evidence, however, suggests that rickets due to vitamin D deficiency may be reemerging as a problem in the U.S. population.**[1,2]

Digital Vision/Getty Images, Inc.

9.1 Fat-Soluble Vitamins in Our Diet

LEARNING OBJECTIVES

- Name the fat-soluble vitamins.
- Discuss factors that impact fat-soluble vitamin status in the United States.

The fat-soluble vitamins—A, D, E, and K—are found along with fats in foods. Because they are fat soluble, they require special handling for absorption into and transport throughout the body. They require bile and dietary fat for absorption. Once absorbed, they are transported with lipids through the lymphatic system in chylomicrons before entering the blood (see Chapter 5). Because excesses of these vitamins can be stored in the liver and fatty tissues, intakes can vary without a risk of deficiency as long as average intake over a period of weeks or months meets needs. Solubility in fat, however, limits their routes of excretion and therefore increases the risk of toxicity.

© Tatiana Emshanova/iStockphoto

Figure 9.1 Leafy greens are a source of fat-soluble vitamins
Leafy green vegetables such as this Swiss chard are good low-calorie sources of vitamin A precursors, as well as vitamin E and vitamin K.

Despite the body's ability to store the fat-soluble vitamins, deficiencies of vitamin A and D are common in the developing world. Elimination of these deficiencies through supplementation and fortification is a focus of public health programs. In the United States and other developed countries, severe deficiencies of these nutrients are rare, but trends in the modern diet have affected the amounts of fat-soluble vitamins people consume. Our rising reliance on fast food has reduced our intake of fruits and vegetables, particularly leafy greens, reducing our intake of vitamins A, E, and K (**Figure 9.1**). We work at indoor jobs, exercise at indoor gyms, and protect ourselves from the sun by using sunscreens when we go out, all of which limit our ability to get adequate vitamin D from sunshine. We limit the number of eggs we eat to reduce our cholesterol intake, but eggs are good sources of vitamin D and vitamin A. We cut fat to reduce calories, but vegetable oils provide much of our vitamin E as well as fat-soluble phytochemicals, some of which serve as precursors to vitamin A. Medications that limit fat absorption also limit fat-soluble vitamin absorption. To avoid the risks of limited intakes, we take supplements of these vitamins and fortify our food with them. This then increases the risks of consuming toxic amounts. It is unclear whether these new dietary patterns will have a long-term impact on our fat-soluble vitamin status.

9.2 Vitamin A

LEARNING OBJECTIVES

- Compare the sources, functions, and potential toxicity of preformed vitamin A and provitamin A.
- Describe the role of vitamin A in night vision.
- Discuss how vitamin A affects gene expression.
- Explain why a vitamin A deficiency can cause eye infections and blindness.

Did you ever hear that eating carrots would help you see in the dark? It turns out to be true. Carrots are high in provitamin A, and vitamin A is important for vision and healthy eyes. This connection between vision and foods that we now know are high in vitamin A has been recognized for centuries. In ancient times, the Egyptians knew that eating liver could improve night vision in those who had difficulty in adjusting from bright light to dim light. In 1968, George Wald earned the Nobel Prize in medicine for identifying the mechanism by which vitamin A is involved in vision. Although this is a key function of vitamin A, attention today is focused more on how vitamin A interacts with genes to regulate growth and **cell differentiation**. Despite our expanding understanding of the functions of vitamin A, deficiency remains a world health problem.

cell differentiation Structural and functional changes that cause cells to mature into specialized cells.

Vitamin A in the Diet

Vitamin A is found preformed and in precursor or provitamin forms in our diet. Preformed vitamin A is found in animal foods, and the provitamin A forms are found in plants (**Figure 9.2**). Both sources can be used to meet vitamin A needs in the body.

Preformed vitamin A compounds are known as **retinoids**. Three retinoids are active in the body: retinol, retinal, and retinoic acid. Animal foods such as liver, fish, egg yolks, and dairy products provide preformed vitamin A, primarily as retinol or retinol attached to a fatty acid (retinyl ester). Margarine and nonfat and reduced-fat milk are fortified with retinol because they are often consumed in place of butter and whole milk, which are good natural sources of this vitamin. Retinal and retinoic acid can be formed in the body from retinol (**Figure 9.3**).

retinoids The chemical forms of preformed vitamin A: retinol, retinal, and retinoic acid.

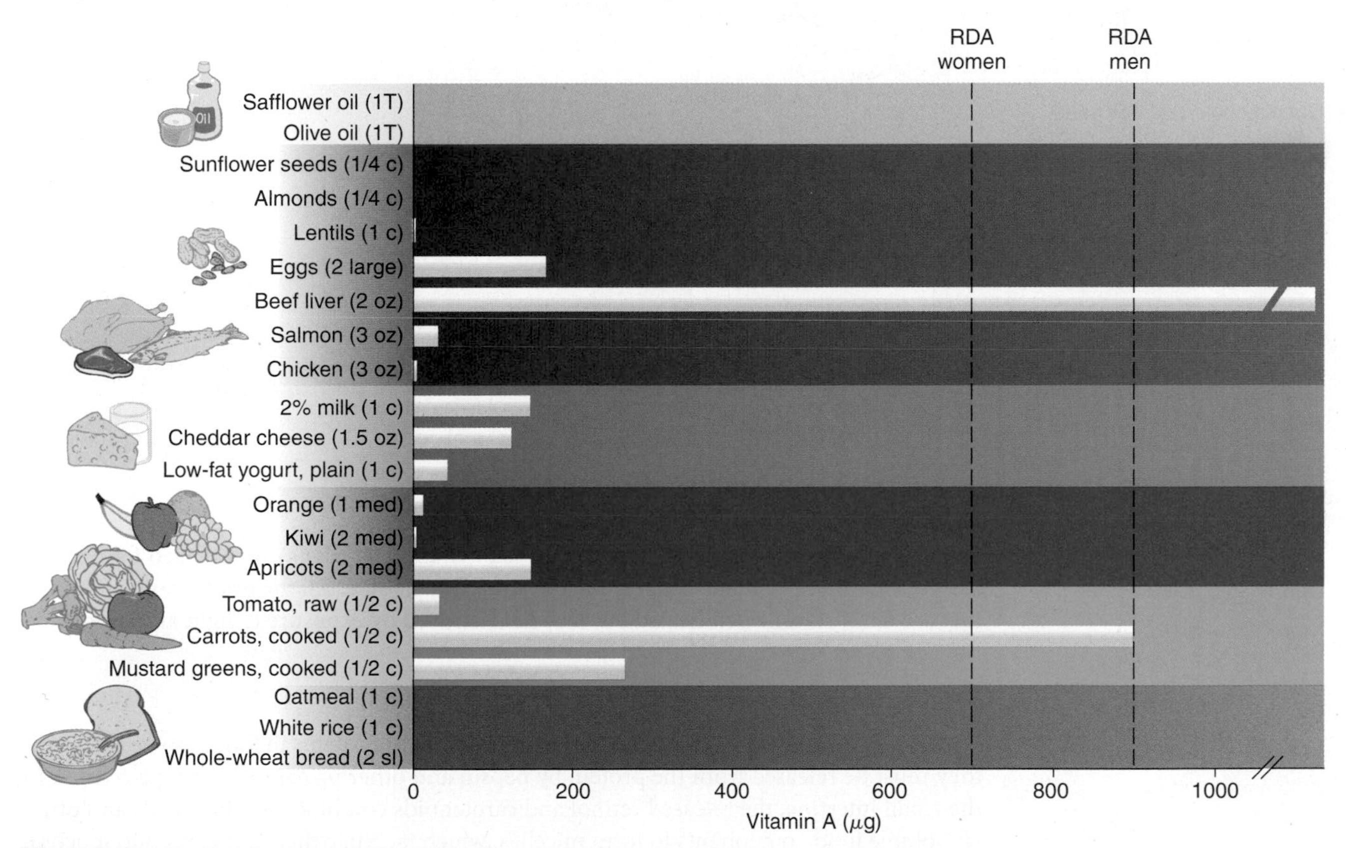

Figure 9.2 **Vitamin A content of MyPlate food groups**
Eggs, liver, and dairy products are good sources of preformed vitamin A, and yellow-orange and leafy green fruits and vegetables are good sources of provitamin A carotenoids; grains are generally poor sources of vitamin A. The dashed lines show the RDAs for men and women.

Plants contain provitamin A compounds called **carotenoids**. Carotenoids are yellow, orange, and red pigments that give these colors to fruits and vegetables. About 50 of the 600 carotenoids that have been isolated provide vitamin A activity. **Beta-carotene (β-carotene)**, the most potent precursor, is plentiful in dark orange fruits and vegetables such as mangos, apricots, cantaloupe, carrots, red peppers, pumpkins, and sweet potatoes as well as in leafy greens where the orange color is masked by the green of chlorophyll. Other carotenoids that provide some provitamin A activity include alpha-carotene (α-carotene), found in carrots, dark green vegetables, and winter squash, and beta-cryptoxanthin (β-cryptoxanthin), found in mangos, papayas, winter squash, and sweet red peppers.[3] Lutein, lycopene, and zeaxanthin are carotenoids with no vitamin A activity (see **Focus on Phytochemicals**).

carotenoids Natural pigments synthesized by plants and many microorganisms. They give yellow and red-orange fruits and vegetables their color.

beta-carotene (β-carotene) A carotenoid that has more provitamin A activity than other carotenoids. It also acts as an antioxidant.

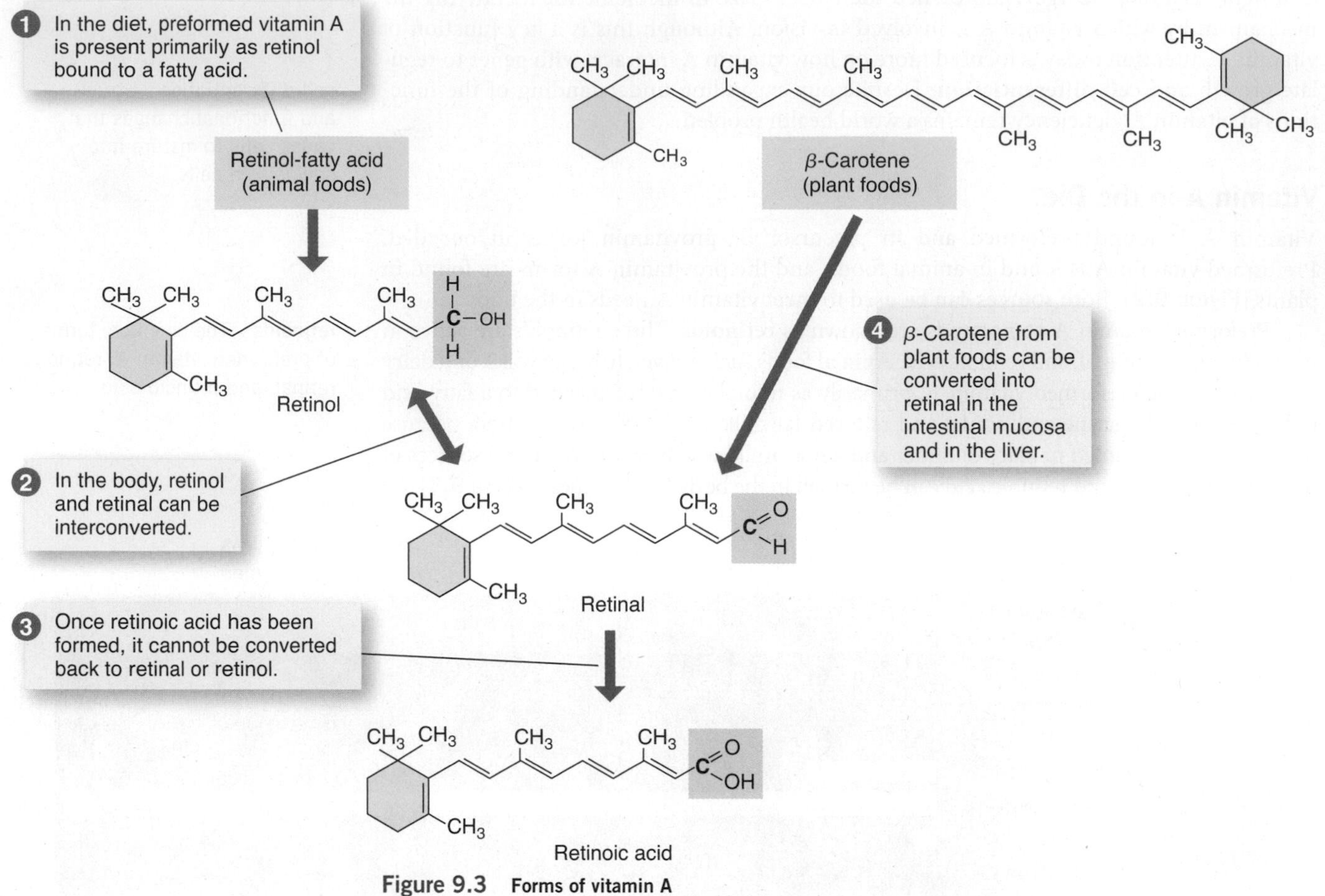

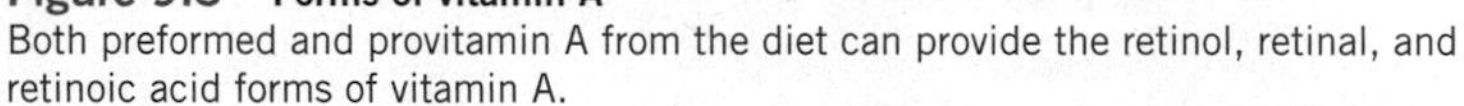

Figure 9.3 Forms of vitamin A
Both preformed and provitamin A from the diet can provide the retinol, retinal, and retinoic acid forms of vitamin A.

To help consumers identify food sources of vitamin A, labels on packaged foods must list the vitamin A content as a percentage of the Daily Value. All forms of vitamin A in the diet are fairly stable when heated but may be destroyed by exposure to light and oxygen.

Vitamin A in the Digestive Tract

Both preformed vitamin A and carotenoids are bound to proteins in foods. To be absorbed, they must be released from the protein by pepsin and other protein-digesting enzymes. In the small intestine, the released retinol and carotenoids combine with bile acids and other fat-soluble food components to form micelles, which facilitate their diffusion into mucosal cells (see Chapter 5). Absorption of preformed vitamin A is efficient—70 to 90% of what is consumed. The provitamin carotenoids are less well absorbed, and absorption decreases as intake increases, so large amounts are not well absorbed.[4] Once inside the mucosal cells, much of the β-carotene is converted to retinal. Cleaving a molecule of β-carotene in half theoretically yields two molecules of retinal; however, because β-carotene is not as well absorbed as preformed vitamin A and may not be efficiently converted to retinal, it takes about 12 μg of dietary β-carotene to yield 1 μg of retinol (see Figure 9.3).

The fat content of the diet and the ability to absorb fat can affect the amount of vitamin A that is absorbed. A diet that is very low in fat (less than 10 g/day) can reduce vitamin A absorption. This is rarely a problem in industrialized countries, where typical fat intake ranges from 50 to 100 grams per day. However, in populations with low dietary fat intakes, vitamin A deficiency may occur due to poor absorption. Diseases that cause fat malabsorption, as well as some medications, can also interfere with vitamin A absorption and cause a deficiency.

How Vitamin A Functions in the Body

Preformed vitamin A and carotenoids absorbed from the diet are transported from the intestine in chylomicrons. These lipoproteins deliver the preformed vitamin A and carotenoids to body tissues such as muscle, kidney, and liver. In the liver, some carotenoids can be converted into retinol and retinol that is not immediately needed is stored. To move from liver stores to the tissues, retinol must be bound to **retinol-binding protein**. There is no specific blood transport protein for carotenoids, but since they are fat soluble, they are incorporated into lipoproteins to travel in the bloodstream.

retinol-binding protein A protein that is necessary to transport vitamin A from the liver to other tissues.

The different forms of vitamin A have different functions. The body can make the retinal and retinoic acid forms from the retinol and carotenoids in the diet (see Figure 9.3). Retinol is the form that circulates in the blood. Retinal is the form that is important for vision. Retinol and retinal can be interconverted from one to the other. Retinoic acid, which is made from retinal, cannot be used in the visual cycle (see the next section) but is the form that affects gene expression and is responsible for vitamin A's role in cell differentiation, growth, and reproduction.[5] Carotenoids that are not converted to retinoids may act as antioxidants or provide other biological functions.

rhodopsin A light-absorbing compound found in the retina of the eye that is composed of the protein opsin loosely bound to retinal.

How Vitamin A Helps Us to See Light Vitamin A is involved in the perception of light. In the eye, the retinal form of the vitamin combines with the protein opsin to form the visual pigment **rhodopsin** (**Figure 9.4**). Rhodopsin helps transform the energy from light into a nerve impulse that is sent to the brain. This nerve impulse allows us to see.

THINK CRITICALLY

Use the diagram to explain why people say eating carrots helps you see in the dark.

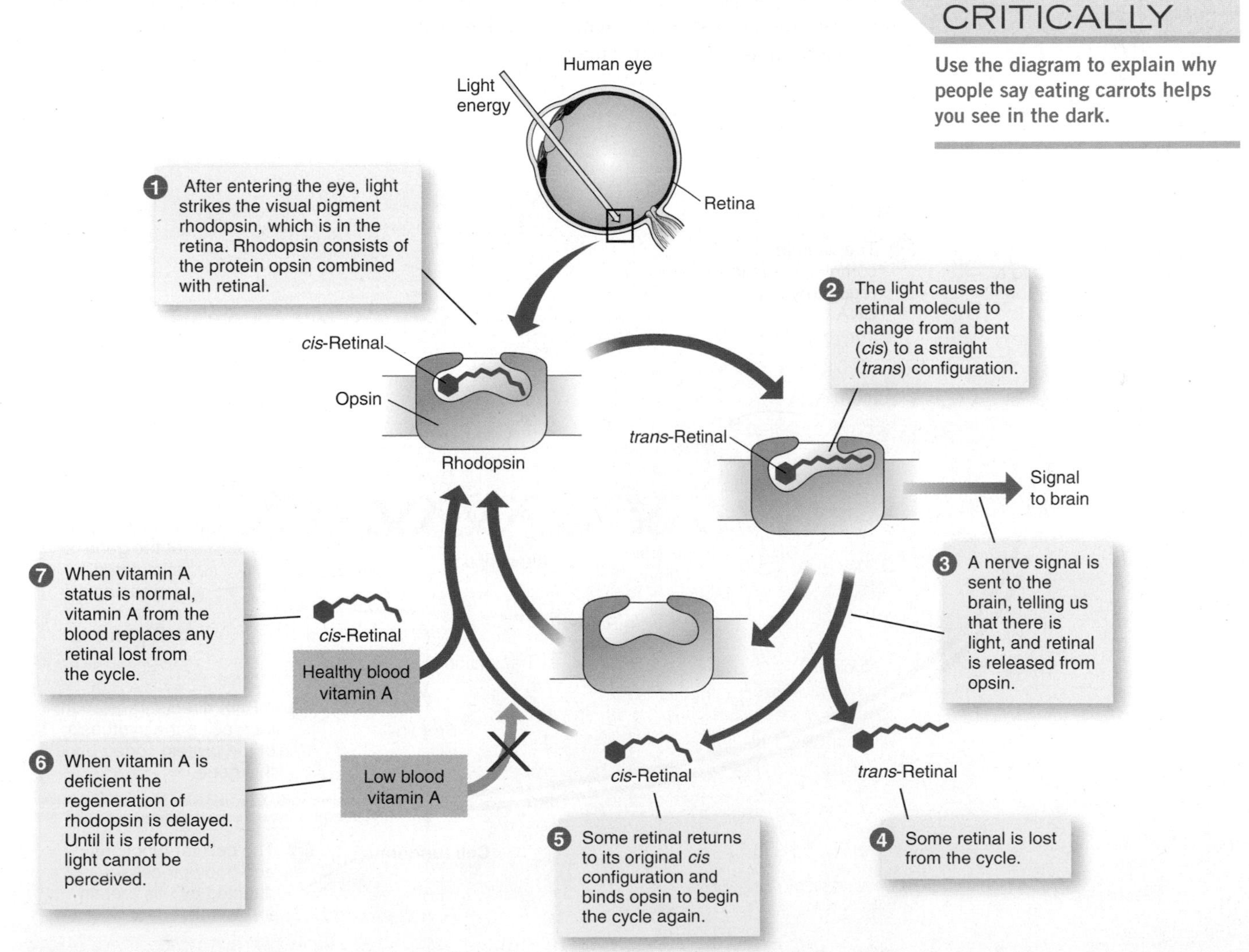

Figure 9.4 The visual cycle
Looking into the bright headlights of an approaching car at night is temporarily blinding for all of us, but for someone with vitamin A deficiency the blindness lasts much longer. This occurs because of the role of vitamin A in the visual cycle.

The visual cycle begins when light passes into the eye and strikes rhodopsin. The light changes the retinal in rhodopsin from a curved molecule to a straight one by converting a *cis* double bond in retinal to a *trans* double bond. This change in shape initiates a series of events that cause a nerve signal to be sent to the brain and retinal to be released from opsin. After the light stimulus has passed, the *trans* retinal is converted back to its original *cis* form and recombined with opsin to regenerate rhodopsin. Each time this cycle occurs, some retinal is lost and must be replaced by retinol from the blood. The retinol is converted into retinal in the eye. When vitamin A is deficient, there is a delay in the regeneration of rhodopsin that causes difficulty seeing in dim light, particularly after exposure to a bright light—a condition called **night blindness** (see Figure 9.4). Night blindness is one of the first and more easily reversible symptoms of vitamin A deficiency.

night blindness The inability of the eye to adapt to reduced light, causing poor vision in dim light.

How Vitamin A Regulates Gene Expression The retinoic acid form of vitamin A is needed for normal gene expression. When a specific gene is expressed, it instructs the cell to make a particular protein. Proteins have structural and regulatory functions within cells and throughout the body. Altering the expression of specific genes increases (or decreases) the production of certain proteins and thereby affects various cellular and body functions. In order to affect gene expression, the retinoic acid form of vitamin A enters the nucleus of specific target cells, where it binds to retinoic acid receptor proteins; the retinoic acid–protein complex then binds to a regulatory region of DNA (**Figure 9.5**). This binding changes the amount of messenger RNA (mRNA) that is made by the gene. The change in mRNA changes the amount of the protein that is produced. For example, vitamin A increases the expression of a gene that makes an enzyme in liver cells. This enzyme enables the liver to make glucose through the process of glyconeogenesis.

THINK CRITICALLY

What would happen to the amount of protein produced in step 4 if vitamin A was absent?

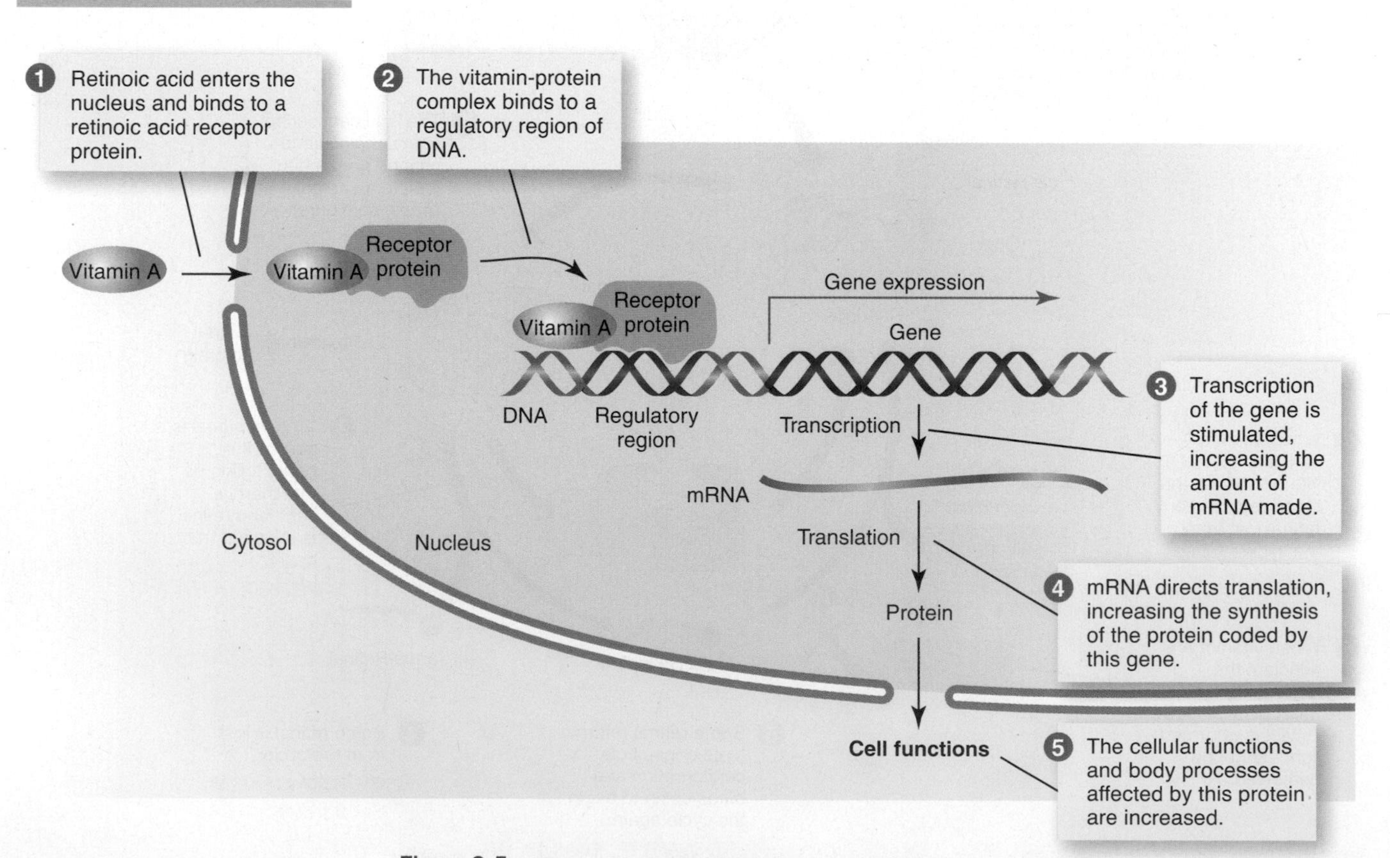

Figure 9.5 Vitamin A and gene expression
The retinoic acid form of vitamin A affects cell function by changing gene expression. The steps illustrate what happens when vitamin A turns on gene expression.

By affecting gene expression, vitamin A can determine what proteins a cell produces and what type of cell a cell will become as it differentiates. Cell differentiation is the process whereby cells change in structure and function to become specialized. For instance, in the bone marrow, some cells differentiate into various white blood cells, whereas others differentiate to form red blood cells.

The role of vitamin A in cell differentiation is particularly important for the maintenance of epithelial tissue. Epithelial tissue is the type of tissue that covers external body surfaces and lines internal cavities and tubes. It includes the skin and the linings of the eyes, intestines, lungs, vagina, and bladder. These cells have a short life span and need to be replaced often. Adequate vitamin A ensures that the old cells are replaced with normal healthy new ones. The ability of vitamin A to regulate the growth and differentiation of cells also makes it essential throughout life for normal reproduction, growth, and immune function.

β-Carotene Functions as a Vitamin A Precursor and an Antioxidant Some carotenoids, particularly β-carotene, can be converted to vitamin A in the intestinal mucosa and liver. Unconverted carotenoids also circulate in the blood and reach tissues where they may function as antioxidants, a role independent of any conversion to vitamin A. β-Carotene and other carotenoids are fat-soluble antioxidants that may play a role in protecting cell membranes from damage by free radicals. The antioxidant properties of carotenoids have stimulated interest in their ability to protect against diseases in which oxidative processes play a role, such as cancer, heart disease, and impaired vision due to macular degeneration or cataracts.

Vitamin A Deficiency

Mild vitamin A deficiency causes night blindness, which is a reversible condition (discussed above). If the deficiency progresses, the ability to maintain epithelial tissue is impaired, particularly in the eye, and growth, reproduction, and immune function become affected.

Why Vitamin A Deficiency Can Cause Blindness When vitamin A is deficient, epithelial cells do not differentiate normally because vitamin A is not there to turn on or turn off the production of particular proteins. For example, the epithelial tissue on many body surfaces contains cells that produce mucus for lubrication. When mucus-secreting cells die, new cells differentiate into mucus-secreting cells to replace them. When vitamin A is deficient, the new cells do not differentiate properly and instead become cells that produce a protein called **keratin** (**Figure 9.6a**). Keratin is the hard protein that makes up hair and fingernails. As the mucus-secreting cells die and are replaced by keratin-producing cells, the epithelial surface becomes hard and dry. This process is known as *keratinization*. The hard, dry surface does not have the protective capabilities of normal epithelium and so the likelihood of infection is increased. The risk of infection is compounded by the fact that vitamin A deficiency also decreases immune function.

keratin A hard protein that makes up hair and nails.

All epithelial tissues are affected by vitamin A deficiency, but the eye is particularly susceptible to damage. The mucus in the eye normally provides lubrication, washes away dirt and other particles, and also contains a protein that helps destroy bacteria. When vitamin A is deficient, the lack of mucus and the buildup of keratin cause the cornea to dry and leave the eye open to infection. A spectrum of eye disorders, known as **xerophthalmia**, is associated with vitamin A deficiency. Xerophthalmia begins as night blindness and extreme dryness and progresses to wrinkling, cloudiness, and softening of the cornea, called **keratomalacia**. If left untreated, it can result in rupture of the cornea, infection, degenerative tissue changes, and permanent blindness (**Figure 9.6b**).

xerophthalmia A spectrum of eye conditions resulting from vitamin A deficiency that may lead to blindness. An early symptom is night blindness, and as deficiency worsens, lack of mucus leaves the eye dry and vulnerable to cracking and infection.

keratomalacia Softening and drying and ulceration of the cornea resulting from vitamin A deficiency

How Vitamin A Deficiency Affects Reproduction, Growth, and Immunity During reproduction, vitamin A is needed to direct cells to differentiate and to form the shapes and patterns needed for the development of a complete organism. Lack of vitamin A during embryonic development results in abnormalities and death. Poor overall growth is an

Figure 9.6 **Vitamin A deficiency and xerophthalmia**

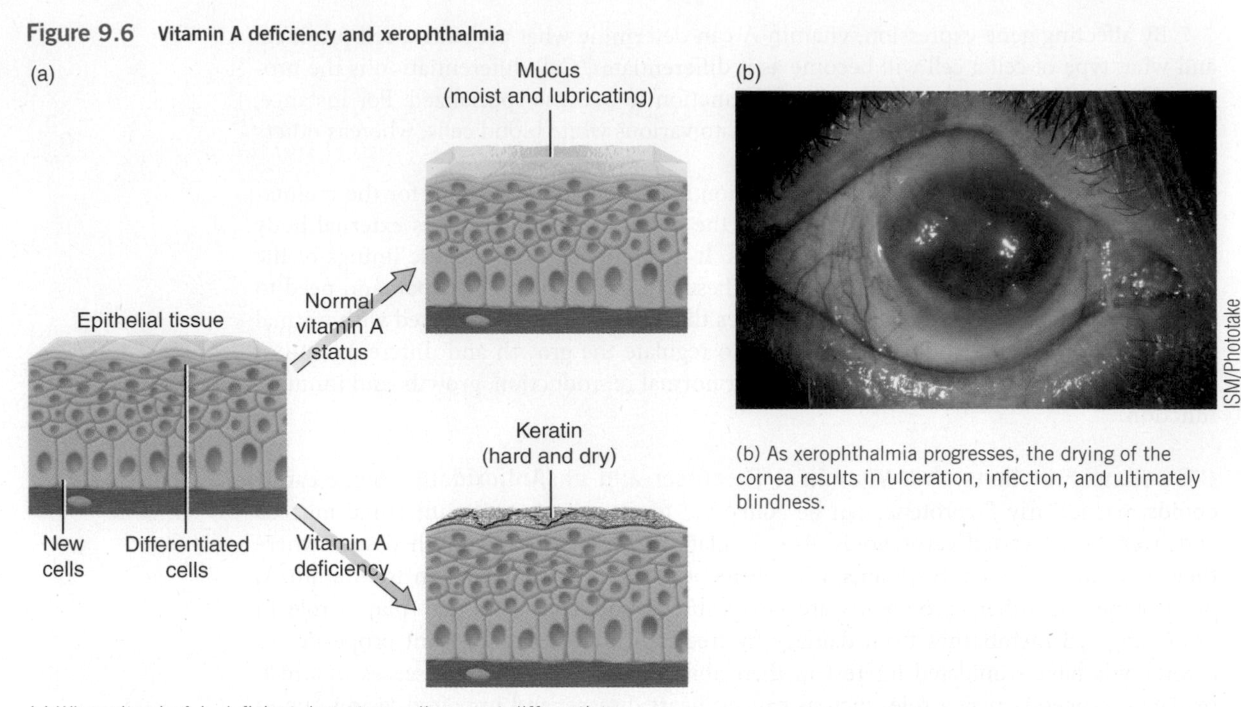

(a) When vitamin A is deficient, immature cells cannot differentiate normally, and instead of mucus-secreting cells, they become cells that produce keratin.

(b) As xerophthalmia progresses, the drying of the cornea results in ulceration, infection, and ultimately blindness.

early sign of vitamin A deficiency in children. In growing children, vitamin A affects the activity of cells that form and break down bone, and a deficiency early in life can cause abnormal jawbone growth, resulting in crooked teeth and poor dental health.

In the immune system, vitamin A is needed for the differentiation that produces the different types of immune cells. Vitamin A deficiency also reduces the function of specific immune cells and hinders the normal regeneration of mucosal barriers damaged by infection.[6] This impaired immune function increases the risk of illness and infection due to defective epithelial tissue barriers (see Chapter 18: Science Applied: Vitamin A: The Anti-Infective Vitamin).

Vitamin A Deficiency Remains a World Health Problem Vitamin A deficiency is a threat to the health, sight, and lives of millions of children in the developing world. Children deficient in vitamin A have poor appetites; are anemic; have an increased susceptibility to infections, including measles; and are more likely to die in childhood. It is estimated that more than 250 million preschool children worldwide are vitamin A deficient and that 250,000 to 500,000 children go blind annually due to vitamin A deficiency.[7] It is most common in India, Africa, Latin America, and the Caribbean (see **Debate: Combating Vitamin A Deficiency with Golden Rice**).

Vitamin A deficiency can be caused by insufficient intakes of vitamin A, fat, protein, or the mineral zinc. As discussed, without fat, vitamin A cannot be absorbed, so a diet very low in fat can cause a deficiency by reducing vitamin A absorption. Protein deficiency can cause vitamin A deficiency because the retinol-binding protein needed to transport vitamin A from the liver cannot be made in sufficient quantities. The importance of zinc for vitamin A utilization is believed to be due to its role in protein synthesis. When zinc is deficient, the proteins needed for vitamin A transport and metabolism are lacking.

Vitamin A deficiency is not common in developed countries, but dietary intake surveys in the United States indicate that many Americans do not meet the recommendations for this vitamin.[4] Intakes below the RDA can be caused by poor food choices even when the food supply is plentiful. In the United States the intake of fruits and vegetables, many of which are excellent sources of provitamin A, does not meet recommendations. A typical fast-food meal of a hamburger and French fries provides almost no vitamin A (see **Critical Thinking: How Much Vitamin A Is in Your Fast-Food Meal?**).

DEBATE

Combating Vitamin A Deficiency with Golden Rice

Each year, vitamin A deficiency takes the sight and lives of hundreds of thousands of children worldwide. The deficiency is most common in impoverished populations where the dietary staple is deficient in vitamin A. Currently this deficiency is addressed using vitamin A supplementation, fortification of the food supply, and interventions that increase the variety of the diet to include foods that are rich in vitamin A. A more controversial solution is Golden Rice. This is rice that has been genetically modified (GM) to synthesize β-carotene, a precursor to vitamin A. Many believe that, if it replaced white rice as a staple, Golden Rice could alleviate vitamin A deficiency. Its development has stimulated debate about whether genetic modification is an effective and safe way to help alleviate malnutrition in the developing world.

Golden Rice has the potential to increase vitamin A intake, but exclusively growing this variety reduces biodiversity.

Initial concerns about Golden Rice focused on whether it would provide enough vitamin A to alleviate deficiency. The original variety provided so little β-carotene that a 2-year-old child would need to eat 3 kg of it each day to get enough vitamin A.[1] A newer variety now provides more than half of the RDA for children in a reasonable serving of half a cup of rice.[2] But even if children eat Golden Rice, many argue that it may not be a solution to malnutrition. The rice will increase vitamin A intake, but deficient populations typically suffer from other nutrient deficiencies; when protein, fat, or zinc is deficient, the body can't efficiently use vitamin A.[3,4]

So should we continue to spend resources developing Golden Rice? Opponents argue that, in the decade since Golden Rice was developed, it has done nothing to prevent vitamin A deficiency, and it has diverted resources from proven programs that address multiple nutrient deficiencies. Proponents contend that the problem is not the rice but rather the regulatory climate that has prevented it from being introduced; Golden Rice was developed in 1999 but will probably not be commercialized until at least 2012.[5,6] The development costs of Golden Rice have been high, but it is predicted to be cost-effective in the long term because, once it is introduced, recurrent costs should be low.[7]

There is also concern about the safety of introducing GM rice. An environmental concern is that its use will decrease biodiversity, which is the variety of plants grown in a region (see photo). Reducing biodiversity increases the risk that the entire rice crop will be destroyed by an infestation of insects or a disease.[3] Proponents of GM crops argue that this issue arises whenever a new crop that is preferred by farmers is introduced.

GM crops such as Golden Rice are not substitutes for traditional solutions to malnutrition but can be used to complement them. Supplementation may be necessary for those in immediate need. Fortification and supplementation may work better in urban settings; Golden Rice may be better for reaching isolated rural populations. In the long term, the goal is to use whatever means are available to solve the problem of vitamin A deficiency and other types of malnutrition.

THINK CRITICALLY: Why is it important to preserve all the varieties of rice that currently exist, even if we do not rely on them as dietary staples?

[1]Enserink, M. Tough lessons from golden rice. *Science* 320:468–471, 2008.

[2]Tang, G., Qin, J., Dolnikowski, G. G., et al. Golden rice is an effective source of vitamin A. *Am J Clin Nutr* 89:1776–1783, 2009.

[3]Greenpeace International. Golden rice's lack of luster: Addressing vitamin A deficiency without genetic engineering. November 9, 2010.

[4]Bienvenido, O. J., Rice in human nutrition.

[5]Potrykus, I. Lessons from the "Humanitarian Golden Rice" project: Regulation prevents development of public good genetically engineered crop products. *N Biotechnol* 27:466–472, July 27, 2010.

[6]GMO Compass. Golden rice: First field tests in the Philippines.

[7]Qaim, M. Benefits of genetically modified crops for the poor: household income, nutrition, and health. *N Biotechnol* 27:552–557, July 17, 2010.

How Much Vitamin A Is in Your Fast-Food Meal?

TNT Magazine/Alamy

John has been living in an apartment all junior year, eating a lot of fast food. He has heard that it is "bad for you" because it is too high in calories and too low in essential nutrients, particularly vitamin A. John looks up the nutrient composition of his favorite fast-food meals.

		ENERGY (kcals)	VITAMIN A (% Daily Value)
McDonald's	Big Mac	540	6
	Medium fries	380	0.2
	Soda (20 oz)	240	0
KFC	Fried chicken leg and breast	710	2.6
	Mashed potatoes and gravy	120	1.5
	Soda (20 oz)	240	0

© Hywit Dimyadi/iStockphoto

iProfile Use iProfile to find out how much vitamin A is in your favorite fast-food meal.

CRITICAL THINKING QUESTIONS

- How does the vitamin A in the McDonald's and KFC meals compare to the RDA?
- If John chooses the Big Mac, fries, and soda, how much more vitamin A does he need to get from his other meals?

CRITICAL THINKING QUESTIONS

- John has two cups of spaghetti with meat sauce, a slice of Italian bread with 1 tsp butter, and a cup of 1% milk for dinner. How much vitamin A does this provide?
- Choose a high–vitamin A vegetable that he could add to his dinner. How much of this vegetable would he need to eat to double the amount of vitamin A in his dinner?
- John needs 2500 kcals to maintain his weight. How many kcals can he eat for breakfast if he has the KFC meal for lunch and the spaghetti meal and vegetable you suggested for dinner?

Recommended Vitamin A Intake

LIFE CYCLE

The recommended intake for vitamin A is based on the amount needed to maintain normal body stores. The RDA is set at 900 μg of vitamin A per day for men and 700 μg/day for women[4] (**Table 9.1**). The RDA is increased in pregnancy to account for the vitamin A that is transferred to the fetus and during lactation to account for the vitamin A lost in milk. The RDA for children is set lower than that for adults based on their smaller body size. For infants, an AI has been set based on the amount of vitamin A consumed by an average healthy breast-fed infant.

Recommendations for vitamin A intake are expressed in micrograms (μg) of retinol. Retinol can be supplied by both preformed vitamin A and carotenoids in the diet. No quantitative recommendations have been made for intakes of β-carotene or other carotenoids.

TABLE 9.1 A Summary of the Fat-Soluble Vitamins (All Values Are RDAs Unless Otherwise Noted)

Vitamin	Sources	Recommended intake for adults	Major functions	Deficiency diseases and symptoms	Groups at risk of deficiency	Toxicity	UL
Vitamin A (retinol, retinal, retinoic acid, vitamin A acetate, vitamin A palmitate, retinyl palmitate, provitamin A, carotene, β-carotene, carotenoids)	Retinol: liver, fish, fortified milk and margarine, butter, eggs; carotenoids: carrots, leafy greens, sweet potatoes, broccoli, apricots, cantaloupe	700–900 μg/day	Vision, health of cornea and other epithelial tissue, cell differen-tiation, reproduction, immune function	Xerophthalmia: night blindness, dry cornea, eye infections; poor growth, dry skin, impaired immune function	People living in poverty (particularly children and pregnant women), people who consume very low-fat or low-protein diets	Headache, vomiting, hair loss, liver damage, skin changes, bone pain, fractures, birth defects	3000 μg/day of preformed vitamin A
Vitamin D (calciferol, cholecalcif-erol, calcitriol, ergocalciferol, dihydroxy vitamin D)	Egg yolk, liver, fish oils, tuna, salmon, fortified milk, synthesis from sunlight	600–800 IU/day (15–20 μg/day)	Absorption of calcium and phosphorus, maintenance of bone	Rickets in children: abnormal growth, misshapen bones, bowed legs, soft bones; osteomalacia in adults: weak bones and bone and muscle pain	Some breast-fed infants; children and elderly people (especially those with dark skin and little exposure to sunlight); people with kidney disease	Calcium deposits in soft tissues, growth retardation, kidney damage	4000 IU/day (100 μg/day)
Vitamin E (tocopherol, α-tocopherol)	Vegetable oils, leafy greens, seeds, nuts, peanuts	15 mg/day	Antioxidant, protects cell membranes	Broken red blood cells, nerve damage	People with poor fat absorption, premature infants	Inhibition of vitamin K activity	1000 mg/day from supple-mental sources
Vitamin K (phylloquinones, menaquinone)	Vegetable oils, leafy greens, synthesis by intestinal bacteria	90–120 μg/day[a]	Synthesis of blood-clotting proteins and proteins in bone	Hemorrhage	Newborns (especially premature), people on long-term antibiotics	Anemia, brain damage	ND

[a]Adequate Intake (AI)

Note: UL, Tolerable Upper Intake Level; ND, not determined due to insufficient evidence.

Their intake is considered only with regard to the amount of retinol they provide. Because carotenoids are less well absorbed and not completely converted to vitamin A, a correction factor, referred to as **retinol activity equivalents (RAE)**, must be applied to carotenoids to determine the amount of usable vitamin A they provide. Twelve micrograms of β-carotene provide 1 RAE of vitamin A, and 24 μg of α-carotene or β-cryptoxanthin provide 1 RAE.[4]

retinol activity equivalent (RAE) The amount of retinol, β-carotene, α-carotene, or β-cryptoxanthin that provides vitamin A activity equal to 1 μg of retinol.

As our understanding of vitamin A has increased, the units in which recommended intakes have been expressed have changed. Prior to 1980, vitamin A was expressed in international units (IUs). The 1989, RDAs used values called retinol equivalents (REs) to account for differences in absorption between preformed vitamin A and carotenoids. These older units are still found in some food-composition databases and tables. Values for converting REs and IUs to micrograms of retinol are given in **Figure 9.7**.

Vitamin A Toxicity and Supplements

Preformed vitamin A is absorbed rapidly and eliminated from the body slowly. Therefore toxicity can result acutely from a high dose in a short period of time or chronically from lower intake over a longer period of time. Acute toxicity causes symptoms such as nausea, vomiting, headache, dizziness, blurred vision, and a lack of muscle coordination. It has

Supplement Facts
Serving Size 1 Tablet

	Amount Per Serving	% Daily Value
Vitamin A (as retinyl acetate and 50% as beta-carotene)	5000 IU	100%
Vitamin C (as ascorbic acid)	60 mg	100%
Vitamin D (as cholecalciferol)	400 IU	100%
Vitamin E (as dl-alpha tocopheryl acetate)	30 IU	100%
Thiamin (as thiamin mononitrate)	1.5 mg	100%
Riboflavin	1.7 mg	100%
Niacin (as niacinamide)	20 mg	100%
Vitamin B_6 (as pyridoxine hydrochloride)	2.0 mg	100%
Folate (as folic acid)	400 mcg	100%
Vitamin B_{12} (as cyanocobalamin)	6 mcg	100%
Biotin	30 mcg	10%
Pantothenic Acid (as calcium pantothenate)	10 mg	100%

Other ingredients: Gelatin, lactose, magnesium stearate, microcrystalline cellulose, FD&C Yellow No. 6, propylene glycol, propylparaben, and sodium benzoate.

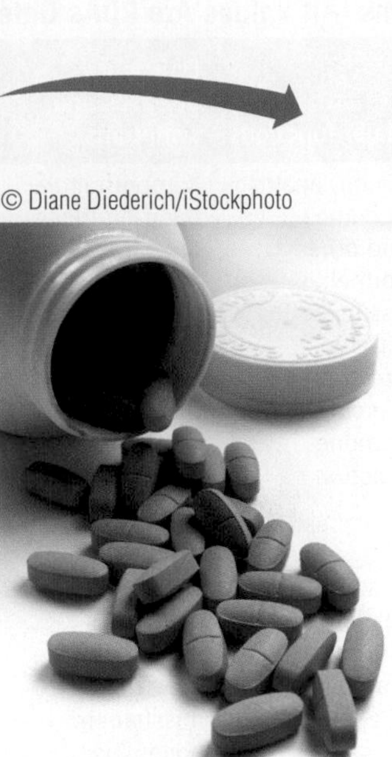

Converting Vitamin A Units

Form and source	Amount equal to 1 μg retinol
Preformed vitamin A in food or supplements	1 μg 1 RAE 1 μg RE 3.3 IU
β-Carotene in food[a]	12 μg 1 RAE 2 μg RE 20 IU
α-Carotene or β-cryptoxanthin in food	24 μg 1 RAE 2 μg RE 40 IU

[a]β-Carotene in supplements may be better absorbed than β-carotene in food and so provides more vitamin A activity. It is estimated that 2 μg of β-carotene dissolved in oil provides 1 μg of vitamin A activity.

Figure 9.7 How much retinol is in your supplement?
The Supplement Facts panel on dietary supplements list the amounts of vitamin A and other fat-soluble vitamins in International Units (IU). The information in the table can be used to convert the amount of vitamin A provided by your supplement or other vitamin A sources to micrograms of retinol.

been reported in Artic explorers who consumed polar bear liver (**Figure 9.8a**). Polar bear liver is not a common dish at most dinner tables, but supplements of preformed vitamin A also have the potential to deliver a toxic dose. Chronic toxicity occurs when preformed vitamin A doses as low as 10 times the RDA are consumed for a period of months to years. The symptoms of chronic toxicity include weight loss, muscle and joint pain, liver damage, bone abnormalities, visual defects, dry scaling lips, and skin rashes.

Excess vitamin A is a particular concern for pregnant women because it may contribute to birth defects.[4,8] High intakes of vitamin A have also been found to cause liver damage and increase the incidence of bone fractures.[9,10] The UL is set at 2800 μg/day of preformed vitamin A for 14- to 18-year-olds and 3000 μg/day for adults.

Carotenoid Toxicity Because of the toxicity of preformed vitamin A, most supplements provide some or all of their vitamin A as carotenoids (**Table 9.2**). Carotenoids are not toxic because their absorption from the diet decreases at high doses, and once in the body,

THINK CRITICALLY

Why is liver from any animal particularly high in vitamin A?

Figure 9.8 High doses of vitamin A and carotenoids

Paul Nicklen/NG Image Collection

(a) Foods generally do not naturally contain large enough amounts of nutrients to be toxic. Polar bear liver is an exception. It contains about 100,000 μg of vitamin A in just 1 ounce and has caused vitamin A toxicity in Arctic explorers who consumed it.

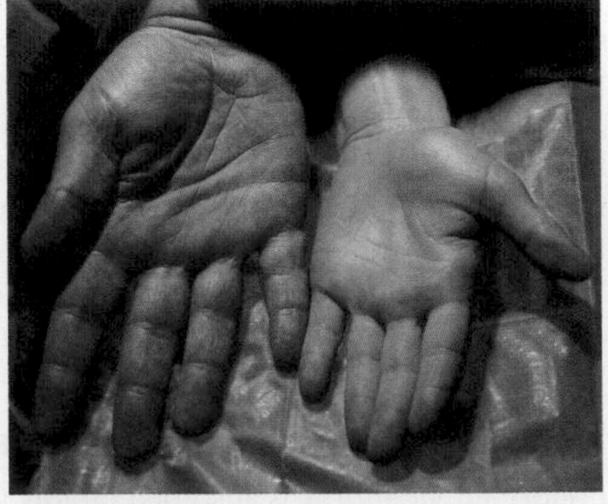

From A. Mazzone and A. Dal Canton, The New England Journal of Medicine, 2002; 347-222-223

(b) The hand on the right illustrates hypercarotenemia, the harmless buildup of carotenoids in the adipose tissue that makes the skin look yellow-orange. The color is particularly apparent on the palms of the hands and the soles of the feet.

TABLE 9.2 Benefits and Risks of Fat-Soluble Vitamin Supplements

Supplement	Marketing claim	Actual benefits or risks
Vitamin A (retinoids)	Improves vision, prevents skin disorders, enhances immunity	Needed for vision and eye health, growth, reproduction, and immunity, but extra as supplements does not provide additional benefits.Toxic at high doses, can cause birth defects and bone loss.
Carotenoids	Needed for vision, prevents skin disorders, antioxidant	Can provide all functions of vitamin A and is an antioxidant, but supplements do not provide additional benefits. High doses can cause orange-colored skin and increase lung cancer risk in smokers.
Vitamin D	Bone health, prevents multiple sclerosis	Needed for calcium absorption and bone maintenance.There is evidence that supplements reduce the risk of autoimmune diseases and cancer. High doses cause heart and kidney damage.
Vitamin E	Prevents heart disease, improves symptoms of fibrocystic breast disease, promotes immune function, reduces scar formation	Antioxidant, protects cell membranes; little evidence that oral supplements reduce risk of heart disease or that topical application reduces scar formation. High doses interfere with anticoagulant medications.

their conversion to retinoids is limited. Large daily intakes of carotenoids—usually from carrot juice or β-carotene supplements—do, however, lead to a condition known as **hypercarotenemia**. In this condition, large amounts of carotenoids stored in the adipose tissue give the skin a yellow-orange color (**Figure 9.8b**). It is not known to be dangerous, and when intake decreases, the skin returns to its normal color.

hypercarotenemia A condition in which carotenoids accumulate in the adipose tissue, causing the skin to appear yellow-orange.

Carotenoid supplements are generally nontoxic and can be used to prevent vitamin A deficiency, but there is concern that large amounts may be harmful to cigarette smokers. A number of human intervention trials found an increased incidence of lung cancer in cigarette smokers who took β-carotene supplements.[11] Not all trials have demonstrated this effect, but smokers are advised to avoid β-carotene supplements and to rely on food sources to obtain carotenoids in their diet. The small amounts found in standard-strength multivitamin supplements are not likely to be harmful for any group. No UL has been determined for carotenoid intake.

Vitamin A as a Drug Derivatives of vitamin A are currently used as drugs. One derivative of retinoic acid, marketed as Retin-A, is used topically to treat acne and to reduce wrinkles due to sun damage. It acts by increasing the turnover of cells. In patients with acne, new cells replace the cells of existing pimples and the rapid turnover of cells prevents new pimples from forming. By a similar mechanism, Retin-A can reduce wrinkles and diminish areas of darkened skin and rough skin. Another vitamin A derivative, 13-*cis*-retinoic acid, marketed as Accutane, is taken orally to treat acne. This medication can have serious side effects, including dry itchy skin and chapped lips, irritated eyes, joint and muscle pain, decreased night vision, depression, and increases in blood lipid levels. In pregnant women, it can cause severe birth defects, including brain damage and physical malformations. Although Retin-A and Accutane are derivatives of vitamin A, they are drugs. Vitamin A supplements cannot be substituted as a treatment for acne, and large doses will cause toxicity symptoms.

9.3 Vitamin D

LEARNING OBJECTIVES

- Explain why vitamin D is known as the sunshine vitamin.
- Relate the functions of vitamin D to the symptoms that occur when it is deficient.
- Discuss why vitamin D deficiency is on the rise.

Vitamin D is known as the "sunshine vitamin" because it can be produced in the skin by exposure to ultraviolet light. Because vitamin D can be made in the body, there is a longstanding debate on whether vitamin D is a vitamin or a hormone. By definition, vitamins are dietary essentials. However, vitamin D can be formed in the skin, so it is essential in the diet only when exposure to sunlight is limited or the body's ability to synthesize the vitamin is reduced. Vitamin D acts like a hormone because it is produced in one organ, the skin, and affects other organs, primarily the intestine, bone, and kidney.

Vitamin D in the Diet

The major source of vitamin D for most humans is exposure to sunlight.[12] Only a few foods are natural sources of vitamin D. These include liver; fatty fish such as salmon, mackerel, and sardines; cod-liver oil; and egg yolks (**Figure 9.9**). These foods contain **cholecalciferol**, also known as vitamin D_3. Vitamin D–fortified foods, such as fortified milk and breakfast cereals, are important sources of vitamin D in the United States. These may contain vitamin D_3 or another active form of the vitamin called vitamin D_2.

cholecalciferol The chemical name for vitamin D_3. It can be formed in the skin of animals by the action of sunlight on 7-dehydrocholesterol.

How Vitamin D Functions in the Body

When sunlight strikes the skin, a compound made from cholesterol, called 7-dehydrocholesterol, is converted to cholecalciferol (vitamin D_3) (**Figure 9.10**). Regardless of whether the vitamin D in your body was synthesized in your skin or consumed in your diet, it is inactive until it is chemically altered in the liver and then the kidney. In the liver, a hydroxyl group (OH) is added to vitamin D to form 25-hydroxy vitamin D_3. This is the form of the vitamin that circulates in the blood and is monitored in the blood as an indicator of patients' vitamin D status. However, 25-hydroxy vitamin D_3 is inactive and must be modified by the kidney, where another hydroxyl group is added to make the active form of vitamin D: 1,25-dihydroxy vitamin D_3 (see Figure 9.10).

Why Vitamin D Is Needed for Healthy Bones The principal function of vitamin D is to maintain levels of calcium and phosphorus in the blood that favor bone mineralization. When blood calcium levels drop too low, the parathyroid gland releases

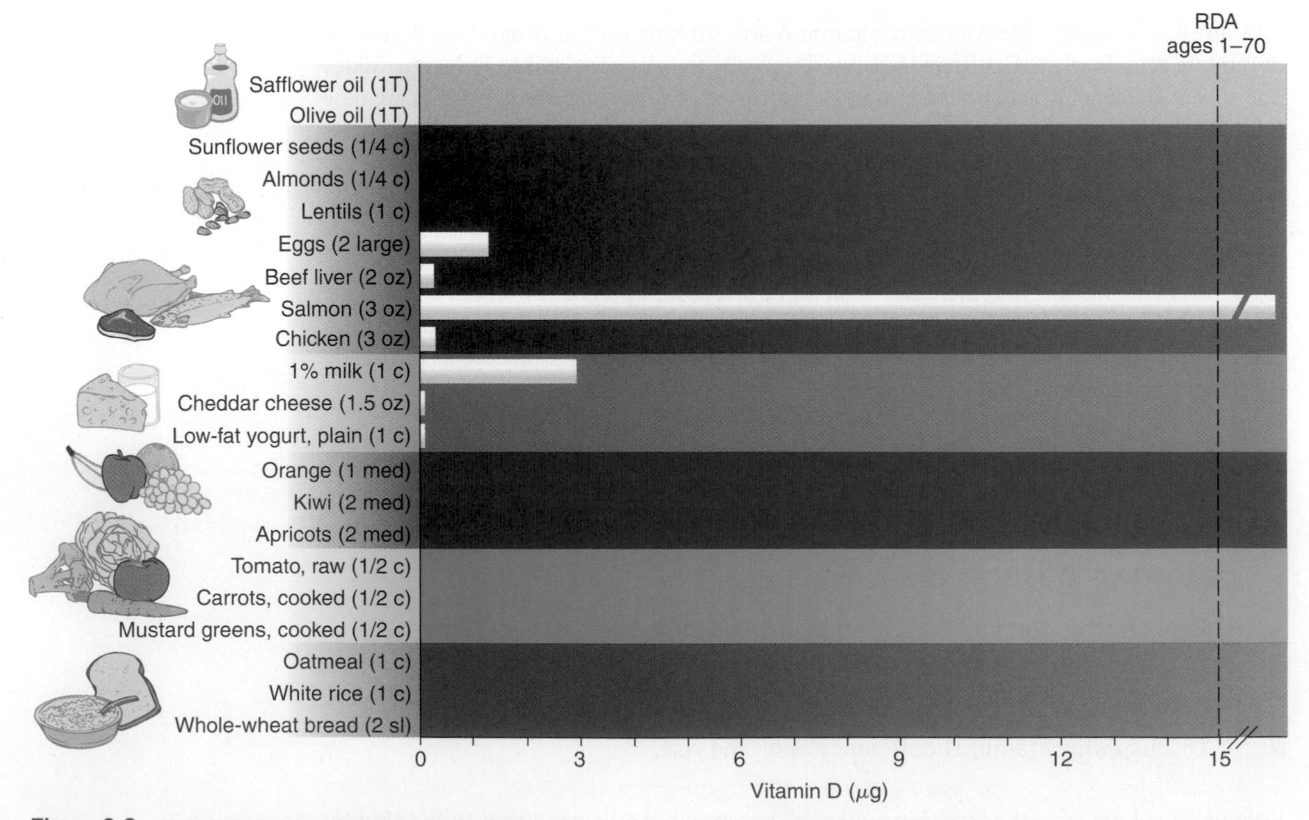

Figure 9.9 Vitamin D content of MyPlate food groups
Only a few foods are natural sources of vitamin D; the dashed line shows the RDA for people ages 1 to 70.

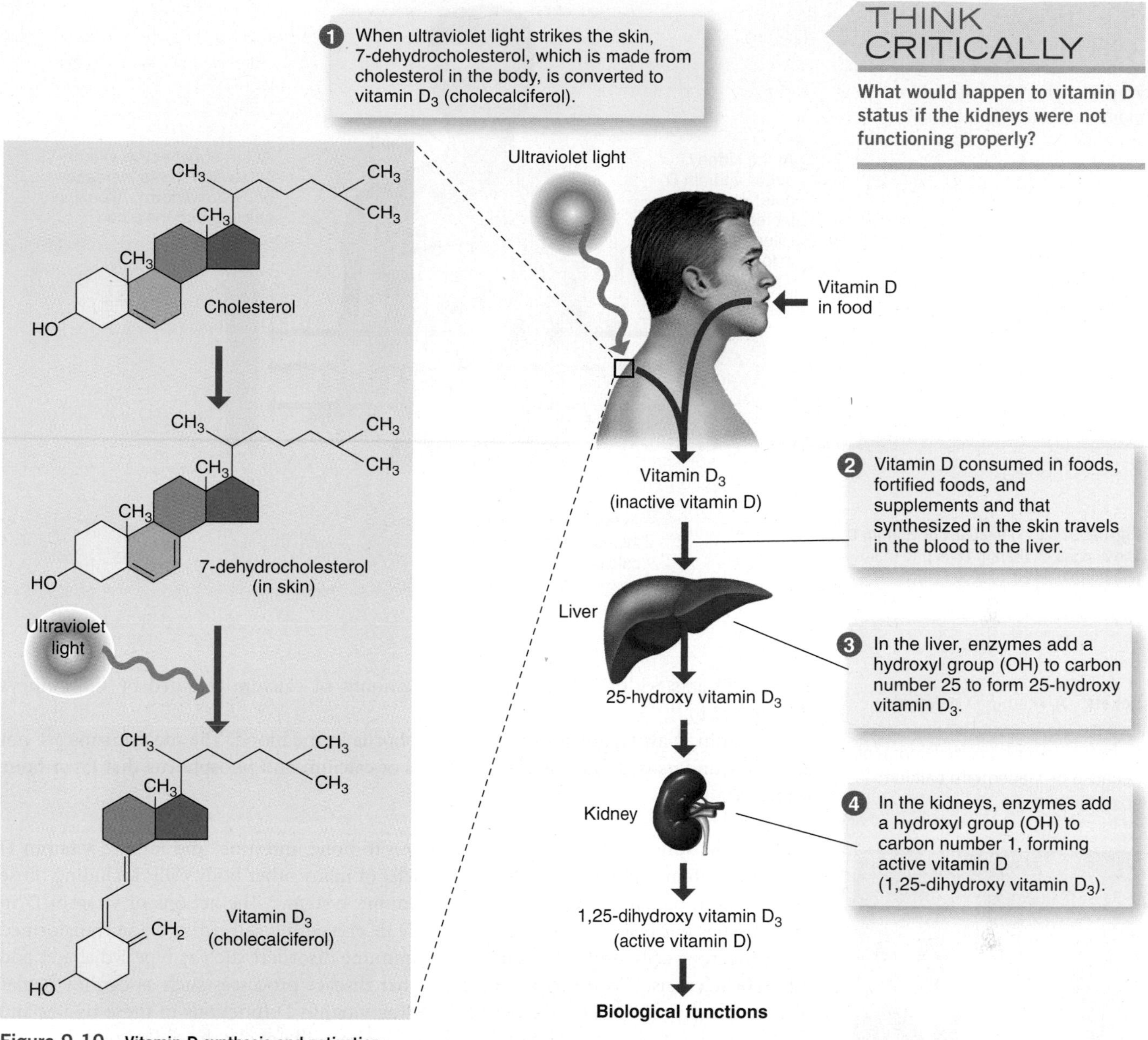

Figure 9.10 Vitamin D synthesis and activation
In order to function, vitamin D from food and from synthesis in the skin must be activated.

THINK CRITICALLY

What would happen to vitamin D status if the kidneys were not functioning properly?

parathyroid hormone (PTH). PTH stimulates enzymes in the kidney to convert 25-hydroxy vitamin D_3 to the active form of the vitamin. Active vitamin D enters the blood and travels to its major target tissues—intestine, bone, and kidney—where it acts to increase calcium levels in the blood. The functions of vitamin D, like those of vitamin A, are due to its role in gene expression. Active vitamin D binds to vitamin D receptor proteins in the nucleus of cells in target tissues and increases the production of certain proteins (see Figure 9.5). Its effect at three different tissues helps to increase blood calcium levels.[13] One target is intestinal cells, where vitamin D increases the expression of genes that code for the production of intestinal calcium transport proteins. This enhances the active transport of dietary calcium from the intestinal lumen into the body. This action requires only very low levels of the active vitamin. If dietary calcium is unavailable, higher levels of vitamin D act at the bone and kidney in conjunction with PTH to return blood calcium to normal. At the bone, vitamin D causes precursor cells to differentiate into cells that break down bone. Bone breakdown releases calcium into the blood. At the kidney,

parathyroid hormone (PTH) A hormone released by the parathyroid gland that acts to increase blood calcium levels.

THINK CRITICALLY

Why is vitamin D included in many calcium supplements?

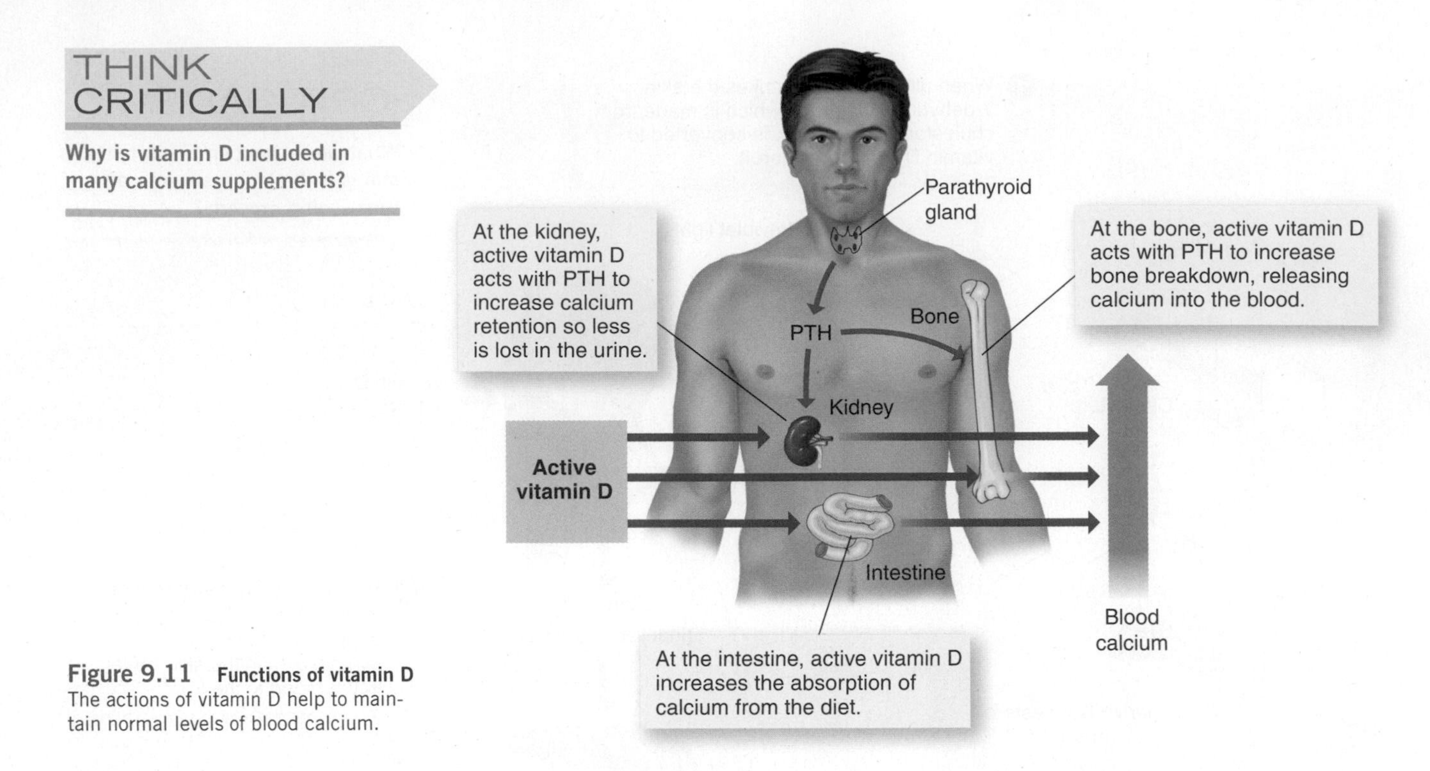

Figure 9.11 Functions of vitamin D The actions of vitamin D help to maintain normal levels of blood calcium.

vitamin D acts with PTH to increase the amount of calcium retained by the kidneys (**Figure 9.11**).[13]

Vitamin D also regulates levels of phosphorus in the blood. The mechanisms are not as clearly understood, but the result is levels of calcium and phosphorus that favor bone mineralization.

rickets A vitamin D deficiency disease in children that is characterized by poor bone development because of inadequate calcium absorption.

Other Functions of Vitamin D In addition to bone, intestine, and kidney, vitamin D receptor proteins have been found in the nuclei of many other body cells, including those of the colon, pancreas, skin, breast, and immune system.[14] The actions of vitamin D in these tissues has been proposed to play a role in preventing cells from being transformed into cancerous cells, protecting against autoimmune disorders such as type 1 diabetes and multiple sclerosis, and protecting against other disease processes such as cardiovascular disease and type 2 diabetes.[14] However, the way vitamin D functions in these tissues and the physiological consequences of these functions are still not clearly understood.[13]

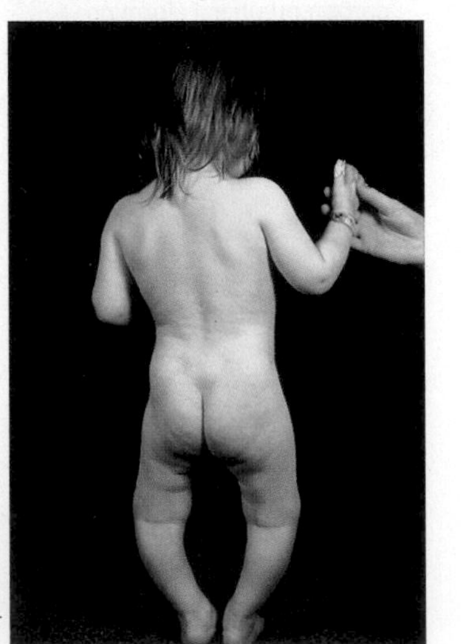
Biophoto Associates/Photo Researchers

Figure 9.12 Rickets The vitamin D deficiency disease rickets causes short stature and bone deformities. The characteristic bowed legs occur because the bones are too weak to support the weight of the body.

Vitamin D Deficiency

When vitamin D is deficient, only about 10 to 15% of the calcium in the diet can be absorbed. As a result, calcium is not available for proper bone mineralization and abnormalities in bone structure occur.

In children, vitamin D deficiency causes **rickets**; it is characterized by weak bones due to inadequate calcium and phosphorus, and bone deformities such as narrow rib cages, known as pigeon breasts, and bowed legs (**Figure 9.12**). Vitamin D deficiency also prevents children from reaching their genetically programmed height and reduces bone mass and causes muscle weakness. Rickets, first recognized in the 1600s, was common during the Industrial Revolution when large numbers of poorly nourished children lived under a layer of smog in the newly industrialized cities. Tall buildings and smog-filled air reduced children's exposure to sunlight. Today, the risk of vitamin D deficiency is increased by having dark skin pigmentation, which prevents UV light rays from penetrating into the layers of the skin where vitamin D is formed, and by using sunscreen, wearing concealing clothing, living in cities with pollution and tall buildings, working and playing indoors, and living too far north or south of the equator, all of which reduce the amount of sunlight that reaches the skin (**Figure 9.13**).

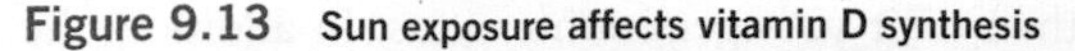

Figure 9.13 Sun exposure affects vitamin D synthesis

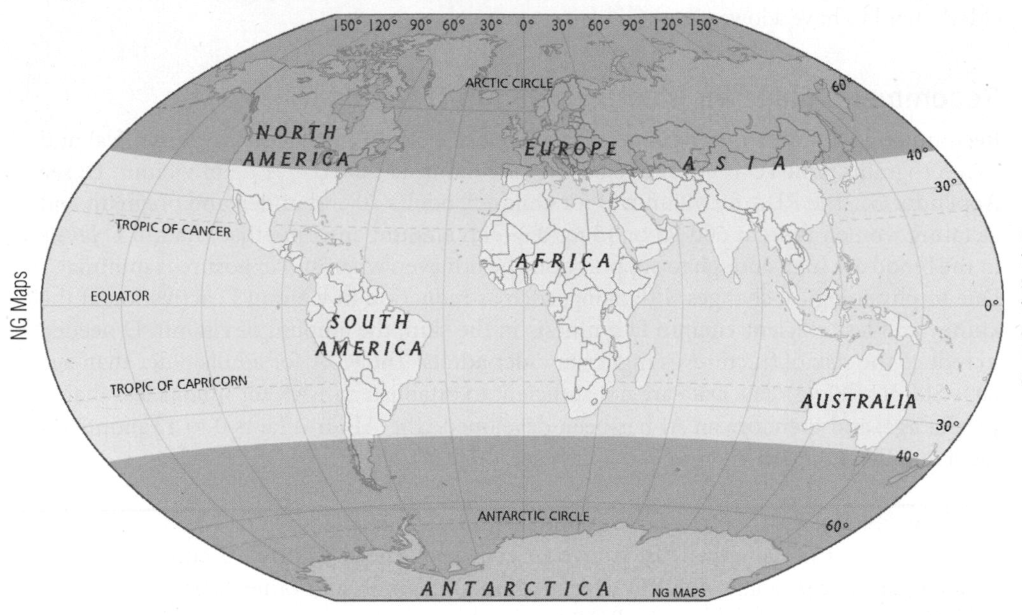

(a) The angle at which the sun strikes the earth affects the body's ability to synthesize vitamin D in the skin. During the winter at latitudes greater than about 40 degrees north or south, there is not enough UV radiation to synthesize adequate amounts.

(b) Sunscreen with an SPF of 15 decreases vitamin D synthesis by more than 95%.[15] Sunscreen is important for reducing the risk of skin cancer, but some time in the sun without sunscreen may be needed to meet vitamin D needs.

(c) Dark skin pigmentation may reduce the body's ability to make vitamin D in the skin by as much as 99%. Concealing clothing worn by certain cultural and religious groups prevents sunlight from striking the skin. This explains why vitamin D deficiency occurs in women and children in some of the sunniest regions of the world.

The fortification of milk with vitamin D has helped greatly to reduce rickets in the United States, but it is still a problem in infants and young children with dark skin, which reduces the formation of vitamin D, and those who are breast-fed because breast milk is low in vitamin D.[2] Rickets is also seen in children with disorders that affect fat absorption and in children who for any number of reasons do not drink milk.

In adults, vitamin D deficiency causes a condition called **osteomalacia**. Because bone growth is complete in adults, osteomalacia does not cause bone deformities, but bones are weakened because not enough calcium is available to form the mineral deposits needed to maintain their health. Insufficient bone mineralization leads to fractures of the weight-bearing bones such as those in the hips and spine. Osteomalacia also causes bone pain and muscle aches and weakness. It can precipitate or exacerbate osteoporosis, which is a loss of total bone mass, not just minerals (see Chapter 10).

osteomalacia A vitamin D deficiency disease in adults characterized by a loss of minerals from bones. It causes bone pain, muscle aches, and an increase in bone fractures.

Osteomalacia is common in adults with kidney failure because the conversion of vitamin D from the inactive to active form is reduced. African Americans are at particular risk for vitamin D deficiency because vitamin D synthesis is low due to dark pigmentation and consumption of milk fortified with vitamin D is low due to the high frequency of lactose intolerance. The elderly are at risk because the ability to synthesize vitamin D in the skin decreases with age and because older adults typically cover more of their skin with clothing and spend less time in the sun than their younger counterparts. In addition, the elderly tend to have a lower intake of dairy products.

Recommendations for Meeting Vitamin D Needs

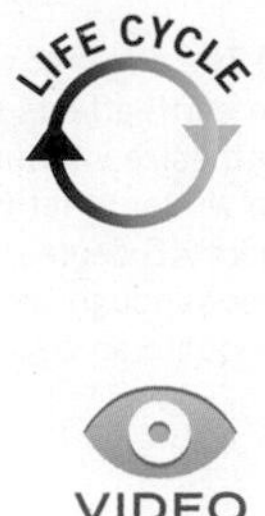

Recommended intakes for vitamin D are expressed both in International Units (IUs) and in micrograms. One IU is equal to 0.025 μg of vitamin D (40 IU = 1 μg of vitamin D; see Appendix G). The RDA for vitamin D for children, adults 70 and under, and pregnant and lactating women is set at 600 IU or 15 μg/day—an amount ensuring that vitamin D levels in the blood are high enough to support bone health even when sun exposure is minimal.[13] Due to physiological changes with aging, such as reduction in vitamin D activation at the kidney and less efficient vitamin D synthesis in the skin, the amount of vitamin D needed to reduce the risk of fractures is higher in older adults. The RDA for adults older than age 70 is 800 IU (20 μg)/day. Data are not sufficient to establish an EAR for infants less than 1 year of age, and therefore an AI has been developed. The AI for infants 0 to 12 months is set at 400 IU (10 μg) of vitamin D per day (see Table 9.1).

Meeting Needs with Food and Sunlight Milk is a good source of vitamin D, but if you synthesized no vitamin D in your skin you would need to drink 5 cups of vitamin D–fortified milk to obtain the RDA for vitamin D. Salmon and other fish are also good sources—3 ounces will provide the RDA.

The recommended intakes for vitamin D are based on the assumption of minimal sun exposure. This assumption is made because of the variation in the extent to which synthesis from sunlight meets the requirement. The amount of vitamin D synthesized in the skin is affected by many variables, so it is not possible to make a single recommendation on the amount of time people need to spend in the sun to meet their vitamin D needs. For example, during the spring, summer, and fall, light-skinned individuals may need to spend only 5 to 10 minutes outdoors, two to three times a week, with their faces, hands, arms, and legs exposed to meet their vitamin D requirement, whereas very dark-skinned individuals (who never sunburn) may need 10 to 50 times more sun exposure to produce the same amount of vitamin D.[16,17] In the summer, children and active adults usually spend enough time outdoors without sunscreens to provide their vitamin D requirement. It has been estimated that more than 90% of the vitamin D requirement for most people comes from casual exposure to sunlight.[18]

When Is Supplemental Vitamin D Recommended? Vitamin D supplements are recommended to a number of groups. Because breast milk is low in vitamin D, it is recommended that breast-fed and partially breast-fed infants be supplemented with 400 IU (10 μg)/day of vitamin D beginning in the first few days of life and continuing until they are consuming about 1 L (4 cups) of vitamin D–fortified formula or milk daily.[19,20] Supplemental vitamin D is also recommended for all non-breast-fed infants who are ingesting less than 4 cups per day of vitamin D-fortified formula or milk and for children and adolescents who do not get regular sunlight exposure and do not ingest at least 4 cups per day of vitamin D–fortified milk. Other groups that might benefit from vitamin D supplements include people who do not drink milk or consume dairy products (see Chapter 6), older adults (see Chapter 16), individuals with dark skin pigmentation, individuals who do not absorb fat normally, and individuals with limited sun exposure because they are homebound, wear robes and head coverings for religious reasons, or work in occupations that prevent sun exposure.[21]

Vitamin D Toxicity

Too much vitamin D in the body can cause high calcium concentrations in the blood and urine, deposition of calcium in soft tissues such as the blood vessels and kidneys, and cardiovascular damage. Synthesis of vitamin D from exposure to sunlight does not produce toxic amounts because the body regulates vitamin D formation. Consumption of unfortified foods is also not a concern, but oversupplementation and overfortification can pose a risk. One case of accidental overfortification of milk resulted in the hospitalization of 56 individuals and the deaths of 2.[22] The UL for ages 9 and older for vitamin D is 4000 IU (100 μg)/day.[13]

9.4 Vitamin E

LEARNING OBJECTIVES

- List two food sources of vitamin E.
- Explain the function of vitamin E.

Vitamin E is a fat-soluble vitamin with an antioxidant function. It was first identified as a fat-soluble component of grains that was necessary for fertility in laboratory rats. It took almost 30 years to isolate this vitamin and to determine that it is also necessary for reproduction in humans. The chemical name for vitamin E, **tocopherol**, is from the Greek *tokos*, meaning childbirth, and *phero*, meaning to bring forth. Although vitamin E has been promoted to slow aging, cure infertility, reduce scarring, and protect against air pollution, research has not shown it to be useful for these purposes. Today we continue to explore the role of this antioxidant in protecting us from chronic disease.

tocopherol The chemical name for vitamin E.

Vitamin E in the Diet

Several naturally occurring forms of vitamin E are found in foods, but only the **alpha-tocopherol (α-tocopherol)** form can meet vitamin E requirements in humans. The other forms do not meet vitamin E needs because they are not converted to α-tocopherol in humans and cannot be transported by the α-tocopherol transfer protein, a liver protein that helps distribute vitamin E to body tissues.

alpha-tocopherol (α-tocopherol) The form of tocopherol that provides vitamin E activity in humans.

Dietary sources of vitamin E include nuts, seeds, and peanuts; plant oils, such as soybean, corn, and sunflower oils; leafy green vegetables; and wheat germ (**Figure 9.14**). Vitamin E is also consumed in fortified foods, such as breakfast cereals, and supplements; however, the vitamin E from these sources may not be as efficient at meeting needs. The

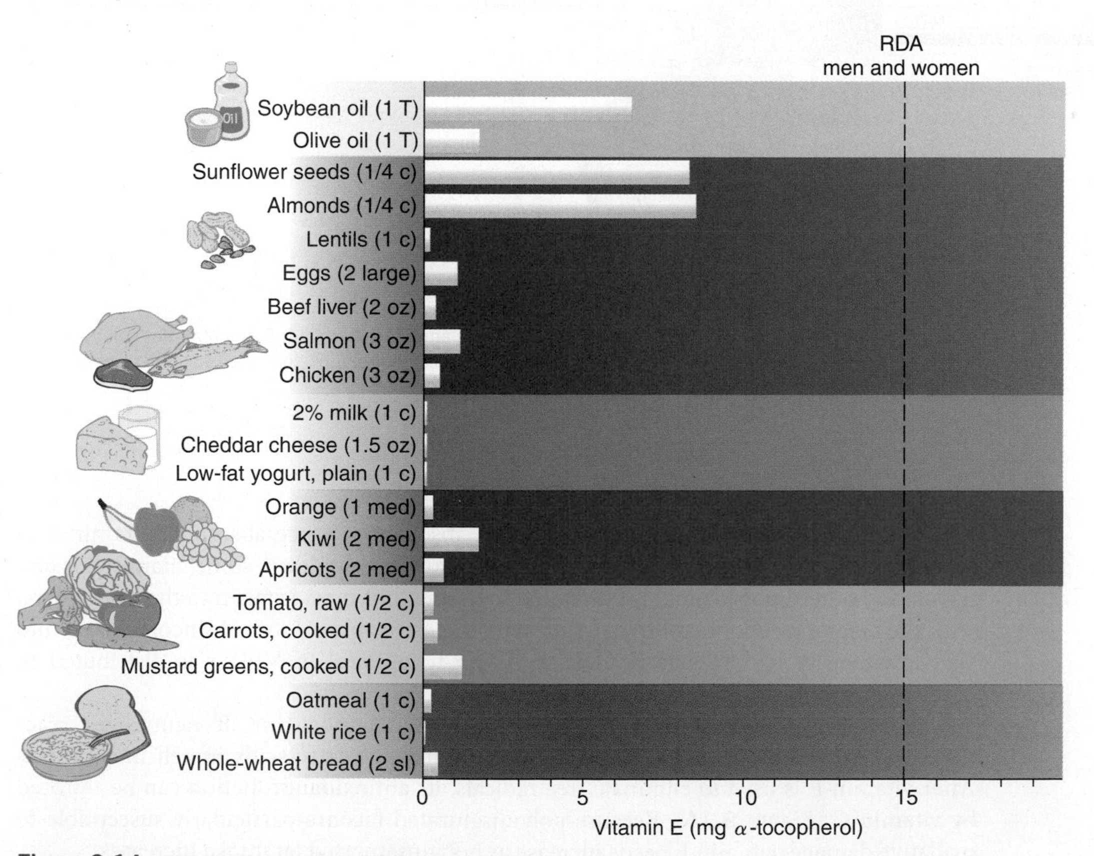

Figure 9.14 Vitamin E content of MyPlate food groups
Adults can obtain their RDA for vitamin E (dashed line) by consuming plant oils, nuts and seeds, and leafy green vegetables.

isomers Molecules with the same molecular formula but a different arrangement of the atoms.

synthetic form of α-tocopherol found in dietary supplements and fortified foods is composed of eight different **isomers**. Only half of these are active in the body. Therefore, synthetic α-tocopherol provides half of the biological activity of natural α-tocopherol; 10 mg of synthetic α-tocopherol provides the function of 5 mg of natural α-tocopherol.

Because the discovery that only the α-tocopherol form of vitamin E provides activity is relatively recent, most nutrient databases and nutrition labels overrepresent the amount of functional vitamin E in foods. Previously, all forms of tocopherol were included when calculating vitamin E content. Vitamin E content was expressed as either IUs, defined as 1 mg of synthetic α-tocopherol, or α-tocopherol equivalents (α-TEs), which considered other tocopherols as fractions of α-tocopherol. To correct for this, formulas have been developed to convert IUs and α-TEs into milligrams of α-tocopherol (**Figure 9.15**).

Because vitamin E is sensitive to destruction by oxygen, metals, light, and heat, some is lost during food processing, cooking, and storage. Although it is relatively stable at normal cooking temperatures, the vitamin E in cooking oils may be destroyed if the oil is repeatedly heated to the high temperatures used for deep-fat frying.

Supplement Facts
Serving Size 1 Tablet

	Amount Per Serving	% Daily Value
Vitamin A (as retinyl acetate and 50% as beta-carotene)	5000 IU	100%
Vitamin C (as ascorbic acid)	60 mg	100%
Vitamin D (as cholecalciferol)	400 IU	100%
Vitamin E (as dl-alpha tocopheryl acetate)	30 IU	100%
Thiamin (as thiamin mononitrate)	1.5 mg	100%
Riboflavin	1.7 mg	100%
Niacin (as niacinamide)	20 mg	100%
Vitamin B_6 (as pyridoxine hydrochloride)	2.0 mg	100%
Folate (as folic acid)	400 mcg	100%
Vitamin B_{12} (as cyanocobalamin)	6 mcg	100%
Biotin	30 mcg	10%
Pantothenic Acid (as calcium pantothenate)	10 mg	100%

Other ingredients: Gelatin, lactose, magnesium stearate, microcrystalline cellulose, FD&C Yellow No. 6, propylene glycol, propylparaben, and sodium benzoate.

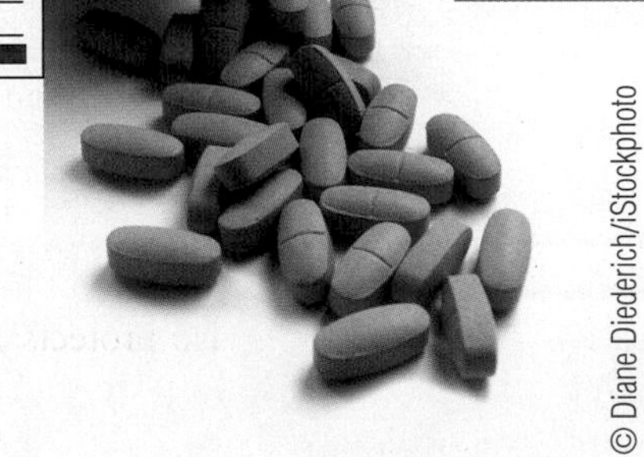

© Diane Diederich/iStockphoto

Converting Vitamin E Units

To estimate the α-tocopherol intake from foods:

- If values are given as mg α-TEs:
  mg α-TE × 0.8 = mg α-tocopherol
- If values are given as IUs:
  First, determine if the source of the α-tocopherol is natural or synthetic.
  - For natural α-tocopherol:
    IU of natural α-tocopherol × 0.67 = mg α-tocopherol
  - For synthetic α-tocopherol (dl-α-tocopherol):
    IU of synthetic α-tocopherol × 0.45 = mg α-tocopherol

Figure 9.15 How much vitamin E is in your supplement?
The Supplement Facts panel on dietary supplements lists the amounts of vitamin E and other fat-soluble vitamins in IUs. The amount of vitamin E provided by your supplement can be converted to milligrams of α-tocopherol using the equations shown in the table.

How Vitamin E Functions in the Body

Vitamin E absorption depends on normal fat absorption. Once absorbed, vitamin E is incorporated into chylomicrons. As chylomicrons are broken down, some vitamin E is distributed to other lipoproteins and delivered to tissues, but most is taken to the liver where, with the help of α-tocopherol transfer protein, the α-tocopherol form is incorporated into very-low-density lipoproteins (VLDLs).[23] The α-tocopherol in VLDLs is distributed to other plasma lipoproteins and delivered to cells.

Vitamin E functions primarily as a fat-soluble antioxidant. It neutralizes reactive oxygen compounds before they damage unsaturated fatty acids in cell membranes. After vitamin E is used to eliminate free radicals, its antioxidant function can be restored by vitamin C (**Figure 9.16**). Because polyunsaturated fats are particularly susceptible to oxidative damage, vitamin E needs increase as polyunsaturated fat intake increases.

By protecting cell membranes, vitamin E is important in maintaining the integrity of red blood cells, cells in nervous tissue, cells of the immune system, and lung cells, where it

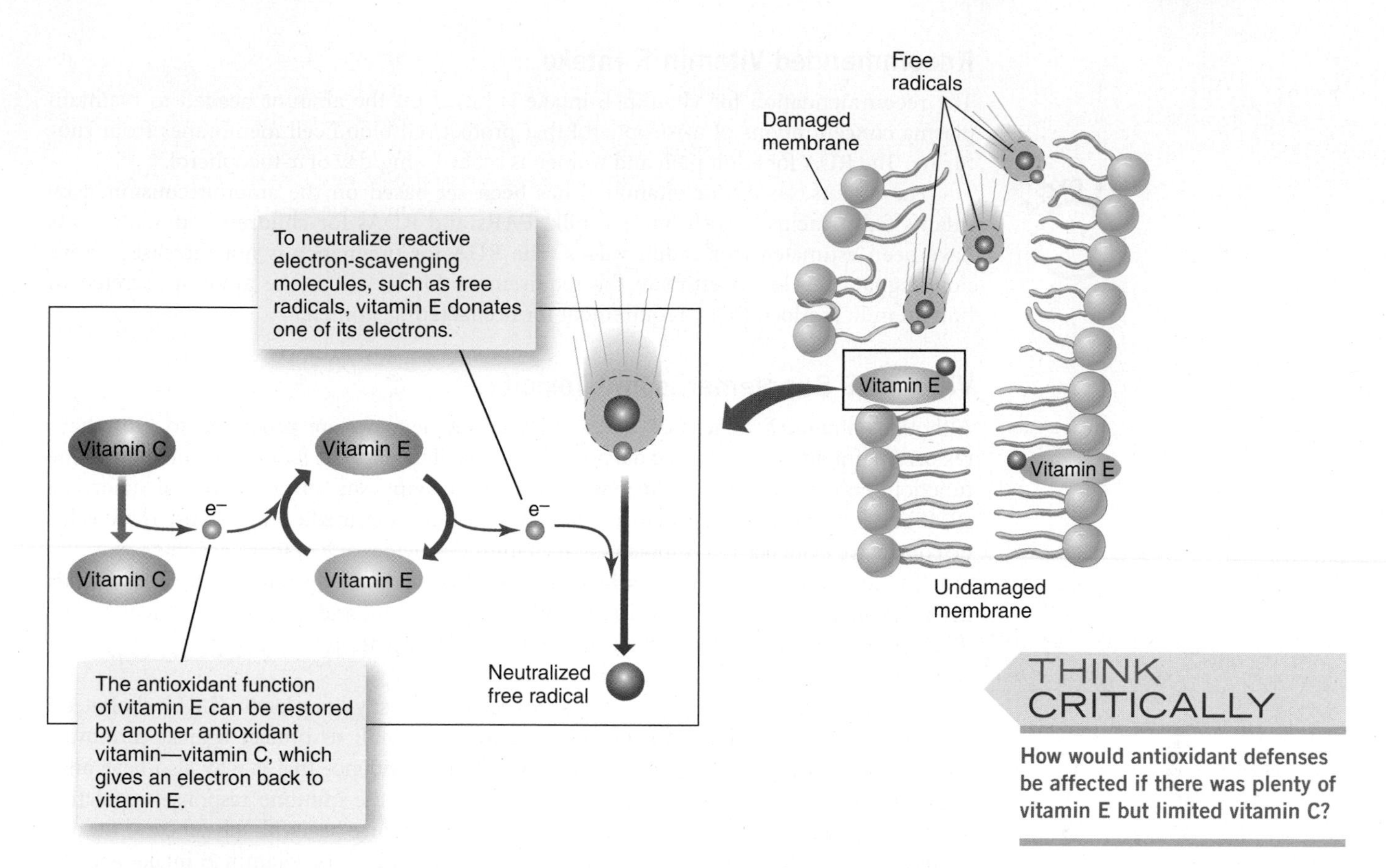

Figure 9.16 Antioxidant mechanism of vitamin E
By neutralizing free radicals, vitamin E guards not only cell membranes, as shown here, but also body proteins, DNA, and cholesterol. Vitamin E must be regenerated by vitamin C to restore it to the form that can act as an antioxidant.

THINK CRITICALLY

How would antioxidant defenses be affected if there was plenty of vitamin E but limited vitamin C?

is particularly important because oxygen concentrations in those cells are high.[24] Vitamin E can also defend cells from damage by heavy metals, such as lead and mercury, and toxins, such as carbon tetrachloride, benzene, and a variety of drugs. It also protects against some environmental pollutants such as ozone. A number of vitamin E's roles are hypothesized to reduce the risk of heart disease and other chronic disorders.

Vitamin E Deficiency

Because vitamin E is needed to protect cell membranes, a deficiency causes those membranes to break down. Red blood cells and nerve cells are particularly susceptible. With a vitamin E deficiency, red blood cell membranes may rupture, causing a type of anemia called **hemolytic anemia**. This is most common in premature infants. All newborn infants have low blood vitamin E levels because there is little transfer of this vitamin from mother to fetus until the last weeks of pregnancy. The levels are lower in premature infants because they are born before much vitamin E has been transferred from the mother. To prevent vitamin E deficiency in premature infants, they can be fed special formulas that contain higher amounts of vitamin E.

hemolytic anemia Anemia that results when red blood cells break open.

Because vitamin E is plentiful in the food supply and is stored in many of the body's tissues, vitamin E deficiency is rare in adults, occurring only when other health problems interfere with fat absorption, which reduces vitamin E absorption. In such cases, the vitamin E deficiency is usually characterized by symptoms associated with nerve degeneration, such as poor muscle coordination, weakness, and impaired vision. For example, in individuals with cystic fibrosis, an inherited condition that reduces fat absorption, deficiency can develop rapidly, causing serious neurological problems that, if untreated, can become permanent.

Recommended Vitamin E Intake

The recommendation for vitamin E intake is based on the amount needed to maintain plasma concentrations of α-tocopherol that protect red blood cell membranes from rupturing. The RDA for adult men and women is set at 15 mg/day of α-tocopherol.[25]

For infants, an AI for vitamin E has been set based on the amount consumed by infants fed principally with human milk. EARs and RDAs for children and adolescents have been estimated from adult values. The RDA for pregnancy is not increased above nonpregnant levels. To estimate the requirement for lactation, the amount secreted in human milk is added to the requirement for nonlactating women.[25]

Vitamin E Supplements and Toxicity

Although vitamin E deficiency is uncommon, supplements are promoted to grow hair; restore, maintain, or increase sexual potency and fertility; alleviate fatigue; maintain immune function; enhance athletic performance; reduce the symptoms of premenstrual syndrome (PMS) and menopause; slow aging; and treat a host of other medical problems. There is little conclusive evidence that supplemental vitamin E provides any of these benefits.

The antioxidant role of vitamin E suggests that it may help reduce the risk of heart disease, cancer, Alzheimer's disease, macular degeneration, and a variety of other chronic diseases associated with oxidative damage. Particular attention has been paid to its potential benefits in guarding against heart disease.

As an antioxidant, vitamin E helps protect low-density-lipoprotein (LDL) cholesterol from oxidation. In addition to the potential for vitamin E to protect against cardiovascular disease through its antioxidant function, there is evidence that it may also have anti-inflammatory functions and be involved in modulating the immune response, regulating genes that affect cell growth and cell death, and detoxifying harmful substances.[26] Studies examining the relationship between blood levels of vitamin E or vitamin E intake and the incidence of cardiovascular disease have been mixed, but studies investigating the effect of vitamin E supplements on the incidence of cardiovascular disease and other chronic diseases have failed to provide evidence of any benefits.[26,27] The best approach is to get plenty of dietary vitamin E; however, most Americans do not consume enough vitamin E to meet the RDA.[28]

Vitamin E is relatively nontoxic. There is no evidence of adverse effects from consuming large amounts from food. The UL is 1000 mg/day from supplemental sources.[25] Vitamin E supplements should not be taken by individuals taking blood-thinning medications because it reduces blood clotting and interferes with the action of vitamin K.

9.5 Vitamin K

LEARNING OBJECTIVES

- Describe how vitamin K is involved in blood clotting.
- Explain why newborns and people taking antibiotics are at risk for vitamin K deficiency.

Vitamin K is one of the few vitamins about which extravagant claims are not made. Like the other fat-soluble vitamins, it was discovered inadvertently by feeding animals a fat-free diet. In this case, researchers in Denmark noted that chicks fed this diet developed a bleeding disorder that was cured by feeding them a fat-soluble extract from green plants. Vitamin K was named for *koagulation*, the Danish word for **coagulation**, or blood clotting.

coagulation The process of blood clotting.

phylloquinone The form of vitamin K found in plants.

menaquinones The forms of vitamin K synthesized by bacteria and found in animals.

Vitamin K in the Diet

Like other fat-soluble vitamins, vitamin K is found in several forms. **Phylloquinone** is the form found in plants and the primary form in the diet. A group of vitamin K compounds, called **menaquinones**, are found in fish oils and meats and are synthesized by bacteria, including those in the human intestine. Menaquinones are the form found in supplements. Only a small number of foods provide significant amounts of vitamin K; however, typical

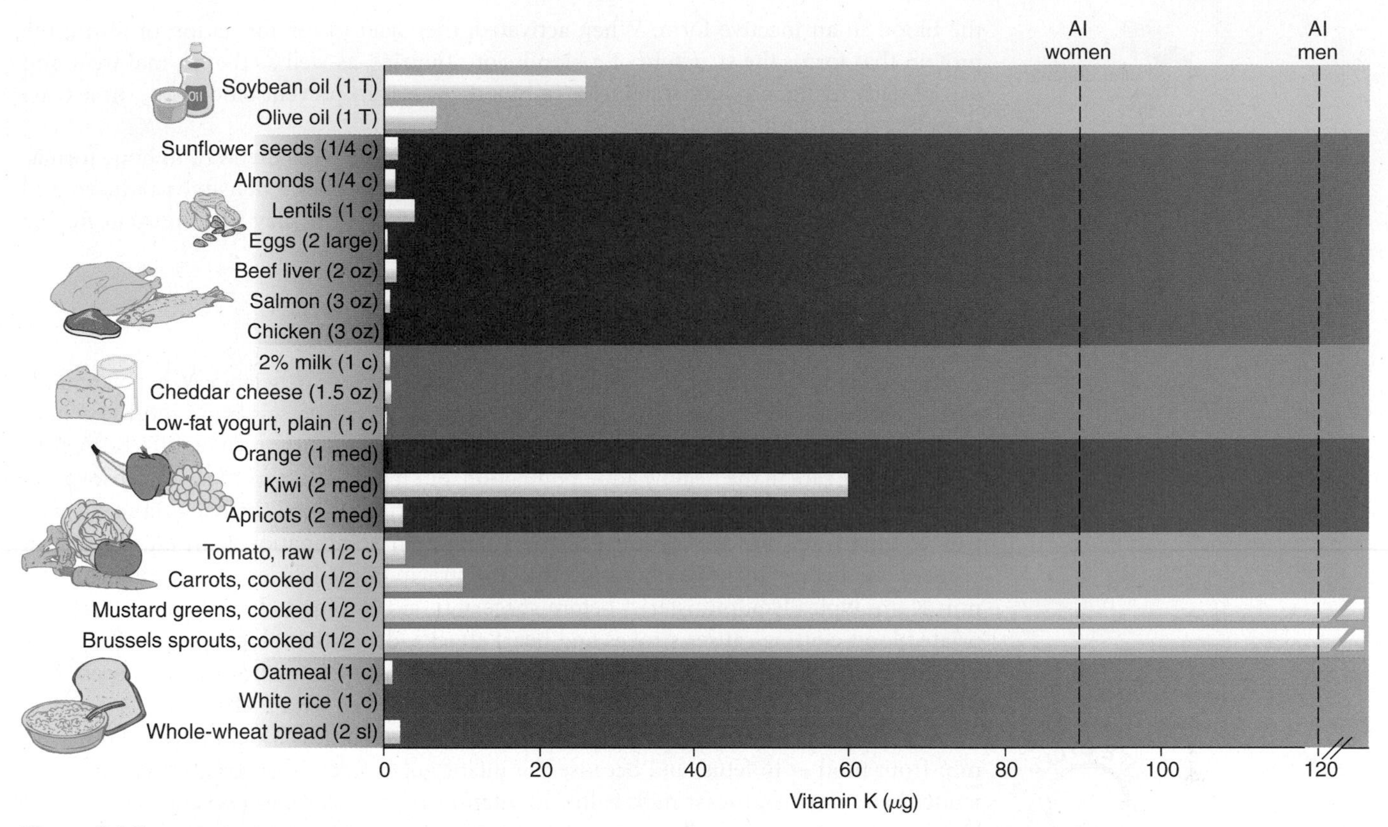

Figure 9.17 Vitamin K content of MyPlate food groups
The best sources of vitamin K are leafy green vegetables and some plant oils; the dashed lines represent the AIs for adult men and women.

intakes in the United States meet recommendations.[4] The best dietary sources are liver and leafy green vegetables such as spinach, broccoli, Brussels sprouts, kale, and turnip greens. Some vegetable oils are also good sources (**Figure 9.17**). Some of the vitamin K produced by bacteria in the human gastrointestinal tract is also absorbed. Vitamin K is destroyed by exposure to light and low- or high-acid conditions.

How Vitamin K Functions in the Body

Vitamin K is a coenzyme needed for the production of **prothrombin** and other specific blood-clotting factors (**Figure 9.18**). Blood-clotting factors are proteins that circulate in

prothrombin A blood protein required for blood clotting.

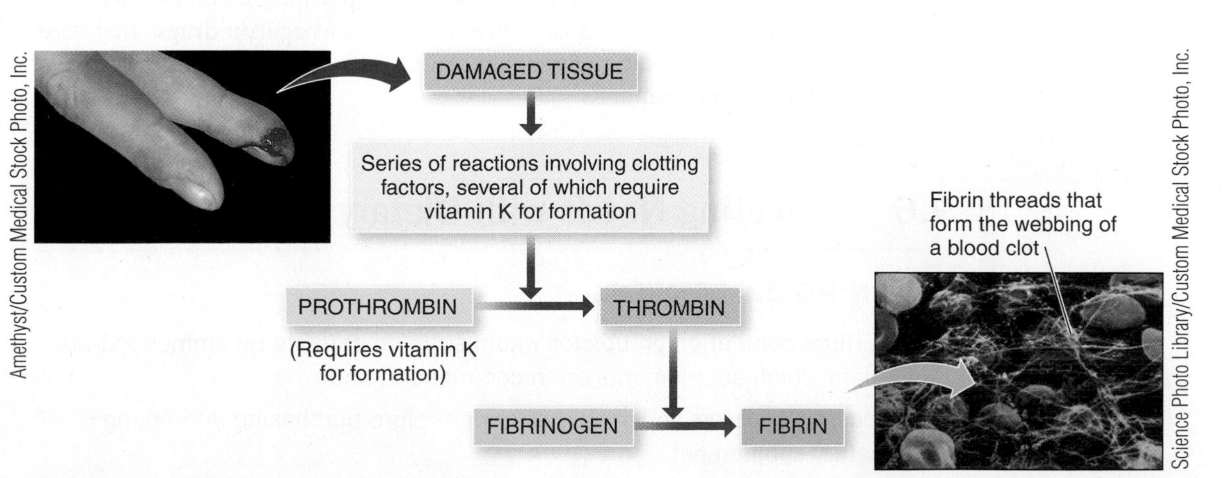

Figure 9.18 Role of vitamin K in blood clotting
Blood clotting requires a series of reactions that result in the formation of a fibrous protein called fibrin. Fibrin fibers form a net that traps platelets and blood cells and forms the structure of a blood clot. Several clotting factors, including prothrombin, require vitamin K for synthesis. If vitamin K is deficient, they are not made correctly, and the blood will not clot.

the blood in an inactive form. When activated, they lead to the formation of fibrin, the protein that forms the structure of a blood clot . Injuries, as well as the normal wear and tear of daily living, produce small tears in blood vessels. To prevent blood loss, these tears must be repaired with blood clots.

Vitamin K is also needed for the synthesis of several proteins involved in bone formation and breakdown. With a vitamin K deficiency, bone mineral density is reduced and the risk of fractures increases.[29] Therefore, adequate vitamin K may be important for the prevention or treatment of osteoporosis.[30]

Vitamin K Deficiency

Abnormal blood coagulation is the major symptom of vitamin K deficiency. When vitamin K is deficient, the blood does not clot to seal ruptured blood vessels and blood loss goes unchecked. If the deficiency is severe, it can eventually cause death from blood loss. A deficiency is very rare in the healthy adult population, but it may result from fat-malabsorption syndromes or the long-term use of antibiotics. The antibiotics kill the bacteria in the gastrointestinal tract that are a source of the vitamin. In combination with an illness that reduces the dietary intake of vitamin K, this may precipitate a deficiency. Injections of vitamin K are typically administered before surgery to aid in blood clotting. Since inappropriate blood clotting causes strokes and heart attacks, drugs that block vitamin K activity have been used to reduce blood-clot formation in patients with cardiovascular disease (see **Science Applied: Saving Cows, Killing Rats, and Surviving Heart Attacks**).

Vitamin K deficiency is most common in newborns. There is little transfer of this vitamin from mother to fetus, and because the infant gut is free of bacteria, no vitamin K is made there. Further, breast milk is low in vitamin K. Therefore, to prevent uncontrolled bleeding, infants are typically given a vitamin K injection within six hours of birth.

Recommended Vitamin K Intake

Unlike other fat-soluble vitamins, the body uses vitamin K rapidly, so a constant supply is necessary. The vitamin K produced by bacteria in the gastrointestinal tract contributes to needs, but this is not well absorbed and alone is not enough to meet requirements. An AI for dietary vitamin K has been set at about 120 μg/day for men and 90 μg/day for women (see Table 9.1). The AI is not increased for pregnancy or lactation. An AI for infants was set based on the amount typically consumed in breast milk.

Vitamin K Toxicity and Supplements

A UL has not been established for vitamin K because there are no well-documented side effects, even with intakes up to 370 μg/day from food and supplements. Because vitamin K functions in blood clotting, high doses can interfere with anticoagulant drugs. Therefore, individuals prescribed these medications should consult with their physicians before taking supplements containing vitamin K.

9.6 Meeting Needs with Dietary Supplements

LEARNING OBJECTIVES

- List three population groups for whom supplements are recommended and Explain which supplements are recommended and why.
- Discuss five factors you should consider before purchasing and taking a dietary supplement.

Currently 66% all adult Americans consider themselves supplement users.[31] People take dietary supplements to energize themselves, to protect themselves from disease, to cure

SCIENCE APPLIED

Saving Cows, Killing Rats, and Surviving Heart Attacks

(Elena Schweitzer/Shutterstock) Victoriano Izquierdo/Getty Images, Inc. (DAJ/Getty Images, Inc.)

In the 1930s, hemorrhagic sweet clover disease was killing cows across the prairies of the midwestern United States and Canada. It occurred in cattle that were fed moldy, spoiled sweet clover hay. These animals died because their blood did not clot. Even a minor scratch from a barbed wire fence could be fatal; once bleeding began, it did not stop.

THE SCIENCE

On a snowy night in 1933, a disgruntled farmer delivered a bale of moldy clover hay, a pail of unclotted blood, and a dead cow to the laboratory of Dr. Carl Link at the University of Wisconsin. Link, the university's first professor of biochemistry, was already making a name for himself in the scientific community when he was presented with this challenge. What was in the clover that killed the man's cows? What could he do to stop the bleeding disease that was killing them? Link had just begun to research hemorrhagic sweet clover disease, but at the time the only advice he could offer the farmer was to find alternative feed and try blood transfusions to save his remaining animals. Link and his colleagues began a line of inquiry that ultimately led to the development of an anti-blood-clotting factor that today saves the lives of hundreds of thousands of people.

Six years after the farmer's challenge, Link and colleagues had isolated the anticoagulant *dicumarol* from moldy clover. Dicumarol is a derivative of coumarin, which gives clover its sweet scent; mold converts coumarin to dicumarol. Cows fed moldy clover consume dicumarol, which interferes with vitamin K activity and consequently prevents normal blood clotting (see figure).

THE APPLICATION

The discovery of dicumarol enhanced our understanding of the blood-clotting mechanism and led to the development of anticoagulant drugs. These drugs help eliminate blood clots and prevent their formation. In carefully regulated doses, they are used to treat heart attacks, which occur when blood clots block blood flow to the heart muscle. Dicumarol, first synthesized in 1940, was the first anticoagulant that could be administered orally to humans. Further work with dicumarol led Link to propose the use of a more potent derivative, called *warfarin*, as rat poison. The name warfarin comes from the initials of the Wisconsin Alumni Research Foundation (WARF), which patented the drug, and from coumarin, the compound from which it is derived. When rats consume the odorless, colorless warfarin, their blood fails to clot, and they bleed to death. Warfarin was used as a rodenticide for nearly a decade before it was introduced into clinical medicine in 1954. Sodium warfarin, also known by the brand name Coumadin, soon became the most widely prescribed anticoagulant drug in the nation. It was used to treat President Dwight D. Eisenhower when he suffered a heart attack in 1955.

Sodium warfarin and dicumarol have been administered to millions of patients to prevent blood clots, which can cause heart attacks and strokes. Thus a substance that killed hundreds of cattle across the Great Plains during the Great Depression and today efficiently kills rodents invading our homes also saves a multitude of human lives.

THINK CRITICALLY: Why should patients taking warfarin avoid supplements that contain vitamin K?

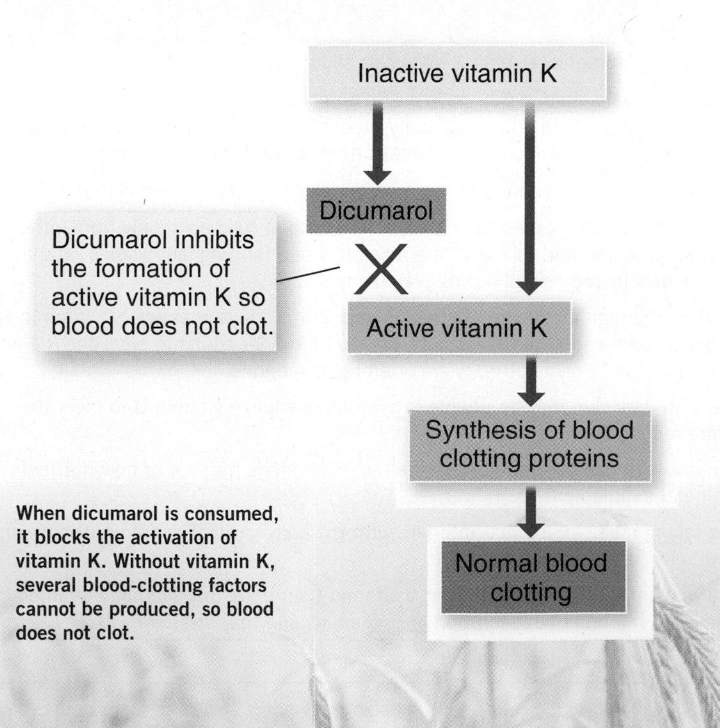

When dicumarol is consumed, it blocks the activation of vitamin K. Without vitamin K, several blood-clotting factors cannot be produced, so blood does not clot.

their illnesses, to enhance what they get in food, and simply to ensure against deficiencies. These products may be beneficial and even necessary under some circumstances for some people, but they also have the potential to cause harm.

Who Needs Vitamin/Mineral Supplements?

Eating a variety of foods is the best way to meet nutrient needs, and most healthy adults who consume a reasonably good diet do not need supplements. In fact, an argument against the use of supplements is that it gives people a false sense of security, causing them to pay less attention to the nutrient content of the foods they choose. For some people, however, supplements may be the only way to meet certain nutrient needs because they have low intakes, increased needs, or excess losses. Groups for whom vitamin and mineral supplements are typically recommended and the type of supplement are given in **Table 9.3**.

Choose Supplements with Care

Despite the benefits of supplements to some individuals, they can also carry risks. Concentrated doses of vitamins and minerals can result in toxicity, and supplements of other substances, such as herbs, may have side effects that outweigh any benefits they may provide. There are people who should not take certain supplements. For example, smokers should not take β-carotene supplements, and people who are unable to regulate iron absorption need to avoid iron supplements. People who tend to develop kidney stones should avoid vitamin C supplements. People taking medications should also be cautious because some interact with supplements. For example, taking supplements of vitamin E or vitamin K can alter the effectiveness of anticoagulant medications. Individuals who routinely take medications should discuss nutrient-drug interactions and the need for specific vitamin and mineral supplementation with their doctor or pharmacist.

Supplements can be part of an effective strategy to promote good health, but they should never be considered a substitute for other good health habits, and they should never be used instead of medical therapy to treat a health problem. If you choose to use dietary supplements, a safe choice is a multivitamin/mineral supplement that does not exceed 100% of the Daily Values (see **Off the Label: Read the Label before You Supplement**).

TABLE 9.3 Groups for Whom Dietary Supplements Are Recommended[32]

Group	Recommendation
Dieters	People who consume fewer than 1200 kcals should take a multivitamin/multimineral supplement.
Vegans and those who eliminate all dairy foods	To obtain adequate vitamin B_{12}, people who do not eat animal products need to take supplements or consume vitamin B_{12}-fortified foods. Because dairy products are an important source of calcium and vitamin D, those who do not consume dairy products may benefit from taking supplements that provide calcium and vitamin D.
Infants and children	Supplemental fluoride, vitamin D, and iron are recommended under certain circumstances.
Young women and pregnant women	Women of childbearing age should consume 400 μg of folic acid daily from either fortified foods or supplements. Supplements of iron and folic acid are recommended for pregnant women, and multivitamin/multimineral supplements are usually prescribed during pregnancy.
Older adults	Because of the high incidence of atrophic gastritis in adults over age 50, vitamin B_{12} supplements or fortified foods are recommended. It may also be difficult for older adults to meet the RDAs for vitamin D and calcium, so supplements of these nutrients are often recommended.
Individuals with dark skin pigmentation	People with dark skin may be unable to synthesize enough vitamin D to meet their needs for this vitamin and may therefore require supplements.
Individuals with restricted diets	Individuals with health conditions that affect what foods they eat or how nutrients are used may require vitamin and mineral supplements.
People taking medications	People taking medications that interfere with the body's use of certain nutrients may require supplements of these nutrients.
Cigarette smokers and alcohol users	People who smoke heavily require more vitamin C and possibly vitamin E than do nonsmokers.[26,33] Alcohol consumption inhibits the absorption of B vitamins and may interfere with B vitamin metabolism.

© iStockphoto

OFF THE LABEL

Read the Label before You Supplement

You can purchase almost any nutrient as an individual supplement or you can choose from a surfeit of combinations of nutrients, herbs, and other components. The labels entice you with descriptions like "all natural," "mega," "advanced formula," "high potency," and "ultra." If you need a supplement or choose to take one, how can you decide which is best?

If you choose to take a dietary supplement, use some common sense and read the label. Supplement labels are required to carry a standardized Supplement Facts panel (see figure and Chapter 2) that can be used to determine if a supplement includes the nutrients you want in appropriate amounts. For example, if you are looking for a vitamin/mineral supplement, the Supplement Facts panel will indicate if it provides minerals as well as vitamins. If you want to increase your calcium intake, check to see if it provides the amount of calcium you want. If you have low iron stores, check to see if it provides iron. If you are over 50 or eat a vegan diet, check to see if it provides enough vitamin B_{12} to meet your needs. Special supplement formulas for men, seniors, and women are available but may not necessarily provide what you need even if you are in the group they target. These formulations are not defined or regulated, so it is up to the company to decide what they contain.

Be sure to look at the serving size. If the serving is one tablet, the supplement facts will be shown for one tablet—if you take two, you'll have to double the numbers. The amounts of nutrients contained in a dose vary depending on the supplement. Some contain less than or equal to 100% of the Daily Value, while others supply two, three, or ten times the Daily Value. High doses of individual nutrients or combinations of several different nutrients or other substances can lead to nutrient imbalances or toxicities.

To assess the safety of the amounts of nutrients in a supplement dose, compare it to the Tolerable Upper Intake Levels (ULs) (see inside cover). If the supplement contains amounts of nutrients in excess of the UL, symptoms of toxicity are more likely to occur. One nutrient to be particularly careful of is vitamin A; too much increases the risk of bone fractures.[1] To minimize your risk, do not take supplements of preformed vitamin A and if you take a multivitamin look for one that contains vitamin A as β-carotene. When checking how much is in your supplement, be aware that the % Daily Value of vitamin A listed on food and supplement labels is based on a Daily Value of 1500 μg (5000 IU) per day for adults. For women, the RDA is only 700 μg (2330 IU) per day, so a supplement containing 100% of the Daily Value provides over twice the RDA. Although the ULs are a good guide, they are not always available. For example, there are no ULs or Daily Values for herbs, carotenoids, and other phytochemicals, so it is difficult to assess whether the amount in a supplement is too low to be of benefit or high enough to be harmful.

THINK CRITICALLY: If you took two tablets of the vitamin/mineral supplement shown here, would it pose a risk for nutrient toxicity?

Supplement Facts

Serving Size 1 Tablet

Each Tablet Contains	% DV
Vitamin A 5000 IU (20% as Beta Carotene)	100%
Vitamin C 60 mg	100%
Vitamin D 400 IU	100%
Vitamin E 30 IU	100%
Vitamin K 25 mcg	31%
Thiamin 1.5 mg	100%
Riboflavin 1.7 mg	100%
Niacin 20 mg	100%
Vitamin B6 2 mg	100%
Folic Acid 400 mcg	100%
Vitamin B12 6 mcg	100%
Biotin 30 mcg	10%
Pantothenic Acid 10 mg	100%
Calcium 162 mg	16%
Iron 18 mg	100%
Phosphorus 109 mg	11%
Iodine 150 mcg	100%
Magnesium 100 mg	25%
Zinc 15 mg	100%
Selenium 20 mcg	29%
Copper 2 mg	100%
Manganese 2 mg	100%

Each Tablet Contains	% DV
Chromium 120 mcg	100%
Molybdenum 75 mcg	100%
Chloride 72 mg	2%
Potassium 80 mg	2%
Boron 150 mcg	*
Nickel 5 mcg	*
Silicon 2 mg	*
Tin 10 mcg	*
Vanadium 10 mcg	*
Lutein 250 mcg	*

*Daily Value (%DV) not established.

SUGGESTED USE:
Adults - One tablet daily

WARNING: Accidental overdose of iron-containing products is a leading cause of fatal poisoning in children under 6. Keep this product out of reach of children. In case of accidental overdose, call a doctor or poison control center immediately.

Keep bottle tightly closed.
Store at room temperature.

George Semple

There are thousands of dietary supplements on the market. They contain a multitude of nutrients and other ingredients, some of which may pose a risk of toxicity.

[1]Genaro Pde, S., and Martini, L. A. Vitamin A supplementation and risk of skeletal fracture. *Nutr Rev* 62:65–67, 2004.

Although there is little evidence that the average person can benefit from such a supplement, there is also little evidence of harm. Here are a number of suggestions that will help you when choosing or using dietary supplements:

- **Consider why you want it**. If you are taking it to ensure good health, does it provide both vitamins and minerals? If you want to supplement specific nutrients, are they contained in the product?
- **Compare product costs**. Just as more isn't always better, more expensive is not always better either.
- **Read the label**. Does the supplement contain potentially toxic levels of any nutrient? Are you taking the amount recommended on the label? For any nutrients that exceed 100% of the Daily Value, check to see if they exceed the UL (see UL table in the back cover of the book). Does the supplement contain any nonvitamin/nonmineral ingredients? If so, have any of them been shown to be toxic to someone like you?
- **Check the expiration date**. Some nutrients degrade over time, so expired products will have a lower nutrient content than is shown on the label. This is particularly true if the product has not been stored properly.
- **Consider your medical history**. Do you have a medical condition that recommends against certain nutrients or other ingredients? Are you taking prescription medication that an ingredient in the supplement may interact with? Check with a physician, dietitian, or pharmacist to help identify these interactions.
- **Approach herbal supplements with caution**. If you are pregnant, ill, or taking medications, consult your physician before taking herbs. Do not give them to children. Do not take combinations of herbs. Do not use herbs for long periods. Stop taking any product that causes side effects.
- **Report harmful effects**. If you suffer a harmful effect or an illness that you think is related to the use of a supplement, seek medical attention and go to the FDA Reporting Web site for information about how to proceed.

OUTCOME

© DNY59/iStockphoto

Digital Vision/Getty Images, Inc.

Ayana is now a happy, healthy 2-year-old. You would never know that a year ago she had been diagnosed with the vitamin D deficiency disease rickets. She developed rickets because breast milk is low in vitamin D and she was not exposed to enough sunlight to synthesize adequate amounts of the vitamin to meet her needs. Without adequate vitamin D, she was unable to absorb sufficient calcium to support bone growth and health. After diagnosing Ayana with rickets, her pediatrician prescribed oral supplements of both vitamin D and calcium and recommended that Ayana consume a calcium-rich diet. After only a week of treatment, Ayana's laboratory values had improved, and X-rays showed that her bones were healing. After a few months, the bony protrusions on her ribs had disappeared, the slight bowing in her legs was corrected, and new teeth began to emerge. If Ayana had not been treated when she was young and still growing, her skeletal deformities would have been permanent.

Ayana's mother is now pregnant with her second child. She now understands that having dark skin and living in a cold climate, where there is little sunshine during the winter months, means that it is important to get enough vitamin D in her diet. She cannot drink milk, an important dietary source of calcium and vitamin D, but she does drink soy milk that is fortified with both these nutrients. **She plans to breast-feed the new baby but now knows that the baby will also need vitamin D supplements.**

APPLICATIONS

ASSESSING YOUR DIET

1. **Do you get enough vitamin A?**
   a. Use iProfile and your food intake record from Chapter 2 to calculate your average daily intake of vitamin A.
   b. How does your vitamin A intake compare to the RDA for someone of your age and gender?
   c. What are three major food sources of vitamin A in your diet?
   d. Do the major food sources of vitamin A in your diet contain preformed vitamin A or carotenoids?
2. **How much vitamin D do you get in your diet?**
   a. Does your diet meet the RDA of 600 IU?
   b. Name three foods you could add to increase your vitamin D intake.
   c. Based on the color of your skin, where you live, and the amount of time you spend outside, do you think you need to be concerned about the amount of vitamin D in your diet?
3. **How much vitamin E is in your diet?**
   a. Use iProfile and the food intake record you kept in Chapter 2 to calculate your average intake of vitamin E.
   b. How does your intake of vitamin E compare with the RDA?
   c. If your diet does not meet the RDA, suggest modifications that will add enough vitamin E to your diet to meet the RDA without increasing your energy intake.

CONSUMER ISSUES

4. **What supplements are your friends taking? Are they safe?**
   a. Do a supplement survey of 10 people. Record all of the vitamin and mineral supplements they take, including the dose and the number of doses taken per day as well as the reason they chose to take each supplement.
   b. Tabulate the total amount of each vitamin and mineral each person takes.
   c. Are any of the survey subjects consuming nutrients in excess of recommendations (RDA or AI)?
   d. Are any nutrients consumed in excess of the UL?
   e. Do you think these supplements will fulfill the expectations of the consumers? Why or why not?

CLINICAL CONCERNS

5. **Who is at risk of vitamin D deficiency in your town?**
   a. Assume you have a friend or relative living in a nursing home. Find out about how much time nursing-home residents spend outdoors without sunscreen or clothing covering their skin. Do you think they are at risk for vitamin D deficiency? Why or why not?
   b. At higher northern and southern latitudes, little vitamin D is synthesized in the winter months. At what latitude is your community located? Does this put you at risk of vitamin D deficiency?
   c. How much time do the schoolchildren at your local elementary school spend outdoors during recess? Do you think they are at risk for vitamin D deficiency? Why or why not?
   d. Do a survey to determine the percentage of people who apply sunscreen daily. Why might sunscreen affect vitamin D status?
6. **Explain why infections are more common and immunization programs are less successful in regions where vitamin A deficiency is prevalent.**
7. **Zachary's grandmother is taking Coumadin. Her doctor has told her that she can eat foods with vitamin K but she must keep her vitamin K Intake consistent from day to day. The foods she typically consumes all contain low or moderate amounts of vitamin K.**
   a. Why does she need to worry about changes in her intake of vitamin K?
   b. Her family is bringing food for a large potluck Christmas dinner. Use the Internet to identify five foods that contain over 400 μg of vitamin K.

SUMMARY

9.1 Fat-Soluble Vitamins in Our Diet

- Vitamins A, D, E, and K are soluble in fat, which affects how they are absorbed, transported, stored, and excreted. Deficiencies of vitamins A and D are common in the developing world. In the United States, the risk of deficiencies of fat-soluble vitamins is increasing due to low intakes of fruits and vegetables and limited sun exposure.

9.2 Vitamin A

- Vitamin A is found both preformed as retinoids and in precursor forms called carotenoids. The major food sources of preformed vitamin A include liver, eggs, fish, and fortified dairy products. Carotenoids are found in plant foods such as yellow-, orange-, and red-colored fruits and vegetables and leafy green vegetables. Some carotenoids are precursors of vitamin A. The most potent is β-carotene.
- In the body the retinoids, which include retinol, retinal, and retinoic acid, are needed for vision and for the growth and differentiation of cells. Retinol is transported in the blood and can be converted into retinal or retinoic acid. Retinal binds to opsin in the eye to form rhodopsin. After light strikes rhodopsin to begin the visual cycle, a nerve impulse is sent to the brain so the light is perceived. Retinoic acid affects cell differentiation by altering gene expression. It is needed for healthy epithelial tissue and normal reproduction

and immune function. β-Carotene functions as an antioxidant, a role that is independent of its conversion to vitamin A.

- Mild vitamin A deficiency causes night blindness. More severe deficiency interferes with cell differentiation. This impairs immune function and growth and causes the epithelial surface of the eye to become hard and dry, leading to infection and blindness. Vitamin A deficiency is a world health problem that increases the frequency of infectious disease and causes blindness and death in hundreds of thousands of children.
- The need for vitamin A can be met with preformed or provitamin A. Preformed vitamin A can be toxic at doses as low as ten times the RDA and can increase the risk of bone fractures and birth defects. Carotenoids are not toxic, but a high intake can give the skin an orange appearance.

9.3 Vitamin D

- Go to WileyPLUS to view a video clip on Vitamin D VIDEO
- Vitamin D can be made in the skin by exposure to sunlight, so dietary needs vary depending on the amount synthesized. Vitamin D is found in fish oils and fortified milk.
- Dietary vitamin D as well as vitamin D synthesized in the skin must be modified by the liver and then the kidney to form active vitamin D. Active vitamin D promotes calcium and phosphorus absorption from the intestines and acts with parathyroid hormone to cause the release of calcium from bone and calcium retention by the kidney. These roles are essential for maintaining proper levels of calcium and phosphorus in the body.
- Vitamin D deficiency in children results in a condition called rickets; in adults, vitamin D deficiency causes osteomalacia. The risk of vitamin D is increased by factors such as living at high latitudes, working and playing indoors, wearing concealing clothing, the presence of pollution and tall buildings, having dark skin, and using sunscreen that reduce the amount of sunlight that reaches the skin.
- The RDA for vitamin D assumes limited sun exposure because the amount that is synthesized in the skin is extremely variable. Vitamin D supplements are recommended for a number of groups including breast-fed babies, people who do not drink milk, older adults and individuals with dark skin pigmentation or limited sun exposure. Too much vitamin D is toxic, but synthesis of vitamin D from exposure to sunlight does not produce toxic amounts.

9.4 Vitamin E

- Vitamin E is found in nuts, plant oils, green vegetables, and fortified cereals.
- Vitamin E functions primarily as a fat-soluble antioxidant. It is necessary for reproduction and protects cell membranes from oxidative damage.
- Vitamin E deficiency can cause hemolytic anemia and neurological problems.
- Many people take vitamin E supplements. There is little risk of toxicity, but there is also little documented evidence of any benefit from supplements.

9.5 Vitamin K

- Vitamin K is found in plants and is synthesized by bacteria in the gastrointestinal tract.
- Vitamin K is a coenzyme essential for the formation of blood clotting factors as well as proteins needed for normal bone mineralization.
- Deficiency causes bleeding and low-bone density. Since vitamin K deficiency is a problem in newborns, they are routinely given vitamin K injections at birth.

9.6 Meeting Needs with Dietary Supplements

- Some groups require supplements to meet the need for certain nutrients, because they have low intakes, increased needs, or excess losses. This includes dieters, vegans, infants and children, pregnant women and women of childbearing age, older adults, individuals with dark skin pigmentation, and people who have restricted diets, are taking medications, or who smoke or use alcohol.
- Appropriate supplement use can promote health, but supplements should never be considered a substitute for other good health habits and they should never be used instead of medical therapy to treat a health problem.

REVIEW QUESTIONS

1. List two food sources of preformed vitamin A and two of provitamin A.
2. How is vitamin A involved in the perception of light?
3. How does vitamin A affect the proteins made by a cell?
4. Why does a deficiency of vitamin A cause night blindness? dry eyes?
5. What is β-carotene?
6. Explain why β-carotene is not toxic but preformed vitamin A is.
7. Why is vitamin D called "the sunshine vitamin?"
8. Name two sources of vitamin D in the diet.
9. What is the primary function of vitamin D?
10. Describe the symptoms of vitamin D deficiency in children.
11. Explain how vitamin D's effect on gene expression alters calcium absorption.
12. What is the function of vitamin E?
13. Name two sources of vitamin E in the diet.
14. What is the main function of vitamin K?
15. What are the symptoms of vitamin K deficiency?
16. Explain why certain groups need supplements to meet their nutrient needs.

REFERENCES

1. Rajakumar, K., and Thomas, S. B. Reemerging nutritional rickets: A historical perspective. *Arch Pediatr Adolesc Med* 159:335–341, 2005.
2. Weisberg, P., Scanlon, K. S., Li, R., and Cogswell, M. E. Nutritional rickets among children in the United States: Review of cases reported between 1986 and 2003. *Am J Clin Nutr* 80(suppl):1697S–1705S, 2004.
3. Holden, J. M., Eldridge, A. L., Beecher, G. R. et al. Carotenoid content of U.S. foods: An update of the database. *J Food Comp and Anal* 12:169–196, 1999.
4. Food and Nutrition Board, Institute of Medicine. *Dietary Reference Intakes: Vitamin A, Vitamin K, Arsenic, Boron, Chromium, Copper, Iodine, Iron, Manganese, Molybdenum, Nickel, Silicon, Vanadium, and Zinc*. Washington, DC: National Academies Press, 2001.
5. Ross, A. C. Vitamin A and carotenoids. In *Modern Nutrition in Health and Disease*, 10th ed. M. E. Shils, M. Shike, A. C. Ross, et al. eds. Philadelphia: Lippincott Williams & Wilkins, 2006, pp. 351–375.
6. Wintergerst, E. S., Maggini, S., and Hornig, D. H. Contribution of selected vitamins and trace elements to immune function. *Ann Nutr Metab* 51:301–323, 2007.
7. World Health Organization. Micronutrient deficiencies. Vitamin A deficiency: The challenge at the WHO micronutrient nutrition website.
8. Adams, J. The neurobehavioral teratology of retinoids: A 50-year history. *Birth Defects Res A Clin Mol Teratol* 88:895–905, 2010.
9. Penniston, K. L., and Tanumihardjo, S. A. The acute and chronic toxic effects of vitamin A. *Am J Clin Nutr* 83:191–201, 2006.
10. Castaño, G., Etchart, C., and Sookoian, S. Vitamin A toxicity in a physical culturist patient: A case report and review of the literature. *Ann Hepatol* 5:293–395, 2006.
11. Goralczyk, R. Beta-carotene and lung cancer in smokers: Review of hypotheses and status of research. *Nutr Cancer* 61:767–774, 2009.
12. Holick, M. F. Resurrection of vitamin D deficiency and rickets. *J Clin Invest* 116:2062–2072, 2006.
13. Institute of Medicine, Food and Nutrition Board. *Dietary Reference Intakes for Calcium and Vitamin D.*Washington, DC: National Academies Press, 2011.
14. Holick, M. F. Vitamin D: Extraskeletal health. *Endocrinol Metab Clin North Am* 39:381–400, 2010.
15. Holick, M. F., and Chen, T. C. Vitamin D deficiency: A worldwide problem with health consequences. *Am J Clin Nutr* 87(suppl.):1080S–1086S, 2008.
16. Holick, M. F. Vitamin D deficiency: What a pain it is. *Mayo Clin Proc* 78(12):1457–1459, 2003.
17. Holick, M. F. Vitamin D: Importance in the prevention of cancers, type 1 diabetes, heart disease, and osteoporosis. *Am J Clin Nutr* 79:362–371, 2004.
18. Holick, M. F. Sunlight and vitamin D for bone health and prevention of autoimmune diseases, cancers, and cardiovascular disease. *Am J Clin Nutr* 80(suppl):1678S–1688S, 2004.
19. American Academy of Pediatrics Section on Breastfeeding, Policy statement. Breastfeeding and the use of human milk. *Pediatrics* 129:e827–e841, 2012.
20. Wagner, C. L., and Greer, F. R. American Academy of Pediatrics Section on Breastfeeding and Committee on Nutrition. Prevention of rickets and vitamin D deficiency in infants, children, and adolescents. *Pediatrics* 122:1142–1152, 2008.
21. Office of Dietary Supplements. Dietary supplement fact sheet: Vitamin D.
22. Blank, S., Scanlon, K. S., Sinks, T. H., and Falk, H. An outbreak of hypervitaminosis D associated with the overfortification of milk from a home-delivery dairy. *Am J Public Health* 85: 656–659, 1995.
23. Traber, M. G., Burton, G. W., and Hamilton, R. L. Vitamin E trafficking. *Ann N Y Acad Sci* 1031:1–12, 2004.
24. Sabat, R., Guthmann, F., and Rüstow, B. Formation of reactive oxygen species in lung alveolar cells: Effect of vitamin E deficiency. *Lung* 186:115–122, 2008.
25. Institute of Medicine, Food and Nutrition Board. *Dietary Reference Intakes for Vitamin C, Vitamin E, Selenium, and Carotenoids*. Washington, DC: National Academies Press, 2000.
26. Cordero, Z., Drogan, D., Weikert, C., and Boeing, H. Vitamin E and risk of cardiovascular diseases: A review of epidemiologic and clinical trial studies. *Crit Rev Food Sci Nutr* 50:420–440, 2010.
27. Galli, F., and Azzi, A. Present trends in vitamin E research. *Biofactors* 36:33–42, 2010.
28. Maras, J. E., Bermudez, O. I., Qiao, N., et al. Intake of alpha-tocopherol is limited among U.S. adults. *J Am Diet Assoc* 104:567–575, 2004.
29. Bügel, S. Vitamin K and bone health in adult humans. *Vitam Horm* 78:393–416, 2008.
30. Prabhoo, R., and Prabhoo, T. R. Vitamin K2: A novel therapy for osteoporosis. *J Indian Med Assoc* 108:253–254, 256–258, 2010.
31. Council for Responsible Nutrition. Supplement usage, consumer confidence remains steady.
32. American Dietetic Association. Position of the American Dietetic Association: Nutrient supplementation. *J Am Diet Assoc* 109:2073–2085, 2009.
33. Bruno, R. S., and Traber, M. G. Cigarette smoke alters human vitamin E requirements. *J Nutr* 135:671–674, 2005.

***To access links to online sources, please go to** www.wiley.com/college/smolin **and select Nutrition: Science and Applications, 3rd edition. From this page, select either the student or instructor companion site. Once on the desired site, select References.**

Focus on Phytochemicals

© Valentyn Volkov/iStockphoto

CHAPTER OUTLINE

functional foods Foods that provide a potential benefit to health beyond that attributed to the nutrients they contain.

phytochemicals Substances found in plant foods (*phyto-* means plant) that are not essential nutrients but may have health-promoting properties.

zoochemicals Substances found in animal foods (*zoo-* means animal) that are not essential nutrients but may have health-promoting properties.

Food presents an unlimited array of tastes, textures, colors, and aromas. With this gastronomic variety and delight come a myriad of nutrient combinations. In addition, food contains substances that have not been identified as nutrients but may promote health and reduce disease risk. Foods have been used in folk medicine for centuries. Today, researchers continue to discover beneficial effects of various food components and explore the relationships among the consumption of specific foods, typical dietary patterns, and health. Foods that provide health benefits beyond basic nutrition are called **functional foods.**[1] **Table F3.1** provides examples of functional foods and their potential benefits. **Health-promoting substances in plant foods are called phytochemicals, while those found in animal foods are called zoochemicals.**

TABLE F3.1 Benefits of Functional Foods

Food	Potential health benefit
Blueberries	May reduce the risk of heart disease and cancer.[2,3]
Breakfast cereal with added flaxseed	Helps reduce blood cholesterol levels and the overall risk of heart disease.[4]
Chocolate	May help reduce blood pressure and other risk factors for heart disease.[5]
Garlic	Helps reduce blood cholesterol levels and the overall risk of heart disease.[6]
Kale	May reduce the risk of age-related blindness (macular degeneration).[7]
Margarine with added plant sterols	Reduces blood cholesterol levels.[8]
Nuts	May reduce the risk of heart disease.[9]
Oatmeal	Helps reduce blood cholesterol.[10]
Orange juice with added calcium	Helps prevent osteoporosis.
Salmon	Reduces the risk of heart disease.[11]
Tea, green and black	May reduce the risk of certain types of cancer.[12]
Whole-grain bread	Helps reduce the risk of cancer, heart disease, obesity, and diabetes.[13]

F3.1 Phytochemicals in Our Food

LEARNING OBJECTIVES

- Distinguish phytochemicals from essential nutrients.
- Discuss how the color of fruits and vegetables is related to their phytochemical content.
- Give some examples of the health benefits of phytochemicals.

Phytochemicals include the hundreds, perhaps thousands of biologically active nonnutritive chemicals found in plants. In the plants themselves, phytochemicals provide biological functions. For example, compounds in onions and garlic are natural pesticides that protect plants from insects. Gardeners often plant these near more vulnerable plant varieties to protect them from insect infestation. Most plant chemicals have no effect on human health, but many promote health, and a few can be toxic. For example, the phytochemicals in soy and tea have been shown to inhibit tumor growth.[12,14] On the other hand, chemicals in rhubarb leaves can cause symptoms ranging from burning in the mouth and throat, abdominal pain, nausea, vomiting, and diarrhea to convulsions, coma, and death from cardiovascular collapse (**Figure F3.1**). Despite the variety of effects these chemicals can have, the term "phytochemical" is generally used to refer to those substances found in plants that have health-promoting properties. Phytochemicals are responsible for the health-promoting properties afforded by a variety of natural functional foods.

antioxidant A substance that decreases the adverse effects of reactive molecules on normal physiological function.

The phytochemicals in our diet protect our health in a variety of ways. Some, such as carotenoids, are **antioxidants** (see Chapters 8 and 9). Others provide benefits because they mimic the structure or function of natural substances in the body. Phytoestrogens, such as those in soy for instance, have structures similar to the hormone estrogen and affect us by blocking or mimicking estrogen action. Phytosterols resemble cholesterol in structure and thus compete with cholesterol for absorption from the gastrointestinal tract. Their presence reduces cholesterol absorption and helps lower blood cholesterol—a major risk factor for cardiovascular disease. Some phytochemicals stimulate the body's natural defenses.

Figure F3.1 Rhubarb leaves are toxic
Not all of the chemicals in plants benefit our health. The leaves of the rhubarb plant are poisonous. The stems have lower levels of the toxin and are safe to eat.

carcinogen A substance that causes cancer.

For example, indoles and isothiocyanates found in broccoli stimulate the activity of enzymes that help deactivate **carcinogens**. Other phytochemicals are health-promoting because they can alter the way in which cells communicate, affect DNA repair mechanisms, or influence other cell processes that may affect cancer development.[15] The phytochemicals described in this chapter are just a few of the ones that have been identified in our food. Many others have been studied and still more have yet to be isolated.

Carotenoids

Carrots, sweet potatoes, acorn squash, apricots, and mangos get their yellow-orange color from the carotenoids they contain. These and other yellow-orange and red fruits and vegetables as well as leafy greens, in which the color of the carotenoids is masked by the green color of chlorophyll, are the major sources of dietary carotenoids. People who eat more of these have higher blood carotenoid levels. A high intake of carotenoid-containing fruits and vegetables has been associated with a reduced risk of certain cancers, cardiovascular disease, and age-related eye diseases.[16]

Carotenoids are phytochemicals that have antioxidant properties; some also have vitamin A activity[17] (**Table F3.2**). The most prevalent carotenoids in the North American diet include β-carotene, α-carotene, β-cryptoxanthin, lycopene, lutein, and zeaxanthin. β-Carotene is the best known and provides the most vitamin A activity, but it may be a less effective antioxidant than others. Lycopene, the carotenoid that gives tomatoes their

TABLE F3.2 Phytochemicals in Foods

Phytochemical name or class	Food sources	Biological activities and possible effects
Carotenoids: α-carotene, β-carotene, β-cryptoxanthin, lutein, zeaxanthin, lycopene	Apricots, carrots, cantaloupe, tomatoes, peppers, sweet potatoes, squash, broccoli, spinach, and other leafy greens	Some are converted to vitamin A, provide antioxidant protection, some decrease the risk of macular degeneration.
Flavonoids: flavonols (quercetin, kaempferol, myricetin), flavones (apigenin), flavanols (catechins), anthrocyanidins (cyanidin, delphinidin)	Berries, citrus fruits, onions, margarine, purple grapes, green tea, red wine, and chocolate	Make capillary blood vessels stronger, block carcinogens and slow the growth of cancer cells.
Phytoestrogens including lignins and isoflavones such as genistein, biochanin A, and daidzein	Tofu, soy milk, soybeans, flax seed, and rye bread	Mimic effect of estrogen, reduce menopause symptoms, slow the growth of cancer cells, reduce blood cholesterol, may reduce risk of osteoporosis.
Phytosterols:β-sitosterol, stigmasterol, and campesterol	Nuts, seeds, and legumes	Decrease cholesterol absorption, reduce the risk of colon cancer by slowing growth of colon cells.
Capsaicin	Hot peppers	Modulates blood clotting.
Glucosinolates, isothiocyanates, indoles	Broccoli, Brussels sprouts, and cabbage	Increase the activity of enzymes that deactivate carcinogens, alter estrogen metabolism, affect the regulation of gene expression.
Sulfides and allium compounds	Onions, garlic, leeks, and chives	Deactivate carcinogens, kill bacteria, protect against heart disease.
Inositol	Sesame seeds and soybeans	Protects against free radicals, protects against cancer.
Saponins	Beans and herbs	Decrease cholesterol absorption, decrease cancer risk, antioxidant.
Ellagic acid	Nuts, grapes, and strawberries	Anticancer properties, prevents the formation of carcinogens in the stomach.
Tannins, catechins	Tea and red wine	Antioxidants, cancer protection.
Curcumin	Turmeric and mustard	Reduces carcinogen formation, antioxidant, anti-inflammatory.
Sulforaphane	Broccoli and other cruciferous vegetables	Detoxifies carcinogens, shown to protect animals from breast cancer.
Limonene	Citrus fruit peels	Inhibits cancer cell growth.

Figure F3.2 **Lycopene in tomatoes**
The red color of tomatoes is due to the carotenoid lycopene, a potent antioxidant. Processing of tomatoes into tomato paste and sauce increases lycopene concentration and bioavailability.

red color, does not provide vitamin A activity, but it is a more potent antioxidant than other dietary carotenoids[18] (**Figure F3.2**). The carotenoids lutein and zeaxanthin are antioxidants that accumulate in the macula, which is the central portion of the retina of the eye. High intakes of these are associated with reduced risk of **macular degeneration**, the leading cause of blindness in older adults (see Chapter 16).[19]

macular degeneration An incurable eye disorder that is caused by deterioration of the retina of the eye. It is the leading cause of blindness in adults over the age of 55 years.

Flavonoids

Like carotenoids, flavonoids are plant pigments that add color to your plate. They are found in fruits, vegetables, wine, grape juice, chocolate, and tea. One of the most abundant types of flavonoids is the anthocyanins, which give the blue and red colors to blueberries, raspberries, and red cabbage. Other types of flavonoids give the pale yellow color to potatoes, onions, and orange rinds. These compounds are strong antioxidants that protect against cancer and cardiovascular disease. Citrus fruits contain about 60 flavonoids that inhibit blood clotting and have antioxidant, anti-inflammatory, and anticancer properties.[20]

Indoles, Isothiocyanates, and Alliums

Cruciferous vegetables such as broccoli, cauliflower, Brussels sprouts, cabbage, and greens such as mustard and collards are particularly good sources of sulfur-containing phytochemicals (**Figure F3.3**). These stimulate the activity of enzymes that detoxify carcinogens. Sulforaphane, a phytochemical in broccoli, is particularly effective at boosting the activity of these enzyme systems and has been shown to protect animals from breast cancer.[21] Crucifers also contain phytochemicals called indoles that inactivate the hormone estrogen. Exposure to estrogen is believed to increase the risk of cancer; therefore, because indoles reduce estrogen exposure, they protect against cancer.

cruciferous A group of vegetables (also called crucifers) named for the cross shape of their four-petal flowers. They include broccoli, Brussels sprouts, cabbage, cauliflower, kale, kohlrabi, mustard greens, rutabagas, and turnips. Their consumption is linked with lower rates of cancer.

Sulfur compounds called allium compounds are found in plants belonging to the species *Allium* such as garlic, onions, leeks, chives, and shallots (**Figure F3.4**). These phytochemicals boost the activity of cancer-destroying enzyme systems, protect against oxidative damage, and defend against heart disease by lowering blood cholesterol, blood pressure, and platelet activity.[6] In addition, these compounds prevent bacteria in the gut from converting nitrates into nitrites, which can form carcinogens.

iStockphoto

Figure F3.3 Cruciferous vegetables
Cruciferous vegetables, such as cabbage and cauliflower, are excellent sources of phytochemicals that may protect against cancer.

Ulrich Kerth/StockFood/Getty Images, Inc.

Figure F3.4 Allium vegetables
Garlic and onions provide sulfur-containing phytochemicals that may help protect against cancer and heart disease.

© 4kodiak/iStockphoto

Figure F3.5 Soy isoflavones
A high intake of soy and soy products, such as this miso soup with tofu, throughout life is one factor that is believed to contribute to the health benefits of a traditional Asian diet.

Phytoestrogens

Human hormones help regulate body processes and maintain homeostasis. Plants have hormones too. When ingested, they can affect human health. Phytoestrogens are plant hormones that are believed to interrupt cancer development and affect health by interfering with the action of the human hormone estrogen. Phytoestrogens include isoflavones and lignans. Soy products are particularly rich in isoflavones and flaxseed and sesame oil are the best sources of lignans (**Figure F3.5**).

Phytoestrogens are modified by the microflora in the intestines to form compounds that are structurally similar to estrogen. In some situations they are believed to mimic estrogen and in others to block estrogen function by tying up estrogen receptors on cells. For example, lignin metabolites have been shown to inhibit the growth of estrogen-stimulated breast cancer cells.[22] There is evidence that consuming isoflavones reduces the symptoms of menopause (including hot flashes), reduces bone loss, and affects the risk and progression of breast cancer. A review of studies on the effects of isoflavones on menopause symptoms suggests that the specific isoflavone genistein may reduce hot flashes.[23] Clinical trials support a role of isoflavones in the prevention of bone loss, but results are inconsistent.[24] Isoflavones have been found to promote the growth of breast tumors in animals, and therefore women who have breast cancer are typically advised to avoid soy, which is high in isoflavones.[25] In women without breast cancer, isoflavones from soy appear to protect against breast cancer when consumed in moderate amounts throughout life, as is common in the Asian diet. Increasing soy consumption after menopause may have no effect.[25]

F3.2 How to Choose a Phytochemical-Rich Diet

LEARNING OBJECTIVES

- List foods that are high in phytochemicals.
- Compare the health benefits of taking phytochemical supplements with those of a diet rich in unrefined plant foods.

TABLE F3.3 How to Choose: Phytochemicals

- Choose five different colors of fruits and vegetables each day.
- Try a new fruit or vegetable each week.
- Spice up your food—herbs and spices are a great source of phytochemicals.
- Add vegetables to your favorite entrees such as spaghetti sauces and casseroles.
- Try baked or dried fruit for dessert.
- Double your typical serving of vegetables.
- Add pesto, spinach, artichokes, or asparagus to pizza.
- Buy jars of chopped garlic, ginger, and basil to make it easy to add more of these to your cooking.
- Snack on whole-grain crackers.
- Add barley or bulgur to casseroles or stews.
- Switch to whole-wheat bread, brown rice, and whole-wheat pasta.
- Add fruit to your cereal or vegetables to your eggs.
- Dice up some tofu and add it to your stir fry.
- Include nuts in stir fries and baked goods.
- Sprinkle flaxseed in your oatmeal.

It has not been possible to make quantitative recommendations for the intake of specific phytochemicals. Many of these substances have not been identified or classified. Many function differently depending on the form in which they are consumed. Their effects may vary if they are removed from the foods in which they are found. Therefore, to benefit the most from phytochemicals, choose a diet based on plant foods. The impact of the total diet is more significant than that of any single phytochemical. For example, in some people a diet high in plant sterols, soy, almonds, and foods such as oats, barley, psyllium, okra, and eggplant that are high in soluble fibers has been shown to be as effective at lowering cholesterol as prescription medications.[26] Likewise, it has been hypothesized that a diet rich in flaxseed, cruciferous vegetables, and fruits and vegetables in general could significantly reduce the risk of breast, colon, prostate, lung, and other cancers.[27] **Table F3.3** includes some suggestions for increasing the intake of foods that provide a variety of phytochemicals.

Eat More Fruits and Vegetables

The 2010 Dietary Guidelines and MyPlate recommend a diet with plenty of fruits and vegetables: two cups of fruit and 2.5 cups of vegetables for a 2000-kcalorie diet. If these fruits and vegetables include a rainbow of colors—dark green vegetables as well as yellow-orange, pale yellow, and deep red and purple fruits and vegetables—you are getting an abundance of carotenoids, flavonoids, and other phytochemicals (**Figure F3.6**).[28]

Unfortunately, most people do not follow these recommendations. Recent surveys of the typical American diet suggest that fewer than one in ten Americans consumes the recommended servings of fruits and vegetables.[29] In addition, white potatoes represent a disproportionately large share of the total vegetable intake (many of these consumed as French fries), while phytochemical- and nutrient-rich

Figure F3.6 Eat a rainbow of fruits and vegetables
Choosing fruits and vegetables with all the colors of the rainbow is a good way to include a variety of phytochemicals in your diet.

dark green and deep yellow and orange vegetables accounted for a disproportionately small share.

Make Half Your Grains Whole

Fruits and vegetables aren't the only source of dietary phytochemicals. In fact, whole grains deliver as many if not more phytochemicals and antioxidants than do fruits and vegetables, and therefore the Dietary Guidelines recommend that we make half our grains whole.[28] Epidemiological studies have found that whole-grain intake is associated with a lower risk of cancer, cardiovascular disease, diabetes, and obesity. These health-promoting properties are believed to be due to the synergistic effects of the wide variety of nutrients and phytochemicals found in whole grains.[13] In addition to fiber, whole grains are rich in antioxidant phytochemicals called phenols as well as phytate, phytoestrogens such as lignan, and plant stanols and sterols. The bran and germ portions of whole-wheat flour contribute more than half of the total phenols, flavonoids, lutein, and zeaxanthin, as well as over 80% of the water-soluble and fat-soluble antioxidant activity.[30] These phytochemicals and vitamins are lost when the bran and germ are removed to make white flour. The Dietary Guidelines recommend Americans consume at least three servings, or half of their grain servings, as whole grains. The average intake of whole grains in the United States today is less than one serving per day.

Choose Plant Proteins

The amounts of health-promoting substances in the diet can further be enhanced by choosing plant sources of protein. Phytochemical-rich soybeans and flaxseed are good sources of protein and, other legumes, nuts, and seeds are high-protein foods that make important fiber and phytochemical contributions. The Dietary Guidelines and MyPlate encourage the consumption of these high-phytochemical protein sources by recommending that Americans choose more beans, peas, nuts, and seeds.

What About Added Phytochemicals?

In addition to foods that are natural sources of phytochemicals, phytochemical supplements are available and many foods are fortified with phytochemicals. For example, special margarines such as Benecol® and Take Control® contain added plant sterols to help lower blood cholesterol, some brands of oatmeal are fortified with soy and flaxseed, and carrot juice is available with added lutein. These modified foods have also been called *nutraceuticals* because they blur the line between nutrients and pharmaceuticals, or *designer foods* because they are designed to provide specific health benefits (see Chapter 8: Debate: Super-Fortified Foods: Are They a Healthy Addition to Your Diet?).

While phytochemical supplements and phytochemical-fortified foods may offer some specific advantages, they do not provide all the benefits obtained from a diet high in natural sources of these compounds. One reason is that most contain only a few of the many phytochemicals contained in foods. In addition, the benefits provided by many foods are believed to be due to the interactions among a variety of phytochemicals and nutrients, and therefore cannot be replicated by a supplement that contains only one or a few of these substances. Finally, the amounts that can be added to supplements or foods may be too small to have an impact on overall health. For example, a multivitamin advertising that it provides lycopene may include only about 0.3 mg per dose, whereas a half-cup serving of spaghetti sauce will give you about 20 mg.

REFERENCES

1. American Dietetic Association. Position of the American Dietetic Association: Functional foods. *J Am Diet Assoc* 109:735–746, 2009.

2. Basu, A., Rhone, M., and Lyons, T. J. Berries: Emerging impact on cardiovascular health. *Nutr Rev* 68:168–177, 2010.

3. Seeram, N. P. Recent trends and advances in berry health benefits research. *J Agric Food Chem* 58:3869–3870, 2010.

4. Prasad, K. Flaxseed and cardiovascular health. *J Cardiovasc Pharmacol* 54:369–377, 2009.

5. Grassi, D., Desideri, G., and Ferri, C. Blood pressure and cardiovascular risk: What about cocoa and chocolate? *Arch Biochem Biophys* 501:112–115, 2010.

6. Butt, M. S., Sultan, M. T., Butt, M. S., and Iqbal, J. Garlic: Nature's protection against physiological threats. *Crit Rev Food Sci Nutr* 49:538–551, 2009.

7. Carpentier, S., Knaus, M., and Suh, M. Associations between lutein, zeaxanthin, and age-related macular degeneration: An overview. *Crit Rev Food Sci Nutr* 49:313–326, 2009.

8. Sanclemente, T., Marques-Lopes, I., Puzo, J., and García-Otín, A. L. Role of naturally-occurring plant sterols on intestinal cholesterol absorption and plasmatic levels. *J Physiol Biochem* 65:87–98, 2009.

9. Bolling, B. W., McKay, D. L., and Blumberg, J. B. The phytochemical composition and antioxidant actions of tree nuts. *Asia Pac J Clin Nutr* 19:117–123, 2010.

10. Sadiq Butt, M., Tahir-Nadeem, M., Khan, M. K., et al. Oat: Unique among the cereals. *Eur J Nutr* 47:68–79, 2008.

11. Manerba, A., Vizzardi, E., Metra, M., and Dei Cas, L. n-3 PUFAs and cardiovascular disease prevention. *Future Cardiol* 6:343–350, 2010.

12. Lambert, J. D., and Elias, R. J. The antioxidant and pro-oxidant activities of green tea polyphenols: A role in cancer prevention. *Arch Biochem Biophys* 501:65–72, 2010.

13. Fardet, A. New hypotheses for the health-protective mechanisms of whole-grain cereals: What is beyond fibre? *Nutr Res Rev* 23:65–134, 2010.

14. Zhou, J. R., Yu, L., Mai, Z., and Blackburn, G. L. Combined inhibition of estrogen-dependent human breast carcinoma by soy and tea bioactive components in mice. *Intl J Cancer* 108:8–14, 2004.

15. Zuzino, S. How plants protect us. *Agriculture Res Mag.* March 2008, pp 9–11.

16. Krinsky, N. I., Johnson, E. J. Carotenoid actions and their relation to health and disease. *Mol Aspects Med* 6:459–516, 2005.

17. Rao, A. V., and Rao, L. G. Carotenoids and human health. *Pharmucol Res* 5:207–216, 2007.

18. Kelkel, M., Schumacher, M., Dicato, M., and Diederich, M. Antioxidant and anti-proliferative properties of lycopene. *Free Radic Res* 45:925–940, 2011.

19. Ma, L., Dou, H. L., Wu, Y. Q., et al. Lutein and zeaxanthin intake and the risk of age-related macular degeneration: A systematic review and meta-analysis. *Br J Nutr* 107:350–359, 2012.

20. Erdman, J. W., Balentine, D., and Arab, L. Flavonoids and heart health: Proceedings of the ILSI North America Flavonoids Workshop, May 31–June 1, 2005, Washington, DC. *J Nutr* 137:718S–737S, 2007.

21. Stan, S. D., Kar, S., Stoner, G. D., and Singh, S. V. Bioactive food components and cancer risk reduction. *J Cell Biochem* 104:339–356, 2008.

22. Rice, S., and Whitehead, S. A. Phytoestrogens oestrogen synthesis and breast cancer. *J Steroid Biochem Mol Biol* 108:186–195, 2008.

23. Williamson-Hughes, P. S, Flickinger, B. D., Messina, M. J., and Empie, M. W. Isoflavone supplements containing predominantly genistein reduce hot flash symptoms: A critical review of published studies. *Menopause* 13:831–839, 2006.

24. Taku, K., Melby, M. K., Takebayashi, J., et al. Effect of soy isoflavone extract supplements on bone mineral density in menopausal women: Meta-analysis of randomized controlled trials. *Asia Pac J Clin Nutr* 19:33–42, 2010.

25. Hilakivi-Clarke, L., Andrade, J. E., and Helferich, W. Is soy consumption good or bad for the breast? *J Nutr* 140:2326S–2334S, 2010.

26. Jenkins, D. J., Kendall, C. W., Marchie, A., et al. Direct comparison of a dietary portfolio of cholesterol-lowering foods with a statin in hypercholesterolemic participants. *Am . Clin Nutr* 81:380–387, 2005.

27. Donaldson, M. S. Nutrition and cancer: A review of the evidence for an anti-cancer diet. *Nutr J* 3:19–30, 2004.

28. U.S. Department of Agriculture and U.S. Department of Health and Human Services. *Dietary Guidelines for Americans*, 2010, 7th ed., Washington DC: U.S Government Printing Office, December, 2010.

29. Kimmons, J., Gillespie, C., Seymour, J., et al. Fruit and vegetable intake among adolescents and adults in the United States: Percentage meeting individualized recommendations. *J Med* 11: 26–36, 2009.

30. Adom, K. K., Sorrells, M. E., and Liu, R. H. Phytochemicals and antioxidant activity of milled fractions of different wheat varieties. *J Agric Food Chem* 53:2297–2306, 2005.

CHAPTER OUTLINE

Eiichi Onodera/Dex Image/Getty Images, Inc.

Water and the Electrolytes

10

CASE STUDY

When Gabriele Andersen-Scheiss emerged from the tunnel into the Olympic stadium in 1984 for the final lap of the women's marathon, she didn't look much like a world-class athlete. The 37-year-old runner staggered around the track as if drunk, her left arm limp, her right leg stiff. She had completed about 26 of the 26.2 miles of the marathon, but it wasn't clear whether she now could negotiate a single lap around the track. What was wrong? Should someone stop her?

Medical officers rushed over to help, but she waved them off. After observing her as she continued around the track, doctors decided to allow her to go on unaided. Andersen-Scheiss was demonstrating the signs of serious dehydration. Because the medics could see that she was sweating profusely, though, they concluded that she was not yet suffering from heat stroke, the most severe and life-threatening form of heat-related illness. The gold medalist in this race, Joan Benoit, had crossed the finish line 20 minutes earlier, so Andersen-Scheiss was struggling simply to finish, not to win. One minute the crowd cheered her on, and the next they pleaded for the medics to stop her. Her final staggering lap around the track to the finish line lasted 5 minutes and 44 seconds. Medics immediately treated her for dehydration and heat exhaustion.

David Tenenbaum/©AP/Wide World Photos

This case illustrates the importance of water in maintaining body systems. Water and the minerals dissolved in it make up the soup in which the reactions that maintain life occur. **Water and mineral imbalances can come on rapidly and be devastating, but they can also be alleviated faster than other nutrient deficiencies.**

10.1 Water: The Internal Sea

LEARNING OBJECTIVES

- Discuss the forces that move water back and forth across cell membranes.
- Describe five functions of water in the body.
- List the sources of body water and describe the routes by which water is lost.
- Discuss the effects of dehydration.

The complex molecules necessary for the emergence of life were forged in the Earth's first seas. These primordial seas supported life because they were rich in inorganic minerals as well as organic substances. As organisms grew in complexity, the water and chemicals critical to their survival were incorporated into an internal sea. Just as the right amounts of water, organic molecules, and minerals were necessary for the beginning of life, the right combination is necessary in the body for the maintenance of life. This internal sea allows the reactions necessary for life to proceed.

How Water Functions in the Body

Water is an essential nutrient that we must consume in our diet to survive. Without water, death occurs in only a few days. In the body, water serves as a medium in which chemical reactions take place; it also transports nutrients and wastes, provides protection, helps regulate temperature, participates in chemical reactions, and helps maintain acid–base balance (**Table 10.1**).

TABLE 10.1 A Summary of Water and the Electrolytes

Nutrient	Sources	Recommended intake for adults	Major functions	Deficiency diseases and symptoms	Groups at risk of deficiency	Toxicity	UL
Water	Drinking water, other beverages, and food	2.7–3.7 L/day[a]	Solvent, reactant, protector, transporter, regulator of temperature and pH	Thirst, dark-colored urine, weakness, poor endurance, confusion, disorientation	Infants; people with fever, vomiting, or diarrhea; elderly individuals; athletes	Confusion, coma, convulsions	ND
Sodium	Table salt, processed foods	<2300 mg; ideally 1500 mg/day[a]	Major positive extracellular ion, nerve transmission, muscle contraction, fluid balance	Muscle cramps	People consuming a severely sodium-restricted diet, those who sweat excessively	High blood pressure in sensitive people	2300 mg/day
Potassium	Fresh fruits and vegetables, legumes, whole grains, milk, meat	4700 mg/day[a] or more	Major positive intracellular ion, nerve transmission, muscle contraction, fluid balance	Irregular heartbeat, fatigue, muscle cramps	People consuming poor diets high in processed foods, people with diarrhea or vomiting, those taking thiazide diuretics	Abnormal heartbeat	ND
Chloride	Table salt, processed foods	<3600 mg/day; ideally 2300 mg/day[a]	Major negative extracellular ion, fluid balance	Unlikely	None	None likely	3600 mg/day

[a]Adequate Intakes (AIs).

Note: UL, Tolerable Upper Intake Level; ND, not determined due to insufficient data.

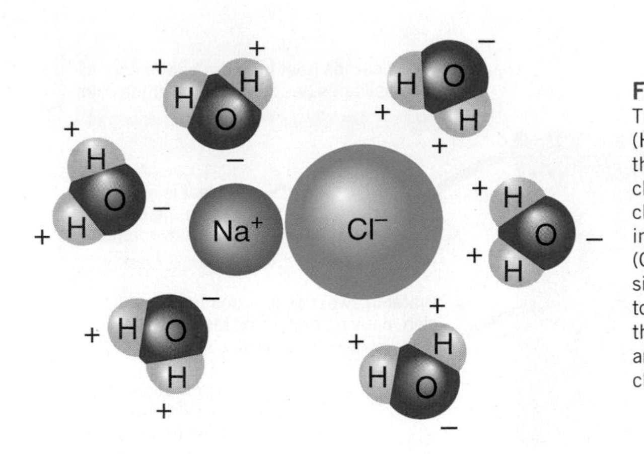

Figure 10.1 **The polar nature of water** The oxygen side of a water molecule (H_2O) has a slightly negative charge and the hydrogen side has a slightly positive charge. This polar nature causes sodium chloride (NaCl) to dissolve when placed in water; the sodium (Na^+) and chloride (Cl^-) dissociate because the negative sides of the water molecules are attracted to the positively charged sodium ions and the positive sides of the water molecules are attracted to the negatively charged chloride ions.

Why Water Is a Good Solvent One of the key functions of water in the body is as a **solvent**. A solvent is a fluid in which **solutes** can dissolve to form a solution. Water is an ideal solvent for some substances because it is **polar**; that is, the two sides or poles of the water molecule have opposite electrical charges. The polar nature of water comes from its structure, which consists of two hydrogen atoms and one oxygen atom (H_2O). These atoms, like all atoms, are made up of a positively charged central core, or nucleus, with negatively charged **electrons** orbiting around it. To form a water molecule, the two hydrogen atoms move close enough to share their electrons with an atom of oxygen. But the sharing is not equal. The shared electrons spend more time around the oxygen atom than around the hydrogen atoms, giving the oxygen side of the molecule a slightly negative charge and the hydrogen side a slightly positive charge. This polar nature of water allows it to surround other charged molecules and disperse them. Table salt, which dissolves in water, consists of a positively charged sodium **ion** (Na^+) bound to a negatively charged chloride ion (Cl^-). When placed in water, the sodium and chloride ions move apart, or **dissociate**, because the positively charged sodium ion is attracted to the negative pole of the water molecule and the negatively charged chloride ion is attracted to the positive pole (**Figure 10.1**).

solvent A fluid in which one or more substances dissolve.

solutes Dissolved substances.

polar Used to describe a molecule that has a positive charge at one end and a negative charge at the other.

electrons Negatively charged particles.

ion An atom or group of atoms that carries an electrical charge.

dissociate To separate two charged ions.

Water as a Transport Medium Blood, which is about 90% water, transports oxygen and nutrients to cells. It then carries carbon dioxide and other waste products away from the cells. Blood also distributes hormones and other regulatory molecules throughout the body so they can reach target cells. Water in urine helps to carry waste products, such as urea and ketones, out of the body.

Water Lubricates and Protects Water functions as a lubricant and cleanser. Watery tears lubricate the eyes and wash away dirt; synovial fluid lubricates the joints; and saliva lubricates the mouth, making it easier to chew, taste, and swallow food. Water inside the eyeballs and spinal cord acts as a cushion against shock. Similarly, during pregnancy, water in the amniotic fluid provides a protective cushion around the fetus.

Water Regulates Body Temperature Body temperature is closely regulated to maintain a normal level of around 98.6°F (37°C). If the body temperature rises above 108°F or falls below 80°F, death is likely. The fact that water holds heat and changes temperature slowly helps keep body temperature constant when the outside temperature fluctuates, but water is also more actively involved in temperature regulation.

The water in blood helps regulate body temperature by increasing or decreasing the amount of heat lost from the surface of the body. When body temperature starts to rise, the blood vessels in the skin dilate, causing blood to flow close to the surface of the body where it

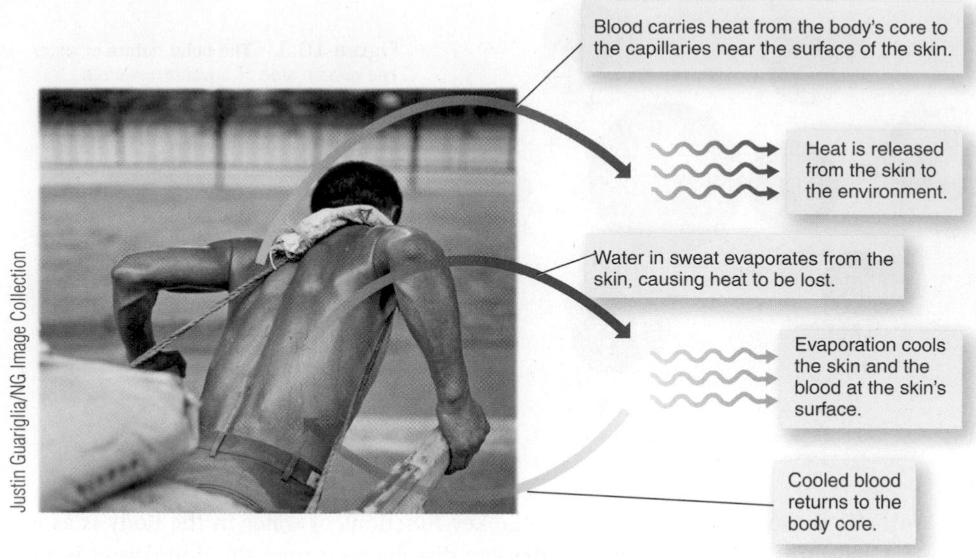

Blood carries heat from the body's core to the capillaries near the surface of the skin.

Heat is released from the skin to the environment.

Water in sweat evaporates from the skin, causing heat to be lost.

Evaporation cools the skin and the blood at the skin's surface.

Cooled blood returns to the body core.

Justin Guariglia/NG Image Collection

Figure 10.2 Water Helps Cool the Body
Shunting blood to the skin allows heat to be transferred from the blood to the surroundings. Evaporation of the water in sweat cools the skin and the blood near the surface of the skin.

can release some of the heat to the environment (**Figure 10.2**). This occurs with fevers as well as when environmental temperature rises. In a cold environment, blood vessels in the skin constrict, restricting the flow of blood near the surface and conserving body heat.

Water also helps regulate body temperature through evaporation. When body temperature increases, the brain triggers the sweat glands in the skin to produce sweat, which is mostly water. As the water evaporates from the skin, heat is lost, cooling the body (see Figure 10.2).

hydrolysis reaction Chemical reaction that breaks large molecules into smaller ones by adding water.

condensation or **dehydration reaction** Chemical reaction that joins two molecules together. During the reaction hydrogen and oxygen are lost from the two molecules and form water.

Water Participates in Metabolic Reactions Water is involved in some of the chemical reactions of metabolism. A **hydrolysis reaction** breaks large molecules into smaller ones by adding water. For example, water is added in the reaction that breaks a molecule of maltose into two glucose molecules (**Figure 10.3**). Water also participates in reactions that join two molecules. This type of reaction is referred to as a **condensation** or **dehydration reaction**. The formation of the disaccharide maltose from two glucose molecules is a condensation reaction and therefore requires the removal of a water molecule.

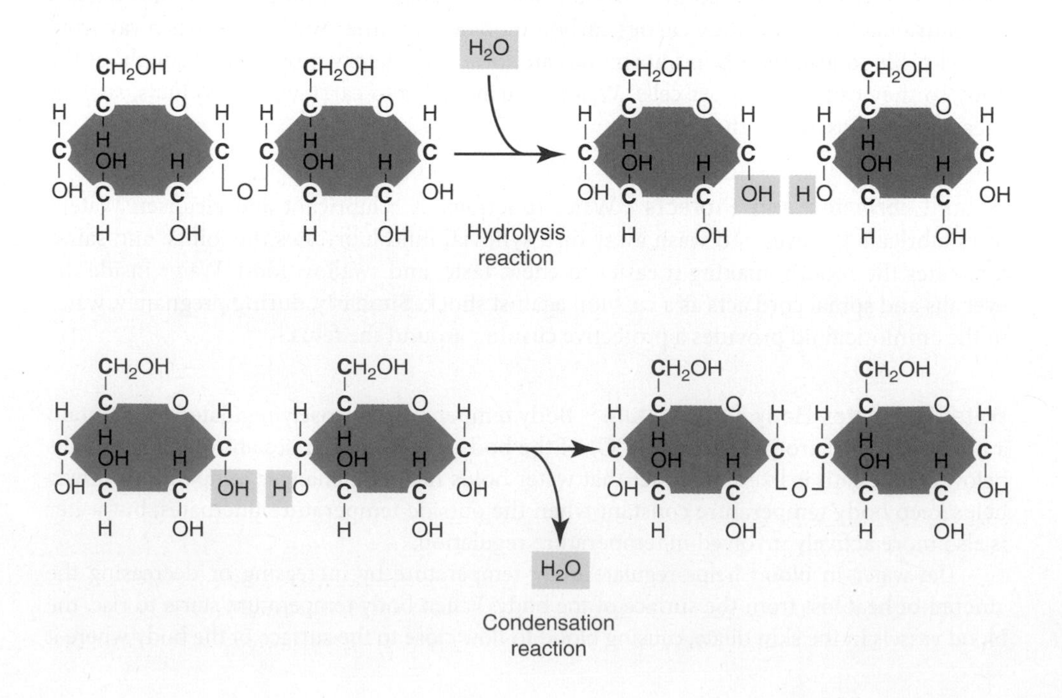

Figure 10.3 Hydrolysis and condensation reactions
The cleavage of the disaccharide maltose into two molecules of glucose involves the addition of a molecule of water and is therefore a hydrolysis reaction. Disaccharide formation releases a molecule of water and is therefore a condensation reaction.

Water Helps Maintain Acid–Base Balance The chemical reactions that occur in the body are very sensitive to acidity. Acidity is expressed in units of **pH**. The range of pH units is from 1 to 14, with 1 being very acidic, 14 being very basic or alkaline, and 7 being neutral (**Figure 10.4**). Most reactions in the body occur in slightly basic solutions, around pH 7.4. If body solutions become too acidic or too basic, chemical reactions cannot proceed efficiently. Water and the dissolved substances it contains are important for maintaining the proper level of acidity. Water serves as a medium for the chemical reactions that prevent changes in pH, and it participates in some of these reactions. Water is also needed as a transport medium to allow the respiratory tract and kidneys to regulate acid–base balance.

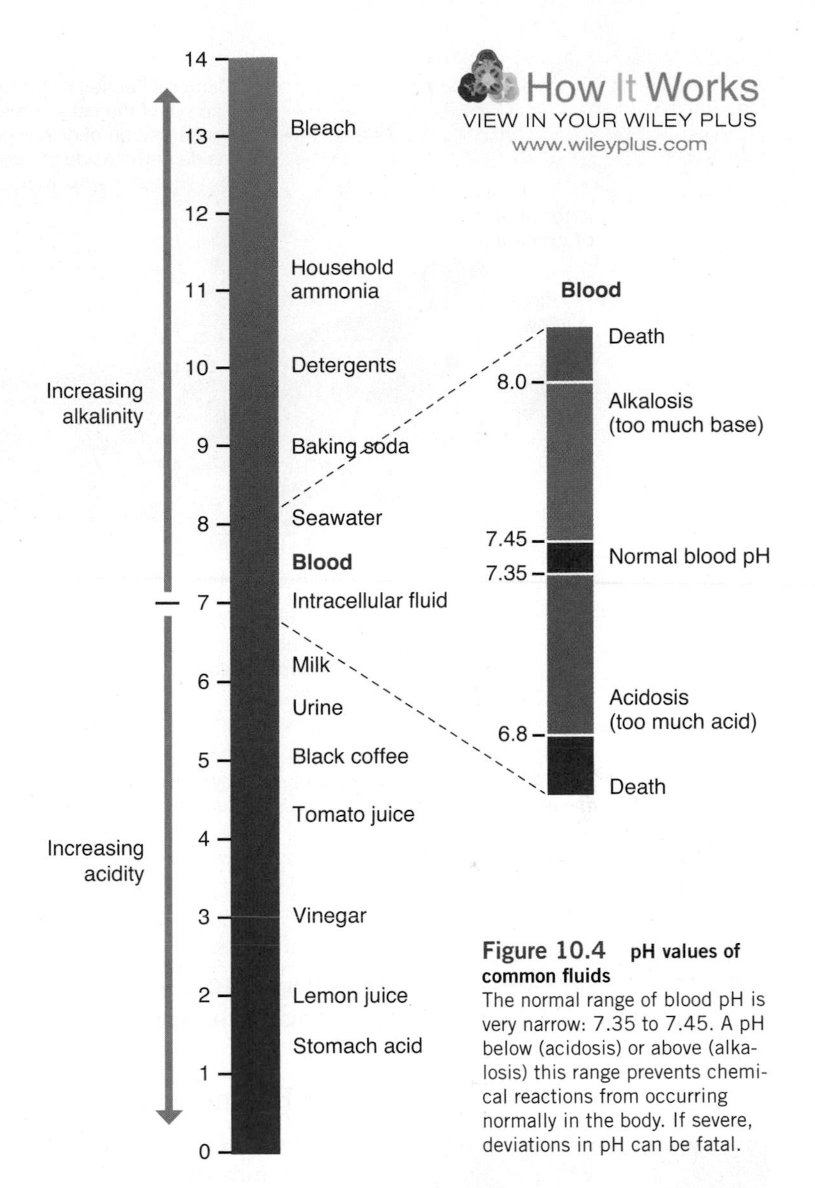

Figure 10.4 **pH values of common fluids**
The normal range of blood pH is very narrow: 7.35 to 7.45. A pH below (acidosis) or above (alkalosis) this range prevents chemical reactions from occurring normally in the body. If severe, deviations in pH can be fatal.

How Water Is Distributed Throughout the Body

Water is found in varying proportions in all the tissues of the body; blood is about 90% water, muscle about 75%, and bone about 25%. Adipocytes have a low water content—only about 10%. In adults, about 60% of total body weight is water, but this varies with age and other factors that affect body composition.

The percentage of water is higher in infants than in adults and decreases as adults age, primarily due to increases in body fat and a loss of muscle mass. Because women typically have a higher percentage of body fat than men, they have less body water. Obese individuals have a lower percentage of body weight as water and a higher percentage as fat than their lean counterparts.

About two-thirds of body water is found inside cells; this is known as **intracellular fluid**. The remaining one-third is outside cells as **extracellular fluid**. Extracellular fluid includes primarily blood plasma, lymph, and the fluid between cells, called **interstitial fluid**. Other extracellular fluids include fluid secreted by glands, such as saliva and other digestive secretions, and fluid in the eyes, joints, and spinal cord. The concentration of substances dissolved in body water varies among these body compartments. The concentration of protein is highest in intracellular fluid, lower in blood plasma, and even lower in interstitial fluid. Extracellular fluid has a higher concentration of sodium and chloride and a lower concentration of potassium, and intracellular fluid is higher in potassium and lower in sodium and chloride.

The movement of water from one compartment to another is affected by fluid pressure and osmosis, which depends on the concentration of solutes in each compartment (**Figure 10.5**). The fluid pressure of blood against the blood vessel walls, or **blood pressure**, causes water to move from the blood into the interstitial space. The difference in the concentration of solutes between the blood in the capillaries and the fluid in the interstitial space causes much of this water to reenter the capillaries by osmosis (see Chapter 3). Osmosis occurs when there is a selectively permeable membrane, such as a cell membrane, that allows water to pass freely but regulates the passage of other substances. Water moves across this membrane in a direction that will equalize the concentration of solutes on both

pH A measure of the level of acidity or alkalinity of a solution.

intracellular fluid The fluid located inside cells.

extracellular fluid The fluid located outside cells. It includes fluid found in the blood plasma, lymph, gastrointestinal tract, spinal column, eyes, and joints and that found between cells.

interstitial fluid The portion of the extracellular fluid located in the spaces between the cells of body tissues.

blood pressure The amount of force exerted by the blood against the artery walls.

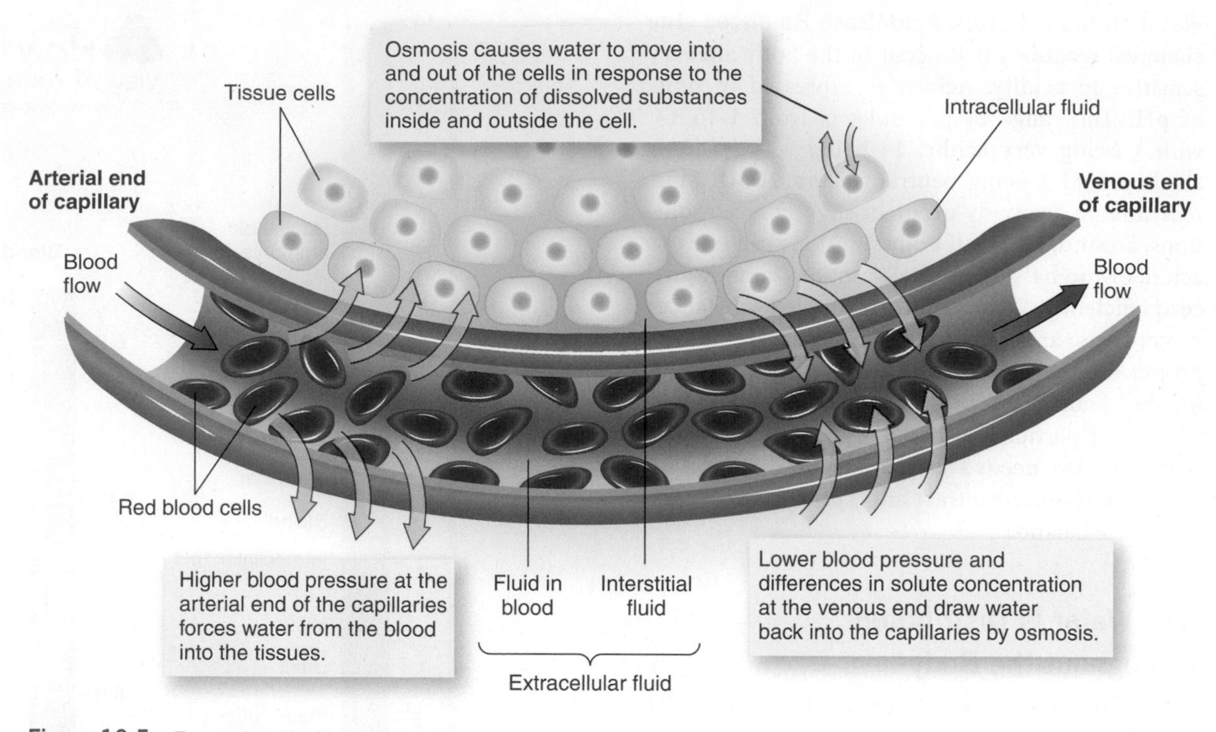

Figure 10.5 Forces that distribute body water
The amount of water in blood and tissues is determined by blood pressure and the force generated by osmosis.

Draw an illustration showing the concentration of water molecules and sugar molecules inside and outside the strawberries in each of these photos.

Figure 10.6 Osmosis
When sugar is sprinkled on fresh strawberries, the water inside the strawberries moves across the skin of the fruit to try to equalize the sugar concentration on each side. This pulls water out of the fruit and causes it to shrink.

sides (**Figure 10.6**). The body can regulate the amount of water in each compartment by adjusting the concentration of solutes and relying on osmosis to move water.

Water Balance

The amount of water in the body remains relatively constant over time. Since the body does not store water, to maintain this homeostasis the water taken into the body must equal the amount of water lost in urine, in feces, and through evaporation (**Figure 10.7**). When water losses are increased, as they are in hot weather and with exercise, intake must increase to keep body water at a healthy level.

Water Intake Most of the water in the body comes from the diet—not only as water we drink but also from the water in other liquids and in solid food (**Figure 10.8**). About 75 to 80% of total water intake comes from fluids, with food providing the remaining 20 to 25%.[1] Milk is about 90% water, apples are about 85% water, and roast beef is about 50% water. A small amount of water is also generated inside the body by metabolism, but this is not significant in meeting body water needs.

Water consumed in the diet is absorbed from the gastrointestinal tract by osmosis. This is possible because of the concentration gradient created by the absorption of nutrients from the lumen into the blood. As nutrients are absorbed, the solute concentration in the lumen decreases. Water therefore moves with the absorbed nutrients toward the area with the highest solute concentration. The rate of water absorption is affected by the volume of water and the concentration of nutrients consumed with it. Consuming a large volume of water increases the rate of water absorption; increasing the nutrients and other solutes it contains decreases the rate of absorption.

Water Losses Water is lost from the body in urine, in feces, through evaporation from the lungs and skin, and in sweat. A typical adult who is not sweating loses about 2.7 to 3.7 liters of water daily (see Figure 10.7).

Urinary Losses Typical urine output is 1 to 2 liters per day, but the amount of water lost in the urine varies with water intake (drinking more proportionately increases urine output) and the amount of waste that needs to be excreted in the urine. The waste products that must be excreted in urine include urea and other nitrogen-containing products from protein breakdown, ketones from fat breakdown, phosphates, sulfates, and other minerals. The amount of urea that must be excreted is increased when dietary protein increases or body protein breakdown increases. Ketone excretion is increased when body fat is broken down. In both cases, the need for water increases in order to produce more urine to excrete the extra wastes.

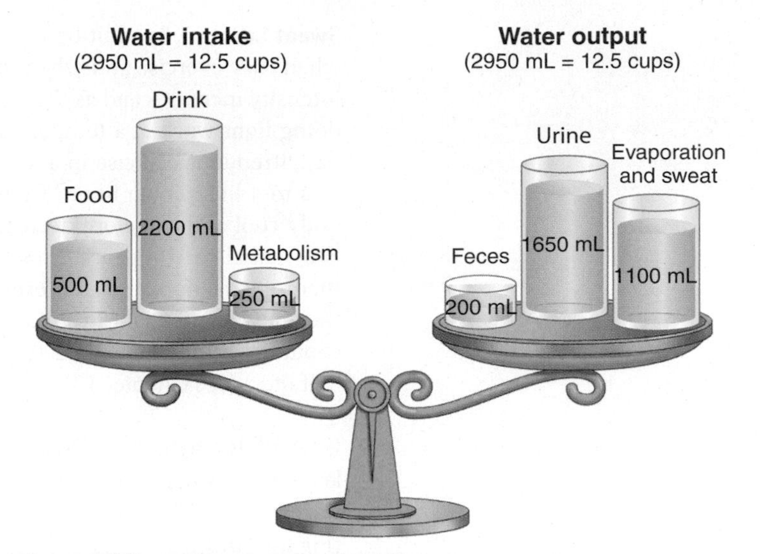

Figure 10.7 Water balance
To maintain water balance, intake from food and drink and water produced by metabolism must equal water output from evaporation, sweat, urine, and feces. This figure illustrates approximate amounts of water that enter and leave the body daily in a typical woman in a temperate enviroment who is in water balance.

Fecal Losses In a healthy person, the amount of water lost in the feces is usually small, only about 200 mL/day (less than a cup). This is remarkable because every day about 9 liters (38 cups) of fluid enter the gastrointestinal tract via food, water, and secretions. Under normal conditions, more than 95% is reabsorbed before the feces are eliminated. However, in cases of severe diarrhea, large amounts of water can be lost through the gastrointestinal tract and can compromise health.

Evaporative Losses People are continuously losing water from their skin and respiratory tract due to evaporation. This occurs without the individual being aware that it is occurring; such losses are therefore referred to as **insensible losses**. The amount of water lost through evaporation varies greatly, depending on activity, temperature, humidity, and body size. In a temperate climate, an inactive person loses about 1 L (4 cups) per day through insensible losses; the amount increases with increases in activity, environmental temperature, and body size, as well as when humidity is low, as in an airplane or in the desert.

insensible losses Fluid losses that are not perceived by the senses, such as evaporation of water through the skin and lungs.

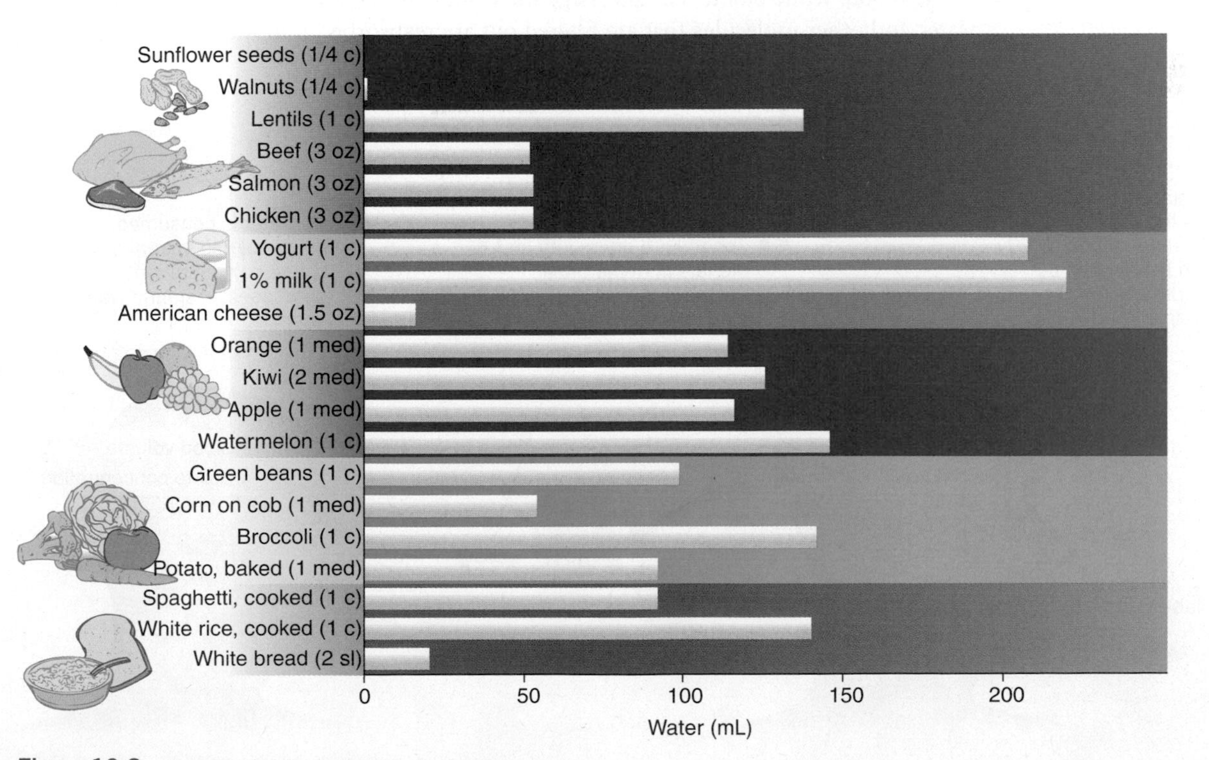

Figure 10.8 Sources of water in MyPlate food groups
Fruits and vegetables, as well as many choices from other food groups, have a high water content.

Sweat Losses In addition to losing water through evaporation, people lose water in sweat when they exercise and when the environment is hot. More sweat is produced as exercise intensity increases and as the environment becomes hotter and more humid. An individual doing light work at a temperature of about 84°F will lose about 2 to 3 liters of sweat per day. Strenuous exercise in a hot environment can cause water losses in sweat to be as high as 2 to 4 liters in an hour.[2] Clothing that permits the evaporation of sweat helps keep the body cool and therefore decreases the amount of sweat produced. Adequate water intake is essential to compensate for these losses. Athletes can estimate water loss by weighing themselves before and after exercise. Lost water can then be replaced by consuming at least the equivalent weight in fluids. For instance, an athlete who loses 2 lbs during a workout should consume an extra 2 to 4 pints or 1 to 2 liters of fluid (1 lb = 1 pint, or about one-half liter) (see Chapter 13).

How Water Intake Is Regulated When water losses increase, intake must increase to keep body water at a healthy level. The need to consume water is triggered by the sensation of thirst. Thirst is caused by dryness in the mouth as well as by signals from the brain (**Figure 10.9**).

Over the course of days and weeks, fluid intake, driven by thirst and that typically consumed as beverages during meals, is adequate to maintain water homeostasis. However, this is not always the case in the short term. People don't or can't always drink when they are thirsty, and thirst is quenched almost as soon as fluid is consumed and long before short-term water balance is restored. Also, the sensation of thirst often lags behind the need for water. For example, athletes exercising in hot weather lose water rapidly but do not experience intense thirst until they have lost so much body water that their physical performance is compromised.[3,4] A person with fever, vomiting, or diarrhea may also be losing water rapidly, and thirst mechanisms may not be adequate to replace the fluid. Thirst is a powerful urge, but to maintain adequate levels of water in the body it is also necessary to regulate water excretion.

How Water Losses Are Regulated To maintain water balance, the kidneys regulate the amount of water excreted in the urine. The kidneys function like a filter. As blood flows through them, water molecules and other small molecules move through the filter and out of the blood, while blood cells and large molecules are retained in the blood. Some of the water and other molecules that are filtered out are reabsorbed into the blood, and the rest are excreted in the urine.

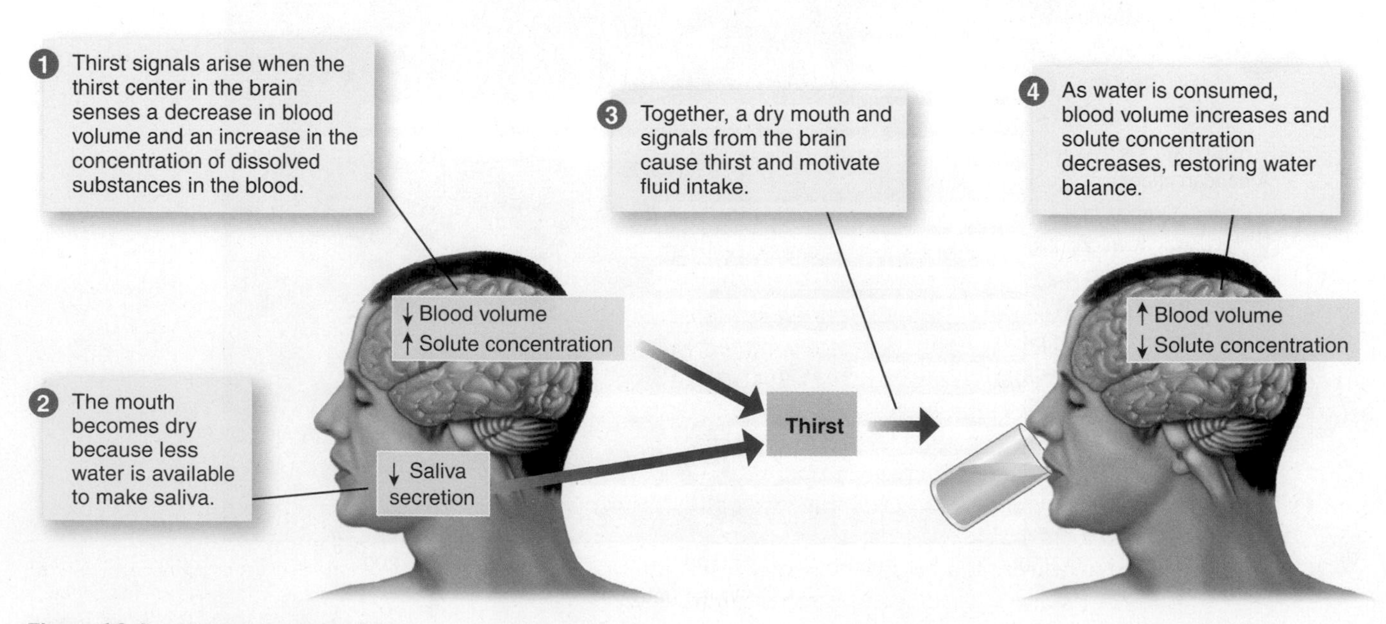

Figure 10.9 **Stimulating water intake**
The sensation of thirst helps motivate fluid intake in order to restore water balance.

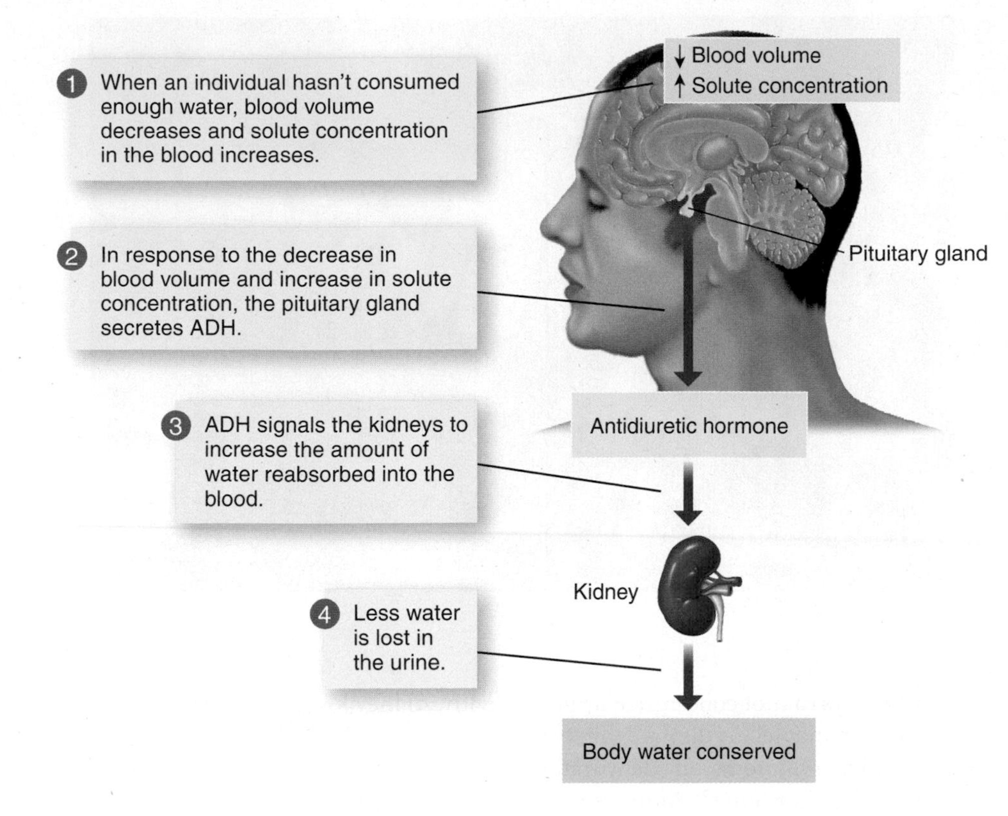

Figure 10.10 Regulating urinary water losses
The kidneys help regulate water balance by adjusting the amount of water lost in the urine in response to the release of antidiuretic hormone (ADH).

What happens to ADH levels and urine volume if someone consumes a lot of extra water?

The amount of water that is reabsorbed into the blood rather than excreted in the urine depends on conditions in the body. When the concentration of solutes in the blood is high, as in someone who has exercised strenuously and not consumed enough water, **antidiuretic hormone (ADH)**, which is secreted from the pituitary gland, signals the kidneys to reabsorb water to reduce the amount lost in the urine. This reabsorbed water is returned to the blood, preventing the solute concentration in the blood from increasing further (**Figure 10.10**). When the solute concentration in the blood is low, as in someone who has just guzzled several glasses of water, ADH levels decrease. Now the kidneys reabsorb less water and more is excreted in the urine, allowing blood solute concentration to increase to normal. The amount of sodium in the blood, blood volume, and blood pressure also play a role in regulating body water.

antidiuretic hormone (ADH) A hormone secreted by the pituitary gland that increases the amount of water reabsorbed by the kidney and therefore retained in the body.

dehydration A condition that results when not enough water is present to meet the body's needs.

Water Deficiency: Dehydration

Even minor changes in the amount and distribution of body water can be life-threatening. Without food, an average individual can live for about eight weeks, but a lack of water reduces survival to only a few days. A deficiency of water causes symptoms more rapidly than any other nutrient deficiency. Likewise, health can be restored in a matter of minutes or hours when fluid is replaced.

Dehydration occurs when water loss exceeds water intake. It causes a reduction in blood volume, which impairs the ability to deliver oxygen and nutrients to cells and remove waste products. Even mild dehydration—a body water loss of 1 to 2% of body weight—can impair physical and cognitive performance.[1] Early symptoms of dehydration include headache, fatigue, loss of appetite, dry eyes and mouth, and dark-colored urine (**Figure 10.11**). A loss of 5% body weight as water can cause nausea and difficulty concentrating. Confusion and disorientation can occur when water loss approaches 7% of body weight. A loss of about 10 to 20% can be fatal.

Athletes are at risk for dehydration because they may lose large amounts of water in sweat. Older adults are at risk because the thirst mechanism becomes less sensitive with age. Infants are at risk because their body surface area relative to their weight is much greater than that of adults, so they lose proportionately more water through evaporation;

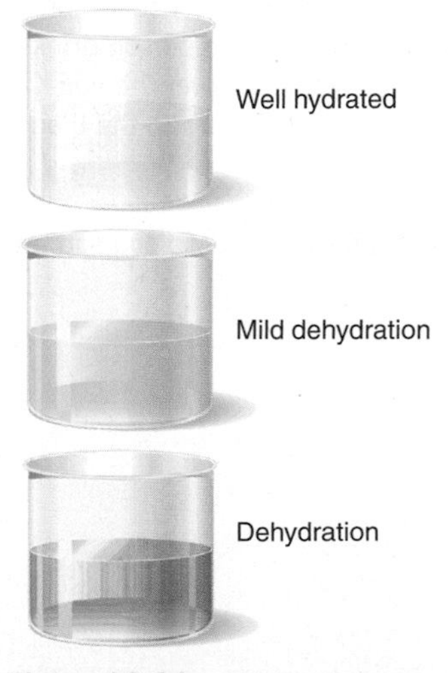

Figure 10.11 Are you at risk for dehydration?
Urine color is an indication of whether you are drinking enough. Pale yellow urine indicates you are well hydrated. The darker the urine, the greater the level of dehydration.

Joe Raedle/Staff/Getty Images, Inc.

Figure 10.12 **Rehydration saves lives**
Oral rehydration solutions maximize the rate of water absorption, but when water losses are too severe intravenous fluids may be necessary. This bypasses the gastrointestinal tract and puts fluid directly into the blood.

Why would a solution of 0.9% sodium chloride in water be preferred over plain water as an intravenous fluid?

also, their kidneys cannot concentrate urine efficiently, so they lose more water in urine. In addition, they cannot tell us they are thirsty.

The milder symptoms of dehydration disappear quickly after water or some other beverage is consumed. More severe dehydration can require medical attention. Dehydration due to diarrhea is a leading cause of death in children under 5 in the developing world. Replacing lost fluids and electrolytes in the right combinations can save lives. In some cases, oral rehydration therapy is sufficient. Drinking mixtures made by simply dissolving a large pinch of salt (½ tsp) and a scoop of sugar (2 T) in 1 L of clean water can restore the body's water balance by promoting the absorption of water and sodium. In severe cases of diarrhea, administration of intravenous fluids is needed to restore hydration (**Figure 10.12**).

Water Intoxication: Overhydration

water intoxication A condition that occurs when a person drinks enough water to lower the concentration of sodium in the blood significantly.

hyponatremia Low blood sodium concentration.

It is difficult to consume too much water under normal circumstances. However, overhydration or **water intoxication** may occur with illness and during exercise, particularly when exercise lasts several hours and fluids are replaced with plain water. An excess of water can affect the distribution of water among body compartments. When there is too much water relative to the amount of sodium in the blood, a condition called **hyponatremia**, water moves out of the blood vessels into the tissues by osmosis, causing them to swell. Swelling in the brain can cause disorientation, convulsions, coma, and death. Water intoxication and hyponatremia are far less common than dehydration but can be just as dangerous.

The early symptoms of water intoxication may be similar to dehydration: nausea, muscle cramps, disorientation, slurred speech, and confusion. It is important to determine if the problem is dehydration or water intoxication because if you assume the symptoms are from dehydration and drink plain water, the symptoms will worsen and can result in seizure, coma, or death. To help avoid water intoxication when exercising for more than an hour, beverages such as sports drinks, containing dilute solutions of sodium as well as sugar, should be used to replace water losses (see Chapter 13).

Recommended Water Intake

We need more water each day than any other nutrient. As discussed above, the actual amount of water needed each day can vary greatly depending on activity level, environmental temperature, and humidity. Despite this variability, AIs have been set for water:

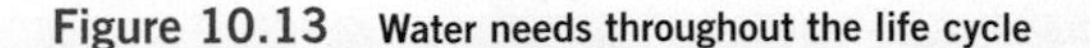
Figure 10.13 Water needs throughout the life cycle

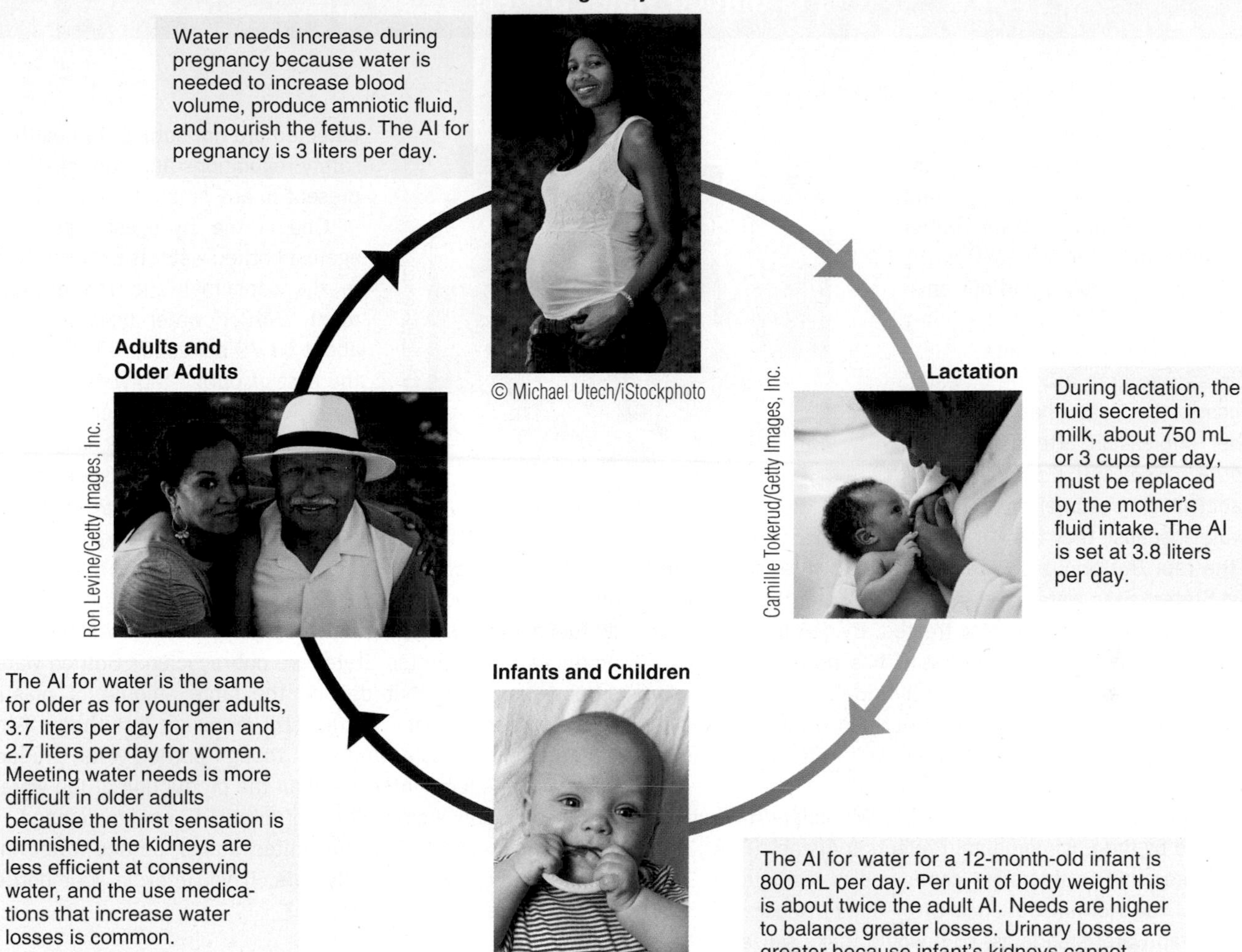

3.7 liters (3700 grams) per day for adult men and 2.7 liters (2700 grams) per day for adult women (**Figure 10.13**).[1]

Diet can also affect water needs. A high-protein diet increases water needs because the urea produced from protein breakdown is excreted in the urine. A low-calorie diet increases water needs because as body fat and protein are broken down to fuel the body, ketones and urea are produced and must be excreted in the urine. A high-sodium diet increases water losses because the excess sodium must be excreted in the urine. A high-fiber diet increases water needs because more water is held in the intestines and lost in the feces. Despite variations in our water needs, most people consume adequate fluids on a day-to-day basis to maintain body water at a normal level.[1]

In the United States, beverages provide about 80% of the body's requirement for water—about 3 L (13 cups) for men and 2.2 L (9 cups) for women.[1] The rest comes from the water consumed in food. Most beverages, whether water, milk, juice, or soda, help meet the overall need for water (see **Debate: Is Bottled Water Better?**).

Beverages containing caffeine, such as coffee, tea, and cola, increase water losses for a short period because caffeine is a diuretic. Over the course of a day, however, the increase in water loss is small, so the net amount of water that caffeinated beverages add to the body is similar to the amount contributed by noncaffeinated beverages. Alcohol is also a diuretic; its overall effect on water balance depends on the relative amounts of water and alcohol in the beverages being consumed.[1]

DEBATE

Is Bottled Water Better?

© Skip ODonnell/iStockphoto

Americans consume almost 28 gallons of bottled water per person per year.[1] We choose it because it is convenient and because we think it tastes better and is safer than tap water. But the cost to our pocketbooks and our environment is high. Should we be filling our bottles at the tap instead?

It is easy to grab a bottle of water, and most bottled water has no chlorine or other unpleasant aftertaste. Words like *pure, crisp*, and *fresh tasting* on the label lead consumers to buy bottled water because they think it is better than water that comes from the tap. But about 25% of the bottled water sold in the United States is tap water, and some is tap water that has been filtered, disinfected, or otherwise treated. By definition, bottled water can be any water, as long as it has no added ingredients (except antimicrobial agents or fluoride). Labels may help you distinguish: Distilled water and purified water are treated tap water; artesian water, spring water, well water, and mineral water come from underground sources.

Is bottled water safer than tap water? Municipal (tap) water is regulated by the Environmental Protection Agency (EPA), and bottled water sold in interstate commerce is regulated by the Food and Drug Administration (FDA). The FDA uses most of the EPA's tap water standards, so it would make sense that tap water and bottled water would be equally safe. However, tap water advocates argue that tap water may actually be safer. A certified outside laboratory tests municipal water supplies every year, while bottled water companies are permitted to do their own tests for purity.[2] Tap water must also be filtered and disinfected, but there are no federal filtration or disinfection requirements for bottled water.

Contamination is a safety concern for both bottled water and tap water. A study of contaminants in bottled water found that 10 popular brands contained a total of 38 chemical pollutants—everything from caffeine and Tylenol to bacteria, radioactive isotopes, and fertilizer residue.[3] Sounds scary, but bottled water advocates argue that public drinking water may also fall short of pollutant standards. Since 2004, testing by water utilities has found 315 pollutants in tap water. Some of these are substances that are regulated but were found at levels above guidelines, but more than half of the chemicals detected are not subject to health or safety regulations and can legally be present in any amount.[4]

One of the strongest arguments against bottled water is the cost, both to the consumer and the environment. Bottled water typically costs about $3.79 per gallon—1900 times the cost of public tap water. Bottled water drinkers may feel that the added cost is worth it because of the advantages in terms of taste and convenience. Opponents argue that even if you can afford it, the planet can't. Globally, bottled water generates 1.5 million tons of plastic waste per year and consumes oil, which is used to produce the bottles and the gasoline and jet fuel to transport it.[5] About three-quarters of the water bottles produced in the United States are not recycled.[2] Bottled water proponents argue that despite the large amount of plastic waste, discarded water bottles still represent less than 1% of total municipal waste. And even though bottled water production is more energy-intensive than the production of tap water, it makes up only a small share of total U.S. energy demand.[2]

So which is better? In the United States, bottled water and tap water are both generally safe. If you recycle your bottle, does it matter which you choose?

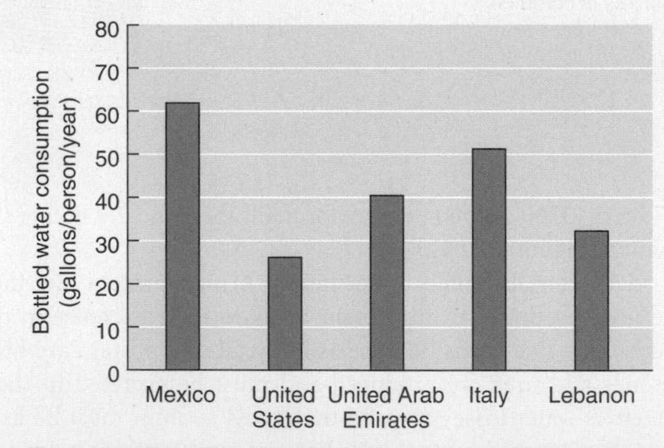

The graph compares the per capita bottled water consumption in the United States to several other countries.

THINK CRITICALLY: What factors might affect a country's per capita consumption of bottled water?

[1]International Bottled Water Association. Bottled water 2009.

[2]U.S. Government Accountability Office. Bottled water: FDA safety and consumer protections are often less stringent than comparable EPA protections for tap water, June 2009.

[3]Environmental Working Group. Bottled water quality investigations: 10 major brands, 38 pollutants, October 2008.

[4]Environmental Working Group. National Drinking Water Database.

[5]Sierra Club. Bottled Water Campaign.

10.2 Electrolytes: Salts of the Internal Sea

LEARNING OBJECTIVES

- Define the term electrolyte in nutrition.
- Contrast the dietary sources of sodium and potassium.
- Describe how electrolytes function in fluid balance, nerve conduction, and muscle contraction.
- Summarize the events that restore low blood pressure to normal.
- Give an example of a situation or condition that might lead to electrolyte imbalance.

The water in the body—the internal sea—contains a variety of mineral salts. The right amounts and combinations of these are necessary for the maintenance of life. The distribution of these minerals affects the distribution of water in different body compartments. The properties of these minerals, including electrical charge, affect nerve and muscle function. Mineral salts dissociate in water to form ions. Positively and negatively charged ions are called **electrolytes**. Although there are many electrolytes in the body, in nutrition the term is typically used to refer to the three principal electrolytes in body fluids: sodium, potassium, and chloride. Sodium and potassium carry a positive charge, and chloride carries a negative charge.

electrolytes Positively and negatively charged ions that conduct an electrical current in solution. Commonly refers to sodium, potassium, and chloride.

Sodium, Potassium, and Chloride in Our Food

Much of the sodium we consume is in the form of sodium chloride, what we call salt or table salt. Salt is 40% sodium and 60% chloride by weight, so 9 g of salt contains 3.6 g of sodium (9 g × 0.4 = 3.6 g) and 5.4 g of chloride (9 g × 0.6 = 5.4 g). The U.S. diet is high in sodium and low in potassium. The reason is that we eat a lot of processed foods, which are high in sodium and generally low in potassium, and too few fresh unprocessed foods such as fruits, vegetables, whole grains, and fresh meats, which are low in sodium and high in potassium (**Figure 10.14**).

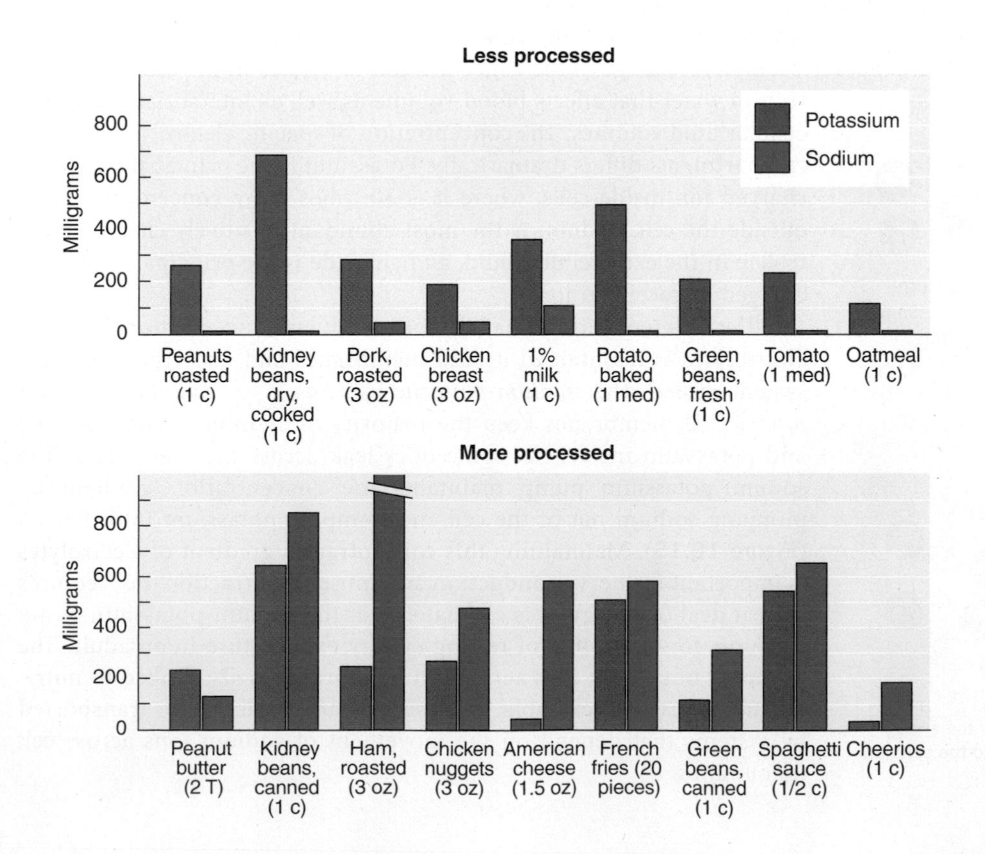

Figure 10.14 Sodium and potassium in processed and unprocessed foods
Less processed foods tend to be low in sodium and good sources of potassium, whereas more processed foods are generally higher in sodium and may also be lower in potassium.

About 77% of the sodium Americans eat is from that added to food during processing and manufacturing. Only about 12% comes from sodium found naturally in food, while 11% is from salt added in cooking and at the table.[1] Less than 1% of the sodium we consume is from tap water.[1] Softened water or mineral water is often higher in sodium than tap water and, if consumed in large quantities, can contribute significantly to daily sodium intake. Some processed foods, such as potato chips, lunchmeats, and canned soups, contain large amounts of added sodium. Others, such as bread, have smaller amounts but add a lot of sodium to the diet because they are consumed frequently. Sodium is added to processed foods for flavoring and as a preservative because it inhibits bacterial growth. In addition to sodium chloride, other sodium salts, such as sodium bicarbonate, sodium citrate, and sodium glutamate, are also used as preservatives.

The human diet has not always been high in salt. Prehistoric diets consisted of plant foods such as nuts, berries, roots, and greens. These foods are high in potassium and low in sodium. Fresh animal foods, such as meat and milk, are also low in sodium and high in potassium. Because of its value as a food preservative and flavor enhancer, salt was highly prized by ancient cultures in Asia, Africa, and Europe, where it was used in rituals as well as in the preservation of food. Roman soldiers were paid in *sal*, the Latin word for salt from which we get our word *salary*. Today, rather than a prized commodity, salt is a substance we attempt to limit in the diet. The reason for restricting salt is that diets high in salt have been implicated as a risk factor for **hypertension**.

hypertension Blood pressure that is consistently elevated to 140/90 mm Hg or greater.

How Sodium, Potassium, and Chloride Function in the Body

Almost all of the sodium, chloride, and potassium consumed in the diet is absorbed. In the body, these electrolytes help regulate fluid balance and are important for nerve conduction and muscle contraction.

How Electrolytes Regulate Fluid Balance The distribution of fluid among body compartments depends on the concentration of electrolytes and other solutes. Water moves by osmosis in response to solute concentration. All body fluids are in osmotic balance. So, for example, a change in the concentration of solutes in the blood causes a shift in water that affects blood volume as well as interstitial and intracellular fluid volumes. The concentration of specific electrolytes in body compartments differs dramatically. Potassium is the principal positively charged ion inside cells, where it is 30 times more concentrated than outside the cell. Sodium is the most abundant positively charged electrolyte in the extracellular fluid, and chloride is the principal negatively charged extracellular ion.

The different intracelluar and extracellular concentrations of these electrolytes is maintained by cell membranes and an active transport system called the *sodium-potassium-ATPase,* or *sodium-potassium pump*. Cell membranes keep the majority of sodium outside the cell and potassium inside, but some does leak across the membrane. The sodium-potassium pump maintains the concentration gradient by pumping sodium out of the cell and pumping potassium into the cell (**Figure 10.15**). Maintaining this concentration gradient of electrolytes is important for nerve conduction and muscle contraction, but requires a great deal of energy: It is estimated that the sodium-potassium pump accounts for 20 to 40% of resting energy expenditure in an adult. The pumping of sodium ions across cell membranes is also linked to nutrient transport. For example, glucose and amino acids are transported by systems that depend on the movement of sodium ions across cell membranes.

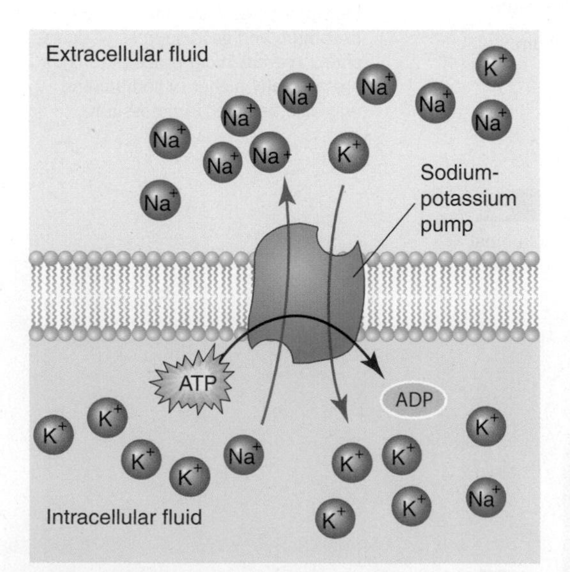

Figure 10.15 Sodium-potassium pump
Each cycle of sodium-potassium pump uses the energy in ATP to pump two potassium ions (K^+) into the cell and expel three sodium ions (Na^+).

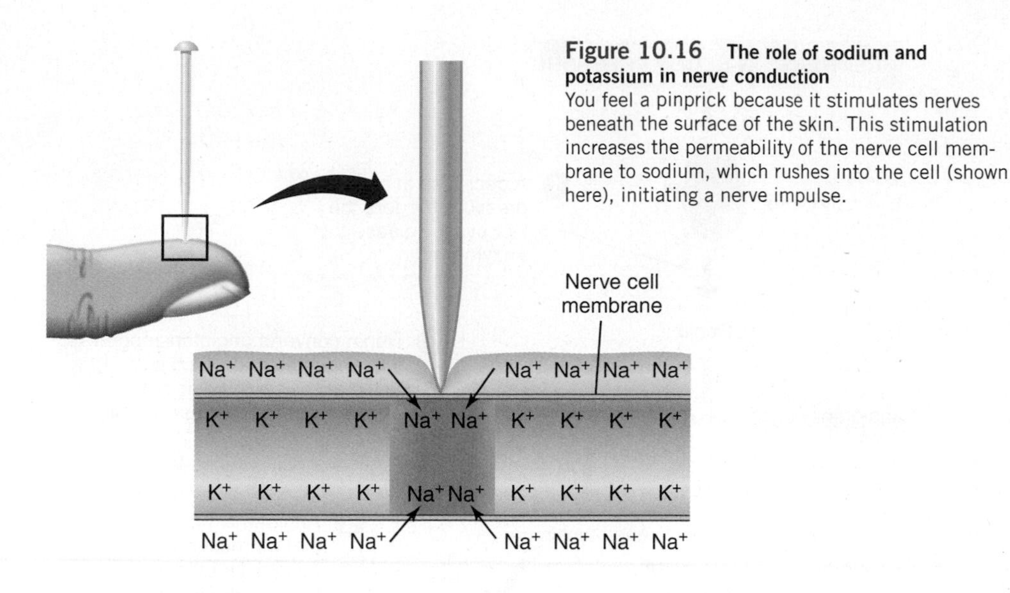

Figure 10.16 **The role of sodium and potassium in nerve conduction**
You feel a pinprick because it stimulates nerves beneath the surface of the skin. This stimulation increases the permeability of the nerve cell membrane to sodium, which rushes into the cell (shown here), initiating a nerve impulse.

Why Electrolytes Are Needed for Nerve Conduction and Muscle Contraction The concentration gradient of sodium and potassium across nerve cell membranes is important for the conduction of nerve impulses (see Table 10.1). An electrical charge, or membrane potential, exists across nerve cell membranes because the number of negative ions just inside the cell membrane is greater than the number outside. This occurs because the cell membrane allows more positively charged ions to leak out of the cell than to leak into the cell. Nerve impulses are created by a change in the electrical charge across cell membranes. Stimuli, such as touch or the presence of neurotransmitters, increase the cell membrane's permeability to sodium, allowing it to rush into the cell (**Figure 10.16**). This reverses, or depolarizes, the charge of the cell membrane at that location, and an electrical current is generated. The nerve impulse travels along the nerve cell as an electrical current. Once the nerve impulse passes, the original membrane potential is rapidly restored by another change in cell membrane permeability; then the original distribution of sodium and potassium ions across the cell membrane is restored by the sodium-potassium pump in the cell membrane. A similar mechanism causes the depolarization of the muscle cell membranes, leading to muscle contraction.

How Electrolyte Balance Is Regulated

In northern China, typical daily sodium chloride intake is greater than 13.9 g; in the Kalahari Desert, it is less than 1.7 g; among the Yanomani Indians of Brazil, consumption may be less than 0.06 g.[1] Despite these variations in intake, homeostatic mechanisms ensure that blood levels of sodium do not differ significantly among these groups.

Sodium and chloride homeostasis is regulated to some extent by the intake of both water and salt. When salt intake is high, thirst is stimulated to increase water intake. When salt intake is very low, a "salt appetite" causes the individual to crave salt. While this salt appetite promotes adequate salt intake, the desire for salt is also a learned preference. For example, the amounts of salt in the U.S. diet are far too high for salt appetite to be activated; the craving that triggers your desire to plunge into a bag of salty chips is due to this learned preference. If you cut back on your salt intake, you will find that your taste buds become more sensitive to the presence of salt, and foods taste saltier.

Thirst and salt appetite help ensure that appropriate proportions of salt and water are taken in, but the kidneys are the primary regulator of sodium, chloride, and potassium balance in the body. Excretion of these electrolytes in the urine is decreased when intake is low and increased when intake is high. Because water follows sodium by osmosis, the ability of the kidneys to conserve sodium provides a mechanism to conserve body water.

Sodium plays a pivotal role in regulating extracellular fluid volume. When the concentration of sodium in the blood increases, water follows by osmosis, causing an increase in blood volume. Changes in blood volume can change blood pressure. Changes in blood

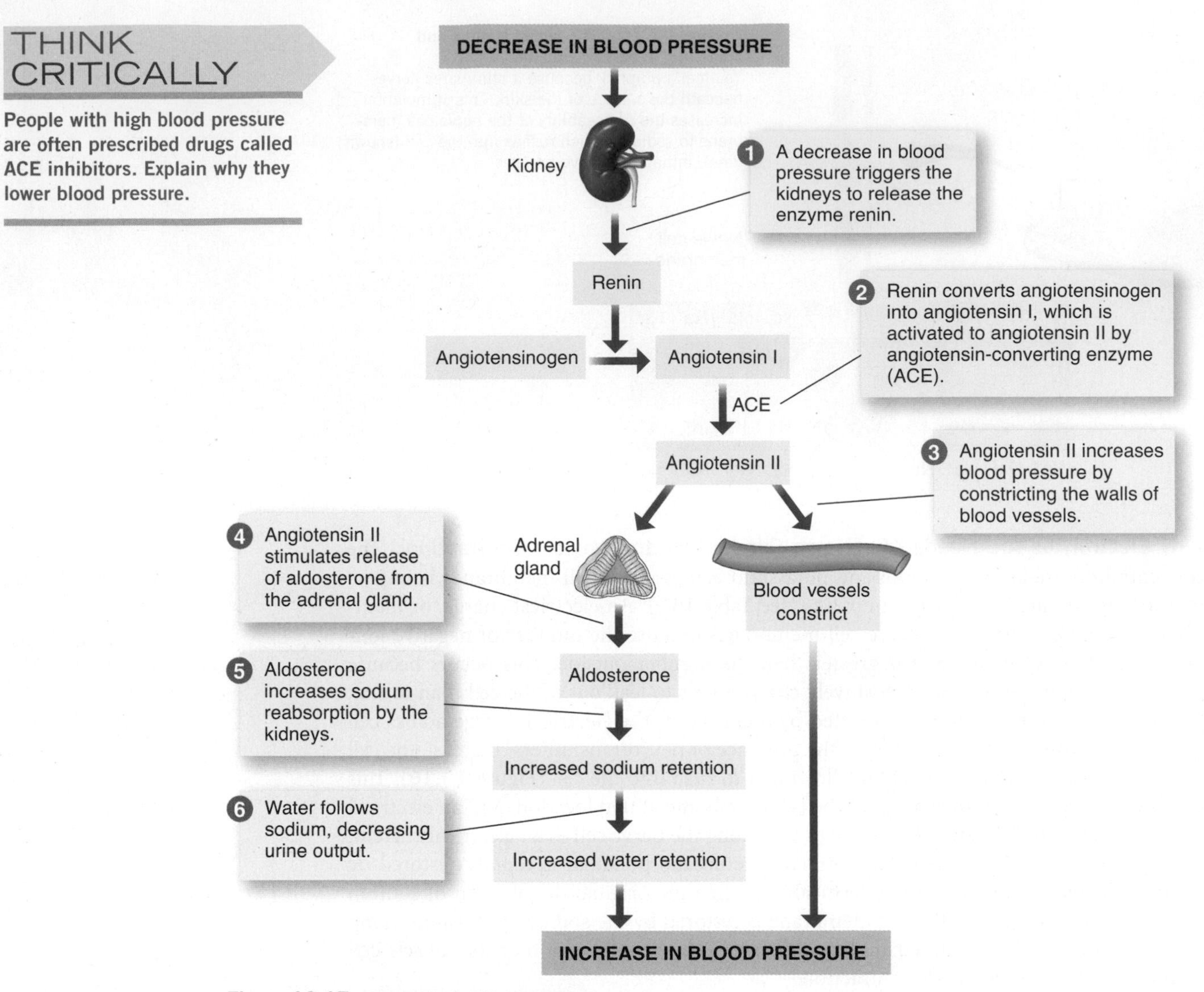

Figure 10.17 Regulation of blood pressure
A drop in blood pressure triggers events that cause blood vessels to constrict and the kidneys to retain water. An increase in blood pressure inhibits these events so that blood pressure does not continue to rise.

renin An enzyme produced by the kidneys that converts angiotensinogen to angiotensin I.

angiotensin II A compound that causes blood vessel walls to constrict and stimulates the release of the hormone aldosterone.

aldosterone A hormone that increases sodium reabsorption by the kidney and therefore enhances water retention.

pressure trigger the production and release of proteins and hormones that affect the amount of sodium, and hence water, retained by the kidneys. For example, when blood pressure decreases, the kidneys release the enzyme **renin**, beginning a series of events leading to the production of **angiotensin II** (**Figure 10.17**). Angiotensin II increases blood pressure both by causing the blood vessel walls to constrict and by stimulating the release of the hormone **aldosterone**, which acts on the kidneys to increase sodium (and chloride) reabsorption. Water follows the reabsorbed sodium, helping to maintain blood volume and, consequently, blood pressure. As blood pressure increases, it inhibits the release of renin and aldosterone so that blood pressure does not continue to rise.

The amount of potassium in the body is also tightly regulated. If blood levels begin to rise, mechanisms are activated to stimulate the cellular uptake of potassium. This short-term regulation prevents the amount of potassium in the extracellular fluid from getting lethally high. The long-term regulation of potassium balance depends on aldosterone release, which causes the kidneys to excrete potassium and retain sodium. Aldosterone release is stimulated by high blood potassium, low blood sodium, or angiotensin II.

Electrolyte Deficiency

The electrolytes are plentiful in the diet, and the kidneys of a healthy individual are efficient at regulating amounts in the body. However, illness and extreme conditions such as strenuous exercise in a hot environment can increase electrolyte losses and affect overall health.

Sodium, chloride, and potassium depletion can occur when losses are increased due to heavy and persistent sweating, chronic diarrhea or vomiting, or kidney disorders that lead to excessive urinary losses. It is obvious that these situations increase water losses, but the water that is lost carries electrolytes with it. If the electrolytes are not replaced along with the water, it can lead to electrolyte imbalances. For example, if athletes lose a lot of sodium in sweat but replace only the water, they may suffer from hyponatremia (see Chapter 13). Likewise, someone who has lost fluid and electrolytes from diarrhea should replace both. These can be replaced by consuming electrolyte-containing liquids, such as chicken broth or commercial oral rehydration fluids. These are often recommended for young children who have had an extended bout of diarrhea or vomiting.

Medications can also interfere with electrolyte balance. For example, thiazide diuretics, which are used to treat hypertension, cause potassium loss. Generally, potassium supplements are prescribed along with or incorporated into medications that cause potassium loss. Deficiencies of any of the electrolytes can lead to electrolyte imbalance, which can cause disturbances in acid–base balance, poor appetite, muscle cramps, confusion, apathy, constipation, and, eventually, an irregular heartbeat. For example, the sudden death that can occur in fasting, anorexia nervosa, or starvation may be due to heart failure caused by potassium deficiency.

Electrolyte Toxicity

Electrolyte toxicity is rare when water needs are met and kidney function is normal. It is not possible for healthy people to consume too much potassium from foods. If, however, potassium supplements are consumed in excess or kidney function is compromised, blood levels of potassium can increase and potentially cause death due to an irregular heartbeat. A high oral dose of potassium generally causes vomiting, but if too much potassium enters the blood, it can cause the heart to stop.

It is difficult to consume more sodium than the body can handle because we usually drink more water when we consume more sodium. Though rare, elevation of blood sodium can result from massive ingestion of salt, such as may occur from drinking seawater or consuming salt tablets. The most common cause of high blood sodium is dehydration, and the symptoms of high blood sodium are similar to those of dehydration.

10.3 Hypertension

LEARNING OBJECTIVES

- Define "hypertension," and list its symptoms and consequences.
- Discuss the effect of dietary salt intake on blood pressure.
- Describe the DASH diet and how it affects blood pressure.

Hypertension, or high blood pressure, has been called "the silent killer" because it has no outward symptoms but can lead to atherosclerosis, heart attack, stroke, kidney disease, and early death. Hypertension is a serious public health concern in the United States: about one in three U.S. adults has hypertension; of these, 20% don't know they have it.[5] Among hypertensive adults, 71% are using medications to control their blood pressure but only 48% have their blood pressure under control.[5]

A certain level of blood pressure is necessary to ensure that blood is delivered to all the body tissues. An optimal blood pressure is less than 120/80 millimeters of mercury (mm Hg). The higher number is systolic pressure, the maximum pressure in the artery, occurring when the heart is contracting. The lower number is diastolic pressure, the minimum pressure in the artery, occurring when the heart is resting between beats. Blood pressures from 120/80 to 139/89 mm Hg are referred to as *prehypertension*. Prehypertension affects 36% of Americans and indicates an increased risk for developing hypertension.[6]

Andersen Ross/Getty Images, Inc.

Figure 10.18 **Assessing blood pressure**
Blood pressure should be monitored regularly. It is measured using a blood pressure meter called a sphygmomanometer. It consists of an inflatable cuff that restricts blood flow and a pressure meter. A stethoscope is used to listen to the sounds of blood flow that indicate systolic and diastolic pressures.

Blood pressure that is consistently 140/90 mm Hg or greater indicates hypertension (**Figure 10.18**) (see Appendix C).[7]

What Causes Hypertension?

Hypertension occurs when there is an increase in the pressure of the blood against the arterial wall. In most cases we don't know the cause. Hypertension with no obvious external cause is referred to as *essential hypertension*. It is a complex disorder, most likely resulting from disturbances in one or more of the mechanisms that control body fluid and electrolyte balance. High blood pressure that occurs as a result of other disorders is referred to as *secondary hypertension*. For example, if atherosclerosis causes a reduction in blood flow to the kidneys, the kidneys respond by releasing renin. This triggers events that lead to an increase in blood volume and a constriction of blood vessels. This raises blood pressure to the kidneys but also raises blood pressure throughout the body.

Factors That Affect the Risk of Hypertension

The risk of developing high blood pressure is affected by genetics, age, existing disease conditions, and lifestyle factors. A family history of high blood pressure increases your risk of developing this disorder. The genetic basis of hypertension is also illustrated by the fact that it is more common in people of certain ethnic and racial backgrounds. For example, 41.4% of African Americans have high blood pressure; compared to Caucasian Americans, in African Americans hypertension develops earlier in life and average blood pressure is higher.[5] The increased incidence among African Americans is reflected in their 1.8 times greater rate of death from stroke, 1.5 times greater rate of death from heart disease, and 4.2 times greater rate of hypertension-related kidney failure compared to whites.[5] The risk of hypertension increases in adults as they age. One reason is that as people age the arteries lose their elasticity. Hypertension is generally uncommon in children and adolescents, but the frequency has been increasing, in part due to the increased prevalence of obesity.[5] Obesity, particularly abdominal obesity, increases the risk of hypertension. Excess adipose tissue adds miles of capillaries through which blood must be pumped. Weight loss can prevent or delay the onset of hypertension in obese individuals. Diabetes also increases the risk of high blood pressure. Kidney damage is generally the cause of high blood pressure in type 1 diabetes. In type 2 diabetes, the higher incidence of hypertension may also be due to the effect of high insulin levels on sodium retention in the kidneys as well as the presence of excess body fat. A lack of physical activity, heavy alcohol consumption, smoking, stress, and a number of dietary factors can also increase blood pressure.[8]

How Diet Affects Blood Pressure

The correlation between sodium and blood pressure is well known. On average, as sodium intake goes up, so does blood pressure, but sodium is not the only nutrient that affects blood pressure. Diets high in potassium, calcium, and magnesium are associated with a lower average blood pressure. Other components of the diet, such as the amount of fiber and the type and amount of fat, may also affect the risk of developing hypertension.

Sodium Intake and Blood Pressure The relationship between sodium intake and blood pressure was first identified by examining the incidence of hypertension in populations with different average dietary salt intakes. It was found that in populations consuming less than 4.5 g of salt per day (1800 mg sodium), average blood pressure was low and hypertension was rare or absent. In populations consuming 5.8 g of salt (2320 mg sodium) or more per day, blood pressure increased with sodium intake.[9] More recent intervention trials have examined the effect of different levels of sodium intake on blood pressure. It was found that the lower the amount of sodium in the diet the lower the blood pressure.[10] When compared to an intake of 3300 mg of sodium (a level slightly less than the average amount consumed by Americans), an intake of 2400 mg of sodium reduced blood pressure in those with and without hypertension, and even more significant reductions were seen when sodium intake was reduced to 1500 mg of sodium per day. These studies and others are the basis for the current DRIs and Dietary Guidelines recommendations, which are discussed below. Despite the general effect of sodium intake on blood pressure, not everyone who consumes more than the recommended daily amount of sodium will develop hypertension. The degree of salt sensitivity an individual has determines how their blood pressure will respond to changes in dietary sodium intake. Individuals with hypertension, diabetes, or chronic kidney disease, as well as older individuals and African Americans, tend to be more sensitive to sodium intake.[1]

Dietary Patterns and Blood Pressure Dietary patterns with high intakes of fiber and the minerals potassium, magnesium, and calcium are associated with lower blood pressure.[11] For example, populations and individuals consuming vegetarian diets, which are high in these nutrients, generally have lower blood pressure than nonvegetarians.[12] Population surveys, like NHANES (see Chapter 2 Science Applied, NHANES: A Snapshot of America's Nutritional Health), have shown that a dietary pattern low in calcium, potassium, and magnesium is associated with hypertension in American adults.[13] Despite these data, studies that have explored the impact of individual nutrients on blood pressure are often inconclusive. This may be because the impact of each individual nutrient is small and other dietary components are also important in blood pressure regulation.

The first intervention trial to look at the effect of a dietary pattern high in potassium, calcium, and magnesium on blood pressure was the DASH trial, which stands for Dietary Approaches to Stop Hypertension[14] (see **Science Applied: A DASH to Heart Health**). In this trial, the amount of dietary sodium was kept constant and different dietary patterns were compared in terms of their effect on blood pressure. The greatest reduction in blood pressure was found with a dietary pattern that is high in fruits and vegetables and includes low-fat dairy products and lean meat, fish, and poultry. This pattern provides plenty of fiber, potassium, magnesium, and calcium; is low in total fat, saturated fat, and cholesterol; and is lower in sodium than the typical American diet (see Appendix D). Consuming a diet that follows the DASH pattern lowers blood pressure in individuals with elevated blood pressure even when sodium levels are not severely restricted. Reductions in blood pressure are greater when sodium intake is lower.[15,16] The results of the DASH trial demonstrate that changing the dietary pattern can lower blood pressure. The DASH dietary pattern may also reduce cancer risk, prevent osteoporosis, and protect against heart disease. The DASH dietary pattern or DASH Eating Plan is included in the recommendations for the 2010 Dietary Guidelines for Americans.[6]

Diet is not the only lifestyle factor that impacts blood pressure. Maintaining body weight in a healthy range, staying active, limiting alcohol consumption, and other lifestyle factors all help keep blood pressure in the normal range (**Table 10.2**).

TABLE 10.2 Lifestyle Choices to Keep Blood Pressure in the Normal Range

- Eat plenty of fruits and vegetables—they are naturally low in salt and calories and rich in potassium.
- Choose and prepare foods with less salt.
- Aim for a healthy weight—blood pressure increases with increases in body weight and decreases when excess weight is reduced.
- Increase physical activity—it helps lower blood pressure, reduce the risk of other chronic diseases, and manage weight.
- If you drink alcoholic beverages, do so in moderation. Excessive alcohol consumption has been associated with high blood pressure.
- Quit smoking.
- Reduce stress—over time, stress can raise blood pressure and contribute to the development of hypertension.

SCIENCE APPLIED

A DASH to Heart Health

For over 30 years, the National Heart, Lung, and Blood Institute (NHLBI) of the National Institutes of Health (NIH) made recommendations for the treatment of high blood pressure through modifications of individual dietary components such as sodium and other minerals as well as fat, cholesterol, fiber, and protein. But the data supporting some of these nutrient-intake recommendations were often conflicting and equivocal. This led the NHLBI to pursue a clinical trial that would evaluate the effects of dietary patterns rather than individual dietary components on blood pressure.[1] The result was the DASH (Dietary Approaches to Stop Hypertension) trial, which provided conclusive evidence that diet does affect blood pressure.[2]

THE SCIENCE

Designing the DASH trial presented several challenges. The study required large numbers of subjects to consume a specific diet for a fairly long period. To accrue enough subjects, the study needed to be run simultaneously from several different study centers. Subjects' diets had to be standardized across these centers.[1] Controlling dietary intake also presented challenges because the subjects had to consume one meal a day, five days a week at the research study centers; the remaining meals were prepared by the research centers and sent home with the subjects to be consumed at home. To control for extra intake, subjects were also required to keep a daily diary of any nonstudy items consumed.

The DASH study population included 459 adults with slightly elevated blood pressure who were not taking blood pressure medications. All study subjects consumed the same control diet for the first three weeks of the study, and their blood pressure was monitored. Subjects were then randomly assigned to one of three dietary groups for eight weeks. During this intervention period, blood pressure was measured by individuals who did not know to which dietary groups the participants had been assigned.

The study diets included a "control" diet, a "fruits and vegetables" diet, and a "combination" diet. Each diet provided about 3000 mg of sodium—slightly lower than the typical U.S. intake—and enough calories to maintain body weight. The "control" diet was a typical American dietary pattern—low in fiber and high in fat and saturated fat. It matched the average U.S. nutrient intake except for potassium, magnesium, and calcium, which were set below average. The "fruits and vegetables" diet increased servings of fruits and vegetables and decreased sweets but was otherwise similar to the control diet. This diet was higher in fiber than the control, and the potassium and magnesium were above the typical American intake. The "combination" diet increased fruits, vegetables, and low-fat dairy products and reduced red meat, sweets, and sugar-sweetened drinks; it was higher in potassium, calcium, magnesium, and fiber than the control diet and was lower in total fat, saturated fat, and cholesterol.

After only two weeks, participants consuming the "fruits and vegetables" diet had a blood pressure reduction of 2.8 mm Hg systolic and 1.1 mm Hg diastolic. Those consuming the "combination" diet, which is now known as the DASH diet or DASH Eating Plan, had a reduction in blood pressure of 5.5 mm Hg systolic and 3 mm Hg diastolic compared with the control group (see graph).[2] If the U.S. population adopted the DASH diet, it is estimated that the incidence of coronary heart disease would be reduced by 15% and stroke by 27%.

10.4 Healthy Electrolyte Intakes

LEARNING OBJECTIVES

- Discuss the rationale for lowering sodium intake.
- Plan a DASH diet that meets your calorie needs.

Our current dietary pattern that is high in processed foods and low in fruits and vegetables has led to high sodium and low potassium intakes. To meet recommendations for a healthy diet, most people in the United States need to reduce their sodium intake and increase their potassium intake.

(Elena Schweitzer/Shutterstock) Victoriano Izquierdo/Getty Images, Inc. (DAJ/Getty Images, Inc.)

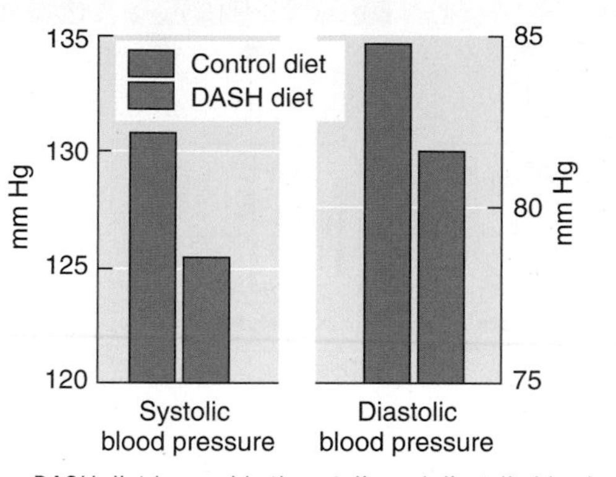

Consuming a DASH diet lowered both systolic and diastolic blood pressure compared to the control diet (typical American diet) even though both dietary groups consumed 3000 mg of sodium per day. Reducing dietary sodium intake causes further reduction in blood pressure.

A second DASH study, called DASH-Sodium, compared the effect of different levels of sodium intake on blood pressure in subjects consuming either the DASH Eating Plan or a control diet (typical American diet).[3] Lowering the sodium lowered blood pressure in subjects consuming both the DASH and control diets. Those consuming the DASH diet and the lowest sodium level, 1500 mg/day, had the greatest reduction in blood pressure—an effect equal to or greater than what would be expected from treatment with a single antihypertensive medication.

THE APPLICATION

When many dietary factors are modified simultaneously, as in the DASH trial, it is impossible to determine which ones cause the effects seen. Therefore, the DASH trial taught scientists less about the physiology of hypertension than do studies that look at individual factors, but it provided more information about what people should eat to prevent or reduce hypertension. It shifted the focus of diet and disease prevention from individual nutrients or foods to dietary patterns. There is no single dietary prescription for health. Healthy dietary patterns can include a multitude of food choices and combinations to satisfy differences in personal tastes and preferences based on culture, ethnicity, and tradition. Healthy eating patterns help reduce the risk of not only hypertension but also a host of other nutrition-related chronic diseases. The Dietary Guidelines for Americans now recommends the DASH Eating Plan as a healthy meal-planning tool for the general public.[4]

THINK CRITICALLY: If you had to recommend one aspect of diet to lower blood pressure, what would it be?

[1]Vogt, T. M., Appel, L. J., Obarzanek, E., et al. Dietary Approaches to Stop Hypertension: Rationale, design and methods. *J Am Diet Assoc* 99:12s–18s, 1999.

[2]Appel, L. J., Moore, T. J., Obarzanek, E., et al. A clinical trial of the effects of dietary patterns on blood pressure. DASH Collaborative Research Group. *N Engl J Med* 336: 1117–1124, 1997.

[3]Sacks, F. M., Svetkey, L. P., Vollmer, W. M., et al. Effects on blood pressure of reduced dietary sodium and the Dietary Approaches to Stop Hypertension (DASH) diet. DASH-Sodium Collaborative Research Group. *N Engl J Med* 344:3–10, 2001.

[4]U.S. Department of Agriculture and U.S. Department of Health and Human Services. *Dietary Guidelines for Americans*, 2010, 7th ed. Washington, DC: U.S. Government Printing Office, December 2010.

Recommended Sodium, Chloride, and Potassium Intakes

The typical daily intake of sodium in the United States is about 3400 mg. The AI for sodium for adults ages 19 to 50 years is 1500 mg/day (see Table 10.1).[1] This amount ensures an adequate intake of other nutrients and accounts for sodium losses in sweat. The UL for sodium is 2300 mg, based on the increase in blood pressure seen with higher sodium intakes. This is the same as the 2010 Dietary Guidelines recommendation of a daily sodium intake of less than 2300 mg.[6] For people who are 51 or older and those of any age who are African American or have hypertension, diabetes, or kidney disease, sodium intake should be reduced to 1500 mg/day, which is equivalent to the AI for sodium for adults ages 19 to 50.

The AI for chloride is 2300 mg/day; the UL for chloride is 3600 mg/day.[1] The AI for sodium and chloride is equivalent to 3.8 g of salt per day, which is less than

Figure 10.19 Dietary patterns to lower blood pressure

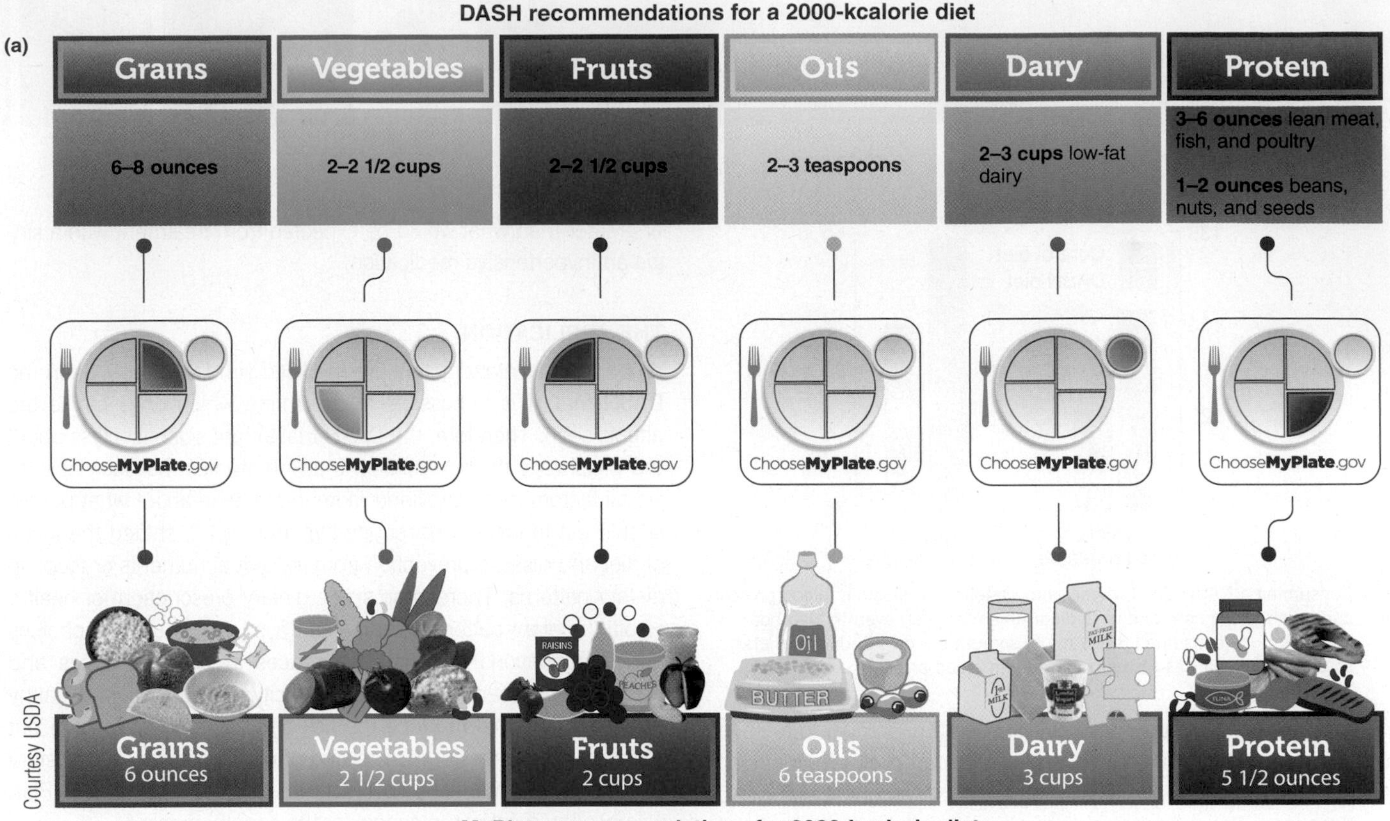

(a) The recommendation of the DASH eating plan (upper portion of the figure) are similar to those of MyPlate (lower portion). Both recommend eating plenty of fruits and vegetables; choosing whole grains; having beans, nuts, seeds, and fish more often; and choosing lean meats and low-fat dairy products. The table provides the number of servings from each of the DASH eating plan food groups for four different energy levels.

half of the typical daily intake of 8.5 g of salt, or about 1.5 teaspoons per day.[6] The Daily Value for sodium used on food labels is no more than 2400 mg of sodium per day (6 g salt).

The AI for potassium is set at 4700 mg/day, a level that will lower blood pressure and reduce the adverse effects of sodium on blood pressure. Few Americans currently consume amounts of potassium that are equal to or greater than this amount. The Dietary Guidelines also recommend an intake of 4700 mg of potassium per day; the Daily Value is at least 3500 mg/day for adults. No UL has been set for potassium because potassium intake from foods is not a risk in healthy people with normal kidney function.

There is no evidence that sodium and chloride requirements differ during pregnancy, so pregnant women are advised to follow the recommendations for the general population[1] (see Chapter 14). At one time, a dietary salt restriction was common during pregnancy to prevent a spectrum of conditions involving high blood pressure known as hypertensive disorders of pregnancy. The cause of these disorders is not known, but salt restriction is no longer recommended. Potassium recommendations are not increased during pregnancy, but during lactation the AI is higher to account for the potassium lost in milk. In infants, sodium, chloride, and potassium needs are estimated from the amount consumed in human milk, which contains more chloride than sodium. This same chloride-to-sodium ratio has been recommended for infant formulas.

Choosing a Dietary Pattern to Lower Blood Pressure

Choosing a dietary pattern that will increase potassium and reduce sodium to meet recommendations will help to control blood pressure. MyPlate and the DASH Eating Plan

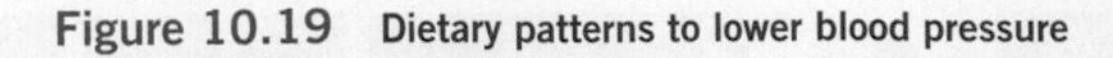

(b)

DASH Eating Plan at Different Energy Levels

		Daily servings per kilocalorie level			
Food group	Serving sizes	1600	2000	2600	3100
Grains[a]	1 slice bread, 1 oz dry cereal, ½ cup cooked rice, pasta, or cereal	6	6–8	10–11	12–13
Vegetables	1 cup raw leafy vegetables, ½ cup raw or cooked vegetables, or vegetable juice	3–4	4–5	5–6	6
Fruits	1 medium fruit, ¼ cup dried fruit, ½ cup fresh, frozen, or canned fruit, or fruit juice	4	4–5	5–6	6
Low-fat dairy	8 oz milk, 1 cup yogurt, 1½ oz cheese	2–3	2–3	3	3–4
Lean meats, fish, poultry	1 oz cooked meat, poultry, or fish	3–4 or less	6 or less	6 or less	6–9
Beans, nuts, and seeds	⅓ cup or 1½ oz nuts, 2 T or ½ oz seeds, ½ cup cooked dry beans or peas	3–4/ week	4–5/ week	1/day	1/day
Fat and oils[b]	1 tsp soft margarine, 1 T low-fat mayonnaise, 1 T light salad dressing, 1 tsp vegetable oil	2	2–3	3	4
Sweets and added sugar	1 T sugar, 1 T jelly or jam, ½ cup sorbet, 8 oz lemonade	≤3/ week	≤5/ week	≤2/ week	≤2/ week

[a]Whole grains are recommended for most servings to meet fiber recommendations.

[b]Fat content changes the number of servings for fats and oils. 1 T regular salad dressing equals one serving; 1 T low-fat dressing equals ½ serving; 1 T of fat-free dressing equals 0 servings.

(c)

Andy Washnik

(c) A diet that is high in fruits and vegetables, which are good sources of potassium, magnesium, and fiber, reduces blood pressure compared to a similar diet containing fewer fruits and vegetables. The amounts in the measuring cups shown here, about 2 cups of fruit and 2½ cups of vegetables, represent the amount recommended for a 2000-kcalorie diet.

recommend dietary patterns that are high in fruits and vegetables and whole grains, and include lean protein sources and low-fat dairy products (**Figure 10.19**). These dietary patterns are high in potassium, calcium, magnesium, and fiber and low in sodium (see **Critical Thinking: DASHing a Diet**).

The typical American diet contains only about 2000 to 3000 mg of potassium per day—well below the recommendation of 4700 mg/day. This is because most Americans do not consume the recommended amounts of fruits and vegetables—the best sources of potassium per calorie. Meat, milk, and cereal products also provide potassium. Those who consume a diet that follows the MyPlate or the DASH Diet Plan recommendations for fruit and vegetable intake will easily meet the potassium recommendation (see Figure 10.19). When diets are high in fruits and vegetables, potassium intakes of 8000 to 11,000 mg/day are not uncommon.[1]

Even a diet high in fruits and vegetables can provide more than the recommended amount of sodium. Americans can lower the amount of sodium in their diet by limiting their intake of processed foods and cutting down on the amount of salt added in cooking and at the table. Food labels can be helpful in selecting lower-sodium foods (see **Off the Label: Pass on the Salt**).

CRITICAL THINKING

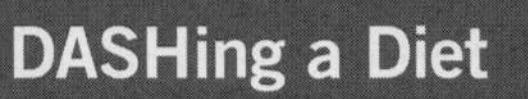

DASHing a Diet

Rashamel has a family history of hypertension. He exercises for about 30 minutes on most days of the week, doesn't smoke, and watches his weight and salt intake. Despite these efforts, at his recent physical his blood pressure was 138/87 mm Hg, which is in the prehypertension category. Rashamel's doctor suggests he see a dietitian to help him manage his blood pressure with a DASH Eating Plan. His current diet is shown here.

CURRENT DIET	SERVING	ENERGY (kcal)	SODIUM (mg)
Breakfast			
Orange juice	¾ cup	84	2
1% low-fat milk	1 cup	102	107
Wheaties w/1 tsp sugar	1 cup	149	253
Whole-wheat bread w/jelly	1 slice	147	161
Margarine	1 tsp	33	44
Lunch			
Tuna salad	¾ cup	288	618
Whole-wheat bread	2 slices	256	259
Chips	1 oz	150	136
Cola	1 can	136	15
Dinner			
Baked chicken	3 oz	140	63
Teriyaki sauce	1 T	30	1220
Fried rice	1 cup	333	822
Salad	1 cup	22	36
Light salad dressing	1 T	30	160
Dinner roll	1	78	138
Margarine	2 tsp	66	87
Cantaloupe	½ cup	27	13
Iced tea (sweetened)	12 oz	90	4
Snacks			
Cookies	2 large	156	110
Dried apricots	5	84	4
Milky Way candy bar	1	260	95
Cola	1 can	136	15

CRITICAL THINKING QUESTIONS

- Compare Rashamel's diet to the recommendations of a 2600-kcal DASH Eating Plan (see Figure 10.19b).
- Suggest foods that Rashamel could add to his diet to meet the recommendations of the DASH Eating Plan.
- How many kcalories does this modification add to Rashamel's diet? Assuming he is maintaining his weight on his current diet, what foods would he need to eliminate to prevent weight gain?
- Compare Rashamel's sodium intake to the AI. Which four foods provide the most sodium?

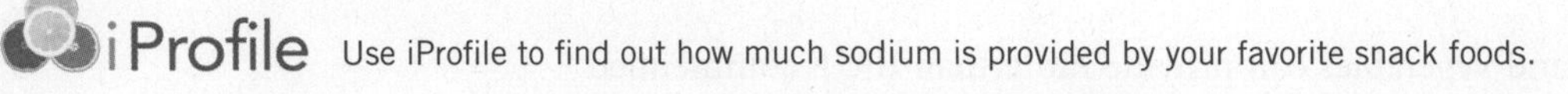

iProfile Use iProfile to find out how much sodium is provided by your favorite snack foods.

© iStockphoto

Pass on the Salt

Most of the sodium in the American diet comes from processed foods. Sodium chloride is the most common form in which sodium is added to foods; other sodium-containing ingredients include sodium hydroxide, sodium salts such as baking soda and baking powder, and monosodium glutamate (MSG).

Food labels list the sodium-containing ingredients in the ingredient list, and the Nutrition Facts panel gives the total amount of sodium in milligrams. To help you assess how the amount of sodium fits into the recommendations for a healthy diet, the sodium content of a serving is given as a percentage of the Daily Value, which is 2400 mg. For example, the Nutrition Facts label here indicates that a serving of this spaghetti sauce contains 250 mg of sodium, or 10% of the Daily Value. In general, a food with 5% or less of the Daily Value is low in sodium, and one with 20% or more is high in sodium.

Food labels also include descriptors relating to salt or sodium content (see table). The spaghetti sauce shown here, labeled "Light in Sodium," contains 50% less sodium than regular spaghetti sauce. Additionally, a health claim that diets low in sodium may reduce the risk of high blood pressure can appear on foods that meet the definition of a low-sodium food and provide less than 20% of the Daily Value for fat, saturated fat, and cholesterol per serving. Some medications, such as pain relievers, antacids, and cough medications, can also contribute a significant amount of sodium. Drug facts labels on over-the-counter medications can help identify those that contain significant amounts of sodium.

Nutrition Facts
Serving Size 1/2 cup (125g)
Servings Per Container about 3½

Amount Per Serving	
Calories 50	Calories from Fat 10
	%Daily Value**
Total Fat 1g	2%
Saturated Fat 0g	0%
Trans Fat 0g	
Cholesterol 0mg	0%
Sodium 250mg	10%
Potassium 530mg	15%
Total Carbohydrate 9g	3%
Dietary Fiber 1g	4%
Sugars 7g	
Protein 2g	
Vitamin A 10% •	Vitamin C 25%
Calcium 2% •	Iron 10%

*Percent Daily Values are based on a 2,000 calorie diet. Your daily values may be higher or lower depending on your calorie needs.

		Calories: 2,000	2,500
Total Fat	Less than	65g	80g
Sat Fat	Less than	20g	25g
Cholesterol	Less than	300mg	300mg
Sodium	Less than	2,400mg	2,400mg
Potassium		3,500mg	3,500mg
Total Carbohydrate		300g	375g
Dietary Fiber		25g	30g

Light Spaghetti Sauce, 250 milligrams (mg) per serving
Regular Spaghetti Sauce, 500mg per serving

SPAGHETTI SAUCE
LIGHT IN SODIUM
50% Less Sodium than our regular spaghetti sauce (See side panel for nutrition information.)

THINK CRITICALLY: If instead of a serving of this spaghetti sauce, which is light in sodium, you ate a serving of regular spaghetti sauce, how much more sodium would you be consuming?

Salt and Sodium on Food Labels

Sodium-free	Contains less than 5 mg of sodium per serving.
Salt-free	Must meet criterion for "sodium-free."
Very low sodium	Contains 35 mg or less of sodium per serving.
Low sodium	Contains 140 mg or less of sodium per serving.
Reduced or less sodium	Contains at least 25% less sodium per serving than a reference food.
Light in sodium	Contains at least 50% less sodium per serving than the average reference amount for the same food with no sodium reduction.
No salt added, without added salt, and unsalted	No salt added during processing, and the food it resembles and for which it substitutes is normally processed with salt. (If the food is not "sodium-free," the statement "not a sodium-free food" or "not for control of sodium in the diet" must appear on the same panel as the Nutrition Facts panel.)
Lightly salted	Contains at least 50% less sodium per serving than a reference food. (If the food is not "low in sodium," the statement "not a low-sodium food" must appear on the same panel as the "Nutrition Facts" panel.)
Low-sodium meal	Contains 140 milligrams or less sodium per 100 grams.

TABLE 10.3 How to Choose: A Low-Sodium Diet

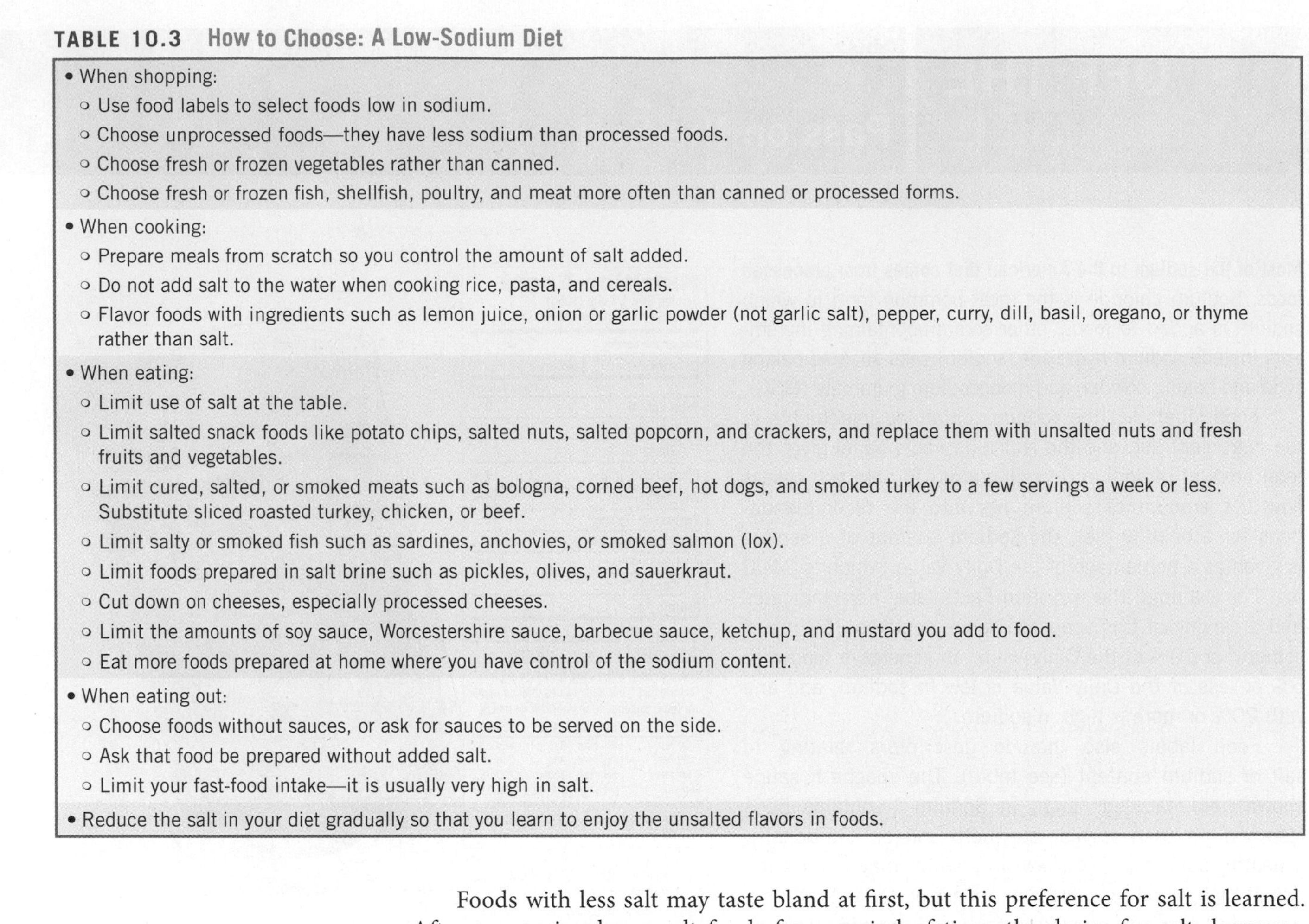

- When shopping:
  - Use food labels to select foods low in sodium.
  - Choose unprocessed foods—they have less sodium than processed foods.
  - Choose fresh or frozen vegetables rather than canned.
  - Choose fresh or frozen fish, shellfish, poultry, and meat more often than canned or processed forms.
- When cooking:
  - Prepare meals from scratch so you control the amount of salt added.
  - Do not add salt to the water when cooking rice, pasta, and cereals.
  - Flavor foods with ingredients such as lemon juice, onion or garlic powder (not garlic salt), pepper, curry, dill, basil, oregano, or thyme rather than salt.
- When eating:
  - Limit use of salt at the table.
  - Limit salted snack foods like potato chips, salted nuts, salted popcorn, and crackers, and replace them with unsalted nuts and fresh fruits and vegetables.
  - Limit cured, salted, or smoked meats such as bologna, corned beef, hot dogs, and smoked turkey to a few servings a week or less. Substitute sliced roasted turkey, chicken, or beef.
  - Limit salty or smoked fish such as sardines, anchovies, or smoked salmon (lox).
  - Limit foods prepared in salt brine such as pickles, olives, and sauerkraut.
  - Cut down on cheeses, especially processed cheeses.
  - Limit the amounts of soy sauce, Worcestershire sauce, barbecue sauce, ketchup, and mustard you add to food.
  - Eat more foods prepared at home where you have control of the sodium content.
- When eating out:
  - Choose foods without sauces, or ask for sauces to be served on the side.
  - Ask that food be prepared without added salt.
  - Limit your fast-food intake—it is usually very high in salt.
- Reduce the salt in your diet gradually so that you learn to enjoy the unsalted flavors in foods.

Foods with less salt may taste bland at first, but this preference for salt is learned. After consuming lower-salt foods for a period of time, the desire for salt decreases. Adding other flavorings such as onions, garlic, lemon juice, vinegar, black pepper, parsley, and other herbs to food may also help satisfy your taste for flavorful food without adding sodium. **Table 10.3** provides some tips for reducing sodium intake.

OUTCOME

Eiichi Onodera/Dex Image/Getty Images, Inc.

David Tenenbaum/©AP/Wide World Photos

Despite her ailing appearance, Gabriele Andersen-Scheiss recovered rapidly. Once she finished her lap, she was given intravenous fluids and sprayed with a hose, and cold towels were placed on her body. After two hours she returned to the Olympic Village where she ate while doctors observed her for another few hours. Her dehydration had been caused by a combination of large fluid losses, a low fluid intake, and the fact that she was not well acclimatized to the southern California heat. As she ran, water evaporated from her respiratory tract and skin, and she produced large amounts of sweat to cool her body. In such hot conditions, an athlete can lose several liters of water every hour. Andersen-Scheiss was not able to replace her fluid losses even with frequent drinks at water stations, so the amount of water in her body began to decline. Her blood volume decreased, and sufficient oxygen and nutrients could not be delivered to her exercising muscles. By the time she entered the stadium for her final lap, the lack of blood to her limbs had caused them to become numb. She observed, "I knew what I wanted to do, but my muscles wouldn't respond anymore." Andersen-Scheiss never ran another Olympic marathon but did go on to become a top Masters runner, setting many distance records. **Her wobbly struggle to finish the women's Olympic marathon in 1984 is a vivid example of what happens when dehydration disrupts the balance of body fluids and electrolytes.**

APPLICATIONS

ASSESSING YOUR DIET

1. **Do you drink enough fluid?**
   a. Keep a log of all the fluids you consume in one day. Calculate your intake by totaling the volume of water, beverages, as well as foods that are liquid at room temperature.
   b. How does your intake on this day compare with the AI?
   c. How should your intake change if you added an hour of jogging or basketball to your day?
2. **How does your diet compare to the DASH Eating Plan?**
   a. Use one day of the food record you kept in Chapter 2 to compare the number of servings you ate from each of the food groups to the number of servings recommended by the DASH Eating Plan for your energy intake, as shown in the table in Figure 10.19.
   b. Suggest modifications to your diet that would allow it to meet the DASH guidelines.
   c. What difficulties or inconveniences do you see with following this dietary pattern?
   d. What other dietary or lifestyle changes might you make if you are at high risk of hypertension?

CONSUMER ISSUES

3. **How much sodium do processed foods add to our diet?**
   a. Survey your cupboard or dorm room for processed foods
   b. Make a list of those that contain more than 10% of the Daily Value for sodium (2400 mg) per serving.
   c. Suggest substitutions for these that provide less sodium.
   d. If you choose these alternatives, would your calorie intake change?

CLINICAL CONCERNS

4. **Virginia's mother has high blood pressure and has had several strokes. A recent physical exam indicates that Virginia also has high blood pressure. Her physician prescribes medication to lower her blood pressure, but he believes that with some changes in diet and lifestyle Virginia's blood pressure can be brought into the normal range without drugs. Virginia works at a desk job. The only exercise she gets is when she takes care of her nieces and nephews one weekend a month. Her typical diet includes a breakfast of cereal, tomato juice, and coffee. She has a snack of doughnuts and coffee at work, and for lunch she joins coworkers for a fast-food cheeseburger, fries, and milkshake. When she gets home she has a soda and snacks on peanuts or chips. Dinner is usually a frozen dinner with milk.**
   a. What dietary changes would you recommend for Virginia?
   b. What lifestyle changes would you recommend?
5. **Brian has high blood pressure. His doctor suggests some dietary and lifestyle changes and prescribes a thiazide diuretic to help lower his blood pressure.**
   a. Why might a diuretic help lower blood pressure?
   b. Thiazide diuretics cause potassium to be lost in the urine. List some foods Brian can increase in his diet to boost his potassium intake.
   c. There are many other types of antihypertensive medications available. Use the Internet and your knowledge of physiology and blood pressure regulation to explain how two other classes of antihypertensive drugs help lower blood pressure.

SUMMARY

10.1 Water: The Internal Sea

- Water is an essential nutrient that provides many functions in the body. The polar structure of water allows it to act as a solvent for the molecules and chemical reactions involved in metabolism. Water helps to transport other nutrients and waste products within the body and to excrete wastes from the body. It also helps to protect the body, regulate body temperature, and lubricate areas such as the eyes and the joints.
- The adult human body is about 60% water by weight. Body water is distributed between intracellular and extracellular compartments. The amount in each compartment depends largely on the concentration of solutes. Since water will diffuse by osmosis from a compartment with a lower concentration of solutes to one with a higher concentration, changes in the concentration of electrolytes and other solutes in each compartment cause changes in the distribution of water.
- Water cannot be stored, so intake must balance losses to maintain homeostasis. Fluid intake is stimulated by the sensation of thirst, which occurs in response to a decrease in blood volume and an increase in the concentration of solutes. Water is lost from the body in urine and feces, through evaporation from the skin and lungs, and in sweat. The kidney is the primary regulator of water output. If water intake is low, antidiuretic hormone will cause the kidney to conserve water. If water intake is high, more water will be excreted in the urine.
- Dehydration can occur if water intake is too low or output is excessive. Mild dehydration can cause headache, fatigue, loss of appetite, dry eyes and mouth, and dark-colored urine. Severe dehydration can be fatal.
- Water intoxication is not a common as dehydration. It causes hyponatremia, which can result in abnormal fluid accumulation in body tissues.

- The recommended intake of water is 2.7 L/day for adult women and 3.7 L/day for adult men; needs vary depending on environmental conditions and activity level.

10.2 Electrolytes: Salts of the Internal Sea

- The U.S. diet is abundant in sodium and chloride from processed foods and table salt but generally low in potassium, which is high in unprocessed foods such as fresh fruits and vegetables.
- The minerals sodium, chloride, and potassium are electrolytes that are important in the maintenance of fluid balance and the formation of membrane potentials essential for nerve transmission and muscle contraction.
- Electrolyte and fluid homeostasis is regulated primarily by the kidneys. A decrease in blood pressure or blood volume signals the release of the enzyme renin, which helps form angiotensin II. Angiotensin II causes blood vessels to constrict and the hormone aldosterone to be released. Aldosterone causes the kidneys to reabsorb sodium and hence water, thereby preventing any further loss in blood volume. Failure of these regulatory mechanisms may be a cause of hypertension.
- Low levels of sodium, chloride, and potassium can occur when water and electrolyte losses are increased due to excessive sweating, chronic diarrhea or vomiting, or kidney disorders. High blood sodium most commonly results from dehydration.

10.3 Hypertension

- A healthy blood pressure is less than 120/80 mm Hg. Blood pressure from 120/80 to 139/89 mm Hg is referred to as prehypertension, and blood pressure that is consistently 140/90 mm Hg or greater indicates hypertension. Hypertension affects about a third of adults in the United States. Essential hypertension has no obvious external cause.
- The risk of developing high blood pressure is increased by a family history of hypertension, aging, obesity, diabetes, and lifestyle factors such as a lack of physical activity, heavy alcohol consumption, smoking, stress, and a poor diet. A diet high in sodium increases blood pressure in most individuals. High intakes of the minerals potassium, magnesium, and calcium help lower blood pressure.
- Diets high in sodium and low in potassium are associated with an increased risk of hypertension. The DASH diet—a dietary pattern moderate in sodium; high in potassium, magnesium, calcium, and fiber; and low in fat, saturated fat, and cholesterol—lowers blood pressure. Blood pressure management also requires maintenance of a healthy weight, an active lifestyle, and a limit on alcohol consumption.

10.4 Healthy Electrolyte Intakes

- Recommendations for health suggest that we increase our intake of potassium and consume less sodium. The recommended salt intake is 3.8 g/day for adults ages 19 to 50. Because salt is 40% sodium and 60% chloride by weight, this represents 1500 mg of sodium and 2300 mg of chloride. The UL for sodium, which is based on the increase in blood pressure seen with higher sodium intakes, is only 2300 mg/day. The AI for potassium is 4700 mg/day, an amount significantly higher than the 2000 to 3000 mg consumed by most Americans. No UL has been set for potassium.
- Public health recommendations including the Dietary Guidelines suggest that Americans consume a dietary pattern that is high in fruits, vegetables, whole grains, legumes, nuts and seeds, and provides low-fat dairy products and lean meat, fish, and poultry. This dietary pattern, which can be achieved by following the recommendations of the DASH Eating Plan or MyPlate, contains less sodium and more potassium than the typical American diet.

REVIEW QUESTIONS

1. Describe the functions of water in the body.
2. How is the total amount of water in the body regulated?
3. How is the amount of water in each body compartment regulated?
4. What is the recommended water intake for adults?
5. List three factors that increase water needs.
6. Define electrolyte. How is the term used in nutrition?
7. How do sodium, potassium, and chloride function in the body?
8. Explain how a drop in blood pressure is returned to normal.
9. What are the consequences of untreated hypertension?
10. What types of foods contribute the most sodium to the North American diet?
11. What types of foods are good sources of potassium?
12. What is the relationship between dietary sodium and blood pressure?
13. What is the DASH diet, and how does it affect blood pressure?

REFERENCES

1. Institute of Medicine, Food and Nutrition Board. *Dietary Reference Intakes for Water. Salt and Potassium*. Washington, DC: National Academies Press, 2004.

2. Shen, H-P. Body fluids and water balance. In *Biochemical and Physiological Aspects of Human Nutrition*, 2nd ed. M. Stipanuk, ed. St. Louis: Saunders Elsevier, 2006, pp. 973–1000.

3. Murray, B. Hydration and physical performance. *J Am Coll Nutr* 26:542S–548S, 2007.

4. Maughan, R .J., and Shirreffs, S. M. Dehydration and rehydration in competative sport. *Scand J Med Sci Sports* 20 (suppl 3): 40–47, 2010.

5. Roger, V. L., Go, A. S., Lloyd-Jones, D. M., et al. Heart disease and stroke statistics—2011 update: a report from the American Heart Association. *Circulation* 123:e18–e209, 2011.

6. U.S. Department of Agriculture and U.S. Department of Health and Human Services. *Dietary Guidelines for Americans, 2010*, 7th ed. Washington, DC: U.S. Government Printing Office, December 2010.

7. Chobanian, A. V., Bakris, G. L., Black, H. R., et al. Seventh report of the Joint National Committee on Prevention, Detection, Evaluation, and Treatment of High Blood Pressure. Joint National Committee on Prevention, Detection, Evaluation, and Treatment of High Blood Pressure. National Heart, Lung, and Blood Institute. *Hypertension* 42: 1206–1252, 2003.

8. American Heart Association. Am I at risk? Factors that contribute to high blood pressure.

9. Carvalho, J. J., Baruzzi, R. G., Howard, P. F., et al. Blood pressure in four remote populations in the Intersalt study. *Hypertension* 14:238–246, 1989.

10. Sacks, F. M., Svetkey, L. P., Vollmer, W. M., et al. Effects on blood pressure of reduced dietary sodium and the Dietary Approaches to Stop Hypertension (DASH) diet. DASH–Sodium Collaborative Research Group. *N Engl J Med* 344:3–10, 2001.

11. Houston, M. C., and Harper, K. J. Potassium, magnesium, and calcium: Their role in both the cause and treatment of hypertension. *J Clin Hypertens (Greenwich)* 10 (7 Suppl. 2):3–11, 2008.

12. Craig, W. J., Mangels, A. R., and American Dietetic Association. Position of the American Dietetic Association: Vegetarian diets. *J Am Diet Assoc* 109:1266–1282, 2009.

13. Townsend, M. S., Fulgoni, V. L. III, Stern, J. S., et al. Low mineral intake is associated with high systolic blood pressure in the Third and Fourth National Health and Nutrition Examination Surveys: Could we all be right? *Am J Hypertens* 18:261–269, 2005.

14. Appel, L. J., Moore, T. J., Obarzanek, E., et al. A clinical trial of the effects of dietary patterns on blood pressure. *N Engl J Med* 336:1117–1124, 1997.

15. Greenland, P. Beating high blood pressure with low sodium DASH. *N Engl J Med* 344:53–55, 2001.

16. Sacks, F. M., Svetkey, L. P., Vollmer, W. M., et al. Effects on blood pressure of reduced dietary sodium and the Dietary Approaches to Stop Hypertension (DASH) diet. DASH-Sodium Collaborative Research Group. *N Engl J Med* 344:3–10, 2001.

***To access links to online sources, please go to www.wiley.com/college/smolin and select Nutrition: Science and Applications, 3rd edition. From this page, select either the student or instructor companion site. Once on the desired site, select References.**

CHAPTER OUTLINE

PhotoDisc, Inc./Getty Images

Major Minerals and Bone Health

11

CASE STUDY

Mika felt a knife-like pain in the top of her left foot. She tried to keep running, but the pain continued. Frustrated, she limped back to see the trainer. Mika is a 22-year-old college student who currently runs outdoors about 40 miles each week. She has been running track since she was a freshman in high school. During her junior year in college, she suffered a stress fracture in her right foot. Now, as a senior, she fears she has a second stress fracture in the left foot.

An evaluation by a physician and some diagnostic tests confirm that Mika has a new stress fracture. In addition to treatment for the stress fracture, the doctor recommends she have a full medical evaluation. Mika is 5 feet 3 inches tall and weighs 102 pounds (BMI = 18.1 kg/m^2). She has lost about five pounds since the start of her spring training. Her body fat is currently 17%. A medical history reveals that she has not menstruated in almost a year. A diet history discloses that she is lactose intolerant, so doesn't consume much dairy, and that she restricts her overall intake to keep her weight low to improve her running performance. A bone scan shows that Mika has low bone mass in her hips and spine and that her overall bone density is similar to that of a woman in her 60s. Mika is shocked. How could a healthy young woman who gets plenty of outdoor exercise have such low bone density? **What can she do to improve her bone health and prevent future fractures?**

Blend Images/Getty Images, Inc.

11.1 What Are Minerals?

LEARNING OBJECTIVES

- Define *mineral* in terms of nutrition.
- Describe how interactions among minerals and other dietary components affect mineral bioavailability.

mineral In nutrition, an element needed by the body in small amounts for structure and to regulate chemical reactions and body processes.

major mineral A mineral needed in the diet in an amount greater than 100 mg/day or present in the body in an amount greater than 0.01% of body weight.

trace mineral or **trace element** A mineral required in the diet in an amount of 100 mg or less per day or present in the body in an amount of 0.01% of body weight or less.

Minerals are inorganic elements needed by the body as structural components and regulators of body processes. Minerals may combine with other elements in the body, but they retain their chemical identity. Unlike vitamins, they are not destroyed by heat, oxygen, or acid. The ash that remains after a food is combusted in a bomb calorimeter contains the minerals that were present in that food.

Humans need to consume more than 20 minerals to stay healthy. Some of these make up a significant portion of our body weight; others are found in minute quantities. If a mineral is needed in the diet in amounts greater than 100 mg/day (an amount equivalent in weight to about two drops of water) or is present in the body in amounts greater than 0.01% of body weight, the mineral is considered a **major mineral**. The major minerals include the electrolytes sodium, chloride, and potassium, which were discussed in Chapter 10, and calcium, phosphorus, magnesium, and sulfur, which are discussed in this chapter. Minerals that are needed in the diet in an amount of 100 mg or less per day or are present in the body in an amount of 0.01% or less of body weight are referred to as **trace minerals** or **trace elements** (**Figure 11.1**). The trace minerals include iron, copper, zinc, selenium, iodine, chromium, fluoride, manganese, molybdenum, and others and are discussed in Chapter 12. Just because we need more of the major minerals than of the trace minerals doesn't mean that one group is more important than the other. A deficiency of a trace mineral is just as damaging to health as a deficiency of a major mineral.

Minerals in Our Diet

Minerals in our diet come from both plant and animal sources (**Figure 11.2**). Most foods naturally contain minerals, and some foods provide minerals that are added intentionally

H																	He
Li	Be											B	C	N	O	F	Ne
Na	Mg											Al	Si	P	S	Cl	Ar
K	Ca	Sc	Ti	V	Cr	Mn	Fe	Co	Ni	Cu	Zn	Ga	Ge	As	Se	Br	Kr
Rb	Sr	Y	Zr	Nb	Mo	Tc	Ru	Rh	Pd	Ag	Cd	In	Sn	Sb	Te	I	Xe
Cs	Ba	La	Hf	Ta	W	Re	Os	Ir	Pt	Au	Hg	Tl	Pb	Bi	Po	At	Rn
Fr	Ra	Ac	Rf	Db	Sg	Bh	Hs	Mt	Ds	Rg	Cn	Uut	Fl	Uup	Lv	Uus	Uuo

Figure 11.1 Major and trace minerals in the periodic table
The major minerals, shown in purple, and the trace minerals, shown in blue, are the same elements found in the periodic table. Although these are essential nutrients, not all the elements in the periodic table are needed by the body.

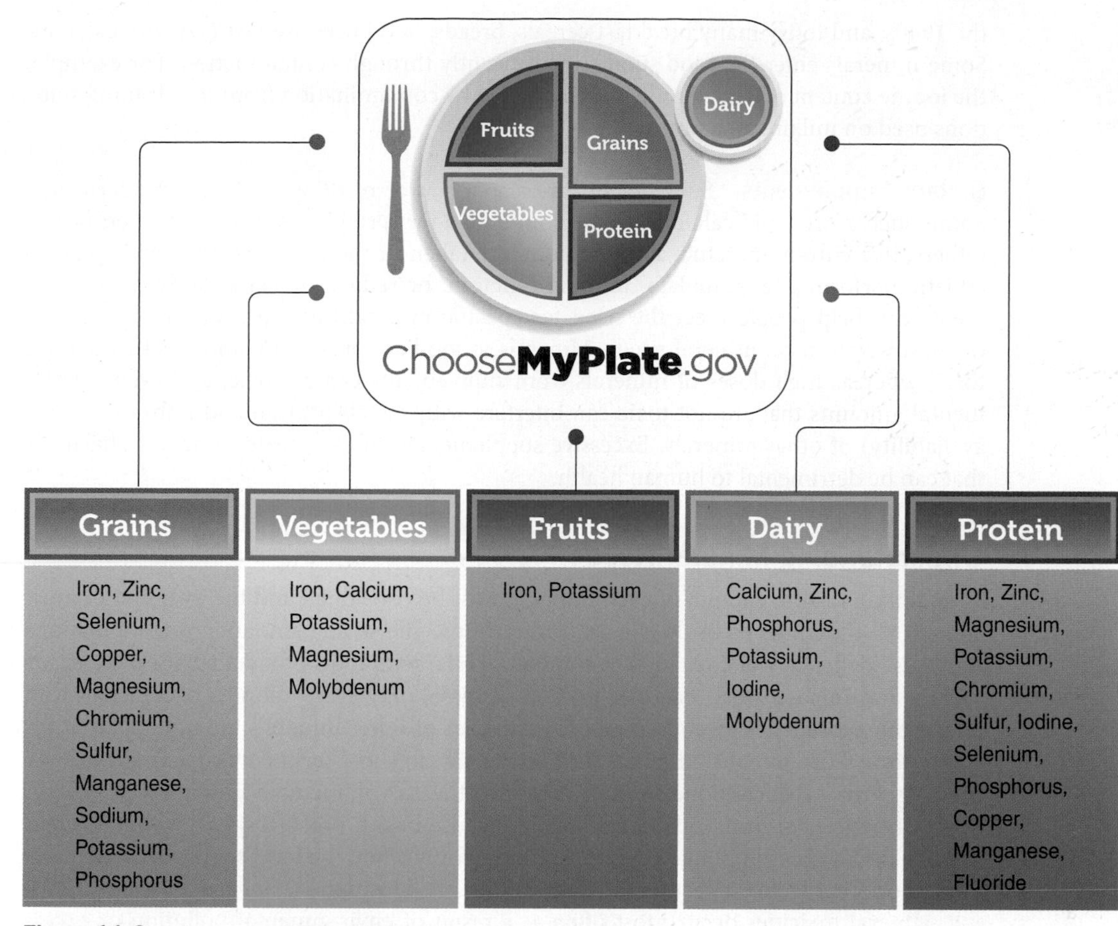

Figure 11.2 **Minerals in MyPlate food groups**
Minerals are found in all the MyPlate food groups; some groups are particularly good sources of specific minerals. Eating a variety of foods, including fresh fruits, vegetables, nuts, legumes, whole grains and cereals, milk, seafood, and lean meats can maximize your diet's mineral content.

by fortification or accidentally through contamination. Dietary supplements are also a source of minerals.

Natural Food Sources of Minerals In some foods, the amounts of minerals naturally present are predictable because the minerals are regulated components of the plant or animal. For instance, calcium is a component of milk; therefore, drinking a glass of milk reliably provides a known amount. Magnesium is a component of chlorophyll, so it is found in consistent amounts in leafy greens. For some minerals, however, the amounts in food vary depending on the mineral concentration in the soil and water at the food's source. For example, the soil content of iodine is high near the ocean but usually quite low in inland areas. Therefore, foods grown near the ocean are better sources of iodine than those grown inland. In developed countries, modern transportation systems make available foods produced in many locations, so the diet is unlikely to be deficient in minerals that vary in concentration depending on where the food is produced. Individual trace-element deficiencies and excesses are more likely to occur in countries where the diet consists predominantly of locally grown foods.

How Processing Affects the Minerals in Food Food processing and refining can affect the mineral content of foods. Processing does not destroy minerals, but it can still cause losses. For example, when vegetables are cooked, the cells are broken down and potassium is lost in the cooking water. When the skins of fruits and vegetables or the bran and germ of grains are detached, magnesium, iron, selenium, zinc, and copper are lost. Processing also adds minerals to foods. Sodium is frequently added for flavor or as a preservative. The enrichment of grains has been adding iron to our breads, baked goods, and rice since the 1940s, and today many breakfast cereals, breads, and juices are fortified with calcium. Some minerals enter the food supply inadvertently through contamination. For example, the iodine content of dairy products is increased by contamination from the cleaning solutions used on milking machines.

Dietary Supplements Supplements are also a source of minerals in the modern diet. Some, such as iron and calcium, are recommended for certain groups to meet their needs. Others, like chromium, zinc, and selenium, are taken in the hope that they will enhance athletic performance, stimulate immune function, or reduce cancer risk. While supplements can help people meet the need for specific minerals, eating a variety of foods is the best way to meet mineral needs. Minerals in food are present in amounts that are not toxic, whereas high doses of minerals from supplements can be toxic, and even supplemental amounts that are not toxic can interfere with the absorption and utilization (bioavailability) of other minerals. Excessive supplementation can create mineral imbalances that can be detrimental to human health.

Understanding Mineral Needs

To maintain health, enough of each mineral must be consumed and the overall diet must contain all the minerals in the correct proportions. The wrong amounts or combinations can cause deficiency or toxicity. For some minerals, symptoms of a deficient intake occur rapidly and impact short-term health. For example, only a few months of deficient iron intake can cause lethargy and fatigue. Deficiencies of other minerals are not apparent for a long time. For example, a low calcium intake has no short-term consequences, but over the long term it reduces bone density, increasing the risk of fractures later in life.

Deficiencies of iron, iodine, and calcium cause health problems worldwide, whereas deficiencies of other minerals are rare, occurring only when the food supply is particularly limited or other factors affect mineral absorption or utilization.

Mineral toxicities occur most often as a result of environmental pollution or excessive use of supplements. For example, lead is toxic. Chronic exposure to lead from old lead paint, lead pipes, and soil and air contamination can cause growth retardation and learning disabilities in children (see Chapter 15). Even minerals such as iron that are essential in small doses can be toxic when consumed in excess.

What Affects Mineral Bioavailability? The bioavailability of the minerals in foods varies. For some minerals, such as sodium, almost all that is present in our food is absorbed, but for others, only a small percentage of what is consumed is absorbed. For instance, only about 25% of the calcium in our diet is absorbed, and iron absorption may be as low as 5%. How much of a particular mineral is absorbed and utilized may vary from food to food, meal to meal, and person to person.

In general, the minerals in animal products are better absorbed than those in plant foods. The difference in absorption is due in part to the fact that plants contain substances such as **phytates** (also called phytic acid), **tannins**, **oxalates**, and fiber that bind minerals in the gastrointestinal tract and can reduce absorption (**Figure 11.3**). The North American diet generally does not contain enough of any of these components to cause a mineral deficiency, but diets in developing countries may. For example, the amount of phytate in three-quarters of a cup of dry, unrefined cereal grains doubles the amount of zinc that needs to be consumed to meet needs.[1,2]

The presence of one mineral can also interfere with the absorption of another. For example, mineral ions that carry the same charge compete for absorption in the gastrointestinal tract. Calcium, magnesium, zinc, copper, and iron all carry a 2+ charge, so a high intake of one may reduce the absorption of another. Although this is generally not a problem when whole foods are consumed, a large dose of one mineral from a dietary supplement may interfere with the absorption of other minerals.

The body's need for a mineral may also affect how much of that mineral is absorbed. For instance, if plenty of iron is stored in your body, you absorb less of the iron you consume. Life stage can also affect absorption; for example, calcium absorption doubles during pregnancy, when the body's needs are high.

phytate or **phytic acid** A phosphorus storage compound found in seeds and grains that can bind minerals and decrease their absorption.

tannin A substance found in tea and some grains that can bind minerals and decrease their absorption.

oxalate An organic acid found in spinach and other leafy green vegetables that can bind minerals and decrease their absorption.

Charles D. Winters

Figure 11.3 Compounds that interfere with mineral absorption
Plant foods such as these contain substances that can reduce mineral absorption when consumed in large amounts.

The ability to transport minerals from intestinal mucosal cells to the rest of the body also affects bioavailability. Some minerals must bind to plasma proteins or specific transport proteins to be transported in the blood. This binding helps regulate their absorption and prevents minerals that are chemically reactive from forming free radicals that could cause oxidative damage. Nutritional status and nutrient intake can affect mineral transport in the body. For instance, when protein intake is deficient, transport proteins (and proteins in general) cannot be synthesized. In this case, therefore, even if a mineral is adequate in the diet, it cannot be transported to the cells where it is needed.

How Minerals Function Minerals contribute to the body's structure and help regulate body processes. Many serve more than one function. For example, we need calcium to keep our bones strong and also to maintain normal blood pressure, allow muscles to contract, and transmit nerve signals from cell to cell. Some minerals help regulate water balance, others help regulate energy metabolism, and some affect growth and development through their role in the expression of certain genes. Many minerals act as **cofactors** needed for enzyme activity (**Figure 11.4**). None of the minerals we require acts in isolation. Instead, they interact with one another as well as with other nutrients and other components of the diet.

cofactor An inorganic ion or coenzyme required for enzyme activity.

Recommended Mineral Intakes As with other nutrients, the DRI recommendations for mineral intakes provide either an RDA value, when sufficient information is available to establish an EAR, or an AI, when population data are used to estimate needs. To help avoid toxic amounts of minerals, ULs have been established when adequate data are available. Recommendations are based on evidence from many types of research studies, ranging from laboratory studies done with animals and clinical trials using human subjects to epidemiological observations. In addition to planned experiments, information about trace mineral needs has come from the study of inherited diseases affecting trace mineral utilization and of individuals fed for long periods of time solely by total parenteral nutrition (TPN) solutions that were inadvertently deficient in specific essential minerals. As with other nutrients, when no other data are available, mineral needs can be estimated by evaluating the intake in a healthy population. It is assumed that if there are no deficiency symptoms, the diet must meet the requirement for that nutrient. One problem with this approach, however, is that deficiency symptoms may become apparent only when the deficiency is severe. Subtle signs of a mineral deficiency in a population may be difficult to detect, and deficiencies that have no symptoms for many years may be hard to identify.

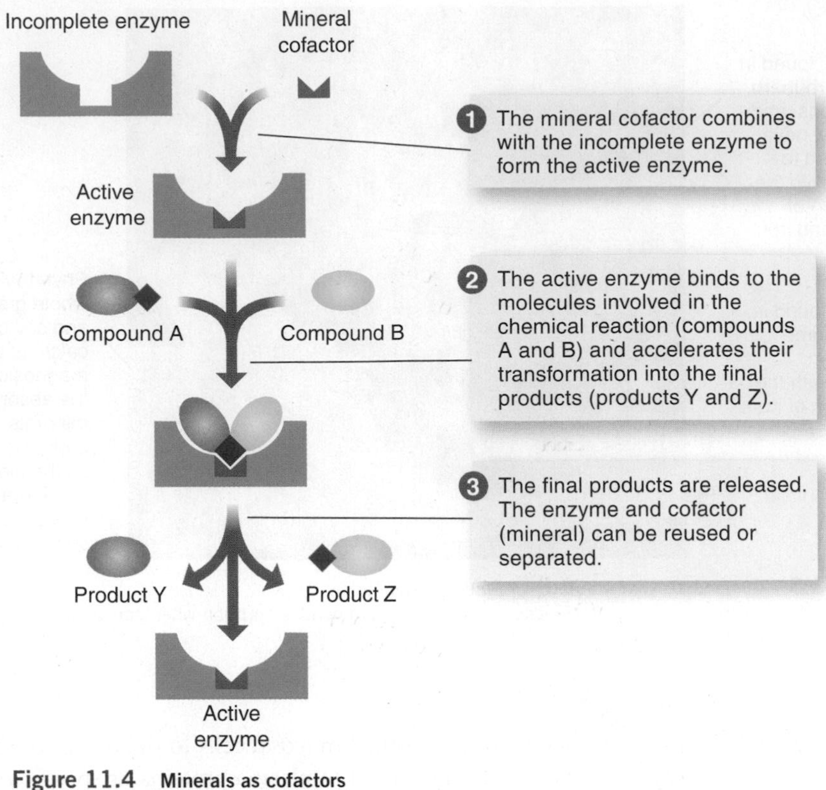

Figure 11.4 **Minerals as cofactors**
The binding of a cofactor to an enzyme activates the enzyme.

11.2 Minerals, Osteoporosis, and Bone Health

LEARNING OBJECTIVES

- Explain why protein, vitamin C, vitamin D, and calcium are needed for bone formation.
- Explain why bone remodeling is important.
- Discuss how the rates of bone formation and breakdown change throughout life.
- Describe dietary and lifestyle factors that affect the risk of osteoporosis.

Bones are the hardest, strongest structures in the human body. They are strong because of the minerals they contain—calcium along with phosphorus, magnesium, sodium, fluoride, and a number of others. Bones support our weight, whether we are walking, running, or jumping rope. However, with age, some bone is lost, causing bone strength to decrease. For many people, the loss of bone is so great that the force of stepping off a curb is enough to cause their bones to fracture. This condition is known as **osteoporosis**.

osteoporosis A bone disorder characterized by a reduction in bone mass, increased bone fragility, and an increased risk of fractures.

hydroxyapatite A crystalline compound composed of calcium and phosphorus that is deposited in the protein matrix of bone to give it strength and rigidity.

Bone Structure, Formation, and Breakdown

Bone consists of a framework or matrix made up primarily of the protein collagen. Imbedded in this protein matrix are solid mineral crystals known as **hydroxyapatite**. Healthy bone requires adequate dietary protein and vitamin C to maintain the collagen and a sufficient supply of calcium and other minerals to ensure solidity. Adequate vitamin D (discussed in Chapter 9) is needed to maintain appropriate levels of calcium and phosphorus. There is also growing evidence of the importance of vitamin K for bone health.[3]

There are two types of bone: **cortical** or **compact bone**, which accounts for about 80% of the skeleton and forms the sturdy, dense outer surface layer, and **trabecular** or **spongy bone**, which forms an inner lattice that supports the cortical shell (**Figure 11.5a**). Trabecular bone is found in the knobby ends of the long bones, the pelvis, wrists, vertebrae, scapulae, and the areas of the bone that surround the bone marrow.

Like other tissues in the body, bone is alive, and it is constantly being broken down and re-formed through a process called **bone remodeling** (**Figure 11.5b**). Bone is formed

cortical or **compact bone** Dense, compact bone that makes up the sturdy outer surface layer of bones.

trabecular or **spongy bone** The type of bone forming the inner spongy lattice that lines the bone marrow cavity and supports the cortical shell.

bone remodeling The process whereby bone is continuously broken down and re-formed to allow for growth and maintenance.

Figure 11.5 **Bone structure and remodeling**

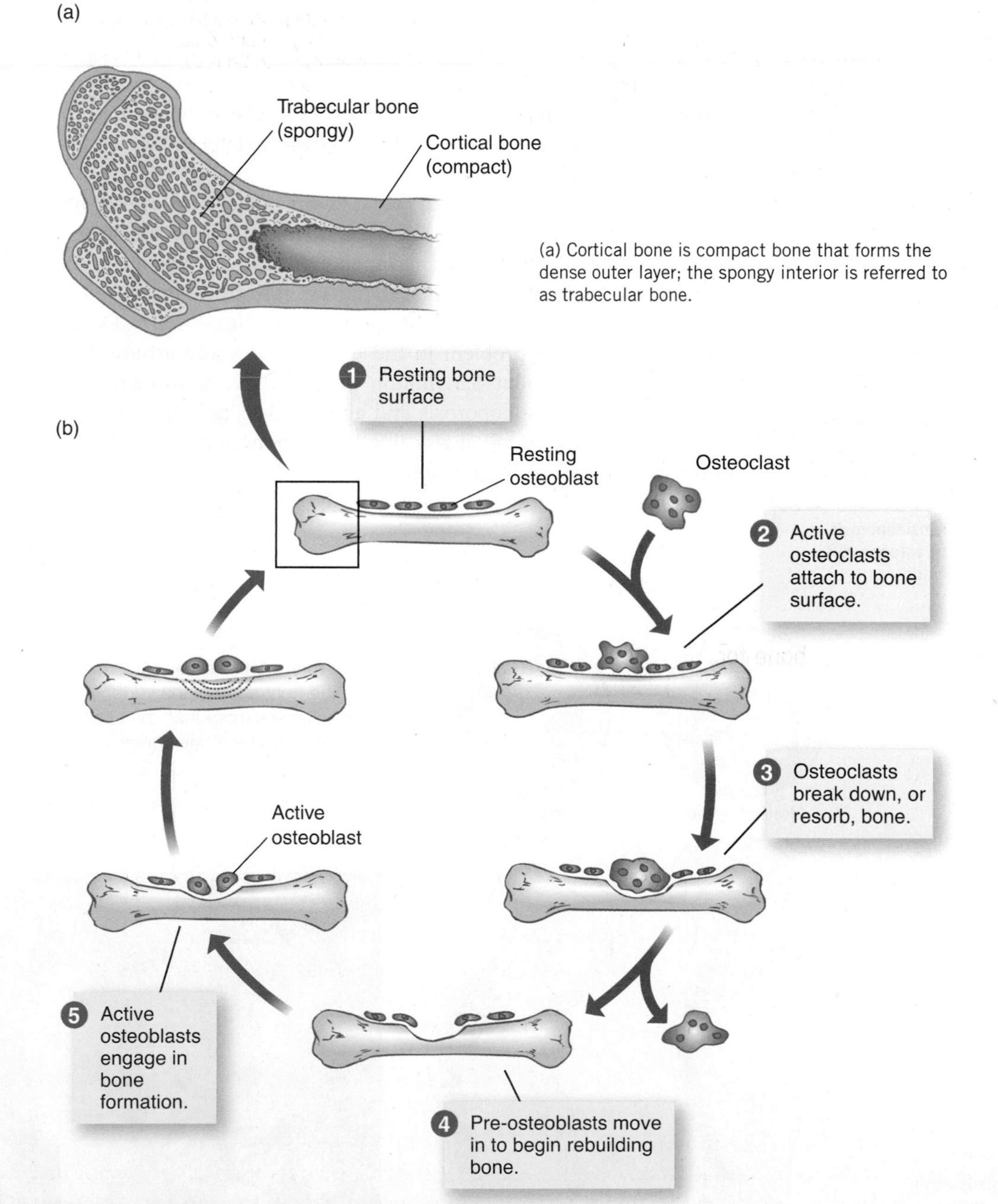

(a) Cortical bone is compact bone that forms the dense outer layer; the spongy interior is referred to as trabecular bone.

(b) During bone remodeling, osteoclasts (pink) break down bone and osteoblasts (blue) build bone.

osteoblast A type of cell responsible for the deposition of bone.

osteoclast A type of cell responsible for bone breakdown.

peak bone mass The maximum bone density attained at any time in life, usually occurring in young adulthood.

by cells called **osteoblasts** and broken down in a process called bone resorption by cells called **osteoclasts**. During bone formation, the activity of the bone-building osteoblasts exceeds that of the osteoclasts. When bone is being broken down, the osteoclasts resorb bone more rapidly than the osteoblasts can rebuild it (see Figure 11.5b).

Most bone is formed early in life. In the growing bones of children, bone formation occurs more rapidly than breakdown. Even after growth stops, bone mass continues to increase into young adulthood when **peak bone mass** is achieved, somewhere between the ages of 16 and 30.[4] When bone breakdown and formation are in balance, bone mass remains constant. After about ages 35 to 45, the amount of bone broken down begins to exceed that which is formed. If enough bone is lost, the skeleton is weakened by osteoporosis and fractures occur easily.

Osteoporosis

Osteoporosis is caused by a loss of both the protein matrix and the mineral deposits of bone, resulting in a decrease in the total amount of bone (**Figure 11.6**). Although both types of bone are lost with age, the greater surface area of the trabecular bone gives it a higher turnover rate than cortical bone and it is therefore more vulnerable to bone loss. As a result, the regions in the skeleton that have higher amounts of trabecular bone, such as the spine and the upper part of the femur, are more susceptible to fracture later in life.

Michael Klein/Peter Arnold/Getty Images, Inc.

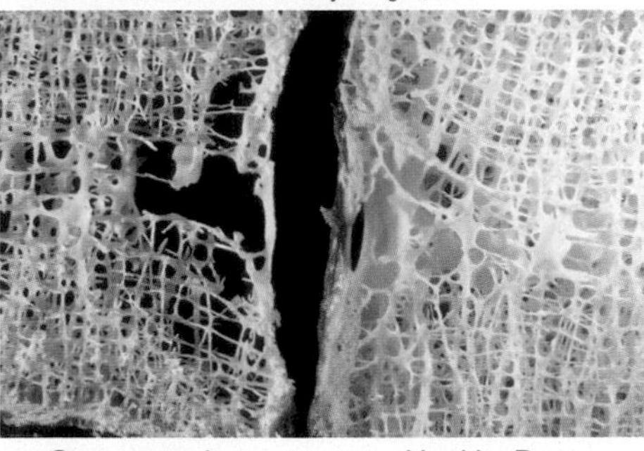

Figure 11.6 Healthy bone versus osteoporosis When trabecular bone is weakened by osteoporosis (left) it appears more porous than healthy bone (right).

How Osteoporosis Impacts Health Osteoporosis is a silent disease because it initially causes no symptoms. By the fifth or sixth decade of life, the bones of individuals with this disorder have weakened enough to cause back pain and bone fractures of the spine, hip, and wrist. Spinal compression fractures may result in loss of height and a stooped posture (called dowager's hump) (**Figure 11.7**). Osteoporosis is a major public health problem in the United States and around the world. In the United States, about 5.3 million people (10% of women and 2% of men) over age 50 have osteoporosis and another 34.5 million (49% of women and 30% of men) over 50 are at risk due to low bone mass, called *osteopenia*.[5] Osteoporosis leads to 1.5 million fractures annually, which account for \$12 to \$18 billion per year in direct medical costs.[6]

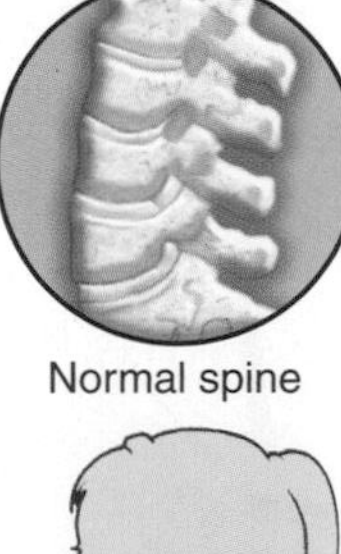

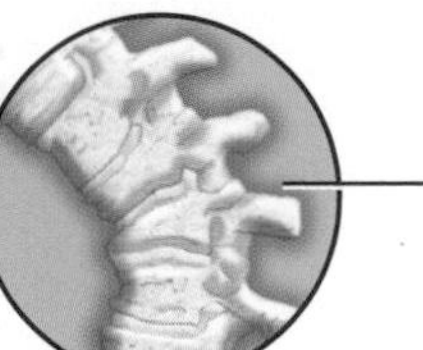

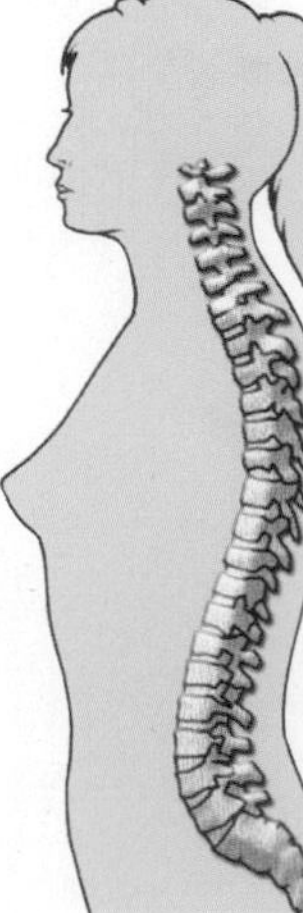

Figure 11.7 Effects of Osteoporosis Spinal compression fractures, shown here, are common with osteoporosis and may result in loss of height and a stooped posture (called "dowager's hump").

TABLE 11.1 Factors Affecting the Risk of Osteoporosis

Risk factor	How it affects risk
Gender	Fractures due to osteoporosis are about twice as common in women as in men. Men are larger and heavier than women and therefore have a greater peak bone mass. Women lose more bone than men due to postmenopausal bone loss.
Age	Bone loss is a normal part of aging, and risk increases with age.
Race	African Americans have denser bones than do Caucasians and Southeast Asians, so their risk of osteoporosis is lower.
Family history	Having a family member with osteoporosis increases risk.
Body size	Individuals who are thin and light have an increased risk because they have less bone mass.
Smoking	Tobacco use weakens bones.
Exercise	Weight-bearing exercise, such as walking and jogging, throughout life strengthens bone, and increasing weight-bearing exercise at any age can increase bone density.
Alcohol abuse	Long-term alcohol abuse reduces bone formation and interferes with the body's ability to absorb calcium.
Diet	A diet that is lacking in calcium and vitamin D plays a major role in the development of osteoporosis. Low calcium intake during the years of bone formation results in a lower peak bone mass, and low calcium intake in adulthood can accelerate bone loss.

What Factors Affect the Risk of Osteoporosis? The causes of osteoporosis are not fully understood, but the risk depends on the level of peak bone mass and the rate at which bone is lost. These are affected by age, gender, hormone levels, genetics, and lifestyle (**Table 11.1**).

Why Age Increases Risk The risk of osteoporosis increases with age. This is because bone density declines after about age 35 when bone breakdown begins to exceed bone formation. This **age-related bone loss** occurs in both men and women (**Figure 11.8**). Bone is lost at a rate of about 0.3 to 0.5% per year. Factors that increase bone loss in older adults include a decline in calcium and vitamin D intake, a decrease in physical activity, a

age-related bone loss The bone loss that occurs in both cortical and trabecular bone of men and women as they advance in age.

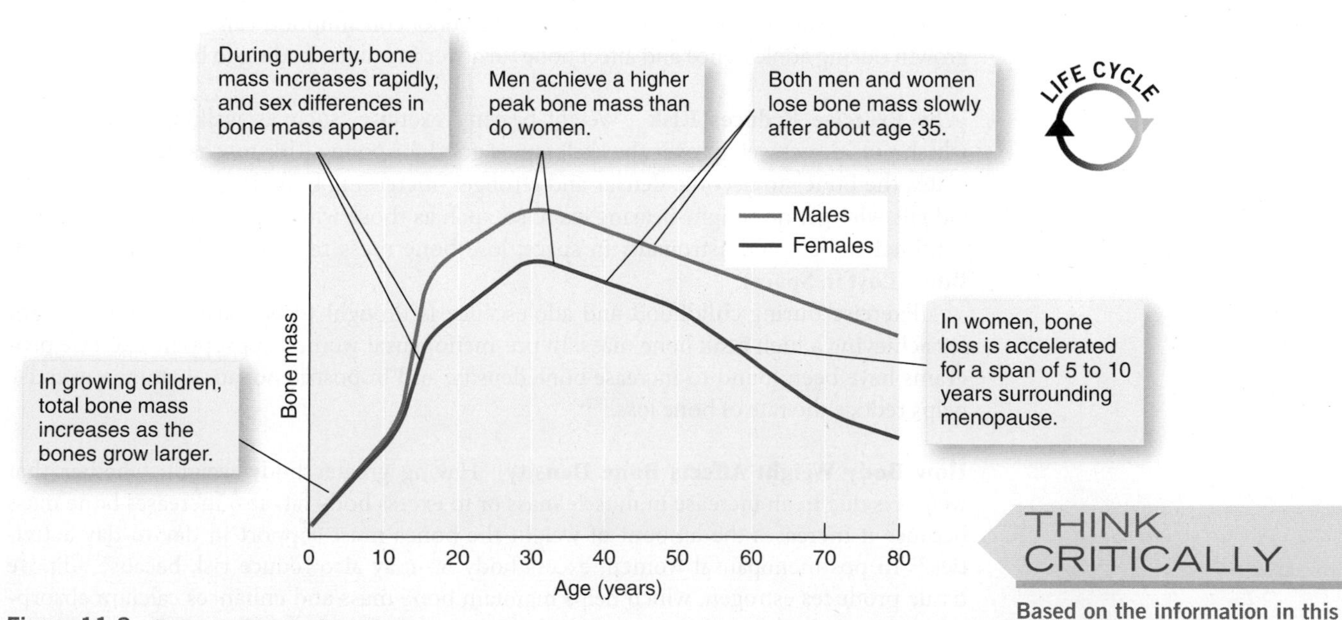

Figure 11.8 Bone mass by gender and age
Changes in the balance between bone formation and bone breakdown cause bone mass to increase in children and adolescents and decrease in adults as they grow older. Although both men and women lose bone after about age 35, women have a lower bone mineral density than men and experience accelerated bone loss after menopause.

THINK CRITICALLY

Based on the information in this graph, why would women over 50 be at greater risk of osteoporosis than men over 50?

decrease in the efficiency of vitamin D activation by the kidney, and a decrease in dietary calcium absorption.[7] In addition, older adults typically spend less time in the sun and wear more clothing to cover the skin when outdoors, which may reduce calcium absorption by decreasing the amount of vitamin D synthesized in the skin.

menopause The physiological changes that mark the end of a woman's capacity to bear children.

postmenopausal bone loss Accelerated bone loss that occurs in women for about 5 to 10 years surrounding menopause.

How Gender Affects Risk About 80% of those affected by osteoporosis are women. Osteoporosis-related fractures occur in one out of every two women over age 50 compared to about one in every eight men over 50.[8] The risk of osteoporosis is greater in women than men because men have a higher peak bone mass to begin with and because bone loss is accelerated in women for a period of about 5 to 10 years surrounding **menopause** (see Figure 11.8). This **postmenopausal bone loss** is related to the drop in estrogen levels that occurs around menopause. Declining estrogen affects bone cells, allowing an increase in bone breakdown. It also decreases intestinal calcium absorption.[9,10] During this 5- to 10-year period, bone loss may be increased 10-fold to a rate of 3 to 5% per year. After the postmenopausal period, women continue to lose bone but at a slower rate.

How Your Genes Affect Bone Density Studies of families and twins have shown that genetic factors are important determinants of bone density, bone size, bone turnover, and osteoporosis risk.[11] Lifestyle factors such as nutrition, activity level, smoking, and alcohol consumption interact with genetic factors over time to determine actual bone density. Genetic differences among racial groups lead to differences in the risk of osteoporosis. For example, the incidence of osteoporosis in African American women is half that of white women even though African American women have lower blood levels of vitamin D. The reason for this lower risk of fractures is that African American women begin menopause with higher bone density and have lower rates of postmenopausal bone loss.[12] Bone density in Asians is generally lower than in non-Hispanic whites and that of Hispanics is similar or slightly lower than non-Hispanic whites, but bone density measures do not always explain racial differences in osteoporosis risk.[13,14]

Smoking and Alcohol Use Cigarette smoking and alcohol consumption both can decrease bone mass and increase the risk of bone fractures. It is estimated that smoking increases the lifetime risk of developing a fracture by 25% and that it increases the risk of a hip fracture by 84% in men.[15] The mechanisms underlying smoking-associated bone loss remain poorly understood.[16] The effect of smoking may be partially reversed by smoking cessation. Although some evidence suggests that moderate drinking may decrease the risk of fracture in postmenopausal women, long-term heavy alcohol consumption can interfere with bone growth during adolescence and affect bone turnover in adults, leading to bone loss.[17,18]

Why Exercise Reduces Risk Weight-bearing exercise, such as walking and jogging, which puts direct weight over the skeleton, is good for bones. This mechanical stress stimulates the bones to become denser and stronger, increasing bone mass. In contrast, individuals who get no weight-bearing exercise, such as those with spinal cord injuries, those confined to bed, and astronauts in space, lose bone mass rapidly (see **Science Applied: Bone: Lost in Space**).

Exercise during childhood and adolescence is thought to be particularly important for achieving a high peak bone mass. In pre-menopausal women appropriate exercise programs have been found to increase bone density, and in postmenopausal women exercise helps reduce the rate of bone loss.[19,20]

How Body Weight Affects Bone Density Having greater body weight, whether that weight is due to an increase in muscle mass or to excess body fat, also increases bone mass because it increases the amount of weight the bones must support in day-to-day activities.[21] In postmenopausal women, excess body fat may also reduce risk because adipose tissue produces estrogen, which helps maintain bone mass and enhances calcium absorption. Therefore, the risk and severity of osteoporosis are decreased in adults with higher body weight and fat. The relationships between body fat and bone density are not necessarily the same during growth, and there is evidence that obesity may have a detrimental rather than a beneficial effect on bone mass in children and adolescents.[22]

SCIENCE APPLIED

Bone: Lost in Space

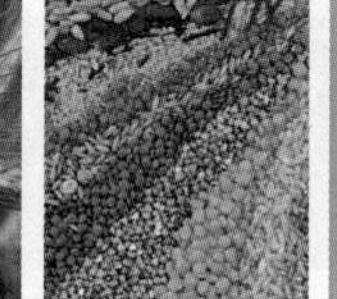

(Elena Schweitzer/ Shutterstock) Victoriano Izquierdo/Getty Images, Inc. (DAJ/Getty Images, Inc.)

When John Glenn became the first American to orbit the Earth in 1962, there was little concern about the effect his five-hour flight would have on his bones. But by 1997, when Shannon Lucid spent 188 days aboard the Russian Mir space station, the effect of weightlessness on bone health had become a serious concern (see photo). Weight-bearing activities such as walking, jogging, and weight training are important for the maintenance of bone health. Under the force of Earth's gravity, these activities mechanically stress the bones, and this stimulates the deposition of calcium. When an astronaut goes into space, weightlessness eliminates this stimulus, calcium no longer accumulates in bone, and that present in bone upon leaving Earth's gravitational field begins to dissipate, elevating calcium levels in the rest of the body. This may lead to kidney stones and calcification of the soft tissues.[1] Unless bone loss in space can be prevented, prolonged space flights may not be possible.

Courtesy NASA

Shannon Lucid exercising on a treadmill aboard the Mir space station.

THE SCIENCE

Information on bone loss in space has been accumulating for 60 years. Studies done in the 1960s and early 1970s during the Gemini and Apollo space missions first showed that astronauts lost calcium from their bones during space travel. Skylab missions then offered an opportunity to study the effects of more extended periods in zero gravity. In one study, nine astronauts maintained a constant dietary intake and made continuous urine and fecal collections for 21 to 31 days before their flight, during their Skylab missions, and for 17 to 18 days after returning to Earth. The results showed that urinary calcium losses were increased during the flight. These losses occurred despite vigorous exercise regimens while in flight and were comparable to losses seen in normal adults subjected to prolonged bed rest.[2]

The examination of bone metabolism in space continued with joint Russian-American studies, many conducted among astronauts who spent long periods on the Mir space station. Techniques used to study bones in space included measuring hormones and other indicators of bone metabolism in blood and urine samples before and after flights; measuring muscle strength and bone density before, during, and after return to Earth's gravity; and taking X-ray scans to determine bone mass.[3] The most significant bone loss was found in weight-bearing parts of the skeleton; the lower vertebrae, hips, and upper femur—the same areas at risk for fracture in osteoporosis.[4] Once the astronauts returned to Earth, calcium loss slowed, but even after six months bone mass in most subjects had not completely recovered.[5] Biochemical measures indicate that the bone loss that occurs during space flight is due to an increase in bone resorption and decreased intestinal calcium absorption.[1,3] The decrease in calcium absorption is likely due to low levels of vitamin D from insufficient dietary intake and lack of ultraviolet light exposure during space flight.[3]

THE APPLICATION

What we learn about preventing bone loss in space can be applied to those on Earth who are at risk for osteoporosis. To counteract the effects of weightlessness, astronauts exercise in space on stationary bikes and treadmills and pull against bungee cords. Exercises that provide higher loads can counteract bone loss to some degree but do not prevent it.[3] In addition to weight-bearing exercise, future studies may address nutritional factors such as vitamin D, calcium, and omega-3 fatty acids intakes as well as appropriate levels of other nutrients that affect bone metabolism such as vitamin K, sodium, and protein.[6] Understanding how to optimize nutrient intake and exercise patterns for bone mineral retention will help not only astronauts during long space flights but also people who are on prolonged bed rest because of surgery, serious illness, or complications of pregnancy and those who are experiencing immobilization of some part of the body because of stroke, fracture, spinal cord injury, or other chronic conditions. These people often experience significant bone loss and are at high risk for developing osteoporosis and fracturing a bone.[7]

---

[1]Okada, A., Ichikawa, J., and Tozawa, K. Kidney stone formation during space. *Clin Calcium* 21:1505–1510, 2011.

[2]Whedon, G. D., Lutwak, L., Rambaut, P., et al. Mineral and nitrogen metabolic studies on Skylab flights and comparison with effects of Earth long-term recumbency. *Life Sci Space Res* 14:119–127, 1976.

[3]Smith, S., Wastney, M., O'Brien, K., et al. Bone markers, calcium metabolism and calcium kinetics during extended-duration space flight on the Mir space station. *J Bone Miner Res* 20:208–218, 2005.

[4]Grigoriev, A. I., Oganov, V. S., Bakulin, A. V., et al. Clinical and physiological evaluation of bone changes among astronauts after long-term space flights. *Aviakos. Ekolog Med* 32:21–25, 1998.

[5]Shackelford, L. C., Oganov, V., LeBlanc, A., et al. Bone mineral loss and recovery after shuttle-Mir flights.

[6]Zwart, S. R., Pierson, D., Mehta, S., et al. Capacity of omega-3 fatty acids or eicosapentaenoic acid to counteract weightlessness-induced bone loss by inhibiting NF-kappa B activation: From cells to bed rest to astronauts. *J Bone Miner Res* 25:1049–1057, 2010.

[7]Grossman, J. M. Osteoporosis prevention. *Curr Opin Rheumatol* 23:203–210, 2010.

CRITICAL THINKING Osteoporosis Risk

Sylvia is worried about her risk for osteoporosis. Her 75-year-old mother recently suffered a fractured hip due to reduced bone density caused by osteoporosis. Her previously independent mother is now living in a nursing home and struggling to return to her former life. Sylvia is frightened that she will face the same future. To assess her risk, she completes an osteoporosis risk-factor questionnaire from a health magazine and also records her food intake for one day.

OSTEOPOROSIS QUESTIONNAIRE

Gender?	_ Male X Female
Age?	_ 12 to 18 years _ 19 to 30 years X 31 to 50 years _ 51 to 70 years _ > 70 years
Have you ever broken a bone?	_ Yes X No
What is your bone density?	X Never been measured _ Normal _ Low density
What is your Body Mass Index?	_ <18.5 kg/m² _ 18.5 to 24.9 X 25.0 to 29.9 _ 30.0 to 34.9 _ > 35
Do you smoke cigarettes?	X Yes _ No
How much alcohol do you drink?	_ > 2 drinks/day _ 1–2 drinks/day X Several drinks per week _ None
How much milk do you drink?	_ None X 2 or fewer glasses a day _ 3 or more glasses a day
How much milk did you drink as a child?	_ None X 2 or fewer glasses a day _ 3 or more glasses a day
How much milk did you drink as an adolescent?	X None _ 2 or fewer glasses a day _ 3 or more glasses a day
How often do you exercise?	X Less than 3 times a week _ 3 or more times a week
What types of activities do you participate in?	_ None X Walking, jogging, tennis _ Swimming, bicycling
What is your exercise history?	_ I have been active all my life. X I was active as a child but no longer exercise often. _ I recently started exercising.
If you are female, are you currently menstruating?	X Yes _ No
If you are postmenopausal, how long ago did menopause occur?	_ Less than 5 years ago _ More than 5 years ago
Do you have a family history of osteoporosis?	X Yes _ No

CRITICAL THINKING QUESTIONS

- Based on this survey how would you describe Sylvia's osteoporosis risk?
- List 4 risk factors Sylvia can change and how each could be changed to reduce her risk.

How Diet Affects Osteoporosis Risk Diet can have a significant effect on osteoporosis risk. Adequate calcium intake is important for bone development and bone health throughout life. During childhood and adolescence, adequate calcium intake is needed to maximize bone density; low calcium intakes during the years of bone formation result in a lower peak bone mass. If calcium intake is low after peak bone mass has been achieved, the rate of bone loss may be increased and, along with it, the risk of osteoporosis.

Despite its importance, calcium intake alone does not predict the risk of osteoporosis. Other dietary components affect bone mass and bone health by affecting calcium absorption, urinary calcium losses, and bone physiology. Low intakes of vitamin D reduce calcium absorption, thereby increasing the risk of osteoporosis. Diets high in phytate, oxalates, and tannins also reduce calcium absorption. High sodium has been

© Elena Elisseeva/iStockphoto

Sylvia's Food Record

FOOD	ENERGY (kcals)	CALCIUM (mg)
Breakfast		
Eggs (2 large)	150	5
Toast with margarine (2 slices)	200	50
Orange juice (¾ cup)	80	15
Coffee with cream (1 cup)	45	35
Lunch		
Bologna sandwich on white bread with mayonnaise	260	60
Lettuce and tomato	10	5
Milk (1 cup)	120	300
Apple (1 medium)	80	10
Snack		
Chips (1 oz)	150	10
Beer (12 fl oz)	140	20
Dinner		
Roast beef (3 oz)	225	5
Mashed potatoes and gravy (1 cup)	350	50
Spinach, steamed (¾ cup)	31	184
Iced tea (12 oz)	4	0
Ice cream (½ cup)	140	70
Total	1985	819

CRITICAL THINKING QUESTIONS

- Compare Sylvia's calcium intake to her RDA for calcium.
- Compare the bioavailability of the two main sources of calcium in Sylvia's diet.
- Suggest changes to her diet that will increase her calcium intake to meet the RDA without increasing her calories.

iProfile Use iProfile to find out how much calcium is provided by the nondairy foods in your diet.

implicated as a risk factor for osteoporosis because high dietary sodium intake increases calcium loss in the urine.[23] Adequate protein is necessary for bone health, but increasing protein intake increases urinary calcium losses. Despite this, high protein intakes are generally not associated with a higher risk of osteoporosis. This is because diets higher in protein are typically higher in calcium and may enhance calcium absorption. Bone mass depends more on the ratio of calcium to protein than the amount of protein alone. When intakes of calcium and vitamin D are adequate, neither high protein nor high sodium is believed to adversely affect bone health and the risk of osteoporosis.[24-26] Higher intakes of zinc, magnesium, potassium, fiber, and vitamin C—nutrients that are plentiful in fruits and vegetables—are associated with greater bone mass[27] (see **Critical Thinking: Osteoporosis Risk**).

DEBATE

Are Supplements a Safe Way to Meet Your Calcium Needs?

Calcium is essential for bone health, but many of us don't consume the recommended amount in our diets; instead, almost 50% of American adults rely on calcium supplements.[1,2] These help us meet the RDA, but now several studies have suggested that calcium supplementation may increase the risk of cardiovascular disease. Will the calcium you take to keep your bones strong cause a heart attack? Should you supplement calcium?

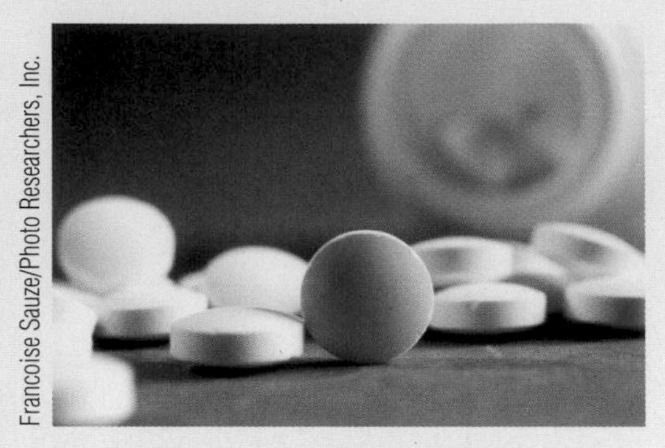

Francoise Sauze/Photo Researchers, Inc.

Calcium supplements are important for some people, but the 2011 DRIs for calcium caution that oversupplementation may be causing many postmenopausal women to be getting too much calcium.

The connection between calcium and cardiovascular disease is related to the fact that calcium supplements cause a rise in blood calcium levels after ingestion. Since the atherosclerotic deposits that cause heart disease are hardened by calcium, elevated blood levels have been hypothesized to cause calcium to be transferred to the artery wall. High blood calcium may also affect blood clotting and the function of endothelial cells that line blood vessels and smooth muscle cells, all key factors in cardiovascular disease. An analysis of a large number of calcium supplementation trials found about a 30% increase in the incidence of heart attacks in patients taking the supplements compared to the controls.[3] Analyses of studies that included both calcium and vitamin D supplementation were less definitive but still suggest that there may be a connection.[4,5]

Not all scientists believe that there is cause for concern. Some argue that flaws in the way calcium supplementation studies were analyzed make the association between calcium supplements and heart disease questionable. One study looked only at calcium supplements, not calcium plus vitamin D, as is typically prescribed.[3] A study that did include both calcium and vitamin D supplements eliminated some study subjects from the analyses.[5] In addition, the trials that were analyzed were not specifically designed to examine the effect of calcium supplements on cardiovascular events. These scientists argue that further support for this hypothesis will need trials specifically designed to examine the effect of calcium and vitamin D supplements on cardiovascular health.[6]

There is strong evidence that adequate calcium throughout life reduces the risk of osteoporosis, but this does not necessarily mean that adding a calcium supplement in adulthood will have the same effect. Calcium supplements have been shown to increase bone density, but most trials do not show a reduction in the incidence of bone fractures.[7-9] In the United States 82.6 million people suffer from cardiovascular disease,[10] whereas only 1.5 million have osteoporosis.[11] Some suggest that if you consider the overall benefits and risks of calcium supplements in the population, the number of cardiovascular

How to Prevent and Treat Osteoporosis

You can't feel your bones weakening, so people with osteoporosis may not know that their bone mass is dangerously low until they are in their 50s or 60s and experience a bone fracture. Once osteoporosis has developed, it is difficult to restore lost bone. Therefore, the best treatment for osteoporosis is to prevent it by achieving a high peak bone mass and slowing the rate of bone loss. During childhood, adolescence, and young adulthood, diet and exercise can help prevent osteoporosis by ensuring maximum peak bone density. A diet that contains adequate amounts of calcium and vitamin D produces greater peak bone mass during the early years and slows bone loss as adults age. Adequate intakes of zinc, magnesium, potassium, fiber, vitamin K, and vitamin C—nutrients that are plentiful in fruits and vegetables—are also important for bone health. Weight-bearing exercise before about age 35 helps to increases peak bone mass, and maintaining an active lifestyle that includes weight-bearing exercise throughout life helps preserve bone density. Limiting smoking and alcohol consumption can also help to increase and maintain bone density.

events caused by calcium supplements may actually be greater than the number of fractures prevented.[9]

So if calcium supplements increase heart attack risk, should you skip them? Are you increasing your osteoporosis risk by not taking supplements? Everyone agrees that getting adequate calcium from food is the best option. It is absorbed more slowly, so it does not cause a rapid rise in blood calcium.[9] Unlike data for supplements, epidemiology does not correlate high calcium intake from foods with heart disease; it may even be associated with a lower risk of cardiovascular disease.[12] But not everyone agrees on how much calcium the average person needs each day to keep bones strong and healthy. The World Health Organization recommends 400 to 500 mg of calcium a day to prevent osteoporosis, while the RDA is 1,000 mg/day for younger adults and 1,200 mg/day for women over 50 and men over 70. In postmenopausal women, the group with the highest risk of osteoporosis, typical calcium intake is about 900 mg and over 35% of this may come from supplements.[1] If these women do not take supplements they are getting only about half of the RDA. Are supplements worth the risk?

THINK CRITICALLY: What dietary changes would you suggest to a teenage girl who is consuming 600 mg of calcium per day?

[1]Mangano, K. M., Walsh, S. J., Insogna, K. L., et al. Calcium intake in the United States from dietary and supplemental sources across adult age groups: New estimates from the National Health and Nutrition Examination Survey 2003–2006. *J Am Diet Assoc* 111:687–695, 2011.

[2]U.S. Department of Agriculture and U.S. Department of Health and Human Services. *Dietary Guidelines for Americans, 2010.* 7th ed., Washington, DC: U.S. Government Printing Office, December 2010.

[3]Bolland, M. J., Avenell, A., Baron, J. A., et al. Effect of calcium supplements on risk of myocardial infarction and cardiovascular events: Meta-analysis. *BMJ* 341:c3691, 2010.

[4]Hsia, J., Heiss, G., Ren, H., et al. Calcium/vitamin D supplementation and cardiovascular events. *Circulation* 115:846–854, 2007.

[5]Bolland, M., Grey, A., Avenell, A., et al. Calcium supplements with or without vitamin D and risk of cardiovascular events: Reanalysis of the Women's Health Initiative limited access dataset and meta-analysis *BMJ*, 342:d2040, 2011.

[6]Hennekens, C. H., and Barice, E. J. Calcium supplements and risk of myocardial infarction: A hypothesis formulated but not yet adequately tested. *Am J Med* 124:1097–1098, 2011.

[7]Tang, B. M., Eslick, G. D., Nowson, C., et al. Use of calcium or calcium in combination with vitamin D supplementation to prevent fractures and bone loss in people aged 50 years and older: A meta-analysis. *Lancet* 370:657–666, 2007.

[8]Lips, P., Bouillon, R., van Schoor, N. M., et al. Reducing fracture risk with calcium and vitamin D. *Clin Endocrinol (Oxf)* 73:277–285, 2010.

[9]Reid, I. R., Bolland, M. J., Sambrook, P. N., and Grey, A. Calcium supplementation: Balancing the cardiovascular risks. *Maturitas* 69:289–295, 2011.

[10]AHA statistical update: Heart disease and stroke statistics—2011.

[11]Bone health and osteoporosis: A report of the surgeon general.

[12]Kaluza, J., Orsini, N., Levitan, E. B., et al. Dietary calcium and magnesium intake and mortality: A prospective study of men. *Am J Epidemiol* 171:801–807, 2010.

Supplements of calcium and vitamin D are currently used by many people, particularly women, to prevent and treat osteoporosis (see **Debate: Are Supplements a Safe Way to Meet Your Calcium Needs?**). In young individuals supplemental calcium can promote the development of a higher peak bone mass. In postmenopausal women, calcium supplements have a small but beneficial effect on bone mass.[28] Supplementation with 700 to 800 IU of vitamin D per day appears to reduce the risk of hip fractures in elderly persons.[29]

Osteoporosis is commonly treated with estrogen and progesterone to reduce bone breakdown and increase calcium absorption, along with supplements of calcium and vitamin D and regular weight-bearing exercise. Replacing the hormones estrogen and progesterone, lost in menopause—known as *hormone replacement therapy*—has been shown to reduce bone loss and restore some lost bone, but this therapy can also increase the risk of heart disease, stroke, blood clots, and breast cancer, so its use should be considered within the context of the individual's other health risks.[30] Treatments also include other hormones and drugs, called bisphosphonates, that inhibit the activity of cells that break down bone. Bisphosphonates have been shown to prevent postmenopausal bone loss, increase bone mineral density, and reduce the risk of fractures.[31]

11.3 Calcium (Ca)

LEARNING OBJECTIVES

- List foods that are good sources of calcium.
- Explain the functions of calcium in the body.
- Compare the roles of parathyroid hormone, calcitonin, and vitamin D in the regulation of blood calcium levels.

Calcium is the most abundant mineral in the body. It provides structure to bones and teeth and has essential regulatory roles. Because of the importance of calcium in regulating body processes, blood levels of this mineral are strictly controlled. However, this occurs at the expense of bone calcium; calcium is released from bone when blood levels drop. As a result, if the diet is not sufficient in calcium, over the long term, the amount of calcium in bone declines and the risk of bone fractures due to osteoporosis is increased.

Calcium in the Diet

The main source of calcium in the U.S. diet is dairy products such as milk, cheese, and yogurt. Fish, such as sardines, that are consumed in their entirety, including the bones, are also a good source, as are legumes and some green vegetables such as broccoli, Chinese cabbage, and kale (**Figure 11.9**). Grains provide smaller amounts of calcium, but because they are consumed in such large quantities they make a significant contribution to dietary calcium intake.

If bioavailability were the same, how much cooked kale would you need to eat to get as much calcium as you do from a cup of milk?

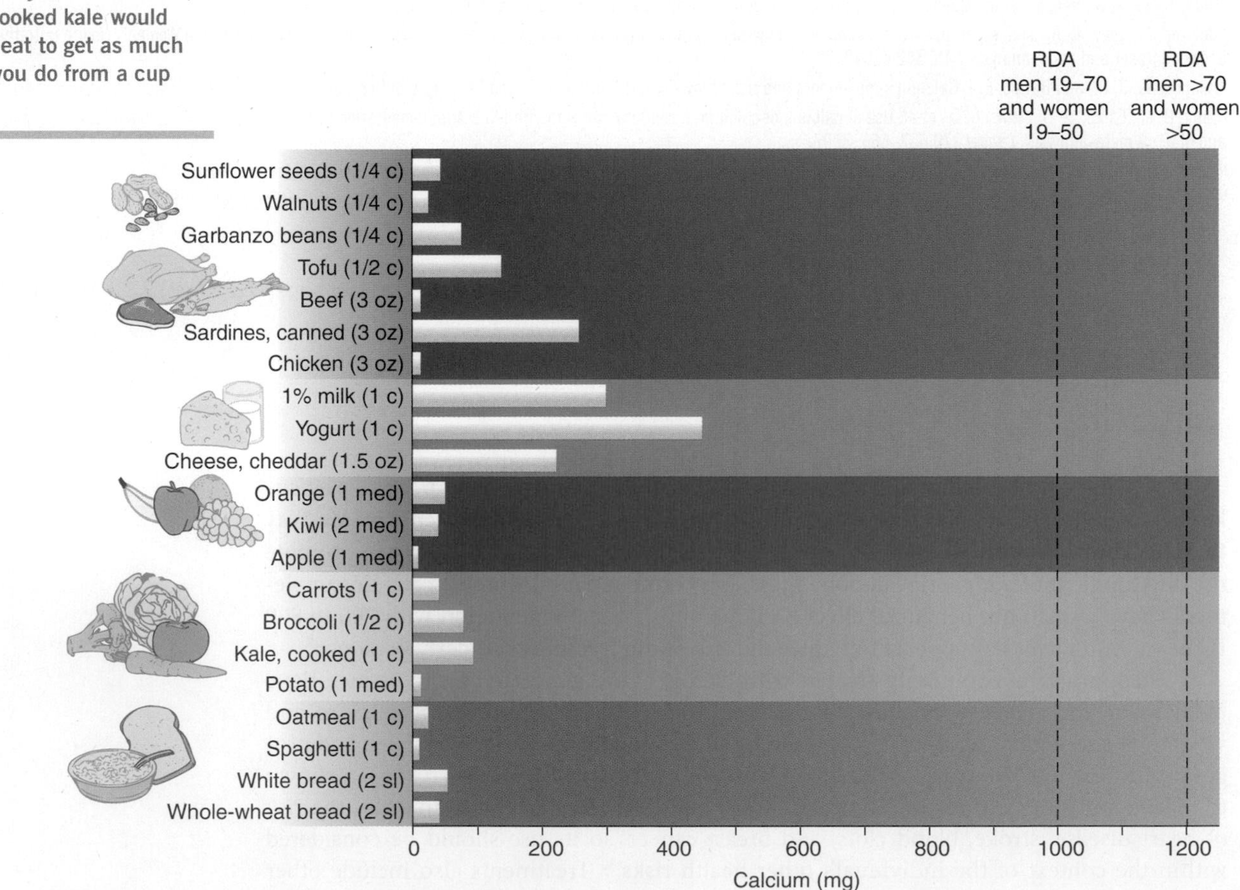

Figure 11.9 Calcium content of MyPlate food groups
Calcium is provided by many different foods, but dairy products and fish consumed with bones are the best sources; the dashed lines represent the RDAs for adult men and women.

Some of the calcium in the diet comes from supplements or is added to foods during food processing. Baked goods such as breads, rolls, and crackers, to which nonfat dry milk powder has been added, provide calcium. Tortillas that are treated with lime water (calcium hydroxide) provide calcium. Tofu is a good source when calcium is used in its processing. In addition, there are numerous products on the market, such as orange juice and breakfast cereals, that are fortified with calcium.

Calcium in the Digestive Tract

Calcium is absorbed by both active transport and passive diffusion (**Figure 11.10**). Active transport depends on the availability of the active form of vitamin D, which turns on the synthesis of calcium transport proteins in the intestine. Active transport accounts for most calcium absorption when intakes are low to moderate. At high calcium intakes, passive diffusion becomes more important. As calcium intake increases, the percentage that is absorbed declines. Both active and passive absorption require stomach acid. The acid allows the calcium to dissolve to form calcium ions (Ca^{2+}) that can be absorbed.[32]

The efficiency of calcium absorption is higher at times when the body's need for calcium is greater. During pregnancy, when calcium is needed for formation of the fetal skeleton, elevated estrogen helps increase calcium absorption to over 50%. Calcium need is also increased during lactation, but some of the calcium needed to make milk appears to come from the mother's bones. After lactation stops, an increase in calcium absorption and retention of calcium by the kidney help restore bone calcium. During infancy, when

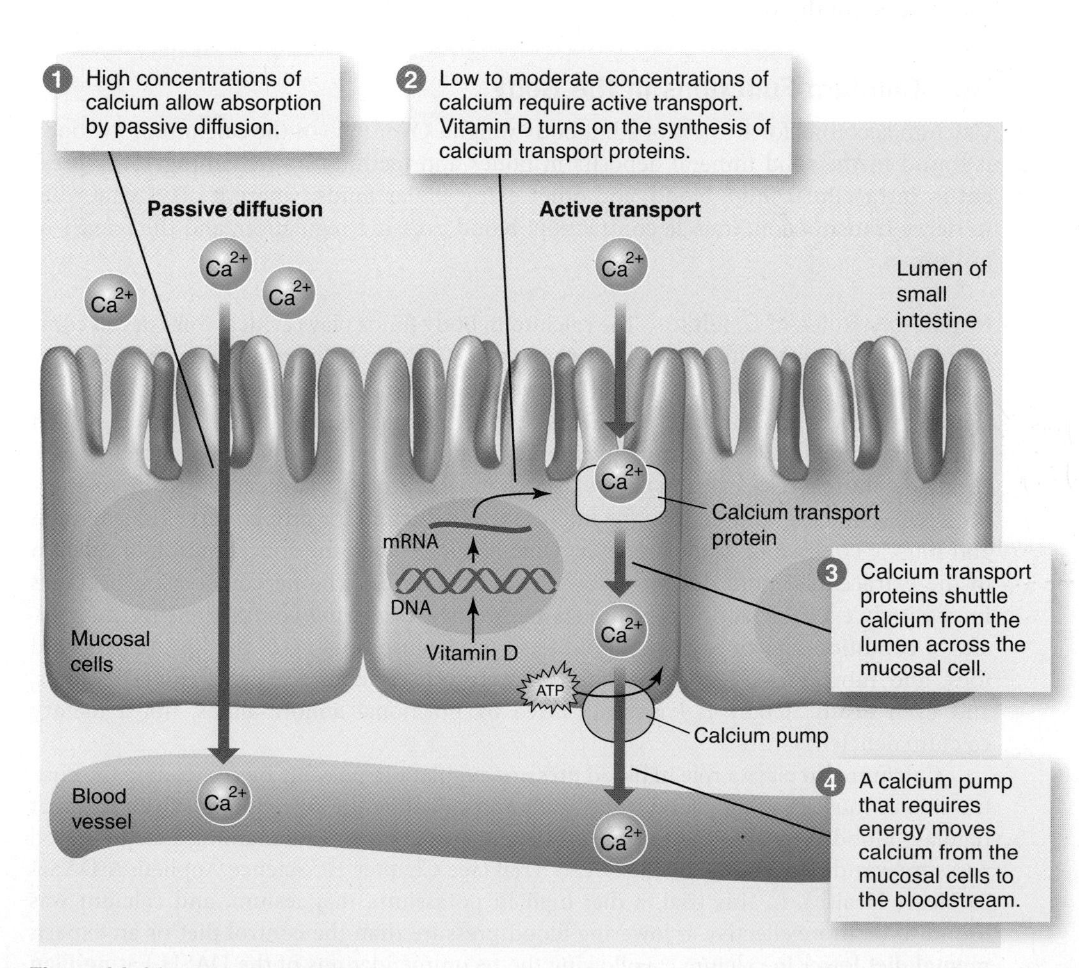

Figure 11.10 Calcium absorption
Some calcium can be absorbed by passive diffusion, particularly when calcium concentrations are high; at lower calcium concentrations, absorption occurs primarily by an active transport mechanism that requires vitamin D.

the skeleton is growing rapidly, about 60% of calcium consumed is absorbed. In young adults, absorption is about 25%. In older adults, absorption declines. This may be due to a reduction in stomach acid, a decrease in blood levels of the active form of vitamin D, or a decrease in responsiveness to vitamin D.[7] An additional decrease in calcium absorption occurs in women after menopause due to the decrease in estrogen.

The bioavailability of calcium is affected by a number of dietary factors, the most significant of which is vitamin D. When vitamin D is deficient, absorption drops to about 10%.[33] Calcium absorption is increased by the presence of acidic foods, lactose, and fat, and decreased by tannins, fiber, phytates, and oxalates. For example, spinach is a high-calcium vegetable but only about 5% of its calcium is absorbed; the rest is bound by oxalates and excreted in the feces.[34] Vegetables such as kale, collard greens, turnip greens, mustard greens, and Chinese cabbage are low in oxalates, so their calcium is well absorbed. Chocolate also contains oxalates, but chocolate milk is still a good source of calcium because the amount of oxalates from the chocolate added to a glass of milk is small. Fiber can also reduce calcium absorption, but with few exceptions, the effect is small. Phytates, however, can have a significant effect on the absorption of calcium from foods such as wheat bran and pinto, red, and white beans. It would take almost 10 servings of red beans to provide the same amount of absorbable calcium as one serving of milk. When calcium intake is low, dietary components that alter absorption have a greater effect on calcium status than when calcium intake is adequate.[35]

Medications that reduce the secretion of stomach acid, such as H2 blockers (e.g., Pepcid AC and Zantac) and proton pump inhibitors (e.g., Prilosec and Prevacid), can also interfere with calcium absorption. These are used to treat heartburn and gastroesophageal reflux disease (GERD) and may be taken over long periods. By blocking stomach acid secretion, they reduce calcium absorption and may contribute to the development of osteoporosis in the long term.[32]

How Calcium Functions in the Body

Calcium accounts for 1 to 2% of adult body weight. Over 99% of the calcium in the body is found in the solid mineral deposits in bones and teeth.[33] The remaining 1% is present in intracellular fluid, blood, and other extracellular fluids, where it plays vital roles in nerve transmission, muscle contraction, blood pressure regulation, and the release of hormones.

Regulatory Roles of Calcium The calcium in body fluids plays critical roles in cell communication and the regulation of body processes. Calcium helps regulate enzyme activity and is necessary in blood clotting. It is involved in transmitting chemical and electrical signals in nerves and muscles. It is necessary for the release of neurotransmitters, which allow nerve impulses to pass from one nerve to another and from nerves to target tissues. Inside the muscle cells, calcium allows the two muscle proteins, actin and myosin, to interact to cause muscle contraction. The importance of calcium for proper nerve transmission and muscle contraction is illustrated by what happens when the concentration of calcium in the extracellular fluid drops too low. When this occurs, the nervous system becomes increasingly excitable and nerves fire spontaneously, triggering contractions of the muscles, a condition known as *tetany*. Mild tetany can cause tingling of the lips, fingers, and toes, and more serious tetany results in severe muscle contractions, tremors, cramps, and even death. Tetany is typically caused by hormonal abnormalities, not a dietary calcium deficiency.

Calcium also plays a role in blood pressure regulation, possibly by controlling the contraction of muscles in the blood vessel walls and signaling the secretion of substances that regulate blood pressure. The impact of adequate calcium on maintaining healthy blood pressure was demonstrated by the DASH Trial (see Chapter 10, Science Applied: A DASH to Heart Health). In this trial, a diet high in potassium, magnesium, and calcium was found to be more effective at lowering blood pressure than the control diet or an experimental diet lower in calcium.[36] Following the recommendations of the DASH Eating Plan can help keep both bones and blood pressure healthy.

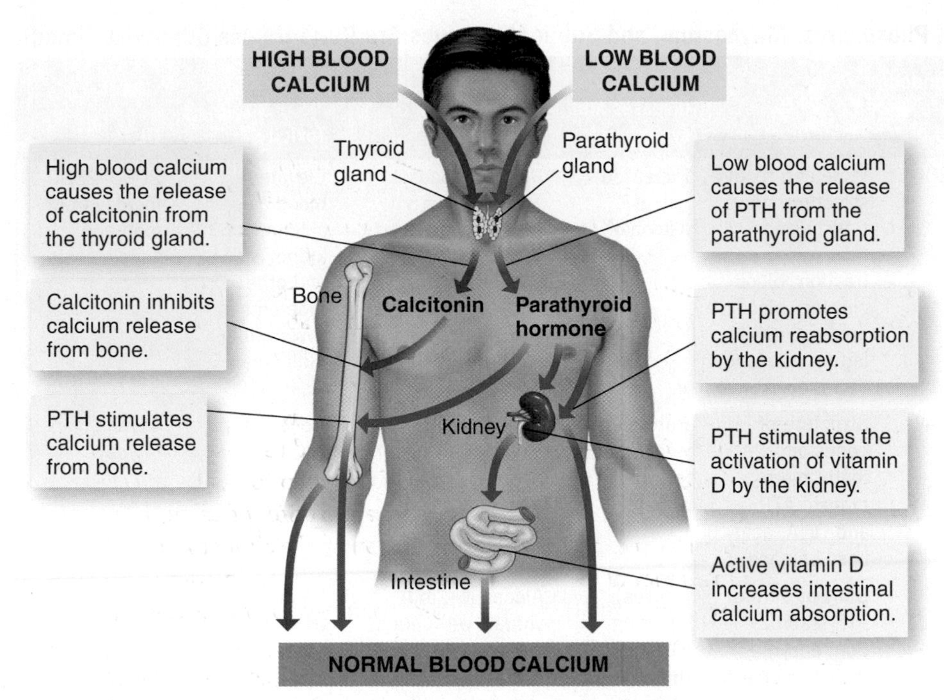

Figure 11.11 **Regulation of blood calcium levels**
Levels of calcium in the blood are very tightly regulated by parathyroid hormone and calcitonin.

THINK CRITICALLY

When blood calcium is normal, PTH levels are low. How does this affect the amount of calcium excreted in the urine and the amount absorbed from the GI tract?

How Blood Calcium Levels Are Regulated The roles of calcium are so vital to survival that powerful regulatory mechanisms ensure that constant intracellular and extracellular concentrations are maintained. Slight changes in blood calcium levels trigger responses that quickly raise or lower them back to normal. The hormones **parathyroid hormone (PTH)**, which raises blood calcium, and **calcitonin**, which lowers blood calcium, help maintain calcium homeostasis (**Figure 11.11**). If the level of blood calcium falls too low, it triggers an increase in the secretion of PTH from the parathyroid glands. PTH stimulates the release of calcium from bone and causes the kidneys to reduce calcium loss in the urine and to activate vitamin D. Activated vitamin D increases the amount of calcium absorbed from the gastrointestinal tract and acts with PTH to stimulate calcium release from the bone and calcium retention by the kidney (see Chapter 9). The overall effect is to rapidly increase blood calcium levels. If blood calcium levels become too high, the secretion of PTH is reduced. This increases excretion of calcium by the kidney, decreases vitamin D activation so less dietary calcium is absorbed, and reduces calcium release from bone. High blood calcium also stimulates the secretion of calcitonin from the thyroid gland. Calcitonin acts primarily on bone to inhibit the release of calcium. Together, low PTH levels and the presence of calcitonin cause blood calcium levels to decrease to normal (see Figure 11.11).

parathyroid hormone (PTH) A hormone secreted by the parathyroid gland that increases blood calcium levels.

calcitonin A hormone secreted by the thyroid gland that reduces blood calcium levels.

Calcium Deficiency

When calcium intake is not adequate, normal blood levels are maintained by resorbing calcium from bone. This provides a steady supply of calcium to support cell communication, nerve transmission, muscle contraction, and other regulatory roles. Although there are no short-term symptoms of a calcium deficiency related to the removal of calcium from bone, a deficient diet affects bone mass. Low calcium intake during the years of bone formation results in lower bone density. Low intake during the adult years increases the rate of bone loss. As discussed previously, both of these effects increase the risk of osteoporosis.

TABLE 11.2 A Summary of Calcium, Phosphorus, Magnesium, and Sulfur (All Values Are RDAs Unless Otherwise Noted)

Mineral	Sources	Recommended intake for adults	Major functions	Deficiency diseases and symptoms	Groups at risk of deficiency	Toxicity	UL
Calcium	Dairy products, fish consumed with bones, leafy green vegetables, fortified foods	1000–1200 mg/day	Bone and tooth structure, nerve transmission, muscle contraction, blood clotting, blood pressure regulation, hormone secretion	Increased risk of osteoporosis	Postmenopausal women; elderly people; those who consume a vegan diet, are lactose intolerant, or have kidney disease	Elevated blood calcium, kidney stones, and other problems in susceptible individuals	2000–2500 mg/day from food and supplements
Phosphorus	Meat, dairy, cereals, baked goods	700 mg/day	Structure of bones and teeth, membranes, ATP, and DNA; acid–base balance	Bone loss, weakness, lack of appetite	Premature infants, alcoholics, elderly people	None likely	4000 mg/day
Magnesium	Greens, whole grains, legumes, nuts, seeds	310–420 mg/day	Bone structure, ATP stabilization, enzyme activity, nerve and muscle function	Nausea, vomiting, weakness, muscle pain, heart changes	Alcoholics, individuals with kidney or gastrointestinal disease	Nausea, vomiting, low blood pressure	350 mg/day from nonfood sources
Sulfur	Protein foods, preservatives	None specified	Part of some amino acids and vitamins, acid–base balance	None when protein needs are met	None	None likely	ND

Note: UL, Tolerable Upper Intake Level, ND, not determined due to insufficient evidence.

Recommendations for Meeting Calcium Needs

For adults ages 19 to 50, 1000 mg/day of calcium is recommended to maintain bone health. Since natural bone loss begins earlier in women than in men due to menopause, the RDA for women 51 to 70 is set at 1200 mg/day to slow bone loss. For men in this age group it remains at 1000 mg/day. Bone loss and resulting osteoporotic fractures are concerns for both men and women 70 and older, so the RDA for both genders in this age group is 1200 mg/day[37] (**Table 11.2**). In children and adolescents the RDA is set at a level that supports bone growth. For adolescents the RDA is higher than for adults—1300 mg/day for boys and girls ages 9 through 18 years.

Infants thrive on the amount of calcium they obtain from human milk. For infants, an AI is set based on the calcium content of human milk. The RDA for calcium during pregnancy is the same as for nonpregnant women. This is because there is an increase in maternal calcium absorption during pregnancy that helps to supply the calcium needed for the fetal skeleton. The RDA during lactation is the same as for nonlactating women even though calcium is secreted in milk. The source of this calcium does appear to be the maternal skeleton, but this bone resorption occurs regardless of calcium intake and the calcium lost appears to be regained following weaning.[38]

Meeting Calcium Needs with Food The best way to meet calcium needs is with food. Food provides you with other nutrients, some of which are important for absorbing and using calcium. Americans typically do not consume enough calcium. Milk is the major source of calcium in the U.S. diet, but teenage boys and girls today drink more soda and other sweetened beverages than milk. Since the 1970s, children and teens (ages 2 to 18 years) have more than doubled the number of calories they consume from soft drinks while cutting the calories they obtain from milk by 34%.[39] This has significantly reduced their calcium intake (**Figure 11.12**). Teenage girls consume only 60% of the recommended amount of calcium, with soda drinkers consuming almost one-fifth less calcium than those who don't drink soda.[40] Both the DASH Eating Plan and MyPlate recommend 3 cups of milk or milk products plus 2 to 5 cups of vegetables daily to ensure adequate calcium intake. Ice cream, puddings, and dishes made with milk or cream are also good calcium sources but contribute empty calories from added sugar and solid fats.

THINK CRITICALLY

How do you think the trend away from milk consumption will affect the incidence of osteoporosis 30 years from now?

Individuals who are lactose intolerant can meet their calcium needs by consuming high-calcium foods that contribute little or no lactose. Those who can tolerate some lactose may be able to consume fermented dairy products such as yogurt and cheese. Lactose-free calcium sources include dark-green vegetables such as kale, broccoli, and mustard greens; soy products processed with calcium; and fish consumed with the bones (**Figure 11.13a**). Drinking milk treated with the lactose-digesting enzyme, lactase (Lactaid milk) or taking lactase pills to help digest the lactose consumed with a meal can also help those with lactose intolerance to meet calcium needs (see Chapter 4). Calcium supplements and fortified foods such as breakfast cereals and juice products also provide calcium without lactose (**Figure 11.13b**).

When Is Supplemental Calcium Recommended? Calcium supplements are currently recommended for individuals who do not consume dairy products, older adults, and anyone else who does not meet their calcium needs with diet alone. The type of calcium supplement and the amount per dose affect how much you absorb. Calcium is present in foods and supplements as a compound with other molecules. During digestion, the calcium separates from the other molecule and becomes available for absorption. Calcium citrate and calcium carbonate are calcium compounds commonly found in supplements. Calcium carbonate is 40% calcium by weight and 60% carbonate. Other calcium compounds such as calcium phosphate,

	Low-fat milk	Cola soft drink
Serving size (oz)	8	12
Energy (kcal)	102	150
Protein (g)	8	0
Calcium (mg)	300	0
Phosphorus (mg)	235	45
Riboflavin (mg)	0.4	0
Vitamin A (μg)	144	0
Vitamin D (μg)	3	0
Caffeine (mg)	0	40

Figure 11.12 **Soda versus milk**
A 12-ounce can of soda contains about 10 teaspoons of sugar and few other nutrients. Replacing a glass of milk with a soda increases the amount of added sugar in the diet by about 40 g and reduces protein, calcium, vitamin A, vitamin D, and riboflavin intake.

Figure 11.13 **Dietary sources of calcium**

Felicia Martinez/PhotoEdit

(a) Dairy products, fish consumed with bones, leafy greens, and legumes are good sources of calcium.

Andy Washnik

(b) A variety of calcium-fortified foods are available to help meet needs. A cup of calcium-fortified orange juice or soymilk provides as much calcium as a glass of milk.

OFF THE LABEL Counting All Your Calcium

To find out if you get enough calcium, you'll need to count all your sources. You can obtain calcium from natural sources such as milk, yogurt, and leafy greens; from foods fortified with the mineral; and from calcium supplements.

You can see if a packaged food is a good source of calcium by looking at the label. The Nutrition Facts panel lists the % Daily Value for calcium. To calculate the milligrams of calcium in that food, multiply the % Daily Value by 1000 mg (the Daily Value for calcium). Descriptors such as "high in calcium" or "a good source of calcium" can help you identify foods that make a significant calcium contribution (see table). Foods high in calcium may include the health claim that a diet high in calcium helps reduce the risk of osteoporosis.

If you rely on supplements to increase your calcium count, be aware that a multivitamin and mineral supplement provides only a small amount of the calcium you need. To maximize absorption, choose a product that has calcium along with vitamin D and provides no more than 500 mg calcium per dose (see figure). The amount of calcium you absorb from supplements also depends on how quickly the pills dissolve. To ensure that the supplement you choose provides calcium that will dissolve quickly, look for the "USP" letters or symbol on the label. This indicates that it meets the U.S. Pharmacopeia's standards for how well the tablet dissolves as well as how much calcium the tablet contains.

Calcium Descriptors on Food Labels

Descriptor	Definition
High calcium, rich in calcium, excellent source of calcium	Contains 200 mg of calcium or more per serving
Good source of calcium	Contains 10 to 190 mg of calcium per serving
More or added calcium	Contains at least 100 mg more calcium per serving than a reference food

Over-the-counter antacids can also be taken to supplement calcium intake. These have a Drug Facts, rather than a Supplement Facts, label. Many of these are safe, effective calcium supplements, such as Tums, which contains calcium carbonate. However, antacids that contain aluminum and magnesium may actually increase calcium loss.

In short, to see if you are getting enough calcium, check the labels on your foods, supplements, and medications; consider the form of calcium in each; and watch for excesses of other nutrients and contaminants. Sound complicated? Maybe an extra glass of milk is easier.

calcium lactate, and calcium gluconate have much smaller percentages of calcium, so you would need to consume more tablets per day to get the same dose. Dolamite, oyster shell, and bone meal are natural sources of calcium that may be contaminated with heavy metals such as lead. The FDA has set an upper limit as to the amount of lead a calcium supplement may contain, but it is up to the manufacturer to determine if their supplement meets FDA standards. Because calcium absorption decreases when large amounts are consumed at one time, calcium bioavailability is better when a lower-dose calcium supplement (no more than 500 mg per dose) is taken twice a day than when a supplement that provides 100% of the RDA is taken once a day (see **Off the Label: Counting All Your Calcium**).

Excessive Calcium

Too much calcium can also cause problems. Elevated blood calcium levels can cause symptoms such as loss of appetite, nausea, vomiting, constipation, abdominal pain, thirst, and frequent urination. Severe elevations may cause confusion, delirium, coma, and even death. Elevated blood calcium is rare and is most often caused by cancer or disorders that increase the secretion of PTH rather than intake from diet or supplements. But it can also result from increases in calcium absorption due to excessive vitamin D intake or a high

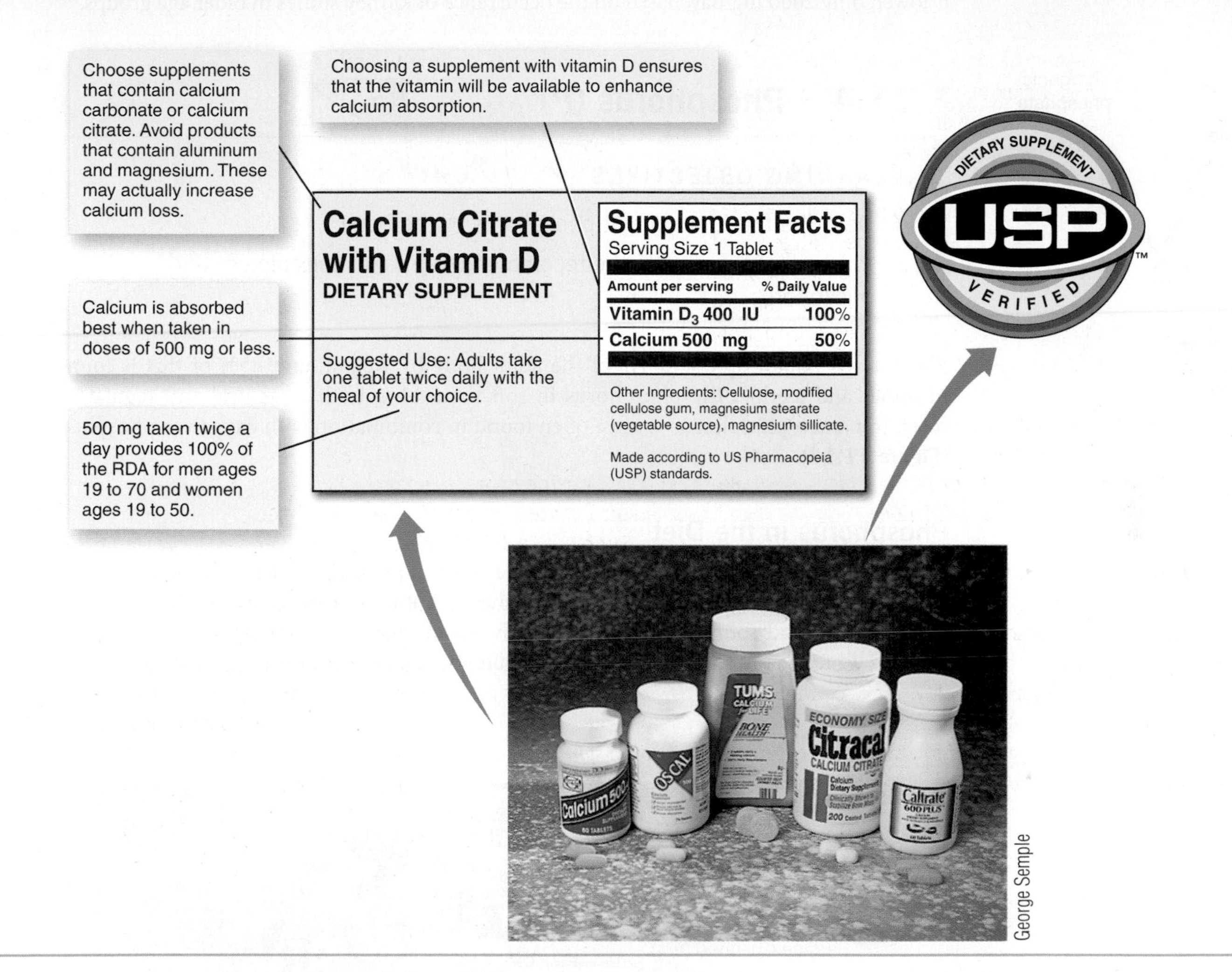

intake of calcium in combination with antacids. Consuming large amounts of milk along with antacids used to be a common treatment for peptic ulcers. This combination is associated with a condition called milk-alkali syndrome, which is characterized by high blood calcium along with calcification of the kidney that can lead to kidney failure. After the treatment of ulcers changed, the incidence of this condition declined, but it has risen again due to the increased use of calcium-containing antacids as calcium supplements and to treat heartburn.[41]

Too much calcium from supplements may also promote the formation of kidney stones. Kidney stones, which are usually composed of calcium oxalate or calcium phosphate, affect approximately 12% of the U.S. population. Although their cause is usually unknown, abnormally elevated urinary calcium, which can result from high doses of supplemental calcium, increases the risk of developing calcium stones.[37] Some postmenopausal women taking supplements may be getting too much calcium and increasing their risk of kidney stones. High calcium intake has also been linked with increased risk of prostate cancer, but it has been difficult to separate the potential effect of dairy products from that of calcium.

Excessive calcium intake can cause constipation and may interfere with the absorption of other minerals such as iron, zinc, magnesium, and phosphorus. Although calcium supplements inhibit iron absorption, there is no evidence that the long-term use of calcium supplements with meals affects iron status.[42] High intakes of calcium from supplements have also been found to reduce zinc absorption and thereby increase the amount of zinc needed in the diet.[43] There is no evidence of depletion of phosphorus or magnesium associated with calcium intake.

The UL for calcium in young adults ages 19 to 50 is 2500 mg/day. In older adults the UL is lower, only 2000 mg/day, based on the occurrence of kidney stones in older age groups.[37]

11.4 Phosphorus (P)

LEARNING OBJECTIVES

- Describe the functions of phosphorus in the body.
- Plan a diet that meets the recommended intakes for calcium and phosphorus.

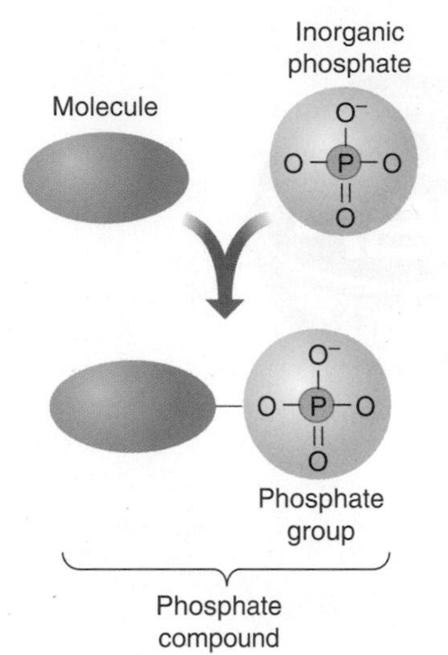

Figure 11.14 Phosphate When inorganic phosphate (phosphorus combined with oxygen) joins with another molecule, it is called a phosphate group.

Phosphorus makes up about 1% of the adult body by weight, and 85% of this is found in bones and teeth.[33] The phosphorus in soft tissues has both structural and regulatory roles. In nature, phosphorus is most often found in combination with oxygen as phosphate (**Figure 11.14**).

Phosphorus in the Diet

Phosphorus is more widely distributed in the diet than calcium. Like calcium, it is found in dairy products such as milk, yogurt, and cheese, but meat, cereals, bran, eggs, nuts, and fish are also good sources (**Figure 11.15**). Food additives used in baked goods, cheese, processed meats, and soft drinks also contribute to dietary phosphorus.

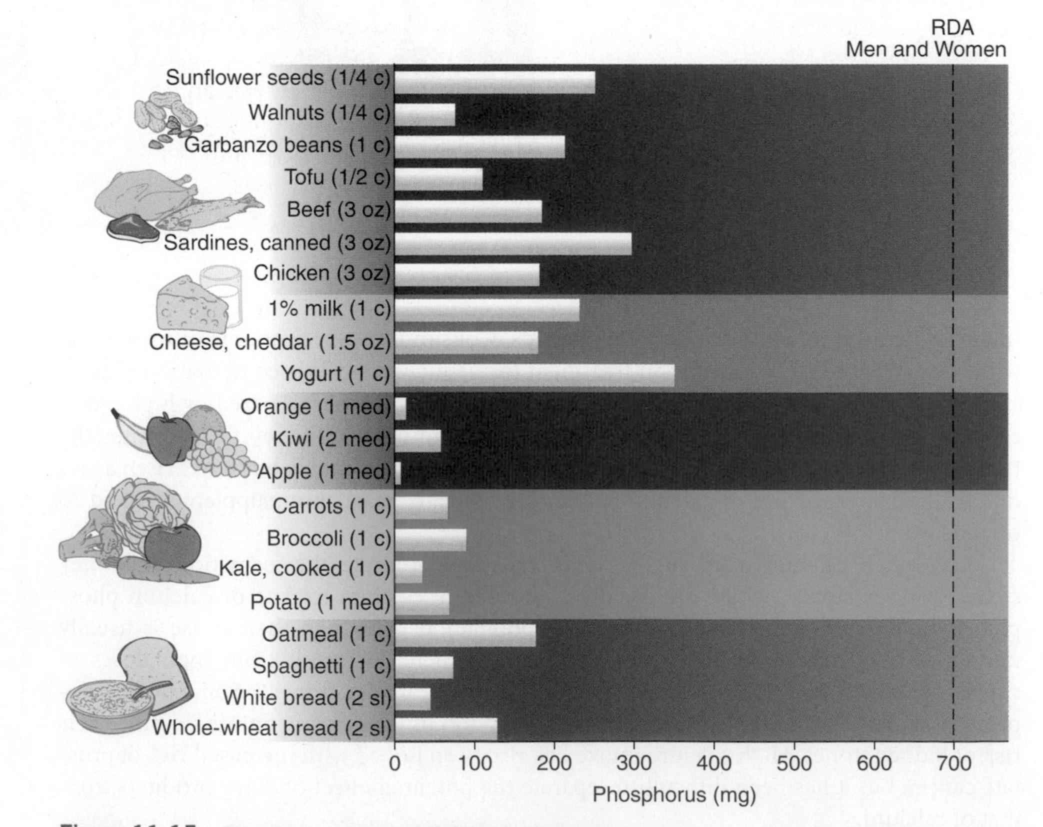

Figure 11.15 Phosphorus content of MyPlate food groups Adults can obtain their RDA for phosphorus (dashed line) by consuming foods in all the food groups.

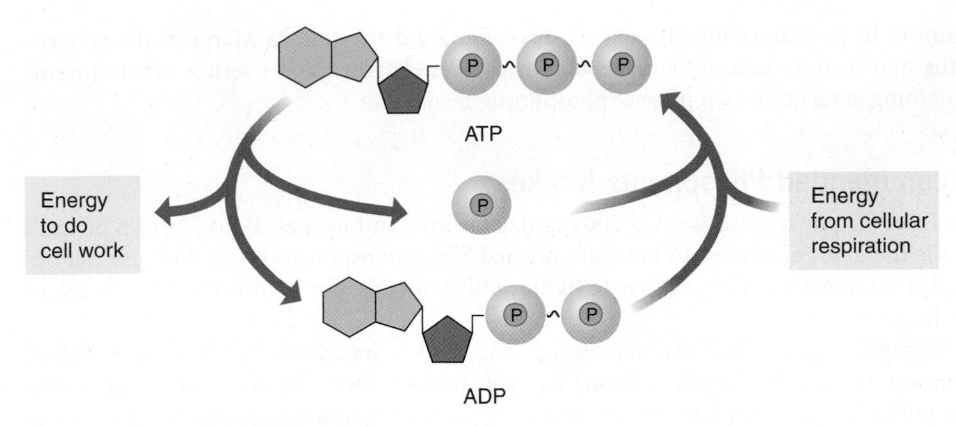

Figure 11.16 **Phosphorus and ATP**
Breaking the high-energy bond between the second and third phosphate groups of ATP releases energy for cellular work. The ADP (adenosine diphosphate) that is formed can be converted back to ATP by using the energy trapped by the electron transport chain of cellular respiration to add a phosphate group.

Phosphorus in the Digestive Tract

Phosphorus is more readily absorbed than calcium. About 60 to 70% is absorbed from a typical diet. There is no evidence that the efficiency of absorption is affected by the amount in the diet. Vitamin D does aid phosphorus absorption via an active mechanism, but most absorption occurs by a mechanism that does not depend on vitamin D. Therefore, when vitamin D is deficient, phosphorus can still be absorbed, but its absorption is reduced.

How Phosphorus Functions in the Body

Phosphorus is an important component of a number of molecules with structural or regulatory roles (see Table 11.2). Phosphorus, along with calcium, forms hydroxyapatite crystals that provide rigidity to bones. Phosphorus is a component of the water-soluble head of phospholipid molecules, which form the structure of cell membranes (see Chapter 5). Phosphorus is a major constituent of the genetic material DNA and RNA, and it is essential for energy metabolism because the high-energy bonds of ATP are formed between phosphate groups (**Figure 11.16**). Phosphorus is also a component of other high-energy compounds, including creatine phosphate, which provides energy to exercising muscles. Phosphorus-containing molecules are important in relaying signals to the interior of cells to mediate hormone action and perform other metabolic activities. Phosphorus is involved in regulating enzyme activity because the addition of a phosphate group can activate or deactivate certain enzymes. It is also part of the phosphate buffer system that helps regulate the pH in the cytosol of all cells so that chemical reactions can proceed normally.

Blood levels of phosphorus are not as strictly controlled as those of calcium, but levels are maintained in a ratio with calcium that allows bone mineralization. When blood levels of phosphorus are low, the active form of vitamin D is synthesized. This increases the absorption of both phosphorus and calcium from the intestine and increases their release from bone. When phosphorus intake is high, more is lost in the urine, so plasma levels rise only slightly. A rise in serum phosphorus indirectly stimulates PTH release, causing phosphorus excretion and calcium retention by the kidney as well as calcium release from bone. When PTH is not secreted, as when calcium levels rise, phosphorus is retained by the kidney and calcium is excreted.

Phosphorus Deficiency

Phosphorus deficiency can lead to bone loss, weakness, and loss of appetite. Deficiency is rare in healthy people because phosphorus is so widely distributed in foods. Most people in the United States easily meet their needs; the average daily intake is about 1400 mg for adult men and 1000 mg for adult women.[33] Marginal phosphorus deficiencies are most common in premature infants, vegans, alcoholics, and the elderly. Marginal phosphorus status may also be caused by losses due to chronic diarrhea and overuse of aluminum-containing antacids, which prevent phosphorus absorption.

Recommended Phosphorus Intake

The RDA for phosphorus is set at 700 mg/day for men and women 19 to 50 years of age.[33] This is the amount needed to maintain normal blood phosphorus levels. Because neither absorption nor urinary losses change significantly with age, the RDA is the same for older adults.

For growing children and adolescents, the RDA is based on the phosphorus intake necessary to meet the needs for bone and soft-tissue growth. There is no evidence that phosphorus requirements are increased during pregnancy; intestinal absorption increases by about 10%, which is sufficient to provide the additional phosphorus needed by the mother and fetus. The RDA is not increased during lactation because the phosphorus in milk is provided by an increase in bone resorption and a decrease in urinary excretion that are independent of dietary intake of either phosphorus or calcium.

Excessive Phosphorus

Toxicity from high phosphorus intake is rare in healthy adults, but excessive intakes can lead to bone resorption. Typical intake in the United States is above recommendations. One reason is the increased use of phosphorus-containing food additives. This has led to concern about its impact on bone health. High phosphorus intake has been found to increase bone resorption, but some of this can be prevented by adequate calcium intake.[44] Therefore levels of phosphorus intake typical in the United States are not believed to affect bone health as long as calcium intake is adequate. Based on the upper level of normal serum phosphate, a UL for phosphorus of 4 g/day has been set for adults ages 19 to 70 years.[33]

11.5 Magnesium (Mg)

LEARNING OBJECTIVES

- Name three foods that are good sources of magnesium.
- Describe the functions of magnesium in the body.

There are approximately 25 g of magnesium in the adult human body. Magnesium is a mineral that affects the metabolism of calcium, sodium, and potassium.

Magnesium in the Diet

Magnesium is found in leafy greens such as spinach and kale because it is a component of chlorophyll. Nuts, seeds, bananas, and the germ and bran of whole grains are also good sources (**Figure 11.17**). Processed foods are generally poor sources. For example, removing the bran and germ of the wheat kernel reduces the magnesium content of a cup of white flour to only 28 mg, compared with the 166 mg in a cup of whole-wheat flour. In areas with hard water, the water supply may provide a significant amount of magnesium.

Magnesium in the Digestive Tract

About 50% of the magnesium in the diet is absorbed, and the percentage decreases as intake increases. The active form of vitamin D can enhance magnesium absorption to a small extent, and the presence of phytate decreases absorption. As calcium in the diet increases, the absorption of magnesium decreases, so the use of calcium supplements can reduce the absorption of magnesium.

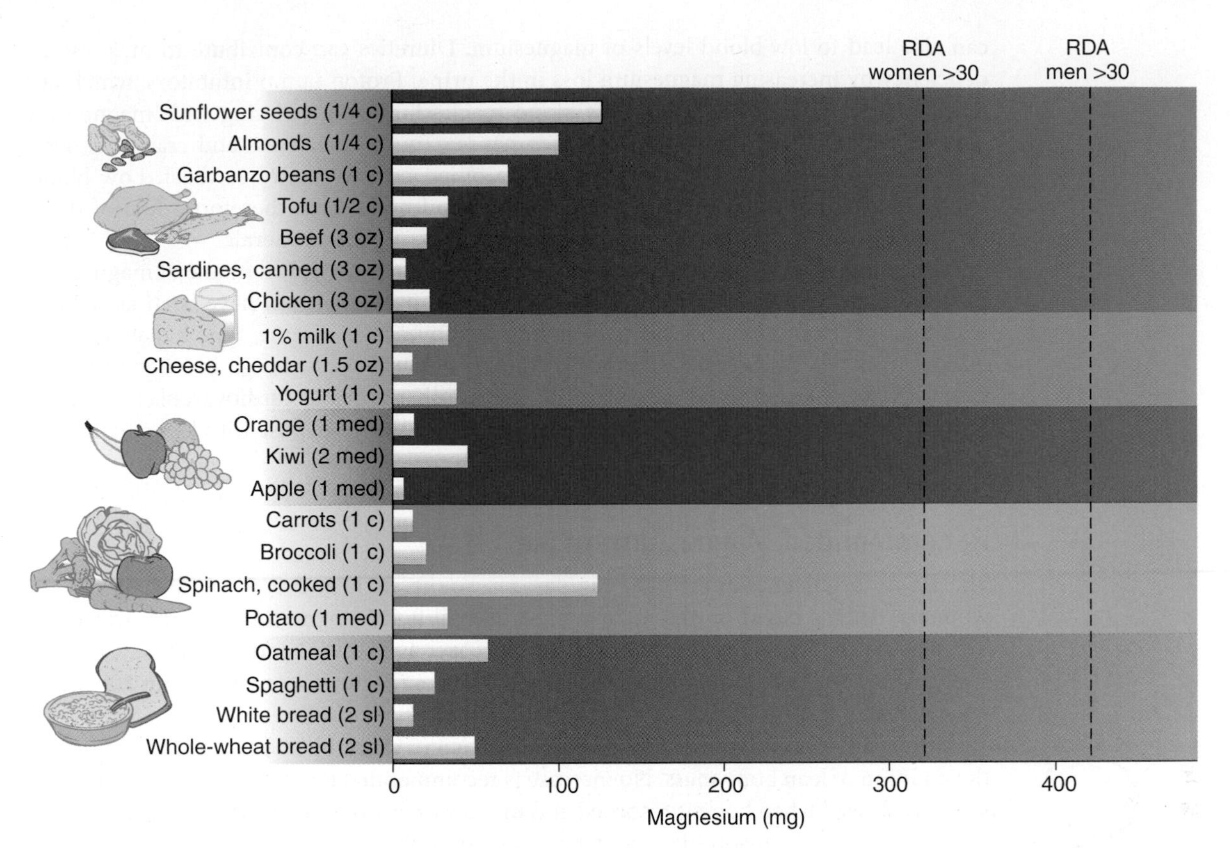

Figure 11.17 **Magnesium content of MyPlate food groups**
Magnesium is found in nuts, seeds, legumes, and leafy greens; the dashed lines represent the RDA for adult men and women over age 30.

How Magnesium Functions in the Body

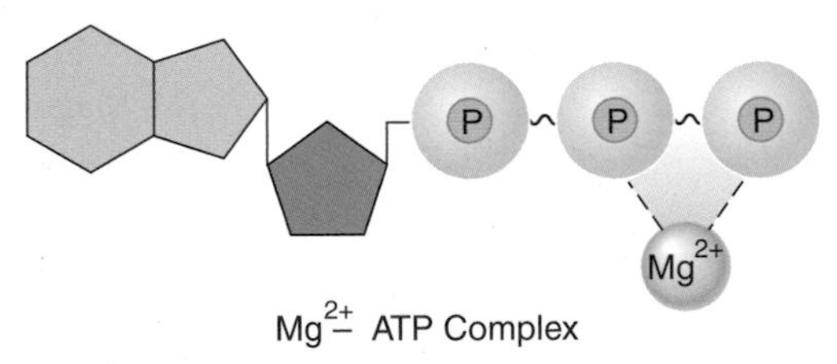

Figure 11.18 **Magnesium stabilizes ATP**
The magnesium ion (Mg^{2+}) stabilizes ATP structure by forming a magnesium-ATP complex. It is therefore important for all reactions that use or generate ATP.

About 50 to 60% of the magnesium in the body is in bone, where it is essential to maintain healthy bone structure. Most of the remaining body magnesium is present inside cells, where it is the second most abundant positively charged intracellular ion (after potassium). Magnesium is associated with the negative charge on phosphate-containing molecules such as ATP (**Figure 11.18**). Magnesium is also involved in regulating calcium homeostasis and is needed for the action of vitamin D and many hormones, including PTH.[45]

Magnesium is a cofactor for over 300 enzymes. It is necessary for the generation of ATP from carbohydrate, fat, and protein (see Table 11.2). In some of these reactions, it is involved indirectly as a stabilizer of ATP and, in others, directly as an enzyme activator. Magnesium is needed for the activity of the sodium-potassium pump, which is responsible for active transport of sodium and potassium across membranes. It is therefore essential for maintenance of electrical potentials across cell membranes and proper functioning of the nerves and muscles, including those in the heart. It is important for DNA and RNA synthesis and for almost every step in protein synthesis. Therefore, magnesium is particularly important for dividing growing cells.

The kidneys closely regulate blood levels of magnesium. When magnesium intake is low, excretion in the urine is decreased. As intake increases, urinary excretion increases to maintain normal blood levels. This efficient regulation permits homeostasis over a wide range of dietary intakes.

Magnesium Deficiency

Magnesium deficiency is rare in the general population. It does occur in those with alcoholism, malnutrition, kidney disease, and gastrointestinal disease. Certain medications can also lead to low blood levels of magnesium. Diuretics can contribute to magnesium deficiency by increasing magnesium loss in the urine. Proton pump inhibitors, which are taken to treat GERD, may cause low blood magnesium by interfering with magnesium absorption.[46] Deficiency symptoms include nausea, muscle weakness and cramping, irritability, mental derangement, and changes in blood pressure and heartbeat. Low blood magnesium levels affect levels of blood calcium and potassium; therefore some of these symptoms may be due to alterations in the levels of these other minerals.

Although symptomatic magnesium deficiency is rare, the typical intake of magnesium in the United States is below the RDA. Low intakes of magnesium have been associated with a number of chronic diseases, including cardiovascular disease, type 2 diabetes, and osteoporosis.[47] As discussed in Chapter 10, dietary patterns that are high in magnesium are associated with lower blood pressure. The risk of other types of cardiovascular disease is also lower for people with adequate magnesium intake than for those with less magnesium in their diet.

Recommended Magnesium Intake

The RDA for magnesium is 400 mg/day for young men and 310 mg/day for young women.[33] This is based on the maintenance of total body magnesium balance over time. The RDA is slightly higher for men and women over age 30: 420 and 320 mg/day, respectively. A cup of whole-grain breakfast cereal, spinach, or legumes contains about 100 mg of magnesium.

The magnesium requirement for pregnancy is increased by 40 mg/day to account for the addition of lean body mass. No increase is recommended for lactation because magnesium is released when bone is resorbed and urinary excretion is decreased. An AI is set for infants based on the magnesium content of human milk.

Magnesium Toxicity and Supplements

No adverse effects have been observed from ingestion of magnesium from food, but toxicity may occur from concentrated sources such as magnesium-containing drugs and supplements. Toxicity has been reported in elderly patients with impaired kidney function who frequently use magnesium-containing laxatives and antacids such as milk of magnesia. Magnesium toxicity is characterized by nausea, vomiting, low blood pressure, and cardiovascular changes. The UL for adults and adolescents over 9 years of age is 350 mg from nonfood sources of magnesium.

11.6 Sulfur (S)

LEARNING OBJECTIVES

- Name sources of sulfur in the diet.
- Discuss the role of sulfur in the body.

Dietary sulfur is found in organic molecules such as the sulfur-containing amino acids in proteins and the sulfur-containing vitamins. It is also found in some inorganic food preservatives such as sulfur dioxide, sodium sulfite, and sodium and potassium bisulfite, which are used as antioxidants.

In the body, the sulfur-containing amino acids methionine and cysteine are needed for protein synthesis. Cysteine is also part of the compound glutathione, which is important in detoxifying drugs and protecting cells from oxidative damage. The vitamins thiamin and biotin, essential for ATP production, contain sulfur. Sulfur-containing ions are important in regulating acid-base balance.

There is no recommended intake for sulfur, and no deficiencies are known when protein needs are met (see Table 11.2).

OUTCOME

PhotoDisc, Inc./Getty Images

Mika's low bone density is due to a number of factors. She doesn't consume enough calcium to maximize her bone density, and she is not menstruating normally, a condition call amenorrhea. Her energy-restricted diet combined with her high level of exercise contributes to a low level of body fat and low levels of estrogen, leading to amenorrhea. Just as low estrogen contributes to bone loss in postmenopausal women, it has a similar impact in young women. The combination of disordered eating, amenorrhea, and low bone density is referred to as the female athlete triad (see **Focus on Eating Disorders** and Chapter 13). To prevent further bone loss and hopefully improve her bone density, Mika needs to make some lifestyle changes. The goals for Mika's recovery plan are to increase her caloric intake, improve the quality of her diet, and cut back on her training so she can gain weight and restore her menstrual cycle. With the help of behavioral and dietary counseling, after about six months she is able to bring her BMI back into the healthy range. **She is now menstruating and is running three to four days a week, rather than six days a week.**

Blend Images/Getty Images, Inc.

APPLICATIONS

ASSESSING YOUR DIET

1. **Do you get enough calcium?**
   a. Using iProfile and the food record you kept in Chapter 2, calculate your average calcium intake.
   b. How does your intake compare with the RDA for calcium for someone of your age and gender?
   c. If your calcium intake is below the RDA, suggest modifications to increase the amount of calcium in your diet without significantly increasing your energy intake.
2. **Evaluate your risk of developing osteoporosis.**
   a. Make list of risk factors for osteoporosis and note which ones affect you.
   b. Note which risk factors you can change and which you cannot.
   c. Suggest changes that would reduce your risk.

CONSUMER ISSUES

3. **How much calcium do fortified foods contribute?**
   a. Choose three foods that are fortified with calcium and use their labels to determine how much calcium is in a serving of each.
   b. How many servings per day of each would you need to consume to meet your calcium needs?
   c. Are there other dietary components in any of these foods that would interfere with calcium absorption or utilization?
4. **Imagine you are a scientist with the U.S. Department of Agriculture and you have been assigned the task of deciding on a food or group of foods that will be fortified with calcium to assure that the population meets its calcium needs.**
   a. What food or group of foods would you recommend? Why?
   b. Does the food or group of foods contain dietary components that interfere with calcium absorption?
   c. Is this a food or group of foods that is consumed by the population groups most at risk for calcium deficiency?
   d. How much calcium would you recommend adding? Would this amount meet recommendations but not put the population at risk of calcium excess?

CLINICAL CONCERNS

5. **Many people in the United States must limit milk consumption due to lactose intolerance. How can they meet their calcium needs?**

a. Use iProfile to plan a day's diet that provides 1000 mg of calcium but does not include any dairy products or calcium-fortified foods.

b. Is this a reasonable diet to follow every day? Would you recommend including fortified foods or calcium supplements? Why or why not?

6. This table shows the incidence of hip fractures in different groups within a population of older adults. Use your knowledge of bone physiology to explain the differences observed among Caucasian women, Caucasian men, African American women, and African American men.

Population group	Annual incidence of hip fracture per 1000
Caucasian women	30
Caucasian men	13
African American women	11
African American men	7

SUMMARY

11.1 What Are Minerals?

- Minerals are elements needed by the body to regulate chemical reactions and provide structure. They are divided into major minerals and trace minerals based on the amount required in the diet and present in the body. They are found in both plant and animal foods. Minerals are added to some foods through fortification and get into others as a result of contamination. Dietary supplements are also a source of minerals.
- Deficiencies of certain minerals are world health problems. Mineral bioavailability is affected by body needs as well as interactions with other minerals, vitamins, and dietary components such as fiber, phytates, oxylates, and tannins. Minerals contribute to the body's structure and help regulate body processes, often as cofactors.

11.2 Minerals, Osteoporosis, and Bone Health

- Bone consists of a protein matrix hardened by mineral deposits. It is a living tissue that is constantly being broken down and reformed in a process known as bone remodeling. Early in life, bone formation occurs more rapidly than bone breakdown to allow bone growth and an increase in bone mass. Loss of bone due to age and other factors increases the risk of developing osteoporosis, a condition in which loss of bone mass increases the risk of bone fractures.
- The risk of osteoporosis is related to the level of peak bone mass and the rate of bone loss. These are affected by age, gender, hormone levels, genetics, smoking, alcohol use, exercise, and diet. Peak bone mass occurs in young adulthood. With age, bone breakdown begins to outpace formation, causing a decrease in bone mass; this is accelerated in women for about 5 to 10 years surrounding menopause.
- Osteoporosis risk can be reduced by an active lifestyle and a diet adequate in calcium and vitamin D. Osteoporosis is treated with supplements of calcium and vitamin D, medications that inhibit bone breakdown, and in some cases hormone replacement therapy.

11.3 Calcium (Ca)

- Sources of calcium in the American diet include dairy products, fish consumed with bones, and leafy green vegetables. Fortified foods and supplements also contribute to calcium intake.
- Sufficient calcium absorption depends on adequate levels of vitamin D. The absorption of calcium is reduced by the presence of tannins, fiber, phytates, and oxalates. Calcium absorption varies with life stage and is highest during infancy and pregnancy, when needs are greatest.
- Most of the calcium in the body is in bone. Calcium not found in bone is essential for cell communication, nerve transmission, muscle contraction, blood clotting, and blood pressure regulation. Blood levels of calcium are regulated by parathyroid hormone (PTH) and calcitonin. PTH stimulates the release of calcium from bone, decreases calcium excretion by the kidney, and activates vitamin D to increase the amount of calcium absorbed from the gastrointestinal tract and released from bone. Calcitonin blocks calcium release from bone.
- Calcium deficiency can reduce bone mass and increase the risk of osteoporosis.
- The RDA for calcium ranges from 1000 to 1200 mg/day for adults and is 1300 mg/day in adolescents and pregnant and lactating women.
- Too much calcium can contribute to the formation of kidney stones, raise blood calcium levels, and interfere with the absorption of other minerals.

11.4 Phosphorus (P)

- Phosphorus is more widely distributed in the diet than calcium. It is found in dairy products such as milk, yogurt, and cheese, but meat, cereals, bran, eggs, nuts, and fish are also good sources. Food additives used in baked goods, cheese, processed meats, and soft drinks also contribute to dietary phosphorus.
- About 60 to 70% of the phosphorus in a typical diet is absorbed. Vitamin D aids phosphorus absorption via an active mechanism, but most absorption occurs by a mechanism that does not depend on vitamin D.
- Most of the phosphorus in the body is found in bones and teeth. In addition to its structural role in these tissues, phosphorus is an essential component of phospholipids, ATP, and DNA. Phosphorus is also part of a buffer system that helps prevent changes in pH.

- Phosphorus deficiency is rare in healthy people because phosphorus is so widely distributed in foods. Deficiency can lead to bone loss, weakness, and loss of appetite.
- The RDA for adults is 700 mg phosphorus per day.
- Symptoms related to high phosphorus intake are rare in healthy adults, but excessive intakes can lead to bone resorption. The levels of phosphorus intake typical in the United States are not believed to affect bone health as long as calcium intake is adequate.

11.5 Magnesium (Mg)

- Magnesium is found in leafy greens such as spinach and kale because it is a component of chlorophyll. Nuts, seeds, bananas, and the germ and bran of whole grains are also good sources.
- About half of the magnesium in the diet is absorbed, and the percentage decreases as intake increases. The active form of vitamin D can enhance magnesium absorption and the presence of phytates decreases absorption.
- Magnesium is important for bone health, and it is needed as a cofactor for numerous reactions throughout the body. In reactions involved in energy metabolism it acts as an enzyme activator and stabilizer of ATP. It is also needed to maintain membrane potentials; thus it is essential for nerve and muscle conductivity. Homeostasis is regulated by the kidney.
- Magnesium deficiency is rare in the general population. It does occur in those with alcoholism, malnutrition, kidney disease, and gastrointestinal disease. Symptoms include nausea, muscle weakness and cramping, irritability, mental derangement, and changes in blood pressure and heartbeat.
- The RDA for magnesium is 400 mg/day for young men and 310 mg/day for young women.
- No adverse effects have been observed from ingestion of magnesium from food, but toxicity may occur from magnesium-containing drugs and supplements.

11.6 Sulfur (S)

- Sulfur is in the diet as preformed organic molecules such as the amino acids methionine and cysteine, which are needed to synthesize proteins and glutathione, and the vitamins thiamin and biotin, needed for energy metabolism. Sulfur is also part of a buffer system that regulates acid–base balance. A dietary deficiency is unknown in the absence of protein malnutrition.

REVIEW QUESTIONS

1. Explain the difference between major minerals and trace elements.
2. List four factors that can affect mineral bioavailability.
3. Describe the structure of bone.
4. What is bone remodeling?
5. How does the rate of bone formation and breakdown change throughout life?
6. How does the level of peak bone mass affect the risk of osteoporosis?
7. How is calcium intake related to the risk of osteoporosis?
8. What factors other than calcium intake are related to the risk of osteoporosis?
9. What is the major source of calcium in the North American diet?
10. What is the function of calcium in bones and teeth?
11. What are the roles of calcium in body fluids?
12. How are blood calcium levels restored when they drop too low? Rise too high?
13. List sources of dietary calcium acceptable to those who are lactose intolerant.
14. Name some food sources of phosphorus.
15. What are the functions of phosphorus in the body?
16. Name some food sources of magnesium.
17. What is the function of magnesium in the body?
18. Where is sulfur found in the body?

REFERENCES

1. Hambidge, K. M., Miller, L. V., Westcott, J. E., et al. Zinc bioavailability and homeostasis. *Am J Clin Nutr* 91:1478S–1483S, 2010.
2. Gibson, R. S., Bailey, K. B., Gibbs, M., and Ferguson, E. L. A review of phytate, iron, zinc, and calcium concentrations in plant-based complementary foods used in low-income countries and implications for bioavailability. *Food Nutr Bull* 31 (2 suppl):S134–S146, 2010.
3. Shea, M. K., and Booth, S. L. Update on the role of vitamin K in skeletal health. *Nutr Rev* 66:549–557, 2008.
4. Wang, Q., and Seeman, E. Skeletal growth and peak bone strength. *Best Pract Res Clin Endocrinol Metab* 22:687–700, 2008.
5. Looker, A. C., Melton, L. J. III, Harris, T. B., et al. Prevalence and trends in low femur bone density among older US adults:

NHANES 2005–2006 compared with NHANES III. *J Bone Miner Res* 25:64–71, 2010.

6. U.S. Department of Health and Human Services. *Bone Health and Osteoporosis: A Report of the Surgeon General*. Rockville, MD: U.S. Department of Health and Human Services, Office of the Surgeon General, 2004.
7. Nordin, B. E. C., Need, A. G., Morris, H. A., et al. Effect of age on calcium absorption in postmenopausal women. *Am J Clin Nutr* 80:998–1002, 2004.
8. Cawthon, P. M. Gender differences in osteoporosis and fractures. *Clin Orthop Relat Res* 469:1900–1905, 2011.
9. Cifuentes, M., Advis, J. P., and Shapses, S. A. Estrogen prevents the reduction in fractional calcium absorption due to energy restriction in mature rats. *J Nutr* 134:1929–1934, 2004.
10. Van Cromphaut, S. J., Rummens, K., Stockmans, I., et al. Intestinal calcium transporter genes are upregulated by estrogens and the reproductive cycle through vitamin D receptor-independent mechanisms. *J Bone Miner Res* 18:1725–1736, 2003.
11. Walker, M. D., Novotny, R., Bilezikian, J. P., and Weaver, C. M. Race and diet interactions in the acquisition, maintenance, and loss of bone. *J Nutr* 138:1256S–1260S, 2008.
12. Aloia, J. F. African Americans, 25-hydroxyvitamin D, and osteoporosis: A paradox. *Am J Clin Nutr* 88:545S–550S, 2008.
13. Ferrari, S. Human genetics of osteoporosis. *Best Prac Res Clin Endocrinol Metab* 22:723–735, 2009.
14. Looker, A. C., Melton, L. J. III, Harris, T., et al. Age, gender, and race/ethnic differences in total body and subregional bone density. *Osteoporos Int* 20:1141–1149, 2009.
15. Kanis, J., Johnell, O., Oden, A., et al. Smoking and fracture risk: A meta-analysis. *Osteoporos Int.* 16: 155–162, 2005.
16. Wong, P. K., Christie, J. J., and Wark, J. D. The effects of smoking on bone health. *Clin Sci* (Lond) 113:233–241, 2007.
17. Chakkalakal, D. A. Alcohol-induced bone loss and deficient bone repair. *Alcohol Clin Exp Res* 29:2077–2090, 2005.
18. Jin, L. H., Chang, S. J., Koh, S. B., et al. Association between alcohol consumption and bone strength in Korean adults: The Korean Genomic Rural Cohort Study. *Metabolism* 60:351–358, 2011.
19. Martyn-St James, M., and Carroll, S. Effects of different impact exercise modalities on bone mineral density in premenopausal women: A meta-analysis. *J Bone Miner Metab* 28:251–267, 2010.
20. Martyn-St James, M., and Carroll, S. A meta-analysis of impact exercise on postmenopausal bone loss: The case for mixed loading exercise programmes. *Br J Sports Med* 43:898–908, 2009.
21. Reid, I. R. Fat and bone. *Arch Biochem Biophys* 503:20–27, 2010.
22. Pollock, N. K., Bernard, P. J., Gutin, B., et al. Adolescent obesity, bone mass, and cardiometabolic risk factors. *J Pediatr* 158:727–734, 2011.
23. Teucher, B., and Fairweather-Tait, S. Dietary sodium as a risk factor for osteoporosis: Where is the evidence? *Proc Nutr Soc* 62:859–866, 2003.
24. Ilich, J. Z., Brownbill, R. A., and Coster, D. C. Higher habitual sodium intake is not detrimental for bones in older women with adequate calcium intake. *Eur J Appl Physiol* 109:745–755, 2010.
25. Kerstetter, J. E., O'Brien, K. O., Caseria, D. M., et al. The impact of dietary protein on calcium absorption and kinetic measures of bone turnover in women. *J Clin Endocrinol Metab* 90:26–31, 2005.
26. Darling, A. L., Millward, D. J., Torgerson, D. J., et al. Dietary protein and bone health: A systematic review and meta-analysis. *Am J Clin Nutr* 90:1674–1692, 2009.
27. Prynne, C. J., Mishra, G. D., O'Connell, M. A., et al. Fruit and vegetable intakes and bone mineral status: A cross-sectional study in five age and sex cohorts. *Am J Clin Nutr* 83:1420–1428, 2006.
28. Shea, B., Wells, G., Cranney, A., et al. Meta-analysis of calcium supplementation for the prevention of postmenopausal osteoporosis. *Endocrine Rev* 23:552–559, 2002.
29. Heike, A., Bischoff-Ferrari, H. A., Willett, W. A., et al. Fracture prevention with vitamin D supplementation: A meta-analysis of randomized controlled trials. *JAMA* 293:2257–2264, 2005.
30. U.S. Preventive Services Task Force. Hormone therapy for the prevention of chronic conditions in postmenopausal women: Recommendations from the U.S. Preventive Services Task Force. *Ann Intern Med* 142:855–860, 2005.
31. Papapoulos, S. E. Use of bisphosphonates in the management of postmenopausal osteoporosis. *Ann N Y Acad Sci* 1218:15–32, 2011.
32. Sipponen, P., and Härkönen, M. Hypochlorhydric stomach: A risk condition for calcium malabsorption and osteoporosis? *Scand J Gastroenterol* 45:133–138, 2010.
33. Institute of Medicine, Food and Nutrition Board. *Dietary Reference Intakes for Calcium, Phosphorus, Magnesium, Vitamin D, and Fluoride*. Washington, DC: National Academies Press, 1997.
34. Heaney, R. P., Weaver, C. M., and Recker, R. R. Calcium absorption from spinach. *Am J Clin Nutr* 47:707–709, 1988.
35. Bronner, F., and Pansu, D. Nutritional aspects of calcium absorption. *J Nutr* 129:9–12, 1999.
36. Conlin, P. R., Chow, D., Miller, E. R. III, et al. The effect of dietary patterns on blood pressure control in hypertensive patients: Results from the Dietary Approaches to Stop Hypertension (DASH) trial. *Am J Hypertens* 13:949–955, 2000.
37. Institute of Medicine, Food and Nutrition Board. *Dietary Reference Intakes for Calcium and Vitamin D*. Washington, DC: National Academies Press, 2011.
38. Prentice, A. Micronutrients and the bone mineral content of the mother, fetus and newborn. *J Nutr* 133(5 suppl 2): 1693S–1699S, 2003.
39. Nielsen, S. J., and Popkin, B. M. Changes in beverage intake between 1977 and 2000. *Am J Prev Med* 27:205–210, 2004.
40. Center for Science in the Public Interest. Liquid candy: How soft drinks are harming America's health.
41. Beall, D. P., Henslee, H. B., Webb, H. R., and Scofield, R. H. Milk-alkali syndrome: A historical review and description of the modern version of the syndrome. *Am J Med Sci* 331: 233–242, 2006.

42. Mølgaard, C., Kaestel, P., and Michaelsen, K. F. Long-term calcium supplementation does not affect the iron status of 12-14-year-old girls. *Am J Clin Nutr* 82:98–102, 2005.

43. Wood, R. J., and Zheng, J. J. High dietary calcium intakes reduce zinc absorption and balance in humans. *Am J Clin Nutr* 65:1803–1809, 1997.

44. Kerni, V. E., Kärkkäinen, M. U., Karp, H. J., et al. Increased calcium intake does not completely counteract the effects of increased phosphorus intake on bone: An acute dose-response study in healthy females. *Br J Nutr* 99:832–839, 2008.

45. Konrad, M., and Schlingmann, K-P. Magnesium. In *Biochemical, Physiological, and Molecular Aspects of Human Nutrition*, 2nd ed. M. H. Stipanuk, ed. St. Louis: Saunders Elsevier, 2006, pp. 921–941.

46. Chen, J., Yuan, Y. C., Leontiadis, G. I., and Howden, C. W. Recent safety concerns with proton pump inhibitors. *J Clin Gastroenterol* 46:93–114, 2012.

47. Rosanoff, A., Weaver, C. M., and Rude, R. K. Suboptimal magnesium status in the United States: Are the health consequences underestimated? *Nutr Rev* 70:153–164, 2012.

***To access links to online sources, please go to www.wiley.com/college/smolin and select Nutrition: Science and Applications, 3rd edition. From this page, select either the student or instructor companion site. Once on the desired site, select References.**

CHAPTER OUTLINE

© amana images Inc./Alamy Limited

The Trace Minerals

12

CASE STUDY

Nissi is a 12-year-old girl who lives in a small village in India. She has been feeling tired and has noticed a lump in her neck that seems to be getting bigger. Her mother takes her to the local clinic, where the doctor determines that her malaise and the swelling in her neck, which is called a goiter, are caused by an iodine deficiency. He gives Nissi an injection of iodine and explains that, if left untreated, Nissi's symptoms would have continued to worsen.

How could an apparently healthy young girl have a nutritional deficiency? Her diet consists almost entirely of homegrown grains, pulses (seeds, beans, and lentils), and vegetables. These locally grown foods are deficient in iodine because the Ganges River valley has flooded repeatedly over the centuries washing the iodine out of the soil. Food grown there is therefore low in iodine so a diet based solely on local foods does not provide enough of this essential mineral to meet needs. Iodine deficiency is a problem not only for Nissi but also for others in her village. Goiter is common and young women who are deficient in iodine during pregnancy are at risk of giving birth to babies with developmental abnormalities.

One way to prevent iodine deficiency is to include foods that are higher in iodine. However, this approach is not practical for Nissi because all of the food grown in the region—and thus her entire diet—is deficient in iodine. **How might Nissi's family and others in her village obtain adequate iodine?**

PhotosIndia.com/Getty Images, Inc.

12.1 Trace Minerals in Our Diet

LEARNING OBJECTIVES

- Distinguish trace minerals from major minerals.
- Explain why bioavailability is particularly important in meeting trace mineral needs.

The trace minerals, which include iron, zinc, copper, manganese, selenium, iodine, chromium, fluoride, and molybdenum, as well as several others, are required in the diet in an amount of 100 mg or less per day or present in the body in an amount of 0.01% or less of body weight. Like the major minerals, they serve a variety of essential structural and regulatory roles. Some of their functions are unique: Iodine is needed to make thyroid hormones, iron is needed to carry oxygen to body cells, and fluoride is needed for strong teeth. Other functions are similar and complementary: Selenium, copper, zinc, iron, and manganese are each cofactors for antioxidant enzyme systems.

Although the trace minerals are distinguished from the major minerals only by the amounts required in the diet and present in the body, the small amounts in which they are found make them difficult to study. Usually nutrient needs are evaluated by feeding a diet devoid of that nutrient. However, with some trace minerals, needs are so small that contamination from the environment and minerals already present in the body can obscure experimental results. Bioavailability is also more a concern with trace minerals because such small amounts are present in the diet. Phytates, tannins, oxalate, and fiber can bind minerals, reducing their absorption. For example, when the diet is based on unleavened grains, the phytate content may be high enough to decrease zinc absorption and cause a zinc deficiency. The interactions among the minerals that can affect their absorption and utilization also have a greater impact on trace mineral status. For example, a deficiency of copper can decrease available iron by reducing the amount of iron that can bind to iron transport proteins in the blood.

Determining the amounts of trace minerals that foods contribute to the diet is difficult and is compromised by the fact that the amounts of some are affected by the soil content where the food is grown or produced. For example, a loaf of bread made from wheat grown in one location may supply a different amount of selenium from a loaf made from wheat grown elsewhere. When modern transportation systems make foods produced in many locations available, this variation is unlikely to affect mineral status. In countries where the diet consists predominantly of locally grown foods, individual trace element deficiencies and excesses are more likely.

12.2 Iron (Fe)

LEARNING OBJECTIVES

- List some dietary sources of heme and nonheme iron.
- Explain how the amount of iron in the body is regulated.
- Describe the primary function of iron and the effects of iron deficiency.
- Discuss who is at risk for iron toxicity and why.

Iron was identified as a major constituent of blood in the eighteenth century. By 1832, iron tablets were used to treat young women in whom "coloring matter" was lacking in the blood. Today we know that the red color in blood is due to the iron-containing protein **hemoglobin** and that a deficiency of iron decreases hemoglobin production. Even though iron is one of the best understood of the trace minerals, iron deficiency remains the most common nutritional deficiency worldwide and is a problem among certain population groups in the United States.[1,2]

hemoglobin An iron-containing protein in red blood cells that binds oxygen and transports it through the bloodstream to cells.

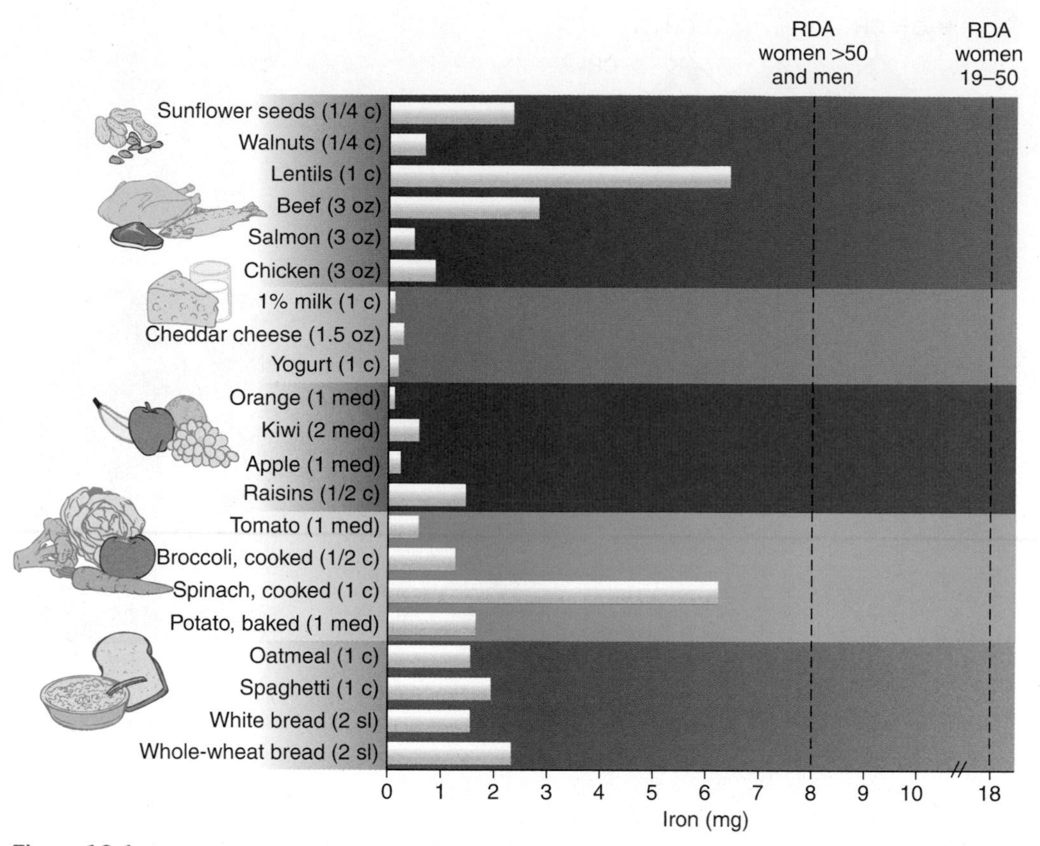

Figure 12.1 **Iron content of MyPlate food groups**
Both plant and animal foods are good sources of iron. Meat, poultry, and fish provide both heme and nonheme iron. Plant foods contain only nonheme iron. The dashed lines represent the RDA for women of childbearing age and for men and postmenopausal women.

Iron in the Diet

Iron in the diet comes from both plant and animal sources (**Figure 12.1**). Much of the iron in animal products is **heme iron**—iron that is part of a chemical complex, called a heme group, found in proteins such as hemoglobin in blood and **myoglobin** in muscle (**Figure 12.2**). Meat, poultry, and fish are good sources of heme iron. Heme iron accounts for about 10 to 15% of the dietary iron in meat-eating populations.[3]

Leafy green vegetables, legumes, and whole and enriched grains are good sources of **nonheme iron**. Another source of nonheme iron in the diet is iron cooking utensils, from which iron leaches into food. Leaching is enhanced by acidic foods. For example, 3 ounces of spaghetti sauce cooked in a glass pan contains about 0.6 mg of iron, but the same sauce cooked in an iron skillet contains about 5.7 mg, depending on how long it is cooked.

heme iron A readily absorbed form of iron found in animal products that is chemically associated with proteins such as hemoglobin and myoglobin.

myoglobin An iron-containing protein in muscle cells that binds oxygen.

nonheme iron A poorly absorbed form of iron found in both plant and animal foods that is not part of the iron complex found in hemoglobin and myoglobin.

Iron in the Digestive Tract

Iron from the diet is absorbed into the intestinal mucosal cells. The amount absorbed depends on whether the iron is heme or nonheme iron as well as on the presence of dietary components that enhance or inhibit iron absorption.

Heme iron is absorbed more efficiently than nonheme iron and is not significantly affected by most dietary factors that affect nonheme iron absorption. Absorption of heme iron ranges from 15 to 35%, compared to only 2 to 20% for the nonheme iron in plant foods.[4] When foods containing heme proteins are consumed, the iron-containing heme group is released from the proteins by protein-digesting enzymes. The heme binds to receptors on the surface of mucosal cells, allowing it to enter the cells, where the iron is released from the heme group.

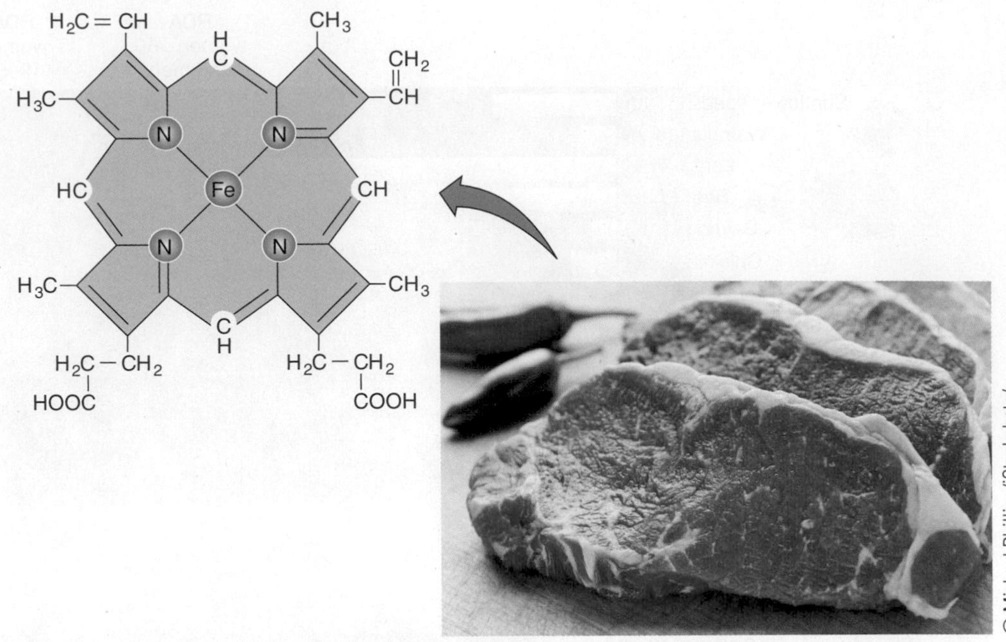

Figure 12.2 Heme iron
The myoglobin in red meat provides heme iron and gives the meat its red color. A heme group contains 4 five-membered nitrogen-containing rings that form a cage around a central iron ion. In hemoglobin and myoglobin the iron ion is in the Fe^{2+} state.

When foods containing nonheme iron are consumed, stomach acid helps convert the ferric form (Fe^{3+}) of iron to the ferrous form (Fe^{2+}). The ferrous form of iron remains more soluble when it enters the intestine and therefore is absorbed into the mucosal cells more readily. When foods containing nonheme iron are consumed with foods containing acids, such as ascorbic acid (vitamin C), citric acid, or lactic acid, iron absorption is enhanced because the acids help to keep iron in the ferrous (Fe^{2+}) form. The best studied of these acids is vitamin C, which enhances nonheme iron absorption both by keeping the iron in its more absorbable form and by forming a complex with iron that remains soluble and more bioavailable.[3] Vitamin C can enhance nonheme iron absorption up to sixfold. Another dietary component that increases the absorption of nonheme iron is meat such as beef, fish, or poultry. For example, a small amount of ground beef in a pot of chili will enhance the body's absorption of nonheme iron from the beans.

Dietary factors that interfere with the absorption of nonheme iron include fiber, phytates found in cereals, tannins found in tea, and oxalates found in some leafy greens such as spinach.[3] These prevent absorption by binding iron in the gastrointestinal tract. They are not consumed in large enough quantities in the United States to cause an iron deficiency. However, these components do contribute to iron deficiency in parts of the developing world where the diet is low in heme iron and factors that enhance nonheme iron absorption and high in foods that limit absorption.[5] The presence of other minerals may also interfere with iron absorption. For instance, calcium consumed in the same meal with iron decreases iron absorption. However, the long-term effect of calcium in a varied diet that includes other inhibitors and enhances of iron absorption, is minimal.[3,6]

Iron in the Body

Iron is essential for life, but in excess it is toxic. To protect against the toxic effects of iron, the body regulates the amount that enters the blood from the mucosal cells of the gastrointestinal tract and has evolved ways to safely transport and store it.

How Iron Homeostasis Is Regulated The amount of iron in the body is controlled primarily at the intestine (**Figure 12.3**). Iron that has entered the mucosal cells of the small

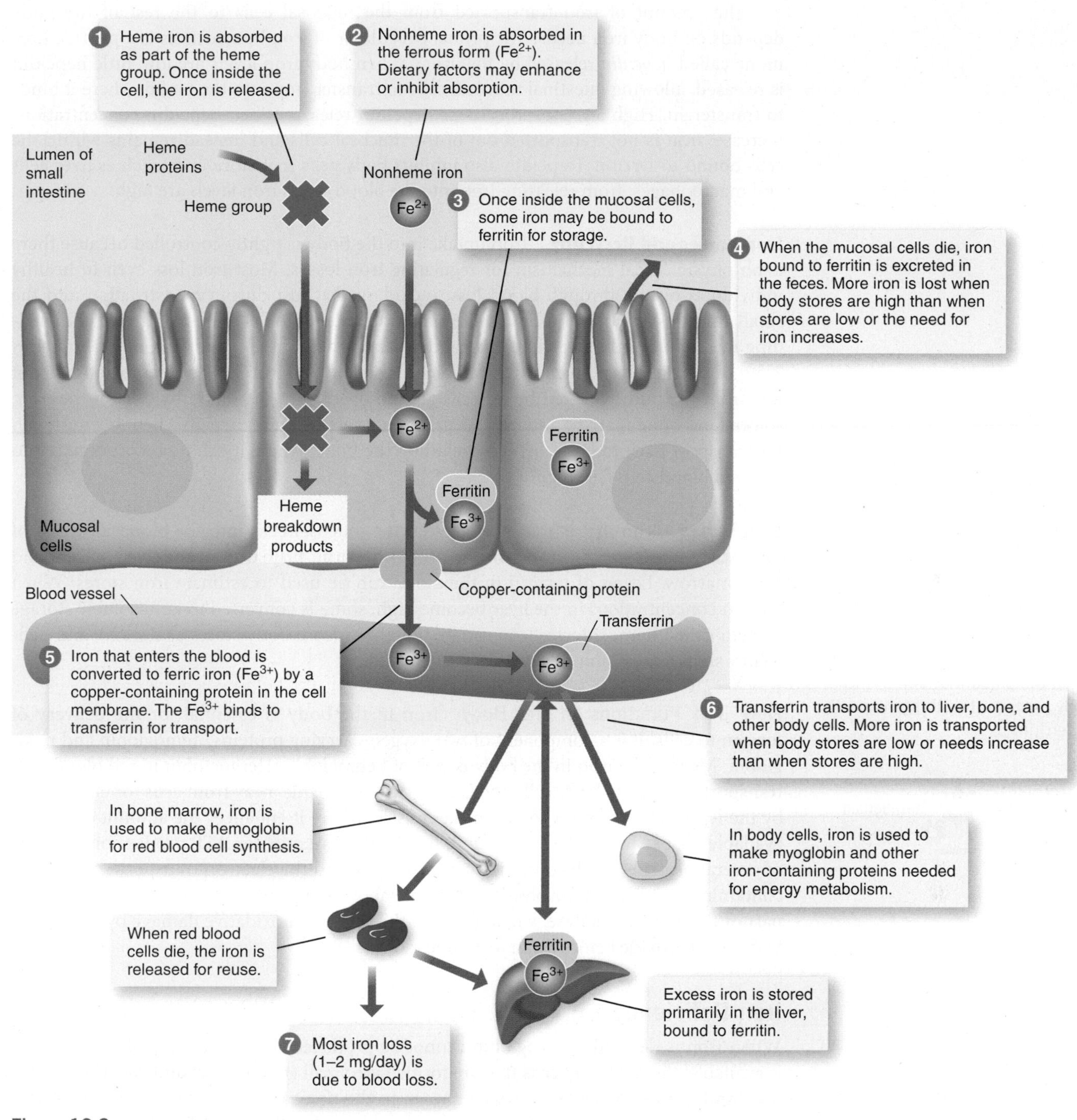

Figure 12.3 **Iron absorption and transport**
The amount of iron available to the body depends on both the amount absorbed into the mucosal cells of the small intestine and the amount transported to the rest of the body. The amount of iron that leaves the mucosal cells for transport to liver, bone, and other tissues is carefully regulated to maintain iron homeostasis. Iron loss is due primarily to blood loss.

intestine can be bound to the iron storage protein **ferritin** or transferred into the blood and picked up by the iron transport protein **transferrin**. Iron that remains bound to ferritin is excreted in the feces when mucosal cells die and are sloughed off into the intestinal lumen (see Figure 12.3). Iron that is picked up by transferrin is transported in the blood to the liver, bones, and other body tissues. Transferrin picks up iron from the intestinal mucosal cells in the small intestine as well as iron released from the breakdown of hemoglobin. Before iron can bind transferrin, it must be converted to the ferric (Fe^{3+}) form. In the intestine, this is accomplished by the action of a copper-containing protein located in the mucosal cell membrane.[7]

ferritin The major iron storage protein.

transferrin An iron transport protein in the blood.

The amount of iron transported from the mucosal cells to the rest of the body depends on body iron needs. The primary regulator of iron homeostasis is a peptide hormone called *hepcidin* released by the liver.[7] When body iron levels are low, little hepcidin is released, allowing intestinal mucosal cells to transfer iron into the blood, where it binds to transferrin. High levels of iron cause hepcidin release. When hepcidin concentrations increase, iron is not transported out of the mucosal cells and instead remains within the cells bound to ferritin. Hepcidin also inhibits body cells that store iron, such as liver cells and macrophages, from releasing iron into the blood when iron levels are high.

Iron Losses and Recycling Iron uptake into the body is tightly controlled because there is no physiological mechanism for regulating iron losses. Most iron loss even in healthy individuals occurs through blood loss, including that lost during menstruation and the small amounts lost from the gastrointestinal tract. Some iron is also lost through the shedding of cells from the intestine, skin, and urinary tract. These losses total only about 1 to 2 mg of iron per day.[7] Even when red blood cells die, the iron in their hemoglobin is not lost from the body. Old red blood cells are removed from the blood by cells in the liver, spleen, and bone marrow and degraded; the recovered iron is then attached to transferrin for transport back to body tissues, including the bone, where it can be incorporated into new red blood cells (see Figure 12.3).

Iron Stores Iron that is transported from the mucosal cells into the blood in excess of immediate needs can be stored in the protein ferritin, primarily in the liver, spleen, and bone marrow. Levels of ferritin in the blood can be used to estimate iron stores. When ferritin concentrations in the liver become high, some is converted to an insoluble storage protein called **hemosiderin**. Iron can be mobilized from body stores as needed, and deficiency signs appear only after stores are depleted.

hemosiderin An insoluble iron storage compound that stores iron when the amount of iron in the body exceeds the storage capacity of ferritin.

How Iron Functions in the Body Iron in the body is essential for the delivery of oxygen to cells. It is a component of two oxygen-carrying proteins, hemoglobin and myoglobin. Most of the iron in the body is part of hemoglobin. Hemoglobin in red blood cells transports oxygen to body cells and carries carbon dioxide away from cells for elimination by the lungs. Myoglobin is found in the muscle, where it enhances the amount of oxygen available for use in muscle contraction. Iron is also essential for ATP production as a part of several proteins involved in the citric acid cycle and the electron transport chain. Iron-containing proteins are involved in drug metabolism and immune function. Iron is also part of the enzyme catalase, which protects the cells from oxidative damage by destroying hydrogen peroxide before it can form free radicals.

Iron Deficiency

When iron is deficient, hemoglobin cannot be produced. When not enough hemoglobin is available, the red blood cells that are formed are small (microcytic) and pale (hypochromic) and unable to deliver adequate oxygen to the tissues. This microcytic, hypochromic anemia is known as **iron deficiency anemia**. Anemia is the last stage of iron deficiency. Earlier stages have no symptoms because they do not affect the amount of iron in red blood cells. Iron depletion can be detected by blood tests that measure indicators of iron levels in the plasma and in body stores (**Figure 12.4**).

iron deficiency anemia An iron deficiency disease that occurs when the oxygen-carrying capacity of the blood is decreased because there is insufficient iron to make hemoglobin.

What Are the Symptoms of Iron Deficiency Anemia? The symptoms of iron deficiency anemia include fatigue, weakness, headache, decreased work capacity, an inability to maintain body temperature in a cold environment, changes in behavior, decreased resistance to infection, adverse pregnancy outcomes, impaired development in infants, and an increased risk of lead poisoning in young children. Many of these symptoms are due to the role of iron in energy metabolism and oxygen transport.

Who Is at Risk for Iron Deficiency? It is estimated that as much as 80% of the world's population may be iron deficient and that 30% (2 billion people) suffer from iron

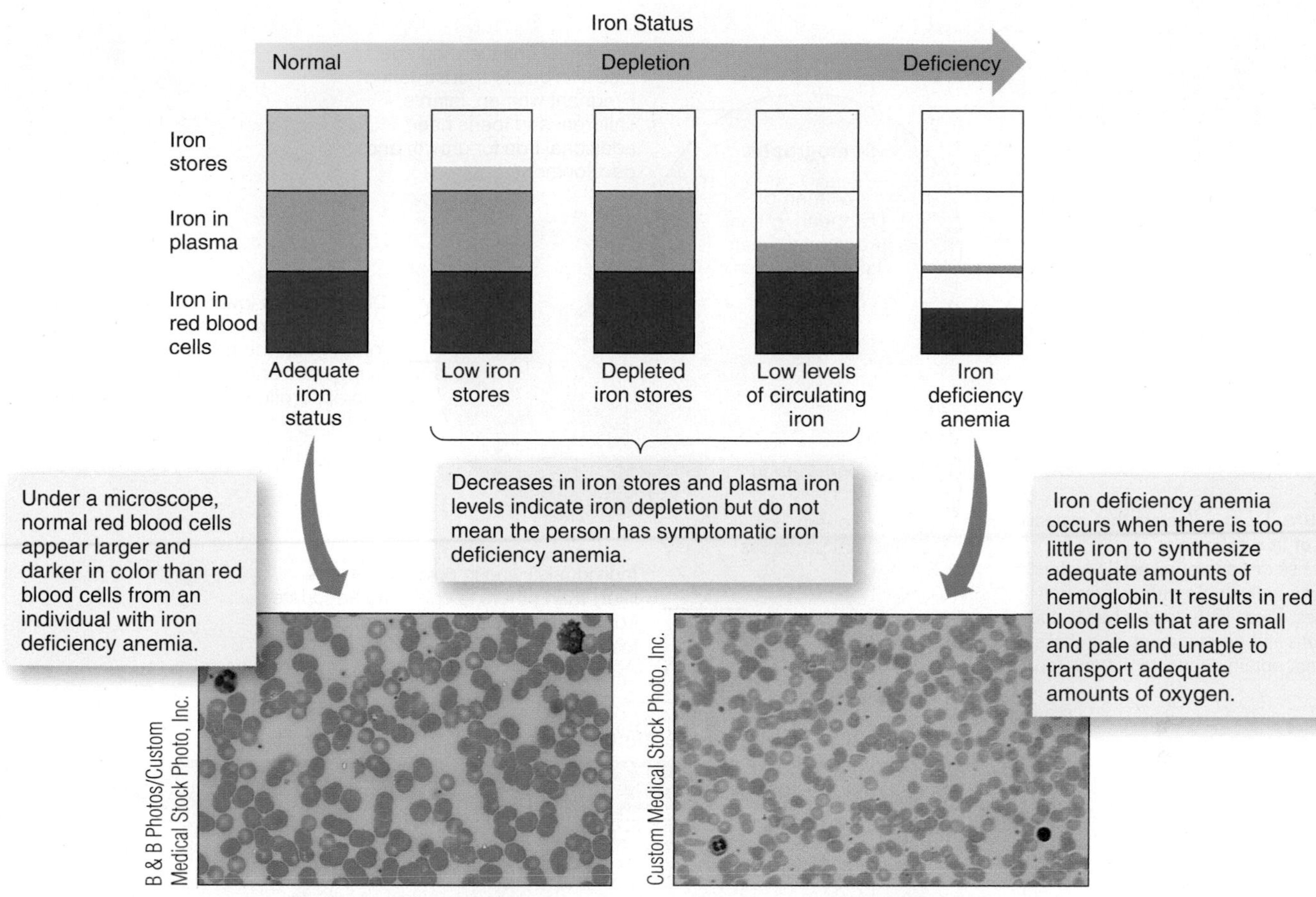

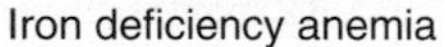

Figure 12.4 Progression of iron deficiency
Inadequate iron first causes a decrease in the amount of stored iron, followed by low iron levels in the plasma. It is only after plasma levels drop that there is no longer enough iron available to maintain adequate hemoglobin in red blood cells, resulting in iron deficiency anemia.

deficiency anemia.[2] In the United States about 2% of toddlers ages 1 to 2 years and 3% of adolescent and adult women are anemic due to iron deficiency.[8] Among low-income and minority women and children, the incidence of iron deficiency anemia is even greater (**Figure 12.5**).

Women of reproductive age are at risk for iron deficiency anemia because of iron loss due to menstruation. Menstruation is absent during pregnancy, but the need for iron is increased because of the expansion of maternal blood volume and the growth of other maternal tissues and the fetus. Iron deficiency is common among pregnant women even in industrialized countries and can lead to premature delivery and greater risk to the mother.

Iron deficiency is common in infants, children, and adolescents. In infants and children, rapid growth increases iron needs. About 7% of infants in the United States have iron deficiency.[9] Toddlers may also be at risk because finicky eating habits often reduce intake (see Chapter 15). In adolescent boys, rapid growth and an increase in muscle mass and blood volume increase iron need. In adolescent girls, iron needs are increased because weight gain is almost as great as in boys and iron losses are increased by the onset of menstruation (see Chapter 15). About 9 to 12% of adolescent girls and women of childbearing age have low iron levels.[9]

Athletes are another group at risk for iron deficiency. This may be due to a low iron intake as well as to increased losses due to prolonged training. Based on the amount lost, the EAR may be 30 to 70% higher for athletes than for the general population[10] (see Chapter 13).

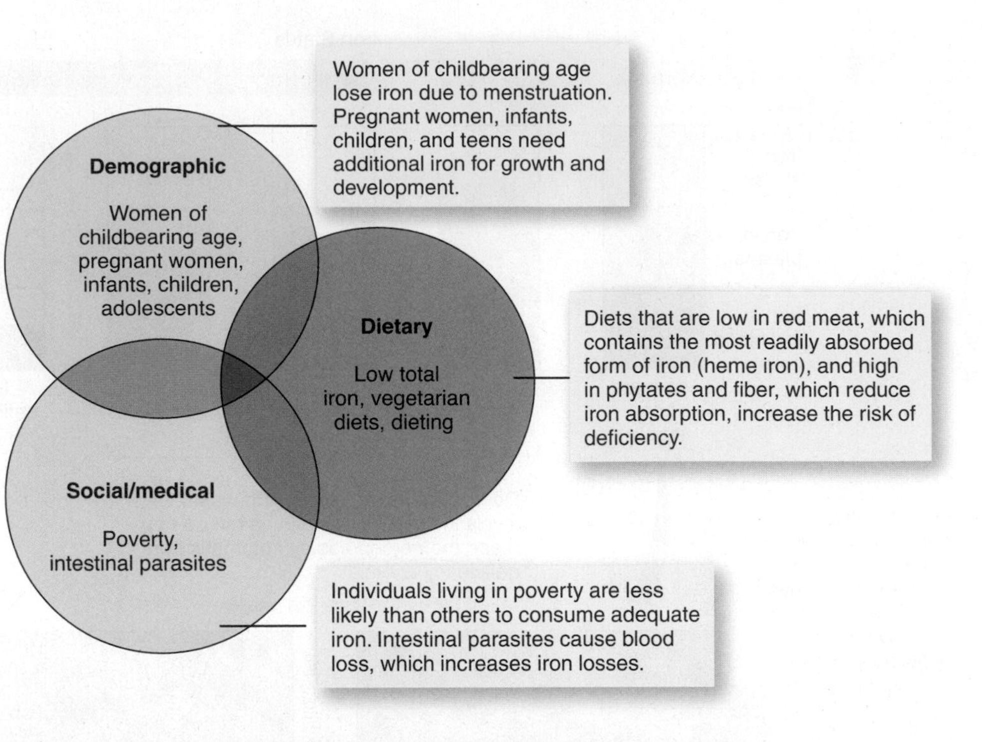

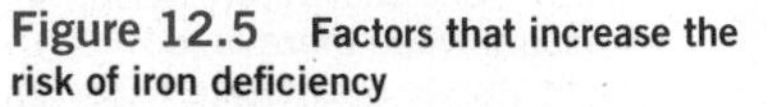

Figure 12.5 Factors that increase the risk of iron deficiency
The risk of iron deficiency is highest among individuals with greater iron losses, those with greater needs due to growth and development, and those who cannot obtain adequate dietary iron.

TABLE 12.1 Dietary Reference Intake Values for Iron

Gender/life stage	Recommended intake
Infants	
0–6 months	0.27 mg/d[a]
7–12 months	11 mg/d
Children	
1–3 years	7 mg/d
4–8 years	10 mg/d
Males	
9–13 years	8 mg/d
14–18 years	11 mg/d
≥19 years	8 mg/d
Females	
9–13 years	8 mg/d
14–18 years	15 mg/d
19–50 years	18 mg/d
≥51 years	8 mg/d
Females taking oral contraceptives	
14–18 years	11.4 mg/d
19–50 years	10.9 mg/d
Pregnant women	27 mg/d
Lactating women	
≤18 years	10 mg/d
19–50 years	9 mg/d
Vegetarians[b]	
Men ≥ 19 years	14 mg/d
Women ≥ 51 years	14 mg/d
Menstruating women (19–50 years)	32 mg/d
Adolescent girls	27 mg/d

[a]This value is an AI; all other values are RDAs.

[b]Value is RDA × 1.8.

Recommended Iron Intake

The RDA for iron is based on the amount needed to maintain normal function but only minimal iron stores. The RDA is set at 8 mg/day for adult men ages 19 and older and for postmenopausal women.[10] The RDA for menstruating women is increased to 18 mg/day to compensate for the iron lost in menstrual blood. Other specific recommendations have been made for each gender and life-stage group by considering the percentage of dietary iron absorbed, iron losses from the body, and conditions that increase needs, such as growth and pregnancy. A separate RDA category has been created for vegetarians because iron is poorly absorbed from plant sources (**Table 12.1**).

The recommended iron intake during pregnancy is increased to 27 mg/day to account for the iron deposited in the fetal and maternal tissues. The RDA during lactation is set lower than the RDA for menstruating women because little iron is lost in milk and menstruation is usually absent (see Table 12.2). The RDA for infants, children, and adolescents considers the additional iron needed for growth. An AI has been set for infants from newborn to 6 months based on the mean iron intake of infants principally fed human milk. Because the iron in human milk is more bioavailable than that in infant formula, it is recommended that infants who are not fed human milk or are only partially nourished with human milk be fed iron-fortified formula.[11]

Iron Toxicity

Iron is essential for cellular metabolism, but too much can be toxic (**Table 12.2**). Iron promotes the formation of free radicals and causes cell death due to excess oxidation of cellular components. Iron toxicity can be acute, resulting from ingestion of a single large dose at one time, or chronic, due to the accumulation of iron in the body over time, referred to as *iron overload*. A UL has been set at 45 mg/day from all sources.[10]

How Acute Toxicity Affects the Body Taking a single large dose of iron can be life-threatening. This acute iron poisoning can damage the intestinal lining and cause abnormalities in body pH, shock, and liver failure. Iron toxicity from supplements is one of the most common forms of poisoning among children under age 6 and is the leading cause of liver transplants in children. To protect children from accidental poisoning from iron-containing drugs and supplements, these products display a warning on the label (**Figure 12.6**).[12]

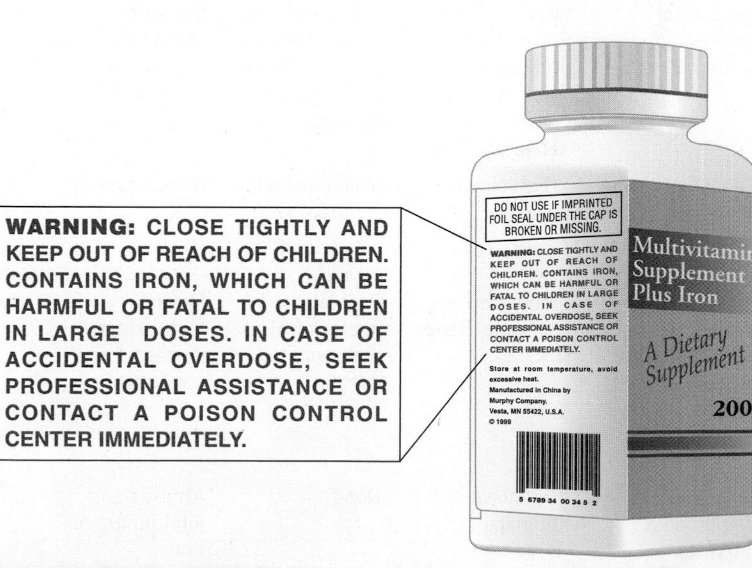

Figure 12.6 Iron toxicity from supplements
Iron-containing products must carry this warning and should be stored out of the reach of children or other individuals who might consume them in excess.

TABLE 12.2 A Summary of the Trace Minerals (All Values Are RDAs Unless Otherwise Noted)

Mineral	Sources	Recommended intake for adults	Major functions	Deficiency diseases and symptoms	Groups at risk of deficiency	Toxicity	UL
Iron	Red meats, leafy greens, dried fruit, whole and enriched grains	8–18 mg/d	Part of hemoglobin, which delivers oxygen to cells; myoglobin, which holds oxygen in muscle; and electron carriers in the electron transport chain; needed for immune function	Iron deficiency anemia: fatigue, weakness, small pale red blood cells, low hemoglobin	Infants and preschool children, adolescents, women of childbearing age, pregnant women, athletes, vegetarians	Gastrointestinal upset, liver damage	45 mg/d
Zinc	Meat, seafood, whole grains, eggs	8–11 mg/d	Regulates protein synthesis; functions in growth, development, wound healing, immunity, and antioxidant protection	Poor growth and development, skin rashes, decreased immune function	Vegetarians, low-income children, elderly	Decreased copper absorption, depressed immune function	40 mg/d
Copper	Organ meats, nuts, seeds, whole grains, seafood, cocoa	900 μg/d	A part of proteins needed for iron absorption, lipid metabolism, collagen synthesis, nerve and immune function, and antioxidant protection	Anemia, poor growth, bone abnormalities	Those who oversupplement zinc	Vomiting	10 mg/d
Manganese	Nuts, legumes, whole grains, tea	1.8–2.3 mg/d[a]	Functions in carbohydrate and lipid metabolism and antioxidant protection	Growth retardation	None	Nerve damage	11 mg/d
Selenium	Organ meats, seafood, eggs, whole grains	55 μg/d	Antioxidant protection as part of glutathione peroxidase, synthesis of thyroid hormones; spares vitamin E	Muscle pain, weakness, Keshan disease	Populations in areas with low-selenium soil	Nausea, diarrhea, vomiting, fatigue, hair and nail changes	400 μg/d
Iodine	Iodized salt, saltwater fish, seafood, dairy products	150 μg/d	Needed for synthesis of thyroid hormones	Goiter, cretinism, intellectual disability, growth and developmental abnormalities	Populations in areas with low-iodine soil and iodized salt is not used	Enlarged thyroid	1110 μg/d
Chromium	Brewer's yeast, nuts, whole grains, mushrooms	25–35 μg/d[a]	Enhances insulin action	High blood glucose	Malnourished children	None reported	ND
Fluoride	Fluoridated water, tea, fish, toothpaste	3–4 mg/d[a]	Strengthens tooth enamel, enhances remineralization of tooth enamel, reduces acid production by bacteria in the mouth	Increased risk of dental caries	Populations in areas with unfluoridated water	Mottled teeth, kidney damage, bone abnormalities	10 mg/d
Molybdenum	Milk, organ meats, grains, legumes	45 μg/d	Cofactor for a number of enzymes	Unknown in humans	None	Arthritis and joint inflammation	2 mg/d

[a]Adequate Intake (AI).

Note: UL = Tolerable Upper Intake Level; ND = Not determined due to insufficient evidence.

Why Iron Overload Is Damaging If too much iron enters the body, over time it accumulates in tissues such as the heart and liver. Iron overload can occur in people with conditions that cause abnormal red blood cell synthesis and in those with diseases requiring frequent blood transfusions, but the most common cause of chronic iron overload is **hemochromatosis.**[13] Hemochromatosis is an inherited condition, caused by a defect in the gene for hepcidin or other proteins involved in iron uptake, that allows excess iron to enter the circulation.[14] Hemochromatosis afflicts about 0.3 to 0.7% of people of northern European ancestry and is the most common genetic disorders in the Caucasian population.[15]

hemochromatosis An inherited condition that results in increased iron absorption.

Hemochromatosis has no symptoms early in life, but in middle age nonspecific symptoms such as weight loss, fatigue, weakness, and abdominal pain typically begin. If allowed to progress, the accumulation of excess iron that occurs in hemochromatosis causes oxidative changes resulting in heart and liver damage, diabetes, certain types of cancer, and other chronic conditions. Iron deposits also darken the skin. To have these symptoms, an individual must inherit the hemochromatosis gene from both parents. People who inherit the gene from only a single parent (about 10% of the population) don't have these serious symptoms but do absorb iron better than people who do not have the gene at all.

The rate at which iron accumulates and leads to serious symptoms in those with hemochromatosis depends on the amount of iron consumed and other dietary factors that affect iron absorption, as well as factors that cause iron loss such as menstruation and blood donations. The public health impact of hemochromatosis is potentially significant. The availability of red meat and the prevalence of iron-fortified foods in the American diet virtually assure that individuals with two genes for hemochromatosis will eventually accumulate damaging levels of iron. If individuals with hemochromatosis can be identified, treatment is simple: regular blood withdrawal. This will prevent the complications of iron overload, but to be effective it must be initiated before organs are damaged.[15] Therefore, genetic screening to identify and treat young healthy individuals is essential in preventing complications.

Overconsumption of iron supplements or a diet high in absorbable iron can also increase iron stores. Although these iron stores are not high enough to cause the serious problems that occur with hemochromatosis, it has been hypothesized that people with high iron stores are at increased risk of the same diseases that occur in hemochromatosis—heart disease, diabetes, and cancer. Iron is hypothesized to promote heart disease by increasing the formation of oxidized LDL cholesterol, which then leads to atherosclerosis. In some studies, elevated levels of the iron-storage protein ferritin have been associated with an increased risk of heart attacks, but the majority of studies do not support this association.[16] A relationship between higher iron stores and an increased risk of diabetes has been found in overweight and obese individuals with impaired glucose tolerance.[17] Iron is hypothesized to contribute to diabetes because it promotes the formation of free radicals, which contribute to insulin resistance and eventually decreased insulin secretion. The increase in free radical formation may also increase cancer risk. To avoid iron overload, iron supplements are not recommended for adult men and postmenopausal women unless prescribed by a physician.[4]

Meeting Iron Needs

Both the amount and the bioavailability of iron from the diet need to be considered when planning a diet to meet iron needs (see **Critical Thinking: Increasing Iron Intake and Uptake**). The best food sources of iron are red meats and organ meats such as liver and kidney because they provide the more absorbable heme iron. Good nonheme iron sources are legumes, dried fruit, leafy greens such as spinach and kale, and fortified grain products. Nonheme iron absorption can be enhanced by including meat, fish, poultry, and foods rich in vitamin C in meals containing iron while decreasing the consumption of dairy products, which are high in calcium, at these meals.[3] Food labels can help identify good sources of iron (**Figure 12.7**).

Although diet is the ideal way to meet iron needs, supplements are often recommended for groups at risk for deficiency such as small children, women of childbearing age, and pregnant women. Iron is commonly available as an individual supplement or as part of multivitamin and mineral supplements. These contain nonheme iron. As with nonheme iron in the diet, to enhance the absorption of iron in a supplement, it should be

Increasing Iron Intake and Uptake

© Mona Makela/iStockphoto/

BACKGROUND

Hanna is a 23-year-old graduate student. She has been feeling tired and run down all semester. She goes to the health center where blood tests indicate that she has iron deficiency anemia. A review of her vegetarian diet is shown here.

TYPICAL DIET

FOOD	AMOUNT	IRON (mg)
Breakfast		
Grits with	1 cup	1.4
butter	1 tsp	0
Plantain	1 (3 oz)	1.8
Whole-wheat toast	1 slice	1.4
Apple juice	¾ cup	0.3
Tea with	1 cup	0
sugar	1 tsp	0
Lunch		
Apple	1 medium	0.2
Cornbread with	1 piece	0.9
Butter	1 tsp	0
Yogurt, low-fat, vanilla	1 cup	0.2
Tomato, raw	1 medium	0.5
Tea with	1 cup	0
sugar	1 tsp	0
Dinner		
Rice, brown	1 cup	0.8
Peanuts	⅓ cup	1.2
Green beans, boiled	1 cup	0.8
Yams, boiled	1 cup	0.7
Apple juice	¾ cup	0.2
Tea with	1 cup	0
sugar	1 tsp	0
TOTAL		10.4

CRITICAL THINKING QUESTIONS

- Based on the diet shown here, what factors contributed to Hanna's iron deficiency?
- Suggest two changes in Hanna's vegetarian diet that would increase the amount of iron she consumes.
- Suggest two changes in Hanna's diet that would help to increase the absorption of the iron in her meals.
- Hanna's diet does not provide enough calcium. Suggest a substitution she could make that would increase her calcium intake by about 200 mg.
- Are there other nutrient deficiencies for which she may be at risk?

iProfile Use iProfile to compare the amount of iron in the dark meat and white meat of poultry.

consumed with foods containing vitamin C, such as orange juice; taken with a meal containing meat, fish, or poultry; and not taken with dairy products, calcium supplements, or substances that bind iron. Iron from supplements that contain the ferrous form (Fe^{2+}) of iron, such as ferrous sulfate, is more readily absorbed than iron from those with the ferric form (Fe^{3+}). Iron supplements can improve iron status, but large intakes of iron from supplements can cause toxicity symptoms and interfere with the absorption of zinc and copper. Iron-containing supplements should be taken only as suggested on the label or prescribed by a physician (**Table 12.3**).

Charles D. Winters

Nutrition Facts
Serving Size 1/4 cup (40g)

Amount Per Serving	
Calories 130	**Calories from Fat** 0
	% Daily Value*
Total Fat 0g	0%
Saturated Fat 0g	0%
Trans Fat 0g	0%
Cholesterol 0mg	0%
Sodium 10mg	0%
Total Carbohydrate 31g	10%
Dietary Fiber 2g	9%
Sugars 29g	
Protein 1g	
Vitamin A	0%
Vitamin C	0%
Calcium	2%
Iron	6%

*Percent Daily Values are based on a 2,000 calorie diet. Your daily values may be higher or lower depending on your calorie needs:

	Calories	2,000	2,500
Total Fat	Less Than	65g	80g
Sat. Fat	Less Than	20g	25g
Cholesterol	Less Than	300mg	300mg
Sodium	Less Than	2,400mg	2,400mg
Total Carbohydrate		300g	375g
Dietary Fiber		25g	30g

Calories per gram:
Fat 9 Carbohydrate 4 Protein 4

Figure 12.7 Iron on food labels
The foods in the photo are good sources of iron. Because iron is a nutrient at risk for deficiency in the American diet, the iron content of packaged foods must be listed on food labels. It is given as a percentage of the Daily Value for iron, which is 18 mg for adults.

THINK CRITICALLY

If you eat the box of raisins shown here, which provides 6% of the Daily Value for iron, what additional foods could you consume throughout the day to meet your iron needs?

TABLE 12.3 Benefits and Risks of Trace Element Supplements

Supplement	Claims	Actual benefits or risks
Iron	Increases energy	Needed to make hemoglobin to deliver oxygen to tissues. Supplements are beneficial if iron is deficient. High doses cause constipation, liver damage, and death.
Zinc	Treats colds, prevents aging, improves immune function, enhances fertility	Needed for enzyme function, protein synthesis, and vitamin and hormone function. Supplements do not enhance these effects. Zinc losenges do not prevent colds but may reduce their duration and severity. High doses cause copper deficiency, nausea, and vomiting.
Copper	Prevents heart disease and osteoporosis. Alleviates arthritis symptoms, maintains healthy skin and hair color, treats hypoglycemia	Supplements are useful for improving bone health and improving blood lipids in those with copper deficiency, but there is no evidence that intakes above recommended levels prevent heart disease or are effective for the treatment of arthritis or skin conditions. High doses can cause vomiting.
Selenium	Protects against cancer. Promotes heart health, stimulates immune function	Antioxidant; evidence that it may protect against cancer in those with low levels. High doses cause loss of hair and nail changes.
Chromium	Controls diabetes, lowers cholesterol, reduces body fat, and increases lean tissue	Needed for insulin action. Supplements may improve blood sugar regulation but do not affect body composition.
Vanadium	Aids insulin action; allows more rapid and intense muscle pumping for body builders	No evidence to support a benefit for body builders. Supplements can reduce insulin requirements, but the dose required exceeds the UL.

12.3 Zinc (Zn)

LEARNING OBJECTIVES

- List some dietary sources of zinc.
- Describe how zinc absorption is regulated.
- Discuss the role of zinc in gene expression.

The essentiality of zinc in the human diet was first recognized about 50 years ago, when a syndrome of growth depression and delayed sexual development, seen in Iranian and Egyptian men consuming diets based on vegetable protein, was alleviated by supplemental zinc.[18] Although their diet was not low in zinc, it was high in grains containing phytates, which interfered with zinc absorption and caused their zinc deficiency.

Zinc in the Diet

Zinc is found in foods from both plant and animal sources. Zinc from animal sources is better absorbed than that from plants because the zinc in plant foods is often bound by phytates. Zinc is abundant in red meat, liver, eggs, dairy products, vegetables, and some seafood (**Figure 12.8**). Whole grains are a good source, but refined grains are not because zinc is lost in milling and not added back in enrichment. Grain products leavened with yeast provide more zinc than unleavened products because the yeast leavening of breads reduces the phytate content.[10]

Zinc in the Digestive Tract

The gastrointestinal tract is the major site for regulation of zinc homeostasis. Both the amount of zinc in the mucosal cells and the amount that leaves these cells for distribution to the rest of the body are regulated.

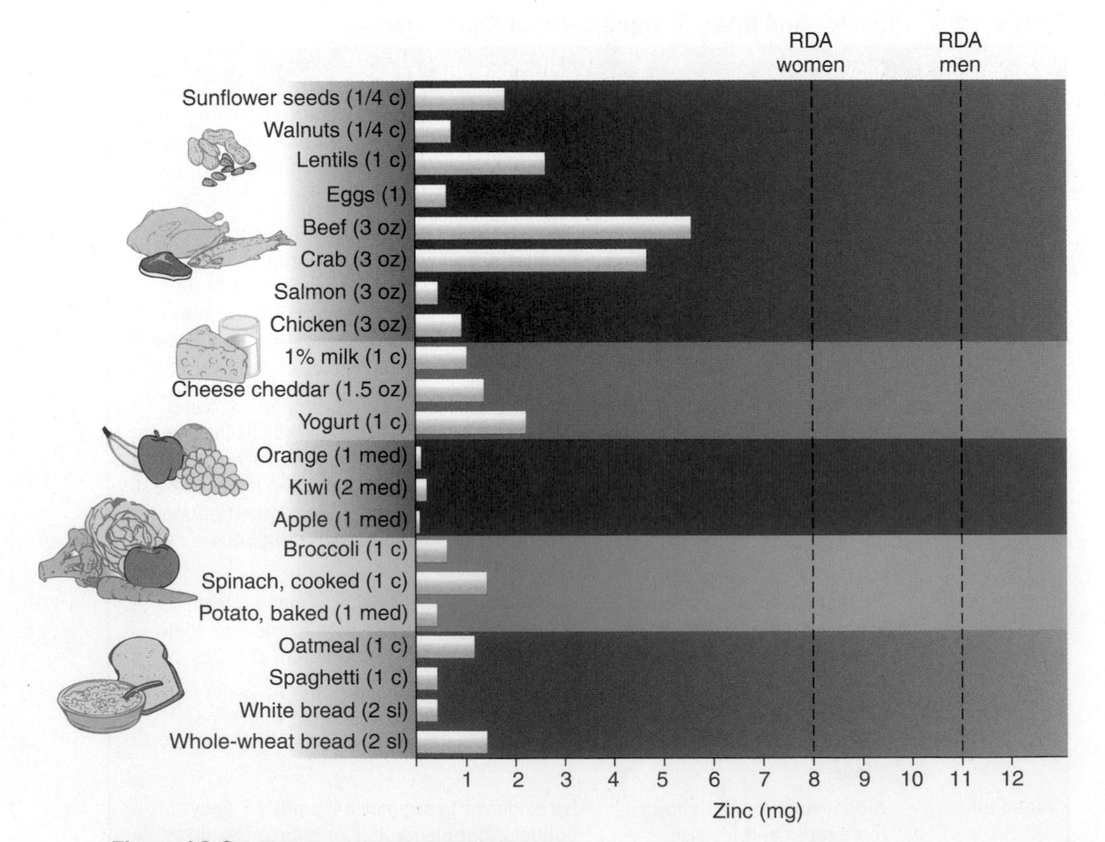

Figure 12.8 Zinc content of MyPlate food groups
Meat, seafood, dairy products, and whole grains are good sources of zinc; the dashed lines represent the RDA for adult men and women.

Figure 12.9 Regulation of zinc absorption

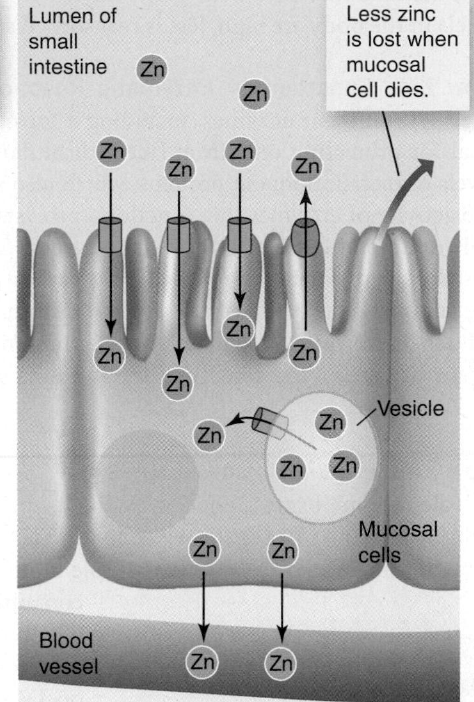

(a) When zinc intake is high, fewer of the zinc transport proteins (shown in blue) are available to move zinc from the lumen into the cells and more zinc transporters (shown in orange) move zinc out of the mucosal cells into the lumen and from the cytosol into storage vesicles. The synthesis of metallothionein, which binds zinc and limits its uptake into the blood, increases. This allows the body to lose zinc when the cells die.

(b) When zinc intake is low, more zinc transport proteins (shown in blue) are available to move zinc from the lumen of the intestine into the mucosal cells and from storage vesicles into the cytosol. Little metallothionein is synthesized, so zinc is not retained in the mucosal cells to be lost when the cells die.

Zinc transport proteins regulate the amount of zinc in the cytosol of the mucosal cells. Some of these proteins increase the amount of zinc absorbed into the mucosal cell by promoting the transport of zinc from the intestinal lumen into the cell. Others reduce the amount of zinc in the cytosol of the mucosal cell by transporting zinc back into the lumen or into storage vesicles in the cell (**Figure 12.9**).[19] The amount of zinc that is in the cytosol and is transported into the blood can be regulated by increasing or decreasing the synthesis of proteins that transport zinc into versus those that transport it back out of the mucosal cells.[20] For example, if zinc intake is high (see Figure 12.9a), expression of zinc transport proteins that move zinc from the lumen into the mucosal cells will decrease relative to the expression of proteins that export zinc out of the mucosa. Low zinc levels will have the opposite effect, decreasing zinc export to the lumen relative to transport into the cell (see Figure 12.9b).

The amount of zinc that passes from the mucosal cell into the blood is also regulated by a metal-binding protein called **metallothionein.**[20] When zinc intake is high, metallothionein synthesis increases. Zinc in the mucosal cell binds to metallothionein, slowing its transfer into the blood; this provides more opportunity for export of zinc back into the lumen when zinc intakes are high or for it to be lost if the mucosal cell dies (see Figure 12.9a). Metallothionein also binds copper, and high levels can inhibit copper absorption.

metallothionein A group of proteins that bind metals. One such protein binds zinc and copper in intestinal cells, limiting their absorption into the blood.

Zinc in the Body

Zinc is the most abundant intracellular trace element. It is found in the cytosol, in cellular organelles, and in the nucleus. Once zinc has been absorbed, homeostasis can be

superoxide dismutase (SOD) An enzyme that protects the cell by neutralizing damaging superoxide free radicals. One form of the enzyme requires zinc and copper for activity, and another form requires manganese.

maintained to some extent by regulating excretion. Zinc is secreted in pancreatic and intestinal juices, which enter the lumen of the intestine. When zinc levels in the body are low, the zinc that enters the gastrointestinal tract can be reabsorbed and recycled. When levels in the body are high, less is reabsorbed and more is therefore eliminated in the feces.

How Zinc Functions in Enzymatic Reactions Zinc is involved in the functioning of over 300 different enzymes, including a form of **superoxide dismutase (SOD)**, which is vital for protecting cells from free radical damage. Zinc is needed to maintain adequate levels of metallothionein proteins, which also scavenge free radicals.[21] Zinc is essential for the activity of enzymes that function in the synthesis of proteins, DNA, and RNA; in carbohydrate metabolism; in acid–base balance; and in a reaction that is necessary for the absorption of folate from food. Zinc plays a role in the storage and release of insulin, the mobilization of vitamin A from the liver, and the stabilization of cell membranes. It influences hormonal regulation of cell division and is therefore needed for the growth and repair of tissues, the activity of the immune system, and the development of sex organs and bone.

How could a zinc deficiency lead to a secondary vitamin A deficiency?

How Zinc Regulates Gene Expression Some of the functions of zinc can be traced to its role in gene expression. For example, zinc stimulates the production of metallothionein by binding to a regulatory factor and activating the transcription of the gene for this protein. Zinc also plays a structural role in proteins essential for gene expression. Proteins containing zinc fold around the zinc atom to form a loop or "finger." These zinc fingers allow nuclear receptor proteins to bind to regulatory regions on DNA, stimulating the transcription of specific genes and therefore the synthesis of the proteins for which they code. For example, the retinoic acid-receptor protein contains zinc fingers. When it binds to vitamin A (retinoic acid), the zinc fingers bind to DNA, allowing gene transcription to be stimulated (**Figure 12.10**). Without zinc, vitamin A, as well as vitamin D and certain hormones, including the thyroid hormones, estrogen, and testosterone, cannot interact with DNA to increase or decrease gene expression and, hence, the synthesis of certain proteins.

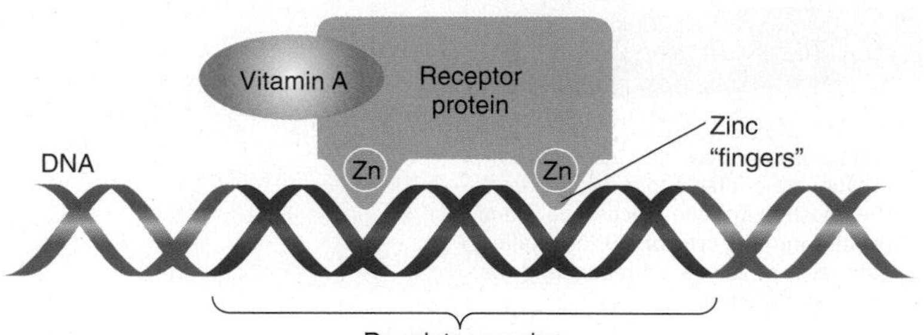

Figure 12.10 Zinc fingers and gene expression
Finger-like structures called zinc fingers allow nuclear receptor proteins that bind to vitamin A to interact with the regulatory region of a gene and thus affect gene expression. Similar nuclear receptor proteins are essential for the activity of vitamin D and certain hormones.

Zinc Deficiency

Zinc deficiency has been seen in individuals with a genetic defect in zinc absorption and metabolism called *acrodermatitis enteropathica*, in those fed total parenteral nutrition (TPN) solutions lacking zinc, and in those consuming diets low in protein and high in phytates. It may also occur in individuals with kidney disease, sickle cell anemia, alcoholism, cancer, or AIDS.

The symptoms of zinc deficiency include poor growth and development, skin rashes, hair loss, diarrhea, neurological changes, impaired reproduction, skeletal abnormalities, and reduced immune function.[22] Many of these symptoms reflect zinc's importance in protein synthesis and gene expression. Because it is needed for the proper functioning of vitamins A and D and the activity of numerous enzymes, some of the zinc deficiency symptoms resemble deficiencies of other essential nutrients. Decreased immune function is one of the main concerns with even moderate zinc deficiency.[22] The impact of zinc deficiency on immune function is rapid and extensive, causing a decrease in the number and function of immune cells in the blood, which can lead to an increased incidence of infections.

Symptomatic zinc deficiency is relatively uncommon in North America. Only about 8% of the U.S. population consumes less than the EAR for zinc, but this percentage would be much higher without the zinc provided by fortified foods and supplements.[23]

The risk of zinc deficiency is greater in areas of the world where the diet is high in phytate, fiber, tannins, and oxalates, as it is in many developing countries. Through its relationship with immune function, zinc deficiency contributes to infection and overall mortality in children in the developing world.[24] Pregnant women, the elderly, and vegans are also at particular risk.

Recommended Zinc Intake

The RDA for zinc is 11 mg/day for adult men and 8 mg/day for adult women. This is based on the amount of zinc needed to replace daily losses from the body. Vegetarians may need to consume as much as 50% more zinc to meet their needs, depending on the phytate content of their diet.[10]

During pregnancy, the recommendation for zinc intake is increased to account for the zinc that accumulates in maternal and fetal tissues. During lactation, the RDA is increased to compensate for zinc secreted in breast milk. For infants from newborn to 6 months, an AI has been established based on the zinc intake of breast-fed infants. RDAs have been established for older infants, children, and adolescents based on the amount of zinc lost from the body, the amount needed for growth, and the absorption of zinc from the diet.

Zinc Toxicity

Zinc can be toxic when consumed in excess of recommendations. A single dose of 1 to 2 g can cause gastrointestinal irritation, vomiting, loss of appetite, diarrhea, abdominal cramps, and headaches. This has occurred with consumption of foods and beverages contaminated with zinc that has leached from galvanized containers. Intakes in the range of 50 to 300 mg/day have been shown to decrease rather than enhance immune function and to reduce HDL cholesterol, the type of cholesterol that has a protective effect against heart disease.[10] Supplements providing 50 mg/day of zinc have been shown to interfere with the absorption of copper. When high zinc intake inhibits copper absorption, it leads to a reduction in the activity of the copper-dependent enzyme copper-zinc superoxide dismutase in red blood cells. A UL has been set at 40 mg/day from all sources based on the adverse effect of excess zinc on copper absorption and metabolism.

Zinc Supplements

Zinc is often marketed as a supplement to improve immune function and enhance fertility and sexual performance. For individuals consuming adequate zinc, there is no evidence that extra is beneficial. In individuals with a mild zinc deficiency, supplementation may result in improved wound healing, immunity, and appetite; in children, it can result in improved growth and learning. In healthy older adults, supplements of zinc have been shown to improve the immune response[25] (see Chapter 16). Zinc supplements in lozenge form are currently popular for preventing and treating colds. Supplemental zinc is also used therapeutically to treat genetic diseases.

THINK CRITICALLY

How many lozenges could you take per day without exceeding the UL for zinc?

Do Zinc Lozenges Relieve Cold Symptoms? Americans suffer from about 62 million colds a year, and the common cold is a leading cause of doctor visits and missed days from school and work.[26] It has been suggested that zinc lozenges, containing either zinc glyconate or zinc acetate, reduce the duration and severity of the common cold by inhibiting the replication of the cold viruses.[27] The zinc swallowed in a mineral supplement will not have any effect because this zinc goes to your stomach and doesn't contact the mucosal surfaces in the nose and throat that are affected by cold viruses. Although clinical trials to assess the efficacy of zinc lozenges have been inconsistent, most support the value of zinc lozenges in reducing the duration and severity of symptoms when administered within 24 hours of the onset of cold symptoms.[28] If zinc lozenges are used as a cold remedy, they should be used cautiously (**Figure 12.11**). Nausea, constipation, diarrhea, abdominal pain, dry mouth, and oral irritation have been

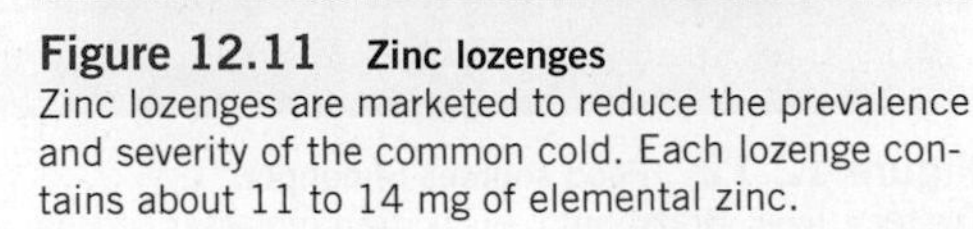

Andy Washnik

Figure 12.11 Zinc lozenges
Zinc lozenges are marketed to reduce the prevalence and severity of the common cold. Each lozenge contains about 11 to 14 mg of elemental zinc.

reported with the use of zinc lozenges.[28] Too much zinc can also suppress the immune system, lower HDL cholesterol levels, and impair copper absorption.

Supplemental Zinc to Treat Genetic Abnormalities Acrodermatitis enteropathica is due to an inherited defect in zinc absorption that results in zinc deficiency. This condition causes skin lesions, damages the eyes, and increases the risk of infection. If untreated, patients with acrodermatitis enteropathica usually die within the first few years of life. Symptoms can be reversed by providing supplemental zinc in amounts greater than 1 to 2 mg per kg per day for life. These large doses bypass the normal absorptive mechanisms, allowing enough zinc to get into the body. This therapy achieves a survival rate of 100%. Because treatment involves consumption of amounts well in excess of the UL, patients with this disorder must be monitored to ensure that the high zinc level does not cause copper deficiency.

Zinc supplements are also used to treat *Wilson's disease*. Wilson's disease is due to an inherited defect in the excretion of copper and causes copper to accumulate in the body, leading to toxicity. A few drugs are available to remove copper from the body, but supplemental zinc acetate, which blocks the absorption of copper, increases copper excretion in the stool, and causes no serious side effects, is considered the best treatment.[29]

12.4 Copper (Cu)

LEARNING OBJECTIVES

- Explain why copper deficiency can lead to anemia.
- Describe how high intakes of zinc affect copper absorption.

The ability of copper to treat certain types of anemia helped establish the essentiality of this mineral in human nutrition.[30] Further understanding of the impact of copper deficiency in humans came from studying individuals who were inadvertently fed intravenous (TPN) solutions deficient in copper and those with a rare genetic disease called *Menkes disease* or *kinky hair disease* in which there is a defect in intestinal copper absorption.

Copper in the Diet

The richest dietary sources of copper are organ meats such as liver and kidney. Seafood, nuts and seeds, whole-grain breads and cereals, and chocolate are also good sources (**Figure 12.12**). As with many other trace minerals, soil content affects the amount of copper in plant foods.

Keith Weller/Courtesy USDA

Figure 12.12 **Food sources of copper**
Oysters, liver, Brazil nuts, blackstrap molasses, cocoa, and black pepper are the highest in copper, but lobster, other nuts, sunflower seeds, green olives, and wheat bran are also good sources.

Copper in the Digestive Tract

About 30 to 40% of the copper in a typical diet is absorbed.[10] The absorption of copper is affected by the presence of other minerals in the diet. As discussed, the zinc content of the diet can have a major impact on copper absorption. When zinc intake is high, it stimulates the synthesis of the protein metallothionein in the mucosal cells. Although metallothionein binds zinc, it binds to copper more tightly. Therefore, when metallothionein is synthesized, it binds copper, preventing it from being moved out of mucosal cells into the blood (**Figure 12.13**). The antagonism between copper and zinc is so great that phytates, which inhibit zinc absorption, actually increase the absorption and utilization of copper. Copper absorption is also reduced by high intakes of iron, manganese, and molybdenum. Other factors that affect copper absorption include vitamin C, which decreases

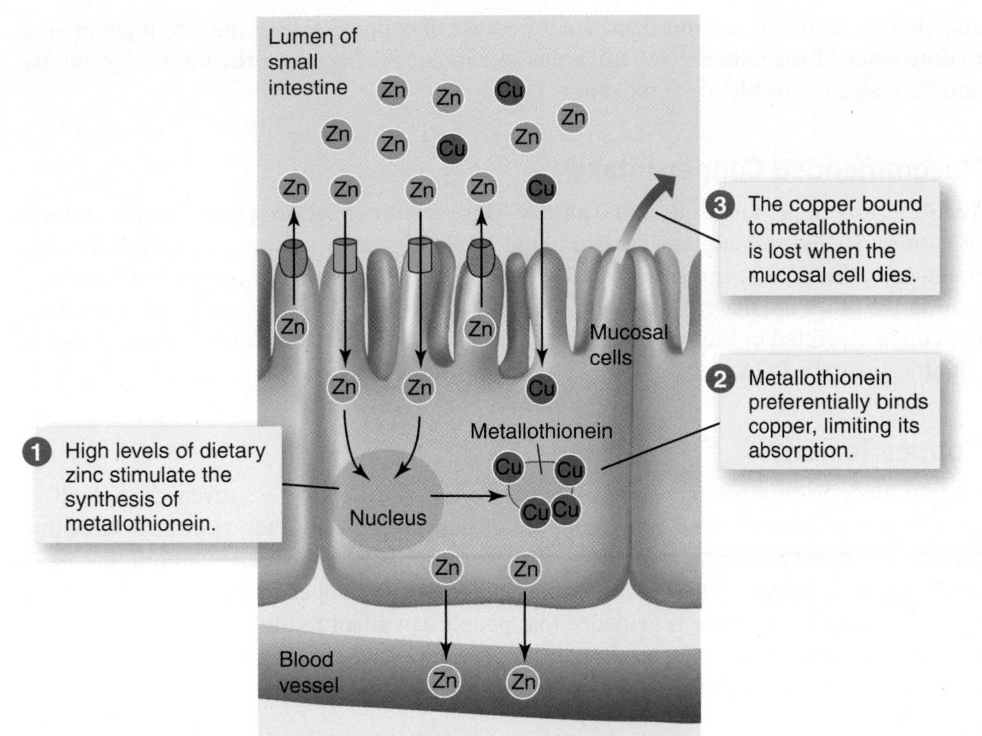

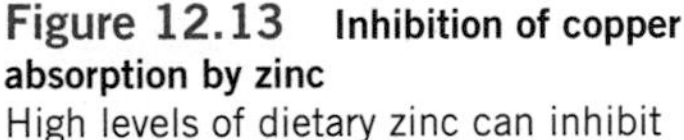

Figure 12.13 Inhibition of copper absorption by zinc
High levels of dietary zinc can inhibit copper absorption by stimulating the synthesis of metallothionein, which then preferentially binds copper and limits its absorption.

copper absorption, and large doses of antacids, which also inhibit copper absorption and, over the long term, can cause copper deficiency.

How Copper Functions in the Body

Once absorbed, copper binds to albumin, a protein in the blood, and travels to the liver, where it binds to the protein **ceruloplasmin** for delivery to other tissues. Copper must be transported bound to proteins such as albumin and ceruloplasmin because free copper ions can trigger oxidation leading to cellular damage. Copper can be removed from the body by secretion in the bile and subsequent elimination in the feces.

ceruloplasmin The major copper-carrying protein in the blood.

Copper functions in a number of important proteins and enzymes that are involved in iron and lipid metabolism, connective tissue synthesis, maintenance of heart muscle, and the functioning of the immune and central nervous systems.[10] The copper-containing plasma protein ceruloplasmin converts iron into a form that can bind to transferrin for transport from body cells, and an analogous copper-containing protein found in intestinal cells is essential for the transport of absorbed iron from the intestine. Copper is an essential component in a form of the antioxidant enzyme superoxide dismutase. Copper also plays a role in cholesterol and glucose metabolism; elevated blood cholesterol levels have been reported in copper deficiency. Copper is needed for the synthesis of the neurotransmitters norepinephrine and dopamine and several blood-clotting factors. It may also be involved in the synthesis of myelin, which is necessary for transmission of nerve signals.

Copper Deficiency

Severe copper deficiency is relatively rare, occurring most often in preterm infants. Marginal copper deficiency may be more prevalent but has been difficult to diagnose. The most common manifestation of copper deficiency is anemia. This is due primarily to the importance of copper-containing proteins in iron transport. In copper deficiency, even if iron is sufficient in the diet, iron cannot be transported out of the intestinal mucosa. Copper deficiency also causes skeletal abnormalities similar to those seen in vitamin C deficiency (scurvy). This is because the enzyme needed for the cross-linking of connective tissue proteins requires copper. Copper deficiency has also been associated with impaired growth, degeneration of the heart muscle, degeneration of the nervous system, and changes in hair color and structure.[10] Because of copper's role in the development and maintenance of the immune system, a diet low in copper decreases the immune response and increases the incidence of infection.[31]

Recommended Copper Intake

LIFE CYCLE

The RDA for copper for adults is 900 μg/day. This recommendation is based on the amount of copper needed to maintain normal blood levels of copper and ceruloplasmin. During pregnancy, the RDA is increased to 1000 μg/day to account for the copper that accumulates in the fetus and maternal tissues. The RDA for lactation is 1300 μg/day to account for the copper secreted in human milk. The amount of copper in the North American diet is slightly above the RDA.

Copper Toxicity

Copper toxicity from dietary sources is extremely rare but has occurred as a result of drinking from contaminated water supplies or consuming acidic foods or beverages that have been stored in copper containers. Excessive copper intake causes abdominal pain, vomiting, and diarrhea. These symptoms may occur with copper intakes of 4.8 mg/day in some individuals, but there is evidence that people can adapt to higher exposures without experiencing any adverse effects. High doses of copper have also been shown to cause liver damage. The UL has been set at 10 mg of copper per day. [10]

12.5 Manganese (Mn)

LEARNING OBJECTIVE

- Discuss the antioxidant function of manganese, copper, and zinc.

The best dietary sources of manganese are whole grains, legumes, and nuts (**Figure 12.14**). Fruits and vegetables are fair sources; meat, dairy products, and refined grains are poor sources.

Manganese in the Body

Manganese homeostasis is maintained by regulating both absorption and excretion. Manganese absorption increases when intake is low and decreases when intake is high. Manganese is eliminated by secretion into the intestinal tract in bile. It is a constituent of some enzymes and an activator of others. Manganese-requiring enzymes are involved in amino acid, carbohydrate, and cholesterol metabolism; cartilage formation; urea synthesis; and antioxidant protection. Like copper and zinc, manganese is needed for the activity of a form of superoxide dismutase. The form of the enzyme requiring manganese is located inside the mitochondria.

Rita Maas/The Image Bank/Getty Images, Inc.

Figure 12.14 Food sources of manganese Legumes, nuts, and whole grains are high in manganese.

Manganese Deficiency and Toxicity

Manganese deficiency in animals results in growth retardation, reproductive problems, congenital malformations in the offspring, and abnormalities in brain function, bone formation, glucose regulation, and lipid metabolism.

Although a naturally occurring manganese deficiency has never been reported in humans, a man participating in a study of vitamin K was inadvertently fed a diet deficient in manganese for six months. He lost weight, his black hair turned a red color, and he developed dermatitis and low blood cholesterol. Manganese deficiency was further studied in young male volunteers fed a manganese-deficient diet for 39 days. These men developed dermatitis and had altered blood levels of cholesterol, calcium, and phosphorus.[32]

Toxic levels of manganese result in damage to the nervous system. In humans, toxicity has been reported in mine workers exposed to high concentrations of inhaled manganese dust. The UL is 11 mg/day from all sources.[10]

Recommended Manganese Intake

There is not sufficient evidence to set an RDA for manganese; the AI is 2.3 mg/day for men and 1.8 mg/day for women. Recommended intakes are higher during pregnancy and lactation.[10]

12.6 Selenium (Se)

LEARNING OBJECTIVES

- Compare the antioxidant functions of selenium and vitamin E.
- Explain why the amount of selenium in the soil may impact the selenium status of certain populations.
- Discuss the relationship between selenium and cancer.

Although selenium was discovered about 180 years ago, its essential role in human nutrition was not recognized until the 1970s when it was found to prevent a heart disorder in children living in regions of China with low soil selenium levels. Today selenium is known to be an important part of the body's antioxidant defenses.

Selenium in the Diet

Seafood, kidney, liver, and eggs are excellent sources of selenium (**Figure 12.15**). Grains, seeds, and dairy products are good sources and fruits, vegetables, and drinking water are

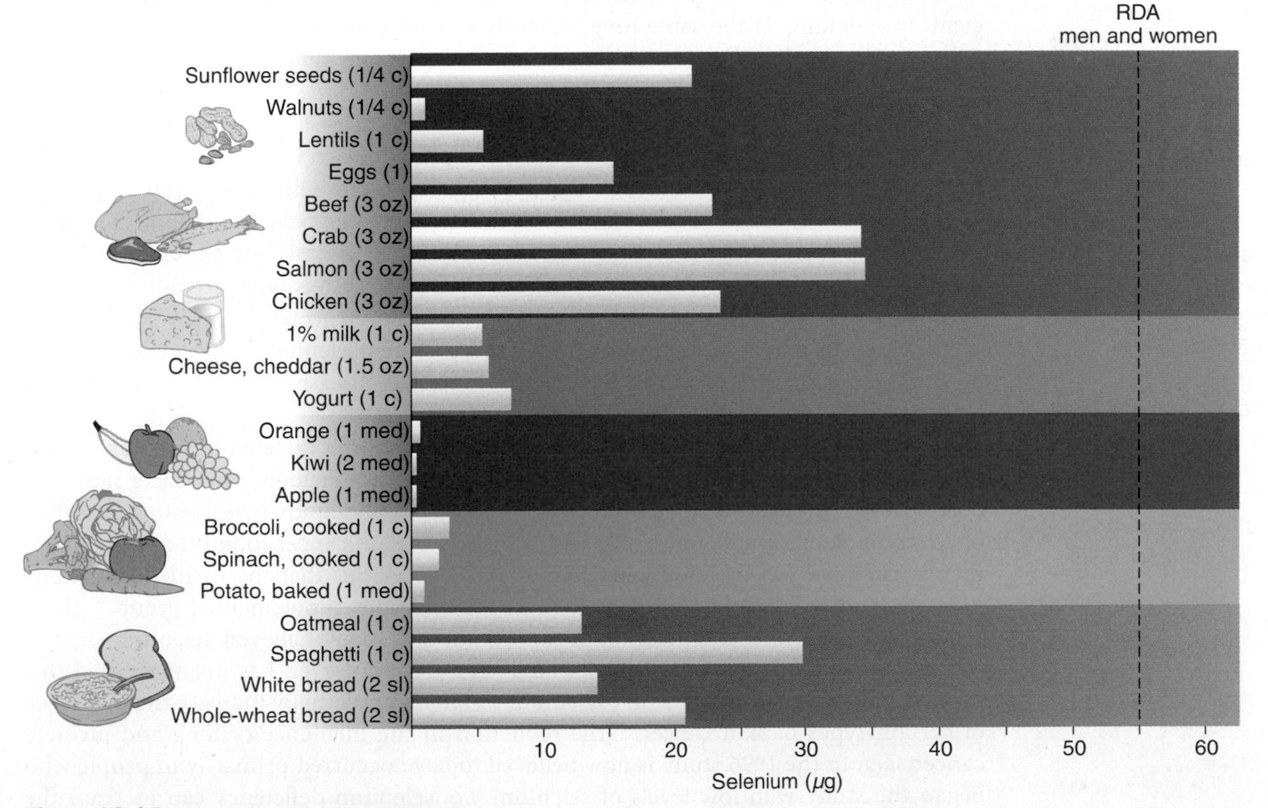

Figure 12.15 Selenium content of MyPlate food groups
Both plant and animal foods are good sources of selenium. The dashed line represents the RDA for adults.

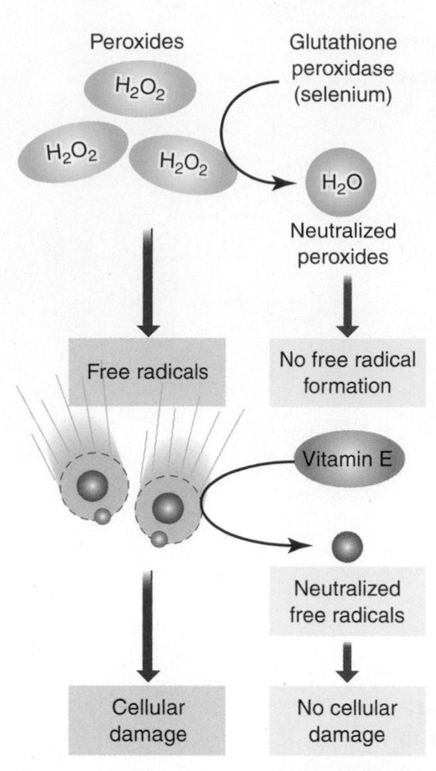

Figure 12.16 **Glutathione peroxidase** Selenium is a part of the enzyme glutathione peroxidase, which neutralizes peroxides before they form free radicals. Fewer free radicals means less vitamin E is needed to eliminate them.

generally poor sources. The selenium content of plant foods depends on the selenium content of the soil where they were grown and the content of animal products is affected by the amount of selenium in their feed.[33] For example, wheat grown in Kansas may have a different selenium content from wheat grown in Michigan. Soil selenium can have a significant impact on the selenium intake of populations consuming primarily locally grown food. Selenium deficiency is not likely to be a problem when the diet includes foods produced in many different locations.

How Selenium Functions in the Body

Selenium absorption is efficient and does not appear to be regulated. Once absorbed, selenium homeostasis is maintained by regulating its excretion in the urine.

Selenium is a mineral that functions mostly through association with proteins called **selenoproteins**. Several of these, including **glutathione peroxidase**, are enzymes that help protect cells from oxidative damage. Glutathione peroxidase neutralizes peroxides so they no longer form free radicals, which cause oxidative damage. By reducing free radical formation, selenium can reduce the need for vitamin E because this vitamin stops the action of free radicals once they are produced (**Figure 12.16**). In addition to its role in glutathione peroxidase, selenium is important for the function of the thyroid gland.[33] Selenoproteins help protect the thyroid gland from reactive oxygen species and selenium-containing enzymes are needed for the synthesis of the thyroid hormones, which regulate basal metabolic rate.

selenoproteins Proteins that contain selenium as a structural component of their amino acids. Selenium is most often found as selenocysteine, which contains an atom of selenium in place of the sulfur atom.

glutathione peroxidase A selenium-containing enzyme that protects cells from oxidative damage by neutralizing peroxides.

Keshan disease A type of heart disease that occurs in areas of China where the soil is very low in selenium. It is believed to be caused by a combination of viral infection and selenium deficiency.

Selenium Deficiency

Symptoms of selenium deficiency include muscular discomfort, weakness, poor immune function, and cognitive decline. Deficiency was not identified in humans until the late 1970s, when it was observed in patients fed TPN solutions inadvertently deficient in selenium. At the same time scientists in China described a disease of the heart muscle called **Keshan disease**, which was linked to selenium deficiency.

Selenium Deficiency and Keshan Disease Keshan disease causes an enlarged heart and poor heart function. It used to be endemic in regions of China where the diet was restricted to locally grown food and the soil was deficient in selenium (**Figure 12.17**). It affected primarily children and women of childbearing age. Selenium supplementation was found to dramatically reduce the incidence of Keshan disease, but it could not reverse heart damage once it had occurred. Although Keshan disease is now virtually eliminated by selenium supplementation, the disease itself is believed to be due to a combination of selenium deficiency and infection with a virus. When selenium is deficient, the virus becomes more virulent, causing the symptoms of Keshan disease.[34]

How Selenium Status Is Related to Cancer The role of selenium in cancer has been under investigation for three decades. An increased incidence of cancer has been observed in regions where selenium intake is low. In 1996 a study investigating the effect of selenium supplements on people with a history of skin cancer found that the supplements had no effect on the recurrence of skin cancer but that the incidence of lung, prostate, and colon cancer all decreased in the selenium-supplemented group.[35] There was a great deal of excitement about this result, and many believed selenium supplements could reduce cancer risk. Subsequent research, however, has not supported this result. Evidence now suggests that selenium supplements actually increase the incidence of certain types of skin cancer.[36] The reduction in the incidence of lung and prostate cancer seen in the 1996 study is now believed to have occurred primarily in people who began the study with low levels of selenium. So, selenium deficiency can increase the risk of cancer, but supplements of selenium have not been shown to be of additional

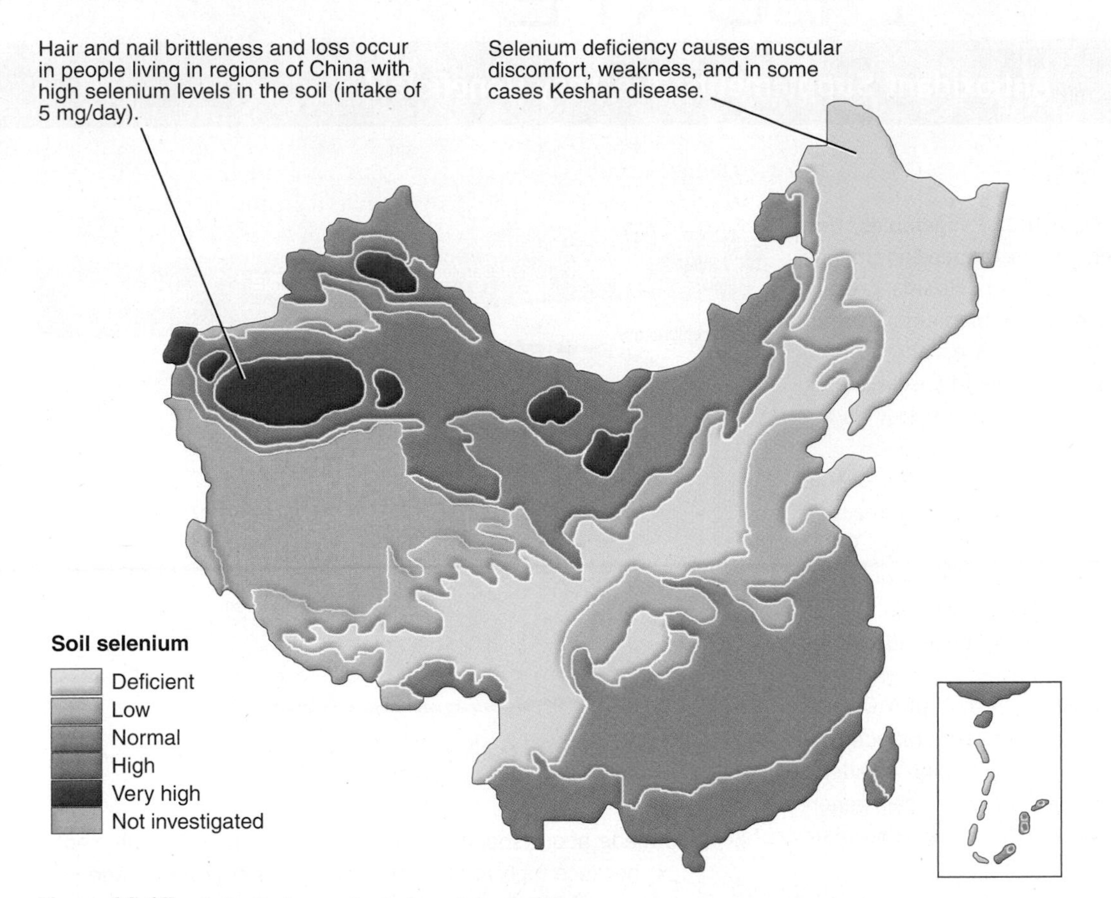

Figure 12.17 **Soil selenium and selenium status in China**
The amount of selenium in the soil affects the selenium content of crops grown in the soil. Keshan disease is most prevalent in the selenium-deficient soil zone that crosses China from the northeast to the southwest. Selenium toxicity symptoms occur in areas of China that have high soil selenium.

benefit in the general population (see **Debate: Antioxidant Supplements: Are They Beneficial?**).

Recommended Selenium Intake

The RDA for selenium for adults is 55 μg/day.[37] This is based on the amount needed to maximize the activity of the enzyme glutathione peroxidase in the blood. The estimated average intake of selenium in the United States meets or nearly meets this recommendation for all age groups.

An increase in selenium intake is recommended during pregnancy based on the amount transferred to the fetus and during lactation to account for the amount of selenium secreted in milk. The AI for infants is based on the amount contained in breast milk.

Selenium Toxicity and Supplements

Selenium toxicity can result from oversupplementation or high levels in the food supply. Symptoms include brittle hair; brittle, thickened finger and toe nails, leading to nail loss in some cases; and a garlic odor on the breath and skin.[33] Hair and fingernail loss from selenium toxicity were reported in a region of China with very high selenium in the soil that led to selenium intake of 5 mg/day (see Figure 12.17). Selenium toxicity has also been reported in the United States because of a manufacturing error that created mineral supplements containing a dose of 27 mg of selenium per day. The individuals who used these

DEBATE

Antioxidant Supplements: Are They Beneficial?

Diets rich in antioxidants from fruits, vegetables, and whole grains are associated with a reduced incidence of chronic diseases. Many of the antioxidants present in these foods are marketed as supplements to fight cancer, protect us from heart disease, and defend against age-related cell damage. Should we be taking these antioxidant supplements to help us stay young and healthy?

The rationale for extra antioxidants is based on the fact that we are constantly bombarded with reactive molecules, such as free radicals, that damage DNA, proteins, and lipids. Free radicals come from air pollution, radiation, and cigarette smoke and are also formed in our bodies. Free radical damage is believed to play a role in aging and diseases such as cancer, cardiovascular disease, and diabetes. However, not all free radicals are harmful. There are free radicals produced by the immune system that help destroy foreign invaders. There must be a balance that allows free radicals generated by the immune system to work while preventing unwanted free radicals from causing damage.

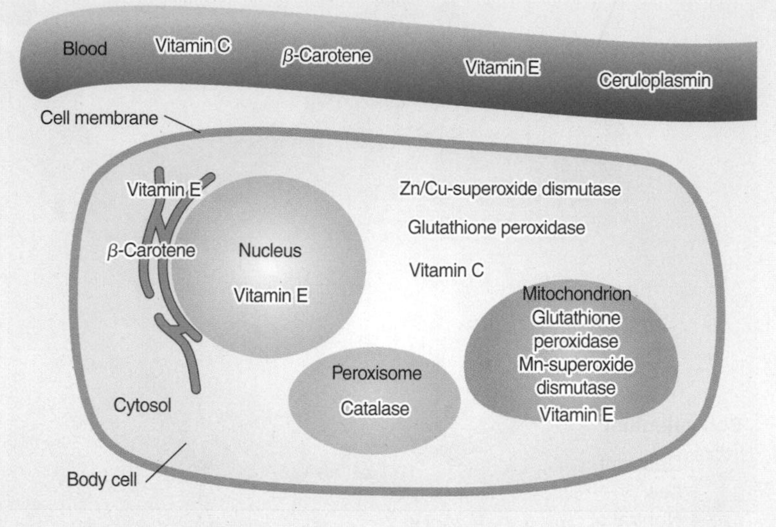

Each of the antioxidant defenses functions in specific locations in the body and protects against specific types of damaging reactions.

If too many free radicals are generated, the body's antioxidant defenses are overwhelmed, a condition referred to as *oxidative stress*, and cell death can result. Those who support the use of antioxidant supplements believe that extra antioxidants can prevent oxidative stress and help us live longer, healthier lives. If your diet is deficient in these nutrients, increasing your intake with a supplement will enhance antioxidant defenses. But supplementing these nutrients to levels above the recommended intake may not be beneficial. A meta-analysis, which is an analysis that combines the results of many studies, has concluded that antioxidant supplementation does not reduce mortality either in healthy people or in those with various diseases and supplements of antioxidants such as β-carotene, vitamin E, and higher doses of vitamin A actually increase mortality.[1]

Not everyone is convinced by these findings. Skeptics argue that combining data on all antioxidants and focusing just on life expectancy may miss the benefits of some individual antioxidant supplements on specific conditions in certain individuals. For example, a combination of dietary antioxidant supplements has been shown to slow the progression of age-related macular degeneration.[2]

Whether you believe the meta-analysis or not, most agree that we need more information about optimum antioxidant doses and the risk of imbalances before recommendations can be made about specific amounts of antioxidant supplements. Just because high intakes of antioxidants from fruits and vegetables reduce the risk of heart disease does not mean the amounts and combinations in supplements will have the same effect.[3] A deficiency of one antioxidant could increase the need for another, and an excess of one may create a deficiency of another. For instance, vitamin C is necessary to regenerate the active form of vitamin E, selenium reduces some of the need for vitamin E, and excesses of zinc can cause copper deficiency. Some antioxidants are harmful to certain individuals. For example, β-carotene supplements increase the risk of lung cancer in smokers, and one analysis found supplemental vitamin E to increase prostate cancer.[4,5]

Antioxidant nutrients are important for good health. Eating more fruits, vegetables, and whole grains will boost the intake of antioxidant nutrients and benefit health, but not everyone agrees that supplements of these antioxidant nutrients are beneficial. Foods provide fiber and phytochemicals that don't come in a supplement, and the antioxidants in food are present in ratios and with other dietary components known to promote rather than harm health. Supplements may be an easy way to increase antioxidant intake, but they also appear to be detrimental to some people.

THINK CRITICALLY: How could a substance that decreases free radical formation be harmful?

---

[1]Bjelakovic, G., Nikolova, D., Gluud, L. L., et al. Antioxidant supplements for prevention of mortality in healthy participants and patients with various diseases. *Cochrane Database Syst Rev* 3:CD007176, 2012.

[2]Olson, J. H., Erie, J. C., and Bakri, S. J. Nutritional supplementation and age-related macular degeneration. *Semin Ophthalmol* 26:131–136, 2012.

[3]Tinkel, J., Hassanain, H., and Khouri, S. J. Cardiovascular antioxidant therapy: A review of supplements, pharmacotherapies, and mechanisms. *Cardiol Rev* 20:77–83, 2012.

[4]The Alpha-Tocopherol, beta-carotene Cancer Prevention Study Group. The effects of vitamin E and beta-carotene on the incidence of lung cancer and other cancers in male smokers. *N Engl J Med* 330:1029–1035, 1994.

[5]Klein, E. A., Thompson, I. M. Jr., Tangen, C. M., et al. Vitamin E and the risk of prostate cancer: The Selenium and Vitamin E Cancer Prevention Trial (SELECT). *JAMA* 306:1549–1556, 2011.

supplements had symptoms that included nausea, diarrhea, and abdominal pain as well as fingernail and hair changes.[32] The UL for adults is 400 μg/day from diet and supplements combined.[37]

Selenium supplements are marketed with claims that they will protect against environmental pollutants, prevent cancer and heart disease, slow the aging process, and improve immune function. Although selenium does play a role in these processes, supplements that increase intake above the RDA will not provide additional benefits and can increase the risk of toxicity.[38]

12.7 Iodine (I)

LEARNING OBJECTIVES

- Describe the function of iodine.
- Explain why iodine deficiency causes the thyroid gland to enlarge.
- Discuss factors that impact the iodine status of a population.

Iodine is needed for the synthesis of thyroid hormones. In the early 1900s, iodine deficiency was common in the central United States and Canada, but it has virtually disappeared due to the addition of iodine to table salt. Iodine deficiency, however, remains a world health problem.

Iodine in the Diet

The iodine content of foods varies depending on the soil where plants are grown or where animals graze. Iodine is found in seawater, so seafood and plants grown near the sea are high in iodine. The soil in inland areas is generally low in iodine, so plants grown inland have lesser amounts.

Most of the iodine in the North American diet comes from salt fortified with iodine, referred to as *iodized salt*. Iodized salt contains about 100 μg of iodine per gram. It is commonplace in the United States and only iodized salt is sold in Canada. Iodized salt should not be confused with sea salt, which is a poor source of iodine because the iodine is lost in the drying process.

Iodine in our diet also comes from contaminants and additives in foods. Dairy products may contain iodine because of the iodine-containing additives used in cattle feed and the use of iodine-containing disinfectants on cows, milking machines, and storage tanks. Iodine-containing sterilizing agents are also used in food-service establishments, and iodine is used in dough conditioners and some food colorings.

How Iodine Functions in the Body

Iodine is absorbed completely and rapidly from the gastrointestinal tract in the form of iodide ions. Iodine can be eliminated from the body by excretion in the urine. More than half of the iodine in the body is located in the thyroid gland in the front of the neck. It is concentrated here because it is an essential component of the thyroid hormones, thyroxine (T_4) and triiodothyronine (T_3), which are made from the amino acid tyrosine (**Figure 12.18**). T_4 is the predominant thyroid hormone in the blood and is converted into the active T_3 form by a selenium-containing enzyme. The thyroid hormones act by affecting gene expression in target cells in a manner similar to vitamins A and D (**Figure 12.19**). Through gene expression, thyroid hormones promote protein synthesis and regulate basal metabolic rate, growth, and development.

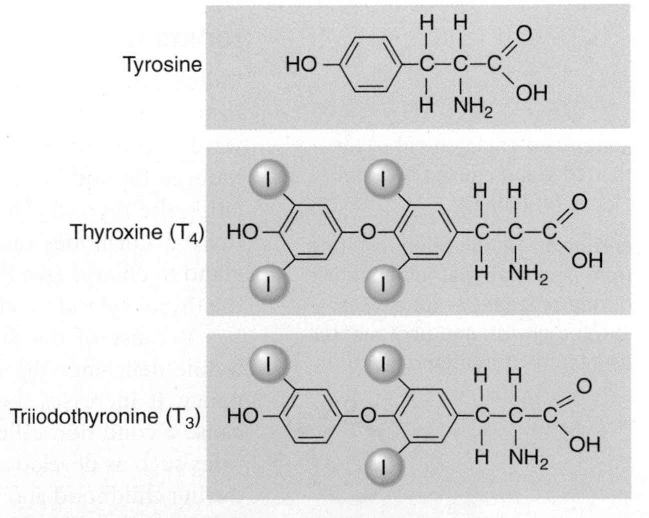

Figure 12.18 Structure of the thyroid hormones
The thyroid hormones thyroxine (T_4) and triiodothyronine (T_3) are made by attaching iodine to the amino acid tyrosine.

Figure 12.19 Role of thyroid hormones in gene expression

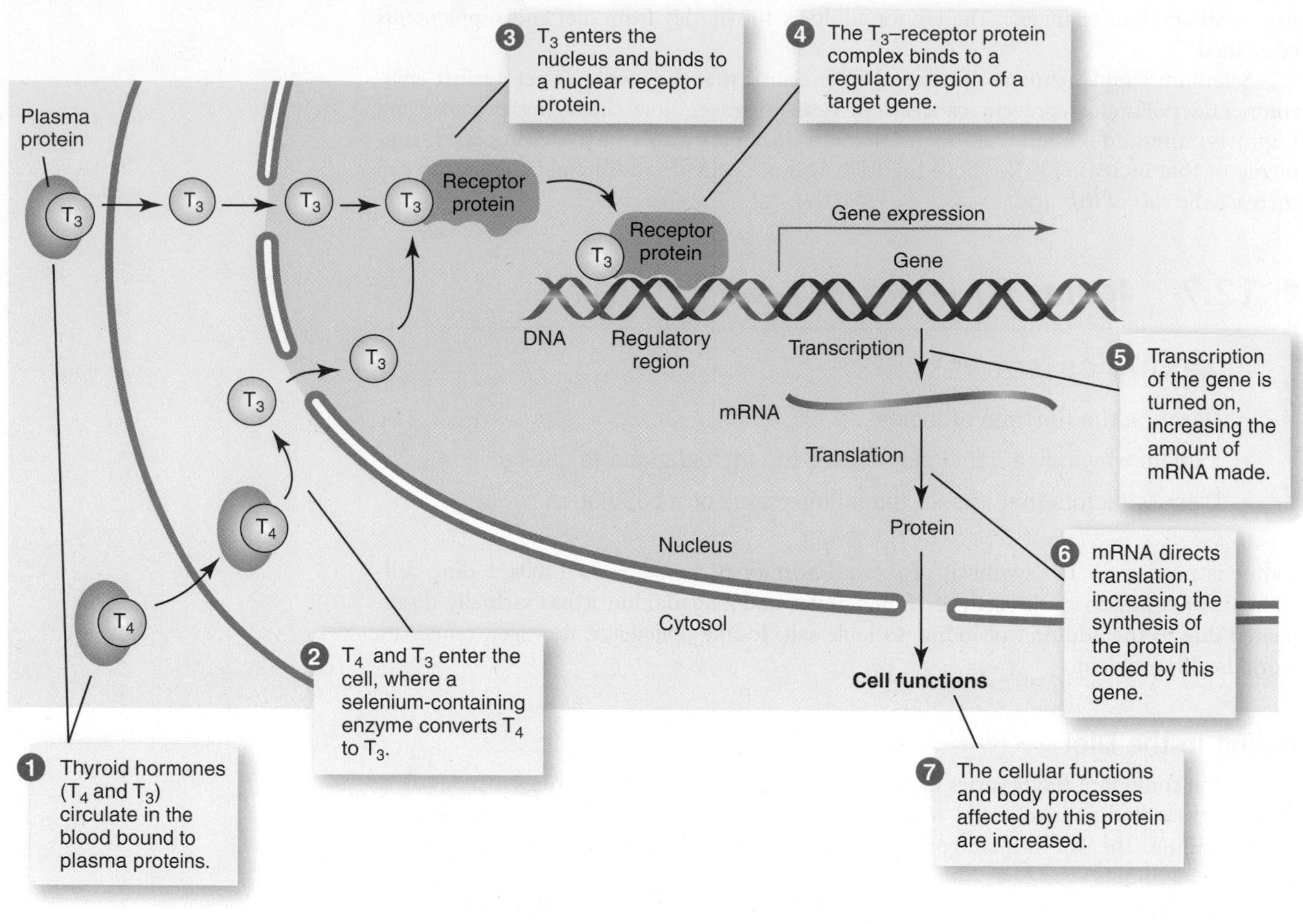

thyroid-stimulating hormone (TSH) A hormone that stimulates the synthesis and secretion of thyroid hormones from the thyroid gland.

Levels of the thyroid hormones are carefully controlled. If blood levels drop, **thyroid-stimulating hormone (TSH)** is released from the anterior pituitary. This hormone signals the thyroid gland to take up iodine and synthesize thyroid hormones. When the supply of iodine is adequate, thyroid hormones can be made and their presence turns off the synthesis of thyroid-stimulating hormone (**Figure 12.20**).

Iodine Deficiency

goiter An enlargement of the thyroid gland caused by a deficiency of iodine.

cretinism A condition resulting from poor maternal iodine intake during pregnancy that causes stunted growth and poor mental development in offspring.

Iodine deficiency reduces the production of thyroid hormones. Metabolic rate slows with insufficient thyroid hormones, causing fatigue and weight gain. The most obvious outward sign of deficiency is an enlarged thyroid gland called a **goiter**. A goiter forms when reduced thyroid hormone levels cause thyroid-stimulating hormone to be released, stimulating the thyroid gland to make more thyroid hormones. Because iodine is unavailable, thyroid hormones cannot be made and the stimulation continues, causing the thyroid gland to enlarge (see Figure 12.20). In milder cases of goiter, treatment with iodine causes the thyroid gland to return to normal size, but in severe cases it may remain enlarged.

Because of the importance of the thyroid hormones for growth and development, iodine deficiency disorders occur at all stages of life. If iodine is deficient during pregnancy, it increases the risk of stillbirth and spontaneous abortion. Deficiency also can cause a condition called **cretinism** in the offspring. Cretinism is characterized by symptoms such as developmental disability, deaf-mutism, and growth failure. Iodine deficiency during childhood and adolescence can also result in goiter and impaired mental function that lowers intellectual capacity. Iodine deficiency is the world's most prevalent, yet easily preventable, cause of brain damage.

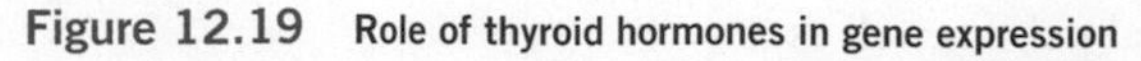

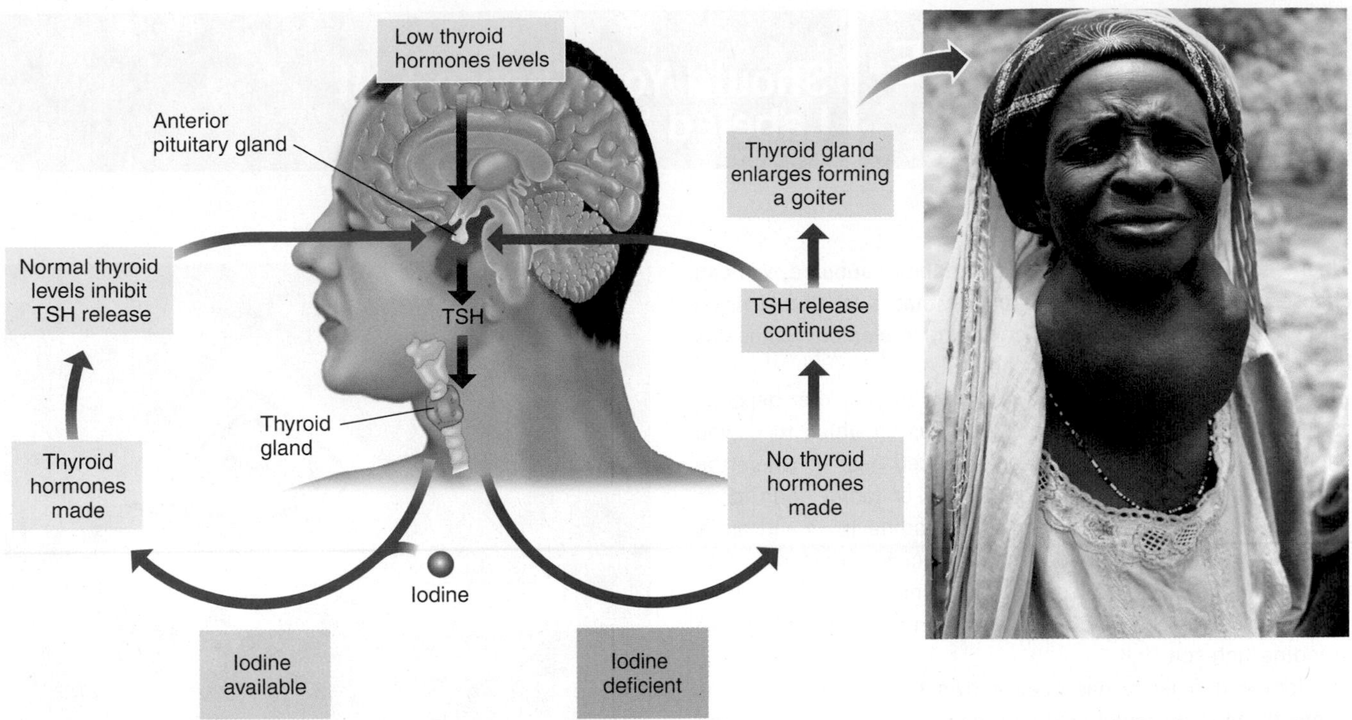

Figure 12.20 **Thyroid hormone regulation and goiter**
When thyroid hormone levels drop too low, thyroid-stimulating hormone (TSH) is released and stimulates the thyroid gland to take up iodine and synthesize more hormones (blue arrows). If iodine is not available (purple arrows), thyroid hormones cannot be made and the stimulation continues. This causes the thyroid gland to enlarge, forming a goiter.

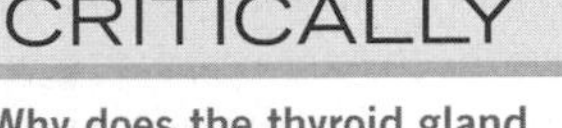

Why does the thyroid gland enlarge when iodine is deficient?

Dietary Factors That Increase Iodine Deficiency Iodine deficiency is most common in regions where the soil is low in iodine and there is little access to fish and seafood, which are good sources of this mineral. The risk of iodine deficiency is also increased by consuming **goitrogens**, substances in food that interfere with the utilization of iodine or with thyroid function. Goitrogens are found in turnips, rutabaga, cabbage, cassava, and millet. When these foods are boiled, the goitrogen content is reduced because some of these compounds leach into the cooking water. Goitrogens contribute to iodine deficiency in African countries where cassava is a dietary staple. In the United States, goitrogens are not a problem because they are present in foods that are typically not consumed in significant quantities.

goitrogens Substances that interfere with the utilization of iodine or the function of the thyroid gland.

Combating Iodine Deficiency Since it was first used in Switzerland in the 1920s, iodized salt has been the major means of combating iodine deficiency (see **Off the Label: Should You Choose Salt Labeled "Iodized"?**). Because of the fortification of table salt with iodine, cretinism and goiter are now rare in North America. Since universal salt iodization was adopted in 1993, the incidence of iodine deficiency worldwide has been cut in half, but it remains a global health issue, particularly among children and pregnant women in low-income countries.[39] About one-third of the world's population is estimated to have a low iodine intake, and poor regulation of iodine fortification has led to excessive iodine intakes in some countries.[40]

Recommended Iodine Intake

The RDA for iodine in adult men and women is 150 μg/day. This is based on the amount needed to maintain normal iodine levels in the thyroid gland. Since the iodization of salt, the intake of iodine in North America has met or exceeded the RDA.

The RDA is higher during pregnancy to account for the iodine that is taken up by the fetus and during lactation to account for the iodine secreted in milk. The recommended intake for infants is based on the amount obtained from breast milk.

© iStockphoto

OFF THE LABEL Should You Choose Salt Labeled "Iodized"?

When selecting a box of salt for the kitchen cupboard, you can choose one that just says "salt" or one that is labeled "iodized salt." Iodized salt is salt to which the trace element iodine has been added. Which should you choose?

The amount of iodine you consume in your diet depends as much on where the foods are grown as on which foods you choose. Foods from the ocean or produced near it, where the soil is rich in iodine, are better sources of iodine than foods produced inland. This is because the iodine in inland areas has been washed into the oceans by glaciers, snow, rain, and floodwaters. The iodine content of plants grown in iodine-deficient soil may be 100 times less than that of plants grown in iodine-rich soil.[1]

Iodine deficiency has been known for centuries in areas where the soil is depleted. In Europe the presence of iodine deficiency was recorded in classical art, which portrayed even the wealthy with goiter and cretinism. Leonardo da Vinci is said to have been more knowledgeable about goiter than medical professors of his time. A century ago, goiter was endemic in the central regions of North America. However, the iodization of salt, which began in the early twentieth century, has virtually eliminated iodine deficiency in North America, Switzerland, and other European countries.

Why fortify salt? Salt was selected as the vehicle for added iodine because it is a food item consistently consumed by the majority of the population at risk. Also, iodine can be added to salt uniformly, inexpensively, and in a form that is well utilized by the body. It can be added in amounts that eliminate deficiency with typical consumption but do not cause toxicity in people consuming larger amounts of iodized salt or individuals meeting iodine needs from other sources.

Since the introduction of iodized salt, iodine intake in the United States has been adequate and iodine deficiency has been rare. However, U.S. surveys have shown a reduction in average iodine status during the past decade.[2] Factors that may have contributed to this include increased consumption of processed foods, which contain noniodized salt; reduced consumption of eggs, which are rich in iodine; and declining amounts of salt added in the home. Also, the iodine content of milk has decreased due to a reduction in the amount of iodine added to cattle feed, and the baking industry has eliminated some of the iodine-containing dough conditioners, thus reducing the iodine content of commercially manufactured breads.[2] If you live on the coast or buy food imported from many locations, you still probably get plenty of iodine. However, if you eat little seafood, live inland where the soil is deficient in iodine, and consume primarily foods grown locally, choose salt labeled "iodized" to ensure that your iodine needs are met.

George Semple

[1]Dunn, J. T. Iodine. In *Modern Nutrition in Health and Disease*, 10th ed. M. E. Shils, M. Shike, A. C. Ross, et al., eds. Philadelphia: Lippincott Williams & Wilkins, 2006, pp. 300–311.
[2]Pearce, E. N. National trends in iodine nutrition: Is everyone getting enough? *Thyroid* 17:823–827, 2007.

Iodine Toxicity

Acute toxicity can occur with very large doses of iodine. Intakes between 200 and 500 μg per kilogram of body weight have caused death in laboratory animals.[32] Chronically high intakes of iodine can cause an enlargement of the thyroid gland that resembles goiter. The UL for adults is 1100 μg of iodine per day from all sources.[10] Goiter from excessive iodine can also occur if iodine intake changes drastically. For example, in a population with a marginal intake, a large increase in intake due to supplementation can cause thyroid enlargement even at levels that would not be toxic in a healthy population.

12.8 Chromium (Cr)

LEARNING OBJECTIVES

- Describe the relationship between blood glucose levels and chromium.
- Explain why chromium supplements are promoted to increase lean body mass.

It has been known since the 1950s that chromium is needed for normal glucose utilization, but only recently have scientists begun to understand the role of chromium in insulin function. The popular supplement chromium picolinate is promoted to increase lean body mass.

Chromium in the Diet

Dietary sources of chromium include liver, brewer's yeast, nuts, and whole grains. Milk, vegetables, and fruit are poor sources. Refined carbohydrates such as white breads, pasta, and white rice are also poor sources because chromium is lost in milling and not added back in the enrichment process. Cooking in stainless steel cookware can increase chromium intake because chromium leaches from the steel into the food.

How Chromium Functions in the Body

After absorption, chromium is bound to the iron transport protein transferrin for transport in the blood. Chromium is involved in carbohydrate and lipid metabolism. When carbohydrate is consumed, insulin is released and binds to receptors in cell membranes. This binding triggers the uptake of glucose by cells, an increase in protein and lipid synthesis, and other effects. Chromium is believed to act as part of a small peptide that binds to the insulin receptor after insulin is bound, enhancing its effect[41] (**Figure 12.21**). When chromium is deficient, it takes more insulin to produce the same effect.

Chromium Deficiency

Overt chromium deficiency is not a problem in the U.S. population. Deficiencies have been reported in patients receiving long-term TPN not containing chromium and in malnourished children. Deficiency symptoms include impaired glucose tolerance with diabetes-like symptoms, such as elevated blood glucose levels and increased insulin levels.[42] Chromium deficiency may also cause elevated blood cholesterol and triglyceride levels, but the role of chromium in lipid metabolism is not fully understood.

Figure 12.21 **Chromium and insulin function**

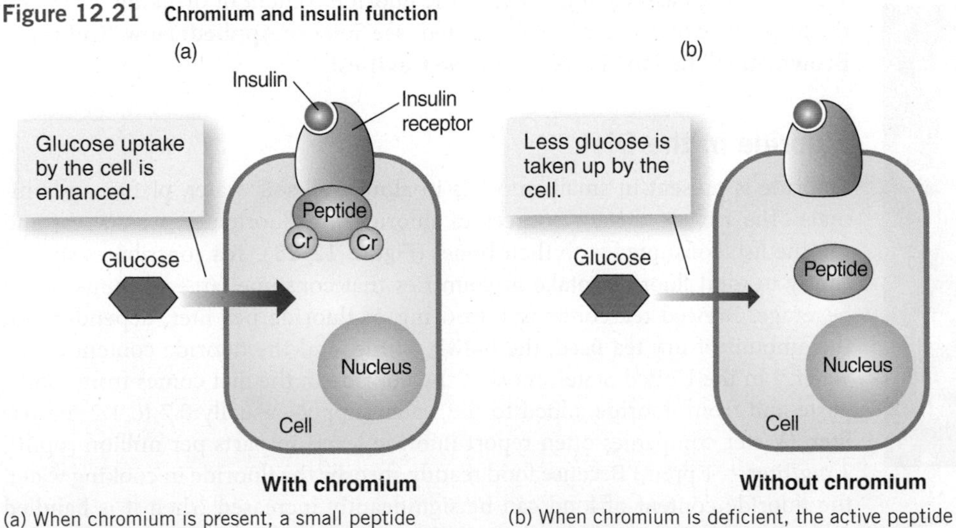

(a) When chromium is present, a small peptide inside cells becomes active and enhances the action of insulin by binding to the insulin receptor, which increases glucose uptake.

(b) When chromium is deficient, the active peptide is not formed and thus cannot bind the insulin receptor. The result is that insulin is less effective and less glucose can enter the cell.

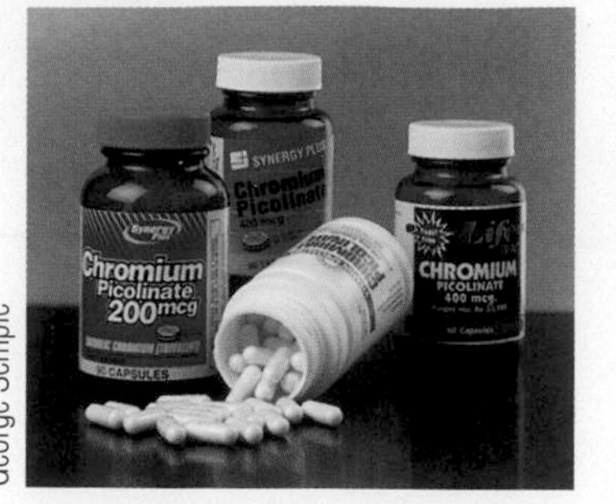

George Semple

Figure 12.22 **Chromium supplements**
Supplements containing chromium picolinate claim to help increase lean body mass and decrease body fat.

Recommended Chromium Intake

Based on the amount of chromium in a balanced diet, an AI has been set at 35 μg/day for men and 25 μg/day for women. The AI is increased during pregnancy and lactation. The AI for older adults is lower because needs decrease as energy intake decreases.

Chromium Supplements and Toxicity

Chromium supplements, particularly chromium picolinate, are marketed to reduce body fat and increase lean body tissue (**Figure 12.22**). These claims appeal to individuals wanting to lose weight as well as to athletes trying to build muscle. Because chromium is needed for insulin action and insulin promotes protein synthesis, it is likely that adequate chromium is necessary to increase lean body mass. However, most recent studies on the effects of chromium picolinate or other chromium supplements in healthy human subjects have found no beneficial effects on muscle strength, body composition, weight loss, or other aspects of health.[43] Chromium supplementation has been shown to have beneficial effects on blood glucose levels in individuals with type 2 diabetes.[44]

Controlled trials have reported no dietary chromium toxicity in humans.[10] Despite the apparent safety of chromium supplements, a few concerns have been raised. Two cases of renal failure have been associated with chromium picolinate supplements, but both of these individuals were taking other drugs known to cause renal toxicity, so it is unclear whether the effect was due to the chromium supplements.[45] The safety of chromium picolinate has also been questioned because of studies in cell culture that suggest it may cause DNA damage.[46] This effect is specific to the picolinate form of chromium and may be due to the ability of this form to generate DNA-damaging free radicals. However, the majority of studies using the standard supplemental doses of chromium picolinate have not found it to cause DNA damage.[47] Despite these concerns, the DRI committee concluded that there was insufficient data to establish a UL for chromium.

12.9 Fluoride (F)

LEARNING OBJECTIVES

- Identify dietary sources of fluoride.
- Discuss the role of fluoride in maintaining dental health.

The importance of fluoride for dental health has been recognized since the 1930s, when an association between the fluoride content of drinking water and the prevalence of dental caries was noted (see **Science Applied: How "Colorado Brown Stain" Led to "Look, Mom! No Cavities!"**).

Charles D. Winters

Figure 12.23 **Dietary sources of fluoride**
Most of the fluoride we consume comes from toothpaste, fluoridated water, tea, and fish eaten with bones.

Fluoride in the Diet

Fluoride is present in small amounts in almost all soil, water, plants, and animals. The richest dietary sources of fluoride are fluoridated water, tea, and marine fish consumed with their bones (**Figure 12.23**). Tea contributes significantly to total fluoride intake in countries that consume large amounts of the beverage. Brewed tea contains 1 to 6 mg of fluoride per liter, depending on the amount of dry tea used, the brewing time, and the fluoride content of the water.[48] In the United States, most of the fluoride in the diet comes from toothpaste and from fluoride added to the water supply—usually 0.7 to 1.2 mg per liter. (Water companies often report fluoride levels in parts per million [ppm]; 1 mg/liter = 1 ppm.) Because food readily absorbs the fluoride in cooking water, the fluoride content of food can be significantly increased when it is handled and prepared using fluoridated water. Cooking utensils also affect food fluoride content. Foods cooked with Teflon utensils can pick up fluoride from the Teflon,

SCIENCE APPLIED

How "Colorado Brown Stain" Led to "Look, Mom! No Cavities!"

(Elena Schweitzer/Shutterstock) Victoriano Izquierdo/Getty Images, Inc. (DAJ/Getty Images, Inc.)

When Dr. Fredrick McKay set up his dental practice in Colorado Springs, Colorado in 1901, he noticed that many of his patients had stained or mottled tooth enamel. McKay noted that those with stained teeth, a condition called "Colorado brown stain," seemed to be less susceptible to tooth decay.[1] At the time, dental caries—tooth decay—were extremely prevalent, there was no known way to prevent the disease, and the most common way to treat it was to extract the affected teeth. McKay believed that Colorado brown stain was due to something in the local water supply.

THE SCIENCE

In 1930, Dr. McKay sent water samples to a chemist for analysis. Using a new methodology called spectrographic analysis, the chemist identified high levels of fluoride, a naturally occurring element, in McKay's samples. This finding led to the establishment of the Dental Hygiene Unit at the National Institutes of Health headed by Dr. H. Trendley Dean. His task was to investigate the association between fluoride intake and mottled enamel, which he termed "fluorosis."

Dean conducted epidemiological surveys to establish the prevalence of fluorosis across the country. He noted a strong inverse relationship between the presence of fluorosis and the presence of dental caries among children.[2] In other words, children with fluorosis had fewer dental caries. Further studies revealed that the protective effect of fluoride against dental caries was seen at water fluoride levels of 1 ppm, a level low enough to cause little fluorosis (see graph). Thus, work designed to identify the harm caused by too much fluoride had discovered the benefits of enough fluoride.

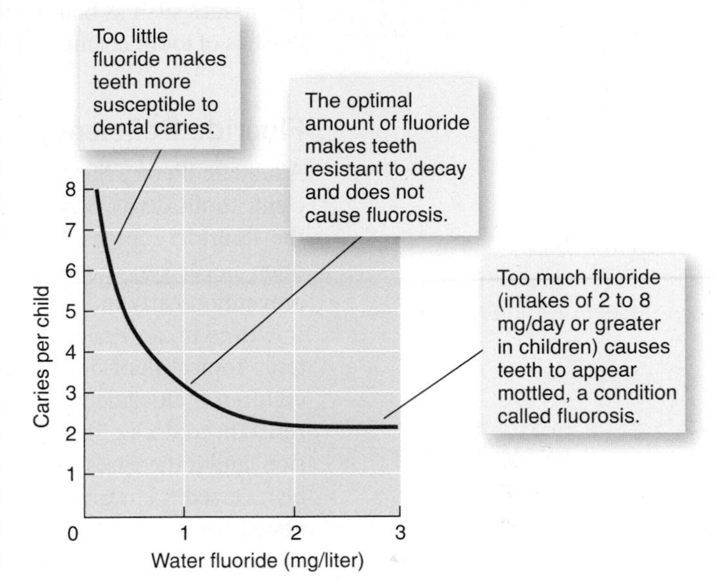

The effect of water fluoridation level on dental caries in children 12 to 14 years of age.

THE APPLICATION

Dr. Dean's research led to an intervention trial to test the effectiveness of community water fluoridation for preventing caries. The trial began in 1945 and included four pairs of cities in Michigan, New York, Illinois, and Ontario, Canada. One city of each pair received the intervention—fluoridated water—and the second city served as a control. Surveys over a 13- to 15-year period found that the incidence of caries was reduced 50 to 70% among children in the communities with the fluoridated water.[3] In 1962, epidemiological studies of water consumption patterns and caries incidence in different climates and geographical locations across the country were used to make the first recommendation for an optimal range of fluoride concentration (0.7–1.2 ppm) in the water supply, with the lower range suggested for warmer climates, where more water is consumed, and the higher range for colder climates.[3]

Although there was some political and medical opposition to fluoridation in the early 1960s that opposition has dwindled to a few voices that still believe that water fluoridation represents a public health hazard and increases the risk of cancer; however, these beliefs are not supported by scientific fact. Based on epidemiological data and available evidence related to the adverse effects of fluoride, the small amounts consumed in drinking water do not pose a health risk.[4]

Today, 72.4% of the U.S. population that is served by a public water system receives fluoridated water.[5] Fluoride intake has also increased due to the widespread use of fluoride toothpaste and fluoridated water in foods and beverages that are distributed in nonfluoridated areas. Although dental caries remain a public health problem, increased fluoride intake, combined with advances in dental care, have dramatically improved the dental health of the American public.

[1]McKay, F. S. Relation of mottled enamel to caries. *J Am Dent Assoc* 15:1429–1437, 1928.

[2]Dean, H. T. Endemic fluorosis and its relation to dental caries. *Public Health Rep* 53:1443–1452, 1938.

[3]Achievements in public health, 1900–1999: Fluoridation of drinking water to prevent dental caries. *Morb Mortal Wkly Rep* 48:933–940, 1999.

[4]National Research Council. *The Health Effects of Ingested Fluoride.* Report of the Subcommittee on the Health Effects of Ingested Fluoride, Committee on Toxicology, Board of Environmental Studies and Toxicology, Commission on Life Sciences. Washington, D.C.: Academy Press, 1993.

[5]Centers for Disease Control and Prevention. Fact sheet: Community water fluordination statistics.

whereas aluminum cookware can decrease fluoride content. Fluoride is absorbed into the body in proportion to its content in the diet.

How Fluoride Functions in the Body

About 80 to 90% of ingested fluoride is absorbed. Fluoride has a high affinity for calcium, so, when consumed with milk or other high-calcium foods, the fluoride forms an insoluble salt with calcium and absorption is reduced. In the body, fluoride is usually associated with calcified tissues such as bones and teeth. When fluoride is incorporated into the hydroxyapatite crystals of tooth enamel, it makes the enamel more resistant to the acid that causes decay.

Fluoride Deficiency

Adequate dietary fluoride is important for bone and dental health. When fluoride is deficient, tooth decay is more common. Because there are few food sources of fluoride and the fluoride content of water is variable, fluoride supplements or water fluoridation is often needed to minimize dental caries. Fluoride has its greatest effect on dental caries prevention early in life, during maximum tooth development up to the age of 13. During this time it can be incorporated into tooth enamel, making the enamel more acid resistant. Topical fluoride also protects the teeth, so fluoride is beneficial for adults as well as children.[49] Fluoride in saliva is incorporated into the surface of the teeth, making it more resistant to decay. Topical fluoride can help prevent root decay in people with gum recession, and it increases enamel remineralization, so early decay does not enlarge and can even be reversed.[49] Fluoride seems to stimulate new bone formation and has therefore been suggested for strengthening bones in adults with osteoporosis. Slow-release fluoride supplements have been shown to increase bone mass and prevent new fractures.[50]

Recommended Fluoride Intake

Epidemiology has confirmed the effectiveness of fluoridated water in reducing dental cavities.[51] The criterion used to establish an AI for fluoride is the estimated intake shown to reduce the occurrence of dental caries maximally without causing unwanted side effects. The AI for fluoride from all sources is set at 0.05 mg per kg per day for everyone 6 months of age and older.[48] Thus, for children age 4 through 8 years, the AI is set at 1.1 mg/day using a reference weight of 22 kg. For adult men ages, 19 and older, the AI is 3.8 mg/day based on a weight of 76 kg; for women, it is 3.1 mg/day based on a weight of 61 kg. The AI is not increased during pregnancy or lactation.

Breast milk is low in fluoride, and ready-made infant formulas are prepared with unfluoridated water. Unless infant formula is prepared at home with fluoridated water, it contains little fluoride. The American Academy of Pediatrics suggests a supplement of 0.25 mg of fluoride per day for children 6 months to 3 years of age, 0.5 mg/day for ages 3 to 6 years, and 1.0 mg/day for ages 6 to 16 who are receiving less than 0.3 mg per liter of fluoride in the water supply.[49] These supplements are available by prescription for children living in areas with low water fluoride concentrations. Swallowed toothpaste is estimated to contribute about 0.6 mg/day of fluoride in young children.[48]

Fluoride Toxicity

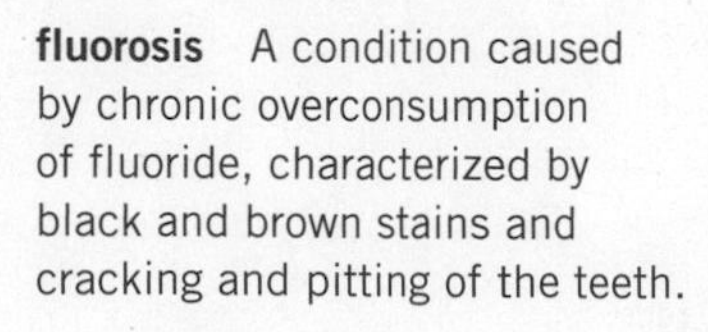

fluorosis A condition caused by chronic overconsumption of fluoride, characterized by black and brown stains and cracking and pitting of the teeth.

Fluoride can cause adverse effects in high doses. In children, fluoride intakes of 2 to 8 mg/day can cause stained, pitted teeth, a condition called **fluorosis** (**Figure 12.24a**). A recent increase in the prevalence of fluorosis in the United States has occurred due to the chronic ingestion of toothpaste containing fluoride.[52] In adults, doses of 20 to 80 mg/day can result in changes in bone that can be crippling, as well as changes in kidney function and possibly nerve and muscle function. Death has been reported with an intake of 5 to 10 grams per day. Due to concern over excess fluoride intake, a warning is now required on fluoride-containing toothpastes (**Figure 12.24b**). The UL for fluoride is set at 0.1 mg per kg per day for infants and children less than 9 years of age and at 10 mg/day for people ages 9 through 70.[48]

Figure 12.24 **Dental fluorosis**

(a)

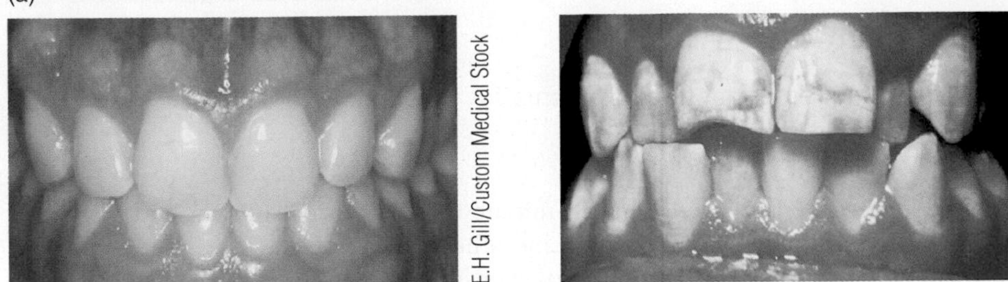

E.H. Gill/Custom Medical Stock

NIH/Custom Medical Stock

(a) Too much dietary fluoride causes staining and pitting of the teeth. These photos compare normal teeth (left) and teeth showing enamel fluorosis (right).

(b)

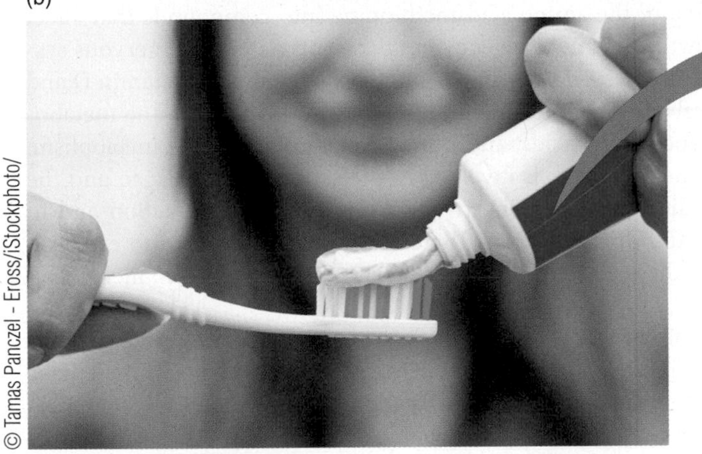

© Tamas Panczel - Eross/iStockphoto/

WARNING: Keep out of the reach of children under 6 years of age. If you accidentally swallow more than used for brushing, get medical help or contact a Poison Control Center right away.

(b) To help reduce the risk of excess fluoride intake, toothpaste labels must include this warning.

12.10 Molybdenum (Mo)

LEARNING OBJECTIVE

- Discuss how the molybdenum content of the soil affects the content of food.

Like many other trace elements, molybdenum is needed to activate enzymes. The molybdenum content of food varies with the molybdenum content of the soil where the food is produced. The most reliable sources include milk, milk products, organ meats, breads, cereals, and legumes.

Molybdenum is readily absorbed from foods. The amount in the body is regulated by excretion in the urine and bile. Molybdenum is a cofactor for enzymes necessary for the metabolism of sulfur-containing amino acids and nitrogen-containing compounds that make up the structure of DNA and RNA, the production of uric acid (a nitrogen-containing waste product), and the oxidation and detoxification of various other compounds.

Although molybdenum deficiency in humans has been reported as a result of long-term TPN, a naturally occurring deficiency has never been reported. Deficiency has been induced in laboratory animals by feeding them high doses of the element tungsten, which inhibits molybdenum absorption. The resulting deficiency caused growth retardation, decreased food intake, impaired reproduction, and decreased life expectancy.

Based on the results of molybdenum balance studies, an RDA has been set at 45 μg/day for adults. The RDA is increased for pregnancy and lactation. An AI has been set for infants based on the amount of molybdenum in breast milk.

There are few data on adverse effects of high intakes of molybdenum in humans. A UL of 2000 μg/day has been set based on impaired growth and reproduction in animals.[10]

12.11 Other Trace Elements

LEARNING OBJECTIVE

- Name four essential trace elements for which no RDA or AI has been established.

Many other trace elements are found in minute amounts in the human body. Some of these may be essential for human health, and others, such as lead (see Chapter 15), may be present only as a result of environmental exposure. Arsenic, boron, nickel, silicon, and vanadium have been reviewed by the DRI committee and found to have a significant role in human health.[10] Arsenic is better known as a poison than an essential nutrient, but the organic forms of arsenic that occur in foods are nontoxic. It is hypothesized that arsenic is involved in the conversion of the amino acid methionine into compounds that affect heart function and cell growth. Arsenic deficiency has been correlated with nervous system disorders, blood vessel diseases, and cancer. Boron may be involved in vitamin D and estrogen metabolism. Nickel is thought to function in enzymes involved in the metabolism of certain fatty acids and amino acids, and it may play a role in folate metabolism. Silicon, the primary constituent of sand, is involved in the synthesis of collagen and the calcification of bone. Vanadium has been shown to have an insulin-like action and to stimulate cell proliferation and differentiation. There are not sufficient data to establish an AI or an RDA for any of these elements, but ULs have been set for boron, nickel, and vanadium.

Other trace elements that have a physiological role include aluminum, bromine, cadmium, germanium, lead, lithium, rubidium, and tin. The specific functions of these have not been defined, and they have not been evaluated by the DRI committee. All the minerals, both those known to be essential and those still being assessed for their role in human health, can be obtained by choosing a variety of foods from each of the food groups of MyPlate.

OUTCOME

© amana images Inc./Alamy Limited

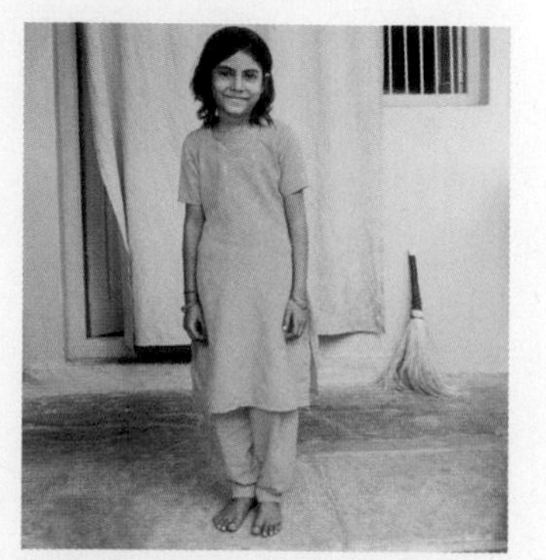

PhotosIndia.com/Getty Images, Inc.

About a year after Nissi was diagnosed with goiter, the Indian government began promoting the use of salt fortified with iodine and banned the production and sale of noniodized salt. Nissi's family and many others in the village are now using iodized salt. Nissi has grown to be an energetic teenager, with no trace of goiter.

Some families in Nissi's village still resist using the new salt, but information and education programs are spreading the message that iodized salt is safe and affordable and that it helps children to achieve their full intellectual capacity. The incidence of iodine deficiency continues to drop.

In the United States, where the diet includes food from many locations, the mineral content of the soil is unlikely to affect the risk of nutrient deficiency. However, in Nissi's village where the only affordable food is locally grown, the risk of deficiency is greater. **Without being able to introduce high-iodine foods from other locations, iodine fortification was needed to solve the problem.**

APPLICATIONS

ASSESSING YOUR DIET

1. **Are you at risk for iron deficiency?**
   a. Use iProfile and the three-day food intake record you kept in Chapter 2 to calculate your average daily intake of iron.
   b. How does your iron intake compare with the recommendation for someone of your age, gender, and life stage?
   c. If your intake is low, suggest modifications to your diet that would increase your iron intake enough to meet the RDA for someone your age and gender.
   d. If your diet already meets the recommendations for iron, make a list of foods you like that are good sources of iron.
   e. Identify the major food sources of iron in your diet and indicate whether they contribute heme iron.
2. **Do you consume enough zinc?**
   a. Use iProfile and your food record to calculate your zinc intake.
   b. If you eliminated meat from your diet, would you meet the RDA for zinc?
   c. What foods could you substitute for meats that are good sources of zinc?

CONSUMER ISSUES

3. **What promises are made about trace element supplements?**
   a. Using the Internet, search for information on a supplement discussed in this chapter—for instance, zinc lozenges or chromium picolinate.
   b. How does the information compare to the discussion in the text?
   c. Who provided the information? Does it promote the sale of a product?
   d. Is the information supported by scientific studies?
4. **What trace minerals are included in fortified foods and supplements?**
   a. Look at the Nutrition Facts label on your breakfast cereal. What trace minerals are listed? What percentage of the Daily Value is included for each?
   b. Look at the Supplement Facts label on a multivitamin/mineral supplement. What trace minerals are listed? What percentage of the Daily Value is included for each?
   c. If you had two servings of the cereal and took this supplement every day, how much iron would you be consuming? How much zinc? What percentage of the Daily Value for iron and zinc would you be consuming? How do the amounts compare to the UL?
   d. Would eating this every day put you at risk for consuming a toxic amount of iron, zinc, or any other trace mineral?

CLINICAL CONCERNS

5. **Fred has hemochromatosis so needs to be careful not to consume foods fortified with iron.**
   a. Find a breakfast cereal that does not contain added iron.
   b. What about bread? Is there bread that is not made from enriched flour? How much iron does it contain?
   c. What could he have for dessert that doesn't have added iron?
6. **Jessica's son has iron-deficiency anemia. He is 2 years old and is a finicky eater. His typical diet is shown here:**

   Breakfast: 1 banana and ½ c milk
   Snack: 1 graham cracker square and ½ c milk
   Lunch: ½ c vegetable soup and ½ c milk
   Snack: ½ c milk
   Dinner: ⅓ c peas, ½ c mashed potatoes, 1 c milk

   a. How much iron does he consume in this diet?
   b. How does his milk intake affect his iron status?
   c. He doesn't like meat. Suggest some food substitutions Jessica could make that would add nonmeat sources of iron to his diet without increasing his calories or decreasing his protein intake?
   d. Suggest some foods high in vitamin C that would improve his iron absorption.
7. **A company is testing a new meal-replacement drink that it will market for weight loss. The subjects in the clinical trial are consuming this product as their sole source of nutrients. After a few weeks, several subjects begin to show the symptoms of anemia. A manufacturing error is suspected. The ingredients intended to be in the drink are:**

Starch	Potassium
Sucrose	Magnesium
Casein (protein)	Chloride
Corn oil	Zinc
Mixed plant fibers	Iron
Vitamin A	Iodine
Vitamin D	Selenium
Vitamin E	Copper
Vitamin K	Manganese
B vitamin mix	Chromium
Calcium	Molybdenum
Sodium	

   a. Omission of which trace minerals could have caused anemia?
   b. Excessive amounts of which trace minerals could have caused anemia?
   c. Omission of which vitamins could have caused anemia?
   d. If a blood test reveals that the anemia is microcytic, what are the possible dietary causes?

SUMMARY

12.1 Trace Minerals in Our Diet

- Trace minerals are required by the body in an amount of 100 mg or less per day or present in the body in an amount of 0.01% or less of body weight. They include iron, zinc, copper, manganese, selenium, iodine, fluoride, chromium, and molybdenum.

12.2 Iron (Fe)

- Iron is found in both plant and animal foods. Heme iron, the easily absorbable form, is found in animal products. Nonheme iron, which is less well absorbed, comes from both animal and plant sources.
- The amount of iron that is absorbed from the diet depends on the type of iron and the presence of other dietary components. The absorption of nonheme iron can be increased by consuming it with meat vitamin C, or other acids; its absorption is decreased by consuming it with foods containing fiber, phytates, oxalates, and tannins.
- Iron homeostasis is regulated at the intestine. If iron stores are low, more iron is transferred from the mucosal cells to the blood and bound to transferrin for transport to body cells. When body stores are high, less iron is transported from the mucosa into the blood and more is bound to ferritin and lost when mucosal cells die.
- Iron functions as part of hemoglobin, which transports oxygen in the blood, and myoglobin, which enhances the amount of oxygen available during muscle contraction. Iron is also a component of proteins involved in ATP production and is needed for activity of the antioxidant enzyme catalase.
- When iron is deficient, adequate hemoglobin cannot be made, resulting in an anemia characterized by small, pale red blood cells. Iron deficiency anemia is the most common nutritional deficiency worldwide.
- The RDA for iron for women ages 19 to 50 is 18 mg/day, more than double the RDA of 8 mg/day for adult men and postmenopausal women.
- Iron can be toxic. Ingestion of a single large dose can be fatal. The accumulation of iron in the body over time causes heart and liver damage and contributes to diabetes and cancer. The most common cause of chronic iron overload is hemochromatosis, a genetic disorder in which too much iron is absorbed.

12.3 Zinc (Zn)

- Good sources of zinc include red meats, eggs, dairy products, and whole grains. Zinc from animal sources is better absorbed than that from plant foods. Phytates inhibit zinc absorption from many plant sources.
- Zinc absorption is regulated by zinc transport proteins that determine how much zinc is in the mucosal cell and by metallothionein, a protein that binds zinc in the mucosal cell. When zinc intake is high, more metallothionein is synthesized and zinc absorption is limited.
- Zinc is needed for the activity of many enzymes, including a form of the antioxidant enzyme superoxide dismutase. Many of the functions of zinc are related to its role in gene expression. Zinc is needed for tissue growth and repair, development of sex organs and bone, proper immune function, storage and release of insulin, mobilization of vitamin A from the liver, and stabilization of cell membranes.
- Zinc deficiency results in poor growth, delayed sexual maturation, skin changes, hair loss, skeletal abnormalities, and depressed immunity.
- The RDA for zinc is 11 mg/day for adult men and 8 mg/day for adult women.
- Since copper binds more tightly to metallothionein than zinc, an excess of zinc, which stimulates metallothionein synthesis, can trap copper in the mucosal cells, causing a copper deficiency.
- When administered within 24 hours of the onset of common cold symptoms, supplemental zinc in lozenge form can reduce the duration and severity of cold symptoms.

12.4 Copper (Cu)

- Good sources of copper in the diet include organ meats, seafood, nuts, and seeds.
- The absorption of copper is affected by the presence of other minerals in the diet. The zinc content of the diet can have a major impact on copper absorption.
- Copper functions in a number of important proteins that affect iron and lipid metabolism, synthesis of connective tissue, and antioxidant protection. Copper is transported in the blood bound to proteins such as ceruloplasmin.
- A copper deficiency can cause anemia and connective tissue abnormalities. Copper toxicity from dietary sources is extremely rare.
- The RDA for copper for adults is 900 μg/day.

12.5 Manganese (Mn)

- Good dietary sources of manganese include whole grains and nuts. The AI is 2.3 mg/day for men and 1.8 mg/day for women.
- Manganese is necessary for the activity of some enzymes, including a form of the antioxidant enzyme superoxide dismutase. Manganese is involved in amino acid, carbohydrate, and lipid metabolism.

12.6 Selenium (Se)

- Excellent dietary sources of selenium include seafood, eggs, and organ meats. The selenium content of plant foods depends on the selenium content of the soil where they are grown, and the selenium content of animal products is affected by the amount of selenium in their feed.
- Selenium protects against oxidative damage as an essential part of the enzyme glutathione peroxidase. Glutathione peroxidase destroys peroxides before they can form free radicals. Adequate dietary selenium reduces the need for vitamin E.

- Severe selenium deficiency is rare except in regions with very low soil selenium content and limited diets. In China, selenium deficiency contributes to the development of a heart condition known as Keshan disease. Low selenium intake has been linked to increased cancer risk.
- The RDA for selenium for adults is 55 μg/day.
- Very high selenium intake (5 mg/day) causes fingernail changes and hair loss.
- Selenium supplements are marketed with claims that they will protect against environmental pollutants, prevent cancer and heart disease, slow the aging process, and improve immune function. Supplements that increase intake above the RDA do not provide additional benefits.

12.7 Iodine (I)

- The best sources of iodine in the diet are seafood, foods grown near the sea, and iodized salt.
- Iodine is an essential component of thyroid hormones, which promote protein synthesis and regulate basal metabolic rate, growth, and development.
- When iodine is deficient, continued release of thyroid-stimulating hormone causes the thyroid gland to enlarge, forming a goiter. Iodine deficiency during pregnancy causes a condition in the offspring known as cretinism, which is characterized by growth failure and developmental disability. Iodine deficiency during childhood and adolescence can impair mental function. Iodized salt has virtually eliminated iodine deficiency in North America. Although salt iodization is being used successfully worldwide to prevent iodine deficiency, iodine deficiency remains a world health problem.
- The RDA for iodine in adult men and women is 150 μg/day
- Acute toxicity can occur with very large doses of iodine. Chronically high intakes of iodine can cause an enlargement of the thyroid gland that resembles goiter.

12.8 Chromium (Cr)

- Chromium is found in liver, brewer's yeast, nuts, and whole grains.
- Chromium is needed for normal insulin action and glucose utilization.
- Overt chromium deficiency is not a problem in the U.S. population.
- Recommended chromium intake is 35 μg/day for men and 25 μg/day for women.
- Chromium supplements are marketed to control blood sugar and increase lean body mass. Controlled trials have reported no dietary chromium toxicity in humans.

12.9 Fluoride (F)

- Most of the fluoride in the diet in the United States comes from fluoridated drinking water and toothpaste.
- Fluoride is necessary for the maintenance of bones and teeth. Adequate dietary fluoride helps prevent dental caries in children and adults.
- Low fluoride intake increases the risk of dental caries. Excess fluoride causes dental fluorosis in children.
- The recommended fluoride intake from all sources is set at 0.05 mg per kg per day for everyone 6 months of age and older because it protects against dental caries with no adverse effects.

12.10 Molybdenum (Mo)

- Molybdenum is a cofactor for enzymes involved in the metabolism of the amino acids methionine and cysteine and nitrogen-containing compounds such as DNA and RNA.

12.11 Other Trace Elements

- There is evidence that boron, arsenic, nickel, silicon, and vanadium may be essential in humans as well as animals. These elements may be necessary in small amounts but can be toxic if consumed in excess.

REVIEW QUESTIONS

1. What are the functions of iron in the body?
2. Why does iron deficiency cause red blood cells to be small and pale?
3. List three life-stage groups at risk for iron deficiency anemia, and explain why they are at risk.
4. List several good sources of iron in the diet, and indicate if they contain heme or nonheme iron.
5. Discuss three dietary factors that affect iron absorption.
6. Explain the roles of ferritin, transferrin, and hepcidin in regulating the amount of iron in the body.
7. What is hemochromatosis?
8. How does zinc affect the synthesis of proteins?
9. How does a high zinc intake reduce the amount of zinc that enters the blood?
10. Why does excess zinc cause a deficiency of copper?
11. Explain why a deficiency of copper can contribute to anemia.
12. What is the role of selenium in the body?
13. Why does selenium decrease the need for vitamin E?
14. What is a goiter, and why does iodine deficiency cause it to form?
15. What is the role of chromium in the body?
16. What do zinc, copper, and manganese have in common?
17. How does adequate fluoride prevent dental caries in children and in adults?

REFERENCES

1. Centers for Disease Control and Prevention. FastStats, Anemia or iron deficiency.
2. World Health Organization. Micronutrient deficiencies, iron deficiency anemia: The challenge.
3. Hurrell, R., and Egli, I. Iron bioavailability and dietary reference values. *Am J Clin Nutr* 91:1461S–1467S, 2010.
4. National Institutes of Health. Office of Dietary Supplements. Dietary supplement fact sheet: Iron.
5. Hurrell, R. F., Lynch, S., Bothwell, T., et al. SUSTAIN Task Force. Enhancing the absorption of fortification iron. A SUSTAIN Task Force report. *Int J Vit Nutr Res* 74:387–401, 2004.
6. Grinder-Pedersen, L., Bukhave, K., Jensen, M., et al. Calcium from milk or calcium-fortified foods does not inhibit nonheme-iron absorption from a whole diet consumed over a 4-d period. *Am J Clin Nutr* 80:404–409, 2004.
7. Zhang, A. S., and Enns, C. A. Molecular mechanisms of normal iron homeostasis. *Hematology* 2009:207–214, 2009.
8. Killip, S., Bennett, J. M., and Chambers, M. D. Iron deficiency anemia. *Am Fam Phys* 75:671–678, 2007.
9. Centers for Disease Control and Prevention. Iron deficiency—United States 1999–2000. *Morb Mortal Wkly Rep* 51:897–899, 2002.
10. Food and Nutrition Board, Institute of Medicine. *Dietary Reference Intakes: Vitamin A, Vitamin K, Arsenic, Boron, Chromium, Copper, Iodine, Iron, Manganese, Molybdenum, Nickel, Silicon, Vanadium, and Zinc.* Washington, DC: National Academies Press, 2001.
11. O'Connor, N. R. Infant formula. *Am Fam Physician* 79: 565–570, 2009.
12. Food and Drug Administration. Iron-containing supplements and drugs; Label warning statements and unit-dose packaging requirements; Removal of regulations for unit-dose packaging requirements for dietary supplements and drugs. *Fed Regist* 68:59714–59715, 2003.
13. Siah, C. W., Trinder, D., and Olynyk, J. K. Iron overload. *Clin Chem Acta* 358:24–36, 2005.
14. Camaschella, C., and Poggiali, E. Inherited disorders of iron metabolism. *Curr Opin Pediatr* 23:14–20, 2011.
15. Utzschneider, K. M., and Kowdley, K. V. Hereditary hemochromatosis and diabetes mellitus: Implications for clinical practice. *Nat Rev Endocrinol* 6:26–33, 2010.
16. Zegrean, M. Association of body iron stores with development of cardiovascular disease in the adult population: A systematic review of the literature. *Can J Cardiovasc Nurs* 19:26–32, 2009.
17. Fowler, C. F. Hereditary hemochromatosis: Pathophysiology, diagnosis, and management. *Crit Care Nurs Clin N Am* 20:191–201, 2008.
18. Prasad, A. S. Discovery of human zinc deficiency and studies in an experimental human model. *Am J Clin Nutr* 53:403–412, 1991.
19. Wang, X., and Zhou, B. Dietary zinc absorption: A play of Zips and ZnTs in the gut. *IUBMB Life* 62:176–182, 2010.
20. Cousins, R. J. Gastrointestinal factors influencing zinc absorption and homeostasis. *Int J Vitam Nutr Res* 80:243–248, 2010.
21. Tapiero, H., and Tew, K. D. Trace elements in human physiology and pathology: Zinc and metallothioneins. *Biomed Pharmacother* 57:399–411, 2003.
22. Prasad, A. S. Zinc in human health: Effect of zinc on immune cells. *Mol Med* 14:353–357, 2008.
23. Fulgoni, V. L., III, Keast, D. R., Bailey, R. L., and Dwyer, J. Foods, fortificants, and supplements: Where do Americans get their nutrients? *J Nutr* 141:1847–1854, 2011.
24. Black, R. E. Zinc deficiency, infectious disease, and mortality in the developing world. *J Nutr* 133:1485S–1489S, 2003.
25. Haase, H., Mocchegiani, E., and Rink, L. Correlation between zinc status and immune function in the elderly. *Biogerontology* 7:421–428, 2006.
26. National Institute of Allergy and Infectious Diseases, National Institutes of Health. Common cold.
27. Eby, G. A., III. Zinc lozenges as cure for the common cold—a review and hypothesis. *Med Hypotheses* 74:482–492, 2010.
28. Singh, M., and Das, R. R. Zinc for the common cold. *Cochrane Database Syst Rev* Feb 16(2):CD001364, 2011.
29. National Institute of Neurological Disorders and Stroke. NINDS Wilson's disease information page.
30. Mills, E. S. The treatment of idiopathic (hypochromic) anemia with iron and copper. *Can Med Assoc J* 22:175–178, 1930.
31. Muñoz, C., Rios, E., Olivos, J., et al. Iron, copper and immunocompetence. *Br J Nutr* 98(suppl 1):S24–S28, 2007.
32. National Research Council, Food and Nutrition Board. *Recommended Dietary Allowances*, 10th ed. Washington, DC: National Academy Press, 1989.
33. Fairweather-Tait, S. J., Bao, Y., Broadley, M. R., et al. Selenium in human health and disease. *Antioxid Redox Signal* 14:1337–1383, 2011.
34. Beck, M. A., Levander, O. A., and Handy, J. Selenium deficiency and viral infection. *J Nutr* 133(5 suppl 1):1463S–1467S, 2003.
35. Clark, L. C., Combs, G. F. Jr., Turnbull, B. W., et al. Effect of selenium supplementation for cancer prevention in patients with carcinoma of the skin. *JAMA* 276:1957–1968, 1996.
36. Duffield-Lillico, A. J., Slate, E. H., Reid, M. E. et al. Selenium supplementation and secondary prevention of nonmelanoma skin cancer in a randomized trial. *J Natl Cancer Inst* 95: 1477–1481, 2003.
37. Institute of Medicine, Food and Nutrition Board. *Dietary Reference Intakes for Vitamin C, Vitamin E, Selenium, and Carotenoids*. Washington, DC: National Academies Press, 2000.
38. Rayman, M. P. Selenium and human health. *Lancet* 379: 1256–1268, 2012.

39. World Health Organization. Micronutrient deficiencies: Iodine deficiency disorders.

40. Andersson, M., de Benoist, B., and Rogers, L. Epidemiology of iodine deficiency: Salt iodisation and iodine status. *Best Pract Res Clin Endocrinol Metab* 24:1–11, 2010.

41. Hua, Y., Clark, S., Ren, J., and Sreejayan, N. Molecular mechanisms of chromium in alleviating insulin resistance. *J Nutr Biochem* 23:313–319, 2012.

42. Vincent, J. B. Recent advances in the nutritional biochemistry of trivalent chromium. *Proc Nutr Soc* 63:41–47, 2004.

43. Di Luigi, L. Supplements and the endocrine system in athletes. *Clin Sports Med* 27:131–151, 2008.

44. Balk, E. M., Tatsioni, A., Lichtenstein, A. H., et al. Effect of chromium supplementation on glucose metabolism and lipids: A systematic review of randomized controlled trials. *Diabetes Care* 30:2154–2163, 2007.

45. Jeejeebhoy, K. N. The role of chromium in nutrition and therapeutics and as a potential toxin. *Nutr Rev* 57:329–335, 1999.

46. Stearns, D. M., Wise, J. P. Sr., Patierno, S. R., and Wetterhahn, K. E. Chromium (III) picolinate produces chromosome damage in Chinese hamster ovary cells. *FASEB J* 9:1643–1648, 1995.

47. EFSA Panel on Food Additives and Nutrient Sources Added to Food (ANS). Scientific opinion on the safety of chromium picolinate as a source of chromium added for nutritional purposes to foodstuff for particular nutritional uses and to foods intended for the general population. *EFSA* 8:1883–1932, 2010.

48. Institute of Medicine, Food and Nutrition Board. *Dietary Reference Intakes for Calcium, Phosphorus, Magnesium, Vitamin D, and Fluoride*. Washington, DC: National Academies Press, 1997.

49. American Dental Association. Fluoridation facts. 2005.

50. Rubin, C. D., Pak, C. Y., Adams-Huet, B. et al. Sustained-release sodium fluoride in the treatment of the elderly with established osteoporosis. *Arch Intern Med* 161:2325–2333, 2001.

51. Parnell, C., Whelton, H., and O'Mullane, D. Water fluoridation. *Eur Arch Paediatr Dent* 10:141–148, 2009.

52. Beltrán-Aguilar, E. D., Barker, L., and Dye, B. A. Prevalence and severity of dental fluorosis in the United States, 1999–2004. *NCHS Data Brief* 53:1–8, 2010.

***To access links to online sources, please go to** www.wiley.com/college/smolin **and select Nutrition: Science and Applications, 3rd edition. From this page, select either the student or instructor companion site. Once on the desired site, select References.**

Focus on Nonvitamin/Nonmineral Supplements

© Selahattin Gezer/iStockphoto

CHAPTER OUTLINE

If you are looking for a dietary supplement, you won't have to look far. There are thousands of different products from which to choose. They come as tablets, capsules, powders, softgels, gelcaps, and liquids. They are sold in health food stores, supermarkets, drug stores, and national discount chain stores, as well as through mail-order catalogs, television, the Internet, and direct sales. Most of these products contain vitamins and minerals, but a large number of the more popular dietary supplements contain nutrients other than vitamins or minerals and substances that are not classified as nutrients at all. Surveys have reported that over 50% of adult Americans use supplements and over 17% use nonvitamin/nonmineral supplements.[1,2] They are taken to improve athletic performance, promote weight loss, prevent and treat disease, slow aging, and a variety of other reasons. Some of these contain protein, amino acids, or fatty acids. Others contain substances, such as hormones, enzymes, and coenzymes, which are made in the body but are not dietary essentials. Some contain plant-derived substances such as phytochemicals and herbs. For many, there is scientific evidence of beneficial effects. **For others, the benefits are more questionable, and for a few, the risk of using them clearly outweighs any benefits they may provide.**

F4.1 What Is a Dietary Supplement?

LEARNING OBJECTIVES

- List the types of substances that are included in dietary supplements.
- Explain the regulatory impact of classifying dietary supplements as foods, rather than drugs.

The term *dietary supplement* traditionally referred to products designed to add essential nutrients to the diet. But the 1994 Dietary Supplement Health and Education Act (DSHEA) broadened this definition to include any product intended for ingestion as a supplement to the diet. These products can include not only vitamins and minerals, but also herbs, botanicals, and other plant-derived substances as well as amino acids and concentrates, metabolites, constituents, and extracts of these substances. Products sold as dietary supplements must carry a standard label that meets the specifications of the DSHEA.[3] But the fact that a supplement is labeled doesn't mean it is safe. Unlike most foods, dietary supplements may contain concentrations of nutrients and other substances that exceed levels normally found in the diet and may therefore present a risk, particularly for people taking medications or with certain medical conditions. Unlike drugs, dietary supplements are not strictly regulated in terms of dose or constituents.

Dietary Supplement Labels

According to FDA regulations, any product intended for ingestion as a supplement to the diet must include the words "dietary supplement" on the label and carry a "Supplement Facts" panel (**Figure F4.1** and Chapter 2). The Supplement Facts panel lists the recommended serving size and the name and quantity of each ingredient per serving. The nutrients for which Daily Values have been established are listed first, followed by other dietary ingredients for which Daily Values have not been established.[3]

To ensure that dietary supplements do not contain contaminants and that the information listed on the supplement label is accurate, the FDA has established regulations regarding current Good Manufacturing Practices (cGMP) for dietary supplements.[4] These require manufacturers to evaluate their products for identity, purity, strength, and composition. The U.S. Pharmacopeia (USP) has also developed a voluntary dietary supplement verification program. Those manufacturers who participate in the program can include the USP-Verified mark on their product label (see Chapter 2, Figure 2.13).

Regulation of Dietary Supplements

Although dietary supplements are often taken as drugs to prevent or treat disease, their safety and marketing are not regulated as strictly as drugs. Drugs are required to undergo testing that must be reviewed by the FDA before they can be marketed to the public. Supplement manufacturers are responsible for ensuring product safety, but these products do not require FDA review. If a problem arises with a specific product, the FDA must prove that the supplement represents a risk before it can be removed from the market. The exception to this rule is products that contain new ingredients. Any supplement containing an ingredient not sold in the United States before October 15, 1994, must provide the FDA with safety data prior to marketing the product. Ingredients sold prior to this date are presumed safe based on their history of use by humans.[3]

Supplement Facts

Serving Size: 2 Capsules
Servings Per Container: 60

	Amount Per Serving	DV%
Vitamin C (as ascorbic acid)	60mg	100%
Pantothenic Acid (as calcium pantothenate)	20mg	200%
Vitamin B_6 (as pyridoxine HCL)	8mg	400%
Niacin	5mg	25%
Folate (as folic acid)	100mcg	25%
Zinc (as zinc gluconate)	5mg	33%
Copper (as copper gluconate)	500mcg	25%
NADH (Nicotinamide Adenine Dinucleotide)	1000mcg	*
Hoodia Gordonii Extract (20:1 Extract-Equal to 2000 mg of whole plant)	100mg	*
5-Hydroxytryptophan (Griffonia Simplicifolia)	25mg	*
N, N Dimethylglycine	50mg	*
Trimethylglycine	75mg	*
L-Phenylalanine	600mg	*
Decaffinated Green Tea Extract (Total Catechins 130mg, Epigaliocatechin Galiate (EGCG) 70mg)	175mg	*
Salvia Scalarea Extract	50mg	*
Choline (as bitartrate)	75mg	*

*Daily Value (DV, not established)

Recommended Use: As a dietary supplement take two capsules before breakfast on a empty stomach (or before exercise) and two capsules at mid-afternoon preferably with 8 oz of water.

This supplement includes a variety of vitamins and minerals. Do the amounts of any exceed 100% of the Daily Value? Do the amounts of any exceed the UL?

This supplement contains many ingredients that are not vitamins or minerals and therefore have no Daily Value or UL. Are they safe when taken in these amounts?

Figure F4.1 Supplement facts labels
This Supplement Facts panel from a supplement marketed to reduce appetite and aid in weight loss indicates that it contains vitamins, as well as amino acids, plant extracts, and some substances made in the body.

F4.2 Macronutrient Supplements

LEARNING OBJECTIVES

- Explain why carbohydrate and protein supplements do not come in pill form.
- Explain why people might take supplements containing protein, amino acids, or fatty acids.

A varied balanced diet will meet nutrient needs for most healthy people (see Chapter 9). However, some people choose to supplement specific nutrients because they don't think they get enough in their diets, or because they are looking for additional beneficial effects such as disease prevention or performance enhancement. For example, carbohydrate and protein supplements are frequently used by athletes. Carbohydrate pills can't provide enough carbohydrate to have a significant effect, so supplemental carbohydrate usually comes in the form of sports beverages, bars, or carbohydrate gels. These are packaged for convenience and easy transport and the carbohydrate is in a form that is absorbed from the GI tract quickly to provide glucose to exercising muscles. Protein supplements generally come as powders, drinks, and bars. They provide no benefit beyond that obtained from the protein in an adequate diet. Individual amino acids and fatty acids are also marketed for their effect on performance and body composition as well as disease prevention.

Amino Acids

Amino acids supplements are popular among athletes and dieters. The amino acids arginine, ornithine, and lysine are taken because they have been found to stimulate the release of growth hormone, which promotes muscle growth. Therefore, these are promoted to people who want to build muscle and lose fat. However, controlled studies have found that oral supplements of these amino acids taken before exercise do not significantly boost growth hormone release, or increase muscle mass or strength above that obtained from training alone.[5,6] Branched-chain amino acids (leucine, isoleucine, and valine) are marketed to athletes to improve athletic performance and aid in recovery from strength exercise. Although studies examining branched-chain amino acid supplements have not found them to enhance performance, there is evidence that they may have some benefits in improving muscle recovery and preventing muscle soreness after intense exercise.[7,8] However, adequate amounts of these amino acids can be obtained from whole-food sources of protein, without amino acid supplements. Glutamine and glycine supplements are also marketed to athletes, but neither has been found to enhance exercise performance. Because some amino acids share transport systems, supplementing one amino acid can cause a deficiency of others that use that same transport system (see Chapter 6).

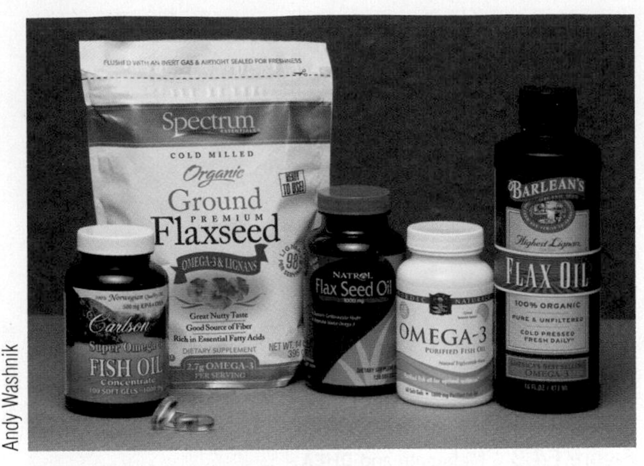

Figure F4.2 Supplements providing omega-3 fatty acids Fish oil and flaxseed oil supplements can increase your omega-3 fatty acid intake, but do not provide all the health benefits of eating a diet that includes fish and flaxseeds.

Fatty Acids

Supplements of omega-3 fatty acids from fish oils and flaxseed are taken to reduce the risk of heart disease (**Figure F4.2**). Scientific evidence shows that the long-chain omega-3 fatty acids EPA and DHA protect against heart disease, but supplements may not be as beneficial as food sources of these fatty acids.[9] Eating fish has been found to provide greater benefits than taking fish oil supplements.[10] Flaxseed oil provides α-linolenic acid. To have heart protective effects, this needs to be converted into the longer-chain omega-3 fatty acids, EPA and DHA. Because the conversion of α-linolenic acid to EPA and DHA is inefficient in humans, flaxseed oil supplements do not provide as much EPA and DHA as fish oil.[10] However, consuming the entire flaxseed, rather than just the oil, adds not only omega-3 fatty acids, but also fiber and phytochemicals to the diet. Eating fish several times a week and sprinkling ground flaxseed on your cereal may be a better option than taking supplemental oils. The risks of taking supplements containing omega-3 fatty acids are minimal for healthy people, but these products are not recommended for those taking blood-thinning medications because they inhibit blood clotting.

F4.3 Substances Made in the Body

LEARNING OBJECTIVES

- Discuss why most enzyme supplements are ineffective.
- Describe the benefits of popular hormone supplements.

Many molecules essential to normal metabolic and physiological function are made in the body in amounts sufficient to meet needs and are therefore not essential in the diet. No deficiency symptoms occur when they are absent from the diet. Nonetheless, supplements of many of these biological molecules are sold to enhance bodily functions. Most of these provide little benefit to healthy, well-nourished individuals.

Enzyme Supplements

Enzymes are proteins. Those needed for normal function are made in the body. When consumed in supplements, like other proteins, they are broken down into amino acids in the GI tract. The enzyme protein itself never reaches cells inside the body, and therefore most enzyme supplements have little effect on body function (see Chapter 6). The exception to this is digestive enzymes that act in the gastrointestinal tract. When supplemented, these can perform their functions before they are broken down. For example, lactase, the enzyme that breaks down the milk sugar lactose, can benefit individuals with lactose intolerance

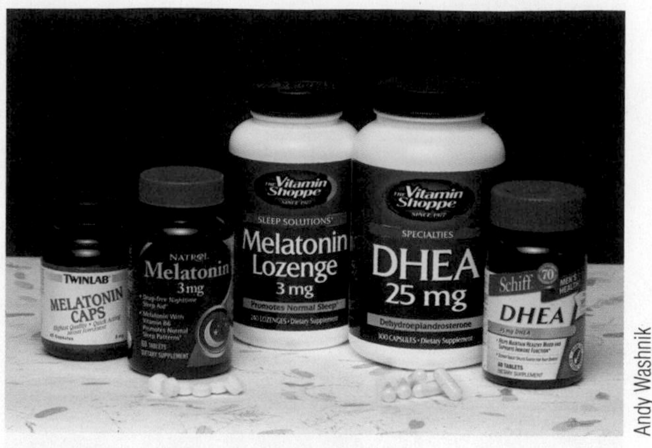

Andy Washnik

Figure F4.3 Melatonin and DHEA
Melatonin and DHEA are hormones that can be purchased in supplement form. DHEA use is banned by the International Olympic committee, the National College Athletic Association, and most major league sporting organizations.

when consumed along with foods containing lactose. The enzyme breaks down lactose that is in the GI tract. Eventually the lactase is also digested.

A number of other digestive enzymes, including proteases, lipases, amylases, and plant enzymes, such as bromelain from pineapple and papain from papaya, are marketed to improve digestion in normal, healthy people. Advertisements for these products claim that the digestive enzymes naturally present in food are destroyed by cooking, leaving partially digested substances in the GI tract and forcing the body to synthesize more of its own digestive enzymes. Despite the popularity of enzyme supplements, there is no evidence that they are beneficial for healthy people.

Hormone Supplements

Hormones are another popular supplement ingredient. Protein or peptide hormones such as growth hormone have the same problem as enzymes—they are proteins, so they are broken down in the gut before they can reach their target and thus cannot function as hormones. Despite this, oral growth hormone supplements are sold. Growth hormone is appealing to athletes because it increases muscle protein synthesis. Despite this physiological effect, even growth hormone injections, which are the only way to get the active hormone into the body, have not been shown to enhance muscle strength, power, or aerobic exercise capacity, although there is evidence that they improve anaerobic exercise capacity.[11]

Hormones that are not proteins can be absorbed into the body intact and supplemental doses may affect body functions (**Figure F4.3**). Melatonin is a hormone made by the pineal gland in the brain. Although it is synthesized from the amino acid tryptophan, it is not a protein and supplemental doses can be absorbed intact from the GI tract. Supplements of melatonin are taken to delay aging, protect against free radical damage, and help with sleep disorders and jet lag. There is little evidence that it slows the aging process or that it reduces oxidative damage, but there is significant evidence linking the hormone melatonin to sleep cycles in humans. It has been suggested that in situations where the body's production of melatonin is reduced (advancing age) or the normal sleep/wake cycle is disrupted, such as with jet lag, supplemental melatonin may improve both sleep duration and quality. Melatonin may help to decrease jet lag symptoms and hasten the return to normal energy levels. Melatonin may also shorten the time it takes for people with insomnia to fall asleep, but it does not appear to improve sleep quality in people whose jobs require them to work on rotating shift schedules.[12]

Like melatonin, steroid hormones can be absorbed in the digestive tract and reach the bloodstream intact, thereby affecting function. DHEA (dehydroepiandrosterone), a steroid hormone made by the adrenal glands, is a popular supplement. It is a precursor to the sex hormones estrogen and testosterone. The production of DHEA in the body peaks in one's mid-twenties, and gradually declines with age in most people. Preventing this drop has been suggested to delay the aging process by maintaining levels of estrogen and testosterone. It has been claimed that DHEA improves energy level, strength, and immunity. Athletes use it as a substitute for anabolic steroids to increase muscle mass and decrease body fat. However, supplementation has not been found effective in increasing muscle size or strength or in treating advancing age, male sexual dysfunction, or menopausal symptoms.[13,14] The long-term effects of DHEA supplements have not been studied, but there is concern that they may interfere with normal levels of the sex hormones and have detrimental effects on the body, including liver damage.

Coenzyme Supplements

Many vitamins have coenzyme functions, but there are also coenzymes that are not dietary essentials. One of these is lipoic acid, which is a coenzyme needed for two reactions necessary for the production of ATP by aerobic metabolism. Although essential to energy metabolism, lipoic acid can be synthesized in adequate amounts by human cells, so supplements do not accelerate the reactions catalyzed by this coenzyme.

Ubiquinone, also called coenzyme Q, is another coenzyme important for the production of ATP from carbohydrate, fat, and protein. It is needed in the electron transport chain, the final stage of aerobic metabolism where ADP is converted to ATP. Because, as its name implies, it is present ubiquitously in animals, plants, and microorganisms, and is synthesized in the human body, supplements are not necessary for most people. However, supplements may be beneficial in individuals with inherited defects that affect coenzyme Q synthesis.[15] Low levels of coenzyme Q are also a concern in those taking cholesterol-lowering statin drugs such as Crestor and Lipitor. Statins inhibit an enzyme needed for the synthesis of both cholesterol and ubiquinone, which can cause depletion of this coenzyme. Depletion of coenzyme Q in the mitochondria of muscle cells has been suggested to be a cause of the muscle pain and weakness that appears as a side effect in some individuals taking statins. The effect of coenzyme Q supplementation on muscle pain and weakness in patients taking statin drugs is equivocal, with some trials showing that supplements reduce pain severity and interference with daily activities and others showing no effect.[16]

Supplements Containing Structural and Regulatory Molecules

A number of structural and regulatory molecules are taken as supplements for a variety of different reasons. For instance, carnitine and creatine are both sold to enhance athletic performance (see Chapter 13). Carnitine is a molecule that is needed to transport fatty acids into the mitochondria, where they are broken down by β-oxidation and aerobic metabolism to produce ATP. Supplements are marketed to increase endurance by improving the use of fat as an energy source during exercise. Supplements have not been shown to improve endurance, but more recent studies focus on its potential to enhance recovery from exercise.[17] Creatine is a small nitrogen-containing molecule found primarily in muscle, where it is used to make a high-energy molecule called creatine phosphate. Creatine phosphate provides energy to the muscle for short bursts of activity. Creatine supplementation has been shown to improve performance in high-intensity exercise lasting 30 seconds or less. It is therefore beneficial for exercise that requires explosive bursts of energy, such as sprinting and weight lifting, but not for long-term endurance activities, such as marathons[18] (see Chapter 13).

Glucosamine, chondroitin, and SAM-e are molecules needed for the maintenance of healthy joints and are sold to alleviate the pain and progression of osteoarthritis. Glucosamine and chondroitin are molecules found in and around the cells of cartilage, the type of connective tissue that cushions joints. They are made in the body and consumed in the diet in meat. Supplements of both glucosamine and chondroitin are reported to reduce arthritis pain, stop cartilage degeneration, and possibly stimulate the repair of damaged joint cartilage (**Figure F4.4**). Clinical studies have not found the effect of these supplements on pain and function to be significant[19] (see Chapter 16).

SAM-e, chemically known as S-adenosylmethionine, is present in the body normally as an intermediate in the metabolism of the amino acid methionine. SAM-e enhances the production of cartilage, and supplements are promoted for the treatment of arthritis; they are also claimed to be effective for treating depression. For patients with arthritis SAM-e, supplements appear to be as effective as aspirin and similar drugs, but they can take much longer to have an effect.[20] There is also evidence that SAM-e may be beneficial in individuals with depression, but further study is needed.[21]

Supplements of inositol can also be found on the shelf. Inositol is a component of phospholipids in cell membranes, where it plays a role in relaying messages to the inside of the cell. Inositol can be synthesized from glucose. There is no evidence that it is essential in the human diet, but it may have some clinical value in treating diseases such as diabetes and kidney failure.[22]

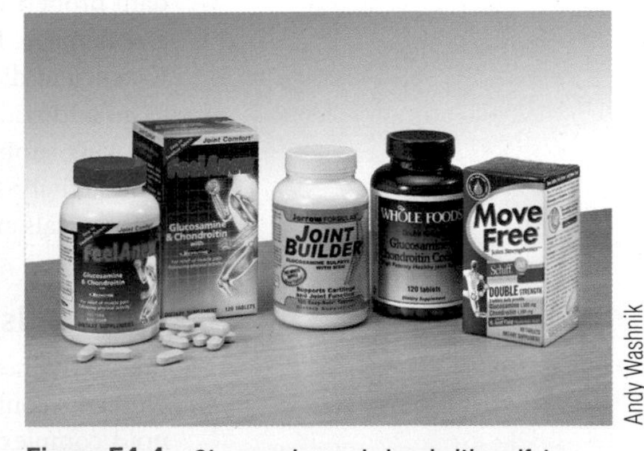

Andy Washnik

Figure F4.4 Glucosamine and chondroitin sulfate supplements
People suffering from arthritis commonly take supplements containing glucosamine and chondroitin sulfate in the hope that they will reduce pain. The pills are large and about three a day are recommended, but the benefits have not been proven.[19]

F4.4 Phytochemical Supplements

LEARNING OBJECTIVES

- Explain why phytochemical supplements may not have the same benefits as foods containing these phytochemicals.
- Discuss how phytochemical supplements may affect antioxidant levels in the body.

A diet rich in phytochemicals from fruits, vegetables, and whole grains has been shown to reduce the risk of heart disease and cancer (see **Focus on Phytochemicals**). Although we have not yet isolated and identified all of these health-promoting substances, some have been extracted, purified, and pressed into pills or capsules. These are advertised as having health-promoting properties, but they do not appear to provide all the benefits obtained from whole foods that are rich sources of these as well as a host of other phytochemicals and nutrients.

Figure F4.5 **Foods rich in carotenoids**
Fruits and vegetables provide most of the carotenoids in our diet.

Carotenoids

Carotenoids are a group of more than 600 yellow, orange, and red compounds found in living organisms. Carotenoids have antioxidant properties, and some also have vitamin A activity. The most prevalent carotenoids in the North American diet include *ß*-carotene, *α*-carotene, *ß*-cryptoxanthin, lycopene, lutein, and zeaxanthin. The major sources of carotenoids in the diet are fruits and vegetables (**Figure F4.5**).

Lutein and zeaxanthin are carotenoids found in leafy green vegetables. They do not have vitamin A activity but are concentrated in the macula of the eye, where they protect against oxidative damage. High intakes of these carotenoids have been associated with a reduced incidence of age-related eye disorders, such as macular degeneration[23] (see Chapter 16).

The most common carotenoid supplement is *ß*-carotene. It is marketed for its antioxidant properties, but under some circumstances it may promote rather than prevent oxidative damage. For example, some studies show that when *ß*-carotene is added to a vitamin E-deficient diet, it promotes oxidation.[24] But in the presence of vitamin E, *ß*-carotene acts as an antioxidant. One explanation for the increase in the incidence of lung cancer among smokers supplemented with *ß*-carotene is that pro-oxidant activity prevails over antioxidant activity. Because supplements do not provide the same amounts and variety of phytochemicals and other substances found in foods, their effect is sometimes unpredictable.

Flavonoids

Like carotenoids, flavonoids, which are often called bioflavonoids, are antioxidants. Supplements containing categories of flavonoids such as rutin and hesperidian and the flavonoid complex pycnogenol are advertised as cures for arthritis, heart disease, high blood pressure, and colds. Hesperidian is often called vitamin P but it does not meet the definition of a vitamin because it has not been shown to be essential in the diet. Bioflavonoids are often included in supplements containing vitamin C because they are purported to promote the action of this vitamin. Although the foods containing these phytochemicals have been shown to have health-promoting properties, supplements have not been shown to have the same health effects.

F4.5 Herbal Supplements

LEARNING OBJECTIVES

- Explain why people choose to take herbal supplements.
- Compare the risks and benefits of herbal supplements.

Technically, an herb is a nonwoody, seed-producing plant that dies at the end of the growing season. However, the term *herb* is generally used to refer to any botanical or plant-derived substance. Throughout history, folk medicine has used herbs to prevent and treat disease. Today, herbs and herbal supplements are more popular than ever. It is estimated that about one in six Americans uses herbs to treat maladies or boost health.[25] The seven most popular herbal medicinal products in the United States today are ginkgo biloba, St. John's wort, ginseng, garlic, echinacea, saw palmetto, and kava.[26]

Herbal supplements are readily available and relatively inexpensive. They can be purchased without a trip to the doctor or a prescription. Although consumers who want to manage their own health may view this as beneficial, it can also cause problems. When a drug is prescribed by a physician, it is assumed that it will have a beneficial effect, that each dose will contain the same amount of drug, that the physician or pharmacist has considered other medications and other medical conditions that may alter the effectiveness of the drug, and that the drug itself will cause no severe side effects. These assumptions cannot be made with herbs. Many botanical compounds are toxic, either alone or in combination with other drugs and herbs being consumed; the FDA has issued warnings about ingredients such as comfrey, kava, and aristolochic acid, and it took ephedra off the market.[27] Because herbal products are made from unpurified plant material, there is also a risk of contamination with pesticides, microbes, metals, and other toxins.[28] And it is difficult to know what dose of an herb you are taking. Even those pressed and packaged into pills may not provide the same dose in each pill. Also, because consumers decide what to treat, herbal remedies can be used inappropriately or instead of necessary medical intervention.

Some herbs may offer benefits, but serious side effects from excessive doses or unusual combinations of herbs and medications are not uncommon. The use of herbal supplements also may be inappropriate at certain times. For example, St. John's wort can prolong and intensify the effects of narcotic drugs and anesthetic agents, so it should not be taken for two to three weeks prior to surgery. Herbal products should also be avoided during pregnancy. Blue cohosh (used to treat menstrual cramps), juniper (used for heartburn), pennyroyal or rosemary (used for digestive problems), sage (used for stomach upset), and thuja (used for respiratory infections), and even raspberry tea (used to treat morning sickness) may stimulate uterine contractions, which can increase the risk of miscarriage or premature labor.[29] Some guidelines to follow if you are considering taking herbs or herbal supplements are included in **Table F4.1**.

Ginkgo Biloba

Ginkgo biloba, also called "maidenhair," is one of the top-selling herbal medicinal products in the United States (**Figure F4.6a**). It has been used for millennia in Chinese medicine

TABLE F4.1 Considerations When Choosing Herbal Supplements

• If you are ill or taking medications, consult your physician before taking herbs.
• Do not take herbs if you are pregnant.
• Do not give herbs to children.
• Do not assume herbal products are safe.
• Do not take herbs with known toxicities.
• Read label ingredients and the list of precautions.
• Start with low doses, and stop taking any product that causes side effects.
• Do not take combinations of herbs.
• Do not use herbs for long periods.
• Do not choose products that claim to be a secret cure and be wary of terms such as "breakthrough," "magical," "miracle cure," and "new discovery."

TABLE F4.2 Potential Benefits and Side Effects of Common Herbal Ingredients

Product	Suggested benefit and uses	Side effects
Astragalus (bei qi, huang qi, ogi, hwang ki, milk vetch)	Enhances the immune system	Can interact with drugs that suppress the immune system and affect blood sugar levels and blood pressure
Bitter orange (Seville orange, sour orange, Zhi shi)	Relieves heartburn and nasal congestion, stimulates appetite, promotes weight loss	Increased heart rate and blood pressure, fainting, heart attack, stroke
Cat's Claw (uña de gato)	Relieves arthritis and stimulates the immune system	Headache, dizziness, vomiting; should not be taken by individuals who are pregnant
Chamomile	Aids gastrointestinal upset, promotes relaxation and sleep	Allergic reactions
Dandelion (lion's tooth, blowball)	Relieves minor digestive problems, increases urine production, supports liver and kidney health	Upset stomach and diarrhea, allergic reactions
Echinacea (purple coneflower, coneflower)	Stimulates the immune system, prevents and treats colds and other upper respiratory infections	Allergic reactions
Ephedra (Ma Huang, Chinese ephedra)	Treats colds and nasal congestion, aids in weight loss, increases energy, and enhances athletic performance	High blood pressure, irregular heartbeat, heart attack, stroke, death; banned by the FDA, but the ban does not apply to traditional Chinese herbal remedies or to products like herbal teas regulated as conventional foods
Ginger	Relieves motion sickness and nausea	Gas, bloating, heartburn, nausea
Ginkgo (Ginkgo biloba, maidenhair tree, fossil tree)	Improves memory and mental function, improves circulation	Gastrointestinal distress, headache, dizziness, allergic skin reactions
Ginseng (Asian ginseng, Chinese ginseng)	Increases sense of well-being and stamina, improves mental and physical performance, enhances immune function, improves sexual function, lowers blood glucose, controls blood pressure	Headache, insomnia, gastrointestinal upset with prolonged use
Hawthorn	Strengthens heart muscle	Possible drug interactions
Hoodia (Kalahari cactus, Xhoba)	Suppresses appetite	Safety unknown; potential risks, side effects, and interactions with medicines and other supplements have not been studied
Kava (kava kava, awa, kava peeper)	Relieves anxiety, stress, insomnia, menopausal symptoms	Liver damage, including hepatitis and liver failure (which can cause death); FDA has issued a warning that using kava supplements has been linked to a rise of severe liver damage
Milk thistle (Mary thistle, holy thistle)	Protects against liver disease, improves liver function	Gastrointestinal upset, allergic reactions, low blood sugar
Red clover (cow clover, meadow clover, wild clover)	Relieves menopausal symptoms, breast pain associated with menstrual cycles, and symptoms of prostate enlargement; lowers blood cholesterol	Headache, nausea, rash
Saw palmetto (American dwarf palm tree, cabbage palm)	Improves urinary flow with enlarges prostate	Mild stomach discomfort, headache, fatigue, decreased libido, rhinitis
St. John's wort (hypericum, Klamath weed, goatweed)	Promotes mental well-being; treats depression, anxiety, and/or sleep disorders	Increased sensitivity to sunlight, anxiety, dry mouth, dizziness, gastrointestinal symptoms, fatigue, headache, sexual dysfunction; interacts with many medications, including antidepressants, birth control pills, digoxin, warfarin, and seizure-control drugs
Valerian (all-heal, garden heliotrope)	Acts as a mild sedative, relieves sleep disorders and anxiety	Gastrointestinal upset, headache, and tiredness possible with prolonged use
Yohimbe (yohimbe bark)	Acts as an aphrodisiac; treats sexual dysfunction, including erectile dysfunction in men	High blood pressure, increased heart rate, headache, anxiety, dizziness, nausea, vomiting, tremors, and sleeplessness

Sources: National Center for Complementary and Alternative Medicine, *Herbs at a Glance,* Office of Dietary Supplements, *Dietary Supplement Fact Sheets,* and Office of Dietary Supplements, *Botanical Supplement Fact Sheets.*

Figure F4.6 Herbal supplements
Herbal supplements are derived from plants, many of which are common in fields, gardens, or kitchens.

(a)

Altrendo/Getty Images, Inc.

(a) The leaves of the ginkgo biloba plant are used to make the top-selling herbal supplement in the United States.

(b)

Michael Gadomski/Photo Researchers

(b) These yellow flowers of St. John's wort grow wild in fields and pastures and are considered a perennial weed.

(c)

Lynn Johnson/NG Image Collection

(c) Ginseng capsules, extracts, and powders are made from the dried root of the ginseng plant.

(d)

© Christine Glade/iStockphoto

(d) Garlic is used to spice food in the kitchen, but extracts are also processed into pills or tablets and marketed as herbal supplements.

(e)

Linda Lewis/Botanica/Getty Images

(e) Flowers of the echinacea plant, like the ones shown here, can be seen growing along the roadside in the western United States.

(f)

© William Mullins/Alamy Limited

(f) Saw palmetto (*Serenoa repens*) is the dominant ground cover in some pine forests in Florida, Louisiana, and North Carolina.

to treat asthma and skin infections, but today it is marketed to enhance memory and to treat a variety of circulatory ailments.[30] Supplements have not been found to reduce the incidence of dementia or protect against cognitive decline in the elderly,[31] but there is evidence that they may benefit mood and attention in healthy adults.[32] Taking *Ginkgo biloba* may cause side effects that include headaches and gastrointestinal symptoms[33] (**Table F4.2**). *Ginkgo biloba* also interacts with a number of medications. It can cause bleeding when combined with warfarin or aspirin, elevated blood pressure when combined with a thiazide diuretic, and coma when combined with the antidepressant trazodone.[34]

St. John's Wort

St. John's wort is taken to promote mental well-being (**Figure F4.6b**). Analysis reveals that it contains low doses of the chemical found in the antidepressant drug fluoxetine (Prozac). The results of clinical trials suggest that it is effective for the treatment of depression.[35] Side effects include nausea and sensitivity to sunlight. St John's wort should not be used in conjunction with prescription antidepressant drugs, and it has been found to interact with anticoagulants, heart medications, birth control pills, immunosupressants, antibiotics, medications used to treat HIV infection, and others.[33,34]

Ginseng

Ginseng has a long history in traditional medicine (**Figure F4.6c**). It has been used in Asia for centuries for its energizing, stress-reducing, and aphrodisiac properties. Today, it is popular in the West for its effects on cardiovascular health, the central nervous system, endocrine function, and sexual function. Although ginseng contains substances that have antioxidant, anti-inflammatory, immunostimulating, and central nervous system effects, controlled trials investigating its health benefits have been equivocal.[36] Despite its history of use in Chinese medicine, ginseng may not be safe for everyone. It can alter bleeding time and therefore should not be used by those taking warfarin, and it has been found to interact with other medications such as estrogens, corticosteroids, antidepressants, and morphine. Even in those not taking other medications, it may cause side effects such as diarrhea, headache, and insomnia.[34]

Garlic

Throughout history garlic has been used for a lot more than to keep vampires away (**Figure F4.6d**). Hippocrates, who is considered the father of Western medicine, recommended garlic to treat pneumonia and other infections, as well as cancer and digestive disorders. Although it is no longer recommended for these purposes, recent research has shown that it may cause a modest reduction in blood cholesterol and triglyceride levels.[37] Eating enough garlic to lower cholesterol will probably keep your friends and family away, as well as the vampires, so supplement manufacturers have provided a way to increase intake without eating this odiferous food at every meal; some preparations contain a deodorized form. Even though we spice our food with it, garlic supplements are not safe for everyone. The National Institutes of Health has concluded that supplements could be harmful for people undergoing treatment for HIV infection and could lead to bleeding in those taking the anticoagulant drug warfarin.[34]

Echinacea

Native Americans used petals of the echinacea plant as a treatment for colds, flu, and infections (**Figure F4.6e**). Today, the echinacea root is popular as an herbal cold remedy. Echinacea is hypothesized to act as an immune system stimulant, but there is little evidence that it is beneficial in either preventing or treating the common cold.[38] Although side effects have not been reported, allergies are possible.

Saw Palmetto

Traditionally, saw palmetto, which comes from the berries of the American dwarf palm (**Figure F4.6f**), has been used to treat problems of the urinary and genital tract; enhance sperm production, breast size, or libido; and as a mild diuretic. Today, it is marketed to treat prostate enlargement and therefore improve urinary flow. A review of the effectiveness of this supplement found that it provides mild to moderate improvement in urinary symptoms and flow measures.[39] No drug interactions have been reported. Available data suggest that saw palmetto is well tolerated by most users. Most side effects, including abdominal pain, decreased libido, headache, fatigue, nausea, and rhinitis, are mild.[40]

Kava

Kava is traditionally served as a special drink at ceremonies such as weddings and coming-out-of-mourning celebrations in the South Pacific. It is used today to relieve stress and anxiety. A review of the effectiveness of kava concluded that administration is effective in reducing anxiety.[41] However, kava may not be the safest way to relieve stress. In 2002 the FDA issued a warning about kava because it may cause liver damage, including hepatitis, cirrhosis, and liver failure. It has been taken off the market in many European countries and in Canada, Australia, and Singapore because of this danger.

REFERENCES

1. Council for Responsible Nutrition. *Supplement Usage, Consumer Confidence Remains Steady.*
2. National Center for Complementary and Alternative Medicine. What is complementary and alterative medicine?
3. FDA, U.S. Department of Health and Human Services. Dietary Supplement Labeling.
4. Food and Drug Administration. Current good manufacturing practice in manufacturing, packaging, labeling, or holding operations for dietary supplements: Final rule. *Fed Regist* 72:34751–34958, June 25, 2007.
5. Kanaley, J. A. Growth hormone, arginine and exercise. *Curr Opin Clin Nutr MetabCare* 11:50–54, 2008.
6. Chromiak, J. A., and Antonio, J. Use of amino acids as growth hormone-releasing agents by athletes. *Nutrition* 18:657–661, 2002.
7. Shimomura, Y., Inaguma, A., Watanabe, S., et al. Branched-chain amino acid supplementation before squat exercise and delayed-onset muscle soreness. *Int J Sport Nutr Exerc Metab.* 20:236–244, 2010.
8. Negro, M., Giardina, S., Marzani, B., and Marzatico, F. Branched-chain amino acid supplementation does not enhance athletic performance but affects muscle recovery and the immune system. *J Sports Med Phys Fitness* 48:347–351, 2008.
9. Anderson, B. M., and Ma, D. W. Are all *n-3* polyunsaturated fatty acids created equal? *Lipids Health Dis* 8:33, 2009.

10. U.S. Department of Agriculture and U.S. Department of Health and Human Services. *Dietary Guidelines for Americans, 2010.* 7th ed., Washington, DC: U. S. Government Printing Office, December 2010.

11. Birzniece, V., Nelson, A. E., and Ho, K. K. Growth hormone and physical performance. *Trends Endocrinol Metab* 22:171–178, 2011.

12. National Institutes of Health, U.S. National Library of Medicine, Medline Plus, Melatonin.

13. Cameron, D. R., and Braunstein, G. D. The use of dehydroepiandrosterone therapy in clinical practice. *Treat Endocrinol* 4:95–114, 2005.

14. Dayal, M., Sammel, M. D., Zhao, J., et al. Supplementation with DHEA: Effect on muscle size, strength, quality of life, and lipids. *J Women's Health* 14:391–400, 2005.

15. Villalba, J. M., Parrado, C., Santos-Gonzalez, M., and Alcain, F. J. Therapeutic use of coenzyme Q10 and coenzyme Q10-related compounds and formulations. *Expert Opin Investig Drugs* 19:535–554, 2010.

16. Littarru, G. P., and Tiano, L. Clinical aspects of coenzyme Q10: an update. *Nutrition* 26:250–254, 2010.

17. Kraemer, W. J., Volek, J. S., and Dunn-Lewis, C. L-carnitine supplementation: Influence upon physiological function. *Curr Sports Med Rep* 7:218–223, 2008.

18. Tarnopolsky, M. A. Caffeine and creatine use in sport. *Ann Nutr Metab* 57(Suppl. 2):1–8, 2011.

19. Wandel, S., Jüni, P., Tendal, B., et al. Effects of glucosamine, chondroitin, or placebo in patients with osteoarthritis of hip or knee: Network meta-analysis. *BMJ* 341:c4675, 2010.

20. Soeken, K. L., Lee, W. L., Bausell, R. B., et al. Safety and efficacy of S-adenosylmethionine (SAMe) for osteoarthritis. *J Fam Pract* 51:425–430, 2002.

21. Williams, A. L., Girard, C., Jui, D., et al. S-adenosylmethionine (SAMe) as treatment for depression: A systematic review. *Clin Invest Med.* 28:132–139, 2005.

22. Ooms, L. M., Horan, K. A., Rahman, P., et al. The role of the inositol polyphosphate 5-phosphatases in cellular function and human disease. *Biochem. J.* 419:29–49, 2009.

23. Age-Related Eye Disease Study Research Group, AREDS Report No. 22, The relationship of dietary carotenoid and vitamin A, E, and C intake with age-related macular degeneration in a case-control study. *Arch Ophthalmol* 125:1225–1232, 2007.

24. Palozza, P. Prooxidant actions of carotenoids in biologic systems. *Nutr Rev* 56:257–265, 1998.

25. Gershwin, M. E., Borchers, A. T., Keen, C. L., et al. Public safety and dietary supplementation. *Ann N Y Acad Sci* 1190:104–117, 2010.

26. Ernst, E. The risk-benefit profile of commonly used herbal therapies: Ginko, St. John's wort, ginseng, echinacea, saw palmetto, and kava. *Ann Intern Med* 136:42–53, 2002.

27. Food and Drug Administration. *Dietary Supplement Warnings and Safety Information.*

28. Saper, R. B., Phillips, R. S., Sehgal, A., et al. Lead, mercury, and arsenic in U.S.- and Indian-manufactured Ayurvedic medicines sold via the Internet. *JAMA* 300:915–923, 2008.

29. Schweitzer, A. Dietary supplements during pregnancy. *J PerinatEduc* 15:44–45, 2006.

30. National Center for Complementary and Alternative Medicine. *Herbs at a Glance: Ginko.*

31. Birks, J., and Grimley Evans, J. Ginkgo biloba for cognitive impairment and dementia. *Cochrane Database Syst Rev* CD003120, 2009.

32. Gorby, H. E., Brownawell, A. M., and Falk, M. C. Do specific dietary constituents and supplements affect mental energy? Review of the evidence. *Nutr Rev* 68:697–718, 2010.

33. National Center for Complementary and Alternative Medicine. *Herbs at a Glance: St. John's Wort.*

34. Izzo, A. A., and Ernst, E. Interactions between herbal medicines and prescribed drugs: an updated systematic review. *Drugs* 69:1777–1798, 2009.

35. Linde, K., Berner, M. M., and Kriston, L. St John's wort for major depression. *Cochrane Database Syst Rev* CD000448, 2008.

36. Xiang, Y. Z., Shang, H. C., Gao, X. M., and Zhang, B. L. A comparison of the ancient use of ginseng in traditional Chinese medicine with modern pharmacological experiments and clinical trials. *Phytother Res* 22:851–858, 2008.

37. Reinhart, K. M., Talati, R., White, C. M., and Coleman, C. I. The impact of garlic on lipid parameters: A systematic review and meta-analysis. *Nutr Res Rev* 22:39–48, 2009.

38. Barrett, B., Brown, R., Rakel, D., et al. Echinacea for treating the common cold: A randomized trial. *Ann Intern Med* 153:769–777, 2010.

39. Suzuki, M., Ito, Y., Fujino, T., et al. Pharmacological effects of saw palmetto extract in the lower urinary tract. *Acta Pharmacol Sin* 30:227–281, 2009.

40. Agbabiaka, T. B., Pittler, M. H., Wider, B., and Ernst, E. Serenoa repens (saw palmetto): A systematic review of adverse events. *Drug Saf* 32:637–647, 2009.

41. Sarris, J., LaPorte, E., and Schweitzer, I. Kava: A comprehensive review of efficacy, safety, and psychopharmacology. *Aust N Z J Psychiatry* 45:27–35, 2011.

CHAPTER OUTLINE

13.1 Exercise, Fitness, and Health
- How Exercise Improves Fitness
- How Exercise Benefits Health

13.2 Exercise Recommendations
- What to Look for in a Fitness Program
- Exercise Recommendations for Children
- How to Create an Active Lifestyle

13.3 Exercise and Energy Metabolism
- How Exercise Duration Affects Metabolism
- How Exercise Intensity Affects Metabolism
- Why Training Increases Aerobic Capacity

13.4 Energy and Nutrient Needs for Physical Activity
- Energy Needs
- Carbohydrate, Fat, and Protein Needs
- Vitamin and Mineral Needs

13.5 Fluid Needs for Physical Activity
- How Dehydration Affects Health and Performance
- Why Heat-Related Illness Is a Concern for Athletes
- Causes and Consequences of Hyponatremia
- Recommended Fluid Intake for Exercise

13.6 Food and Drink to Maximize Performance
- Maximizing Glycogen Stores
- What to Eat Before Exercise
- What to Eat During Exercise
- What to Eat After Exercise

13.7 Ergogenic Aids: Do Supplements Enhance Athletic Performance?
- Vitamin and Mineral Supplements
- Supplements to Build Muscle
- Supplements to Enhance Performance in Short, Intense Activities
- Supplements to Enhance Endurance
- How Diet and Supplements Affect Performance

© mbbirdy/iStockphoto

Nutrition and Physical Activity

13

CASE STUDY

Jason bought a road bike a year ago. He lives in a rural area where he can ride away from the traffic and enjoy the countryside. He typically goes for a 15- or 20-mile ride a few evenings after work, and on the weekends he often rides as far as 30 miles. A friend suggested they sign up for a weeklong ride in Oregon in the fall. The route would require riding 60 to 80 miles a day, often over mountain passes in the Cascades. Jason is thrilled by the challenge. He trains over the summer by finding hillier routes and increasing his distances. By the start of the Oregon ride he is fit and ready. During the ride, water, sports drinks, and snacks or lunch are provided every 20 miles. Jason and his friend do great the first day. They stop and eat and fill their water bottles at all the rest stops and complete the 75-mile ride by 3:30 in the afternoon. The next day the route takes them to Crater Lake. Jason is feeling great and is excited about visiting this extinct volcano, so he convinces his friend to ride past the second rest stop and hurry up to the lake. His friend is carrying an extra water bottle and has some snack bars in his pocket, which he eats soon after the skipped rest stop. Jason is less prepared. He feels fine for about 10 miles and then the road to Crater Lake turns steeply uphill for about 12 miles. Jason's water bottle is now empty and so is his stomach. His legs feel weak and he has no energy. How could he have felt so great yesterday and so awful today? With his friend's encouragement, he struggles slowly on to the lunch stop at the rim of the crater. **Why was he so fatigued?**

© technotr/iStockphoto

fitness The ability to perform routine physical activity without undue fatigue.

overload principle The concept that the body will adapt to the stresses placed on it.

cardiorespiratory endurance The efficiency with which the body delivers to cells the oxygen and nutrients needed for muscular activity and transports waste products from cells.

Figure 13.1 Cardiorespiratory endurance and aerobic capacity

© Joe Michl/iStockphoto/

(a) The cardiovascular and respiratory systems are strengthened by aerobic exercise, such as jogging, bicycling, and swimming. A guideline to assess whether an activity is aerobic is that the intensity is low enough for you to carry on a conversation while exercising but high enough that you cannot sing.

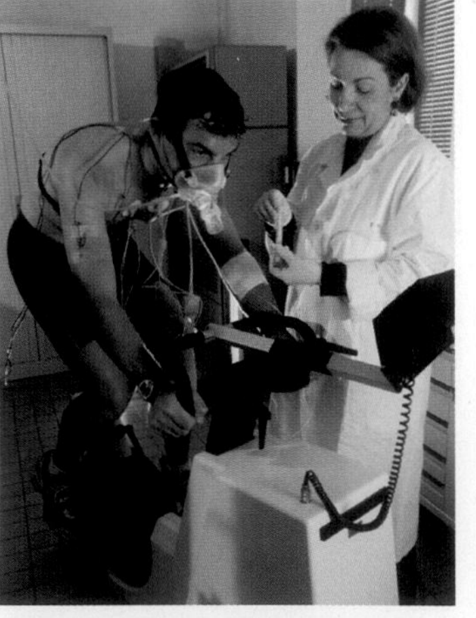

Philippe Psaila/Photo Researchers, Inc.

(b) To measure this athlete's aerobic capacity he rides a stationary bicycle while his oxygen consumption (oxygen inhaled – oxygen exhaled) is measured. His exercise workload is increased gradually by increasing resistance on the bike until he can no longer continue. The amount of oxygen consumed at the highest workload achieved is his aerobic capacity.

13.1 Exercise, Fitness, and Health

LEARNING OBJECTIVES

- Describe the characteristics of a fit individual.
- Explain what is meant by the overload principle.
- Evaluate the impact of exercise on health.
- Discuss the role of exercise in weight management.

Exercise improves **fitness** and overall health. For some people, fitness means being able easily to walk around the block, mow the lawn, or play with their children. For more seasoned athletes, fitness means optimal performance of strenuous exercise. For everyone, fitness reduces the risk of chronic diseases such as cardiovascular disease, diabetes, and obesity. You don't need to run marathons or compete in the Olympics to be physically fit. Even a small amount of exercise is better than none, and, within reason, more exercise is better than less. Whether you are 8, 18, 80, or older, fitness through regular exercise can improve your overall health.

How Exercise Improves Fitness

Fitness level is defined by endurance, strength, flexibility, and body composition. Engaging in regular exercise improves these parameters. When you exercise, changes occur in your body: You breathe harder, your heart beats faster, and your muscles stretch and strain. If you exercise regularly, your body adapts to the exercise you perform and as a result can continue for a few minutes longer, lift a heavier weight, or stretch a millimeter farther. This is known as the **overload principle**: the more you do, the more you are capable of doing. For example, if you run a given distance three times a week, in a few weeks you can run farther; if you lift weights a few days a week, in a few weeks you will have more muscle and will be able to lift more weight more easily. These adaptations improve overall fitness.

Cardiorespiratory Endurance How long you can jog or ride your bike depends on the ability of your cardiovascular and respiratory systems, referred to jointly as the cardiorespiratory system, to deliver oxygen and nutrients to your tissues and remove wastes. **Cardiorespiratory endurance** is increased by **aerobic exercise**, the type of exercise that increases heart rate and uses oxygen. Aerobic activities include walking, dancing, jogging, cross-country skiing, cycling, and swimming (**Figure 13.1a**).

Regular aerobic exercise increases cardiorespiratory endurance because it strengthens the heart muscle and increases **stroke volume**, which is the amount of blood pumped with each beat of the heart. This in turn decreases **resting heart rate**, the rate at which the heart must beat to supply blood to the tissues at rest. The more fit a person is, the lower their resting heart rate and the more blood their heart can pump to muscles during exercise. In addition to increasing the amount of oxygen-rich blood that is pumped to muscles, regular aerobic exercise increases the muscle's ability to use oxygen to produce ATP. The body's maximum ability to generate ATP by aerobic metabolism during exercise is called **aerobic capacity** or **VO_2 max**.

Aerobic capacity is dependent on the ability of the cardiorespiratory system to deliver oxygen to the cells and the ability of the cells to use oxygen to produce ATP. The greater a person's aerobic capacity, the more intense activity he or she can perform before a lack of oxygen affects performance. A trained athlete will have a greater aerobic capacity than an untrained individual. Aerobic capacity can be determined in an exercise laboratory by measuring oxygen consumption or uptake during maximal exercise on a stationary bicycle or treadmill (**Figure 13.1b**).

Muscle Strength and Endurance Greater muscle strength enhances the ability to perform tasks such as pushing or lifting. In daily life, this could mean lifting a gallon of milk off the top shelf of the refrigerator with one hand, carrying a full trash can out to the curb, or moving a couch into your new apartment. Greater muscle endurance enhances your ability to continue repetitive muscle activity, such as shoveling snow or raking leaves. Muscle strength and endurance are increased by repeatedly using muscles in activities that require moving against a resisting force. This type of exercise is called **muscle-strengthening exercise**, strength-training, or resistance exercise, and includes activities such as weight lifting and calisthenics. Lifting a heavy weight stresses muscles. This stress or overload causes muscles to adapt by increasing in size and strength—a process called **hypertrophy**. The larger, stronger muscles can now lift the same weight more easily (**Figure 13.2**).

When muscles are not used due to a lapse in training, an injury, or illness, they become smaller and weaker. This process is called **atrophy**. For example, when individuals are bedridden and unable to move about, their muscles atrophy. Once they are up and active again, their muscles regain strength and size. There is truth in the expression "use it or lose it."

Flexibility Fitness is not just about bulging muscles; it also involves flexibility. Flexibility determines range of motion—how far you can bend and stretch muscles and ligaments. Regularly moving the limbs, neck, and torso through their full ranges of motion helps increase and maintain flexibility. If flexibility is poor, people cannot easily bend to tie their shoes or stretch to remove packages from the car. Being flexible can enhance postural stability and balance.[1] Stretching has not been shown to reduce injury, but an exercise warm-up that includes both cardiorespiratory activities and stretching has benefits for certain types of sports such as dancing. As seen in **Figure 13.3**, stretching exercises can be dynamic, meaning they involve motion, or static, meaning they involve no motion. What combination is best depends on the person and their sport.[2]

How Exercise Affects Body Composition Exercise builds and maintains muscle. Individuals who are physically fit have a greater proportion of lean body tissue than unfit individuals of the same body weight (**Figure 13.4**). Not everyone who is fit is thin, but in a fit person who carries extra pounds, more of the weight is from muscle. How much body fat a person has is also affected by gender and age. In general, women have more stored body fat than men. For young adult women, the desirable percentage of body fat is 21 to 32% of total weight; in adult men, the desirable percentage is about 8 to 19%.[3] With aging,

aerobic exercise Endurance exercise such as jogging, swimming, or cycling that increases heart rate and requires oxygen in metabolism.

stroke volume The volume of blood pumped by each beat of the heart.

resting heart rate The number of times that the heart beats per minute while a person is at rest.

aerobic capacity or **VO_2 max** The maximum amount of oxygen that can be consumed by the tissues during exercise. This is also called maximal oxygen consumption.

Wang Leng/Asia Images/Getty Images, Inc.

Figure 13.2 Increasing muscle size and strength
Building muscle requires an increase in the amount of resistance exercise. Progressively increasing the amount of weight at each exercise session slowly causes the muscle to hypertrophy.

Figure 13.3 Types of stretching

(a) To most of us, stretching means static stretching. In a static stretch you get into a position that stretches a muscle or group of muscles to its farthest point and then hold that position.

RW Photographic/Masterfile

© Kemter/iStockphoto

(b) Most athletes incorporate dynamic stretching into their warm-up. This uses controlled leg and arm swings to extend muscles gently to the limits of their range of motion. Controlled torso twists or side lunges, shown here, are examples of dynamic stretching.

muscle-strengthening exercise Activities that are specifically designed to increase muscle strength, endurance, and size; also called strength-training or resistance exercise.

hypertrophy An increase in the size of a muscle or organ.

atrophy Wasting or decrease in the size of a muscle or other tissue caused by lack of use.

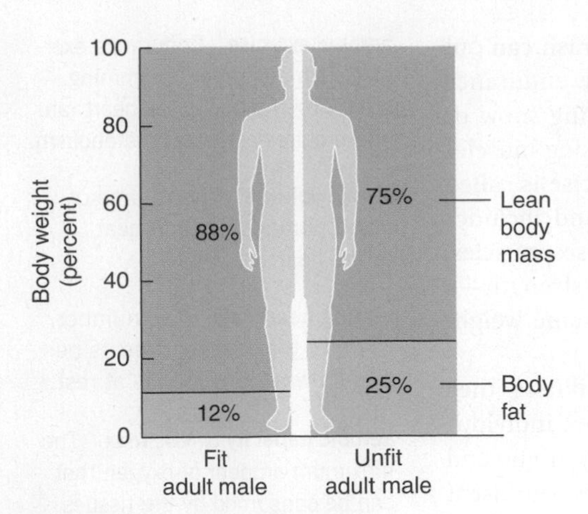

Figure 13.4 Body composition
Body composition, which refers to the percentage of fat versus non-fat or lean tissue (muscle, bones, cartilage, skin, nerves, and internal organs), is an indicator of health and fitness. Engaging in aerobic exercise and muscle-strengthening exercise has a positive impact on body composition, reducing body fat and increasing the proportion of lean tissue.

lean body mass decreases in both men and women, and the percentage of body fat increases even if body weight remains the same. Some of this change may be prevented by staying physically active (see Chapter 16).

How Exercise Benefits Health

In addition to making the tasks of everyday life easier, maintaining fitness through regular activity offers many health benefits. Regular exercise helps maintain muscles, bones, and joints; and reduces the risk of osteoporosis. It can help to prevent or delay the onset of cardiovascular disease, hypertension, stroke, diabetes, and colon and breast cancer (**Figure 13.5**).[1,4] A regular exercise program makes it easier to maintain a healthy body weight and it can also prevent depression and improve mood, sleep patterns, and overall outlook on life.

Cardiovascular Health Maintaining fitness through regular exercise reduces the risk of cardiovascular disease.[1] Because aerobic exercise strengthens the heart muscle, it reduces the number of times the heart must beat to deliver blood to the tissues at rest and during exercise. The changes that occur with exercise also help to lower blood pressure and increase HDL cholesterol levels in the blood.[5] All these effects help to reduce the risk of cardiovascular events such as heart attack and stroke.

VIDEO

Diabetes Prevention and Management People with excess body fat are more likely to develop diabetes. By keeping body fat within the normal range, aerobic exercise can decrease the risk of developing type 2 diabetes.[6] Physical activity that includes both aerobic exercise and strength training is also important in the treatment of diabetes because exercise increases the sensitivity of tissues to insulin. Maintaining an exercise program can reduce or eliminate the need for medication to maintain normal blood glucose levels.[6] Therefore, to prevent low blood glucose, people with diabetes should develop exercise programs with the help of their physicians and dietitians.

Bone and Joint Health Just as lifting weights helps maintain muscle size and strength, weight-bearing exercise stimulates bones to become denser and stronger. One of the causes of bone loss, like muscle loss, is lack of use; therefore, weight-bearing exercise such

Courtesy Lori Smolin

Figure 13.5 Health benefits of physical activity
Maintaining fitness through regular exercise helps you live longer by promoting a healthy weight and reducing the risk of chronic disease, and it helps you feel better by improving sleep patterns and self-image and relieving stress, anxiety, and depression.

as walking, running, and aerobic dance can increase peak bone mass and prevent bone loss, and therefore reduce the risk of osteoporosis (see Chapter 11). An exercise program can also benefit individuals with arthritis because the strength and flexibility promoted by exercise help manage pain and allow arthritic joints to move more easily.[7]

Cancer Risk Individuals who exercise regularly may be reducing their cancer risk by as much as 40%.[8] There is evidence that exercise reduces breast cancer risk; the risk reduction is related to exercise intensity, duration, and the age at which the exercise is performed.[9] The evidence that exercise reduces colon cancer risk is also strong: active individuals are less likely to develop colon cancer than their sedentary counterparts.[8] When evaluating the impact of exercise on cancer risk, diet and other lifestyle factors also must be considered. It is possible that some of the effect is due to the fact that people who exercise regularly are more likely to have healthier overall diets and lifestyles.

Weight Management People who exercise regularly are more likely to maintain a healthy body weight. Exercise makes weight management easier because it increases energy needs so more calories can be consumed without weight gain. During exercise, energy expenditure can rise well above the resting rate and some of this increase persists for many hours after activity slows.[10] Exercise can also boost energy expenditure through its effect on body composition. Exercise increases lean tissue mass and because, even at rest, lean tissue uses more energy than fat tissue, this increases basal energy needs. The combination of increased energy output during exercise, the rise in expenditure that persists after exercise, and the increase in basal needs can have a major impact on total energy expenditure (**Figure 13.6**). Besides increasing energy needs, exercise also promotes the loss of body fat and slows the loss of lean tissue that occurs with energy restriction. This makes exercise an essential component of any weight-reduction program.

Overall Well-Being Physical activity improves mood, boosts self-esteem, and increases overall well-being.[11] It has been shown to reduce depression and anxiety, and improve the quality of life.[12] The exact mechanisms involved are not clear, but one hypothesis has to do with the production of **endorphins**. Exercise stimulates the release of these chemicals, which are thought to be natural mood enhancers that play a role in triggering what athletes describe as an "exercise high." In addition to causing this state of exercise euphoria, endorphins are thought to aid in relaxation, pain tolerance, and appetite control. Regular exercise is also associated with a lower risk of cognitive decline and dementia.[1] It may benefit mental well-being by affecting the levels of certain mood-enhancing neurotransmitters in the brain, releasing muscle tension, improving sleep patterns, and reducing levels of the stress hormone cortisol. Exercise also raises body temperature, which is believed to have a calming effect. These changes in both the brain and the body can reduce anxiety, irritability, stress, fatigue, anger, self-doubt, and hopelessness.[12]

endorphins Compounds that cause a natural euphoria and reduce the perception of pain under certain stressful conditions.

Why does regular exercise cause an increase in basal energy expenditure?

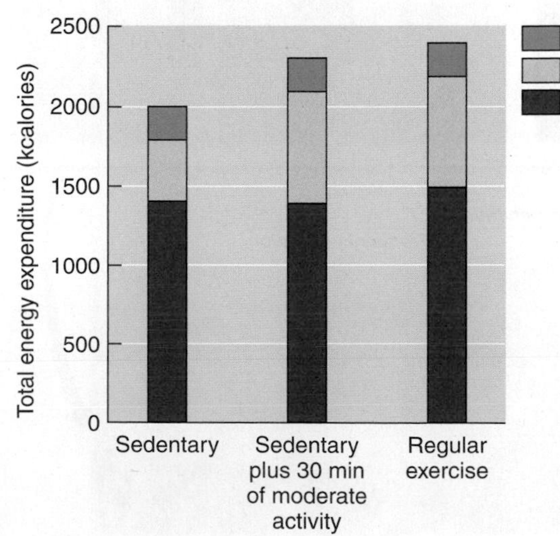

Figure 13.6 Exercise and total energy expenditure
Total energy expenditure is the sum of the energy used for basal energy expenditure (BEE), physical activity, and the thermic effect of food (TEF). Adding 30 minutes of moderate activity can increase total energy expenditure by as much as 300 kcalories. Regular exercise increases muscle mass, which increases basal energy expenditure, further increasing total energy expenditure.

13.2 Exercise Recommendations

LEARNING OBJECTIVES

- Describe the amounts and types of exercise recommended to improve health.
- Classify activities as aerobic or anaerobic.
- Plan a fitness program that can be integrated into your daily routine.
- Explain overtraining syndrome.

Most Americans do not exercise regularly, and 32% of American adults get no physical activity at all during their leisure time.[13] To reduce the risk of chronic disease, public health guidelines, including the 2010 Dietary Guidelines, advise at least 150 minutes of moderate-intensity or 75 minutes of vigorous-intensity aerobic physical activity each week (or an equivalent combination of both).[1,4,14] Greater health benefits can be obtained by exercising more vigorously or for a longer duration. As a rule of thumb, an activity with an intensity rating of 5 to 6 on a scale of 1 to 10, with 1 being lying on the couch and 10 being maximal activity, is considered moderate. For most people, moderate-intensity exercise is the equivalent of walking 3 miles in about an hour or bicycling 8 miles in about an hour. Vigorous-intensity exercise is equivalent to jogging at a rate of 5 miles per hour or faster or bicycling at 10 miles per hour or faster.[4] Adults should also include muscle-strengthening activities on

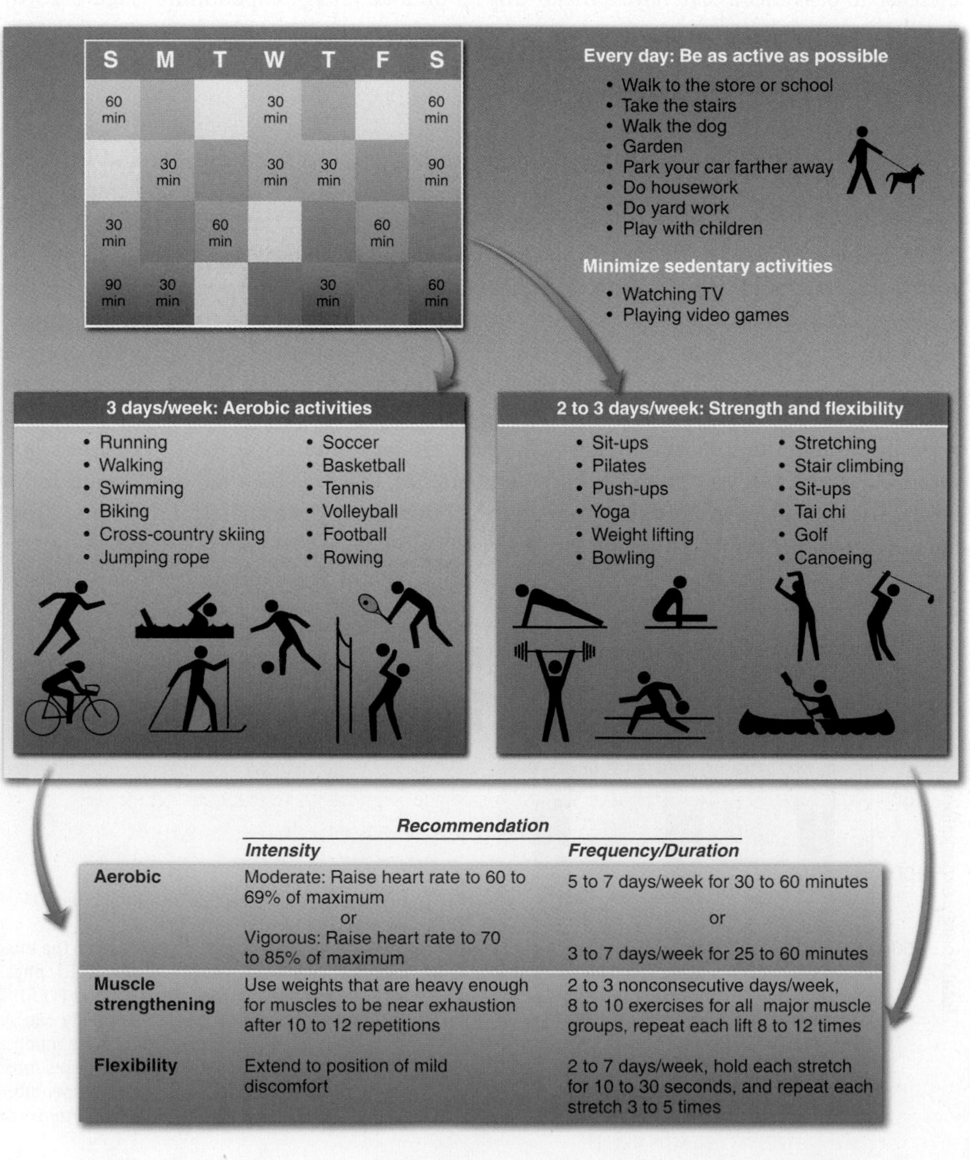

	Recommendation	
	Intensity	*Frequency/Duration*
Aerobic	Moderate: Raise heart rate to 60 to 69% of maximum or Vigorous: Raise heart rate to 70 to 85% of maximum	5 to 7 days/week for 30 to 60 minutes or 3 to 7 days/week for 25 to 60 minutes
Muscle strengthening	Use weights that are heavy enough for muscles to be near exhaustion after 10 to 12 repetitions	2 to 3 nonconsecutive days/week, 8 to 10 exercises for all major muscle groups, repeat each lift 8 to 12 times
Flexibility	Extend to position of mild discomfort	2 to 7 days/week, hold each stretch for 10 to 30 seconds, and repeat each stretch 3 to 5 times

Figure 13.7 Exercise recommendations A healthy lifestyle minimizes sedentary activities and includes a variety of everyday activities as well as aerobic exercise and exercises that improve muscle strength, flexibility, and balance. Achieving the recommended amounts of these activities will help improve and maintain fitness and health. In general, more exercise is better than less.

Figure 13.8 The aerobic zone
(a) You can calculate your aerobic zone by multiplying your maximum heart rate by 0.6 and 0.85. Maximum heart rate depends on age; it can be estimated for men and women by using the following equations:

Men: Maximum heart rate = 220 − age
Women: Maximum heart rate = 206 − (0.88 × age)

For example, a 20-year-old male would have a maximum heart rate of 200 (220 − 20) beats per minute. If he exercises at a pace that keeps his heart rate between 120 (0.6 × 200) and 170 (0.85 × 200) beats per minute, he is in his aerobic zone.

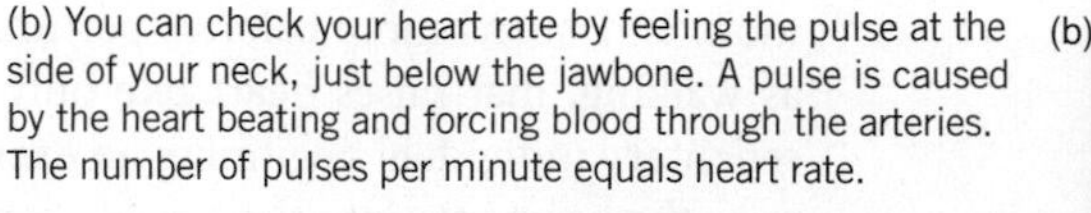

(b) You can check your heart rate by feeling the pulse at the side of your neck, just below the jawbone. A pulse is caused by the heart beating and forcing blood through the arteries. The number of pulses per minute equals heart rate.

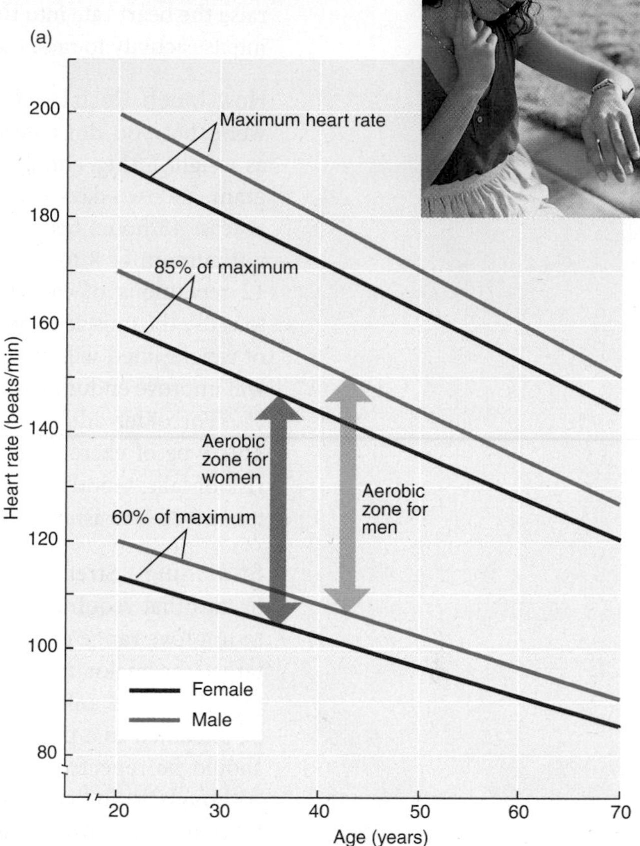

THINK CRITICALLY

What is the aerobic zone for a 30-year-old woman? What happens to this range when she turns 40? Why do you think this change occurs?

two or more days per week, but time spent in muscle-strengthening activities does not count toward meeting the aerobic activity guidelines.

Exercise cannot compensate for extended periods of time spent in sedentary pastimes. Therefore, even people who meet exercise guidelines may be at increased risk of cardiovascular disease, depression, increased waist circumference, and other adverse effects if they spend long periods sitting in a car, at a desk, or in front of the television.[1] Reducing total time spent in sedentary pursuits and breaking up periods of sedentary activity with short bouts of physical activity and standing, which can attenuate the adverse effects of sedentary behavior, should be a goal for all adults, irrespective of their exercise habits.

What to Look for in a Fitness Program

A well-planned fitness regimen includes aerobic exercise, which raises heart rate and therefore improves cardiorespiratory fitness; resistance exercises, which increase muscle strength and endurance and maintain or increase muscle mass; stretching, which promotes and maintains flexibility; and exercises that focus on balance, agility, and coordination, referred to as neuromotor exercise.[1,4,15] Exercise should be integrated into an active lifestyle that includes a variety of everyday activities, enjoyable recreational activities, and a minimum amount of time spent in sedentary activities (**Figure 13.7**).

How Much Aerobic Activity? Aerobic exercise, such as walking, bicycling, skating, swimming, or jogging, should be performed for at least 30 to 60 minutes per day, depending on intensity, most days of the week. An activity is in the aerobic zone if it raises heart rate to 60 to 85% of its maximum (**Figure 13.8**). **Maximum heart rate** is the maximum number of beats per minute that the heart can attain. It is dependent on age and can be estimated in men by subtracting age from 220 and in women by subtracting 88% of age from 206.[16] Each exercise session should begin with a warm-up, such as mild stretching and easy jogging, to increase blood flow to the muscles, and end with a cool-down period, such as walking or stretching, to help prevent muscle cramps and slowly reduce heart rate.

maximum heart rate The maximum number of beats per minute that the heart can attain. It declines with age.

Aerobic activities of different intensities can be combined to meet recommendations and achieve health benefits. The total amount of energy expended in physical activity depends on the intensity, duration, and frequency of the activity. Vigorous physical activity, such as jogging, that raises heart rate to the high end of the aerobic zone (70–85%) improves fitness more and burns more calories per unit of time than does moderate-intensity activity, such as walking, that raises heart rate only to the low end of the aerobic zone (60–69%). For a sedentary individual beginning an exercise program, even mild exercise such as walking can raise the heart rate into the aerobic zone. As fitness improves, exercisers must perform more intense activity to raise their heart rates to this level.

How Much Resistance Exercise? Developing and maintaining strong muscles requires work, but you don't need to lift weights every day. Muscle-strengthening exercises, such as weight lifting, should be done two to three days a week at the start of an exercise program, and two days a week after the desired strength has been achieved. Adults should wait at least 48 hours between each muscle-strengthening session. Each session should include a minimum of 8 to 10 exercises that train the major muscle groups. One to 3 sets of 8 to 12 repetitions of each exercise is recommended.[4] The weights should be heavy enough to cause the muscle to be near exhaustion after the 8 to 12 repetitions. Increasing the amount of weight lifted will increase muscle strength, whereas increasing the number of repetitions will improve endurance.

For older adults, neuromotor exercise should be included in an exercise program.[1] This type of exercise helps improve balance, agility, and muscle strength and reduce the risk of falls. Good options include tai chi and yoga, which use combinations of neuromotor exercise, resistance exercise, and flexibility exercise.

Stretching Stretching may improve performance in activities such as dancing and gymnastics that require flexibility.[4] A combination of static and active stretching can be used to improve range of motion. To improve and maintain flexibility, stretching exercises that target the major muscles and tendons of the shoulders, chest, neck, trunk, lower back, hips, legs, and ankles should be done at least two to three days a week.[1] Muscles should be stretched to a position of mild discomfort and held for 10 to 30 seconds. Each stretch should be repeated three to five times to achieve a total stretching time of 60 seconds. Although flexibility is an important component of fitness, time spent stretching should not be counted towards the goal of at least 150 minutes of aerobic activity per week.[4]

Exercise Recommendations for Children

Children and adolescents should spend at least 60 minutes per day in developmentally appropriate physical activity.[4,14] Aerobic, muscle-strengthening, and bone-strengthening activities should be included in this 60 minutes. Most of the 60 minutes will be from moderate- and vigorous-intensity aerobic exercise, with vigorous-intensity exercise at least three days per week. Muscle-strengthening and bone-strengthening activities should also be included at least three days per week. Muscle-strengthening exercise can be unstructured play activities such as climbing on playground equipment or playing tug of war. Bone-strengthening activities include weight-bearing play such as hopscotch, jumping rope, basketball, and running. Activity for young children should be intermittent, with periods of moderate to vigorous activity lasting 10 to 15 minutes or more along with periods of rest and recovery. To promote the recommended amount of exercise, a variety of enjoyable activities should be stressed and competition deemphasized.

Modern lifestyles do not promote activity in children and adolescents; television, computers, video games, and cell phone activities are often chosen over physical activity. Reducing the amount of time spent in these sedentary activities can increase fitness, lower BMI, and improve blood pressure and cholesterol levels.[17] Children who learn to enjoy physical activity are more likely to be active adults who maintain a healthy body weight and have a lower risk of cardiovascular disease, diabetes, osteoporosis, and certain types of cancer. Learning by example is always best. Children who have physically active parents are the leanest and the fittest (**Figure 13.9**).

Figure 13.9 Exercise for children Creating active environments for children at home, in schools, and in communities can help children participate in and enjoy exercise. Community parks, bike paths, and hiking trails promote active adults, children, and families.

How to Create an Active Lifestyle

Almost everyone can participate in some form of exercise, no matter where they live, how old they are, or what physical limitations they have. Exercise classes are taught in nursing homes. People with heart disease, visual impairments, and physical disabilities compete in athletic events. You are never too old to exercise, and it is never too late to start.

TABLE 13.1 Suggestions for Starting and Maintaining an Exercise Program

Start slowly. Set specific attainable goals. Once you have met them, add more.
• Walk around the block after dinner.
• Get off the bus or subway one stop early.
• Use half of your lunch break to exercise.
• Do a few biceps curls each time you take the milk out of the refrigerator.
Make your exercise fun and convenient.
• Opt for activities you enjoy—bowling and dancing are more fun than a treadmill in the basement.
• Find a partner to exercise with you.
• Choose times that fit your schedule.
Stay motivated.
• Vary your routine—swim one day and mountain bike the next.
• Challenge your strength or endurance once or twice a week and do moderate workouts on other days.
• Track your progress by recording your activity.
• Reward your success with a new book, movie, or some workout clothes.
Keep your exercise safe.
• Warm-up before you start and cool-down when you are done.
• Wear light-colored or reflective clothing that is appropriate for the environmental conditions.
• Don't overdo it—alternate hard days with easy days and take a day off when you need it.
• Listen to your body so you stop before an injury occurs.

Find Convenient, Enjoyable Activities Incorporating exercise into day-to-day life may require a behavior change, and changing behavior is not easy. The first step in beginning an exercise program is to recognize the reasons for not exercising and identify ways to overcome them. Many people avoid exercise because they do not enjoy it, feel they have to join an expensive health club, have little motivation to do it alone, or find it inconvenient and uncomfortable. Finding a type of exercise that is enjoyable, a time that is realistic and convenient, and a place that is appropriate and safe are important first steps in adopting an exercise program. Riding your bike to class or work rather than driving or taking the bus, taking a walk during your lunch break, and enjoying a game of catch or tag with your children are all effective ways to increase your everyday activity level. The goal is to make gradual lifestyle changes that increase physical activity. Behavioral strategies such as those listed in **Table 13.1** may help promote regular exercise.

Keep Exercise Safe Safety should be a consideration in planning any fitness regimen. Before beginning, everyone should check with their physician to be sure that their plans are safe based on their medical history. Then the location and environment for exercise can be considered. Busy work schedules often force people to exercise in the dark, early morning or evening hours. Exercisers who use the street for walking or jogging should wear light-colored, reflective clothing so they can be seen by motorists. Exercising with a partner is safer and more enjoyable.

Weather conditions can also be a safety concern. Physical activity produces heat, which normally is dissipated to the environment, partly by the evaporation of sweat. When the environmental temperature is high, heat is not efficiently transferred to the environment, and when humidity is high, sweat evaporates slowly, making it difficult to cool the body. Thus, exercise should be reduced or curtailed in hot and humid conditions. Cold environments can also pose problems for the outdoor exerciser. In general, cold does not impair exercise capacity, but the numbing of exposed flesh and the bulk of extra clothing can cause problems for joggers and bicyclists. Clothing must allow the body to dissipate heat while providing protection from the cold. For swimmers, cold water can cause performance to deteriorate.

Tailor Exercise Frequency, Intensity, and Duration Individuals should structure their fitness program based on their needs, goals, and abilities. For example, some people might prefer a short, intense workout such as 30 minutes of running, while others would rather work out for a longer time, at a lower intensity, such as a one-hour walk. Some may choose to complete all their exercise during the same session, while others may spread their exercise throughout the day, in shorter bouts. Three short bouts of 10-minute duration can be as effective as a continuous bout of 30 minutes in reducing the risk of chronic disease.[4,15] A combination of intensities, such as a brisk 30-minute walk twice during the week in addition to a 20-minute jog on two other days, can meet recommendations. Also, what is best for a middle-aged man trying to reduce his risk of chronic disease is different from what is best for a 19-year-old college basketball player, and different still from what is best for an octogenarian trying to continue living independently. Young healthy athletes may require very intense activity to obtain a training effect. Older adults and those who have not previously been active can increase their fitness by exercising at a lower intensity if the duration and frequency of exercise are increased.

Don't Overdo It To improve cardiorespiratory fitness and muscle strength, the body must be stressed and respond to the stress by increasing aerobic capacity and muscle size and strength. Initially, training can cause fatigue and weakness, but during rest the body rebuilds to become stronger. If not enough rest occurs between exercise sessions, there is no time to regenerate, so fitness and performance do not improve. In athletes, excessive training can lead to **overtraining syndrome**, which involves emotional, behavioral, and physical symptoms that persist for weeks to months. It is caused by repeated training without sufficient rest to allow for recovery. The most common symptom of overtraining syndrome is fatigue that limits workouts and is felt even at rest. Some athletes experience a decrease in appetite and weight loss as well as muscle soreness, increased frequency of viral illnesses, and increased incidence of injuries. They may become moody, easily irritated, depressed, have altered sleep patterns, or lose their competitive desire and enthusiasm. Overtraining syndrome occurs only in serious athletes who are training extensively, but rest is essential for anyone working to increase fitness.

overtraining syndrome A collection of emotional, behavioral, and physical symptoms that occurs when training without sufficient rest persists for weeks to months.

13.3 Exercise and Energy Metabolism

LEARNING OBJECTIVES

- Compare the fuels used to generate ATP by anaerobic and aerobic metabolism.
- Discuss the effect of exercise duration and intensity on the type of fuel used.
- Describe the physiological changes that occur in response to exercise.

Whether your goal is maintaining health or competing in athletic events, nutrition provides a launching pad from which physical fitness can be improved. Just as an automobile engine runs on energy from gasoline, the body machine runs on energy from the carbohydrate, fat, and protein in food and body stores. These fuels are needed whether you are writing a letter, walking around the block, or running a marathon. But before nutrients can be used to fuel activity, their energy must be converted into the high-energy compound ATP. ATP provides an immediate source of energy for all body functions, including muscle contraction. ATP can be generated both in the presence of oxygen by **aerobic metabolism** and in the absence of oxygen by **anaerobic metabolism** or **anaerobic glycolysis**. The way ATP is produced during activity depends on how long an activity is performed, the intensity of the activity, and the physical conditioning of the exerciser. This in turn affects how much carbohydrate, fat, and protein are used to produce this ATP.

aerobic metabolism Metabolism in the presence of oxygen. In aerobic metabolism, glucose, fatty acids, and amino acids are completely broken down to form carbon dioxide and water and produce ATP.

anaerobic metabolism or **anaerobic glycolysis** Metabolism in the absence of oxygen. Each molecule of glucose generates two molecules of ATP. Glucose is metabolized in this way when oxygen cannot be delivered to the tissues quickly enough to support aerobic metabolism.

How Exercise Duration Affects Metabolism

Resting muscles do not need much energy. At rest, the heart and lungs are able to deliver enough oxygen to meet energy demands using aerobic metabolism. During exercise, to increase the amount of energy provided by aerobic metabolism, the amount of oxygen

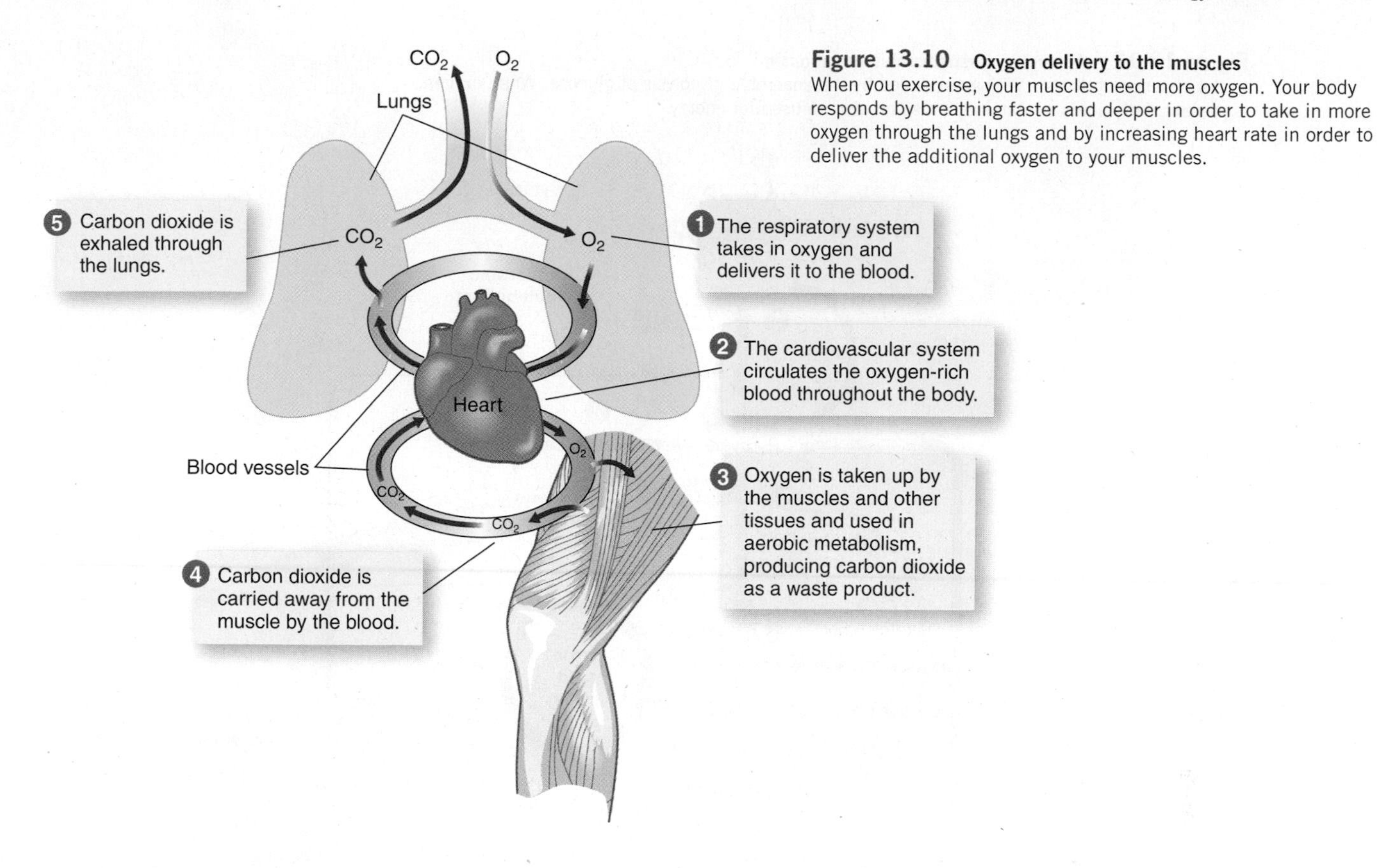

Figure 13.10 Oxygen delivery to the muscles
When you exercise, your muscles need more oxygen. Your body responds by breathing faster and deeper in order to take in more oxygen through the lungs and by increasing heart rate in order to deliver the additional oxygen to your muscles.

available at the muscle must be increased (**Figure 13.10**). To do this, breathing and heart rate are increased, but this takes time. When exercise first begins, breathing and heart rate have not yet had enough time to increase the amount of oxygen available at the muscle.

Instant Energy: Stored ATP and Creatine Phosphate When you jump up to answer the phone or take the first steps of your morning jog, your muscles increase their activity but your heart and lungs have not had time to step up oxygen delivery to them. To get the needed energy, the muscles rely on small amounts of ATP that are stored in resting muscle. It is enough to sustain activity for a few seconds. As the ATP in muscle is used, enzymes break down another high-energy compound, called **creatine phosphate** or **phosphocreatine**, to replenish the ATP supply and allow activity to continue. But, like ATP, the amount of creatine phosphate stored in the muscle at any time is small. It will fuel muscle activity for about an additional 8 to 10 seconds before it, too, is used up. So, during the first 10 to 15 seconds of exercise, the muscles rely on energy from the ATP and creatine phosphate that is stored in them (**Figure 13.11**).

creatine phosphate or **phosphocreatine** A compound found in muscle that can be broken down quickly to make ATP.

Short-Term Energy: Anaerobic Metabolism As exercise continues beyond 10 to 15 seconds, the ATP and creatine phosphate in the muscles are used up but the heart and lungs have still not had time to increase oxygen delivery to the muscles. Therefore, the additional ATP needed to fuel muscle contraction must be produced without oxygen.

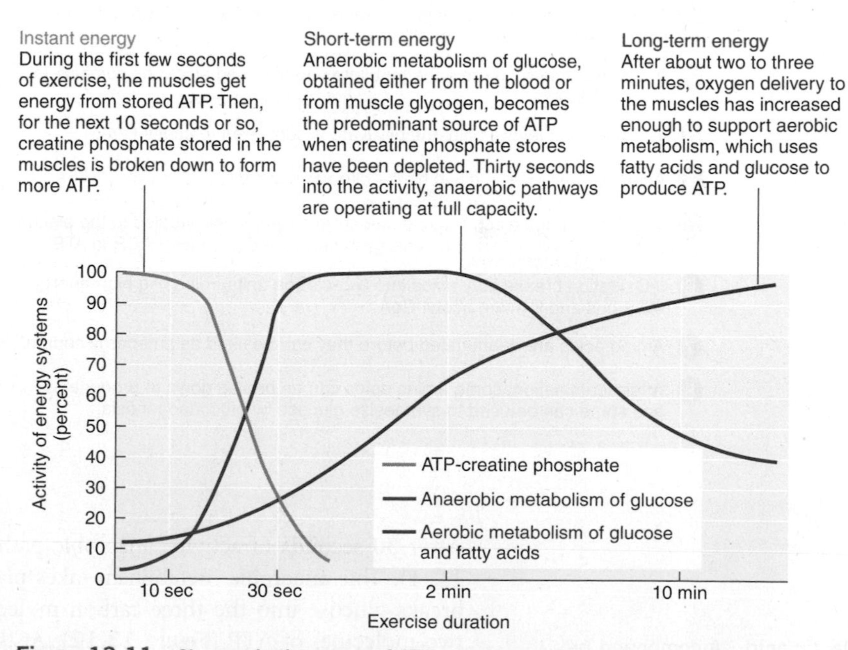

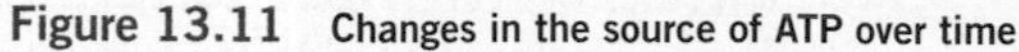
Figure 13.11 Changes in the source of ATP over time
The source of the ATP that fuels muscle contraction changes over the first few minutes of exercise.

Figure 13.12 Anaerobic versus aerobic metabolism
In the absence of oxygen, ATP is produced by the anaerobic glycolysis of glucose. When oxygen is present, fatty acids and amino acids can also be used for energy.

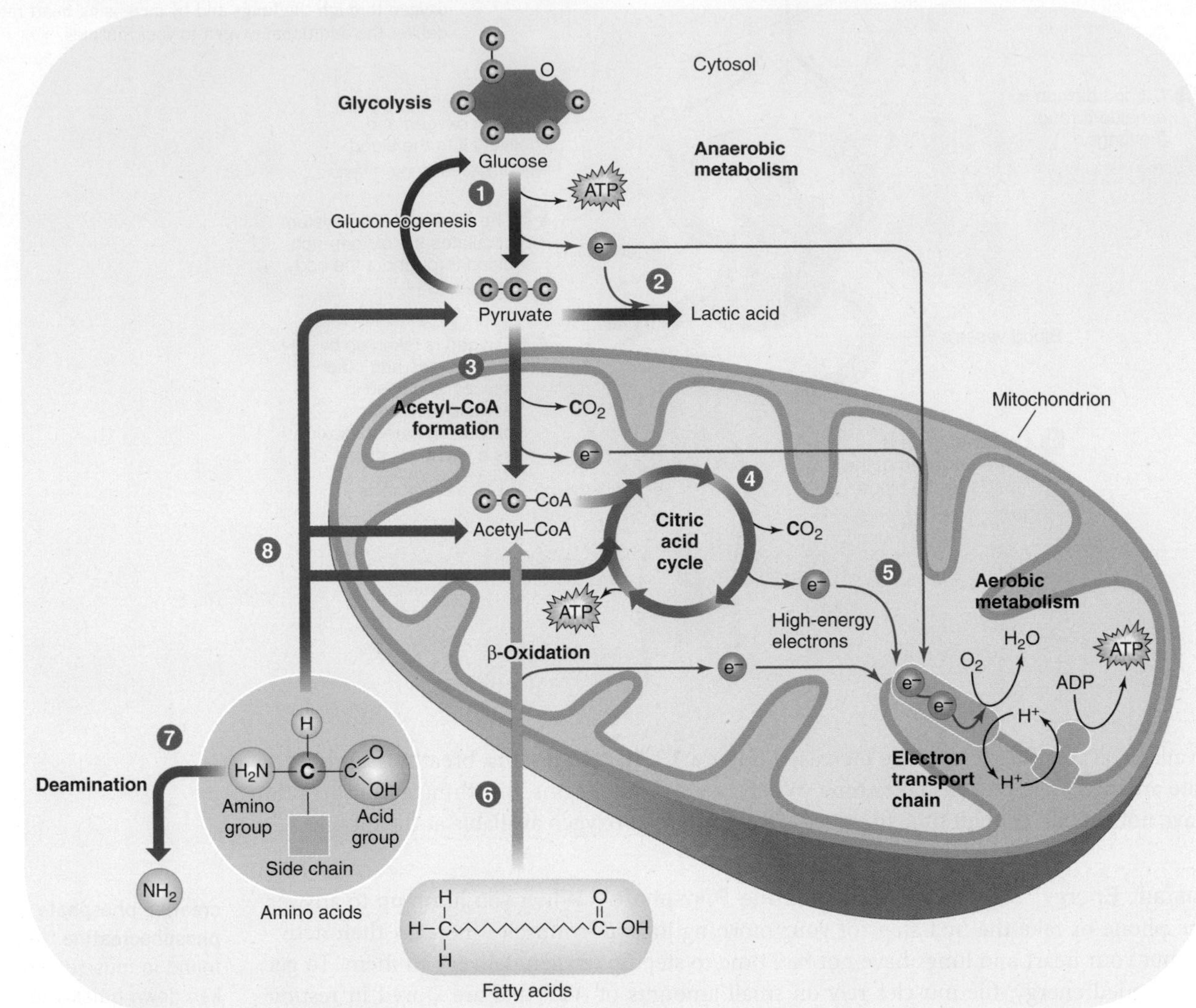

1. Glucose is split into two molecules of pyruvate, releasing electrons, and producing two molecules of ATP.
2. In the absence of oxygen, the pyruvate and released electrons combine to form lactic acid.
3. When oxygen is available, the pyruvate is converted to acetyl-CoA.
4. Acetyl-CoA is broken down by the citric acid cycle.
5. The high-energy electrons released from all steps are shuttled to the electron transport chain, where their energy is harnessed to convert ADP to ATP.
6. β-Oxidation breaks fatty acids into two-carbon units, releasing high-energy electrons and forming acetyl-CoA.
7. Amino acids are deaminated before they can be used as an energy source.
8. After deamination, some amino acids can be broken down to produce ATP and some can be used to synthesize glucose by gluconeogenesis.

After 30 seconds of activity, anaerobic pathways are operating at full capacity (see Figure 13.11). This anaerobic metabolism takes place in the cytosol. It includes glycolysis, which breaks glucose into the three-carbon molecule pyruvate, releases electrons, and produces two molecules of ATP (**Figure 13.12**). At this point, if oxygen is unavailable, the pyruvate and released electrons combine to form **lactic acid**. The lactic acid is transported out of the muscle for use in other tissues.

lactic acid A compound produced from the breakdown of glucose in the absence of oxygen.

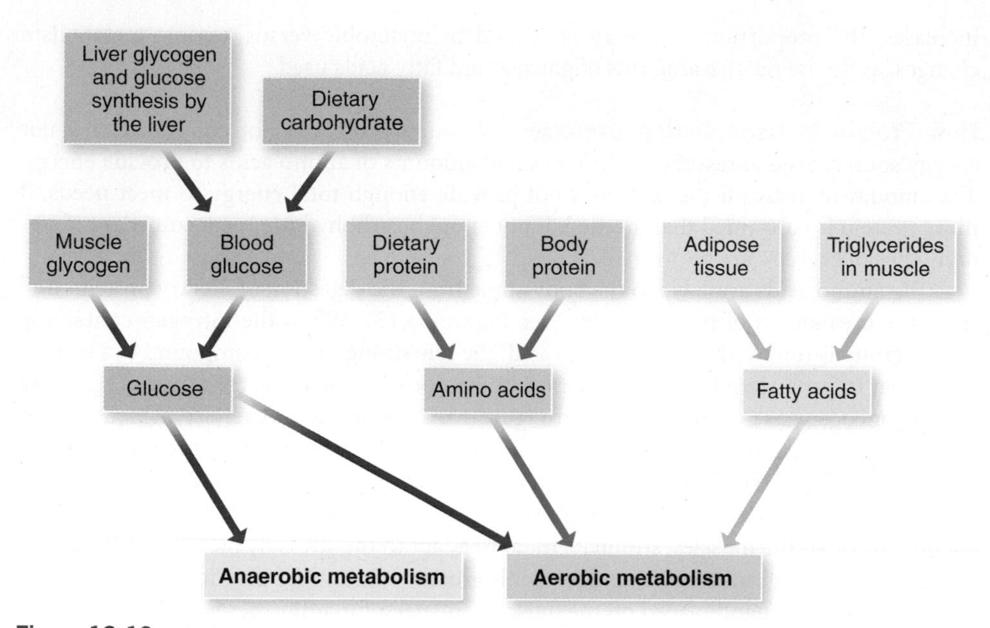

Figure 13.13 **Fuels for anaerobic and aerobic metabolism**
Glucose for anaerobic and aerobic metabolism can come from muscle glycogen or blood glucose. Aerobic metabolism uses fatty acids and amino acids as well as glucose to produce ATP.

Anaerobic metabolism can produce ATP very rapidly, but can only use glucose as a fuel. This glucose may come from the breakdown of glycogen inside the muscle or from glucose delivered via the bloodstream. The glucose delivered in the blood comes from the breakdown of liver glycogen, the synthesis of glucose by the liver, or the ingestion of carbohydrate during exercise (**Figure 13.13**). Anaerobic metabolism predominates during the first few minutes of exercise (see Figure 13.11) and is also important during periods of intense exercise because oxygen cannot be delivered to the cells quickly enough to meet energy demands. Anaerobic metabolism uses glucose rapidly. Because the amount of glucose available is limited, if activity is to continue, the body must use its glucose more efficiently and find a more plentiful fuel source.

Long-Term Energy: Aerobic Metabolism After two to three minutes of exercise, breathing and heart rate have increased to supply more oxygen to the muscles. When oxygen is available, ATP can be produced by aerobic metabolism. The reactions of aerobic metabolism take place in the mitochondria (see Figure 13.12). When glucose is broken down by aerobic metabolism, the pyruvate produced by glycolysis is converted to acetyl-CoA, so lactic acid is not formed. Acetyl-CoA is broken down by the citric acid cycle, producing carbon dioxide and some ATP and releasing high-energy electrons. The electrons are shuttled to the electron transport chain, where their energy is harnessed to convert ADP to ATP and water is formed.

Aerobic metabolism produces ATP more slowly than anaerobic metabolism but is much more efficient, producing about 18 times more ATP for each molecule of glucose. In addition, aerobic metabolism can use fatty acids, and sometimes amino acids from protein, to generate ATP (see Figure 13.12 and 13.13). Fatty acids to fuel muscle contraction come from triglycerides stored in adipose tissue as well as small amounts stored in the muscle itself. During exercise, triglycerides are broken down into fatty acids and glycerol. Fatty acids from adipose tissue are released into the blood and are then taken up by muscle cells. Inside muscle cells, fatty acids from triglycerides within the muscle and those delivered by the blood need to be transported into the mitochondria to produce ATP. To enter the mitochondria, fatty acids must be activated with the help of **carnitine**. Inside the mitochondria, fatty acids are broken into two-carbon units by β-oxidation to form acetyl-CoA (see Figure 13.12). Acetyl-CoA is metabolized via the citric acid cycle and electron transport chain to produce ATP, carbon dioxide, and water.

carnitine A molecule synthesized in the body that is needed to transport fatty acids and some amino acids into the mitochondria for metabolism.

When exercise continues at a low to moderate intensity, aerobic metabolism predominates and fat becomes the principal fuel source for exercising muscles. If exercise intensity increases, the proportion of energy generated by anaerobic versus aerobic metabolism changes, as do the relative amounts of glucose and fatty acids used.

How Protein Is Used During Exercise Although protein is not considered a major energy source, even at rest the body uses small amounts of amino acids to provide energy. The amount increases if the diet does not provide enough total energy to meet needs, if more protein is consumed than needed, if not enough carbohydrate is consumed, or if certain types of exercise are performed.

The amino acids available to the body come from the digestion of dietary proteins and from the breakdown of body proteins (see Figure 13.13). When the nitrogen-containing amino group is removed from an amino acid, the remaining carbon compound can be broken down to produce ATP by aerobic metabolism or in some cases used to make glucose via gluconeogenesis (see Chapter 4 and Figure 13.12). Endurance exercise, which continues for many hours, increases the use of amino acids both as an energy source and as a raw material for glucose synthesis. When exercise stops, amino acid use increases because protein synthesis is accelerated to build and repair muscle. Muscle-building activities, which slightly overload the muscles, stimulate them to adapt to the stress by breaking down existing muscle proteins and replacing them with greater amounts of new muscle proteins to meet the higher demand placed on the muscle. The need for amino acids for muscle building and repair is greater in strength athletes because they are actively overloading their muscles to stimulate the synthesis of new muscle tissue.

How Exercise Intensity Affects Metabolism

During exercise, ATP is produced by both anaerobic and aerobic metabolism. Together, the contributions made by each of these systems ensure that muscles get enough ATP to meet the demand placed on them. The relative contribution of anaerobic versus aerobic metabolism depends on how intense the activity is. With very intense activity, the ability to deliver and use oxygen at the muscle becomes limiting. When the amount of ATP that can be produced by aerobic metabolism cannot meet the demand, the proportion of ATP produced anaerobically from glucose increases. Generally, the more intense the exercise, the more the muscles must rely on glucose to provide energy (**Figure 13.14a**). When intensity reaches the aerobic capacity of the athlete, most energy is derived from anaerobic metabolism of glucose. When the exercise is lower in intensity, the cardiorespiratory system can deliver enough oxygen to the muscles to allow aerobic metabolism to predominate, so fatty acids as well as some glucose are used as fuel. Thus, exercise intensity determines the contributions that carbohydrate and fat make as fuels for ATP production as well as the total number of calories burned (**Figure 13.14b**). In turn, which fuels are used affects how long exercise can continue before **fatigue** occurs.

fatigue The inability to continue an activity at an optimal level.

Why High-Intensity Exercise Contributes to Fatigue If you run faster, you tire sooner. Fatigue occurs much more quickly with high-intensity exercise than with lower-intensity exercise because more intense exercise relies more on anaerobic metabolism, which can use only glucose for fuel. Glycogen stores thus are rapidly depleted. Anaerobic metabolism also produces lactic acid. With low-intensity exercise, the small amounts of lactic acid produced are carried away from the muscles and used by other tissues as an energy source or converted back into glucose by the liver. During high-intensity exercise, the amount of lactic acid produced exceeds what can be used by other tissues, and the lactic acid builds up in the muscle and subsequently in the blood. Until recently, it was assumed that lactic acid buildup was the cause of muscle fatigue, but we now know that although lactic acid buildup occurs with high-intensity exercise, it is not a major factor in muscle fatigue.[18] When exercise stops and oxygen is available again, lactic acid can be either carried away by the blood to other tissues to be broken down or metabolized aerobically in the muscle. Fatigue most likely has many causes, including glycogen depletion and changes in the muscle cells and the concentrations of molecules involved in muscle energy metabolism.

When athletes run out of glycogen, they experience a feeling of overwhelming fatigue that is sometimes referred to as "hitting the wall" or "bonking." Glycogen depletion is a concern for athletes because the amount of glycogen available to produce glucose during

Figure 13.14 The effect of exercise intensity on fuel use and energy expenditure

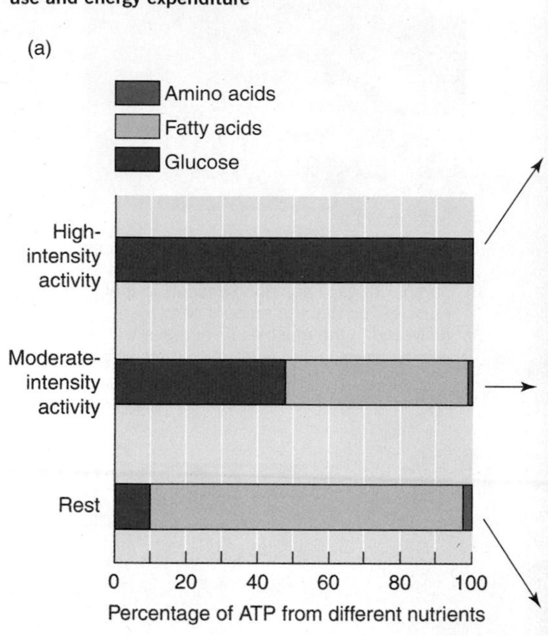

(b)
Kilocalories from carbohydrate
Kilocalories from fat
Kilocalories/hour
0
100
200
300
400
500
600
50% fat
40% fat
Fat-burning zone (lower intensity)
Cardio zone (higher intensity)

(a) Exercise intensity determines the contributions of carbohydrate, fat, and protein as fuels for ATP production. At rest and during low- to moderate-intensity exercise, aerobic metabolism predominates, so fatty acids are an important fuel source. As exercise intensity increases, the proportion of energy supplied by anaerobic metabolism increases, so glucose becomes the predominant fuel. Keep in mind, however, that the total amount of energy expended per unit of time is greater as the intensity of exercise increases.

(b) Aerobic exercise equipment typically includes workout programs that target your "fat-burning zone" and "cardio zone." Both are aerobic, but the fat-burning program is lower in intensity so a higher proportion of fat is used for fuel. The higher-intensity cardio workout uses proportionately less fat, but the actual amount of fat used is the same and the total number of calories burned is greater.

exercise is limited. There are between 60 and 120 g of glycogen stored in the liver; stores are highest just after a meal. Liver glycogen is used to maintain blood glucose between meals and during the night. Eating breakfast replenishes the liver glycogen used overnight. There are about 200 to 500 g of glycogen in the muscles of a 70-kg person. The glycogen in a muscle is used to fuel the activity of that muscle. Muscle glycogen levels can be increased by a combination of rest and a very high-carbohydrate diet.

THINK CRITICALLY

Which workout will help you lose the most weight: 30 minutes in the cardio zone or 30 minutes in the fat-burning zone? Why?

Why Low-Intensity Exercise Can Continue Longer Low-intensity exercise can be continued for longer periods than high-intensity exercise because it relies on aerobic metabolism, which is more efficient than anaerobic metabolism and uses both glucose and fatty acids for energy. The body's fat reserves are almost unlimited, so that if fat is the fuel, exercise can theoretically continue for a very long time. For example, it is estimated that a 130-lb woman has enough energy stored as body fat to run 1000 miles.[19] However, even aerobic activity uses some glucose, so if exercise continues long enough, glycogen stores will eventually be depleted and contribute to fatigue.

Why Training Increases Aerobic Capacity

Training with repeated bouts of aerobic exercise causes physiological changes that increase aerobic capacity—the amount of oxygen that can be delivered to and used by the muscle cells. This in turn affects the proportion of glucose versus fatty acids that can be used by the exercising muscle cells.

Regular aerobic exercise causes adaptations in the cardiorespiratory system. The heart becomes larger and stronger so that the stroke volume is increased. The number of capillary blood vessels in the muscles increases so that blood is delivered to muscles more efficiently. And the total blood volume and number of red blood cells expands, increasing the amount of hemoglobin so more oxygen can be transported to the muscle cells (**Figure 13.15**).

Figure 13.15 Effect of exercise training
The physiological effects of training allow trained athletes to sustain aerobic exercise for longer periods at higher intensities than untrained individuals.

Training also causes changes at the cellular level that affect the ability of cells to use different types of fuel to produce ATP. There is an increase in the ability to store glycogen, and there is an increase in the number and size of muscle-cell mitochondria (see Figure 13.15). Because aerobic metabolism occurs in the mitochondria, this increases the cell's capacity to burn fatty acids to produce ATP. The use of fatty acids spares glycogen, which delays the onset of fatigue. Because trained athletes store more glycogen and use it more slowly, they can sustain aerobic exercise for longer periods at higher intensities than can untrained individuals. Conditioned athletes can also exercise at a higher percentage of their aerobic capacity before lactic acid begins to accumulate.

Living and working at high altitudes, where the atmosphere contains less oxygen, also causes adaptations that improve the ability of the cardiorespiratory system to deliver oxygen. Therefore, endurance athletes often train at high altitudes to enhance their aerobic capacity.

13.4 Energy and Nutrient Needs for Physical Activity

LEARNING OBJECTIVES

- Compare the energy and protein needs of athletes and nonathletes.
- List micronutrients that are at risk for deficiency in athletes.
- Explain why athletes are at risk for iron deficiency.
- Describe the female athlete triad.

Good nutrition is essential to performance, whether you are a marathon runner or a mall walker. The diet must provide sufficient energy from the appropriate sources to fuel activity, protein to maintain muscle mass, micronutrients to allow utilization of the energy-yielding nutrients, and water to transport nutrients and cool the body. The major difference between the nutritional needs of a serious athlete and those of a casual exerciser is the amount of energy and water required.

Energy Needs

The amount of energy needed for activity depends on the intensity, duration, and frequency of the activity, as well as on the characteristics of the exerciser, and even his or her location. For a casual exerciser, the energy needed for activity may increase energy expenditure by only a few hundred kcalories a day. For an endurance athlete, such as a marathon runner, the energy needed for training may increase expenditure by 2000 to 3000 kcalories per day. Some athletes require 6000 kcalories a day to maintain body weight. In general, the more intense the activity, the more energy it requires, and the more time spent exercising, the more energy it burns (**Table 13.2** and Appendix F). For example, walking for 30 minutes involves less work than running for the same amount of time and therefore requires less energy. Riding a bicycle for 60 minutes requires six times the energy needed to ride for 10 minutes. The body weight of the exerciser is another factor in determining energy needs. Moving a heavier body requires more energy than moving a lighter one. Therefore, it requires less energy for a 120-lb woman to walk for 30 minutes than for a 250-lb woman.

The DRIs have developed equations to estimate energy requirements based on an individual's age, gender, size, and physical activity (PA) level (see Chapter 7 and inside cover). For the purposes of calculating estimated energy requirement (EER), an individual who performs no exercise other than the activities of daily living is in the "sedentary" PA category and one who performs less than an hour of moderate activity fits into the "low-active" PA category. Someone who engages in 60 minutes of moderate exercise each day is considered to be in the "active" PA category (see Chapter 7, Figure 7.14). An "active" activity level can be achieved with less than 60 minutes of exercise per day if the exercise is more intense, for example, jogging at 5 mph or greater or swimming at a moderate to fast pace. Individuals who perform moderate exercise for more than 2.5 hours per day or more intense exercise for more than 1.25 hours per day are in the "very active" PA category.

TABLE 13.2 Energy Needs for Various Activities

Body Weight		100 lbs	125 lbs	140 lbs	155 lbs	170 lbs	185 lbs	200 lbs
Activity	**Rate**	**Energy (kcal/hr)**						
Bicycling	< 10 mph	137	171	191	212	233	252	273
	10–11.9 mph	228	285	318	353	388	420	455
	12–13.9 mph	319	399	445	494	543	588	637
	14–15.9 mph	410	513	572	635	698	756	819
	16–19 mph	501	627	699	776	853	924	1001
Running	12 min/mile	319	399	445	494	543	588	637
	10 min/mile	410	513	572	635	698	756	819
	8 min/mile	523	713	730	811	891	966	1047
	6 min/mile	683	855	953	1058	1163	1260	1365
Skiing, cross-country	2.5 mph	273	342	381	423	465	504	546
	4–4.9 mph	319	399	445	494	543	588	637
	5–7.9 mph	364	456	508	564	620	672	728
Swimming	leisurely	228	285	318	353	388	420	455
	50 yds/min	319	399	445	494	543	588	637
	75 yds/min	455	570	635	705	775	840	910
Walking	2 mph	68	86	95	106	116	126	137
	3 mph	105	131	146	162	178	193	209
	4 mph	182	228	254	282	310	336	364

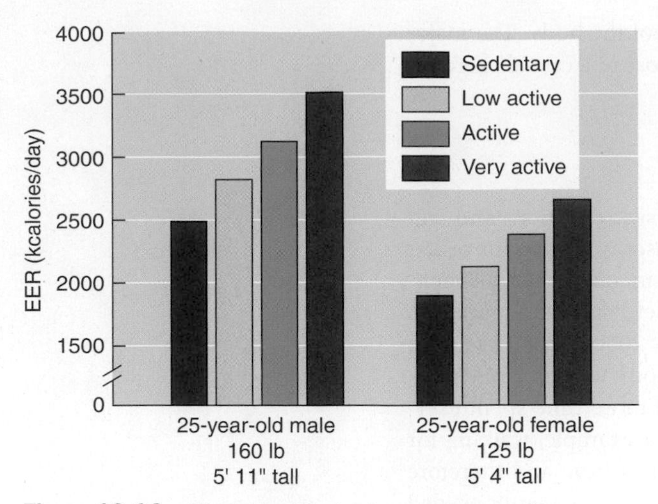

Figure 13.16 **Effect of activity level on energy expenditure** Increasing the amount of activity you routinely engage in can significantly increase the number of calories you need to consume each day to maintain body weight.

Calculating EER at different activity levels can demonstrate the dramatic impact activity can have on energy needs. For example, the EER for a 25-year-old, sedentary, 5′11″ tall, 160-lb man is about 2500 kcalories. If this same person becomes a runner and trains several hours per day, his energy needs may increase to 3500 kcalories per day or more (**Figure 13.16**).

There are also some special considerations that affect the energy needed for activity. For example, because of the buoyancy of adipose tissue, the energy required for an individual with excess body fat to swim may be less than that of his lean counterpart. If a lean individual and an obese individual were in the weightlessness of space, it would require no more energy for one to leap across the room than the other. There are also special circumstances that affect the amount of energy an individual needs for daily activity. A paraplegic in a wheelchair may have lower energy needs because many of the major muscles in the body are always inactive. At the other extreme, a person with the form of cerebral palsy that causes uncontrolled muscle movements may have higher energy needs because the muscles never stop moving.

Losing or Gaining Weight Body weight and composition can affect exercise performance. In sports such as ballet, gymnastics, and certain running events, small, light bodies offer an advantage, so athletes may restrict energy intake in order to maintain a low body weight. In sports such as football and weight lifting, having a large amount of muscle is advantageous, and athletes may try to build muscle and increase body weight.

Healthy Weight Loss Athletes should follow the general guidelines for healthy weight loss—reduce energy intake, increase activity, and change the behaviors that led to weight gain (see Chapter 7). To preserve lean body mass and enhance fat loss, weight should be lost at a rate of about 0.5 to 2 lb/week. This can be accomplished by reducing total energy intake by 200 to 500 kcalories per day and increasing exercise. An athlete who needs to lose weight should do so in advance of the competitive season to prevent the restricted diet from affecting performance.[20]

Unhealthy Weight-Loss Practices Athletes are often under extreme pressure to achieve and maintain a body weight that optimizes their performance. Failure to meet weight-loss goals may have serious consequences such as being cut from the team or restricted from competition. This pressure may compel some athletes to use strict diets and maintain body weights that are so low that health and performance are threatened. The motivation and self-discipline characteristic of successful athletes contribute to their increased risk of eating disorders.[21] In athletes with anorexia, restricted food intake causes a deficiency of energy and nutrients, which can affect growth and maturation and impair exercise performance. In athletes with bulimia, purging can cause dehydration and electrolyte imbalance, which affect performance and put overall health at risk. In addition to using restricted food intake or purging to keep body weight low, some athletes focus on the other side of the energy balance equation by exercising compulsively to burn calories (see **Focus on Eating Disorders**).

Athletes involved in sports with weight classes, such as wrestling and boxing, are at particular risk for unhealthy weight-loss practices because they are under pressure to lose weight before a competition so they can compete in lower weight classes. Competing at the high end of a weight class is thought to give one an advantage over smaller opponents. To lose weight rapidly, these athletes may use sporadic diets that severely restrict energy intake or dehydrate themselves through such practices as vigorous exercise, fluid restriction, wearing plastic vapor-impermeable suits, and using hot environments such as saunas and steam rooms. They may also resort to even more extreme measures such as self-induced vomiting and the use of diuretics and laxatives. These practices can be dangerous and even fatal (**Figure 13.17**). They may impair performance and can adversely affect heart and kidney function, temperature regulation, and electrolyte balance.

Suggestions for Weight Gain Adding weight and muscle mass can enhance performance and help to protect athletes in contact sports like football and hockey. Weight gain, like weight loss takes time. A weight gain of 0.5 to 1 pound per week can be accomplished by slowly increasing portion sizes and adding frequent snacks to increase calorie intake by 500 to 1000 kcalories per day. These extra calories can include healthy choices from each food group of MyPlate. Increasing calories from salty, fatty, or sweet snacks and sugary beverages will increase not only calories, but also the risk of chronic disease.

Strength training is an essential part of weight gain for athletes. It will stimulate appetite as well as promote an increase in lean tissue. Additional calories along with muscle-building exercise will help maximize lean weight gain and minimize weight gain from fat. Adequate protein intake is essential for increasing muscle mass, but most of the energy needed to fuel muscle growth comes from carbohydrate and fat, so protein supplements and very high-protein diets are unnecessary (see **Critical Thinking: Getting Enough Protein**).

Enigma/Alamy

Figure 13.17 Making weight
After three young wrestlers died while exercising in plastic suits in order to sweat off water weight, wrestling guidelines were changed to improve safety.[22] Weight classes were altered to eliminate the lightest class, plastic sweat suits were banned, weigh-ins were moved to one hour before competition, and mandatory weight-loss rules were instituted. The percentage of body fat can be no less than 5% for college wrestlers and 7% for high school wrestlers.

Carbohydrate, Fat, and Protein Needs

The source of dietary energy can be as important as the amount of energy in an athlete's diet. In general, the diets of physically active individuals should contain the same proportion of carbohydrate, fat, and protein as is recommended to the general public—about 45 to 65% of total energy as carbohydrate, 20 to 35% of energy as fat, and 10 to 35% of energy as protein (**Figure 13.18**).

Carbohydrate Carbohydrate is needed to maintain blood glucose levels during exercise and to replace glycogen stores after exercise. The amount recommended for athletes depends on the total energy expenditure, type of sport, gender, and environmental conditions, but ranges from 6 to 10 g per kilogram of body weight per day.[20] For a 150-lb person burning 3000 kcalories per day, this would be about 60% of calories, or about 450 g of carbohydrate. Most of the carbohydrate in the diet should be complex carbohydrates from

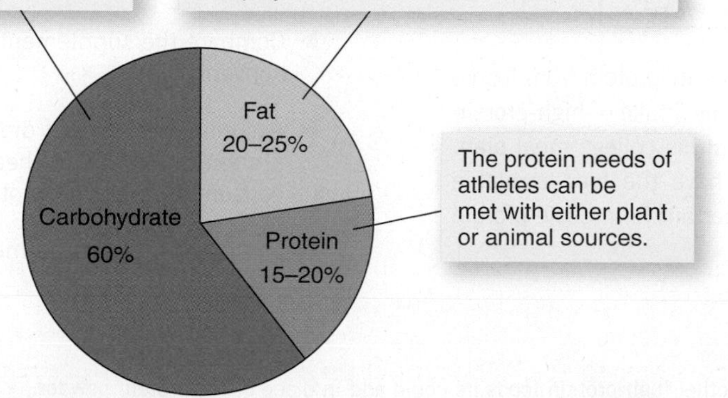

Figure 13.18 Proportions of energy-yielding nutrients in an athlete's diet
The proportions of carbohydrate, fat, and protein recommended in the diets of athletes, shown in the pie chart, are within the ranges recommended for the general public.

Getting Enough Protein

© Leigh Schindler/iStockphoto

BACKGROUND

Paulo is a football player who weighs 240 lb and is trying to put on about 20 lb of muscle before the season begins. He has read that eating protein immediately after exercise and at other key times during the day enhances muscle protein synthesis, so he adds three protein shakes to his daily diet, one first thing in the morning, one right after practice, and one before going to bed. The label from his protein shake is shown below.

Powdered Protein Shake Mix

Nutrition Facts

Serving Size 1 Scoop (28.5g)
Servings Per Container 30

Amount Per Serving

Calories 110 **Calories from Fat** 20

	% Daily Value**
Total Fat 2g	3%
Saturated Fat 1g	5%
Cholesterol 30mg	10%
Sodium 40mg	2%
Potassium 197mg	6%
Total Carbohydrate 4g	1 %
Dietary Fiber 1g	4%
Sugars 2g	†
Protein 18g	36%

Not a significant source of vitamin A, vitamin C, calcium, or iron.

*Percent Daily Values are based on a 2,000 calorie diet. Your daily values may be higher or lower depending on your calorie needs:
†Daily Value (DV) not established

	Calories:	2,000	2,500
Total Fat	Less than	65g	80g
Sat. Fat	Less than	20g	25g
Cholesterol	Less than	300mg	300mg
Sodium	Less than	2,400mg	2,400mg
Potassium		3,500mg	3,500mg
Total Carbohydrate		300g	375g
Dietary Fiber		25g	30g
Protein		50g	65g

Calories per gram:
Fat 9 • Carbohydrate 4 • Protein 4

Paulo recognizes that he can get high-quality protein from foods. It doesn't cost more for him to increase his intake of high-protein foods because all his food is included in his college meal plan. However, it is sometimes difficult to have the food available right after practice, when protein is important for muscle protein synthesis.

CRITICAL THINKING QUESTIONS

- Based on the RDA for a strength athlete, how much protein should Paulo consume per day?

CRITICAL THINKING QUESTIONS

- How much protein will three scoops add to his diet? How many calories will it add?
- If he used low-fat milk instead of the protein powder, how much milk would he have to drink to get the amount of protein in three scoops?
- How many ounces of chicken breast would he need to eat to equal the protein in three scoops? How many calories would this provide?
- The protein powder costs $20 per jar. Is it cheaper to use the supplement or get the protein from milk or chicken?
- Compare the supplement to food in terms of convenience.
- Assuming his current diet provides 100 g of protein, are there any health risks associated with consuming the extra protein from the powder?
- What would you recommend to Paulo?

iProfile Use iProfile to find other high-protein foods he could add in place of the protein powder.

whole grains and starchy vegetables, with some naturally occurring simple sugars from fruit and milk. These foods provide vitamins, minerals, phytochemicals, and fiber as well as energy. Snacks and meals consumed before or during exercise should be lower in fiber to avoid cramping and gastrointestinal distress.

Fat Dietary fat supplies essential fatty acids, ensures the absorption of fat-soluble vitamins, and provides an important source of energy. Body stores of fat provide enough energy to support the demands of even the longest endurance events. For physically active individuals, diets providing 20 to 25% of energy as fat, primarily from unsaturated vegetable oils, have been recommended.[20] Diets higher in fat may not contain enough carbohydrate to maximize glycogen stores and optimize performance. Excess dietary fat is unnecessary and excess energy consumed as fat, carbohydrate, or protein can cause an increase in body fat. Diets very low in fat (less than 20% of calories) have not been found to benefit performance.

Protein Protein is not a significant energy source, accounting for only about 5 to 10% of energy expended, but dietary protein is needed to maintain and repair lean tissues, including muscle. Enough protein is essential to maintain muscle mass and strength, but eating extra protein does not produce bigger muscles. A diet in which 15 to 20% of calories come from protein will meet the needs of most athletes. This protein can come from plant sources such as legumes, grains, and nuts, and/or animal sources such as lean meat, fish, eggs, and low-fat dairy products.

A diet that contains the RDA for protein (0.8 g/kg/day) provides adequate protein for most active individuals. As discussed in Chapter 6, competitive athletes participating in endurance and strength sports may require more protein.[20] In endurance events such as marathons, protein is used for energy and to maintain blood glucose, so these athletes may benefit from 1.2 to 1.4 g of protein per kg per day. Strength athletes who require amino acids to synthesize new muscle proteins may benefit from 1.2 to 1.7 g per kilogram per day. While this amount is greater than the RDA, it is not greater than the amount of protein habitually consumed by athletes.[23] For example, an 85-kg man consuming 3000 kcalories per day, 18% of which is from protein, would be consuming 135 grams of protein, or 1.6 g/kg of body weight.

Vitamin and Mineral Needs

An adequate intake of vitamins and minerals is essential for optimal performance. These nutrients are needed for energy metabolism, oxygen delivery, antioxidant protection, and repair and maintenance of body structures. During exercise, the amounts of many vitamins and minerals used in energy metabolism are increased and after exercise the amounts of those needed to repair tissue damage are increased. Exercise may also increase the losses of some nutrients. Nonetheless, most athletes can meet their needs by consuming the amounts of vitamins and minerals recommended for the general population. In addition, because athletes must eat more food to satisfy their higher energy needs, they consume extra vitamins and minerals in these foods, particularly if nutrient-dense choices are made. Athletes who restrict their intake to maintain low body weight may be at risk for vitamin or mineral deficiencies.

B Vitamins B vitamins such as thiamin, riboflavin, and niacin are important for the production of ATP from carbohydrates and fat. Vitamin B_6, folate, and vitamin B_{12} are needed for proper synthesis of red blood cells, which deliver oxygen to body tissues. Vitamin B_6 is needed to break down glycogen to release glucose and to make the protein hemoglobin, which carries oxygen in red blood cells. Despite the importance of all of these roles during exercise, the recommended intake of B vitamins is no greater for athletes than for the rest of the population.

Antioxidant Nutrients Exercise increases the amount of oxygen at the muscle and the rate of metabolic reactions that produce ATP; oxygen utilization in active muscles can rise as much as 200-fold above resting levels.[24] This increased oxygen use increases the production of free radicals that can lead to oxidative stress and contribute to muscle fatigue.[25] To protect the body from damage caused by free radicals, muscle cells contain antioxidant defenses, some of which may interact with dietary antioxidants such as vitamin C, vitamin E, β-carotene, and selenium. Despite the increase in free radical production that occurs during exercise, there is no evidence that supplementation of antioxidants improves performance or that athletes require more of these nutrients than the general public.[26]

Iron and Anemia The body requires iron to form hemoglobin and myoglobin and a number of iron-containing proteins that are essential for the production of ATP by aerobic metabolism. Exercise increases the need for a number of these proteins and may thereby increase iron needs. For example, exercise stimulates the production of red blood cells, so more iron is needed for hemoglobin synthesis. Prolonged training may also contribute to iron deficiency because iron losses in feces, urine, and sweat increase.[20] Although a specific RDA has not been set for athletes, the DRIs acknowledge that, based on iron losses, the EAR may be 30 to 70% higher for athletes than for the general population.[27]

Another cause of iron loss is foot-strike hemolysis. This is the breaking of red blood cells due to the contraction of large muscles or impact in events such as running. Foot-strike hemolysis rarely causes anemia because most of the iron from these cells is recycled, and the breaking of red blood cells stimulates the production of new ones.

Reduced iron stores are not uncommon in athletes.[20] Female athletes are at particular risk; their needs are higher than those of male athletes because they need to replace the iron lost in menstrual blood.[28] In athletes of both sexes, inadequate iron intake often contributes to low iron stores. Iron intake may be low in athletes who are attempting to keep their body weight down and in those who do not eat meat, which is an excellent source of readily absorbable heme iron. If iron deficiency progresses to anemia, the body's ability both to transport oxygen and to provide energy by aerobic metabolism is reduced, impairing exercise performance as well as overall health.

Iron deficiency anemia (see Chapter 12) should not be confused with **sports anemia**, which is a temporary decrease in hemoglobin concentration that occurs during exercise training. This is an adaptation to training that does not seem to impair delivery of oxygen to tissues. It occurs when blood volume expands to increase oxygen delivery, but the synthesis of red blood cells lags behind the increase in plasma volume (**Figure 13.19**).

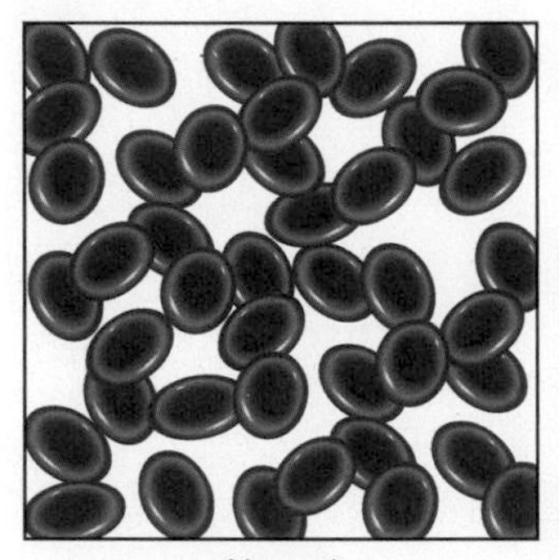

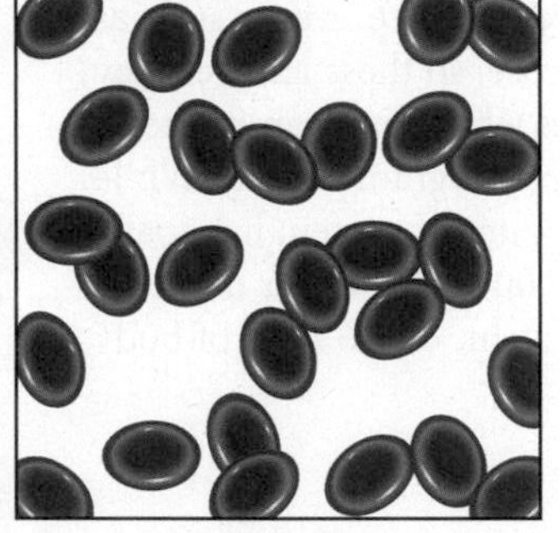

Figure 13.19 **Sports anemia**
Training causes a decrease in the percentage of blood volume that is red blood cells. As training progresses, the number of red blood cells increases to catch up with the increase in total blood volume.

sports anemia Reduced hemoglobin levels that occur as part of a beneficial adaptation to aerobic exercise in which expanded plasma volume dilutes red blood cells.

Calcium and Vitamin D Calcium and vitamin D are needed to maintain blood calcium levels and promote and maintain healthy bone density, which in turn reduces the risk of osteoporosis. In general, exercise—particularly weight-bearing exercise—increases bone density, thereby reducing the risk of osteoporosis, but this occurs only when levels of vitamins, minerals, and hormones are adequate.

An athlete's diet should include adequate sources of calcium to maintain bone health. The additional energy needs of athletes allow calcium sources, such as a slice of cheese on a sandwich or a milk shake, to be added without exceeding energy needs. Even if the diet is adequate in calcium, it can't be absorbed efficiently without sufficient vitamin D (see Chapter 9). Vitamin D deficiency is a concern in athletes because it can contribute to stress fractures and possibly impair muscle function and athletic performance.[29] Since vitamin D can be synthesized in the skin from sun exposure, status is generally higher in outdoor than indoor athletes. Outdoor athletes can usually achieve adequate vitamin D status during the summer, spring, and fall, but during the winter vitamin D supplements may be needed to attain adequate blood levels of vitamin D.[30]

female athlete triad The combination of disordered eating, amenorrhea, and osteoporosis that occurs in some female athletes, particularly those involved in sports in which low body weight and appearance are important.

In female athletes, too much exercise combined with restricted food intake can cause hormonal abnormalities that affect calcium and bone metabolism and can lead to bone loss. This **female athlete triad** is a combination of restrictive eating patterns that can lead to eating disorders, abnormalities in hormone levels that cause **amenorrhea**, and disturbances in bone formation and breakdown that contribute to osteoporosis (see **Focus on Eating Disorders**). Hormonal abnormalities occur when extreme energy restriction and exercise create a physiological condition similar to starvation. Estrogen levels drop, causing amenorrhea. Because estrogen is needed for calcium homeostasis in the bone and calcium absorption in the intestines, low levels lead to premature bone loss, low peak bone mass, and an increased risk of stress fractures.[31] Neither adequate dietary calcium nor the increase in bone mass caused by weight-bearing exercise can compensate for bone loss due to low estrogen levels. Treatment for female athlete triad involves increasing energy intake and reducing activity so that menstrual cycles resume. This is essential for preserving long-term bone health.

amenorrhea Delayed onset of menstruation or the absence of three or more consecutive menstrual cycles.

13.5 Fluid Needs for Physical Activity

LEARNING OBJECTIVES

- Discuss dehydration in relation to performance and heat-related illness.
- Describe an exercise scenario that might lead to hyponatremia.
- Explain why the types of fluid recommended for a 30-minute workout and a two-hour workout are different.

During exercise, water is needed to eliminate heat and to transport both oxygen and nutrients to the muscles and waste products away from the muscles. The ability to dissipate the heat generated during exercise is affected by the hydration status of the exerciser as well as by environmental conditions. At rest in a temperate environment, an individual loses about 1.2 L (about 4.5 cups) of water per day, or 50 mL/hour, through evaporation from the skin and lungs. Exercise in a hot environment can increase losses more than tenfold. Even when fluids are consumed at regular intervals during exercise, it may not be possible to drink enough to compensate for losses from sweat and evaporation through the lungs. Failure to consume adequate fluids to replace water lost can be critical for even the most casual exerciser. If heat cannot be lost from the body, body temperature rises and health as well as exercise performance may be jeopardized.

How Dehydration Affects Health and Performance

Dehydration occurs when water loss is great enough for blood volume to decrease, thereby reducing the ability to deliver oxygen and nutrients to exercising muscles. Dehydration hastens the onset of fatigue and makes a given exercise intensity seem more difficult. Even mild dehydration—a body water loss of 2 to 3% of body weight—can impair exercise performance (**Figure 13.20**).[20] A 3% reduction in body weight can significantly reduce the

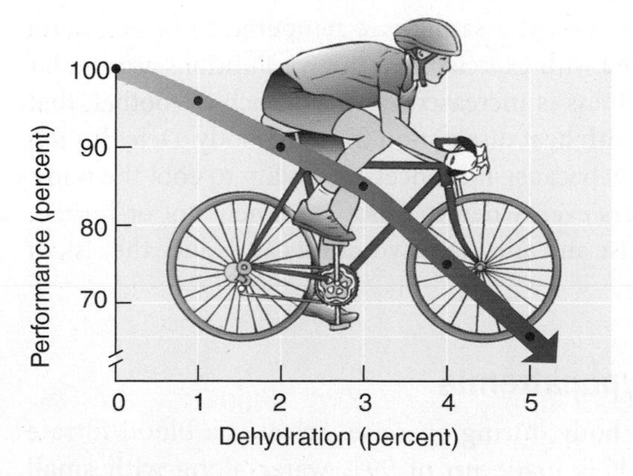

Figure 13.20 Effect of dehydration on exercise performance
As the severity of dehydration increases, exercise performance declines.[32]

THINK CRITICALLY

If a person loses 4% of his body weight as water during a competition, by what percentage will his performance be decreased by the end of the event?

heat-related illnesses Conditions, including heat cramps, heat exhaustion, and heat stroke, that can occur due to an unfavorable combination of exercise, hydration status, and climatic conditions.

amount of blood pumped with each heartbeat. This reduces the ability of the circulatory system to deliver oxygen and nutrients to cells and remove waste products. The decrease in blood volume that occurs with dehydration reduces blood flow to the skin and sweat production, which limits the body's ability to sweat and cool itself. Core body temperature can then increase and with it the risk of various **heat-related illnesses**.

The risk of dehydration is greater in hot environments, but it may also occur when exercising in the cold. Cold air tends to be dry air, so evaporative losses from the lungs are greater. Insulated clothing may increase sweat losses and fluid intake may be reduced because a chilled athlete may be reluctant to drink a cold beverage. Female athletes tend to limit fluid intake in the cold to avoid the inconvenience of removing clothing to urinate.[20]

Children, older athletes, and obese individuals are at a greater risk for dehydration and heat-related illness. Children produce more heat, are less able to transfer heat from muscles to the skin, take longer to acclimatize to heat, and sweat less than adults. To reduce risks on hot days, children should rest periodically in the shade, consume fluids frequently, and limit the intensity and duration of activities. Also, children lose more heat in cold environments than adults because they have a greater surface area per unit of body weight. Therefore, they are more prone to **hypothermia**. Older athletes are at greater risk of dehydration because the thirst sensation decreases with age, and the kidneys may be less able to concentrate urine, thus increasing the amount of fluid lost in urine. Excess weight increases the risk of heat-related illness because it increases the amount of work and therefore the amount of heat produced in a given activity. The fat also acts as an insulator, retarding the conduction of heat to the body surface. Obese individuals also have a smaller surface-area-to-body-mass ratio than lean people, so they are less efficient at dissipating heat through blood flow to the surface and the evaporation of sweat.

hypothermia A condition in which body temperature drops below normal. Hypothermia depresses the central nervous system, resulting in the inability to shiver, sleepiness, and eventually coma and death.

Why Heat-Related Illness Is a Concern for Athletes

Exercising in hot weather can lead to heat-related illness. Heat-related illnesses include heat cramps, heat exhaustion, and heat stroke. Heat cramps are involuntary muscle spasms that occur during or after intense exercise, usually in the muscles involved in exercise. They are caused by an imbalance of the electrolytes sodium and potassium at the muscle cell membranes and can occur when water and salt are lost during extended exercise. Heat exhaustion occurs when fluid loss causes blood volume to decrease so much that it is not possible to both cool the body and deliver oxygen to active muscles. It is characterized by a rapid weak pulse, low blood pressure, fainting, profuse sweating, and disorientation. Someone experiencing the symptoms of heat exhaustion should stop exercising and move to a cooler environment. If exercise continues, heat exhaustion may progress to heat stroke. Heat stroke, the most serious form of heat-related illness, occurs when the temperature regulatory center of the brain fails due to a very high core body temperature (greater than 105°F). Heat stroke is characterized by elevated body temperature, hot dry skin, extreme confusion, and unconsciousness. It requires immediate medical attention.

Both temperature and humidity greatly affect the risk of heat-related illness. As environmental temperature rises, it becomes more difficult to dissipate heat and as humidity rises, the ability to cool the body by evaporation declines.[33] When the humidity is high, the same air temperature feels hotter than when the humidity is lower. For example, when the humidity is 100%, a temperature of 82°F feels the same as a temperature of 90°F and a humidity of only 50%. The risks associated with exercising in these conditions are similar (**Figure 13.21**). The risk of heat-related illness is increased in sports such as football that require protective clothing that interferes with heat dissipation from the body. Dehydration also increases the risk of heat-related illness because it reduces the ability to cool the body. Dehydration can precipitate these disorders even when it is not extremely hot or humid. Conditioning with repeated bouts of exercise and adequate hydration can reduce the risk of heat-related illness.

Causes and Consequences of Hyponatremia

The evaporation of sweat helps cool the body during exercise. Sweat is a blood filtrate produced by sweat glands in the skin. It is made up of 99% water along with small

Relative humidity (%) / Air temperature

°F	40	45	50	55	60	65	70	75	80	85	90	95	100
110	136												
108	130	137											
106	124	130	137										
104	119	124	131	137									
102	114	119	124	130	137								
100	109	114	118	124	129	136							
98	105	109	113	117	123	128	134						
96	101	104	108	112	116	121	126	132					
94	97	100	102	106	110	114	119	124	129	135			
92	94	96	99	101	105	108	112	116	121	126	131		
90	91	93	95	97	100	103	106	109	113	117	122	127	132
88	88	89	91	93	95	98	100	103	106	110	113	117	121
86	85	87	88	89	91	93	95	97	100	102	105	108	112
84	83	84	85	86	88	89	90	92	94	96	98	100	103
82	81	82	83	84	84	85	86	88	89	90	91	93	95
80	80	80	81	81	82	82	83	84	84	85	86	86	87

Heat index (apparent temperature)

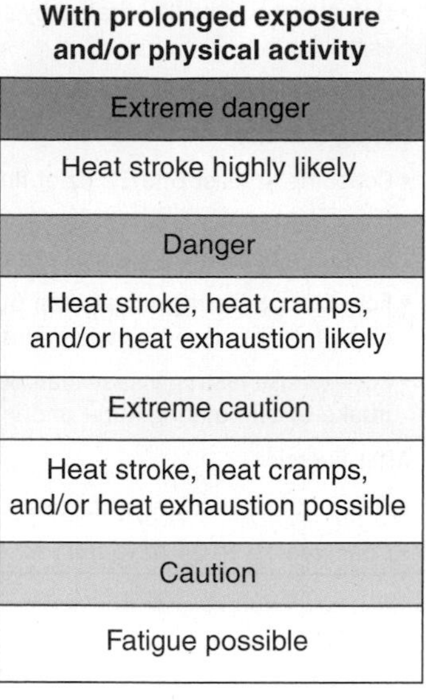

Figure 13.21 Heat index and the risk of heat-related illness[33]
The "heat index" or "apparent temperature" is a measure of how hot it feels when the relative humidity is added to the actual air temperature. To find the heat index, find the intersection of the temperature on the left side of the table and the relative humidity across the top of the table. The shaded zones indicate the relative risk of heat-related illness with continued exposure and/or physical activity.

amounts of minerals, primarily sodium and chloride, acids, and trace amounts of other substances. Because sweat is mostly water, for most activities sweat losses can be replaced with plain water. However, when sweating continues for more than four hours, as it may during endurance events such as marathons, enough sodium may be lost to affect electrolyte balance. A reduction in the concentration of sodium in the blood is referred to as **hyponatremia**. Hyponatremia can occur if an athlete loses large amounts of water and salt in sweat and replaces the loss with water alone. This causes the sodium that remains in blood to be diluted, so that the amount of water is too great for the amount of sodium. This is analogous to taking a full glass of salt water, dumping out half, and replacing what was poured out with plain water. The sodium in the glass is now more dilute (**Figure 13.22**). It is also possible to develop hyponatremia when salt losses from sweating are not excessive. This can occur if athletes drink more water than is lost in sweat, diluting the sodium in their system. A study of runners in the Boston marathon found that 13% of those tested had hyponatremia and 0.6% had serum sodium concentrations low enough for this condition to be considered critical.[34] It is the concentration of sodium that is important, not the absolute amount.

hyponatremia Abnormally low concentration of sodium in the blood.

Hyponatremia is dangerous because the sodium in the blood helps hold fluid in the blood vessels. As sodium concentration drops, water will leave the bloodstream by osmosis and accumulate in the tissues, causing swelling. Fluid accumulation in the lungs interferes with gas exchange, and fluid accumulation in the brain causes disorientation, seizure, coma, and death. The early symptoms of hyponatremia may be similar to dehydration: nausea, muscle cramps, disorientation, slurred speech, and

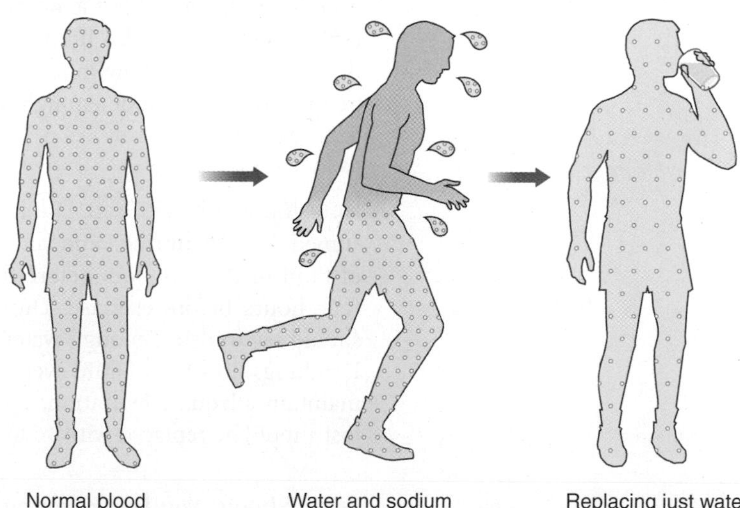

Figure 13.22 Hyponatremia
Hyponatremia can occur when an athlete loses water and sodium in sweat but replaces these losses with plain water, diluting the remaining sodium in the blood.

Figure 13.23 **Recommended fluid intake**
Athletes should be well hydrated before they start exercising. To maintain adequate hydration during extended exercise, it is important to carry water. Hydration packs such as the one worn by this runner can be used for biking, hiking, and jogging.

Universal Images Group/SuperStock

Before Exercise

- Begin exercise well hydrated by consuming generous amounts of fluid in the 24 hours before exercise.
- Consume about 2 cups of fluid at least 4 hours before exercise.

During Exercise

- Consume at least 6 to 12 oz of fluid every 15 to 20 minutes.
- For exercise lasting 60 minutes or less, plain water is the only fluid needed but beverages containing carbohydrate and electrolytes will not hurt performance.
- For exercise lasting longer than 60 minutes,consuming a fluid containing about 6 to 8% carbohydrate may improve endurance.
- For exercise lasting longer than 60 minutes, a fluid containing electrolytes can increase fluid intake by stimulating thirst and increasing absorption.

After Exercise

- Begin fluid replacement immediately after exercise.
- Consume 16 to 24 oz of fluid for each pound of weight lost.

confusion. Drinking water alone will make the problem worse and can result in seizure, coma, or death.

For most exercise, hyponatremia is not a concern and lost electrolytes can be replaced during the meals following exercise. During long-distance events when hyponatremia is more likely, risk can be reduced by increasing sodium intake several days prior to competition, consuming sodium-containing sports drinks during the event, and avoiding Tylenol, aspirin, ibuprofen, and other nonsteroidal anti-inflammatory agents. These medications interfere with kidney function and may contribute to the development of hyponatremia. Mild symptoms of hyponatremia can be treated by eating salty foods or drinking sodium-containing beverages such as sports drinks. More severe symptoms require medical attention.

Recommended Fluid Intake for Exercise

Anyone exercising should consume extra fluids. Good hydration is important before exercise and, since thirst is not a reliable indicator of immediate fluid needs, it is important to schedule regular fluid breaks during exercise. The amount and type of fluid that is best depends on how much water you lose and how long you exercise. Because many people do not consume enough during exercise, beverages consumed after exercise must restore hydration.

How Much Should You Drink? To ensure hydration, adequate fluids should be consumed before, during, and after exercise. Exercisers should drink generous amounts of fluid in the 24 hours before the exercise session and about two cups of fluid at least four hours before exercise. During exercise, whether casual or competitive, exercisers should try to drink enough water to prevent weight loss in excess of 2% of body weight.[20] Drinking 6 to 12 oz of fluid every 15 to 20 minutes beginning at the start of exercise should maintain adequate hydration. To restore lost water after exercise, each pound of weight lost should be replaced with 16 to 24 oz (2 to 3 cups) of fluid (**Figure 13.23**).

What Should You Drink During Short Workouts? For exercise lasting an hour or less, water is the only fluid needed. For a 20-minute jog, 40 minutes at the gym, or a brisk walk through the park, sports drinks offer no advantage over a water bottle filled at the drinking fountain. Sports drinks will not hurt your performance in a short workout, but they may be counterproductive if the goal of exercise is weight loss. A typical sports drink contains about 50 kcalories per cup, so drinking a 16-oz bottle at the gym will replace about half of the 200 kcalories expended during your 40-minute ride on the stationary bicycle.

What Should You Drink During Long Workouts? For exercise lasting more than 60 minutes, beverages containing a small amount of carbohydrate and electrolytes are recommended. Consuming carbohydrate in a beverage helps to maintain blood glucose levels, thus providing a source of glucose for the muscle and delaying fatigue. The right proportion of carbohydrate to water is important. If the concentration of carbohydrate is too low, it will not help performance; if it is too high, it will delay stomach emptying (see **Debate: Energy Drinks for Athletic Performance?**). Beverages containing 15 to 20 g of carbohydrate per cup (6 to 8%) are best. This is the amount of carbohydrate found in popular sports beverages such as Gatorade and Powerade. Beverages containing larger amounts of carbohydrate, such as fruit juices and soft drinks, are not recommended unless they are diluted with an equal volume of water. Water and carbohydrate trapped in the stomach do not benefit the athlete. Choosing flavored beverages over plain water may also be advantageous because they tempt athletes to drink more, helping to ensure adequate hydration.

The amount of sodium lost during exercise lasting less than three to four hours is small. Even though there may not be a physiological need to replace sodium, a beverage containing 500 to 700 mg of sodium per liter (around 150 mg/cup) is recommended for exercise lasting more than an hour.[20] This is because the sodium enhances palatability and the drive to drink, so it may cause an increase in fluid intake. The presence of small amounts of sodium and glucose also slightly increase the rate of water absorption. A sodium-containing beverage will also help prevent hyponatremia in athletes who overhydrate and in those participating in endurance events, such as ultramarathons or Ironman Triathlons, in which significant amounts of sodium may be lost in sweat.

13.6 Food and Drink to Maximize Performance

LEARNING OBJECTIVES

- Discuss the advantages and disadvantages of carbohydrate loading.
- Explain the recommendations for food and drink during extended exercise.
- Plan pre- and post-competition meals for a marathon runner.

glycogen supercompensation or **carbohydrate loading** A regimen designed to maximize muscle glycogen stores before an athletic event.

For most of us, a trip to the gym requires no special nutritional planning, but for competitive athletes, when they eat and what they eat before, during, and after competition is as important as a balanced overall diet. Food eaten at these times may give or take away the extra seconds that can mean victory or defeat.

Maximizing Glycogen Stores

Glycogen provides a source of stored glucose. Larger glycogen stores allow exercise to continue for longer periods. Glycogen stores and hence endurance are increased by increasing carbohydrate intake (**Figure 13.24**).

Serious endurance athletes who want to increase their glycogen stores substantially before a competition may choose to follow a dietary regimen referred to as **glycogen supercompensation** or **carbohydrate loading**. This involves resting for one to three days before competition while consuming a very high-carbohydrate diet.[35,36] The diet should provide 10 to 12 g of carbohydrate per kg of body weight per day. For a 150-pound person, this is equivalent to about 700 g of carbohydrate per day. Having a stack of pancakes with syrup and a glass of milk or a plate of pasta with garlic bread and a glass of juice provides more than 200 g of carbohydrate.

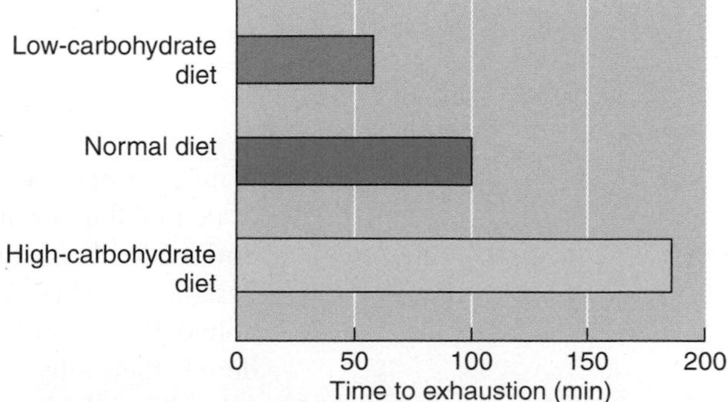

Figure 13.24 Effect of carbohydrate consumption on endurance
The amount of carbohydrate consumed in the diet affects the level of muscle glycogen and hence an athlete's endurance. This graph shows endurance capacity during cycling exercise after three days of a very-low-carbohydrate diet (less than 5% of energy from carbohydrate), a normal diet (about 55% carbohydrate), and a high-carbohydrate diet (82% carbohydrate).[37]

DEBATE

Energy Drinks for Athletic Performance?

Tony Cenicola/Redux Pictures

The popularity of energy drinks with names like Monster, Red Bull, and Full Throttle has soared over the past decade. They promise to keep you alert to study, work, drive, party all night, and perhaps excel at your next athletic competition. They are sold alongside sports drinks, and manufacturers of these beverages often sponsor athletes and athletic events. Should they be used as ergogenic aids? Is drinking them a safe way to improve your game?

The main ingredients in these drinks are sugar and caffeine. Sugar is an important fuel for exercise. A traditional sports drink, like Gatorade, contains about 28 g of sugar in 16 oz; a typical energy drink provides twice this much (55 to 60 g, or about 14 teaspoons) (see table). Since glucose fuels activity, the additional sugar could provide the extra energy needed for prolonged exercise. However, providing more sugar does not increase the rate at which it is absorbed, so energy drinks don't get any more glucose to the muscles than sports drinks do. The unabsorbed sugar in the stomach and intestines can cause GI distress and slow fluid absorption. So more sugar may not always be better during activity.

The caffeine content of energy drinks ranges from 50 to 505 mg per can or bottle. Caffeine is an effective ergogenic aid that enhances endurance when consumed before or during exercise.[1] So, the caffeine in energy drinks could provide an ergogenic benefit. However, too much caffeine, referred to as *caffeine intoxication*, causes nervousness, anxiety, restlessness, insomnia, gastrointestinal upset, tremors, increased blood pressure, and rapid heartbeat. This is a particular concern in young athletes because their lower body weight means that the energy drink supplies more caffeine per kg than in an adult. A number of cases of caffeine-associated cardiac arrest, seizure, and death have occurred after consumption of energy

[1]Ganio, M. S., Klau, J. F., Casa, D. J., et al. Effect of caffeine on sport-specific endurance performance: A systematic review. *J Strength Cond Res* 23:315–324, 2009.
[2]Berger, A. J., and Alford, K. Cardiac arrest in a young man following excess consumption of caffeinated "energy drinks." *Med J Aust* 190:41–43, 2009.
[3]Clauson, K. A., Sheilds, K. M., McQueen, C. E., and Persad, N. Safety issues associated with commercially available energy drinks. *J Am Pharm Assoc* 48:e55–e63, 2008.
[4]Ballard, S. L., Wellborn-Kim, J. J., and Clauson, K. A. Effects of commercial energy drink consumption on athletic performance and body composition. *Phys Sportsmen* 38:107–117, 2010.
[5]Higgins, J. P., Tuttle, T. D., and Higgins, C. L. Energy beverages: Content and safety. *Mayo Clin Proc* 85:1033–1041, 2010.
[6]Duchan, E., Patel, N. D., and Feucht, C. Energy drinks: A review of use and safety. *Phys Sportsmed* 38:171–179, 2010.

A number of commercial high-carbohydrate beverages, containing 50 to 60 g of carbohydrate in 8 fluid oz, are available to help athletes consume the amount of carbohydrate recommended to maximize glycogen stores. These should not be confused with sports drinks designed to be consumed during competition, which contain only about 15 to 20 g of carbohydrate in 8 fluid oz. Trained athletes who follow a carbohydrate-loading regimen can double their muscle glycogen content.[35]

Although glycogen supercompensation is beneficial to endurance athletes, it will provide no benefit and even has some disadvantages for people exercising for periods less than 90 minutes. For every gram of glycogen in the muscle, 3 g of water are also deposited. This water can produce a 2- to 7-lb weight gain and may cause some muscle stiffness. The extra weight is a disadvantage for those competing in events of short duration. As glycogen is used, the water is released; this can be an advantage when exercising in hot weather.

Comparison of caffeinated beverages

Beverage	Serving (fluid oz)	Caffeine (mg)	Sugar (g)	Energy (kcalories)
Coffee	8	100–200	0	0
Espresso with sugar	2.0	100	7–15	30–60
Coca-Cola Classic	12	35	39	140
Mountain Dew	12	54	46	170
Monster	16	160	54	200
Jolt Cola	8	80	30	120
Arizona Caution Extreme Energy Shot	8	100	33	130
Red Bull	8	80	28	110
Rockstar	16	160	62	280
Monster	8	80	27	100
Full Throttle	16	160	57	220

drinks.[2,3,4] Even if the caffeine in an energy drink increases endurance, depending on when it is consumed, it can affect timing and coordination and hurt overall performance. Caffeine is also a diuretic; at the levels contained in these drinks, it may contribute to dehydration, particularly in first-time users.[5] The FDA limits the amount of caffeine in soft drinks to 0.02% (about 71 mg in 12 oz), but energy drinks are considered dietary supplements, so the caffeine content is not regulated.

Energy drinks often also contain other ingredients that promise to improve performance, such as B vitamins, taurine, guarana, and ginseng. B vitamins are needed to produce ATP, so they are marketed to enhance energy production from sugar. But unless you are deficient in these vitamins, getting them from an energy drink will not enhance your ATP production. Taurine is an amino acid that may reduce the amount of muscle damage and improve exercise performance and capacity, but not all research supports these claims.[4] Guarana is an herbal ingredient that contains caffeine as well as small amounts of the stimulants theobromine and theophylline. The extra caffeine from guarana (not included in the caffeine listed for these beverages) may contribute to caffeine intoxication. Ginseng is also claimed to have performance-enhancing effects, but these effects have not been demonstrated scientifically.[3,6] Even if these ingredients are ergogenic, the amounts contained in energy drinks may be too small to have an effect, and the safety of consuming them in combination with caffeine prior to or during exercise has yet to be established.[3]

So should you down an energy drink before your next competition? They do provide a caffeine boost, but is it so much caffeine that you risk dehydration, high blood pressure, and heart problems? Energy drinks provide sugar to fuel activity, but will they upset your stomach? What about the herbal ingredients—do they offer a benefit you are looking for?

THINK CRITICALLY: Use the table above to assess the advantages and disadvantages of consuming an 8-oz can of Red Bull versus a 12-oz can of Coca-Cola Classic before your 30-minute run.

What to Eat Before Exercise

An athlete who is hungry will not perform at his or her best, but the wrong meal can hinder performance more than the right one can enhance it. The size, composition, and timing of the preexercise meal are all important. The goal of meals eaten before exercise is to maximize glycogen stores and provide adequate hydration while minimizing hunger and any undigested food in the stomach that can lead to gastric distress.

Ideally, a preexercise meal should provide enough fluid to maintain hydration and be high in carbohydrate (60 to 70% of calories). This will help to maintain blood glucose and maximize glycogen stores. Muscle glycogen is only depleted by exercise, but liver glycogen is used to supply blood glucose, so it is depleted even during rest if no food is ingested. So, first thing in the morning, liver glycogen stores have been reduced by the overnight fast. A high-carbohydrate meal eaten two to four hours before the event will fill liver glycogen stores. In addition to being high in carbohydrate, the preexercise meal should be moderate in protein (10 to 20%) and low in fat (10 to 25%) and fiber to minimize gastrointestinal distress and bloating during competition. A cup of pasta with tomato sauce and a slice of bread, or a turkey sandwich and a cup of juice, are good choices. Spicy foods that could cause heartburn, and large amounts of simple sugars that could cause diarrhea, should be avoided unless the athlete is accustomed to eating these foods.

In addition to providing nutritional clout, a meal that includes "lucky" foods may provide some athletes with an added psychological advantage. Some athletes find that in addition to a precompetition meal, a small high-carbohydrate snack or beverage consumed shortly before an event may enhance endurance. Because foods affect people differently, athletes should test the effect of these meals and snacks during training, not during competition.

What to Eat During Exercise

Regardless of the type or duration of exercise, maintaining adequate fluid intake is important while exercising (see Figure 13.23). For exercise that lasts more than an hour, consuming about 30 to 60 g of carbohydrate per hour (the amount in a banana or an energy bar) can enhance endurance.[20] Consuming carbohydrate during exercise is particularly important for athletes who exercise in the morning, when liver glycogen levels are low.

Carbohydrate intake should begin shortly after exercise commences and regular amounts should be consumed every 15 to 20 minutes during exercise. The carbohydrate should provide glucose, glucose polymers (chains of glucose molecules), or a combination of glucose and fructose.[20] Fructose alone is not as effective and may cause diarrhea. Some athletes may prefer to obtain this carbohydrate from a sports drink, while others prefer a high-carbohydrate snack or energy gel consumed with water (see **Off the Label: What Are You Getting from That Sports Bar?**). Energy gels consist of a thick carbohydrate syrup packaged in a palm-sized packet. The contents can be sucked out of the packet, providing about 20 g of carbohydrate.

During exercise, sodium and other minerals are lost in sweat. Although the amounts lost during exercise lasting less than three hours are usually not enough to affect health or performance, a snack or beverage containing sodium is recommended for exercise lasting one hour or more. As discussed above, sodium enhances the palatability of beverages and increases the drive to drink so even if sodium losses are small, consuming it during exercise may cause an increase in fluid intake.

What to Eat After Exercise

When exercise ends, the body must shift from the catabolic state of breaking down glycogen, triglycerides, and muscle proteins for fuel to the anabolic state of restoring muscle and liver glycogen, depositing lipids, and synthesizing muscle proteins. The goal for meals after exercise is to replenish fluid, electrolyte, and glycogen losses and to provide amino acids for muscle protein synthesis and repair. For example, a mixed meal such as pancakes and a glass of milk consumed soon after a strenuous competition or training session will help the athlete prepare for the next exercise session.

The first priority for all exercisers is to replace fluid losses. For serious athletes, it may also be important to replenish glycogen stores rapidly. To maximize glycogen replacement, a high-carbohydrate meal or drink should be consumed within 30 minutes of completion of the competition and again every two hours for six hours after the event.[20] Ideally, the meals should provide about 1.0 to 1.5 g of carbohydrate per kg of body weight, which is about 70 to 100 g of carbohydrate for a 70-kg (154-lb) person—the equivalent of 1.5 cups of pasta and 8 oz of chocolate milk.[38] Consuming foods such as these that contain both carbohydrate and protein enhances glycogen synthesis even more than does consuming carbohydrate alone.[39,40] Including protein with carbohydrate in postexercise meals also stimulates muscle protein synthesis and provides the amino acids needed for muscle protein synthesis and repair.[41]

© iStockphoto

OFF THE LABEL What Are You Getting from That Sports Bar?

Looking for a convenient snack that can give you an energy boost during your bike ride or day of cross-country skiing? A sports bar may be the answer, but which should you choose? There are hundreds of varieties. Some are high in protein and low in carbohydrate; others are high in carbohydrate and low in fat; and some claim to have just the right balance of everything. They promise to optimize performance, build lean muscle, reduce body fat, increase strength, and speed recovery.

Nutrition Facts
Serving Size 1 bar (2 oz.) (57g)

Amount Per Serving	
Calories 271	Calories from Fat 122
	% Daily Value*
Total Fat 14g	**21%**
Saturated Fat 5g	**26%**
Trans Fat 0g	
Cholesterol 5mg	**2%**
Sodium 140mg	**6%**
Total Carbohydrates 35g	**12%**
Dietery Fiber 1g	**5%**
Sugars 30g	
Protein 4g	

Vitamin A 2% • Vitamin C 0%
Calcium 5% • Iron 2%

*Based on a 2,000 calorie diet

Andy Washnik

Carbohydrate is the fuel that becomes limiting during prolonged exercise, so if you want to have the energy to keep pedaling or skiing, choose high-carbohydrate bars, called energy or endurance bars. They have the carbohydrate needed to prevent hunger and maintain blood glucose during extended sports activities. Use the label to check out the amount of carbohydrate, fat, protein, and energy in different bars. A bar that provides about 45 g of carbohydrate (70% of calories) will help maintain your blood glucose level during exercise. Watch the fat and protein; in 300 kcalories you want no more than about 8 g of fat and 16 g of protein. Bars higher in fat or protein or lower in carbohydrate will not give you the blood glucose boost that you need to continue exercising.

Nutrition Facts	Amount/Serving	% DV*	Amount/Serving	% DV*
Serving Size 1 bar (65g)	**Total Fat** 2g	3%	**Total Carb** 45g	15%
Calories 230	Saturated Fat 0.5g	3%	Dietary Fiber 3g	12%
Calories from Fat 20	*Trans* Fat 0g		Sugars 14g	
Calories from Sat Fat 5	**Cholesterol** 0mg	0%	Other Carb 28g	
*Percent Daily Values (DV) are based on a 2,000 calorie diet.	**Sodium** 90mg	4%	**Protein** 10g	
	Potassium 145mg	4%		

Vitamin A 0% • Vitamin C 100% • Calcium 30% • Iron 35% • Vitamin E 100% Thiamin 100% • Riboflavin 100% • Niacin 100% • Vitamin B_6 100% Folate 100% • Vitamin B_{12} 100% • Biotin 100% • Pantothenic Acid 100% Phosphorus 35% • Magnesium 35% • Zinc 35% • Copper 35% • Chromium 20%

George Semple

Is a sports bar any better for you than a candy bar? A look at the labels shows that there is a difference (see figures). Typically, sports bars are lower in fat, higher in fiber, and contain more vitamins and minerals than candy bars. Many contain vitamin C, vitamin E, calcium, iron, magnesium, copper, zinc, and a host of B vitamins. They must list the % Daily Value for calcium, iron, and vitamins A and C. The Daily Values of other vitamins and minerals do not need to be included in the Nutrition Facts portion of the label, but all the nutrients added will appear in the ingredient list.

So, should you be packing a sports bar on your next outing? It won't take the place of the whole grains, fresh vegetables and fruits, low-fat dairy products, and lean meats or other high-protein foods that make up a healthy diet. But, if having a compact, individually wrapped bar that can travel with you means the difference between consuming a snack or no food at all, sports bars can be beneficial. They may also provide a psychological edge if you believe they will enhance your performance.

If you choose to use these bars, wash them down with plenty of water. They don't provide fluid—an essential during any activity. Also remember that one sports bar provides around 200 to 300 kcalories. Even though they are eaten to support activity, they still add to your overall energy intake and can contribute to weight gain if consumed in excess.

THINK CRITICALLY: Suggest a snack that provides about the same amount of carbohydrate and calories as an energy bar, but is less expensive.

The glycogen-restoring regimen just described can replenish muscle and liver glycogen within 24 hours of an athletic event and is critical for athletes who must perform again the following day. Athletes not competing again the next day can replenish their glycogen stores more slowly by consuming high-carbohydrate foods for the next day or so. A diet providing about 65% of calories from carbohydrate, or about 400 g of carbohydrate in a 2500-kcalorie diet, should provide sufficient carbohydrate during the recovery period.[20] More than one-third of this amount could be provided by a six-inch sub, 12 oz of low-fat chocolate milk, a banana, and some pretzels.

Most people do not need a special glycogen replacement strategy to ensure glycogen stores are full by their next gym visit. Eating a typical diet that provides about 55% of calories as carbohydrate will quickly replace the glycogen used during a 30- to 60-minute workout at the gym.

13.7 Ergogenic Aids: Do Supplements Enhance Athletic Performance?

LEARNING OBJECTIVES

- Assess the health risks associated with anabolic steroids.
- Explain why creatine supplements affect sprint performance.
- Describe one way in which a supplement might improve endurance.

Citius, altius, fortius—faster, higher, stronger—the Olympic motto. For as long as there have been competitions, athletes have yearned for something—anything—that would give them the competitive edge. Anything designed to enhance performance can be considered an **ergogenic aid**: running shoes are mechanical aids; psychotherapy is a psychological aid; drugs are pharmacological aids. Many dietary supplements are also used as ergogenic aids. Although these supplements are often expensive and most have not been shown to improve performance, athletes are vulnerable to their enticements. When considering the use of an ergogenic supplement, an athlete should first weigh the health risks against potential benefits: dietary supplements do not have to be proven safe or effective before they can be sold (see Chapter 2 and **Focus on Nonvitamin Nonmineral Supplements** as well as **Table 13.3**). The following sections discuss some of the more popular products. Others are reviewed in **Table 13.4**.

ergogenic aid Anything designed to increase work or improve performance.

TABLE 13.3 Evaluating the Benefits and Risks of Ergogenic Supplements

Does the supplement meet your needs?
• Does the product contain the nutrient or other ingredient you are looking for?
• Has it been shown to provide the benefits you want?
Are the ingredients safe for you?
• Does it contain any ingredients that have been shown to be toxic to someone like you?
• Do you have a medical condition that would make it dangerous to take this product?
• Are you taking prescription medication that might interact with the supplement?
Is the dose safe?
• Does it contain potentially toxic levels of any nutrient? Check the % Daily Value for any nutrients that exceed 100%. If they do, do they exceed the UL?
• Follow the recommended dose on the package. More isn't always better and may cause side effects.
How much does it cost?
• More expensive is not always better.
• Compare costs and ingredients before you buy.

TABLE 13.4 Claims, Benefits, and Risks of Popular Ergogenic Aids

Ergogenic aid	Promoter claims	Proven benefits	Potential risks
Arginine, ornithine, and lysine	Causes the release of growth hormone, which stimulates muscle development and decreases body fat.	No increase in lean body mass observed with supplementation.	Reduced absorption of other amino acids. Diarrhea at high doses.
β-Alanine	Increases strength, power, and endurance during high-intensity exercise.	Increases muscle carnosine, which prevents changes in muscle acidity. Delays fatigue and increases power output during short intense activity.	Safe at moderate doses. High doses can cause flushing and tingling.
Bicarbonate (sodium bicarbonate, baking soda)	Helps buffer lactic acid produced during exercise and delays fatigue.	Increases blood pH and may enhance performance and strength during intense anaerobic activities.	Causes bloating, diarrhea, and high blood pH.
Branched chain amino acids (leucine, isoleucine, and valine)	Improves recovery from strength exercise.	No performance benefit, but may decrease muscle soreness after intense exercise.	No toxicity reported.
Caffeine	Increases the release of fatty acids from adipose tissue, spares glycogen, and enhances endurance.	Increases endurance in some individuals.	Nervousness, anxiety, insomnia, digestive discomfort, abnormal heartbeat.
Carnitine	Enhances the utilization of fatty acids and spares glycogen.	No increase in fatty acid utilization or improvement in exercise performance found.	D,L-carnitine and D-carnitine forms can be toxic.
Chromium (chromium picolinate)	Increases lean body mass, decreases body fat, delays fatigue.	No effect on protein or lipid metabolism unless a chromium deficiency exists.	No toxicity reported in humans.
Coenzyme Q10	Increases mitochondrial ATP production, acts as an antioxidant, and may combat fatigue.	No effect on exercise performance observed.	No toxicity reported.
Creatine (creatine monohydrate)	Increases ATP production and speeds recovery after high-intensity exercise.	Increases muscle creatine and creatine phosphate synthesis after exercise. Enhances strength, performance, and recovery from high-intensity exercise.	5 g/day for up to a year appears to be safe.
DHEA (dehydroepiandrosterone)	Builds muscles, burns fat, and delays chronic diseases associated with aging.	No proven benefits.	Acne, oily skin, facial hair, voice deepening, hair loss, mood changes, liver damage, and stimulation of existing cancers.
Glutamine	Increases muscle glycogen deposition following intense exercise, enhances immune function, and prevents the adverse effects of overtraining.	Little evidence that glutamine increases immune function, prevents the symptoms of overtraining, or increases glycogen synthesis.	No evidence of toxicity.
Glycerol	Improves hydration and endurance.	Evidence of an effect is equivocal.	May cause cellular dehydration, nausea, vomiting, diarrhea.
HMB (β-hydroxy-β-methylbutyrate)	Increases ability to build muscle in response to exercise.	Has not been found to enhance muscle size, strength, or endurance.	No toxicity in animals, but little information in humans.
Medium-chain triglycerides (MCT)	Provides energy without promoting fat deposition; increases the availability of fatty acids for aerobic metabolism.	Increase blood fatty acid levels but do not increase endurance or other aspects of performance.	None known.
Vanadium (vanadyl sulfate)	Aids insulin action; allows more rapid and intense muscle pumping for body builders.	No evidence to support a benefit for body builders.	Reduces insulin production.

Vitamin and Mineral Supplements

Many of the promises made to athletes about the benefits of vitamin and mineral supplements are extrapolated from their biochemical functions. For example, B-vitamin supplements are promoted to enhance ATP production because of the roles B vitamins play in muscle energy metabolism. Vitamins B_6, B_{12}, and folic acid are promoted for aerobic exercise because they are involved in the transport of oxygen to exercising muscles. These vitamins are indeed needed for energy metabolism, and a deficiency of one or more of these can interfere with ATP production and impair athletic performance; however, providing more than the recommended amount does not deliver more oxygen to the muscle, cause more ATP to be produced, or enhance athletic performance.

Supplements of vitamin E, vitamin C, β-carotene, and selenium are promoted to athletes because of their antioxidant functions. As discussed earlier, exercise increases oxidative processes and therefore increases the production of free radicals, which cause cellular damage and have been associated with fatigue during exercise. It has been suggested that antioxidant supplements reduce the levels of free radicals and hence delay fatigue. Research, however, has not found that supplementation of antioxidant nutrients improves performance or reduces oxidative stress (see Chapter 12 Debate: Antioxidant Supplements: Are They Beneficial?).[26] The best way to ensure adequate antioxidant protection is to consume a diet that includes plenty of whole grains, fruits, and vegetables, which are rich in antioxidant nutrients as well as antioxidant phytochemicals.

Supplements of chromium, vanadium, selenium, zinc, and iron are marketed to strengthen muscles or enhance endurance. As with vitamin supplements, many of the claims made about these minerals are based on their physiological functions. And as with vitamins, there is little evidence that consuming more than the recommended amount provides any benefits.[42]

Chromium supplements, as chromium picolinate, claim to increase lean body mass and decrease body fat. Chromium is needed for insulin action and insulin promotes protein synthesis. Therefore, adequate chromium status is likely to be important for lean tissue synthesis. The picolinate form is typically used because it is believed to be absorbed better than other forms of chromium. However, studies in humans have not consistently demonstrated an effect of supplemental chromium picolinate on muscle strength, body composition, body weight, or other aspects of health.[43]

Vanadium, usually as vanadyl sulfate, is another mineral marketed for its ability to promote the action of insulin. Vanadium supplements promise to increase lean body mass, but there is no evidence that they have an anabolic effect, and toxicity is a concern.[44]

Selenium is marketed for its antioxidant properties and zinc for its role in protein synthesis and tissue repair, but neither of these supplements has been found to improve athletic performance in individuals with adequate mineral status. Iron is also marketed as an ergogenic mineral because it is needed for hemoglobin synthesis. If an iron deficiency exists, as it frequently does in female athletes, supplements can be of benefit.

Supplements to Build Muscle

Hundreds of millions of dollars are spent each year on supplements to build muscle size, strength, and power. The most popular are protein supplements, but supplements of certain amino acids, as well as muscle-building hormones with dangerous side effects, are also used.

Do Protein Supplements Help Build Muscle? Protein supplements—from powders you mix in your beverage to bars you put in your backpack– are marketed to athletes with the promise of enhancing muscle growth or improving performance. Muscles enlarge in response to exercise. Resistance training stimulates muscle protein synthesis. Adequate protein is necessary for this to occur. Consuming high-quality proteins such as milk and soy proteins after intense resistance training has been found to maximize muscle growth stimulated by exercise.[45] This effect can be obtained by consuming protein from a protein supplement or a glass of chocolate milk. However, overall protein intake in excess of recommended amounts either as food or supplements does not increase muscle growth or strength.[44] If an athlete's diet provides enough energy, it usually provides enough protein without adding a supplement (see Critical Thinking: Getting Enough Protein?).

Growth Hormone and Growth Hormone Releasers Growth hormone, although banned by the World Anti-Doping Agency, is appealing to athletes because it increases muscle protein synthesis. Despite this physiological effect, however, it has not been shown to enhance muscle strength, power, or aerobic exercise capacity, but there is evidence that it improves anaerobic exercise capacity.[46] Prolonged use of growth hormone can cause heart dysfunction, high blood pressure, and excessive growth of some body parts, such as hands, feet, and facial features.

Supplements of the amino acids ornithine, arginine, and lysine are marketed with the promise that they will stimulate the release of growth hormone, which increases the growth of muscles. Large doses of these amino acids have been shown to stimulate the release of growth hormone, and some studies have found that supplements of these amino acids cause growth hormone levels to increase more than they do from exercise alone, but the effect this has on exercise performance needs further investigation.[47,48,49]

Anabolic Steroids and Steroid Precursors Anabolic steroids are synthetic versions of the human steroid hormone testosterone. Natural testosterone stimulates and maintains the male sexual organs and promotes the development of bones and muscles and the growth of skin and hair. The synthetic testosterone in anabolic steroids has a greater effect on muscle, bone, skin, and hair than it does on sexual organs. When taken in conjunction with exercise and an adequate diet, anabolic steroids cause increases in muscle size and strength. However, they have extremely dangerous side effects and they are controlled substances that are banned by the International Olympic Committee, the National Collegiate Athletic Association (NCAA), and most other sporting organizations.[50] Some of the side effects occur because these drugs make the body think natural testosterone is being produced, so the body shuts down its own testosterone production (**Figure 13.25**). Without natural testosterone, the sexual organs are not maintained; this leads to shrinkage of the testicles and a decrease in sperm production.[51]

LIFE CYCLE

In adolescents, the use of anabolic steroids causes cessation of bone growth and stunted height. Anabolic steroid use may also cause oily skin and acne, water retention in the tissues, yellowing of the eyes and skin, coronary artery disease, liver disease, and sometimes death. Users may experience psychological and behavioral side effects such as violent outbursts and depression, possibly leading to suicide.[51]

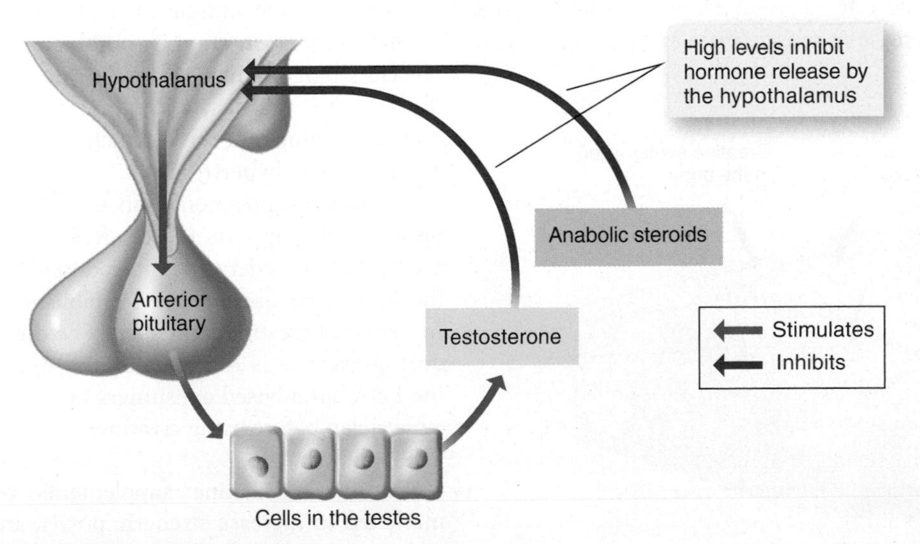

Figure 13.25 Anabolic steroids
When testosterone levels are low, the hypothalamus releases a hormone that stimulates the anterior pituitary to secrete a hormone that increases the production of testosterone by the testes. High levels of either natural or synthetic testosterone inhibit the release of the stimulatory hormone from the hypothalamus, shutting down the synthesis of natural testosterone.

THINK CRITICALLY

Why does anabolic steroid use promote muscle development while causing the testes to shrink?

Steroid precursors, which are compounds that can be converted into steroid hormones in the body, are also controlled substances. The best known of these is androstenedione, often referred to as "andro." It was launched into public prominence when professional baseball player Mark McGwire announced his use of it during the 1998 Major League Baseball season, when he hit 70 home runs, breaking the league's single-season home-run record. Contrary to marketing claims, the use of andro or other steroid precursors has not been found to increase testosterone levels or produce any ergogenic effects, and they may cause some of the same side effects as anabolic steroids.[52]

β-Hydroxy-β-Methylbutyrate β-Hydroxy-β-methylbutyrate, known as HMB, is a metabolite of the branched-chain amino acid leucine that is promoted as a muscle builder. Supplements of HMB are hypothesized to increase strength and muscle mass by decreasing muscle breakdown and speeding muscle repair. Early studies found that oral HMB supplements decreased muscle damage and increased muscle strength in response to resistance training.[53] However, more recent studies have not found these supplements to have an ergogenic effect on muscular strength, size, or endurance.[54]

Supplements to Enhance Performance in Short, Intense Activities

A number of supplements are marketed to athletes who seek to improve performance in sports that depend on quick bursts of intense activity. Some have benefits because they help the body cope with the metabolic stress of very intense activities.

Creatine Supplements Creatine is a nitrogen-containing compound found primarily in muscle, where it is used to make creatine phosphate, a source of energy for short-term exercise. It can be synthesized in the liver, kidneys, and pancreas from the amino acids arginine, glycine, and methionine. It is also consumed in the diet in meat and milk. The more creatine in the diet, the greater the amount of creatine stored in muscle. Creatine supplements increase levels of both creatine and creatine phosphate in muscle (**Figure 13.26**). Higher levels of these provide muscles with more quick energy for short-term maximal exercise. Creatine supplementation has been shown to improve performance in high-intensity exercise lasting 30 seconds or less. It is therefore beneficial for exercise that requires explosive bursts of energy, such as sprinting and weight lifting, but not for long-term endurance activities such as marathons.[55]

Creatine supplements are also taken by athletes to increase muscle mass and strength. Some of the increase in lean body mass that occurs when supplements are taken is believed to be due to water retention related to creatine uptake in the muscle. In addition, an increase in muscle mass and strength may occur when supplementation is combined with muscle-strengthening exercises because the increase in muscle creatine permits higher training intensity, which leads to greater muscle hypertrophy.[56]

Creatine supplementation at intakes of up to 5 g/day appears to be safe for up to a year, but the safety of higher doses over the long term has not been established.[57] Ingestion of creatine immediately before or during exercise is not recommended, and the FDA has advised consumers to consult a physician before using creatine.

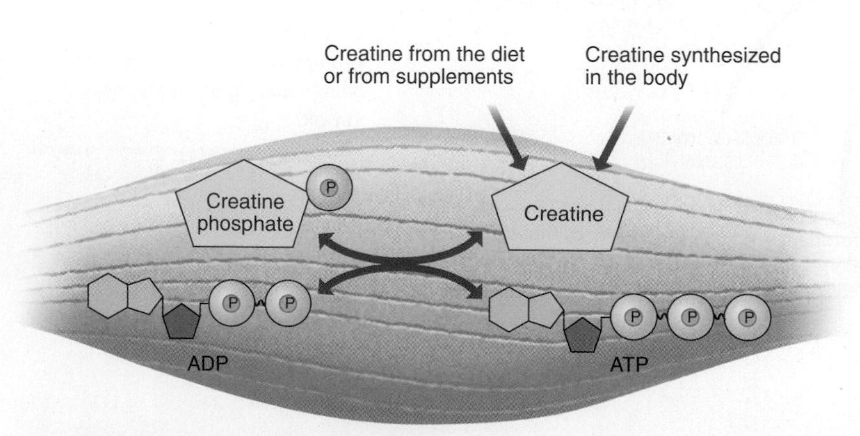

Figure 13.26 Creatine and creatine phosphate
The more creatine consumed, the greater the amount of creatine and creatine phosphate stored in the muscles. During short bursts of intense activity, creatine phosphate can transfer a phosphate group to ADP, forming creatine, and ATP that can be used for muscle contraction.

β-Alanine β-Alanine supplements are marketed to increase strength, power, and endurance during high-intensity exercise. β-Alanine is a nonessential amino acid that is used in the muscle to synthesize a compound called carnosine. Carnosine is a buffer that prevents changes in acidity within the muscle. During high-intensity exercise, lactic acid is produced, causing muscle pH to decline. The increase in acidity is believed to interfere with muscle function and contribute to fatigue. Supplements of β-alanine increase muscle carnosine concentrations, which is hypothesized to delay the onset of fatigue by preventing the decrease in muscle pH. Studies of β-alanine supplementation have found that they can delay fatigue and increase the ability to sustain muscle power output during high-intensity exercise lasting between one and four minutes.[58,59]

Bicarbonate Bicarbonate also acts as a buffer in the body. Supplementing it is thought to neutralize acid and thus delay fatigue and allow improved performance. Taking sodium bicarbonate, which is just baking soda from the kitchen cupboard, before exercise has been found to improve performance and delay exhaustion in sports, such as sprint cycling, that entail intense exercise lasting only one to seven minutes, but it is of no benefit for lower-intensity aerobic exercise.[60] Just because baking soda is an ingredient in your cookies does not mean that it is without risk. Many people experience abdominal cramps and diarrhea after taking sodium bicarbonate, and other possible side effects have not been carefully researched.

Branched-Chain Amino Acids Supplements of the branched-chain amino acids (leucine, isoleucine, and valine) are promoted to improve athletic performance by reducing protein breakdown and increasing protein synthesis in skeletal muscle during recovery from strength exercise. Although studies examining branched-chain amino acid supplements have not found them to enhance performance, there is evidence that they may have some benefits in improving muscle recovery and preventing muscle soreness after intense exercise.[61,62]

Supplements to Enhance Endurance

A primary performance concern for long-distance runners and cyclists is running out of glycogen. Glycogen is spared and endurance enhanced when aerobic metabolism predominates and fat is used as an energy source. A number of substances, including carnitine, medium-chain triglycerides, and caffeine, are taken to increase endurance by increasing the amount of fat available to the muscle cells. Athletes also use the banned hormone erythropoietin, known as EPO, to increase aerobic capacity (see **Science Applied: EPO: Lifesaver and Scandal**).

Carnitine Carnitine supplements are marketed as fat burners—substances that increase the utilization of fat during exercise. Carnitine is produced in the body from the amino acids lysine and methionine. It is needed to transport fatty acids into the mitochondria, where they are used to produce ATP by aerobic metabolism. Supplements promise to speed up the delivery of fatty acids to muscle mitochondria and therefore enhance the utilization of fat during exercise. Carnitine is made in the body, so it does not need to be supplied in the diet to ensure the efficient use of fatty acids. Studies that examined the effect of carnitine supplements found that they did not affect the utilization of fat as fuel during exercise or improve exercise endurance.[63]

Medium-Chain Triglycerides Most of the fatty acids used by the muscle to generate ATP are delivered in the blood, so, theoretically, higher levels in the blood increase the availability of fatty acids as a fuel for exercise. Most of the fatty acids consumed in the diet are incorporated into chylomicrons, which enter the lymphatic system before they appear in the blood. However, the fatty acids provided by medium-chain triglycerides (MCT) are only 8 to 10 carbons in length and are therefore soluble in water. They are absorbed directly into the blood without being incorporated into chylomicrons. In addition to being absorbed much more quickly, these fatty acids cross cell membranes more easily and can enter the mitochondria for oxidation without the help of carnitine. When MCTs are ingested, blood fatty acid levels increase. Despite this, research has not found that supplementation with MCT increases endurance, spares glycogen, or enhances performance.[64]

Caffeine Caffeine is a stimulant found in coffee, tea, and some soft drinks (see Debate: Energy Drinks for Athletic Performance?). Consuming 3 to 6 mg of caffeine per kg

SCIENCE APPLIED

EPO: Lifesaver and Scandal

Since the 19th century, it has been known that people who live at high altitudes have "thicker" blood (that is, more red blood cells) than people who live at low altitudes, and that moving to a higher altitude can increase red blood cell production.[1] More red blood cells means the blood can carry more oxygen to support aerobic metabolism. Understanding how the thin air at high altitudes stimulates red blood cell production would have significant impact on the treatment of anemia and, interestingly enough, sports performance and sports scandal.

THE SCIENCE

In an attempt to identify the substance responsible for stimulating red blood cell production, French researchers in the early 20th century examined the effects of bloodletting on rabbits. When the blood taken from bled rabbits was given to unbled ones, red blood cell production increased in these animals, indicating the presence of the substance that stimulated red blood cell synthesis. Scientists ultimately called this hormone erythropoietin after erythropoiesis, the process of red blood cell formation.

Interest in erythropoietin increased in the 1950s, when the cold war focused research priorities on ways to protect the human body from the effects of nuclear radiation, including anemia. In 1955 the Atomic Energy Commission awarded Dr. Eugene Goldwasser of the University of Chicago a grant to isolate and characterize erythropoietin. To find out where erythropoietin was produced, Dr. Goldwasser systematically removed the organs of laboratory rats. He concluded that the kidneys produced erythropoietin because blood withdrawal did not stimulate red blood synthesis when the kidneys were removed, but it did after removal of other organs. To isolate the compound he needed large amounts. A breakthrough came in 1975, when Dr. Takaji Mijake of Japan's Kumamoto University arrived in Chicago and met Dr. Goldwasser with a vial containing the dried concentrate of 2,550 liters, or about 674 gallons, of urine he had collected from patients with aplastic anemia.[1] This anemia causes excess erythropoietin to be produced and excreted in the urine. Mijake, Goldwasser, and their colleagues were able to isolate 8 mg of human erythropoietin from this concentrate.[2] The 20-year quest to isolate the hormone had come to an end.

THE APPLICATION

In 1980 Dr. Goldwasser gave a sample of purified erythropoietin to a small new California-based biotech company, which became known as Amgen. Scientists at Amgen went on to clone the gene that codes for erythropoietin, and use it to produce the protein through genetic engineering. Amgen was interested in erythropoietin, which became known as EPO, because it believed the hormone could cure various types of anemia, including that which occurs in patients with kidney failure. When kidney function declines, natural erythropoietin is no longer produced, resulting in anemia (see figure). In clinical trials, administration of genetically engineered EPO was shown to correct anemia in dialysis patients.[3] This led to the approval of the first recombinant human erythropoietin product in 1989. FDA approval of EPO launched Amgen product sales from $2.8 million in 1989 to roughly $140 million in 1990.[4]

The usefulness of genetically engineered EPO did not stop with kidney patients. It is now used to treat HIV patients who have anemia associated with antiviral drug treatments, cancer patients who are suffering from anemia associated with chemotherapy, and to boost blood production in patients who wish to donate their own blood for use during surgery.

[1]Fisher, J. W. Landmark advances in the development of erythropoietin. *Exp Biol Med* 235: 1398–1411, 2010.

[2]Miyake, T., Kung, C. K.-H., and Goldwasser, E., Purification of human erythropoietin. *J Biol Chem* 252:5558–5564, 1977.

[3]Eschbach, J. W., Egrie, J. C., Downing, M. R., et al. Correction of the anemia of end stage renal disease with recombinant human erythropoietin. Results of a combined phase I and II clinical trial. *N Eng J Med* 316:73–78, 1987.

[4]Amgen, Inc. Company History and Profiles. Available online at Amgen website.

[5]Noakes, T. D. Tainted glory—doping and athletic performance. *N Engl J Med* 351:847–849, 2004.

[6]World Anti-Doping Agency.

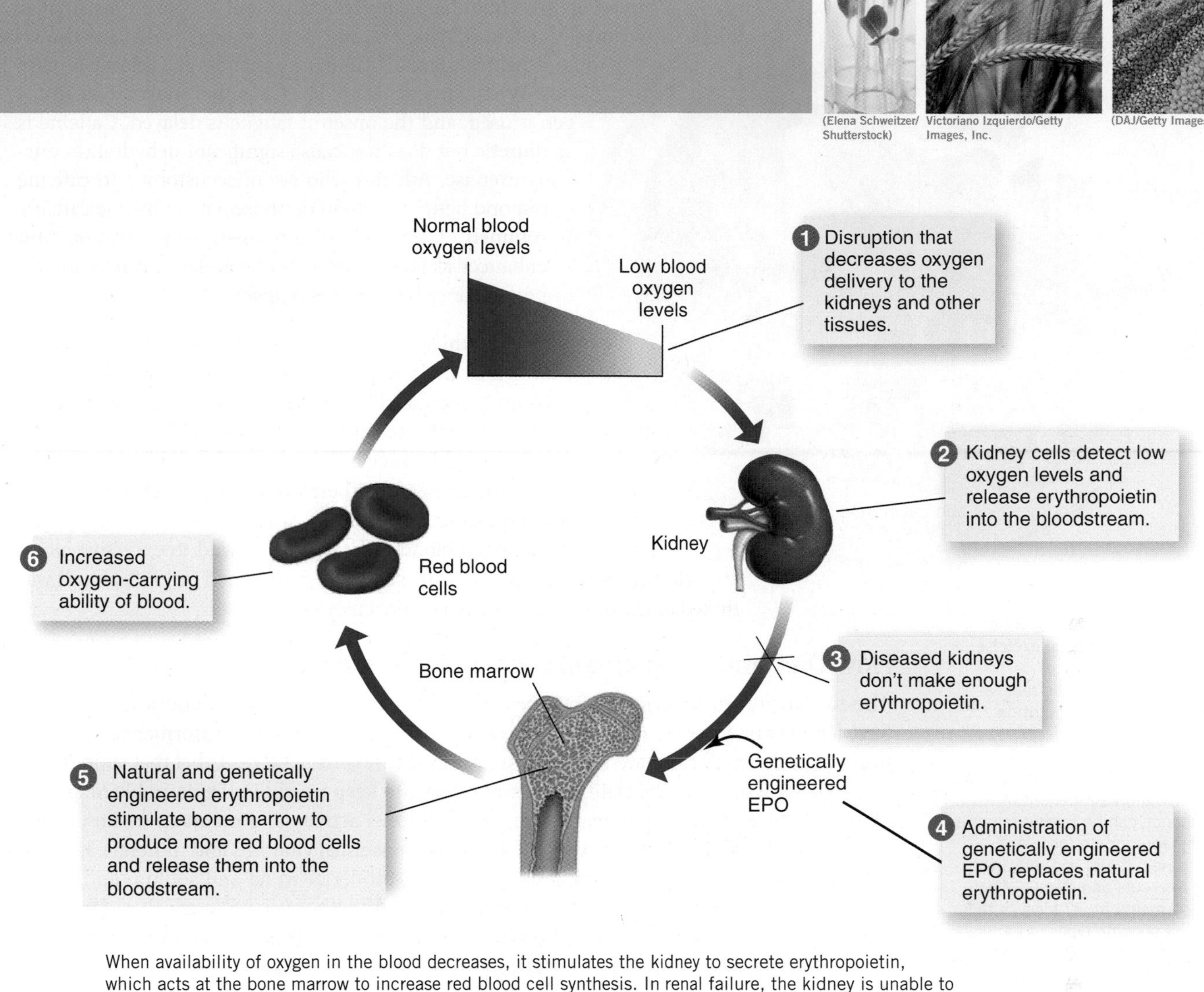

When availability of oxygen in the blood decreases, it stimulates the kidney to secrete erythropoietin, which acts at the bone marrow to increase red blood cell synthesis. In renal failure, the kidney is unable to produce erythropoietin, so red cell production declines, resulting in anemia. Administration of genetically engineered erythropoietin prevents or cures the anemia.

Athletes also became aware of the potential benefits of EPO. To increase aerobic capacity, athletes often train at high altitudes where low oxygen concentrations stimulate the production of red blood cells; upon returning to lower elevations their blood then delivers more oxygen to their exercising muscles. The use of oxygen deprivation tents also simulated this effect. Yet all these solutions are temporary. EPO provided a convenient, more lasting solution. Injections of EPO increase the total circulating red cell volume and therefore increase aerobic capacity and delay fatigue. Unfortunately, the benefits do not come without risks. In the late 1980s, 18 young professional cyclists died from apparent unknown causes when erythropoietin was first introduced into the world of cycling, and since then many other deaths have been linked to its use. It was determined that EPO can endanger athletes because the high concentration of red blood cells can clog arteries and lead to stroke and heart attack. EPO can also cause sudden death by reducing heart rate, usually at night, and users can develop antierythropoietin antibodies, which may cause a reduction in the red-cell mass.[5] The use of EPO by athletes has been banned since the early 1990s.[6]

Today EPO saves the lives of renal, cancer, and HIV patients, but has also led to scandal in the sports world as athletes are tempted by this drug that boosts their ability to produce ATP aerobically. Research now focuses not just on the clinical applications of EPO, but also on ways to detect the illegal use of EPO among athletes.

of body weight, an amount equivalent to about 2.5 cups of percolated coffee, up to an hour before exercising, as well as consuming smaller doses of caffeine during exercise (1 to 2 mg/kg), have been shown to improve endurance.[55] Caffeine enhances the release of fatty acids. When fatty acids are used as a fuel source, less glycogen is used, and the onset of fatigue is delayed. Caffeine is a diuretic but does not cause significant dehydration during exercise. Athletes who are unaccustomed to caffeine respond better to it than do those who consume caffeine routinely. Caffeine also improves concentration and enhances alertness, but in some athletes, it may impair performance by causing GI upset.

EPO Athletes also use the hormone erythropoietin, known as EPO, to enhance endurance. Natural erythropoietin is produced by the kidneys and stimulates cells in the bone marrow to differentiate into red blood cells. EPO can enhance endurance by increasing the ability to transport oxygen to the muscles. It therefore increases aerobic capacity and spares glycogen. However, too much EPO can cause production of too many red blood cells, which can lead to excessive blood clotting, heart attacks, and strokes. EPO was banned in 1990, after it was linked to the deaths of more than a dozen cyclists.[65]

Figure 13.27 The impact of diet and supplements on exercise performance
This figure illustrates the relative importance of various nutrition strategies for exercise performance. Along with talent and hard work, eating a healthy overall diet provides the most significant benefit. Performance can be further improved by using appropriate foods and fluids to help refuel and rehydrate during workouts and events. Where do supplements fit in? Most of them don't, but specific types of athletes in certain events will receive an additional small benefit from a few select supplements.

How Diet and Supplements Affect Performance

The foundation of superior athletic performance is talent, hard work, and a healthy diet. Supplements may garner most of the press when it comes to exercise performance, but their impact is very small compared to that of a healthy diet (Figure 13.27). A diet that supports an active lifestyle provides the right number of calories to keep weight in the desirable range; the proper balance of carbohydrate, protein, and fat to fuel activity and maintain tissues; plenty of water; and sufficient but not excessive amounts of essential vitamins and minerals. It is rich in whole grains, fruits and vegetables, high in fiber, moderate in fat and sodium and low in saturated fat, cholesterol, *trans* fat, and added sugars. Whether you are a couch potato or an Olympic hopeful, the recommendations of MyPlate and the Dietary Guidelines can help you choose a diet that provides the foundation for optimal performance.

OUTCOME

© mbbirdy/iStockphoto

© technotr/iStockphoto

After reaching the rim of Crater Lake, Jason takes his time at the lunch stop, not only to enjoy the gorgeous vistas, but also to eat a good lunch and drink plenty of fluids. After refueling and rehydrating, Jason feels good for the rest of the day's ride. Jason ran into trouble because he didn't feel hungry, so he passed on the opportunity to pick up food and water. His friend knew to anticipate his needs by eating every hour or two and drinking often, even if he didn't feel hungry or thirsty, because it takes time for dietary carbohydrate to boost blood glucose levels. By not eating, Jason had cut off the supply of glucose from his diet. On the uphill ride to Crater Lake, his body was forced to use stored glycogen faster, depleting his muscle and liver stores. His liver could not supply blood glucose fast enough to meet his needs. He was also mildly dehydrated, further contributing to his fatigue. The next day had more hard climbs, but Jason made sure his water bottle was filled and he ate some food at every rest stop and took additional snacks with him. **Jason and his friend completed the ride and are hoping to do a similar ride next year.**

APPLICATIONS

ASSESSING YOUR DIET

1. **How much exercise do you get?**
   a. Keep a log of your activity for one day.
   b. Refer to Chapter 7, Figure 7.14 to help you determine your physical activity level and PA value.
   c. Use Table 7.3 to determine your estimated energy expenditure (EER).
   d. If you increased your exercise enough to move to the "active" physical activity level, what would your new EER be? (If you are already active, what would your EER be in the "very active" level?)
   e. Use iProfile to find foods that you could add to your diet to balance the added expenditure of this increase in activity.

2. **What types and amounts of exercise work for you?**
   a. Taking into consideration your typical weekly schedule of activities and events, design a reasonable exercise program for yourself. Include the types of activities, the times during the week you will be involved in each activity, and the length of time you will engage in each activity. Choose activities you enjoy and schedule them for practical lengths of time and at reasonable frequencies.
   b. Which activities are aerobic, which are for muscle strengthening? Are any of your activities bone-strengthening?
   c. Can each of these activities be performed year-round? If not, suggest alternative activities and locations for inclement weather.
   d. What everyday changes could you make that would increase the energy expended in day-to-day activities?

CONSUMER ISSUES

3. **David is beginning an exercise program. He plans to run before lunch and then play racquetball every night after dinner. Once he begins his exercise program, he finds that he feels lethargic and hungry before his late-morning run. After running, he doesn't have much of an appetite, so he saves his lunch until midafternoon. He is still hungry enough to eat dinner at home with his family, but finds that he is getting stomach cramps and is too full when he goes to play racquetball. His typical diet is**

   Breakfast: Orange juice, coffee

   Lunch: Ham and cheese sandwich, potato chips, soft drink, cookies

   Dinner: Steak, baked potato with sour cream and butter, green beans in butter sauce, salad with Italian dressing, whole milk

   a. How might David change his diet so it is better suited to his exercise program?
   b. Does his exercise program include both an aerobic and a strength-training component?
   c. Do you think David will be able to stick with this exercise program? Why or why not?
   d. Suggest some changes that would make David's exercise program more balanced.

4. **Do a risk-benefit analysis of an ergogenic aid. List the risks and benefits and then write a conclusion stating why you would or would not recommend this substance.**

CLINICAL CONCERNS

5. **Two friends are running a marathon together. One has participated in an intensive training program. The other was too busy and trained only a few hours a week. About five minutes into the marathon, they have settled into a slow, steady pace and are able to carry on a conversation. After an hour, the well-trained individual is feeling good and so increases her pace. The untrained person tries to keep up but is no longer able to talk, and after about 15 minutes is fatigued and needs to stop. Why does the untrained person tire faster?**

6. **Diana is undergoing chemotherapy for breast cancer. She is extremely tired and breathless. A blood sample reveals that she is anemic. Her doctor stops the chemotherapy and prescribes several injections of erythropoietin. Her son, who is a long-distance runner, tells her that erythropoietin, or EPO, is a banned substance for sports and has dangerous side effects.**
   a. Why was EPO prescribed for Diana?
   b. Why might athletes want to take this drug? What type of athlete?
   c. Compare the risks and benefits for Diana to those of a healthy athlete using the drug.

SUMMARY

13.1 Exercise, Fitness, and Health

- How fit an individual is depends on his or her cardiorespiratory endurance, muscle strength, muscle endurance, flexibility, and body composition. Regular aerobic exercise increases aerobic capacity and resistance exercise increases muscle strength.
- Go to WileyPLUS to view a video clip on the health benefits of exercise VIDEO
- Regular exercise can reduce the risk of chronic diseases such as obesity, heart disease, diabetes, osteoporosis, and certain types of cancer. Exercise helps manage body weight by increasing energy expenditure and by increasing the proportion of body weight that is lean tissue. Exercise also improves sleep patterns and promotes well-being by improving mood, boosting self-esteem, and reducing depression and anxiety.

13.2 Exercise Recommendations

- Current exercise recommendations suggest a minimum of 150 minutes per week of moderate intensity aerobic activity. In addition to aerobic exercise, an exercise program should include muscle-strengthening exercise and exercises to promote flexibility, balance, and agility. Time spent in strength and flexibility training does not count towards the goal of 150 minutes of aerobic exercise per week.

- Exercise recommendations for children suggest a minimum of 60 minutes of activity per day, some of which is muscle-strengthening and bone-strengthening. Activities should be age appropriate.
- An exercise program should include activities that are enjoyable, convenient, and safe. Various combinations of exercise frequency, intensity, and duration can meet exercise goals. Rest is important to allow the body to recover and rebuild. In serious athletes, inadequate rest can lead to overtraining syndrome.

13.3 Exercise and Energy Metabolism

- During the first 10 to 15 seconds of exercise, ATP and creatine phosphate stored in the muscle provide energy to fuel activity. During the next two to three minutes, the amount of oxygen at the muscle remains limited, so ATP is generated by the anaerobic metabolism of glucose. After a few minutes, the delivery of oxygen at the muscle increases, and ATP can be generated by aerobic metabolism. Aerobic metabolism is more efficient than anaerobic metabolism and can utilize glucose, fatty acids, and amino acids as energy sources. The use of protein as an energy source increases when exercise continues for many hours.
- For short-term, high-intensity activity, ATP is generated primarily from the anaerobic metabolism of glucose from muscle glycogen stores. Exercise that relies on anaerobic metabolism depletes glycogen rapidly, resulting in fatigue. For lower-intensity exercise of longer duration, aerobic metabolism predominates, and both glucose and fatty acids are important fuel sources.
- Fitness training causes changes in the cardiovascular system and muscles that improve oxygen delivery and utilization and increase glycogen stores, allowing aerobic exercise to be sustained for longer periods at higher intensity.

13.4 Energy and Nutrient Needs for Physical Activity

- The amount of energy needed for activity depends on the intensity, duration, and frequency of the activity, as well as the characteristics of the exerciser. The EER is increased by increases in activity level. The desire for weight loss to improve performance can lead to unhealthy weight-loss practices and eating disorders. A combination of adequate protein, resistance exercise, and increased energy intake can help promote an increase in weight and muscle mass.
- To maximize glycogen stores, optimize performance, and maintain and repair lean tissue, a diet providing about 60% of energy from carbohydrate, 20 to 25% of energy from fat, and 15 to 20% of energy from protein is recommended.
- Sufficient vitamins and minerals are needed to generate ATP from macronutrients, to maintain and repair tissues, and to transport oxygen and wastes to and from the cells. Most athletes who consume a varied diet that meets their energy needs also meet their vitamin and mineral needs. Those who restrict their food intake may be at risk for deficiencies. Increased iron needs and greater iron losses due to fitness training put athletes, particularly female athletes, at risk of iron deficiency. Athletes are also at risk of deficiencies of calcium and vitamin D. Female athletes may develop the female athlete triad, a combination of eating disorders, amenorrhea, and osteoporosis.

13.5 Fluid Needs for Physical Activity

- Water is needed to ensure that the body can be cooled and that nutrients and oxygen can be delivered to body tissues. If water intake is inadequate, dehydration can lead to a decline in exercise performance and increase the risk of heat-related illness. The risk of dehydration is increased when environmental temperature and humidity are elevated.
- Exercising in hot weather can lead to heat-related illnesses, including heat cramps, heat exhaustion, and heat stroke.
- Drinking plain water during extended exercise increases the risk of hyponatremia, a low concentration of sodium in the blood.
- Adequate fluid intake before exercise ensures that athletes begin exercise well hydrated. Fluid intake during and after exercise must replace water lost in sweat and from evaporation through the lungs. Plain water is an appropriate fluid to consume for most exercise. Beverages containing carbohydrate and sodium are recommended for exercise lasting more than an hour.

13.6 Food and Drink to Maximize Performance

- Competitive endurance athletes may benefit from a dietary regimen called glycogen supercompensation (carbohydrate loading), which maximizes glycogen stores before an event.
- Meals eaten before competition should help ensure adequate hydration, provide moderate amounts of protein, be high enough in carbohydrate to maximize glycogen stores, be low in fat and fiber to speed gastric emptying, and satisfy the psychological needs of the athlete.
- During exercise, athletes need beverages and food to replace lost fluid and provide carbohydrate and sodium.
- Postcompetition meals should replace lost fluids and electrolytes, provide carbohydrate to restore muscle and liver glycogen, and provide protein for muscle protein synthesis and repair.

13.7 Ergogenic Aids: Do Supplements Enhance Athletic Performance?

- Many types of ergogenic aids are marketed to improve athletic performance. An individual risk–benefit analysis should be used to determine whether a supplement is appropriate for you.
- Vitamin and mineral supplements are usually not necessary for athletes who are consuming a diet that meets their energy needs.
- Protein supplements are popular for athletes trying to build muscle. Adequate protein is necessary to support muscle growth, but dietary protein can meet this need as effectively as the protein provided by supplements. Anabolic steroids combined with resistance-training exercise increase muscle size and strength, but these supplements are illegal and have dangerous side effects.
- Creatine and β-alanine supplements have been shown to have benefits for improving performance in short-duration, high-intensity exercise.

- Caffeine use can improve performance in endurance activities. EPO increases endurance by boosting the number of red blood cells, but it is a banned substance that can contribute to excessive blood clotting, heart attacks, and strokes.
- A healthy diet is the base for successful athletic performance. Beverages and foods that supply fluids and energy during exercise can enhance performance, and there are a few supplements that benefit specific activities.

REVIEW QUESTIONS

1. What characterizes a fit individual?
2. What is aerobic exercise?
3. How does aerobic exercise affect resting heart rate?
4. What is strength training?
5. What causes muscle hypertrophy? Muscle atrophy?
6. List five of the health benefits of exercise.
7. How much of what types of exercise is recommended?
8. What is aerobic capacity? How is it affected by training?
9. From where does the ATP to fuel the first few minutes of exercise come?
10. What fuels are used to produce ATP in anaerobic metabolism?
11. Why is aerobic metabolism more efficient than anaerobic metabolism?
12. What factors affect the availability of oxygen at the muscle cell and the type of fuel used during exercise?
13. What fuels are used in exercise of long duration, such as marathon running?
14. What are the recommendations for fluid intake before, during, and after exercise?
15. How does exercise affect protein needs?
16. What is glycogen supercompensation or carbohydrate loading?
17. Explain why supplements of creatine and of β-alanine are ergogenic. What types of athletes would benefit?

REFERENCES

1. Garber, C. E., Blissmer, B., Deschenes, M. R., et al. American College of Sports Medicine. American College of Sports Medicine position stand: Quantity and quality of exercise for developing and maintaining cardiorespiratory, musculoskeletal, and neuromotor fitness in apparently healthy adults: guidance for prescribing exercise. *Med Sci Sports Exerc* 43:1334–1359, 2011.
2. Behm, D. G., and Chaouachi, A. A review of the acute effects of static and dynamic stretching on performance. *Eur J Appl Physiol* 111:2633–2651, 2011.
3. Gallagher, D., Heymsfield, S., Heo, M. et al. Healthy percentage body fat ranges: an approach for developing guidelines based on body mass index. *Am J Clin Nutr* 72:694–701, 2000.
4. United States Department of Health and Human Services. *2008 Physical Activity Guidelines for Americans.*
5. Ahmed, H. M., Blaha, M. J., Nasir, K., et al. Effects of physical activity on cardiovascular disease. *Am J Cardiol* 109:288–295, 2012.
6. Teixeira-Lemos, E., Nunes, S., Teixeira, F., and Reis, F. Regular physical exercise training assists in preventing type 2 diabetes development: Focus on its antioxidant and anti-inflammatory properties. *Cardiovasc Diabetol* 10:12, 2011.
7. Jones, G., Schultz, M. G., and Dore, D. Physical activity and osteoarthritis of the knee: Can MRI scans shed more light on this issue? *Phys Sportsmed* 39:55–61, 2011.
8. Newton, R. U., and Galvão, D. A. Exercise in prevention and management of cancer. *Curr Treat Options Oncol* 9:135–146, 2008.
9. Reigle, B. S., and Wonders, K. Breast cancer and the role of exercise in women. *Methods Mol. Biol* 472:169–189, 2009.
10. Institute of Medicine, Food and Nutrition Board. *Dietary Reference Intakes for Energy, Carbohydrates, Fiber, Fat, Protein and Amino Acids.* Washington, DC: National Academies Press, 2002.
11. Fontaine, K. R. Physical activity improves mental health. *Phys Sportsmed* 28:83–84, 2000.
12. Aan het Rot, M., Collins, K. A., and Fitterling, H. L. Physical exercise and depression. *Mt Sinai J Med* 76:204–214, 2009.
13. Healthy People 2020. *Physical Activity Objectives.*
14. U.S. Department of Agriculture and U.S. Department of Health and Human Services. *Dietary Guidelines for Americans, 2010*, 7th ed., Washington, DC: U.S. Government Printing Office, December 2010.
15. Haskell, W. L., Lee, I-M., Pate, R. R., et al. Physical activity and public health: updated recommendations for adults from the American College of Sports Medicine and the American Heart Association. *Circulation* 116:1081–1093, 2007.
16. Gulati, M., Shaw, L. J., Thisted, R. A., et al. Heart rate response to exercise stress testing in asymptomatic women: The St. James Women Take Heart project. *Circulation* 122:130–137, 2010.

17. Swinburn, B., and Shelly, A. Effect of TV time and other sedentary pursuits. *Int J Obes (Lond)* 32:S5132–S5136, 2008.

18. Allen, D. G., Lamb, G. D., and Westerblad, H. Skeletal muscle fatigue: Cellular mechanisms. *Physiol Rev* 88:287–332, 2008.

19. Manore, M., and Thompson, J. *Sport Nutrition for Health and Performance*. Champaign, IL: Human Kinetics, 2000.

20. Rodriguez, N. R., DiMarco, N. M., and Langley, S. Position of the American Dietetic Association, Dietitians of Canada, and the American College of Sports Medicine: Nutrition and athletic performance. *J Am Diet Assoc* 109:509–527, 2009.

21. Sundgot-Borgen, J., and Torstveit, M. K. Aspects of disordered eating continuum in elite high-intensity sports. *Scand J Med Sci Sports* 20 (suppl 2):112–121, 2010.

22. Remick, D., Chancellor, K., Pederson, J. et al. Hyperthermia and dehydration-related deaths associated with intentional rapid weight loss in three collegiate wrestlers—North Carolina, Wisconsin, and Michigan, November–December, 1997. *MMWR Morb Mortal Wkly Rep* 47:105–108, 1998.

23. Tipton, K .D. Efficacy and consequences of very-high-protein diets for athletes and exercisers. *Proc Nutr Soc* 7:1–10, 2011.

24. Jackson, M. J., Khassaf, M., Vasilaki, F., et al. Vitamin E and the oxidative stress of exercise. *Ann NY Acad Sci* 1031:158–168, 2004.

25. Powers, S. K., and Jackson, M. J. Exercise-induced oxidative stress: Cellular mechanisms and impact on muscle force production. *Physiol Rev* 88:1243–1276, 2008.

26. Margaritis, I., and Rousseau, A. S. Does physical exercise modify antioxidant requirements? *Nutr Res Rev* 21:3–12, 2008.

27. Food and Nutrition Board, Institute of Medicine. *Dietary Reference Intakes: Vitamin A, Vitamin K, Arsenic, Boron, Chromium, Copper, Iodine, Iron, Manganese, Molybdenum, Nickel, Silicon, Vanadium, and Zinc*. Washington, DC: National Academies Press, 2001.

28. Di Santolo, M., Stel, G., Banfi, G., et al. Anemia and iron status in young fertile non-professional female athletes. *Eur J Appl Physiol* 102:703–709, 2008.

29. Bartoszewska, M., Kamboj, M., and Patel, D. R. Vitamin D, muscle function, and exercise performance. *Pediatr Clin North Am* 57:849–861, 2010.

30. Halliday, T. M., Peterson, N. J., Thomas, J. J., et al. Vitamin D status relative to diet, lifestyle, injury, and illness in college athletes. *Med Sci Sports Exerc* 43: 335–343, 2011.

31. Lambrinoudaki, I., and Papadimitriou, D. Pathophysiology of bone loss in the female athlete. *Ann N Y Acad Sci* 1205:45–50, 2010.

32. Saltin, B., and Castill, D. I. Fluid and electrolyte balance during prolonged exercise. In *Exercise, Nutrition, and Energy Metabolism*, E. S. Horton and R. I. Tergung, eds. New York: Macmillan, 1988.

33. NOAA's National Weather Service: *Heat Index*.

34. Almond, C. S. D., Shin, A. Y., Fortescue, E. B., et al. Hyponatremia among runners in the Boston marathon. *N Engl J Med* 352:1550–1556, 2005.

35. Bussar, V. A., Fairchild, T. J., Rao, A., et al. Carbohydrate loading in human muscle: An improved 1-day protocol. *Eur J Appl Physiol* 87:290–295, 2002.

36. Burke, L. M. Nutrition strategies for the marathon: Fuel for training and racing. *Sports Med* 37:344–347, 2007.

37. Bergstrom, J., Hermansen, L., Hultman, E., and Saltin, B. Diet, muscle glycogen and physical performance. *Acta Physiol Scand* 71:140–150, 1967.

38. Karp, J. R., Johnston, J. D., Tecklenburg, T. D., et al. Chocolate milk as a post-exercise recovery aid. *Int J Sport Nutr Exerc Metab* 16:78–91, 2006.

39. Beelen, M., Burke, L. M., Gibala, M. J., and van Loon, L. J. C. Nutritional strategies to promote postexercise recovery. *Int J Sport Nutr Exerc Metab* 20:515–532, 2010.

40. Berardi, J. M., Price, T. B., Noreen, E. E., and Lemon, P. W. Postexercise muscle glycogen recovery enhanced with a carbohydrate-protein supplement. *Med Sci Sports Exerc* 38: 1106–1113, 2006.

41. Howarth, K. R., Moreau, N. A., Phillips, S. M., and Gibala, M. J. Coingestion of protein with carbohydrate during recovery from endurance exercise stimulates skeletal muscle protein synthesis in humans. *J Appl Physiol* 106:1394–1402, 2009.

42. Williams, M. H. Dietary supplements and sports performance: Minerals. *J Internat Soc Sport Nutr* 2:43–49, 2005.

43. Di Luigi, L. Supplements and the endocrine system in athletes. *Clin Sports Med* 27:131–151, 2008.

44. Nissen, S. L. and Sharp, R. L. Effect of dietary supplements on lean mass and strength gains with resistance exercise: A meta-analysis. *J Appl Physiol* 94:651–659, 2003.

45. Phillips, S. M., Tang, J. E., and Moore, D. R. The role of milk- and soy-based protein in support of muscle protein synthesis and muscle protein accretion in young and elderly persons. *J Am Coll Nutr* 28:343–354, 2009.

46. Birzniece, V., Nelson, A. E., and Ho, K. K. Growth hormone and physical performance. *Trends Endocrinol Metab* 22: 171–178, 2011.

47. Kanaley, J. A., Growth hormone, arginine and exercise. *Curr Opin Clin Nutr Metab Care* 11:50–54, 2008.

48. Chromiak, J. A., and Antonio, J. Use of amino acids as growth hormone-releasing agents by athletes. *Nutrition* 18:657–661, 2002.

49. Zajac, A., Poprzecki, S., Zebrowska, A., et al. Arginine and ornithine supplementation increases growth hormone and insulin-like growth factor-1 serum levels after heavy-resistance exercise in strength-trained athletes. *J Strength Cond Res.* 24:1082–1090, 2010.

50. NutraBio.com. *Congress Passes Steroid Control Act.*

51. Van Amsterdam, J., Opperhuizen, A., and Hartgens, F. Adverse health effects of anabolic-androgenic steroids. *Regul Toxicol Pharmacol* 57:117–123, 2010.

52. Brown, G. A., Vukovich, M., and King, D. S. Testosterone prohormone supplements. *Med Sci Sports Exerc* 38:1451–1461, 2006.

53. Nissen, S., Sharp, R., Ray, M., et al. Effect of leucine metabolite beta-hydroxy-beta-methylbutyrate on muscle metabolism

during resistance-exercise training. *J Appl Physiol* 81: 2095–2104, 1996.

54. O'Connor, D. M., and Crowe, M. J. Effects of six weeks of beta-hydroxy-beta-methylbutyrate (HMB) and HMB/creatine supplementation on strength, power, and anthropometry of highly trained athletes. *J Strength Cond Res* 21:419–423, 2007.

55. Tarnopolsky, M. A. Caffeine and creatine use in sport. *Ann Nutr Metab* 57(suppl. 2):1–8, 2011.

56. Volek, J. S., and Rawson, E. S. Scientific basis and practical aspects of creatine supplementation for athletes. *Nutrition* 20:609–614, 2004.

57. Shao, A., and Hathcock, J. N. Risk assessment for creatine monohydrate. *Regul Toxicol Pharmacol* 45:242–251, 2006.

58. Hobson, R. M., Saunders, B., Ball, G., et al. Effects of β-alanine supplementation on exercise performance: A meta-analysis. *Amino Acids* 43:25–37, 2012.

59. Culbertson, J. Y., Kreider, R. B., Greenwood, M., and Cooke, M. Effects of beta-alanine on muscle carnosine and exercise performance: A review of the current literature. *Nutrients* 2:75–98, 2010.

60. Raymer, G. H., Marsh, G. D., Kowalchuk, J. M., and Thompson, R. T. Metabolic effects of induced alkalosis during progressive forearm exercise to fatigue. *J Appl Physiol* 96:2050–2056, 2004.

61. Shimomura, Y., Inaguma, A., Watanabe, S., et al. Branched-chain amino acid supplementation before squat exercise and delayed-onset muscle soreness. *Int J Sport Nutr Exerc Metab* 20:236–244, 2010.

62. Negro, M., Giardina, S., Marzani, B., and Marzatico, F. Branched-chain amino acid supplementation does not enhance athletic performance but affects muscle recovery and the immune system. *J Sports Med Phys Fitness* 48:347–351, 2008.

63. Smith, W. A., Fry, A. C., Tschume, L. C., and Bloomer, R .J. Effect of glycine propionyl-L-carnitine on aerobic and anaerobic exercise performance. *Int J Sport Nutr Exerc Metab* 18: 19–36, 2008.

64. Clegg, M. E. Medium-chain triglycerides are advantageous in promoting weight loss although not beneficial to exercise performance. *Int J Food Sci Nutr* 61:653–679, 2010.

65. Birkeland, K. I., Stray-Gundersen, J., Hemmersbach, P., et al. Effect of rhEPO administration on serum levels of sTfR and cycling performance. *Med Sci Sports Exerc* 32:1238–1243, 2000.

***To access links to online sources, please go to www.wiley.com/college/smolin and select Nutrition: Science and Applications, 3rd edition. From this page, select either the student or instructor companion site. Once on the desired site, select References.**

CHAPTER OUTLINE

UpperCut Images/Getty Images, Inc.

Nutrition During Pregnancy and Lactation

14

CASE STUDY

Jasmine is 26 years old and expecting her first baby. She is at a healthy prepregnancy weight and exercises regularly by jogging or bicycling. At her initial prenatal visit, the obstetrician prescribes a prenatal supplement and gives her a pamphlet on nutrition during pregnancy. The pamphlet recommends that she gain 25 to 35 pounds over the course of her pregnancy. Jasmine is more concerned about this weight gain than her nutrient intake.

Jasmine has a sister with three young children who is now 25 lbs heavier than she was before the birth of her first child eight years ago. Her sister also developed gestational diabetes during her last pregnancy. She had a difficult delivery because the baby was very large and she is still struggling to keep her blood sugar under control. Jasmine doesn't want to end up like that. She decides to continue to follow her usual exercise and eating patterns. After all, she reasons, the prenatal supplement will give her all the nutrients her baby needs.

By her next prenatal visit, during the fourth month of her pregnancy, Jasmine has gained only 1 lb. She tells the doctor that she has been limiting what she eats because she is afraid of gaining too much weight. He explains that a deficiency of nutrients or energy may cause birth defects or increase the risk of having a baby that is born too soon or too small. Her prenatal supplement is designed to be taken in addition to a healthy diet. Eating a balanced diet is essential to supply all of the energy and protein necessary for the changes that occur in her physiology during pregnancy, as well as for the growth and development of the infant. Exercising regularly will help Jasmine avoid excess weight gain, but as her pregnancy progresses she may need to modify her exercise activities to ones that minimize the risk of injury. When Jasmine expresses her concern about gestational diabetes, the doctor tells her that keeping her weight gain within recommendations, limiting foods high in sugar, and exercising regularly will help to keep her blood sugar healthy. **He assures Jasmine that they will check for diabetes in a few weeks.**

© Miroslav Ferkuniak/iStockphoto

14.1 The Physiology of Pregnancy

LEARNING OBJECTIVES

- Explain why the mother's nutrient intake during pregnancy is so important.
- Discuss the risks associated with gaining too little or too much weight during pregnancy.
- List some types of exercise that are appropriate for pregnant women.
- Describe some of the nutrition-related discomforts and complications of pregnancy.

conception The union of sperm and egg (ovum) that results in pregnancy.

lactation Milk production and secretion.

fertilization The union of sperm and egg (ovum).

oviduct or **fallopian tube** Narrow duct leading from the ovary to the uterus.

zygote The cell produced by the union of sperm and ovum during fertilization.

Pregnancy, from **conception** to birth, usually lasts 40 weeks in humans. During pregnancy, a single cell grows and develops into an infant that is ready for life outside the womb. This development requires a safe environment to which oxygen and nutrients are provided in the right amounts and from which waste products are removed. Many physiological changes take place in the mother to support prenatal development and prepare her for **lactation**.

Prenatal Growth and Development

Reproduction requires the **fertilization** of an egg, or ovum, from the mother by a sperm from the father. Fertilization, which occurs in the **oviduct** or **fallopian tube**, produces a single-celled **zygote**. The zygote travels down the mother's fallopian tube into the uterus.

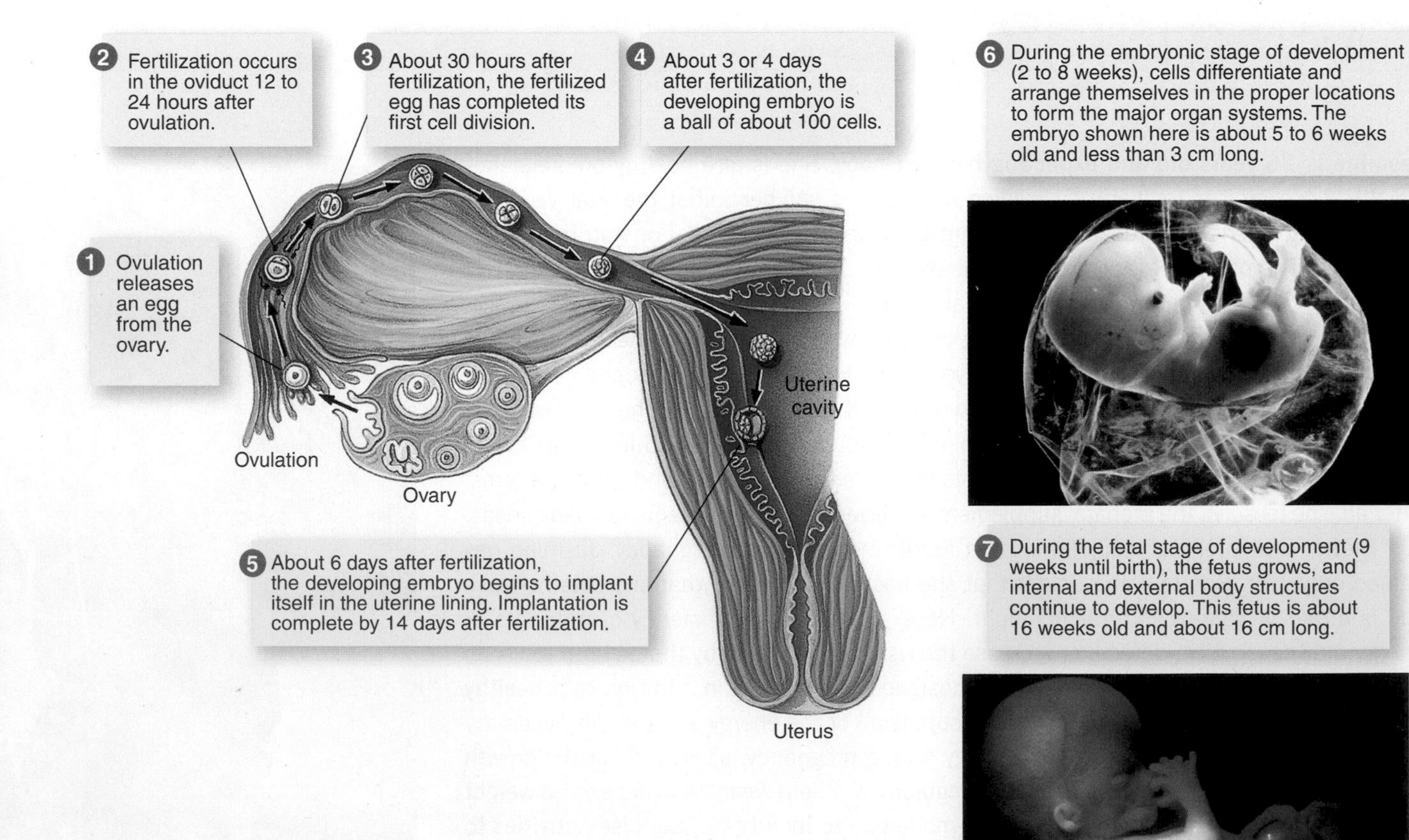

Figure 14.1 Prenatal development
This cross section of the female reproductive tract shows the path of the egg and developing embryo from the ovary, where the egg is produced, through the oviduct to the uterus, where most prenatal development occurs.

Along the way, it divides many times to form a ball of smaller cells. In the uterus, the developing embryo imbeds in the uterine lining in a process known as **implantation** (**Figure 14.1**).

implantation The process by which the developing embryo embeds in the uterine lining.

Prenatal growth and development continue after implantation. The cells differentiate into the multitude of specialized cell types that make up the human body, and arrange themselves in the proper shapes and locations to form organs and other structures. About two weeks after fertilization, implantation is complete and the developing offspring is known as an **embryo** (see Figure 14.1). The embryonic stage of development lasts until the eighth week after fertilization, when rudimentary organ systems have been formed. The embryo at this point is approximately 3 centimeters long (a little more than an inch) and has a beating heart. All major external and internal structures have been formed. Beginning at the ninth week of development and continuing until birth, the developing offspring is known as a **fetus** (see Figure 14.1). During the fetal period of development, structures that appeared during the embryonic period continue to grow and mature. Anything that interferes with development can cause birth defects. If the birth defects are severe, they may result in a **spontaneous abortion** or **miscarriage**.

embryo The developing human from two to eight weeks after fertilization. All organ systems are formed during this time.

fetus The developing human from the ninth week to birth. Growth and refinement of structures occur during this time.

spontaneous abortion or **miscarriage** Interruption of pregnancy prior to the seventh month.

The early embryo gets its nourishment by breaking down the lining of the uterus, but soon this source is inadequate to meet its growing needs. After about five weeks, the **placenta** takes over the role of nourishing the embryo (**Figure 14.2**). The placenta is a network of blood vessels and tissues that allows nutrients and oxygen to be transferred from mother to fetus and waste products to be transferred from the fetus to the mother's blood for elimination.

placenta An organ produced from both maternal and embryonic tissues. It secretes hormones, transfers nutrients and oxygen from the mother's blood to the fetus, and removes wastes.

The placenta is made up of tissue from both the mother and the fetus. The maternal portion of the placenta develops from the uterine lining. The fetal portion of the placenta develops from the outer layer of pre-embryonic cells. These cells divide to form branch-like projections that grow into the lining of the uterus, where they are surrounded by pools of maternal blood. The projections contain blood vessels that supply the developing fetus (see Figure 14.2). Although maternal and fetal blood do not mix, the close proximity of fetal blood vessels to maternal blood allows nutrients and oxygen to easily pass from mother to baby, and allows carbon dioxide and other wastes to pass from baby to mother for elimination. The placenta also secretes hormones that are necessary to maintain pregnancy.

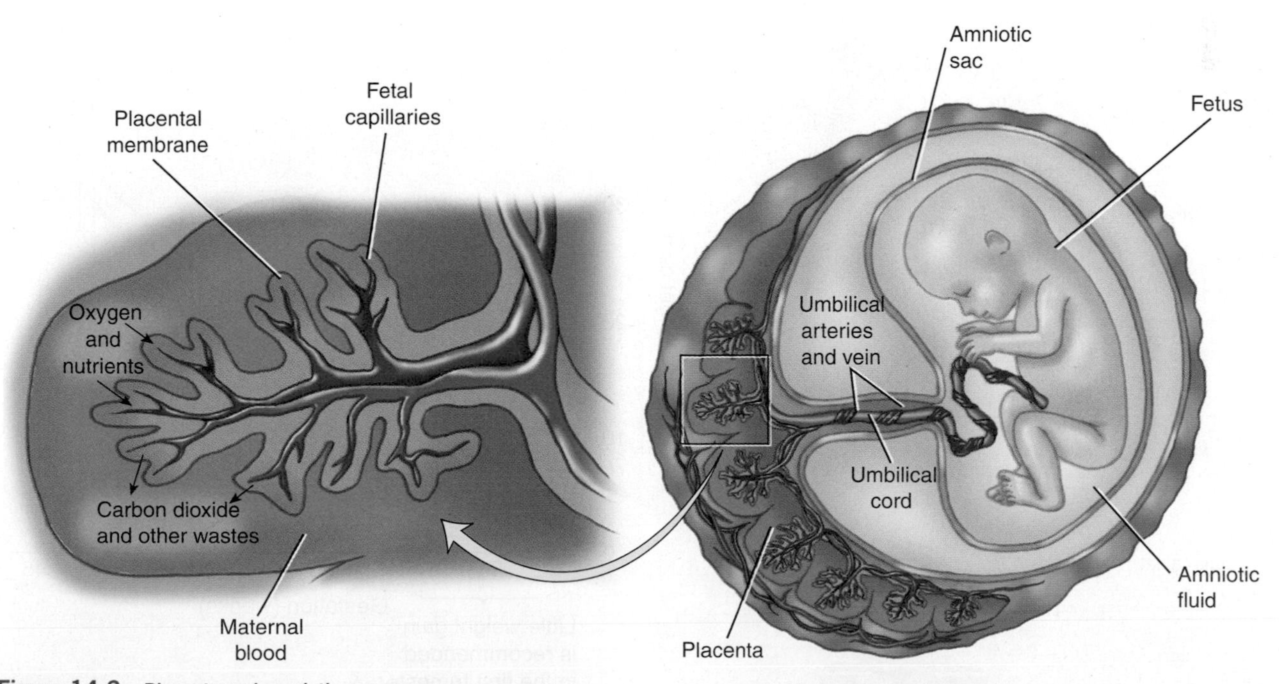

Figure 14.2 Placenta and amniotic sac
The placenta is made of branch-like projections that extend from the embryo into the uterine lining, placing maternal and fetal blood in close proximity. The placenta allows the transfer of nutrients and wastes between mother and baby. Fetal blood travels to and from the placenta via the umbilical cord. The amniotic sac and the amniotic fluid, in which the baby floats, provide protection.

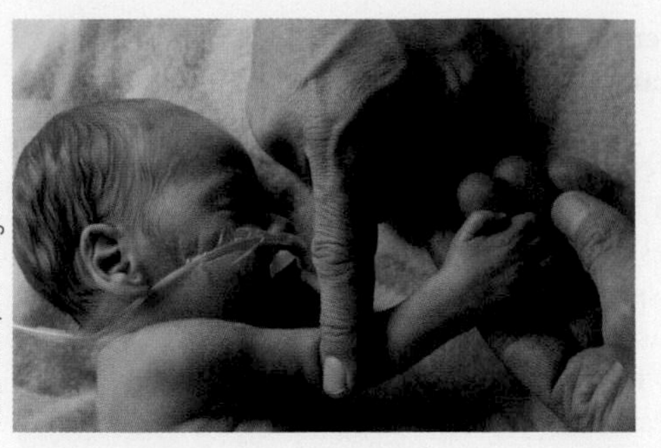
Sarah Leen/NG Image Collection

Figure 14.3 **Premature infants**
Low-birth-weight and very-low-birth-weight infants require special care and a special diet in order to grow and develop.

gestation The time between conception and birth, which lasts about nine months (or about 40 weeks) in humans.

small for gestational age An infant born at term weighing less than 2.5 kg (5.5 lbs).

preterm or **premature** An infant born before 37 weeks of gestation.

low-birth-weight A birth weight less than 2.5 kg (5.5 lbs).

very-low-birth-weight A birth weight less than 1.5 kg (3.3 lbs).

Pregnancy usually ends after 40 weeks of **gestation** with the birth of an infant weighing about 3 to 4 kilograms (6.6 to 8.8 lbs).[1] Infants who are born on time but have failed to grow normally in the uterus are said to be **small for gestational age**. Those born before 37 weeks of gestation are said to be **preterm** or **premature**. Whether born too soon or just too small, **low-birth-weight** infants (those weighing less than 2.5 kg [5.5 lbs] at birth) and **very-low-birth-weight** infants (those weighing less than 1.5 kg [3.3 lbs]), are at increased risk for illness and early death.[2] They often require special care and a special diet in order to continue to grow and develop successfully. Survival improves with increasing gestational age and birth weight. Today, with advances in medical and nutritional care, infants born as early as 25 weeks of gestation and those weighing as little as 1 kg (2.2 lbs) can survive[3] (**Figure 14.3**).

What Changes Occur in the Mother

A woman's body undergoes many changes during pregnancy to support the growth and development of the embryo and fetus. These continuous physiological adjustments affect the metabolism and distribution of nutrients in her body. Maternal blood volume increases by 50%, and the heart, lungs, and kidneys work harder to deliver nutrients and oxygen and remove wastes. The placenta develops and the hormones it produces orchestrate other changes: They promote uterine growth; they relax muscles and ligaments to accommodate the growing fetus and allow for childbirth; they promote breast development; and they increase fat deposition to provide the energy stores that will be needed during late pregnancy and lactation. These changes result in weight gain and can affect the type and level of physical activity that is safe for the pregnant woman.

Figure 14.4 **Composition and rate of weight gain in pregnancy**

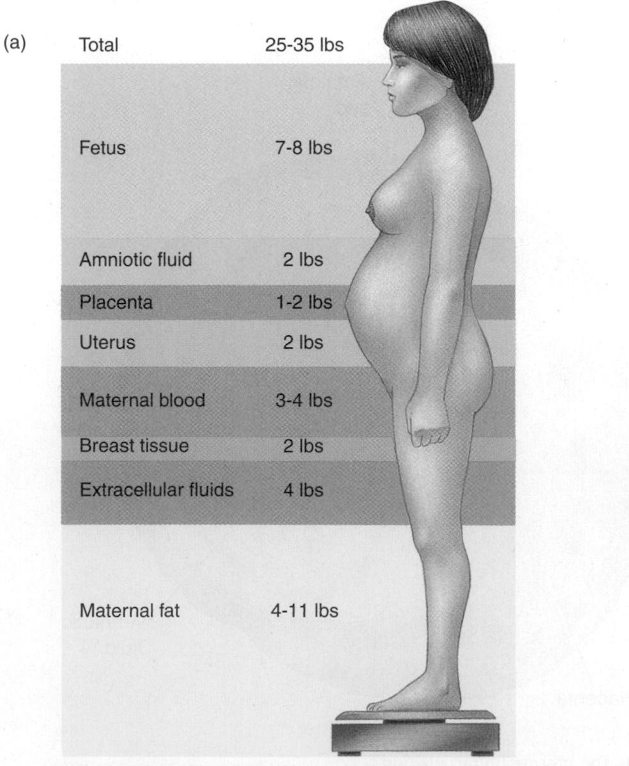

(a) Weight gained during pregnancy is due to increases in the weight of the mother's tissues, as well as the weight of the fetus, placenta, and amniotic fluid.

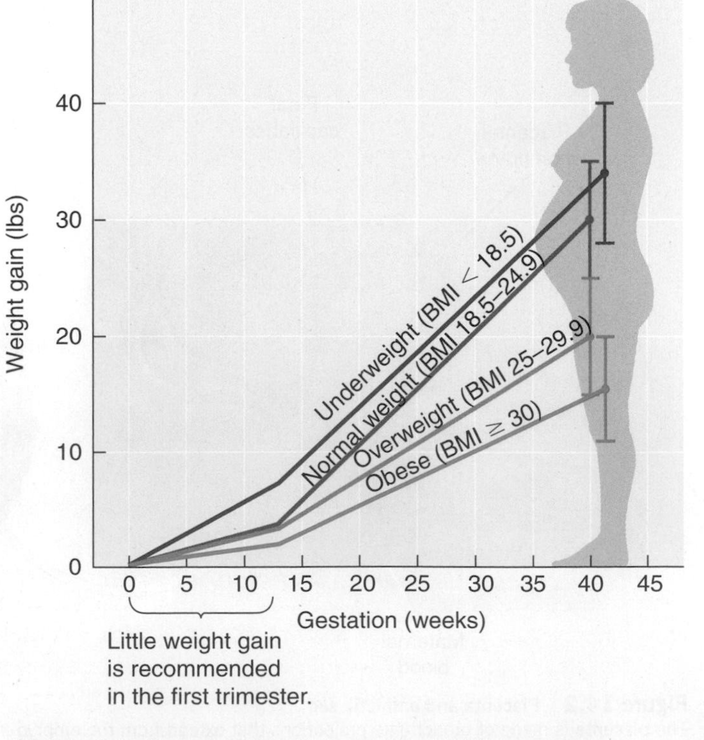

(b) A similar pattern of weight gain is recommended for women who are normal weight, underweight, overweight, or obese at the start of pregnancy, but total weight gain recommendations differ.

TABLE 14.1 Recommended Weight Gain During Pregnancy[1]

	Total Weight Gain		Rates of Weight Gain[a] (2nd and 3rd Trimester)	
Prepregnancy BMI	(kg)	(lbs)	Mean (range) (kg/week)	Mean (range) (lbs/week)
Underweight (< 18.5 kg/m^2)	12.5–18	28–40	0.51 (0.44–0.58)	1 (1–1.3)
Normal weight (18.5–24.9 kg/m^2)	11.5–16	25–35	0.42 (0.35–0.50)	1 (0.8–1)
Overweight (25.0–29.9 kg/m^2)	7–11.5	15–25	0.28 (0.23–0.33)	0.6 (0.5–0.7)
Obese (≥ 30.0 kg/m^2)	5–9	11–20	0.22 (0.17–0.27)	0.5 (0.4–0.6)
Twin pregnancy[b] (18.5–24.9 kg/m^2)	17–25	37–54		

[a]Calculations assume a 0.5–2 kg (1.1–4.4 lbs) weight gain in the first trimester (based on Siega-Riz et al., 1994; Abrams et al., 1995; Carmichael et al., 1997).
[b]Provisional recommendation

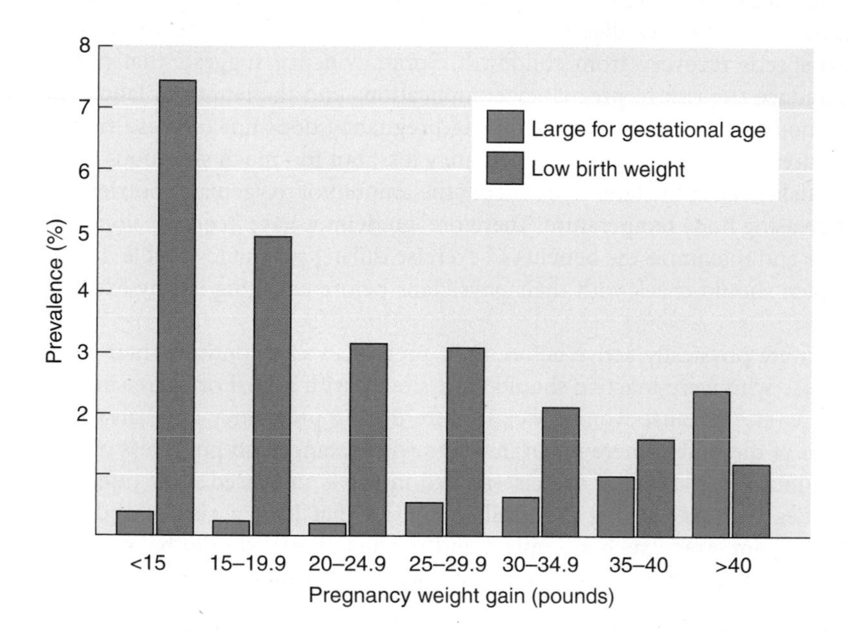

THINK CRITICALLY

Based on the information in this graph, how much weight would you recommend that a woman with a BMI of 24 gain during pregnancy?

Figure 14.5 Relationship between maternal weight gain and birth weight Gaining less or more than the recommended amount of weight during pregnancy increases the incidence of low-birth-weight and large-for-gestational-age babies, respectively.

Weight Gain During Pregnancy Gaining the right amount of weight during pregnancy is essential to the health of both mother and fetus. The recommended weight gain for healthy, normal-weight women is 25 to 35 lbs (11.4 to 15.9 kg). Typically, the weight of the infant at birth is about 25% of the total pregnancy weight gain. The placenta, amniotic fluid, and changes in maternal tissues account for the rest. This includes increases in the size of the uterus and breasts, expansion of blood and extracellular fluid volume, and deposition of fat stores (**Figure 14.4a**). The rate of weight gain is as important as the total weight gain (**Table 14.1**). Little gain is expected in the first three months, or **trimester**, of pregnancy—usually about 2 to 4 lbs (0.9 to 1.8 kg). In the second and third trimesters, when the fetus grows from less than 1 lb to 6 to 8 lbs, the recommended maternal weight gain is about 1 lb (0.45 kg)/week. Women who are underweight and women who are overweight or obese at conception should still gain weight at a slow, steady rate (**Figure 14.4b**). Weight gains of up to 40 lbs (18 kg) are recommended for women who begin pregnancy underweight. Overweight and obese women should gain less. Additional weight gain is recommended for multiple-birth pregnancies (see Table 14.1).[1]

trimester A term used to describe each third, or 3-month period, of a pregnancy.

Being underweight by 10% or more at the onset of pregnancy or gaining too little weight during pregnancy increases the risk of producing a low-birth-weight baby or preterm baby (**Figure 14.5**). It can also increase the child's risk of developing heart disease or diabetes later in life.[2] Excess weight, whether present before conception or gained during pregnancy, can also compromise the outcome of the pregnancy. The mother's risks for high blood pressure, diabetes, difficult delivery, and **cesarean section** are increased by excess weight, as is the risk of having a **large-for-gestational-age** baby (see Figure 14.5). Maternal obesity may also increase the risk of neural tube defects and fetal death.[2,4] Despite this,

cesarean section The surgical removal of the fetus from the uterus.

large-for-gestational-age An infant weighing more than 4 kg (8.8 lbs) at birth.

dieting during pregnancy is not advised even for obese women. If possible, excess weight should be lost before the pregnancy begins or, alternatively, after the child is born and weaned.

Some women are concerned that weight gained during pregnancy will be permanent, but most women lose all but about 2 lbs within a year of delivery. Approximately 10 lbs are lost at birth from the weight of the baby, amniotic fluid, and placenta. In the week after delivery, another 5 lbs of fluid are typically lost. Once this initial fluid and tissue weight is lost, further weight loss requires that energy intake be less than energy output. After the mother has recovered from delivery, a balanced, reduced-calorie diet combined with moderate exercise will promote weight loss and the return of muscle tone.

Physical Activity During Pregnancy For healthy, well-nourished women, carefully chosen moderate exercise is recommended during and after pregnancy.[2] Physical activity during pregnancy improves overall fitness, reduces stress, prevents excess weight gain, prevents low back pain, improves digestion, reduces constipation, improves mood and body image, and speeds recovery from childbirth. Some evidence suggests that physical activity may reduce the risk of pregnancy complications and the length of labor.[5] In healthy women, moderate-intensity activity during pregnancy does not increase risk of low birth weight, preterm delivery, or early pregnancy loss, but too much strenuous exercise has the potential to harm the fetus by reducing the amount of oxygen and nutrients it receives or by increasing body temperature. Therefore, guidelines have been developed to minimize the risks and maximize the benefits of exercise during pregnancy (**Table 14.2**). All pregnant women should check with their physicians before engaging in any exercise program.

Women who were physically active before their pregnancy can continue their exercise programs; those who were inactive should start slowly with a goal of exercising for 150 minutes each week.[5] Because women weigh more during pregnancy and carry that weight in the front of the body, where it can interfere with balance and put stress on the bones, joints, and muscles, the risk of exercise-related injury is increased. Low-intensity, low-impact activities such as walking are ideal. Activities that have a risk of abdominal trauma, falls, or joint stress, such as contact and racquet sports, should be avoided.[5] Exercise in the water is recommended because the body's buoyancy in water compensates for the changes in weight distribution. To ensure adequate delivery of oxygen and nutrients to the fetus, intense exercise should be limited during pregnancy. To prevent overheating, plenty of fluids should be consumed, and exercise should be carried out in a well-ventilated environment.

Discomforts of Pregnancy

The physiological changes that occur during pregnancy can cause uncomfortable side effects in the mother. Some are caused by changes in fluid distribution, others by hormonal

TABLE 14.2 Guidelines for Physical Activity During Pregnancy

Do . . .	Don't . . .
Obtain permission from your health care provider before beginning an exercise program.	Exercise strenuously during the first trimester.
Increase activity gradually if you were inactive before pregnancy.	Exercise strenuously for more than 15 minutes at a time during the second and third trimesters.
Exercise regularly rather than intermittently.	Exercise to the point of exhaustion.
Stop exercising when fatigued.	Exercise lying on your back after the first trimester.
Choose non-weight-bearing activities, such as swimming, that entail minimal risk of falls or abdominal injury.	Scuba dive or engage in activities that entail risk of abdominal trauma, falls, or joint stress.
Drink plenty of fluids before, during, and after exercise.	Exercise in hot or humid environments.

Tracy Frankel/The Image Bank/Getty Images, Inc.

changes that affect the digestive tract. Most of these problems are minor, but in some cases they may endanger the mother and the fetus.

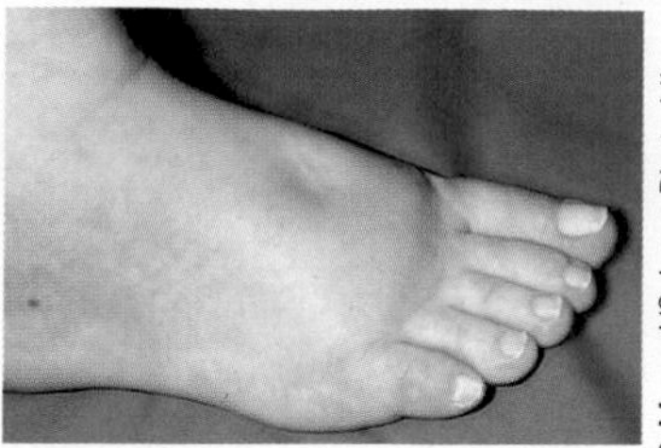

P. Marazzi/Science Photo Library/ Photo Researchers, Inc.

Figure 14.6 Edema
Edema in the feet and ankles is common during pregnancy. The swelling may be reduced by elevating the feet.

edema Swelling due to the buildup of extracellular fluid in the tissues.

morning sickness Nausea and vomiting that affects many women during the first few months of pregnancy and in some women can continue throughout the pregnancy.

Why Edema is Common During Pregnancy During pregnancy, blood volume expands to nourish the fetus, but this expansion may also cause the accumulation of extracellular fluid in the tissues, known as **edema**. Edema is common in the feet and ankles during pregnancy because the growing uterus puts pressure on the veins that return blood from the legs to the heart. This causes blood to pool in the legs, forcing fluid from the veins into the tissues of the feet and ankles (**Figure 14.6**). Edema can be uncomfortable, but does not increase medical risks unless it is accompanied by a rise in blood pressure. Reducing fluid and sodium intake below the amount recommended for the general population is not recommended.

How Morning Sickness Affects Health **Morning sickness** is a syndrome of nausea and vomiting that occurs in about 90% of women during pregnancy.[6] The incidence peaks at about 8 to 12 weeks of pregnancy and symptoms usually resolve by week 20. The term "morning sickness" is somewhat of a misnomer, because symptoms can occur anytime during the day or night. Although the cause is unknown, this nausea and vomiting is hypothesized to be related to the hormonal changes of pregnancy. The symptoms may be alleviated to some extent by eating small frequent snacks of dry starchy foods, such as plain crackers or bread. In most women, symptoms are mild and do not affect maternal or fetal health. Up to 2% of pregnant women, however, have a severe and intractable form of nausea and vomiting called *hyperemesis gravidarum* that may result in weight loss, nutritional deficiencies, and abnormalities in fluids, electrolyte levels, and acid-base balance. Treatment may require intravenous nutrition to assure that needs are met, and medications to reduce nausea.

Why Heartburn is a Problem for Pregnant Women Heartburn, a burning sensation caused by stomach acid leaking up into the esophagus, is another common digestive complaint during pregnancy because the hormones produced to relax the muscles of the uterus also relax the muscles of the gastrointestinal tract. This involuntary relaxation of the gastroesophageal sphincter allows the acidic stomach contents to back up into the esophagus, causing irritation. The problem gets more severe as pregnancy progresses because the growing baby crowds the stomach (**Figure 14.7**). The fuller the stomach, the more likely its contents are to back up into the esophagus. Heartburn can be reduced by avoiding substances that are known to cause it, such as caffeine and chocolate, and by consuming many small meals throughout the day rather than a few large meals. Limiting intake of high-fat foods that leave the stomach slowly, such as fried foods, rich sauces, and desserts, can also help reduce heartburn. Because a reclining position makes it easier for acidic juices to flow into the esophagus, remaining upright after eating, limiting eating in the hours before bedtime, and sleeping with extra pillows to produce a semi-reclining sleep position can also reduce heartburn.

How Pregnancy Contributes to Constipation and Hemorrhoids Constipation is a frequent complaint during pregnancy. The pregnancy-related hormones that cause muscles to relax also decrease intestinal motility and slow transit time. Constipation is a more common problem late in pregnancy, when the enlarging uterus puts pressure on the gastrointestinal tract (see Figure 14.7). Iron supplements prescribed during pregnancy also contribute to constipation. Maintaining a moderate level of physical activity and consuming plenty of water and other fluids along with high-fiber foods, such as whole grains, vegetables, and fruits, are recommended to prevent constipation. Hemorrhoids are also more common during pregnancy, as a result of both constipation and physiological changes in blood flow.

Complications of Pregnancy

While most pregnancies are problem-free, some women do experience complications that increase risks to both mother and baby.[7,8] Some, such as anemia, can be prevented

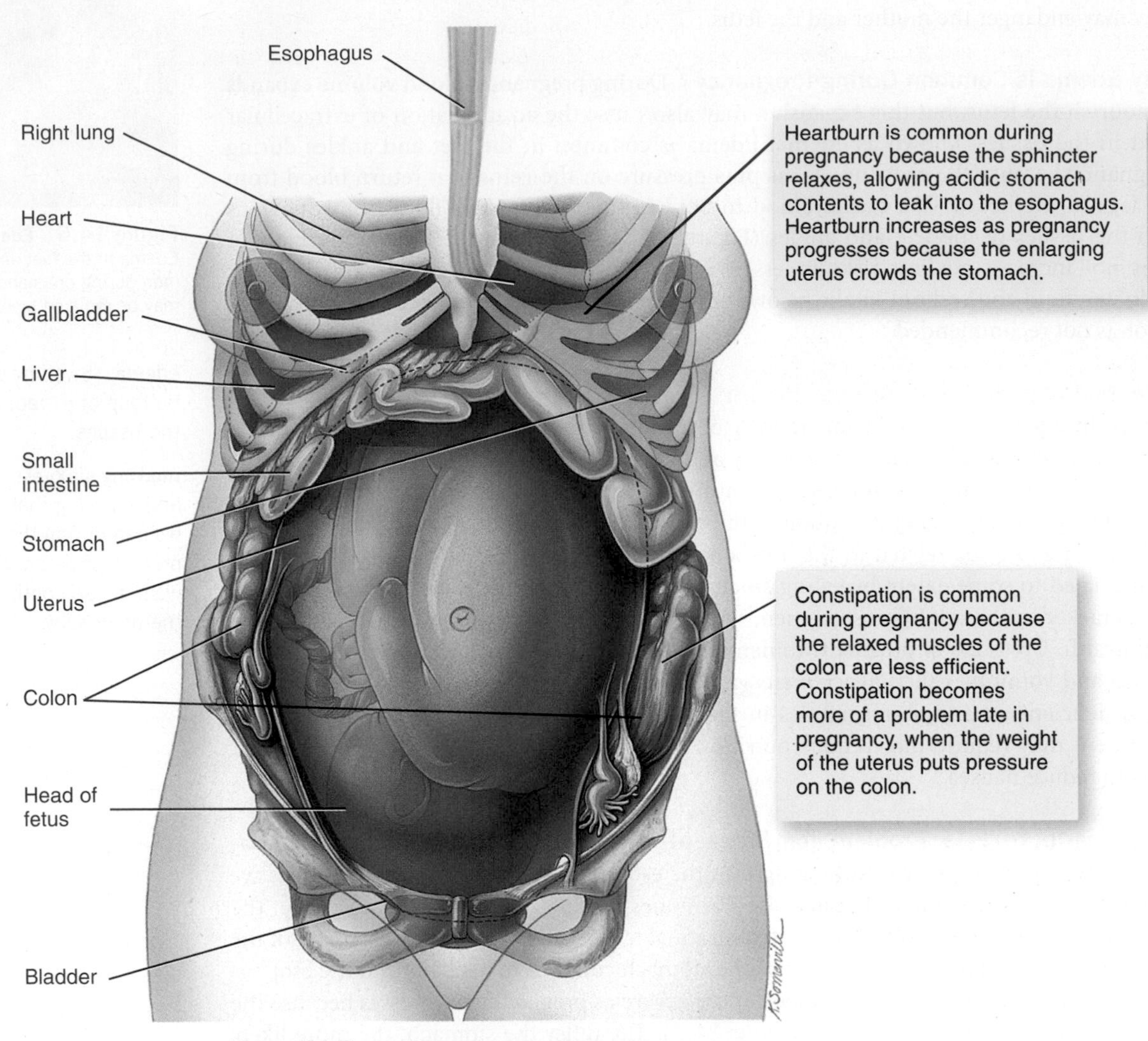

Figure 14.7 **Crowding of the gastrointestinal tract**
During pregnancy the uterus enlarges and pushes higher into the abdominal cavity, exerting pressure on the stomach and intestines. This contributes to both heartburn and constipation.

hypertensive disorders of pregnancy High blood pressure during pregnancy that is due to chronic hypertension, gestational hypertension, preeclampsia-eclampsia, or preeclampsia superimposed on chronic hypertension.

gestational hypertension The development of hypertension after the twentieth week of pregnancy.

preeclampsia A condition characterized by an increase in body weight, elevated blood pressure, protein in the urine, and edema. It can progress to **eclampsia**, which can be life-threatening to mother and fetus.

by proper nutrition and good prenatal care. For others, such as hypertension and diabetes, the cause is not as well understood, making prevention difficult. However, if caught early, these complications can usually be managed, allowing a healthy delivery (**Figure 14.8**).

High Blood Pressure About 6 to 8% of pregnant women experience high blood pressure during pregnancy. **Hypertensive disorders of pregnancy** refers to a spectrum of conditions involving elevated blood pressure during pregnancy. It accounts for more than 12% of pregnancy-related maternal deaths in the United States.[9,10] It is especially common in mothers under 20 and over 35 years of age, low-income mothers, and mothers with chronic hypertension or kidney disease.

About one-third of the cases of hypertensive disorders of pregnancy are due to chronic hypertension that was present before the pregnancy, but the remainder are related to the pregnancy. The least problematic of these is **gestational hypertension**, an abnormal rise in blood pressure that occurs after the 20th week of pregnancy. Gestational hypertension may signal the potential for a more serious condition called **preeclampsia**. Preeclampsia is characterized by high blood pressure along with fluid retention and excretion of protein in the urine; it can result in weight gain of several pounds within a few days. It is dangerous to the baby because it reduces blood flow to the placenta, and it is dangerous to the mother because it can progress to a condition called **eclampsia**, in which life-threatening seizures occur. Women with preeclampsia require bed rest and careful medical monitoring. The condition usually resolves after delivery.

The causes of preeclampsia are not fully understood. At one time, low-sodium diets were prescribed to prevent it, but studies have not found them to be effective.[11] Calcium may play a role in preventing the hypertensive disorders of pregnancy; calcium supplements have been found to reduce the risk of high blood pressure and preeclampsia.[12] Although calcium supplements are not routinely recommended for healthy pregnant women, pregnant teens, individuals with inadequate calcium intake, and women who are known to be at risk of developing hypertension during pregnancy may benefit from additional dietary calcium.

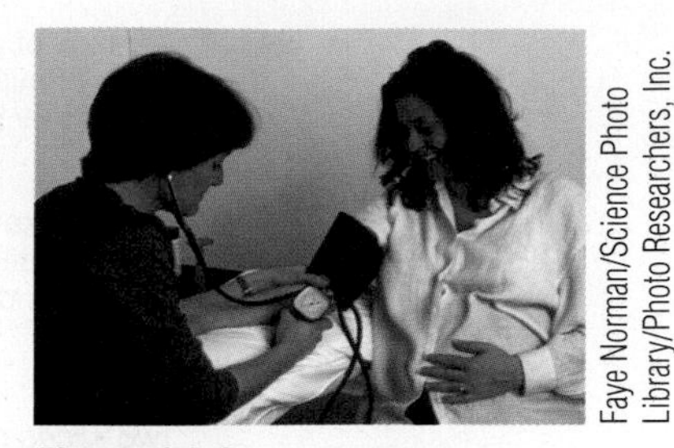

Faye Norman/Science Photo Library/Photo Researchers, Inc.

Figure 14.8 **Prenatal care** Adequate prenatal care, which monitors the mother's blood pressure, weight, and blood sugar and the baby's size and heartbeat, can allow early identification and treatment of pregnancy complications.

Gestational Diabetes Mellitus Consistently elevated blood glucose levels during pregnancy in a woman without previously diagnosed diabetes is known as **gestational diabetes mellitus**. It occurs in about 2 to 10% of all pregnancies and is most common in obese women and those with a family history of type 2 diabetes.[2,13] It occurs more frequently among African American, Hispanic/Latino American, and Native American women than among Caucasian women.[14] This form of diabetes usually disappears when the pregnancy is completed, but women who have had gestational diabetes have a 35 to 60% chance of developing type 2 diabetes over the next 10 to 20 years (see Chapter 4).[13] In addition to its impact on the mother's health, gestational diabetes increases risks to the baby. Because glucose in the mother's blood passes freely across the placenta, when the mother's blood levels are high, the growing fetus receives extra glucose calories. This extra energy promotes rapid growth, resulting in babies who are large for gestational age and consequently at increased risk of complications during delivery. Babies born to mothers with gestational diabetes are also at increased risk of developing diabetes as adults.[15] As with other types of diabetes, the treatment of gestational diabetes involves consuming a carefully planned diet, maintaining moderate daily exercise, and in some cases using medications to control blood glucose.

gestational diabetes mellitus A consistently elevated blood glucose level that develops during pregnancy and returns to normal after delivery.

14.2 The Nutritional Needs of Pregnancy

LEARNING OBJECTIVES

- Discuss the recommendations for energy intake during each trimester.
- Compare the protein and micronutrient needs of pregnant women with those of nonpregnant women.
- Discuss the possible effects of too little folic acid during pregnancy.

In order to produce a healthy baby, maternal intake must supply all the nutrients needed to provide for the growth and development of the fetus while continuing to meet the mother's needs. Because the increased need for energy is proportionately smaller than the increased need for protein, vitamins, and minerals, a well-balanced, nutrient-dense diet is required.

Energy Needs During Pregnancy

Energy needs increase during pregnancy to deposit and maintain the new fetal and maternal tissues. The estimated energy requirement (EER) for pregnancy is calculated by totaling the energy needs of nonpregnant women, the increase in energy needs due to pregnancy, and the energy deposited in tissues.[16] During the first trimester, total energy expenditure changes little, so the EER is not increased above nonpregnant levels. During

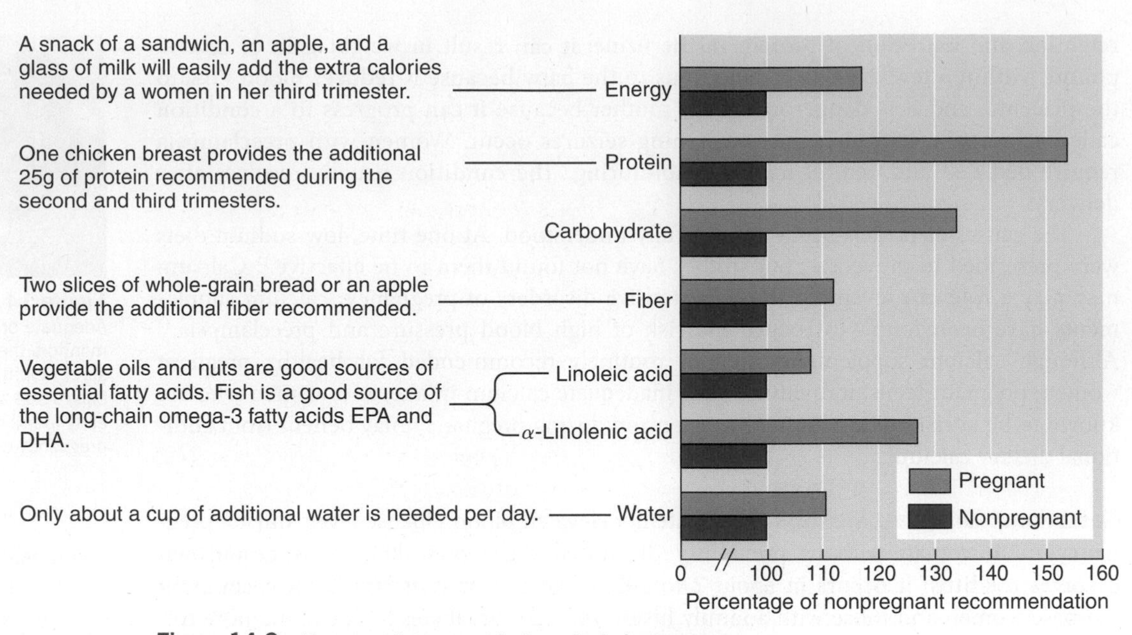

Figure 14.9 Energy and macronutrient needs during pregnancy
This graph illustrates the differences between the recommended daily intake of energy, protein, carbohydrate, fiber, essential fatty acids, and water for nonpregnant 25-year-old women and pregnant 25-year-old women during the third trimester of pregnancy.

the second and third trimesters, an additional 340 and 452 kcalories/day, respectively, are recommended (**Figure 14.9**).

Protein, Carbohydrate, and Fat Recommendations

The RDA for protein is increased during pregnancy (see Figure 14.9). The additional protein is needed because protein is essential for the formation and growth of new cells. During pregnancy, the placenta develops and grows, the uterus and breasts enlarge, and a single cell develops into a fully formed infant. An additional 25 g of protein per day above the RDA for nonpregnant women, or 1.1 g/kg of body weight per day, is recommended in the second and third trimesters of pregnancy.[16] For example, for a woman weighing 136 lbs (62 kg), this increases protein needs to a total of about 75 g/day. This amount of protein would be provided by a diet that includes three cups of milk or yogurt and seven ounce equivalents of meat or beans.

The RDA for carbohydrate is increased by 45 g/day during pregnancy to provide sufficient glucose to fuel the fetal and maternal brains. Therefore the RDA for carbohydrate during pregnancy is 175 g/day. This is well below the typical intake of about 300 g/day, and therefore most women do not need to consciously increase carbohydrate intake.

Although total fat intake does not need to increase during pregnancy, more of the essential fatty acids linoleic and α-linolenic acid are recommended because these are incorporated into the placenta and the fetal tissues. The long-chain polyunsaturated fatty acids docosahexanoic acid (DHA) and arachidonic acid are important because they not only support maternal health but are essential for the development of the eyes and nervous system in the fetus.[17] Since there is insufficient data to determine how much is required to meet these needs, AIs for the essential fatty acids have been established based on the median intake in the United States.

Despite increases in the recommended intakes of protein, carbohydrate, and specific fatty acids during pregnancy, the macronutrient distribution of the diet should be about the same as that recommended for the general population. If the additional energy and protein needed during pregnancy come from nutrient-dense choices, the diet will provide the additional carbohydrate and fatty acids needed for a healthy pregnancy.

Water and Electrolyte Needs

The need for water is increased during pregnancy because of the increase in blood volume, the production of amniotic fluid, and the needs of the fetus. Throughout pregnancy a woman will accumulate about 6 to 9 liters of water. Some is intracellular, but most is due to increases in the volume of blood and interstitial fluid. The need for water from food and beverages is therefore increased from 2.7 L/day in nonpregnant women to 3 L/day.[11] This is equivalent to drinking a little more than an extra cup a day. Despite changes in the amount and distribution of body water during pregnancy, there is no evidence that the requirements for potassium, sodium, or chloride are different from those of nonpregnant women.[11]

Micronutrient Needs During Pregnancy

The need for many vitamins and minerals is increased during pregnancy (**Figure 14.10**). To form new maternal and fetal cells and to meet the needs for protein synthesis in fetal and maternal tissues, the requirements for folate, vitamin B_{12}, vitamin B_6, zinc, and iron increase. To provide for the rise in energy utilization, the requirements for thiamin, niacin, and riboflavin increase. Adequate calcium, vitamin D, and vitamin C are needed for the growth and development of bone and connective tissue. For many of these nutrients, intake is easily increased when energy intake rises to meet needs, but for others there is a risk that inadequate amounts will be consumed.

Calcium The fetus retains about 30 g of calcium over the course of gestation. Most of the calcium is deposited in the last trimester when the fetal skeleton is growing most rapidly and the teeth are forming. Many women have trouble getting enough calcium to meet their own needs, let alone enough to provide this amount for the fetus. Fortunately, they don't need to consume any more than is recommended for nonpregnant women because calcium absorption increases during pregnancy. This increase is believed to be

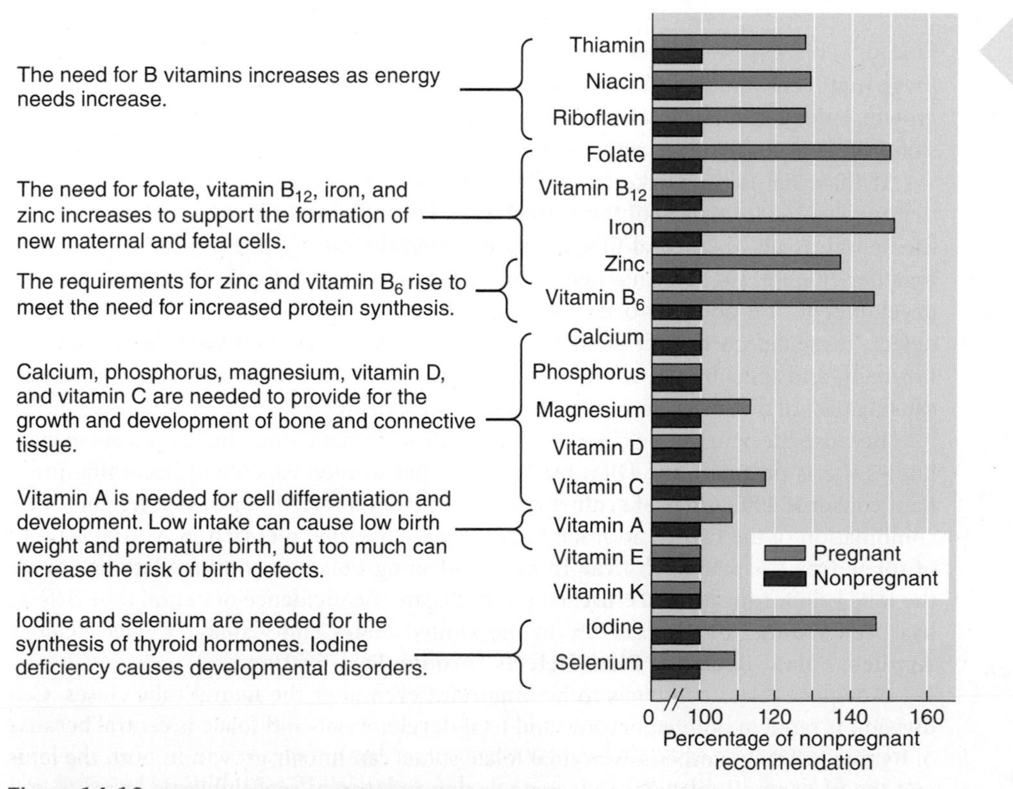

Figure 14.10 **Vitamin and mineral needs during pregnancy**
This graph compares the recommended micronutrient intakes for 25-year-old nonpregnant women and 25-year-old women during the third trimester of pregnancy.

THINK CRITICALLY

If the RDA for calcium is not increased during pregnancy, where does the calcium for the fetal skeleton come from?

due in part to the rise in estrogen that occurs during pregnancy as well as an increase in the concentration of active vitamin D in the blood. At one time there was concern that the calcium needed by the fetus would come from maternal bones if intake was not increased. It is now known that the increased need for calcium does not increase maternal bone resorption, and studies have found no correlation between the number of pregnancies a woman has had and the density of her bones.[18] The RDA for calcium for pregnant women aged 19 and older—1000 mg/day—is not increased above nonpregnant needs.[19] The RDA can be met by consuming three to four servings of milk or other dairy products daily. Women who are lactose intolerant can meet their calcium needs with yogurt, cheese, reduced-lactose milk, calcium-rich vegetables, fish consumed with bones, calcium-fortified foods, or calcium supplements. Low calcium intake may increase the risk of preeclampsia.[20]

Vitamin D Adequate vitamin D is essential to ensure efficient calcium absorption. The RDA for vitamin D during pregnancy is 600 IU (15 μg)/day, the same as for nonpregnant women.[19] This is based on the assumption of minimal sun exposure. Pregnant women who receive regular exposure to sunlight can synthesize sufficient vitamin D. If exposure to sunlight is limited, dietary sources such as milk must supply the needed amounts. A recent study has found that almost 70% of pregnant women have low blood levels of vitamin D and the percentage is much higher among African-American women.[21] African-American women are at increased risk of deficiency not only because their darker pigmentation reduces the synthesis of vitamin D in the skin, but also because milk intake is frequently reduced due to lactose intolerance. Most prenatal supplements provide 200 IU (10 μg) of vitamin D, which is only a third of the RDA.

Vitamin C Vitamin C is important for bone and connective tissue formation because it is needed for the synthesis of collagen, which gives structure to skin, tendons, and the protein matrix of bones. Vitamin C deficiency during pregnancy increases the risk for premature birth and preeclampsia. The RDA is increased by 10 mg/day during pregnancy.[22] The requirement for vitamin C can easily be met by including citrus fruits and juices in the diet, and supplements are generally not necessary.

neural tube The portion of the embryo that develops into the brain and spinal cord.

Folate Folate is needed for the synthesis of DNA and thus for cell division. During pregnancy, cells multiply to form the placenta, expand maternal blood, and allow fetal growth. Adequate folate intake is crucial even before conception because rapid cell division occurs in the first days and weeks of pregnancy.

If maternal folate intake is low, there is an increased risk of fetal abnormalities that involve the formation of the **neural tube**. During development, a groove forms in the neural tissue. The neural tube is created when the sides of the groove rise and fold together (**Figure 14.11**). This neural tube closure occurs between 21 and 28 days of development. If it does not occur normally, the infant will be born with a neural tube defect. These defects include anencephaly, in which the brain and skull do not develop normally, and spina bifida, a condition in which the vertebrae do not close completely, causing part of the spinal cord to be exposed (see Chapter 8).

Because the neural tube closes so early in development, often before a woman even knows she is pregnant, the DRIs recommend that women capable of becoming pregnant consume 400 μg/day of synthetic folic acid from fortified foods, supplements, or a combination of the two, in addition to consuming a varied diet rich in natural sources of folate (see Chapter 8, Critical Thinking: Meeting Folate Recommendations). Since the folic acid fortification of enriched grains began, the incidence of neural tube defects has been reduced by almost 50% in the United States and Canada[23,24] (see **Science Applied: Folate: Reducing Birth Defects Through Fortification**).

Adequate folate continues to be important even after the neural tube closes. Cell division is rapid in both embryonic and fetal development, and folate is central because of its role in DNA synthesis. Marginal folate status can impair growth in both the fetus and the placenta. If folate is inadequate during pregnancy, megaloblastic anemia—the type of anemia in which blood cells do not mature properly—may result (see Chapter 8).

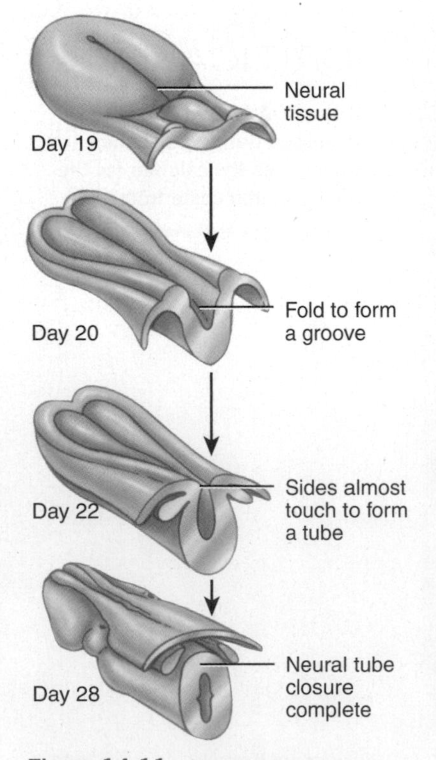

Figure 14.11 Neural tube formation The neural tube develops from a flat plate of neural tissue.

SCIENCE APPLIED

Folate: Reducing Birth Defects Through Fortification

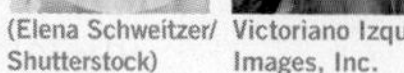

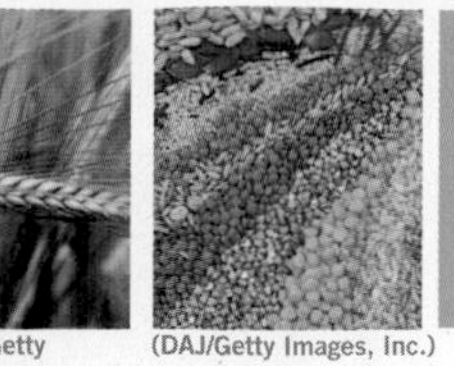

(Elena Schweitzer/ Shutterstock) Victoriano Izquierdo/Getty Images, Inc. (DAJ/Getty Images, Inc.)

Until the late 1990s, neural tube defects (NTDs), such as spina bifida and anencephaly, affected approximately 3900 pregnancies in the United States each year. For decades, researchers had suspected that NTDs might be related to the mother's dietary intake. Today, public health policies mandate the fortification of certain foods with folic acid, the synthetic form of folate, preventing birth defects in thousands of babies.

THE SCIENCE

The link between maternal folate status and NTDs was identified in the 1970s when lower first-trimester red-blood-cell folate concentrations were found in women who later gave birth to NTD-affected babies.[1] An intervention study completed in 1980 demonstrated that folic acid supplementation in early pregnancy reduced the incidence of NTDs in women who had previously given birth to a baby with an NTD.[2] These findings were supported by a second, large-scale study, which randomly assigned nonpregnant women to receive folic acid, other supplemental vitamins, or a placebo. This trial was stopped early when researchers concluded that folic acid supplementation alone reduced NTD recurrence by 71%.[3]

A link between NTDs and folic acid was now clear, but early trials used 800 μg of folic acid. Could a smaller amount be as effective? Also, would folic acid supplementation prevent the first occurrence of NTDs? To answer these questions, women who had no previous NTD–affected pregnancies and were planning a pregnancy were randomly assigned to receive folic acid or a placebo for at least one month before conception and until at least the date of the second missed menstrual cycle.[4] The study evaluated 5453 pregnancies. In the 2391 women receiving the placebo, six babies were born with NTDs; in the 2471 women in the supplement group, there were no NTDs.[5] This and other studies helped determine that 400 μg of supplemental folic acid dramatically reduced the incidence of NTDs.[6,7]

THE APPLICATION

These results led the Centers for Disease Control and Prevention and the U.S. Public Health Service to recommend that all women of childbearing age who are capable of becoming pregnant consume 400 μg/day of folic acid. But how could the population's folate intake be increased? Educating women to consume foods high in folate would be costly and ineffective.

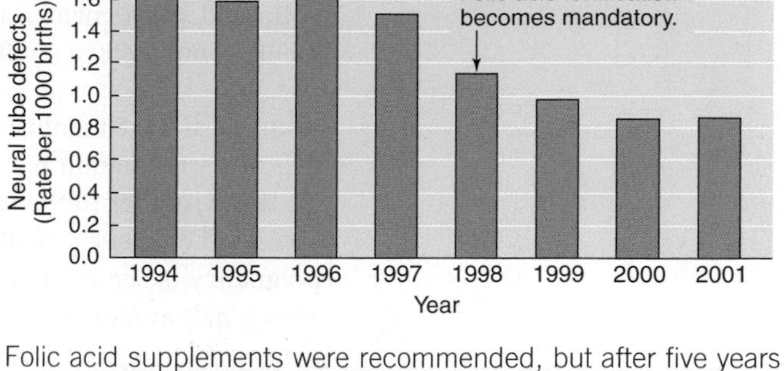

Folic acid supplements were recommended, but after five years only one-third of women of childbearing age were consuming a supplement containing the recommended amount.[8] Therefore, it was concluded that the most reliable way to increase folate consumption was through food fortification.

Fortification of the food supply requires a careful analysis to determine how much of the nutrient to add to provide the needed benefit without undue risk. The right amount of supplemental folic acid could reduce the incidence of NTD-affected pregnancies and the risk of heart disease in the general population. But because a high folic acid intake can mask the symptoms of vitamin B_{12} deficiency, a level of fortification needed to be chosen that would reduce NTDs without compromising the elderly population at risk for B_{12} deficiency.

The foods chosen for fortification were enriched grain products such as bread, flour, cornmeal, pasta, grits, and rice. These products are regularly consumed by the target population—women of childbearing age of all races and cultures. The folic acid could be added to these grains while they were being enriched with other nutrients. The amount added, 140 μg per 100 g of grain product, was chosen because it balanced the need to provide enough folic acid to reduce the risk of NTDs with the possibility of masking vitamin B_{12} deficiency. The success of the folic acid fortification program can be seen in the decline in the estimated number of NTD-affected pregnancies that has occurred since the fortification of grains with folic acid[9,10] (see graph).

THINK CRITICALLY: Suggest some possible reasons why fortifying grains with folic acid has not completely eliminated neural tube defects.

[1]Smithells, R. W., Sheppard, S., and Schorah, C. J. Vitamin deficiencies and neural tube defects. *Arch Dis Child* 51:944–949, 1976.

[2]Smithells, R. W., Sheppard, S., Schorah, C. J., et al. Possible prevention of neural tube defects by periconceptional vitamin supplementation. *Lancet* 1:339–340, 1980.

[3]MRC Vitamin Study Research Group. Prevention of neural tube defects: Results of the MRC vitamin study. *Lancet* 338:131–137, 1991.

[4]Czeizel, A. E., and Dudás, I. Prevention of the first occurrence of neural-tube defects by periconceptional vitamin supplementation. *N Engl J Med.* 327:1832–1835, 1992.

[5]Czeizel, A. E. Folic acid and the prevention of neural tube defects. *J. Pediatr. Gastroenterol. Nutr.* 20:4–16, 1995.

[6]Werler, M. M., Shapiro, S., and Mitchell, A. A. Periconceptional folic acid exposure and the risk of occurrent neural tube defects. *JAMA* 269:1257–1261, 1993.

[7]Daly, S., Mills, J. L., Molloy, A. M., et al. Minimum effective dose of folic acid for food fortification to prevent neural-tube defects. *Lancet* 350:1666–1669, 1997.

[8]Use of folic acid-containing supplements among women of childbearing age—United States, 1997. *MMWR* 47:131–134, 1998.

[9]Berry, R. J., Bailey, L., Mulinare, J., and Bower, C. Fortification of flour with folic acid. *Food Nutr Bull* 31(1 Suppl.):S22–S35, 2010.

[10]De Wals, P., Tairou, F., Van Allen, M. I., et al. Reduction in neural-tube defects after folic acid fortification in Canada. *N Engl J Med* 357:135–142, 2007.

Low dietary folate intakes and low circulating folate levels are associated with increased risk of preterm delivery, low birth weight, and fetal growth retardation.[25] Thus, to maintain red blood cell folate levels in pregnant women, the RDA for folate is set at 600 μg dietary folate equivalents per day.[26] Natural sources of folate include orange juice, legumes, leafy green vegetables, and organ meats. A half cup of legumes plus a cup of raw spinach provide about a third of the RDA. Fortified sources include enriched breads, cereals, and other grain products; a serving of fortified cereal provides from 25 to 100% of the RDA. Folic acid supplements can also be used to meet recommendations. Most prenatal supplements contain 400 μg of folic acid.

Vitamin B_{12} Vitamin B_{12} is essential for the regeneration of active forms of folate, so a deficiency of vitamin B_{12} can also result in megaloblastic anemia. Vitamin B_{12} is transferred from the mother to the fetus during pregnancy. Based on the amount transferred and the increased efficiency of vitamin B_{12} absorption that occurs during pregnancy, the RDA for pregnancy is set at 2.6 μg/day.[26] This recommendation is easily met by a diet containing even small amounts of animal products. Vegetarian diets are generally safe for pregnant women, but vegans must consume foods fortified with vitamin B_{12} or take vitamin B_{12} supplements daily to meet the needs of mother and fetus.

Zinc Zinc is involved in the synthesis of DNA, RNA, and proteins. It is therefore extremely important for growth and development. Zinc deficiency during pregnancy is associated with an increased risk of fetal malformations, prematurity, and low birth weight.[27] Because zinc absorption is inhibited by high iron intakes, iron supplements may compromise zinc status if the diet is low in zinc. The RDA is 12 mg/day for pregnant women age 18 and younger and 11 mg/day for pregnant women 19 years of age and older.[28] The zinc in red meat is more absorbable than from other sources; a 3-oz serving of lean ground beef provides about 2 mg of zinc.

Iron Iron needs are high during pregnancy to provide for the synthesis of hemoglobin and other iron-containing proteins in both maternal and fetal tissues. The physiological changes of pregnancy allow increased iron absorption, and iron losses are decreased due to the cessation of menstruation. Nonetheless, iron-deficiency anemia is common during pregnancy. Part of the reason for this is that low iron stores are common among women of childbearing age, so many women start pregnancy with diminished iron stores and quickly become deficient.

The RDA for iron during pregnancy is 27 mg/day compared with 18 mg/day for nonpregnant women.[28] It takes an exceptionally well-planned diet to meet iron needs during pregnancy. Red meats, leafy green vegetables, and fortified cereals are good sources of iron. Foods that enhance iron absorption, such as citrus fruit and meat, should also be included in the diet. Iron supplements are typically recommended during the second and third trimesters of pregnancy (see **Critical Thinking: Nutrient Needs for a Successful Pregnancy**).

When iron needs are not met, iron-deficiency anemia may occur. Iron-deficiency anemia during pregnancy has been associated with an increased risk of low birth weight and preterm delivery.[29] The fetus draws iron from the mother to ensure adequate fetal hemoglobin production, mostly during the last trimester. Babies born prematurely may not have had time to accumulate sufficient iron, but babies born at term usually have adequate iron stores even if the mother is deficient.

Meeting Nutrient Needs with Food and Supplements

The energy and nutrient needs of pregnancy can be met by following the recommendations of the 2010 Dietary Guidelines and MyPlate for Pregnancy. Additional grains, vegetables, and fruits provide energy, protein, folate, vitamin C, and fiber, particularly if whole grains are chosen. An extra serving of dairy provides energy, protein, calcium, vitamin D, and riboflavin. Additional lean meat provides energy, protein, vitamin B_6, vitamin B_{12}, iron, and zinc. For example, adding a snack such as a turkey sandwich on whole-grain

CRITICAL THINKING

Nutrient Needs for a Successful Pregnancy

Mark Burstyn/Masterfile

BACKGROUND

Tina is three months pregnant. From the start, she has been careful about her nutritional health. She has been eating well and taking her prenatal supplement. Now that she is approaching her second trimester, her doctor is concerned about her intake of iron. Even though there is iron in her supplement, she also needs to increase the iron in her diet.

Tina is 26 years old, 5'4" tall, and weighed 126 lbs before she became pregnant. She exercises for about 40 minutes per day. Her typical intake is shown in the table.

FOOD	IRON (mg)	PROTEIN (g)	ENERGY (kcals)
Breakfast			
1 cup corn flakes	0.4	2	130
with 1 cup reduced-fat milk	0.1	8	121
¾ cup orange juice	1.1	1.3	105
1 cup decaffeinated coffee with sugar and cream	0.1	0	51
Lunch			
Tuna sandwich			
3 oz tuna	1.2	21.7	170
2 tsp mayonnaise	0.5	0.1	66
2 slices white bread	2.5	6.6	200
20 French fries	1.3	3.2	203
1 can orange soda	0.2	0	180
3 chocolate chip cookies	0.8	1.7	144
1 apple	0.4	0	80
Dinner			
3 oz chicken leg	1.5	29.6	250
½ cup peas	1.2	7.9	67
1 piece corn bread	0.8	4	152
1 tsp margarine	0	0	33
1 cup lettuce and tomato salad	1.3	5	25
1 T Italian dressing	0	0	69
1 cup reduced-fat milk	0.1	8	121
TOTAL	13.5	99.1	2167

CRITICAL THINKING QUESTIONS

- By how much will Tina need to increase her energy and protein intake during her second trimester? Third trimester?
- Tina's diet meets the MyPlate recommendations for her first trimester. From which groups does she need to add servings during her second and third trimester?
- Plan a snack for her second trimester that will provide the appropriate food group servings and the recommended increase in calories and protein.
- What about iron? Does Tina's current diet meet her needs without the supplement? How much iron does the snack you suggested provide?

iProfile Use iProfile to find good food sources of heme and nonheme iron.

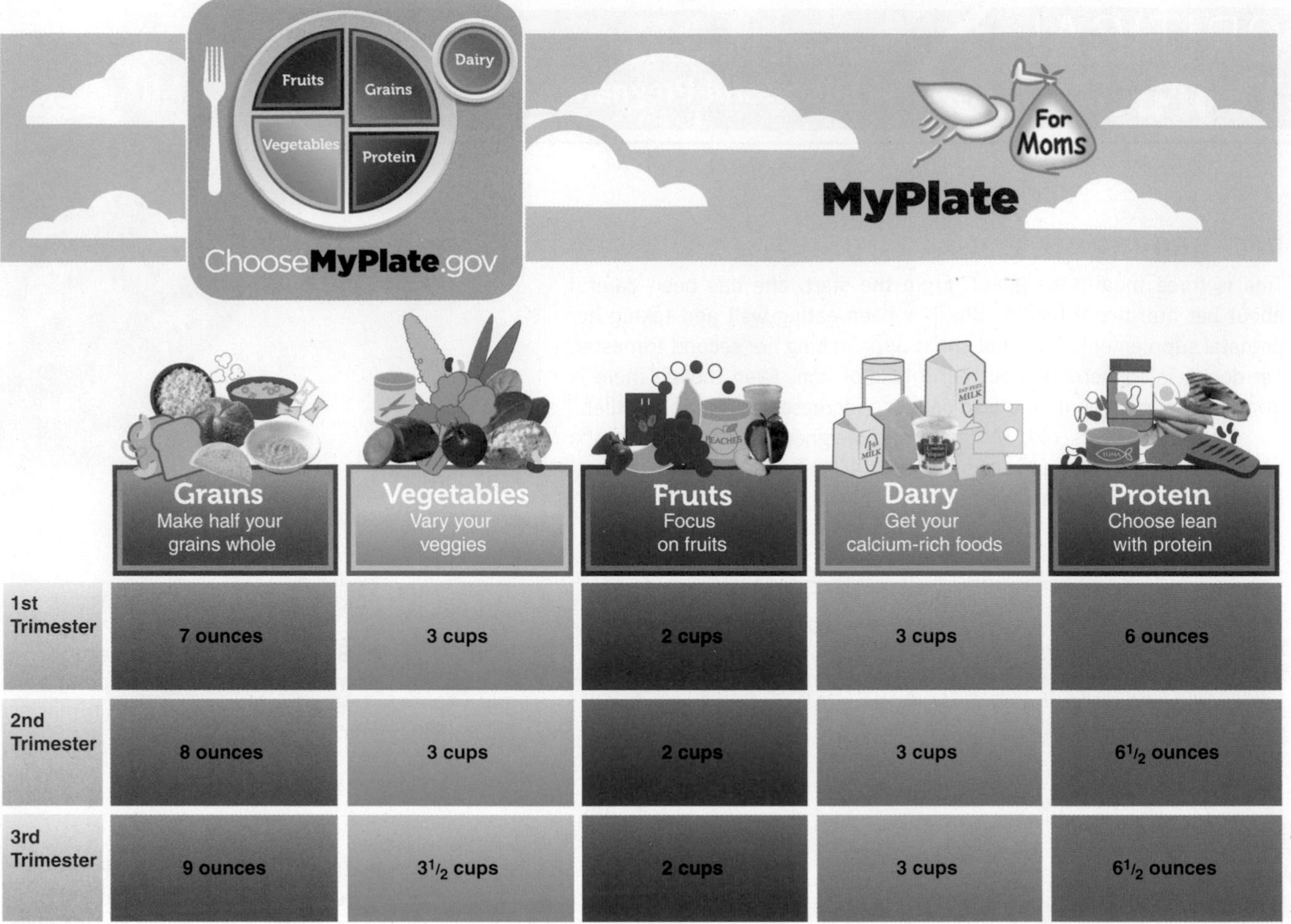

	Grains Make half your grains whole	Vegetables Vary your veggies	Fruits Focus on fruits	Dairy Get your calcium-rich foods	Protein Choose lean with protein
1st Trimester	7 ounces	3 cups	2 cups	3 cups	6 ounces
2nd Trimester	8 ounces	3 cups	2 cups	3 cups	$6\frac{1}{2}$ ounces
3rd Trimester	9 ounces	$3\frac{1}{2}$ cups	2 cups	3 cups	$6\frac{1}{2}$ ounces

Figure 14.12 **MyPlate for pregnancy**
This Daily Food Plan for Moms is for a 26-year-old woman who is 5 feet 4 inches tall, gets 30 to 60 minutes of exercise a day, and weighed 125 lbs before she became pregnant. Energy needs are not increased during the first trimester, so the recommended amounts from each group for the first trimester are the same as for a nonpregnant woman.

bread, an apple, and a glass of low-fat milk will provide all the extra energy and nutrients needed daily for a healthy pregnancy (**Figure 14.12**).

Dietary Supplements Even when a healthy diet based on a MyPlate Daily Food Plan for Moms is consumed, it is difficult to meet all vitamin and mineral needs. Generally, supplements of folic acid are recommended before and during pregnancy, and iron supplements are recommended during the second and third trimesters.[2] A multivitamin and mineral supplement may also be necessary in those whose food choices are limited, such as vegetarians, or in those whose needs are very high, such as pregnant teenagers. A prenatal supplement, however, must be taken in conjunction with, not in place of, a carefully planned diet (see **Off the Label: What's in a Prenatal Supplement?**).

Food Cravings and Aversions Most women change their diets during pregnancy. Some changes are made in an effort to improve nutrition to ensure a healthy infant, but others are based on cravings, aversions, or cultural or family traditions. Foods that are commonly craved include fruit and fruit juices, sweets, candy (particularly chocolate), and dairy products. Common aversions include coffee and other caffeinated drinks, alcohol, meat, fish, poultry, eggs, highly seasoned foods, or fried foods. It has been suggested that hormonal or physiological changes during pregnancy—in particular, changes in taste and smell—may be the cause of such cravings and aversions.

© iStockphoto

OFF THE LABEL

What's in a Prenatal Supplement?

Most pregnant women leave their first prenatal doctor's visit with a prescription for a prenatal vitamin and mineral supplement. What's in these supplements? Do they meet all the needs of pregnancy?

A look at a label shows that a typical prenatal supplement contains more than 15 vitamins and minerals. It supplies enough folate, iron, and many other micronutrients to meet recommendations. But taking the supplement does not mean that pregnant women can ignore their diet. Some nutrients in these supplements are present in amounts that do not meet the needs of pregnancy, and others are missing altogether. For example, the tablet whose ingredients are shown in the Supplements Facts panel here does not provide any magnesium and contains only 200 mg of calcium, which is only 20% of the recommended intake for a pregnant woman aged 19 to 50 years of age. A pregnant woman taking this tablet would still need to consume plenty of whole grains, nuts, seeds, and leafy greens to meet her magnesium needs and enough milk or other high-calcium foods to meet the recommendation of 1000 mg/day of calcium.

Even if all the calcium and magnesium needed for pregnancy could be packed into one little pill, it still would not provide everything needed in a healthy prenatal diet. Prenatal supplements do not contain the protein needed for tissue synthesis or the carbohydrates required for energy. They lack fiber, which helps prevent constipation, and they do not contain the fluid needed for expanding blood volume and maintaining normal bowel function. They also don't contain other food components such as the phytochemicals supplied by a diet rich in whole grains, fruits, and vegetables. Thus, taking a multivitamin and mineral supplement during pregnancy can be beneficial if the recommended dosage is not exceeded and the supplement does not take the place of a carefully planned diet.

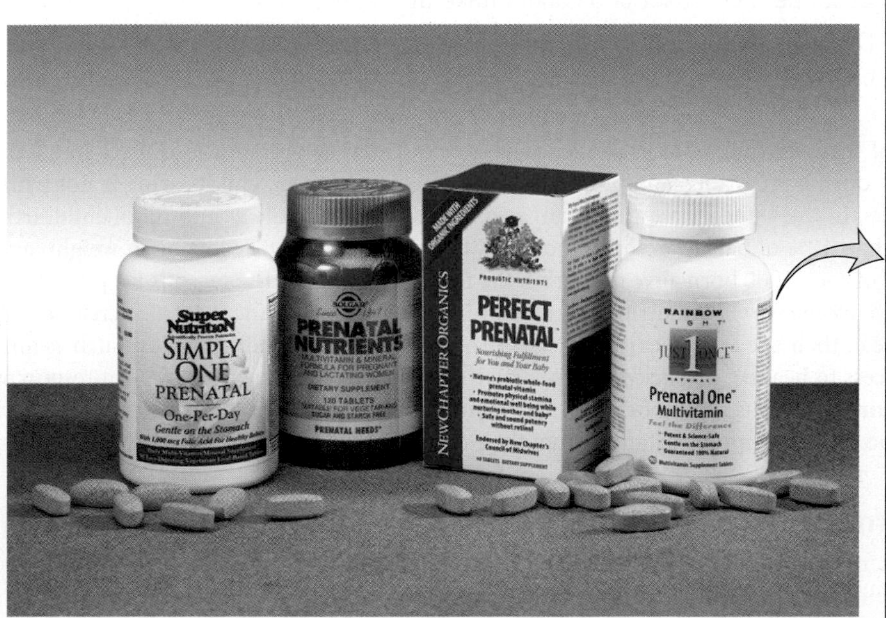

Andy Washnik

Supplement Facts

Serving Size 1 Tablet
Servings Per Container 60

Amount Per 1 Tablet	% Daily Value*
Vitamin A (as beta carotene) 5000 IU	63%
Vitamin C (as ascorbic acid) 85 mg	100%
Vitamin D (as cholecalciferol) 400 IU	200%
Vitamin E (as d-alpha tocopheryl) acetate (Covitol™) 22 IU	67%
Vitamin K 90 mcg	100%
Thiamin 1.4 mg	100%
Riboflavin 1.6 mg	100%
Niacin (as niacinamide) 17 mg	100%
Vitamin B_6 (as pyridoxine HCl) 2.6 mg	137%
Folic acid 1000 mcg	167%
Vitamin B_{12} (as cyanocobalamin) 2.6 mg	100%
Pantothenic Acid (as as d-calcium pantothenate) 6 mg	100%
Iron (as iron fumarate) 27 mg	100%
Iodine (kelp) 220 mcg	100%
Zinc (as monomethionine & gluconate) 11mg	100%
Selenium (as sodium selenate) 60 mcg	100%
Copper (as copper sulfate) 1000 mcg	100%
Calcium (as calcium carbonate) 200 mg	20%

*** Daily Values based on RDAs for pregnant women ages 19-50**

Other ingredients: stearic acid, vegetable stearate, silicon dioxide, croscarmellose sodium, microcrystalline cellulose, natural coating (contains hydroxypropyl methylcellulose, titanium dioxide, riboflavin, polyethylene glycol and polysorbate)

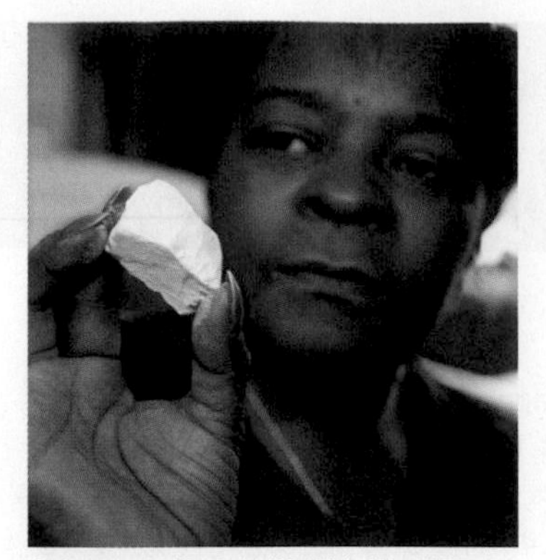

© Michael DiBari, Jr./AP Wide World Photos

Figure 14.13 **Pica and cultural beliefs**
This African-American woman in Georgia is eating a white clay called kaolin, which some women crave during pregnancy. Eating kaolin is also a traditional remedy for morning sickness. This example of pica may be related to cultural beliefs and traditions.

pica An abnormal craving for and ingestion of unusual food and nonfood substances.

Usually the food cravings of pregnant women can be indulged within reason with no harmful effects. But abnormal cravings leading to consumption of nonfood substances can have serious consequences. **Pica** is an abnormal craving for and ingestion of nonfood substances having little or no nutritional value (see **Focus on Eating Disorders**). Pica has been described since antiquity but its cause is still a mystery. It was once thought that pica was an attempt to meet micronutrient needs. It is now believed that it may be more related to cultural factors than the need for micronutrients. It is more common in African-American women and in those with a family or personal history of the practice (**Figure 14.13**).[30] Women with pica commonly consume clay, laundry starch, ice and freezer frost, baking soda, cornstarch, and ashes. Consuming large amounts of these can reduce the intake of nutrient-dense foods, reduce nutrient absorption from food, increase the risk of consuming toxins and harmful microorganisms, and even cause intestinal obstructions. Complications of pica include iron-deficiency anemia, lead poisoning, and parasitic infections.[30] Anemia and hypertensive disorders of pregnancy are more common in mothers who practice pica, but it is not clear if pica is a result of these conditions or a cause. In newborns, anemia and low birth weight are often related to pica in the mother.

14.3 Factors That Increase the Risks of Pregnancy

LEARNING OBJECTIVES

- Explain how maternal nutritional status, health status, age, and income level affect the risks of pregnancy.
- Discuss why deficiencies or excesses of nutrients affect fetal health.
- List three environmental hazards that can affect pregnancy
- Describe the impact of alcohol intake during pregnancy.
- Describe the effects of maternal drug and alcohol use on pregnancy outcome.

Most of the over four million women who give birth every year in the United States are healthy during pregnancy and produce healthy babies. However, childbearing is not without risks. In the United States, 12 out of every 100,000 women die as a result of childbirth. More than 12% of babies are born too soon, 8% are of low or very low birth weight, and 6.4 out of each 1000 born alive die within the first year of life.[8,31] The reasons for poor pregnancy outcome vary. Malnutrition is a factor in some women. Others are at increased risk because of their age and preexisting health problems or socioeconomic factors such as limited access to health care, lack of a supportive home environment, or insufficient resources to acquire nutritious foods (**Table 14.3**). Some women and babies are at risk because they are exposed to harmful substances from their diet or environment.

Maternal Nutritional Status

Proper nutrition is important before pregnancy to support conception and maximize the likelihood of a healthy pregnancy. At any time during pregnancy, maternal malnutrition due to a deficiency or excess of energy or individual nutrients can affect pregnancy outcome.

Nutritional Status Before Pregnancy A woman's nutritional status before she becomes pregnant may affect her ability to conceive and successfully complete a pregnancy. Starvation diets, anorexia nervosa, and excessive exercise, such as marathon running, can reduce body fat and affect hormone levels. If hormone levels are too low, ovulation does not occur and conception is not possible. Too much body fat can also reduce fertility by

TABLE 14.3 Factors That Increase Pregnancy Risks

Maternal factor	Maternal risk	Infant/Fetal risk
Prepregnancy BMI < 18.5 or gaining too little weight during pregnancy	Anemia, premature rupture of the membranes, hemorrhage after delivery	Low birth weight, preterm birth
Prepregnancy BMI ≥ 30 or gaining too much weight during pregnancy	Hypertensive disorders of pregnancy, gestational diabetes, difficult delivery, cesarean section	Large-for-gestational-age, low Apgar scores (a score used to assess the health of a baby in the first minutes after birth), and neural tube defects
Malnutrition	Decreased ability to conceive, anemia	Fetal growth retardation, low birth weight, birth defects, preterm birth, spontaneous abortion, stillbirth, increased risk of chronic disease later in life
Phenylketonuria	High blood levels of phenylketones	Brain damage leading to intellectual disabiity if low phenylalanine diet is not carefully followed by mother
Hypertension	Stroke, heart attack, premature separation of the placenta from the uterine wall	Low birth weight, fetal death
Diabetes	Difficulty adjusting insulin dose, preeclampsia, cesarean section	Large-for-gestational-age, congenital abnormalities, fetal death
Frequent pregnancies: 3 or more during a 2-year period	Malnutrition	Low birth weight, preterm birth
History of poor obstetric or fetal outcome	Recurrence of problem in subsequent pregnancy	Birth defects, death
Age:		
Younger than 20	Malnutrition, hypertensive disorders of pregnancy	Low birth weight
Older than 35	Hypertensive disorders of pregnancy, gestational diabetes	Down syndrome and other chromosomal abnormalities
Alcohol consumption	Poor nutritional status	Fetal alcohol spectrum disorders, including fetal alcohol syndrome
Cigarette smoking	Lung cancer and other lung diseases, miscarriage	Low birth weight, stillbirth, preterm birth, sudden infant death syndrome, respiratory problems
Cocaine use	Hypertension, miscarriage, premature labor and delivery	Intrauterine growth retardation, low birth weight, preterm birth, birth defects, sudden infant death syndrome

altering hormone levels. Deficiencies or excesses of nutrients can affect pregnancy outcome. For instance, a deficiency of folate or an excess of vitamin A early in pregnancy can cause birth defects.

Nutritional status can be affected by some birth-control methods and these can therefore have an impact on a subsequent pregnancy. For example, oral contraceptives are associated with reduced blood levels of vitamins B_6 and B_{12}.[26] If conception occurs soon after oral contraceptive use stops, these levels will not have had time to return to normal before pregnancy begins.

Malnutrition During Pregnancy

Maternal malnutrition can cause fetal growth retardation, low infant birth weight, birth defects, premature birth, spontaneous abortion, and stillbirth. The effect of malnutrition depends on how severe the nutrient deficiency or excess is and when during the pregnancy it occurs. In general, poor nutrition early in pregnancy affects embryonic development and the potential of the embryo to survive, and poor nutrition in the latter part of pregnancy affects fetal growth.

Immediate Effects of Maternal Malnutrition A low energy intake during early pregnancy is not likely to interfere with fetal growth because the energy demands of the embryo are small. However, if the embryo does not receive adequate amounts of the nutrients needed for cell division and differentiation, such as folate and vitamin A, malformations or death can result. Inadequate folate intake in the first few weeks of pregnancy may affect neural tube development.[26] Too much vitamin A is of particular concern because the risk of kidney problems and central nervous system abnormalities in the offspring increases even when maternal intake is not extremely high. High intakes early in pregnancy are the most damaging. A UL of 3000 μg/day has been established for pregnant women ages 19 to 50 years.[28] Supplements consumed during pregnancy should therefore contain β-carotene, which is not damaging to the fetus.

Malnutrition is most likely to cause developmental defects (malformations) during the first trimester. After the first trimester, nutrient deficiencies or excesses are less likely to cause malformations because most organs and structures have already formed. However, undernutrition in the mother after the first trimester can interfere with fetal growth. Even a mild energy restriction during the last trimester, when the fetus is growing rapidly, can affect birth weight. Malnutrition also interferes with the growth and function of the placenta. Then, in turn, a poorly developed placenta cannot deliver sufficient nutrients to the fetus, and the result is a small infant who may also have other developmental abnormalities.

Long-Term Effects of Maternal Malnutrition It has been proposed that problems in maternal nutrition can cause adaptations that change fetal structure, physiology, and metabolism and can affect the child's risk of developing chronic diseases later in life. Evidence for this comes from epidemiological studies that suggest that individuals who were small at birth or disproportionately thin or short have higher rates of heart disease, high blood pressure, high blood cholesterol, and diabetes in middle age.[32]

Maternal Health Status

The general health as well as nutritional status of the mother affects the outcome of pregnancy. Women who begin pregnancy with chronic diseases such as hypertension, diabetes, and phenylketonuria (PKU) must manage their health carefully to assure a healthy pregnancy. The effect of hypertension depends on when it develops and how severe it is. As in nonpregnant individuals, high blood pressure in pregnant women increases the risk of stroke and heart attack, but in pregnancy it also increases the risk of low birth weight and premature separation of the placenta from the wall of the uterus, resulting in fetal death.

Women with diabetes must carefully manage diet and medication to ensure that glucose levels stay in the normal range throughout pregnancy. Uncontrolled diabetes early in pregnancy increases the risk of birth defects. When normal blood glucose is maintained throughout pregnancy, the risk of complications is greatly reduced. The need for insulin increases during the second and third trimesters, so women with preexisting diabetes may need to adjust their medication dosage. When maternal blood glucose is elevated, it provides extra calories to the growing fetus, resulting in large-for-gestational-age newborns who are at increased risk.

Reproductive History Frequent pregnancies, with little time between, increase the risk of poor pregnancy outcomes. One reason is that the mother may not have replenished nutrient stores depleted in the first pregnancy when she becomes pregnant again. An interval of less than 18 months increases the risk of delivering a full-term but small-for-gestational-age infant. An interval of only three months has been shown to increase the risk of a preterm or small-for-gestational-age infant as well as neonatal death.[33] Women with a history of poor pregnancy outcome are also at increased risk. For example, a woman who has had a number of miscarriages is more likely to have another, and a woman who has had one child with a birth defect has an increased risk for defects in subsequent children.

The Pregnant Teenager Pregnancy places a stress on the body at any age, but this is compounded when the mother herself is still growing. Although the rate of teen pregnancy has been decreasing, from 62 babies per 1000 teens in 1991 to 39 per thousand in 2009, it remains a major public health problem.[34] Pregnant teens are at greater risk of developing hypertensive disorders of pregnancy and are more likely to deliver preterm and low-birth-weight babies. To produce a healthy baby, a pregnant teenager needs early medical intervention and nutritional counseling.

Adolescent girls continue to grow and mature physically for about four to seven years after menstruation begins. Therefore the diet of a pregnant teen must provide both for her growth and that of her baby. Even teenagers who deliver normal-birth-weight infants may stop growing themselves.[35] Because the nutrient needs of a pregnant teen may be higher than those of a pregnant adult, the DRIs include a special set of nutrient recommendations for pregnant teens (**Figure 14.14**). Consuming a diet that meets these needs can be challenging. Even nonpregnant teens often fall short of meeting their nutrient needs. Nutrients that are commonly low in the diets of pregnant teens are calcium, iron, zinc, magnesium, vitamin D, folate, and vitamin B_6.[22,26]

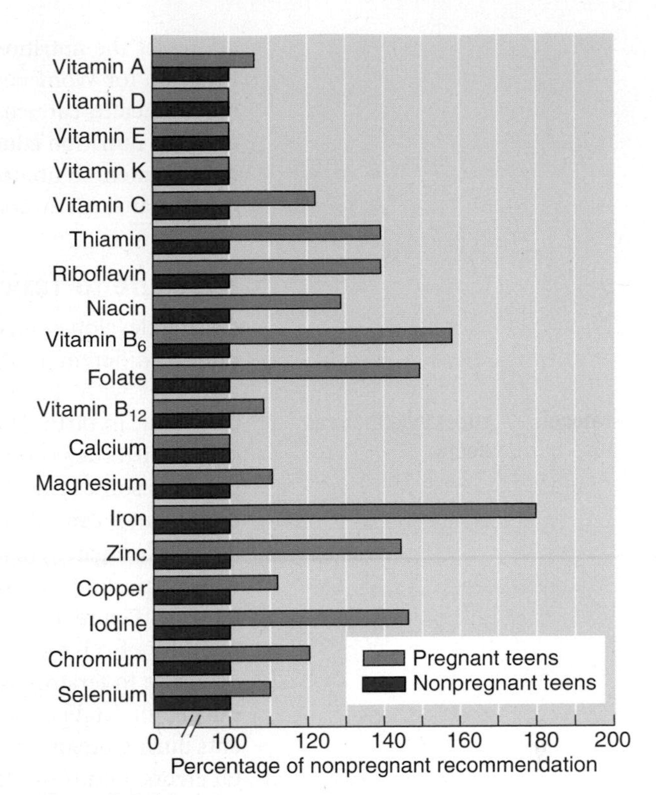

Figure 14.14 **Nutrient needs of pregnant teens**
Because the nutrient needs of pregnant teens are different from those of pregnant adults, the DRIs include a special set of recommendations for this group. This graph compares the micronutrient needs of 14- to 18-year-old pregnant and nonpregnant teens.

The Older Mother The nutritional requirements for older women during pregnancy are no different than those for women in their twenties, but pregnancy after the age of 35 does carry additional risks because older women are more likely to start pregnancy with medical conditions such as cardiovascular disease, kidney disease, obesity, and diabetes. During pregnancy, they also are more likely to develop gestational diabetes, hypertensive disorders of pregnancy, and other complications. They have a higher incidence of low-birth-weight deliveries and of chromosomal abnormalities, especially **Down syndrome** (**Figure 14.15**). Today, careful medical monitoring throughout pregnancy is reducing the risks to older mothers and their babies.

Socioeconomic Factors

One of the greatest risk factors for poor pregnancy outcome is low-income level. Poverty limits access to food, education, and health care. Low-income women have a higher incidence of low-birth-weight and preterm infants.[37] Many low-income women do not receive any medical care until late in pregnancy. Women who begin prenatal care after the first trimester are at a higher risk for poor pregnancy outcomes such as premature birth, low birth weight, or growth retardation. One federally funded program that

THINK CRITICALLY

Why are the micronutrient needs of a pregnant teen higher than those of a 25-year-old pregnant woman of the same weight?

Down syndrome A disorder caused by extra genetic material that results in cognitive disabilities, distinctive facial characteristics, and other abnormalities.

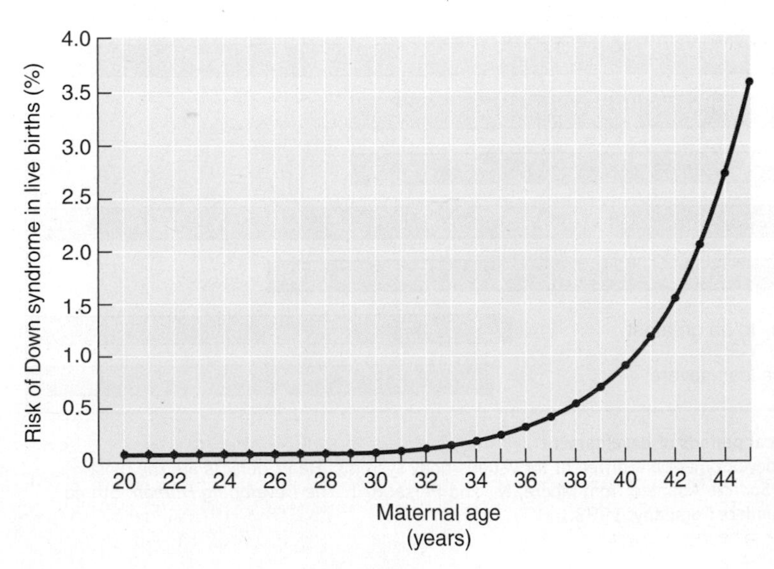

Figure 14.15 **Incidence of Down syndrome**
Down syndrome is caused by a defect during egg or sperm formation that results in an extra copy of chromosome 21. It causes developmental and cognitive delays, distinctive facial characteristics, and other health problems. The incidence of Down syndrome rises with maternal age of 35 years and older. However, over 80% of Down syndrome births are to women under the age of 35, as younger women have more babies overall.[36]

addresses the nutritional needs of pregnant women is the Special Supplemental Nutrition Program for Women, Infants, and Children (WIC). WIC participation has been shown to reduce health-care costs by providing preventative care to low-income pregnant women through nutrition education and food vouchers.[38] This program provides services to pregnant women, to nonlactating women for six months after birth, to lactating women for 12 months after birth, and to infants and children up to 5 years of age.

Exposure to Toxic Substances

teratogen A substance that can cause birth defects.

During development, cells are particularly vulnerable to damage because they are dividing rapidly, differentiating, and moving to form organs and other structures. Anything that interferes with this development can cause a baby to be born too soon or too small or can result in birth defects. A **teratogen** is any chemical, biological, or physical agent that causes birth defects. Even some vitamins have been found to be teratogens. The placenta prevents some teratogens from passing from the mother's blood to the embryonic or fetal blood, but it cannot prevent the passage of all hazardous substances.

Each organ system develops at a different rate and time, so each has a critical period when exposure to a teratogen or other insult can disrupt development and cause irreversible damage (**Figure 14.16**). If the damage is severe, it may result in a miscarriage. Because the majority of cell differentiation occurs during the embryonic period, this is the time when exposure to teratogens can do the most damage, but vital body organs can still be affected during the fetal period. As discussed previously, deficiencies or excesses of energy or nutrients during pregnancy can affect the health of the embryo and fetus and cause developmental errors. Other substances present in the environment, consumed in the diet, or taken as medications or recreational drugs can also act as teratogens.

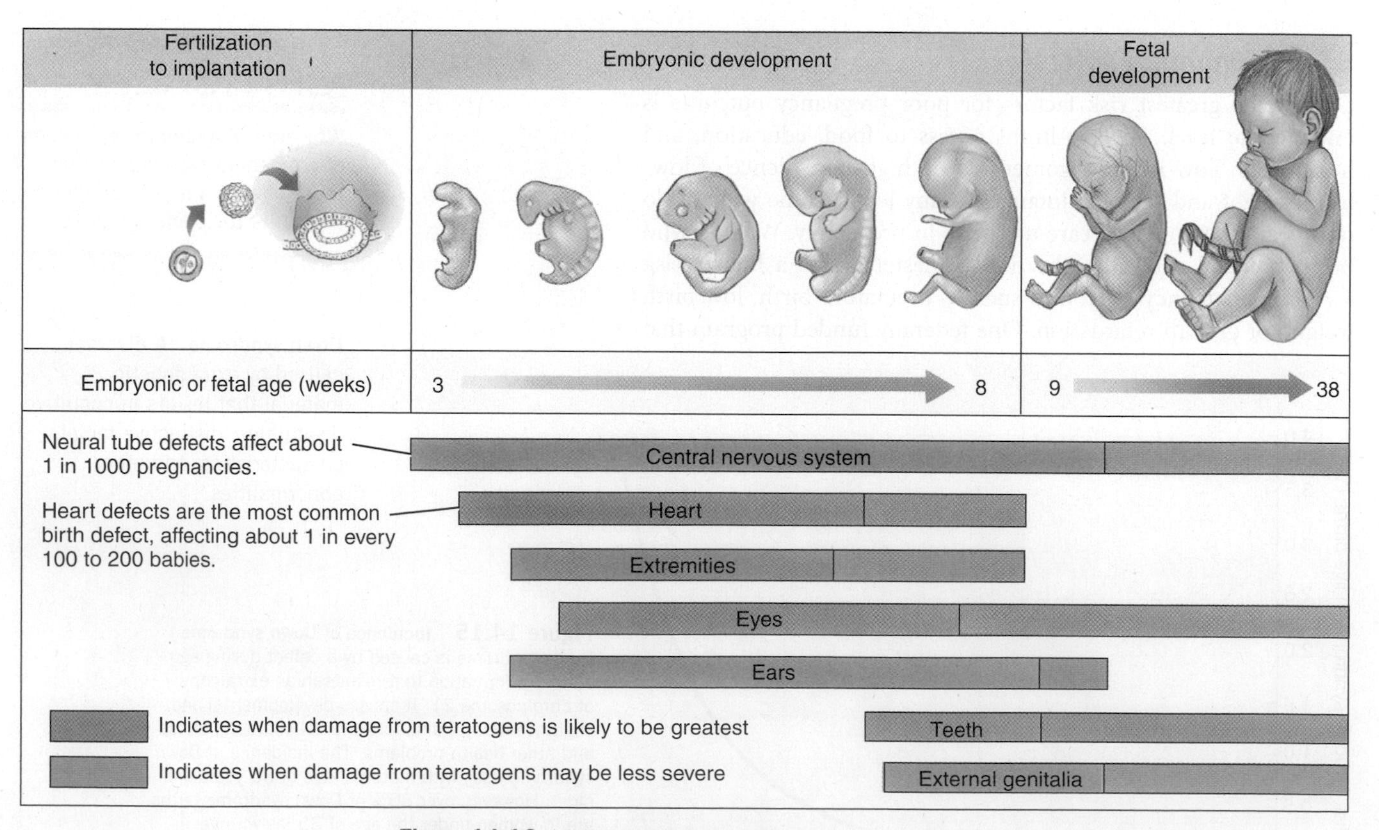

Figure 14.16 Critical periods of development
The critical periods of development are different for various body systems. Heart defects are the most common birth defect. (*Source*: Adapted from Moore, K., and Persaud, T. *The Developing Human*, 5th ed. Philadelphia: W. B. Saunders Company, 1993.)

Environmental Toxins In a pregnant woman, exposure to environmental toxins—such as cleaning solvents, lead, mercury, some insecticides, and paint—can affect her developing child. Therefore, pregnant women need to be aware of the potential toxins in their food, water, and environment. Fish has both benefits and risks for pregnant women. It is a source of lean protein for tissue growth and of omega-3 fatty acids and iodine needed for brain development, but if it is contaminated with mercury, consumption during pregnancy can cause developmental delays and brain damage in the baby. Rather than avoid fish, pregnant women should be informed consumers. Exposure to mercury can be controlled by avoiding fish such as shark, swordfish, king mackerel, or tilefish, which can be high in mercury, and by limiting tuna intake to no more than 6 oz of albacore per week. Up to 12 oz/week of varieties of fish and shellfish that are lower in mercury, such as shrimp, canned light tuna, salmon, pollock, and catfish, can be safely consumed.[39]

Caffeine and Herbs Caffeine-containing beverages like coffee are a part of our typical diet, but consuming too much caffeine during pregnancy has been associated with reductions in birth weight and an increased risk of miscarriage.[40] It is recommended that pregnant women avoid consuming more than 200 mg of caffeine per day. This is the amount in about two cups of regular coffee or five cups of tea or cola beverages. Herbal teas are also popular beverages. Some are consumed to treat the discomforts of pregnancy. For example, ginger and raspberry leaves are often used to treat morning sickness. However, because there is not a great deal of research on herbal teas and other herbal products, pregnant women should avoid them until they are shown to be safe during pregnancy.[2]

Food-Borne Illness The immune system is weakened during pregnancy, increasing susceptibility to and severity of certain food-borne illnesses. *Listeria* infections are about 20 times more likely during pregnancy and are especially dangerous for pregnant women, often resulting in miscarriage, premature delivery, stillbirth, or infection of the fetus.[41] About one-quarter of babies born with *Listeria* infections do not survive. The bacteria are commonly found in unpasteurized milk, soft cheeses, and uncooked hot dogs and lunch meats.

Toxoplasmosis is an infection caused by a parasite. If a pregnant woman becomes infected, there is about a 40% chance that she will pass the infection to her unborn baby.[42] Some infected babies develop vision and hearing loss, intellectual disability, and/or seizures. The toxoplasmosis parasite is found in cat feces, soil, and undercooked infected meat. Pregnant women should follow the safe food-handling recommendations discussed in Chapter 17.

How Alcohol Affects Pregnancy Outcome Alcohol consumption during pregnancy is one of the leading causes of preventable birth defects (see **Focus on Alcohol**). Alcohol is a teratogen that is particularly damaging to the developing central nervous system.[43] It also indirectly affects fetal growth and development because it is a toxin that reduces blood flow to the placenta, thereby decreasing the delivery of oxygen and nutrients to the fetus. The use of alcohol can also impair maternal nutritional status, further increasing the risk to the embryo or fetus. Despite this, about 12% of women report drinking alcohol during pregnancy.[2]

Prenatal exposure to alcohol can cause a spectrum of disorders depending on the dose, timing, and duration of the exposure. One of the most severe outcomes of drinking alcohol during pregnancy is a baby with **fetal alcohol syndrome (FAS)**. FAS is characterized by facial deformities, growth retardation, and permanent brain damage (**Figure 14.17**). Newborns with the syndrome may be shaky and irritable, with poor muscle tone and alcohol withdrawal symptoms. Other problems include heart and urinary-tract defects, impaired vision and hearing, and delayed language development. Intellectual disability is the most common and most serious effect. Not all babies who are exposed to alcohol while in the uterus have FAS. The term **fetal alcohol spectrum disorders (FASDs)** is used to refer to all the physical or behavioral disorders or conditions and functional or mental impairments linked to prenatal alcohol exposure.[44] As many as 2 to 5% of young school children in the United States are affected by FASDs.[45]

fetal alcohol syndrome A characteristic group of physical and mental abnormalities in an infant resulting from maternal alcohol consumption during pregnancy.

fetal alcohol spectrum disorders (FASDs) A continuum of permanent birth defects in a child due to maternal alcohol consumption during pregnancy.

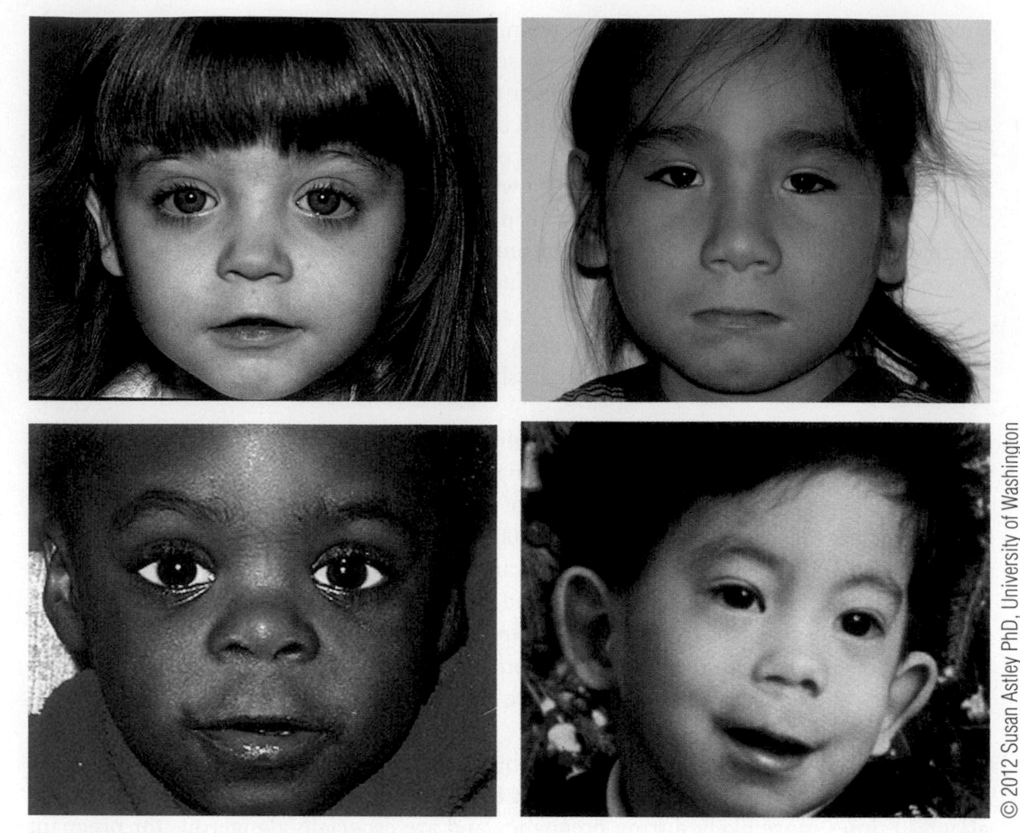

Figure 14.17 Fetal alcohol syndrome
Facial characteristics shared by children with fetal alcohol syndrome include a low nasal bridge, a short nose, distinct eyelids, and a thin upper lip.

Because alcohol consumption in each trimester has been associated with abnormalities and because there is no level of alcohol consumption that is known to be safe, complete abstinence from alcohol is recommended during pregnancy. Warning labels that appear on containers of beer, wine, and hard liquor state that "According to the Surgeon General, women should not drink alcoholic beverages during pregnancy because of the risk of birth defects."

Cigarette Smoke If a woman smokes cigarettes during pregnancy, her baby will be affected before birth and throughout life. The carbon monoxide in tobacco smoke binds to hemoglobin, reducing oxygen delivery to fetal tissues. The nicotine absorbed from cigarette smoke is a teratogen that can affect brain development.[46] Nicotine also constricts arteries and limits blood flow, reducing both oxygen and nutrient delivery to the fetus.[47] Cigarette smoking during pregnancy reduces birth weight and increases the risk of preterm delivery, stillbirth, neurobehavioral problems, and early death.[2,46] Even exposure to cigarette smoke from the environment has been found to increase the risk of low birth weight. The risk of **sudden infant death syndrome (SIDS),** or **crib death**, and respiratory problems are also increased in children exposed to cigarette smoke both in the uterus and after birth. The effects of maternal smoking follow children throughout life; they are more likely to have frequent colds and develop lung problems later in life.[48]

sudden infant death syndrome (SIDS) or **crib death** The unexplained death of an infant, usually during sleep.

Legal and Illicit Drug Use The use of drugs—whether over-the-counter, prescribed, or illicit—can also affect both fertility and pregnancy outcome. For example, the acne medications Accutane and Retin-A, which are derivatives of vitamin A, can cause birth defects if used during pregnancy. A woman who is considering pregnancy should discuss her plans with her physician in order to determine the risks associated with any medication she is taking.

Substance abuse during pregnancy is a national health issue. It is estimated that from 1 to 11% of babies born each year have been exposed to drugs during the prenatal period. These numbers include only the use of illicit drugs and would be much larger if alcohol and nicotine were included.[49]

Marijuana and cocaine are drugs that are commonly used during pregnancy. Both cross the placenta and enter the fetal blood. There is little evidence that marijuana affects fetal outcome, but cocaine use increases the risk of complications to the mother and creates problems for the infant before, during, and after delivery. Cocaine is a central nervous system stimulant, but many of its effects during pregnancy occur because it constricts blood vessels, thereby reducing the flow of oxygen and nutrients to the rapidly dividing fetal cells. Cocaine use during pregnancy is associated with a high rate of miscarriage, intrauterine growth retardation, spontaneous abortion, premature labor and delivery, low birth weight, and birth defects.[50] Exposure to cocaine, opiates, or amphetamines has been shown to affect infant behavior and impact learning and attention span during childhood.[51]

14.4 Lactation

LEARNING OBJECTIVES

- Compare the nutritional needs of lactating women with those of nonpregnant nonlactating women of childbearing age.
- Explain the relationship between suckling and milk production and letdown.

colostrum The first milk, which is secreted in late pregnancy and up to a week after birth. It is rich in protein and immune factors.

prolactin A hormone released by the anterior pituitary that acts on the milk-producing glands in the breast to stimulate and sustain milk production.

The nutrient requirements of pregnancy include those needed to prepare for lactation. After childbirth, the breast-feeding mother's nutrient intake must support milk production and can influence the nutrient composition of her milk.

Milk Production and Letdown

Lactation involves both the synthesis of the milk components, such as milk proteins, lactose, and milk lipids, and the movement of these milk components through the milk ducts to the nipple. Throughout pregnancy, hormones prepare the breasts for lactation by stimulating the enlargement and development of the milk ducts and the milk-producing glands, called *alveoli* (**Figure 14.18**). During the first few days after childbirth, the breasts produce and secrete a small amount of a clear yellow fluid called **colostrum**. Colostrum is immature milk. It is rich in protein, including immune factors that help protect the newborn from disease. Within about a week of childbirth, there is a rapid increase in milk secretion, and its composition changes from colostrum to that of mature milk.

Milk production and release is triggered by hormones that are released in response to the suckling of the infant. The pituitary hormone **prolactin** stimulates milk production; the more the infant suckles, the more milk is produced. The release of milk from the milk-producing glands and its movement

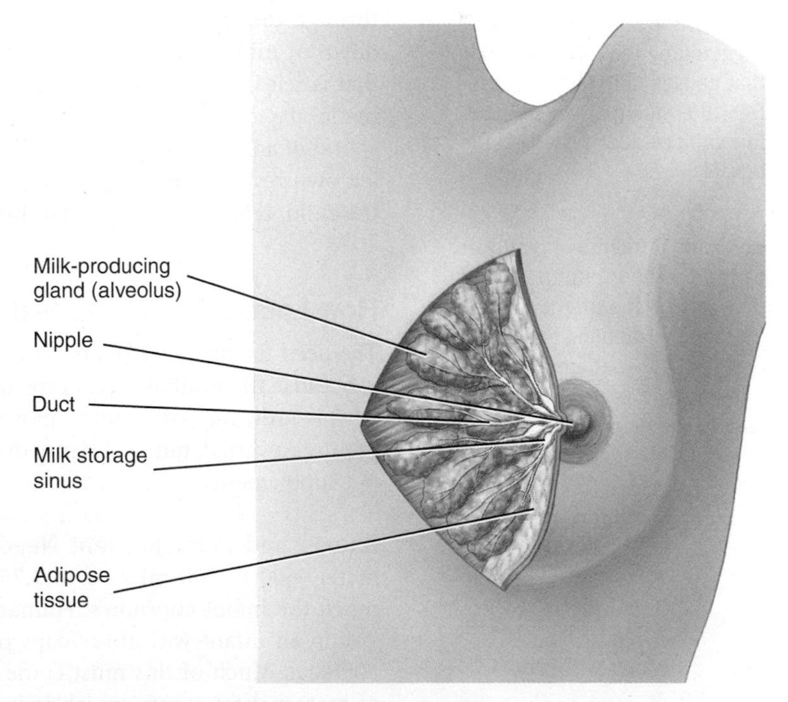

Figure 14.18 **Anatomy of lactation**
During lactation, milk travels from the milk-producing glands through the ducts to milk storage sinuses and then to the nipple.

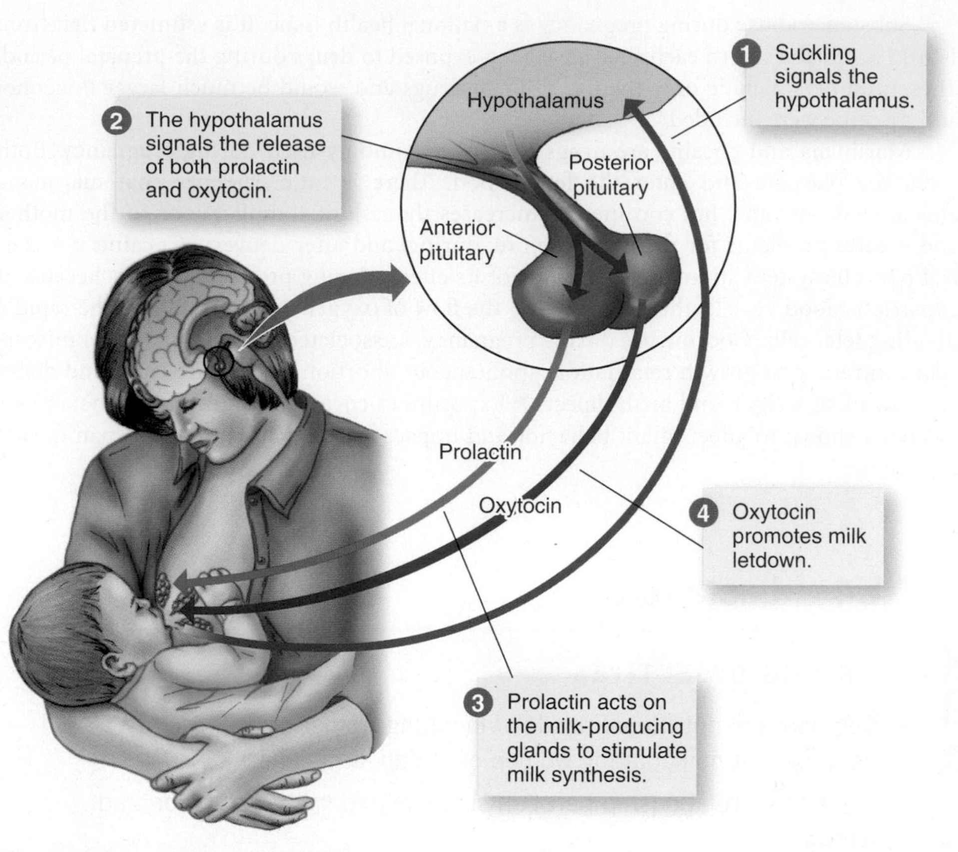

Figure 14.19 **Hormones of lactation**
When the infant suckles, nerve receptors in the nipple send signals to the hypothalamus, which trigger the release of prolactin and oxytocin.

letdown A hormonal reflex triggered by the infant's suckling that causes milk to be released from the milk glands and flow through the duct system to the nipple.

oxytocin A hormone produced by the posterior pituitary gland that acts on the uterus to cause uterine contractions and on the breast to cause the movement of milk into the secretory ducts that lead to the nipple.

through the ducts and storage sinuses is referred to as **letdown** (**Figure 14.19**). The letdown of milk is caused by **oxytocin**, another hormone produced by the pituitary gland that is released in response to the suckling of the infant. As nursing becomes more automatic, oxytocin release and the letdown of milk may occur in response to the sight or sound of an infant. It can be inhibited by nervous tension, fatigue, or embarrassment. The letdown response is essential for successful breast-feeding and makes suckling easier for the child. If letdown is slow, the child can become frustrated and difficult to feed.

How Lactation Affects Maternal Nutrient Needs

The need for many nutrients is even greater during lactation than during pregnancy. This is because the mother is still providing for all the energy and nutrient needs of the infant, who is growing faster and is more active than the fetus. Meeting the needs of lactation requires a varied, nutrient-dense diet. Most lactating women can meet all their needs without supplements.

Energy and Macronutrient Needs During the first six months of lactation, approximately 600 to 900 mL (2.5 to 3.75 cups) of milk is produced daily, depending on how much the infant consumes. Human milk contains about 160 kcal/cup (240 mL), so providing an infant with three cups of milk requires the mother to expend approximately 500 kcal. Much of this must come from the diet, but some can come from mobilization of maternal fat stores, which increase during pregnancy. The EER for lactation is estimated by adding the total energy expenditure of nonlactating women and the energy in milk and then subtracting the energy supplied by maternal fat stores.[16] This is equal to an additional 330 kcal/day during the first six months of lactation, and 400 kcal/day

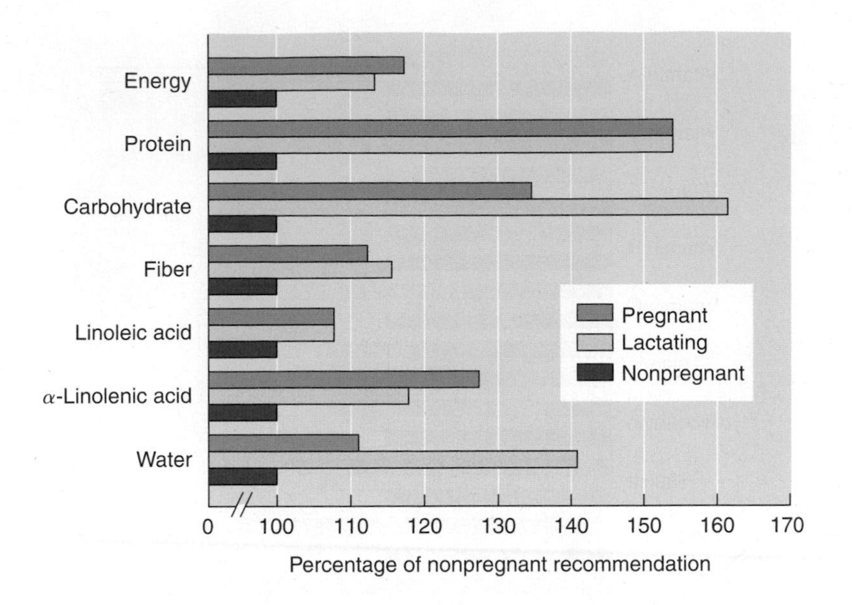

Why are water needs higher during lactation than during pregnancy?

Figure 14.20 Energy and macronutrient needs during lactation This graph compares the energy and macronutrient recommendations for 25-year-old nonpregnant women and 25-year-old women during the third trimester of pregnancy and the first six months of lactation.

during the second six months (**Figure 14.20**). To ensure adequate protein for milk production, the RDA for lactation is 25 g/day higher than for nonlactating women. The RDA for carbohydrate and the AIs for linoleic and α-linolenic acids are also higher during lactation.[16]

Even though some of the energy for lactation comes from maternal fat stores, the impact of lactation on maternal weight loss is variable. Beginning one month after birth, most lactating women lose 1 to 2 lbs (0.5 to 1 kg)/month for six months. Some women will lose more, and others may maintain or even gain weight regardless of whether or not they breast-feed. Rapid weight loss is not recommended during lactation because it can decrease milk production. However, regular exercise can make weight loss easier and does not impair milk production.

Water Needs During Lactation The amount of milk a woman produces depends on how much her baby demands. The more the infant suckles, the more milk is produced. To avoid dehydration and ensure adequate milk production, lactating women need to consume about 1 liter of additional water per day. The AI of 3.8 L/day, of which about 3.1 L is from drinking water and other beverages, is based on typical intake during lactation.[11] Consuming an extra glass of milk, juice, or water at every meal and whenever the infant nurses can help ensure adequate fluid intake.

Micronutrient Needs During Lactation The recommended intakes for several vitamins and minerals are increased during lactation to meet the metabolic needs of synthesizing milk and to replace the nutrients secreted in the milk itself. Maternal intake of some vitamins, including B_6, B_{12}, C, A, and D, can affect milk composition. When maternal intake is low, the amounts in milk are decreased. The recommended intakes of vitamin B_6, vitamin B_{12}, other B vitamins, and vitamins C, A, and E are increased above nonlactating levels (**Figure 14.21**). Because vitamin B_{12} may be deficient in the breast milk of vegan mothers, their infants should be supplemented with vitamin B_{12}.

For other nutrients, levels in the milk are maintained at the expense of maternal stores. For example, much of the calcium secreted in human milk comes from an increase in maternal bone resorption. However, the RDA for calcium is not increased above nonlactating levels because the loss of calcium from maternal bones is not prevented by increases in dietary calcium. In addition, the lost bone calcium is fully restored within a few months of weaning, and therefore women who breast-feed have not been found to have a long-term deficit in bone mineral density.[52]

Iron needs are not increased during lactation because little iron is lost in milk, and, in most women, losses are decreased because menstruation is absent. The RDA for lactation is 9 mg/day, half that of nonlactating women.

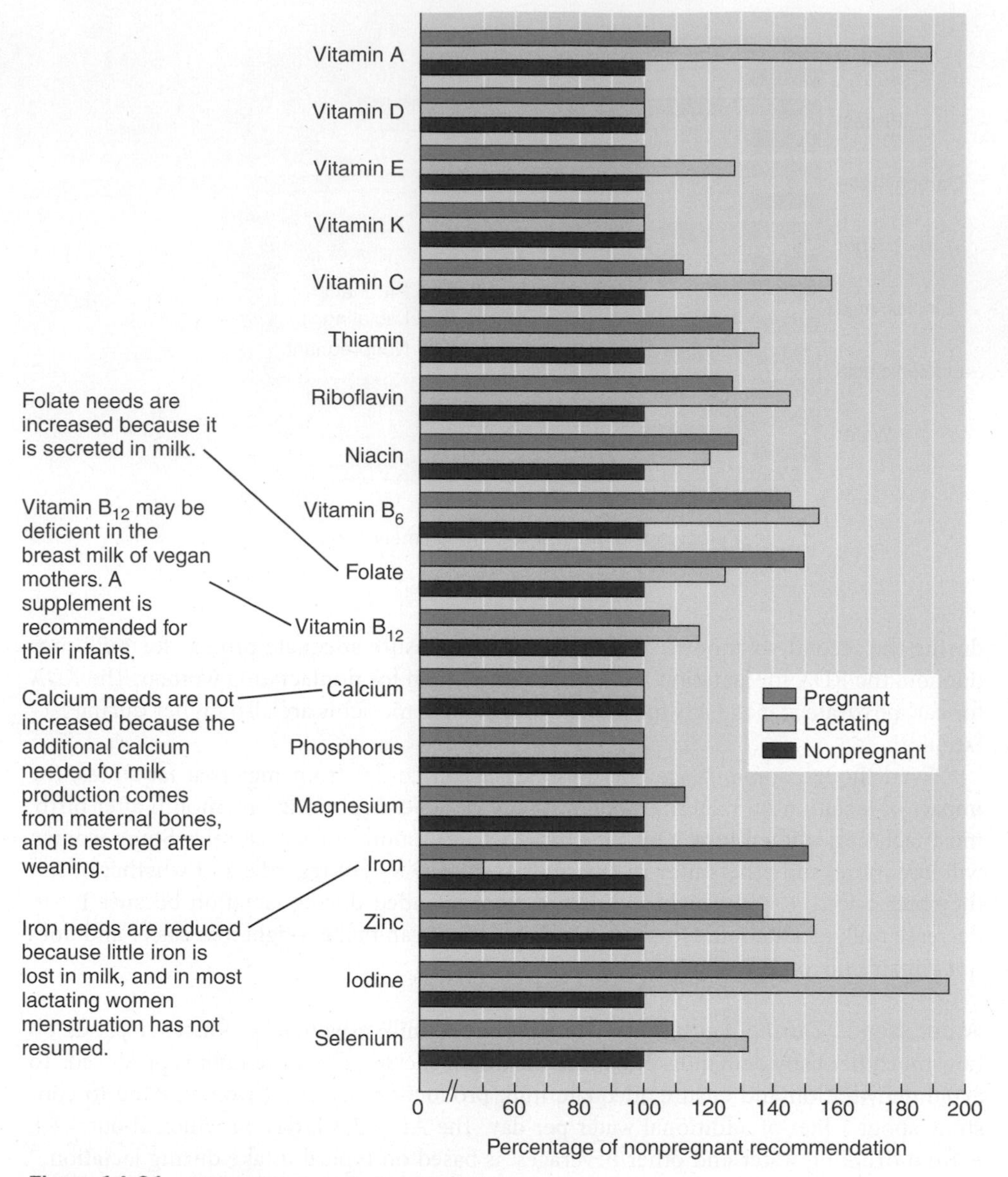

Figure 14.21 Micronutrient needs during lactation
This graph compares the vitamin and mineral recommendations for 25-year-old nonpregnant women and 25-year-old women during the third trimester of pregnancy and the first six months of lactation.

14.5 The Nutritional Needs of Infancy

LEARNING OBJECTIVES

- Compare the energy needs per kg of body weight of infants with those of adults.
- Compare the carbohydrate, protein, and fat recommendations for infants with those for adults.
- List micronutrients that are at risk for deficiency in infants.
- Explain the best way to monitor the adequacy of an infant's dietary intake.

When a child is born and the umbilical cord is cut, he or she suddenly becomes actively involved in obtaining nutrients rather than being passively fed through the placenta. The

TABLE 14.4 Energy and Nutrient Needs of Infants Compared to Adults

Nutrient/Energy	Infant recommendation (0–6 mo)	Adult recommendation
Energy[a]	493–606 kcal/day (~100 kcal/kg/day)	2403–3067 kcal/day (~30 kcal/kg/day)
Protein	9.1 g/day 1.52 g/kg/d	46–56 g/day 0.8 g/kg/day
Carbohydrate	at least 60 g/day 40% of energy intake[b]	at least 130 g/day 45%–65% of energy intake
Fat	50% of energy[b]	20%–35% of energy
Linoleic acid	4.4 g/day[c]	12–17 g/day
α-Linolenic acid	0.5 g/day[d]	1.1–1.6 g/day
Fluid	0.7 L	2.7–3.7 L

[a]The energy values are based on EER prediction equations for infants 0–6 months of age and for adults ≥19 years of age.

[b]Based on the composition of human milk.

[c]Refers to all omega-6 polyunsaturated fatty acids.

[d]Refers to all omega-3 polyunsaturated fatty acids.

energy and nutrients the infant consumes must support his or her continuing growth and development and increasing level of activity (**Table 14.4**).

Energy Needs During Infancy

During the first few months after birth, growth is more rapid than at any other time of life. As a result, newborn infants need more calories per kg of body weight than at any other time (**Figure 14.22a**). EERs for infants are calculated from total energy expenditure

Figure 14.22 Energy and macronutrient needs of infants and adults

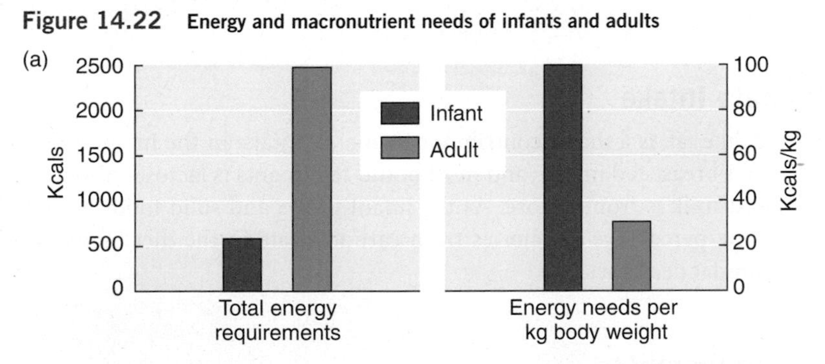

(a) The total amount of energy required by a newborn is less than the amount needed by an adult. When energy needs are expressed per kg of body weight, however, we see that infants require about three times more energy than an adult male.

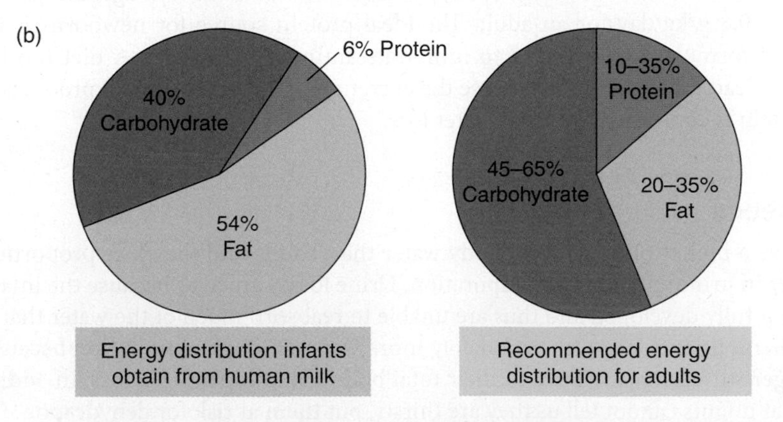

(b) A comparison of the distribution of calories from carbohydrate, fat, and protein in human milk with that recommended for an adult illustrates proportionally how much more fat infants need.

Which macronutrient makes up the largest percent of calories in breast milk? Why is this important for infant health?

plus the energy deposited in tissues due to growth[16] (see inside cover). After the first few months, growth slows some and activity increases. Differences in growth rates and activity levels are reflected in the separate EER prediction equations for infants 0 to 3 months, 4 to 6 months, and 7 to 12 months.

Fat Recommendations

Healthy infants consume about 55% of their energy as fat during the first six months of life, and 40% during the second six months. This is far greater than the 20 to 35% of energy from fat recommended in the adult diet (**Figure 14.22b**). The high proportion of fat increases the energy density of the diet, allowing the infant's small stomach to hold enough to meet energy needs. An AI for total fat has been set at 31 g/day for infants from birth to six months of age and at 30 g/day for infants seven to 12 months of age.

In addition to getting enough fat, infants need the right kinds of fat. A sufficient supply of the long-chain polyunsaturated fatty acids DHA (an omega-3 fatty acid) and arachidonic acid (an omega-6 fatty acid) are important for nervous system development. These fatty acids are constituents of cell membranes and are incorporated into brain tissue and the retina of the eye. Breast milk contains both of these fatty acids. Infant formulas supplemented with DHA and arachidonic acid are available, but the addition of these fatty acids to infant formulas is not required in the United States.[53] Infants can synthesize these fatty acids from linoleic and α-linolenic acid, but since breast-fed infants have higher plasma concentrations of these long-chain polyunsaturated fatty acids than infants fed non-fortified formula, it is hypothesized that the rate of conversion may not be optimal. Evidence that inclusion of these fatty acids in infant formulas benefits growth, visual function, and cognitive development is equivocal (see **Debate: DHA-Fortified Infant Formulas**).[54] AIs for infants have been set for total omega-3 and total omega-6 fatty acids based on the amounts of these types of fatty acids in human milk.[16]

Carbohydrate Intake

Carbohydrate, like fat, is a major contributor to energy intake in the infant. The source of carbohydrate for breast-fed infants and most bottle-fed infants is lactose. About 40% of the energy in breast milk is from lactose. As the infant grows and solid foods are introduced into the diet, the percentage of calories from carbohydrate in the diet increases and the percentage from fat decreases.

Protein Requirements

The infant's protein requirement per unit of body weight is very high compared with the adult requirement: The AI is 1.52 g/kg/day from birth to six months of age, compared with the RDA of 0.8 g/kg/day for an adult. The ideal protein source for newborns is human milk. Infant formulas are designed to mimic its amino acid pattern. A diet too high in protein may lead to dehydration because the excretion of metabolic wastes produced when excess protein is consumed increases water loss.

Water Needs

Infants have a higher proportion of body water than adults and they lose proportionately more water in urine and through evaporation. Urine losses are high because the infant kidneys are not fully developed and thus are unable to reabsorb much of the water that passes through them. Infants lose proportionately more water through evaporation because they have a larger surface area relative to their total body volume. These factors, in addition to the fact that infants cannot tell us they are thirsty, put them at risk for dehydration. Despite this, infants who are exclusively breast-fed do not require additional water. The AI is based on the volume of human milk consumed and the water content of the milk. It is set at

DEBATE

DHA-Fortified Infant Formulas

DHA and arachidonic acid (ARA) are polyunsaturated fatty acids that can be made in the body from the essential omega-3 fatty acid α-linolenic acid and the essential omega-6 fatty acid linoleic acid, respectively. Studies in animals indicate DHA and ARA are found in high concentrations in the brain and retina and are needed for normal vision and brain development.[1] Accumulation of these fatty acids in the brain and retina occurs most rapidly between the third trimester of pregnancy and 24 months of age, so adequate amounts are crucial during this developmental period. Some infant formulas in the U.S. are fortified with DHA and ARA, and advertisements suggest that they provide an advantage for infant development. Should all babies be fed fortified formula?

Science Source/Photo Researchers, Inc.

DHA and ARA are found in breast milk, so it seems logical that they should be added to infant formula. But, unlike most nutrients, the amounts in breast milk are variable, depending on maternal diet, so it is unclear what constitutes optimal levels. The amounts of these fatty acids transferred to the fetus by the placenta during the last trimester may be enough to ensure adequate brain deposition.[1] Infants born at term are also capable of synthesizing DHA and ARA, so those fed unfortified formula may be able to meet their needs if they have enough α-linolenic acid and linoleic acid in their diet. But there is wide individual variation in the ability to convert α-linolenic acid to DHA, and in some infants conversion may be too low for optimal brain and visual development.[1] Infants born before term cannot synthesize enough of these fatty acids, so they must be included in formula for premature infants.

So will higher intakes of DHA and ARA make children smarter? Numerous studies have explored the impact that postnatal intake of these fatty acids has on intelligence and vision. Some have found that higher intakes of DHA increase blood levels of DHA in infants and are associated with improvements in cognitive development or vision compared to infants with lower intakes.[1,2,3] But not all studies agree. A study of 18-month-old babies found that feeding fortified formulas did not have a significant effect on mental or psychomotor development.[4] A review concluded that feeding infant formulas fortified with DHA and ARA has not been proven to yield benefits with regard to vision or cognition.[5]

There are a number of reasons why the reported effects of fortified formulas have been ambiguous. The effects could be different due to differences in the DHA content of the formula, duration of formula feeding, and the methods used to assess visual acuity and cognitive development. Differences in children's intelligence may be due more to maternal or family characteristics, such as maternal IQ, than the type of milk they are fed.[6]

When it comes to brain and eye development, no one knows exactly how much DHA or ARA an infant needs. A direct link between fortified formula and better vision or higher IQ than with unfortified formula has yet to be established.[1,2] Fortified formulas may not make your baby smarter, but published literature has not found them to be harmful for infants.[7] Breast milk is always best, so the biggest downside to fortified formulas is that advertising may make new mothers believe that the formulas are as good as or better for their babies than breast milk.

THINK CRITICALLY: Why is breast milk still better than these fortified formulas?

[1]Guesnet, P., and Alessandri, J. M. Docosahexaenoic acid (DHA) and the developing central nervous system (CNS): Implications for dietary recommendations. *Biochimie* 93:7–12, 2011.

[2]Hoffman, D. R., Boettcher, J. A., and Diersen-Schade, D. A. Toward optimizing vision and cognition in term infants by dietary docosahexaenoic and arachidonic acid supplementation: A review of randomized controlled trials. *Prostaglandins, Leukot Essent Fatty Acids* 81:151–158, 2009.

[3]Birch, E. E., Garfield, S., Castañeda, Y., et al. Visual acuity and cognitive outcomes at 4 years of age in a double-blind, randomized trial of long-chain polyunsaturated fatty acid-supplemented infant formula. *Early Hum Dev* 83:279–284, 2007.

[4]Beyerlein, A., Hadders-Algra, M., Kennedy, K., et al. Infant formula supplementation with long-chain polyunsaturated fatty acids has no effect on Bayley developmental scores at 18 months of age—IPD meta-analysis of four large clinical trials. *J Pediatr Gastroenterol Nutr* 50:79–84, 2010.

[5]Simmer, K., Patole, S. K., and Rao, S .C. Long-chain polyunsaturated fatty acid supplementation in infants born at term. *Cochrane Database Syst Rev* Dec 7; (12):CD000376, 2011.

[6]Gale, C. R., Marriott, L. D., Martyn, C. N., and Group for Southampton Women's Survey Study. Breastfeeding, the use of docosahexaenoic acid-fortified formulas in infancy and neuropsychological function in childhood. *Arch Dis Child* 95:174–179, 2009.

[7]Makrides, M., Smithers, L. G., and Gibson, R. A. Role of long-chain polyunsaturated fatty acids in neurodevelopment and growth. *Nestle Nutr Workshop Ser Pediatr Program* 65:123–133, 2010.

0.7 L/day for infants zero to six months and at 0.8 L/day for older infants (seven to 12 months).[11] In older infants, some fluid is obtained from other beverages and foods. Although breast milk can meet fluid needs in healthy infants, additional fluids may be needed when water losses are increased by diarrhea or vomiting.

In the developing world, dehydration from diarrhea is the most common cause of infant death, and in the United States it kills one child each day. The cause of the diarrhea is usually a bacterial or viral infection. If an infant has diarrhea, fluid intake should be monitored carefully and a pediatrician should be contacted. Mixtures of sugar, water, and electrolytes are available to replace fluid loses.

Micronutrient Needs of Infants

Human milk and formula are designed to meet the nutrient needs of young infants. Nonetheless, infants may still be at risk for deficiencies of iron, vitamin D, and vitamin K, and suboptimal levels of fluoride. The breast-fed infants of vegan mothers may also be at risk of vitamin B_{12} deficiency.

Iron Iron is the nutrient most commonly deficient in infants who are consuming adequate energy and protein. Iron deficiency is usually not a problem during the first four to six months of life because infants have iron stores at birth and the iron in human milk, though not particularly abundant, is very well absorbed. The AI for iron from birth to six months is only 0.27 mg/day.[28] After 4 to 6 months, iron stores are depleted but iron needs remain high to provide for hemoglobin synthesis, tissue growth, and iron storage. The RDA for infants seven to 12 months jumps to 11 mg/day. To meet needs after four to six months the diets of breast-fed infants should contain other sources of iron, such as iron-fortified rice cereal. Formula-fed infants can obtain iron from fortified formula.

Vitamin D Newborns are also potentially at risk for vitamin D deficiency. Breast milk is relatively low in vitamin D, so breast-fed infants who do not receive adequate exposure to sunlight, such as those living in cold climates, may not obtain enough vitamin D. It is recommended that breast-fed and partially breast-fed infants be supplemented with 400 IU (10 μg)/day of vitamin D beginning in the first few days of life and continuing until they are consuming about 1 L (4 cups) of vitamin D-fortified formula or milk daily.[19,55] Infant formulas contain at least 400 IU (10 μg)/L of vitamin D. Most formula-fed infants consume nearly 1 L of formula per day after the first month of life, so they obtain enough vitamin D from the formula; those consuming less than 1L/day should receive a vitamin D supplement of 400 IU (10 μg)/day. The amount of sun exposure necessary to maintain adequate levels of vitamin D in any given infant at any point in time is not easy to determine; light-skinned infants are more likely than darker-skinned babies to meet their vitamin D needs from sunlight (**Figure 14.23**).[55]

Photodisc/Getty Images, Inc.

Figure 14.23 Vitamin D and skin pigmentation
Dark skin pigmentation reduces the amount of vitamin D that can be synthesized in the skin, putting darker-skinned babies at greater risk for deficiency.

Vitamin K Vitamin K, which is essential for normal blood clotting, is another nutrient for which newborns are at risk of deficiency. Little of this vitamin crosses the placenta from mother to fetus, and because the gut is sterile at birth, no microbial vitamin K synthesis occurs. Breast milk is also low in vitamin K, so breast-fed infants are at risk of hemorrhage due to vitamin K deficiency. To prevent this, it is recommended that all breast-fed infants receive a single intramuscular injection containing 0.5 to 1.0 mg of vitamin K after the first feeding is completed and within the first six hours of life.[56] This provides them with enough vitamin K to last until their intestines are colonized with the bacteria that synthesize it.

Fluoride Fluoride is important in the development of teeth even before they erupt. Breast milk is low in fluoride, and formula manufacturers use unfluoridated water in preparing liquid formula. Therefore, breast-fed infants, infants fed premixed formula, and those fed formula mixed with low-fluoride water are often supplemented beginning at six months of age. In areas where the drinking water is fluoridated, infants fed formula reconstituted with tap water should not be given fluoride supplements.

Vitamin B_{12} Breast milk and infant formula typically contain enough vitamin B_{12} to meet the infant's needs. An exception is the breast milk of vegan mothers, which may be deficient in vitamin B_{12}. Therefore, breast fed-infants of vegan mothers should be given vitamin B_{12} supplements.

How Infant Growth Is Assessed

Although nutrient needs for infants are fairly well defined, it is difficult to calculate an infant's actual nutrient intake—particularly if they are breast-feeding. The best indicator of adequate nourishment is normal growth. Most healthy infants follow standard patterns of growth, so their growth can be monitored using growth charts[57] (see Appendix B).

Growth Charts Growth charts plot typical growth patterns of infants, children, and adolescents in the United States (**Figure 14.24**). They can be used to monitor an infant's pattern of growth and to compare length, weight, and head circumference to standards for infants of the same age. The resulting ranking, or percentile, indicates where the infant's growth falls in relation to population standards. For example, if a newborn boy is at the 20th percentile for weight, it means that 19% of newborn boys weigh less and 80% weigh more. Children usually continue at the same percentiles as they grow.

Whether an infant is 6 lbs or 8 lbs at birth, the pattern of growth should be approximately the same—rapid initially and slowing slightly as the infant approaches 1 year of age.

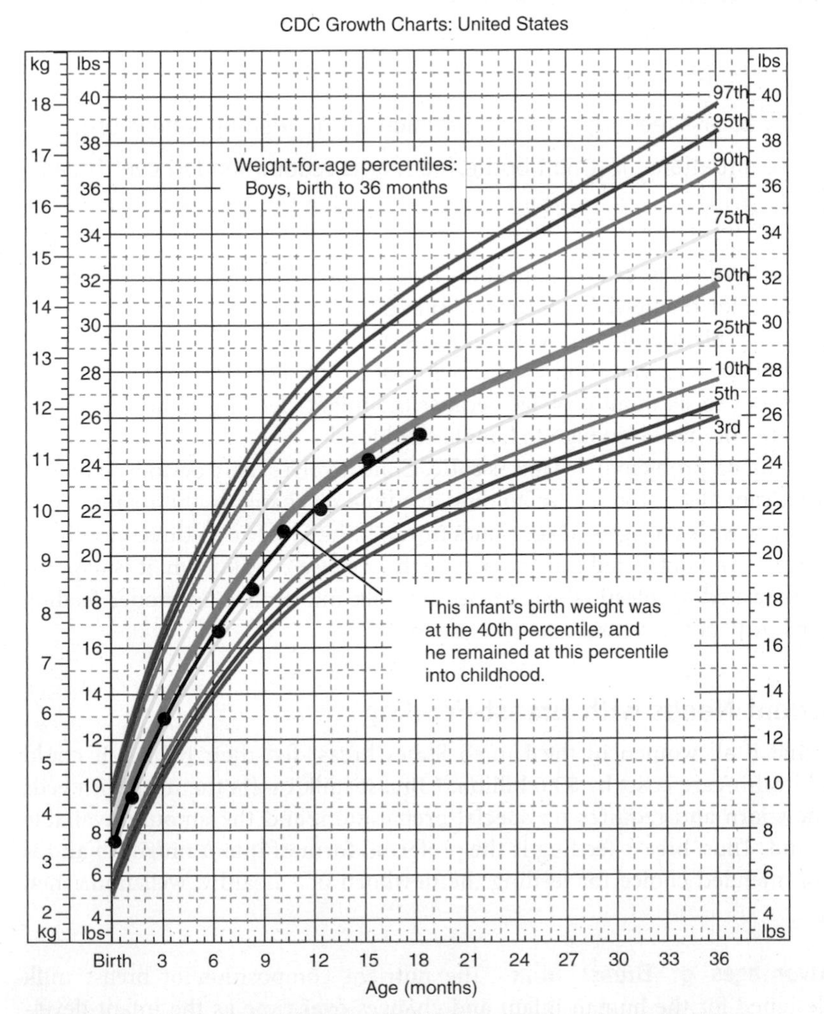

Source: Developed by the National Center for Health Statistics in collaboration with the National Center for Chronic Disease Prevention and Health Promotion (2000).

Figure 14.24 Monitoring infant growth
The black line on this weight-for-age growth chart for boys from birth to 36 months shows the pattern of growth for an infant whose birth weight was at the 40th percentile.

THINK CRITICALLY

How much would a 15-month-old boy weigh if he was at the 50th percentile?

A rule of thumb is that an infant's birth weight should double by 4 months and triple by 1 year of age. In the first year of life, most infants increase their length by 50%. Small infants and premature infants often follow a pattern parallel to but below the growth curve for a period of time, and then experience catch-up growth that brings them onto the growth curve in a place compatible with their genetic growth potential.

Abnormal Growth and Nutritional Status Slight fluctuations in growth rate are normal, but a consistent pattern of not following the growth curve or a sudden change in growth pattern is cause for concern and could indicate overnutrition or undernutrition. A rapid increase in weight without an increase in height may be an indicator that the infant is being overfed. Because overweight children grow into overweight adults, this pattern of weight increase should be addressed early in life.

failure to thrive The inability of a child's growth to keep up with normal growth curves.

Growth that is slower than the predicted pattern indicates **failure to thrive**. This is a catch-all term for any type of growth failure in a young child. The cause may be a congenital condition, the presence of disease, poor nutrition, neglect, abuse, or psychosocial problems. The treatment is usually an individualized plan that includes adequate nutrition and careful monitoring by physicians, dietitians, and other health-care professionals. Just as there are critical periods in fetal life, there are critical periods for growth and development during infancy when undernutrition can permanently affect development.

14.6 Feeding the Newborn

LEARNING OBJECTIVES

- Explain why breast-feeding is the first choice for nourishing a newborn.
- Discuss two situations in which bottle feeding is recommended over breast-feeding.

Newborns have small stomachs, can consume only liquids, and have high nutrient requirements. The ideal food for the newborn is breast milk, but infant formula can also meet a newborn's needs. From birth until four to six months of age, infants don't need anything other than breast milk or formula. Solid food should not be introduced into the diet until the child is at least four to six months of age because the infant's feeding abilities and gastrointestinal tract are not mature enough to handle solid foods.

colic Inconsolable crying that is believed to be due to pain from gas buildup in the gastrointestinal tract or immaturity of the central nervous system.

A relatively common problem in infants is **colic**. Colic involves daily periods of inconsolable crying that cannot be stopped by holding, feeding, or changing the infant. Colic usually begins at a few weeks of age and continues through the first two to three months. It occurs in both breast- and bottle-fed infants. Although its cause is unknown, it is hypothesized that colic is related to intestinal gas caused by milk intolerance, improper feeding practices, or immaturity of the central nervous system.

Meeting Nutrient Needs with Breast-Feeding

It is estimated that if all women in the United States breast-fed their babies, it could decrease annual health-care costs by $3.6 billion.[58] Breast milk meets the nutrient needs of the human newborn and requires no special preparation, and the amount available varies with demand. Thus, breast-feeding is the preferred form of infant nutrition and is usually the recommended choice for feeding the newborn of a healthy, well-nourished mother.[59]

Nutritional Advantages of Breast Milk The nutrient composition of breast milk is specifically designed for the human infant and changes over time as the infant develops, meeting the nutrient needs of the child for up to the first year of life. The first fluid that is produced by the breast after delivery is colostrum. This yellowish fluid is higher in water, protein, immune factors, minerals, and some vitamins than mature breast milk.

It is produced for up to a week after delivery. While colostrum is produced, it may seem that the newborn is not receiving enough to eat; however, supplemental bottle feedings are not necessary. The nutrients in colostrum meet infant needs until mature milk production begins. Colostrum also has beneficial effects on the gastrointestinal tract, acting as a laxative that helps the baby excrete the thick, mucusy stool produced during life in the womb.

Lactalbumin is the predominant protein in human milk. In the infant's stomach it forms a soft, easily digested curd. The amino acids methionine and phenylalanine, which are difficult for the infant to metabolize, are present in lower amounts in human milk proteins than in cow's milk proteins. Human milk is also a good source of taurine, an amino acid needed for bile salt formation and eye and brain function.

The lipids in human milk are easily digested. They are high in cholesterol and the fatty acids linoleic acid, arachidonic acid, and DHA, which are essential for normal brain development, eyesight, and growth. The fat content of breast milk changes throughout a feeding, gradually increasing during the nursing session. Thus, for the baby to attain satiety and obtain adequate energy, it is important for nursing to continue long enough for the infant to obtain the higher-fat milk.

Lactose is the primary carbohydrate in human milk. It is digested slowly so it stimulates the growth of acid-producing bacteria. It also promotes the absorption of calcium and other minerals and provides a source of the sugar galactose for nervous system development.

Breast milk is low in sodium and the zinc, iron, and calcium present are in forms that are easily absorbed. About 50% of the iron in human milk is absorbed, compared with only 2 to 30% from many other foods.

Other Advantages of Breast-Feeding Breast-feeding can be a relaxing, emotionally enjoyable interaction for both mother and infant. In addition to its nutritional advantages, breast-feeding is convenient, inexpensive, and has immunological and physiological benefits for both mother and child (**Figure 14.25**).

Figure 14.25 **The benefits of breast feeding**[60]

Benefits for infants

- Provides optimum nutrition
- Enables strong bonding with mother
- Enhances immune protection
- Reduces allergies
- Decreases ear infections, respiratory illnesses, and asthma
- Reduces the likelihood of constipation, diarrhea, or chronic digestive diseases
- Reduces risk for SIDS
- Lowers risk for obesity, type 1 and 2 diabetes, heart disease, hypertension, high blood cholesterol, and childhood leukemia
- Aids in the development of the facial muscles, speech development, and correct formation of the teeth
- Lessens the risk of overfeeding because the amount of milk consumed cannot be monitored visually

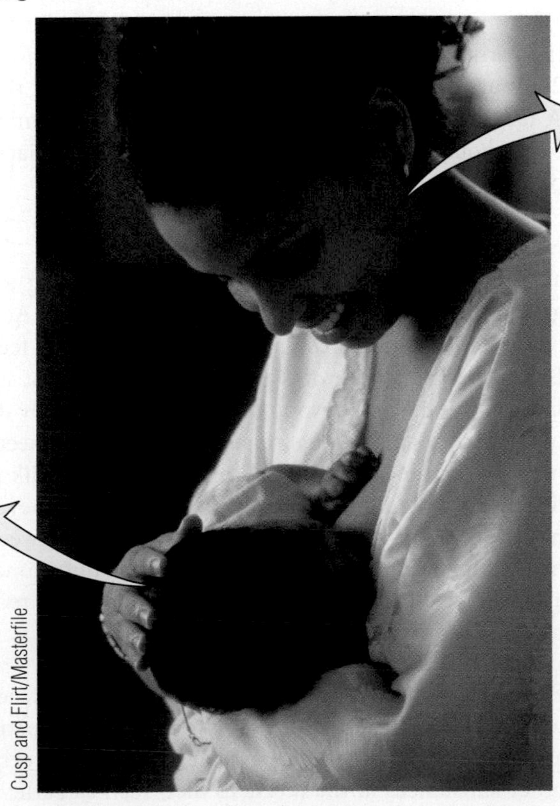

Cusp and Flirt/Masterfile

Benefits for mothers

- Provides relaxing, emotionally enjoyable interaction; strengthens bonding with infant
- Reduces financial costs
- Requires less preparation and clean-up time; always available
- Causes uterine contractions that help the uterus return to its normal size more quickly after delivery
- Increases energy expenditure, which may speed return to prepregnancy weight
- Lowers risk of developing type 2 diabetes and breast and ovarian cancers
- Improves bone density and decreases risk of fractures
- Inhibits ovulation, lengthening the time between pregnancies; however, breast-feeding cannot be relied on for birth control
- Decreases the risk of postpartum depression
- Enhances self-esteem in the maternal role

During the first few months of life, the immune factors provided, first by colostrum, and later by mature milk, compensate for the infant's immature immune system. These include antibody proteins and immune system cells that pass from the mother into her milk. Breast milk also contains a number of enzymes and other proteins that prevent the growth of harmful microorganisms. Several carbohydrates have been identified that protect against disease-causing microorganisms, including viruses that cause diarrhea. One carbohydrate favors the growth of the beneficial bacterium *Lactobacillus bifidus* in the infant's colon, which inhibits the growth of harmful bacteria. Breast-fed babies have fewer allergies, ear infections, respiratory illnesses, and urinary tract infections than formula-fed babies, and have fewer problems with constipation and diarrhea. There is also evidence that breast-feeding protects against sudden infant death syndrome, diabetes, and chronic digestive diseases.[59]

In addition to providing disease protection, the strong suckling required by breast-feeding aids in the development of facial muscles, which help in speech development and the correct formation of teeth. Breast-fed babies are also less likely to be overfed, because the amount of milk consumed cannot be monitored visually. In bottle-feeding, it is often tempting to encourage the baby to finish the entire bottle whether or not he or she is hungry.

For the mother, breast-feeding has the advantage of providing a readily available and inexpensive source of nourishment for her infant (see Figure 14.25). It requires no preparation or bottles and nipples that must be washed. It is more ecological because it doesn't require energy for manufacture or generate waste from discarded packaging. Physiologically, breast-feeding causes contractions that help the mother's uterus return to normal size more quickly and may promote weight loss in some women. Women who breast-feed may have a lower risk of developing osteoporosis and breast and ovarian cancer. Lactation also inhibits ovulation, lengthening the time between pregnancies; however, it does not reliably prevent ovulation and so cannot be effectively used for birth control. Oral contraceptives can be used immediately postpartum, but those containing only progestin are preferable because they do not affect milk volume or composition. Oral contraceptives containing estrogen may decrease milk volume.

How Much Is Enough? A strong, healthy baby will be able to suckle shortly after birth. Within a week, milk production and breast-feeding are usually fully established. During the early weeks of breast-feeding the infant should be fed about eight to 12 times every 24 hours (every two to three hours) or whenever the infant shows early signs of hunger. A feeding should last approximately eight to 12 minutes at each breast. A well-fed newborn should urinate enough to soak six to eight diapers a day and gain about one-third to one-half pound per week.

How Long Should Breast-Feeding Continue? Physiologically, lactation can continue as long as suckling is maintained. Breast-feeding alone is sufficient to support optimal growth for about six months, and the American Academy of Pediatrics and many other health organizations recommend exclusive breast-feeding for the first six months of life.[59] Breast-feeding along with supplemental feeding of solids is recommended for at least the first year of life and beyond for as long as mutually desired by mother and child. After 12 months, the baby no longer needs breast milk to meet nutrient needs. As the infant obtains more and more of its energy from solid foods, milk production decreases due to reduced demand by the infant. However, breast-feeding beyond 12 months continues to provide nutrition, comfort, and an emotional bond between mother and child. In developing nations it may continue to give children the nutritional advantage needed to fight infection and stay healthy. The World Health Organization recommends that infants in developing nations be breast-fed for two years or more.[61]

Practical Aspects of Breast-Feeding Breast-feeding does not always come naturally to mother or infant and can require practice and patience. Effective suckling by the infant and relaxation of the mother are essential to successful breast-feeding. Some foods and other substances in the mother's diet, such as garlic or spicy foods, contain chemicals or flavors that pass into breast milk and cause adverse reactions in some babies. These reactions seem to be individual to the mother and child. As long as a food does not affect the infant's response to feeding, it can be included in the mother's diet. Caffeine in the mother's diet can make the infant jittery and excitable, so large amounts should be avoided while breast-feeding. Alcohol, which is harmful for infants, passes into breast milk. It is most concentrated an hour to an hour and a half after consumption and is cleared from the milk at about the same rate at which it disappears from the bloodstream. Therefore, occasional limited alcohol consumption while breast-feeding is probably not harmful if intake is timed to minimize the amount present in milk when the infant is fed.

Breast-feeding does not mean that a mother must be available for every feeding. Milk may be pumped from the breast and stored for later feedings (**Figure 14.26**). Since pumped milk is exposed to pumps and bottles, care must be taken to avoid contamination. If pumped milk is not immediately fed to the baby, it should be refrigerated. It can be kept refrigerated for 24 to 48 hours, but if it will not be used within that period, it should be frozen in clean containers. Warming breast milk in a microwave is not recommended because this destroys some of its immune properties and it may result in dangerously hot portions of the milk. The best way to warm milk is by running warm water over the bottle.

Figure 14.26 Pumped breast milk
Freezing pumped breast milk allows other caregivers to feed the baby breast milk from a bottle if the mother is not available.

Meeting Nutrient Needs with Bottle-Feeding

A hundred years ago, a baby who could not be breast-fed had little chance of survival. Today, infants who cannot breast-feed can still thrive. There are many commercially available infant formulas modeled after the nutrient content of breast milk.

When Is Bottle-Feeding Best? There are some situations when breast-feeding is not the best choice. An infant who is small or weak may not have the strength to receive adequate nutrition from breast-feeding. In this case, formula, which provides almost the same nutrients as breast milk, can be used, or pumped breast milk can be offered to the infant in a bottle.

Bottle-feeding prevents the transmission of drugs and disease via breast milk. Women who are taking medications should check with their physician on whether it is safe to breast-feed. If prescription drugs are taken by the mother for only a short time, a breast pump may be used to maintain milk production and the milk discarded until the medication is no longer needed. Because alcohol and drugs such as cocaine and marijuana can be passed to the baby in breast milk, alcoholic and drug-addicted mothers are counseled not to breast-feed. Nicotine from cigarette smoke is also rapidly transferred from maternal blood to milk, and heavy smoking may decrease the supply of milk. Tuberculosis and certain viral infections can be transmitted to the infant in breast milk, but common illnesses such as colds, flu, and skin infections should not interfere with breast-feeding. Human immunodeficiency virus (HIV), the virus that causes AIDS, can be transmitted to the infant in breast milk. In the United States, women who are infected with HIV are advised not to breast-feed, but in developing nations, the risk of malnutrition associated with not breast-feeding often outweighs the risk of passing this infection on to the infant.[61]

How Much Is Enough? As with breast-fed infants, formula-fed infants should be fed on demand every few hours. Newborns have small stomachs, so at each feeding they may consume only a few ounces of formula. As the infant grows, the amount consumed at each feeding will increase to 4 to 8 oz. Caregivers should respond to cues from the infant that hunger is satisfied, even if a bottle of formula is not finished. Encouraging infants to finish every bottle can result in overfeeding and excess weight gain. As with breast-fed infants, adequate intake can be judged from the amount of urine produced and the amount of weight gained.

Practical Aspects of Bottle-Feeding Infant formula must be prepared carefully in order to avoid mixing errors and contamination. If the proper measurements are not used in preparing formula, the child can receive an excess or deficiency of nutrients and

© Sally and Richard Greenhill/Alamy Limited

Figure 14.27 Preparing infant formula Formulas are marketed in three basic forms: ready-to-feed, liquid concentrates, and powdered. Ready-to-feed formulas require no preparation and are packaged in many sizes. Liquid concentrates and powdered formulas are prepared for use by mixing them with specific amounts of water. When properly prepared, all provide the needed nutrients in appropriate concentrations.

an improper ratio of nutrients to fluids (**Figure 14.27**). If the water and all the equipment used in preparing formula are not clean or if the prepared formula is left unrefrigerated, food-borne illness may result. Because sanitation is often a problem in developing nations, infections that lead to diarrhea and dehydration occur more commonly in formula-fed than in breast-fed infants. Commercially prepared formulas are sterile and powdered formulas contain no harmful microorganisms. To avoid introducing harmful microorganisms, the water used to mix powdered formula should be boiled for one to two minutes and allowed to cool before mixing. Hands should be washed before preparing formula, and bottles and nipples should be washed in a dishwasher or placed in a pan of boiling water for five minutes. Formula should be prepared immediately before a feeding, and any excess should be discarded. Opened cans of ready-to-feed and liquid concentrate formula should be covered and refrigerated and used within the time indicated on the can. Formula may be fed either warm or cold, but the temperature should be consistent.

The position of the child is important during feeding. The infant's head should be higher than his or her stomach, and the bottle should be tilted so that there is no air in the nipple. If the hole in the nipple is too large, the infant may feel full before receiving adequate nutrition. If the hole is too small, the infant may tire before nutrient needs are met. Just as breast-fed infants alternate breasts, bottle-fed infants should be held alternately between the left and right arms to promote equal development of the head and neck muscles.

Infants should never be put to bed with a bottle of formula or other sweetened liquids. At night, while the child sleeps, the flow of saliva is decreased and the sugary formula is allowed to remain in contact with the teeth for many hours. This causes the rapid and serious decay of the upper teeth referred to as **nursing bottle syndrome**. Usually, the lower teeth are protected by the tongue and are unaffected (**Figure 14.28**).

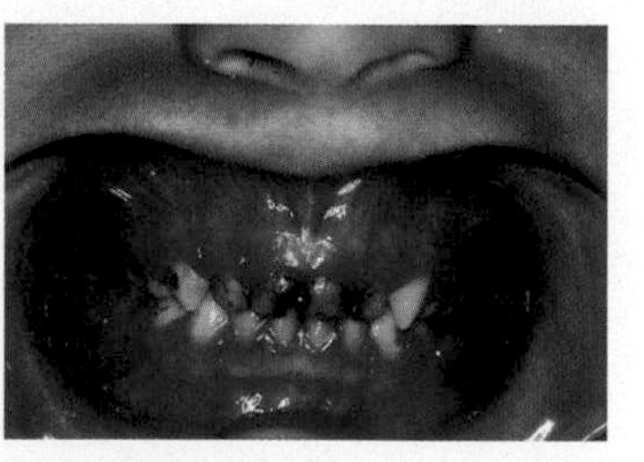

K.L. Boyd, D.D.S./Custom Medical Stock Photo, Inc.

Figure 14.28 Nursing bottle syndrome Putting infants to bed with a bottle causes rapid and serious decay of the upper teeth, referred to as nursing bottle syndrome.

nursing bottle syndrome Extreme tooth decay in the upper teeth resulting from putting a child to bed with a bottle containing milk or other sweet liquids.

Formula Choices Infant formulas can never duplicate the living cells, active hormones, enzymes, and immune system molecules in human milk, but formulas today try to replicate human milk as closely as possible in order to match the growth, nutrient absorption, and other parameters obtained with breast-feeding. Commercially prepared infant formulas are required by the FDA to provide minimum amounts of 29 nutrients based on levels in breast milk. Because the American Academy of Pediatrics recommends iron-fortified formulas, most infant products are fortified with iron. Some formulas are supplemented with the long-chain polyunsaturated fatty acids DHA and arachidonic acid. These are found in breast milk and are thought to help infant development. Formulas supplemented with fatty acids are more expensive, and it is still not clear whether they improve growth or visual or neurological development (see Debate: DHA-Fortified Infant Formulas).

Most infant formula is based on cow's milk, but unmodified cows' milk should never be fed to infants. Cow's milk is difficult for infants to digest and has a higher protein and mineral content than breast milk or formula, so it taxes the infant kidneys and predisposes the infant to dehydration. Young infants may also become anemic if fed cow's milk because it contains little absorbable iron and can lead to iron loss by causing small amounts of gastrointestinal bleeding. Unmodified goat's milk is also not recommended for infants. Although it is less allergenic and more easily digested than cow's milk, it is low in vitamin D as well as iron, vitamin B_{12}, and folate, which can lead to an iron deficiency or megaloblastic anemia. Cow's milk and goat's milk can be introduced at about 12 months of age when organ systems are more mature and missing nutrients can be provided by other foods.

Special formulas are available for infants with allergies, premature infants, and those with genetic abnormalities that alter dietary needs. For infants who cannot tolerate human milk or cow's milk-based formula, soy protein formulas are available. And for those who are allergic to soy protein, formulas made from predigested proteins (elemental formulas), called protein hydrolysates, are an option.

Premature infants have special needs because they do not have fully developed organ systems or metabolic pathways. If they are too small or weak to nurse or take a bottle, pumped breast milk or formula can be fed through a tube. Some nutrients that are produced in the bodies of full-term infants are essential in the diets of premature infants. For example, preterm infants are less able to synthesize the amino acids tyrosine and cysteine and the fatty acid DHA. These and other substances, such as taurine and carnitine, are needed in greater amounts in the diets of premature babies. The energy, protein, and micronutrient requirements of preterm infants are also higher due to their rapid growth and development. Preterm infant formulas are available to meet the needs of premature babies.

Genetic abnormalities that prevent the normal metabolism of specific nutrients may alter dietary needs. For instance, infants with the genetic disease PKU lack an enzyme needed to metabolize the amino acid phenylalanine (see Chapter 6). If a child with PKU is fed breast milk or a formula that contains too much phenylalanine, the by-products of phenylalanine metabolism accumulate and cause brain damage. This can be prevented by feeding infants with PKU a special formula that provides only enough phenylalanine to meet the need for protein synthesis. Because this special diet must be started as soon as possible, infants born in the United States and Canada are tested for PKU at birth.

OUTCOME

UpperCut Images/Getty Images, Inc.

After her doctor's visit, Jasmine realized that the health of her baby was more important to her than whether or not she needed to lose a few pounds after the delivery. Her doctor helped her understand that she needed to eat a diet higher in energy, protein, and many micronutrients than the one she consumed before she became pregnant. Jasmine took a prenatal supplement to help meet her micronutrient needs, but a pill can't provide the calories and protein she needs to build new tissues or fiber to prevent constipation. To construct her diet, Jasmine looked up her MyPlate Plan for Moms. It called for an increase in her intake of grains and vegetables, which added energy, fiber, protein, folate, and vitamin C to her diet, and extra-lean meat, which provided iron, protein, zinc, and B vitamins. She continued to exercise regularly but switched to walking and swimming, which were safer and more comfortable when her belly was large. She skipped the sugary sodas and was able to keep her blood sugar normal throughout her pregnancy. After nine months, Jasmine had gained 32 pounds and gave birth to a healthy 7 lb 8 oz baby boy. She breast-fed him for a year. While she was lactating, her need for energy, fluid, and certain micronutrients continued to be higher than they were before her pregnancy. **Although she felt as if she was eating all the time, after six months she was back to her prepregnancy weight.**

© Miroslav Ferkuniak/iStockphoto

APPLICATIONS

ASSESSING YOUR DIET

1. **If you were a 25-year-old pregnant woman, would your diet meet your needs?**
   a. Pick one day of the food record you kept in Chapter 2. Use iProfile to determine whether this diet meets the third trimester energy and protein recommendations for a 5'5" tall, 25-year-old sedentary pregnant woman who weighed 130 lbs at the beginning of her pregnancy. If it doesn't, what foods would you add to the diet to meet the needs of pregnancy?
   b. Does this diet meet the iron and calcium needs of a 25-year-old pregnant woman? List three foods that are good sources of each.
   c. Does this diet meet the folate needs of a 25-year-old pregnant woman? What foods could you add to a diet that is low in folate to meet needs without supplements? What foods in this diet are fortified with folic acid?

CONSUMER ISSUES

2. Use the Internet to find out about the WIC program in your area.

a. What services would this program provide for you if you were a pregnant or lactating woman or had a young child?

b. What income levels does it serve? Would you be eligible?

c. List four foods that you could purchase using WIC.

CLINICAL CONCERNS

3. Marina is 16 years old and is pregnant with her first child. She is 5'4" and weighs 110 lbs. She eats breakfast at home with her mother and two brothers and has lunch in the school cafeteria. After school she often has a snack with friends and then has dinner at home. A day's sample diet is listed here:

SAMPLE DIET

FOOD	SERVING SIZE	FOOD	SERVING SIZE
Breakfast		*Snack*	
Pastry	1	Ice cream cone	1
Fruit punch	6 oz	*Dinner*	
Lunch		Chicken	1 drumstick
Hamburger	1	Rice	1 cup
on bun	1	Refried beans	½ cup
Canned peaches	½ cup	Tortillas	3
Diet cola	12 oz	Fruit punch	12 oz

a. What is the RDA for folate, vitamin D, calcium, iron, and zinc for a 16-year-old pregnant female?

b. Does Marina's diet meet the recommendations for these nutrients?

c. Use iProfile to determine how many kcalories her diet provides?

d. Will Marina gain the recommended amount of weight if she consumes this diet throughout her pregnancy?

e. What dietary changes would you suggest to meet the needs of the second trimester of pregnancy and to ensure a healthy pregnancy for both Marina and her baby?

4. How do the energy and nutrient needs of nonpregnant, pregnant, and lactating women differ?

a. For each of the following, describe any differences between the needs of nonpregnant, pregnant, and lactating women of similar age and size.

- Energy
- Protein
- Calcium
- Iron
- Folate

b. For each of the above, explain why the requirements for pregnancy and lactation do or do not differ from those for the nonpregnant state.

SUMMARY

14.1 The Physiology of Pregnancy

- Pregnancy begins with the fertilization of an egg by a sperm. About two weeks after fertilization, implantation is complete. The embryo grows, and the cells differentiate to form the organs and structures of the body. Growth and maturation continue in the fetal period, which begins at nine weeks, and continues until birth, about 38 weeks after fertilization.
- During pregnancy, the placenta develops; maternal blood volume increases; the uterus and supporting muscles expand; body fat is deposited; the heart, lungs, and kidneys work harder; the breasts enlarge; and total body weight increases. Changes in the mother and growth of the fetus contribute to weight gain. Recommended weight gain during pregnancy is 25 to 35 lbs (11 to 16 kg) for normal-weight women. Too little or too much weight gain can place both mother and baby at risk. Weight loss should never be attempted during pregnancy. Normal-weight, underweight, overweight, and obese mothers should all gain weight at a steady rate during pregnancy, but the total amount of weight gain recommended depends on prepregnancy weight.
- During healthy pregnancies, a carefully planned program of moderate-intensity exercise can be safe and beneficial.
- Changes in blood volume and hormone levels and the enlargement of the uterus can result in edema, morning sickness, heartburn, constipation, and hemorrhoids during pregnancy.
- The hypertensive disorders of pregnancy include chronic hypertension, gestational hypertension, and preeclampsia, which can lead to life-threatening eclampsia. Gestational diabetes can result in a large-for-gestational-age baby.

14.2 The Nutritional Needs of Pregnancy

- During pregnancy, the requirements for energy, protein, water, and some vitamins, and minerals rise above levels for nonpregnant women. The need for B vitamins is increased to support increased energy and protein metabolism. The RDA for calcium is not increased because the greater need during pregnancy is met by an increase in absorption. Vitamin D deficiency is a concern, particularly for darker-skinned women. Adequate folic acid early in pregnancy reduces the risk of neural tube defects. Iron is needed for red blood cell synthesis, so requirements are high and deficiency is common during pregnancy.
- Even with a nutrient-dense diet that follows a MyPlate Plan for Moms, supplements of folic acid and iron are recommended during pregnancy.

14.3 Factors That Increase the Risks of Pregnancy

- Nutritional status is important before, during, and after pregnancy. Poor nutrition before pregnancy can decrease fertility or lead to a poor pregnancy outcome. Malnutrition during pregnancy can affect fetal growth and development and the risk that the child will develop chronic disease later in life.
- Poor maternal health status, age that is under 20 or over 35 years, a short interval between pregnancies, a history of poor reproductive outcomes, and poverty all increase the risk of complications for the mother and baby.
- Because the embryo and fetus are developing and growing rapidly, they are susceptible to damage from physical, chemical, or other environmental teratogens. Mercury in food, food-borne pathogens, cigarette smoking, alcohol use, and certain prescription and illegal drugs can interfere with growth and development of the embryo and fetus.

14.4 Lactation

- Milk production and letdown are triggered by the hormones prolactin and oxytocin, respectively. They are released in response to the suckling of the infant.
- Lactation requires energy and nutrients from the mother to produce adequate milk. The energy for milk production comes from the diet and maternal fat stores. During lactation, the need for water, and many vitamins and minerals is even greater than during pregnancy.

14.5 The Nutritional Needs of Infancy

- Newborns grow more rapidly and require more energy and protein per kg of body weight than at any other time in life. Fat and water needs are also proportionately higher than in adults.
- A diet that meets energy, protein, and fat needs may not necessarily meet the needs for iron, fluoride, and vitamins D and K.
- Growth, which is assessed using growth charts, is the best indicator of adequate nutrition in an infant. Growth charts can be used to compare an infant's growth with the growth of other infants of the same age.

14.6 Feeding the Newborn

- Breast milk is the ideal food for new babies. It is designed specifically for the human newborn; is always available; requires no special equipment, mixing, or sterilization; and provides immune protection.
- There are many infant formulas on the market that are patterned after human milk and provide adequate nutrition to the baby. Infant formulas are the best option when the mother is ill or is taking prescription or illicit drugs, or when the infant has special nutritional needs. The major disadvantages of formula feeding are the potential for bacterial contamination, overfeeding, and the possibility of errors in mixing formula.

REVIEW QUESTIONS

1. List three physiological changes that occur in the mother's body during pregnancy.
2. How much weight should a woman gain during pregnancy?
3. How do the recommendations for weight gain differ for underweight, overweight, and obese women?
4. What kind of exercise is safe during pregnancy?
5. List three common digestive-system discomforts that afflict pregnant women and explain why they occur.
6. Explain why the hypertensive disorders of pregnancy can be a risk to the mother and baby.
7. Why does gestational diabetes increase the risk of having a large-for-gestational-age baby?
8. How do energy and protein requirements change during pregnancy?
9. Why does the recommendation for iron intake increase during pregnancy and decrease during lactation?
10. Why are folic acid supplements recommended even before pregnancy for women of childbearing age?
11. Are vegetarian diets safe for pregnant women? Why or why not?
12. Why does malnutrition early in pregnancy have different effects from malnutrition during the last trimester?
13. How does maternal age affect nutrient requirements during pregnancy?
14. How does alcohol consumed by a woman during pregnancy affect the child?
15. Why is the need for some nutrients greater during lactation than pregnancy?
16. What is the best indicator of adequate nutrition in an infant?
17. What are the advantages of breast-feeding?
18. When is bottle-feeding a better choice?

REFERENCES

1. Committee to Reexamine IOM Pregnancy Weight Guidelines, Institute of Medicine, National Research Council. *Weight Gain During Pregnancy: Reexamining the Guidelines*. Washington, DC: National Academies Press, 2009.
2. Kaiser, L., and Allen, L. A. Position of the American Dietetic Association: Nutrition and lifestyle for a healthy pregnancy outcome. *J Am Diet Assoc* 108:553–561, 2008.
3. Iacovidou, N., Varsami, M., and Syggellou, A. Neonatal outcome of preterm delivery. *Ann NY Acad Sci* 1205:130–134, 2010.
4. Rasmussen, S. A., Chu, S. Y., Kim, S. Y., et al. Maternal obesity and risk of neural tube defects: a metaanalysis. *Am J Obstet Gynecol* 198:611–619, 2008.
5. United States Department of Health and Human Services. 2008 Physical Activity Guidelines for Americans.
6. Ebrahimi, N., Maltepe, C., and Einarson, A. Optimal management of nausea and vomiting of pregnancy. *Int J Women's Health* 2:241–248, 2010.
7. National Center for Health Statistics, Centers for Disease Control and Prevention. *FastStats: Infant Health*.
8. National Center for Health Statistics, Centers for Disease Control and Prevention. *Deaths: Final Data for 2004*.
9. Leeman, L., and Fontaine, P. Hypertensive disorders of pregnancy. *Am Fam Physician* 78:93–100, 2008.
10. Berg, C. J., Callahan, W. M., Syverson, C., and Henderson, Z. Pregnancy-related mortality in the United States 1998–2005, *Obstet Gynecol* 166:1302–1309, 2010.
11. Food and Nutrition Board, Institute of Medicine. *Dietary Reference Intakes Water, Potassium, Sodium, Chloride, and Sulfate*. Washington, DC: National Academies Press, 2004.
12. Kumar, A., Devi, S. G., Batra, S., et al. Calcium supplementation for the prevention of pre-eclampsia. *Int J Gynaecol Obstet.* 104:32–36, 2009.
13. Centers for Disease Control and Prevention. National Diabetes Fact Sheet, 2011.
14. Office of Minority Health and Health Disparities, Centers for Disease Control and Prevention. *Eliminate Disparities in Diabetes*.
15. Damm, P. Future risk of diabetes in mother and child after gestational diabetes mellitus. *Int J Gynaecol Obstet* 104(Suppl. 1): S25–S26, 2009.
16. Food and Nutrition Board, Institute of Medicine. *Dietary Reference Intakes for Energy, Carbohydrate, Fiber, Fat, Fatty Acids, Cholesterol, Protein, and Amino Acids*. Washington, DC: National Academies Press, 2002, 2005.
17. Koletzko, B., Lien, E., Agostoni, C., et al. The roles of long chain polyunsaturated fatty acids in pregnancy, lactation and infancy: Review of current knowledge and consensus recommendations. *J Perinat Med* 36:5–14, 2008.
18. Kalkwarf, H. J., and Specker, B. L. Bone mineral changes during pregnancy and lactation. *Endocrine* 17:49–53, 2002.
19. Food and Nutrition Board, Institute of Medicine. *Dietary Reference Intakes for Calcium and Vitamin D*. Washington, DC: National Academies Press, 2011.
20. Thangaratinam, S., Langenveld, J., Mol, B. W., and Khan, K. S. Prediction and primary prevention of pre-eclampsia. *Best Pract Res Clin Obstet Gynaecol.* 25:419–433, 2011.
21. Ginde, A. A., Sullivan, A. F., Mansbach, J. M., and Camargo, C. A. Jr. Vitamin D insufficiency in pregnant and nonpregnant women of childbearing age in the United States. *Am J Obstet Gynecol* 202:436.e1–436.e8, 2010.
22. Food and Nutrition Board, Institute of Medicine. *Dietary Reference Intakes for Vitamin C, Vitamin E, Selenium, and Carotenoids*. Washington, DC: National Academies Press, 2000.
23. Berry, R. J., Bailey, L., Mulinare, J., and Bower, C. Fortification of flour with folic acid. *Food Nutr Bull* 31(1 Suppl.):S22–S35, 2010.
24. De Wals, P., Tairou, F., Van Allen, M. I., et al. Reduction in neural-tube defects after folic acid fortification in Canada. *N Engl J Med* 357:135–142, 2007.
25. Scholl, T. O., and Johnson, W. G. Folic acid: influence on the outcome of pregnancy. *Am J Clin Nutr* 71(Suppl.):1295S–1303S, 2000.
26. Food and Nutrition Board, Institute of Medicine. *Dietary Reference Intakes for Thiamin, Riboflavin, Niacin, Vitamin B-6, Folate, Vitamin B-12, Pantothenic Acid, Biotin, and Choline*. Washington, DC: National Academies Press, 1998.
27. Hess, S. Y., and King, J. C. Effects of maternal zinc supplementation on pregnancy and lactation outcomes. *Food Nutr Bull* 30(1 Suppl):S60–S78, 2009.
28. Food and Nutrition Board, Institute of Medicine. *Dietary Reference Intakes for Vitamin A, Vitamin K, Arsenic, Boron, Chromium, Copper, Iodine, Iron, Manganese, Molybdenum, Nickel, Silicon, Vanadium, and Zinc*. Washington, DC: National Academies Press, 2001.
29. Scholl, T. O. Iron status during pregnancy: setting the stage for mother and infant. *Am J Clin Nutr* 81:1218S–1222S, 2005.
30. Mills, M. E. Craving more than food: the implications of pica in pregnancy. *Nurs Women's Health* 11:266–273, 2007.
31. Hoyert, D. L. Maternal mortality and related concepts. National Center for Health Statistics. *Vital Health Stat* 3(33), 2007.
32. Gluckman, P. D., Hanson, M. A., Cooper, C., and Thornburg, K. L. Effect of in utero and early-life conditions on adult health and disease. *N Engl J Med* 359:61–73, 2008.
33. Centers for Disease Control and Prevention. Pediatric and Pregnancy Nutrition Surveillance System, Health Indicators.
34. National Center for Health Statistics, Centers for Disease Control and Prevention. *FastStats: Teen Births*.
35. Casanueva, E., Roselló-Soberón, M. E., and De-Regil, L. M. Adolescents with adequate birth weight newborns diminish

energy expenditure and cease growth. *J Nutr* 136:2498–2501, 2006.

36. National Down Syndrome Congress. Facts About Down Syndrome.

37. Reinold, C., Dalenius, K., Brindley, P., et al. Pregnancy Nutrition Surveillance 2009 Report. Atlanta: U.S. Department of Health and Human Services, Centers for Disease Control and Prevention, 2011.

38. USDA Food and Nutrition Service, How WIC helps.

39. U.S. Food and Drug Administration, Center for Food Safety and Applied Nutrition. Food Safety for Moms-to-Be.

40. Weng, X., Odouli, R., and Li, D. K. Maternal caffeine consumption during pregnancy and the risk of miscarriage: A prospective cohort study. *Am J Obstet Gynecol* 198:279e1–279e8, 2008.

41. U.S. Food and Drug Administration, Center for Food Safety and Applied Nutrition. Food Safety for Moms-to-Be. While You're Pregnant—Listeria.

42. U.S. Food and Drug Administration, Center for Food Safety and Applied Nutrition. Food Safety for Moms-to-Be. While You're Pregnant—Toxoplasmosis.

43. Goodlett, C. R., and Horn, K. H. Mechanisms of alcohol-induced damage to the developing nervous system. *Alcohol Res Health* 25:175–184, 2001.

44. Riley, E. P., Infante, M. A., and Warren, K. R. Fetal alcohol spectrum disorders: An overview. *Neuropsychol Rev* 99:298–302, 2011.

45. May, P. A., Gossage, J. P., Kalberg, W. O., et al. Prevalence and epidemiologic characteristics of FASD from various research methods with an emphasis on recent in-school studies. *Devel Disabil Res Rev* 15:176–192, 2009.

46. Rogers, J. M. Tobacco and pregnancy: Overview of exposures and effects. *Birth Defects Res. C. Embryo Today* 84:1–15, 2008.

47. Xiao, D., Huang, X., Yang, S., and Zhang, L. Direct effects of nicotine on contractility of the uterine artery in pregnancy. *J Pharmacol Exp Ther* 322:180–185, 2007.

48. National Center for Chronic Disease Prevention and Health Promotion. *Women and Smoking: A Report of the Surgeon General*, March 2001.

49. The National Council on Alcoholism and Drug Dependence. Alcohol- and Other Drug-Related Birth Defects, Facts and Information.

50. Fajemirokun-Odudeyi, O., and Lindow, S. W. Obstetric implications of cocaine use in pregnancy: A literature review. *Eur J Obstet Gyncol Reprod Biol* 112:2–8, 2004.

51. Singer, L. T., Nelson, S., Short, E., et al. Prenatal cocaine exposure: Drug and environmental effects at 9 years. *J Pediatr* 153:105–111, 2008.

52. Kovacs, C. S. Calcium and bone metabolism during pregnancy and lactation. *J Mammary Gland Biol Neoplasia* 10:105–118, 2005.

53. American Academy of Pediatric Committee on Nutrition. New infant formula additives approved by FDA. *AAP News* 20:209, 2002.

54. Wright, K., Coverston, C., Tiedeman, M., and Abegglen, J. A. Formula supplemented with docosahexaenoic acid (DHA) and arachidonic acid (ARA): A critical review of the research. *J Spec Pediatr Nurs* 11:100–112, 2006.

55. Wagner, C. L., and Greer, F. R. American Academy of Pediatrics Section on Breast-Feeding, American Academy of Pediatrics Committee on Nutrition. Prevention of rickets and vitamin D deficiency in infants, children, and adolescents. *Pediatrics* 122:1142–1152, 2008.

56. American Academy of Pediatrics, Committee on Fetus and Newborn. Controversies concerning vitamin K and the newborn. *Pediatrics* 112:191–192, 2003.

57. U.S. Department of Health and Human Services, Centers for Disease Control and Prevention, National Center for Health Statistics. CDC Growth Charts: United States, Advance Data, No. 314, June 8, 2000 (revised).

58. Weimer, J. The Economic Benefits of Breast Feeding: A Review and Analysis. Food Assistance and Nutrition Research Report No. 13. Washington, DC: Food and Rural Economics Division, Economic Research Service, US Department of Agriculture; 2001.

59. American Academy of Pediatrics, Section on Breast-Feeding. Executive Summary: Breastfeeding and the Use of Human Milk.

60. James. D. C., and Lessen, R. American Dietetic Association. Position of the American Dietetic Association: Promoting and supporting breastfeeding. *J Am Diet Assoc* 109:1926–1942, 2009.

61. WHO Fact Sheet No. 342: Infant and young child feeding, July 2010.

***To access links to online sources, please go to** www.wiley.com/college/smolin **and select Nutrition: Science and Applications, 3rd edition. From this page, select either the student or instructor companion site. Once on the desired site, select References.**

CHAPTER OUTLINE

LWA/The Image Bank/Getty Images, Inc.

Nutrition from Infancy to Adolescence

15

CASE STUDY

Thirteen-year-old Felicia has been overweight since she was 3. When she started eighth grade, she was 5' tall and weighed 140 lb. At a recent checkup, her physician noted that her blood pressure was elevated and her cholesterol levels were at the high end of normal. The physician recommended that Felicia and her mother meet with a dietitian to discuss ways to manage Felicia's weight and improve her diet.

At that meeting, Felicia explained that she knows she should eat less and exercise more but can't find a way to make those changes. She spends most of her school day sitting at a desk. Physical education class meets only twice a week. She has to eat what is on the school menu because she gets a reduced-cost lunch as part of the National School Lunch Program. It is difficult for her to exercise after school because she goes to the library study program until her mother gets out of work at 5 pm. Felicia gets a little more exercise on weekends by seeing friends, but they often go out for ice cream or fast food and Felicia feels left out if she doesn't join them.

The dietitian helps Felicia figure out some changes that will help reduce her calorie intake. She also works with Felicia's mother to plan low-calorie, nutritious dinners that the entire family will enjoy. Finally, Felicia and her mother decide to get some regular exercise by walking after dinner. These evening walks soon progress to activities that include the whole family, such as skating and bike riding. **Felicia is on the right track, as good nutrition and exercise patterns learned early in life are key to maintaining long-term health.**

Blend Images/SUPERSTOCK

15.1 Starting Right for a Healthy Life

LEARNING OBJECTIVES

- Discuss the healthfulness of children's diets in the U.S.
- Explain the impact of diet and lifestyle during childhood on the risk of chronic disease.

Nutrient intake during childhood helps shape the adult that a child will become. A nutritious well-balanced eating pattern and active lifestyle allow children to grow to their potential and can prevent or delay the onset of the chronic diseases that plague American adults. Therefore, teaching healthy eating and exercise habits will benefit not only today's children, but tomorrow's adults (**Figure 15.1**).

What Are American Children Eating?

The diets of children today are not as healthy as they could be. Most children and adolescents in the United States consume more saturated fat and sodium and less calcium and fiber than is recommended. The typical diet contains too few fruits and vegetables and whole grains and too many processed foods that are high in fat, salt, or sugar.[1] When analyzed using the Healthy Eating Index, which is the government's diet quality scoring system, 64 to 88% of children between 2 and 9 years of age consumed diets that "need improvement." As children get older, the quality of their diet gets worse: they drink less milk and eat less fruit (**Figure 15.2**).[2]

Comparing the diets of children today with those of 25 years ago shows that intake of milk, vegetables, eggs, and grains has decreased while intake of cheese, fruits, juices, and sweetened beverages has increased. Children today eat more meals away from home and consume larger portion sizes. Studies in adults as well as children suggest that the larger the portion served, the more the individual will eat. For example, when given lunch portions double the age-appropriate standard, children ages 3 to 5 consumed 15% more calories than when served the standard portion.[3] The amount of energy children consume from snacks has also increased, and eating more calories from snacks is associated with being overweight.[4]

© Hongqi Zhang/iStockphoto

Figure 15.1 **Developing healthy eating habits**
The foods and nutrients this child eats influence the eating habits and health of the adult he will become.

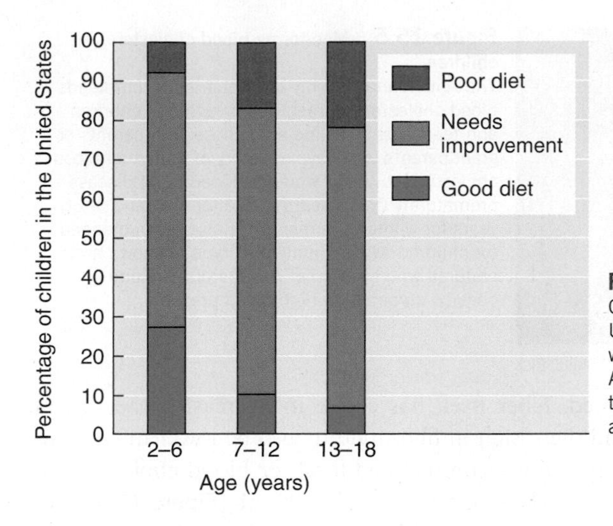

Figure 15.2 **How good are children's diets?** Only a small percentage of children in the United States have "good" diets, based on how well they meet dietary recommendations. Most American children consume too few fruits, vegetables, and whole grains and too much fat, salt, and sugar. As children grow, their diets worsen.

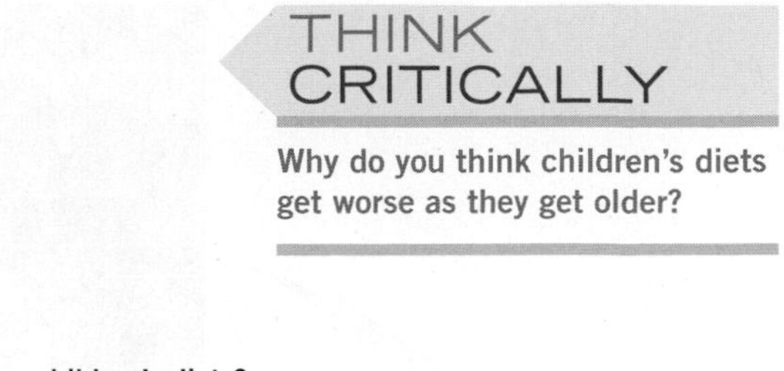

Why do you think children's diets get worse as they get older?

How Poor Nutrition Is Impacting Children's Health

The high-calorie, high-saturated-fat, high-salt diet and low-activity lifestyle that contribute to obesity and chronic disease in adults is having the same effect in kids. Like adult obesity, childhood obesity increases the risks of chronic disease. Inactive, overweight children may have high blood cholesterol and glucose levels and elevated blood pressure. These factors increase the risk of developing heart disease, diabetes, and hypertension.[5,6]

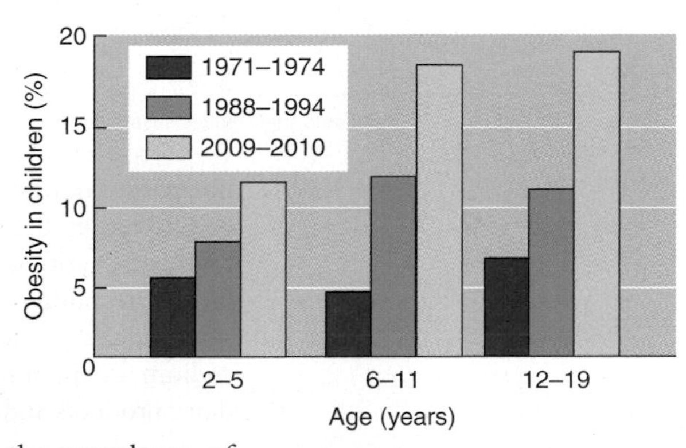

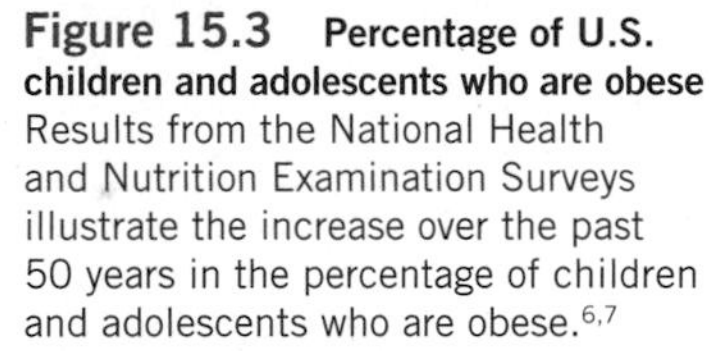

Figure 15.3 **Percentage of U.S. children and adolescents who are obese** Results from the National Health and Nutrition Examination Surveys illustrate the increase over the past 50 years in the percentage of children and adolescents who are obese.[6,7]

Obesity Overweight and obesity are major problems in children in the United States. Over the past 25 years the total amount of energy consumed by American children has increased, and along with it the prevalence of obesity.[2] More than twice as many children and almost three times as many teens are overweight or obese today as in 1980. About 17% of children and teens ages 2 to 19 are now obese (**Figure 15.3**).[6]

In addition to the health issues, obese children and adolescents in the United States face social and psychological challenges. They are less well accepted by their peers than normal-weight children and are frequently ridiculed and teased. They often have poor self-image and low self-esteem, particularly during the teenage years. Obese adolescents may be discriminated against by adults as well as by their peers. This contributes to feelings of rejection, social isolation, and low self-esteem. The isolation of obese adolescents from teen society results in boredom, depression, inactivity, and withdrawal—all of which can cause an increase in eating and a decrease in energy output, worsening the problem (**Figure 15.4**).

Robert E. Daemmrich/Getty Images, Inc.

Figure 15.4 **Social isolation** The social isolation experienced by many overweight teens contributes to inactivity and overeating, further exacerbating their weight problems.

Type 2 Diabetes Until recently, type 2 diabetes was considered a disease that affected primarily adults over 40, but it is now on the rise among America's youth. About 3700 U.S. children and adolescents are diagnosed with type 2 diabetes each year. Most of these cases are in children 10 to 19 years, and the incidence is higher among minorities.[8,9] The typical picture of type 2 diabetes in this population is a child from age 10 to mid-puberty, overweight, with a family history of the disease. Little is known about type 2 diabetes in children, but it is a progressive disease that increases in severity with time from diagnosis. The longer an individual has diabetes, the greater the risk of complications that involve the circulatory system or nervous system and that can lead to blindness, amputations, kidney failure, and heart disease (see Chapter 4). Keeping children's weight in the healthy range and maintaining an active lifestyle can prevent type 2 diabetes.

Elevated Blood Cholesterol and Heart Disease Children in the U.S. currently consume more than the recommended maximum of 10% of their energy from saturated fat.[10] This can lead to elevated blood cholesterol levels. Children's diets are also lower in fiber

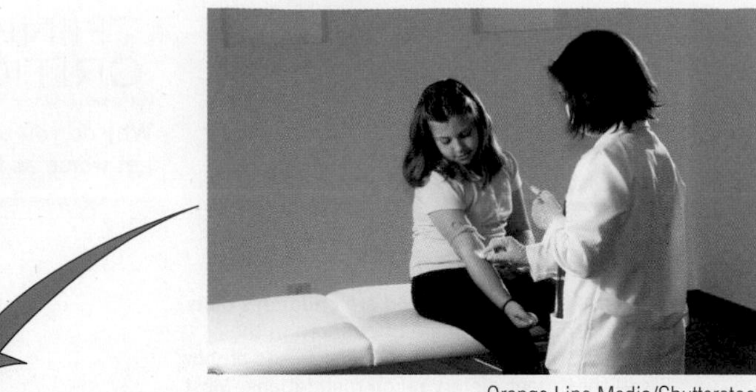

Orange Line Media/Shutterstock

Figure 15.5 Monitoring blood cholesterol in children
The American Academy of Pediatrics recommends blood cholesterol screening for high-risk children and teenagers. This includes those with parents or grandparents who have a history of abnormal blood cholesterol levels or who developed heart disease prematurely (≤ 55 years of age for men and ≤ 65 years for women). Screening is also recommended for children whose family history is unknown and children who have other risk factors, including obesity, diabetes, or high blood pressure.[11]

Cholesterol levels in children and adolescents 2 to 19 years old*

	Total cholesterol (mg/100 mL)	LDL cholesterol (mg/100 mL)
Acceptable	< 170	< 110
Borderline	170–199	110–129
High	≥200	≥130

*HDL levels should be ≥ 35 mg/dL and triglycerides should be ≤ 150 mg/dL.

than recommended. Fiber itself has a role in decreasing the risk of heart disease, and diets high in fiber tend to also be lower in fat, cholesterol, and energy. The recommended level for blood cholesterol in children ages 2 to 19 is less than 170 mg per 100 mL (**Figure 15.5**).[12] In the United States, many children have blood cholesterol levels higher than this. Elevated blood cholesterol levels during childhood and adolescence are associated with higher blood cholesterol and higher mortality rates from cardiovascular disease in adulthood.[13]

Hypertension People who have blood pressure at the high end of normal as youngsters are more likely to develop hypertenion as adults.[14] Blood pressure can be affected by the amount of body fat, activity level, and sodium intake, as well as by the total pattern of dietary intake, so attention should be paid to these nutritional and lifestyle factors in children. This is particularly important if there is a family history of hypertension. As in adults, an active lifestyle, maintenance of a healthy body weight, and a diet low in sodium and high in whole grains, fruits, and vegetables, with moderate amounts of low-fat dairy products and lean meats, is recommended to maintain normal blood pressure.[15]

Healthy Eating Is Learned for Life

Much of what we choose to eat as adults depends on what we learned to eat as children. Children learn by example, and therefore the eating patterns, attitudes, and feeding styles of their caregivers influence what they learn to eat. When caregivers drink milk, choose whole grains, and eat plenty of fruits and vegetables, children follow their example. Likewise, if a child's role models eat a diet high in fat and low in fruits and vegetables, the child will follow suit. The eating and exercise habits developed during childhood and adolescence are important because they establish a pattern that may last a lifetime and affect how healthy people will be as they get older.

15.2 Nourishing Infants, Toddlers, and Young Children

LEARNING OBJECTIVES

- Explain why growth is the best indicator of nutrient intake in children.
- Compare the energy and protein requirements of infants and children with those of adults.
- Give examples of how environment influences children's nutrient intake.
- Summarize the recommendations for preventing and managing food allergies.

Nourishing a growing child is not always an easy task. The foods offered must supply enough energy and nutrients to meet the needs of maintenance, activity, and growth, as well as suit children's tastes. This can be a challenge to caregivers, whether they are feeding infants (ages

4 months to 1 year) sampling solid foods for the first time, toddlers (ages 1 to 3 years) experimenting with new foods, or young children (ages 4 to 8 years) eating meals at school or with friends.

Monitoring Growth

The best indicator that a child is receiving adequate nourishment, neither too little nor too much, is a normal growth pattern. If a child does not get enough to eat, growth may be slowed. If intake is excessive, the child is at risk for becoming obese and developing the chronic diseases that are increasingly common in American adults.

The ultimate size (height and weight) that a child will attain is affected by genetic, environmental, and lifestyle factors. A child whose parents are 5 feet tall may not have the genetic potential to grow to 6 feet, but when adequately nourished, most children follow standard patterns of growth. Growth is most rapid in the first year of life, when an infant's length increases by 50%, or about 10 inches. In the second year of life, children generally grow about 5 in.; in the third year, 4 in.; and thereafter, about 2 to 3 in./year. During adolescence, there is a period of growth that is almost as rapid as that of infancy. Growth can be monitored by comparing a child's growth pattern to standard patterns using growth charts (see Appendix B).[16] For infants, charts are available to monitor weight-for-age, length-for-age, and head circumference-for-age. For children and adolescents ages 2 to 20 years, weight-for-age, length-for-age, and body mass index (BMI)-for-age charts are available. The BMI-for-age growth chart is recommended for identifying children who are underweight, overweight, or obese (**Figure 15.6**).

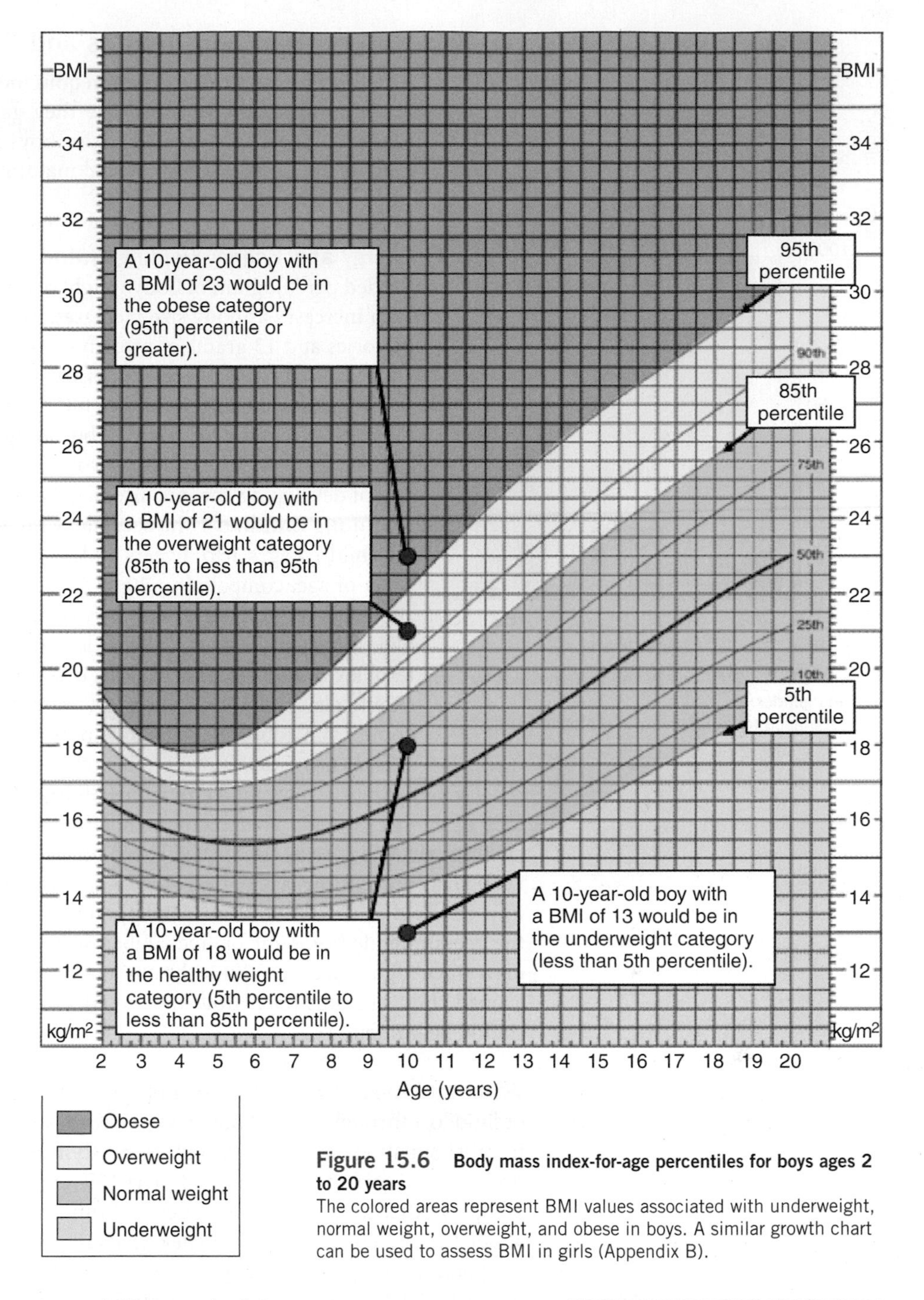

Figure 15.6 **Body mass index-for-age percentiles for boys ages 2 to 20 years**
The colored areas represent BMI values associated with underweight, normal weight, overweight, and obese in boys. A similar growth chart can be used to assess BMI in girls (Appendix B).

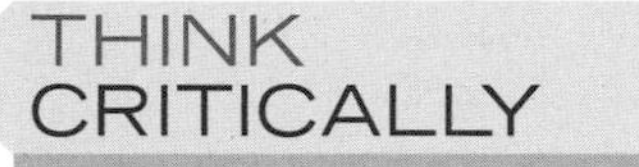

If a 10-year-old boy who is obese continues to grow in height but not gain weight, what will happen to his BMI?

Although growth often occurs in spurts and plateaus, overall growth patterns are predictable. If a child's overall pattern of growth changes, his or her dietary intake should be evaluated to determine the reason for the sudden change. Children who fall below the fifth percentile of the BMI-for-age distribution are considered underweight and their intake should be evaluated to be sure they are meeting their needs. Malnutrition during childhood can cause lasting damage for which adequate nutrition later on may not be able to compensate. Children are considered overweight when their BMI is greater than or equal to the 85th percentile and less than the 95th percentile and are considered obese when their BMI is greater than or equal to the 95th percentile (see Figure 15.6). An increase in activity and/or decrease in energy intake may be needed to keep a child's weight in the healthy range.

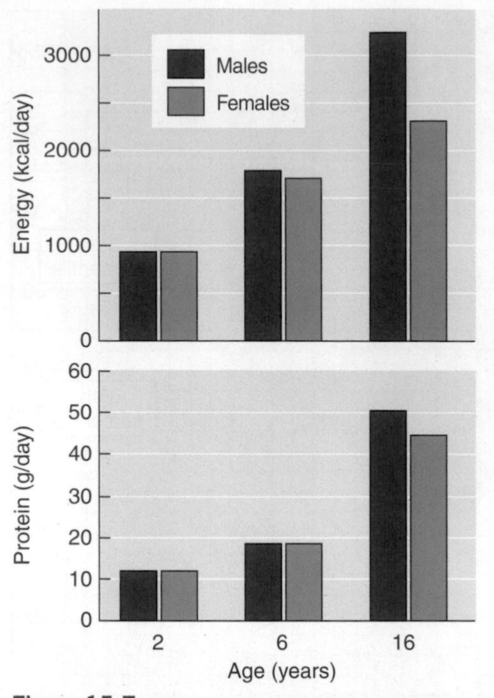

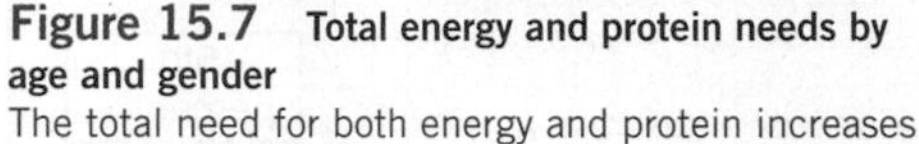
Figure 15.7 Total energy and protein needs by age and gender
The total need for both energy and protein increases as children get older.

Nutrient Needs of Infants and Children

As children grow, their nutrient requirements per unit of body weight decrease, but total needs increase because they gain weight and become more active. Recommended nutrient intakes for boys and girls do not differ until about 9 years of age, at which time sexual maturation causes differences in their nutrient requirements.

Energy and Energy-Yielding Nutrients The amount of energy and protein needed per kilogram of body weight decreases with age, but the total amount of each increases as body size increases. The average 2-year-old needs about 1000 kcalories and 13 grams of protein per day. By age 6, that child will need about 1600 kcals and 19 g of protein per day (**Figure 15.7**).[17]

Infants need a high-fat diet (40 to 55% of energy intake) to support their rapid growth and development, but by age 4, the recommended proportion of calories from fat is reduced to provide adequate energy without increasing the risk of developing chronic disease (see **Off the Label: Reading Labels on Food for Young Children**). The acceptable range for fat intake is 30 to 40% of energy for children ages 1 to 3 years and 25 to 35% of energy for those 4 through 18 years of age, compared to 20 to 35% for adults. Children's diets must provide adequate amounts of essential fatty acids; specific AIs have been set for linoleic and α-linolenic acid. To reduce the risk of developing high blood cholesterol levels and subsequent heart disease, the diets of children over the age of 3 should also be low in cholesterol, saturated fat, and *trans* fat. Most fats should come from foods rich in polyunsaturated and monounsaturated fatty acids, such as fish, nuts, and vegetable oils.[15]

Carbohydrate recommendations for children are the same as those for adults: 45 to 65% of energy. Specific fiber recommendations have not been made for infants, but for children an AI has been established based on data that show an intake of 14 g of fiber per 1000 kcals reduces the risk of heart disease. As in the adult diet, most of the carbohydrate in a child's diet should be from whole grains, fruits, and vegetables. These will provide the recommended amount of fiber. Fiber supplements are not recommended for children since high intakes can limit the amount of food and, consequently, the nutrients that a small child can consume. Foods high in added sugars, such as cookies, candy, and soda, should be limited.

Water and Electrolytes By 1 year of age, a child's kidneys have matured and the amount of fluid lost through evaporation has decreased, so total fluid losses decline. As with adults, in most situations drinking enough to satisfy thirst will provide sufficient water. In children 1 to 3 years of age, about 1.3 liters (5½ cups) of water daily will meet needs; about 4 cups of this should be from water and other fluids. Older children, ages 4 to 8, need about 1.7 L (7 cups) per day.[18] Water needs increase with illness, when the environmental temperature is high, and when activity increases sweat losses.

A UL of 2.3 g of sodium per day has been set for adults and teens 14 to 18 years of age because a high sodium intake is associated with elevated blood pressure. The UL is somewhat lower in children and younger teens (see inside cover). The typical sodium intake in children and teens currently exceeds the UL.[18]

Micronutrients Children are smaller than adolescents and adults, and for the most part the recommended amounts of micronutrients are also smaller (see inside cover). Generally, a nutrient-dense diet that follows the MyPlate recommendations for children will meet needs. Consuming the recommended amounts of meats and whole and enriched grains helps ensure enough B vitamins. Adequate fruits and vegetables provide vitamin C and vitamin A. Milk provides calcium and vitamins A and D. Fortified breakfast cereals help compensate for poorer diets by providing the recommended amounts of a variety of vitamins and minerals in a single serving. However, despite the relative abundance of vitamins and minerals in the modern diet, poor food choices put many children in the United States today at risk for deficiencies of calcium, vitamin D, and iron.

OFF THE LABEL Reading Labels on Food for Young Children

© iStockphoto

Children have different nutrient needs from adults. Therefore, the labels on foods designed for young children must follow different rules. The most obvious difference is how fat is listed in the Nutrition Facts. Labels for foods intended for children younger than 2 years list total fat but do not list the amount of saturated fat, polyunsaturated fat, monounsaturated fat, cholesterol, calories from fat, or calories from saturated fat.[1] These labels also may not carry most claims about a food's nutrient content or health effects. The reason is that dietary fat is needed for brain development and meeting energy needs during the rapid growth and development of infancy and early childhood. It is hoped that excluding information about fat content on the label will prevent caregivers from restricting fats in the diets of young children.

As children develop, the amount of fat in their diet can safely be reduced. Therefore, labels on foods designed for 2- to 4-year-olds must include information on the amount of cholesterol and saturated fat per serving and may provide information on calories from fat and saturated fat, as well as polyunsaturated and monounsaturated fat per serving. The serving sizes listed must be based on servings appropriate for small children.

Another difference between standard food labels and those for foods designed for children under age 4 is the absence of % Daily Values for total fat, saturated fat, cholesterol, total carbohydrate, fiber, and sodium.[1] For children under 4 years, the FDA has set Daily Values only for vitamins, minerals, and protein. Labels include the % Daily Values for these nutrients when they are present in significant amounts.

Nutrient and health claims allowed on young children's foods include claims that describe the percentage of vitamins or minerals as they apply to the Daily Values for children under age 2, such as "provides 50% of the Daily Value for vitamin C." Also, for children under 2, the descriptors "unsweetened" and "unsalted" are allowed. "No sugar added" and "sugar free" are approved only for use on dietary supplements for children.

Many of the foods consumed by young children do not have special labels because they are also adult foods. When selecting these foods, keep in mind that the needs of young children, especially for fat, are different from the needs of adults.

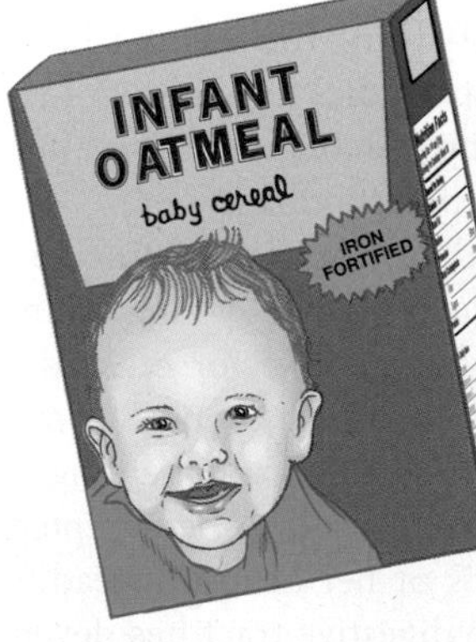

Nutrition Facts

Serving Size 1/4 cup (15g)
Servings Per Container About 30

Amount Per Serving		
Calories 60		
Total Fat		1g
Sodium		0mg
Potassium		50mg
Total Carbohydrate		10g
Fiber		1g
Sugars		0g
Protein		2g
Daily Value	Infants 0-1	Children 1-4
Protein	7%	6%
Vitamin A	0%	0%
Vitamin C	0%	0%
Calcium	15%	10%
Iron	45%	60%
Vitamin E	15%	8%
Thiamin	45%	30%
Riboflavin	45%	30%
Niacin	25%	20%
Phosphorus	15%	10%

Nutrition facts label for foods for children under age 2.

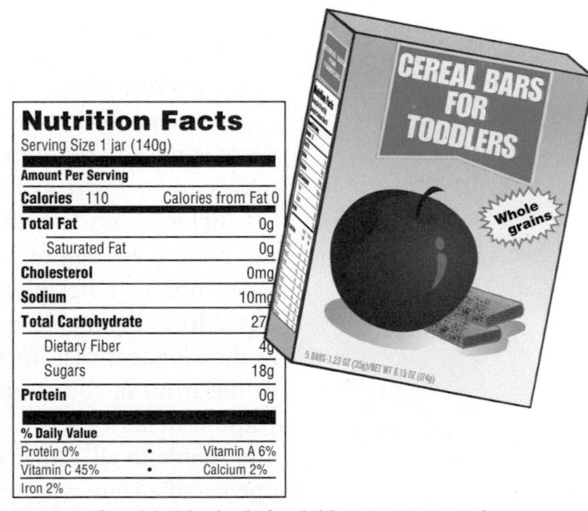

Nutrition Facts

Serving Size 1 jar (140g)

Amount Per Serving	
Calories 110	Calories from Fat 0
Total Fat	0g
Saturated Fat	0g
Cholesterol	0mg
Sodium	10m
Total Carbohydrate	27
Dietary Fiber	4
Sugars	18g
Protein	0g
% Daily Value	
Protein 0% •	Vitamin A 6%
Vitamin C 45% •	Calcium 2%
Iron 2%	

Nutrition facts label for foods for children 2 to 4 years of age.

[1]U.S. Food and Drug Administration. *Food Labeling Guide*, September 1994; revised April 2008; revised October 2009.

If the sweetened beverages in the graph were replaced with 100% fruit juice, what would happen to the nutrient density of the diet?

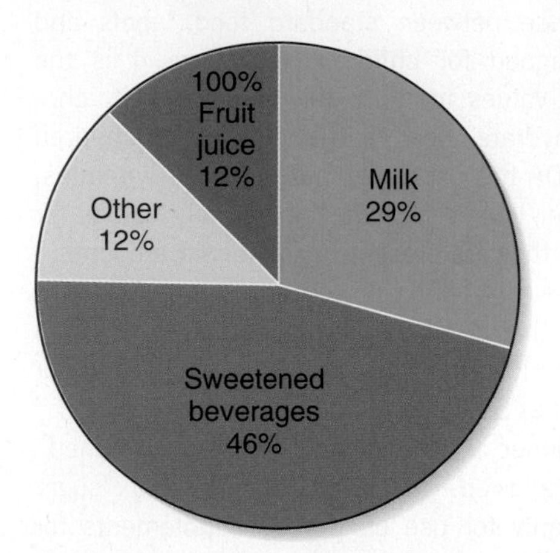

Figure 15.8 Beverage choices in children and teens
Consumption of soft drinks by children and adolescents has increased since the 1970s, and milk consumption has decreased. Today, 46% of their beverage intake by weight is soft drinks and other sweetened beverages, while only 29% is milk.[21] This pattern has increased the intake of added sweeteners and decreased the amounts of calcium, phosphorus, magnesium, potassium, protein, riboflavin, vitamin A, and zinc in their diets.

Calcium, Vitamin D, and Bone Health Calcium intake in school-age children has been declining, primarily due to a decrease in the consumption of dairy products (**Figure 15.8**). Adequate calcium intake during childhood is essential for achieving maximum peak bone mass, which is important for preventing osteoporosis later in life (see Chapter 11). The RDA for calcium is 700 mg/day for toddlers (ages 1 to 3) and 1000 mg/day for young children (ages 4 to 8).[19]

Vitamin D, which is needed for calcium absorption, is also essential for bone health. Low intakes of milk combined with limited sun exposure may put many children at risk for vitamin D deficiency (see Chapter 9).[20] The RDA is 600 IU (15 μg)/day for children, adolescents, and young adults (see **Science Applied: Vitamin D: From Cod Liver Oil to Fortified Milk**).

Iron and Anemia The RDA for iron for infants 7 to 12 months of age is 11 mg/day. This drops to 7 mg/day for toddlers, but for young children ages 4 to 8 it is 10 mg/day, which is higher than the RDA for adult men. Milk is a poor source of iron and high in calcium, which inhibits iron absorption. Children, especially those ages 1 through 5, who consume large amounts of milk may not consume enough iron to meet their needs. Children in this age group should be limited to no more than 24 ounces of milk daily and encouraged to eat good sources of iron such as fortified grains and breakfast cereals, raisins, eggs, and lean meats, and to consume fruits and vegetables high in vitamin C to promote iron absorption.[22] Iron deficiency can lower resistance to illness and slow recovery time. It can affect learning ability, intellectual performance, stamina, and mood.[23,24] If anemia is diagnosed, iron supplements are usually prescribed until iron stores are repleted. These supplements should be kept out of the reach of children. Overdoses of iron-containing supplements are the leading cause of poisoning deaths among children under 6 years.[25] To help protect children, products containing iron include a warning about the hazards to children of ingesting large amounts of iron.

Meeting Infants' Nutrient Needs

Although breast milk or infant formula meets most nutritional needs until 1 year of age, semisolid and solid foods can be gradually introduced into the infant's diet starting between 4 and 6 months. Introducing solid foods earlier provides no nutritional or developmental advantages. Despite this, some parents offer semisolid foods before this age, often by adding infant cereal to the baby's bottle because they think the infant is hungry or the added food will help the child sleep through the night. Studies have shown that there is no difference in sleeping patterns based on such feeding practices and that they may increase the risk of developing food allergies and obesity by preschool age.[26,27,28]

Before 4 to 6 months of age, the infant's feeding abilities and gastrointestinal tract are not mature enough to handle foods other than breast milk or formula. The young infant takes milk by a licking motion of the tongue called suckling, which strokes or milks the liquid from the nipple. Solid food placed in the mouth at an early age is usually pushed out as the tongue thrusts forward. By 4 to 6 months of age, the early reflex to bring the tongue to the front of the mouth to suckle has diminished, allowing solid food to be accepted without being expelled. By this age, the infant also can hold his or her head up steadily and is able to sit, either with or without support. Internally, the digestive tract has developed, and enzymes are present for starch digestion. The kidneys are more mature and better able to concentrate urine. With all of these changes, the child is ready to begin a new approach to eating.

What Foods to Introduce First? The most commonly recommended food to be given first to a child is iron-fortified infant rice cereal mixed with formula or breast milk. Rice cereal is the recommended first food because it is easily digested and rarely causes allergic reactions. After rice has been successfully included in the diet, other grains can be introduced, with wheat cereal offered last because it is most likely to cause an allergic reaction. After cereals are introduced, pureed vegetables or fruits can be tried; some suggest that vegetables be offered before fruits so that the child will learn to enjoy food that is not

SCIENCE APPLIED

Vitamin D: From Cod Liver Oil to Fortified Milk

(Elena Schweitzer/Shutterstock) Victoriano Izquierdo/Getty Images, Inc. (DAJ/Getty Images, Inc.)

Rickets, a vitamin D deficiency disease that results in bone deformities, has plagued infants for centuries. Soranus, a late-first-century Roman physician, was one of the first to describe these bone abnormalities in infants.[1] In the 17th century, the English physician Daniel Whistler published the first description of the clinical features and symptoms of this malady. At the beginning of the 20th century, when air pollution from the Industrial Revolution darkened the skies, rickets affected more than 90% of children in Northern Europe and 80% of children in Boston and New York City.[2] At the time, the cause and cure for the disease were unknown.

THE SCIENCE

The relationship between rickets and sunlight did not go unnoticed by scientists. In the late 1800s it was observed that children living in northern European cities such as London and Glasgow were plagued with rickets, while even impoverished children in sunny regions of the world were free of the disease.[2] The idea that sun exposure could help bones seemed preposterous to many, so interest also focused on the curative properties of cod-liver oil, which had for centuries been a folk remedy for rickets. When lion cubs at the London Zoo developed rickets in 1889, cod-liver oil and crushed bones added to the diet caused a complete recovery.[1] Since cod-liver oil was known to be a source of vitamin A, a vitamin-A deficiency was proposed as the cause of rickets, and foods rich in vitamin A, such as cod-liver oil, butter, and whole milk were suggested as a remedy. The true anti-rachitic factor was identified by Dr. Elmer McCollom and his colleagues at Johns Hopkins University. They experimented with cod-liver oil that had been treated in a way that destroyed the vitamin A. The treated oil was not able to cure vitamin-A deficiency, but still retained its ability to cure rickets in rats. They concluded that the anti-rachitic substance found in the oil was distinct from vitamin A and called the substance vitamin D.[1]

Mary Evans/Photo Researchers, Inc.

Cod-liver oil was marketed as a cure for many ailments in the early 20th century. Today children get their vitamin D from fortified milk rather than a spoonful of cod liver oil.

A complete understanding of vitamin D and rickets was complicated by the fact that it can be obtained both from the diet and from sunlight, but soon researchers began to recognize that vitamin D was both a nutrient and a hormone. Studies conducted in rodents with experimentally induced rickets demonstrated that rickets could be cured both by exposure to a particular spectrum of ultraviolet light and by supplementation with cod-liver oil.[3] Experiments in human infants conducted in Vienna and at Yale University were able to confirm that both exposure to sunlight and the vitamin D in cod-liver oil were able to prevent rickets.[1,3]

THE APPLICATION

Understanding the sources and function of vitamin D gave public health officials the tools to eliminate a scourge that had afflicted children since Roman times. By the late 1920s and '30s, public health agencies in the U.S. were promoting sensible sun exposure, vitamin D supplements, and vitamin D fortification of foods. The use of cod-liver oil in the treatment and prevention of rickets became commonplace and in some states capsules of the oil were distributed to residents. But cod-liver oil was not an ideal solution because people had to remember to take it and some preparations were much lower in vitamin D than others.[4] The solution emerged from the discovery, by Harry Steenbock at the University of Wisconsin, that irradiating foods and feeding them to rats with rickets cured the rickets.[3] Steenbock's finding led to an inexpensive way to enhance the vitamin D content of foods, including milk. The introduction and consumption of vitamin D-fortified foods led to the virtual eradication of rickets in the United States. People no longer needed to remember to take cod-liver oil because science had found a way to put sunshine into their food.

[1]Rajakumar, K. Vitamin D, Cod-liver oil, sunlight, and rickets: A historical perspective. *Pediatrics* 112: e132–e135, 2003.
[2]Holick, M. F. The vitamin D deficiency pandemic: A forgotten hormone important for health. *Public Health Rev* 32:267–283, 2010.
[3]Rajakumar, K., Greenspan, S. L., Thomas, S. B., and Holick, M. F. Solar ultraviolet radiation and vitamin D: A historical perspective. *Am J Public Health* 97:1746–1754, 2007.
[4]Veick, M. T. A history of rickets in the United States. *Am J Clin Nutr* 20: 1234–1244, 1967.

Figure 15.9 **Nourishing a developing infant**

Raul Touzon/NG Image Collection | Andy Lim/Shutterstock | Bubbles Photolibrary/Alamy | Ashok Rodrigues/iStockphoto

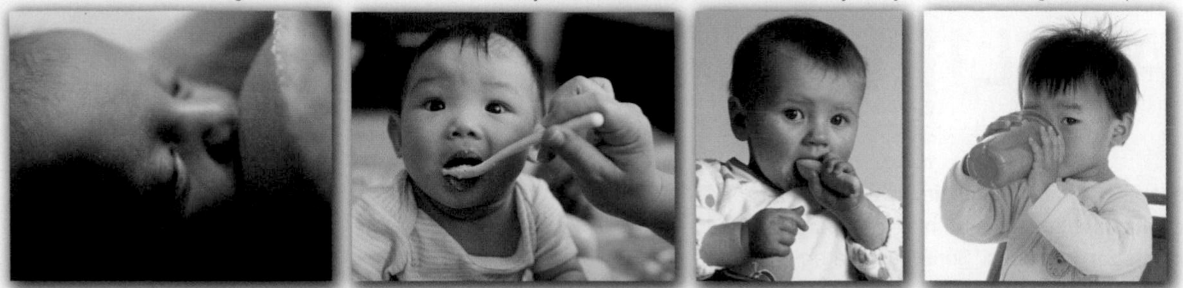

Age	Birth to 4 months	4 to 6 months	6 to 9 months	9 to 12 months
Developmental milestones	The infant takes milk from the nipple by suckling, Solid food placed in the mouth is usually pushed out as the tongue thrusts forward.	The tongue is held farther back in the mouth, allowing solid food to be accepted without being expelled. The infant can hold his or her head up and is able to sit, with or without support.	The infant can sit without support, chew, hold food, and easily move hand to mouth.	The infant can drink from a cup and feed him/herself.
Foods	Breast milk or iron-fortified infant formula.	Breast milk or formula, iron-fortified infant cereal; rice cereal is usually the first solid food introduced because it is easily digested and less likely than other grains to cause allergies. After cereals, pureed vegetables and fruits can be introduced.	Breast milk or formula, iron-fortified infant cereal, pureed or strained vegetables, fruits, meats and beans, limited finger foods.	Breast milk or formula, iron-fortified infant cereal, chopped vegetables, soft fruits, meats and beans, fruit juice, nonchoking finger foods such as dry cereal, cooked pasta, and well-cooked vegetables.

sweet before being introduced to sweet foods. Once teeth have erupted, foods with more texture can be added (**Figure 15.9**). For the 6- to 12-month-old child, small pieces of soft or ground fruits, vegetables, and meats are appropriate (**Table 15.1**).

Honey should not be fed to children less than a year old because it may contain spores of *Clostridium botulinum*, the bacterium that causes botulism poisoning (see Chapter 17). Older children and adults are not at risk from botulism spores because the environment in a mature gastrointestinal tract prevents the bacterium from growing.

Increasing Variety with Developmentally Appropriate Choices As the child becomes familiar with more variety, food choices should be made from each of the food groups in MyPlate. At 1 year of age, whole cow's milk should be offered and continued until 2 years of age, after which reduced-fat milks can be used. Cow's milk should not be used before 1 year of age because it is difficult to digest and the infant's kidneys are too immature to handle

TABLE 15.1 Typical Meal Patterns for Infants

	Amounts per day		
	4–6 months	6–9 months	9–12 months
Breast milk or formula[a]	32 oz	32 oz	32 oz
Cereal	4 T infant rice cereal	½ cup Cheerios	½ cup breakfast cereal, rice, or small pasta
Vegetables		4–6 T pureed or baby food	⅓–½ cup cooked vegetables such as peas, diced green beans or cooked carrots
Fruits		4 T mashed, pureed, or baby food	½ cup soft diced fruits
Meats, fish, poultry, egg yolk		2–4 T strained or pureed	4–6 T, chopped or ground; ¼ cup cottage cheese
Finger foods		Include finger foods, advancing in size and texture with development: teething biscuits, dry toast, crackers, bananas	Table foods except foods in shapes and sizes likely to cause choking, such as large pieces of meat, whole grapes, or hot dogs or carrots cut in circular slices

[a]Includes formula or breast milk added to cereal.

its higher protein and mineral content. To avoid choking, foods that can easily lodge in the throat, such as carrots, grapes, and hot dogs, should not be offered to infants or toddlers.

As children become more independent, they will want to feed themselves. Although this is not always a neat and clean process, it is important for development. By the age of 8 or 9 months, infants can hold a bottle and self-feed finger foods such as crackers. By 10 months, most infants can drink from a cup, so water and fruit juices can be offered (see Figure 15.9). The amount of juice offered to children should be limited. The American Academy of Pediatrics recommends 4 to 6 ounces per day of 100% juice for children ages 1 to 6, and 8 to 12 oz for children age 7 and older.[29] More than half the fruit children consume is as juice rather than whole fruit. Although 100% fruit juice can be part of a healthy diet, too much can cause diarrhea, over- or undernutrition, and dental caries.

A Caution About Food Allergies Although true food allergies are relatively rare, they are more common in infants.[30] Their immature digestive tracts allow incompletely digested proteins to be absorbed. Exposure to an allergen for the first time causes the immune system to produce antibodies to that allergen (see Chapters 3 and 6). When the allergen is encountered again by eating the same food, symptoms such as vomiting, diarrhea, asthma, hives, eczema, runny nose and swelling of respiratory tissues, hay fever, and general cramps and aches may result as the immune system battles the allergen. The symptoms may occur almost immediately or take up to 24 hours to appear, and can vary from mild to severe and life threatening. Foods that commonly cause allergies include wheat, peanuts, eggs, milk, nuts, fish, shellfish, and soy.

To monitor for food allergies when solid foods are given to infants, it is important to introduce new foods one at a time. Each new food should be offered for a few days without adding any other new foods. If an allergic reaction occurs, it is most likely due to the newly introduced food. Foods that cause allergy symptoms such as rashes, digestive upsets, or respiratory problems should be discontinued before any other new foods are added. After an infant is about 3 months of age, the risk of developing food allergies is reduced because incompletely digested proteins are less likely to be absorbed. Many children who develop food allergies before the age of 3 years will outgrow them. For example, most children allergic to eggs at 1 year of age are no longer allergic by age 5. Allergies that appear after 3 years of age are more likely to be a problem for life.

Adverse reactions to foods are also caused by food intolerances. In contrast to food allergies, food intolerances do not involve antibody production by the immune system. Rather, they are caused by foods that create problems during digestion. Food intolerances can be caused by chemical components in foods, by toxins that occur naturally in foods, by substances added to foods during processing or preparation, or simply by large amounts of foods, such as onions or prunes, that cause local gastrointestinal irritation. Lactose intolerance is an example of a food intolerance caused by a reduced ability to digest milk sugar. It is not an allergy to milk proteins.

Diagnosing Food Allergies Several laboratory methods are available to identify foods that are likely to cause an individual's allergic reaction, but they cannot determine the source of the problem with 100% reliability. The cause of a food allergy can be confirmed by using an **elimination diet** and **food challenge**. This involves eliminating all foods suspected of causing an allergic reaction from the diet. Once a diet that causes no symptoms has been established, it should be consumed for two to four weeks. Then, in the food challenge, small amounts of a food suspected of causing a reaction are reintroduced under a doctor's supervision. If no reaction to the food occurs, then increasing amounts are introduced until a normal portion is offered. If there is still no reaction, then the food can be ruled out as an allergen.

elimination diet and **food challenge** A regimen that eliminates potential allergy-causing foods from an individual's diet and then systematically adds them back to identify any foods that cause an allergic reaction.

Preventing and Managing Food Allergies Preventing the development of food allergies is not always possible. Breast-feeding can reduce the risk of food allergies and is recommended for infants from families with a history of allergies.[31] Infants who are breast-fed are less likely to be exposed to foreign proteins that cause food allergies. In addition, their gut matures earlier and they are protected by antibodies and other components of human milk. There is no evidence that delaying the introduction of solid food, including foods thought to be highly allergenic, past 6 months protects against the development of food allergies.[31]

Once a food allergy has developed, the best way to manage it is to avoid consuming the offending food. The information on food labels can be helpful in identifying foods that contain allergy-causing ingredients (see Chapter 6, Off the Label: Is It Safe for You?).

Meeting Children's Nutrient Needs

The development of nutritious eating habits begins in infancy and childhood. Parents and caregivers influence which foods children learn to eat through the examples they set and the amounts and types of foods offered. Children are more likely to eat foods that are available and easily accessible, and they tend to eat greater quantities when larger portions are provided.[32] Caregivers are responsible for deciding what foods and portions should be offered to a child, when they should be offered, and where they should be eaten. The child must then decide whether to eat, what foods to eat, and how much of what they are offered to consume.[33] As children get older their choices are affected more by social activities, what they see at school, and what their friends are eating.

What to Offer? Children should be offered a balanced and varied diet adequate in energy and essential nutrients and appropriate to the child's developmental needs. A healthy meal plan is based on whole grains, vegetables, and fruits; is adequate in milk and other high-protein foods; and is moderate in fat and sodium. Food choices to meet these goals can be determined by following the MyPlate recommendations and spreading these amounts throughout the meals and snacks served each day (**Figure 15.10**).

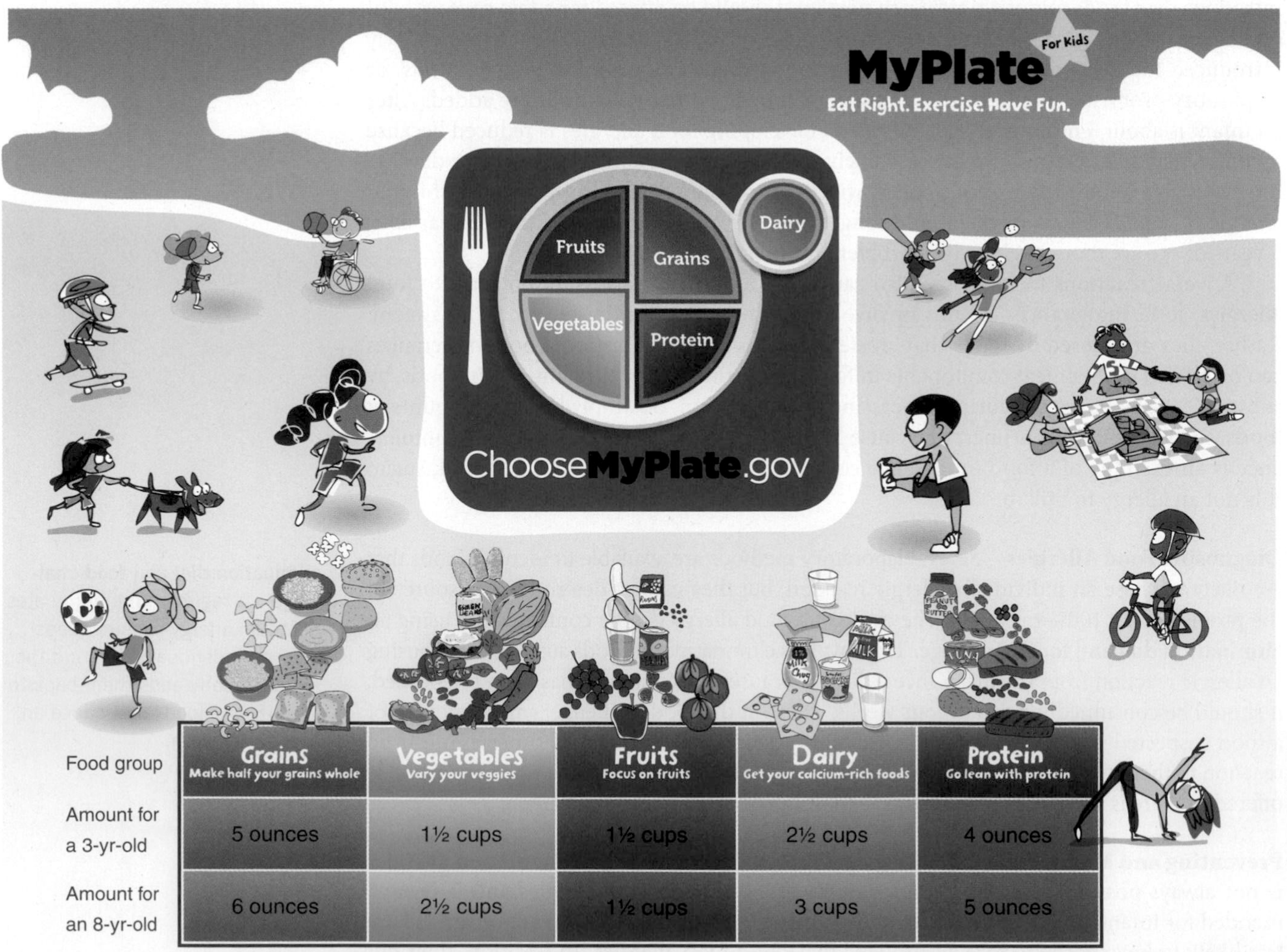

Food group	Grains Make half your grains whole	Vegetables Vary your veggies	Fruits Focus on fruits	Dairy Get your calcium-rich foods	Protein Go lean with protein
Amount for a 3-yr-old	5 ounces	1½ cups	1½ cups	2½ cups	4 ounces
Amount for an 8-yr-old	6 ounces	2½ cups	1½ cups	3 cups	5 ounces

Figure 15.10 MyPlate for kids
MyPlate for Kids recommends amounts of food from each group that are appropriate for young children. The recommendations shown here are for a 3-year-old and an 8-year-old girl who engage in more than 60 minutes of activity daily. Food choices to meet these amounts should be spread throughout the meals and snacks served each day.

TABLE 15.2 A Typical Meal and Snack Pattern for 3- and 8-Year-Old Children

	Amount	
Food	3-year-old	8-year-old
Breakfast		
Whole grain cereal	½ cup	1 cup
Milk, 2%	½ cup	½ cup
Banana	½ medium	1 medium
Snack		
Peanut butter	2 T	2 T
Whole wheat crackers	5	5
Milk, 2%	½ cup	1 cup
Lunch		
Vegetable soup	½ cup	1 cup
Grilled tuna sandwich	½	1
Tomato	¼ medium	½ medium
Orange	½ medium	1 medium
Milk, 2%	½ cup	1 cup
Snack		
Broccoli crowns	4	6
Ranch salad dressing	1 tsp	2 tsp
Dinner		
Chicken drumsticks	1	2
Baked sweet potato	½ cup	1 cup
Green beans	¼ cup	½ cup
Milk, 2%	½ cup	1 cup
Graham crackers	1	2
Snack		
Yogurt	½ cup	1 cup
Berries	½ cup	¾ cup

For example, a 6-year-old, moderately active boy would only need 5 oz of grains. These could include a half cup of cereal at breakfast, some crackers for a snack, half a sandwich at lunch, and a half cup of rice at dinner. This child would need 2½ cups of milk a day, but these may be consumed in smaller portions throughout the day (**Table 15.2**). The need for nutrient-dense choices from all food groups is just as important for children as it is for adults.

Getting a child to eat all the foods recommended by MyPlate may not always be easy. To increase variety, new foods should regularly be introduced into a child's diet. Children's food preferences are learned through repeated exposure to foods; children develop an increased preference for a food if it is offered a minimum of 8 to 10 times.[2] Children are also more likely to try a new food if it is introduced at the beginning of a meal, when the child is hungry, and if the child sees his or her parents or peers eating it. If a new food becomes associated with a bad experience, such as burning the mouth, the child will be unlikely to try it again. Incorporating refused foods into familiar dishes can also increase the variety of the diet. If vegetables are refused, they can be added to soups and casseroles. Fruit can be served on cereal or in milkshakes. Cheese can be included in recipes such as macaroni and cheese, cheese sauce, and pizza. Milk can be added to hot cereal, cream soups, puddings, and custards, and powdered milk can be used in baking. Meats can be added to spaghetti sauce, stews, casseroles, burritos, or pizza.

Children often have periods, known as **food jags**, when they will eat only certain foods and nothing else. For example, a child may refuse to eat anything other than peanut butter and jelly sandwiches for breakfast, lunch, and dinner. The general guideline is to

food jag When a child will only eat one food item meal after meal.

Courtesy Mary Grosvenor

Figure 15.11 **Food choices**
Children should be allowed to select what and how much they will eat from a variety of healthy choices.

Baerbel Schmidt/Stone/Getty Images, Inc.

Figure 15.12 **School lunches**
Participation in the National School Lunch Program reduces the intake of salty snacks and beverages other than milk and 100% fruit juice.[35]

continue to offer other foods along with those the child is focused on. What a child will not touch at one meal, he or she may eat the next day or the next week. No matter how erratic children's food intake may be, caregivers should offer a variety of appropriate healthy food choices at each meal and let their children select what and how much they will eat (**Figure 15.11**).

How Often Should Meals and Snacks Be Offered? Children have small stomachs and high nutrient needs; therefore they should consume nutritious meals and snacks throughout the day. For young children, a meal or snack should be offered every two to three hours and, because children thrive on routines and feel secure in knowing what to expect, a consistent pattern should be maintained from day to day. A missed meal or snack can leave a child without sufficient energy to perform optimally at school or play. A good breakfast is particularly important for ensuring optimal performance at school. Snacks should be as nutritious as meals and should focus on fruits, vegetables, low-fat dairy products, lean meats, and whole-grain products, not soda and chips (**Table 15.3**).

Meals at Day Care or School All meals need to contribute to a child's nutrient intake, but parents may have little input into what children eat while at day care or school. Ensuring that meals eaten away from home are nutritious is not easy because there is no guarantee that what is served or brought from home will be eaten. A packed lunch should contain foods the child likes and that do not require refrigeration (even if a refrigerator is available, the child is likely to forget to put the lunch in it). Even the most carefully planned lunch is not nutritious if it is not eaten.

For children who buy their lunch at school, the National School Lunch Program provides low-cost meals designed to meet nutrient needs and promote healthy diets. The goals of this program are to improve the dietary intake and nutritional health of America's children, and to promote nutrition education by teaching children to make appropriate food choices.[34] Each lunch meal must provide one-third of the RDA for protein, vitamin A, vitamin C, iron, calcium, and energy, and provide no more than 30% of energy from fat and 10% from saturated fat. Within these guidelines, each school or school district can decide which foods to serve and how they are prepared. An analysis of the eating patterns and body weight of children who participated in the school lunch program found that participants ate fewer desserts, salty snacks, and fats and more milk, meat, and beans than nonparticipants (**Figure 15.12**).[35]

In addition to lunches, federal guidelines regulate foods sold in snack bars and vending machines that compete with school lunch programs. These must provide at least 5% of the RDA for one or more of the following: protein, vitamin A, vitamin C, niacin, riboflavin, calcium, and iron.

TABLE 15.3 Healthy Snacks for Young Children

Focus on Fruit—Make popsicles from fruit juice, cut fruit into interesting shapes, try frozen grapes, make trail mix with raisins, dried cranberries, and other dried fruit.
Vary Your Veggies—Dip baby carrots in ranch dressing, add peanut butter to celery, serve salsa with baked tortilla chips.
Get Your Calcium-Rich Foods—Include milk with after-school cookies, try melted cheese on a tortilla, top some fruit with yogurt, snack on string cheese, make a yogurt and fruit shake.
Make Half Your Grains Whole—Try whole-grain breadsticks or pretzels, have graham crackers instead of cookies, pop some corn.
Go Lean with Protein—Put peanut butter on apple slices or crackers, snack on some turkey slices, roll some black beans into a tortilla, spread hummus on crackers.

Vitamin and Mineral Supplements Like adults, children who consume a well-selected, varied diet can meet all their vitamin and mineral requirements with food. In fact, average intakes of most vitamins and minerals for children 2 to 11 years of age exceed 100% of the recommended amounts.[1] Occasional skipped meals and unfinished dinners are a normal part of most children's eating behavior. However, children with particularly erratic eating habits, those on regimens to manage obesity, those with limited food availability, and those who consume a vegan diet may benefit from supplements that provide no more than 100% of the Daily Values. To avoid misuse, children's supplements should be stored out of reach and administered by caregivers.

Eating Environment Factors such as whether families eat together, watch TV during meals, and choose food prepared at home or at a restaurant influence children's eating patterns. To develop sound eating habits, children need companionship, conversation, and a pleasant location at mealtimes. Caregivers should sit with children and eat what they eat (**Figure 15.13**). Children should be given plenty of time to finish eating. Slow eaters are unlikely to finish eating if they are abandoned by siblings who run off to play and adults who leave to wash dishes.

To make mealtime a nutritious, educational, and enjoyable experience, it should not be a battle zone. Threats and bribes are counterproductive and can create a problem where none had previously existed. Food is not a reward or a punishment: It is simply nutrition.

Masterfile

Figure 15.13 Family meals Companionship and conversation at meals help create a positive eating environment and foster good eating patterns.

15.3 Nutrition and Health Concerns in Children

LEARNING OBJECTIVES

- Discuss the relationship between food intake patterns and dental caries.
- Describe the impact of the modern American lifestyle on childhood obesity.
- Discuss the recommendations for preventing and treating childhood obesity.

A number of diet and lifestyle factors put children at risk for illness and malnutrition. Some are a greater risk in young children because of their size and stage of development and others are problems that may continue into adolescence.

Dental Caries

Children and teens typically eat more sugar than is recommended. Added sugars reduce the nutrient density of foods, so excessive consumption of foods high in added sugars makes it difficult to meet nutrient needs. In addition, a diet high in sugary foods promotes the growth of bacteria on the surface of the teeth, increasing the risk of dental caries, commonly known as tooth decay or cavities. Decay occurs when bacteria metabolize sugar, producing acids, which damage the structure of the teeth. Because the primary teeth guide the growth of the permanent teeth, maintaining healthy primary teeth is just as important as preserving permanent ones.

Much of the added sugar in children's diets comes from soft drinks and other sweetened beverages; when these are sipped slowly between meals, the contact time between sugar and teeth is prolonged, hence increasing the risk of tooth decay (**Figure 15.14**). Changing the types of beverages consumed by children may impact the incidence of dental caries. A study found fewer cavities in children who drank more milk and more in those who consumed soda and juice.[36] Although sugary foods are the most cavity-promoting, any

Tim Laman/NG Image Collection

Figure 15.14 Limiting sweetened beverages Offering beverages that can be carried around encourages continuous sipping and is not recommended.[15,29] When juice is sipped slowly over a long period, it provides a continuous supply of sugar to feed cavity-causing bacteria.

carbohydrate-containing food can cause tooth decay, especially if the food sticks to the teeth (see Chapter 4).

Both diet and dental hygiene can affect the risk of dental caries. Preventing tooth decay involves limiting high-carbohydrate snacks, especially those that stick to teeth; brushing teeth frequently to remove sticky sweets; and consuming adequate fluoride. It is recommended that children consume three servings of dairy foods daily, limit intake of 100% juice, and restrict other sugared beverages to occasional treats.[29,36] Children's teeth should be brushed as soon as they erupt and children 3 years of age and older should be examined by a dentist regularly.

Diet and Hyperactivity

Hyperactivity is a problem in 5 to 10% of school-age children, occurring more frequently in boys than in girls. This syndrome involves extreme physical activity, excitability, impulsiveness, distractibility, short attention span, and a low tolerance for frustration. Hyperactive children have more difficulty learning but usually are of normal or above-average intelligence. Hyperactivity is now considered part of a larger syndrome known as **attention deficit hyperactivity disorder (ADHD)**.

attention deficit hyperactivity disorder (ADHD) A condition characterized by a short attention span, acting without thinking, and a high level of activity, excitability, and distractibility.

One popular misconception is that hyperactivity is caused by a high sugar intake, but research on sugar intake and behavior has failed to support this hypothesis.[37,38] Hyperactive behavior that is observed after sugar consumption is likely the result of other circumstances in that child's life. For example, the excitement of a birthday party rather than the cake is most likely the cause of hyperactive behavior. Other situations that might cause hyperactivity include lack of sleep, overstimulation, the desire for more attention, or lack of physical activity.

Another possible cause of hyperactive behavior in children is caffeine. Caffeine is a stimulant that can cause sleeplessness, restlessness, and irregular heartbeats. Beverages, foods, and medicines containing caffeine are often a part of children's diets. For example, caffeinated beverages such as Coke and Mountain Dew are commonly included in children's fast-food meals.

Lead Toxicity

Lead is an environmental contaminant that can be toxic, especially in children under age 6. Lead is found in soil contaminated with lead paint dust; it enters drinking water from old corroded lead plumbing, lead solder on copper pipes, or brass faucets. It is found in polluted air, in leaded glass, and in glazes used on imported and antique pottery. Children are at particular risk for lead toxicity because they absorb lead much more efficiently than do adults. It is estimated that infants and young children may absorb as much as 50% of ingested lead, whereas adults absorb only 10 to 15%.[39] Lead is better absorbed from an empty stomach and when other minerals such as calcium, zinc, and iron are deficient, so malnourished children are at particular risk.

Once lead has been absorbed from the gastrointestinal tract, it circulates in the bloodstream and then accumulates in the bones and, to a lesser extent, the brain, teeth, and kidneys. Lead disrupts the functioning of neurotransmitters and thus interferes with the functioning of the nervous system. In young children, lead poisoning can cause learning disabilities and behavior problems. There is evidence that even blood lead levels below 10 μg/100 mL, which were once thought to be safe, may impair IQ in young children.[40] Higher levels of lead can contribute to iron-deficiency anemia, changes in kidney function, nervous system changes, and even seizures, coma, and death.

Because of the risks of lead toxicity from environmental contamination, state legislatures began to prohibit the use of lead in house paint, gasoline, and solder. As a result, the number of children ages 1 to 5 with elevated blood lead levels decreased from 88.2% in 1976–1980 to only 0.9% in 2003–2008.[41] Despite these gains, low-income children remain at higher risk because low-income families are more likely to live in older buildings that still have lead paint and lead plumbing. Typical blood lead levels for non-Hispanic black children remain slightly higher than those for Mexican-American and non-Hispanic white children, in part reflecting differences in socioeconomic status.[42]

The effects of lead poisoning are permanent, but if high levels are detected early, the lead can be removed with medical treatment. The economic benefit of lowering lead levels among children by preventing lead exposure is estimated at $213 billion per year.[41]

Childhood Obesity

As discussed previously, the incidence of obesity in children ages 2 to 19 has increased from 5% in 1971–1974 to almost 17% in 2009–2010.[6] Changes in lifestyle that have led to decreases in activity and increases in energy intake among children are the major reasons for this increase in obesity. Reversing this trend requires changes in the way children play and eat.

How Television Affects Body Weight Many children today spend more time watching television than they do in any activity other than sleep. Television affects nutritional status in a number of ways: it introduces children to foods they might otherwise not be exposed to, it promotes snacking, and it reduces physical activity.

Through advertising, television has a strong influence on the foods selected by young children; children exposed to advertising choose advertised food products more often than children who are not exposed.[43] Food is the most frequently advertised product category on children's TV. The majority of these ads target products high in fat, sodium, and added refined sugars, such as sweetened breakfast cereals, candy, cookies, doughnuts, and other desserts; snacks; sweetened beverages; and fast-food (Figure 15.15a).[44] The more hours of TV a child watches, the more he or she asks for advertised food items and the more likely these items are to be in the home. Television also promotes snacking behavior. Although snacks are an important part of a growing child's diet, while watching TV many children snack on sweet and salty foods that are low in nutrient density. In terms of overall diet, more hours of TV watching have been associated with higher intakes of energy, fat, sweet and salty snacks, and carbonated beverages and lower intakes of fruits and vegetables.[43]

Perhaps the most important nutritional impact of television is that it reduces activity (Figure 15.15b). Hours spent watching television are hours when physical activity is at a minimum. Generally the more physically active a child is, the less TV he or she watches. Children and adolescents today have replaced physically demanding activities with sedentary pastimes including watching television, playing video games, and networking on cell phones and computers.

Should the types of advertising allowed on children's TV be regulated?

Figure 15.15 Television affects food intake and activity level

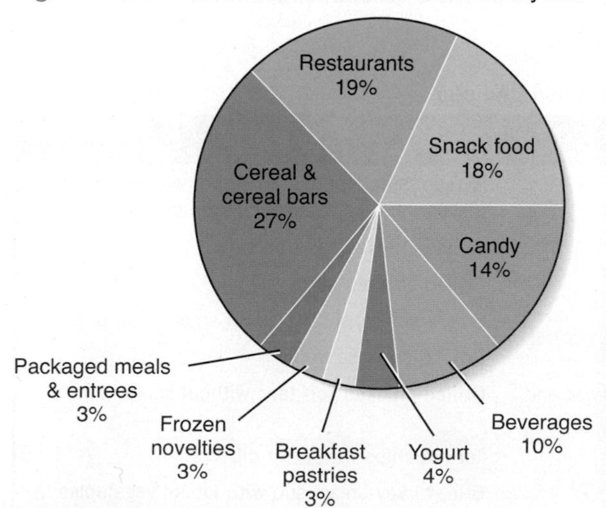

(a) During children's television programming, food is the most frequently advertised product category. This chart, which illustrates the types of food advertised on Saturday morning children's television, shows that almost half of the commercials advertise candy, snack foods, beverages, and pastries.[44]

Donna Day/Getty Images, Inc.

(b) Watching television is a sedentary pastime. Children who watch four or more hours of TV per day are 40% more likely to be overweight than those who watch an hour or less a day.[45]

Robert Ginn/Photolibrary/Getty Images, Inc.

Figure 15.16 **Fast food** Consuming too much fast food and making poor fast-food choices can result in a diet high in calories and low in fruits and vegetables.

Why Fast Food Impacts Body Weight Children and teens generally love fast food, and there is nothing wrong with an occasional fast-food meal. But a steady diet of burgers, fries, and soda will likely contribute to an overall diet that is high in energy, fat, sugar, and salt and low in calcium, fiber, and vitamins A and C (**Figure 15.16**).

The portions served at fast-food restaurants are a major contributor to children's increased energy intake. A meal that includes a hamburger, small fries, and a small drink will provide about 600 kcals, 12 g of saturated fat, and 36 g of sugar. But a meal with a Big Mac, a large order of fries, and a large drink will increase the numbers to 1390 kcals, 15 g of saturated fat, and 94 g of sugar. A steady diet of such meals can significantly increase energy intake and consequently body weight. Choosing carefully from the old fast-food standbys such as plain, single-patty hamburgers or newer low-fat options such as grilled chicken sandwiches can keep fat and energy intake reasonable (see **Table 15.4**).

Some food groups are limited in typical fast-food meals. For example, the few pieces of shredded lettuce and the slice of tomato that garnish a burger or taco make only a small contribution to the 2½ cups of vegetables that should be included in the diet of an average 8-year-old. Typically fast-food meals are also excessive in calories and lacking in milk and fruits. Many fast-food franchises now offer vegetables, salads, yogurt, fruit, and milk, which are moderate in calories and can increase calcium and vitamin intake if selected. Even if these foods are not chosen, if the missing milk, fruits, and vegetables are consumed at other times during the day and overall energy intake balances output, the total diet can still be a healthy one.

Preventing and Treating Obesity The Institute of Medicine Committee on Prevention of Obesity in Children and Youth has developed a prevention-focused plan to decrease the prevalence of childhood obesity in the United States.[46] The committee recommends action by federal, state, and local governments; industry and the media; health-care professionals; community and nonprofit organizations; schools; and parents and families. One recommendation is that the food, beverage, restaurant, entertainment, leisure, and recreation industries help encourage healthy, appropriately portioned foods and develop products and opportunities that stimulate physical activity. The report also recommends changes in the food label and the development of advertising and marketing guidelines that help minimize the risk of obesity. At the community level, recommendations focus on programs that encourage healthy eating behaviors and regular exercise by providing children with safe places to walk, bike, play games and sports, and engage in other physical activities. The report targets schools because they are a place where children can be reached with information about energy balance, good nutrition, and physical activity. Schools can offer foods and beverages that meet nutritional standards. In addition, children should achieve at least 60 minutes of physical activity daily.[5] These recommendations apply at home as

TABLE 15.4 Make Healthier Fast-Food Choices

Instead of . . .	Choose . . .
Double-patty hamburger with cheese, mayonnaise, special sauce, and bacon	Regular single-patty hamburger without mayonnaise, special sauce, and bacon
Breaded and fried chicken sandwich	Grilled chicken sandwich
Chicken nuggets or tenders	Grilled chicken strips
Large order of French fries	Baked potato, side salad, or small order of fries
Fried chicken wings	Broiled skinless wings
Crispy-shell chicken taco with extra cheese and sour cream	Grilled-chicken soft taco without sour cream
Nachos with cheese sauce	Tortilla chips with bean dip
12-in. meatball sub	6-in. turkey breast sub with lots of vegetables
Thick-crust pizza with extra cheese and meat toppings	Thin-crust pizza with extra veggies
Doughnut	Cinnamon and raisin bagel with low-fat cream cheese

well as at school because parents make daily decisions regarding food choices and opportunities for recreation, they determine the setting for foods eaten in the home, and they are responsible for other rules and policies that influence the extent to which various members of the family engage in healthful eating and physical activity.

These strategies will help reduce the prevalence of obesity in the population, but weight problems must also be addressed in individuals. Because children are still growing, weight loss is rarely recommended. Rather, overweight children should be encouraged to slow their weight gain while they continue to grow taller. This allows them to "grow into" their weight. A child who is at the 85th percentile for BMI at age 7 and gains only a few pounds a year can be at the 75th percentile by age 9. This requires that the child's behavior be modified to reduce energy intake to a moderate level and to increase activity.

Changing Behavior As in adults, weight gain in children is related to patterns of eating and exercise. Any permanent change in weight requires a permanent change in lifestyle. Changing eating patterns and activity is key to developing and maintaining a healthy weight. Because children, like adults, may overeat for comfort, self-reward, or out of boredom, parent involvement in helping the child find other sources of gratification can be vital.

Reducing Intake Modifying a child's food consumption patterns can be difficult. Denying food may promote further overeating by making the child feel that there will not be enough to satisfy hunger. The child may then overeat whenever there is a chance. Thus, energy intake restrictions should be relatively mild, and the focus instead should be on offering nutrient-dense foods such as whole grains, fruits and vegetables, lean meats, and reduced-fat dairy products. Planning ahead can help manage eating at social events. For example, a teenage girl with a weight problem could plan how much she will eat at a pizza party and then increase her exercise to compensate for the excess calories consumed.

Increasing Activity Whether or not a child is overweight, he or she should be physically active for at least an hour per day.[5] Children have short attention spans, so their activities should be intermittent. Periods of moderate to vigorous activity lasting 10 to 15 minutes or more each day should be interspersed with periods of rest and recovery. Children and adolescents should be exposed to a variety of different types of activities that are of various levels of intensity and include both muscle-strengthening and bone-strengthening activities (**Figure 15.17**). Learning to enjoy sports and exercise in childhood will set the stage for an active lifestyle in adulthood. For inactive children, physical activity should be increased gradually in order to make exercise a positive experience. A good way to start is to

Masterfile

Exercise Guidelines for Children and Adolescents[5]

- Children and adolescents should do 60 minutes or more of physical activity daily.
  - Aerobic: Most of the 60 or more minutes a day should be either moderate- or vigorous-intensity aerobic physical activity, and should include vigorous-intensity physical activity at least 3 days a week.
  - Muscle-strengthening: As part of their 60 or more minutes of daily physical activity, children and adolescents should include muscle-strengthening physical activity on at least 3 days of the week.
  - Bone-strengthening: As part of their 60 or more minutes of daily physical activity, children and adolescents should include bone-strengthening physical activity on at least 3 days of the week.
- It is important to encourage young people to participate in physical activities that are appropriate for their age, that are enjoyable, and that offer variety.

Figure 15.17 Meeting activity goals Exercise should be fun. Riding bikes, using rollerblades and skateboards, and playing basketball and Frisbee all count toward fulfilling exercise goals.

encourage activities such as games, walks after dinner, bike rides, hikes, swimming, and volleyball that can be enjoyed by the whole family. This sends a positive message to "be more active." Again, involvement of the whole family is key to increasing activity.[47]

15.4 Adolescents

LEARNING OBJECTIVES

- Describe how growth and body composition are affected by puberty.
- Compare the energy needs of adolescents with those of children and adults.
- Explain why iron and calcium are of particular concern during the teen years.
- Use MyPlate to plan a day's diet that would appeal to a teenager.

Once a child has reached about 9 to 12 years of age, the physical changes associated with sexual maturation begin to occur. These physical changes, along with the social and psychological changes that accompany them, have a significant impact on the nutritional needs and nutrient intakes of adolescents. The DRIs divide recommended intakes for adolescence into ages 9 through 13 and ages 14 through 18.

The Changing Body: Sexual Maturation

puberty A period of rapid growth and physical changes that ends in the attainment of sexual maturity.

adolescent growth spurt An 18- to 24-month period of peak growth velocity that begins at about ages 10 to 13 in girls and 12 to 15 in boys.

menarche The onset of menstruation, which occurs normally between the ages of 10 and 15 years.

During adolescence, organ systems develop and grow, **puberty** occurs, body composition changes, and the growth rates and nutritional requirements of boys and girls diverge. During the teenage years, boys and girls grow about 11 inches and gain about 40% of their eventual skeletal mass. From ages 10 to 17, girls gain about 53 lbs and boys about 70 lbs. During adolescence, there is an 18- to 24-month period of peak growth velocity, called the **adolescent growth spurt** (**Figure 15.18a**).

The hormonal changes that occur with sexual development orchestrate growth and affect body composition. During the growth spurt, boys tend to grow taller and heavier than girls and do so at a faster rate. Boys gain fat but also add so much lean mass as muscle and bone that their percentage of body fat actually decreases. In girls, **menarche**, the onset of menstruation, is typically followed by a deceleration in growth rate and an increase in fat deposition. By age 20, females have about twice as much adipose tissue as males and only about two-thirds as much lean tissue (**Figure 15.18b**).

Nutrient intake during childhood and adolescence can affect sexual development. Nutritional deficiencies can cause poor growth and delayed sexual maturation. Taller, heavier children usually enter puberty sooner than shorter, lighter ones.[48]

Adolescent Nutrient Needs

The physiological changes that occur during adolescence affect nutrient needs. Total nutrient needs are greater during adolescence than at any other time of life. Because of the large individual variation in the age at which these growth and developmental changes occur, the stage of maturation is often a better indicator of nutritional requirements than actual chronological age. The best indicators of adequate intake are satiety and growth that follows the curve of the growth charts.

Energy and Energy-Yielding Nutrients The proportion of energy from carbohydrate, fat, and protein recommended for adolescents is similar to that of adults, but the total amount of energy needed by teenagers exceeds adult needs. Energy requirements for boys are greater than those for girls because boys have more muscle and a greater body size. Adolescent girls need 2100 to 2400 kcals/day, whereas boys require about 2200 to 3150 kcals/day (see Figure 15.7). Protein requirements for both groups reach the adult

Figure 15.18 **Adolescent growth**

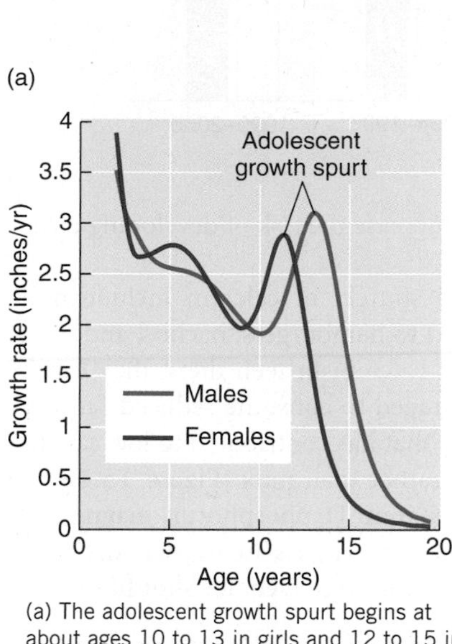

(a) The adolescent growth spurt begins at about ages 10 to 13 in girls and 12 to 15 in boys. During a one-year growth spurt, boys can gain 4 in. in height and girls 3.5 in.

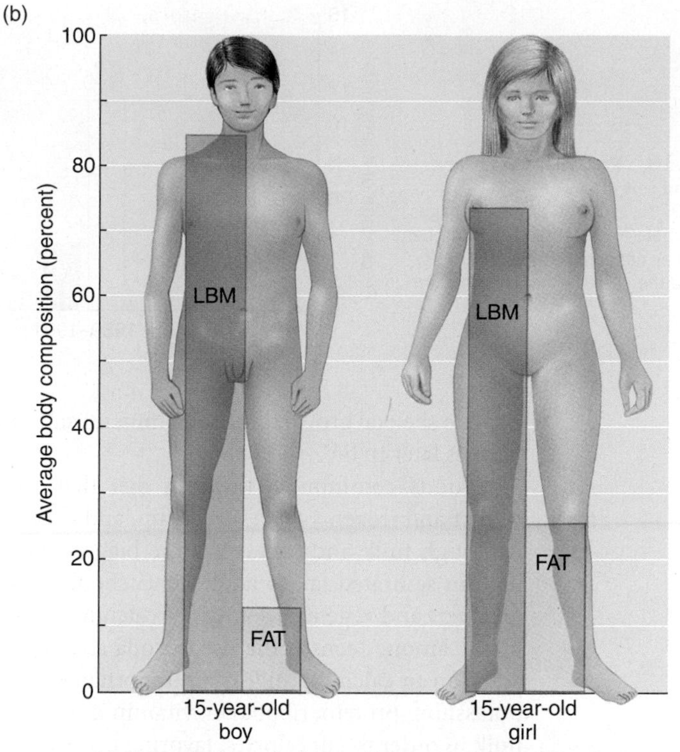

(b) After puberty, males have a higher percentage of lean body mass (LBM) and less body fat than females. (*Source*: Adapted from Forbes, G. B. Body composition. In *Present Knowledge in Nutrition*, 6th ed. M. L. Brown, ed. Washington, DC: International Life Sciences Institute–Nutrition Foundation, 1990.)

recommendation of 0.8 g/kg/day by about age 19, but since boys are generally heavier, they require more total protein than girls. These higher requirements for males continue throughout life.

Micronutrients The requirements for many vitamins and minerals are greater during adolescence than childhood. Some of these are increased because of increased energy needs and others are increased because of their roles in growth and maturation.

Vitamins The need for most of the vitamins rises to adult levels during adolescence. The requirement for B vitamins, which are involved in energy metabolism, is much higher in adolescence than in childhood because of higher energy needs. The rapid growth of adolescence further increases the need for vitamin B_6, which is important for protein synthesis, and for folate and vitamin B_{12}, which are essential for cell division. In general, the fact that teens have high energy needs allows them to meet most of their vitamin needs because eating more food provides more vitamins. However, intakes of riboflavin and vitamin D are frequently low in teens' diets, especially in those of girls, possibly due to low milk intake. Low vitamin-D intake results in low blood levels when synthesis from sunlight is limited due to dark skin pigmentation or inadequate exposure to sunlight during the winter months. Vitamin D is important to support the rapid skeletal growth that occurs during adolescence (see Chapter 7).[49]

Calcium During the adolescent growth spurt, both the length and the mass of bones increases. Adequate calcium is essential to form healthy bone. Calcium retention varies with growth rate, with the fastest-growing adolescents retaining the most calcium. The RDA for calcium during adolescence is 1300 mg/day for everyone between the ages of 9 and 18, but intake is typically below this amount in both sexes.[19] Fewer than 10% of girls and only 25% of boys ages 9 to 13 consume the recommended amount of calcium.[50] This low intake, in combination with a sedentary lifestyle in childhood and adolescence, can

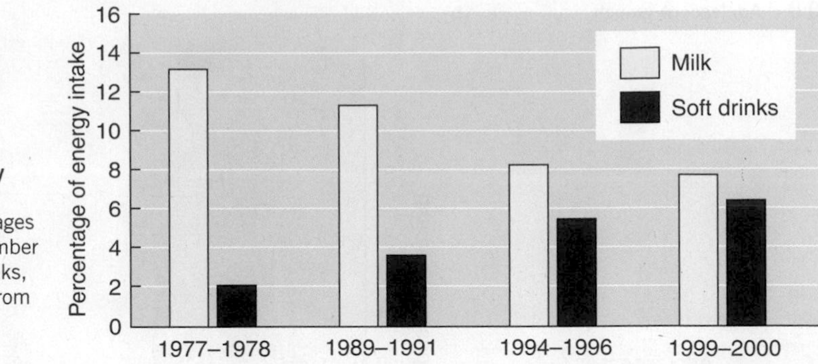

Figure 15.19 Percentage of energy from milk and soft drinks[51]
Since the 1970s, children and teens (ages 2–18) have more than doubled the number of calories they consume from soft drinks, while cutting the calories they obtain from milk by 34%.

impede skeletal growth and bone mineralization and increase the risk of developing osteoporosis later in life.

Foods common in the teen diet that are good sources of calcium include milk, yogurt and frozen yogurt, ice cream, and cheese added to hamburgers, nachos, and pizza. Although milk and cheese are the biggest sources of calcium in teen diets, they can be high in saturated fat, so adolescents should be encouraged to consume reduced-fat dairy products and vegetable sources of calcium. One factor that has contributed to low calcium intake among teens is the use of soda as a beverage, rather than milk (**Figure 15.19**). In addition to calcium, milk is an important source of vitamin D, phosphorus, magnesium, potassium, protein, riboflavin, vitamin A, and zinc. Adolescent girls are likely to drink less milk in order to cut calories, favoring low-calorie soft drinks (see **Debate: Should the Sale of Soda Be Banned from Schools?**).

Iron Iron deficiency is common in adolescent females, affecting about 9% of girls ages 12 to 15 and 16% of young women ages 16 to 19.[52] The deficiency progresses to anemia in about 3% of adolescent girls.[53] This mineral is needed to synthesize hemoglobin for the expansion of blood volume and myoglobin for the increase in muscle mass. Because blood volume expands at a faster rate in boys than in girls, boys require more iron for tissue synthesis than girls. However, the iron loss due to menstruation makes total needs greater in young women. The RDA is set at 11 mg/day for boys (this is greater than the 8 mg RDA for adult men) and 15 mg/day for girls ages 14 to 18.[54] Girls are at a higher risk for iron deficiency because they require more iron and they tend to eat fewer iron-rich foods so are less likely to consume the recommended amount.

Zinc During adolescence, the increase in protein synthesis required for the growth of skeletal muscle and the development of organs increases the need for zinc. The RDA is 11 mg/day for boys and 9 mg/day for girls ages 14 to 18. A long-term deficiency results in growth retardation and altered sexual development. Although severe zinc deficiency is rare in developed countries, even mild deficiency can cause poor growth, affect appetite and taste, impair immune response, and interfere with vitamin A metabolism. Since adolescents are growing rapidly and maturing sexually, adequate zinc is essential for this age group. Good sources of zinc include meats and whole grains. The fortification of many breakfast cereals with zinc has increased the average zinc intake in the United States.[55]

Meeting Adolescent Nutrient Needs

During adolescence, physiological changes dictate nutritional needs but peer pressure may dictate food choices. Parents often have little control over what adolescents eat, and skipped meals and meals away from home are common. A food is more likely to be selected because it tastes good, it is easy to grab on the go, or friends are eating it than because it is healthy.

Filling MyPlate No matter when foods are consumed during the day, an adolescent's diet should follow the recommendations of MyPlate for the appropriate age, gender, and activity level. For example, an 18-year-old boy who exercises more than an hour a day would need a diet containing about 3200 kcals a day that includes 10 oz of grains. The diet

DEBATE

Should the Sale of Soda Be Banned from Schools?

Obesity is a complex and growing crisis among children in the United States. One factor believed to contribute to childhood obesity is consumption of sugar-sweetened beverages, especially soda.[1] While both the American Academy of Pediatrics and the 2010 Dietary Guidelines recommend that children limit consumption of sugar-sweetened beverages, soft-drink vending machines line the halls in many American schools. Should the sale of soda in our schools be banned?

Supporters of a soda ban stress that eliminating soda at school is an important step toward meeting nutrition recommendations and improving children's health. They point out that these beverages add empty calories and caffeine to the diet and are linked to obesity and a variety of other health problems. There is evidence that children who drink more sugar-sweetened soft drinks have greater body weights than those who drink less and that limiting soda consumption contributes to a modest amount of weight reduction.[2,3] Soda consumption increases the risk of osteoporosis by displacing milk and thus reducing calcium intake. It contributes to dental caries by providing a supply of sugar to bacteria in the mouth.[1,4]

No one will argue that fact that soda is bad for America's youth, but many contend that removing soda from schools will do little to decrease consumption, promote healthy diets, or reduce obesity. They believe that it is too small an intervention to have an impact and that simply eliminating soda does not make the diet healthy. Limiting the sale of soda in the schools appears to reduce in-school access and purchase of sugar-sweetened beverages, but does not significantly reduce overall consumption.[5] It has been suggested that being deprived of soda in the school may even encourage children to drink more when they leave school.

Many feel it is the schools' responsibility to eliminate this unhealthy choice from children's environment.[4,6] Schools should not be endorsing the sale of beverages that public health officials recommend we limit. Others argue that banning soda infringes on the individual's right to make choices. A ban implies that parents and children cannot make responsible choices and some fear it will lead to bans on other products and foods. They point out that removing soda does nothing to educate children about healthy diets and lifestyle, and some believe it actually undermines children's ability to learn how to make healthy decisions for themselves.

Paul J. Richards/NewsCom

Another consideration is that the sale of these beverages generates revenue for schools. This money is used for everything from textbook purchases to funding extracurricular activities such as sports programs, field trips, and proms. Some argue that eliminating this source of revenue would result in cuts in sports programs and actually contribute to obesity by reducing physical activity. Supporters of the ban say that revenue should not be generated at the expense of children's health. They point out that schools would never consider selling cigarettes to raise money, so why would they promote soda sales?

Everyone will agree that banning soda will not solve the obesity epidemic, but that schools can be part of the solution.[7] Schools can have a positive impact on students' diets by providing nutrition education and improving the quality of school lunches.[8] Children need to have access to healthy choices and understand what foods to choose. Changing what children eat at school can be important in preventing obesity, but it is still unclear whether allowing the sale of soda in vending machines undermines this goal.

THINK CRITICALLY: Do you think banning soda but allowing the sale of 100% juice and sports drinks is an appropriate compromise? Why or why not?

---

[1]U.S. Department of Health and Human Services, U.S. Department of Agriculture. *Dietary Guidelines for Americans, 2010*, 7th ed. Washington, DC, U.S. Government Printing Office, December, 2010.

[2]Malik, V. S., Schulze, M. B., and Hu, F. B. Intake of sugar-sweetened beverages and weight gain: A systematic review. *Am J Clin Nutr* 84:274–288, 2006.

[3]Coffey, S. Pepsi: Removing its sodas from schools worldwide.

[4]American Academy of Pediatrics, Expert Committee Recommendations Regarding the Prevention, Assessment, and Treatment of Child and Adolescent Overweight and Obesity: Summary Report. *Pediatrics* 120:S164–S192, 2007.

[5]Taber, D. R., Chriqui, J. F., Powell, L. M., et al. Banning all sugar-sweetened beverages in middle schools. *Arch Pediatr Adolesc Med* 166:256–262, 2012.

[6]American Dietetic Association. Position of the American Dietetic Association: Local support for nutrition integrity in schools. *J Am Diet Assoc* 110:1244–1245, 2010.

[7]American Dietetic Association. Joint Position of ADA, Society for Nutrition Education and School Nutrition Association: Comprehensive school nutrition services. *J Am Diet Assoc* 110:1738–1749, 2010.

[8]American Heart Association. Understanding childhood obesity.

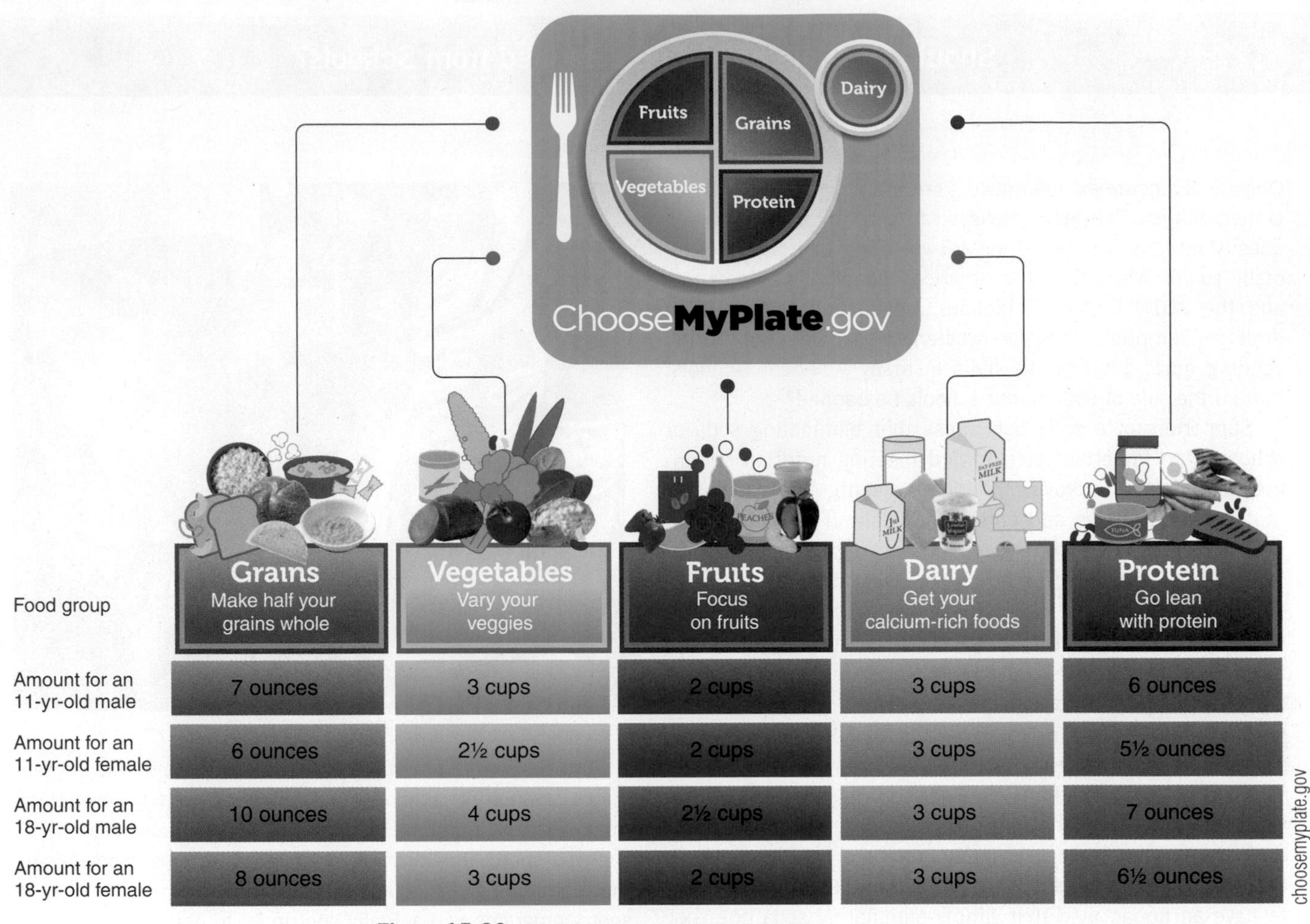

Food group	Grains Make half your grains whole	Vegetables Vary your veggies	Fruits Focus on fruits	Dairy Get your calcium-rich foods	Protein Go lean with protein
Amount for an 11-yr-old male	7 ounces	3 cups	2 cups	3 cups	6 ounces
Amount for an 11-yr-old female	6 ounces	2½ cups	2 cups	3 cups	5½ ounces
Amount for an 18-yr-old male	10 ounces	4 cups	2½ cups	3 cups	7 ounces
Amount for an 18-yr-old female	8 ounces	3 cups	2 cups	3 cups	6½ ounces

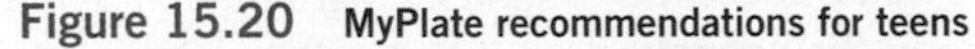

Figure 15.20 MyPlate recommendations for teens
The recommendations shown here are for 11- and 18-year-old males and females who engage in more than 60 minutes of activity daily. The 10 oz equivalents of grains recommended for an active 18-year-old boy may seem like a huge amount. But when spread over the course of a day (in the form of, perhaps, a large bowl of cereal and toast for breakfast, two tacos for lunch, and a dinner that includes spaghetti and garlic bread), it is an amount that a teenage male can easily consume.

should also provide 4 cups of vegetables and 2½ cups of fruit. Fruits and vegetables are the food groups most likely to be lacking in the teen diet. An analysis of fruit and vegetables consumption among teens showed that 29% of high school students consumed fruit less than once daily, and 33% consumed vegetables less than once daily (**Figure 15.20**).[56]

Healthy Breakfasts and Snacks The nutrients and energy provided by a healthy breakfast affect nutritional status and school performance, yet children and teens are more likely to skip breakfast than to skip any other meal.[57] Skipping breakfast may result in a span of 15 or more hours without food. Because breakfast provides energy and nutrients to the brain, children who skip it are more likely to have academic, emotional, and behavioral problems than those who eat breakfast.[58] Studies have found that, compared with nonbreakfast eaters, children who participate in school breakfast programs have better nutrient intakes, and that these improvements in nutrient intakes are associated with improvements in academic performance, reductions in hyperactivity, better psychosocial behaviors, and less absence and tardiness.[59]

Snacks are also an important part of the adolescent diet; they provide about a quarter of the calories for a typical teen. Unfortunately, many of these snacks are burgers, fries, potato chips, candy, cookies, and other high-fat, high-sodium, high-sugar foods. To balance these, the meals and snacks offered to adolescents at home should be high in dairy products, vegetables, and fruits. Sources of fruits and vegetables acceptable to teens include fruit juice, salads, and tomato sauce and vegetables on pizza and spaghetti.

15.5 Special Concerns of Teenagers

LEARNING OBJECTIVES

- Explain how a vegetarian diet could be high in saturated fat.
- Discuss how teens' concerns for appearance and performance can affect their nutritional status.
- Describe why peer pressure affects nutrition in teenagers.

As teens grow and become more independent, they are responsible for decisions about their diet and lifestyle. Choices they make regarding the foods they eat, their appearance, and their physical activities as well as the use of tobacco, alcohol, and illegal drugs can affect their nutrition as well as their overall health.

How Vegetarian Diets Affect Nutritional Health

Some children and adolescents consume vegetarian diets because their families are vegetarians, but teens may also decide to consume a vegetarian diet even if the rest of the family does not. Some do it for health reasons or to lose weight, but most give up meat because they are concerned about animals and the environment. About 3% of children and teens ages 6 to 17, or about a million school-aged youth, are vegetarian.[60]

We typically think of vegetarian diets as healthful alternatives and for many teens, they are. However, as with any diet, vegetarian foods must be carefully chosen to meet nutrient needs and avoid excesses.[15] Meatless diets can be low in iron and zinc, and vegan diets, which contain no animal products, may put teens at risk of vitamin B_{12} deficiency and inadequate calcium and vitamin D intake. If the diet includes high-fat dairy products, it can be high in saturated fat and cholesterol. If only refined grains are chosen and fruit and vegetable intake goals are not met, the diet can be low in fiber (**Figure 15.21**). The key to a healthy vegetarian diet, like all healthy diets, is to choose a variety of nutrient-dense vegetables, fruits, grains, legumes, and dairy products (or other calcium sources).

Bill Aron/PhotoEdit

Robyn Mackenzie/iStockphoto

Figure 15.21 Making healthy vegetarian choices Cheese pizza and ice cream are high-saturated fat, high-cholesterol, low-fiber vegetarian choices. In contrast, whole-grain pita bread stuffed with chickpeas, corn, spinach, and tomatoes, served with reduced-fat milk, is low in saturated fat and cholesterol, high in fiber, and a good source of calcium and iron.

Eating for Appearance and Performance

Appearance is probably of more concern during adolescence than at any other time of life. Many girls want to lose weight even if they are not overweight. Some boys also want to reduce their weight, but more want to gain weight to achieve a muscular, strong appearance and enhance their athletic abilities (see **Critical Thinking: Less Food May Not Mean Fewer Calories**).

Eating Disorders Although eating disorders are usually not diagnosed until adolescence, the excessive concern about weight and body image that characterizes these conditions may begin as early as the preschool years. As children grow, the pressure of taking on the responsibilities of adulthood, combined with pressure from peers and society to be thin, may contribute to the development of eating disorders. It is estimated that 0.5% of adolescent females have anorexia nervosa and 1 to 2% have bulimia nervosa.[61] Disordered eating is often hidden by other eating patterns. For example, many women choose vegetarian diets for weight control and vegetarian college women have been found to be at greater risk of disordered eating than nonvegetarians.[62]

CRITICAL THINKING

Less Food May Not Mean Fewer Calories

© Fuat Kose/iStockphoto

Jenny is a busy 16-year-old high school junior who has never paid much attention to her diet. Now that she has a part-time job, she eats meals on the run. Despite eating less often, she is gaining weight on her new diet. She is also feeling run down, which is a concern because she has been anemic in the past.

JENNY'S ORIGINAL DIET			JENNY'S NEW DIET		
FOOD	**SERVING**	**ENERGY (kcals)**	**FOOD**	**SERVING**	**ENERY (kcals)**
Breakfast			**Breakfast**		
Cornflakes	1 cup	97	Bagel	1 (4" diameter)	225
Reduced-fat milk	1 cup	120			
Orange juice	1 cup	110			
Wheat toast	2 slices	140			
Margarine	2 tsp	68			
Lunch			**Lunch**		
Hamburger	3 oz patty 1 bun 1 tsp ketchup 2 pickles slices	360	Potato chips	2.5 oz	400
Reduced-fat milk	1 cup	120	Diet coke	20 oz	0
Apple	1	81			
Corn chips	1 oz	153			
Snack			**Snack**		
Cheese pizza	1 slice	200	Frozen yogurt	1 cup	288
Cola	16 oz	185	Hershey bar	1.5 oz	210
Dinner			**Dinner**		
Skinless chicken breast	3 oz	110	Wendy's cheeseburger	1	580
Broccoli	½ cup	25	Fries	Medium	420
Brown rice	½ cup cooked	115	Vanilla shake	Medium	400
Reduced-fat milk	1 cup	120			
Fruit salad	1 cup	120			
TOTALS		**1904**			**2523**

CRITICAL THINKING QUESTIONS

- Assuming Jenny needs 2000 kcals to maintain her weight, how much weight will she gain in a month if she continues with her new diet? (one lb = 3500 kcals)
- How much iron does Jenny's new meal pattern provide? Is she meeting the RDA for a 16-year-old female?
- Compare the calcium content of Jenny's original diet with her new diet.
- Suggest modifications for each of the meals and snacks in Jenny's new diet to reduce her calories and increase her iron intake. Don't forget to consider the fact she is eating lunch at school and needs snack and dinner options she can eat in the car or purchase at the food court where she works.

iProfile Use iProfile to find snack choices that are nutrient-dense and filling, but still low in calories.

The nutritional consequences of an eating disorder can affect growth and development and have a lifelong impact on health (see **Focus on Eating Disorders**).

How Athletics Affect Nutritional Health Despite all the benefits of exercise, nutrition misinformation and the desire to excel in a sport can cause adolescent athletes to take dietary supplements, use anabolic steroids, or consume inappropriate training diets and experiment with fad diets, all of which can impact health (see Chapter 13).

Dennis MacDonald/Age Fotostock America, Inc.

Figure 15.22 Bulking up
High school football players often use dietary supplements and illegal ergogenic aids to gain muscle mass and strength.

Teen athletes may require more water, energy, protein, carbohydrate, and micronutrients than their less active peers, but supplements are rarely needed to meet these needs. If the extra energy needs of teen athletes are met with whole grains, fresh fruits and vegetables, reduced-fat dairy products, and lean meats, their protein, carbohydrate, and micronutrient needs will easily be met. An exception is iron, which may need to be supplemented, particularly in female athletes. The combination of poor iron intake, iron losses from menstruation and sweat, and increased needs for building new lean tissue puts many female athletes at risk for iron-deficiency anemia.[63]

Many of the most dangerous practices associated with adolescent sports are those that attempt to control body weight. Some sports such as football demand that the athlete be large and muscular (**Figure 15.22**). In order to "bulk up," high school athletes may experiment with anabolic steroids, androstenedione (andro), and creatine. Anabolic steroids are illegal, and although they do increase muscle mass, the risks far outweigh the benefits. Androstenedione is a testosterone precursor that is also a banned substance. The use of andro and other steroid precursors has not been found to increase testosterone levels or produce any ergogenic effects, and they may cause some of the same side effects as anabolic steroids.[64] Creatine is sold as a dietary supplement. It improves exercise performance in sports requiring short bursts of activity and has not been associated with serious side effects.[65] Nonetheless, the best and safest way for young athletes to increase muscle mass is the hard way: lift weights and eat more. Lifting weights three times a week will stimulate the muscles to enlarge and adding snacks such as milkshakes and peanut-butter sandwiches will provide the energy and protein needed to support muscle growth (see Chapter 13).

Female athletes involved in sports that require lean, light bodies, such as gymnastics and ballet, are likely to abuse weight-loss diets. Sexual maturation, which causes an increase in body fat and changes in weight distribution, can be disturbing to young women involved in such sports. The combination of hard training and weight restriction can lead to a syndrome known as the female athlete triad that includes disordered eating, amenorrhea, and osteoporosis[66] (see **Focus on Eating Disorders** and Chapter 13).

Weight loss is also a concern for adolescents participating in sports such as wrestling that require athletes to fit into a specific weight class on the day of the event. In these athletes, dangerous methods of quick weight loss—such as severe energy intake restriction, water deprivation, self-induced vomiting, and diuretic and laxative abuse—are common practice. Low-energy diets can interfere with normal growth and may be too limited in variety to meet these athletes' needs for vitamins and minerals. Even more of a danger is the practice of restricting water intake and encouraging sweat loss to decrease body weight. This may achieve the temporary weight loss necessary to put the athlete in a lower weight class, but dehydration is dangerous and can impair athletic performance (see Chapters 10 and 13).

Oral Contraceptive Use

Oral contraceptive hormones may be prescribed to adolescent girls for a number of reasons and can change nutritional status because they affect nutrient metabolism. Oral contraceptives may cause a rise in fasting blood sugar and a tendency toward abnormal glucose tolerance in those with a family history of diabetes. They may also cause changes in body composition, including weight gain due to water retention and an increase in lean body mass. Oral contraceptives may reduce the need for iron by reducing menstrual flow and increasing iron absorption. Therefore, a special RDA for iron of 11.4 mg/day has been established for those taking oral contraceptives.[54]

Teenage Pregnancy

Because adolescent girls continue to grow and mature for several years after menstruation starts, the diet of a pregnant teenager must meet her own nutrient needs for growth and development as well as the needs of pregnancy. These elevated needs put the pregnant adolescent at nutritional risk. In order for the mother and fetus to remain healthy, special attention must be paid to all aspects of prenatal care, including nutrient intake (see Chapter 14). Due to the special nutrient needs of this group, the DRIs have included a life-stage group for pregnant girls age 18 or younger.

Tobacco Use

Approximately 20% of high school students in the United States smoke cigarettes.[67] Smoking affects hunger, body weight, and nutritional status, and increases the risk of cardiovascular disease and lung cancer. Many teens start smoking in order to promote weight loss or maintenance. Because smoking is associated with lower body weights in adult women, some teens believe smoking will curb their appetite and help them stay thin or lose weight.[68] Smokers often do not want to quit because they fear that they will gain weight. Smoking impacts nutrient intake. A comparison of the diets of smokers and nonsmokers found that smokers consumed fewer fruits and vegetables, leading to lower intakes of folate, vitamin C, and fiber.[69] This dietary pattern can affect nutritional status and the risk of developing chronic disease. In addition to being associated with lower intakes of vitamin C, the increase in oxidative stress caused by smoking increases the requirement for vitamin C; thus the DRIs recommend that smokers consume an extra 35 mg/day.[70]

Alcohol Use

Although it is illegal to sell alcohol to adolescents, alcoholic beverages are commonly available at teen social gatherings, and the peer pressure to consume them is strong. Surveys of American youth suggest that almost 45% of students in 9th through 12th grades have had at least one drink during the previous month; of these, 26% have engaged in **binge drinking**.[71] Binge drinking is a problem on college campuses that causes about 50 deaths and hundreds of cases of alcohol poisoning annually. It is estimated that about 40% of college students "binge" on alcohol at least once during a two-week period.[72] Alcohol is a drug that has short-term effects that occur soon after ingestion and long-term health consequences that are associated with overuse. It provides 7 kcals/g but no nutrients. The alcohol in the diet displaces foods that are nutritious. Once ingested, alcohol alters nutrient absorption and metabolism. The metabolism of alcohol as well as its impact on nutritional and overall health are discussed in **Focus on Alcohol**.

binge drinking The consumption of five or more drinks in a row for males or four or more for females.

OUTCOME

LWA/The Image Bank/Getty Images, Inc.

After a few months of consuming healthy, low-calorie meals and exercising regularly, Felicia had lost two pounds. More important, she began to feel better about herself and was even more motivated to change her old habits to improve her health and appearance. Her fitness improved enough for her to ride her bike to school. She began helping her mother find healthy recipes and learned to make lower-calorie choices when she ate out or snacked with friends. For example, instead of ice cream she chose sorbet or low-fat frozen yogurt, and she passed on burgers and fries in favor of a grilled chicken sandwich and a salad. When she was hungry, she reached for fruit or raw vegetables. Felicia began taking a multivitamin and mineral supplement to make sure she got enough iron, and she drank fat-free milk with her meals for a low-calorie source of calcium. She still ate lunch in the school cafeteria, but learned to make better choices from the foods offered.

Felicia, now 15 years old, has grown 3 inches and gained only 5 pounds over the last two years. Her BMI is at about the 90th percentile. Her self-confidence has increased so much that she is considering trying out for the field hockey team next year. **If she grows another 2 inches and gains weight slowly, by the time she is 17 she will have a BMI in the healthy range.**

Blend Images/SUPERSTOCK

APPLICATIONS

ASSESSING YOUR DIET

1. **What nutrients are in your favorite fast-food meal?**
   - **a.** Use iProfile to look up the nutrient composition of your favorite fast-food meal.
   - **b.** Compare the percentage of calories from carbohydrate, fat, and protein in the meal with the percentages recommended for a 10-year-old boy. Assuming that he exercises for 30 to 60 minutes a day, would he be able to eat this meal and still keep his day's intake within the recommended percentages? Why or why not?
   - **c.** Compare the amount of iron, calcium, vitamin C, and vitamin A in this meal to the recommended amounts of each for an 18-year-old male. If he eats this meal for lunch, what can he eat at his at other meals to ensure he meets the RDA for these nutrients?
2. **How does a fast-food lunch fit on MyPlate?**
   - **a.** How much food from each MyPlate food group is provided by a Big Mac, a small order of fries, and a 16-oz cola?
   - **b.** Estimate how many empty calories this meal provides.
   - **c.** Go to ChooseMyPlate.gov to see how much food from each food group is recommended for you and how many empty calories you can allow each day.
   - **d.** If you consumed this fast-food lunch, how much more food from each group would you need to consume to satisfy the daily recommendations of MyPlate?
   - **e.** Select foods from each group that are low in empty calories to complete your intake for the day.
   - **f.** Answer parts c, d, and e again, assuming you are 8 years old.

CONSUMER ISSUES

3. **Use MyPlate to design a day's menu for a 15-year-old boy who spends two hours a day playing soccer.**
4. **What recommendations would you make for each of the following young athletes?**
   - **a.** Frank is a wrestler. His usual weight is just at the low end of a weight class and he wants to lose enough weight to be in the next lower class.
   - **b.** Sam is a football player. He has been getting hit hard in the last few games and wants to bulk up. He is looking into taking supplements to speed up the process.
   - **c.** Talia is a dancer who has recently gained a few pounds. Her coach notices that she no longer drinks or snacks during practice.

CLINICAL CONCERNS

5. **Is this girl growing normally?**

Age	Height (in.)	Weight (lb)
6	45	44
7	48	53
8	50	77
9	52	97

   - **a.** Use the height and weight measurements recorded in the table above for a girl from age 6 to age 9 to calculate her BMI at each age, and plot these values on the appropriate BMI-for-age growth chart in Appendix B.
   - **b.** What recommendations do you have about this girl's weight?
6. **Jared is 8 years old. He refuses to eat breakfast before leaving for school. What can Jared's parents do to ensure that he has a nutritious meal before school starts?**

SUMMARY

15.1 Starting Right for a Healthy Life

- The current diet of American children is low in fruits and vegetables and whole grains and high in processed foods that are excessive in fat, salt, and sugar.
- The poor diet and low activity level of children contributes to obesity, diabetes, elevated blood pressure, and unhealthy blood lipid levels.
- Healthy eating habits learned in childhood set the stage for nutrition and health in the adult years.

15.2 Nourishing Infants, Toddlers, and Young Children

- Growth that follows standard patterns is the best indicator of adequate nutrition. BMI-for-age growth charts can be used to evaluate whether a child is normal weight, underweight, overweight, or obese.
- Energy and protein needs per kg of body weight decrease as children grow, but total needs increase because of the increase in total body weight and activity level. Children need a greater percentage of fat and the same percentage of carbohydrates as adults. Dietary carbohydrates should come primarily from whole grains, vegetables, fruits, and milk. Baked goods, candy, and soda should be limited.
- Under most situations, water needs in children can be met by consuming enough fluid to alleviate thirst. Activity and a hot environment increase water needs. As with adults, the typical sodium intake in children and teens currently exceeds the recommended amounts.
- Calcium and iron intakes are often low in children's diets. Adequate calcium is important for preventing osteoporosis later in life but calcium intake is declining due to reductions in the consumption of dairy products. Excessive milk consumption can contribute to iron deficiency.
- Introducing solid foods between 4 and 6 months of age adds iron and other nutrients to the diet and aids in muscle development. Newly introduced foods should be appropriate to the child's stage of development and offered one at a time to monitor for food allergies.
- Food allergies are caused by the absorption of allergens, most of which are proteins. Food allergies involve antibody production by the immune system and are more common in infancy because the infant's immature gastrointestinal tract is more likely to absorb incompletely digested proteins. Specific foods that cause allergies can be identified by an elimination diet and food challenge. Unlike food allergies, food intolerances do not involve the immune system.
- In order for children to meet nutrient needs and develop nutritious habits, a variety of healthy foods should be offered at meals and snacks throughout the day. The National School Lunch Program provides low-cost school lunches designed to meet nutrient needs and promote healthy diets. Some children can benefit from vitamin/mineral supplements, but children who consume a well-selected, varied diet can meet all their vitamin and mineral requirements with food.

15.3 Nutrition and Health Concerns in Children

- Dietary patterns high in sugars combined with poor dental hygiene put children at risk of dental caries. Sugar intake has been blamed for hyperactive behavior, but there is little evidence that it is the cause. Lack of sleep, overstimulation, the desire for more attention, lack of physical activity, and caffeine consumption may contribute to hyperactivity.
- Children's health is harmed by lead. Lead disrupts the activity of neurotransmitters and thus interferes with the functioning of the nervous system. Due to reductions in the use of lead, the number of children with high blood lead levels has declined dramatically.
- Obesity is a growing problem among children. Watching television contributes to childhood obesity by promoting high-calorie foods and by reducing the amount of exercise children get. Fast food can also add excess calories to a child's diet unless nutrient-dense choices are made throughout the rest of the day. Solutions to the problem of childhood obesity involve action from government, industry, health-care providers, communities, schools, and parents and families. Individuals who are overweight need to change behaviors to decrease food intake and increase activity.

15.4 Adolescents

- During adolescence, body composition changes and the nutritional requirements of boys and girls diverge. Males gain more lean body tissue, while females have a proportionally greater increase in body fat.
- During adolescence, accelerated growth and sexual maturation have an impact on nutrient requirements. Total energy and protein requirements are higher than at any other time of life. Vitamin requirements increase to meet the needs of rapid growth. Vitamin D, calcium, iron, and zinc are likely to be low in the adolescent diet. Vitamin D and calcium intake is often low because teens drink soda instead of milk. Iron-deficiency anemia is common, especially in girls as they begin losing iron through menstruation.
- Adolescent nutrient needs can be met by following the recommendations of MyPlate. Since meals are frequently missed, healthy snacks should be included in the diet.

15.5 Special Concerns of Teenagers

- Teens often turn to vegetarian diets, sometimes for environmental reasons and sometimes for weight control. Poorly planned vegetarian diets can put teens at risk of iron, zinc, calcium, vitamin D, or vitamin B_{12} deficiency. Poor food choices can result in vegetarian diets that are high in saturated fat and cholesterol and low in fiber.
- Psychosocial changes occurring during the adolescent years make physical appearance of great concern. Eating disorders are more common in adolescence than at any other time. Adolescent athletes are susceptible to nutrition misinformation, and they may experiment with dangerous practices such as using anabolic steroids to increase muscle mass or fad diets and fluid restriction to lose weight.
- During the teen years, pregnancy, the use of oral contraceptives, and tobacco may affect nutritional status. Alcohol consumption and over-consumption often start in the teen years and can have negative nutritional and social consequences.

REVIEW QUESTIONS

1. How does nutrient intake during childhood affect health later in life?
2. What improvements should be made in the diets of American children?
3. How can parents and caregivers influence children's food choices?
4. What is the best way to determine if a child is eating enough?
5. What factors influence the maximum height a child will reach?
6. What nutritional problems can be signaled by sudden changes in weight patterns?
7. How do the recommendations for fat intake change as a child gets older?
8. When should solid and semisolid foods be introduced into an infant's diet?
9. How should new foods be introduced to monitor for the development of food allergies?
10. Why is anemia a problem in young children? In teenage girls?
11. Why are snacks an important part of children's diets?
12. Why are malnourished children at greater risk of lead toxicity than adults or well-nourished children?
13. How does watching TV impact the nutritional status of children?
14. How can fast foods be incorporated into a healthy diet?
15. How does the treatment of obesity in children differ from treatment in adults?
16. What is the adolescent growth spurt? How does it affect nutrient requirements?
17. Describe two physiological differences between males and females after puberty that affect their nutrient needs.
18. Explain why soda intake among teens may be contributing to osteoporosis.
19. Why are vegetarian diets not always healthier than diets that include meat?
20. Why is eating breakfast important for children and adolescents?
21. Why might participation in athletics contribute to the development of eating disorders in adolescents?

REFERENCES

1. U.S. Department of Agriculture, Center for Nutrition Policy and Promotion. The quality of children's diets in 2003–04 as measured by the Healthy Eating Index—2005. Nutrition Insight 43, April 2009.
2. American Dietetic Association. Position of the American Dietetic Association: Nutrition guidance for healthy children 2 to 11 years. *J Am Diet Assoc.* 108:1038–1047, 2008.
3. Orlet Fisher, J., Rolls, B. J., and Birch, L. L. Children's bite size and intake of an entree are greater with large portions than with age-appropriate or self-selected portions. *Am J Clin Nutr* 77:1164–1170, 2003.
4. Johnson, L., Mander, A. P., Jones, L. R., et al. Energy-dense, low-fiber, high-fat dietary pattern is associated with increased fatness in childhood. *Am J Clin Nutr* 87:846–854, 2008.
5. DHHS, 2008 Physical Activity Guidelines for Americans.
6. Ogden, C. L., Carroll, M. D., Kit, B. K., and Flegal, K. M., Prevalence of obesity and trends in body mass index among U.S. children and adolescents, 1999–2010. *JAMA* 307:483–490, 2012.
7. Ogden, C., and Carroll, M., NCHS Health E-Stat. *Prevalence of Obesity Among Children and Adolescents: United States, Trends 1963–1965 Through 2007–2008.*
8. National Diabetes Information Clearinghouse. *National Diabetes Statistics*, 2007.
9. Mayer-Davis, E. J. Type 2 diabetes in youth: Epidemiology and current research toward prevention and treatment. *J Am Diet Assoc* 108:S45–S51, 2008.
10. Center for Disease Control, National Health and Nutrition Examination Survey. Intake of Calories and Selected Nutrients for the United States Population, 1999–2000.
11. Daniels, S. R., Greer, F. R., and the Committee on Nutrition. Lipid screening for cardiovascular health in childhood. *Pediatr* 122:198–208, 2008.
12. American Heart Association. *Cholesterol and Atherosclerosis in Children.*
13. Academy of Pediatrics, Committee on Nutrition, Lipid Screening and Cardiovascular Health in Childhood.
14. Luma, G. B., and Spiotta, R. T. Hypertension in children and adolescents. *Am Fam Physician* 73:1558–1568, 2006.
15. U.S. Department of Agriculture and U.S. Department of Health and Human Services. *Dietary Guidelines for Americans*, 2010, 7th ed., Washington D.C: U.S. Government Printing Office, December, 2010.
16. U.S. Department of Health and Human Services, Centers for Disease Control and Prevention, National Center for Health Statistics. *CDC Growth Charts: United States.* Advance Data, No. 314, June 8, 2000 (revised).
17. Institute of Medicine, Food and Nutrition Board. *Dietary Reference Intakes for Energy, Carbohydrate, Fiber, Fat, Protein and Amino Acids.* Washington, DC: National Academies Press, 2002.
18. Institute of Medicine, Food and Nutrition Board. *Dietary Reference Intakes for Water, Potassium, Sodium, Chloride, and Sulfate.* Washington, DC: National Academies Press, 2004.

19. Food and Nutrition Board, Institute of Medicine. *Dietary Reference Intakes for Calcium and Vitamin D.* Washington, DC: National Academies Press, 2011.

20. Rovner, A. J., and O'Brien, K. O. Hypovitaminosis D among healthy children in the United States: A review of the current evidence. *Arch Pediatr Adolesc Med* 162:513–519, 2008.

21. Risk Factor Monitoring and Methods Branch, National Cancer Institute. *Sources of Beverage Intakes Among the US Population, 2005–06.*

22. Centers for Diseases Control and Prevention. *Nutrition for Everyone. Iron and Iron Deficiency.*

23. Haas, J. D., and Brownlie, T., IV. Iron deficiency and reduced work capacity: A critical review of the research to determine a causal relationship. *J. Nutr.* 131:676S–690S, 2001.

24. WHO, Iron Deficiency Anemia Assessment, Prevention and Control.

25. Food and Drug Administration. Iron-Containing Supplements and Drugs; Label Warning Statements and Unit-Dose Packaging Requirements; Removal of Regulations for Unit-Dose Packaging Requirements for Dietary Supplements and Drugs. *Fed Regist* 68:59714–59715, 2003.

26. Macknin, M. L., Medendorp, S. V., and Maier, M. C. Infant sleep and bedtime cereal. *Am J Dis Child* 143:1066–1068, 1989.

27. American Academy of Pediatrics Section on Breastfeeding (2005). Breastfeeding and the use of human milk. *Pediatrics* 115:496–506, 2005.

28. Huh, S. Y., Rifas-Shiman, S. L., Tavares, E. M., et al. Timing of solid food introduction and the risk of obesity in preschool-aged children. *Pediatrics* 127:544–551, 2011.

29. American Academy of Pediatrics. The use and misuse of fruit juice in pediatrics. *Pediatrics* 107:1210–1213, 2001.

30. Formanek, R. Food allergies. When food becomes the enemy. *FDA Consumer,* July/Aug, 35:40, 2001.

31. Greer, F. R., Sicherer, S. H., and Burks, A. W. Effects of early nutritional interventions on the development of atopic disease in infants and children: The role of maternal dietary restriction, breastfeeding, timing of introduction of complementary foods, and hydrolyzed formulas. *Pediatrics* 121:183–191, 2008.

32. Fisher, J. O., Liu, J., Birch, L. L. and Rolls, B. J. Effects of portion size and energy density on young children's intake at a meal. *Am J Clin Nutr* 86:174–179, 2007.

33. Slatter, E. *Child of Mine.* Palo Alto, CA: Bull Publishing, 2000.

34. U.S. Department of Agriculture, Food and Nutrition Service. *National School Lunch Program.*

35. Gleason, P., Briefel, R., Wilson, A. and Dodd, A. H. *School Meal Program Participation and Its Association with Dietary Patterns and Childhood Obesity.* Contractor and Cooperator Report No. 55, July 2009.

36. Marshall, T. A., Levy, S. M., Broffitt, B., et al. Dental caries and beverage consumption in young children. *Pediatrics* 112:e184–e191, 2003.

37. Cormier, E., and Elder, J. H. Diet and child behavior problems: Fact or fiction? *Pediatr Nurs* 33:138–143, 2007.

38. Rojas, N. L., and Chan, E. Old and new controversies in the alternative treatment of attention-deficit hyperactivity disorder. *Ment Retard Dev Disabil Res Rev* 11:116–130, 2005.

39. Farley, D. Dangers of lead still linger. *FDA Consum* 32:16–21, 1998.

40. Wengrovitz, A. M., and Brown, M. J. Recommendations for blood lead screening of Medicaid-eligible children aged 1–5 years: An updated approach to targeting a group at high risk. *MMWR Morb Mortal Wkly Rep* 58:1–11, 2009.

41. Centers for Disease Control and Prevention (CDC). Ten great public health achievements—United States, 2001–2010. *MMWR Morb Mortal Wkly Rep* 60:619–623, 2011.

42. Centers for Disease Control and Prevention (CDC). Blood lead level.

43. Coon, K. A., and Tucker, K. L. Television and children's consumption patterns. A review of the literature. *Minerva Pediatr.* 54:423–436, 2002.

44. Batada, A., Seitz, M., Woxan, M. et al. Nine out of ten food advertisements shown during Saturday morning children's television programming are for food high in fat, sodium, or added sugars, or low in nutrients. *J Am Diet Assoc* 108:673–678, 2008.

45. Eisenmann, J. C., Bartee, R. T., and Wang, M. Q. Physical activity, TV viewing, and weight in U.S. youth: 1999 Youth Risk Behavior Survey. *Obes Res* 10:379–385, 2002.

46. Institute of Medicine, Food and Nutrition Board.*Preventing Childhood Obesity: Health in the Balance.* Washington, DC: National Academies Press, 2005.

47. Centers for Disease Control and Prevention (CDC). *Making Physical Activity a Part of a Child's Life.*

48. Mandel, D., Zimlichman, E., Mimouni, F. B., et al. Age at menarche and body mass index: A population study. *J Pediatr Endocrinol Metab* 17:1507–1510, 2004.

49. Weng, F. L., Shults, J., Leonard, M. B., et al. Risk factors for low serum 25-hydroxyvitamin D concentrations in otherwise healthy children and adolescents. *Am J Clin Nutr* 86:150–158, 2007.

50. National Institutes of Health, National Institute of Child Health and Human Development. Milk Matters, *About Milk Matters.*

51. Nielsen, S. J., and Popkin, B. M. Changes in beverage intake between 1977 and 2001. *Am J Prev Med* 27:205–210, 2004.

52. Centers for Disease Control and Prevention. Iron deficiency—United States, 1999–2000. *MMWR Morb Mortal Wkly Rep* 51:897–899, 2002.

53. Killip, S., Bennett, J. M., and Chambers, M. D. Iron deficiency anemia. *Am Fam Phys* 75:671–678, 2007.

54. Institute of Medicine, Food and Nutrition Board. *Dietary Reference Intakes for Vitamin A, Vitamin K, Arsenic, Boron, Chromium, Copper, Iodine, Iron, Manganese, Molybdenum, Nickel, Silicon, Vanadium, and Zinc.* Washington, DC: National Academies Press, 2001.

55. Arsenault, J. E., and Brown, K. H. Zinc intake of US preschool children exceeds new dietary reference intakes. *Am J Clin Nutr* 78:1011–1017, 2003.

56. Centers for Disease Control and Prevention (CDC). Fruit and vegetable consumption among high school students–United States, 2010. *MMWR Morb Mortal Wkly Rep* 60: 1583–1586, 2011.

57. Rampersaud, G. C., Pereira, M. A., Girard, B. L., et al. Breakfast habits, nutritional status, body weight, and academic performance in children and adolescents. *J Am Diet Assoc* 105:743–760, 2005.

58. Alaimo, K., Olson, C., and Frongillo, E. Food insufficiency and American school-aged children's cognitive, academic, and psychosocial development. *Pediatrics* 108:44–53, 2001.

59. Kleinman, R. E., Hall, S., Green, H., et al. Diet, breakfast, and academic performance in children. *Ann Nutr Metab* 46(Suppl. 1):24–30, 2002.

60. The Vegetarian Resource Group. *How Many Teens Are Vegetarian? 2010 National Poll.*

61. Rosen, D.S.; American Academy of Pediatrics Committee on Adolescence. Identification and management of eating disorders in children and adolescents. *Pediatrics* 12:1240–1253, 2010.

62. Klopp, S. A., Heiss, C. J., and Smith, H. S. Self-reported vegetarianism may be a marker for college women at risk for disordered eating. *J Am Diet Assoc* 103:745–747, 2003.

63. Rodriguiez, N. R., Dimarco, N. M., and Langley, S. Position of the American Dietetic Association, Dietitians of Canada, and the American College of Sports Medicine: Nutrition and athletic performance. *J Am Diet Assoc* 109:509–527, 2009.

64. Brown, G. A., Vukovich, M., and King, D. S. Testosterone prohormone supplements. *Med Sci Sports Exerc* 38:1451–1461, 2006.

65. Tarnopolsky, M. A. Caffeine and creatine use in sport. *Ann Nutr Metab* 57(Suppl. 2):1–8, 2011.

66. Kazis, K., and Iglesias, E. The female athlete triad. *Adolesc Med* 14:87–95, 2003.

67. CDC *Trends in Current Cigarette Smoking Among High School Students and Adults, United States, 1965–2010.*

68. Facchini, M., Rozensztejn, R., and Gonzalez, C. Smoking and weight control behaviors. *Eat Weight Disord* 10:1–7, 2005.

69. McClure, J. B., Divine, G., Alexander, G., Tolsma, D., et al. A comparison of smokers' and nonsmokers' fruit and vegetable intake and relevant psychosocial factors. *Behav Med* 35:14–22, 2009.

70. Food and Nutrition Board, Institute of Medicine. *Dietary Reference Intakes for Vitamin C, Vitamin E, Selenium and Carotenoids.* Washington, DC: National Academies Press, 2000.

71. YRBSS, 2007 *National Youth Risk Behavior Survey Overview.*

72. National Council on Alcoholism and Drug Dependence (NCAAD). Youth, Alcohol and Other Drugs.

***To access links to online sources, please go to** www.wiley.com/college/smolin **and select Nutrition: Science and Applications, 3rd edition. From this page, select either the student or instructor companion site. Once on the desired site, select References.**

CHAPTER OUTLINE

Robert E Daemmrich/Stone/Getty Images, Inc.

Nutrition and Aging: The Adult Years

16

CASE STUDY

Cheng is 70 years old and has always been healthy. He and his wife exercise regularly, watch their weight, and eat a healthy diet. Recently Cheng experienced a few episodes of forgetfulness, but he laughed it off as old age. Then he began feeling tired and having tingling in his hands and feet, difficulty walking, and diarrhea. Fearing the worst, he made an appointment to see his doctor.

Hiroshi Yagi/Getty Images, Inc.

Laboratory tests showed that Cheng had low levels of vitamin B_{12}. A diet history revealed that his diet didn't provide much of the vitamin. He eats lots of grains, fruits, and vegetables, which do not provide vitamin B_{12}, and very little meat or dairy, which are sources of this vitamin. The doctor explained that Cheng's symptoms were due to a vitamin B_{12} deficiency. The deficiency was probably caused by a condition common in older adults that reduces the ability to absorb the vitamin B_{12} from food. Cheng's doctor gave him an injection of vitamin B_{12} and recommended that he start taking a daily supplement containing the vitamin.

Because of our plentiful food supply and the availability of vitamin supplements, Americans tend to think of vitamin deficiencies as a thing of the past, or perhaps as conditions that afflict people only in the developing world. For the most part that is true. **But, as Cheng's case illustrates, older adults are a segment of the U.S. population that is at risk of malnutrition due to physical, psychological, and social changes that may limit intakes, or medical conditions or medications that interfere with nutrient absorption or utilization.**

Figure 16.1 Aging
The five generations of women in this family illustrate that the process of aging occurs continuously in individuals of all ages.

16.1 What Is Aging?

LEARNING OBJECTIVES

- Define aging.
- Compare life expectancy with healthy life expectancy.
- Explain what is meant by compression of morbidity.

Biologically, **aging** is not something that begins at age 55, 65, or 75; it is a process that begins with conception and continues throughout life (**Figure 16.1**). Aging involves the inevitable accumulation of changes with time that are associated with and responsible for an ever-increasing susceptibility to disease and death. The maximum age to which any human can live—the **life span**—is about 100 to 120 years. Life span is a characteristic of a species; dogs can live about 20 years, gorillas 39 years, and mice only about three years.

SCIENCE APPLIED

Eat Less—Live Longer?

Throughout history, humans have searched for ways to stay young, avoid the ravages of old age, and extend life. Ponce de Leon came up empty in his search for the fountain of youth, but an Italian nobleman named Luigi Cornaro may have come closer. Born in 1464, Cornaro lived a life of excess until the age of 35, when a physician advised him to adopt a more frugal existence. His response was to "eat as little as possible," which he promoted in a series of books translated as "The Art of Living Long."[1] He lived to 102 at a time when average life expectancy was only about 30. Had he found the fountain of youth?

THE SCIENCE

Since the 1930s, research studies have described the life-extending effect of caloric restriction in rats and other organisms.[2,3] Consuming a calorie-restricted diet (one that provides 20 to 40% less than the recommended intake) that meets the need for all essential nutrients has been shown to extend the maximal life span of organisms as diverse as worms, insects, and rodents by as much as 50%.[3] It also reduces age-related chronic disease, improves immune function, increases resistance to numerous stressors and toxins, and maintains function later into life.

To explore the possibility that caloric restriction would lengthen life and reduce chronic disease in animals more closely related to humans, in the late 1980s researchers began to study the effect of this diet in rhesus monkeys. Adult monkeys were fed nutritionally adequate diets containing 30% fewer calories than a control group. These studies are still ongoing, but results suggest that caloric restriction does extend lifespan: the animals have lower body fat and less muscle loss, as well as a lower incidence of cancer, heart disease, and type 2 diabetes.[4] Studying caloric restriction in humans is fraught with problems, but the National Institute on Aging has taken on the challenge by sponsoring a randomized controlled trial that is currently underway.[5] Long-term data is not yet available, but caloric restriction for 6 to 12 months causes a reduction in body weight, increased insulin sensitivity, and improved indices of oxidative stress.[1] Currently, the only long-term evidence of the benefits of human caloric restriction comes from epidemiology. The people of Okinawa,

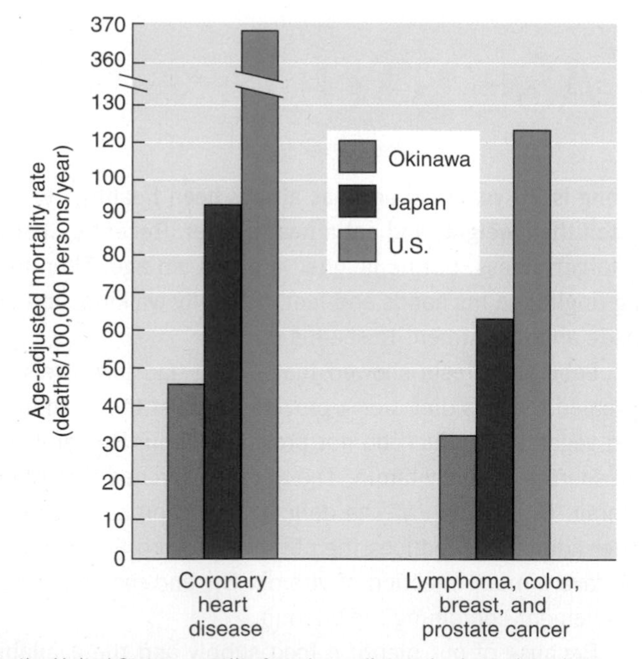

In the United States mortality from heart disease is about eight time higher, and from certain forms of cancer about four time higher, than it is in Okinawa.

Life span is believed to be genetically determined. The only experimental treatment that has been found to extend the life span of a variety of mammalian species is a nutritional manipulation called *caloric restriction*.[1] Restricting the energy intake of animals to about 70% of typical intake slows and/or delays the aging processes. Although the underlying biological mechanisms responsible for the life-span-extending effects of this diet are still under investigation, changes in four regulatory pathways are believed to play an important role.[2] Despite promising data from animal studies, caloric restriction is not currently considered a viable option for extending life in humans (see **Science Applied: Eat Less—Live Longer?**).

aging The process of becoming older. It is genetically determined and modulated by environmental and lifestyle factors.

life span The maximum age to which members of a species can live.

How Long Can Americans Expect to Live?

Although humans can live to be 120 years, most do not. **Life expectancy**, the average length of time that a person can be expected to live, varies between and within populations. It is affected by genetics, lifestyle, and environmental factors. In the United States

life expectancy The average length of life for a population of individuals.

(Elena Schweitzer/Shutterstock) Victoriano Izquierdo/Getty Images, Inc. (DAJ/Getty Images, Inc.)

a series of islands lying between mainland Japan and Taiwan, enjoy not only the longest life expectancy in the world (81.2 years) but also the longest healthy life expectancy.[6] Traditionally, Okinawans practice caloric restriction by eating until they are only 80% full. In addition, most lead active lives, working as farmers, and consume a nutrient-dense diet that is high in vegetables and fish and low in meat, refined grains, saturated fat, salt, and sugar.[7] They maintain a low BMI and have a lower incidence of chronic diseases than people living in the United States or on mainland Japan (see graph).[7]

THE APPLICATION

The health benefits of caloric restriction on experimental animals have led some people apply it to their own lives. This requires a diet providing 20 to 40% fewer calories than would typically be consumed, but containing all the necessary nutrients to support life. When individuals who follow calorie-restricted diets were studied, their health status was found to be consistent with the beneficial effects of caloric restriction seen in animals.[1] Unfortunately, this is a self-selected population, with no controls, making the results questionable. Even if caloric restriction in humans provides all the benefits it does in animals, following a calorie-restricted diet is not easy. Caloric restriction would mean that a person who typically eats about 2000 kcalories per day could eat only 1200 to 1600 kcalories per day. This would cause food cravings, weight loss, a lack of energy, and in some cases psychological consequences. To eat this little while meeting nutrient needs requires carefully planned meals and snacks; a poorly planned diet will lead to malnutrition.

While a group of dedicated supporters continue to restrict their intake in attempts to extend their lives, research on caloric restriction has discovered that the effects are mediated via a relatively small number of regulatory pathways. This has led to research into developing drugs that mimic the effects of caloric restriction. The goal is a drug that provides the positive health effects of caloric restriction without the negative side effects. A number of drugs have been developed, but thus far none have shown consistent benefits in animal testing.[1]

THINK CRITICALLY: If a promising anti-aging drug is developed in animals, what problems might arise in trying to test it in humans?

[1]Speakman, J. R., and Mitchell, S. E. Caloric restriction. *Mol Aspects Med* 32:159–221, 2011.

[2]McCay, C. M., Crowell, M. F., and Maynard, L. A. The effect of retarded growth upon the length of life and upon ultimate size. *J Nutr* 10, 63–79, 1935.

[3]Park, H. W. Longevity, aging, and caloric restriction: Clive Maine McCay and the construction of a multidisciplinary research program. *Hist Stud Nat Sci* 40:79–124, 2010.

[4]Kemnitz, J. W. Calorie restriction and aging in nonhuman primates. *ILAR J* 52:66–77, 2011.

[5]Rickman, A. D., Williamson, D. A., Martin, C. K., et al. The CALERIE Study: Design and methods of an innovative 25% caloric restriction intervention. *Contemp Clin Trials* 32:874–881, 2011.

[6]Willcox, B. J., Willcox, D. C., Todoriki, H., et al. Caloric restriction, the traditional Okinawan diet, and healthy aging: The diet of the world's longest-lived people and its potential impact on morbidity and life span. *Ann N Y Acad Sci* 1114:434–455, 2007.

[7]Willcox, D. C., Willcox, B. J., Todoriki, H., and Suzuki, M. The Okinawan diet: Health implications of a low-calorie, nutrient-dense, antioxidant-rich dietary pattern low in glycemic load. *J Am Coll Nutr* 28(Suppl.):500S–516S, 2009.

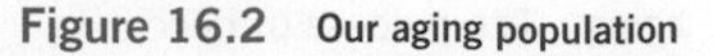
Figure 16.2 **Our aging population**

(a)

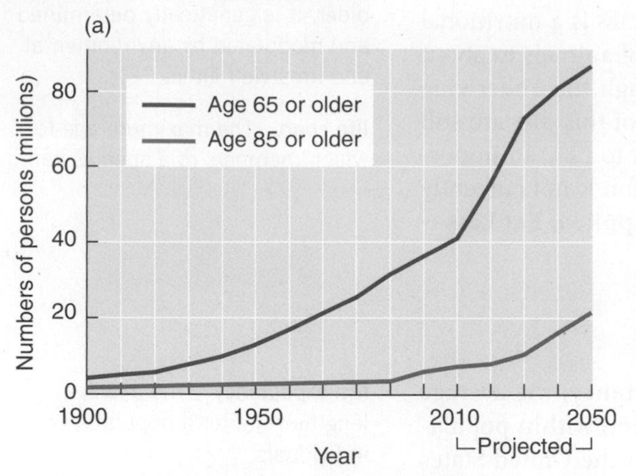

(a) Projections indicate that in the next few decades there will be almost 80 million people in the United States who are 65 or older.[5]

(b)

Purestock/Getty Images, Inc.

(b) Chronological age is not always the best indicator of a person's health. A person who is 75 may have the vigor and health of someone who is 55, or vice versa. Some older adults are healthy, independent, and active, while others are chronically ill, dependent, and at high risk for malnutrition.

in 1900, life expectancy was 50 years, but today, with advances in technology and improved nutrition and health care, the average is about 78.7 years.[3] With this increase in life expectancy, has come an increase in the number of older adults. During the 20th century, the population of older adults, those 65 years of age and older, grew from 3 million in 1900 to 35 million in 2000, or about 12% of the population (**Figure 16.2a**). Over the same time period, the oldest-old population, those over the age of 85 years, grew from just over 100,000 to 4.2 million. This increase in the number of older adults is expected to continue. By 2030, it is projected that nearly 20% of the total U.S. population, or 71.5 million Americans, will be older than 65 years of age.[4]

How Long Can Americans Expect to Be Healthy?

Even though average life expectancy in the United States has increased to over 78 years, the average healthy life expectancy is only about 69 years.[6] This means that, on average, the last 9 years of life are restricted by disease and disability. The goal of successful aging is to increase not only life expectancy but the number of years of healthy life that an individual can expect (**Figure 16.2b**). This is achieved by slowing the physiological changes that accumulate over time and postponing the diseases of aging long enough to approach or reach the limits of life span before any adverse symptoms appear. This is referred to as **compression of morbidity**. When applied to the population as a whole, this term means that people are healthier and living longer (**Figure 16.3**); applied to the individual, it means staying healthy through the later years of life.

Because the incidence of disease and disability increases with increasing age, the older population accounts for a large part of the public health budget. Thus, keeping older adults healthy is beneficial not only for the aging individuals themselves, but for the health-care system as well. Postponing the changes that occur with age is an important public health goal.

How will the increase in the elderly population affect health care costs?

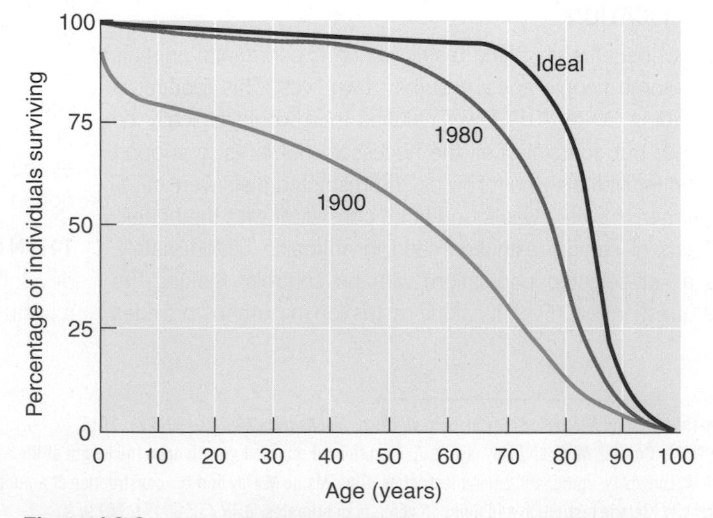

Figure 16.3 **Compression of morbidity**[7]
The decline in deaths from infectious diseases between 1900 and 1980 allowed more people to survive into adulthood. Delaying the onset of chronic diseases will allow more people to remain healthy and survive into their seventies, eighties, and nineties. This is illustrated by the curve labeled "ideal."

compression of morbidity The postponement of the onset of chronic disease so that disability occupies a smaller and smaller proportion of the life span.

16.2 What Causes Aging?

LEARNING OBJECTIVES

- Describe the biological causes of aging.
- Discuss genetic, environmental, and lifestyle factors that affect the aging process.

Although universal to all living things, aging is a process we don't fully understand. We do know that as organisms become older, the number of cells they contain decreases and the functioning of the remaining cells declines. As tissues and organs lose cells, the ability of the organism to perform the physiological functions necessary to maintain homeostasis decreases; disease becomes increasingly common and the risk of malnutrition increases. This loss of cells and cell function occurs throughout life, but the effects are not felt for many years because organisms begin life with extra functional capacity, or **reserve capacity**. Reserve capacity allows an organism to continue functioning normally despite a decrease in the number and function of cells. In young adults, the reserve capacity of organs is four to ten times that required to sustain life. As a person ages and reserve capacity decreases, the effects of aging become evident in all body systems.

reserve capacity The amount of functional capacity that an organ has above and beyond what is needed to sustain life.

There are two major hypotheses to explain why aging occurs. One favors the idea of a genetic clock and argues that genetically programmed cell death leads to aging. The other views the events of aging as the result of cellular wear and tear. The actual cause of the cell death associated with aging is probably some combination of both of these, and the rate at which cell death occurs and at which aging proceeds is determined by the interplay among our genetic makeup, our lifestyle, and the environment in which we spend our years.

Programmed Cell Death

Programmed cell death refers to the selective, orderly death of individual cells or groups of cells that is triggered when genes that disrupt cell function are activated. It allows changes in body structure and shape to occur during growth and development and helps maintain homeostasis at any age by getting rid of old or damaged cells. Both too much and too little programmed cell death are hypothesized to contribute to aging.[8] For example, too much cell death decreases the total number of cells, resulting in a loss of organ function and a decrease in life span. Too little cell death will allow damaged cells such as cancer cells to proliferate.

programmed cell death The death of cells at specific predictable times.

Wear and Tear

Aging is also hypothesized to be caused by an accumulation of cellular damage. This wear and tear may result from errors in DNA synthesis, increases in glucose levels, or damage caused by free radicals. Free radicals are reactive chemical substances that are generated from both normal metabolic processes and exposure to environmental factors (see **Chapter 8 and the Online Appendix: Focus on Metabolism**). They cause oxidative damage to proteins, lipids, carbohydrates, and DNA, and may also indirectly harm cells by producing toxic products. For example, age spots—brown spots that appear on the skin with age—are caused by the oxidation of lipids, which produces a pigment called lipofuscin, or age pigment. The damage done to cells by free radicals is associated with aging and has been implicated in the development of a number of chronic diseases common among older adults, including cardiovascular disease, and cancer.

How Genetics, Environment, and Lifestyle Interact

The rate at which the changes associated with aging accumulate depends on the genes people inherit, the environment they live in, and their lifestyle. Genes determine the efficiency with which cells are maintained and repaired. Individuals with less cellular repair capacity will lose cells more readily and consequently age more quickly. Likewise, genes determine

Figure 16.4 Factors that affect the rate of aging
The rate at which individuals age is affected by their genetic makeup, the environment in which they live, and the lifestyle choices they make.

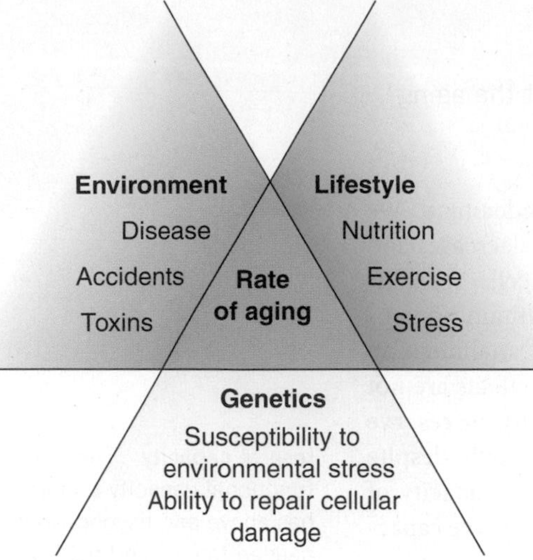

THINK CRITICALLY

Based on the information in this figure, what could someone do to slow his or her rate of aging?

susceptibility to age-related diseases such as cardiovascular disease and cancer. However, individuals who inherit a low capacity to repair cellular damage may still live long lives if they reside in an environment with few factors that damage cells and if they eat well and exercise regularly. In contrast, individuals with exceptional cellular repair ability may accumulate cellular damage rapidly and die young if they smoke cigarettes, consume a diet high in saturated fat and low in antioxidant nutrients, and live sedentary lives. No matter what individuals' genes predict about how long they will live, their actual **longevity** will also be affected by lifestyle factors and the extent to which they are able to avoid accidents and disease (**Figure 16.4**).

longevity The duration of an individual's life.

16.3 Aging and the Risk of Malnutrition

LEARNING OBJECTIVES

- Discuss how the physiological changes of aging can affect nutritional status.
- Explain how changes in body composition affect energy needs.
- Explain how the nutrient needs of older adults are affected by disease and medication use.
- Discuss social and economic factors that increase the risk of malnutrition.

The aging process usually does not cause malnutrition in healthy, active adults, but nutritional health can be compromised by the physical changes that occur with age, the presence of disease, and economic, psychological, and social circumstances (**Table 16.1**). These

TABLE 16.1 Factors That Increase the Risk of Malnutrition Among the Elderly

• **Reduced food intake** due to:
Decreased appetite caused by lack of exercise, depression, or social isolation
Changes in taste, smell, and vision
Dental problems
Limitations in mobility
Medications that restrict meal times or affect appetite
Lack of money to buy food
Lack of nutrition knowledge
• **Reduced nutrient absorption and utilization** due to:
Gastrointestinal changes
Medications that affect absorption
Diseases such as diabetes, kidney disease, alcoholism, and gastrointestinal disease
• **Increased nutrient requirements** due to:
Illness with fever or infection
Injury or surgery
• **Increased nutrient losses** due to:
Medications that increase excretion of nutrients
Diseases such as gastrointestinal and kidney disease

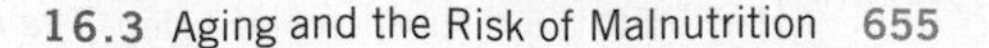

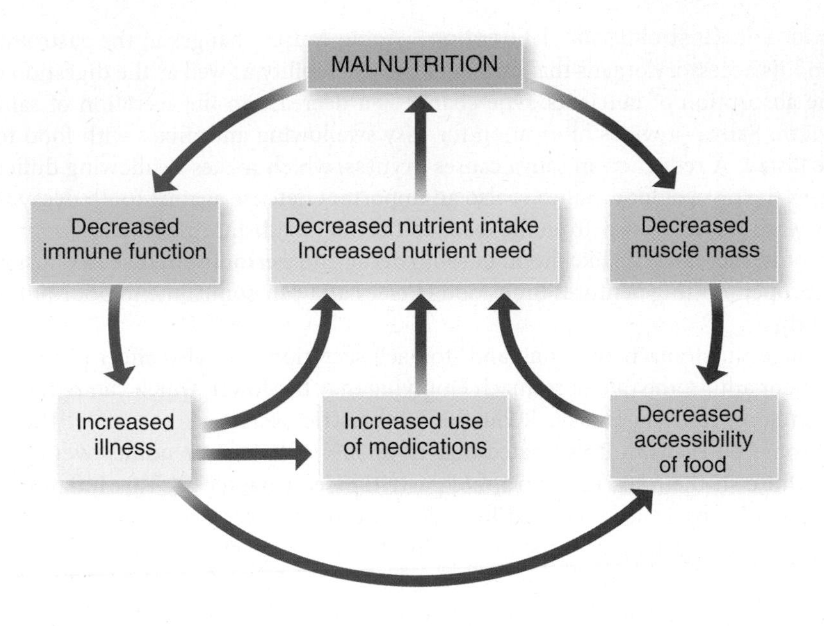

Figure 16.5 Causes and consequences of malnutrition
Many of the physiological changes and health problems associated with aging can affect nutritional status. This illustration shows that decreases in muscle mass and immune function and an increase in medication use contribute to malnutrition and that, in turn, malnutrition reduces immunity and muscle mass.

factors can increase the risk of malnutrition by altering nutrient needs and decreasing the motivation to eat and the ability to acquire and enjoy food. Malnutrition then exacerbates some of these factors, contributing to a downward health spiral from which it is difficult to recover (**Figure 16.5**).

How Physiological Changes Affect the Risk of Malnutrition

It is difficult to determine which of the changes that occur with aging are inevitable consequences of the aging process and which are the result of disease states. But whether caused by disease or normal aging, the changes that occur in organs and organ systems can affect nutritional status by altering the appeal of food, the ability to obtain food, and the digestion, absorption, metabolism, and excretion of nutrients.

Sensory Decline Beginning around age 60, there is a progressive decline in the ability to taste and smell; this becomes more severe in persons over 70. The deterioration of these senses can contribute to impaired nutritional status by decreasing the appeal and enjoyment of food.[9] Some studies suggest that the decrease in taste acuity is due to a reduction in the number of taste buds on the tongue; others suggest that it is the result of changes in sensitivity to specific flavors such as salty and sweet. The appeal of food is also affected by the decline in the sense of smell. Odors provide important clues to food acceptability before food enters the mouth and, once in the mouth, some molecules reach the nasal cavity where their odor is detected. It is the blending of the odor message from the nasal cavity and the taste message from the tongue that provides the overall food flavor. When the sense of smell is diminished, food is not as flavorful.

Vision also typically declines with age, making shopping for and preparation of food difficult. **Macular degeneration** is the most common cause of blindness in older Americans. The macula is a small area of the retina of the eye that distinguishes fine detail. With age, oxidative damage reduces the number of viable cells in the macula. As the macula degenerates, visual acuity declines, ultimately resulting in blindness. **Cataracts** are another common reason for declining sight (**Figure 16.6**). Of people who live to age 85, half will have cataracts that impair vision. Oxidative damage is believed to cause both macular degeneration and cataracts. Therefore, a diet high in foods containing antioxidant nutrients might slow or prevent these eye disorders.

macular degeneration Degeneration of a portion of the retina that results in a loss of visual detail and blindness.

cataract A disease of the eye that results in cloudy spots on the lens (and sometimes the cornea) that obscure vision.

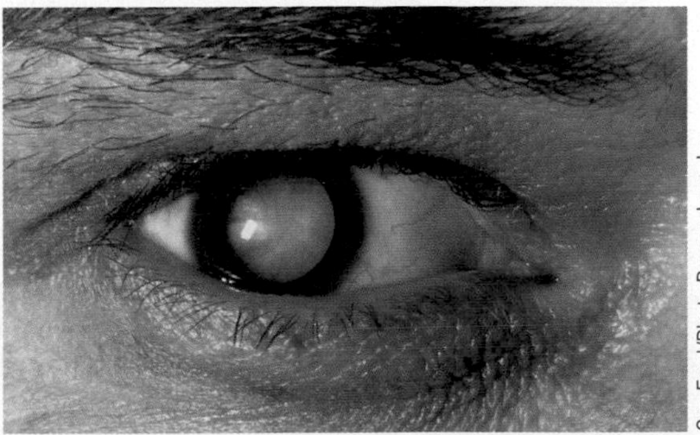

Sue Ford/Photo Researchers, Inc.

Figure 16.6 Cataracts
Cataracts cause the lens of the eye to become cloudy and impair vision. When cataracts obscure vision, the affected lens can be removed and replaced with an artificial plastic lens.

periodontal disease A degeneration of the area surrounding the teeth, specifically the gum and supporting bone.

atrophic gastritis An inflammation of the stomach lining that causes a reduction in stomach acid and allows bacterial overgrowth.

Alterations in Gastrointestinal Function Aging causes changes in the gastrointestinal tract and its accessory organs that may alter the palatability as well as the digestion of food and the absorption of nutrients. One change is a decrease in the secretion of saliva into the mouth. Saliva provides lubrication for easy swallowing and mixes with food to allow it to be tasted. A reduction in saliva causes dryness, which makes swallowing difficult and decreases the taste of food. Saliva is also an important defense against tooth decay because it helps wash material away from the teeth and contains substances that kill bacteria. Thus, a dry mouth increases the likelihood of tooth decay and **periodontal disease**. Loss of teeth and improperly fitting dentures limit food choices and can contribute to poor nutrition in the elderly.

Changes in stomach emptying and stomach secretions can also affect nutritional status. In older adults, the rate of stomach emptying may be slower, which can reduce hunger and, therefore, nutrient intake. Reductions in gastric secretions can affect the absorption of some nutrients. It is estimated that 10 to 30% of American adults over age 50 and 40% of those in their 80s have **atrophic gastritis** (see Chapter 3). This inflammation of the stomach lining is accompanied by a decrease in the secretion of stomach acid and, in severe cases, a reduction in the production of intrinsic factor (see Chapter 8).[10] When stomach acid is reduced, the enzymes that release vitamin B_{12} from food do not function properly and the vitamin B_{12} naturally present in food cannot be absorbed. Absorption of iron, folate, calcium, and vitamin K may also be reduced. Reduced stomach acid secretion also allows microbial overgrowth in the stomach and small intestine.[10] The increase in the number of gut microbes that use vitamin B_{12} further reduces the amount of the vitamin available for absorption into the body.

With age, there is a reduction in digestive enzymes from the pancreas and small intestine, but there is enough reserve capacity that digestion and absorption are rarely significantly impaired. In the colon there are functional changes, including decreased motility and elasticity, weakened abdominal and pelvic muscles, and decreased sensory perception, which can lead to constipation. Low fiber and fluid intakes and lack of activity also contribute to constipation. Constipation affects about 50% of elderly individuals living in the community and more than 75% of elderly patients in hospitals and nursing homes.[11] Maintaining regular exercise and consuming adequate fluid and fiber are safe ways to prevent constipation.

Changes in Other Organs Age-related changes in organs other than those of the gastrointestinal tract may also affect nutrient metabolism. Most absorbed nutrients travel from the intestine to the liver for metabolism or storage. The liver has a greater regenerative capacity than most organs, but with age, there is a decrease in liver size and blood flow and an increase in fat accumulation, which eventually decrease the liver's ability to metabolize nutrients and break down drugs and alcohol. With age, the pancreas may become less responsive to blood glucose levels, and the body cells may become more resistant to insulin, resulting in diabetes. Changes in the heart and blood vessels reduce blood flow to the kidneys, making waste removal less efficient. The kidneys themselves become smaller and their ability to filter blood and to excrete the products of protein breakdown declines.[12] In some individuals, blood urea levels may increase if protein intake is too high. The ability of the kidneys to concentrate urine also decreases with age, as does the sensation of thirst, increasing the risk of dehydration.

Changes in Body Weight and Composition Although obesity is a problem among older adults, after the age of 60 the incidence of obesity decreases. In those over 80 years of age, obesity is only half as common as in 50- to 59-year-olds.[13] In older adults, loss of muscle and bone mass, extreme thinness, and unintentional weight loss are important health problems that increase the risk of malnutrition. Numerous studies have demonstrated that in older adults low BMI is associated with a higher risk of mortality.[14]

Obesity The incidence of obesity in older age groups, as in younger adults, has increased over the past 25 years.[13] Although the risk of death associated with obesity is lower in older than in younger adults, obesity still increases the risk of health complications that reduce the quality of life. It contributes to a higher risk of cardiovascular disease, certain cancers,

THINK CRITICALLY

By what percentage does muscle mass decrease between age 20 and age 70?

hypertension, stroke, sleep apnea, type 2 diabetes, and arthritis. Obesity also contributes to suboptimal physical functioning. As with younger adults, moderate weight loss can decrease obesity-related complications in older adults and, if combined with physical activity, improve physical function and quality of life. The approach to weight loss in the older population must place more emphasis on preventing the loss of muscle and bone mass that occurs with age because it can be further accelerated with weight loss. Including exercise as part of a weight-loss regimen can improve strength, endurance, and overall well-being.

Figure 16.7 Changes in the proportion of muscle and fat with age
(a) With age, the proportion of muscle mass decreases and that of body fat increases.[17]

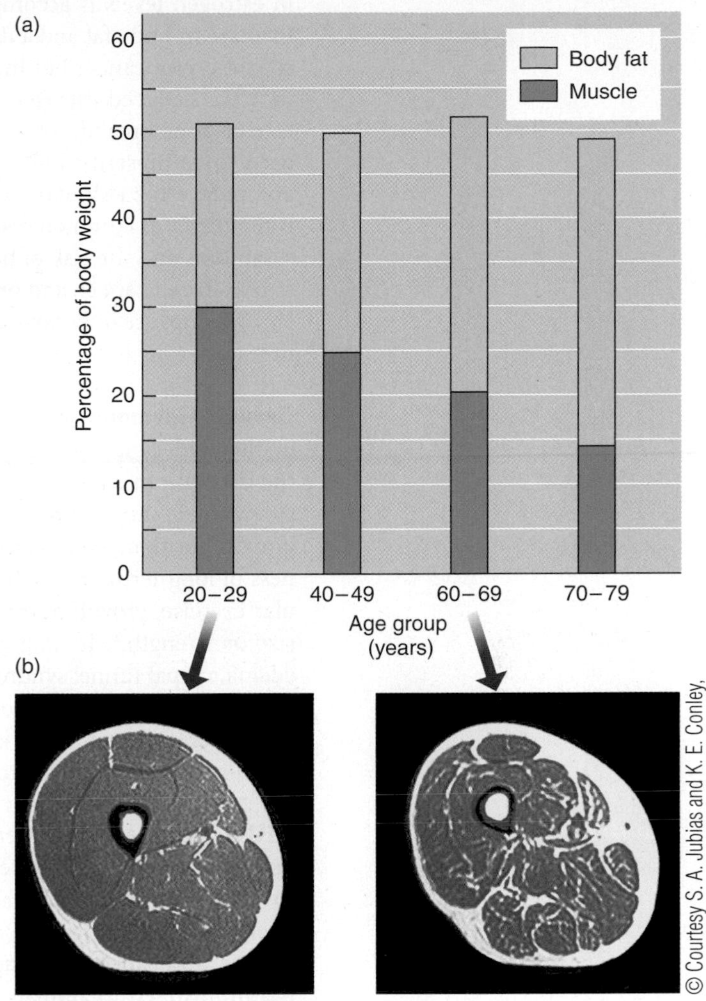

(b) These magnetic resonance images of a thigh cross-section from a 25-year-old man (left) and a 65-year-old man (right) show, that although the thighs are of similar size, the thigh of the older man has more fat (shown in white) around and through the muscle, indicating significant muscle loss.

Changes in Body Composition Even when body weight is in a healthy range, the changes in body composition that occur with age can affect nutritional and overall health. With age, there is an increase in body fat, especially in the abdomen, and a decrease in lean tissue, including a loss of muscle mass and strength, referred to as **sarcopenia** (**Figure 16.7**).[15,16] In older adults, the wasting of lean tissue can result in a high percentage of body fat even when body weight is low or stable.

The decline in muscle size and strength affects both the skeletal muscles needed to move the body and the heart and respiratory muscles needed to deliver oxygen to the tissues (see Figure 16.7). Therefore, both strength and endurance are decreased, making the tasks of day-to-day life more difficult. The loss of muscle strength contributes to **physical frailty**, which is characterized by general weakness, decreased endurance, impaired mobility, and poor balance; this increases the risk of falls and fractures. In the oldest old (those 85 years of age and older), loss of muscle strength becomes the limiting factor determining whether they can continue to live independently. Some of the reduction in muscle strength and mass is due to changes in hormone levels and in muscle protein synthesis, but a lack of exercise is also an important contributor.

sarcopenia Progressive decrease in skeletal muscle mass and strength that occurs with age.

physical frailty Impairment in function and reduction in physiological reserves severe enough to cause limitations in the basic activities of daily living.

Aging is also accompanied by a decrease in bone mass, often resulting in osteoporosis, which further increases the risk of fractures. Although obesity makes many chronic conditions worse, the loss of bone and the risk of osteoporosis is reduced in individuals who weigh more. This is partly due to the added mechanical stress on the bones caused by carrying excess body weight and partly due to the release of estrogen by body fat (see Chapter 11).

Changes in Hormone Levels Some of the hormonal changes that occur with age are considered part of the normal aging process. Others may be a symptom of a disease process. For example, about 4 to 8.5% of adults in the United States have thyroid levels that are below normal. In many individuals this causes no symptoms, and it is not clear whether administering thyroid hormones to these people offers any benefits.[18] If the decrease causes symptoms, the patient is treated by administering thyroid hormones.[19]

menopause Physiological changes that mark the end of a woman's capacity to bear children.

Estrogen The most striking and rapidly occurring age-related hormonal change is **menopause**. Menopause normally occurs in women between the ages of 45 and 55. During menopause, the cyclical release of the female hormones estrogen and progesterone slows and eventually stops, causing ovulation and menstruation to cease. The period of decline in estrogen levels is accompanied by changes in mood, skin, and body composition (an increase in body fat and a decrease in lean tissue). The reduction in estrogen decreases the risk of breast cancer but increases the risk of heart disease to a level more similar to that in men. Reduced estrogen levels also increase the risk of osteoporosis by increasing the rate of bone breakdown and decreasing calcium absorption from the intestine. Estrogen used to be prescribed liberally to older women to alleviate the symptoms of menopause and reduce the risk of osteoporosis and heart disease. This hormone replacement therapy is no longer as common because studies have found that, while it does reduce menopausal symptoms and the risk of bone fractures, it increases the risk of heart disease, blood clots, stroke, breast cancer, and problems with memory and thinking.[20]

Menopause does not occur in men, but with age men do experience a gradual decrease in testosterone levels, which may contribute to a decrease in muscle mass and strength.

Growth Hormone Growth hormone stimulates growth and protein synthesis. Levels gradually decline with age in both men and women and may be responsible for some of the decrease in lean body mass, increase in fat mass, and bone loss that occurs with age. A few small, short-term studies have demonstrated improvements in body composition with growth hormone treatments, but there is little data on the benefits, safety, or cost effectiveness of long-term growth hormone administration.[21] When compared to a program of regular exercise, growth hormone injections did not produce any greater increases in muscle size or strength.[22] In addition, growth hormone administration has side effects including edema, carpal tunnel syndrome, and decreases in insulin sensitivity. Despite the promotion of growth hormone and products that supposedly increase growth hormone release as anti-aging therapies, until more is known about the long-term effects of growth hormone administration, the use of these products for anti-aging purposes is not recommended.[21]

DHEA DHEA (dehydroepiandrosterone) is a precursor to the sex hormones, testosterone, estrogen, and progesterone. Even though low levels of this hormone are not known to cause age-associated disorders, DHEA supplements sold over the counter claim to strengthen bones, muscles, and the immune system, and to prevent diabetes, obesity, heart disease, and cancer. Although some of these effects have been demonstrated when DHEA is administered to animals, beneficial effects of DHEA supplementation in humans have not been clearly established.[23]

Melatonin Melatonin is a hormone that is secreted by the pineal gland. It is involved in regulating the body's cycles of sleep and wakefulness. A decline in melatonin is hypothesized to influence aging by affecting body rhythms and triggering genetically programmed aging at a cellular level. Melatonin is also an antioxidant and may enhance immune function and reduce inflammation in the brain.[24] It is sold as a dietary supplement, but its ability to extend normal longevity in humans has not been determined.[25]

Insulin One of the most common hormone-related changes that occurs with aging is elevated blood glucose. This is due to both a decrease in the amount of insulin released by the pancreas and a decrease in the sensitivity of body tissues to insulin. Insulin resistance is related to poor diet, inactivity, increased abdominal fat mass, and decreased lean mass. Almost 27% of individuals 60 years of age or older have diabetes, and many of these cases are undiagnosed.[26] Individuals with diabetes are treated with diet and lifestyle prescriptions, medications to reduce blood glucose, and, when necessary, administration of the hormone insulin.

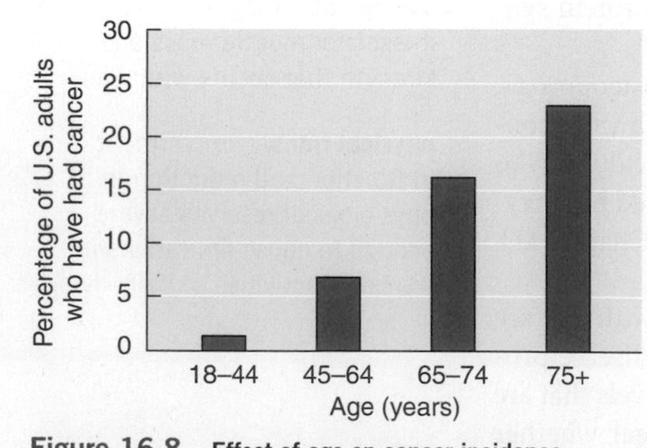

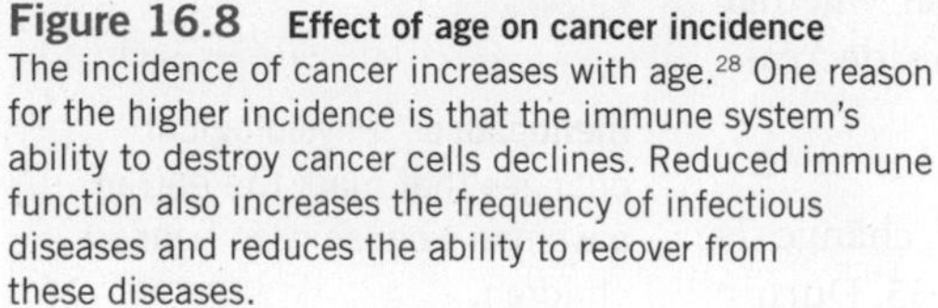
Figure 16.8 Effect of age on cancer incidence
The incidence of cancer increases with age.[28] One reason for the higher incidence is that the immune system's ability to destroy cancer cells declines. Reduced immune function also increases the frequency of infectious diseases and reduces the ability to recover from these diseases.

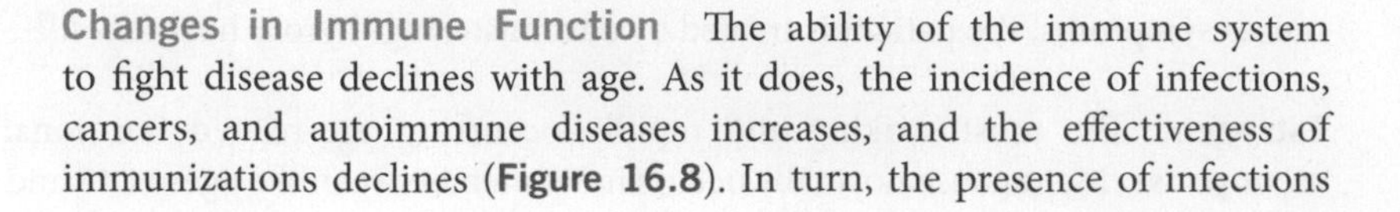
Changes in Immune Function The ability of the immune system to fight disease declines with age. As it does, the incidence of infections, cancers, and autoimmune diseases increases, and the effectiveness of immunizations declines (**Figure 16.8**). In turn, the presence of infections

and chronic disease contributes to malnutrition (see Figure 16.5). Some of the decrease in immune function seen in older adults may be due to nutritional deficiencies.[27] The immune response depends on the ability of cells to differentiate, divide rapidly, and secrete immune factors, so nutrients that are involved in cell differentiation, cell division, and protein synthesis can influence the immune response. Nutrient deficiencies are common in older adults, including deficiencies of zinc, iron, β-carotene, folic acid, and vitamins B_6, B_{12}, C, D, and E. Supplements of some of these individual nutrients have been shown to increase certain aspects of the immune response, but have not been shown to reduce mortality from infections. When a multivitamin/mineral supplement is used, enhanced immune responses and a reduction in the occurrence of common infections are seen. These effects are greater among individuals showing evidence of nutrient deficiencies before supplementation.[27]

High doses of some nutrients, including zinc, copper, and iron, depress immune function, so supplements should not contain more than 100% of the Daily Value.

Why Medical Conditions Contribute to Malnutrition

The reduction in reserve capacity and decline in immune function that occur with age make infectious disease more frequent and more serious in the elderly. In addition, most older adults have at least one chronic medical condition. The incidence of cardiovascular disease, diabetes, osteoporosis, hypertension, cancer, arthritis, and Alzheimer's disease all increases with age. Some of these diseases change nutrient requirements, some decrease the appeal of food, and some impair the ability to obtain and prepare an adequate diet by affecting mobility and mental status. All of these can increase the risk of **food insecurity** and malnutrition.

food insecurity An inability to consistently acquire foods that are nutritionally adequate and individually, socially, and culturally acceptable.

Conditions that Decrease Mobility More than half of the older population suffers from some form of physical disability, and the incidence increases with increasing age. Twenty-seven percent of older adults have disabilities that make it difficult to carry out the activities of daily life.[4,29] These limitations affect the ability to maintain good nutritional health by making it hard to shop, prepare food, get around the house, or go out to eat. Arthritis, a condition that causes pain upon movement, is the most common cause of disability in older individuals, affecting 50% of adults over the age of 65 years (**Figure 16.9**).[30] Half of

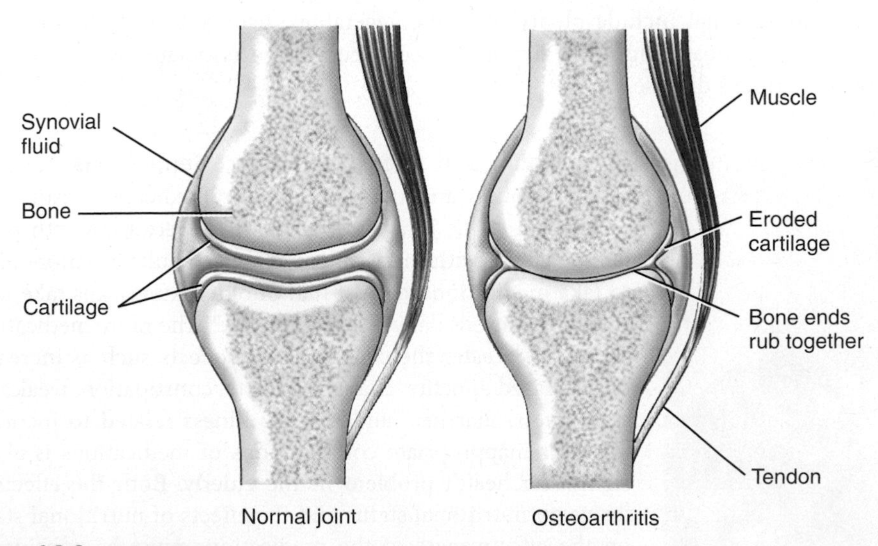

Figure 16.9 Osteoarthritis
Osteoarthritis, the most common form of arthritis, occurs when the cartilage that cushions the joints degenerates, allowing the bones to rub together and cause pain. Anti-inflammatory medications help reduce the pain. Supplements of glucosamine and chondroitin are marketed to improve symptoms and slow the progression of osteoarthritis, but clinical studies have not found the effect of these supplements on pain and function to be significant.[32]

individuals 70 years of age and older with arthritis need help with the activities of daily living, including preparing and eating meals.[31] Osteoporosis and its associated fractures can also affect mobility, which in turn affects the ability to acquire and consume a healthy diet.

Conditions that Impair Mental Status Altered mental status can affect nutrition by interfering with the response to hunger and the ability to eat and to obtain and prepare food. Although many individuals maintain adequate nervous system function into old age, the incidence of dementia increases with age. **Dementia** refers to an impairment in memory, thinking, and/or judgment that is severe enough to cause personality changes and affect daily activities and relationships with others. Causes of dementia include multiple strokes, alcoholism, dehydration, medication side effects and interactions, and **Alzheimer's disease**.

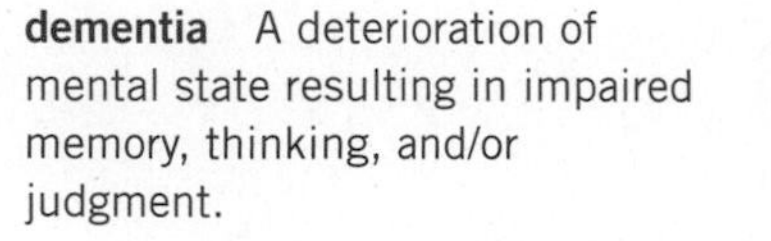

dementia A deterioration of mental state resulting in impaired memory, thinking, and/or judgment.

Alzheimer's disease A disease that causes the relentless and irreversible loss of mental function.

Low levels of vitamin B_{12} and vitamin E have also been suggested to affect mental function in the elderly. With aging, there is a decrease in blood vitamin B_{12} levels and a rise in metabolites indicative of poor vitamin B_{12} status. In most cases, vitamin B_{12} supplements do not improve neurological function; however, in some elderly patients with mild dementia and low blood levels of vitamin B_{12}, supplementation does improve mental function.[33]

The relationship between vitamin E and cognitive impairment is also controversial. In the elderly, lower blood levels of vitamin E have been associated with poor memory and mental functioning, and those who suffer from dementia have lower plasma levels of vitamin E. The effect of vitamin E and vitamin C supplements on preventing dementia is inconsistent, but some evidence suggests that vitamin E supplements may slow the progression of the disease.[34]

Over half of the cases of dementia in the elderly are due to Alzheimer's disease, a progressive, incurable loss of mental function. The brains of patients with Alzheimer's disease are characterized by the accumulation of an abnormal protein and a loss of certain types of nerve cells. Its cause is unknown, but there does appear to be a genetic component in some cases. One controversial theory is that accumulation of aluminum in the brain contributes to Alzheimer's disease. While the brains of patients with Alzheimer's do contain high levels of aluminum, it is not clear whether this is a cause of Alzheimer's. Nonetheless, some recommend reducing aluminum exposure from food, skin care products, antiperspirants, and pharmaceuticals.[35] The brains of patients with Alzheimer's also show damage from oxidative stress, and antioxidants are hypothesized to play a role in preventing and delaying the progression of Alzheimer's disease.[36] Dietary patterns that include plenty of fruits, vegetables, fish, nuts and legumes, and lower intakes of meats, high-fat dairy, and sweets seem to be associated with a reduced risk of Alzheimer's disease.[37]

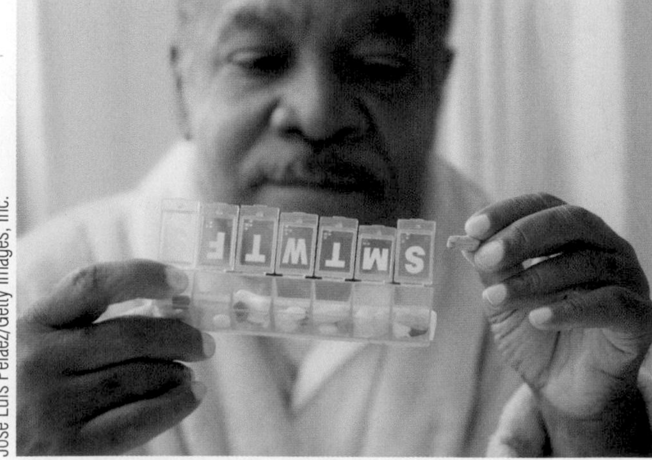

Jose Luis Pelaez/Getty Images, Inc.

Figure 16.10 Multiple medications
It is common for older adults to take multiple medications daily.

Prescription and Over-the-Counter Medications The use of prescription and over-the-counter medications can affect nutritional status in a number of ways. Because health problems increase with increasing age, older adults are more likely to take medications; almost half of older Americans take multiple medications daily (**Figure 16.10**).[38] The more medications taken, the greater the chance of side effects such as increased or decreased appetite, changes in taste, constipation, weakness, drowsiness, diarrhea, and nausea. Illness related to incorrect doses or inappropriate combinations of medications is also a significant health problem in the elderly. Both the effects of drugs on nutritional status and the effects of nutritional status on the effectiveness of the medications must be considered. Many older adults also take a variety of dietary supplements.[39] Some are harmful alone or in combination with prescription or over-the-counter medications (see **DEBATE: Are Herbal Supplements Helpful or Harmful?**).[40]

DEBATE

Are Herbal Supplements Helpful or Harmful?

Herbal supplements are popular as "natural" remedies. These products are commonly used by older adults; one study found that 88% of respondents used herbal products.[1] Are these products a helpful addition to traditional health care, or are they a health risk?

Catherine Karnow/NG Image Collection

Many cultures, including the ancient Greeks and Native Americans, have used herbal products to treat everything from coughs, constipation, and poison ivy to arthritis and heart ailments. Today, about one in six Americans uses herbs to treat maladies or boost health.[2] These products are affordable and widely available, and they do not require prescriptions. Advocates of herbal supplements feel that this availability allows people more control of their own health care. Opponents fear that self-dosing with herbs may lead to toxic reactions and prevent people from seeking traditional medical care and proven treatments.

No one will argue that herbs have demonstrated physiological effects. In fact, some of the prescription drugs used today were derived from plants. Aspirin comes from willow bark; digitalis, a drug prescribed for certain heart conditions, comes from foxglove flowers. Herbs are made from all or part of the plant, so the amounts of active ingredients are affected by growing conditions, harvesting, and processing, and they vary with the brand and batch.[3] In contrast, drugs are purified compounds and are tested to ensure consistent amounts in each pill. Advocates believe that having a mixture of compounds is an advantage because some ingredients can enhance the effects of others. Opponents argue that these interactions may be negative, diminishing the effect of one of the active ingredients.[3]

Proponents of herbal supplements argue that most herbal medicines are well tolerated and have fewer unintended consequences than prescription drugs. Prescription drugs are certainly not without side effects, but because of the levels of regulation, we are assured that a prescribed drug is an effective treatment, and any side effects are documented. Herbal products, like other dietary supplements, do not require FDA approval before they are marketed. Opponents believe that this lack of regulation allows dangerous products to enter the marketplace. Many botanical compounds are toxic; the FDA has issued warnings about ingredients such as comfrey, kava, and aristolochic acid, and it took ephedra off the market.[4] Because herbal products are made from unpurified plant material, there is also a risk of contamination with pesticides, microbes, metals, and other toxins.[5]

Another concern when people self-dose with herbs is that the herbs may interact with drugs and cause unwanted side effects.[2] For example, gingko biloba and garlic can interfere with blood clotting, and so should not be used with blood-thinning medication or before surgery. St. John's wort may interact with anesthetics and antidepressants, and echinacea can limit the effect of some steroids.[6] Prescription drugs come with instructions and warnings about potential side effects; in addition, a physician or pharmacist considers the other medications and conditions that may alter the safety and effectiveness of the drug, and a prescribing physician monitors patient health. This is not the case with dietary supplements.

Herbs are medicines and, like taking other medications, taking herbs has some advantages and risks. Whether herbal supplements are helpful or harmful depends on the supplement, the dose, and the consumer.

THINK CRITICALLY: Would you purchase a remedy from the selection of herbs at the shop shown in this photo? Why or why not?

---

[1]Ness, J., Cirillo, D. J., Weir, D. R., et al. Use of complementary medicine in older Americans: Results from the health and retirement study. *Gerontologist* 45:516–524, 2005.

[2]Gershwin, M. E., Borchers, A. T., Keen, C. L., et al. Public safety and dietary supplementation. *Ann N Y Acad Sci* 1190:104–117, 2010.

[3]Ribnicky, D. M., Poulev, A., Schmidt, B., et al. Evaluation of botanicals for improving human health. *Am J Clin Nutr* 87:472S–475S, 2008.

[4]Food and Drug Administration. *Dietary Supplement Warnings and Safety Information.*

[5]Saper, R. B., Phillips, R. S., Sehgal, A., et al. Lead, mercury, and arsenic in U.S.- and Indian-manufactured Ayurvedic medicines sold via the Internet. *JAMA* 300:915–923, 2008.

[6]Ulbricht, C., Chao, W., Costa, D., et al. Clinical evidence of herb–drug interactions: A systematic review by the Natural Standard Research Collaboration. *Curr Drug Metab* 9: 1063–1120, 2008.

TABLE 16.2 Commonly Used Drugs That May Contribute to Nutritional Deficiencies[41]

Drug Group	Drug	Potential Deficiency
Antacids and other acid-reducing agents	Sodium bicarbonate	Folate, iron, potassium
	Aluminum hydroxide	Calcium, phosphorus
	Ranitidine (Zantac)	Iron, zinc
	Cimetidine (Tagamet)	Iron, zinc, folic acid, vitamin B_{12}, vitamin D
Antimicrobial agents	Tetracycline	Calcium, potassium, magnesium, folate, riboflavin, vitamins B_6, B_{12}, C, K
	Gentamicin	Calcium, potassium, magnesium, vitamin B_6
	Neomycin	Fat, folic acid, iron, calcium, magnesium, potassium, vitamins B_6, B_{12}, A, D, K
	Trimethoprim	Calcium, folate, magnesium, vitamins B_6, B_{12}, K
Anti-inflammatory agents	Sulfasalazine	Folate
	Prednisone	Calcium
	Aspirin	Vitamin C, folate, iron, vitamin E, potassium
Anticancer drugs	Colchine	Fat, β-carotene, potassium, vitamin B_{12}
	Methotrexate	Folate, calcium
Anticoagulant drugs	Warfarin	Vitamin K
Antihypertensive drugs	Hydralazine	Vitamin B_6
Diuretics	Thiazides	Potassium, magnesium, zinc
	Furosemide	Potassium, calcium, magnesium, thiamin, vitamins B_6, C
Hypocholesterolemic agents	Cholestyramine	Fat, fat-soluble vitamins, β-carotene, iron, calcium, magnesium, phosphorus, zinc, folate, vitamin B_{12}
	Statins	Selenium, vitamin E, β-carotene
Laxatives	General	Potassium
	Mineral oil	Fat-soluble vitamins, β-carotene, calcium, phosphorus, potassium
Tranquilizers	Chlorpromazine	Riboflavin

How Medications Affect Nutritional Status Medications can affect nutritional status by altering appetite, nutrient absorption, metabolism, or nutrient excretion (**Table 16.2**). The impact is greatest in individuals who must take medications for extended periods, those who take multiple medications, and those whose nutritional status is already marginal.

Some medications directly affect the gastrointestinal tract. More than 250 drugs, including blood pressure medications, antidepressants, decongestants, and the pain reliever ibuprofen (found in Advil, Motrin, and Nuprin), can cause mouth dryness, which can decrease interest in eating by interfering with taste, chewing, and swallowing. Aspirin is a stomach irritant and can cause small amounts of painless bleeding in the gastrointestinal tract, resulting in iron loss. Digoxin, which is a heart stimulant, can cause gastrointestinal upset, loss of appetite, and nausea. Narcotic pain medications such as codeine can lead to constipation, nausea, and vomiting.

Other drugs can decrease nutrient absorption. Cholestyramine (Questran), which is used to reduce blood cholesterol, and Orlistat and Alli, used to promote weight loss, can decrease the absorption of fat-soluble vitamins, vitamin B_{12}, iron, and folate. Antacids that contain aluminum or magnesium hydroxide (Rolaids or Maalox) combine with phosphorus in the gut to form compounds that cannot be absorbed; chronic use can result in loss of phosphorus from bone and possibly accelerate osteoporosis. Repeated use of stimulant laxatives can deplete calcium and potassium. Mineral oil laxatives prevent the absorption of fat-soluble vitamins. If it is not possible to prevent constipation by consuming a diet high in fiber and fluid, bulk-forming laxatives such as Metamucil and Citrucel are a safer choice.

The metabolism of drugs can also affect nutritional status. For example, anticonvulsive drugs (used to prevent seizures) increase the liver's capacity to metabolize and eliminate vitamin D, therefore increasing the need for vitamin D.

Some drugs affect nutrient excretion. Diuretics, which are used to treat hypertension and edema, cause water loss, but some types (thiazides) also increase the excretion of potassium. People taking thiazide diuretics are advised to include several good sources of potassium in their diet each day or are prescribed supplements.

How Food and Nutritional Status Affect Drug Absorption and Metabolism Food components can either enhance or retard the absorption and metabolism of drugs. Some drugs, such as the pain medication Darvon, are absorbed better or faster if taken with food. Other drugs, such as aspirin and ibuprofen, should be taken with food because they are irritating to the gastrointestinal tract. Because food can delay how quickly drugs leave the stomach, some medications are best taken with just water. Other drugs interact with specific foods. For instance, the antibiotic tetracycline should not be taken with milk because it binds with calcium, making both unavailable. The metabolism of atorvastatin (Lipitor), taken to lower cholesterol, is blocked by a compound in grapefruit, so eating grapefruit or drinking grapefruit juice can result in drug toxicity.[42]

Nutritional status can also affect drug metabolism. If nutritional status is poor, the body's ability to detoxify drugs may be altered. For example, in a malnourished individual, theophylline, used to treat asthma, is metabolized slowly, resulting in high blood levels of the drug that can cause loss of appetite, nausea, and vomiting.

Specific nutrients can also affect the metabolism of drugs. High-protein diets enhance drug metabolism in general, and low-protein diets slow it. Vitamin K hinders the action of anticoagulants taken to reduce the risk of blood clots. On the other hand, omega-3 fatty acids, such as those in fish oils, inhibit blood clotting and may intensify the effect of an anticoagulant drug and cause bleeding. It is safe to eat fish while taking anticoagulant drugs; however, the use of fish oil supplements is not recommended. Drugs can also interact with one another. For example, alcohol affects the metabolism of over 100 medications. Drug interactions can exaggerate or, in some cases, diminish the effect of a medication. Individuals taking any medication should consult their doctor, pharmacist, or dietitian regarding how the drug could affect the action of other drugs they may be taking, how the drug could affect their nutrition, and how their nutrition could affect the action of the drug.

How Economic, Social, and Psychological Factors Affect Nutritional Status

A variety of social and economic changes accompany aging. Many of these factors are interrelated and impair nutritional health by decreasing the motivation to eat and the ability to acquire and enjoy food.

Low Income Approximately 3.5 million older adults live below the poverty level.[43] Poverty is an even greater problem for African-American and Hispanic elderly. About 7% of elderly whites are poor compared to 18% of elderly African Americans and 18% of elderly Hispanics. Many older adults must live on a fixed income when they retire from their jobs, making it difficult to afford health care, especially medications, and a healthy diet. Food is often the most flexible expense in one's budget, so limiting the types and amounts of foods consumed may be the only option available for older adults trying to meet expenses. As a result, almost 15% of elderly Americans are food insecure.[44] Substandard housing and inadequate food preparation facilities can make the situation worse because food cannot easily be prepared and eaten at home.

Dependent Living Although many older adults continue to live independently in their own homes, the physical decline and psychological issues associated with aging cause some to eventually require assistance in living (**Figure 16.11**). Poor eyesight and other physical restrictions can limit the ability to drive a car. Without help, many older adults may be unable to get to markets and food programs, restricting the types of food available to them. While a social support system consisting of family members, friends, and other caregivers can help many people stay in their own homes, others may require assisted-living facilities, where they have their own apartments but can obtain assistance around the clock. For some, however, a nursing home is required to obtain the appropriate care.

THINK CRITICALLY

Why can limited mobility decrease nutrient intake?

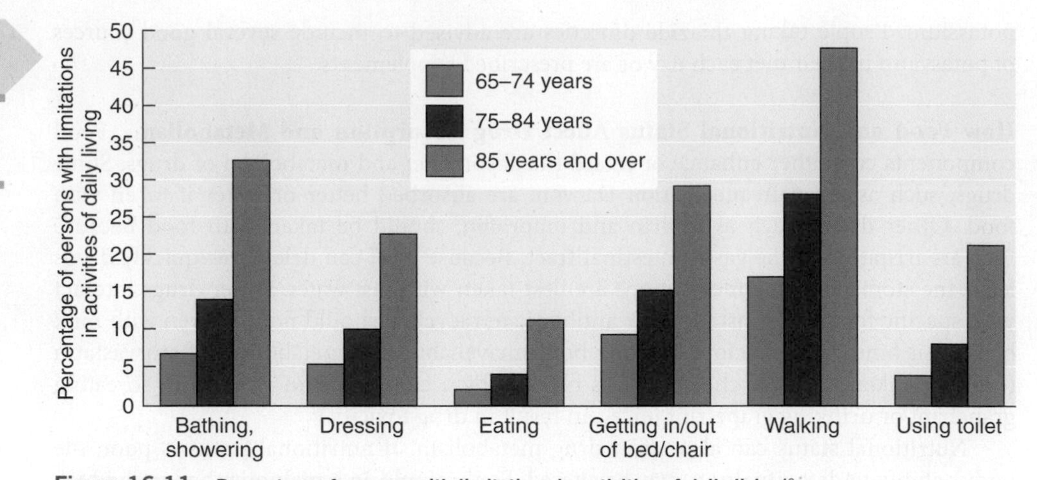

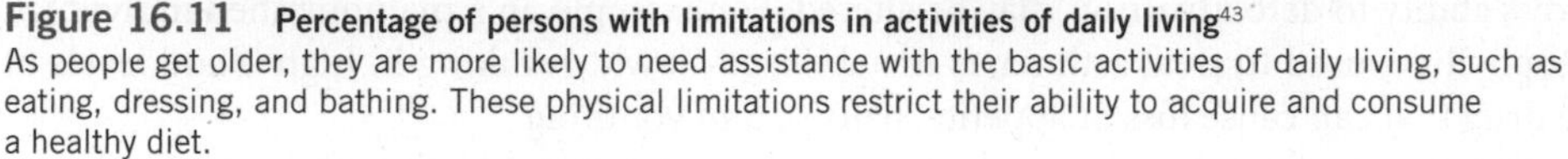

Figure 16.11 Percentage of persons with limitations in activities of daily living[43]
As people get older, they are more likely to need assistance with the basic activities of daily living, such as eating, dressing, and bathing. These physical limitations restrict their ability to acquire and consume a healthy diet.

Those in nursing homes are at increased risk for malnutrition because they are more likely to have medical conditions that increase nutrient needs or that interfere with food intake or nutrient absorption, and because they are dependent on others to provide for their care. In addition, 50% of institutionalized elderly suffer from some form of disorientation or confusion, which further increases the likelihood of decreased nutrient intake. Even when adequate meals are provided, many nursing home residents require assistance in eating and frequently do not consume all of the food served, increasing the likelihood of fluid and energy deficits.[45]

Depression Social, psychological, and physical factors all contribute to depression in the elderly; although the elderly comprise only 12% of the population, 16% of suicides occur in this age group.[46] Social factors such as retirement and the death or relocation of friends and family can cause social isolation and depression. Physical factors such as disability cause loss of independence. This reduces the ability to engage in normal daily activities, visit with friends and family easily, and provide for personal needs, further contributing to depression. Depression can make meals less appetizing and decrease the quantity and quality of foods consumed, thereby increasing the risk of malnutrition.

16.4 Nutritional Needs and Concerns of Older Adults

LEARNING OBJECTIVES

- Compare the energy needs of older adults to those of young adults.
- Describe why older adults are at risk for dehydration.
- List the vitamins and minerals for which older adults are at risk for deficiency.

Older adults are a diverse population. There are 70-year-olds running marathons while others are confined to wheelchairs. The physiological and health changes that accompany aging affect the requirements for some nutrients, how nutrient needs must be met, and the risk of malnutrition. In order best to address the nutrient needs of adults, the DRIs divide adulthood into four age categories: young adulthood, ages 19 through 30; middle age, 31 through 50 years; adulthood, ages 51 through 70; and older adults, those over 70 years of age. Recommendations have been developed to meet the needs of the majority of healthy individuals in each age group. Although the incidence of chronic diseases and disabilities increases with advancing age, these conditions are not considered when making general nutrient intake recommendations.

Energy and Macronutrient Needs

Adult energy needs typically decline with age. This is due to a decrease in all components of total energy expenditure. Basal metabolic rate (BMR) decreases as adults get older, in part due to a decrease in lean body mass. There is a 2 to 3% drop in BMR every decade after about age 20. The thermic effect of food also declines with age and is about 20% lower in older than in younger adults. These decreases are reflected in the energy needs of older adults. For example, the estimated energy requirement (EER) for an 80-year-old man is almost 600 kcalories per day less than that for a 20-year-old man of the same height, weight, and physical activity level.[33] The decrease in physical activity that typically occurs with age is estimated to account for about one-half of the decrease in total energy expenditure.[47,48] Decreased activity also contributes to the reduction in lean body mass and BMR.

Protein Protein is needed at all ages to repair and maintain tissues. Therefore, unlike energy requirements, the requirement for protein does not decline with age. The RDA for older as well as younger adults is 0.8 g/kg of body weight per day. As a result, an adequate diet for older adults must be somewhat higher in protein relative to energy intake. Due to the diversity of older adults, actual need depends on the individual. In some, the protein requirement may be less than the RDA because there is less lean body mass to maintain, whereas in others it may be greater than the RDA because protein absorption or utilization is reduced.

Fat The digestion and absorption of fat do not change with advancing age. Therefore, although total fat intake may be lower in older adults due to lower energy needs, the recommendations regarding the proportion and types of dietary fat apply to older as well as younger adults. A diet with 20 to 35% of energy from fat that contains adequate amounts of the essential fatty acids and limits saturated fat, *trans* fat, and cholesterol is recommended. Following these recommendations will allow older adults to meet their nutrient needs without exceeding their energy requirements and may delay the onset of chronic disease. However, in certain situations, such as being underweight, greater fat intake may be warranted.

Carbohydrates and Fiber As with fat, the recommended proportion of energy from carbohydrate (45 to 65% of energy) does not change with age, but the total amount needed may be lower in older adults due to lower energy needs. Dietary carbohydrate should come from whole grains, fruits, vegetables, and low-fat dairy products, and foods high in added sugars should be limited. This pattern will help assure adequate nutrients without excess energy.

The recommendations for fiber intake for older adults are slightly lower than for younger adults because the AIs for fiber are based on total energy needs.[33] Fiber, from whole grains, fruits, vegetables, and legumes, when consumed with adequate fluid, helps prevent constipation, hemorrhoids, and diverticulosis—conditions that are common in older adults. High-fiber diets may also be beneficial in the prevention and management of diabetes, cardiovascular disease, and obesity.

Water The recommended water intake for older adults is the same as that for younger adults;[49] however, changes in the homeostatic mechanisms that regulate water balance may make meeting these needs more challenging. With age there is a reduction in the sense of thirst, which can decrease fluid intake. In addition, the kidneys are no longer as efficient at conserving water, so water loss increases. Other physical and psychological changes also increase the risk of dehydration. For instance, difficulty in swallowing and restricted mobility may limit access to water even in the presence of thirst. Depression, which decreases water intake, and medications such as laxatives and diuretics, which increase water loss, also contribute to dehydration in the elderly. The elderly may also voluntarily restrict fluid intake to avoid accidents due to incontinence or because numerous trips to the bathroom increase pain from arthritis. In addition to impairing organ function, inadequate fluid intake contributes to the development of constipation.

Micronutrient Needs

Although the recommended intake for many of the micronutrients is no different for older adults than for younger adults, the decrease in energy intake that occurs with age causes a decline in the intakes of micronutrients, especially the B vitamins, calcium, iron, and zinc (**Figure 16.12**).[9] Changes in digestion, absorption, and metabolism also affect micronutrient status. In turn, inadequate levels of certain micronutrients contribute to the development of some of the disorders that are common in older adults.

B Vitamins The only B vitamins for which recommendations differ between older and younger adults are vitamins B_6 and B_{12}. The RDA for vitamin B_6 is greater in people ages 51 and older than for younger adults because higher dietary intakes are needed to maintain the same functional levels in the body. The RDA for vitamin B_{12} is not increased for older adults, but it is a nutrient of concern because of both reduced absorption and low

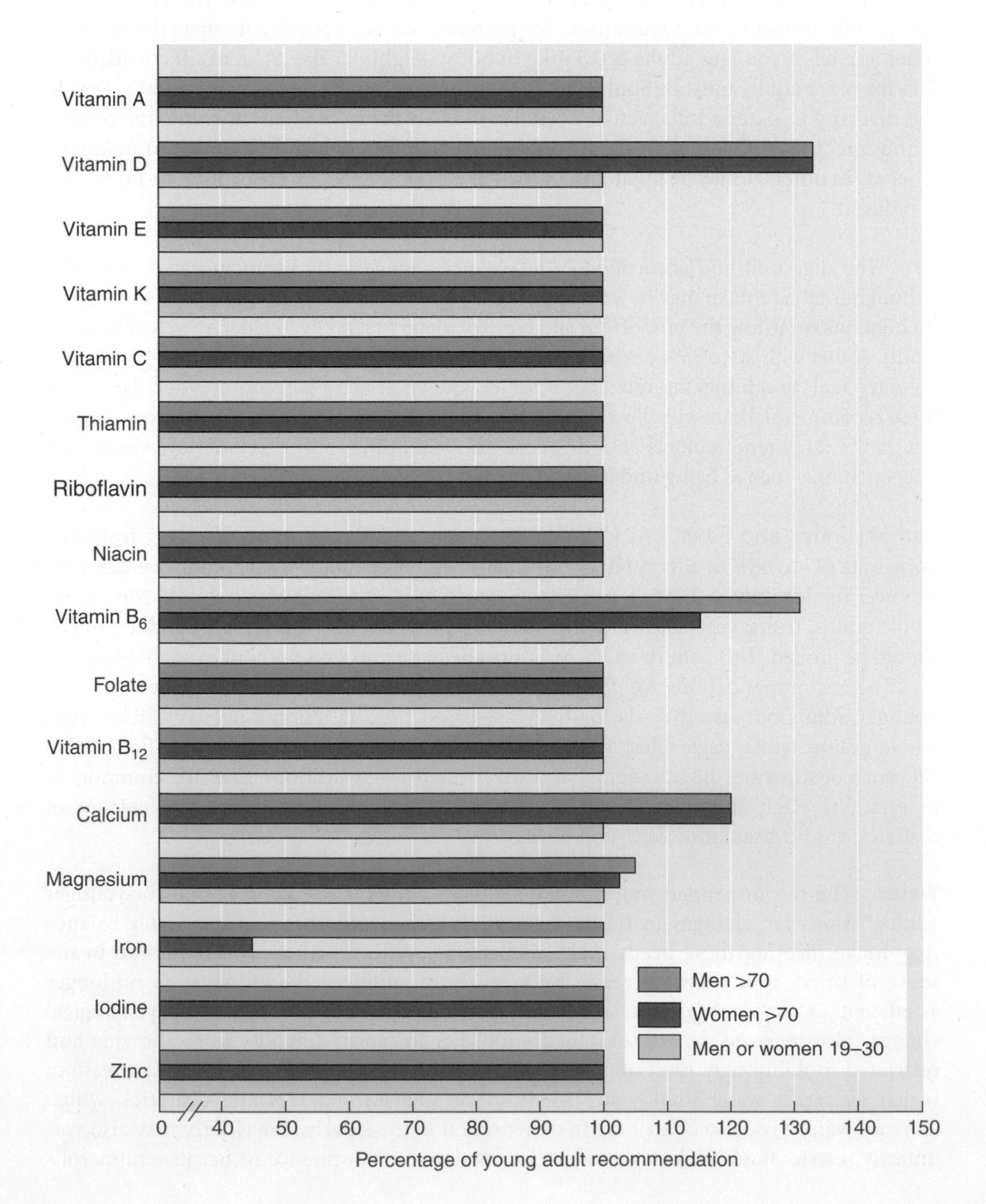

Figure 16.12 Vitamin and mineral needs of older adults This graph compares the recommended micronutrients intakes for young adults ages 19 through 30 to those of adults over age 70. The iron needs of older women decrease by over 50% because iron is no longer lost thorough menstruation.

dietary intakes. Absorption of vitamin B_{12} naturally present in food is reduced in many older adults due to atrophic gastritis. It is therefore recommended that individuals over the age of 50 meet their RDA for vitamin B_{12} by consuming foods fortified with vitamin B_{12}, such as breakfast cereals or soy-based products, or by taking a supplement containing vitamin B_{12}.[10] The vitamin B_{12} in fortified foods and supplements is not bound to proteins, so it is absorbed even when stomach acid is low.

The RDA for folate is the same for adults of all ages, but folate intake is a concern in older adults for several reasons. Deficiencies of folate and vitamin B_{12} contribute to anemia, which is common in older adults. In addition, low folate, along with inadequate levels of vitamin B_6 and B_{12}, may result in an elevated homocysteine level, which increases the risk of cardiovascular disease. Due to the importance of folate in DNA synthesis, low folate intake has also been hypothesized to cause DNA changes that contribute to cancer development.[50] The fortification of enriched grain products with folate, which began in 1998, has increased intake of this vitamin. However, when folate is consumed in excess it can mask the symptoms of vitamin B_{12} deficiency. Diets that include highly fortified products, such as breakfast cereals, and folate supplements may exceed the UL for folate. Levels above the UL increase the risk that vitamin B_{12} deficiency will be masked and therefore go untreated (see Chapter 8).[50]

Antioxidant Vitamins The recommended intake of antioxidant vitamins is not increased in older adults, but low dietary intakes of vitamins C and E and carotenoids are a concern due to low fruit and vegetable consumption among the elderly. Only 28% of individuals 65 years of age and older consume five or more servings of fruits and vegetables daily.[51] As discussed above, eye disorders as well as mental impairment have been correlated with low levels of antioxidants in the elderly.

Calcium and Vitamin D Low intakes of both calcium and vitamin D contribute to osteoporosis in the elderly. Calcium status is a problem in elderly people because calcium intake is low and intestinal absorption decreases with age. Without sufficient calcium, bone mass decreases, and the risk of bone fractures due to osteoporosis increases. The reduction in estrogen that occurs with menopause further increases bone loss in women by increasing the rate of bone breakdown and decreasing the absorption of calcium from the intestine. The RDA for adult men 51 years and older is 1000 mg/day. Because of the accelerated bone loss in women during the postmenopausal period, the RDA for women 51 years and older is increased to 1200 mg/day. To reduce age-related bone loss, the RDA for both men and women over 70 years of age is 1200 mg/day.[52]

Vitamin D, which is necessary for adequate calcium absorption, is also a concern in elderly people. Intake is often low, and synthesis in the skin is reduced due to limited exposure to sunlight and because the capacity to synthesize vitamin D in the skin decreases with age. The RDA for people ages 51 and older is 600 IU (15 μg)/day, the same as for other age groups. For individuals over age 70 years, the RDA is increased to 800 IU (20 μg)/day.[52]

Iron The iron needs of women decline sharply at menopause when blood loss through menstruation stops. The RDA for iron for women over 50 years of age is less than half of that of menstruating women (see Figure 16.12). The iron needs of men do not change. Nonetheless, iron-deficiency anemia is common in the elderly, often due to chronic blood loss from disease and medications and poor iron absorption due to low stomach acid and antacid use.[53]

Zinc The RDA for zinc is not changed in older adults, but lower energy intakes as well as malabsorption, physiological stress, trauma, muscle wasting, and prescription and over-the-counter medications can all contribute to poor zinc status.[54] The consequences of poor zinc status may include loss of taste acuity and impaired immune function and wound healing. Loss of taste acuity can contribute to malnutrition by reducing food intake. Reduction in immune function and wound healing increases the risk of infection, which can also impair nutritional status.

16.5 Keeping Older Adults Healthy

LEARNING OBJECTIVES

- Explain how exercise and a nutritious diet affect the degenerative changes of aging.
- Plan a menu that would meet your MyPlate recommendations if you were 80 years old.
- List nutrients for which supplements are recommended to older adults.
- Describe the purpose of the DETERMINE checklist.

There is no secret dietary factor that will bestow immortality, but good nutrition and an active lifestyle are major determinants of successful aging. A good diet can extend an individual's healthy life span by preventing malnutrition and delaying the onset of chronic diseases. The diseases that are the major causes of disability in older adults—cardiovascular disease, hypertension, diabetes, cancer, and osteoporosis—are all nutrition-related. Exercise and a lifetime of healthy eating will not necessarily prevent these diseases, but they may slow the changes that accumulate over time, postponing the onset of disease symptoms. For example, the risk of developing cardiovascular disease can be decreased by exercise and a diet low in saturated fat, *trans* fat, and cholesterol and high in whole grains, fruits, and vegetables. The risk of osteoporosis may be reduced by adequate calcium and vitamin D intake and exercise throughout life. And the likelihood of developing certain types of cancer can be reduced by consuming a diet low in saturated fat and high in whole grains, vegetables, and fruits. Regular exercise can slow the loss of lean body mass, maintain fitness and independence, and allow an increase in food intake without weight gain, so micronutrient needs are more easily met.

Despite the fact that the nutrient needs of older adults are not drastically different from those of young adults, it is more challenging to meet these needs. Some of this challenge is due to changes in health and social and economic conditions that are more common in this population. Meeting needs requires consideration of each person's medical, psychological, social, and economic circumstances. For some, government aid is needed to assure adequate nutrition.

How to Plan a Healthy Diet for Older Adults

The first step toward meeting the nutrient needs of the elderly is to plan a healthy diet. Because older adults need less energy but the same amounts of most micronutrients, their food choices must be nutrient dense. Meals and snacks should include plenty of liquids because dehydration is a common problem. In some cases, nutrient supplements may be necessary to meet needs.

A MyPlate Food Plan A MyPlate food plan can be used to determine the recommended amounts of food from each group needed to meet, but not exceed, energy needs of older as well as younger adults (**Figure 16.13**). To meet micronutrient needs without exceeding calorie needs, older adults need to choose nutrient-dense foods from each food group. To ensure adequate fiber, whole grains should be chosen from the grains group. To maximize vitamin and phytochemical intake, a variety of fruits and orange, dark green, and starchy vegetables should be included. To maximize nutrient density, low-fat dairy products and lean meats or other protein sources should be chosen. To ensure foods are eaten, those offered should be easy to prepare and well seasoned to enhance appeal. In older adults, there is room for even fewer empty calories than in younger adults (see Figure 16.13).

Dietary Supplements Many older adults may benefit from supplementing particular nutrients. Vitamin D supplements may be advantageous because production of this vitamin in the skin is decreased in the elderly and exposure to sunlight may be limited. A calcium supplement may be necessary to meet needs, particularly in older women, because it can be difficult to consume 1200 mg of calcium from food without exceeding energy

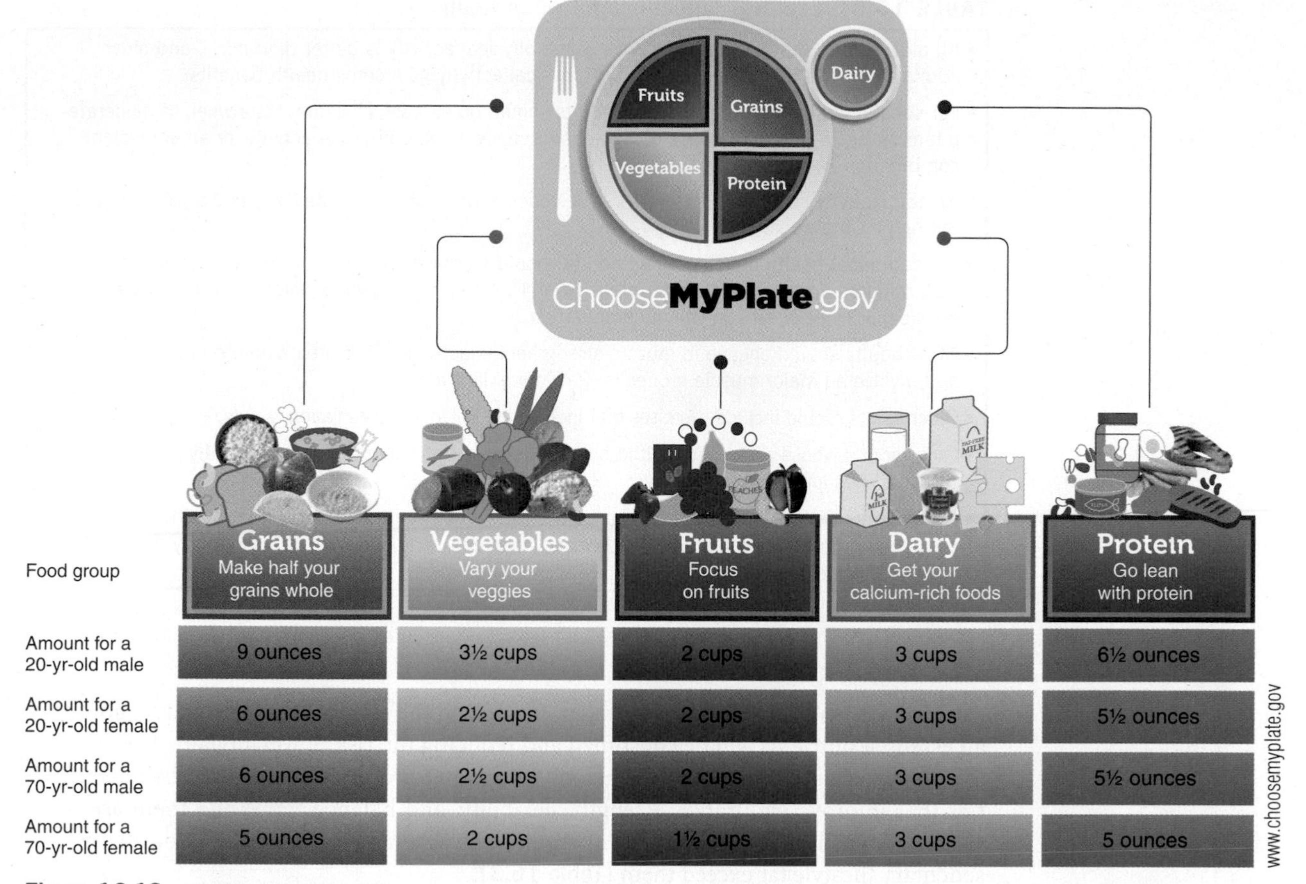

Food group	Grains Make half your grains whole	Vegetables Vary your veggies	Fruits Focus on fruits	Dairy Get your calcium-rich foods	Protein Go lean with protein
Amount for a 20-yr-old male	9 ounces	3½ cups	2 cups	3 cups	6½ ounces
Amount for a 20-yr-old female	6 ounces	2½ cups	2 cups	3 cups	5½ ounces
Amount for a 70-yr-old male	6 ounces	2½ cups	2 cups	3 cups	5½ ounces
Amount for a 70-yr-old female	5 ounces	2 cups	1½ cups	3 cups	5 ounces

Figure 16.13 **MyPlate recommendations**
The MyPlate recommendations shown here are for 20- and 70-year-old men and women who get less than 30 minutes of activity per day. The older adults need to consume fewer servings from most food groups and are allotted fewer empty calories: only 120 kcal/day for a 70-year-old woman and 260 kcal/day for a 70-year-old-man compared to 260 and 360 kcal/day, respectively for their 20-year-old counterparts.

THINK CRITICALLY

Why do 70-year-olds need to eat a more nutrient-dense diet than 20-year-olds?

needs. Supplemental vitamin B_{12} from pills or fortified foods is recommended for older adults because the absorption of vitamin B_{12} often decreases with age. However, supplements should not take the place of a balanced, nutrient-dense diet high in whole grains, fruits, and vegetables. These foods also contain phytochemicals and other substances that may protect against disease. Older adults should be cautious to avoid overdoses, and the resulting toxicities, when selecting nutrient supplements. Supplements of nonnutrient substances should be taken with care. Most of these provide no proven benefit, many are costly, and some can be toxic.

The hypothesis that aging is caused by oxidative damage has contributed to the popularity of antioxidant supplements. Although there is evidence that diets high in antioxidants from fruits and vegetables and other plant foods are associated with a reduced incidence of various chronic diseases, including cardiovascular disease and some types of cancer, intervention trials have not consistently found that antioxidant supplements reduce the risk of chronic disease or decrease mortality (see Chapter 12 Debate: Antioxidant Supplements: Are they Beneficial?).[55] So, rather than taking an antioxidant supplement, older adults, like everyone else, should consume a diet plentiful in plant foods high in these nutrients. When antioxidant nutrients are obtained from foods, they bring with them phytochemicals, some of which offer additional antioxidant protection and others of which protect us from chronic disease in other ways (**Figure 16.14**).

Figure 16.14 **Foods high in antioxidants**
Older adults can avoid antioxidant deficiencies by consuming a variety of foods that are good sources of antioxidants. These also provide fiber, energy, other micronutrients, and phytochemicals.

Physical Activity for Older Adults

Regular physical activity can extend the years of active independent life, reduce disability, and improve the quality of life for older adults. It helps to maintain muscle mass, bone strength, cardiorespiratory function, and balance. Exercise also increases energy

TABLE 16.3 Exercise Guidelines for Older Adults[56,57]

• All older adults should avoid inactivity. Some physical activity is better than none, and older adults who participate in any amount of physical activity gain some health benefits.
• For substantial health benefits, older adults should do at least 150 minutes a week of moderate-intensity, or 75 minutes a week of vigorous-intensity aerobic physical activity, or an equivalent combination.
• Aerobic activity should be performed in episodes of at least 10 minutes and preferably spread throughout the week.
• For additional health benefits, older adults should increase their aerobic physical activity to 300 minutes a week of moderate-intensity, or 150 minutes a week of vigorous-intensity aerobic physical activity, or an equivalent combination.
• Older adults should engage in muscle-strengthening activities that are moderate or high intensity and involve all major muscle groups on 2 or more days a week.
• Older adults should include exercises that incorporate balance, coordination, agility, and proprioception.
• Older adults whose activity is limited by chronic conditions should be as physically active as their abilities and conditions allow.
• Older adults with chronic conditions should discuss their limitations with their health care providers in order to understand whether and how their conditions affect their ability to do regular physical activity safely.

expenditure, so more food can be eaten, increasing the chances that adequate amounts of all essential nutrients will be consumed and reducing the risk of weight gain. To maximize the benefits of exercise, a physical activity program for older adults should include activities that improve endurance, strength, flexibility, and balance.[56,57] While there are some risks associated with participation in regular physical activity, the risks associated with a sedentary lifestyle far exceed them (**Table 16.3**).

How to Get Started Before starting an exercise program or increasing the level of physical activity, older adults should check with their physician. Those who have been inactive should begin with very low-intensity physical activities. To maximize enjoyment and optimize adherence, activities should be tailored to the individual's needs and interests. Exercise classes or other group-based activities can be a good way for older adults to start an exercise program. Classes enhance social interaction and improve compliance by fostering a mutual commitment to continuing the activity. Classes also provide instruction on proper technique, which can minimize injuries, and supervision in the event of an injury or other emergency.

Endurance Endurance activities appropriate for older adults include low-impact aerobic activities such as walking, biking, and swimming. These aerobic activities provide protection against chronic disease; the benefit increases if intensity increases from low to moderate. Water activities such as water aerobics and swimming do not stress the joints, so they can be used to improve endurance in those with arthritis or other bone and joint disorders (**Figure 16.15**). Some weight-bearing exercise, such as walking, is encouraged to promote bone health. Safe walking environments include shopping malls or lighted sidewalks. Lifestyle activities such as vacuuming, sweeping, mopping, and gardening can also enhance endurance. To provide benefits, endurance activities need to be performed for at least 10 minutes without rest.

Photo and Co/Getty Images, Inc.

Figure 16.15 Low-impact activities
Water aerobics classes provide older adults with a low-impact aerobic activity and social interaction.

Strength Resistance training has been shown to increase strength and lean body mass as well as to slow the decrease in bone density common in the elderly. Muscle-strengthening exercise should be performed two or more days per week.[57] Lifting small weights or stretching elastic bands at a level that requires some physical effort can provide strength training (**Figure 16.16**). Weight machines and calisthenics can also be used to increase strength. Both upper- and lower-body muscles should be included in a strengthening regimen. Muscles of the lower body are particularly important for maintaining mobility and independence.

©.Ingram Publishing/Alamy Images

Figure 16.16 Muscle-strengthening exercise
Resistance training at any age improves muscle strength and endurance.

Flexibility Flexibility decreases with age but can be improved with exercises. Better flexibility enhances postural stability and balance, makes the tasks of everyday life easier, and may reduce the risk of injuries.[56,57] Flexibility exercises should include active stretching, which moves the muscle through a full range of motion, as well static stretches, which extend a muscle its full length and hold it for 10 to 30 seconds. The movements should be smooth and stretches held steady without bouncing. Stretching can be incorporated into lifestyle activities such as putting away the dishes on high and low shelves and tying shoes.

Balance Improvements in strength and endurance also improve balance, but specific neuromotor exercise should be included in an exercise program for older adults.[56] This type of exercise helps improve physical function and reduce the risk of falls by improving balance, as well as agility, coordination, and proprioception, which is the ability to sense the position, orientation, and movement of the body and its parts. Neuromotor exercises should include those that improve balance and coordination while standing still, such as balancing on one leg or standing upright with the eyes closed, as well as those that improve balance, agility, and coordination while moving around, such as walking along a straight line or heel-to-toe walking. Balance activities can also be incorporated into daily routines, for example, standing on one foot while waiting in line at the grocery store. Yoga and Tai chi are examples of activities that combine neuromotor exercise, resistance exercise, and flexibility exercise.

How to Prevent Food Insecurity

Once a diet that will meet needs has been developed, steps must be taken to assure that the elderly individual is capable of obtaining and consuming this diet. Preventing food insecurity and ensuring adequate nutrient intake may involve providing nutrient-dense meals and instruction regarding nutrient needs, economical food choices, and how to prepare food. It may also require assistance with shopping and food preparation.

Overcoming Economic and Social Limitations Affording a healthy diet is a problem for many older individuals. Reduced-cost food and meals at senior centers, food banks, and soup kitchens are available to people on limited incomes, as are coupons and debit cards that can be used to purchase food. Programs that provide education about low-cost, nutritious food choices can also help reduce food costs.

Another problem that contributes to poor nutrient intake is loneliness (**Figure 16.17**). Living, cooking, and eating alone can decrease interest in food. This can be a problem not only for the elderly but for anyone who typically eats alone.

Cooking for one is also a challenge. Buying single servings of food is an option, although it can be expensive. To avoid spoilage of perishable items, grocers can be asked to break up packages of meat, eggs, fruits, and vegetables so small amounts can be purchased. Alternatively, large packages can be purchased and shared among friends. Cooking larger portions and freezing foods in meal-size batches can be helpful not only with cost but also to relieve the boredom of eating the same leftovers several days in a row. Creativity and flexibility in what defines a meal can also help. An easy single meal can be prepared by topping a potato with cooked vegetables and cheese, or with leftover chili or spaghetti sauce (**Figure 16.18**). Yogurt or a bowl of cereal with fruit and milk is also a nutritious dinner option.

Thinkstock/Getty Images, Inc.

Figure 16.17 Congregate meals
The social interaction provided by congregate meals is as beneficial to older adults as the meals provided.

Overcoming Physical Limitations Difficulty in cooking due to limited mobility can also reduce food intake. Precooked foods, frozen dinners, canned foods, or salad-bar items,

Tony Freeman/PhotoEdit

Figure 16.18 Easy meals
Selecting acceptable foods that are nutritious and easy to prepare is important in meeting the needs of the elderly.

as well as instant foods such as cereals, rice and noodle dishes, and soups that just require adding water, can provide a meal with almost no preparation. Medical nutritional products such as Ensure or Boost can also be used to supplement intake. These canned, fortified products have a long shelf life and can meet nutrient needs with a small volume. Food can also be ordered by phone if this is affordable. Eating out at senior centers or low-cost restaurants or sharing shopping and cooking chores with a friend can reduce cooking demands and increase social interaction. Home health services can help with cooking and feeding, and most senior centers, health departments, and social service agencies offer meals, rides, and in-home care.

Overcoming Medical Limitations Medical conditions and the use of medications often affect food choices. Meals need to be appealing and easy to prepare and consume, as well as compatible with medical conditions. For instance, an individual with dental problems may not be able to chew fresh fruits and vegetables. Therefore, texture modification is required. Fully cooked, canned, or soft fruit or fruit juices can be substituted for hard-to-chew fruits, and cooked vegetables can replace raw ones. Eggs and stewed meats can provide easy-to-chew protein sources. To overcome changes in the sense of taste and smell, spicy or acidic foods may be limited or emphasized, depending on individual tastes.

Special diets are typically prescribed for individuals with medical conditions such as hypertension, heart disease, diabetes, and kidney disease. These diets may contribute to malnutrition if they restrict favorite foods and if individuals prescribed the diet are not given enough information about how to substitute foods that will provide adequate energy, nutrients, and eating pleasure.[9] Education about what foods are appropriate and how to read food labels can help identify products that fit within dietary restrictions.

The use of prescription or over-the-counter medications to treat medical conditions can also affect eating habits. Physicians, pharmacists, and dietitians can provide information about possible effects on food intake. Purchasing all prescription medications from the same pharmacy will ensure that the pharmacist is aware of all medications taken and can advise of possible interactions. Health-care providers also need to be informed about all nonprescription medications and vitamin, mineral, or other dietary supplements used and whether or not medications are taken according to the prescription instructions.

Nutrition Programs for the Elderly

To address concerns over the nutritional health of the elderly, the federal Nutrition Screening Initiative was developed to promote screening for and intervention in nutrition-related problems in older adults. This program is working to increase the awareness of nutritional problems in the elderly by involving practitioners and community organizations as well as relatives, friends, and others caring for the elderly in evaluating the nutritional status of the aging population.[58] This program developed the DETERMINE checklist (**Table 16.4**); the name is based on an acronym for the physiological, medical,

TABLE 16.4 DETERMINE: A Checklist of the Warning Signs of Malnutrition

Disease	Any disease, illness, or condition that causes changes in eating can predispose a person to malnutrition. Memory loss and depression can also interfere with nutrition if they affect food intake.
Eating poorly	Eating either too little or too much can lead to poor health.
Tooth loss/mouth pain	Poor health of the mouth, teeth, and gums interferes with the ability to eat.
Economic hardship	Having to, or choosing to, spend less than $25 to $30 per person per week on food interferes with nutrient intake.
Reduced social support	Not having contact with people on a daily basis has a negative effect on morale, well-being, and eating.
Multiple medicines	The more medicines a person takes, the greater the chances of side effects such as weakness, drowsiness, diarrhea, changes in taste and appetite, nausea, and constipation.
Involuntary weight loss/gain	Unintentionally losing or gaining weight is a warning sign that should not be ignored. Being overweight or underweight also increases the risk of malnutrition.
Needs assistance in self-care	Difficulty walking, shopping, and cooking increases the risk of malnutrition.
Elder above age 80	The risks of frailty and health problems increase with increasing age.

CRITICAL THINKING Averting Malnutrition

Shirley is an 82-year-old woman who lives alone. She has periodontal disease, which has caused her to lose teeth, so she eats foods that are easy to chew, as well as easy to prepare and easy to carry home on the bus.

iStockphoto

Shirley's Diet

FOOD	AMOUNT
Breakfast	
Bran flakes	1 cup
Reduced-fat milk	1 cup
Coffee	1 cup
Reduced-fat milk	1 T
Sugar	2 tsp
Lunch	
Chicken soup	1 cup
Saltine crackers	6 crackers
Applesauce	1 small
Dinner	
Reduced-fat milk	1 cup
Instant white rice	1 cup
Ground beef	3 oz
Peas	1 cup

CRITICAL THINKING QUESTIONS

- Which of the warning signs of malnutrition in the DETERMINE checklist apply to Shirley?
- Compare Shirley's intake to the MyPlate Daily Food Plan for an 82-year-old woman who gets less than 30 min of activity per day. How many servings from each food group does Shirley need to add?
- Suggest some soft foods she can add to her diet to meet her MyPlate recommendations. How many kcalories will your suggestions add?
- If she is maintaining her weight on her current diet, how much weight will she gain in a month if she follows your suggestions?

iProfile Use iProfile to find snacks that are nutrient dense and easy to chew.

and socioeconomic situations that increase the risk of malnutrition among the elderly. The elderly themselves, family members, and caregivers can use this tool to identify when malnutrition is a potential problem (see **Critical Thinking: Averting Malnutrition**).

To help meet the nutrient needs of older adults, the federal Older Americans Act provides nutrition services to older individuals who are in economic need.[59] Programs that provide nutritious meals in communal settings promote social interaction and can improve nutrient intake. The Elderly Nutrition Programs established by the Older Americans Act provide congregate meals at locations such as senior centers, community centers, schools, and churches. For those who are unable to attend congregate meals, home-delivered meals are available.

Although such programs are a first step in meeting nutritional needs, currently most provide only one meal a day for five days a week. Each meal served must provide at least a third of the RDA. In practice, participants in these elder nutrition programs are receiving 40% to 50% of their daily intake from this one meal.[60] These and other programs addressing the nutritional needs of older adults are described in **Table 16.5**.

TABLE 16.5 National Programs Promoting Better Nutrition Among Older Americans

Program
• Older Americans Act—Title III Congregate and Home-Delivered Nutrition Programs Serves at least one meal five days a week to persons 60 years and older. Meals are served at home or in churches, schools, senior centers, or other facilities.
• Older Americans Act—Title VI Congregate and Home-Delivered Nutrition Programs Provides home-delivered and congregate meals to Native-American organizations.
• Older Americans Act—Title III Health Promotion and Disease Prevention Program Provides health-promotion and disease-prevention services in areas where there are large numbers of economically needy older adults.
• Nutrition Screening Initiative Promotes nutritional screening and more attention to nutrition in all health-care and social-service settings that provide for older adults.
• Supplemental Nutrition Assistance Program (SNAP) Provides coupons or debit cards to low-income individuals including the elderly.These can be used instead of cash to purchase food.
• Nutrition Program for the Elderly Provides grants, cash, and commodity foods to states and tribes to supplement congregate and home-delivered meal programs.
• Commodity Supplemental Food Program—Elderly Provides food, nutrition education, and health-service referrals to individuals with low incomes, including the elderly.
• Child and Adult Care Food Program (Adult Day Care) Provides cash reimbursements and food commodities to community day-care centers that serve meals and snacks to children and elderly with special needs.
• Food Distribution Program on Indian Reservations Distributes commodity foods to low-income persons, including the elderly, living on or near Indian reservations.

OUTCOME

Robert E Daemmrich/Stone/Getty Images, Inc.

Hiroshi Yagi/Getty Images, Inc.

Cheng and his wife work hard to maintain a healthy diet and lifestyle. They were shocked, then, to find that the fruits, vegetables, and whole grains in their diet were not providing all of the nutrients Cheng needs. Because of Cheng's low-meat, low-dairy diet, he was consuming little vitamin B_{12}. In addition, he had developed a condition common in older adults called atrophic gastritis, which reduces the amount of stomach acid. Stomach acid is essential for releasing vitamin B_{12} from the proteins it is bound to in food and allowing it to be absorbed. Without sufficient vitamin B_{12}, the myelin coating on nerves cannot be maintained, and nerve function is disrupted. The vitamin B_{12} injections Cheng's doctor gave him helped to immediately boost his vitamin B_{12} status. To ensure he continued to get enough vitamin B_{12}, he started taking a supplement containing vitamin B_{12}. The vitamin B_{12} in supplements is not bound to proteins, so it is more easily absorbed when stomach acid is low. Cheng is now back to his old self. He continues his healthy diet and lifestyle and takes his supplement. **He is grateful he discovered his condition early, because an untreated vitamin B_{12} deficiency might have resulted in permanent neurological damage.**

APPLICATIONS

ASSESSING YOUR DIET

1. **How does age affect energy needs?**
   a. How does your average energy intake from the food record you kept in Chapter 2 compare to the EER for a person who is your height, weight, and activity level but is 75 years old?
   b. Use iProfile to suggest modifications in your food choices that would allow you to meet the nutrient needs of a 75-year-old without exceeding the energy needs.
   c. Use the ChooseMyPlate Website to determine the amounts of food recommended from each food group if you were 75 years old and at your current activity level.

CONSUMER ISSUES

2. **What do seniors need to know about nutrition?**
   a. Prepare an outline for a 20-minute lecture on nutrition and aging that could be given at a senior center in your area.

b. Define two goals for the lecture.
c. What are five main points that you should discuss?

3. What information and resources are available for the elderly and their families?

a. Use the Internet to determine what kind of nutrition information is available to individuals planning for the care of their elderly parents or relatives. What are the costs?
b. Assume that you have an elderly friend who lives and eats alone. Find resources in your area that would be able to provide meals and other services for your friend.

CLINICAL CONCERNS

4. How do medical conditions and dietary restrictions affect food choices?

a. How might you modify your food choices to accommodate a low-sodium diet?
b. How might you modify your food choices to accommodate a restriction of protein to 0.6 g/kg of body weight?
c. How might you modify your food choices to accommodate a loss of smell and taste?
d. How might you modify your food choices to accommodate a dry mouth and poorly fitting dentures?

SUMMARY

16.1 What Is Aging?

- Aging is the accumulation of changes over time that results in an ever-increasing susceptibility to disease and death. The longest an organism can live, or life span, is a characteristic of a species. The average age to which people in a population live, or life expectancy, is a characteristic of a population.
- As a population, we are living longer but not necessarily healthier lives. The elderly are the fastest-growing segment of the U.S. population. Compression of morbidity, that is, increasing the number of healthy years, is an important public health goal. A healthy diet and lifestyle cannot stop aging but can postpone the onset of many of the physiological changes and diseases that are common in older adults.

16.2 What Causes Aging?

- As organisms become older, the number of cells they contain decreases and the function of the remaining cells declines. This reduces reserve capacity, lowering the organism's ability to maintain homeostasis and increasing the risk of disease.
- The loss of cells and cell function that causes aging is believed to be due to both genetic factors that limit cell life and the accumulation of cellular damage over time.
- How long we live and how long we remain healthy is determined by a combination of genetic, environmental, and lifestyle factors.

16.3 Aging and the Risk of Malnutrition

- The risk of malnutrition increases with age due to the physiological changes that accompany aging. There are changes in the sense of smell that affect the appeal of food, changes in vision that affect the ability to prepare food, changes in digestion and absorption that decrease the intake and absorption of nutrients, changes in metabolism that affect nutrient utilization, changes in weight and body composition that increase health risks and reduce independence, changes in hormonal patterns that affect body function, and changes in immune function that increase the risk of infectious and chronic disease.
- Both infectious and chronic diseases affect nutrient requirements and the ability to consume a nutritious diet. Aging increases the incidence of diseases that reduce mobility and mental capacity, limiting the ability to acquire, prepare, and consume food. The medications used to treat disease also affect nutrition, especially when the medications are taken over long periods of time and when multiple medications are taken simultaneously.
- Low income levels increase the risk of malnutrition among the elderly by limiting the ability to purchase food. Loss of independence contributes to depression, which makes meals less appetizing and decreases the quantity and quality of foods consumed.

16.4 Nutritional Needs and Concerns of Older Adults

- Energy needs are lower in older adults due to a decrease in basal metabolic rate and the thermic effect of food and a reduction in physical activity, but the needs for protein, water, fiber, and most micronutrients remain the same. Intakes of fiber and water are often less than recommended, increasing the risk of dehydration and constipation.
- Low intakes, reduced absorption, and changes in the metabolism of certain micronutrients including vitamin B_{12}, vitamin D, and calcium puts older adults at risk of deficiency. Iron requirements decrease in women after menopause, but many older adults are at risk of iron deficiency due to blood loss from disease or medications.

16.5 Keeping Older Adults Healthy

- A healthy diet and regular exercise can prevent malnutrition, delay the onset of chronic conditions, and increase independence in older adults. MyPlate can be used to determine the amounts of food from each food group needed by adults of all ages. Meals for the elderly must be nutrient-dense, provide plenty of fluid and fiber, and consider individual medical, psychological, social, and economic circumstances. Supplements of calcium, vitamin D, and vitamin B_{12} may be beneficial. In some cases, assistance with shopping and meal preparation may be needed.
- A physical activity program for older adults should include activities that improve endurance, strength, flexibility, and balance. Activities should be tailored to the individual's needs and likes. Exercise classes are advantageous for older adults because they provide both social interaction and professional support and instruction. A well-planned exercise program can reduce the risk of chronic disease, improve mobility, increase independence, and reduce the risk of falls and injuries.
- The DETERMINE checklist helps identify older adults who are at risk for malnutrition.
- The federal Older Americans Act includes programs that provide older adults with low-cost or free meals in their homes or in a social setting. Although these programs are helpful, they do not ensure adequate nutrition for all elderly.

REVIEW QUESTIONS

1. What is life expectancy? How does it differ from healthy life expectancy? How does it differ from life span?
2. What is meant by compression of morbidity?
3. What factors determine at what age the consequences of aging become apparent?
4. Why are older adults at risk for malnutrition?
5. List three physiological changes that occur with aging.
6. How can nutrition affect the risk of developing macular degeneration?
7. Should obese adults over 70 years of age lose weight? Why or why not?
8. Explain how physical disabilities and mental illness affect nutritional status.
9. List three ways in which medication use and nutrition interact.
10. What social and economic factors increase nutritional risk among the elderly?
11. Why are the energy needs of older adults reduced?
12. Why are older adults at risk for vitamin B_{12} deficiency? Vitamin D deficiency?
13. Why is it important that elderly individuals consume a nutrient-dense diet?
14. Compare the MyPlate recommendations for a 65-year-old male and an 85-year-old male.
15. List some activities that are appropriate for older adults to improve endurance, strength, flexibility, and balance.
16. What is the purpose of the DETERMINE checklist?

REFERENCES

1. Anderson, R. M., and Weindruch, R. The caloric restriction paradigm: Implications for healthy human aging. *Am J Hum Biol* 24:101–106, 2012.
2. Speakman, J. R., and Mitchell, S. E. Caloric restriction. *Mol Aspects Med* 32:159–221, 2011.
3. Centers for Disease Control and Prevention. Deaths: Final data for 2010. *Natl Vital Stat Rep* 60:1–68, 2010.
4. Federal Interagency Forum on Aging-Related Statistics. *Older Americans 2008: Key Indicators of Wellbeing, Population.*
5. U.S. Census Bureau, Decennial Census Data and Population Projections.
6. World Health Organization. *The World Health Report, 2004—Changing History.*
7. Fries, J. F. Aging, natural death, and the compression of morbidity. *N Engl J Med* 303:130–135, 1980.
8. Shen, J., and Tower, J. Programmed cell death and apoptosis in aging and life span regulation. *Discov Med* 8:223–226, 2009.
9. Dorner, B., Friedrich, E. K., and Posthauer, M. E. Position of the American Dietetic Association: Individualized nutrition approaches for older adults in health care communities. *J Am Diet Assoc* 110:1549–1553, 2010.
10. Institute of Medicine, Food and Nutrition Board. *Dietary Reference Intakes for Thiamin, Riboflavin, Niacin, Vitamin B_6, Folate, Vitamin B_{12}, Pantothenic Acid, Biotin, and Choline.* Washington, D.C.: National Academies Press, 1998.
11. Satish, S. C., and Jorge, T. G. Update on the management of constipation in the elderly: New treatment options. *Clin Interv Aging* 5:163–171, 2010.
12. Esposito, C., Plati, A., Mazzullo, T., et al. Renal function and functional reserve in healthy elderly individuals. *J Nephrol* 20:617–625, 2007.
13. Villareal, D. T., Apovian, C. M., Kushner, R. F., and Klein, S.; American Society for Nutrition; NAASO, The Obesity Society. Obesity in older adults: Technical review and position statement of the American Society for Nutrition and NAASO, The Obesity Society. *Obes Res* 13:1849–1863, 2005.
14. Thomas, D. R. Weight loss in older adults. *Rev Endocr Metab Disord* 6:129–136, 2005.
15. Thomas, D. R. Loss of skeletal muscle mass in aging: Examining the relationship of starvation, sarcopenia and cachexia. *Clin Nutr* 26:389–399, 2007.
16. Wroblewski, A. P, Amati, F., Smiley, M. A., et al. Chronic exercise preserves lean muscle mass in Masters athletes. *Phys Sports Med* 39:172–178, 2011.
17. Cohn, S. H., Vartsky, D., Yasumura, S., et al., Compartmental body composition based on total-body nitrogen, potassium, and calcium. *Am J Physiol* 239:192–200, 1980.
18. Col, N. F., Surks, M. I., and Daniels, G. H. Subclinical thyroid disease: Clinical applications. *JAMA* 291:239–243, 2004.
19. Surks, M. I., Ortiz, E., Daniels, G. H., et al. Subclinical thyroid disease: Scientific review and guidelines for diagnosis and management. *JAMA* 291:228–238, 2004.
20. U.S. Preventive Services Task Force. Hormone therapy for the prevention of chronic conditions in postmenopausal women: Recommendations from the U.S. Preventive Services Task Force. *Ann Intern Med* 142:855–860, 2005.
21. Bartke, A. Growth hormone and aging: a challenging controversy. *Clin Interv Aging* 3:659–665, 2008.
22. Thorner, M. O. Statement by the Growth Hormone Research Society on the GH/IGF-I axis in extending health span. *J Gerontol A Biol Sci Med Sci* 64:1039–1044, 2009.
23. Cameron, D. R., and Braunstein, G. D. The use of dehydroepiandrosterone therapy in clinical practice. *Treat Endocrinol* 4:95–114, 2005.

24. Bondy, S. C., Lahiri, D. K., Perreau, V. M., et al. Retardation of brain aging by chronic treatment with melatonin. *Ann NY Acad Sci* 1035:197–215, 2004.

25. Karasek, M. Does melatonin play a role in the aging process? *J Physiol Pharmacol* 58:105–113, 2007.

26. National Institute of Diabetes and Digestive and Kidney Diseases. *National Diabetes Information Clearinghouse, National Diabetes Statistics, 2011.*

27. Chandra, R. K. Impact of nutritional status and nutrient supplements on immune responses and incidence of infection in older individuals. *Aging Res Rev* 3:91–104, 2004.

28. U.S. Department of Health and Human Services. *Securing the Benefits of Medical Innovation for Seniors: The Role of Prescription Drugs and Drug Coverage.*

29. Administration on Aging. *A Profile of Older Americans: 2007.*

30. Centers for Disease Control and Prevention (CDC). Prevalence of doctor-diagnosed arthritis and arthritis-attributable activity limitation—United States, 2007–2009. *MMWR Morb Mortal Wkly Rep* 59:1261–1265, 2010.

31. American Dietetic Association. Position of the American Dietetic Association: Nutrition across the spectrum of aging. *J Am Diet Assoc* 105:616–633, 2005.

32. Wandel, S., Jüni, P., Tendal, B., et al. Effects of glucosamine, chondroitin, or placebo in patients with osteoarthritis of hip or knee: Network meta-analysis. *BMJ* 341:c4675, 2010.

33. Institute of Medicine, Food and Nutrition Board. *Dietary Reference Intakes for Energy, Carbohydrate, Fiber, Fat, Protein, and Amino Acids.* Washington, D.C.: National Academies Press, 2002.

34. Cherubini, A., Martin, A., Andres-Lacueva, C., et al. Vitamin E levels, cognitive impairment and dementia in older persons: The InCHIANTI study. *Neurobiol Aging* 26:987–994, 2005.

35. Tomljenovic, L. Aluminum and Alzheimer's disease: After a century of controversy, is there a plausible link? *J Alzheimer's Dis* 23:567–598, 2011.

36. Pocernich, C. B., Lange, M. L., Sultana, R., and Butterfield, D. A. Nutritional approaches to modulate oxidative stress in Alzheimer's disease. *Curr Alzheimer Res* 8:452–469, 2011.

37. Gu, Y., and Scarmeas, N. Dietary patterns in Alzheimer's disease and cognitive aging. *Curr Alzheimer Res.* 8:510–519, 2011.

38. Hajjar, E. R., Cafiero, A. C., and Hanlon, J. T. Polypharmacy in elderly patients. *Am J Geriatr Pharmacother* 5:314–316, 2007.

39. Ness, J., Cirillo, D. J., Weir, D. R., et al. Use of complementary medicine in older americans: Results from the health and retirement study. *Gerontologist* 45:516–524, 2005.

40. Graham, R. E., Gandhi, T. K., Borus, J., et al. *Risk of Concurrent Use of Prescription Drugs with Herbal and Dietary Supplements in Ambulatory Care.*

41. Moss, M. Drugs as anti-nutrients. *J Nutr Environ Med* 16: 149–166, 2007.

42. Bailey, D .G. Fruit juice inhibition of uptake transport: A new type of food-drug interaction. *Br J Clin Pharmacol* 70:645–655, 2010.

43. Administration on Aging. *A Profile of Older Americans: 2011.*

44. Coleman-Jensen, A., Nord, M., Andrews, M., and Carlson, S. *Household Food Security in the United States, 2010.* September 2011.

45. Dorner, B., Niedert, K. C., and Welch, P. K., American Dietetic Association. Position of the American Dietetic Association: Liberalized diets for older adults in long-term care. *J Am Diet Assoc* 102:1316–1323, 2002.

46. National Institute of Mental Health, National Institutes of Health. *Older Adults: Depression and Suicide Facts (Fact Sheet).*

47. Villareal, D. T., Apovian, C. M., Kushner, R. F., and Klein, S. American Society for Nutrition; NAASO, the Obesity Society. Obesity in older adults: Technical review and position statement of the American Society for Nutrition and NAASO, the Obesity Society. *Am J Clin Nutr* 82:923–934, 2005.

48. Roberts, S., B., and Dallal, G. E. Energy requirements and aging. *Public Health Nutr* 8:1028–1036, 2005.

49. Food and Nutrition Board, Institute of Medicine. *Dietary Reference Intakes Water, Potassium, Sodium, Chloride, and Sulfate.* Washington, DC: National Academies Press, 2004.

50. Rampersaud, G. C., Kauwell, G. P., and Bailey, L. B. Folate: A key to optimizing health and reducing disease risk in the elderly. *J Am Coll Nutr* 22:1–8, 2003.

51. Centers for Disease Control and Prevention. *5 a Day.*

52. Food and Nutrition Board, Institute of Medicine. *Dietary Reference Intakes for Calcium and Vitamin D.* Washington, D.C.: National Academies Press, 2011.

53. Price, E. A., Mehra, R., Holmes, T. H., and Schrier, S. L. Anemia in older persons: Etiology and evaluation. *Blood Cells Mol Dis* 46:159–163, 2011.

54. Chernoff, R. Micronutrient requirements in older women. *Am J Clin Nutr* 81:1240S–1245S, 2005.

55. Bjelakovic, G., Nikolova, D., Gluud, L. L., et al. Antioxidant supplements for prevention of mortality in healthy participants and patients with various diseases. *Cochrane Database Syst Rev.* 3:CD007176, 2012.

56. Garber, C. E., Blissmer, B., Deschenes, M. R., et al; American College of Sports Medicine. American College of Sports Medicine position stand: Quantity and quality of exercise for developing and maintaining cardiorespiratory, musculoskeletal, and neuromotor fitness in apparently healthy adults: Guidance for prescribing exercise. *Med Sci Sports Exerc* 43:1334–1359, 2011.

57. United States Department of Health and Human Services, *2008 Physical Activity Guidelines for Americans.*

58. Brunt, A. R. The ability of the DETERMINE checklist to predict continued community-dwelling in rural, white women. *J Nutr Elder* 25:41–59, 2006.

59. DHHS, NIH, Administration on Aging. *Elderly Nutrition Program.*

60. U.S. Department of Health and Human Services, Administration on Aging. Nutrition Services (OAA Title IIIC).

***To access links to online sources, please go to www.wiley.com/college/smolin and select Nutrition: Science and Applications, 3rd edition. From this page, select either the student or instructor companion site. Once on the desired site, select References.**

CHAPTER OUTLINE

© GMVozd/iStockphoto

Food Safety

17

CASE STUDY

One hundred children at Loghill Elementary School were absent or went home sick on Friday. Forty of them vomited at school. It might have been the stomach flu, but only a few of the teachers and parents became ill, even though they were in contact with the children and presumably would have been just as likely as the children to catch the flu. When this many people in one place become ill at the same time, food-borne illness is always suspected.

Even though almost everyone had recovered by Monday, the local health department was notified. Inspectors came to the school to investigate the cause of the illnesses and were able to trace the source to the "Welcome Back, Spring" celebration held the day before the illness struck. For this event, the first graders made cupcakes, the second graders made cookies, the third graders made fruit salad, and the fourth graders made frozen custard. After interviewing the children and adults who were sick, the inspectors determined that only those who ate frozen custard became ill and that they all began having symptoms within 48 hours of consuming it. Symptoms included nausea and vomiting, diarrhea, abdominal pain, and fever. Many of the sick children were seen by physicians, and the organism *Salmonella enteritidis* was isolated from their stool samples. **Further investigation revealed that the recipe the fourth graders used to make the frozen custard included six grade-A raw eggs. The students had cranked the custard by hand, and it had taken about two hours to harden.**

Masterfile

17.1 How Can Food Make Us Sick?

LEARNING OBJECTIVES

- Name the primary cause of food-borne illness in the United States.
- Discuss how and where food can get contaminated.
- Explain why a contaminated food does not cause illness in everyone who eats it.

The American food supply is very safe but it is not risk-free. *Salmonella* bacteria contaminate chickens sold in the United States, industrial waste has polluted some of our waterways, pesticide residues are found on our fruit, and concerns about Mad Cow disease are rising. Headlines announce *Listeria* in cantaloupe; *Escherichia coli* (*E. coli*) in meat and apple juice; *Salmonella* in eggs, on vegetables, and in cereal; *Cyclospora* on fruit; *Cryptosporidium* in drinking water; and hepatitis A in frozen strawberries. The Centers for Disease Control and Prevention estimates that each year about one in six Americans, or 48 million people, get sick, 128,000 are hospitalized, and 3,000 die from food-borne diseases.[1]

If given a choice, most people would elect to consume food that contains no harmful substances. However, choosing a diet that is free of all potential hazards is nearly impossible. Food has always carried the risks of bacterial contamination and naturally occurring toxins. Today, modern agricultural technology, trade patterns, food processing, and changes in dietary habits have increased the risks associated with bacterial contamination and introduced new risks. Regulatory agencies, food manufacturers and retailers, as well as consumers, must work together to maximize the safety of the food supply.

food-borne illness An illness caused by consumption of food containing a toxin or disease-causing microorganism.

pathogen A biological agent that causes disease.

toxin A substance that can cause harm at some level of exposure.

What Is Food-Borne Illness?

Food-borne illness in the broadest sense is any illness that is related to the consumption of food, or contaminants or toxins in food. However, most of the food-borne illness in the United States is caused by the contamination of food with **pathogens**, that is, microorganisms or microbes that can cause disease. **Toxins** produced by these microorganisms, as well as chemical and physical contaminants from the environment and those used in the processing and packaging of food, can also cause food-borne illness. Chemical contaminants include substances such as drugs used in raising cattle and producing milk, pesticides and fertilizers used in growing crops, and wastes from industry that accumulate in the environment. Physical contaminants include substances as diverse as broken glass and insect wings. Even substances we use to protect the food supply, such as packaging materials and preservatives, have the potential to cause harm if not properly used.

How Does Food Get Contaminated?

Food can be contaminated at any point in its journey from production to when it is prepared in your kitchen (**Figure 17.1**). Often food is contaminated where it is grown or produced. *E. coli* from improperly treated manure fertilizer can contaminate strawberries in the field. *Salmonella enteritidis* may enter eggs directly from hens infected with the bacteria. Fish and seafood may be contaminated by agricultural runoff, sewage, and other toxins in the waters where they live. Molds may grow on grains during unusually wet or dry growing seasons. Food can also be contaminated during processing or storage, at retail facilities, during transport, and even at home. This often occurs by **cross-contamination** from a contaminated food or piece of equipment to an uninfected one. For example, *E. coli* from a single cow can contaminate processing equipment and be transferred to thousands of pounds of hamburger. Careful sanitation and handling can control most of these sources of food-borne illness.

cross-contamination The transfer of contaminants from one food or object to another.

Masterfile

1 Farm
Crops can be contaminated with bacteria before they are even harvested.

Thinkstock Images/Getty Images, Inc.

2 Processing
Cross-contamination from food or equipment can transfer microbes to large volumes of food.

David Freund/iStockphoto

3 Transportation
During transport, poor sanitation and inadequate refrigeration can contaminate food and allow microbes to grow.

4 Retail
Food can become contaminated during handling or storage in grocery stores or during preparation in restaurants.

Digital Vision/Getty Images, Inc.

5 Table
Even a safe food can be contaminated in the home if it is not handled, stored, and prepared safely.

Masterfile

Figure 17.1 **Sources of food contamination**
Keeping food safe involves identifying possible points of contamination along its journey from the farm to the dinner table.

When Do Contaminants Make Us Sick?

Even when a food is contaminated, it does not cause every individual who consumes it to become ill. The potential of a substance to cause harm depends on how potent it is, the amount or dose that is consumed, how frequently it is consumed, and who consumes it. Some contaminants in food cause harm even when minute amounts are consumed, and almost any substance can be toxic if a large enough amount is consumed. Many substances have a **threshold effect**; that is, they are harmless up to a certain dose or threshold, after which negative effects increase with increasing intake. Body size, nutritional status, and how a substance is metabolized by the body can also affect toxicity. Small doses are more likely to cause harm in children and small adults because the amount of toxin per unit of body weight is greater. Poor nutritional or health status may decrease the body's natural ability to detoxify harmful substances. Substances that are stored in the body are more likely to be toxic because they accumulate over time. They are deposited in bones, adipose tissue, liver, or other tissues eventually causing toxicity symptoms. Substances that are easily excreted when consumed in excess are less likely to cause toxicity. The interaction of toxins with one another and with other dietary factors also affects toxicity. For example, mercury, which is extremely toxic, is not absorbed well if the diet is high in selenium, and the absorption of lead is decreased by the presence of iron and calcium in the diet.

THINK CRITICALLY

Why does contamination that occurs during processing have the potential to make more people sick than contamination that occurs at home?

threshold effect A reaction that occurs at a certain level of ingestion and increases as the dose increases. Below that level there is no reaction.

17.2 Keeping Food Safe

LEARNING OBJECTIVES

- Explain how a HACCP system helps prevent food-borne illness.
- Discuss the roles of the federal agencies responsible for the safety of the U.S. food supply.
- Discuss the role of the consumer in keeping food safe.

Ensuring the safety of the food supply is a responsibility shared by the government, food manufacturers and retail establishments, and consumers. The steps taken to avoid food-borne illness require weighing the benefits a food provides against the potential risks it presents. This type of risk-benefit analysis is done by regulatory agencies when they evaluate the safety and sanitation of food-processing methods and food service establishments. Not every potential contaminant is harmful, nor can all be avoided, but those that have a great potential for harm need to be addressed. Consumers should also consider the risks and benefits when they choose which foods to buy, which to eat, and how to handle, store, and cook these foods.

What Is the Government's Role?

The safety of the food supply is monitored by agencies at the international, federal, state, and local levels (**Table 17.1**). Federal agencies set standards and establish regulations for the safe and sanitary handling of food and water and for the information on food labels. They regulate the use of agricultural chemicals, additives, and packaging materials; inspect

TABLE 17.1 Agencies that Monitor the Food Supply

Agency	Role
International Organizations	
Food and Agriculture Organization of the United Nations (FAO)	Promotes and shares knowledge in all aspects of food quality and safety and in all stages of food production: harvest, post-harvest handling, storage, transport, processing, and distribution.
World Health Organization (WHO)	Develops international food safety policies, food inspection programs, and standards for hygienic food preparation; promotes technologies that improve food safety and consumer education about safe food practices. Works closely with the FAO.
Federal Organizations	
U.S. Food and Drug Administration (FDA)	Ensures the safety and quality of all foods sold across state lines with the exception of red meat, poultry, and egg products; inspects food processing plants; inspects imported foods with the exception of red meat, poultry, and egg products; sets standards for food composition; oversees use of drugs and feed in food-producing animals; enforces regulations for food labeling, food and color additives, and food sanitation.
U.S. Department of Agriculture (USDA) Food Safety and Inspection Service (FSIS)	Enforces standards for the wholesomeness and quality of red meat, poultry, and egg products produced in the United States and imported from other countries. If an imported food is suspect, it can be tested for contamination and denied entry into the country.
U.S. Environmental Protection Agency (EPA)	Regulates pesticide levels and must approve all pesticides before they can be sold in the United States; establishes water quality standards.
National Marine Fisheries Service	Oversees the management of fisheries and fish harvesting; operates a voluntary program of inspection and grading of fish products.
National Oceanic and Atmospheric Administration (NOAA)	Oversees fish and seafood products. Its Seafood Inspection Program inspects and certifies fishing vessels, seafood processing plants, and retail facilities for compliance with federal sanitation standards.
Centers for Disease Control and Prevention (CDC)	Monitors and investigates the incidence and causes of food-borne illnesses.
State and Local Governments	
Oversee all food within their jurisdiction; also inspect restaurants, grocery stores, and other retail food establishments, as well as dairy farms and milk processing plants, grain mills, and food manufacturing plants within local jurisdictions.	

food-processing and storage facilities; monitor both domestic and imported foods for contamination; and investigate outbreaks of food-borne illness. State agencies have the primary responsibility for milk safety and the inspection of restaurants, retail food stores, dairies, grain mills, and other food-related establishments within their borders. As a result, regulations vary from state to state. To provide guidance, the U.S. Food and Drug Administration (FDA) publishes the **Food Code**, which offers recommendations for safeguarding public health when food is served to the consumer.[2]

Food Code A set of recommendations published by the FDA for the handling and service of food sold in restaurants and other establishments that serve food.

Concern about the prevalence of food-borne illness led to the establishment of the National Food Safety Initiative. The goal of this program is to reduce the incidence of food-borne illness by improving food safety practices and policies throughout the United States. The focus is on reducing the risk of microbial food-borne illness.[3] In 2011, in response to the continued threat from our food supply, the FDA Food Safety Modernization Act was passed. This legislation shifted the focus of regulation to preventing food-borne illness, not just reacting to problems as they occur. It gives the FDA an inspection mandate and new legal powers to ensure that companies are doing their part to stop potentially unsafe food from entering the marketplace.[4]

Hazard Analysis Critical Control Point (HACCP) A food safety system that focuses on identifying, preventing, and eliminating hazards that could cause food-borne illness.

Identifying Potential Problems Food safety used to be monitored by conducting spot-checks of manufacturing conditions and products. These checks often relied on visual inspection to detect contamination and typically did not find a problem until after it had already occurred. The current system for safeguarding the food supply is called **Hazard Analysis Critical Control Point (HACCP)**. This is a science-based approach designed to prevent food contamination rather than catch it after it occurs.

The HACCP approach to food safety involves establishing standardized procedures to prevent, control, or eliminate contamination before food reaches consumers (**Table 17.2**).

TABLE 17.2 Seven Principles of HACCP

Principle	Description
1. Conduct a hazard analysis.	Analyze the processes associated with the production of a food to identify the potential biological, chemical, and physical hazards and determine what type of preventive measures, such as changes in temperature, pH, or moisture level, could be used to control or avoid these hazards.
2. Identify critical control points.	Identify steps in a food's production called critical control points, at which the potential hazard can be prevented, controlled, or eliminated—for example, cooking, cooling, packaging, and metal detection.
3. Set critical limits.	Establish preventive procedures with measurable limits for all critical control points. For example, for a cooked food this might be a minimum cooking time and temperature required to ensure elimination of harmful microbes. If these critical limits are not met, the food safety hazards are not being prevented, eliminated, or reduced to acceptable levels.
4. Monitor the critical control points	Develop procedures to monitor the critical limits. For example, how and by whom will the cooking temperature be monitored? Adjustments can be made while continuing the process.
5. Establish corrective actions.	Establish plans to discard the potentially hazardous product and to correct the out-of-control process when monitoring shows that a critical limit has not been met—for example, reprocessing or discarding food if the minimum cooking temperature is not met.
6. Maintain record-keeping and documentation procedures.	Prepare and maintain a written HACCP plan.This would include records of hazards and their control methods, the monitoring of each critical control point, and notations of corrective actions taken. Each principle must be backed by sound scientific knowledge—for example, published studies on the time and temperatures needed to control specific food-borne pathogens.
7. Institute verification procedures.	Institute procedures to verify the scientific or technical validity of the hazard analysis, the adequacy of the critical control points, and the effectiveness of the HACCP plan. An example of verification is the testing of time and temperature recording devices to verify that a cooking unit is working properly.

Source: U.S. Food and Drug Administration.

critical control points Possible points in food production, manufacturing, and transportation at which controls can be applied and contamination be prevented or eliminated.

It focuses on identifying points in the handling of food, called **critical control points**, where chemical, physical, or microbial contamination can be prevented, controlled, or eliminated. The HACCP system requires the food manufacturing and food service industries to anticipate where contamination might occur. It also establishes record-keeping procedures to verify that the system is working consistently. The advantages of the HACCP system over standard inspections by the FDA are that it is preventive rather than punitive, it is easier to manage, and it puts the responsibility for food safety on the manufacturer, not the regulatory agencies.

Tracking Food-Borne Illness In addition to requiring the application of HACCP principles to identify and prevent potential and actual food hazards, the government has established a system for tracking food-borne illness once it has occurred. Rapid identification of the source of the contaminant that caused an outbreak of food-borne illness can help stop its spread. A national computer network linking public health laboratories enables epidemiologists quickly to respond to serious and widespread food contamination problems.[5] With this system, the distinctive DNA of a pathogenic strain of a microorganism can be tracked. For example, if outbreaks of food-borne illness in Ohio and Minnesota are both caused by the same strain of an organism, epidemiologists know that the outbreaks were caused by the same food source. They can focus their search for the source of contamination on foods distributed to both locations. To confirm the source, the DNA fingerprint isolated from the organisms found in people who became ill can be matched to the DNA fingerprint from a contaminated food source.

What Is the Role of Food Manufacturers and Retailers?

The responsibility of providing safe food to the marketplace falls on the shoulders of food manufacturers. It is their job to establish and implement a HAACP system for their particular business. Once in place, it allows the company to anticipate where contamination might occur and then to prevent hazardous food from reaching the consumer. **Figure 17.2** provides an example of how a HACCP system is used to ensure the safety of liquid egg products. To produce these products, eggs are removed from their shells; mixed together in large vats; heated, in a process called **pasteurization**, to kill *Salmonella* and other microbial contaminants; packaged; and then refrigerated or frozen.

pasteurization The process of heating food products to kill disease-causing organisms.

Food manufacturers are also responsible for proper labeling of their products. In addition to nutritional labeling, some products also contain safe-handling labels as well as some type of product dating. A "use by" date, also called a "best if used by," "freshness," or "quality assurance" date refers to the last date the product is likely to be at peak flavor, freshness, and texture. Beyond this date, the product's quality may diminish, but the food may still be safe if it has been handled and stored properly. A "sell by" or "pull by" date indicates when the grocery store should take the product off the shelf. You should buy the product before the date, but if the food has been handled and stored properly, it is usually still safe for consumption after it. For example, milk usually can be used for up to about seven days after the sell-by date. Some foods also have an "expiration date." This is used to specify the last date that the food should be eaten. State governments regulate these dates for perishable items such as milk and eggs. The FDA regulates only the expiration dates of infant formulas.

Once food has left the manufacturer, it goes to restaurants and other retail establishments. These businesses are responsible for preventing contaminated food from reaching the consumer. They must monitor the food that enters the establishment and prevent infected food, utensils, and employees from cross-contaminating food served to customers. Food in retail establishments has many opportunities to be contaminated because of the large volume of food that is handled and the large number of people involved in food preparation. Although most of the food-borne illness in the United States is caused by food prepared in private homes, an outbreak in a commercial or institutional establishment usually involves more people at a time and is more likely to be reported.

Even when a restaurant uses extreme care in food preparation, customers can be a source of contamination. Because customers serve themselves at salad bars, cross-contamination

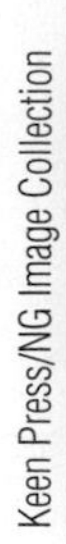

1 Conduct a hazard analysis.
The manufacturer analyzes its processing steps for potential hazards and determines preventive measures that can be taken. Contamination of eggs with the bacterium *Salmonella* is a potential hazard. Adequate heating is a preventive measure that can eliminate this hazard.

2 Identify the critical control points.
The critical control point in egg processing is pasteurization of the shelled egg mixture.

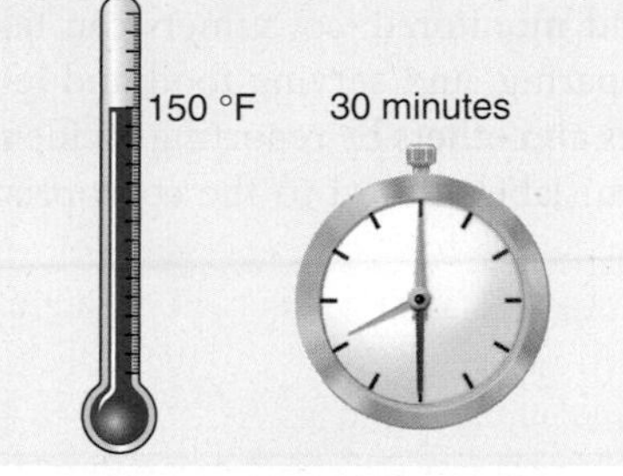

3 Set critical limits.
The critical limits for egg pasteurization are sufficient heating time and temperature to ensure that *Salmonella* bacteria are killed.

Hank Morgan/Photo Researchers, Inc.

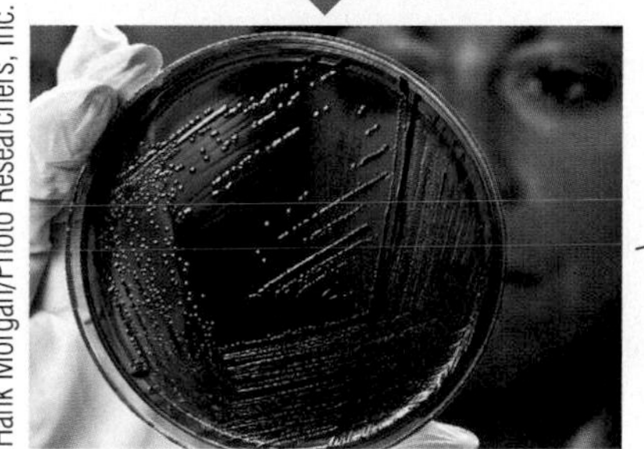

4 Monitor the critical control points.
Each batch of pasteurized eggs is tested for *Salmonella.* If the temperature is not high enough or the heating is not continued long enough, *Salmonella* can survive, as shown here by the growing bacterial colonies.

5 Establish corrective action.
Batches of eggs that are contaminated with *Salmonella* are discarded and the temperature of the heat chamber is adjusted to ensure that *Salmonella* in the next batch will be killed.

6 Maintain record keeping procedures.
Extensive records are kept, documenting the monitoring of critical control points and corrective actions taken. This enables the source of a problem to be traced in the event of an outbreak of food-borne illness.

Dorling Kindersley/Getty Images, Inc.

7 Institute verification procedures.
Plans and records are continually reviewed to ensure that the HACCP plan is working and only safe eggs are reaching consumers.

Figure 17.2 HACCP in liquid egg production
The scrambled eggs served in your cafeteria at school or work most likely came out of a carton rather than a shell. This example shows how a HACCP system might be used to prevent contaminated liquid egg products from reaching consumers.

Catherine Karnow/© Corbis

Figure 17.3 **Sneeze guards protect food**
Clear plastic shields placed above salad bars and buffets prevent customers from contaminating food with microorganisms transmitted by coughs and sneezes.

from one customer to another is a risk. To limit this, salad and dessert bars in restaurants are usually equipped with "sneeze guards" (**Figure 17.3**).

What Is the Consumer's Role?

Consumers should be actively involved in preventing food-borne illness. Individuals must decide what foods they will consume and evaluate the risks involved. A food that has been manufactured, packaged, and transported with the greatest care can still cause food-borne illness if it is not carefully handled at home. For example, contaminated eggs, chicken, or hamburger can cause microbial food-borne illness if they are not thoroughly cooked. Just as manufacturers are asked to identify critical control points in food handling where contamination can be prevented, eliminated, and monitored, consumers can take a similar approach in selecting, storing, preparing, and serving food and leftovers. Consumers can also protect themselves and others by reporting incidents involving unsanitary, unsafe, deceptive, or mislabeled food to the appropriate agencies (**Table 17.3**).

TABLE 17.3 How to Report Food-Related Issues

Before reporting a suspected case of food contamination, get all the facts. Determine whether you have used the product as intended and according to the manufacturer's instruction. Check to see if the item is past its expiration date. After these steps have been taken, report the incident to the appropriate agency:
• **Problems related to any food except meat and poultry, including adverse reactions:** Report emergencies to the FDA's main emergency number, which is staffed 24 hours a day: 301–443–1240. Nonemergencies can be reported to the FDA consumer complaint coordinator in your area, which you can find at www.fda.gov/safety/reportaproblem.
• **Issues related to meat and poultry:** Report first to your state department of agriculture and then to the USDA Meat and Poultry Hotline (1–888-MPHotline or mphotline.fsis@usda.gov).
• **Restaurant food and sanitation problems:** Report directly to your local or state health department.
• **Issues related to alcoholic beverages:** Report to the U.S. Department of the Treasury's Bureau of Alcohol, Tobacco, Firearms, and Explosives.
• **Pesticide, air, and water pollution:** Report first to your state environmental protection department and then to the U.S. EPA.
• **Products purchased at the grocery store:** Return to the store. Grocery stores are concerned with the safety of the foods they sell, and they will take responsibility for tracking down and correcting the problem. They will either refund your money or replace the product.

17.3 Pathogens in Food

LEARNING OBJECTIVES

- Distinguish food-borne infection from food-borne intoxication.
- Discuss three types of bacteria that commonly cause food-borne illness.
- Explain how viruses, molds, parasites, and prions can make us sick.
- Describe how careful food handling can prevent food-borne illness.

food-borne infection Illness produced by the ingestion of food containing microorganisms that can multiply inside the body and cause injurious effects.

Food-borne intoxication Illness caused by consuming a food containing a toxin.

Most of the food-borne illness in the United States is caused by consuming food contaminated with pathogens (**Table 17.4**). Bacteria, viruses, molds, and parasites are pathogens that affect the food supply.[6] An illness caused by consuming food contaminated with pathogens that multiply in the gastrointestinal tract or other parts of the body is called a **food-borne infection**. An illness caused by consuming food containing toxins produced by a pathogen is referred to as a **food-borne intoxication**. Unlike food-borne infections, which are usually caused by ingesting large numbers of pathogens, intoxication can be

TABLE 17.4 Summary of Bacterial, Viral, and Parasitic Food-borne Illnesses

Microbe	Sources	Symptoms	Onset (time after consumption)	Duration
Bacteria				
Salmonella	Fecal contamination, raw or undercooked eggs and meat, especially poultry	Nausea, abdominal pain, diarrhea, headache, fever	6–48 hours	1–2 days
Campylobacter jejuni	Unpasteurized milk, untreated water, undercooked meat and poultry	Fever, headache, diarrhea, abdominal pain	2–5 days	1–2 weeks
Listeria monocytogenes	Raw milk products; soft ripened cheeses; deli meats and cold cuts, raw and undercooked poultry; meats; raw and smoked fish; raw vegetables	Fever, headache, stiff neck, chills, nausea, vomiting. May cause spontaneous abortion or stillbirth in pregnant women and meningitis and blood infections in the fetus.	Days to weeks	Days to weeks
Vibrio vulnificus	Raw seafood from contaminated water	Cramps, abdominal pain, weakness, watery diarrhea, fever, chills	15–24 hours	2–4 days
Staphylococcus aureus	Human contamination from coughs and sneezes; egg, poultry, and meat products; salads such as tuna, potato, and macaroni	Severe nausea, vomiting, diarrhea	2–8 hours	2–3 days
Escherichia coli O157:H7	Fecal contamination, undercooked ground beef	Abdominal pain, bloody diarrhea, kidney failure	5–48 hours	3 days–2 weeks or longer
Clostridium perfringens	Fecal contamination, deep-dish casseroles	Nausea, diarrhea, abdominal pain	8–22 hours	6–24 hours
Clostridium botulinum	Improperly canned foods, deep-dish casseroles, honey	Lassitude, weakness, vertigo, respiratory failure, paralysis	18–36 hours	10 days or longer (must administer antitoxin)
Shigella	Fecal contamination of water or foods, especially salads such as chicken, tuna, shrimp, and potato salads	Diarrhea, abdominal pain, fever, vomiting	12–50 hours	5–6 days
Yersinia enterocolitica	Pork, dairy products, and produce	Diarrhea, vomiting, fever, abdominal pain; often mistaken for appendicitis	24–48 hours	Weeks
Viruses				
Norovirus	Fecal contamination of water or foods, especially shellfish and salad ingredients	Diarrhea, nausea, vomiting	1–2 days	2–6 days
Hepatitis A	Human fecal contamination of food or water, raw shellfish	Jaundice, liver inflammation, fatigue, fever, nausea, anorexia, abdominal discomfort	10–50 days	1–2 weeks to several months
Parasites				
Giardia lamblia	Fecal contamination of water and uncooked foods	Diarrhea, abdominal pain, gas, anorexia, nausea, vomiting	5–25 days	1–2 weeks, but may be chronic
Cryptosporidium parvum	Fecal contamination of food or water	Severe watery diarrhea	Hours	2–4 days, but sometimes weeks
Trichinella spiralis	Undercooked pork, game meat	Muscle weakness, flulike symptoms	Weeks	Months
Anisakis simplex	Raw fish	Severe abdominal pain	1 hour–2 weeks	3 weeks
Toxoplasma gondii	Meat, primarily pork	Toxoplasmosis (can cause central nervous system disorders, flulike symptoms, and birth defects in the offspring of women exposed during pregnancy)	10–23 days	May become chronic

caused by only a few microorganisms that have produced a toxin. These food toxins may be difficult to destroy.

In most cases, the symptoms of a microbial food-borne illness include abdominal pain, nausea, diarrhea, and vomiting. These symptoms are often mistaken for the flu. Food-borne illness can also cause more severe symptoms such as spontaneous abortion, hemolytic uremic syndrome, arthritis, and Guillain-Barré syndrome. Young children, pregnant women, elderly persons, and individuals with compromised immune systems, such as patients with AIDS and cancer, are most susceptible to severe reactions. Avoiding microbial food-borne illness requires knowledge of how contamination occurs and how to handle, store, and prepare food safely.

Bacteria

Bacteria are present in the soil, on our skin, on most surfaces in our homes, and in the food we eat. Most of the bacteria in our environment are harmless, some are beneficial, and some are pathogenic, causing disease either by growing in the body or by producing toxins in food. Pathogenic bacteria may also produce toxins within the body. Usually a large number of bacteria must be consumed to cause illness. Some common causes of bacterial infections include *Salmonella, E. coli, Campylobacter jejuni, Listeria monocytogenes*, and *Vibrio vulnificus*. *Staphylococcus aureus, Clostridium perfringens*, and *Clostridium botulinum* cause food-borne intoxication.

Figure 17.4 The spread of *Salmonella*
Salmonella can infect the ovaries of hens and contaminate the eggs before the shells are formed, so that the bacteria are present inside the shell when the eggs are laid. Therefore, eggs should never be eaten raw.

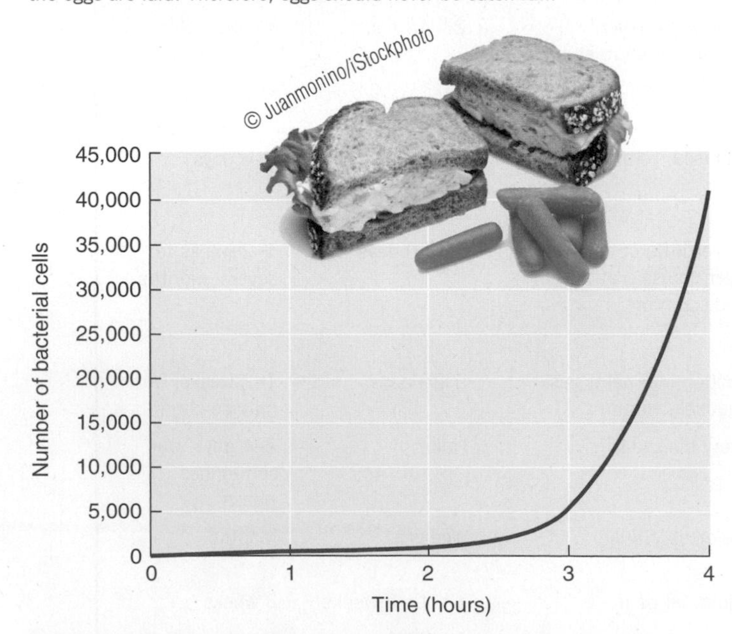

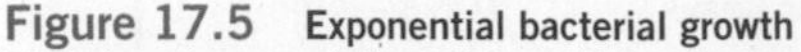

Figure 17.5 Exponential bacterial growth
The size of a population of bacterial cells doubles each time the cells divide; thus, if 10 bacterial cells contaminate an egg-salad sandwich during preparation and it sits in your warm car for four hours, during which the cells divide every 20 minutes, there will be 40,960 bacterial cells in the sandwich by the time you eat it.

Salmonella It is estimated that two to four million people in the United States become infected with *Salmonella* each year. Most of these people just experience abdominal pain and diarrhea, but some infections are more serious.[6] *Salmonella* is found in animal and human feces and infects food through contaminated water or improper handling. *Salmonella* outbreaks have been caused by contaminated meat, meat products, dairy products, seafood, fresh vegetables, and cereal, but poultry and eggs are the most common food sources. Poultry products are often contaminated because poultry farms house large numbers of chickens in close proximity, allowing one infected chicken to infect thousands of others (**Figure 17.4**). One way to reduce infection is to spray chicks with beneficial bacteria. The chicks ingest the bacteria when they preen their feathers and the beneficial bacteria colonize the digestive tract, suppressing the growth of pathogens.

Even if food contaminated with *Salmonella* is brought into the kitchen, careful handling and cooking of the food can prevent the organisms from causing illness. Washing hands, cutting boards, and utensils can prevent cross-contamination. If a contaminated food is stored in the refrigerator, the multiplication of the *Salmonella* will be slowed. In contrast, if a contaminated food is left at room temperature, the *Salmonella* will multiply rapidly, and when the food is eaten, large numbers of bacteria will be ingested with it (**Figure 17.5**). *Salmonella* is killed by heat—so foods likely to be contaminated, such as poultry and eggs, should be cooked thoroughly.

Escherichia coli (E. coli) *E. coli* is a bacterium that inhabits the gastrointestinal tracts of humans and other animals. It comes in contact with food through fecal contamination of water or unsanitary handling of food. Transmission of

E. coli is also a risk at day care centers from cross-contamination if caregivers do not carefully wash their hands after diaper changes. Some strains of *E. coli* are harmless, but others can cause serious food-borne illness. One strain of *E. coli*, found in water contaminated by human or animal feces, is the cause of "travelers' diarrhea." Another strain, *E. coli* O157:H7, produces a toxin that causes abdominal pain, bloody diarrhea, and, in severe cases, hemolytic uremic syndrome, which can lead to kidney failure and even death. *E. coli* O157:H7 can live in the intestines of healthy cattle and contaminate the meat after slaughter.

E. coli O157:H7 entered the public spotlight in 1993, when it caused the deaths of several children who had consumed undercooked, contaminated hamburgers from a fast-food restaurant. Thorough cooking of the hamburgers would have killed the bacteria that caused these deaths. *E. coli* can also contaminate produce such as lettuce, spinach, and green onions and cause illness if the produce is eaten raw. In 2010, *E. coli* contamination of lettuce sickened people in five states and in 2011 infected bologna, another food that we don't typically cook, sickening people in five different states.[7,8] In 2012, raw clover sprouts contaminated with *E. coli* infected 29 persons from 11 different states.[9] In addition to concerns about *E. coli* O157:H7 in our meat and produce, a new, even more deadly strain of *E. coli* (*E. coli* 0104:H4) emerged in Germany in 2011, sickening over 4000 people and killing 50.[10]

E. coli on food can multiply, even at refrigerator temperatures, but if a contaminated food is thoroughly cooked to 160°F, both the bacteria and the toxin are destroyed. Ground beef contaminated with *E. coli* O157:H7 is a particular risk because, unlike solid pieces of meat that are only contaminated on the surface, grinding mixes the bacteria throughout the meat (**Figure 17.6**). The *E. coli* on the outside of the meat are quickly killed during cooking, but those in the interior survive if the meat is not cooked thoroughly.

John A. Rizzo/Photodisc/Getty Images, Inc.

Figure 17.6 Ground beef and *E. coli* contamination During grinding, the bacteria on the surface of the meat is mixed throughout. *E. coli*-contaminated meat that comes into contact with a grinder contaminates the grinder and may transfer the bacteria to hundreds of pounds of beef that subsequently passes through it.

Campylobacter There are several species of *Campylobacter* that cause food-borne infections: *Campylobacter jejuni* is the leading cause of acute bacterial diarrhea in developed countries.[11] Common sources are undercooked chicken, unpasteurized milk, and untreated water. A sampling of raw chicken from supermarkets found that the majority of samples (70.1%) were contaminated with *Campylobacter*.[12] This organism grows slowly in cold temperatures and is killed by heat, so careful storage and thorough cooking help prevent infection.

Listeria Another cause of bacterial food-borne infection is *Listeria monocytogenes*. Although most cases of *Listeria* infection result in flu-like symptoms, in high-risk groups such as pregnant women, children, the elderly, and the immunocompromised, it has one of the highest fatality rates of all food-borne illnesses. *Listeria* infection is 18 times more common in pregnant women than in the general population. Infection during pregnancy is associated with an increased risk of spontaneous abortion and stillbirth, and it can be transmitted to the fetus, causing meningitis and serious blood infections.[13] Each year in the United States, an estimated 1,600 persons become seriously ill due to infection with *Listeria monocytogenes*, and of these, 260 die.[14] *Listeria* is a very resistant organism that survives at higher and lower temperatures than most bacteria; it can survive and grow at refrigerator temperatures. *Listeria* frequently contaminates dairy products, but it is destroyed by pasteurization. Because it can grow at cool temperatures, it is found in processed ready-to-eat foods such as hot dogs and lunch meats. To prevent infection, ready-to-eat meats like hot dogs should be heated to steaming and unpasteurized dairy products should be avoided.

Vibrio *Vibrio vulnificus* and *Vibrio parahaemolyticus* are two species of *Vibrio* bacteria that cause vomiting, diarrhea, and abdominal pain in healthy people and can be deadly

THINK CRITICALLY

Why are *Clostridium botulinum* spores in honey dangerous only to babies under 1 year old?

in people with compromised immune systems. The most common way people become infected is by eating raw or undercooked shellfish, particularly oysters. *Vibrio* bacteria grow in warm seawater, so the incidence of *Vibrio* infection is higher during the summer months, when warm water favors growth.

Staphylococcus aureus *Staphylococcus aureus* is a common cause of microbial food-borne intoxication. These bacteria live in human nasal passages and can be transferred through coughing or sneezing when handling food. The bacteria then produce toxins as they grow on the food. When ingested, the toxin causes symptoms that include nausea, vomiting, diarrhea, abdominal cramps, and headache. Ham, salads, bakery products, and dairy products are common sources of *Staphylococcus aureus*.

Figure 17.7 Honey may contain botulism After 1 year of age, the gut microflora have matured enough to prevent *Clostridium botulinum* spores from germinating so it is safe to include honey in the diet.

Clostridium perfringens The bacterium *Clostridium perfringens* may cause illness by both infection and intoxication. It is found in soil and in the intestines of animals and humans. It thrives in conditions with little oxygen (anaerobic conditions) and is difficult to kill because it forms heat-resistant **spores**. Spores are a stage of bacterial life that remains dormant until environmental conditions favor growth. *Clostridium perfringens* is often called the "cafeteria germ" because foods stored in large containers, such as those used to serve food in cafeteria lines, have anaerobic centers that provide an excellent growth environment. Sources include improperly prepared roast beef, turkey, pork, chicken, and ground beef.

Clostridium botulinum Another strain of *Clostridium, Clostridium botulinum*, produces the deadliest bacterial food-borne toxin. Although the bacteria themselves are not harmful, the toxin, produced as the bacteria begin to grow and develop, blocks nerve function, resulting in vomiting, abdominal pain, double vision, dizziness, and paralysis causing respiratory failure. If untreated, botulism poisoning is often fatal, but today modern detection methods and rapid administration of antitoxin have reduced mortality. Low-acid foods such as potatoes or stews that are held in large containers where anaerobic conditions prevail, provide an optimal environment for botulism spores to germinate. Canned foods, particularly improperly home-canned foods, can also be a source of botulism. Canned foods should be discarded if the can is bulging because this indicates the presence of gas produced by bacteria as they grow. Once formed, botulism toxin can be destroyed by boiling, but if the safety of a food is in question, it should be discarded; even a taste of botulism toxin can be deadly.

Infant botulism, though rare, occurs worldwide and is the most common form of botulism in the United States.[15] It occurs when botulism spores germinate in the gastrointestinal tract and produce toxin that is absorbed into the bloodstream. It causes weakness, paralysis, and respiratory problems. In the absence of complications, infants generally recover. The reason infants get botulism from ingesting spores but adults do not is because infants do not have competing intestinal microflora. In adults these healthy bacteria prevent botulism spores from germinating. Botulism spores can contaminate honey, so it should never be fed to infants under 1 year of age (**Figure 17.7**).

spore A dormant state of some bacteria that is resistant to heat but can germinate and produce a new organism when environmental conditions are favorable.

Viruses

Unlike bacteria, the viruses that cause human diseases cannot grow and reproduce in foods. Human viruses can reproduce only inside human cells. When food or water contaminated with virus particles is ingested, the virus makes us ill by entering our cells and converting them into virus-making factories (**Figure 17.8**).

Noroviruses Noroviruses are a group of viruses that cause gastroenteritis, or what we commonly think of as the "stomach flu." Symptoms, which include stomach pain, nausea, vomiting, and diarrhea, typically resolve within one to three days. Noroviruses are

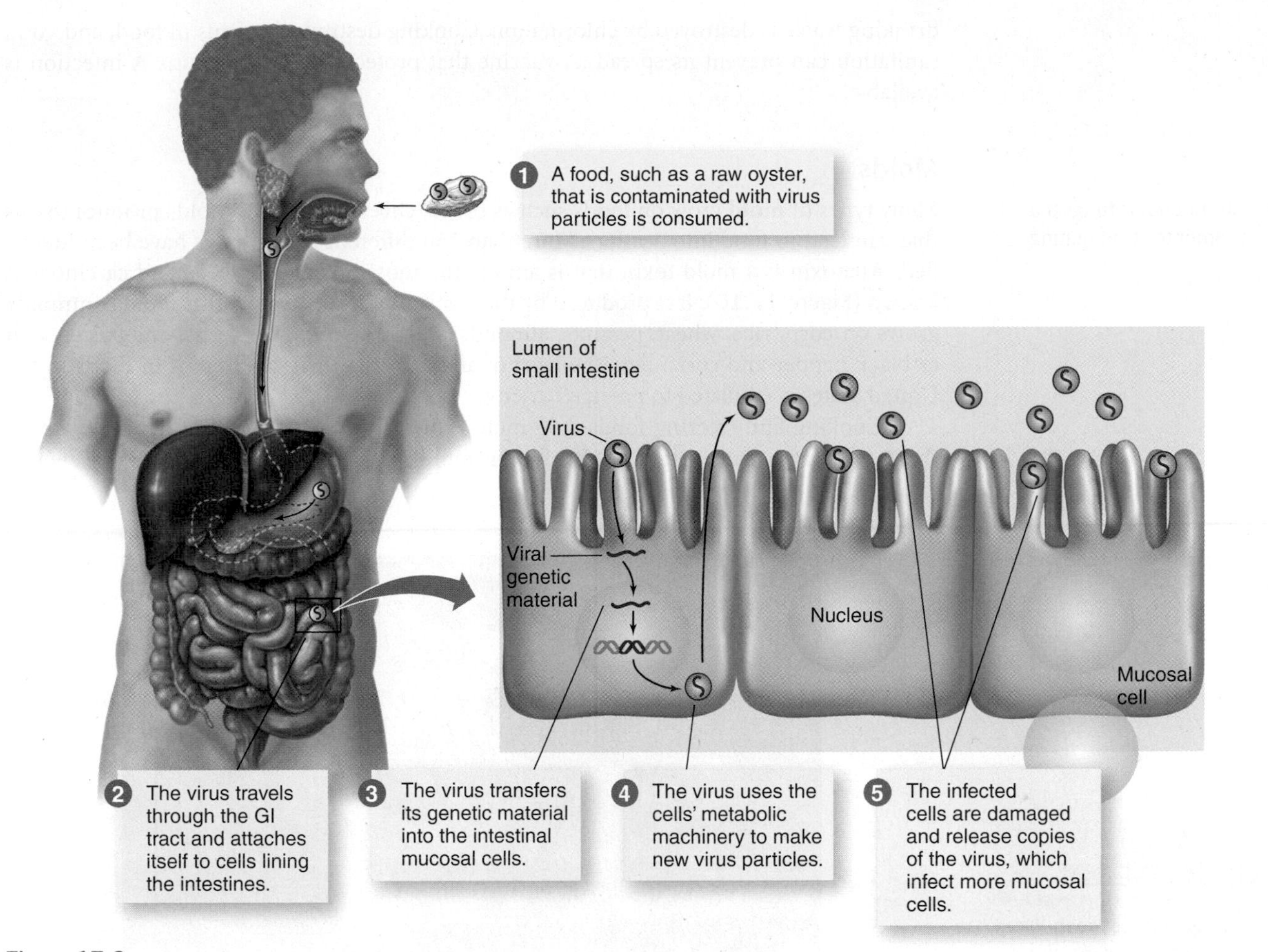

Figure 17.8 How viruses make us sick
Viruses make us sick by reproducing inside our cells. Viruses that cause food-borne illness enter the body through the gastrointestinal tract. Other types of viruses may enter the body through open cuts, the respiratory tract, or the genital tract.

THINK CRITICALLY

Why are people on cruise ships susceptible to outbreaks of norovirus?

now recognized as the most common cause of infectious gastroenteritis among persons of all ages (**Figure 17.9**). Each year, they cause about 21 million people to become ill and lead to about 70,000 hospitalizations and 800 deaths.[16] Noroviruses spread primarily from one infected person to another, but people can also be infected by eating food that is contaminated with the virus or by touching a contaminated surface and then putting their fingers to their face or mouth. Epidemics aboard cruise ships often make headlines, but noroviruses spread rapidly in any confined population, so outbreaks are also common in health care facilities, schools, and dormitories.

Noroviruses are destroyed by cooking, so the most common causes of norovirus food-borne illness are uncooked foods such as raw shellfish and fresh fruits and vegetables. Food-borne norovirus infections can be avoided by carefully washing fresh produce and thoroughly cooking seafood.

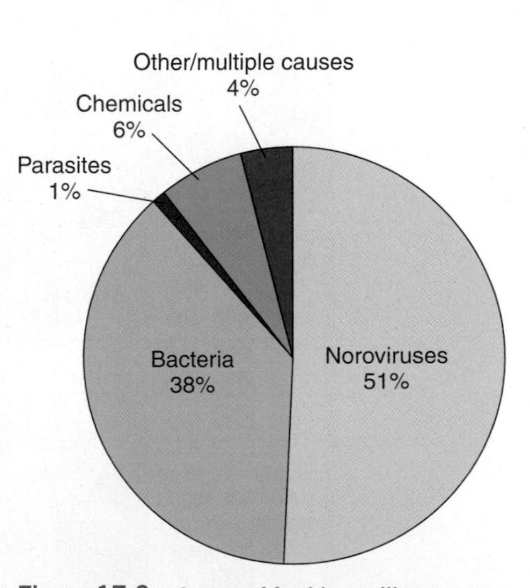

Figure 17.9 Causes of food-borne illness
Over half of the outbreaks of food-borne disease with known causes that were reported to CDC between 2006 and 2008 were due to norovirus infections.[17]

Hepatitis A Hepatitis A is a highly contagious viral disorder that causes inflammation of the liver, jaundice, fever, nausea, fatigue, and abdominal pain. It can be contracted from food contaminated by unsanitary handling or from eating raw or undercooked shellfish caught in sewage-contaminated waters. Hepatitis A can require a recovery period of several months, but it usually does not need treatment and does not cause permanent liver damage. Hepatitis in drinking water is destroyed by chlorination. Cooking destroys the virus in food, and good sanitation can prevent its spread. A vaccine that protects against hepatitis A infection is available.

Molds

mold Multicellular fungi that form a filamentous branching growth.

Many types of **mold** grow on foods such as bread, cheese, and fruit. Molds produce toxins that can lead to food intoxication. More than 250 different mold toxins have been identified. Aflatoxin is a mold toxin that is among the most potent mutagens and carcinogens known (**Figure 17.10**). It is produced by the mold *Aspergillus flavus*. This mold commonly grows on corn, rice, wheat, peanuts, almonds, walnuts, sunflower seeds, and spices such as black pepper and coriander. The level of aflatoxin that may be present in foods in the United States is regulated to prevent toxicity.

Cooking and freezing foods stop mold growth but do not destroy the mold toxins that have already been produced. If a food is moldy, it should be discarded, the area where

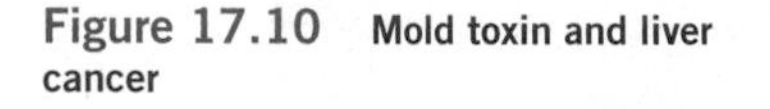

Figure 17.10 Mold toxin and liver cancer

(a)

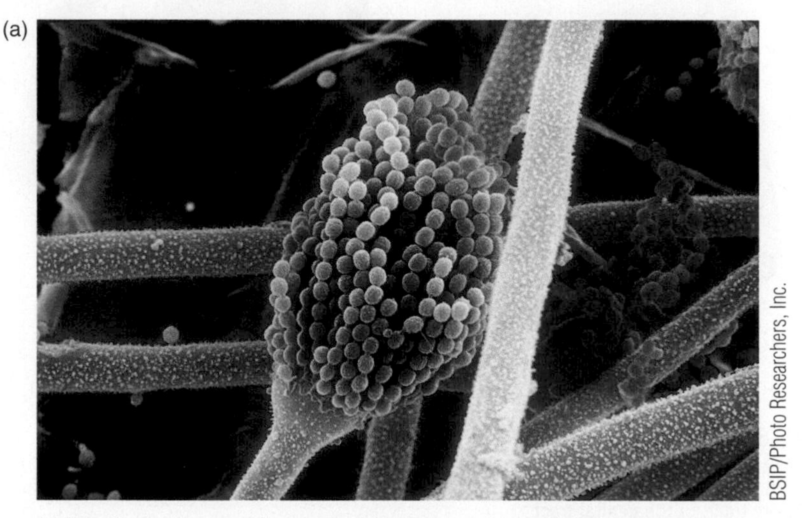

BSIP/Photo Researchers, Inc.

(a) The filamentous growths seen in this electron micrograph belong to the mold *Aspergillus flavus*, which produces aflatoxin.

THINK CRITICALLY

Is the relationship between liver cancer and aflatoxin exposure, shown here, proof that aflatoxin causes liver cancer? Why or why not?

(b)

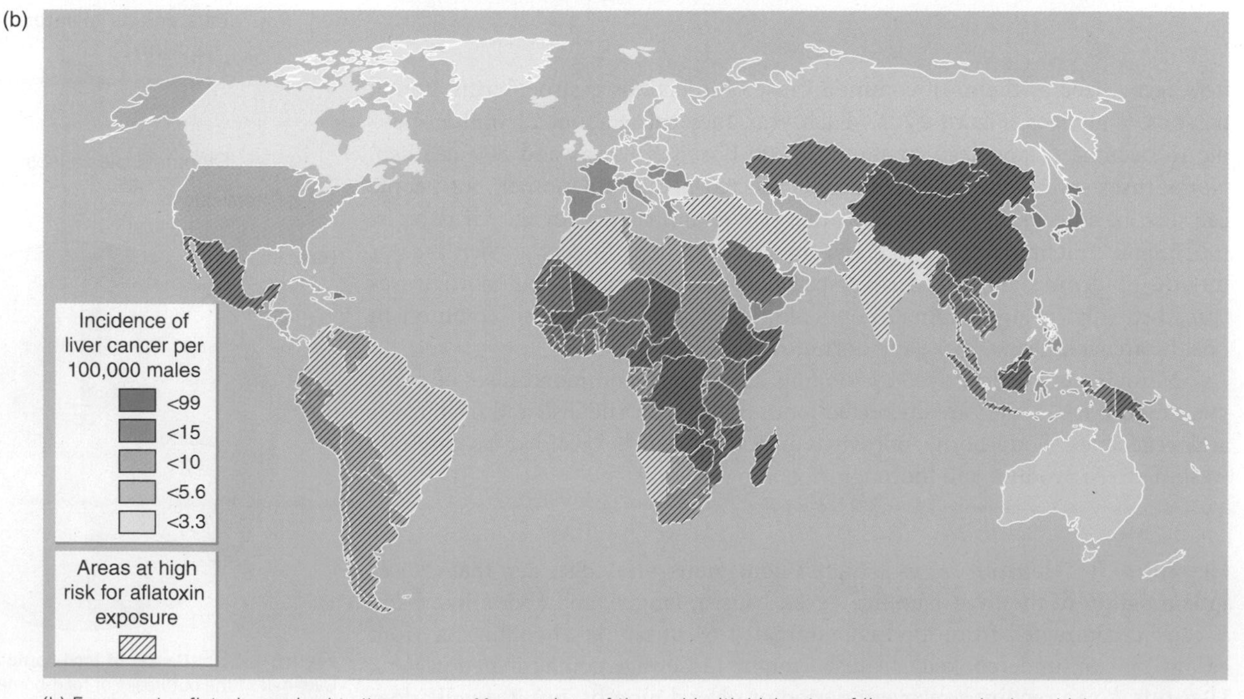

(b) Exposure to aflatoxin can lead to liver cancer. Many regions of the world with high rates of liver cancer also have high exposure to aflatoxin.[18]

it was stored should be cleaned, and neighboring foods should be checked to see if they have also become contaminated.

Parasites

Some **parasites** are tiny single-celled animals, while others are worms that are easily seen with the naked eye. Parasites are killed by thorough cooking. They may be transmitted through consumption of contaminated food or water. *Giardia lamblia* (also called *Giardia duodenalis*) is a single-celled parasite that can infect the gastrointestinal tract through water or food contaminated with human or animal feces (**Figure 17.11**). It is the most frequent cause of diarrhea not due to bacteria or viruses. *Giardia* is sometimes contracted by hikers who drink untreated water from streams contaminated with animal feces. *Giardia* infection is also becoming a problem in day care centers where diapers are changed and hands and surfaces are not thoroughly washed.[6] *Cryptosporidium parvum* is another single-celled parasite. It can be contracted by consuming water or food contaminated with the feces of humans or animals infected with the parasite.[19]

Trichinella spiralis is a parasite that was once commonly found in raw and undercooked pork and pork products. Once ingested, these small, worm-like organisms find their way to the intestine and eventually into the muscles, where they grow, causing flu-like symptoms, muscle weakness, fever, and fluid retention. Trichinosis, the disease caused by *Trichinella* infection, can be prevented by thoroughly cooking meat to kill the parasites before the meat is ingested. The parasites are also destroyed by curing, smoking, canning, or freezing. Current animal feeding practices have decreased the incidence of *Trichinella* in pork, but meat from game like bear and rabbit is still a source of this parasite.

Fish are another common source of parasitic infections. Fish can carry the larvae (the worm-like stage of an organism's life cycle) of parasites such as roundworms, flatworms, flukes, and tapeworms. One such infection, Anisakis disease, is caused by the larval form of the small roundworm *Anisakis simplex*, or herring worm, found in raw fish.[6] Once consumed, these parasites invade the stomach and intestinal tract, causing severe abdominal pain. The fresher the fish is when it is eviscerated, the less likely it is to cause this disease, because the larvae move from the fish's stomach to its flesh only after the fish dies. Parasitic infections from fish can be avoided by consuming cooked fish or freezing fish for 72 hours before consumption. If raw fish is consumed, it should be very fresh. (**Figure 17.12**).

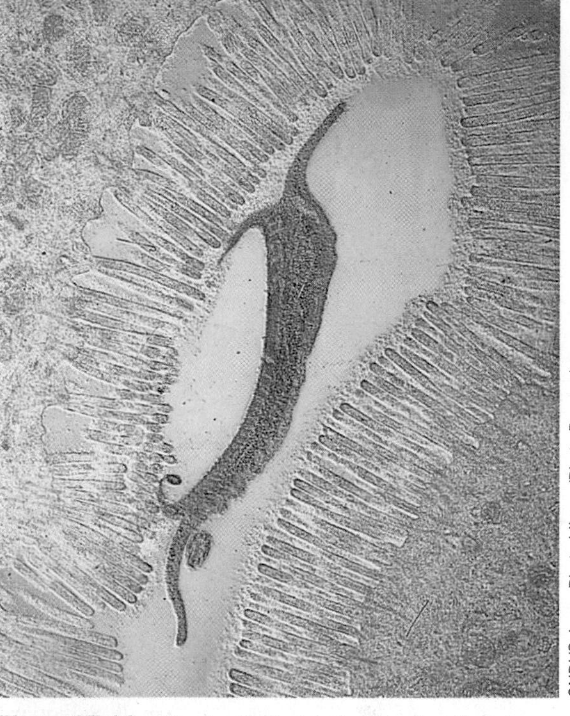

CNRI/Science Photo Library/Photo Researchers

Figure 17.11 Giardia attached to the intestinal lining This electron micrograph shows *Giardia* (green) attached to the microvilli of the human small intestine.

Hoi Fung Tsoi/Getty Images, Inc.

Figure 17.12 Raw fish can be hazardous Although rare, the incidence of parasitic infections from fish has increased along with the popularity of eating raw fish, such as this sushi.

Prions

One of the strangest, most disturbing, though rarest food-borne illness is caused not by a microbe but by a protein, called a **prion**, that has folded improperly. Abnormal prions are believed to be the cause of Mad Cow disease, or bovine spongiform encephalopathy (BSE), a deadly degenerative neurological disease that affects cattle. BSE is thought to have originated from sheep that carried a similar disease called *scrapie*. It moved into cattle in Britain when they were fed protein supplements containing the remains of slaughtered, diseased sheep. People contract the human form of this disease, which is called *variant Creutzfeldt-Jakob disease* (vCJD), by consuming the brain and nervous tissue, intestines, eyes, or tonsils from a cow infected with BSE.[20] Symptoms of vCJD begin as mood swings and numbness, and within about 14 months the nervous tissue damage progresses to dementia and death.

The abnormal prions that cause BSE and vCJD differ from normal prion proteins in the way they are folded—that is, in their three-dimensional structure. These rogue proteins

parasites Organisms that live at the expense of others.

prion A pathogenic protein that is the cause of degenerative brain diseases called *spongiform encephalopathies*. Prion is short for proteinaceous infectious particle.

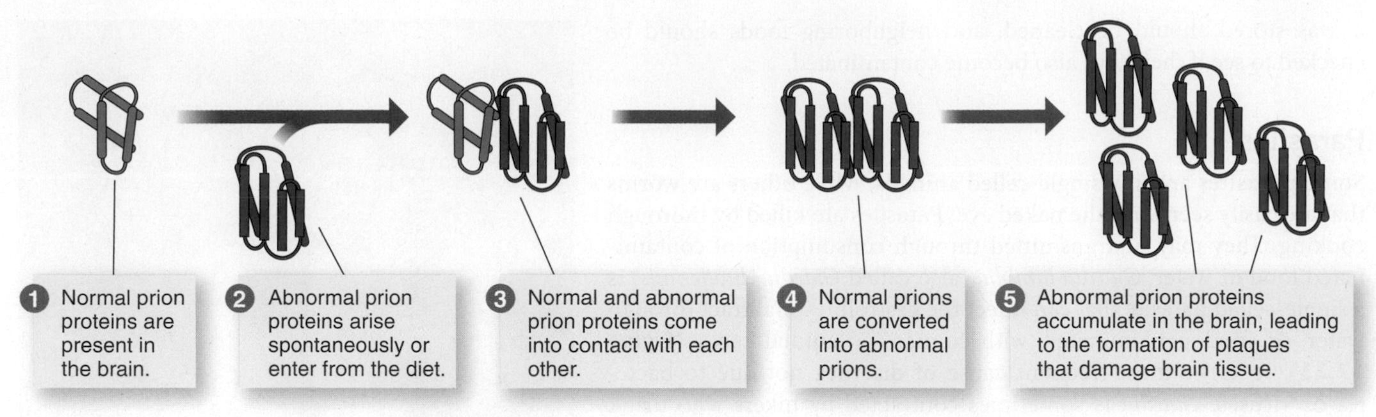

Figure 17.13 How prions multiply
When abnormal prions are introduced into the brain after a person has eaten contaminated tissue, they reproduce by corrupting neighboring proteins, essentially changing their shape so that they, too, become abnormal prions.

reproduce by corrupting neighboring proteins, essentially changing their shape, so they too become abnormal prions (**Figure 17.13**). Because the abnormal prion proteins are not degraded normally, they accumulate and form clumps called *plaques*. These plaques cause the deadly nervous tissue damage.

Even though cooking does not destroy prions, the risk of acquiring vCJD is extremely small. Safeguards are in place to prevent cattle in the United States from contracting BSE. These include restrictions on the import of animals and animal products from countries where BSE has occurred, restrictions on what can be included in cattle feed, and testing for BSE before meat is released into the food supply.[21] Several cows with BSE have been identified in the United States, but meat from these animals did not enter the food supply. Thus far there has been no known instance of U.S. beef causing a case of vCJD. Even in Britain, where over 180,000 cows were infected, only about 167 definite or probable cases of vCJD have been identified.[22]

How to Reduce the Risk of Microbial Food-Borne Illness

Despite the variety of organisms that can cause food-borne illness, most cases can be avoided if food is handled properly (**Figure 17.14**). The first critical control point in preventing food-borne illness is making safe selections at the store to reduce the contaminants that are brought into the home. Food should come from reputable vendors and appear fresh. Foods that are discolored or have an off smell and those in damaged packages should not be purchased or consumed. Tips for the safe selection and handling of food at home are summarized in **Table 17.5**.

Figure 17.14 Fight Bac!
The Fight Bac! educational campaign recommends that consumers follow four steps—clean, separate, cook, and chill—to prevent food-borne illness.

Store Food Properly Once selected, foods need to be stored appropriately. The goal is to keep foods from remaining at temperatures that promote bacterial growth (**Figure 17.15a**). Cold foods should be kept cold, at 40°F or less, and hot foods should be kept hot, at more than 135°F. Refrigerator temperature should be set between 38 and 40 °F and freezers at 0°F. Most produce should be stored in the refrigerator. Fresh meat, poultry, and fish should be frozen immediately if it will not be used within a day or two. Processed meats such as hot dogs and bologna must also be kept refrigerated but can be kept longer than fresh meat[23] (**Figure 17.15b**).

Prevent Cross-Contamination The next critical control point for food in the home is preparation. A clean kitchen is essential for safe food preparation. Hands, countertops, cutting boards, and utensils should be washed with warm soapy water before each food preparation step. Food should be thawed in the refrigerator, in the

TABLE 17.5 Tips for the Safe Selection and Handling of Food

Choose foods wisely
Voluntary freshness dates should be checked and foods with expired dates avoided. Jars should be closed and seals unbroken. Cans that are rusted, dented, or bulging should be rejected. Frozen foods should not contain frost or ice crystals, and food packaging should be secure. Frozen foods should be selected from below the frost line in the freezer. When shopping, cold foods should be purchased last.
Store foods properly
Fresh or frozen foods brought from the store should be refrigerated or frozen immediately at the proper temperature. Food that has been in your refrigerator for longer than is safe should be discarded.
Wash
Hands, cooking utensils, and surfaces should be washed with warm soapy water before each food preparation step. This will prevent cross-contamination.
Thaw
Foods should be thawed in the refrigerator or microwave or under cool running water.
Cook thoroughly
Cooking temperatures should be checked. Thorough cooking destroys most bacteria, toxins, viruses, and parasites.
Refrigerate promptly
Cooked food can be recontaminated, so it should be refrigerated as soon as possible after it is served.
Reheat thoroughly
Foods should be reheated to 165°F to destroy microorganisms that have recontaminated cooked foods and toxins that have been produced.
When in doubt, throw it out.

microwave oven, or under cool running water—not at room temperature. Foods that are eaten raw should not be prepared on the same surfaces as foods that are going to be cooked. For example, if a chicken contaminated with *Salmonella* is cut up on a cutting board and the unwashed cutting board is then used to chop vegetables for a salad, the vegetables will become contaminated with *Salmonella*. When the chicken is cooked, the bacteria will be killed, but the contaminated vegetables are not cooked, so the bacteria can grow and cause food-borne illness. Cross-contamination can also occur when uncooked foods containing live microbes come in contact with foods that have already been cooked. Therefore, cooked meat should never be returned to the same dish that held the raw meat, and sauces used to marinate uncooked foods should never be used as a sauce on cooked food. To remind consumers how to handle meat, the packaging is labeled with safe handling guidelines.

Cook Food Thoroughly Cooking is one of the most important critical control points in preventing food-borne illness. Heat will destroy most harmful microorganisms. To assure that meat is cooked thoroughly, a meat thermometer should be used. Fish should be cooked until the flesh is opaque and separates easily with a fork. Eggs should not be eaten raw, since *Salmonella* can contaminate the inside of the shell; they should be cooked until the yolk and white are firm[24] (**Figure 17.15c**).

Cooked food should be refrigerated as soon as possible after serving. The temperatures conducive to bacterial growth are the temperatures at which food is usually kept between service and storage. Large portions of food should be divided before refrigeration so they will cool quickly. Most leftovers should only be kept for a few days. For example, cooked pasta can be kept refrigerated for three to five days; cooked beef, poultry, pork, vegetables, soup, and stews for three to four days; and stuffing and meat in gravy for one to two days (see Figure 17.15b). When leftovers are reheated, they should be heated to 165 °F to destroy any bacteria that may have grown in them.

Food Safety Away from Home When you prepare your own food, you can monitor the safety of the ingredients and food preparation steps, but at restaurants, picnics, and

Figure 17.15 **Temperature and bacterial growth**

(a) Bacterial growth is most rapid in the danger zone, between 40 and 135 °F.

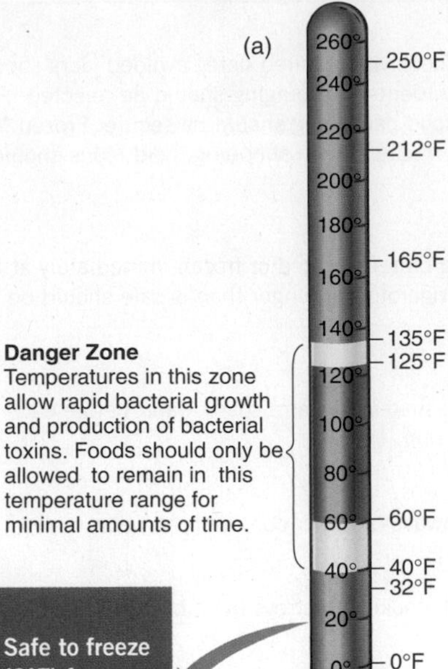

Safe Food Storage Times[23]

(b)

Food	Safe to refrigerate (40°F) for	Safe to freeze (0°F) for
Eggs, Fresh, in shell	3 to 5 wks	Do not freeze
Frozen Dinners and Entrees	Keep frozen until ready to heat	3 to 4 mos
Soups & Stews	3 to 4 days	2 to 3 mos
Hot dogs		
Opened package	1 wk	1 to 2 mos
Unopened package	2 wks	1 to 2 mos
Luncheon Meats		
Opened package	3 to 5 days	1 to 2 mos
Unopened package	2 wks	1 to 2 mos
Bacon and Sausage		
Bacon	7 days	1 mo
Sausage, raw—from chicken, turkey, pork or beef	1 to 2 days	1 to 2 mos
Smoked breakfast links, patties	7 days	1 to 2 mos
Ground meat and poultry	1 to 2 days	3 to 4 mos
Fresh Beef, Veal, Lamb, Pork		
Steaks	3 to 5 days	6 to 12 mos
Chops	3 to 5 days	4 to 6 mos
Roasts	3 to 5 days	4 to 12 mos
Fresh poultry		
Chicken or turkey, whole	1 to 2 days	1 yr
Chicken or turkey, pieces	1 to 2 days	9 mos
Cooked Meat & Poultry Leftovers		
Cooked meat & meat casseroles	3 to 4 days	2 to 3 months
Fried chicken	3 to 4 days	4 mos
Cooked poultry casseroles	3 to 4 days	4 to 6 mos
Poultry pieces, plain	3 to 4 days	4 mos
Chicken nuggets, patties	1 to 2 days	1 to 3 mos
Pizza, cooked	3 to 4 days	1 to 2 mos

Safe Cooking Temperatures[24]

(c)

Food item	Internal temperature (°F) or description
Beef roasts and steaks	145 (allow meat to rest 3 min before carving or consuming)
Ground meat	160
Pork	145 (allow meat to rest 3 min before carving or consuming)
Poultry	165
Fish	145 or until the flesh is opaque and separates easily with a fork
Eggs	Until the yolk and white are firm, not runny (egg dishes 160)
Leftovers and casseroles	165

(c) To ensure that microbes have been killed, meats and casseroles should be cooked, and leftovers reheated, until they reach the internal temperatures shown here.

(b) Storing food in the refrigerator will delay microbial growth, but both refrigerated and frozen food should be discarded if not used by the times shown in the table.

potlucks, you must rely on others to keep your food safe. Consumers should choose restaurants with safety in mind. Restaurants should be clean, and cooked foods should be served hot. Cafeteria steam tables should be kept hot enough that the water is steaming and food is kept above 135°F. Cold foods such as salad-bar items should be kept refrigerated or on ice to keep food at 41°F or colder.

Picnics, potlucks, and other large events where food is served provide a prime opportunity for microbes to flourish because food is often left at room temperature or in the sun for hours before it is consumed. Foods that last well without refrigeration, such as fresh fruits and vegetables, breads, and crackers, should be selected for these occasions (see **Critical Thinking: Safe Picnic Choices**).

CRITICAL THINKING Safe Picnic Choices

© Edward ONeil Photography Inc./iStockphoto

Tamika is planning a potluck picnic. She is worried about food safety, so she tries to apply the HAACP principles.

Tamika decides to find out what people are bringing and perhaps suggest some different choices if the foods seem to present a significant risk.

PICNIC MENU

Chicken salad

Fried chicken

Tamales

Raw vegetables and onion dip

Chips and salsa

Cheese and crackers

Apple pie

Cookies

Mushrooms stuffed with crab meat

Fruit salad

© Michael Valdez/iStockphoto

© Paul Johnson/iStockphoto

© Shawn Gearhart/iStockphoto

CRITICAL THINKING QUESTIONS

- List five foods on the picnic menu that could pose a risk of food-borne illness?
- Describe scenarios by which each of these might become contaminated?
- What steps could be taken during food preparation to reduce the number of bacteria contaminating these foods?
- What could Tamika do at the picnic to reduce the growth of bacteria already present in the food?
- After the picnic, which foods will be safe for Tamika to eat as leftovers?

iProfile Use iProfile to find the number of calories in your favorite picnic foods.

Food safety is also a concern when food must be carried out of the home. Any food that is transported should be kept cold. Lunches should be transported to and from work or school in a cooler or an insulated bag. They should be refrigerated upon arrival or kept cold with ice packs. Most foods that are returned uneaten from work or school should be thrown out and not saved for another day.

17.4 Agricultural and Industrial Chemicals in Food

LEARNING OBJECTIVES

- Illustrate how contaminants move through the food chain and into our foods.
- Discuss how pesticide use is regulated in the United States.
- Compare the risks and benefits of using pesticides with those of growing food organically.
- Describe how consumers can minimize the risks of exposure to chemical contaminants.

Chemicals used in agricultural production and industrial wastes contaminate the environment and can find their way into the food supply. How harmful these chemicals are depends on how long they persist in the environment and whether they accumulate in the organisms that consume them or can be broken down and excreted by these organisms. Some contaminants are eliminated from the environment quickly because they are broken down by microorganisms or chemical reactions. Others remain in the environment for very long periods, and when taken up by plants and small animals, they are not metabolized or excreted. For example, fat-soluble contaminants concentrate in body fat and cannot be excreted. When contaminated plants or small animals are consumed by larger animals that are in turn eaten by still larger animals, the contaminants accumulate, reaching higher concentrations at each level of the food chain (**Figure 17.16**). This process is called **bioaccumulation**. Because the toxins are not eliminated from the body, the greater the amount consumed, the greater the amount present in the body.

bioaccumulation The process by which compounds accumulate or build up in an organism faster than they can be broken down or excreted.

Risks and Benefits of Pesticides

Pesticides are applied to crops growing in the field to prevent plant diseases and insect infestations and to produce after harvesting to prevent spoilage and extend shelf life. Crops grown using pesticides generally produce higher yields and look more appealing because insect damage is limited. Residues of these substances remain on the fruits and vegetables that reach consumers. Pesticides can also travel from the fields where they are applied into water supplies, soil, and other parts of the environment, so pesticide residues are found not only on treated produce but also in meat, poultry, fish, and dairy products.[25]

pesticide Any substance or mixture of substances intended to prevent, destroy, repel, or mitigate insect, animal, plant, fungal, or microbial pests.

Controlling Pesticide Exposure The Environmental Protection Agency (EPA), the FDA, and the U.S. Department of Agriculture (USDA) share the responsibility for limiting consumers' exposure to pesticides. The EPA determines the safety and effectiveness of pesticides and sets limits on the amount of pesticide residues that may remain in or on feed crops, as well as in raw and processed foods marketed in the U.S. These limits or **pesticide tolerances** are normally set at least 100 times below the level that might cause harm to infants, children, adults, or the environment.[25,26]

pesticide tolerances The maximum amount of pesticide residues that may legally remain in food, set by the EPA.

While the EPA sets tolerances, the FDA and USDA monitor pesticide residues in both domestic and imported foods. Domestic samples are collected from growers,

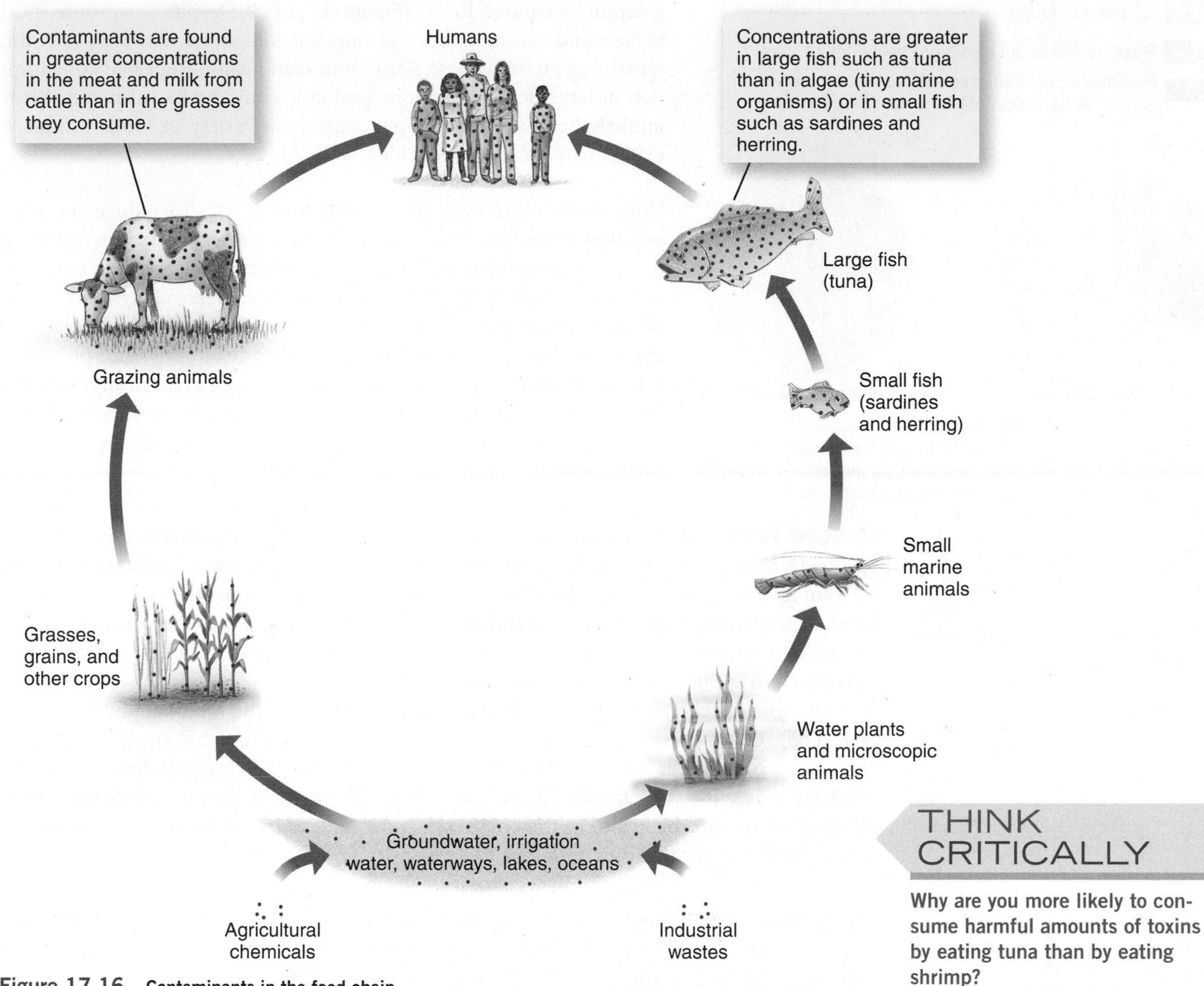

Figure 17.16 **Contaminants in the food chain**
Industrial pollutants and agricultural chemicals that contaminate the water supply enter the food chain and accumulate as they are passed through the chain. An animal later in the food chain has higher concentrations of these contaminants because it consumes all the contaminants that have been eaten by organisms at lower feeding levels.

THINK CRITICALLY

Why are you more likely to consume harmful amounts of toxins by eating tuna than by eating shrimp?

packers, and distributers; import samples are collected at the point of entry into the U.S. market. If domestic samples are found to have illegal residues at a level above an EPA tolerance, the entire lot of food is removed from the market. If imported foods are found to have illegal residues, they are refused entry into U.S. commerce.[27] In addition to this regulatory monitoring that samples raw, unaltered foods, the FDA's Total Diet Study measures pesticide residues in foods that are prepared for consumption. Samples are washed, peeled, and/or cooked before analysis, to simulate how the consumer would handle the food before consumption. The Total Diet Study samples food as "market baskets." Each market basket is made up of about 300 different foods selected to represent the average U.S. consumer's diet. In addition to being analyzed for pesticide residues, this market basket of foods is also analyzed for nutrients, toxins, industrial chemicals, and other chemical contaminants.[27]

In general, the amounts of pesticides to which people are exposed through foods are small. The USDA's Pesticide Data Program has found no more than 1% of samples with residues above established EPA tolerances.[28] The FDA's Pesticide Monitoring Program has found pesticide residue levels in both domestic and imported foods to be well below

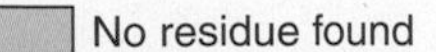

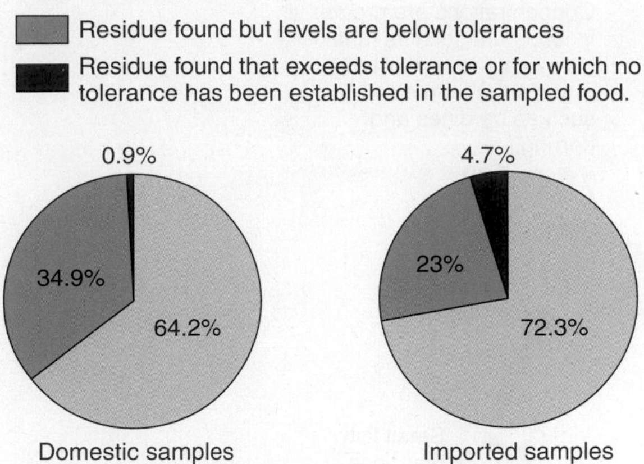

Figure 17.17 **Pesticide tolerances**
An analysis of samples of domestically produced food and imported food from 100 countries found no pesticide residues in 64.2% of domestic and 72.3% of imported samples.[27]

federally permitted limits (**Figure 17.17**).[27] Despite this, some individuals and special-interest groups are concerned that the pesticides remaining on foods pose a risk to human health. Repeated consumption of large doses of any one pesticide could be harmful, but this is unlikely because most people consume a variety of foods produced using many different pesticides.

How Pesticide Risks Are Being Reduced To reduce the risks posed by pesticides, new, more effective chemical pesticides are being developed, and the use of older, more toxic products is decreasing in the United States. One class of pesticide that is less likely to be harmful is biopesticides. These include naturally occurring substances that control pests, microorganisms that control pests, and pesticidal substances that are produced by plants through **genetic engineering**. In addition to developing safer pesticides, production methods are being implemented to make low-pesticide and pesticide-free produce available to the consumer.

genetic engineering A set of techniques used to manipulate DNA for the purpose of changing the characteristics of an organism or creating a new product.

Natural Toxins Many toxins that occur in plants function as natural pesticides that offer protection from bacteria, molds, and insect pests. These naturally pest-resistant crops are advantageous in developing countries because they thrive without the use of expensive added pesticides. Plants high in natural pesticides are being produced through special breeding programs as well as genetic engineering. The natural toxins in plants can also be isolated and applied to crops like synthetic pesticides.

As with all chemical toxins, natural toxins move through the food supply. For example, a cow that has foraged on toxic plants can pass the toxin into her milk and poison the consumer of the milk. Abraham Lincoln's mother died from drinking milk from a cow that had eaten poisonous snakeroot plants. The potential for toxicity, however, depends on the dose of toxin and the health of the consumer. Most natural toxins in the food supply are consumed in doses that pose little risk to the consumer.

integrated pest management (IPM) A method of agricultural pest control that integrates nonchemical and chemical techniques.

Integrated Pest Management **Integrated pest management (IPM)** is a method of agricultural pest control that combines chemical and nonchemical methods and emphasizes the use of natural toxins and effective pesticide application. For example, increasing the use of naturally pest-resistant crop varieties that thrive without the use of added pesticides can reduce costs and do less environmental damage. Integrated pest management programs use information about the life cycles of pests and their interactions with the environment to manage pest damage economically, and with the least possible hazard to people, property, and the environment.

organic food Food that is produced, processed, and handled in accordance with the standards of the USDA National Organic Program.

Organic Methods **Organic food** production methods emphasize the recycling of resources and the conservation of soil and water to protect the environment. Organic food is produced without using most conventional pesticides, fertilizers made with synthetic ingredients, antibiotics, or growth hormones. The USDA's National Organic Program has developed standards for organic foods. These national standards define both substances approved for and prohibited from use in organic food production. For example, an organic food may not include ingredients that are treated with irradiation, produced by genetic modification, or grown using sewage sludge.[29] Certain natural pesticides and other manufactured agents are permitted. Farming and processing operations that produce and handle foods labeled as organic must be certified by the USDA (**Figure 17.18**).

Labeling term	Meaning
100% organic	Contains 100% organically produced raw or processed ingredients.
Organic	Contains at least 95% organically produced raw or processed ingredients.
Made with organic ingredients	Contains at least 70% organically produced ingredients.

Figure 17.18 **Labeling organic foods**
Agricultural products meeting the definition of "100% organic" or "organic" may display the USDA organic seal shown here.

Organic farming techniques reduce the exposure of farm workers to pesticides and decrease the quantity of pesticides introduced into the food supply and the environment. However, organic foods are not risk-free. Animal manure is often used for fertilizer and runoff can pollute lakes and streams. To reduce any risk to consumers, most manure is composted, which reduces the number of pathogens. Regulations prohibit the application of uncomposted manure within 90 days of harvest for crops whose edible portion does not contact the soil, and 120 day for those with edible portions that do. Traces of synthetic pesticides and other agricultural chemicals not approved for organic use can be introduced into organically grown foods by irrigation water, rain, and a variety of other sources. The threshold for pesticide residues in organic foods is set at 5% of the EPA's pesticide-residue tolerance[29] (see **Debate: Should You Go Organic?**).

Courtesy Lori Smolin

Figure 17.19 **Bovine somatotropin** Milk from cows treated with genetically engineered bST is indistinguishable from other milk, but dairies that do not use bST may choose to indicate this on their labels.

Antibiotics and Hormones in Food

Animals, like humans, are treated with antibiotics when they are sick. In addition, some animals are given antibiotics to prevent disease and to promote growth. This increases the amount of meat produced and reduces costs, but if used improperly, residues of these drugs can remain in the meat. To prevent passing these chemicals on to consumers, the FDA regulates which drugs can be used to treat animals used for food production and when they can be administered. The USDA monitors tissue samples for drug residues.[30]

Antibiotic use in animals may also contribute to the creation of antibiotic-resistant bacteria. When bacteria are exposed to an antibiotic, those that are resistant to that antibiotic survive and produce offspring that also carry the antibiotic resistance trait. If these resistant bacteria infect humans, the resulting illness cannot be treated with that antibiotic. Because nearly half the antibiotics produced in the United States are used to prevent disease in animals, this use is suspected of being a major contributor to the development of antibiotic-resistant strains of bacteria.[31]

Hormones are used to increase weight gain in sheep and cattle and milk production in dairy cows. Some naturally occurring hormones such as estrogen and testosterone are used in slow-release form, and levels in the treated animals are no higher than in untreated animals. Before synthetic hormones can be used, it must be demonstrated that residues in meat are within the safe limits. A synthetic hormone that has created public concern is genetically engineered bovine somatotropin (bST). Cows naturally produce somatotropin, a hormone that stimulates milk production. Genetically engineered bST is produced by bacteria and injected into cows to increase milk production. Consumer groups contend that genetically engineered bST causes health problems for the cows and for humans who consume milk or meat from the cows. An FDA review of the effect of bST concluded that it causes no serious long-term health effects in cows, and that milk and meat from bST-treated cows are not health risks to consumers.[32] The FDA does not require milk from bST-treated cows to be specially labeled, but companies may voluntarily label their products as long as the labeling is truthful and not misleading (**Figure 17.19**).

Contamination from Industrial Wastes

One group of industrial chemicals that pollutes the environment is **polychlorinated biphenyls (PCBs)**. These were used in the past in manufacturing electrical capacitors and transformers, plasticizers, waxes, and paper. PCBs in runoff from manufacturing plants contaminated water, particularly near the Great Lakes. Although they are no longer produced, these compounds do not degrade and so are still found in the environment. Fish that live and feed in contaminated waters accumulate PCBs in their adipose tissue; humans who consume large quantities of contaminated fish accumulate PCBs in their adipose tissue.

polychlorinated biphenyls (PCBs) Carcinogenic industrial compounds that have found their way into the environment and, subsequently, the food supply. Repeated ingestion causes them to accumulate in biological tissues over time.

PCB exposure can cause skin conditions, liver damage, and certain kinds of cancer. It is a particular problem for pregnant and lactating women because prenatal exposure and consumption of contaminated breast milk can damage the nervous system and cause learning deficits in children. The American Academy of Pediatrics has concluded that the benefits of breast-feeding outweigh the risks of low levels of PCBs.[33] They recommend that breast-feeding women in areas where high exposures of PCBs have occurred check with the local health department for recommendations on fish consumption.

DEBATE

Should You Go Organic?

The sale of organic foods grew from $1 billion in 1990 to $26.7 billion in 2010. Organic fruits and vegetables now represent 11.4% of all fruit and vegetable sales in the United States.[1] This growing trend is fueled by concern about food safety, nutritional health, and the environment. Is organic food safer, more nutritious, and more environmentally friendly than conventionally produced food?

People often choose organic fruits and vegetables because they believe they are safer than traditional products. If *safer* is defined as containing fewer pesticides, then their choice is the right one. Organic foods are significantly lower in nitrates and pesticide residues than traditionally grown foods.[2] But it can be argued that there is little evidence that the current levels of pesticide exposure from conventional produce are a risk to human health. And there are other contamination risks associated with organic foods. Manure that has been composted to kill bacteria is often used for fertilizer. If the manure is not properly composted, it may contain pathogenic bacteria that can contaminate the food grown in it.[3]

Many consumers believe that organically produced food is not only safer but also more nutritious. *Nutritious* can mean that it is more effective at preventing nutrition-related diseases.[4] When consumption of organic food is compared with consumption of conventionally produced foods, the majority of studies do not show organic foods to be beneficial in terms preventing nutrition-related diseases.[4] An organically produced food may also be considered more nutritious if it provides more nutrients. Research data on the nutrient content of organically produced food are ambiguous. Some studies report that organically produced foods contain more nutrients than conventionally produced food,[2] while others have found no consistent differences in nutrient content.[5] This confusion is not surprising because many factors—including growing conditions, season, and the fertilizer regime and the methods used for crop protection (for example, use of pesticides and herbicides)—affect the nutritional composition of fruits and vegetables. Nutrient content is also affected by how the food is stored, transported, and processed prior to consumption.

What about the environment? It is hard to argue that organic farming is not better for the environment. Instead of chemical pesticides and fertilizers, it relies on crop rotation, compost, and cover crops to maintain the soil. The result is preservation of the soil, so crops can be grown far into the future, and reduction in the amounts of chemicals released into the environment. But organic growing still impacts the environment because manure runoff can pollute waterways and because organic food is often shipped long distances. The shipping uses energy and pollutes the environment.

Is a diet based on organic foods safer, more nutritious, and better for the environment? We can assume that both conventional and organic foods sold in the United States are generally safe. Whether organic is more nutritious depends not only on individual foods but on the diet as a whole. Organic foods are usually more expensive and available in less variety than conventionally grown foods. If your choices of organic foods are limited by availability or cost, then choosing only organic may limit nutrient intake. Are organic foods better for the environment? They reduce pesticide use, but if organic foods are not available locally, the environmental cost of transporting them may outweigh those savings.

THINK CRITICALLY: Are these organic onions that were shipped across the country a better environmental choice than conventionally grown onions from a farm across town?

Drew Rush/NG Image Collection

---

[1]Organic Trade Association. *Industry Statistics and Projected Growth.*

[2]Crinnion, W. J. Organic foods contain higher levels of certain nutrients, lower levels of pesticides, and may provide health benefits for the consumer. *Altern Med Rev.* 1:4–12, 2010.

[3]Mukherjee, A., Speh, D., Dyck, E., et al. Preharvest evaluation of coliforms, *E. coli, Salmonella* and *E. coli* 0157:H7 in organic and conventional produce grown by Minnesota farmers. *J Food Prot* 67:894–900, 2004.

[4]Dangour, A. D., Lock, K., Hayter, A., et al. Nutrition-related health effects of organic foods: A systematic review. *Am J Clin Nutr* 92:203–210, 2010.

[5]Dangour, A. D., Dodhia, S. K., Hayter, A., et al. Nutritional quality of organic foods: A systematic review. *Am J Clin Nutr* 90:680–685, 2009.

Other contaminants from manufacturing, such as chlordane (used to control termites), radioactive substances such as strontium-90, and toxic metals such as cadmium, lead, arsenic, and mercury, have also found their way into fish and shellfish. Cadmium and lead can interfere with the absorption of other minerals, as well as have a direct toxic effect: cadmium can cause kidney damage; lead can impair brain development. Arsenic is believed to contribute to cancer development, and mercury, which has been found in large fish, particularly swordfish, king mackerel, tilefish, and shark, damages nerve cells (**Figure 17.20**).[34] Large fish at the top of the food chain are more likely to contain high levels of industrial contaminants, but shellfish also accumulate contaminants because they feed by passing large volumes of water through their bodies.

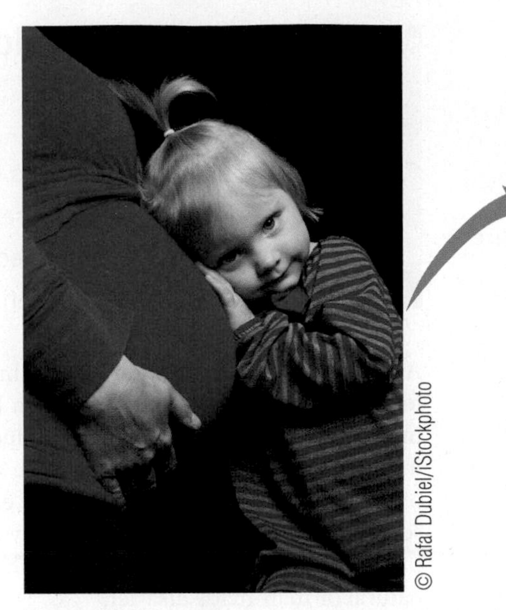

Don't eat swordfish, shark, king mackerel, or tilefish, which can be high in mercury.

Eat up to 6 oz per week of canned albacore or chunk white tuna and up to 12 oz per week of fish and shellfish that are lower in mercury, such as salmon, shrimp, canned light tuna, pollock, catfish, and cod.

Check local advisories about the safety of fish caught in local waters. If no advice is available, eat up to 6 oz per week, but don't consume any other fish during that week.

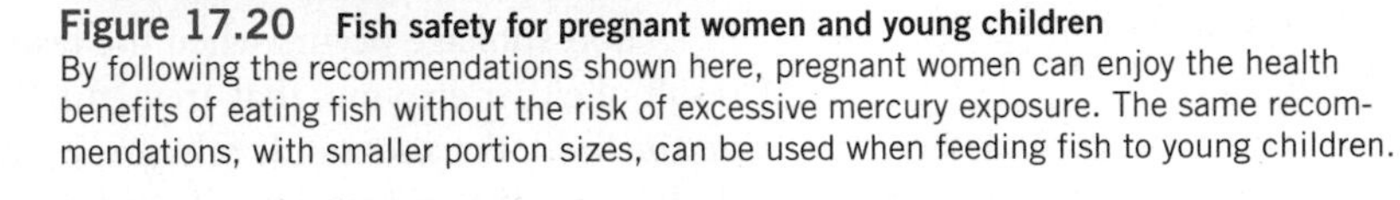

Figure 17.20 Fish safety for pregnant women and young children
By following the recommendations shown here, pregnant women can enjoy the health benefits of eating fish without the risk of excessive mercury exposure. The same recommendations, with smaller portion sizes, can be used when feeding fish to young children.

How to Choose Wisely to Minimize Agricultural and Industrial Contaminants

Even though individual consumers cannot detect chemicals in food, care in selection and preparation can reduce the amounts that are consumed. One of the easiest ways to reduce risk is to choose a wide variety of foods, thus avoiding excessive consumption of any one food. Although some consumers are concerned about the consumption of pesticide residues, the health risk of eliminating foods from the diet that may contain pesticide residues, such as fresh fruits and vegetables, is probably greater than that of the pesticide exposure. To reduce exposure, consumers can choose foods produced organically or using IPM. Locally grown produce will also have fewer pesticide residues because it does not contain pesticides applied to prevent spoilage and extend shelf life during shipping.

Pesticides on conventionally grown produce can be removed or reduced by peeling or washing with tap water and scrubbing with a brush if appropriate. For leafy vegetables such as lettuce and cabbage, the outer leaves can be removed and discarded. Some produce, such as cucumbers, apples, eggplant, squash, and tomatoes, is coated with wax to maintain freshness by sealing in moisture, but wax also seals in pesticides. Much of the wax can be removed by rinsing produce in warm water and scrubbing it with a brush, but to eliminate all wax, the produce must be peeled. Although peeling fruits and vegetables eliminates some pesticides, it also eliminates some fiber and micronutrients.

The risk of ingesting chemical pollutants from fish can be minimized by choosing wisely (see Figure 17.20). Smaller species of fish are safer because they are earlier in the food chain, and smaller fish within a species are safer because they are younger and hence have had less time to accumulate contaminants. The safest fish are saltwater varieties caught well offshore, away from polluted waters. Freshwater fish and saltwater fish that live near shore or spend part of their life cycle in fresh water are more likely to contain contaminants. Migratory fish such as striped bass and bluefish are a problem because they may contain contaminants even when they are caught in clean water well offshore. Consuming a variety of fish rather than just one or two kinds can also reduce the risk of ingesting dangerous amounts of contaminants.

Most toxins concentrate in adipose tissue, so amounts can be reduced by removing the skin, fatty material, and dark meat from fish (and trimming fat from meat and removing skin from poultry). Using cooking methods such as broiling, poaching, boiling, and baking, allows the fatty portions of the fish and other meats to drain out. Do not eat the "tomalley" in lobster. Tomalley, a green paste inside the abdominal cavity of a cooked lobster, serves as the liver and pancreas and is the organ in which toxins accumulate. The analogous organ in blue crabs, called the "mustard" because of its yellow color, should also be avoided.

17.5 Food Technology

LEARNING OBJECTIVES

- Describe how temperature is used to prevent food spoilage.
- Discuss how irradiation preserves food.
- Explain how packaging protects food.
- Compare the risks and benefits of food additives.

fermentation A process in which microorganisms metabolize components of a food and thus change the composition, taste, and storage properties of the food.

food additive A substance that is intentionally added to or can reasonably be expected to become a component of a food during processing.

accidental contaminant A substance not regulated by the FDA that unexpectedly enters the food supply.

aseptic processing A method that places sterilized food in a sterilized package using a sterile process.

Advances in food and agricultural technology have improved the safety and availability of foods. This technology allows food to be stored for long periods, adds nutrients lacking in the diet, and creates new food products. It has helped ensure that food is available even if the local growing season is not ideal. Without technology, the food supply would include only locally grown foods that must be eaten soon after harvest or slaughter. While this has some appeal, it would limit the variety of foods in the diet, particularly during the winter months, and increase the risk for malnutrition if food production were interrupted by a natural or man-made disaster.

Food spoilage occurs when the taste, texture, or nutritional value of food changes as a result of either enzymes that are naturally present in the food or bacteria or mold that grow on the food. For thousands of years, humans have treated food in order to prevent spoilage. Techniques that preserve food work by destroying enzymes present in the food, by killing microbes, or by slowing microbial growth. As shown in **Table 17.6**, the acronym FAT TOM reminds us of the factors that affect microbial growth. Most food preservation techniques modify one or more of these factors to stop or slow microbial growth. Many of the oldest methods of food preservation—including drying, smoking, **fermentation**, adding sugar or salt, and heating or cooling—are still used today. In addition, new methods of improving food quality and preventing spoilage and contamination, such as irradiation and specialized packaging, have been developed. While all these technologies offer benefits, they can also create risks. Some risk arises when substances find their way into food, either accidentally or as a normal part of the production process. The FDA considers any substance that can be expected to become part of a food a **food additive** and regulates its use. Substances that enter food unexpectedly are considered **accidental contaminants** and are not regulated.

George Semple

Figure 17.21 **Aseptic packaging** If unopened, milk and juice in aseptic packages can be stored for long periods without refrigeration.

How Temperature Keeps Food Safe

Cooking food is one of the oldest methods of ensuring food safety. It kills disease-causing organisms and destroys toxins. Cooling food with refrigeration or freezing protects us by slowing or stopping microbial growth. Other preservation techniques that rely on temperature include canning, pasteurization, sterilization, and **aseptic processing**. Aseptic processing heats foods to temperatures that result in sterilization. The sterilized foods are then placed in sterilized packages using sterilized packaging equipment. Aseptic processing is currently used to produce boxes of sterile milk and juices (**Figure 17.21**). These

TABLE 17.6 FAT TOM

Food	Provides a growth media for bacteria.
Acidity	Most bacteria grow best at a pH near neutral. Some food additives, such as citric acid and ascorbic acid (vitamin C), are acids, which prevent microbial growth by lowering the pH of food.
Time	The longer a food sits at an optimum growth temperature, the more bacteria it will contain. Preservation methods such as canning and pasteurization kill microbes by heating food to an appropriate temperature for the right amount of time.
Temperature	The high temperatures of canning, cooking, and pasteurization kill microbes, and the low temperatures of freezing and refrigeration slow or stop microbial growth.
Oxygen	In order to grow, most bacteria need oxygen, so packaging that eliminates oxygen prevents their growth.
Moisture	Bacteria need water to grow, so preservation methods such as drying or the use of high concentrations of salt or sugar, which draw water away by osmosis, prevent bacteria from growing.

Source: Adapted from Food Safety Information from Iowa State University.

can remain free of microbial contamination at room temperature for years (see **Science Applied: Pasteurization: From Spoiled Wine to Safe Milk**).

Preservation techniques that rely on temperature benefit us by providing appealing, safe food, but they are not risk-free, particularly if used incorrectly. If foods are not heated long enough or to a high enough temperature, or if they are not kept cold enough, there is a risk of microbial food-borne illness. In addition, some types of cooking can also generate hazardous chemicals. These are considered accidental contaminants, so they are not regulated by the FDA. The most familiar group of chemicals produced during cooking is the **polycyclic aromatic hydrocarbons (PAHs)**. These carcinogenic substances are formed when fat drips onto the flame of a grill and burns. They rise with the smoke and are deposited on the surface of the food. Grilled meats are therefore high in PAHs. PAH formation can be minimized by selecting lower-fat meats and using a layer of aluminum foil to prevent fat from dripping on the coals.

Broiled foods, which are cooked with the heat source at the top, are low in PAHs. However, broiling and other methods of high-temperature cooking can result in another potential hazard—**heterocyclic amines (HCAs)**, such as benzopyrene. HCAs are formed from the burning of amino acids and other substances in meats. Well-done meat and meat cooked using hotter temperatures contain greater amounts. HCA formation can be reduced by precooking meat, marinating meat before cooking, cooking at lower temperatures, and reducing cooking time by using smaller pieces of meat and avoiding overcooking. The cooking temperatures recommended by the FDA are designed to prevent microbial food-borne illness and minimize the production of PAHs and HCAs (see Figure 17.15).

Another contaminant formed during food preparation is acrylamide. It forms as a result of chemical reactions during high-temperature baking or frying, particularly in carbohydrate-rich foods. The highest levels of acrylamide are found in French fries and snack chips, foods that people should already be eating less of because they are low in nutrients and high in calories. High doses of acrylamide have been found to cause cancer and reproductive problems in animals and to act as a neurotoxin in humans. Thus far, dietary exposure to acrylamide has not been associated with cancer in humans, and more research is needed to determine whether long-term, low-level exposure has any cumulative effects.[35] Methods for reducing the amounts and potential toxicity of acrylamide in foods are being investigated.[36]

polycyclic aromatic hydrocarbons (PAHs) A class of mutagenic substances produced during cooking when there is incomplete combustion of organic materials—such as when fat drips on a grill.

heterocyclic amines (HCAs) A class of mutagenic substances produced when there is incomplete combustion of amino acids during the cooking of meats—such as when meat is charred.

irradiation A process, also called *cold pasteurization*, that exposes foods to radiation to kill contaminating organisms and retard ripening and spoilage of fruits and vegetables.

How Irradiation Preserves and Protects Food

Irradiation, also called *cold pasteurization*, exposes food to a high dose of X-rays, gamma radiation, or high-energy electrons to kill microorganisms and insects and inactivate enzymes that cause germination and ripening of fruits and vegetables. Although irradiated foods must be labeled with the radura symbol shown in **Figure 17.22a** and the statement "treated with radiation" or "treated by irradiation," foods that contain irradiated spices or other irradiated ingredients do not need to display this symbol.

The FDA has approved irradiation to destroy pathogens in red meat and poultry and contaminants in spices; prevent insect infestation in flour and spices; increase the shelf life of potatoes; eliminate *Trichinella* in pork; control insects in fruits, vegetables, and grains; and slow the ripening and spoilage of some produce[37] (**Figure 17.22b**). Because irradiation produces unique compounds in irradiated foods, it is treated as a food additive, and the level of

Figure 17.22 Food irradiation

(a)

TREATED BY
IRRADIATION

(a) This symbol is used to identify foods that have been treated with irradiation.

(b)

Cordelia Molloy/Photo Researchers, Inc.

(b) After two weeks in cold storage, strawberries treated by irradiation remain free of mold (left), whereas untreated strawberries picked at the same time are covered with mold (right).

SCIENCE APPLIED

Pasteurization: From Spoiled Wine to Safe Milk

(Elena Schweitzer/Shutterstock) Victoriano Izquierdo/Getty Images, Inc. (DAJ/Getty Images, Inc.)

In 1857, French sailors were in mutiny because wine supplies were spoiling after only a few weeks at sea. Napoleon recognized this problem as a threat to his hopes for world conquest. He turned to Louis Pasteur for help.

THE SCIENCE

To study the problem, Pasteur traveled to a vineyard in Arbois, France, where spoilage was causing considerable economic losses for the wine industry (see Figure). By examining the spoiled wine under a microscope, Pasteur was able to demonstrate the presence of certain strains of microorganisms. He suggested that spoilage could be prevented by heating the wine to a temperature high enough to kill the harmful microbes but low enough that it did not affect the flavor. Experimentation with heating wine for various times and temperatures revealed that, as Pasteur had predicted, the microorganisms could be killed without damaging the flavor of the wine. This became the foundation for the modern treatment of bottled liquids to prevent their spoilage, a process known as pasteurization.

Pasteurization was eventually used to treat milk. By 1900, pasteurized milk was commonly available, but raw milk was still more popular because the high-temperature pasteurization process used at the time left the milk with a cooked taste. In 1906, Milton J. Rosenau, director of the U.S. Marine Hospital Service Hygienic Laboratory, in Staten Island, NY, established a low-temperature, slow pasteurization process (140°F for 20 minutes) that killed pathogens without changing the taste of the milk. This discovery eliminated the primary obstacle to public acceptance of pasteurized milk.

THE APPLICATION

Pasteurization had a major public health impact, particularly when applied to the milk industry. Between 1880 and 1907, 500 outbreaks of milk-borne diseases occurred in the United States. Milk from cows infected with *Brucella* bacteria caused brucellosis, or undulant fever, in humans. Brucellosis is a chronic debilitating disease characterized by intermittent fever, headache, joint pain, and weight loss. Contaminated milk also caused typhoid, scarlet fever, diphtheria, and tuberculosis. Pasteurization eliminated the pathogenic organisms that caused these diseases. More progress toward improving milk safety and recognizing safe levels of microbial content came when the Public Health Service created a document to assist Alabama in

Jean Loup Charmet/SPL/Photo Researchers

This painting depicts Louis Pasteur studying the souring of wine.

developing a statewide milk sanitation program. This publication ultimately evolved into the Grade A Pasteurized Milk Ordinance. This voluntary agreement established uniform sanitation standards for the interstate shipment of Grade A milk. It now is the basis of milk safety laws in all 50 states and Puerto Rico.[1,2]

Although pasteurization kills pathogenic bacteria and reduces the total number of microorganisms present, it allows many microbes to survive. Today, milk is graded on the basis of bacterial count. A maximum of 20,000 bacteria per milliliter is allowed in Grade A milk. The multiplication of the bacteria that remain eventually causes the milk to spoil. To prevent rapid multiplication after pasteurization, the milk must be cooled immediately and remain refrigerated. In addition to killing microorganisms, the heat also destroys enzymes in the milk; the inactivation of lipase extends the shelf life of homogenized milk by preventing it from becoming rancid. The inactivation of another enzyme called phosphatase is used to gauge the adequacy of pasteurization. Lack of phosphatase activity indicates that the heat treatment has been sufficient, but if phosphatase activity remains, it indicates that the milk was not treated adequately and may not be free of pathogens.

Today almost all milk sold in the United States is pasteurized, and the same techniques are used to prevent spoilage in many other foods. Thanks to the work of Louis Pasteur and Milton J. Rosenau, we now enjoy the nutritional benefits of milk without the risk of infection with disease-causing organisms.

[1]U.S. Public Health Service. 1924. United States Proposed Standard Milk Ordinance, Public Health Reports. Washington, D.C.: Public Health Service, November 7, 1924.

[2]U.S. Food and Drug Administration. Grade "A" Pasteurized Milk Ordinance (2007 Revision).

radiation that may be used is regulated. At the allowed levels of radiation, the amounts of these unique compounds produced are almost negligible and have not been found to be a risk to consumers. Irradiated foods may cost more because of the cost of the extra processing step, but in the future this may be offset by a longer shelf life. Irradiation should be used to complement, not replace, proper food handling by producers, processors, and consumers.

Food irradiation is used in more than 40 countries to treat everything from frogs' legs to rice. However, it is used relatively infrequently in the United States. Part of the reason for this is lack of irradiation facilities, but public fear and suspicion of the technology also limit its use. The word "irradiation" fosters the belief that the food itself becomes radioactive. Opponents to food irradiation claim that it introduces carcinogens, depletes the nutritional value of food, and is used to allow the sale of previously contaminated foods. In fact, irradiated food is not radioactive and scientific studies conducted over the past 50 years have found that the benefits of irradiation outweigh the potential risks.[38] It increases the safety and shelf life of foods and does not noticeably change food texture, taste, or appearance as long as it is properly applied to a suitable product. Irradiation can decrease the amounts of certain nutrients, but losses are similar to those that occur with canning or cold storage.[39] Irradiation can be used in place of chemical treatments, so it benefits consumers and the environment by reducing exposure to chemical pesticides and preservatives.

How Packaging Protects Food

Packaging plays an important role in food preservation: it keeps molds and bacteria out, keeps moisture in, and protects food from physical damage. An open package of cheddar cheese will grow mold in the refrigerator after only a few days. An unopened package will stay fresh for weeks.

Food packaging is continuously being improved. In the past two decades, for instance, consumer demand for fresh, easy-to-prepare foods has led manufacturers to offer partially cooked pasta, vegetables, seafood, fresh and cured meats, and dry products such as whole-bean and ground coffee in packaging that, if unopened, will keep perishable food fresh much longer than will conventional packaging. Vacuum packaging and **modified-atmosphere packaging (MAP)** use plastics or other packaging materials that are impermeable to oxygen. In vacuum packaging the air inside the package is removed prior to sealing in order to remove the oxygen. In modified atmosphere packaging the air is flushed out and replaced with a gas, such as carbon dioxide or nitrogen. In both types of packaging the low oxygen level prevents the growth of aerobic bacteria, slows the ripening of fruits and vegetables, and slows down oxidation reactions, which cause discoloration in fruits and vegetables and rancidity in fats.

modified atmosphere packaging (MAP) A preservation technique used to prolong the shelf life of processed or fresh food by changing the gases surrounding the food in the package.

MAP is often used to package cooked entrees such as pasta primavera or beef teriyaki. The raw ingredients are sealed in a plastic pouch, the air is flushed out, and the pouch and its contents are partially precooked and immediately refrigerated. This processing eliminates the need for the extreme cold of freezing or the extreme heat of canning, so flavor and nutrients are better preserved. Because these products are not heated to temperatures high enough to kill all bacteria and are not stored at temperatures low enough to prevent all bacteria from growing, they could pose a food safety risk. To ensure safety, fresh refrigerated foods should be purchased only from reputable vendors, used before the expiration date printed on the package, refrigerated until use, and heated according to the time and temperature directions on the package.

Packaging can protect food from spoilage, but even the best packaging can introduce risk if it becomes a part of the food. A variety of substances leach into foods from plastics, paper, and even dishes. One potential contaminant from plastics containers is bisphenol A (BPA). BPA is an industrial chemical that has been used since the 1960s to manufacture many hard plastic bottles and the coating inside metal food and beverage cans. The transfer of BPA to food or drinks is a concern for pregnant women, infants, and

Alaska Stock Images/NG Image Collection

Figure 17.23 Does your water bottle contain bisphenol A? To identify plastic containers containing BPA, look for those marked with recycle codes 3 or 7. Some, though not all, of these are made with BPA. To reduce BPA exposure, discard all BPA-containing bottles with scratches and do not use these containers for very hot or boiling liquid that you intend to consume.[41]

young children (**Figure 17.23**). A review by the U.S. National Toxicology Program found that, although there is negligible risk of adverse effects in most adults, exposure of pregnant women to BPA could affect fetal brain development, and exposure of infants and children to the chemical could affect the nervous system and accelerate puberty.[40] To reduce exposure, the FDA supports efforts to eliminate the use of BPA in baby bottles and infant feeding cups and to replace BPA or minimize BPA levels in food can linings.

Substances such as BPA that unintentionally enter our food from sources such as packaging and food processing equipment are referred to as **indirect food additives** and the amounts and types are regulated by EPA tolerance levels and FDA inspections. However, these regulations apply only to the intended use of the product. When used improperly, additional packaging materials can migrate into food. These accidental contaminants are not regulated by the FDA. It is the consumer's responsibility to prevent these substances from entering food. For instance, some plastics migrate into food when heated in a microwave oven. Thus, only packages designed for microwave cooking should be used.

George Semple

Figure 17.24 Additives are everywhere The additives in these foods prevent the bread from molding, the fruit snacks from hardening, and the powdered sugar from clumping; they also smooth the texture of the pudding and give color and flavor to soft drinks and candy.

Food Additives

Food additives keep bread from molding, give margarine its yellow color, and keep Parmesan cheese from clumping in the shaker (**Figure 17.24**). Substances that are intentionally added to foods are regulated by the FDA and are called **direct food additives.**

Food additives are used to make food safer; maintain palatability and wholesomeness; improve color, flavor, or texture; aid in processing; and enhance nutritional value (**Table 17.7**). Their use ensures the availability of wholesome, appetizing, and affordable foods that meet consumer demands throughout the year. The FDA's database "Everything Added to Food in the United States" lists more than 3000 additives.[42]

Additives That Prevent Spoilage

Many substances are added to prevent bacteria and molds from causing food spoilage, to extend shelf life, or to protect the natural color and flavor of food. Sugar and salt are two of the oldest **preservatives.** They prevent microbial growth by decreasing the water availability in the product through osmosis; without adequate water, microbes cannot grow. For example, the high concentration of sugar in jams and jellies draws water away from the microbial cells and prevents them from growing. Antioxidants such as sulfites and BHT (butylatedhydroxytoluene) are also used as preservatives. They prevent fats and oils in baked goods and other foods from becoming rancid or developing an off-flavor and prevent cut fruits such as apples from turning brown when exposed to air. Other preservatives act by blocking the natural ripening and enzymatic processes that continue to occur in foods even after harvest.

indirect food additive A substance that is expected to enter food unintentionally during manufacturing or from packaging. Indirect food additives are regulated by the FDA.

direct food additive A substance that is intentionally added to food. Direct food additives are regulated by the FDA.

preservative A compound that extends the shelf life of a product by retarding chemical, physical, or microbiological changes.

Additives to Improve Nutrient Content, Color, Texture, and Flavor

Additives are not just used to make food safer and last longer. They are also used to enhance the nutrient content, texture, color, and flavor of foods.

Additives to Maintain or Improve Nutritional Quality Nutrients that are added to foods are considered additives. As discussed in Chapter 8, refined grains are enriched with iron and some of the B vitamins that are lost in processing; these are considered additives. Food is also fortified with nutrients, such as calcium and vitamin D, that are typically lacking in the diet. Although the addition of these additives to foods benefits the population by increasing the nutrient content of the diet, it can also increase the risk of nutrient toxicities.

Additives to Improve and Maintain Texture Many different types of additives are used to enhance and maintain desired texture and consistency. Emulsifiers improve the

TABLE 17.7 Common Food Additives[43]

Type of additive	What's on the label	What they do	Where they are used
Preservatives	Ascorbic acid, citric acid, sodium benzoate, calcium propionate, sodium erythorbate, sodium nitrite, calcium sorbate, potassium sorbate, BHA, BHT, EDTA, tocopherols	Maintain freshness; prevent spoilage caused by bacteria, molds, fungi, or yeast; slow or prevent changes in color, flavor, or texture; delay rancidity	Jellies, beverages, baked goods, cured meats, oils and margarines, cereals, dressings, snack foods, fruits and vegetables
Sweeteners	Sucrose, glucose, fructose, sorbitol, mannitol, corn syrup, high-fructose corn syrup, saccharin, aspartame, sucralose, acesulfame potassium (acesulfame-K), neotame	Add sweetness with or without extra calories	Beverages, baked goods, table-top sweeteners, many processed foods
Color additives	FD&C blue nos. 1 and 2, FD&C green no. 3, FD&C red nos. 3 and 40, FD&C yellow nos. 5 and 6, orange B, citrus red no. 2, annatto extract, β-carotene, grapeskin extract, cochineal extract or carmine, paprika oleoresin, caramel color, fruit and vegetable juices, saffron, colorings or color added	Prevent color loss due to exposure to light, air, temperature extremes, and moisture; enhance colors; give color to colorless and "fun" foods	Processed foods, candies, snack foods, margarine, cheese, soft drinks, jellies, puddings and pie fillings
Flavors, spices, and flavor enhancers	Natural flavoring, artificial flavor, spices, monosodium glutamate (MSG), hydrolyzed soy protein, autolyzed yeast extract, disodium guanylate or inosinate	Add specific flavors or enhance flavors already present in foods	Many processed foods, puddings and pie fillings, gelatin mixes, cake mixes, salad dressings, candies, soft drinks, ice cream, BBQ sauce
Nutrients	Thiamine hydrochloride, riboflavin (vitamin B_2), niacin, niacinamide, folate or folic acid, β-carotene, potassium iodide, iron or ferrous sulfate, α-tocopherols, ascorbic acid, vitamin D, amino acids (L-tryptophan, L-lysine, L-leucine, L-methionine)	Replace vitamins and minerals lost in processing; add nutrients that may be lacking in the diet	Flour, breads, cereals, rice, pasta, margarine, salt, milk, fruit beverages, energy bars, breakfast drinks
Emulsifiers	Soy lecithin, mono- and diglycerides, egg yolks, polysorbates, sorbitan monostearate	Allow smooth mixing and prevent separation; reduce stickiness; control crystallization; keep ingredients dispersed	Salad dressings, peanut butter, chocolate, margarine, frozen desserts
Stabilizers and thickeners, binders, and texturizers	Gelatin, pectin, guar gum, carrageenan, xanthan gum, whey	Produce uniform texture, improve "mouth-feel"	Frozen desserts, dairy products, cakes, pudding and gelatin mixes, dressings, jams and jellies, sauces
pH control agents and acidulants	Lactic acid, citric acid, ammonium hydroxide, sodium carbonate	Control acidity and alkalinity, prevent spoilage	Beverages, frozen desserts, chocolate, low-acid canned foods, baking powder
Leavening agents	Baking soda, monocalcium phosphate, calcium carbonate	Promote rising of baked goods	Breads and other baked goods
Anti-caking agents	Calcium silicate, iron ammonium citrate, silicon dioxide	Keep powdered foods free-flowing, prevent moisture absorption	Salt, baking powder, confectioners' sugar
Humectants	Glycerin, sorbitol	Retain moisture	Shredded coconut, marshmallows, soft candies, confections

homogeneity, stability, and uniformity of products such as ice cream. Stabilizers, thickeners, and texturizers, such as pectins and gums, are used to improve consistency or texture in pudding and to stabilize emulsions in foods such as salad dressing. Leavening agents are added to incorporate gas into breads and cakes, causing them to rise. Humectants, such as propylene glycol, cause moisture to be retained so products stay fresh. Anti-caking agents prevent crystalline products such as powdered sugar from absorbing moisture and caking or lumping.

Additives to Affect Flavor and Color Additives are also used to enhance the flavor and color of foods. For example, both natural and alternative sweeteners are added to sweeten foods such as yogurt and fruit drinks. Color additives enhance the appearance of foods. A color additive is defined as any dye, pigment, or substance that can impart color when added or applied to a food, drug, cosmetic, or to the human body. Color additives are used in foods for many reasons, including to balance color loss due to storage or processing and to even out natural variations in food color. Color additives can be used to make foods appear more appetizing; however, they are not allowed to be used as deception to conceal inferiority.

Colors permitted for use in foods are classified as certified or exempt from certification. Certified food colors are synthetic dyes that have been tested to ensure safety, quality, consistency, and strength of color. There are nine certified color additives approved for use in the United States. Colors derived from plant, animal, and certain mineral sources are exempt from certification. They must still meet safety standards before they are approved for use in foods. Examples include annatto extract (yellow), dehydrated beets (bluish-red to brown), caramel (yellow to tan), β-carotene (yellow to orange), and grape skin extract (red, green).

Regulating Food Additives Food additives improve food quality and help protect us from disease, but if the wrong additive is used or if the wrong amount is added, it can do more harm than good. To prevent this, the federal Food, Drug, and Cosmetic (FD&C) Act of 1938 gave the FDA authority over food and food ingredients and defined requirements for truthful labeling of ingredients. This act provided exemptions and safe tolerance levels for additives that were necessary or unavoidable in production and established what are called **standards of identity** for certain foods. Standards of identity define exactly the ingredients that can be contained in certain foods such as mayonnaise, jelly, and orange juice. The FD&C Act also gave the FDA the responsibility of testing food additives for safety. Because the FDA could not possibly test all additives, the 1958 Food Additives Amendment transferred the responsibility for testing from the FDA to the manufacturer. Today, when a manufacturer wants to use a new food additive, a petition must be submitted to the FDA. The petition describes the chemical composition of the additive, how it is manufactured, and how it is detected and measured in food. The manufacturer must prove that the additive will be effective for its intended purpose at the proposed levels, that it is safe for its intended use, and that its use is necessary. Additives may not be used to disguise inferior products or deceive consumers. They cannot be used if they significantly destroy nutrients or where the same effect can be achieved by sound manufacturing processes.

standards of identity Regulations that define the allowable ingredients, composition, and other characteristics of foods.

The safety of food additives is of primary concern. Substances that are toxic at some level of consumption may be harmless at a lower level. To ensure that additives are safe, most of those that are allowed in foods can be added only at levels 100 times below the highest level that has been shown to have no harmful effects (see **Off the Label: Are All Food Additives Safe for You?**). This is a greater margin of safety than exists for many vitamins and other naturally occurring substances.

The regulations for substances that cause cancer are far more rigid because of the **Delaney Clause**, part of the 1958 Food Additives Amendment. It states that a substance that induces cancer in either an animal species or humans, at any dosage, may not be added to food. Debate continues regarding whether the Delaney Clause should be liberalized to allow the use of substances that are added at a level so low that they would not represent a significant health risk.

Delaney Clause A clause added to the 1958 Food Additives Amendment of the Pure Food and Drug Act that prohibits the intentional addition to foods of any compound that has been shown to induce cancer in animals or humans at any dose.

Additives Used Before 1958 When the 1958 Food Additives Amendment was passed, over 600 chemicals defined as food additives were already in common use. To accommodate these substances, the amendment exempted two groups of substances from the food additive regulation process. One group included substances that the FDA or the USDA had determined were safe; these were designated as **prior-sanctioned substances**. The nitrates and nitrites used to retard the growth of *Clostridium botulinum* in cured meats such as ham and hot dogs are on the prior-sanctioned list. However, the use of these has been controversial because they form carcinogenic **nitrosamines** in the digestive tract. They are still allowed in foods because there is little evidence that they pose a serious risk in the amounts consumed in the human diet.[44] To minimize any risk posed by nitrosamines without increasing the risk of bacterial illness, the FDA has limited the amount of nitrate and nitrite that can be added to food and has required the addition of antioxidants, which reduce nitrosamine formation, to foods containing these additives. Consumers can

prior-sanctioned substances Refers to substances that the FDA or the USDA had determined were safe for use in a specific food prior to the 1958 Food Additives Amendment.

nitrosamines Carcinogenic compounds produced by reactions between nitrites and amino acids.

OFF THE LABEL Are All Food Additives Safe for You?

© iStockphoto

Sometimes the ingredients listed on food labels sound like a chemical soup. Calcium propionate is added to bread, disodium EDTA is added to canned kidney beans, and BHA is in potato chips. Most of us ignore this list of chemicals and for most of us that is fine because these additives cause no problems. The FDA does not approve food additives unless they are safe for most consumers, but this doesn't mean they are safe for everyone.

For individuals who are sensitive or allergic to certain additives such as preservatives or colors, the ingredient list provides information that can be lifesaving. For example, in sensitive individuals, sulfites can cause symptoms that range from a stomachache and hives to severe asthmatic reactions. Sulfite sensitivity is most common in individuals with asthma; consuming sulfites causes symptoms in up to 10% of people with asthma.[1] Sulfites are used to preserve foods such as baked goods, canned vegetables, condiments, and maraschino cherries. Sensitive individuals can identify foods that contain sulfites by checking food labels. The forms of sulfites allowed in packaged foods include sulfur dioxide, sodium sulfite, sodium and potassium bisulfite, and sodium and potassium metabisulfite. Foods served in restaurants are also a concern because sulfites are sometimes used in food preparation. For example, a potato dish on the menu may be prepared using potatoes that were peeled and soaked in a sulfite solution before cooking.

Food colors can also cause reactions in sensitive individuals. The color additive FD&C Yellow No. 5, which is listed as tartrazine on medicine labels, may cause itching and hives in sensitive people. It is found in beverages, desserts, and processed vegetables. Sensitivity to FD&C Yellow No. 5 occurs in fewer than 1 in 10,000 people.[2] All foods that contain FDA-certified color additives, such as FD&C Yellow No. 5, must list them by name in the ingredient list. Colors that are exempt from certification, such as dehydrated beets and carotenoids, do not have to be specifically identified and may be listed on the label collectively as "artificial color" (see figure).[3]

© Jack Puccio/iStockphoto

Additives add color and flavor to many foods but can be hazardous to sensitive individuals.

[1]Vally, H., Misso, N. L., and Madan, V. Clinical effects of sulphite additives. *Clin Exp Allergy* 39:1643–51, 2009.
[2]U.S. Food and Drug Administration, Food Additives, FDA/IFIC Brochure: January 1992.
[3]U.S. Food and Drug Administration. *Color Additives*.

reduce nitrosamine exposure by limiting cured meat consumption to three to four ounces per week and maintaining adequate intakes of the antioxidant vitamins C and E.

A second category of substances excluded from the food additive regulation process is **generally recognized as safe (GRAS)** substances. GRAS substances are those whose use is generally recognized as safe, based on their extensive history of use in food before 1958 or based on published scientific evidence.

generally recognized as safe (GRAS) A group of chemical additives that are generally recognized as safe based on their long-standing presence in the food supply without obvious harmful effects.

Just because a substance is on the prior-sanctioned or GRAS list doesn't mean it is safe or that it will stay on these lists. If new evidence emerges that suggests that a substance in either of these categories is unsafe, the FDA may take action to remove the substance from food products.

OUTCOME

© GMVozd/iStockphoto

Masterfile

How could homemade frozen custard, which sounds so wholesome, make so many people sick? The frozen custard that the fourth graders made contained raw eggs, which sometimes harbor the bacterium *Salmonella enteritidis*. Thoroughly cooking eggs kills the bacteria, but any food that contains uncooked eggs is a risk. Since the ingredients for all the frozen custard were mixed as one large batch, bacteria from even a single infected egg could have contaminated the entire recipe. The low temperatures used for making frozen custard would retard bacterial growth but not kill the organisms. The fourth graders were slow in measuring and mixing ingredients, so the mixture of raw eggs, milk, sugar, and cream sat at room temperature for almost an hour before it was chilled and placed in the ice cream maker. When the inspectors sampled the frozen custard, they found it contained large numbers of *Salmonella*. Tracking the ingredients used to make the frozen custard revealed that the school had purchased the eggs from a distributor in Florida, who had bought them from a farm in Maryland. At next year's celebration the school decided to have homemade ice cream, which is not made with raw eggs. **Fortunately, all of the individuals who became ill recovered, but *Salmonella* infection can cause serious illness and even death in elderly people, infants, and persons with impaired immune function.**

APPLICATIONS

ASSESSING YOUR DIET

1. **Do a risk/benefit analysis of your diet.**
   a. All of us consume some foods that present a risk of food-borne illness. For each of the foods listed in the table below, enter any food safety risks associated with consumption and the nutritional benefits these foods provide.

Food	Risk	Nutritional Benefit
Hamburger		
Chicken		
Eggs		
Fish		
Raw fish and shellfish		
Fresh fruits and vegetables		

   b. Are there any foods on this list that you would consider eliminating from your diet?

2. **What food additives are in your favorite snacks?**
   a. Choose a packaged product you like to snack on and write out all the ingredients it contains.
   b. Which of the ingredients are food additives?
   c. Use Table 17.7 to describe the function of each additive.
   d. If these additives could not be used, how would the product differ?

CONSUMER ISSUES

3. **What are the risks and benefits for each scenario described below?**
   a. A restaurant decides not to replace its old dishwasher even though it no longer heats the water to above 140°F.
   b. A town decides that it can improve the health of its citizens and the environment by banning the production and sale of all but organically produced foods.

4. **A train crash spills a load of industrial waste into a river that feeds into a local reservoir. How would this spill affect:**
   a. The safety of the drinking water?
   b. The milk from dairy cattle grazing nearby?
   c. The fish that swim in the river?
   d. The crops irrigated by this water?

CLINICAL CONCERNS

5. **Sixty-seven people became ill after consuming food at a company picnic. Testing of food samples revealed that the tossed salad, the egg salad, and the turkey slices were all contaminated with *Salmonella*. Invent a scenario that would explain how all three became contaminated.**

6. **The graph shows data collected from the people in the Case Study who became ill after the "Welcome Back, Spring" celebration at the Loghill Elementary School.**
   - **a.** Use the graph to explain why health inspectors concluded that the frozen custard was the cause of the illness.
   - **b.** How can eggs become contaminated with *Salmonella*?
   - **c.** The cookie and cupcake recipes also called for eggs. Why are these unlikely to be the cause of the problem?
   - **d.** Suggest a reason why 20 children who consumed the frozen custard did not get ill.

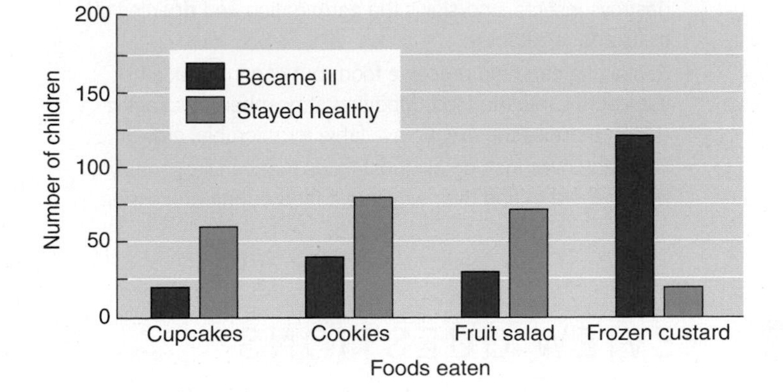

SUMMARY

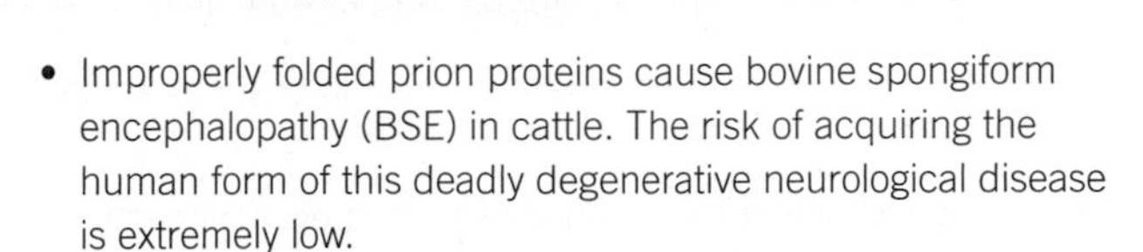

17.1 How Can Food Make Us Sick?

- Food-borne illness is any illness that is related to the consumption of food or contaminants or toxins in food.
- Food may become contaminated where it is grown or produced, during processing or storage, or even in the home. Once contaminated, food preparation utensils and surfaces can cross-contaminate other foods.
- The harm caused by contaminants in the food supply depends on the type of toxin, the dose, the length of time over which it is consumed, and the size and health status of the consumer.

17.2 Keeping Food Safe

- The safety of the food supply is monitored by agencies at the international, federal, state, and local levels. The government promotes the use of HACCP (Hazard Analysis Critical Control Point) principles to ensure food safety. HACCP systems help prevent or eliminate food contamination and track contaminated foods to prevent food-borne illness.
- It is the job of the food manufacturer to label food products properly and establish and implement a HAACP system for its particular business.
- Consumers play an important role in limiting the risks of developing food-borne illness through the way they store, handle, and prepare food.

17.3 Pathogens in Food

- The pathogens that most often affect the food supply include bacteria, viruses, molds, and parasites. Some bacteria cause food-borne infection because they are able to grow in the gastrointestinal tract when ingested. Others produce toxins in food, and consumption of the toxin causes food-borne intoxication.
- Viruses do not grow on food, but when consumed in food, they can enter human cells where they can reproduce and cause food-borne illness.
- Molds that grow on foods produce toxins that can harm consumers.
- Parasites include microscopic single-celled animals, as well as worms that can be seen with the naked eye. They are consumed in contaminated water or food.
- Improperly folded prion proteins cause bovine spongiform encephalopathy (BSE) in cattle. The risk of acquiring the human form of this deadly degenerative neurological disease is extremely low.
- The risk of food-borne illness can be decreased through proper food selection, preparation, and storage. Food should be refrigerated or frozen to slow or stop microbial growth and cooked to recommended temperatures to kill pathogens. Food should not be held between 40° and 135°F because this temperature range favors bacterial growth.

17.4 Agricultural and Industrial Chemicals in Food

- Contaminants such as pesticides applied to crops and industrial wastes that leach into water may find their way into the food supply. To decrease the potential risk of pesticides, safer ones are being developed and U.S. farmers are reducing the amounts applied by using integrated pest management and organic methods.
- Antibiotics are given to animals to prevent disease and increase growth. The widespread use of these drugs may be contributing to the emergence of antibiotic-resistant strains of bacteria.
- Industrial pollutants such as PCBs, radioactive substances, and toxic metals have contaminated some waterways and the fish that live in them. As these contaminants move through the food chain, their concentrations increase.
- Consumers can reduce the amounts of pesticides and other environmental contaminants in food by careful selection and handling of produce; selection of low-fat saltwater varieties of fish caught well offshore in unpolluted waters; and trimming fat from meat, poultry, and fish before cooking.

17.5 Food Technology

- High and low temperatures are used to prevent food spoilage. Cold temperatures slow or prevent microbial growth. High temperatures used in canning, pasteurization, sterilization, and cooking kill microorganisms. However, cooking can also introduce hazardous substances such as polycyclic aromatic hydrocarbons, heterocyclic amines, and acrylamide.
- Irradiation preserves food by exposing it to X-rays, gamma radiation, or high-energy electrons. It kills microorganisms,

destroys insects, and slows the germination and ripening of fruits and vegetables.

- Packaging can help preserve food, but can introduce risk if it leaches into the food. Modified atmosphere packaging (MAP) reduces the oxygen available for microbial growth.
- Food additives include all substances that can reasonably be expected to find their way into a food during processing. This includes direct food additives, which are used to preserve or enhance the appeal of food, and indirect food additives, which are substances known to find their way into food during cooking, processing, and packaging. Direct and indirect food additives are regulated by the FDA. Accidental contaminants that enter food when it is used or prepared incorrectly are not regulated by the FDA.

REVIEW QUESTIONS

1. Discuss 5 pathogens that are common causes of food-borne illness in the United States today.
2. List three factors that determine the likelihood that a contaminant will cause food-borne illness in an individual.
3. How is the federal government involved in ensuring a safe food supply?
4. Explain what HACCP is and how it can prevent or eliminate food contamination.
5. What is the difference between a food-borne infection and a food-borne intoxication?
6. List three common bacterial food contaminants. What can be done to avoid the food-borne illnesses caused by them?
7. What temperature range allows the most rapid bacterial growth?
8. Explain how cross-contamination can occur in home kitchens.
9. How do pesticides applied to crops find their way into animal products?
10. How are pesticides regulated to ensure safety?
11. How does organic food production differ from traditional farming?
12. List some food-processing and packaging techniques that reduce food-borne illnesses.
13. What is food irradiation? What are the benefits and risks?
14. What is the GRAS list?
15. List four reasons for using food additives.

REFERENCES

1. Centers for Disease Control and Prevention. CDC Estimates of Food-Borne Illness in the United States: 2011.
2. U.S. Department of Health and Human Services, USFDA, Food Code, 2005.
3. U.S. Food and Drug Administration, Department of Agriculture, and Environmental Protection Agency. *Food Safety from Farm to Table: A New Strategy for the 21st Century.* February 21, 1997.
4. Taylor, M. R. *The FDA Food Safety Modernization Act: Putting Ideas into Action,* January 27, 2011.
5. Centers for Disease Control and Prevention, U.S. Department of Health and Human Services. *FoodNet—Foodborne Diseases Active Surveillance Network.*
6. U.S. Food and Drug Administration, Center for Food Safety and Nutrition. *Foodborne Pathogenic Microorganisms and Natural Toxins Handbook: The "Bad Bug Book."*
7. Centers for Disease Control and Prevention. *Investigation Update: Multistate Outbreak of Human* E. coli *O145 Infections Linked to Shredded Romaine Lettuce from a Single Processing Facility.*
8. Centers for Disease Control and Prevention. *Investigation Announcement: Multistate Outbreak of* E. coli *O157:H7 Infections Associated with Lebanon Bologna.*
9. Centers for Disease Control and Prevention. *Current Multistate Outbreak:* E. coli *O26 Infections Linked to Raw Clover Sprouts at Jimmy John's Restaurants.*
10. World Health Organization. *Outbreaks of* E. coli *O104:H4 Infection: Update 30,* July 22, 2011.
11. Samuel, M. C., Vugia, D. J., Shallow, S., et al. Epidemiology of sporadic *Campylobacter* infection in the United States and declining trend in incidence, FoodNet 1996–1999. *Clin Infect Dis* 38(Suppl. 3):S165–S174, 2004.
12. Zhao, C., Ge, B., De Villena, J., et al. Prevalence of *Campylobacter* spp., *Escherichia coli,* and *Salmonella* serovars in retail chicken, turkey, pork, and beef from the greater Washington, D.C., area. *Appl Environ Microbiol* 67:5431–5436, 2001.
13. Lamont, R. F., Sobel, J., Mazaki-Tovi, S., et al. Listeriosis in human pregnancy: A systematic review. *J Perinat Med* 39: 227–236, 2011.
14. Centers for Disease Control and Prevention, Division of Foodborne, Bacterial, and Mycotic Disease. *Listeriosis.*
15. Koepke, R., Sobel, J., and Arnon, S. S. Global occurrence of infant botulism, 1976–2006. *Pediatrics* 122:73–82, 2008.
16. Centers for Disease Control and Prevention (CDC). *Norovirus Overview.*

17. Center for Disease Control and Prevention (CDC). Surveillance of Norovirus Outbreaks.

18. National Institute of Environmental Health Sciences, National Institutes of Health. *Aflatoxin and Liver Cancer.*

19. Centers for Disease Control and Prevention. *Cryptospiridium.*

20. Centers for Disease Control and Prevention. *vCJD (Variant Creutzfeldt-Jakob Disease).*

21. U.S. Food and Drug Administration. *BSE (Bovine Spongiform Encephalopathy, or Mad Cow Disease).*

22. The National Creutzfeldt-Jakob Disease Surveillance Unit. University of Edinburgh.

23. FoodSaftey.gov, Charts: Food Safety at a Glance, Storage Times for the Refrigerator and Freezer.

24. FoodSaftey.gov, Charts: Food Safety at a Glance, Safe Minimum Cooking Temperatures.

25. U.S. Environmental Protection Agency. *Setting Tolerances for Pesticide Residues in* Foods.

26. U.S. Environmental Protection Agency. *Assessing Health Risks from Pesticides.*

27. U.S. Food and Drug Administration, Center for Food Safety and Applied Nutrition. *Pesticide Monitoring Program FY 2008.*

28. Punzi, J. S., Lamont, M., Haynes, D., and Epstein, R. L. USDA Pesticide Data Program: Pesticide residues on fresh milk and processed fruit and vegetables, grains, meats, milk, and drinking water. *Outlook on Pest Management*, June, 2005.

29. U.S. Department of Agriculture, Agricultural Marketing Service. *National Organic Program.*

30. Animal Health Institute. *Animal Antibiotics: Keeping Animals Healthy and Our Food Safe.*

31. Pew Commission on Industrial Farm Animal Production. *Final Report: Putting Meat on The Table: Industrial Farm Animal Production in America.*

32. U.S. Food and Drug Administration, Center for Veterinary Medicine. *Report on the Food and Drug Administration's Review of the Safety of Recombinant Bovine Somatotropin.*

33. American Academy of Pediatrics, Committee on Environmental Health. PCBs in breast milk. *Pediatrics* 84:122–123, 1994.

34. U.S. Department of Health and Human Services and U.S. Environmental Protection Agency. *What You Need to Know About Mercury in Fish and Shellfish.* EPA-823-R-04-005, March 2004.

35. Exon, J. H. A review of the toxicology of acrylamide. *J Toxicol Environ Health B Crit Rev* 9:397–412, 2006.

36. Friedman, M., and Levin, C. E. Review of methods for the reduction of dietary content and toxicity of acrylamide *J Agric Food Chem* 56:6113–6140, 2008.

37. U.S. Food and Drug Administration. *Irradiation and Food Safety.*

38. Osterholm, M. T., and Norgan, A. P. The role of irradiation in food safety. *N Engl J Med* 350:1898–1901, 2004.

39. U.S. Food and Drug Administration. *Irradiation of Food and Packaging: An Overview.* 2004.

40. U.S. Food and Drug Administration. *Bisphenol A (BPA): Use in Food Contact Application.*

41. U.S. Food and Drug Administration. *FDA Continues to Study BPA.*

42. U.S. Food and Drug Administration, Center for Food Safety and Applied Nutrition. *Everything Added to Food in the United States (EAFUS).*

43. International Food Information Council (IFIC) and U.S. Food and Drug Administration. *Food Ingredients and Colors.* November 2004; revised April, 2010.

44. U.S. Environmental Protection Agency. *Toxicity and Exposure Assessment for Children's Health: Nitrates and Nitrites.*

***To access links to online sources, please go to** www.wiley.com/college/smolin **and select Nutrition: Science and Applications, 3rd edition. From this page, select either the student or instructor companion site. Once on the desired site, select References.**

Focus on Biotechnology

© kkgas/iStockphoto

CHAPTER OUTLINE

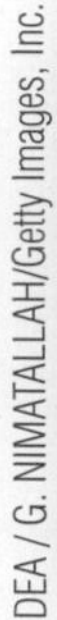

Figure F5.1 **Chimera**
Boyer and Cohen called the small loops of bacterial DNA that contain DNA from another organism *chimeras*, after a mythical fire-breathing beast with the head of a lion and a goat, the body of a goat, and a serpent for a tail.

In 1909, British physician Archibald Garrod hypothesized that genes might be involved in creating proteins. By 1966, investigators had deciphered the genetic code, which links the information in DNA to the synthesis of proteins. The proteins made affect the traits that an organism exhibits. Then, in 1972, a discussion between Dr. Stanley Cohen of Stanford University and Dr. Herbert Boyer of the University of California at San Francisco led to the birth of genetic engineering. They brought together the information needed to allow genetic instructions from one organism to be inserted into another. Cohen's laboratory had been studying how bacterial cells take up small loops of DNA. Boyer had been studying enzymes that could cut DNA at specific locations and paste it back together again. Cohen and Boyer realized that fragments of DNA could be introduced into bacterial cells using the procedure developed in Cohen's lab (**Figure F5.1**). As the bacteria multiplied, so would the new piece of DNA—making copies, or clones. **These techniques for recombining DNA from different sources and cloning it are the basis for all genetic engineering.**[1]

F5.1 How Does Biotechnology Work?

LEARNING OBJECTIVES

- Explain how genetic engineering introduces new traits into plants.
- Compare and contrast traditional breeding methods and modern biotechnology.

Modern **biotechnology** is possible because of the emergence of techniques that allow scientists to alter the DNA of plants, animals, yeast, and bacteria. By modifying DNA, these techniques, referred to as **genetic engineering**, allow scientists to change the proteins that a cell or organism can make, introducing new traits, enhancing desirable ones, or eliminating undesirable ones. Genetic engineering often involves taking the gene for a desired trait from one organism and transferring it to another. It has allowed researchers to create bacteria that make medicines for humans, plants that are disease-resistant, and foods that provide a healthier mix of nutrients. As this new technology has evolved, so has the vocabulary to describe it (**Table F5.1**). Crops and other organisms and food products produced using these techniques are often referred to as *genetically modified* or *GM*.

biotechnology The process of manipulating life forms via genetic engineering in order to provide desirable products for human use.

genetic engineering A set of techniques used to manipulate DNA for the purpose of changing the characteristics of an organism or creating a new product.

Genetics: From Genes to Traits

The characteristics of a plant or animal are carried in its genes. These genes, which are segments of DNA, are passed from generation to generation. Genes contain the information that directs the synthesis of proteins. The specific proteins that are made then determine the traits that an individual organism displays.

DNA Structure DNA is a long, threadlike molecule consisting of two strands that twist around each other, forming a double helix. Each strand has a backbone made up of alternating units of the five-carbon sugar deoxyribose and phosphate groups. Each deoxyribose

TABLE F5.1 Terms Used in Genetic Engineering

Term	Definition
Biotechnology	Manipulating life forms via genetic engineering to provide desirable products for human use.
Bioengineered foods	Foods that have been produced using biotechnology.
Chimera	A DNA molecule composed of DNA from two different species, or an organism consisting of tissues of diverse genetic constitution.
Cloning	Producing an exact duplicate of a gene or an organism.
DNA	A long, thread-like molecule that carries the genetic information of an organism.
Gene	A unit of DNA that provides genetic information coding for a trait. It is the physical basis for the transmission of the characteristics of living organisms from one generation to another.
Gene splicing	The precise joining of DNA from different sources to create a new gene structure.
Genetic engineering	Technology used to selectively alter genes. It can be used to manipulate the genetic material of an organism in such a way as to allow it to produce new and different types of proteins. Other terms applicable to the same techniques are gene splicing, gene manipulation, genetic modification, or recombinant DNA technology.
Genome	The total complement of genetic information in an organism.
GM crops	Genetically modified crops.
GMO	Genetically modified organism.
Hybridization	The mating of different plants to enhance favorable characteristics.
Plasmid	An independent, stable, self-replicating piece of circular DNA in bacterial cells. It is not a part of the normal cell genome.
Recombinant DNA	DNA that has been formed by joining the DNA from different sources.
Transgenic	An organism with a gene or group of genes intentionally transferred from another species or breed.

sugar is attached to a molecule called a base. The four different bases that occur in DNA—adenine, thymine, cytosine, and guanine—are usually abbreviated as A, T, C, and G, respectively. The bases on adjacent strands bind to one another, connecting the two strands. Each base binds only to its complementary base: adenine to thymine and cytosine to guanine (**Figure F5.2**).

The DNA of all organisms—plants, bacteria, and animals, including humans—is made up of the same DNA bases. The differences in the sequence of bases in DNA are responsible for the genetic differences among living things, whether they are differences between species or differences between individuals of the same species. Different individuals of the same species have small differences in the base sequence of their DNA; only identical twins share the same base sequence. Organisms of different species, such as humans and corn plants, have larger differences in the sequences of DNA bases.

Genes to Proteins to Traits The sequence of bases present in a gene specifies the sequence of amino acids that will be present in a protein (see Chapter 6). In the genetic code, three bases code for a single amino acid. The same three bases always code for the same amino acid: for example, the DNA base sequence ACC corresponds to the amino acid tryptophan. Because the code is universal among all life on Earth, a particular three-base sequence codes for the same amino acid whether it is in a bacterium, a mosquito, a corn plant, or a human being.

When a gene is expressed, the protein it codes for is made, providing certain characteristics to the organism. For example, if a protein that stimulates growth is made, the organism will grow bigger, and if a protein pigment is made, it will affect the color of the plant or animal. The presence or absence of specific proteins determines the traits that an individual organism displays (**Figure F5.3**).

Even though organisms of different species are very different, they may have genes with similar base sequences if they both need to make the same protein. For example, humans and pigs both rely on the protein hemoglobin to carry oxygen in the blood, so both humans and pigs have a gene with a very similar sequence of bases that codes for the

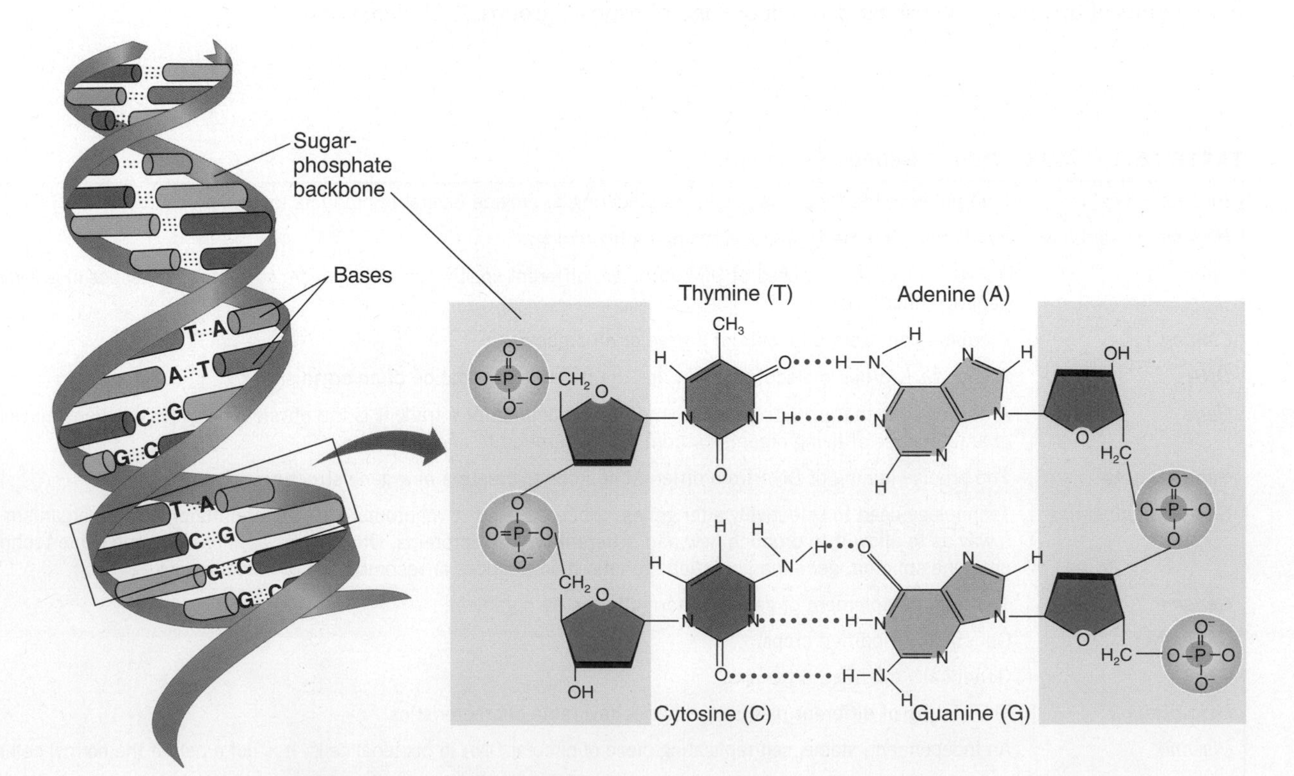

Figure F5.2 DNA: the double helix
DNA is a double-stranded helical molecule. Each strand has a backbone made up of phosphate groups and sugar molecules. Each sugar is attached to adenine (A), thymine (T), cytosine (C), or guanine (G). The two strands of the DNA molecule are bonded together by these bases: A bonds to T, and C bonds to G.

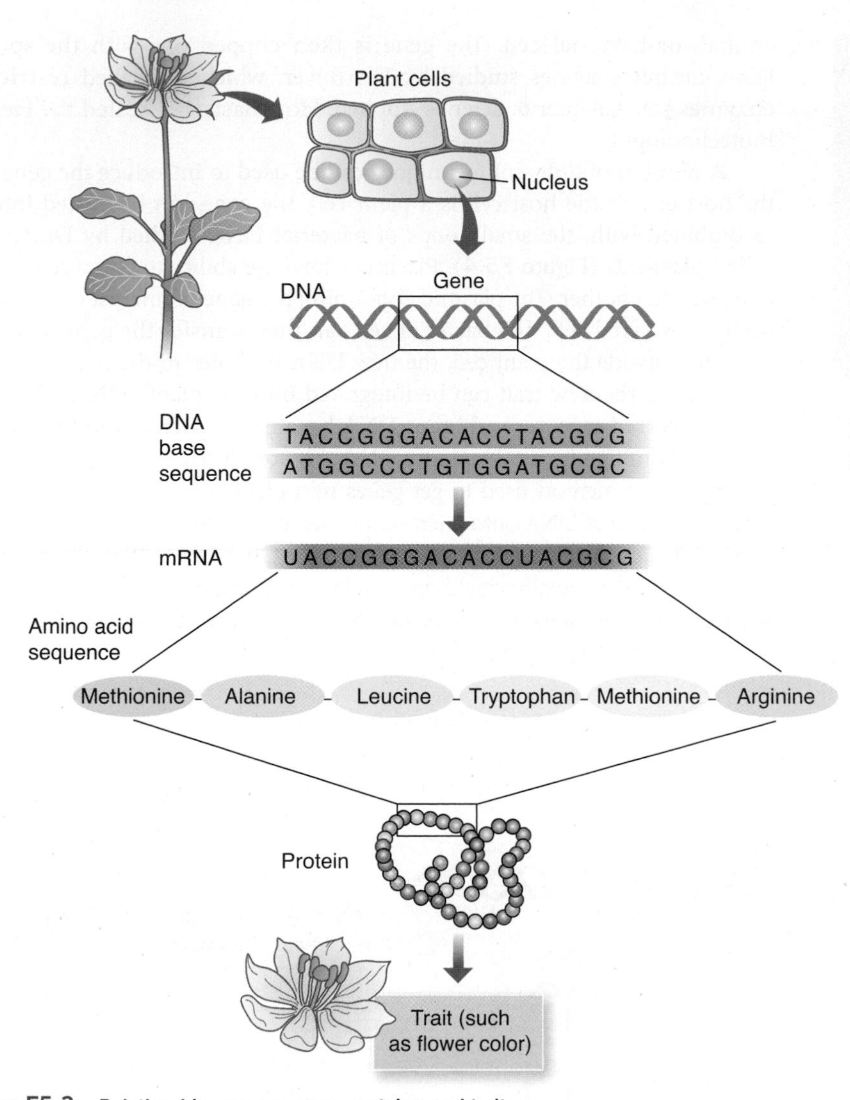

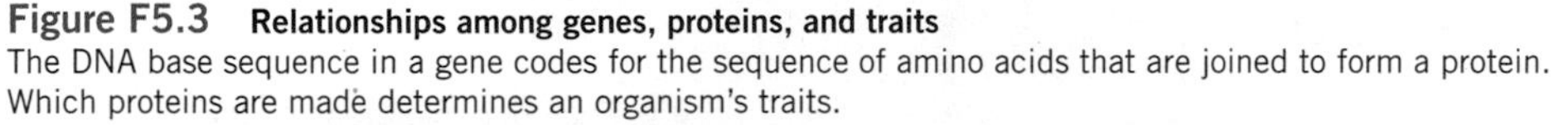

Figure F5.3 Relationships among genes, proteins, and traits
The DNA base sequence in a gene codes for the sequence of amino acids that are joined to form a protein. Which proteins are made determines an organism's traits.

protein hemoglobin. It is estimated that 25% of the genes found in plants are also present in humans, presumably because they code for proteins needed by both organisms.[2]

Passing Traits from Parent to Offspring When an organism reproduces, it passes its genes—and thus the instructions to make the proteins coded for by these genes—to the next generation. When two organisms breed, some genes from each are passed to the offspring. The result is a new combination of genes and the traits for which they code. Over time, mutations can occur in the sequence of bases that make up the genes. Some of these mutations are harmful, reducing the organism's ability to survive and reproduce. Because the organism cannot reproduce, these harmful mutations are not passed on to the next generation. Other mutations are beneficial and result in traits that allow the organism to survive more easily and thus to reproduce and pass on the altered genes. Over millions of years, many genes have been changed by mutations, and those that code for beneficial traits have been passed on because the organisms carrying them thrive and reproduce. The idea behind biotechnology is to speed up the process of introducing beneficial traits that can then be passed from generation to generation.

Methods of Biotechnology

The first step in genetic modification is to identify a stretch of DNA, or gene, for a desired trait, such as resistance to a particular disease. This gene of interest can be from a plant,

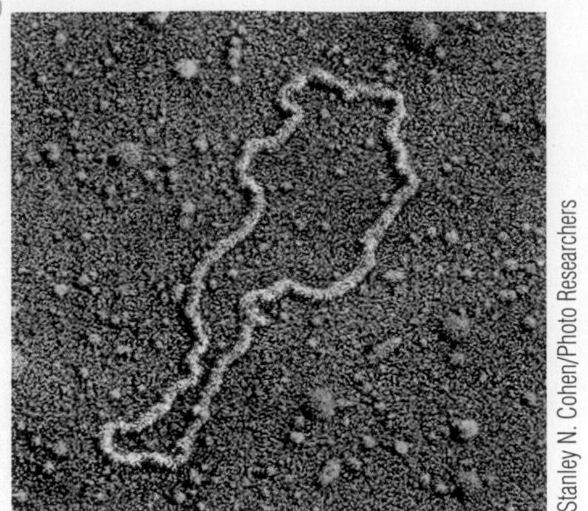

Stanley N. Cohen/Photo Researchers

Figure F5.4 Bacterial Plasmid
This electron micrograph shows a plasmid, which is a small loop of DNA found in bacterial cells.

animal, or bacterial cell. The gene is then clipped out with the specific DNA-cutting enzymes studied by Dr. Boyer, which are called **restriction enzymes** (see Chapter 6: Science Applied: How Bacteria Created the Field of Biotechnology).

A number of different techniques can be used to introduce the gene into the host cell. If the host cell is a plant cell, the gene can be pasted into, or recombined with, the small loops of bacterial DNA studied by Dr. Cohen, called **plasmids** (**Figure F5.4**). Plasmids have the ability to carry genes from one place to another. The plasmid containing the gene of interest can be taken up by a bacterial cell. The bacterial cell can then transfer the gene to a plant cell. Once inside the plant cell, the new DNA migrates to the nucleus, where the gene for the new trait can be integrated into the plant's DNA. The DNA is then referred to as **recombinant DNA** because the DNA from the plasmid has been combined with the plant's DNA (**Figure F5.5**).

A second method used to get genes into plant cells involves painting the desired segment of DNA onto microscopic metal particles. These are then loaded into a "gene gun" and shot into the plant cells. Once inside the cells, the DNA is washed off the metal particles by cellular fluids and migrates to the nucleus, where it is incorporated into the plant's DNA, forming recombinant DNA.

restriction enzyme A bacterial enzyme used in genetic engineering that has the ability to cut DNA in a specific location.

plasmid A loop of bacterial DNA that is independent of the bacterial chromosome.

recombinant DNA DNA that has been formed by joining DNA from different sources.

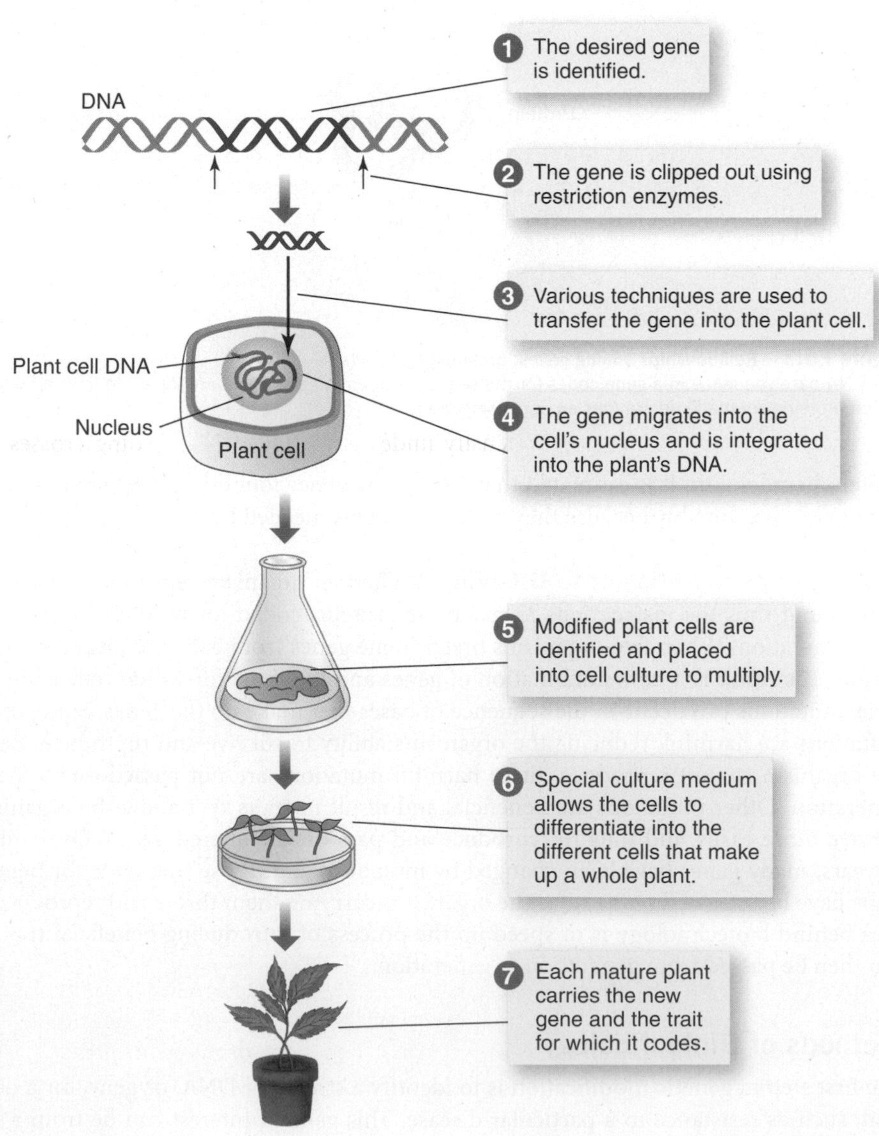

Figure F5.5 Engineering genetically modified plants
Crops developed using the genetic engineering steps shown here are grown all over the world.

The modified plant cells produced by either technique are then allowed to multiply. As they do, the new gene is reproduced with them. The cells are then allowed to divide into more and more cells and eventually differentiate into the various types of cells that make up a whole plant. The new plant is a **transgenic** organism. Each cell in the new plant contains the transferred gene for the desired trait, such as disease resistance (see Figure F5.5).

transgenic An organism with a gene or group of genes intentionally transferred from another species or breed.

Genetic engineering is more difficult in animals because animal cells do not take up genes as easily as plant cells do, and making copies of these cells (clones) is also more difficult. However, these techniques have been used to produce cows that yield more milk, cattle and pigs with more meat on them, and sheep that grow more wool.[3]

Is Genetic Modification Really New?

Modern genetic engineering technology is new, but humans have been directing the genetic modification of plants and animals for about 10,000 years. Farmers thousands of years ago didn't use gene guns or bacterial plasmids, but they selected seeds from plants with the most desirable characteristics to plant for the next year's crop, bred the animals that grew fastest or produced the most milk to improve the productivity of the next generation, and cross-bred plant varieties to combine the desired traits of each. Almost every fruit, vegetable, or crop grown today has been in some way genetically modified using traditional **selective breeding** techniques. Some of these crops, such as pumpkins, potatoes, sugar beets, and varieties of corn, oats, and rice, would not have developed without human intervention. These interventions have allowed modern farmers to grow plants that produce more food that is more nutritious and can better withstand harsh environments and resist disease.

selective breeding Techniques to selectively control mating in plants and animals for the purpose of producing organisms that better serve human needs.

Traditional Breeding Technology

Traditional breeding begins when farmers and ranchers select plants or animals with desired characteristics, such as high yield, palatability, resistance to disease and insects, or aesthetic characteristics. The traits can then be brought together by controlled mating. Plant breeders use a process called **hybridization**, in which two related plants are cross-pollinated or cross-fertilized (**Figure F5.6**). The resulting offspring has characteristics from both parent plants. For example, a breeder who wanted to produce a variety of wheat that was high-yielding and resistant to cool temperatures would cross plants with these two traits. The breeder would then select the offspring that acquired both of the desired traits.

hybridization The process of cross-fertilizing two related plants with the goal of producing an offspring that has the desirable characteristics of both parent plants.

Cross-breeding of animals has a similar goal. Animal breeders use both inbreeding and outbreeding. Inbreeding involves crosses between closely related animals. It can intensify desirable traits but may also intensify undesirable traits. Outbreeding crosses unrelated animals to reduce undesirable traits, increase variability, and introduce new traits. Today, inbreeding and outbreeding are carried out by artificial insemination.

Advantages of Modern Biotechnology

Traditional breeding techniques work well, but they have limitations in terms of both time and outcome. Breeding generations of plants or animals to consistently produce a desired trait is time-consuming; a new trait can only be produced once in the reproductive cycle of the plant or animal. Only plants or animals of the same or closely related species can be interbred. Not every offspring will inherit the desirable traits. Because cross-breeding transfers a set of genes from the parents to the offspring, both desirable and undesirable traits are transferred. Eliminating the undesirable genes while keeping the desirable ones can require many crosses.

Biotechnology, which selects specific genes in the laboratory, has significantly sped up this process and removed certain limitations. It enables breeders to select, modify, and transfer single genes. This speeds up the process by reducing the time and cost of breeding the crosses that are needed to select out the undesirable traits (**Figure F5.7**). In addition, biotechnology is not limited by whether two animals are capable of cross-breeding but, rather, can select desirable genes from any species, since the way DNA codes for proteins in plants, animals, yeast, and bacteria is the same. This allows traits from different species and completely different organisms to be used.

Figure F5.6 Hybridization
Cross-pollination or cross-fertilization, as illustrated with these oat seedlings, is a traditional method for breeding new plant varieties.

Figure F5.7 Traditional plant breeding versus biotechnology

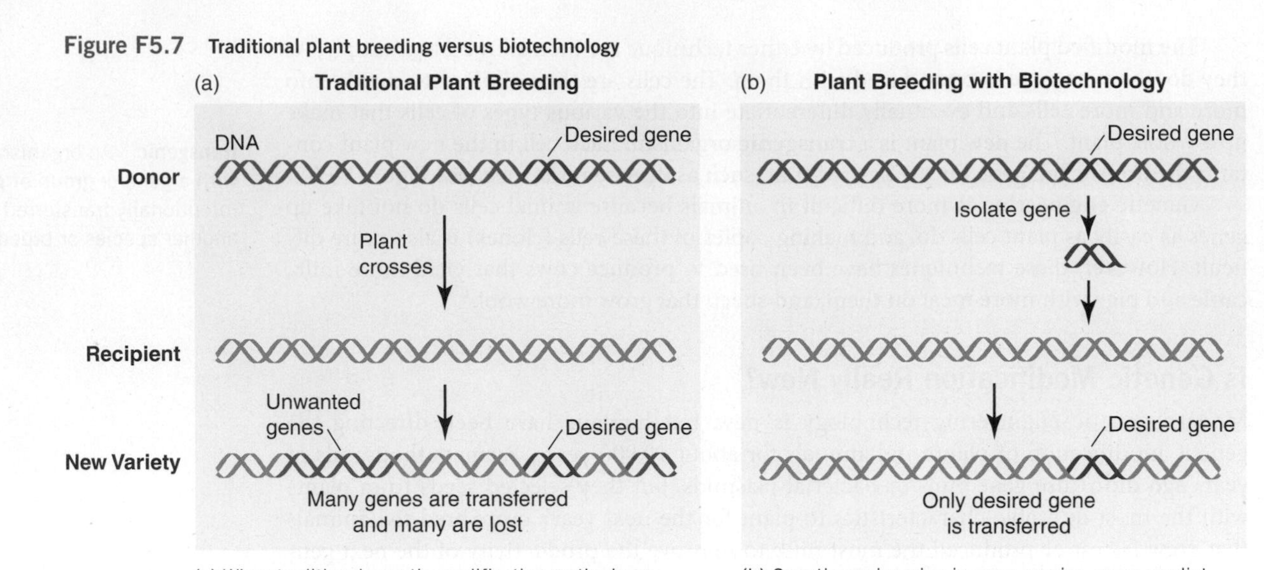

(a) When traditional genetic modification methods are used, thousands of genes are mixed. Not all the new varieties contain the desired gene, and many attempts over many years may be required, to remove the unwanted traits.

(b) Genetic engineering is more precise, more predictable, and faster. It allows the insertion of one or two genes into a plant without the transfer of genes coding for undesirable traits.

F5.2 Applications of Modern Biotechnology

LEARNING OBJECTIVES

- Discuss how genetic engineering is used to increase crop yields.
- Describe how GM crops are used to improve nutrient intake.
- Explain how genetic engineering can be used to combat human disease.

The techniques of biotechnology can be used in a variety of ways in both production and processing to alter the quantity, quality, cost, safety, and shelf life of the food supply (**Table F5.2**). This technology also has great potential for addressing the problem of world hunger and malnutrition.

How Biotechnology Can Increase Crop Yields

Genetic modification of crops can increase yields either directly, by inserting genes that improve the efficiency with which plants convert sunlight into food, or indirectly, by creating plants that are resistant to herbicides, pesticides, and plant diseases, thus reducing crop losses. Scientists are also working to develop plants that can withstand drought, freezing,

TABLE F5.2 Examples of Characteristics Introduced Using Genetic Engineering

Food	Characteristic
Cherry tomato	Better taste, color, texture
Flavr Savr tomato	Delayed ripening
Tomato	Thicker skin, altered pectin content
Corn	Insect protection, herbicide resistance
Cotton	Insect protection, herbicide resistance
Squash	Virus resistance
Papaya	Virus resistance
Potatoes	Potato beetle resistance, virus resistance
Soybeans	Herbicide resistance, high oleic acid content to reduce need for hydrogenation
Sugar beets	Herbicide resistance
Sunflower	High oleic acid content to reduce need for hydrogenation

and high salt concentrations. Many attempts in the last century to increase crop yields in developing countries have failed because they required expensive machinery and chemicals. With genetically engineered crops, simply providing a new type of seed has the potential to increase food production.

Scott Camazine/Photo Researchers/Getty Images, Inc.

Figure F5.8 Bt corn
Genetically engineered corn that produces the Bt protein is toxic to this European corn borer.

Herbicide Resistance Herbicides are chemicals that are sprayed on crops in the field to control weeds. Effective herbicides kill weeds but do not harm crops. A crop such as soybeans is not harmed by a particular herbicide because it contains enzymes that inactivate the herbicide. These enzymes are crop specific, so only certain herbicides can be used with certain crops. In some cases the weeds and the crops are resistant to the same herbicide so the farmer has few or no choices for weed control. Genetic engineering allows researchers to transfer genes for herbicide resistance to plants, making them resistant to new types of herbicides. Farmers then have more options for weed control so they can use herbicides that are less expensive, more effective, and less environmentally damaging.

Insect Resistance Genetic engineering techniques can increase insect resistance in plants. For example, a gene from the bacterium *Bacillus thuringiensis* produces a protein that is toxic to certain insects but safe for humans and other animals. By inserting the gene for this protein into plant cells, scientists have created plants that manufacture their own insecticide. The protein produced by this gene, known as Bt, has been used as an insecticide for over 30 years. But when plants manufacture their own insecticide, Bt does not need to be sprayed on the plants. This saves the farmer money, fuel, and time. It also benefits the environment because the Bt affects only insects feeding on the crop of interest and is not spread to surrounding foliage. Corn, potatoes, and cotton have been genetically modified to produce the Bt protein (**Figure F5.8**). Over the first 13 years of commercial availability, two Bt crops, corn and cotton, reduced chemical insecticide use by about 64 million pounds.[4]

In addition to creating insect-resistant plants, genetic engineering has been used to create environmentally friendly pesticides. These pesticides are produced by bacteria. The bacteria are then killed and sprayed on plants.

Disease Resistance Potato, squash, cucumber, watermelon, and papaya have been modified to resist viral infections. The benefits of this technology are illustrated by how it helped to preserve Hawaii's papaya crop. In the mid-1990s the papaya ring-spot virus threatened to wipe out Hawaii's second largest crop. Traditional plant breeding was not able to produce a virus-resistant strain of papaya, but by inserting a gene that acted like a vaccine into the papaya plant DNA, scientists were able to produce papaya plants that were immune to the virus (**Figure F5.9**).[5]

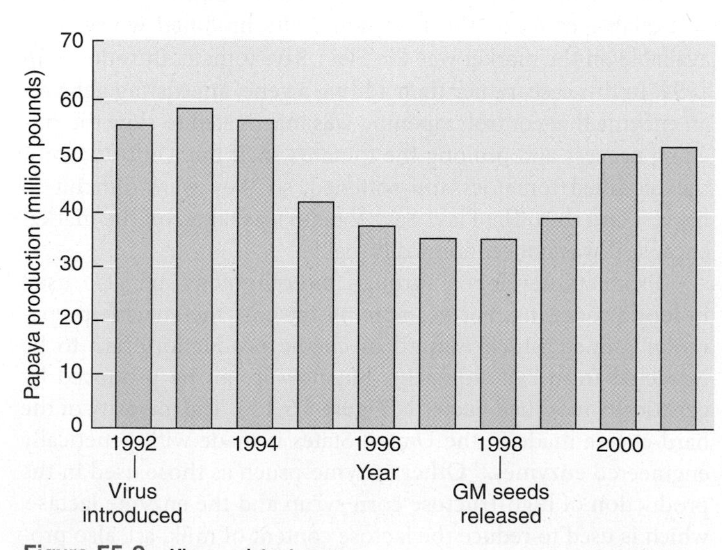

Figure F5.9 Virus-resistant papaya
Seeds for papaya that are resistant to the papaya ring-spot virus were released for commercialization in 1998, allowing Hawaiian papaya production to rebound. This was the first genetically enhanced fruit crop on the market.

How Biotechnology Can Improve Nutritional Quality

Although the causes of malnutrition are rooted in political, economic, and cultural issues that cannot be resolved by agricultural technology alone, genetically modified crops that target some of the major nutritional deficiencies worldwide are being developed. To address protein deficiency, varieties of corn, soybeans, and sweet potatoes with higher levels of essential amino acids are being created. To address iron deficiency, rice has been engineered to contain more iron.[6] To address vitamin A deficiency, genes that code for the production of enzymes needed for the synthesis of the vitamin A precursor β-carotene have been inserted into rice.[7] Half of the world's population depends on rice as a dietary staple, but it is a poor source of vitamin A. The genetically modified rice is called Golden Rice for the color imparted by the β-carotene pigment (**Figure F5.10**). One variety of Golden Rice that has been developed contains enough provitamin A in ½ cup of dry rice (about 1 cup cooked) to provide over 50% of the RDA for a child.[8] Golden Rice was developed in 1999 and the first field trial was completed in 2004, but it has not yet been commercialized.[9] Concerns about the safety and effectiveness of Golden Rice have contributed to the delay in commercialization.[10] One issue is that Golden Rice addresses only one of many nutrient deficiencies that threaten lives in developing countries (see Chapter 9, Debate: Combating Vitamin A Deficiency with Golden Rice). To address multiple nutrient deficiencies, the BioCassava Plus program is using biotechnology to develop cassava with increased zinc, iron, protein, and vitamin A levels.[11]

Golden Rice Humanitarian Board. www.goldenrice.org

Figure F5.10 **Golden rice**
The white rice shown on the left does not contain the vitamin A precursor β-carotene. The rice shown on the right, called Golden Rice, has been genetically modified to produce β-carotene, which gives it a yellow-orange color.

Biotechnology is also being used to provide foods that may help prevent chronic disease. For example, genetic modification has produced a variety of soybeans that has a higher percentage of oleic acid—a monounsaturated fatty acid that may help reduce the risk of cardiovascular disease. Vegetable oils that have lower amounts of saturated fat are also being developed, as are soybeans that contain more of the antioxidant vitamin E than the traditional varieties.[12]

Other qualities that affect a food's role in the diet can also be changed by genetic engineering. For example, potatoes that are denser and have less water have been developed because they absorb less oil when fried—lowering the fat content of an order of French fries.

How Biotechnology Impacts Food Quality and Processing

Genetic engineering is also being used to improve the appeal and quality of food. The first genetically modified whole food available on the market was the Flavr Savr tomato, introduced in 1994. In this case, rather than adding a gene, an existing gene for an enzyme that controls ripening was inactivated to slow the ripening process and prolong the tomato's shelf life. Unfortunately, the modified tomatoes still softened, so they were difficult to harvest and ship. The Flavr Savr tomato was taken off the market because it was not economically viable.

Products developed through biotechnology are also used in food processing. For example, in the past the enzyme preparation rennet, which is used in cheese production, had to be extracted from calf stomachs, but now it can be produced by genetically modified bacteria (**Figure F5.11**). The majority of the hard cheese made in the United States is made with genetically engineered enzymes.[13] Other enzymes, such as those used in the production of high-fructose corn syrup and the enzyme lactase, which is used to reduce the lactose content of milk, are also produced by genetically modified microorganisms. Many food color

Rosenfeld Images Ltd./Photo Researchers

Figure F5.11 **Cheese production**
During cheese production, an enzyme preparation known as rennet is added to clot the milk. Much of the hard cheese produced in the United States today relies on rennet produced by genetically modified bacteria.

and flavor additives are also produced in the laboratory. For example, vanilla can now be produced by plant cells grown in culture, rather than harvested from vanilla orchid plants, which grow only in tropical climates.

How Biotechnology Is Used in Animal Food Production

Genetic engineering is being used to improve methods of preventing, diagnosing, and treating animal disease as well as to enhance growth efficiency and fertility in food animals. The first FDA-approved application of biotechnology in animal production was the use of recombinant bovine somatotropin (bST) in dairy cows. To produce recombinant bST, scientists isolated the gene from cow cells and then inserted it into bacterial cells. The bST, which is produced in large amounts by the bacterial cells, was then isolated and purified. The resulting product can be injected into dairy cows to increase milk production.

Biotechnology is also being applied to aquaculture. Aquaculture, also known as known as fish or shellfish farming, involves the breeding, rearing, and harvesting of plants and animals in ponds, rivers, lakes, and oceans. The nutritional benefits of seafood and the dwindling natural supplies have made aquaculture the fastest growing sector of animal production worldwide. Today, most of the salmon consumed in the United States is farm-raised. Biotechnology allows scientists to increase the nutritional value of fish feed and identify and combine traits in fish and shellfish to increase productivity and improve quality.[14]

How Biotechnology Is Used to Combat Disease

Biotechnology is being used to produce vaccines and medications to treat human disease. For example, before 1978, when bacteria were engineered to produce human insulin, diabetics relied on insulin extracted from pigs or cows (see Chapter 4: Science Applied: How the Discovery of Insulin Made Diabetes Treatable). Other engineered proteins used to treat human disease include tissue plasminogen activator to dissolve blood clots in heart-attack victims, growth factors to stimulate cell replication in bone-marrow transplants, erythropoietin to treat anemia, hepatitis B vaccine, and interferon to attack viruses and stimulate the immune system.

Biotechnology is also being used to prevent allergic reactions by reducing or eliminating naturally occurring allergens from food products. For example, people who are allergic to peanuts may someday enjoy peanut butter sandwiches made with peanuts that have been genetically modified to eliminate the proteins or portions of the proteins to which they are allergic.

F5.3 Safety and Regulation of Genetically Modified Foods

LEARNING OBJECTIVES

- Explain how genetic engineering might create risks for people with allergies.
- Discuss the potential impact of genetic engineering on the environment.
- Describe how genetically modified food and crops are regulated.

The rapid advance in biotechnology during the past decade has created the potential for health problems and environmental damage (**Figure F5.12**). Regulations are in place to control the use of genetic engineering and GM products. Despite these precautions, many consumers and scientists believe that the impact of this booming technology has not yet become apparent. They urge that this technology be used with caution to avoid health or environmental impacts that outweigh the benefits.

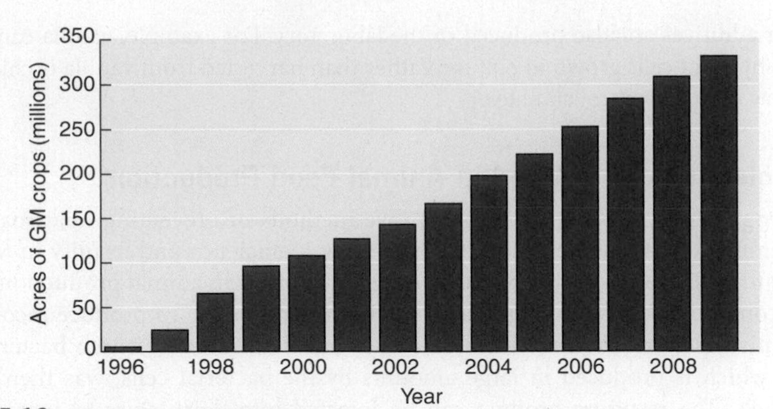

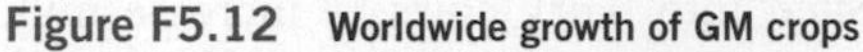

Figure F5.12 **Worldwide growth of GM crops**
Despite concerns about the impact of GM crops, the number of acres planted with them has risen steadily. The most common GM crops are soybeans, corn, cotton, and rapeseed (for canola oil).

Consumer Safety

Consumer safety concerns related to bioengineered foods include the possibility that an allergen or toxin has been inadvertently be introduced into a previously safe food, or that the nutrient content of a food has been negatively affected. Although foods containing ingredients derived from plant biotechnology are not generally required to carry special labels, those that contain potentially harmful allergens or toxins, or that are altered nutritionally, must be labeled. Another potential health concern is that the antibiotic resistance genes used in biotechnology may promote the development of antibiotic-resistant strains of bacteria.

Allergens and Toxins Genes code for proteins, so when a new gene is introduced into a food product, a new protein is made. These new proteins may in some cases be allergens or toxins. If a food contains a new protein that is an allergen, allergic individuals will now react to this food that may previously have been safe for them. For example, if DNA from fish or peanuts—foods that commonly cause allergic reactions—were introduced into tomatoes or corn, these foods would then be dangerous to allergic individuals. To prevent this from happening unintentionally, biotechnology companies have established systems for monitoring the allergenic potential of proteins used for plant genetic engineering.[15] Testing for the transfer of allergens has already proved to be valuable. In 1996, allergy testing successfully prevented soybeans containing a gene from a Brazil nut from entering the market.[16] If these soybeans had entered the food supply, they could have caused allergic reactions in people allergic to nuts. However, despite mandatory testing programs, individuals with food allergies cannot assume that new foods are safe.

Likewise, when a product is created using either a donor or recipient organism that is known to produce a toxin, the manufacturer must verify that the resulting product does not have high levels of the toxin. Toxins are also an issue when plants are produced by traditional breeding; toxic varieties of celery and potatoes have resulted from traditional breeding methods.

Changes in Nutrient Content Changing the proteins made by a plant or animal could also affect the nutrient content of foods. For example, tomatoes are an excellent source of vitamin C. If tomatoes were modified to have no vitamin C, people who rely on tomatoes for this vitamin might no longer get enough in their diets. As with foods containing potentially harmful allergens or toxins, foods with altered nutrient content must be labeled to disclose this information.

Antibiotic Resistance There is a concern that the use of genetic engineering will spread antibiotic-resistance traits to bacteria in the environment or in the gastrointestinal tract. If pathogenic bacteria were to acquire this trait, it would make some of the antibiotics used to treat disease ineffective. The reason for this concern is that genetic engineering techniques use genes that code for antibiotic-resistance as marker genes. By inserting a marker gene along with the gene they want to transfer, scientists are able to check to see whether the gene transfer was successful. Antibiotic-resistant genes are used for this purpose because it is easy to verify that the transfer has occurred by exposing the bacteria to antibiotics—those that survive the antibiotic contain the transferred genes. Current techniques are unlikely to cause problems because the marker genes used are already widespread in the normal bacteria that inhabit the gut and in harmless environmental bacteria. In addition, markers are not used if they confer resistance to clinically important antibiotics. Newer techniques that use alternative selection systems or remove the antibiotic resistance markers before the plants leave the laboratory are becoming available.[17]

Environmental Concerns

Some of the arguments against the use of genetically modified crops are that they will reduce biodiversity, promote the evolution of pesticide-resistant insects, or create "superweeds" that will overgrow our agricultural and forest lands.

Reduced Biodiversity The ability of populations of organisms to adapt to new conditions, diseases, or other hazards depends on the presence of many different species that provide a diversity of genes. When there is biodiversity, a harmful event, such as the emergence of a new disease strain, does not eliminate all the plants—it might kill one species, but others have genes that protect them. Likewise, a drought might be harmful to some species, but others would have genes that make them drought-resistant.

A concern with biotechnology is that farmers will prefer new, resistant, high-yielding crop varieties and stop planting others, causing certain varieties eventually to become extinct. For example, if every farmer began to use only genetically modified rice, then the varieties of rice that carry other traits would become extinct and those traits would be lost. If a new insect or virus emerged that killed the genetically modified rice, breeders and genetic engineers would not have the other varieties available to search for a gene that allows survival.

Concerns about reduced biodiversity are not unique to genetically engineered crops. Biodiversity is also reduced when plants produced by traditional plant breeding are used to the exclusion of others. To preserve biodiversity and ensure a large supply of traits for use in future breeding, biotechnology techniques are being applied to establish gene banks and seed banks and to identify and characterize the genes in many species.[18] These precautions are designed to help prevent genes from being lost, but they will not prevent our agricultural land from being dominated by only a few varieties of crops.

Superweeds and Superbugs Other environmental concerns with regard to genetically modified crops are that they will promote the development of superweeds and superbugs. A superweed might arise if a plant that has been modified to grow faster or to better survive begins growing in areas beyond the farmer's field. Most experts do not feel this is a major concern because domesticated crops depend on a managed agricultural environment and carry traits that make them unable to compete in the wild.

It has also been suggested that the genes inserted to produce hardy, high-yield, fast-growing crops could be transferred to wild relatives by natural cross-breeding. This could result in fast-growing superweeds. Although a possibility, this scenario is unlikely for a number of reasons. First, the probability is small that a weed growing near a genetically modified plant is closely enough related to cross-breed. Even if this does occur, the chances that the new plant will survive and have inherited traits that enhance its survival is even smaller. As a further safeguard to prevent environmental risks, most developers avoid adding traits that could increase the competitiveness or other undesirable properties in weedy relatives.

There is also concern that crops engineered to produce their own pesticides will promote the evolution of pesticide-resistant insects. Although this is an important concern, the risk of it occurring is no greater when pesticides are engineered into the crops than it is when pesticides are sprayed on crops. An illustration of this problem involves insects that are resistant to the Bt toxin.[18] As more and more of the insects' food supply is made up of plants that produce this pesticide or are sprayed with it, only insects that carry genes making them resistant to Bt can survive and reproduce. This increases the number of Bt-resistant insects and therefore reduces the effectiveness of Bt as a method of pest control. To address this problem, strategies are being developed to prevent the number of Bt-resistant insects from increasing. Farmers who grow pesticide-resistant crops are required to grow nonmodified plants in adjacent fields. This provides a food supply for—and encourages the continued existence of—nonresistant insect pests, thereby reducing the likelihood that the number of pesticide-resistant insects will increase.

Chris Knapton/Photo Researchers

Figure F5.13 Field testing GM crops
Field tests of genetically modified crops attempt to determine the impact of the new crops on the environment and how well the plants function. There is concern, however, that the tests themselves may pose a risk to the environment.

Regulation of Genetically Engineered Food Products

Although the United States government does not scrutinize every step in the development of new plant varieties, it is involved in overseeing the process. The government sets guidelines to help researchers address safety and environmental issues at all stages, from the early development of genetically engineered plants through field testing and ultimately commercialization. Companies that develop new plant varieties must provide data to support the safety and wholesomeness of the product. Crops created by both traditional breeding and biotechnology methods must be field tested for several seasons to make sure only desirable changes have been made (**Figure F5.13**). Plants are examined to ensure that they look right, grow right, and produce food that is safe and tastes right. Analytical tests must be performed to determine if the levels of nutrients in the new variety are different and if the food is safe to eat. The FDA, the USDA, and the EPA are all involved in the oversight of plant biotechnology.[19]

The Food and Drug Administration (FDA)

The FDA has jurisdiction over the safety of foods in the marketplace and therefore regulates the safety and labeling of all foods and animal feeds derived from crops, including genetically modified crops.

What Determines the Safety of GM Foods? The FDA policy is that the safety of a food product should be determined on the basis of the characteristics of the food or food product, not the method used to produce it. Foods developed using biotechnology are therefore evaluated to determine their equivalence to foods produced by traditional plant breeding. Emphasis is placed on whether the food creates a new or increased allergenic risk, has an increased level of a naturally occurring toxin, contains a substance not previously present in the food supply, or is nutritionally different from the traditional plant. Currently, premarket approval is required only when the new food contains substances not commonly found in foods or contains a substance that does not have a history of safe use in foods.[20]

THINK CRITICALLY

Do you think GM ingredients are different enough from ingredients produced without genetic engineering to warrant the economic and logistical costs involved in providing this information on food labels?

Labeling GM Foods As with food safety, the FDA's food labeling policies focus on the characteristics of a food, not on how it is produced. For example, virtually all plants have been genetically modified through traditional plant breeding, but the FDA does not require declarations about those modifications. Under this precedent, labeling decisions are based on whether there are differences between the GM food and the traditional food. Foods containing GM ingredients must be labeled as such only if the nutritional composition of the food has been altered; if it contains potentially harmful allergens, toxins, pesticides, or herbicides; if it contains ingredients that are new to the food supply; or if the food has been changed significantly enough that its traditional name no longer applies.

Some people believe that all foods that contain GM ingredients should be labeled as such so that consumers who want to avoid these foods would be able to do so. For instance, a vegan may want to know if the tomato they are about to consume contains some DNA from an animal. However, the FDA stance is that plants and animals already share many of the same DNA segments and such labeling would be misleading, since foods that are not different in quality, nutrient composition, and safety would be viewed as somehow different from traditional foods (**Figure F5.14**).

© Vanessa Miles/Alamy Limited

Figure F5.14 Labeling GM foods
Whether the label is voluntary or mandatory, designing an appropriate label is a complex issue. Some companies are voluntarily labeling foods as "Grown from genetically modified seed." The FDA feels that these simple labels will imply to some people that the product is inherently better than other products, while other people will feel this means that the food is more dangerous than others. However, a more detailed label that describes how a particular product was modified would be too lengthy and would not be understood by many consumers.

Cost is also a consideration. If labeling regulations change, manufacturers would need to add labels to the hundreds, if not thousands, of food products containing GM ingredients. DNA-modified plants would need to be segregated during planting and harvesting, and products containing ingredients from them would require different manufacturing, transport, and storage facilities. Despite these costs, however, the need to sell to markets such as the European Union that do not want GM crops, is already forcing U.S. farmers to separate and label GM crops.

U.S. Department of Agriculture (USDA) The USDA regulates agricultural products and research on the development of new plant varieties. The Animal and Plant Health Inspection Service (APHIS) of the USDA helps to ensure that the cultivation of a new plant variety poses no risk to agricultural production or to the environment. For example, if there is a high probability that a new plant variety will cross-breed with a weed and that the transfer of the new trait could allow the weed plant to survive better, APHIS will not allow further development of this plant. If a plant has been studied and tested and does not pose environmental risks, field-testing is allowed. APHIS continues to oversee the testing until it is determined that the plant is safe.

Environmental Protection Agency (EPA) The EPA regulates any pesticides that may be present in foods and sets tolerance levels for these pesticides. This includes genetically modified plants that can protect themselves from insects or disease. The EPA assesses the safety of the protein that confers the insect or disease resistance for human consumption, for other organisms, and for the environment.

REFERENCES

1. Cohen, S. N., Chang, A. C., Boyer, H. W., and Helling, R. B. Construction of biologically functional bacterial plasmids *in vitro*. *Proc Natl Acad Sci U.S.A.* 70:3240–3244, 1973.
2. Cook, J. R. Testimony before the U.S. House of Representatives Subcommittee on Basic Research hearing on "Plant Genome Research: From the Lab to the Field to the Market." October 5, 1999, Serial No. 106–60. Washington, DC: Government Printing Office, 1999.
3. Margawati, E. T. *Transgenic Animals: Their Benefits to Human Welfare.*
4. The Organic Center. *The First 13 Years of GE Crops: Highlights of a Critical Issue Report*, March, 2011.
5. Gonsalves, D., and Ferreira, S. *Transgenic Papaya: A Case for Managing Risks of Papaya Ringspot Virus in Hawaii.* Plant Management Network, 2003.
6. Sautter, C., Poletti, S., Zhang, P., and Gruissem, W. Biofortification of essential nutritional compounds and trace elements in rice and cassava. *Proc Nutr Soc* 65:153–159, 2006.
7. Ye, X., Al-Babili, S., Kloti, A., et al. Engineering the provitamin A (beta-carotene) biosynthetic pathway into (carotenoid-free) rice endosperm. *Science* 287:303–305, 2000.
8. Muzhingi, T., Gadaga, T. H., Siwela, A. H., et al. Yellow maize with high β-carotene is an effective source of vitamin A in healthy Zimbabwean men. *Am J Clin Nutr* 94:510–519, 2011.
9. Golden Rice. History of Programme: The Road to the Farm Is Bumpy.
10. Potrykus, I. Lessons from the 'Humanitarian Golden Rice' project: Regulation prevents development of public good genetically engineered crop products. *N Biotechnol* 27:466–472, 2010.
11. Sayre, R., Beeching, J. R., Cahoon, E. B., et al. The BioCassava Plus program: Biofortification of cassava for Sub-Saharan Africa. *Ann Rev Plant Biol* 62:251–272, 2011.
12. Hunter, S. C., and Cahoon, E. B. Enhancing vitamin E in oilseeds: Unraveling tocopheroland tocotrienol biosynthesis. *Lipids* 42:97–108, 2007.
13. Lemaux, P. G. Genetically Engineered Plants and Foods: A Scientist's Analysis of the Issues (Part I) *Ann Rev Plant Biol* 59:771–812, 2008.
14. Food and Agriculture Organization of the United Nations, Fisheries and Aquaculture Department. *Biotechnology.*
15. Goodman, R. E., Vieths, S., Sampson, H. A., et al. Allergenicity assessment of genetically modified crops—what makes sense? *Nat Biotechnol* 26:73–81, 2008.
16. Nordlee, J. A., Taylor, S. L., Townsend, J. A., et al. Identification of a Brazil-nut allergen in transgenic soybeans. *N Engl J Med* 334:688–692, 1996.
17. Manimaran, P., Ramkumar, G., Sakthivel, K., et al. Suitability of non-lethal marker and marker-free systems for development of transgenic crop plants: Present status and future prospects. *Biotechnol Adv* 29:703–714, 2011.
18. Lemaux, P. G. Genetically engineered plants and foods: A scientist's analysis of the issues (Part II). *Ann Rev Plant Biol* 60:511–559, 2009.
19. U.S. Food and Drug Administration. Plant Biotechnology for Food and Feed: FDAs Biotechnology Policy.
20. U.S. Food and Drug Administration. *Bioengineered Foods: Statement of Robert E. Brackett, Ph.D.* Center for Food Safety and Applied Nutrition.

CHAPTER OUTLINE

AFP/Getty Images, Inc.

World Hunger and Malnutrition

18

CASE STUDY

Teresa, a Peace Corps volunteer, recently arrived in Rwanda to help care for children in a rural health clinic. She is struck by the number of children suffering from diarrhea and intestinal parasites. Most of the children appear small for their age and are underweight. She often sees children she thinks are 5 or 6 years old, only to discover that they are 10 or 11. She is told that the growth retardation is the result of chronic undernutrition. A small 2-year-old boy named Songe catches her eye. His hair is an odd color, his legs are scrawny, and his belly is bloated, but he is not as emaciated as some of the malnourished children she sees. His mother says that his symptoms began shortly after his younger brother was born. After a bout of diarrhea, he became lethargic and eventually stopped growing.

© MissHibiscus/iStockphoto

The clinic nurse explains to Teresa that Songe is suffering from a form of protein-energy malnutrition. When his brother was born, Songe stopped nursing at his mother's breast and started eating the adult diet, which is high in starch and fiber. He was not getting enough protein to support the needs of his growing body. His belly is bloated because his abdomen is filling with fluid and his liver is filling with fat. There is not enough protein in his blood to hold the fluid there, so it seeps into his abdomen; nor is there enough protein to transport fat, so it accumulates in his liver. Songe is also experiencing health problems that are less visible. Without adequate protein, his immune system cannot function normally, increasing his susceptibility to infections and making the immunizations provided in the clinic less effective.

To treat his malnutrition, the clinic provides Songe with a special high-protein drink to consume along with his regular diet. **Teresa is concerned that when Songe stops consuming the protein supplement, his malnutrition will return.**

18.1 The Two Faces of Malnutrition

LEARNING OBJECTIVES

- Compare the problems of under- and overnutrition in the world today.
- Discuss the impact of malnutrition throughout the life cycle.
- Describe the relationship between malnutrition and infectious disease.
- Explain how nutrition transition affects the incidence of obesity.

hunger Recurrent involuntary lack of food that over time may lead to malnutrition.

starvation A severe reduction in nutrient and energy intake that impairs health and eventually causes death. It is the most extreme form of malnutrition.

cycle of malnutrition A cycle in which malnutrition is perpetuated by an inability to meet nutrient needs at all life stages.

For most of us, the image that comes to mind when we think of malnutrition around the world is one of **hunger** and **starvation**. The nations at risk change, but the soulful eyes and bloated bellies of the starving children remain. About 925 million people around the world are chronically undernourished; over a third of all deaths in children under 5 years of age are due to undernutrition, which kills nearly 6 million children each year.[1,2,3,4] At the same time that global health organizations are struggling with issues of undernutrition, the incidence of illnesses related to overcomsumption is soaring. The overweight and the undernourished both suffer from malnutrition and experience high levels of sickness and disability, shorter life expectancies, and lower levels of productivity. These two faces of malnutrition exist together, complicating the goal of solving the problem of malnutrition worldwide.

The Impact of Undernutrition

In populations where undernutrition is a chronic problem, there is a **cycle of malnutrition** (**Figure 18.1**). The cycle begins when women consume a deficient diet during pregnancy.

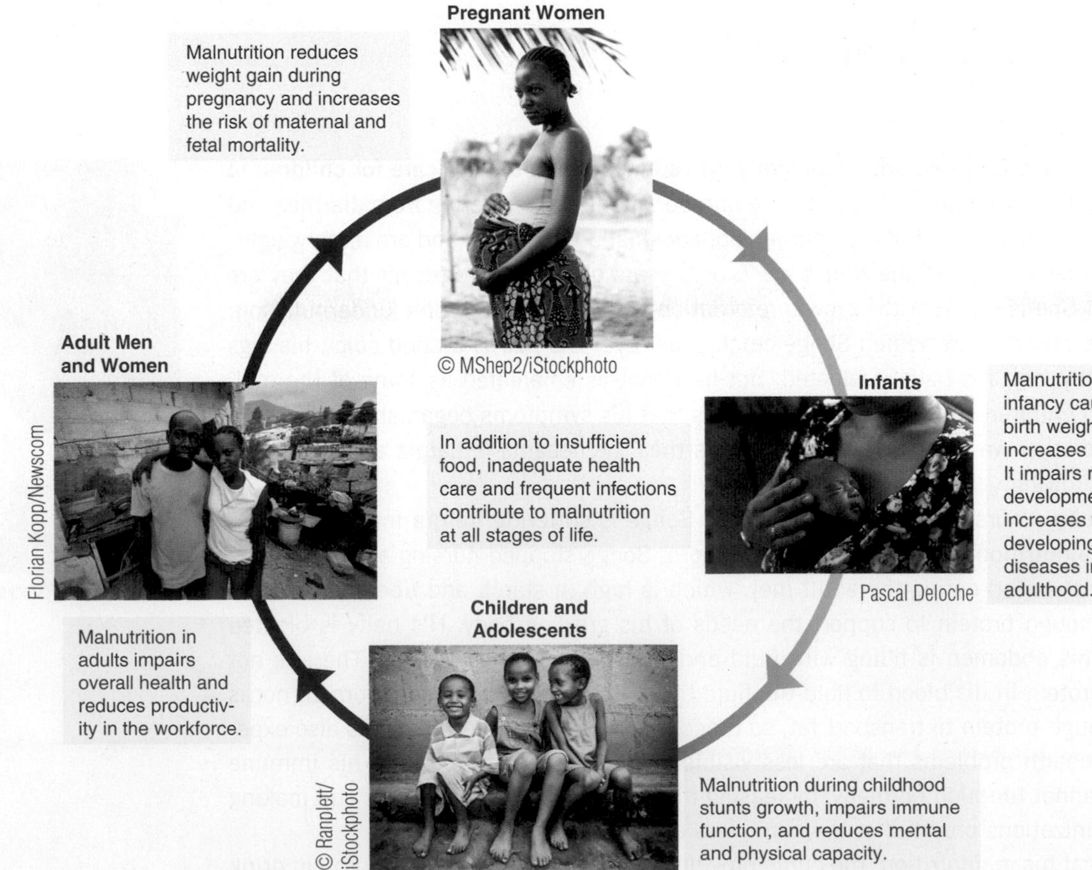

Figure 18.1 Malnutrition through the lifecycle
Malnutrition impairs the health of individuals at every stage of life. It often begins in the womb, continues through infancy and childhood, and extends into adolescent and adult life where it reduces productivity and the ability to live a healthy life.

These women are more likely to give birth to low-birth-weight infants who are susceptible to illness and early death. The children who do survive may be small and weakened physically and mentally. They grow into undernourished adults, who are also susceptible to disease and unable to contribute optimally to economic and social development. The women in this next generation also begin their pregnancies poorly nourished and are therefore likely to give birth to low-birth-weight infants. Interruption of this cycle of malnutrition at any point can benefit the individuals and the society. Healthy children can then grow into healthy adults who produce healthy offspring and can contribute fully to society.

infant mortality rate The number of deaths during the first year of life per 1000 live births.

stunting A decrease in linear growth rate that is an indicator of nutritional well-being in populations of children.

Low Birth Weight and High Infant Mortality Low-birth-weight infants—those weighing less than 5.5 pounds (2.5 kilograms) at birth—are at greater risk of complications, illness, and early death. More low-birth-weight infants means a higher **infant mortality rate**, the number of deaths per 1000 live births in a population. The infant mortality rate and the number of low-birth-weight births are indicators of the health and nutritional status of a population. In industrialized countries like Sweden, the United States, and Japan, the infant mortality rate is less than 7 per 1000 live births; in developing countries like Angola, Somalia, and Afghanistan, the rate is over 80 per 1000 live births (**Table 18.1**).[5] Low-birth-weight infants who do survive require extra nutrients, which are usually not available. Malnutrition in infancy and childhood has a profound effect on mental and physical growth and development as well as on susceptibility to infectious disease.

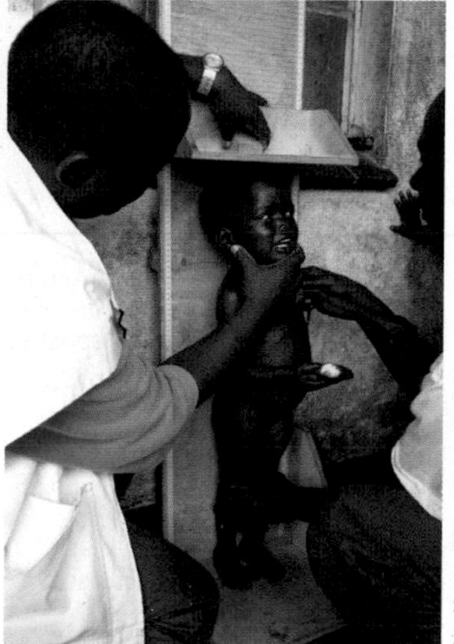

Ton Koene/Zuma Press/Newscom

Figure 18.2 Stunting is an indicator of malnutrition
Stunted children may never regain the height lost as a result of malnutrition. Stunting also impairs the development of vital organs, reducing life expectancy.

Stunted Growth Malnourished children grow poorly. The prevalence of decreased growth in height, referred to as **stunting**, is used as an indicator of the well-being of a population's children (**Figure 18.2**). It is estimated that over 30% of children under 5 years of age in developing countries are stunted.[6] Deficiencies of energy, protein, iron, and zinc, as well as prolonged infections, have been implicated as causes. Stunting in childhood produces smaller adults who have a reduced work capacity. Stunted women are more likely to give birth to low-birth-weight babies. In addition, those who had lower birth weights and early childhood stunting are more likely to have abdominal obesity in adulthood.[7] Abdominal obesity increases the risk of morbidity from cardiovascular disease, hypertension, and diabetes.

TABLE 18.1 Indicators of Poverty and Malnutrition[5]

	Infant mortality (deaths/1000 live births)	Life expectancy (years)	Literacy (percentage of population)	Access to medical care (physicians/1000 people)
More developed countries				
United States	6	78.7	99	2.67
Australia	4.6	81.9	99	2.99
Iceland	3.2	81	99	3.93
Japan	2.2	83.9	99	2.06
Israel	4.1	81.1	97.1	3.63
Germany	3.5	80.2	99	3.53
Less developed countries				
Afghanistan	121.6	49.7	28.1	0.21
Somalia	103.7	50.8	37.8	0.04
Haiti	52.4	62.5	52.9	0.25
Angola	83.5	54.6	70.1	0.08
Ghana	47.3	61.4	67.3	0.08
India	46.1	67.1	61	0.60
Cambodia	54.1	63.0	73.6	0.23

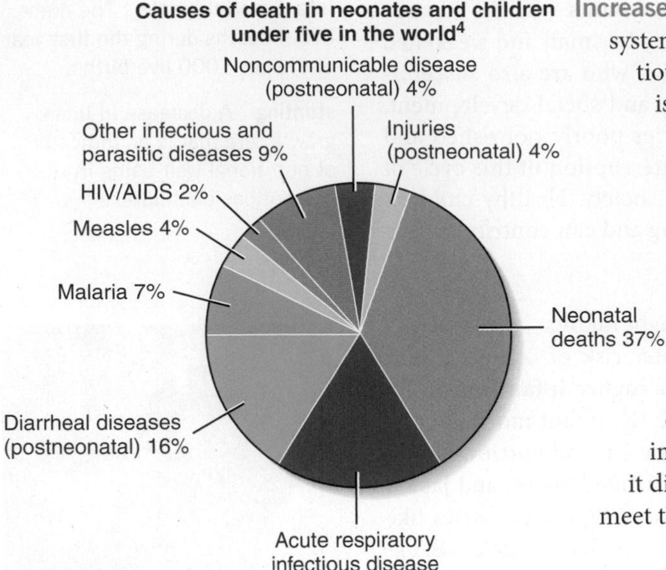

Figure 18.3 Malnutrition increases the risk of infection Infections are the leading cause of death in children under five. Mortality from infectious diseases such as measles, diarrheal diseases, and respiratory infections, is increased in malnourished children.

Increased Infections Undernourished children have depressed immune systems, which reduces their ability to resist infection. Therefore infectious diseases are more common and more severe in undernourished children. Mortality from infections is increased among malnourished children; they may die of infectious diseases that would not be life-threatening in well-nourished children. Well over half of all deaths in children under 5 years are due to infectious disease (**Figure 18.3**).[4] The presence of undernutrition is an underlying cause in an estimated 35% of deaths in children under age 5.[3] Even immunization programs, designed to reduce the incidence of infectious disease, may be ineffective because the immune systems of undernourished individuals cannot respond normally. Infection also contributes to and worsens malnutrition. The infective process increases nutrient needs as well as nutrient losses, thereby making it difficult even for adequately nourished children with infections to meet their needs.[8]

Overnutrition: A World Health Problem

For the first time in human history, the number of overweight people rivals the number who are underweight.[9] There are now more than 1.5 billion adults worldwide who are overweight or obese. One estimate predicts that by 2030, 2.16 billion adults worldwide will be overweight and 1.12 billion will be obese.[10] The prevalence of obesity around the world ranges from less than 5% in rural China, Japan, and some African countries to as high as 75% of the adult population in urban Samoa.[9] In Argentina, Colombia, Mexico, Paraguay, Peru, and Uruguay, more than half of the population is overweight, and more than 15% are obese.[11] Countries such as China and India that have historically been plagued by undernutrition must now also contend with overnutrition. This has been called a *double burden* of disease; as developing countries continue to struggle with problems of undernutrition, they are faced with rising rates of cardiovascular disease, hypertension, type 2 diabetes and other conditions associated with obesity. It is not uncommon in these countries to find undernutrition and obesity existing side by side within the same country, the same community, and even within the same household.

The growing prevalence of obesity among children is also a major concern. Latest estimates suggest that 42 million children under 5 years of age are overweight, and 35 million of these live in developing countries.[12] In some countries a high prevalence of overweight children now exists alongside a high frequency of undernourished children. Overweight and obese children are likely to stay obese into adulthood and are more likely to develop diseases such as diabetes and cardiovascular disease at a younger age.

Why Undernutrition and Overnutrition Exist Side by Side

We see the problems of undernutrition and overnutrition existing side by side because diets and lifestyles change as economic conditions improve (**Figure 18.4a**). Traditional diets in developing countries are based on a limited number of foods—primarily starchy grains and root vegetables. As incomes increase and food availability improves, the diet becomes more varied, and energy intake increases with the addition of meat, milk, and other more energy-dense foods. Along with this **nutrition transition**, there is a decrease in activity due to less physically demanding occupations, greater access to transportation, more labor-saving technology, and more passive leisure time (**Figure 18.4b**).

nutrition transition A series of changes in diet, physical activity, health and nutrition that occurs as poor countries become more prosperous.

Some of the effects of this nutrition transition are positive: Life expectancy increases, and the frequencies of low birth weight, infectious diseases, and nutrient deficiencies decrease. However, at the same time, rates of heart disease, cancer, diabetes, obesity, and childhood obesity increase (Figure 18.4a).[14,15] Transition to a diet high in animal protein and refined foods also increases the use of natural resources and in the long term may deplete nonrenewable resources.

Figure 18.4 Economic development and nutrition transition

(a)

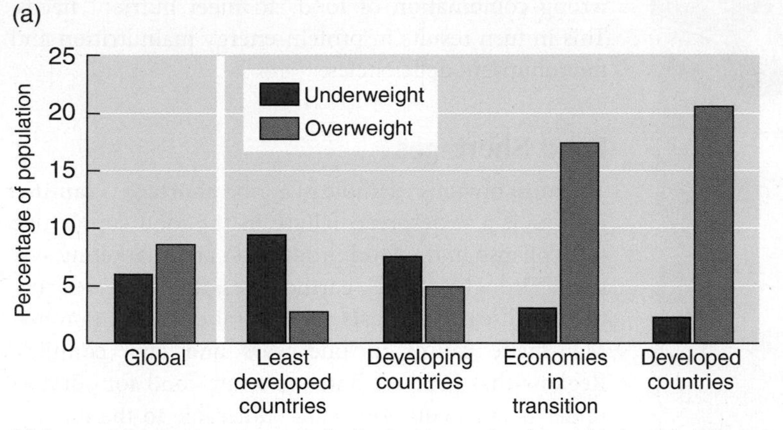

(a) As countries develop economically, the incidence of underweight drops, but the percentage of the population that is overweight increases.

(b)

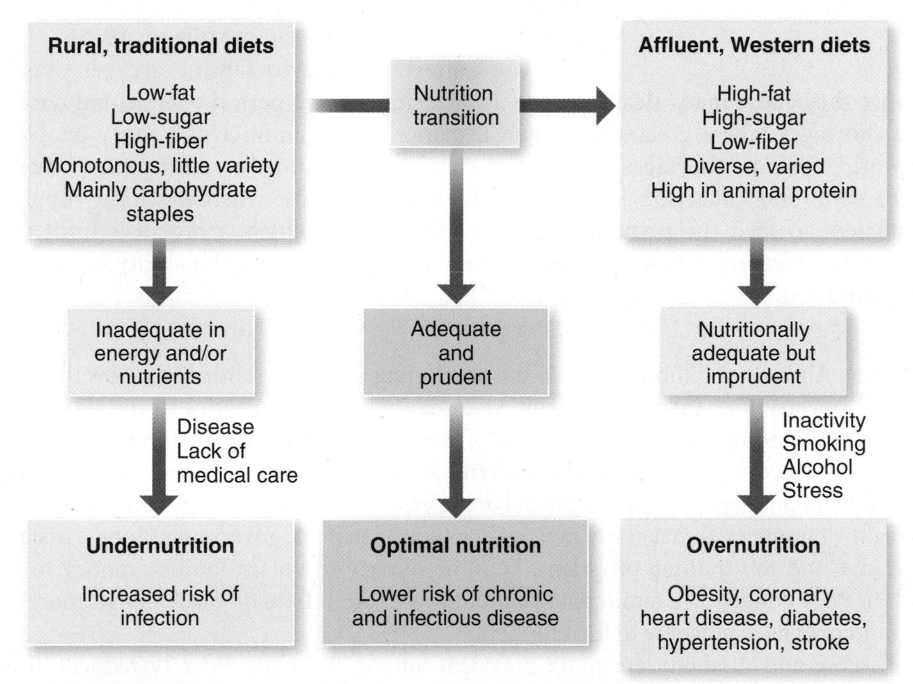

(b) This schematic shows the dietary changes and nutritional consequences of nutrition transition.[13] The traditional rural diet is often inadequate in energy, protein, or micronutrients. The affluent Western diet meets nutrient needs but is high in fat and sugar and low in fiber. A diet that falls somewhere between these extremes is optimal for health.

Infectious disease is the leading cause of death in developing countries. After a country has undergone nutrition transition, what types of diseases are the leading causes of death? Why does this change occur?

18.2 Causes of Hunger and Undernutrition

LEARNING OBJECTIVES

- Discuss the factors that cause food shortages for populations and for individuals.
- Explain the concept of food insecurity.
- List three common nutrient deficiencies worldwide and explain the consequences of these deficiencies.

The specific reasons for hunger vary with the time and place, but the underlying cause is that the food available in the world is not distributed equitably. This inequitable

Alison Wright/NG Image Collection

Figure 18.5 Causes of famine
After the earthquakes in Haiti in 2010, survivors lived in makeshift camps such as this one. This natural disaster destroyed the infrastructure that once distributed food throughout the country, resulting in famine.

distribution results in either not enough food or the wrong combination of foods to meet nutrient needs. This in turn results in protein-energy malnutrition and micronutrient deficiencies.

Food Shortages

The most obvious example of a food shortage is **famine**. Famine is a widespread failure in the food supply due to a collapse in the food production and marketing systems. Drought, floods, earthquakes, and crop destruction by diseases or pests are natural causes of famines. Man-made causes include wars and civil conflicts. Regions that produce barely enough food for survival under normal conditions are vulnerable to the disaster of famine. This situation is analogous to a man standing in water up to his nostrils: if all is calm, he can breathe, but if there is a ripple, he will drown. When a ripple such as a natural or civil disaster occurs, it cuts the margin of survival and creates famine (**Figure 18.5**).

famine A widespread lack of access to food due to a disaster that causes a collapse in the food production and marketing systems.

Food shortages due to famine are very visible because they cause many deaths in one area during a short period of time, but chronic food shortages take a greater toll when it comes to the number of hungry people in the world. Chronic shortages occur when economic inequities result in lack of money, health care, and education for individuals or populations; when the food supply is insufficient to feed the population; when cultural and religious practices limit food choices; or when environmental resources are misused, limiting the ability to continue to produce food.

Poverty Almost 1.3 billion people in the developing world currently live below the international poverty line, earning less than $1.25/day.[16] Poverty is central to the problem of hunger and undernutrition; in most parts of the world their incidence is almost identical (**Figure 18.6**). Poverty creates **food insecurity**, or the limited ability to acquire nutritious, safe foods. Food insecurity can occur in countries, in households, and among individuals. In wealthy countries, social safety nets, such as soup kitchens, government food assistance programs, and job training programs, help the hungry to obtain food or money to buy food. In poor countries, a family that cannot grow enough food or earn enough money to buy food may have nowhere to turn for help.[17]

food insecurity A situation in which people lack adequate physical, social, or economic access to sufficient, safe, nutritious food that meets their dietary needs and food preferences to lead an active and healthy life.

Disease and disability are more prevalent among the poor. Poverty reduces access to health care, so disease goes untreated (see Table 18.1). When left untreated, illness increases nutrient needs and further limits the ability to obtain an adequate diet, contributing to malnutrition. Lack of immunizations and medical treatment increases the incidence of and morbidity from infectious disease and decreases survival rates from chronic diseases such as cancer. Lack of health care also increases infant mortality and the incidence of low-birth-weight births.

The poor have less access to education, which reduces the opportunities to escape poverty. It also increases the risk of undernutrition and disease because lack of education leads to inadequate care for infants, children, and pregnant women. A lack of education about food preparation and storage can affect food safety and the health of the household—unsanitary food preparation increases the incidence of gastrointestinal diseases, which contribute to malnutrition.

Overpopulation Overpopulation exists when a region has more people than its natural resources can support. A fertile river valley can support more people per acre than can a desert environment. But even in fertile regions of the world, if the number of people increases too much, resources are overwhelmed and food shortages occur.

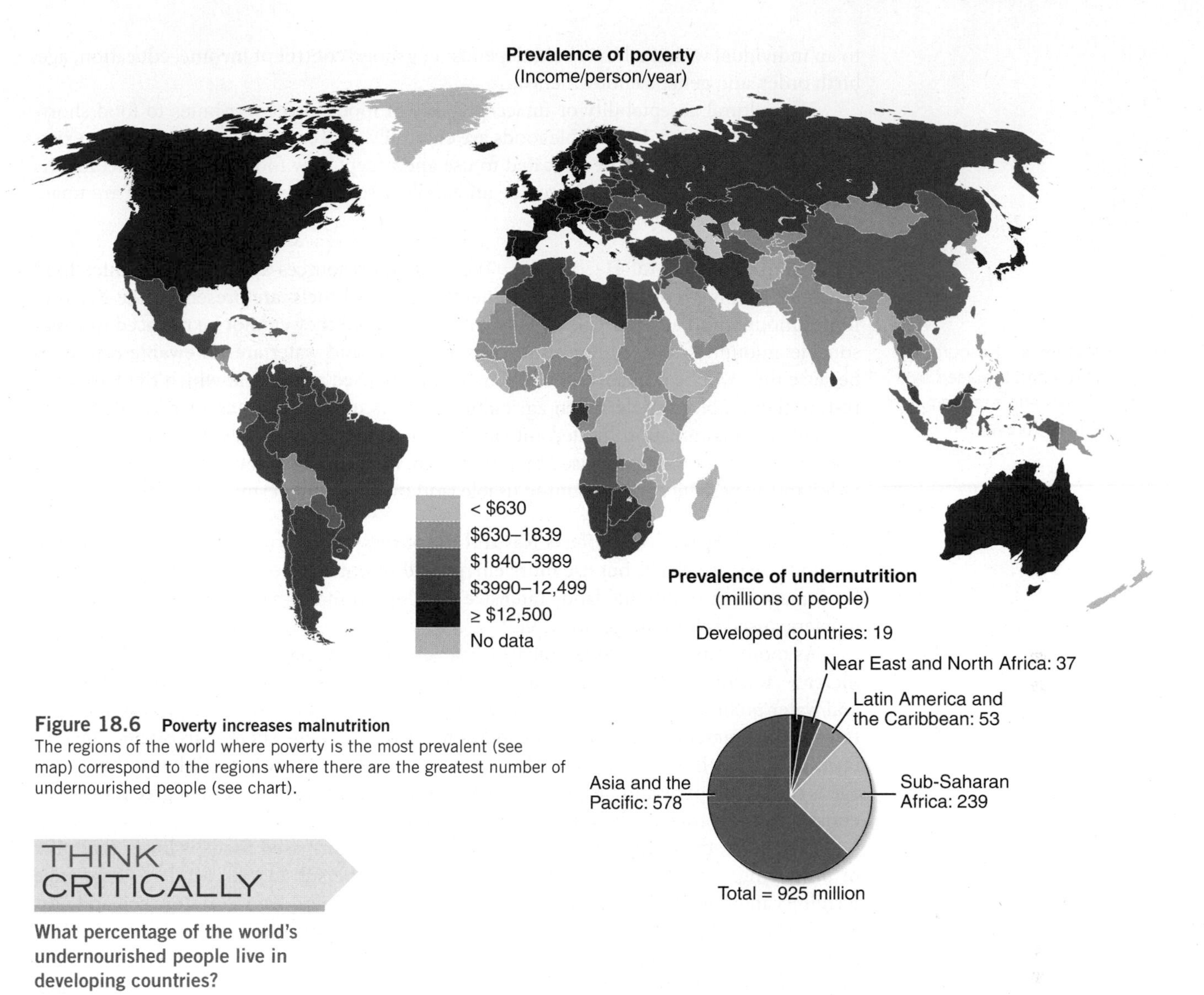

Figure 18.6 **Poverty increases malnutrition**
The regions of the world where poverty is the most prevalent (see map) correspond to the regions where there are the greatest number of undernourished people (see chart).

THINK CRITICALLY

What percentage of the world's undernourished people live in developing countries?

The human population is currently growing at a rate of more than 83 million persons per year, and most of this growth is occurring in developing countries (**Figure 18.7**).[18] These countries cannot escape from poverty because their economy cannot keep pace with the rate of population growth. Efforts to produce enough food can damage the soil and deplete environmental resources, further reducing the capacity to produce food in the future. The problem of hunger today is due primarily to the unequal distribution of resources, but it is estimated that, worldwide, food production has begun to lag behind population growth. If this trend continues, there will soon be too little food in the world to feed the population.

Cultural Practices In some cultures, access to food may be limited for certain individuals within households. For example, women and girls may receive less food than men and boys, because culturally they are viewed as less important. How much food is available

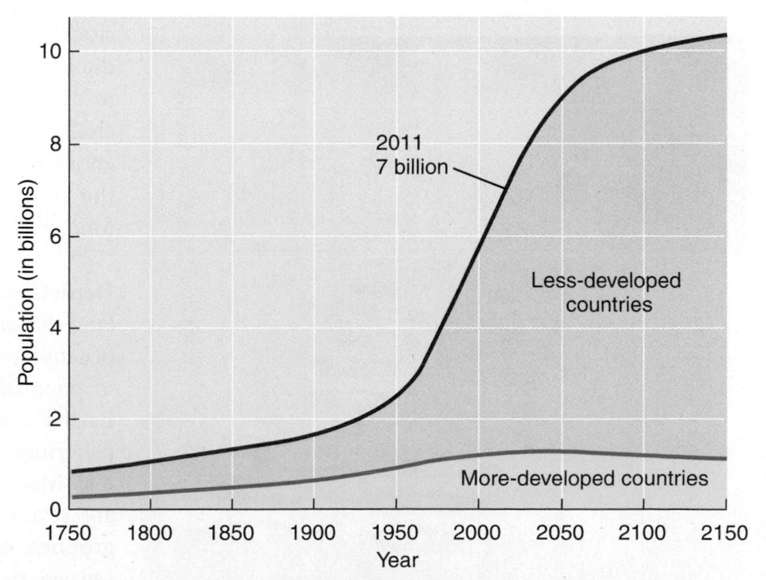

Figure 18.7 **World population growth**
Since about 1950, most of the increase in world population has occurred in developing countries, and this trend is expected to continue.

to an individual within a household depends on gender, control of income, education, age, birth order, and genetic endowments.

The cultural acceptability or unacceptability of foods also contributes to food shortages and malnutrition. If available foods are culturally unacceptable, a food shortage exists unless the population can be educated to use and accept new foods. For example, insects are eaten in some cultures and provide an excellent source of protein, but they are unacceptable to people in other cultures.

Limited Environmental Resources The land and resources available to produce food are limited. Some resources, such as minerals and fossil fuels, are present in the Earth in finite amounts and are nonrenewable—that is, once used they cannot be replaced in a reasonable amount of time. Other resources such as soil and water are **renewable resources** because they will be available indefinitely if they are used at a rate at which the Earth can restore them. For example, when agricultural land is used wisely—crops rotated, erosion prevented, contamination limited—it can be reused almost endlessly. However, if land is not used carefully, damage caused by soil erosion, nutrient depletion, and accumulation of pollutants may reduce the amount of usable land over the long term.

renewable resources Resources that are restored and replaced by natural processes and that can therefore be used forever.

How Modern Agriculture Affects Natural Resources Modern mechanized agricultural methods increase yields, but use more energy and resources and cause more environmental damage than traditional labor-intensive farming. In industrialized nations, about 17% of energy used is for food production.[19]

As more countries undergo nutrition transition, the demand for meat-based diets will increase. Raising cattle uses rangeland, grain, water, and fossil fuels, and creates both air and water pollution. The animals themselves produce methane gas in their gastrointestinal tracts, which enters the air and contributes to the greenhouse effect. Large-scale "factory farming" makes the problem worse because more methane is produced when animal sewage is stored in ponds and heaps. Livestock is responsible for about 18% of greenhouse gas emissions, more than all the cars in the world combined.[20]

In general, the environmental cost of producing plant-based foods is lower than that of producing animal products, but the cost may still be substantial. Modern large-scale farming can erode the soil and deplete its nutrients. Fertilizers used to restore soil and pesticides used to kill insects can contaminate groundwater and eventually waterways. And if the plant products are shipped long distances, require refrigeration or freezing, or need other types of processing, the environmental costs are increased even more (see **Debate: Are Vegetarian Diets Better for People and for the Environment?**).

In addition to an increase in the demand for grain to feed livestock, there has been a sharp acceleration in the use of grain to produce ethanol to fuel cars.[21] The increased demand has contributed to dramatic increases in food prices, which have made it even more challenging for low- and middle-income families worldwide to obtain enough food. The rising price of grain and fuel oil has also reduced the amount of food aid, widening the gap between the amount of food available and the amount needed to meet nutritional needs.[22]

Depletion of Ocean Resources It is not only the resources of the land that are at risk. Throughout human history, fish from the world's oceans have been an important source of protein. However, increases in population have increased demand for fish to the point that the Earth's oceans are being depleted.[23] Because the ocean is open to fishermen from around the world, its use has been difficult to control. Many marine species have been harvested until their numbers are severely reduced. Pollution also threatens the world's fishing grounds. Oil spills and deliberate dumping can occur offshore, and sewage, pesticides, organic pollutants, and sediments from erosion wash into coastal waters where most fish spend at least part of their lives. Even aquaculture, which has the potential to meet the world's demand for seafood, creates pollution, depletes resources, and damages the environment if not managed wisely (**Figure 18.8**).[24]

Figure 18.8 Environmental impact of aquaculture Aquaculture has great potential to alleviate hunger, but can also damage the environment if not managed carefully.

DEBATE

Are Vegetarian Diets Better for People and for the Environment?

It takes only 14 gallons of water to produce a pound of wheat, but 441 gallons to produce a pound of meat. Animal agriculture produces more greenhouse gases than the combined number of cars and trucks worldwide.[1] So it makes sense that eliminating animals from our diet would reduce the environmental impact of human food production. But what if the vegetarian diet relies on foods like frozen pizza and breaded, extruded soy nuggets that are individually wrapped in plastic and shipped across the country? Is the elimination of animal foods the answer to saving the planet and feeding the world?

USDA/Photo Researchers

Grazing cattle on land that is unfit for growing crops preserves arable land for human food production.

Those who promote a vegetarian world argue that producing animal products consumes large amounts of energy, destroys forests and grazing lands, and pollutes the air and water. These problems, in turn, limit the amount of food that can be produced and thus contribute to world hunger. Vegetarian advocates also point out the reduction in chronic disease risk that is associated with a vegetarian diet. Although these arguments are valid in some instances, the impact of food animals on the environment and nutrient intake depends on how these animals are integrated into the ecosystem and what foods are included in the diet.

On a small farm, animals are typically part of the ecosystem. They are fed crop wastes, kitchen scraps, and grasses that people cannot eat and produce meat, eggs, and milk that can nourish people. Animal manure is used for fertilizer in local fields. However, much of the meat that humans eat is produced in large agribusinesses where animals are fed grain grown on land that could be used to grow human food; crops grown to feed animals use 33% of total arable land worldwide.[2] Humans who eat these animals then get back only a fraction of the food energy they could have gotten from eating the grain that was fed to the animals. Animal waste on these huge farms builds up rapidly. It causes runoff that can pollute nearby waterways, and releases gases into the atmosphere that contribute to acid rain and global warming.

The vegetarian foods people choose are also a consideration in the impact of a vegetarian diet. A vegetarian diet can be made up of soy nuggets and French fries or it can be based on locally grown, minimally processed grains, legumes, nuts, and fruits and vegetables. The soy-nugget meal is likely to be lacking in essential nutrients and could actually take more resources to produce than a meat-based meal. A recent study suggested that switching from a beef- and milk-based diet to a diet relying on highly refined protein substitutes, such as texturized soy protein, could actually increase the amount of arable land needed to produce food for humans.[3] Packaging, processing, and transportation have an important environmental impact and need to be factored into the overall impact of vegetarian diets.

Even if everyone made environmentally sound vegetarian choices, eliminating animal products still may not be the best nutritional choice. In the developing world, small amounts of meat and milk in the diet can mean the difference between survival and starvation. In the United States, animal foods provide important sources of vitamin B_{12}, calcium, and highly absorbable forms of iron and zinc. Eliminating animal products would reduce both the variety and nutrient content of the human diet.

So should everyone eat a vegetarian diet? Is it best for nutritional health and the health of the planet? Whatever your view, finding a diet that supports health and the environment is not simple.

THINK CRITICALLY: If you stopped eating meat tomorrow, what would be the main source of protein in your diet?

[1]UN News Centre, *Rearing Cattle Produces More Greenhouse Gasses than Cars*, Nov. 2006.

[2]Steinfeld, H., Gerber, P., Wassenaar, T., et al. *Livestock's Long Shadow: Environmental Issues and Options 2006.*

[3]Audsley, E., Brander, M., Chatterton, J., et al. *How Low Can We Go? An Assessment Of Greenhouse Gas Emissions from the UK Food System and the Scope for Reduction by 2050.*

Poor Quality Diets

Even when there is enough food, undernutrition can occur if the quality of the diet is poor. The typical diet in developing countries is based on high-fiber grain products and has little variety. Adults who are able to consume a relatively large amount of this diet may be able to meet their nutrient needs. But those with increased needs or a limited capacity to consume these foods are at risk for nutrient deficiencies. Children, pregnant women, and the elderly, especially if they are ill, may not be able to eat enough of this bulky grain diet to meet their needs. Deficiencies of protein, iron, iodine, and vitamin A are common because of poor-quality diets (**Table 18.2**).

TABLE 18.2 The Consequences of Malnutrition Throughout Life

Life stage	Common deficiencies	Consequence
In utero	Energy	Low birth weight
	Iodine	Brain damage
	Folate	Neural tube defects
Infant/young child	Protein, energy	Growth retardation, increased risk of infection
	Iron	Anemia
	Iodine	Developmental retardation, goiter
	Vitamin A	Infection, blindness
Adolescent	Protein, energy	Stunting, delayed growth
	Iron	Anemia
	Iodine	Impaired intellectual development, goiter
	Vitamin A	Infection, blindness
	Calcium	Inadequate bone mineralization
Pregnant women	Protein, energy	Intrauterine growth retardation, increased mortality of mother and fetus
	Folate	Maternal anemia, neural tube defects in infants
	Iron	Maternal anemia
	Iodine	Cretinism in infant, goiter in mother
	Vitamin A	Infection, blindness
Adult	Energy, protein	Thinness, lethargy
	Iron	Anemia
Elderly adult	Energy	Thinness, lethargy
	Calcium and protein	Osteoporosis fractures, falls

Source: World Health Organization.

THINK CRITICALLY

Why are children who consume a starchy, low-protein diet more likely to develop kwashiorkor than adults consuming the same diet?

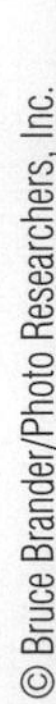

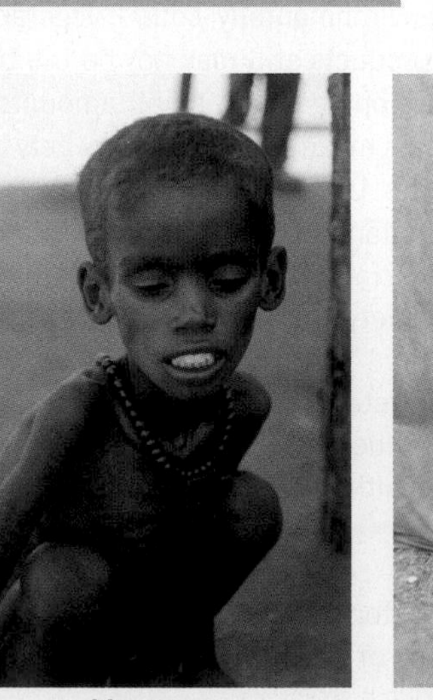

© Alison Wright/NG Images

Marasmus   Kwashiorkor

Figure 18.9 Marasmus and kwashiorkor
A general lack of food causes the wasting associated with marasmus; too little protein to meet needs causes kwashiorkor, which is characterized by a bloated belly.

Protein-Energy Malnutrition Protein and energy deficiencies usually occur together and are most common in children. When there is a general lack of food marasmus results, but when the diet is limited to starchy grains and vegetables, protein deficiency can predominate, particularly in individuals with high protein needs—those who are growing, developing, suffering from infection, or healing (see Chapter 6). Kwashiorkor, a deficiency of protein but not energy, occurs when protein needs are high and the available diet is low in protein, not as a result of a general lack of food. It is common in children over 18 months of age when the main food is a bulky cereal grain low in high-quality protein. Children have high protein needs and small stomachs, so they are not able to consume enough of this grain to meet their protein needs (**Figure 18.9**). Other factors, such as metabolic changes caused by infection, may also play a role in the development of kwashiorkor.

Iron Deficiency Iron deficiency is the most common nutritional deficiency worldwide; two billion people have iron-deficiency anemia, the most severe stage of iron deficiency.[25] The causes of iron deficiency include a limited dietary iron intake and increased losses due to infection. The diet in most developing countries is based on plant foods, which only provide poorly absorbed nonheme iron. The risk and severity of dietary iron deficiency is further increased in developing countries by the high prevalence of intestinal parasites, which cause gastrointestinal blood loss, and acute and chronic infections, such as malaria (**Figure 18.10**).

Iron deficiency can have a major impact on the health and productivity of a population. Anemia during pregnancy increases the risk of maternal and fetal mortality, premature delivery, and low birth weight. Iron deficiency in infants and children can stunt growth, retard mental development, decrease resistance to infection, and increase morbidity due to disease. In older children and adults it causes fatigue and decreases productivity.

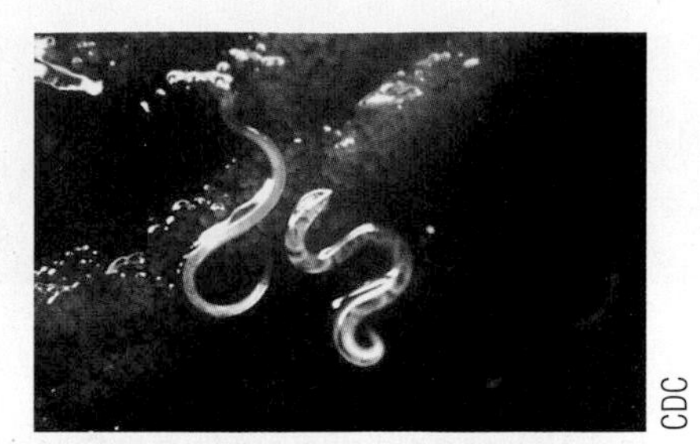
CDC

Figure 18.10 **Hookworm infection**
Hookworm larvae penetrate the skin, infecting people when they walk barefoot in contaminated soil. The parasites attach to the intestinal lining and feed on blood. Hookworm infection affects as many as 740 million people.[26]

Iodine Deficiency Iodine is an essential trace element that is a constituent of the thyroid hormones (see Chapter 12). Globally, over 1.9 billion people, including 285 million school-aged children, have inadequate iodine intake as defined by low urinary iodine excretion.[27] A deficiency of iodine in the food supply affects virtually all members of the community. During pregnancy, iodine deficiency increases the incidence of stillbirths, spontaneous abortions, and developmental abnormalities such as cretinism in the offspring.[28] Although goiter is the most visible symptom of iodine deficiency (**Figure 18.11**), cretinism is the most severe manifestation. Cretinism is characterized by impaired cognitive and physical development. It is devastating to individuals and families, but the more subtle effects of iodine deficiency on mental performance and work capacity may have a greater impact on the population as a whole. Iodine-deficient children have lower IQs and impaired school performance. Iodine deficiency in children and adults is associated with apathy and decreased initiative and decision-making capabilities. Worldwide, iodine deficiency diseases are believed to be one of the main causes of impaired cognitive development in children.[28]

Iodine deficiency occurs in regions with iodine-deficient soil that rely extensively on locally produced food. The eastern Mediterranean region and Africa have the highest incidence of iodine-deficiency disorders. Because soil iodine is low in regions where deficiency is common, the problem can be solved only by importing foods high in iodine or by adding iodine to the local diet through fortification or supplementation. As discussed later, the introduction of iodized salt is the most common way that iodine intake is increased in populations at risk of deficiency.

Figure 18.11 **Goiter**
Goiter, which is an enlargement of the thyroid gland, is the most visible symptom of iodine deficiency.

Vitamin A Deficiency It is estimated that more than 250 million preschool children worldwide suffer from vitamin A deficiency.[29] It causes blindness; depresses immune function, which increases the risk of infections; retards growth; and is often accompanied by anemia (see Chapter 9). It is the leading cause of preventable blindness in children (**Figure 18.12**). In communities where vitamin A deficiency exists, supplementation has been shown to significantly reduce childhood deaths due to infection (see **Science Applied: Vitamin A: The Anti-Infective Vitamin**).

Obtaining sufficient vitamin A is a particular problem during periods of rapid growth and development such as infancy, early childhood, pregnancy, and lactation. Need is increased by frequent infections, such as those causing diarrhea, and illnesses such as measles. Deficiencies of other nutrients, including fat, protein, and zinc, can contribute to vitamin A deficiency because they are needed to absorb and transport the vitamin in the body.

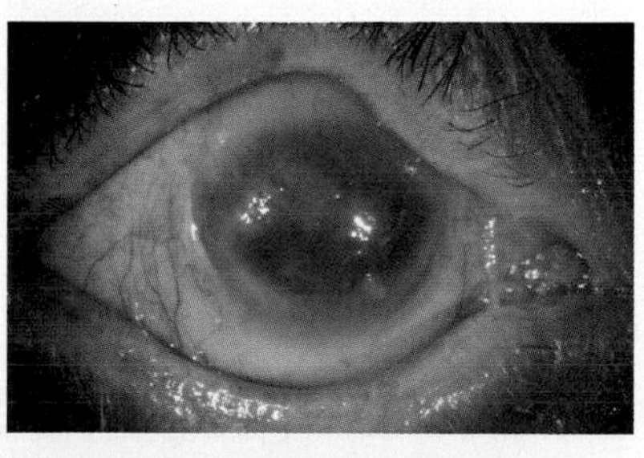
ISM/Phototake

Figure 18.12 **Xerophthalmia due to vitamin A deficiency**
Vitamin A deficiency leads to xerophthalmia, shown here. It is estimated that 250,000 to 500,000 children go blind from vitamin A deficiency every year; half die within a year of losing their sight.[29]

Other Nutrients of Concern In addition to deficiencies of protein, iron, iodine, and vitamin A, there are several vitamin and mineral deficiencies that have recently emerged or reemerged as problems throughout the world. Beriberi, pellagra, and scurvy, caused by deficiencies of thiamin, niacin, and vitamin C, respectively, are rare in the developed world but still occur among the extremely poor and underprivileged and in large refugee populations. Folate deficiency is also a problem in many parts of

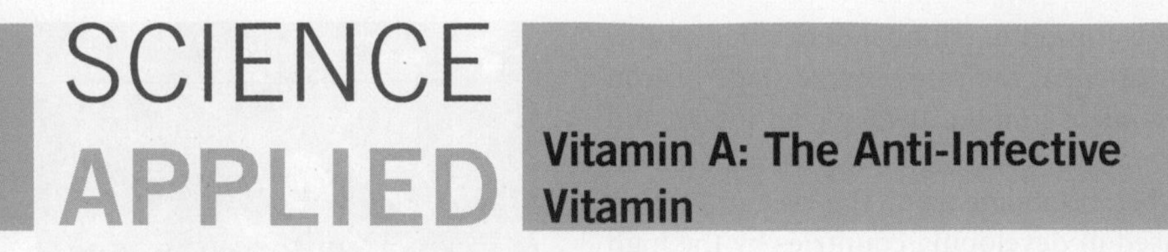

SCIENCE APPLIED

Vitamin A: The Anti-Infective Vitamin

In 1910, one out of every four U.S. infants died before one year of age. The major causes were infectious diseases: epidemics of diarrhea during the summer and respiratory infections during the winter.[1] Similar problems affected children in Europe. In the early part of the twentieth century, the discovery of vitamins led to observations that nutritional deficiency, particularly vitamin A deficiency, was associated with an increased incidence and severity of infectious disease.[2]

THE SCIENCE

In 1925, an epidemic of pneumonia swept through a colony of dogs in a research laboratory in England. The pneumonia occurred almost exclusively in vitamin A-deficient animals. It was hypothesized that the deficiency increased susceptibility to respiratory infections and that this might be relevant to infections in children.[3] Animal experiments confirmed this hypothesis and vitamin A was dubbed the "anti-infective vitamin."[4]

The theory that vitamin A could prevent infections triggered 20 years of clinical investigations. Clinical trials were conducted to determine if vitamin A, given as cod liver oil, could reduce morbidity and mortality from respiratory diseases, measles, and other infections. Although results were mixed, the pharmaceutical industry began promoting cod liver oil to decrease severity and recovery time in ailments such as whooping cough, measles, mumps, chicken pox, and scarlet fever.[1] The administration of cod liver oil became routine for millions of children in the United States and Europe in the 1940s.

In 1959, the World Health Organization (WHO) published a paper that reviewed the mounting evidence of a relationship between nutritional status and infection.[5] It recognized that poor nutritional status leads to more frequent and more severe infections, and infection triggers metabolic responses that cause nutrient losses. For example, a vitamin A deficiency reduces the ability of the immune system to defend itself against measles infection, and the infection itself causes loss of vitamin A that can precipitate acute vitamin A deficiency and blindness (see graph).[6] To a well-nourished child, measles is usually a passing illness, whereas to a malnourished child, it can result in life-long disabilities or death.

THE APPLICATION

Today, antibiotics, vaccinations, and a nutritious and varied diet have reduced infant morbidity and mortality in the United States and other developed nations. Worldwide, though, infections are still responsible for over half of the deaths in children under age 5 every year, and about 35% of these deaths are associated with malnutrition.[7,8] Improved vitamin A status can help reduce the number of child deaths.

Before a vaccine was developed, measles claimed 7 to 8 million lives a year. Today, it has nearly been forgotten in many developed countries, where immunization is universal, but it remains a major problem in developing nations. Because of the interrelationship between vitamin A status and measles, the WHO and UNICEF currently advise that large doses of vitamin A be provided to children with measles and that vitamin A be supplemented at the time of measles vaccination.[9] Providing vitamin-A supplements is a short-term answer that can accompany long-term solutions to vitamin-A deficiency and malnutrition, such as changes in dietary intake patterns and fortification of appropriate foods with vitamin A.

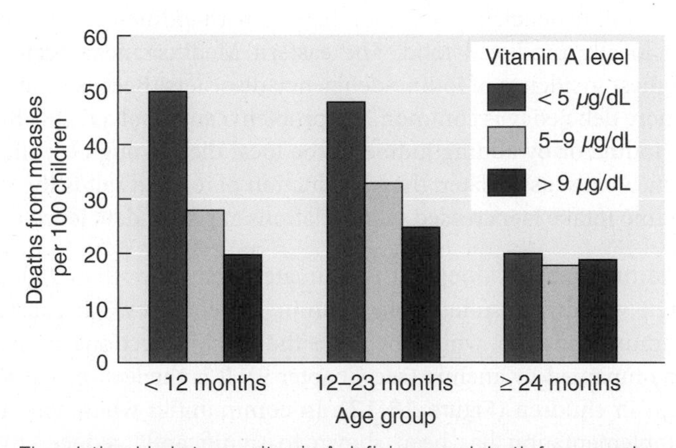

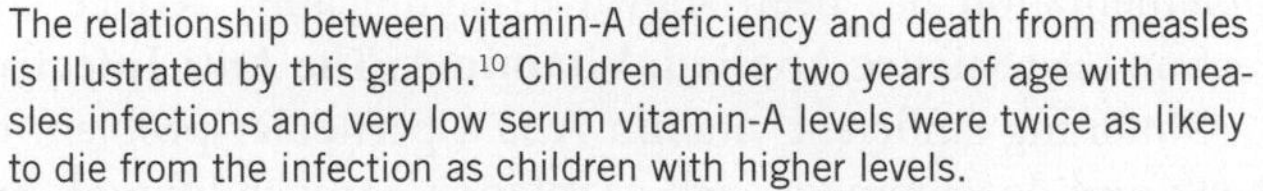
The relationship between vitamin-A deficiency and death from measles is illustrated by this graph.[10] Children under two years of age with measles infections and very low serum vitamin-A levels were twice as likely to die from the infection as children with higher levels.

---

[1] Semba, R. D. Vitamin A as "anti-infective" therapy, 1920–1940. *J Nutr* 129:783–791, 1999.
[2] Brundtland, G. H. Nutrition and infection: Malnutrition and mortality in public health. *Nutr Rev* 58(II):S1–S4, 2000.
[3] Mellanby, E. Diet and disease, with special reference to the teeth, lungs, and prenatal feeding. *Lancet* 1:151–519, 1926.
[4] Green, H. N., and Mellanby, E. Vitamin A as an anti-infective agent. *Br Med J* 2:691–696, 1928.
[5] Scrimshaw, N. S., Taylor, C. E., and Gordon, J. E. Interaction of nutrition and infection. *Am J Med Sci* 237:367–403, 1959.
[6] West, C. E. Vitamin A and measles. *Nutr Rev* 58(II):S46–S54, 2000.
[7] World Health Organization. *The Global Burden of Disease: 2004 Update*, 2008.
[8] Black, R. E., Allen, L. H., Bhutta, Z. A., et al. Maternal and child undernutrition: Global and regional exposures and health consequences. *Lancet* 371:243–260, 2008.
[9] WHO/UNICEF. Joint Statement. *Reducing Measles Mortality.*
[10] World Health Organization. Health Library for Disasters. Vitamin A deficiency and severe complicated measles.

the world. It causes megaloblastic anemia during pregnancy and often compounds existing iron-deficiency anemia. In women of childbearing age, low folate intake increases the risk of having a baby with a neural tube defect. In the United States, enriched grain products are fortified with folic acid to assure adequate intake and to reduce the incidence of neural tube defects. The World Health Organization recommends iron and folic acid supplementation for all menstruating women.[30]

Deficiencies of the minerals zinc, selenium, and calcium are also of concern.[31] Zinc deficiency affects about one-third of the world's population and is believed to cause as many deaths as vitamin A or iron deficiency.[31] Zinc deficiency can cause growth retardation or failure, diarrhea, immune deficiencies, skin and eye lesions, delayed sexual maturation, night blindness, and behavioral changes. It may also contribute to intrauterine growth retardation and neural tube defects in the fetus, and in the elderly it may affect taste acuity and cause dermatitis and impaired immune function. Selenium deficiency has been identified in population groups in China, New Zealand, and the Russian Federation. Selenium deficiency is associated with an increased incidence of Keshan disease (see Chapter 12), a type of heart disease that affects mainly children and young women.[32] Inadequate calcium intake is also a concern worldwide due to its association with the occurrence of osteoporosis. Although factors other than low calcium intake, such as hormone levels and exercise, play a role in the development of osteoporosis, calcium supplementation has been proposed as a means of combating the high prevalence of spine and hip fractures due to osteoporosis, particularly in postmenopausal women.[31]

18.3 Eliminating World Hunger

LEARNING OBJECTIVES

- Discuss two strategies that can help reduce population growth.
- Explain how international trade can help eliminate hunger.
- Discuss the role of sustainable agriculture in maintaining the food supply.
- List three factors that should be considered when fortifying the food supply.

Solving the problem of world hunger is a daunting task. It involves eliminating poverty, controlling population growth, meeting the nutritional needs of a large and diverse population with culturally acceptable foods, increasing production of nutrient-dense foods, and maintaining the global ecosystem. It requires international cooperation, commitment from national and local governments, and the involvement of local populations. The solutions involve economic policies, technical advancement, education, and legislative measures. They require input from politicians, nutrition scientists, economists, and the food industry. Programs and policies must be in place to provide food in the short term and in the long term establish sustainable programs to allow continued production and distribution of food.

Providing Short-Term Food Aid

When people are starving, short-term food and medical aid must be provided right away. The standard approach has been to bring food into the stricken area (**Figure 18.13**). These foods generally consist of agricultural surpluses from other countries and often are not well planned in terms of their nutrient content. Although this type of relief is necessary for a population to survive an immediate crisis such as famine, it does little to prevent future hunger.

Many international, national, and private organizations are working toward the goal of relieving world hunger. The Food and Agriculture Organization (FAO) works to improve the production, intake, and distribution of food worldwide. The World

© AP/Wide World Photos

Figure 18.13 Emergency food relief
Many international relief organizations provide food to hungry people throughout the world.

Health Organization (WHO) targets community health centers and emphasizes the prevention of nutrition problems, such as micronutrient deficiencies. The World Bank finances projects such as supplementation and fortification to foster economic development. The United Nations Children's Fund (UNICEF), which relies on volunteer support, distributes food to all countries in need with a goal of assisting developing countries that occasionally suffer periods of starvation. The Red Cross, the UN Disaster Relief Organization, and the UN High Commissioner for Refugees concentrate on famine relief. The Peace Corps focuses more on fostering long-range development. More and more agencies are engaging in both development and relief. A few examples include the U.S. Agency for International Development, Oxfam, the Hunger Project, and Catholic Relief Services.

Eliminating Poverty

Hunger will exist as long as there is poverty. Even when food is plentiful in a region, the poor do not have access to enough of the right foods to maintain their nutritional health. Economic development that guarantees safe and sanitary housing, access to health care and education, and the resources to acquire enough food are essential to eliminate hunger. Government policies can help to eliminate poverty by increasing the population's income, lowering food prices, or offering feeding programs for the poor, all of which can improve **food security**. International cooperation, such as occurred at the United Nations (UN) Millennium Summit in September 2000, is also important in addressing hunger and poverty. The 189 nations who signed onto the Millennium Declaration recognized, among other goals, the importance of eliminating poverty (**Figure 18.14**).[33]

food security Access by all people at all times to sufficient food required for a healthy and active life.

Providing Enough Food

In the long term, solving the problem of world hunger requires balancing the number of people with the amount of food available globally and locally. Inequities within populations must be addressed by eliminating poverty and providing opportunities for education—both to help people escape poverty and to teach them what constitutes a healthy diet and how to prepare it safely. Long-term solutions need to be based on the cultural and economic needs of the local population.

Controlling Population Growth World population has increased dramatically since the middle of the 20th century, reaching a staggering 7 billion people today; however, population growth has recently begun to slow. The birth rate worldwide has declined—from 5 children per woman in 1950 to 2.46 in 2012.[34] This downward trend in population growth must continue to ensure that food production and natural resources can support the population. Changes in cultural and economic factors as well as family planning and government policies can be used to influence the birth rate.

Population growth can be slowed directly through family planning. To be successful, family-planning efforts must be acceptable to the population and compatible with their cultural and religious needs. A number of approaches, such as provision of contraceptives, education, and economic incentives, have been used to decrease population growth. In Singapore, Thailand, Colombia, and Costa Rica, programs that provide contraceptive information, services, and supplies have been somewhat successful in slowing population growth. In some countries, population-control education is being integrated into the school curriculum, and family-planning messages are carried by popular television programs.

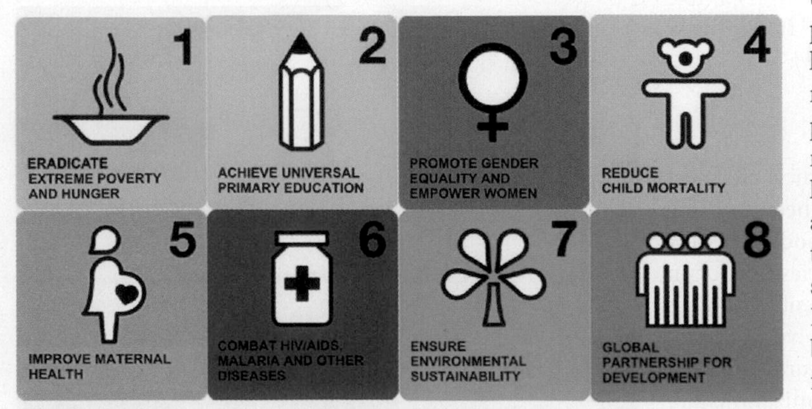

Figure 18.14 End poverty by 2015
The eight Millennium development goals shown here are the world's main development challenges.[33]

An indirect way to reduce population growth is to increase the general level of education and provide economic security. Birthrates decrease when the educational level and economic status of women is improved (**Figure 18.15**).[35] Women with more education tend to marry later and have fewer children. Education also increases the likelihood that women will have control over their fertility, provides knowledge to improve family health, decreases infant and child mortality rates, and offers options other than having numerous children.

Changes in economic policies can help reduce population growth. In some developing countries, higher birth rates are due to the economic and societal roles of children as well as high infant mortality rates. Children are needed to work the farms, support the elders, and otherwise contribute to the economic survival of the family. When infant mortality rates are high, people choose to have many children to ensure that some will survive. Programs that foster economic development and ensure access to food, shelter, and medical care have been shown to cause a decline in birthrates because people feel secure having fewer children and because economic development reduces child mortality and the need for children as workers.

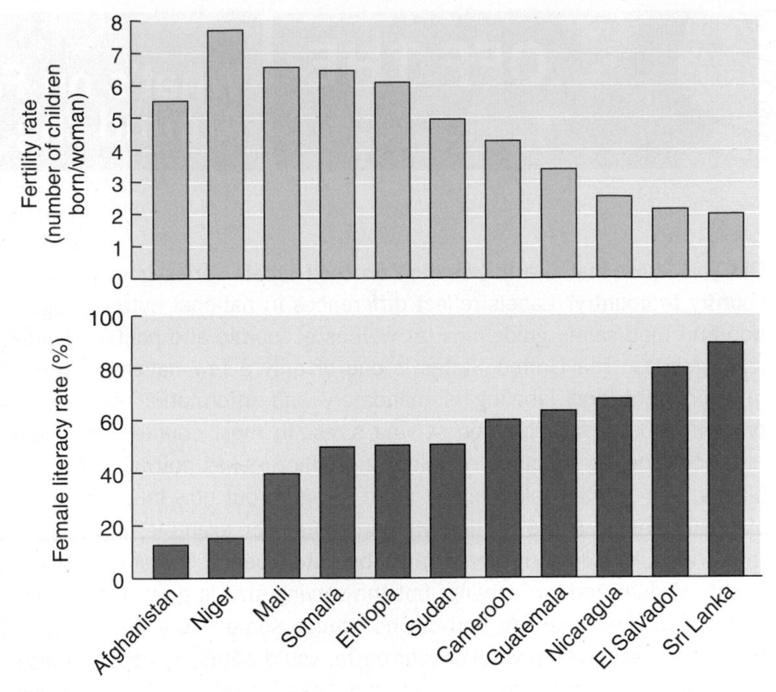

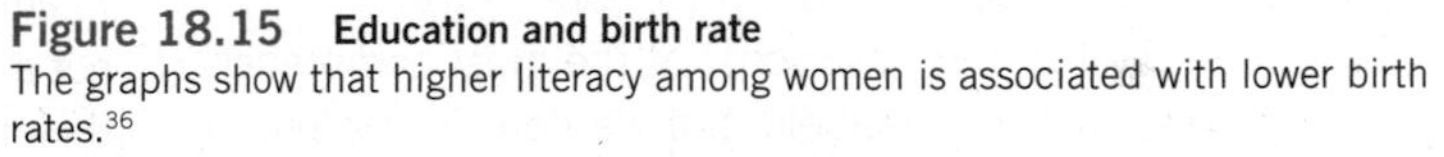

Figure 18.15 Education and birth rate
The graphs show that higher literacy among women is associated with lower birth rates.[36]

Growing and Importing Adequate Food **Food self-sufficiency** is a country's capacity to feed its population. Developing manageable systems for producing acceptable, sustainable sources of food can increase the level of food self-sufficiency. In countries with limited agricultural resources, imports can increase the food supply and help reduce hunger.

Food self-sufficiency The ability of an area to produce enough food to feed its population.

How Agricultural Technology Increases Food Production Technological advancements in agriculture can help a country boost food production. These include newer varieties of plants, better agricultural techniques, and improvements in irrigation. One type of technology being used to increase the quantity and improve the quality of food is genetic engineering (see **Focus on Biotechnology**). Crop yields can be increased either directly, by inserting genes that improve plant growth, or indirectly, by creating plants that are resistant to herbicides, insects, and plant diseases, thus reducing crop losses. Genes that impart insect and herbicide resistance also help the environment because they allow farmers to achieve insect-free crops and weed-free fields with fewer pesticides and safer herbicides. Biotechnology can also be used to modify plant characteristics that are of importance after harvest, such as ease of transport, shelf life, and rate of ripening. The availability of older technologies such as freezing and refrigeration and better storage facilities can reduce food losses due to insects and rodents and thus also increase the amount of food available to the population.

How International Trade Increases Food Availability Some countries have the resources to grow enough food to feed their population and others do not. When a country has few natural resources, access to international trade systems can help provide for their population. The newly industrialized countries of Asia, such as Thailand and Korea, are examples of how an increase in food imports can decrease the number of hungry people. In general, the countries of the world are becoming more interdependent on food imports and on exports to pay for this food. This interdependence can increase the availability of food for the world population (see **Off the Label: What's on Food Labels Around the World?**).

Whether a country's agricultural emphasis focuses on producing **subsistence crops** for local consumption or on producing **cash crops**, which can be sold on the national and international market, influences the availability of food for its people. Shifting to cash crops improves the cash flow of the country but uses local resources to produce crops for

subsistence crop A crop that is grown as food for a farmer's family, with little or nothing left over to sell.

cash crop A crop that is grown to be sold for monetary return, rather than as food for food the local population.

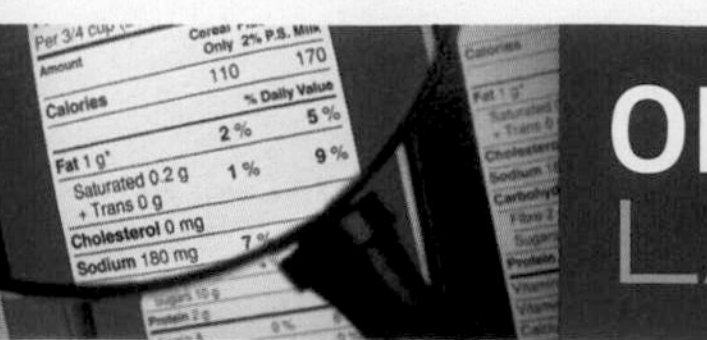
© iStockphoto

OFF THE LABEL What's on Food Labels Around the World?

Did you know that the information on food labels varies from country to country? Labels reflect differences in national nutrition and food-safety guidelines, as well as economic and political agendas. The United States is one of only a few nations in which nutrition labeling is mandatory and information is presented based on common serving sizes. In most countries nutrition labeling is voluntary unless a product makes nutrition claims, and it often takes higher math to figure out how much of a nutrient is in the portion you consume. For example, in the United Kingdom nutrients must be listed per 100 grams of the product and per serving, but the serving size is given in grams. So, if you want to find out how much sodium is in the Bolognese sauce you pour over your pasta, you'd better find out how many grams of sauce is on your plate (see figure).

The United States may have some of the most comprehensive nutrition information on labels, but we don't come out on top when it comes to other types of information. For example, a can of tomatoes labeled in the European Union (EU) would show that the product is 80% tomatoes and 20% water.[1] In the United States, we would only know that tomatoes were the most abundant ingredient by weight. The United States also lags behind in freshness dating, which is not mandatory in the United States but is required in most other developed countries. The information provided about how a food is produced also varies among countries. If you want to know if your food is organically produced, irradiated, or made using genetically modified ingredients, you need to research the labeling guidelines of the country from which you are purchasing the food.[2]

The best label would provide consumers with the information they need to make informed choices. In the United States, consumers are concerned with overconsumption of saturated fat, cholesterol, sodium, and sugar. They can find information about these nutrients on all food labels, but the amounts of niacin, thiamin, and riboflavin are not required because deficiencies are not a concern in the U.S. population. In countries where niacin, thiamin, and riboflavin deficiencies are still prevalent, however, food labels would ideally include the content of these nutrients. In today's global economy, countries should learn from one another and incorporate the best label components from around the world to help their consumers choose wisely.

Lori Smolin

These Bolognese pasta sauces are labeled for sale in the United Kingdom.

Lori Smolin

Amounts of energy and nutrients are given per 100 g of food.

Energy is given in kjoules and kcalories.

This is the EU organic farming logo and indicates that this product is in compliance with the EU organic farming regulations.

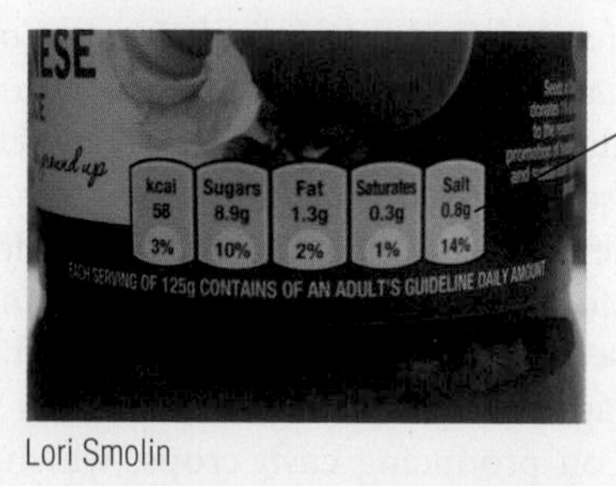

Lori Smolin

Amounts per serving are given as a % of the Guideline Daily Amount (GDA), a guide to how much people should consume each day.

[1]Center for Science in the Public Interest. *Food Labeling for the 21st Century: A Global Agenda for Action*, 1998.

[2]Gruère, G. P., and Rao, S. R. A review of international labeling policies of genetically modified food to evaluate India's proposed rule. *AgBioForum* 10:51–64, 2007.

export and limits the ability of the local people to produce enough food to feed their families. For example, if a large portion of the arable land in West Africa is used to grow cash crops such as coffee and cotton, little agricultural land remains to grow grains and vegetables that nourish the local population. If, however, the cash from the crop is used to purchase nutritious foods for the local people, this decision may help alleviate undernutrition.

Maintaining the Environment The resources needed to support food production depend on the methods used. In developing nations, the resources used by a single person are small but the number of people is large, so it is difficult to produce sufficient food without depleting natural resources such as soil, forests, and water supplies. In developed nations, the population is less dense but the resource demands made by each individual are far greater because of lifestyles and the methods of food production and distribution. A child born in the United States uses 10 to 1000 times more resources daily than the average child born in Chile, Ghana, or Yemen.[37] To allow continued food production for future generations, solutions to the problem of providing enough food must assure that natural resources are conserved in both industrialized and developing nations.

Sustainable Agriculture **Sustainable agriculture** uses food production methods that prevent damage to the environment and allow the land to restore itself, so that food can be produced indefinitely. For example, contour plowing and terracing help prevent soil erosion, keeping the soil available for future crops. Rotating the crops grown in a specific field prevents the depletion of nutrients in the soil, reducing the need for added fertilizers. Sustainable agriculture uses environmentally friendly chemicals that degrade quickly and do not persist as residues in the environment. It also relies on diversification. This approach to farming maximizes natural methods of pest control and fertilization and protects farmers from changes in the marketplace (**Figure 18.16**).

sustainable agriculture Agricultural methods that maintain soil productivity and a healthy ecological balance while having minimal long-term impacts.

Sustainable agriculture is not a single program but involves choosing options that mesh well with the local soil, climate, and farming techniques. In some cases, organic farming, which does not use synthetic pesticides, herbicides, and fertilizers (see Chapter 17), may be the more sustainable option. Organic techniques have less environmental impact because they reduce the use of agricultural chemicals and the release of pollutants into the environment. Organic farming is also advantageous in terms of soil quality and biodiversity, but it is a disadvantage when it comes to land use because crop yields are often lower. A combination of organic and conventional techniques, as is used with integrated

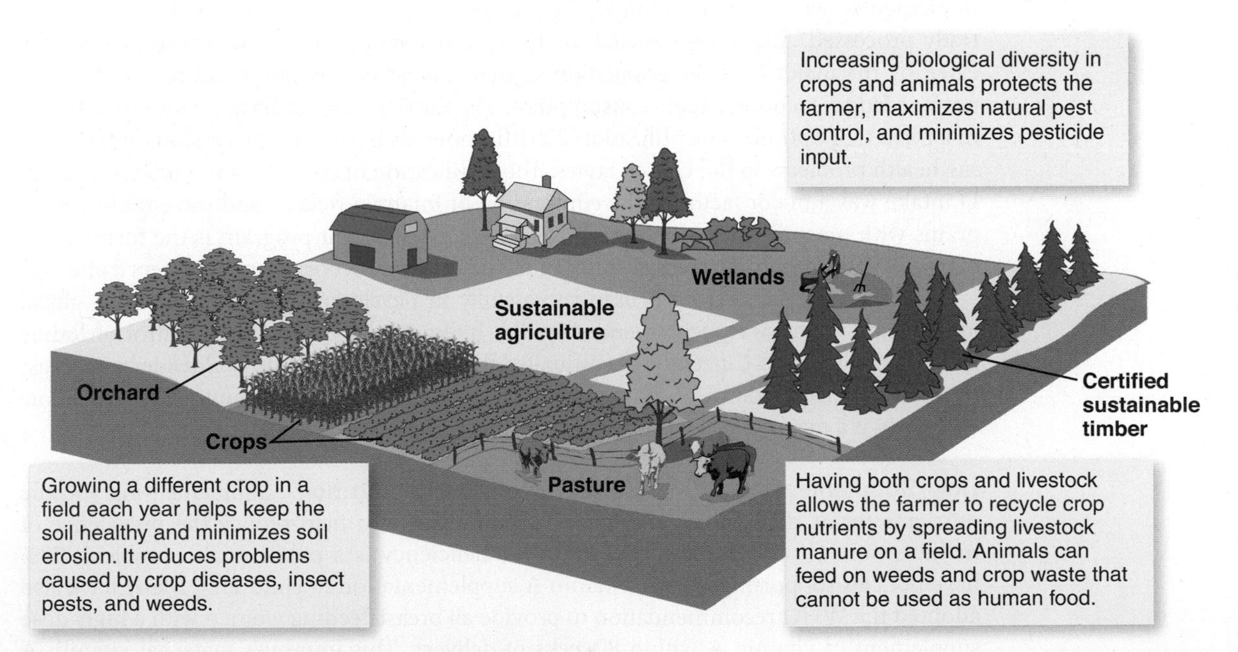

Figure 18.16 A sustainable farm
A sustainable farm consists of a total agricultural ecosystem rather than a single crop. It may include field crops, fruit- and nut-bearing trees, herds of livestock, and forests.

pest management (see Chapter 17), might be best for improving land use and protecting the environment.

Other sustainable programs include agroforestry, in which techniques from forestry and agriculture are used together to restore degraded areas; natural systems agriculture, which attempts to develop agricultural systems that include many types of plants and therefore function like natural ecosystems; and the technique of reducing fertilizer use by matching nutrient resources with the demands of the particular crop being grown.[38]

Sustainable Choices The choices individuals make can also influence the environmental impact of food production. Choosing a diet that is primarily plant-based, with smaller amounts of animal products, can help to minimize the ecological impact of food production. Choosing locally grown foods in season can minimize the energy costs and pollution due to food transport. Choosing organically grown, minimally processed, and ecologically packaged foods can further reduce the environmental impact of food production (see **Critical Thinking: Can You Have an Impact?**).

Ensuring a Nutritious Food Supply

In addition to sufficient energy in the diet, the right mix of nutrients is necessary to ensure the nutritional health of a population. If the foods and crops that are grown or imported do not meet all nutrient needs, the quality of the diet will be poor and malnutrition will occur. If the diet does not provide the right mix of nutrients, deficient nutrients can be added to the diet by changing dietary patterns, fortifying existing foods, or including dietary supplements. Genetic engineering can also address nutrient deficiencies by changing the nutrient content of foods (see Chapter 9 Debate: Combating Vitamin A Deficiency with Golden Rice).

How Food Fortification Alleviates Malnutrition Fortification is the process of adding one or more nutrients to commonly consumed foods with the goal of increasing the nutrient intake of a population. Food fortification will not provide energy to a hungry population, but it can increase the protein quality of the diet and eliminate micronutrient deficiencies. Fortification programs have been created by partnerships among industry, academia, and government. Industry and academia can provide the technology for adding nutrients to foods, and government public health policies can promote the consumption of these fortified foods.

In order for fortification to solve a nutritional problem in a population, it must be implemented wisely. Fortification works if vulnerable groups consume foods that are centrally processed. The foods selected for fortification should be among those consistently eaten by the majority of the population so that extensive promotion and reeducation are not needed to encourage their consumption. The nutrient should be added uniformly and in a form that optimizes its utilization. Fortification has been used successfully in preventing health problems in the United States: The fortification of cow's milk to increase vitamin D intake was a major factor in the elimination of infantile rickets, and the enrichment of grains with niacin helped eliminate pellagra. The most recent program is the fortification of grains with folic acid to reduce neural tube defects in newborns (see Chapters 8 and 14).

Fortification has also been used successfully in developing countries. In 1993, global iodinization of salt was recommended by the International Council for Control of Iodine Deficiency Disorders. Currently, it is estimated that 66% of households worldwide have access to this inexpensive source of iodine. Nonetheless, there are still 54 countries where iodine deficiency is a major public health problem[28] (**Figure 18.17**).

How Supplementation Is Used to Eliminate Malnutrition Supplementing specific nutrients for at-risk segments of the population can also help reduce the prevalence of malnutrition. Of countries where vitamin A deficiency is a public health problem, 78% have policies supporting regular vitamin A supplementation in children.[39] Many have also adopted the WHO recommendation to provide all breast-feeding women with a high-dose supplement of vitamin A within 8 weeks of delivery. This improves maternal vitamin A status and increases the amount of vitamin A that is in breast milk and therefore passed to the infant. Many countries have adopted programs to supplement children older than 6 months with iron and pregnant women with iron and folate.

Figure 18.17 **Iodized salt**

(a)

Percentage of households using iodized salt

100
80
60
40
20
0

Western Pacific | Eastern Mediter-ranean | Europe | Southeast Asia | Americas | Africa

(a) Over the past decade, the number of countries around the world with salt iodization programs has increased dramatically.

(b)

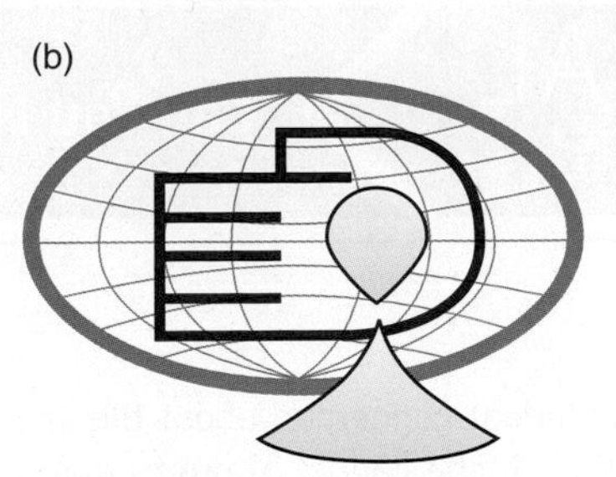

(b) The global iodized salt logo is used around the world as an indicator of iodized salt.

THINK CRITICALLY

What steps would you suggest to convince a population to replace white yams with sweet potatoes?

Providing Education

Education can help improve nutrient intake by teaching consumers which foods are good nutrient sources, so that choices made when purchasing foods or growing vegetables at home can meet nutrient needs. Education about how to prepare foods safely is also critical so food-borne illness doesn't compromise nutritional and overall health. Policies to reduce micronutrient deficiencies also need to include education and strategies to control other infectious and parasitic diseases.

Education is particularly important when introducing a new crop. No matter how nutritious, a new plant variety is not beneficial unless local farmers know how to grow it and the population accepts it as part of their diet and knows how to prepare it. For instance, white yams are common in some regions but are a poor source of β-carotene, which the body can use to make vitamin A. If sweet potatoes, which are rich in β-carotene, became an acceptable choice, the vitamin A available to the population would increase (**Figure 18.18**). Food safety is also a concern when changing traditional dietary practices. For example, introducing papaya to the diet as a source of vitamin A will not improve nutritional status if the fruit is washed in unsanitary water and causes dysentery among the people it is meant to nourish.

© YinYang/iStockphoto

Figure 18.18 **Sweet potatoes**
Replacing yams, which are a poor source of β-carotene, with sweet potatoes, which contain enough β-carotene to meet the requirement for vitamin A in a single serving, can improve the quality of the diet.

Education to encourage breast-feeding can also improve nutritional status and health. Breast-feeding reduces the risk of infectious diseases in infants. When infants are not breast-fed, education about nutritious breast milk substitutes and their safe preparation is essential.

18.4 Hunger at Home

LEARNING OBJECTIVES

- Discuss the causes of food insecurity in the United States.
- Describe factors that prevent people from escaping poverty.
- List the population groups at greatest risk for undernutrition in the United States.
- Describe five federal programs designed to alleviate hunger in the United States.

In the United States, most of the nutritional problems are related to overnutrition. Over 68% of adults in the United States are overweight or obese.[40] Heart disease, hypertension,

CRITICAL THINKING: Can You Have an Impact?

Keesha is a college student concerned about the environment and problems of world hunger. Although she cannot afford to make monetary contributions to relief organizations, she would like her everyday choices to have a minimal impact on the environment. She eats a lot of salad but is not sure which of the options shown here is best.

Waltraud Ingerl/iStockphoto; Justin Sullivan/Getty Images, Inc.

Keesha likes fish but has heard that some fish are endangered. A few of her favorite varieties are listed below.

Atlantic cod

Chilean sea bass

Orange roughy

The following are some inexpensive changes Keesha can make to reduce her impact on the environment.

Bike instead of drive on short trips around town.

Buy a canvas bag to carry groceries.

Bring a water bottle rather than buying bottled water.

Compost vegetable scraps.

Buy locally grown produce.

Buy organically grown produce.

© Todd Bates/iStockphoto

CRITICAL THINKING QUESTIONS

- List one advantage and one disadvantage in terms of convenience, of food safety, and of environmental impact, for the packaged salad and for the fresh produce.
- Use the Internet to find some ocean-friendly substitutes for Keesha's seafood favorites.
- List one advantage and one disadvantage for each of Keesha's changes.

iProfile Use iProfile to lookup the omega-3 fatty acid content of your favorite seafood.

and cancer—all related to obesity—are the leading causes of death. While much of the population is concerned with consuming a diet to lower the risks for these chronic diseases, hungry families are standing in line at soup kitchens. Problems such as poverty and unemployment lead to food insecurity in a land of plenty. Government food and nutrition policy must be concerned with improving economic security as well as providing food to the hungry and maintaining the food supply at an affordable level—and at the same time, that policy must promote healthy diets to reduce diseases related to overconsumption.

Causes of Food Insecurity

In the United States, general food shortage is not the cause of undernutrition, but food insecurity, hunger, and undernutrition are growing problems. Today about 17.2 million

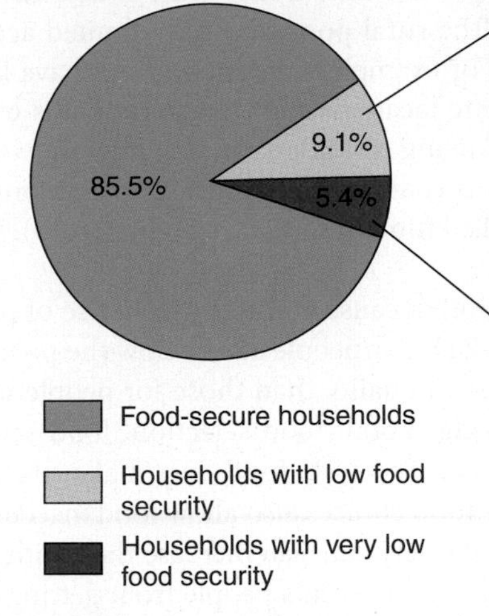

Figure 18.19 **Food insecurity in the United States**
In 2010, approximately 85.5% of U.S. households were food secure; the other 14.5% experienced food insecurity at some point during the year.[41]

American households experience food insecurity due to a lack of money or other resources (**Figure 18.19**).[41] About 6.7 million households, including 3.9 million households with children, experience very low food security. This means that food intake is reduced in one or more household members. Over time, this can lead to malnutrition. The incidence of hunger and food insecurity is higher in the poor; the homeless; women, infants, and children; and the elderly; but illness, disability, a sudden decrease in income, or an increase in living expenses can put anyone at risk for food insecurity.

Poverty Poverty, the main cause of food insecurity, reduces access to food, education, and health care. About 15% of Americans (46.9 million people) live at or below the poverty level (**Figure 18.20**).[42] High poverty and unemployment rates among Native Americans and

2012 Poverty Guidelines*

Persons in family/household	Poverty guideline (dollars/year)
1	11,170
2	15,130
3	19,090
4	23,050
5	27,010
6	30,970
7	34,930
8	38,890

For families/households with more than 8 persons, add $3,960 for each additional person.

*Includes the 48 contiguous states and the District of Columbia

Urban food desert

Richard B. Levine/Newscom

Rural food desert

Clay Peterson/The Californian/©AP/Wide World Photos

Figure 18.20 **Food deserts**
Fresh produce and other nutritious foods are available in American cities, but high prices make them inaccessible to the poor. Poor rural areas can also be food deserts but the reasons differ: even if produce is available, the nearest grocery stores may be 40 miles away.

Alaska Natives contribute to food insecurity.[43] The poor have less money to spend on food and often have less access to affordable food. The high price of real estate in cities has driven supermarkets into the suburbs, and because many low-income city families do not own cars, they must shop at small, expensive corner stores or pay cab fares if they wish to take advantage of lower prices at more distant, larger stores. The rural poor may have limited access to food because they live far from grocery stores. For example, migrant workers have limited access to food because labor camps are in remote locations and transportation is often unavailable. Low incomes and difficult working and living conditions among migrant workers further limit their ability to purchase food and prepare adequate meals. Conditions in both urban and rural areas have created what are called **food deserts** (see Figure 18.20).[44]

food desert An area that lacks access to affordable fruits, vegetables, whole grains, low-fat milk, and other foods that make up a healthy diet.

Lack of Education Lack of education, which is both a cause and a consequence of poverty, also contributes to food insecurity (**Figure 18.21**). For people at or below the poverty level, educational opportunities are fewer and lower in quality than those for people with higher incomes. In the short term, lack of knowledge about food selection, food safety, and home economics can contribute to malnutrition. Too little food may make the diet deficient in energy or particular nutrients, but poor food choices also allow food insecurity to coexist with obesity. Lack of education about food safety can also increase the incidence of food-borne illness. In the long term, lack of education prevents people from getting the higher-paying jobs that could allow them to escape poverty.

Limited Health Care Poverty also limits access to health care, leading to poorer health status. Iron deficiency is more than twice as frequent in low-income children, and the incidence of heart disease, cancer, hypertension, and obesity increases with decreasing income. As in developing nations, poverty is reflected in infant mortality rates. Average infant mortality in the U.S. population is about 6.4 per 1000 live births.[46] However, there are groups within the population that have infant mortality rates as high as those in impoverished nations. Among African Americans, the infant mortality rate is 11.6 per 1000 live births—almost double that of the general population. This difference mirrors the higher poverty rate in this group.

Homelessness The poor must use most of their income to pay for shelter, which seriously reduces the chances that their families will be adequately fed. The high cost of housing not only limits food budgets but also has created a growing problem of homelessness in

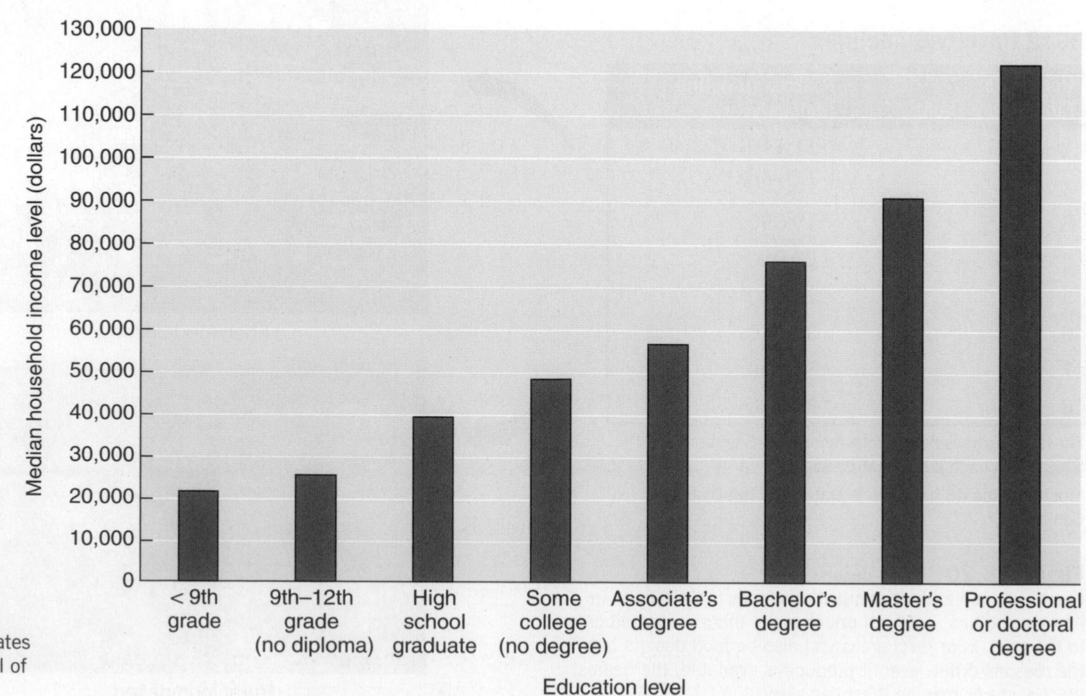

Figure 18.21 Education increases income[45]
Income level in the United States is directly correlated with level of education.

the United States. It is estimated that between 656,000 and 3.5 million Americans are homeless.[47] The homeless are at high risk of malnutrition and food insecurity because they lack not only money but also cooking and food-storage facilities. Without cooking facilities, they must rely on ready-to-eat foods. Without storage facilities, they cannot use less expensive staples such as rice and dried beans, which can be purchased in bulk. Homeless individuals often rely on soup kitchens and shelters to obtain adequate food.

A Cycle of Poverty Many Americans find themselves trapped in a cycle of poverty (**Figure 18.22**). As the U.S. economy has shifted from manufacturing to service-based, factories have closed and manufacturing facilities have moved abroad where labor costs are lower. Former employees often lack the experience or education to move on to other types of work. Unable to find well-paying jobs, they must work longer hours at lower-paying jobs. Low incomes reduce access to transportation and childcare, which can also limit access to better jobs. Long work hours reduce the amount of time available to pursue the additional education or training necessary to find better-paying jobs. Limited income and transportation prohibit relocation to areas where better jobs are available.

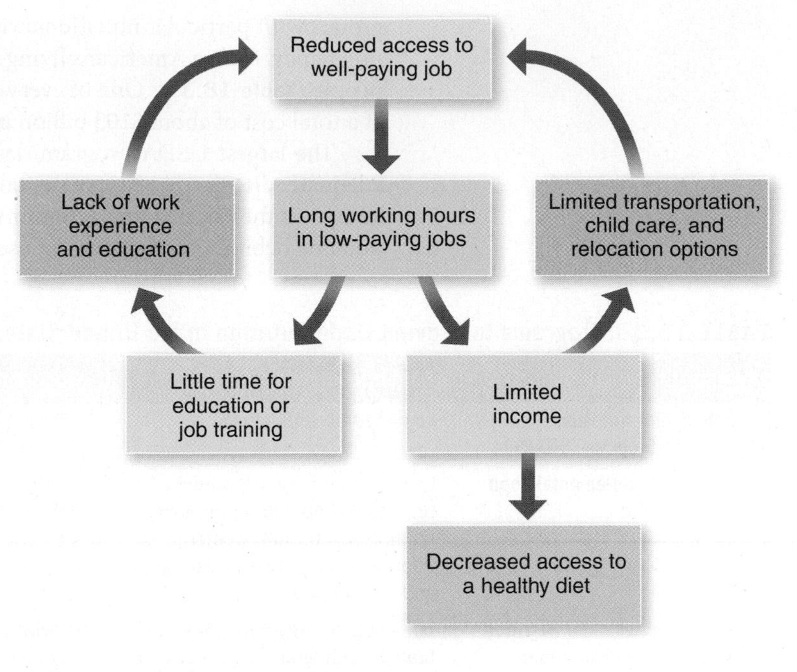

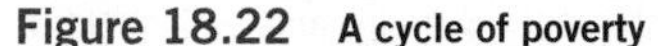

Figure 18.22 A cycle of poverty
Some Americans are trapped in a cycle of poverty because they are unable to acquire the education, training, or resources necessary to obtain better-paying jobs.

High Nutrient Needs The high nutrient needs of pregnant and lactating women and small children put them at particular risk for undernutrition. Almost a third of households with children headed by single women live below the poverty line.[42] Poverty and food insecurity place these women and children at risk of malnutrition, and their special nutritional needs magnify this risk. Because of their increased need for some nutrients, malnutrition may occur in pregnant women, infants, and children even when the rest of the household is adequately fed. For example, the amount of iron in the family diet may be enough to prevent anemia in all but a pregnant teenager.

Disease and Disability Diseases and disabilities limit the ability to purchase, prepare, and physically ingest food. These conditions are more common in the elderly, especially those who are poor, putting them at risk for malnutrition. Greater nutritional risk among older adults is associated with more hospital admissions, and hence greater health-care costs. The number of individuals over age 85 is expected to quadruple by the year 2050; as the number of elderly increases, so will the number at risk of food insecurity.[48] Thus, providing food security for older adults both improves their quality of life and reduces costs for the public health–care system (see Chapter 16).

Federal Programs to Prevent Malnutrition

Solving the problem of undernutrition in the United States involves improving economic security, keeping food affordable, providing food aid to the hungry, and offering education about healthy diets that will meet nutrient needs and reduce diseases related to overconsumption. Historically, many approaches have been attempted to meet this goal. Some have met with success and others have done little to increase access to a nutritious diet for all. Programs that provide access to affordable food and promote healthy eating have been referred to as a *nutrition safety net* for the American population.

The Nutrition Safety Net The nutrition assistance programs in the United States include a combination of general nutrition assistance and specialized programs targeted to groups with particular nutritional risks: children, seniors, infants, women during and after pregnancy, Native Americans living on reservations, people with disabilities, and homeless people (**Table 18.3**).[49] One of every four Americans receives some kind of food assistance, at a total cost of about $103 billion per year.[50]

The largest USDA program designed to make sure that all people have access to an adequate diet is the Supplemental Nutrition Assistance Program (SNAP) (previously known as the Food Stamp Program). SNAP provides monthly benefits in the form of coupons or debit cards that can be used to purchase food, thereby supplementing the food

TABLE 18.3 Programs to Prevent Undernutrition in the United States

Program	Target population	Goals and methods
Supplemental Nutrition Assistance Program (SNAP)	Low-income individuals	Increases access to food by providing coupons or debit cards that can be used to purchase food at a grocery store
Commodity Supplemental Food Program (CSFP)	Low-income pregnant women, breast-feeding and non-breast-feeding postpartum women, infants and children under age 6, and elderly people	Provides food by distributing U.S. Department of Agriculture (USDA) commodity foods
Special Supplemental Nutrition Program for Women, Infants, and Children (WIC)	Low-income pregnant women, breast-feeding and non-breast-feeding postpartum women, and infants and children under age 5	Provides vouchers for the purchase of foods (including infant formula and infant cereal) high in nutrients that are typically lacking in the program's target population; provides nutrition education and referrals for health care
WIC Farmers' Market Nutrition Program	WIC participants	Increases access to fresh produce by providing vouchers that can be used to purchase produce at authorized local farmers' markets
National School Breakfast Program	Low-income children	Provides free or low-cost breakfasts at school to improve the nutritional status of children
National School Lunch Program	Low-income children	Provides free or low-cost lunches at school to improve the nutritional status of children
Special Milk Program	Low-income children	Provides milk for children in schools, camps, and child-care institutions with no federally supported meal program
Summer Food Service Program	Low-income children	Provides free meals and snacks for children when school is not in session
Child and Adult Care Food Program	Children up to age 12 and elderly and disabled adults	Provides nutritious meals to children and adults in day-care settings
Team Nutrition	School-age children	Provides nutrition education, training and technical assistance, and resources to participating schools, with the goal of improving children's lifelong eating and physical activity habits
Head Start	Low-income preschool children and their families	Provides meals and education, including nutrition education
Nutrition Program for the Elderly	Individuals age 60 and over and their spouses	Provides free congregate meals in churches, schools, senior centers, or other facilities and delivers food to homebound people
Senior Farmers' Market Program	Low-income seniors	Provides coupons that can be exchanged for eligible foods at farmers' markets, roadside stands, and community-supported agricultural programs
Homeless Children Nutrition Program	Preschoolers living in shelters	Reimburses providers for meals served
Emergency Food Assistance Program	Low-income people	Provides commodities to soup kitchens, food banks, and individuals for home use
Healthy People 2020	U.S. population	Sets national health promotion objectives to improve the health of the U.S. population through health-care system and industry involvement, as well as individual actions
Expanded Food and Nutrition Education Program (EFNEP)	Low-income families	Provides education in all aspects of food preparation and nutrition
Temporary Assistance for Needy Families (TANF)	Low-income households	Provides assistance and work opportunities to needy families by granting states federal funds to implement welfare programs
Food Distribution Program on Indian Reservations	Low-income households living on reservations and Native Americans living near reservations	Provides food by distributing USDA commodity foods

budgets of low-income individuals. Together with SNAP, four other programs that target high-risk populations—the National School Lunch Program, the Special Supplemental Nutrition Program for Women, Infants, and Children (WIC), the Child and Adult Care Food Program, and the National School Breakfast Program—account for 96% of the USDA's expenditure for food assistance.[50]

In addition to federal nutrition assistance programs, church, community, and charitable emergency food shelters provide for the basic nutritional needs of many Americans. In the United States, about 150,000 nonprofit food distribution programs help direct food to those in need.[51] The leading hunger-relief charity in the United States is Feeding America, which provides food assistance to over 25 million low-income people per year. It includes a network of food banks across the country and supports thousands of local charitable organizations, such as food pantries and soup kitchens, that distribute food directly to hungry Americans.

Virtually all these food distribution programs use food obtained through food recovery, which involves collecting food that is wasted in fields, commercial kitchens, restaurants, and grocery stores and distributing it to those in need. Field **gleaning** is a type of food recovery that involves collecting crops that are not harvested because it is not economically profitable to harvest them or that remain in fields after mechanical harvesting (**Figure 18.23**). It is estimated that over 20% of America's food—enough to feed 49 million people—goes to waste each year.[52]

gleaning The act of collecting the crops left in farmers' fields after they have been commercially harvested or collecting those that were not harvested for economic reasons.

Providing Nutrition Education The link between nutrition education and diet quality is strong. People with more nutrition information and more awareness of the relationship between diet and health consume healthier diets.[53] Healthy diets not only improve current health by optimizing growth, productivity, and well-being, but are essential for preventing chronic diseases in the future. Increasing nutrition knowledge can reduce medical care costs and improve the quality of life.

Education can help individuals with lower incomes stretch limited food dollars by making wise choices at the store and reducing food waste at home. Education can promote community gardens to increase the availability of seasonal vegetables. It can teach people how to prepare foods that become available through commodity distribution and food banks. It can explain safe food handling and food preparation methods. Knowing which foods to choose and how to handle them safely is as important in preventing malnutrition as having the money to buy enough food.

A number of government programs are designed to provide nutrition education. One of the goals of Healthy People 2020 is to increase the nutrition education provided by schools as well as by work sites. The Expanded Food and Nutrition Education Program (EFNEP) provides education in all aspects of food preparation and nutrition to low-income families. In addition, the Dietary Guidelines for Americans, MyPlate, and standardized food labels educate the general public about making wise food choices.

Figure 18.23 Field gleaning
These oranges were harvested in California as part of a local gleaning program.

OUTCOME

AFP/Getty Images, Inc.

When Teresa joined the Peace Corps, she quickly learned the huge effect malnutrition and infectious disease have on the health of children in Rwanda and around the world. Meeting Songe introduced her to the impact of and treatment for protein-energy malnutrition (PEM). After a few months of consuming the high-protein supplement Songe was given at the clinic, his belly had shrunk and his arms and legs began filling out with muscle. Because his malnutrition was treated early, Songe is now a healthy, playful 3-year-old. Others are not as lucky. Many children with short stature due to malnutrition also have cognitive delays that impair their education and job potential throughout their lives. Teresa has seen children with severe anemia from iron deficiency and those who have lost their eyesight due to the complications of vitamin A deficiency. Some children are so ill by the time they reach the clinic that they cannot be saved.

© MissHibiscus/iStockphoto

In addition to her other roles, Teresa helps educate families about the right combinations of foods needed to prevent malnutrition as well as clean water and hygiene to prevent infections. She is learning all she can about the foods available locally so she can recommend meals that provide the nutrients that are at risk of deficiency in the population. **Teresa's suggestions will help, but she recognizes that with the scarcity of certain types of food in the typical local diet, and the high incidence of infectious diseases in the population, she will likely see many other young children with protein and micronutrient deficiencies.**

APPLICATIONS

ASSESSING YOUR DIET

1. Fill in the table below to help you decide the least expensive way to buy these products:

	COST		
PRODUCT	CORNER STORE ($)	SUPER-MARKET ($)	DISCOUNT/ WAREHOUSE STORE ($)
Small orange juice (8 fl oz)			
Large orange juice (1/2 gal)			
Brand-name cereal (e.g., Kix or Corn Flakes) (15 oz box)			
Generic brand (similar cereal) (15 oz box)			
White bread (1 loaf)			
Whole-wheat bread (1 loaf)			
Fresh apples (price/lb)			
Small bag of chips (1 oz)			
Large bag of chips (same brand) (14 to 16 oz)			

a. Calculate the cost per oz for each item at each type of store.

b. How does the size of the package affect cost?

c. How does the brand name affect cost?

d. Will following the MyPlate recommendation to make half your grains whole affect food costs?

2. How much money do you spend on food?

a. Keep a record of how much money you spend on food in a day and use it to estimate your monthly food costs.

b. Suggest two or three changes in the foods you choose that will reduce your food costs.

c. How do these changes affect the nutrient content of your diet?

d. What could you eat if your food budget for the day was only $3?

e. Would the $3-a-day diet you put together meet your nutrient needs? Which nutrients are deficient? Which are excessive?

CONSUMER ISSUES

3. What could you do for World Food Day (October 16)? List some ideas for campus-wide programs to increase awareness of global nutrition issues.

4. **What do you know about hunger in various parts of the world?**
   a. Use the Internet to locate Web sites for organizations such as Worldwatch or Bread for the World Institute.
   b. Choose one area of the world where hunger and undernutrition are a major problem and explain the cause of undernutrition in this area.
   c. What solutions are in place or proposed to solve these problems?

CLINICAL CONCERNS

5. **Imagine that the diet in a developing country is deficient in one of the micronutrients. Most of the food is grown and prepared locally. To solve the problem, the government distributes dietary supplements. When micronutrient deficiency continues to be a problem, interviews reveal that the supplements are not being taken. The government is considering a food fortification program.**

   The following foods are typically consumed:

   Corn and corn tortillas
   Dried beans
   White rice
   Fresh tomatoes
   Fresh vegetables—greens and squash
   Pork
   Fresh fruits

   A typical breakfast is cornmeal and fruit; lunch is a hot meal, usually with some kind of vegetable soup served with tortillas; and dinner is meat, beans, rice, tortillas, and fruit.
   a. Use the information above to suggest a food that might be appropriate to fortify.
   b. What problems must be considered when selecting the food to fortify and the amount to include in the fortified food?
6. **Hookworm is a parasite that attaches to the intestinal lining and feeds on blood. The larvae penetrate the skin, infecting people when they walk barefoot on contaminated soil.**
   a. Based on this information, suggest why hookworm infection is more common in poor tropical and subtropical regions of the world.
   b. What nutrient deficiencies are most likely to result from hookworm infection?
   c. Explain why hookworm infection has such a dramatic impact on the health of marginally nourished individuals.

SUMMARY

18.1 The Two Faces of Malnutrition

- In poorly nourished populations, a cycle of malnutrition exists in which undernourished women give birth to low-birth-weight infants at risk of disease and early death. Children who survive grow poorly and become adults who are physically unable to contribute fully to society. Malnutrition in children causes stunting. The prevalence of stunting is used as an indicator of nutritional well-being in populations of children. Malnutrition depresses immune function at all stages of life, causing an increase in the frequency and severity of infectious diseases.
- Overnutrition is now a global health problem that affects as many people as undernutrition. Rising obesity rates increase the incidence of chronic diseases such as cardiovascular disease and type 2 diabetes worldwide. Obesity now coexists with undernutrition in both developed and developing nations around the world.
- Changes in diet and lifestyle occur as economic conditions in a country improve. Traditional rural diets that contribute to undernutrition are replaced by western-style diets that increase the risk of overnutrition.

18.2 Causes of Hunger and Undernutrition

- World hunger exists due to the inequitable distribution of available food. Natural and manmade disasters can temporarily disrupt food production and distribution and cause famine.
- Chronic food shortage occurs when economic inequities result in lack of money, health care, and education for individuals or populations; when overpopulation and limited natural resources create a situation in which there are more people than food; when cultural practices limit food choices; and when environmental resources are misused, limiting the ability to continue to produce food.
- Go to WileyPLUS to view a video clip on famine in Ethiopia

- Malnutrition occurs when the quality of the diet is poor. High-risk groups with special nutrient needs, such as pregnant women, children, the elderly, and the ill, may not be able to meet their nutrient needs with the available diet. Deficiencies of protein, iron, iodine, vitamin A, and zinc are common worldwide.

18.3 Eliminating World Hunger

- Short-term solutions to undernutrition provide food through relief at the local, national, and international levels.
- Long-term solutions to undernutrition must ensure the availability of food by improving economic conditions to eliminate poverty, controlling population growth, increasing the food supply through agricultural technology or importation, and developing sustainable systems that provide food without damaging the environment.
- Long-term solutions to undernutrition must ensure a nutritionally adequate food supply. Food fortification and dietary supplementation can be used to increase protein quality and eliminate micronutrient deficiencies and improve the overall quality of the diet.
- Education about what to eat and how to prepare food safely can help eliminate malnutrition.

18.4 Hunger at Home

- Both undernutrition and overnutrition are problems in the United States. Food insecurity is associated with poverty, which limits education and access to health care and adequate housing. High nutrient needs increase the risk of malnutrition in women and children, and disease and disability increase risk in the elderly. Limited access to food increases risk in certain segments of the population.

- Nutrition programs in the United States focus on maintaining a nutrition safety net that will provide access to affordable food and promote healthy eating. Some programs designed to help feed the hungry address the general population, whereas others focus on specific high-risk groups. Most programs provide access to food and some provide nutrition education.

REVIEW QUESTIONS

1. What is meant by the statement, "world nutrition policies must address the two faces of malnutrition"?
2. What is the cycle of malnutrition?
3. What is meant by nutrition transition?
4. How does overpopulation contribute to food shortage?
5. How does poverty contribute to world hunger?
6. How are economic growth and population growth related?
7. What segments of the world population are at greatest risk for undernutrition?
8. List three micronutrient deficiencies that are world health problems.
9. Why are environmental issues important in maintaining the world's food supply?
10. How can sustainable agriculture reduce environmental damage?
11. How can food fortification be used to help eliminate malnutrition?
12. List four population groups in the United States that are at risk for undernutrition.
13. List three federal programs that address malnutrition in the United States.

REFERENCES

1. Food and Agriculture Organization of the United Nations. *The State of Food Insecurity in the World 2011.*
2. Caulfield, L. E., de Onis, M., Blössner, M., and Black, R. E. Undernutrition as an underlying cause of child deaths associated with diarrhea, pneumonia, malaria, and measles. *Am J Clin Nutr* 80:193–198, 2004.
3. Black, R. E., Allen, L. H., Bhutta, Z. A., et al. Maternal and child undernutrition: Global and regional exposures and health consequences. *Lancet* 371:243–260, 2008.
4. World Health Organization. *The Global Burden of Disease: 2004 Update, 2008.*
5. Central Intelligence Agency. *The World Factbook.*
6. UNICEF Progress for Children. *A World Fit for Children. Statistical Review.*
7. Martins, V. J., Toledo Florêncio, T. M., Grillo, L. P., et al. Long-lasting effects of undernutrition. *Int J Environ Res Public Health* 8:1817–1846, 2011.
8. Schaible, U. E., and Kaufmann, S. H. E. Malnutrition and infection: Complex mechanisms and global impacts. *PLoS Med* 4:e115, 2007.
9. World Health Organization. *Global Strategy on Diet Physical Activity and Health: Obesity and Overweight.*
10. Popkin, B. M., Adair, L. S., and Ng, S. W. Global nutrition transition and the pandemic of obesity in developing countries. *Nutr Rev* 70:3–21, 2012.
11. Eberwine, D., Pan American Health Organization. Globesity: A crisis of growing proportions. *Perspectives in Health* 7(3):2002.
12. Lobstein, T., Baur, L., and Uauy, R., IASO International Obesity Task Force. Obesity in children and young people: A crisis in public health. *Obes Rev* 5(Suppl 1):4–104, 2004.
13. Vorster, H. H., Bourne, L. T., Venter, C. S., and Oosthuizen, W. Contribution of nutrition to the health transition in developing countries: A framework for research and intervention. *Nutr Rev* 57:341–349, 1999.
14. Kennedy, E. T. The global face of nutrition: What can governments and industry do? *J Nutr* 135:913–915, 2005.
15. World Health Organization, FAO. *The nutrition transition and obesity.*
16. The World Bank. *Poverty and equity data.*
17. Bread for the World Institute. *Are We on Track to End Hunger?* Hunger Report 2004.
18. Population Reference Bureau. World population highlights: Key findings from PRB's 2011 world population data sheet. *Population Bulletin* 63:1–16, 2008.
19. Reijnders, L., and Soret, S. Quantification of the environmental impact of different dietary protein choices. *Am J Clin Nutr* 78(Suppl):664S–668S, 2003.
20. Steinfeld, H., Gerber, P., Wassenaar, T., et al. Livestock's long shadow: Environmental issues and options. Food and Agricultural Organization of the United Nations, Rome, 2006.
21. Brown, L. R. *World Facing Huge New Challenge on Food Front—Business-as-Usual Not a Viable Option.* April 16, 2008.
22. Rosen, S., and Shapouri, S. Rising food prices intensify food insecurity in developing countries. *Amber Waves* 6(1):16–21, February 2008.

23. Worm, B., Hilborn, R., Baum, J. K., et al. Rebuilding global fisheries. *Science* 325:578–585, 2009.

24. Food and Agriculture Organization of the United Nations, Fisheries and Aquaculture Department. *Impact of Aquaculture on Enviroment.*

25. World Health Organization. *Micronutrient Deficiencies. Iron Deficiency Anaemia.*

26. World Health Organization. *Initiative for Vaccine Research. Hookworm Disease.*

27. World Health Organization. *Iodine Status Worldwide.*

28. World Health Organization. *Micronutrient Deficiencies. Iodine Deficiency Disorders.*

29. World Health Organization. *Micronutrient Deficiencies. Vitamin A Deficiency.*

30. World Health Organization. *Intermittent Iron and Folic Acid Supplementation in Menstruating Women.*

31. World Health Organization. *The World Health Report, 1998: Life in the 21st Century—A Vision for All.* Geneva: World Health Organization, 1998.

32. Beck, M. A., Levander, O. A., and Handy, J. Selenium deficiency and viral infection. *J. Nutr* 133(5 suppl 1):1463S–1467S, 2003.

33. United Nations. End Poverty: Millennium Development Goals 2015.

34. Central Intelligence Agency. *World Fact Book. Country Comparisons: Total Fertility Rate.*

35. Berg, L. R., and Hager, M. C. *Visualizing Environmental Science.* Hoboken: John Wiley & Sons, 2009.

36. Association of American Geographers. *How Does Education Affect Fertility Rates in Different Places?*

37. Raloff, J. The human numbers crunch. *Sci News* 149:396–397, 1996.

38. USAID. *Sustainable Agriculture.*

39. United Nations Administrative Committee on Coordination/ Subcommittee on Nutrition. *Third Report on the World Nutrition Situation.*

40. Flegal, K. M., Carroll, M. D., Kit, B. K., and Ogden, C. L. Prevalence of obesity and trends in the distribution of body mass index among U.S. adults, 1999–2010. *JAMA* 307:491–497, 2012.

41. Coleman-Jensen, A., Nord, M., Andrews, M., and Carlson, S. Household Food Security in the United States in 2010. *Economic Research Report* No. (ERR-125), September 2011.

42. DeNavas-Walt, C., Proctor, B. D., and Smith, J. C. *Current Population Reports, Income, Poverty, and Health Insurance Coverage in the United States: 2010.* Washington, DC: U.S. Government Printing Office, 2011.

43. DHHS Office of Minority Health. *American Indian/Alaska Native Profile.*

44. Centers for Disease Control and Prevention. *Food Desert.*

45. U.S. Census Bureau. The 2012 Statistical Abstract. Table 692: Money income of households—distribution by income level and selected characteristics: 2009.

46. Centers for Disease Control and Prevention. *National Center for Healthy Statistics, 2010 Deaths and Mortality.*

47. National Alliance to End Homelessness. *State of Homelessness in America 2011.*

48. Federal Interagency Forum on Aging. *Older Americans 2004: Key Indicators of Well Being.*

49. U.S. Department of Agriculture. Food and Nutrition Service. *The National Nutrition Safety Net: Tools for Community Food Security.* FNS-314.

50. Economic Research Service, U.S. Department of Agriculture. The Food Assistance Landscape: FY 2011 Annual Report. *Economic Bulletin* EIB 93, March 2012.

51. Feeding America. *How We Work.*

52. U.S. Department of Agriculture. *Food Recovery and Gleaning Initiative.* A Citizen's Guide to Food Recovery.

53. U.S. Department of Agriculture, Economic Research Service, USDA Center for Nutrition Policy and Promotion. *USDA's Healthy Eating Index and Nutrition Information.*

***To access links to online sources, please go to www.wiley.com/college/smolin and select Nutrition: Science and Applications, 3rd edition. From this page, select either the student or instructor companion site. Once on the desired site, select References.**

Appendices

Appendix A

Additional DRI Tables

DRI tables for Vitamins and for Minerals are on the front and back covers of this text.

Dietary Reference Intake Values for Energy: Total Energy Expenditure (TEE) Equations for Overweight and Obese Individuals

Life Stage Group	*TEE Prediction Equation(Cal/day)*	*PA Values*
Overweight boys aged 3–18 years	TEE = 114 − (50.9 × age in yrs) + PA[(19.5 × weight in kg) + (1161.4 × height in m)]	Sedentary = 1.00 Low active = 1.12 Active = 1.24 Very active = 1.45
Overweight girls aged 3–18 years	TEE = 389 − (41.2 × age in yrs) + PA[(15.0 × weight in kg) + (701.6 × height in m)]	Sedentary = 1.00 Low active = 1.18 Active = 1.35 Very active = 1.60
Overweight and obese men aged 19 years and older	TEE = 1086 − (10.1 × age in yrs) + PA[(13.7 × weight in kg) + (416 × height in m)]	Sedentary = 1.00 Low active = 1.12 Active = 1.29 Very active = 1.59
Overweight and obese women aged 19 years and older	TEE = 448 − (7.95 × age in yrs) + PA[(11.4 × weight in kg) + (619 × height in m)]	Sedentary = 1.00 Low active = 1.16 Active = 1.27 Very active = 1.44

Source: Institute of Medicine, Food and Nutrition Board, "Dietary Reference Intakes for Energy, Carbohydrate, Fiber, Fat, Fatty Acids, Cholesterol, Protein, and Amino Acids," Washington, D.C.: National Academy Press, 2002, 2005.

Appendix B

Standards for Body Size

Body Mass Index (BMI) and Associated Risk

Body mass index (BMI) is the measurement of choice for determining health risks associated with body weight. To use the table, find the appropriate height in the left-hand column. Move across the row to the given weight. The number at the top of the column is the BMI for that height and weight. Use the table below to determine the risks associated with BMI and waist circumference.

BMI (kg/m^2)	**19**	**20**	**21**	**22**	**23**	**24**	**25**	**26**	**27**	**28**	**29**	**30**	**35**	**40**
Height (in.)	**Weight (lb.)**													
58	91	96	100	105	110	115	119	124	129	134	138	143	167	191
59	94	99	104	109	114	119	124	128	133	138	143	148	173	198
60	97	102	107	112	118	123	128	133	138	143	148	153	179	204
61	100	106	111	116	122	127	132	137	143	148	153	158	185	211
62	104	109	115	120	126	131	136	142	147	153	158	164	191	218
63	107	113	118	124	130	135	141	146	152	158	163	169	197	225
64	110	116	122	128	134	140	145	151	157	163	169	174	204	232
65	114	120	126	132	138	144	150	156	162	168	174	180	210	240
66	118	124	130	136	142	148	155	161	167	173	179	186	216	247
67	121	127	134	140	146	153	159	166	172	178	185	191	223	255
68	125	131	138	144	151	158	164	171	177	184	190	197	230	262
69	128	135	142	149	155	162	169	176	182	189	196	203	236	270
70	132	139	146	153	160	167	174	181	188	195	202	207	243	278
71	136	143	150	157	165	172	179	186	193	200	208	215	250	286
72	140	147	154	162	169	177	184	191	199	206	213	221	258	294
73	144	151	159	166	174	182	189	197	204	212	219	227	265	302
74	148	155	163	171	179	186	194	202	210	218	225	233	272	311
75	152	160	168	176	184	192	200	208	216	224	232	240	279	319
76	156	164	172	180	189	197	205	213	221	230	238	246	287	328

BMI (kg/m^2)		***Waist less than or equal to 40 in. (men) or 35 in. (women)***	***Waist greater than 40 in. (men) or 35 in. (women)***
18.5 or less	Underweight	—	N/A
18.5–24.9	Normal	—	N/A
25.0–29.9	Overweight	Increased	High
30.0–34.9	Obese	High	Very High
35.0–39.9	Obese	Very High	Very High
40 or greater	Extremely Obese	Extremely High	Extremely High

Source: Adapted from Partnership for Healthy Weight Management http://www.consumer.gov/weightloss/bmi.htm

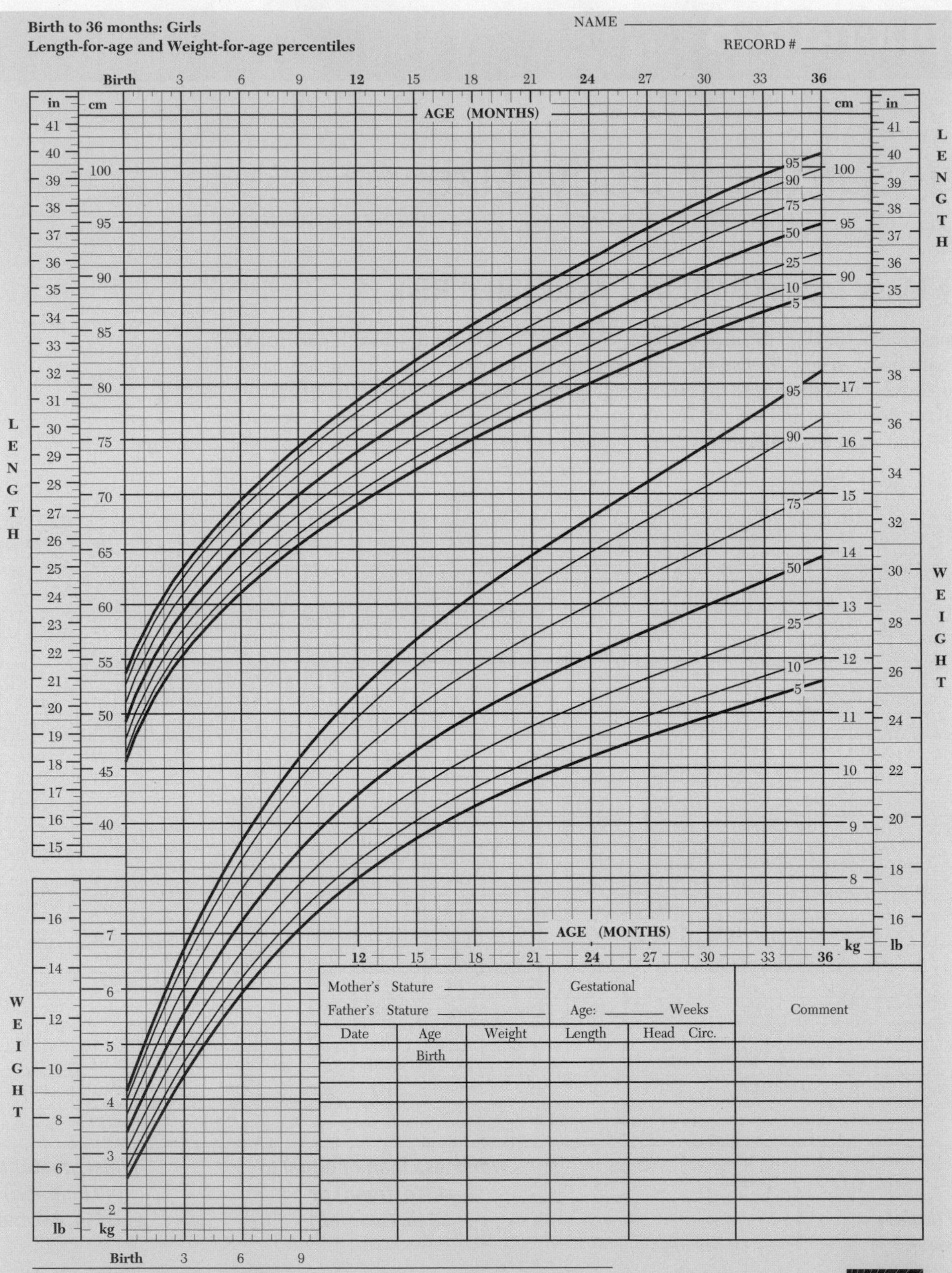

Published May 30, 2000 (modified 4/20/01).

Source: Developed by the National Center for Health Statistics in collaboration with the National Center for Chronic Disease Prevention and Health Promotion (2000). www.cdc.gov/growthcharts

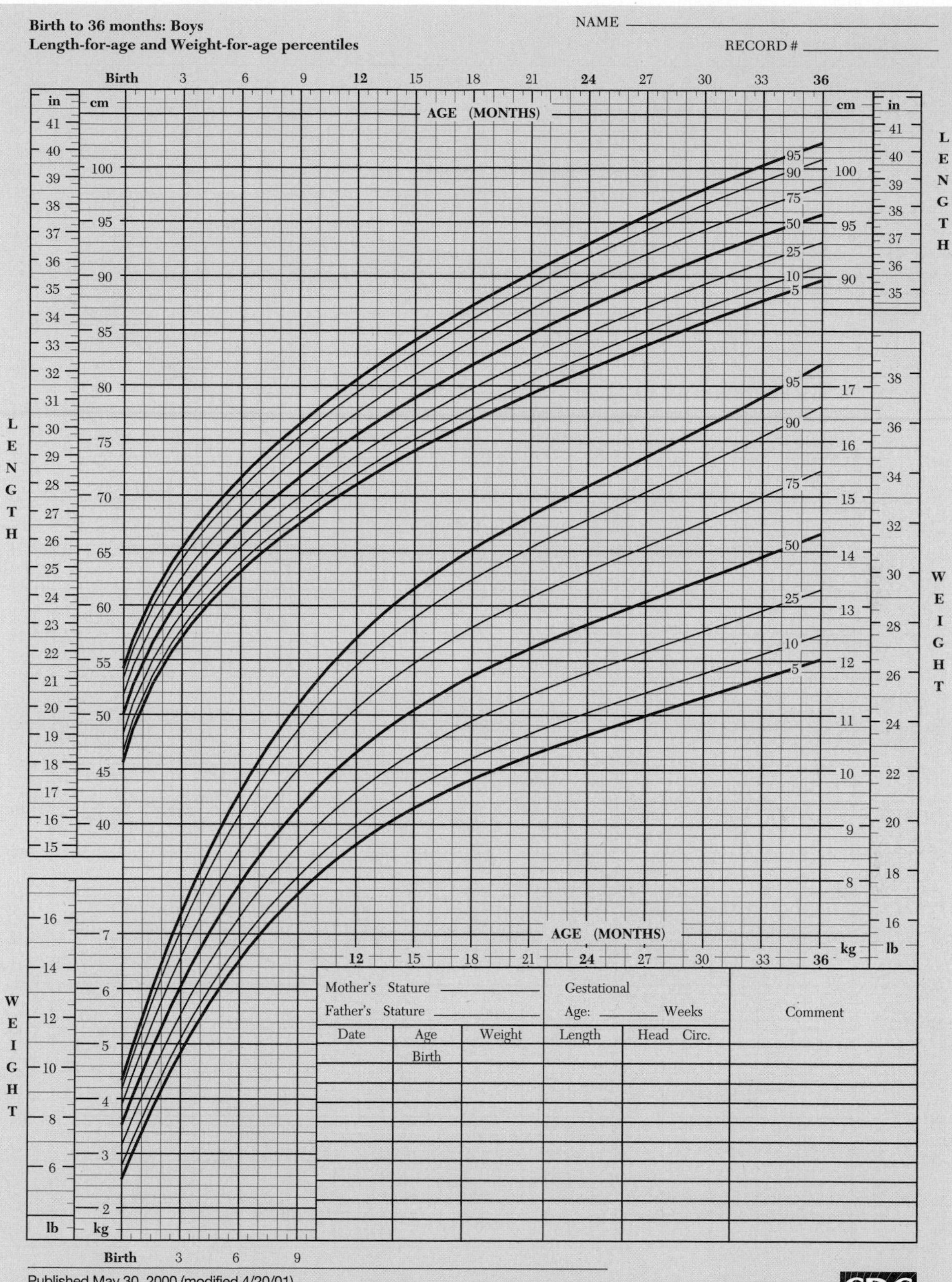

Published May 30, 2000 (modified 4/20/01).

Source: Developed by the National Center for Health Statistics in collaboration with the National Center for Chronic Disease Prevention and Health Promotion (2000). www.cdc.gov/growthcharts

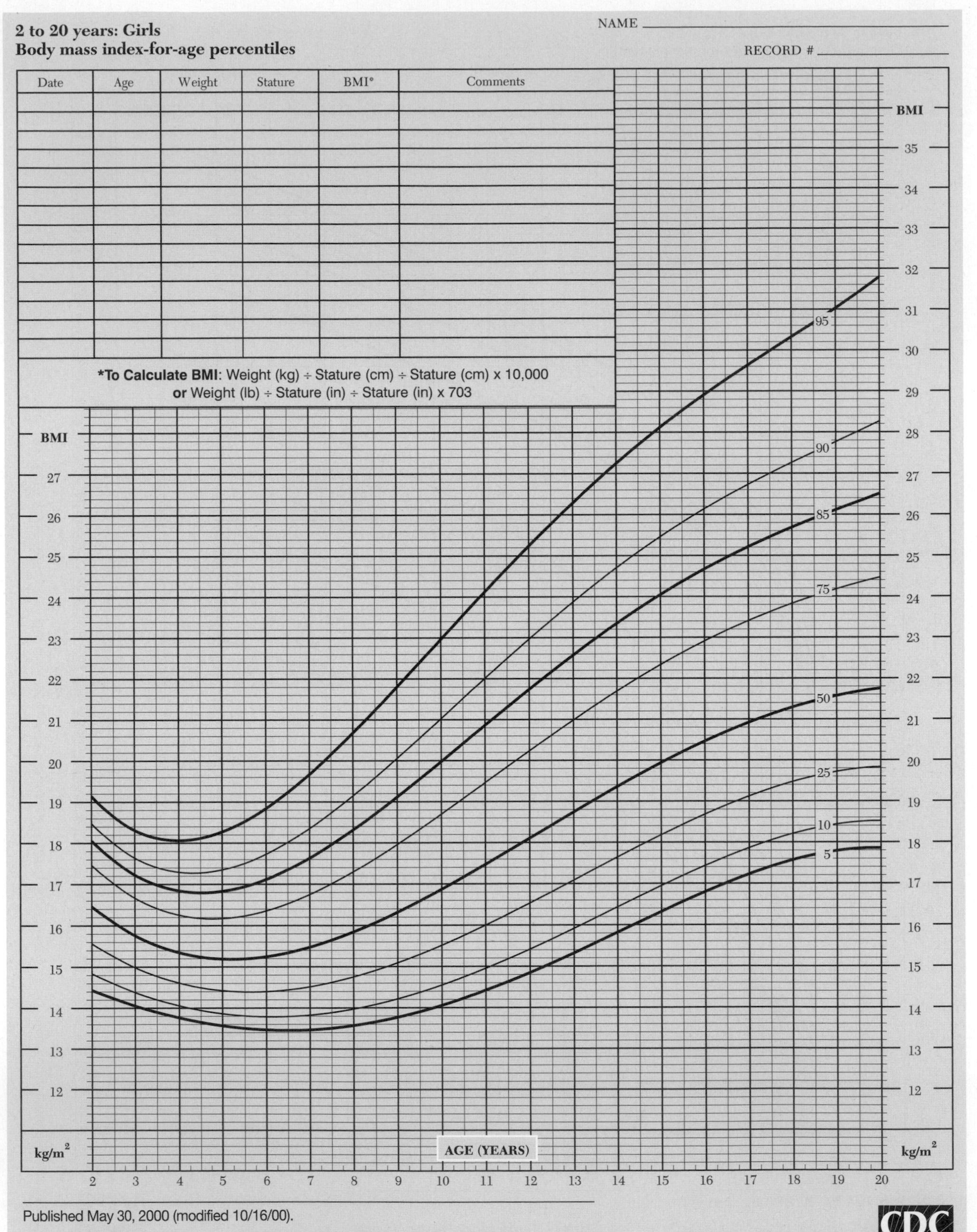

Published May 30, 2000 (modified 10/16/00).

Source: Developed by the National Center for Health Statistics in collaboration with the National Center for Chronic Disease Prevention and Health Promotion (2000). www.cdc.gov/growthcharts

CDC
SAFER • HEALTHIER • PEOPLE™

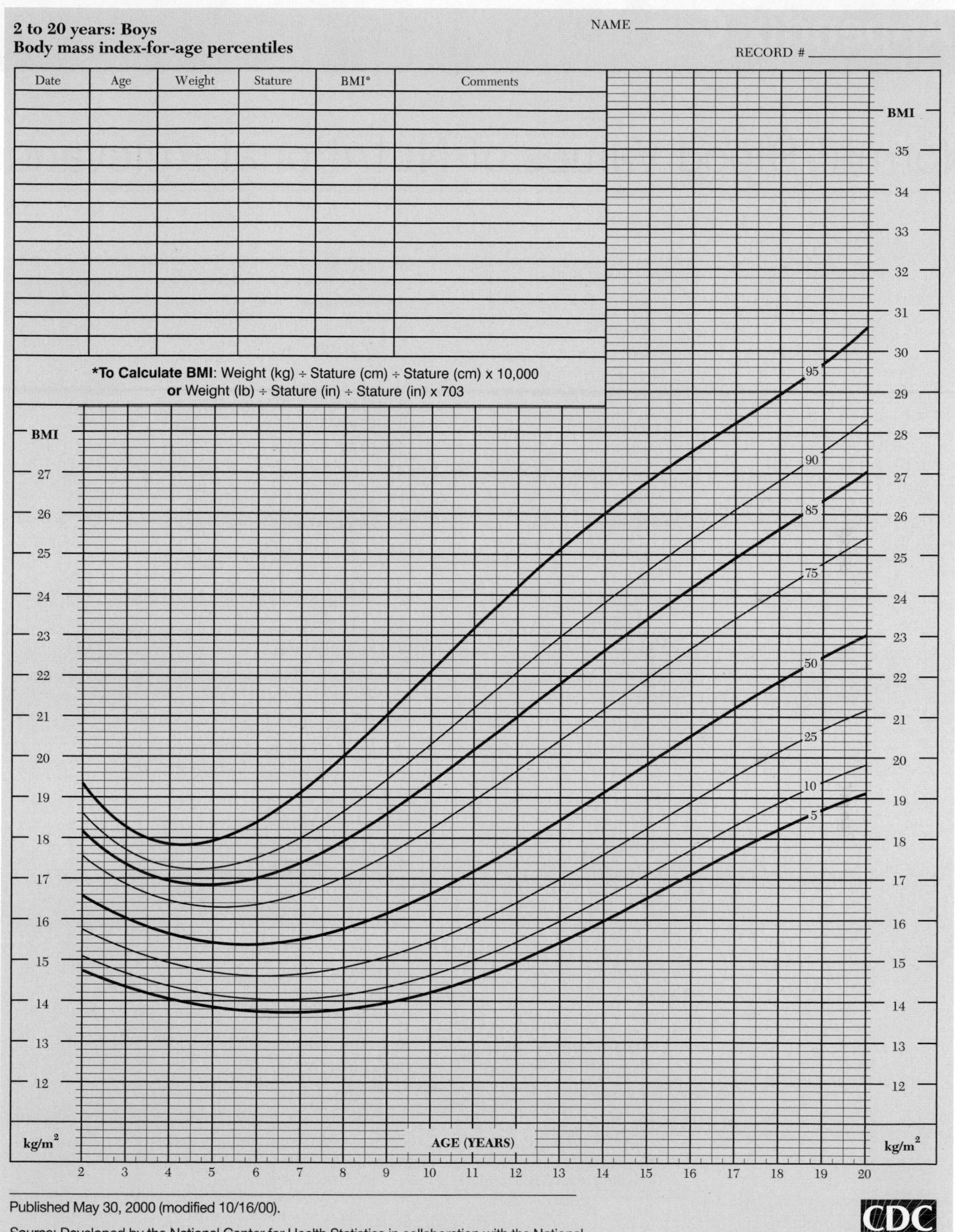

Published May 30, 2000 (modified 10/16/00).

Source: Developed by the National Center for Health Statistics in collaboration with the National Center for Chronic Disease Prevention and Health Promotion (2000). www.cdc.gov/growthcharts

CDC
SAFER • HEALTHIER • PEOPLE™

Appendix C

Normal Blood Values of Nutritional Relevance

Red blood cells	
Men	4.6–6.2 million/mm^3
Women	4.2–5.2 million/mm^3
White blood cells	5,000–10,000/mm^3
Hematocrit	
Men	40–54 ml/100 ml
Women	36–47 ml/100 ml
Children	35–49 ml/100 ml
Hemoglobin	
Men	14–18 g/100 ml
Women	12–16 g/100 ml
Children	11.2–16.5 g/100 ml
Ferritin	
Men	20–300 ng/ml
Women	20–120 ng/ml
Calcium	9–11 mg/100 ml
Iodine	3.8–8 μg/100 ml
Iron	
Men	75–175 μg/100 ml
Women	65–165 μg/100 ml
Zinc	0.75–1.4 μg/ml
Magnesium	1.8–3.0 mg/100 ml
Potassium	3.5–5.0 mEq/liter
Sodium	136–145 mEq/liter
Chloride	100–108 mEq/liter
Vitamin A	20–80 μg/100 ml
Vitamin B_{12}	200–800 pg/100 ml
Vitamin C	0.6–2.0 mg/100 ml
Carotene	48–200 μg/liter
Folate	2–20 ng/ml
pH	7.35–7.45
Total protein	6.6–8.0 g/100 ml
Albumin	3.0–4.0 g/100 ml
Cholesterol	<200 mg/100 ml
Glucose	60–100 mg/100 ml blood, 70–120 mg/100 ml serum

Source: Handbook of Clinical Dietetics, American Dietetic Association, © 1981 by Yale University Press (New Haven, Conn.); Committee on Dietetics of the Mayo Clinic, *Mayo Clinic Diet Manual* (Philadelphia: W. B. Saunders Company, 1981), pp. 275–277.

Guidelines for Determining Healthy Lipid Levels

Classification of LDL, Total, and HDL Cholesterol and Triglycerides (mg/dL)

LDL Cholesterol	
<100	Optimal*
100–129	Near optimal/above optimal
130–159	Borderline high
160–189	High
≥190	Very high
Total Cholesterol	
<200	Desirable
200–239	Borderline high
≥240	High
HDL Cholesterol	
<40	Increases risk
≥60	Decreases risk
Triglycerides	
<150	Normal
150–190	Borderline high
200–499	High
≥500	Very high

*For very high-risk people, LDL cholesterol should be <70.

Source: National Cholesterol Education Program, ATP III, Quick Reference Guide, 2004 update. Available online at http://www.nhlbi.nih.gov/guidlines/cholesterol/atglance.pdf

High Blood Pressure (Hypertension) Guidelines

Classification

High	140/90 or above
Prehypertension	120–139 / 80–89
Normal	119/79 or below

Source: Joint National Committee on Prevention, Detection, Evaluation, and Treatment of High Blood Pressure (2003). *The Seventh Report of the Joint National Committee on Prevention, Detection, Evaluation, and Treatment of High Blood Pressure.* NIH Publication No. 03-5233. Bethesda, MD: U.S. Department of Health and Human Services.

Appendix D

U.S. Nutrition Recommendations and Guidelines

Dietary Guidelines for Americans, 2010

Balancing Calories to Manage Weight
• Prevent and/or reduce overweight and obesity through improved eating and physical activity behaviors. • Control total calorie intake to manage body weight. For people who are overweight or obese, this will mean consuming fewer calories from foods and beverages. • Increase physical activity and reduce time spent in sedentary behaviors. • Maintain appropriate calorie balance during each stage of life—childhood, adolescence, adulthood, pregnancy and breastfeeding, and older age.

Foods and Food Components to Reduce
• Reduce daily sodium intake to less than 2,300 milligrams (mg) and further reduce intake to 1,500 mg among persons who are 51 and older and those of any age who are African American or have hypertension, diabetes, or chronic kidney disease. The 1,500 mg recommendation applies to about half of the U.S. population, including children, and the majority of adults. • Consume less than 10 percent of calories from saturated fatty acids by replacing them with monounsaturated and polyunsaturated fatty acids. • Consume less than 300 mg per day of dietary cholesterol. • Keep *trans* fatty acid consumption as low as possible by limiting foods that contain synthetic sources of *trans* fats, such as partially hydrogenated oils, and by limiting other solid fats. • Reduce the intake of calories from solid fats and added sugars. • Limit the consumption of foods that contain refined grains, especially refined grain foods that contain solid fats, added sugars, and sodium. • If alcohol is consumed it should be consumed in moderation—up to one drink per day for women and two drinks per day for men—and only by adults of legal drinking age.

Foods and Nutrients to Increase

Individuals should meet the following recommendations as part of a healthy eating pattern while staying within their calorie needs.

- Increase vegetable and fruit intake.
- Eat a variety of vegetables, especially dark-green and red and orange vegetables and beans and peas.
- Consume at least half of all grains as whole grains. Increase whole grain intake by replacing refined grains with whole grains.
- Increase intake of fat-free or low-fat milk and milk products, such as milk, yogurt, cheese, or fortified soy beverages.[1]
- Choose a variety of protein foods, which include seafood, lean meat and poultry, eggs, beans and peas, soy products and unsalted nuts and seeds.
- Increase the amount and variety of seafood consumed by choosing seafood in place of some meat and poultry.
- Replace protein foods that are higher in solid fats with choices that are lower in solid fats and calories and/or are sources of oils.
- Use oils to replace solid fats where possible
- Choose foods that provide more potassium, dietary fiber, calcium, and vitamin D, which are nutrients of concern in American diets. These foods include vegetables, fruits, whole grains, and milk and milk products.

Recommendations for Specific Population Groups

Women capable of becoming pregnant[2]

- Choose foods that supply heme iron, which is more readily absorbed by the body, additional iron sources, and enhancers of iron absorption such as vitamin C-rich foods.
- Consume 400 micrograms (mcg) per day of synthetic folic acid (from fortified foods and/or supplements) in addition to food forms of folate from a varied diet.[3]

Women who are pregnant or breastfeeding[2]

- Consume 8 to 12 ounces of seafood per week form a variety of seafood types.
- Due to their high methyl mercury content, limit white (albacore) tuna to 6 ounces per week and do not eat the following four types of fish: tilefish, shark, swordfish, and king mackerel.
- If pregnant, take an iron supplement, as recommended by an obstetrician or other health care provider.

Individuals ages 50 years and older

- Consume foods fortified with vitamin B_{12}, such as fortified cereals, or dietary supplements.

Building Healthy Eating Patterns

- Select an eating pattern that meets nutrient needs over time at an appropriate calorie level.
- Account for all foods and beverages consumed and assess how they fit within a total healthy eating pattern.
- Follow food safety recommendations when preparing and eating foods to reduce the risk of foodborne illnesses.

[1]Fortified soy beverages have been marketed as "soymilk", a product name consumers could see in supermarkets and consumer materials. However, FDA regulations do not contain provisions for the use of the term soymilk. Therefore, in this document, the term "fortified soy beverage" includes products that may be marketed as soymilk.

[2]Includes adolescent girls.

[3]"Folic acid" is the synthetic form of the nutrient, whereas "folate" is the form found naturally in foods.

USDA Food Patterns

For each food group or subgroup,[a] recommended average daily intake amounts[b] at all calorie levels. Recommended intakes from vegetable and protein foods subgroups are per week. For more information and tools for application go to MyPlate.gov.

Calorie level of pattern	1,000	1,200	1,400	1,600	1,800	2,000	2,200	2,400	2,600	2,800	3,000	3,200
	1 c	1 c	1½c	1½ c	1½ c	2 c	2 c	2 c	2 c	2 ½ c	2 ½ c	2 ½ c
	1 c	1½ c	1½ c	2 c	2½ c	2 ½ c	3 c	3 c	3 ½ c	3 ½ c	4 c	4 c
Dark-green vegetables	½ c/wk	1 c/wk	1 c/wk	1½ c/wk	1½ c/wk	1½ c/wk	2 c/wk	2 c/wk	2½ c/wk	2½ c/wk	2½ c/wk	2½ c/wk
Red and orange vegetables	2½ c/wk	3 c/wk	3 c/wk	4 c/wk	5½ c/wk	5½ c/wk	6 c/wk	6 c/wk	7 c/wk	7 c/wk	7½ c/wk	7½ c/wk
Beans and peas (legumes)	½ c/wk	½ c/wk	½ c/wk	1 c/wk	1½ c/wk	1½ c/wk	2 c/wk	2 c/wk	2 ½ c/wk	2 ½ c/wk	3 c/wk	3 c/wk
Starchy vegetables	2 c/wk	3 ½ c/wk	3 ½ c/wk	4 c/wk	5 c/wk	5 c/wk	6 c/wk	6 c/wk	7 c/wk	7 c/wk	8 c/wk	8 c/wk
Other vegetables	1½ c/wk	2½ c/wk	2½ c/wk	3½ c/wk	4 c/wk	4 c/wk	5 c/wk	5 c/wk	5½c/wk	5½ c/wk	7 c/wk	7 c/wk
	3 oz-eq	4 oz-eq	5 oz-eq	5 oz-eq	6 oz-eq	6 oz-eq	7 oz-eq	8 oz-eq	9 oz-eq	10 oz-eq	10 oz-eq	10 oz-eq
Whole grains	1 ½ oz-eq	2 oz-eq	2 ½ oz-eq	3 oz-eq	3 oz-eq	3 oz-eq	3 ½ oz-eq	4 oz-eq	4 ½ oz-eq	5 oz-eq	5 oz-eq	5 oz-eq
Enriched grains	1 ½ oz-eq	2 oz-eq	2 ½ oz-eq	2 oz-eq	3 oz-eq	3 oz-eq	3 ½ oz-eq	4 oz-eq	4 ½ oz-eq	5 oz-eq	5 oz-eq	5 oz-eq
	2 oz-eq	3 oz-eq	4 oz-eq	5 oz-eq	5 oz-eq	5 ½ oz-eq	6 oz-eq	6 ½ oz-eq	6 ½ oz-eq	7 oz-eq	7 oz-eq	7 oz-eq
Seafood	3 oz/wk	5 oz/wk	6 oz/wk	8 oz/wk	8 oz/wk	8 oz/wk	9 oz/wk	10 oz/wk	10 oz/wk	11 oz/wk	11 oz/wk	11 oz/wk
Meat, poultry, eggs	10 oz/wk	14 oz/wk	19 oz/wk	24 oz/wk	24 oz/wk	26 oz/wk	29 oz/wk	31 oz/wk	31 oz/wk	34 oz/wk	34 oz/wk	34 oz/wk
Nuts, seeds, soy products	1 oz/wk	2 oz/wk	3 oz/wk	4 oz/wk	4 oz/wk	4 oz/wk	4 oz/wk	5 oz/wk	5 oz/wk	5 oz/wk	5 oz/wk	5 oz/wk
	2 c	2 ½ c	2 ½ c	3 c	3 c	3 c	3 c	3 c	3 c	3 c	3 c	3 c
	15 g	17 g	17 g	22 g	24 g	27 g	29 g	31 g	34 g	36 g	44 g	51 g
c	137 (14%)	121 (10%)	121 (9%)	121 (8%)	161 (9%)	258 (13%)	266 (12%)	330 (14%)	362 (14%)	395 (14%)	459 (15%)	596 (19%)

[a]All foods are assumed to be in nutrient-dense forms, lean or low-fat and prepared without added fats, sugars, or salt. Solid fats and added sugars may be included up to the daily maximum limit identified in the table.

[b]Food group amounts are shown in cup- (c) or ounce-equivalents (oz-eq). Oils are shown in grams (g). Quantity equivalents for each food group are:

- Grains, 1 ounce-equivalent is 1 one-ounce slice bread; 1 ounce uncooked pasta or rice: 1/2 cup cooked rice, pasta, or cereal; 1 tortilla (6" diameter); 1 pancake (5" diameter); 1 ounce ready-to-eat cereal (about 1 cup cereal flakes).
- Vegetables and fruits, 1 cup equivalent is: 1 cup raw or cooked vegetable or fruit; 1/2 cup dried vegetable or fruit; 1 cup vegetable or fruit juice; 2 cups leafy salad greens.
- Protein foods, 1 ounce-equivalent is: 1 ounce lean meat, poultry, seafood; 1 egg; 1 Tbsp peanut butter; 1/2 ounce nuts or seeds. Also, 1/4 cup cooked beans or peas may also be counted as 1 ounce-equivalent.
- Dairy, 1 cup-equivalent is: 1 cup milk, fortified soy beverage, or yogurt; 1 1/2 ounce natural cheese (e.g., cheddar); 2 ounces of processed cheese.

[c]SoFAS are calories from solid fats and added sugars. The limit for SoFAS is the remaining amount of calories in each food pattern after selecting the specified amount in each food group in nutrient-dense forms (forms that are fat-free or low-fat and with no added sugars). The number of SoFAS is lower in the 1,200, 1,400, and 1,600 calorie patterns than in the 1,000 calorie pattern. The nutrient goals for the 1,200 to 1,600 calorie patterns are higher and require that more calories be used for nutrient-dense foods from the food groups.

American Institute for Cancer Research Recommendations for Cancer Prevention

Ten recommendations for cancer prevention

1. Be as lean as possible without becoming underweight.
2. Be physically active for at least 30 minutes every day.
3. Avoid sugary drinks. Limit consumption of energy-dense foods.
4. Eat more of a variety of vegetables, fruits, whole grains and legumes such as beans.
5. Limit consumption of red meats (such as beef, pork and lamb) and avoid processed meats.
6. If consumed at all, limit alcoholic drinks to 2 for men and 1 for women a day.
7. Limit consumption of salty foods and foods processed with salt (sodium).
8. Don't use supplements to protect against cancer.
9. It is best for mothers to breastfeed exclusively for up to 6 months and then add other liquids and foods.
10. After treatment, cancer survivors should follow the recommendations for cancer prevention.

And always remember—do not smoke or chew tobacco.

Source: American Institute of Cancer Research. Available online at http://www.aicr.org/reduce-your-cancer-risk. Reprinted with permission from the American Institute for Cancer Research.

Dietary Approaches to Stop Hypertension: DASH Diet Recommendations

Type of food	*Daily servings for 1600–3100 calorie diets*	*Daily servings for a 2000 calorie diet*
Grains and grain products (include at least 3 whole-grain foods each day)	6–12	7–8
Fruits	4–6	4–5
Vegetables	4–6	4–5
Low-fat or non-fat dairy foods	2–4	2–3
Lean meats, fish, poultry	1.5–2.5	2 or less
Nuts, seeds, and legumes	3–6 per week	4–5 per week
Fats and sweets	2–4	limited

Source: The DASH Diet Eating Plan. Available online at http://dashdiet.org/

Nutrition and Physical Fitness Recommendations from the American Cancer Society

Maintain a healthy weight throughout life.

- Balance calorie intake with physical activity.
- Avoid excessive weight gain throughout life.
- Achieve and maintain a healthy weight if currently overweight or obese.

Adopt a physically active lifestyle.

- **Adults:** Engage in at least 30 minutes of moderate to vigorous physical activity, above usual activities, on 5 or more days of the week; 45 to 60 minutes of intentional physical activity are preferable.
- **Children and adolescents:** Engage in at least 60 minutes per day of moderate to vigorous physical activity at least 5 days per week.

Eat a healthy diet, with an emphasis on plant sources.

- Choose foods and drinks in amounts that help achieve and maintain a healthy weight.
- Eat 5 or more servings of a variety of vegetables and fruits every day.
- Choose whole grains over processed (refined) grains.
- Limit intake of processed and red meats.

If you drink alcoholic beverages, limit your intake.

- Drink no more than 1 drink per day for women or 2 per day for men.

Source: Some adapted from Complete Guide—Nutrition and Physical Activity for Cancer Prevention. Available online at http://www.cancer.org/docroot/PED/content/PED_3_2X_Diet_and_Activity_Factors_That_Affect_Risks.asp?sitearea=PED. Reprinted by the permission of the American Cancer Society, Inc.

Healthy People 2020

Topics	Indicators	Objectives
Access to care	Proportion of the population with access to healthcare services	1. Increase the proportion of persons with health insurance (AHS 1). 2. Increase proportion of persons with a usual primary care provider (AHS 3). 3. (Developmental) Increase the proportion of persons who receive appropriate evidence-based clinical preventive services (AHS 7).
Healthy Behaviors	Proportion of the population engaged in healthy behaviors	4. Increase the proportion of adults who meet current federal physical activity guidelines for aerobic physical activity and for muscle- strengthing activity (PA 2). 5. Reduce the proportion of children and adolescents who are considered obese (NWS 10). 6. Reduce consumption of calories from solid fats and added sugars in the population aged 2 years and older (NWS 17). 7. Increase the proportion of adults who get sufficient sleep (SH 4).
Chronic Disease	Prevalence and mortality of chronic disease	8. Reduce coronary heart disease deaths (HDS 2). 9. Reduce the proportion of persons in the population with hypertension (HDS 5). 10. Reduce the overall cancer death rate (C 1).
Environmental Determinants	Proportion of the population experiencing a healthy physical environment	11. Reduce the number of days the Air Quality Index (AQI) exceeds 100 (EH 1).
Social Determinants	Proportion of the population experiencing a healthy social environment	12. (Developmental) Improve the health literacy of the population (HC/HIT 1). 13. (Developmental) Increase the proportion of children who are ready for school in all five domains of healthy development: physical development, social-emotional development, approaches to learning, language, and cognitive development (EMC 1). 14. Increase educational achievement of adolescents and young adults (AH 5).
Injury	Proportion of the population that experiences injury	15. Reduce fatal and nonfatal injuries (IVP 1).
Mental Health	Proportion of the population experiencing positive mental health	16. Reduce the proportion of persons who experience major depressive episodes (MDE) (MHMD 4).
Maternal and Infant Health	Proportion of healthy births	17. Reduce low birth weight (LBW) and very low birth weight (VLBW) (MICH 8).
Responsible Sexual Behavior	Proportion of the population engaged in responsible sexual behavior	18. Reduce pregnancy rates among adolescent females (FP 8). 19. Increase the proportion of sexually active persons who use condoms (HIV 17).
Substance Abuse	Proportion of the population engaged in substance abuse	20. Reduce past-month use of illicit substances (SA 13). 21. Reduce the proportion of persons engaging in binge drinking of alcoholic beverages (SA 14).
Tobacco	Proportion of the population using tobacco	22. Reduce tobacco use by adults (TU 1). 23. Reduce the initiation of tobacco use among children, adolescents, and young adults (TU 3).
Quality of Care	Proportion of the population receiving quality health care services	24. Reduce central line-associated bloodstream infections (CLABSI) (HA 1).

Appendix E

Food and Supplement Labeling Information

Sample Food Label for a Granola Bar

Nutrition Facts
Serving Size 1 bar (24g)
Servings Per Container 12

Amount Per Serving

Calories 120	Calories from Fat 45
	% **Daily Value***
Total Fat 5g	**8%**
Saturated Fat 1g	**5%**
Trans Fat	**0%**
Cholesterol 0mg	**0%**
Sodium 65mg	**3%**
Total Carbohydrate 17g	**6%**
Dietary Fiber 1g	**4%**
Sugars 6g	
Protein 2g	

Vitamin A 0% • Vitamin C 0%
Calcium 0% • Iron 4%

*Percent Daily Values are based on a 2,000 calorie diet. Your daily values may be higher or lower depending on your calorie needs:

	Calories:	2,000	2,500
Total Fat	Less than	65g	80g
Sat Fat	Less than	20g	25g
Cholesterol	Less than	300mg	300mg
Sodium	Less than	2,400mg	2,400mg
Total Carbohydrate		300g	375g
Dietary Fiber		25g	30g

Caloies per gram
Fat 9 • Carbohydrate 4 • Protein 4

Ingredients: Rolled oats, sugar, sunflower oil, brown sugar syrup, honey, salt, soy lecithin

Daily Reference Values Used to Establish Daily Values

Food Component	*Daily Reference Value (2000 Kcal)*
Total fat	Less than 65 g (30% of energy)
Saturated fat	Less than 20 g (10% of energy)
Cholesterol	Less than 300 mg
Total carbohydrate	300 g (60% of energy)
Dietary fiber	25 g (11.5 g/1000 Kcal)
Sodium	Less than 2400 mg
Potassium	3500 mg
Protein	50 g (10% of energy)

Source: USDA A Food Labeling Guide Appendix F. Available online at http://www.cfsan.fda.gov/~dms/2lg-xf.html

Sample Supplement Label

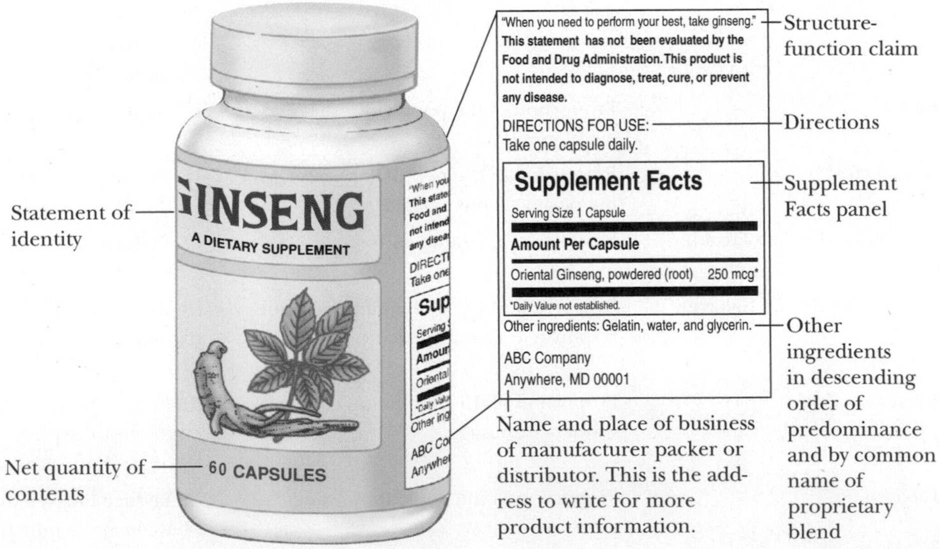

GINSENG
A DIETARY SUPPLEMENT
60 CAPSULES

"When you need to perform your best, take ginseng."
This statement has not been evaluated by the Food and Drug Administration. This product is not intended to diagnose, treat, cure, or prevent any disease.
DIRECTIONS FOR USE:
Take one capsule daily.

Supplement Facts
Serving Size 1 Capsule
Amount Per Capsule
Oriental Ginseng, powdered (root) 250 mcg*
*Daily Value not established.

Other ingredients: Gelatin, water, and glycerin.
ABC Company
Anywhere, MD 00001

Structure-function claim
Directions
Supplement Facts panel
Other ingredients in descending order of predominance and by common name of proprietary blend
Statement of identity
Net quantity of contents
Name and place of business of manufacturer packer or distributor. This is the address to write for more product information.

Recommended Dietary Intakes (RDIs)* Used to Establish Daily Values

Vitamins and Minerals	*Units of Measurement*	*Adults and Children 4 or more Years of Age*	*Infants*	*Children Under 4 Years of Age*	*Pregnant or Lactating Women*
Vitamin A	International Units†	5000 (1000 μg)	1500	2500	8000
Vitamin D	International Units†	400 (10 μg)	400	400	400
Vitamin E	International Units†	30 (10 μg)	5	10	30
Vitamin C	Milligrams	60	35	40	60
Folic acid	Micrograms	400	0.1	0.2	0.8
Thiamin	Milligrams	1.5	0.5	0.7	1.7
Riboflavin	Milligrams	1.7	0.6	0.8	2.0
Niacin	Milligrams	20	8	9	20
Vitamin B_6	Milligrams	2.0	0.4	0.7	2.5
Vitamin B_{12}	Micrograms	6.0	2	3	8
Biotin	Micrograms	300	0.05	0.15	0.30
Pantothenic acid	Milligrams	10	3	5	10
Calcium	Milligrams	1000	0.6	0.8	1.3
Phosphorous	Milligrams	1000	0.5	0.8	1.3
Iodine	Micrograms	150	45	70	150
Iron	Milligrams	18	15	10	18
Magnesium	Milligrams	400	70	200	450
Copper	Milligrams	2.0	0.6	1.0	2.0
Zinc	Milligrams	15	5	8	15
Vitamin K	Micrograms	80	—‡	—‡	—‡
Chromium	Micrograms	120	—	—	—
Selenium	Micrograms	70	—	—	—
Molybdenum	Micrograms	75	—	—	—
Manganese	Milligrams	2	—	—	—
Chloride	Milligrams	3400	—	—	—

*Based on National Academy of Sciences' 1968 Recommended Dietary Allowances.

†The RDIs for fat-soluble vitamins are expressed in International Units (IU). Values that are approximately equivalent in micrograms are given in parentheses.

‡No values yet established for vitamin K, chromium, selenium, molybdenum, manganese, or chloride for this population.

Source: USDA Food Labeling Guide. Available online at http://www.cfsan.fda.gov/~dms/2lg-xf.htm

Health Claims Permitted on Food or Supplement Labels

Health Claims That Meet Significant Scientific Agreement

These are authorized by the FDA based on an extensive review of the scientific literature or information from a scientific body of the U.S. government or the National Academy of Sciences that supports the nutrient/disease relationship.

- Calcium and osteoporosis
- Dietary fat and cancer
- Dietary saturated fat and cholesterol and risk of coronary heart disease
- Dietary non-carcinogenic carbohydrate sweeteners and dental caries
- Fiber-containing grain products, fruits, and vegetables and cancer
- Folic acid and neural tube defects
- Fluoridated water and reduced risk of dental caries
- Fruits, vegetables, and grain products and cancer
- Fruits, vegetables, and grain products that contain fiber, particularly soluble fiber, and risk of coronary heart disease
- Plant sterol/stanol and coronary heart disease
- Potassium and the risk of high blood pressure and stroke
- Saturated fat, cholesterol, and *trans* fat and reduced risk of heart disease
- Sodium and hypertension
- Soluble fiber from certain foods and risk of coronary heart disease
- Soy protein and risk of coronary heart disease
- Stanols/sterols and risk of coronary heart disease
- Whole grain foods and the risk of heart disease and certain cancers

Qualified Health Claims

These are used when there is emerging evidence for a relationship between a food, food component, or dietary supplement and reduced risk of a disease or health-related condition but there is not enough scientific support for the FDA to issue an authorizing regulation.

- Qualified claims about cancer risk: selenium and cancer, antioxidant vitamins and cancer, green tea and cancer, tomatoes and cancer, calcium and colorectal cancer and/recurrent colon rectal polyps.
- Qualified claims about cardiovascular disease risk: nuts and heart disease, walnuts and heart disease, omega-3 fatty acids and coronary heart disease, B vitamins and vascular disease, monounsaturated fatty acids from olive oil and reduced coronary heart disease risk, unsaturated fatty acids from canola oil and coronary heart disease risk, corn oil and coronary heart disease.
- Qualified claims about diabetes: chromium picolinate and diabetes
- Qualified claims about hypertension: calcium and hypertension, pregnancy-induced hypertension and preeclampsia
- Qualified claims about cognitive function: phosphatidylserine and cognitive dysfunction and dementia
- Qualified claims about neural tube birth defects: 0.8 mg folic acid and neural tube birth defects

Source: U.S. Food and Drug Administration. Center for Food Safety and Applied Nutrition. Label Claims. Available online at http://www.cfsan.fda.gov/~dms/2lg-xc.html

Nutrient Content Descriptors Commonly Used on Food Labels

Free	Means that a product contains no amount of, or a trivial amount of, fat, saturated fat, cholesterol, sodium, sugars, or calories. For example,"sugar free" and "fat free" both mean less than 0.5 g per serving. Synonyms for "free" include "without," "no," and "zero."
Low	Used for foods that can be eaten frequently without exceeding the Daily Value for fat, saturated fat, cholesterol, sodium, or calories. Specific definitions have been established for each of these nutrients. For example, "low fat" means that the food contains 3 g or less per serving, and "low cholesterol" means that the food contains less than 20 mg of cholesterol per serving. Synonyms for "low" include "little," "few," and "low source of."
Lean and extra lean	Used to describe the fat content of meat, poultry, seafood, and game meats. "Lean" means that the food contains less than 10 g fat, less than 4.5 g saturated fat, and less than 95 mg of cholesterol per serving and per 100 g. "Extra lean" means that the food contains less than 5 g fat, less than 2 g saturated fat, and less than 95 mg of cholesterol per serving and per 100 g.
High	Can be used if a food contains 20% or more of the Daily Value for a particular nutrient. Synonyms for "high" include "rich in" and "excellent source of."
Good source	Means that a food contains 10 to 19% of the Daily Value for a particular nutrient per serving.
Reduced	Means that a nutritionally altered product contains 25% less of a nutrient or of energy than the regular or reference product.
Less	Means that a food, whether altered or not, contains 25% less of a nutrient or of energy than the reference food. For example, pretzels may claim to have "less fat" than potato chips. "Fewer" may be used as a synonym for "less."
Light	May be used in different ways. First, it can be used on a nutritionally altered product that contains one-third fewer kcalories or half the fat of a reference food. Second, it can be used when the sodium content of a low-calorie, low-fat food has been reduced by 50%. The term "light" can be used to describe properties such as texture and color as long as the label explains the intent—for example, "light and fluffy."
More	Means that a serving of food, whether altered or not, contains a nutrient that is at least 10% of the Daily Value more than the reference food. This definition also applies to foods using the terms "fortified," "enriched," or "added."
Healthy	May be used to describe foods that are low in fat and saturated fat and contain no more than 360 mg of sodium and no more than 60 mg of cholesterol per serving and provide at least 10% of the Daily Value for vitamins A or C, or iron, calcium, protein, or fiber.
Fresh	May be used on foods that are raw and have never been frozen or heated and contain no preservatives.

Source: USDA. A Food Labeling Guide Appendix A. Available online at http://www.cfsan.fda.gov/~dms/2lg-xa.html

Appendix F

Energy Expenditure for Various Activities

Energy Expenditure for Various Activities

	Calories per Hour (by body weight)				
Type of Activity	**100 lb**	**120 lb**	**150 lb**	**180 lb**	**200 lb**
Aerobics (heavy)	363	435	544	653	726
Aerobics (medium)	227	272	340	408	454
Aerobics (light)	136	163	204	245	272
Archery	159	190	238	286	317
Backpacking	408	490	612	735	816
Badminton (doubles)	181	218	272	327	363
Badminton (singles)	231	278	347	416	463
Basketball (nonvigorous)	431	517	646	776	862
Basketball (vigorous)	499	599	748	898	998
Bicycling (6 mph)	159	190	238	286	317
Bicycling (10 mph)	249	299	374	449	499
Bicycling (11 mph)	295	354	442	531	590
Bicycling (12 mph)	340	408	510	612	680
Bicycling (13 mph)	385	463	578	694	771
Billiards	91	109	136	163	181
Bowling	177	212	265	318	354
Boxing—competition	603	724	905	1086	1206
Boxing—sparring	376	452	565	678	753
Calisthenics (heavy)	363	435	544	653	726
Calisthenics (light)	181	218	272	327	363
Canoeing (2.5 mph)	150	180	224	269	299
Canoeing (5 mph)	340	408	510	612	680
Carpentry	227	272	340	408	454
Climbing (mountain)	454	544	680	816	907
Disco dancing	272	327	408	490	544
Ditch digging (hand)	263	316	395	473	526
Fencing	340	408	510	612	680
Fishing (bank/boat)	159	190	238	286	317
Fishing (in waders)	249	299	374	449	499
Football (touch)	340	408	510	612	680
Gardening	145	174	218	261	290
Golf (carry clubs)	227	272	340	408	454
Golf (pull cart)	163	196	245	294	327
Golf (ride in cart)	113	136	170	204	227
Handball (vigorous)	454	544	680	816	907
Hiking (X-country)	249	299	374	449	499
Hiking (mountain)	340	408	510	612	680
Horseback trotting	231	278	347	416	463

(continued)

Energy Expenditure for Various Activities *(continued)*

Type of Activity	*Calories per Hour (by body weight)*				
	100 lb	**120 lb**	**150 lb**	**180 lb**	**200 lb**
Housework	181	218	272	327	363
Hunting (carry load)	272	327	408	490	544
Ice hockey (vigorous)	454	544	680	816	907
Ice skating (10 mph)	263	316	395	473	526
Jazzercise (heavy)	363	435	544	653	726
Jazzercise (medium)	227	272	340	408	454
Jazzercise (light)	136	163	204	245	272
Jog (9 min/mile)	499	599	748	898	998
Jog (10 min/mile)	454	544	680	816	907
Jog (12 min/mile)	385	463	578	694	771
Jog (13 min/mile)	317	381	476	571	635
Jog (14 min/mile)	272	327	408	490	544
Jog (15 min/mile)	227	272	340	408	454
Jog (17 min/mile)	181	218	272	327	363
Lawn mowing (hand)	295	354	442	531	590
Lawn mowing (power)	163	196	245	294	327
Musical instrument playing	113	136	170	204	227
Racquetball (social)	385	463	578	694	771
Racquetball (vigorous)	454	544	680	816	907
Roller skating	231	278	347	416	463
Rowboating (2.5 mph)	200	239	299	359	399
Rowing (11 mph)	590	707	884	1061	1179
Run (5 min/mile)	816	980	1224	1469	1633
Run (6 min/mile)	703	844	1054	1265	1406
Run (7 min/mile)	612	735	918	1102	1224
Run (8 min/mile)	544	653	816	980	1088
Sailing	159	190	238	286	317
Shuffleboard/skeet	136	163	204	245	272
Skiing (X-country)	454	544	680	816	907
Skiing (downhill)	363	435	544	653	726
Square dancing	272	327	407	490	544
Swimming (competitive)	680	816	1020	1224	1361
Swimming (fast)	426	512	639	767	853
Swimming (slow)	349	419	524	629	698
Table tennis	236	283	354	424	472
Tennis (doubles)	227	272	340	408	454
Tennis (singles)	295	354	442	531	590
Tennis (vigorous)	385	463	578	694	771
Volleyball	231	278	347	416	463
Walking (20 min/mile)	159	190	238	286	317
Walking (26 min/mile)	136	163	204	245	272
Water skiing	317	381	476	571	635
Weight lifting (heavy)	408	490	612	735	816
Weight lifting (light)	181	218	272	327	363
Wood chopping (sawing)	295	354	442	531	590

Source: Data reprinted with permission from N-Squared Computing. First Databank Division of the Hearst Corporation.

Appendix G

Chemistry and Structures

A Review of Basic Chemistry

Chemistry is the science of the structure and interactions of matter. All living and nonliving things consist of matter, which is anything that occupies space and has mass. Mass is the amount of matter in any object, which does not change.

Chemical Elements

All forms of matter—both living and nonliving—are made up of a limited number of building blocks called chemical elements. Each element is a substance that cannot be split into a simpler substance by ordinary chemical means. Scientists now recognize 112 elements. Of these, 92 occur naturally on Earth. The rest have been produced from the natural elements using particle accelerators or nuclear reactors. Each element is designated by a chemical symbol, one or two letters of the element's name in English, Latin, or another language. Examples of chemical symbols are H for hydrogen, C for carbon, O for oxygen, N for nitrogen, Ca for calcium, and Na for sodium.

Twenty-six different elements normally are present in the human body. Just four elements, called the major elements, constitute about 96% of the body's mass: oxygen (O), carbon (C), hydrogen (H), and nitrogen (N). Eight others, the lesser elements, contribute 3.8% of the body's mass: calcium, phosphorus (P), potassium (K), sulfur (S), sodium, chlorine (Cl), magnesium (Mg), and iron (Fe). An additional 14 elements—the trace elements—are present in tiny amounts. Together, they account for the remaining 0.2% of the body's mass.

Structure of Atoms

Each element is made up of atoms, the smallest units of matter that retain the properties and characteristics of the element. Atoms are extremely small. Hydrogen atoms, the smallest atoms, have a diameter less than 0.1 nanometer, and the largest atoms are only five times larger.

Dozens of different subatomic particles compose individual atoms. However, only three types of subatomic particles are important for understanding the chemical reactions in the human body: protons, neutrons, and electrons (**Figure G.1**). The dense central core of an atom is its nucleus. Within the nucleus are positively charged protons and uncharged (neutral) neutrons. The tiny, negatively charged electrons move about in a large space surrounding the nucleus. They do not follow a fixed path or orbit but instead form a negatively charged "cloud" that envelops the nucleus.

Even though their exact positions cannot be predicted, specific groups of electrons are most likely to move about within certain regions around the nucleus. These regions, called **electron shells**, are depicted as simple circles around the nucleus. Because each electron shell can hold a specific number of electrons, the electron shell model best conveys this aspect of atomic structure (see Figure G.1). The first electron shell (nearest the nucleus) never holds more than 2 electrons. The second shell holds a maximum of 8 electrons, and the third can hold up to 18 electrons. The electron shells fill with electrons in a specific order, beginning with the first shell. For example, notice in **Figure G.2** that sodium (Na), which has 11 electrons total, contains 2 electrons in the first shell, 8 electrons in the second shell, and 1 electron in the third shell. The number of electrons in an atom of an element always equals the number of protons. Because each electron and proton carries one charge, the negatively charged electrons and the positively charged protons balance each other. Thus, each atom is electrically neural; its total charge is zero.

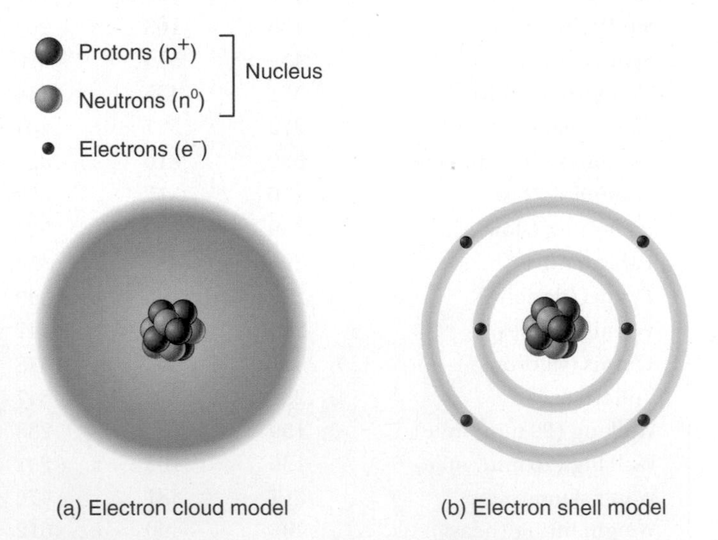

Figure G.1 Two representations of the structure of an atom. Electrons move about the nucleus, which contains neutrons and protons.
(a) In the electron cloud model of an atom, the shading represents the chance of finding an electron in regions outside the nucleus.
(b) In the electron shell model, filled circles represent individual electrons, which are grouped into concentric circles according to the shells they occupy. Both models depict a carbon atom, with six protons, six neutrons, and six electrons.

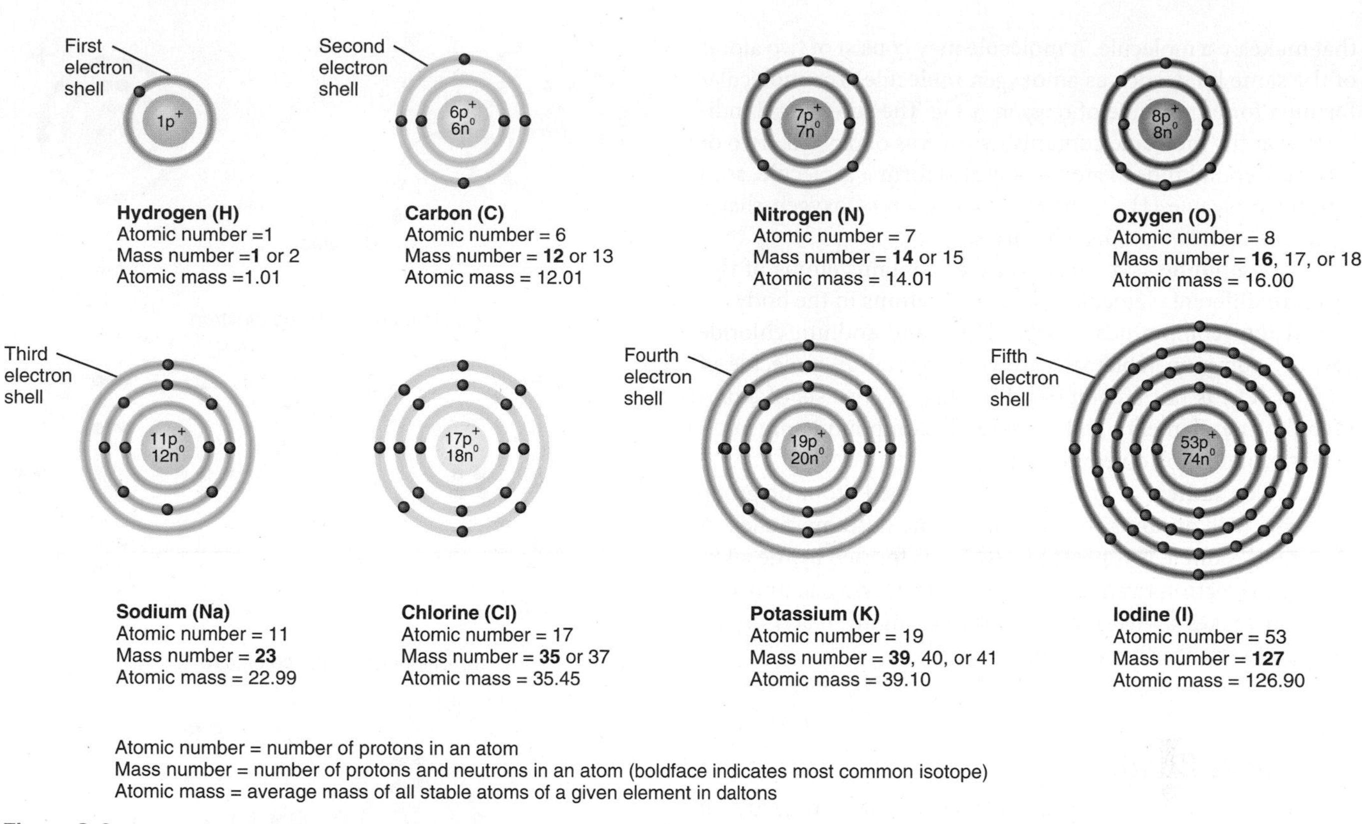

Figure G.2 Atomic structures of several stable atoms.
The atoms of different elements have different atomic numbers because they have different numbers of protons.

The *number of protons* in the nucleus of an atom is an atom's **atomic number**. For example, hydrogen has an atomic number of 1 because its nucleus has one proton, whereas sodium has an atomic number of 11 because its nucleus has 11 protons (see Figure G.2).

The **mass number** of an atom is the sum of its protons and neutrons. Because sodium has 11 protons and 12 neutrons, its mass number is 23 (see Figure G.2). Although all atoms of one element have the same number of protons, they may have different numbers of neutrons and thus different mass numbers. **Isotopes** are atoms of an element that have different numbers of neutrons and therefore different mass numbers. In a sample of oxygen, for example, most atoms have 8 neutrons, and a few have 9 or 10, but all have 8 protons and 8 electrons. Most isotopes are stable, which means that their nuclear structure does not change over time. The stable isotopes of oxygen are designated ^{16}O, ^{17}O, and ^{18}O (or O-16, O-17, and O-18). As you may already have determined, the numbers indicate the mass number of each isotope. As you will discover shortly, the number of electrons of an atom determines its chemical properties. Although the isotopes of an element have different numbers of neutrons, they have identical chemical properties because they have the same number of electrons.

Certain isotopes, called **radioactive isotopes**, are unstable; their nuclei decay into a stable configuration. As they decay, these atoms emit radiation and in the process often transform into a different element. For example, the radioactive isotope of carbon, C-14, decays to N-14. The decay of a radioisotope may be as fast as a fraction of a second or as slow as millions of years. The **half-life** of an isotope is the time required for half of the radioactive atoms in a sample of that isotope to decay into a more stable form. The half-life of C-14, which is used to determine the age of organic samples, is 5600 years, whereas the half-life of I-131, an important clinical tool, is 8 days.

Ions, Molecules, and Compounds

The atoms of each element have a characteristic way of losing, gaining, or sharing their electrons when interacting with other atoms to achieve stability. The way that electrons behave enables atoms in the body to exist in electrically charged forms called ions, or to join with each other into the complex combinations called molecules. If an atom either *gives up* or *gains* electrons, it becomes an ion. An **ion** is an atom that has a positive or negative charge because it has unequal numbers of protons and electrons. An ion of an atom is symbolized by writing its chemical symbol followed by the number of its positive or negative charges. Thus, Ca^{2+} stands for a calcium ion that has two positive charges because it has lost two electrons.

When two or more atoms *share* electrons, the resulting combination is called a **molecule**. A *molecular formula* indicates the elements and the number of atoms of each element that make up a molecule. A molecule may consist of two atoms of the same kind, such as an oxygen molecule. The molecular formula for a molecule of oxygen is O_2. The subscript 2 indicates that the molecule contains two atoms of oxygen. Two or more different kinds of atoms may also form a molecule, as in a water molecule (H_2O). In H_2O one atom of oxygen shares electrons with two atoms of hydrogen.

A **compound** is a substance that contains atoms of two or more different elements. Most of the atoms in the body are joined into compounds. Water (H_2O) and sodium chloride (NaCl), common table salt, are compounds. A molecule of oxygen (O_2) is not a compound because it consists of atoms of only one element. Thus, while all compounds are molecules, not all molecules are compounds.

A **free radical** is an electrically charged atom or group of atoms with an unpaired electron in the outermost shell. A common example is superoxide, which is formed by the addition of an electron to an oxygen molecule. Having an unpaired electron makes a free radical unstable, highly reactive, and destructive to nearby molecules.

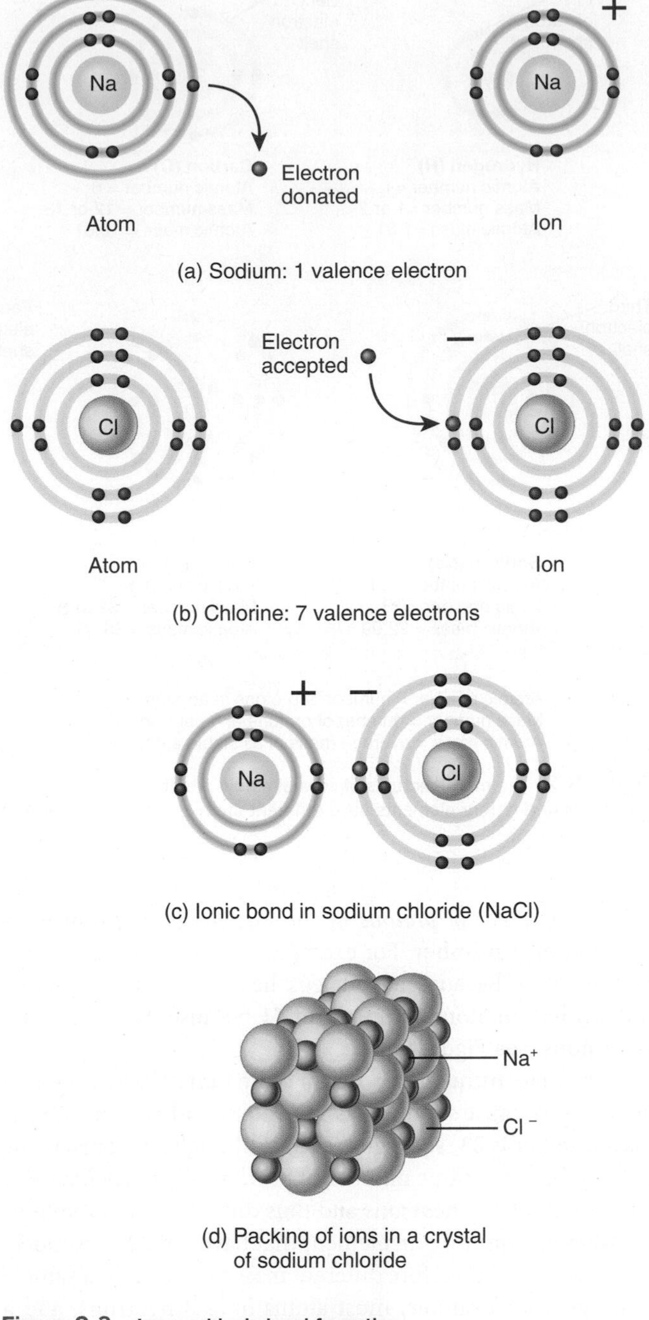

Figure G.3 **Ions and ionic bond formation.**
(a) A sodium atom can have a complete octet of electrons in its outermost shell by losing one electron.
(b) A chlorine atom can have a complete octet by gaining one electron.
(c) An ionic bond may form between oppositely charged ions.

Chemical Bonds

The forces that hold together the atoms of a molecule or a compound are **chemical bonds**. The likelihood that an atom will form a chemical bond with another atom depends on the number of electrons in its outermost shell, also called the **valence shell**. An atom with a valence shell holding eight electrons is *chemically stable*, which means it is unlikely to form chemical bonds with other atoms. Two or more atoms that do not have 8 electrons in their valence shells can interact in ways that produce a chemically stable arrangement of eight valence electrons for each atom. For this to happen, an atom either empties its partially filled valence shell, fills it with donated electrons, or shares electrons with other atoms. The way that valence electrons are distributed determines what kind of chemical bond results. An **ionic bond** is formed when positively and negatively charged ions are attracted to one another. As shown in **Figure G.3**, sodium has one valence electron and chlorine has seven valence electrons. When an atom of sodium donates its sole valence electron to an atom of chlorine, the resulting positive and negative charges pull both ions tightly together, forming an ionic bond. The resulting compound is sodium chloride, written NaCl.

A **covalent bond** forms when two or more atoms *share* electrons rather than gaining or losing them. Atoms form a covalently bonded molecule by sharing one, two, or three pairs of valence electrons (**Figure G.4**). They are the most common chemical bonds in the body, and the compounds that result from them form most of the body's structures. In a **polar covalent bond**, the sharing of electrons between two atoms is unequal: the nucleus of one atom attracts the shared electrons more strongly than the nucleus of the other atom. When polar covalent bonds form, the resulting molecule has a partial negative charge near the atom that attracts electrons more strongly (**Figure G.5**).

Chemical Reactions

A **chemical reaction** occurs when new bonds form or old bonds break between atoms. Chemical reactions are the foundation of all life processes. Each chemical reaction involves energy changes. **Chemical energy** is a form of energy that is stored in the bonds of compounds and molecules. The total amount of energy present at the beginning and end of a chemical reaction is the same. Although energy can be neither created nor destroyed, it may be converted from one form to another.

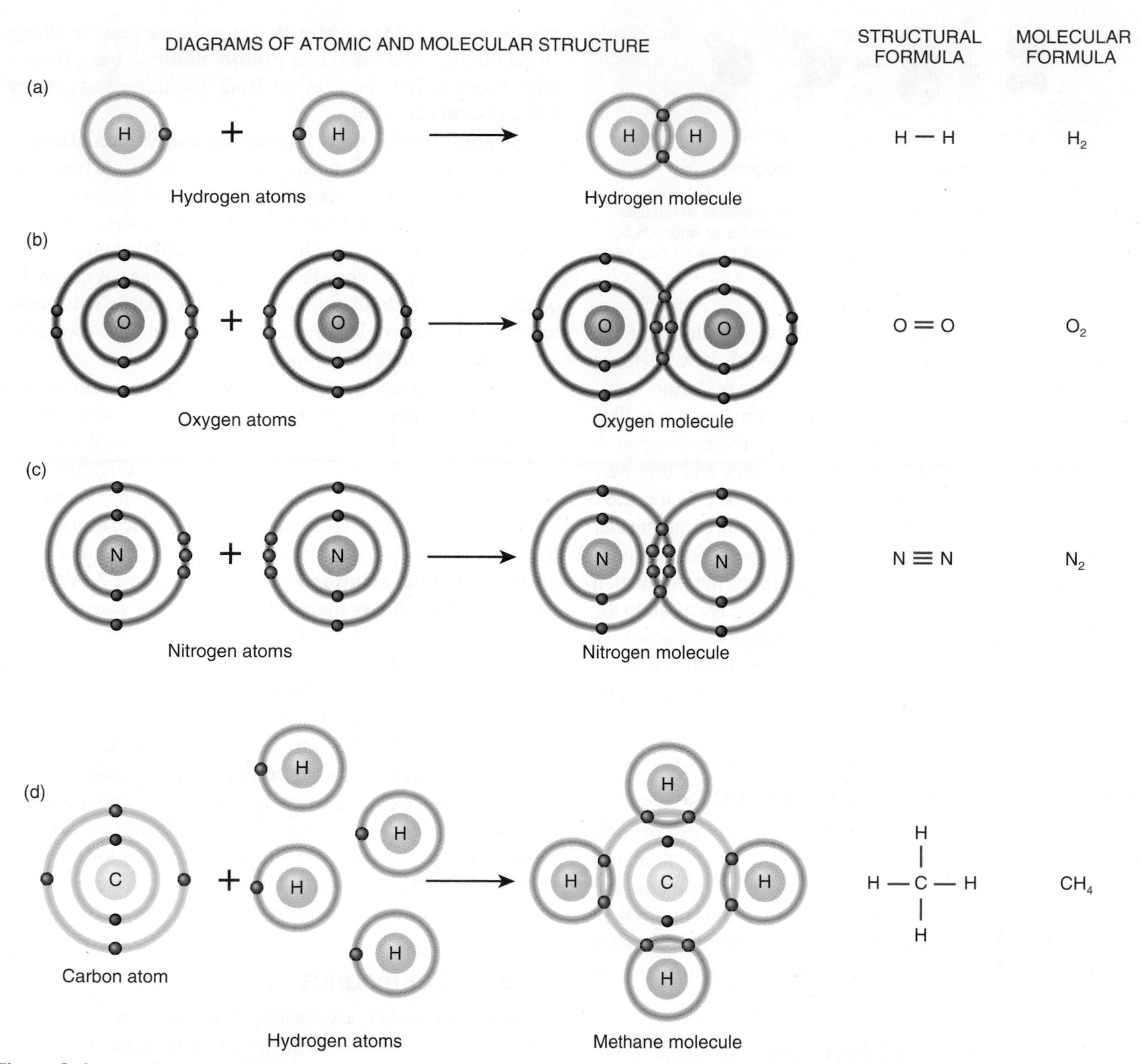

Figure G.4 Covalent bond formation.
The red electrons are shared equally. In writing the structural formula of a covalently bonded molecule, each straight line between the chemical symbols for two atoms denotes a pair of shared electrons. In molecular formulas, the number of atoms in each molecule is noted by subscripts.

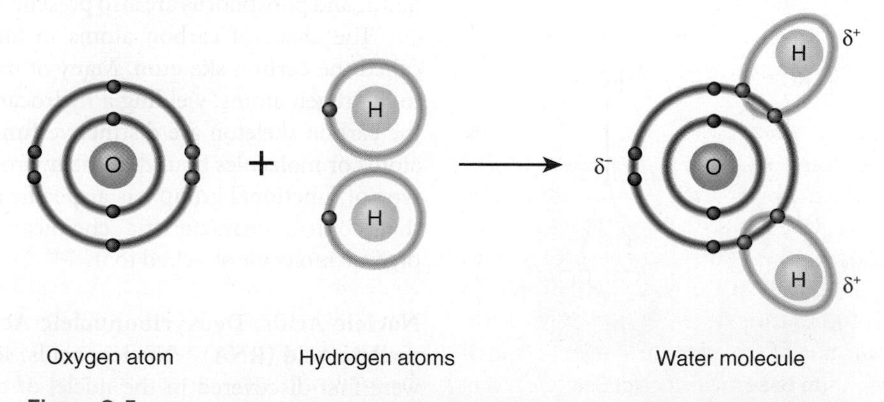

Figure G.5 Polar covalent bonds between oxygen and hydrogen atoms in a water molecule.
The red electrons are shared unequally. Because the oxygen nucleus attracts the shared electrons more strongly, the oxygen end of a water molecule has a partial negative charge, written δ^-, and the hydrogen ends have partial positive charges, written δ^+.

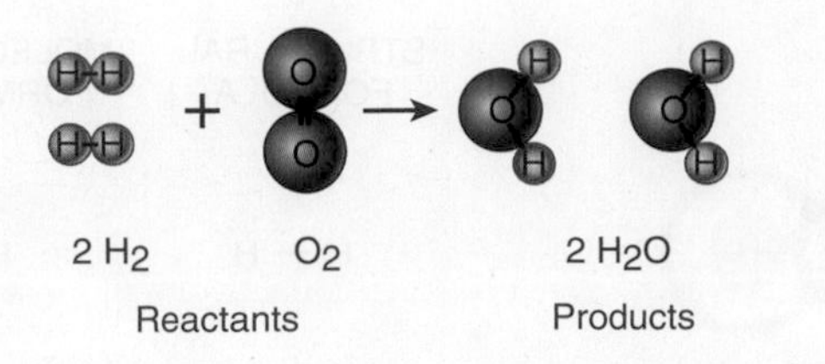

Figure G.6 **The chemical reaction between two hydrogen molecules (H_2) and one oxygen molecule (O_2) to form two molecules of water (H_2O).** Note that the reaction occurs by breaking old bonds and making new ones.

After a chemical reaction takes place, the atoms of the reactants are rearranged to yield products with new chemical properties. When two or more atoms, ions, or molecules combine to form new and larger molecules, the processes are called **synthesis reactions**. One example of a synthesis reaction is the reaction between two hydrogen molecules and one oxygen molecule to form two molecules of water (see **Figure G.6**). **Decomposition reactions** split up large molecules into smaller atoms, ions, or molecules. For instance, the series of reactions that break down glucose to pyruvic acid, with the net production of two molecules of ATP, are important catabolic reactions in the body. Many reactions in the body are **exchange reactions**; they consist of both synthesis and decomposition reactions. Some chemical reactions may be reversible. In a **reversible reaction**, the products can revert to the original reactants.

Inorganic Compounds and Solutions

Most of the chemicals in your body exist in the form of compounds. **Inorganic compounds** contain no more than one carbon atom. They include water and many salts, acids, and bases. Inorganic compounds may have either ionic or covalent bonds. **Organic compounds** always contain carbon and always have covalent bonds. Most are large molecules and many are made up of long chains of carbon atoms.

Inorganic Acids, Bases, and Salts When inorganic acids, bases, or salts dissolve in water, they **dissociate**: they separate into ions and become surrounded by water molecules. An **acid** (**Figure G.7a**) is a substance that dissociates into one or more **hydrogen ions** (**H**) and one or more negatively charged

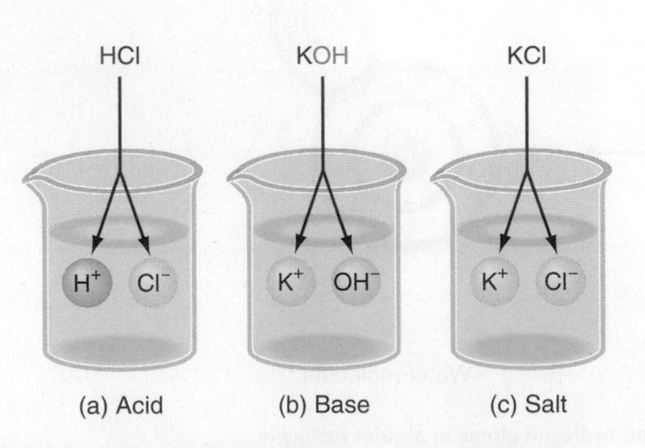

Figure G.7 **Dissociation of inorganic acids, bases, and salts.** Dissociation is the separation of inorganic acids, bases, and salts into ions in a solution.

anions. Because H is a single proton with one positive charge, an acid is also referred to as a **proton donor**. A **base**, by contrast (**Figure G.7b**), removes H+ from a solution and is therefore a **proton acceptor**.

To ensure homeostasis, intracellular and extracellular fluids must contain almost balanced quantities of acids and bases. The more hydrogen ions (H^+) dissolved in a solution, the more acidic the solution; the more hydroxide ions (OH^-), the more basic (alkaline) the solution. The chemical reactions that take place in the body are very sensitive to even small changes in the acidity or alkalinity of the body fluids in which they occur. Any departure from the narrow limits of normal H^+ and OH^- concentrations greatly disrupts body functions.

A solution's acidity or alkalinity is expressed on the **pH scale**, which extends from 0 to 14. This scale is based on the concentration of H^+ in moles per liter. The midpoint of the pH scale is 7, where the concentrations of H^+ and OH^- are equal. A substance with a pH of 7, such as pure water, is neutral. A solution that has more H^+ than OH^- is an **acidic solution** and has a pH below 7. A solution that has more OH^- than H^+ is a **basic** (**alkaline**) **solution** and has a pH above 7. Although the pH of body fluids may differ, as we have discussed, the normal limits for each fluid are quite narrow. Homeostatic mechanisms maintain the pH of blood between 7.35 and 7.45, which is slightly more basic than pure water. If the pH of blood falls below 7.35, a condition called acidosis occurs, and if the pH rises above 7.45, it results in a condition called alkalosis; both conditions can seriously compromise homeostasis. Even though strong acids and bases are continually taken into and formed by the body, the pH of fluids inside and outside cells remains almost constant. One important reason for this is the presence of **buffer systems**, which function to convert strong acids or bases into weak acids or bases.

Organic Compounds

Organic compounds are usually held together by covalent bonds. Carbon has four electrons in its outermost (valence) shell. It can bond covalently with a variety of atoms, including other carbon atoms, to form rings and straight or branched chains. Other elements that most often bond with carbon in organic compounds are hydrogen, oxygen, and nitrogen. Sulfur and phosphorus are also present in organic compounds.

The chain of carbon atoms in an organic molecule is called the **carbon skeleton**. Many of the carbons are bonded to hydrogen atoms, yielding a hydrocarbon. Also attached to the carbon skeleton are distinctive **functional groups**, other atoms or molecules bound to the hydrocarbon skeleton. Each type of functional group has a specific arrangement of atoms that confers characteristic chemical properties upon the organic molecule attached to it.

Nucleic Acids: Deoxyribonucleic Acid (DNA) and Ribonucleic Acid (RNA) **Nucleic acids**, so named because they were first discovered in the nuclei of cells, are huge organic molecules that contain carbon, hydrogen, oxygen, nitrogen, and phosphorus. Nucleic acids are of two varieties. The first, **deoxyribonucleic acid** (**DNA**), forms the inherited genetic

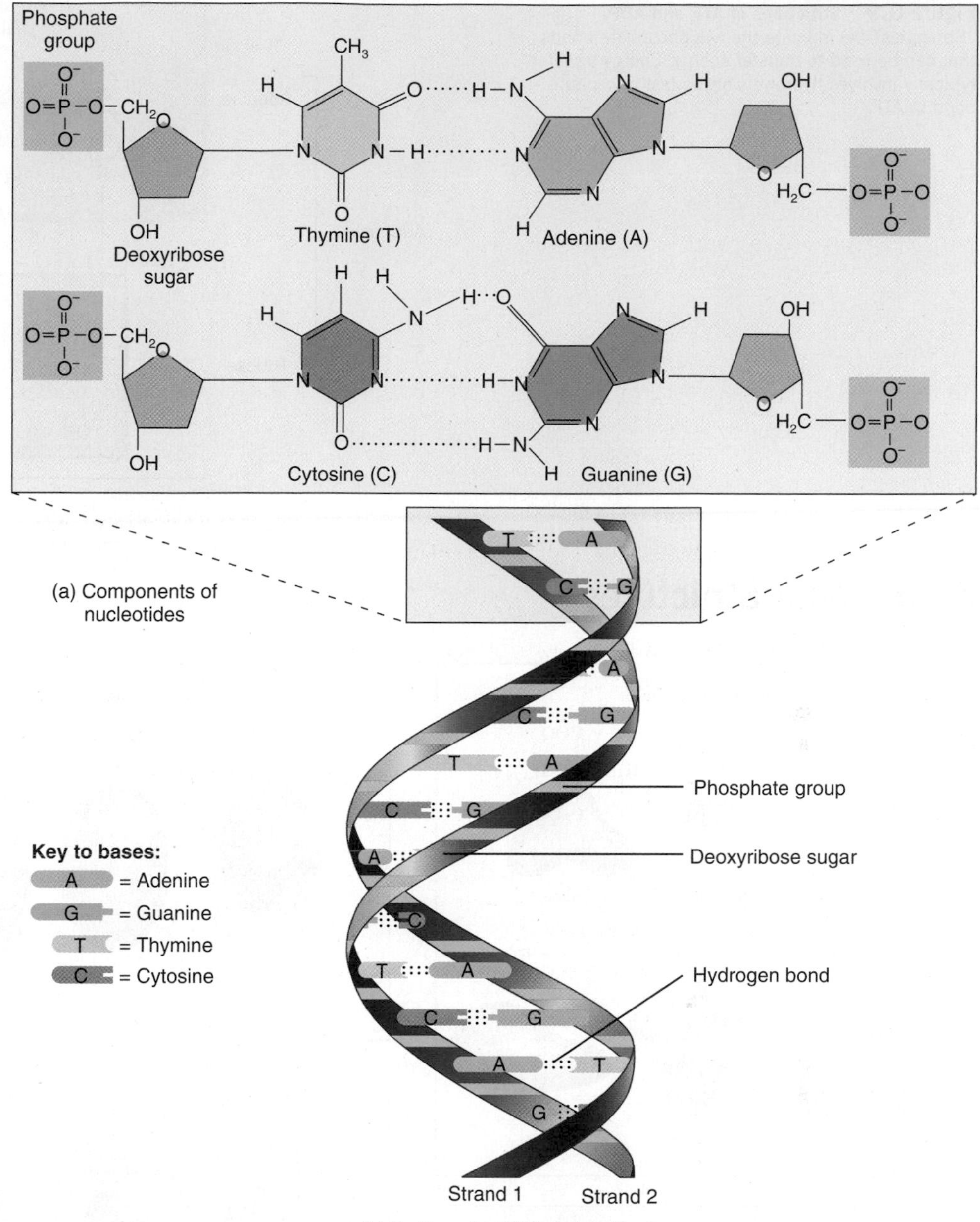

Figure G.8 DNA molecule. (a) A nucleotide consists of a base, a pentose sugar, and a phosphate group. (b) The paired bases project toward the center of the double helix. The structure is stabilized by hydrogen bonds (dotted lines) between each base pair. There are two hydrogen bonds between adenine and thymine and three between cytosine and guanine. Hydrogen bonds result from the attraction of oppositely charged parts of molecules and are weak compared to ionic or covalent bonds.

material inside each human cell (**Figure G.8**). In humans, each **gene** is a segment of a DNA molecule. Our genes determine the traits we inherit, and by controlling protein synthesis they regulate most of the activities that take place in body cells throughout a lifetime. When a cell divides, its hereditary information passes on to the next generation of cells. **Ribonucleic acid (RNA)**, the second type of nucleic acid, relays instructions from the genes to guide each cell's synthesis of proteins from amino acids.

DNA and RNA molecules consist of a chain of repeating **nucleotides**. Each nucleotide consists of three parts: a nitrogenous base, a pentose sugar, and a phosphate group. DNA contains the sugar deoxyribose and four different nitrogenous bases: adenine (A), thymine (T), cytosine (C), and guanine (G). Adenine and guanine are larger, double-ring bases called **purines**; thymine and cytosine are smaller, single-ring bases called **pyrimidines**. RNA contains the sugar ribose and instead of thymine it contains the pyrimidine base uracil (U).

Adenosine Triphosphate **Adenosine triphosphate** or **ATP** is the "energy currency" of living systems (**Figure G.9**). ATP transfers the energy liberated in exergonic catabolic reactions to power cellular activities that require energy (endergonic reactions). Among these cellular activities are muscular contractions, movement of chromosomes during cell division, movement of structures within cells, transport of substances across cell membranes, and synthesis of larger molecules from smaller ones. As its name implies, ATP consists of three phosphate groups attached to adenosine, a unit composed of adenine and the five-carbon sugar ribose.

Figure G.9 Structures of ATP and ADP. "Squiggles" (~) indicate the two phosphate bonds that can be used to transfer energy. Energy transfer typically involves hydrolysis of the last phosphate bond of ATP.

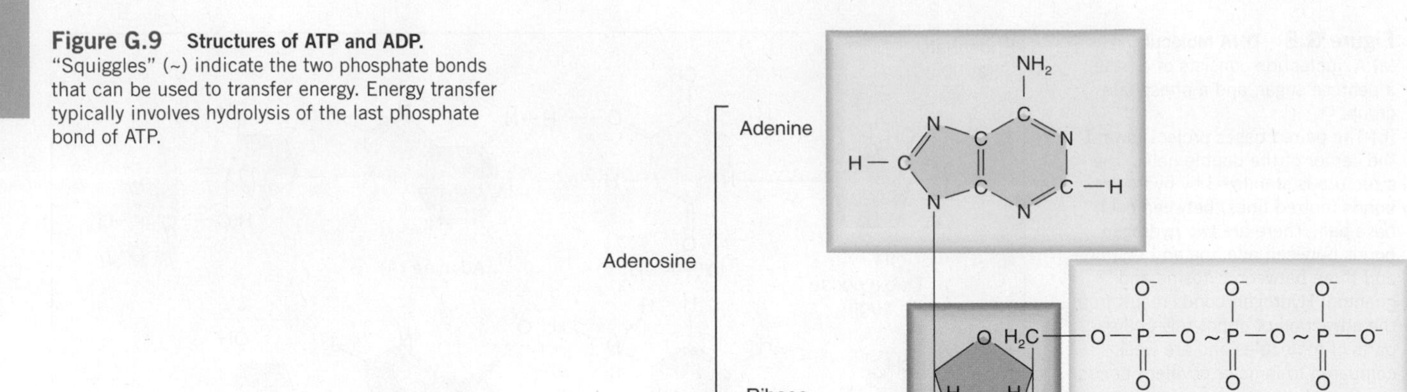

Amino Acid Structures

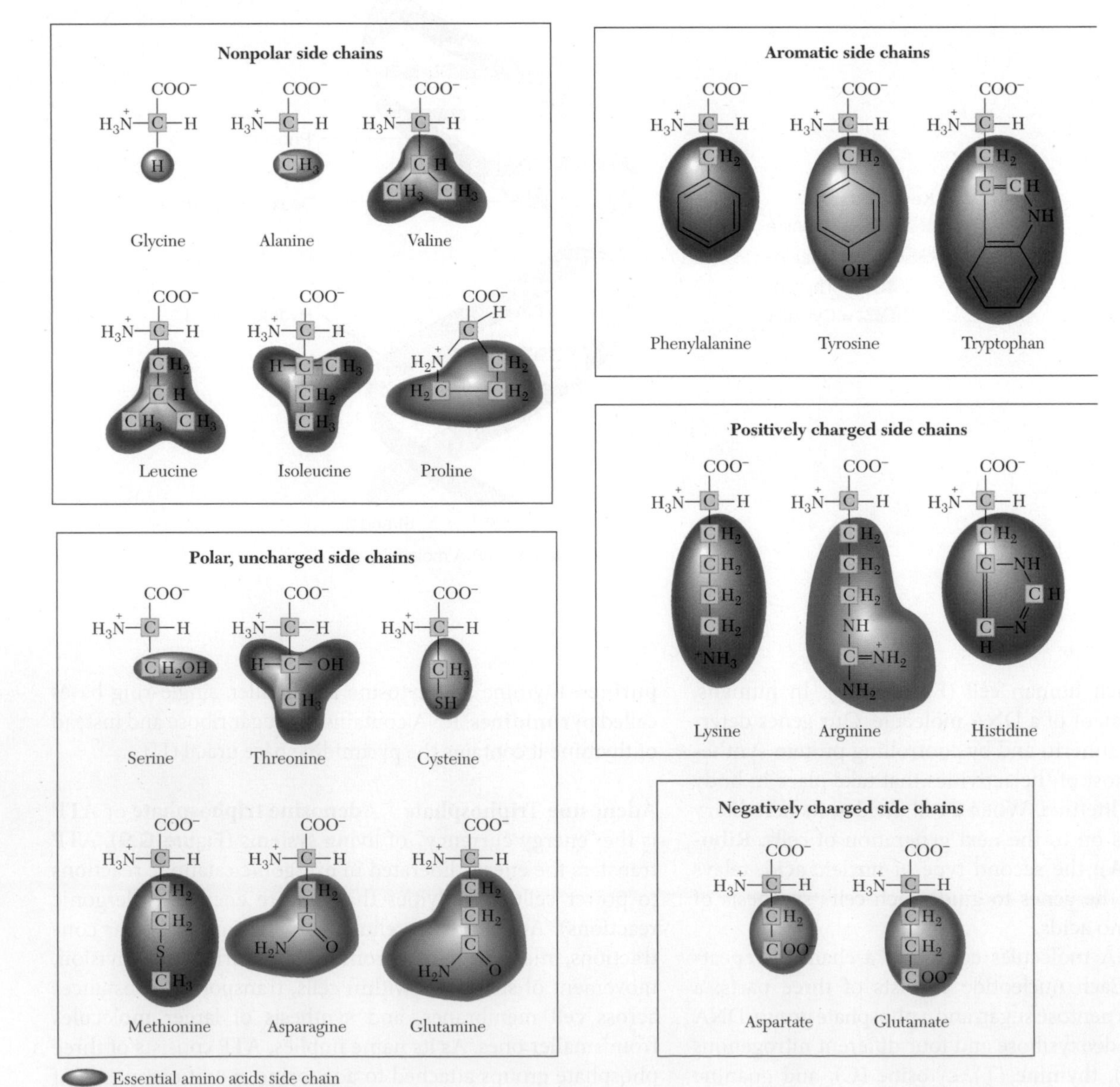

Vitamin Structures

Water-Soluble Vitamins

Thiamin

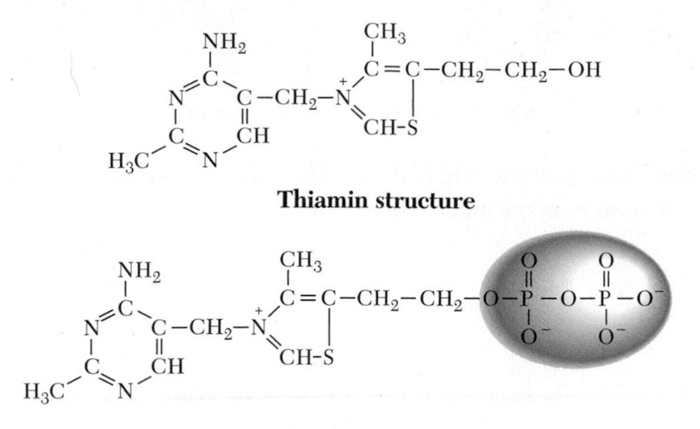

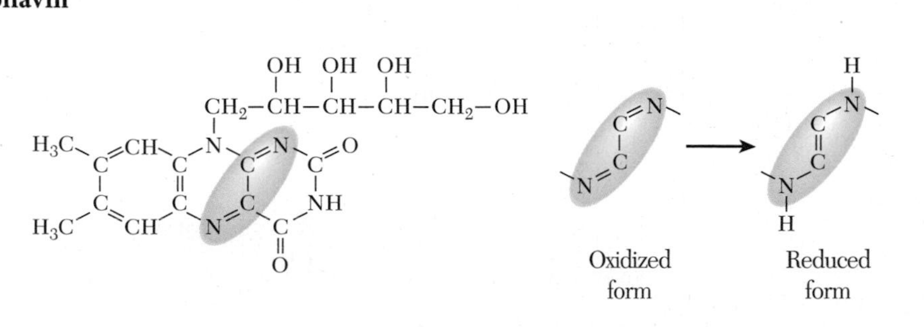

Thiamin pyrophosphate (TPP): The active coenzyme form of thiamin.

Riboflavin

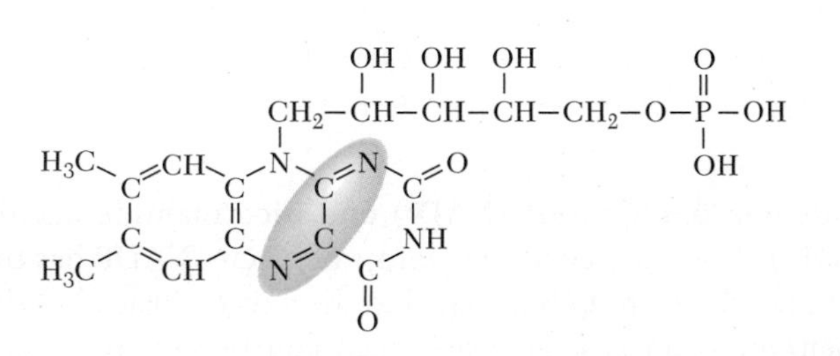

Riboflavin structure: In riboflavin coenzymes the nitrogens can pick up hydrogen atoms.

Flavin mononucleotide (FMN): One of the active coenzyme forms of riboflavin.

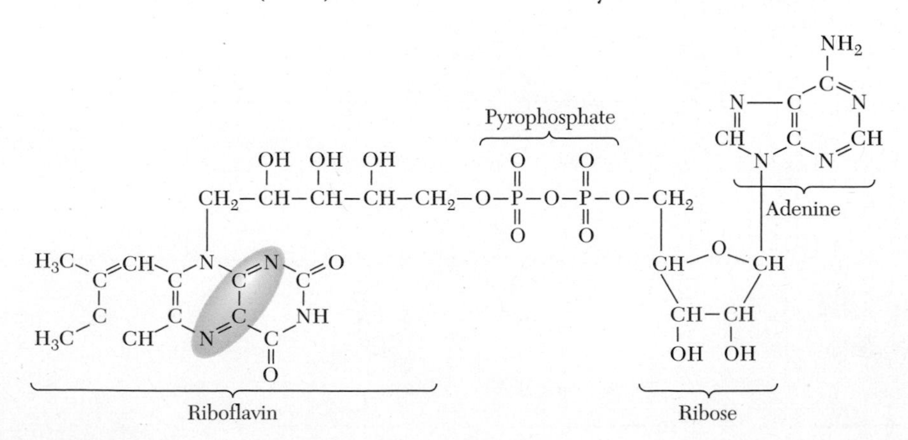

Flavin adenine dinucleotide (FAD): One of the active coenzyme forms of riboflavin.

Niacin

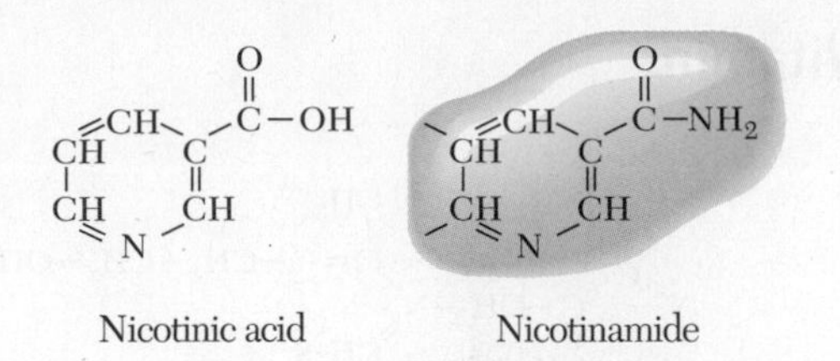

Nicotinic acid Nicotinamide

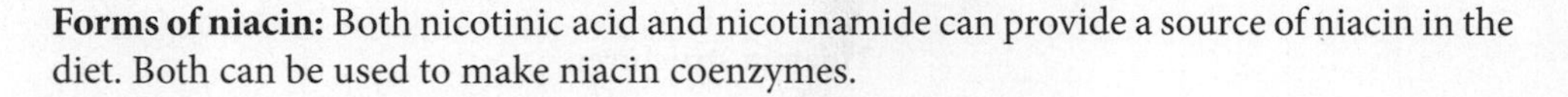

Forms of niacin: Both nicotinic acid and nicotinamide can provide a source of niacin in the diet. Both can be used to make niacin coenzymes.

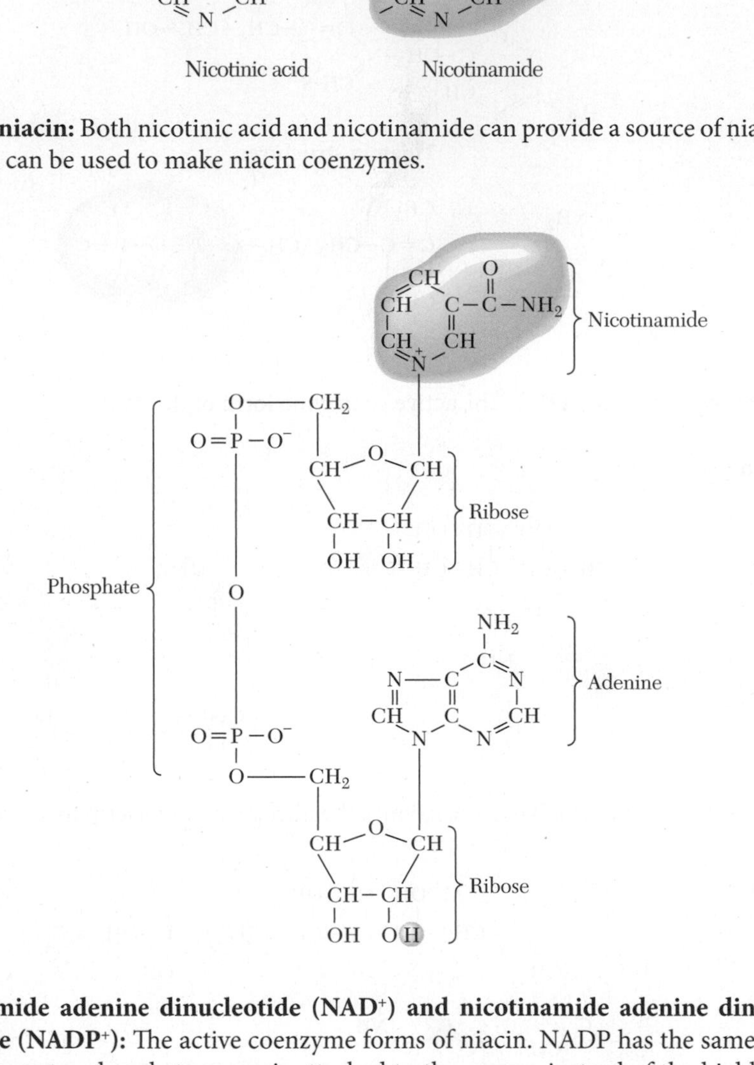

Nicotinamide adenine dinucleotide (NAD^+) and nicotinamide adenine dinucleotide phosphate ($NADP^+$): The active coenzyme forms of niacin. NADP has the same structure as NAD except a phosphate group is attached to the oxygen instead of the highlighted H. These niacin coenzymes can pick up a hydrogen and two electrons to form NADH or NADPH.

Biotin

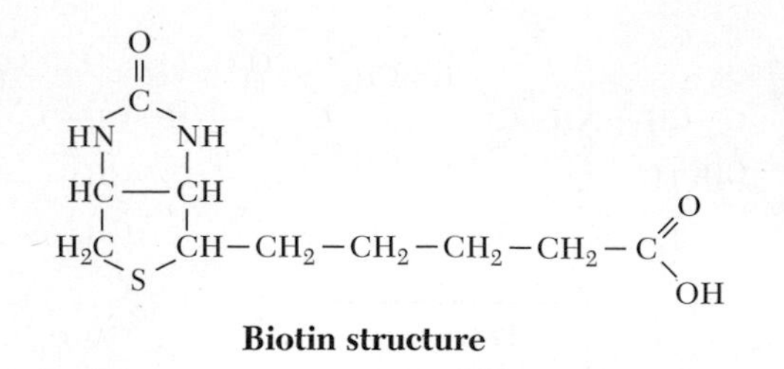

Biotin structure

Pantothenic acid

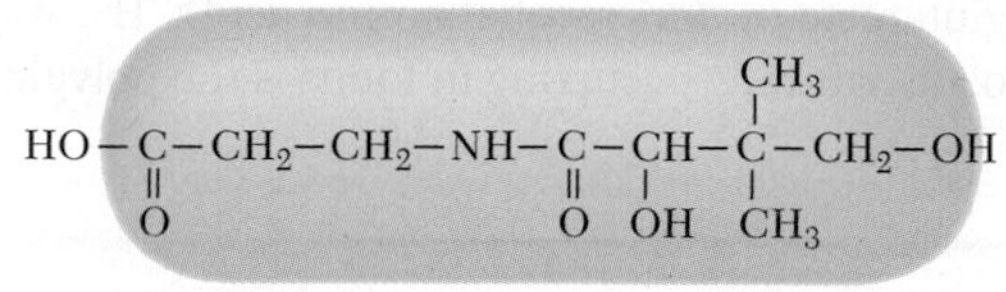

Pantothenic acid: This molecule is part of the structure of coenzyme A (CoA).

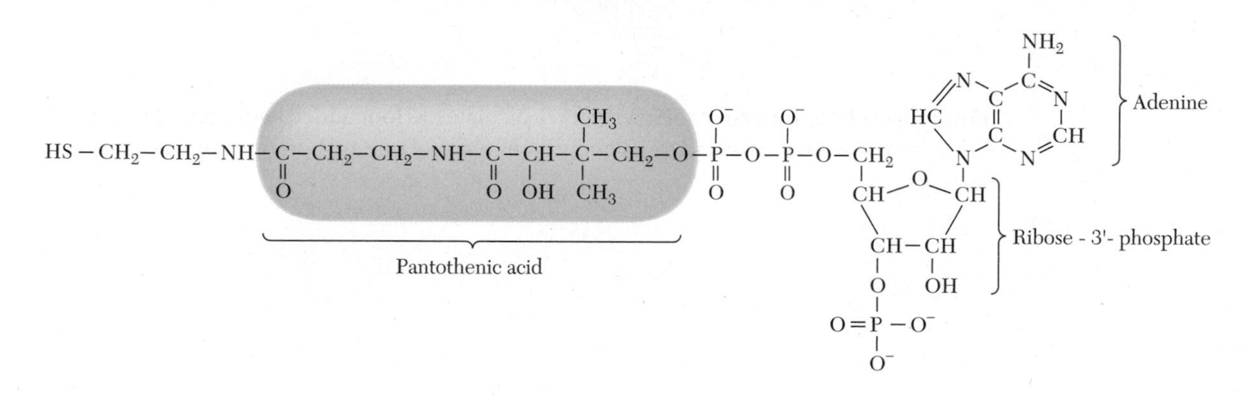

Coenzyme A: This coenzyme includes pantothenic acid as part of its structure.

Vitamin B_6

Pyridoxine | Pyridoxal | Pyridoxamine

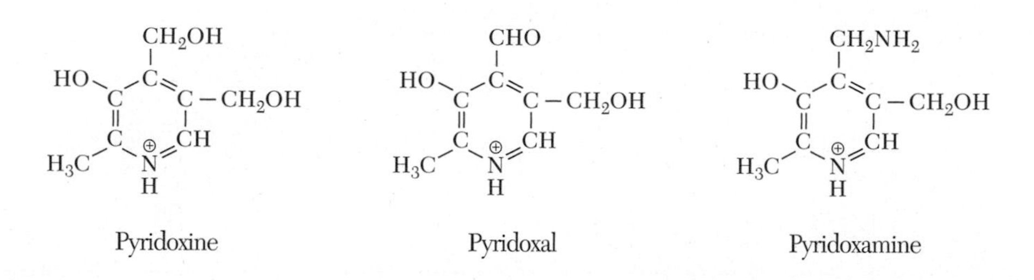

Forms of vitamin B_6: Pyridoxine, pyridoxal, and pyridoxamine are all converted into the active form of vitamin B_6.

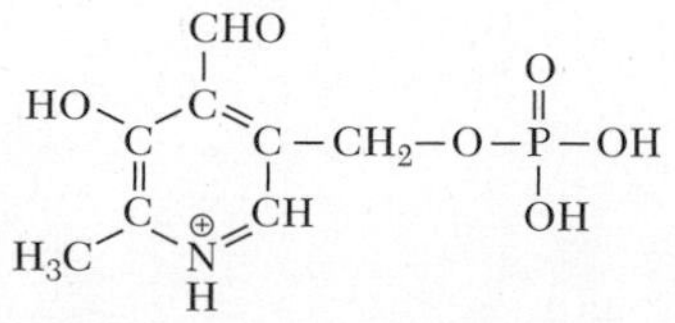

Pyridoxal Phosphate: The active coenzyme form of vitamin B_6.

Folate

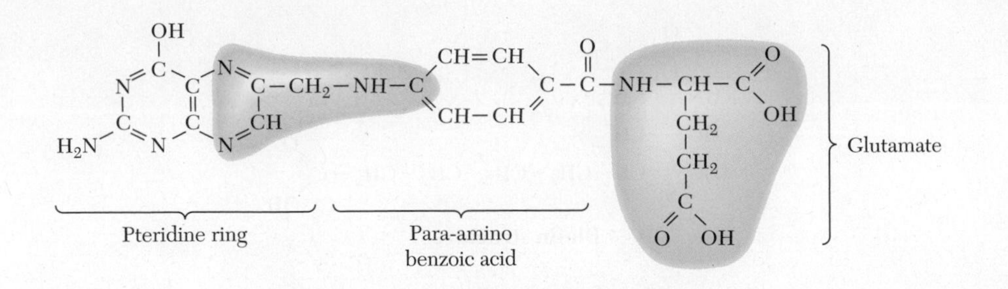

Folate structure: Folate consists of a pteridine ring combined with para-amino benzoic acid and at least one glutamate (a nonessential amino acid). The monoglutamate form is called folic acid. The folate naturally occurring in foods is the polyglutamate form.

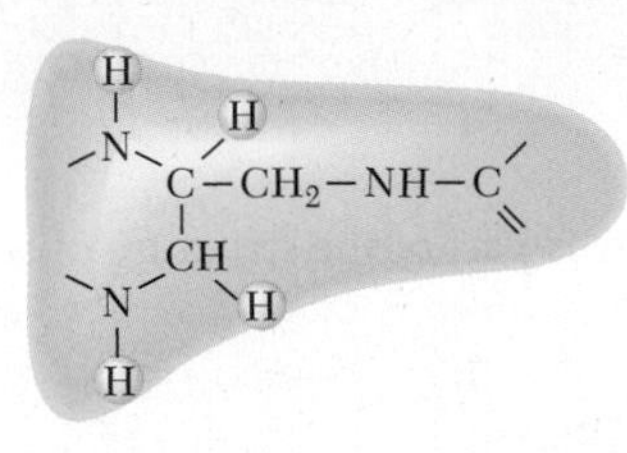

Tetrahydrofolate: The active coenzyme form of folate has four added hydrogens. Derivatives of this form of folate carry and transfer different types of one-carbon units during synthetic reactions.

Vitamin B_{12}

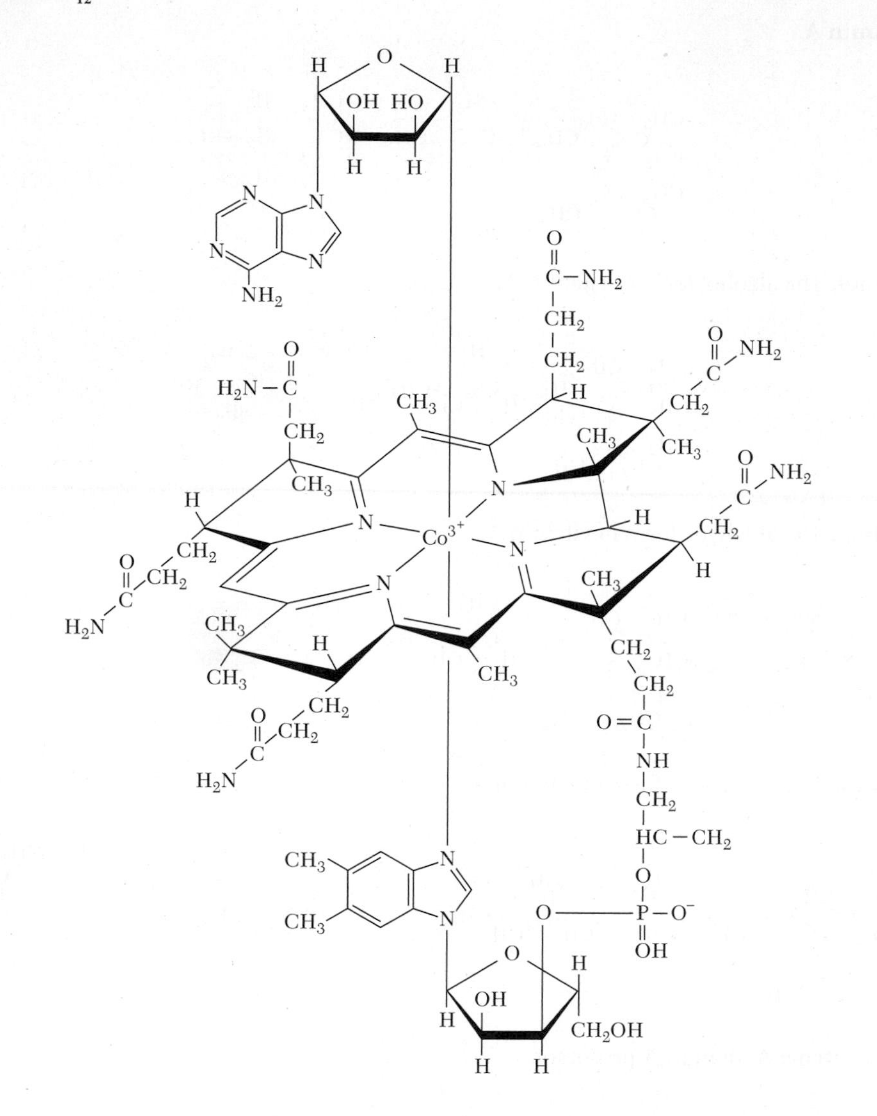

Vitamin B_{12} structure: Cobalamin, commonly known as vitamin B_{12}, is composed of a complex ring structure with a cobalt ion in the center.

Vitamin C

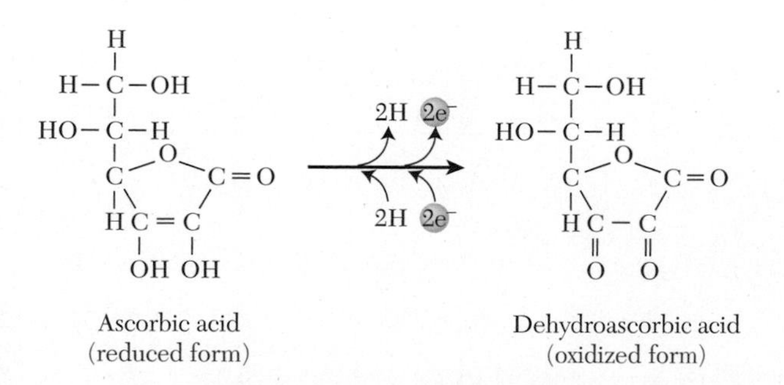

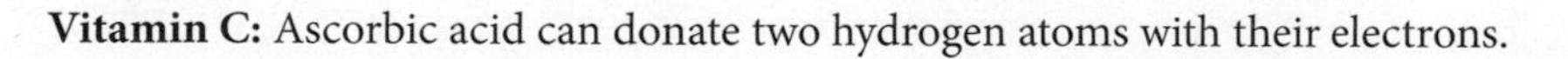

Vitamin C: Ascorbic acid can donate two hydrogen atoms with their electrons.

Fat-Soluble Vitamins

Vitamin A

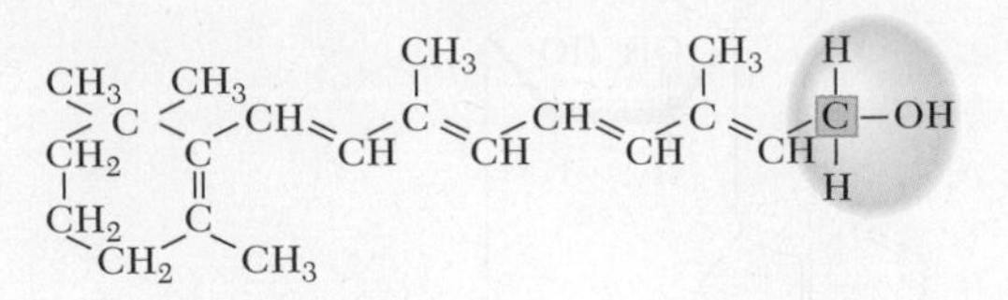

Retinol: The alcohol form of vitamin A.

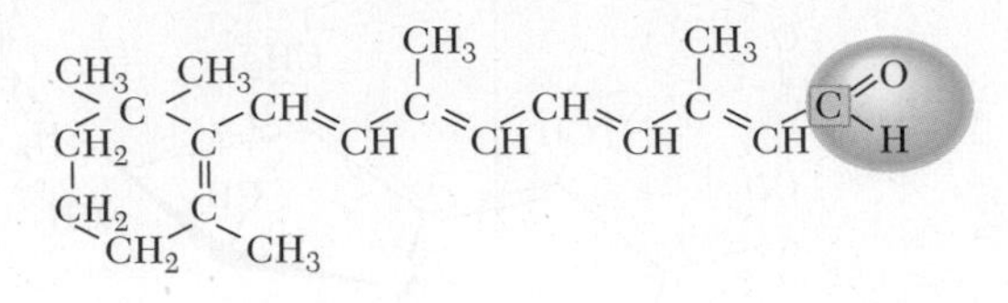

Retinal: The aldehyde form of vitamin A.

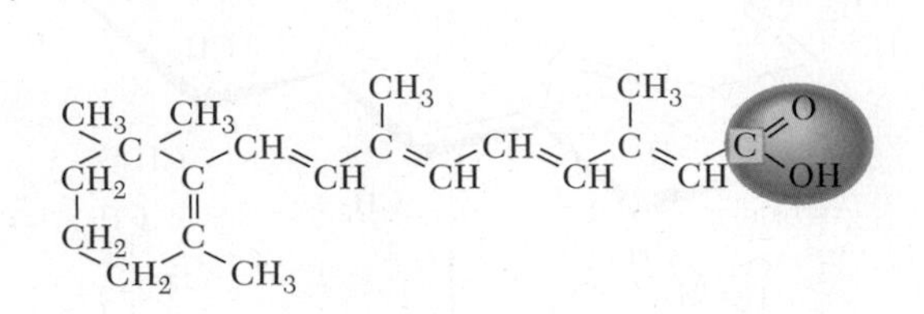

Retinoic acid: The acid form of vitamin A.

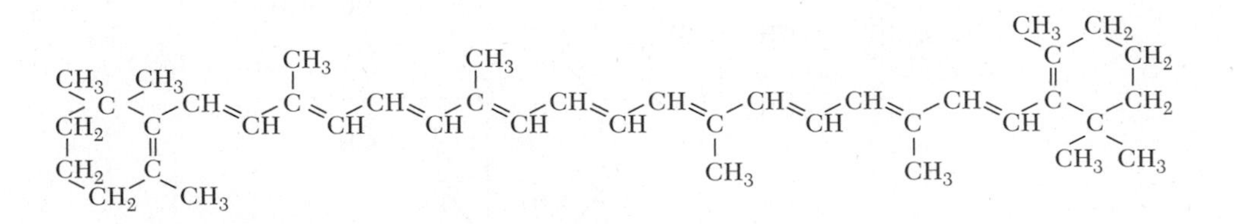

***β*-Carotene:** A vitamin A precursor.

Vitamin D

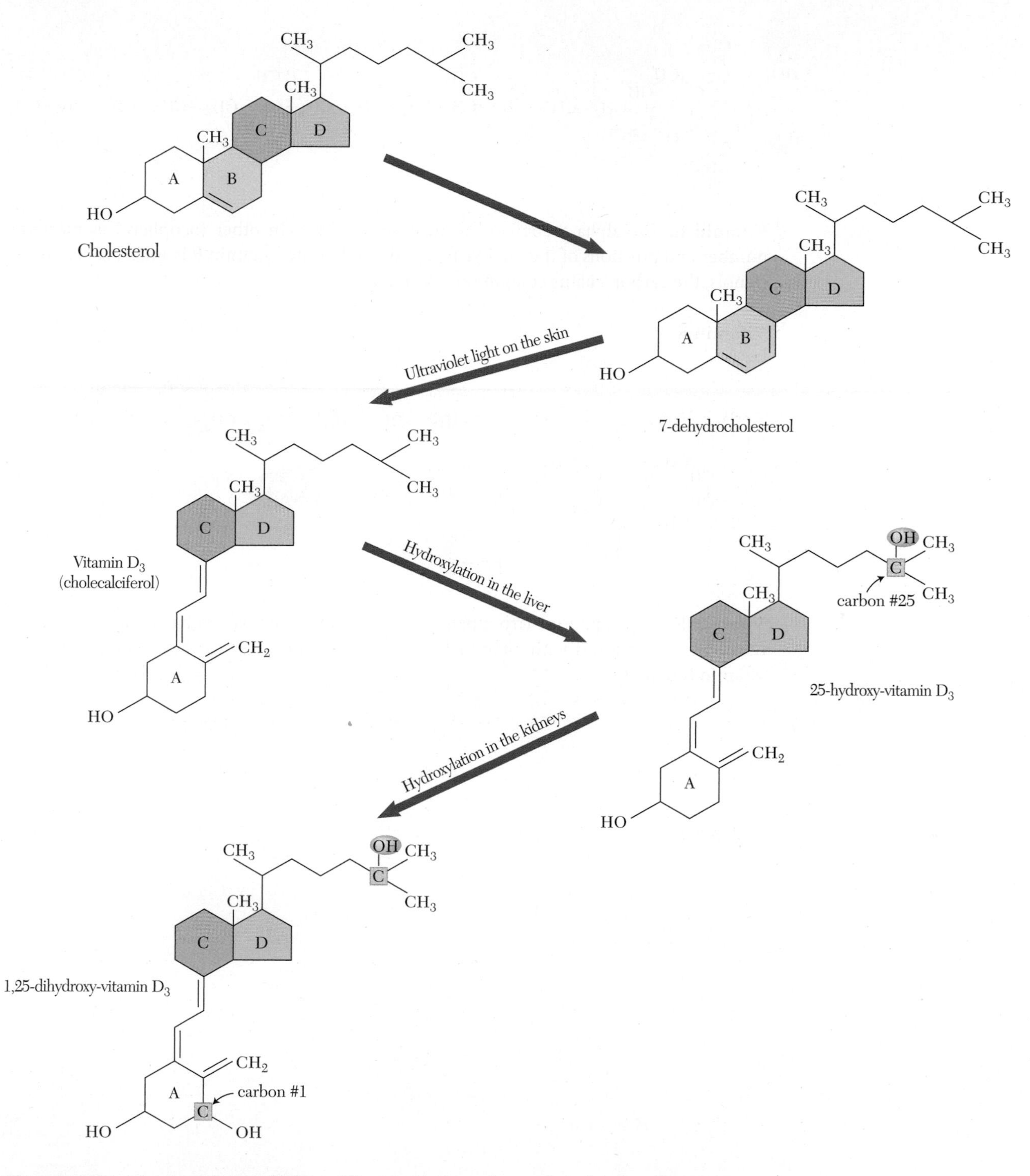

Vitamin D synthesis and metabolism: Vitamin D is synthesized in the skin during exposure to sunlight. Hydroxylation in the liver and kidneys converts it to its active form, 1,25-dihydroxy-vitamin D_3.

Vitamin E

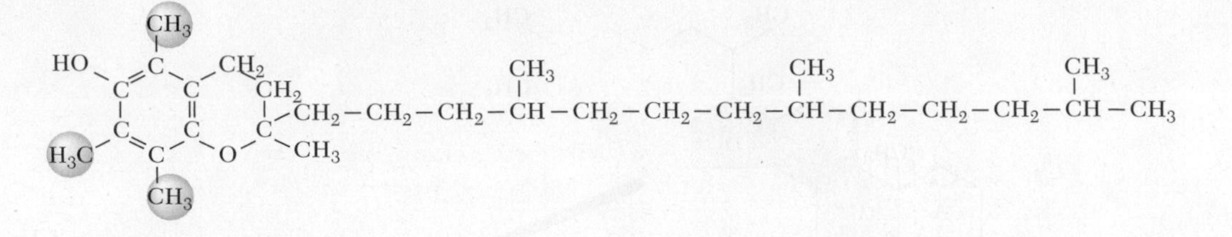

Vitamin E: The alpha-tocopherol form is shown here. In other tocopherol isomers the number and positions of the methyl groups differ. In other vitamin E isomers, called tocotrienols, the carbon chains contain double bonds.

Vitamin K

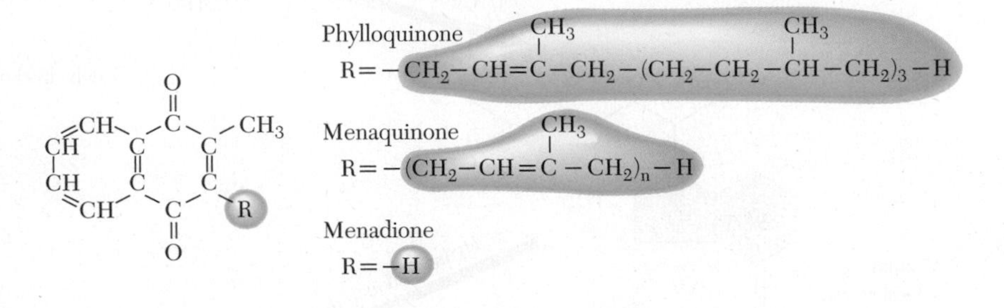

Vitamin K: Phylloquinone (from plants) and menaquinones (from bacteria) are naturally occurring compounds with vitamin K activity. Menadion is a synthetic compound with vitamin K activity.

Appendix H

Calculations and Conversions

Weights and Measures

Measure	*Abbreviation*	*Equivalent*
1 gram	g	1000 milligrams
1 milligram	mg	1000 micrograms
1 microgram	μg	1/1000000 of a gram
1 nanogram	ng	1/1000000000 of a gram
1 picogram	pg	1/1000000000000 of a gram
1 kilogram	kg	1000 grams
		2.2 lb
1 pound	lb	454 grams
		16 ounces
1 teaspoon	tsp	approximately 5 grams
1 tablespoon	Tbsp	3 teaspoons
1 ounce	oz	28.4 grams
1 cup	c	8 fluid ounces
		16 tablespoons
1 pint	pt	2 cups
		16 fluid ounces
1 quart	qt	2 pints
		32 fluid ounces
1 gallon	gal	128 fluid ounces
		4 quarts
1 liter	l	1.06 quarts
		1000 milliliters
1 milliliter	ml	1000 microliters
1 deciliter	dl	100 milliliters
1 kcalorie	kcal, Cal	1000 calories
		4.167 kilojoules
1 kilojoule	kJ	1000 joules

Converting Vitamin A Units

Form and Source	Amount Equal to 1 μg Retinol
Preformed vitamin A in food or supplements	1 μg
	1 RAE
	1 μg RE
	3.3 IU
β-carotene in foods*	12 μg
	1 RAE
	2 μg RE
	20 IU
α-carotene or β-cryptoxanthin in food	24 μg
	1 RAE
	2 μg RE
	40 IU

*β-carotene in supplements may be better absorbed than β-carotene in food and so provide more vitamin A activity. It is estimated that 2 μg of β-carotene dissolved in oil provides 1 μg of vitamin A activity.

Converting Vitamin E Values into mg α-Tocopherol

To estimate the α-tocopherol intake from foods:

- If values are given as mg α-TEs:

  mg α-TE × 0.8 = mg α-tocopherol

- If values are given in IUs:

  First, determine if the source of the α-tocopherol is natural or synthetic.

- For natural α-tocopherol:

  IU of natural α-tocopherol × 0.67 = mg α-tocopherol

- For synthetic α-tocopherol (dl-α-tocopherol):

  IU of synthetic α-tocopherol × 0.45 = mg α-tocopherol

Source: Institute of Medicine, Food and Nutrition Board, http://www.nap.edu/catalog.php?record_id=9810 *Dietary Reference Intakes for Vitamin C, Vitamin E, Selenium, and Carotenoids* (2000). Available online at http://books.nap.edu/openbook.php?record_id=9810&page=192

Calculating Dietary Folate Equivalents in Fortified Foods

The folate listed on labels of fortified foods is primarily folic acid, which is more available than natural forms of folate. In order to compare the folate content of these foods to recommendations, the amount of folic acid must be converted to dietary folate equivalents, expressed as μg DFE. This calculation assumes that all of the folate in these foods is from added folic acid.

Determine the amount (μg) of folic acid in the fortified food:

- Multiply the Daily Value for folate by the % Daily Values listed on the label.
- Daily Value is 400 μg.

Convert the μg folic acid into μg DFE:

- Multiply the μg folic acid by 1.7.
- Folic acid added to a food in fortification provides 1.7 times more available folate per μg than folate naturally present in foods.

For example:

- A serving of English muffins provides 6% of the Daily Value for folate:

  To find the μg folic acid: 400 μg × 6% = 24 μg folic acid

  To convert to μg DFE: 24 μg folic acid × 1.7 = 41 μg DFE

Source: Institute of Medicine, Food and Nutrition Board, http://www.nap.edu/catalog.php?record_id=6015 *Dietary Reference Intakes for Thiamin, Riboflavin, Niacin, Vitamin B6, Folate, Vitamin B12, Pantothenic Acid, Biotin, and Choline* (1998). Available online at http://books.nap.edu/openbook.php?record_id=6015&page=210

Glossary

A

absorption The process of taking substances into the interior of the body.
acceptable Daily Intake (ADI) The amount of a substance in food or drinking water that can be safely consumed daily over a lifetime without adverse effects.
acceptable macronutrient distribution ranges (AMDRs) Ranges of intake for energy-yielding nutrients, expressed as a percentage of total energy intake, that are associated with reduced risk of chronic disease while providing adequate intakes of essential nutrients.
accidental contaminant A substance not regulated by the FDA that unexpectedly enters the food supply.
Accutane A drug that is used orally to treat severe acne. It is a derivative of vitamin A.
acesulfame K (Acesulfame potassium) An alternative sweetener that contains no energy and is 200 times as sweet as sugar.
acetyl-CoA A metabolic intermediate formed during the breakdown of glucose, fatty acids, and amino acids. It is a two-carbon compound attached to a molecule of CoA.
acetylcholine A neurotransmitter that functions in the brain and other parts of the nervous system.
acid A substance that releases hydrogen ions (H^+) in solution.
acrodermatitis enteropathica An inherited defect in zinc absorption and metabolism that leads to severe zinc deficiency.
active transport The transport of substances across a cell membrane with the aid of a carrier molecule and the expenditure of energy. This may occur against a concentration gradient.
acute Refers to effects that develop rapidly.
adaptive thermogenesis The change in energy expenditure induced by factors such as changes in ambient temperature and food intake.
added sugars Sugars and syrups that have been added to foods during processing or preparation.
adequate intakes (AIs) DRI values used as a goal for intake when no RDA can be determined. These values are an approximation of the average nutrient intake that appears to sustain a desired indicator of health.
adipocyte A fat-storing cell.
adipose tissue Tissue found under the skin and around body organs that is composed of fat-storing cells.
adolescent growth spurt An 18- to 24-month period of peak growth velocity that begins at about ages 10 to 13 in girls and 12 to 15 in boys.
adrenaline A hormone secreted by the adrenal gland in response to stress that causes changes, such as an increase in heart rate, that prepare the body for "fight or flight"; also called epinephrine.
aerobic capacity or **VO_2 max** The maximum amount of oxygen that can be consumed by the tissues during exercise. This is also called maximal oxygen consumption.
aerobic exercise Endurance exercise such as jogging, swimming, or cycling that increases heart rate and requires oxygen in metabolism.
aerobic metabolism Metabolism in the presence of oxygen. In aerobic metabolism, glucose, fatty acids, and amino acids are completely broken down to form carbon dioxide and water and produce ATP.
aflatoxin An extremely potent carcinogen that is produced by a mold that grows on peanuts, corn, and grains.
age-related bone loss The bone loss that occurs in both cortical and trabecular bone of men and women as they advance in age.
aging The process of becoming older. It is genetically determined and modulated by environmental and lifestyle factors.
AIDS (acquired immune deficiency syndrome) The syndrome caused by HIV infection that causes the immune system to fail, resulting in frequent recurrent infections that ultimately result in death.
alcohol dehydrogenase (ADH) An enzyme found primarily in the liver and stomach that helps break down alcohol into acetaldehyde, which is then converted to acetyl-CoA.
alcohol poisoning A condition that occurs when too much alcohol is consumed too quickly. It can impair breathing, heart rate, and the gag reflex and can cause unconsciousness, choking, coma, and death.
alcoholic hepatitis Inflammation of the liver caused by alcohol consumption.
alcoholism A chronic disorder characterized by dependence on alcohol and development of withdrawal symptoms when alcohol intake is reduced.
aldosterone A hormone that increases sodium reabsorption by the kidney and therefore enhances water retention.
alimentary canal See gastrointestinal tract.
allergen A substance, usually a protein, that stimulates an immune response causing an allergic reaction.
allergy An adverse reaction involving the immune system that results from exposure to a specific allergen.
alpha-carotene (α-carotene) A carotenoid, some of which can be converted into vitamin A, that is found in leafy green vegetables, carrots, and squash.
alpha-linolenic acid (α-linolenic acid) An 18-carbon omega-3 polyunsaturated fatty acid known to be essential in humans.
alpha-tocopherol (α-tocopherol) The form of tocopherol that provides vitamin E activity in humans.
alveoli (singular, alveolus) Milk-producing glands in the breast.
Alzheimer's disease A disease that causes the relentless and irreversible loss of mental function.
amenorrhea Delayed onset of menstruation or the absence of three or more consecutive menstrual cycles.
amino acid pool All of the amino acids in body tissues and fluids that are available for use by the body.
amino acid score See chemical score.
amino acids The building blocks of proteins. Each contains a central carbon atom bound to a hydrogen atom, an amino group, an acid group, and a side chain.
amniotic fluid The liquid in the amniotic sac that surrounds and protects the fetus during development.
amniotic sac A membrane surrounding the fetus that contains the amniotic fluid.
amylopectin A plant starch that is composed of long, branched chains of glucose molecules.
amylase A plant starch that is composed of long unbranched chains of glucose molecules.
anabolic pathways Energy-requiring biochemical reactions in which simpler molecules are combined to form more complex substances.
anabolic steroids Synthetic fat-soluble hormones that mimic testosterone and are used by some athletes to increase muscle strength and mass.
anabolism Energy-requiring processes in which simpler molecules are combined to form more complex substances.
anaerobic metabolism or **anaerobic glycolysis** Metabolism in the absence of oxygen. Each molecule of glucose generates two molecules of ATP. Glucose is metabolized in this way when the blood cannot deliver oxygen to the tissues quickly enough to support aerobic metabolism.
anaphylaxis An immediate and severe allergic reaction to a substance (e.g., food or drugs). Symptoms include breathing difficulty, loss of consciousness, and a drop in blood pressure and can be fatal.
androstenedione A compound (known as andro) that can be converted into testosterone and estrogen inside the body. It is a controlled substance used by athletes to increase muscle mass and strength.

anecdotal Information based on a story of personal experience.

anemia A condition in which there is a reduced number of red blood cells or a reduced amount of hemoglobin, which reduces the oxygen-carrying capacity of the blood.

anencephaly A birth defect due to failure of the neural tube to close that results in the absence of a major portion of the brain, skull, and scalp.

angiotensin II A compound that causes blood vessel walls to constrict and stimulates the release of the hormone aldosterone.

Anisakis disease A disease caused by infection of the gastrointestinal tract with the Anisakis roundworm, which can contaminate raw fish.

anorexia athletica A type of eating disorder that is characterized by engaging in compulsive exercise to eliminate excess calories.

anorexia nervosa An eating disorder characterized by self-starvation, a distorted body image, and below normal body weight.

antacid A drug used to neutralize acidity in the gastrointestinal tract.

anthropometric measurements External measurements of the body, such as height, weight, limb circumference, and skinfold thickness.

antibiotic A substance that inhibits the growth of or destroys microorganisms; used to treat or prevent infection.

antibiotic resistance When bacteria or other microbes that cause disease evolve into forms that can no longer be killed by antibiotics.

antibody A protein produced by the body's immune system that recognizes foreign substances in the body and helps destroy them.

anticaking agent A substance added to dry food products to prevent clumping.

anticarcinogen A compound that can counteract the effect of cancer-causing substances.

anticoagulant A substance that delays or prevents blood clotting.

antidiuretic hormone (ADH) A hormone secreted by the pituitary gland that increases the amount of water reabsorbed by the kidney and therefore retained in the body.

antigen A foreign substance (almost always a protein) that, when introduced into the body, stimulates an immune response.

antioxidant A substance that can neutralize reactive oxygen molecules and thereby reduce oxidative damage.

antithiamin factor A substance in food that destroys the vitamin thiamin. Some are enzymes and are destroyed by cooking; others are not inactivated by cooking.

anus The outlet of the rectum through which feces are expelled.

apolipoprotein B A protein embedded in the outer shell of low-density lipoprotein (LDL) particles that binds to LDL receptor proteins on body cells.

appetite A desire to consume specific foods that is independent of hunger.

aquaculture The controlled cultivation and harvest of aquatic plants or animals.

arachidonic acid A 20-carbon omega-6 polyunsaturated fatty acid that can be synthesized from linoleic acid.

ariboflavinosis The condition resulting from a deficiency of riboflavin.

arteries Vessels that carry blood away from the heart.

arteriole A small artery that carries blood to capillaries.

arthritis A disease characterized by inflammation of the joints, joint pain, and sometimes changes in joint structure.

ascorbic acid or **ascorbate** The chemical term for vitamin C.

aseptic processing A method that places sterilized food in a sterilized package using a sterile process.

aspartame An alternative sweetener that is 200 times as sweet as sugar and is composed of the amino acids phenylalanine and aspartic acid.

atherosclerosis A type of cardiovascular disease that involves the buildup of fatty material in the artery walls.

atherosclerotic plaque The cholesterol-rich material that is deposited in the arteries of individuals with atherosclerosis. It consists of cholesterol, smooth muscle cells, fibrous tissue, and eventually calcium.

atom The smallest unit of an element that still retains the properties of that element.

ATP (adenosine triphosphate) The high-energy molecule used by the body to perform energy-requiring activities.

atrophic gastritis An inflammation of the stomach lining that causes a reduction in stomach acid and allows bacterial overgrowth.

atrophy Wasting or decrease in the size of a muscle or other tissue caused by lack of use.

attention deficit hyperactivity disorder (ADHD) A condition characterized by a short attention span, acting without thinking, and a high level of activity, excitability, and distractibility.

autoimmune disease A disease that results from immune reactions that destroy normal body cells.

avidin A protein found in raw egg whites that binds biotin, preventing its absorption.

B

bacteria (singular, bacterium) Tiny single-celled organisms found throughout the environment. Most are harmless or beneficial, but a few types can cause disease in humans.

balance study A study that compares the total amount of a nutrient that enters the body with the total amount that leaves the body.

basal energy expenditure (BEE) The energy expended to maintain an awake resting body that is not digesting food.

basal metabolic rate (BMR) The rate of energy expenditure under resting conditions. BMR measurements are performed in a warm room in the morning before the subject rises, and at least 12 hours after the last food or activity.

basal metabolism The energy expended to maintain an awake resting body that is not digesting food.

base A substance that accepts hydrogen ions in solution.

behavior modification A process used to gradually and permanently change habitual behaviors.

beriberi The disease resulting from a deficiency of thiamin.

beta-carotene (β-carotene) A carotenoid that has more provitamin A activity than other carotenoids. It also acts as an antioxidant.

beta-cryptoxanthin (β-cryptoxanthin) A carotenoid found in corn, green peppers, and lemons that can provide some vitamin A activity.

beta-oxidation (β-oxidation) The first step in the production of ATP from fatty acids. This pathway breaks the carbon chain of fatty acids into two-carbon units that form acetyl-CoA and releases high-energy electrons that are passed to the electron transport chain.

bile A substance made in the liver and stored in the gallbladder. It is released into the small intestine to aid in fat digestion and absorption.

bile acids Emulsifiers present in bile that are synthesized by the liver from cholesterol.

binge drinking The consumption, within two hours, of five or more drinks for men or four or more drinks for women.

binge-eating disorder An eating disorder characterized by recurrent episodes of binge eating in the absence of purging behavior.

bingeing or **binge eating** The rapid consumption of a large amount of food in a discrete period of time associated with a feeling that eating is out of control.

bioaccumulation The process by which compounds accumulate or build up in an organism faster than they can be broken down or excreted.

bioavailability A general term describing how well a nutrient can be absorbed and used by the body.

bioelectric impedance analysis A technique for estimating body composition that measures body fat by directing a low-energy electric current through the body and calculating resistance to flow.

biological value A measure of protein quality determined by comparing the amount of nitrogen retained in the body with the amount absorbed from the diet.

biotechnology The process of manipulating life forms via genetic engineering in order to provide desirable products for human use.

biotin A B vitamin needed in energy metabolism.

blackout drinking Amnesia following a period of excess alcohol consumption.

blood pressure The amount of force exerted by the blood against the artery walls.
body image The way a person perceives and imagines his or her body.
body mass index (BMI) A measure of body weight in relation to height that is used to compare body size with a standard.
bomb calorimeter An instrument used to determine the energy content of food. It measures the heat energy released when a dried food is combusted.
bone remodeling The process whereby bone is continuously broken down and re-formed to allow for growth and maintenance.
botulism A severe food-borne intoxication that results from consuming the toxin produced by the bacterium *Colostridium botulinum*.
bovine somatotropin (bST) A hormone naturally produced by cows that stimulates the production of milk. A synthetic version of this hormone is now being produced by genetic engineering.
bovine spongiform encephalopathy (BSE) A fatal neurological disease, also known as mad cow disease, that affects cattle and may be transmitted to humans by consuming beef by-products.
bran The protective outer layers of whole grains. It is a concentrated source of dietary fiber.
Brewer's yeast The type of yeast used in brewing beer; a good source of B vitamins and often used as a nutritional supplement.
brown adipose tissue A type of fat tissue that has a greater number of mitochondria than the more common white adipose tissue. It can waste energy by producing heat.
brush border Refers to the microvilli surface of the intestinal mucosa, which contains some digestive enzymes.
buffer A substance that reacts with an acid or base by picking up or releasing hydrogen ions to prevent changes in pH.
bulimia nervosa An eating disorder characterized by the consumption of large amounts of food at one time (binge eating), followed by purging behaviors such as vomiting or the use of laxatives to eliminate calories from the body.

C

calcitonin A hormone secreted by the thyroid gland that reduces blood calcium levels.
calorie The amount of heat required to raise the temperature of 1 g of water 1 degree Celsius; equal to 4.18 joules. Typically used to refer to the number of kilocalories in food or expended by the body.
calorimetry A technique for measuring energy expenditure.
Campylobacter jejuni A bacterium common in raw milk and undercooked meat that causes food-borne illness.
cancer A disease characterized by cells that grow and divide without restraint and have the ability to grow in areas of the body reserved for other cells.
capillary A small, thin-walled blood vessel where the exchange of gases and nutrients between blood and body cells occurs.
carbohydrate A compound containing carbon, hydrogen, and oxygen in the same proportions as in water; includes sugars, starches, and most fibers.
carbohydrate loading See Glycogen supercompensation.
carbon dioxide A waste product produced by cellular respiration that is eliminated from the body by the lungs.
carcinogen A substance that causes cancer.
cardiorespiratory endurance The efficiency with which the body delivers to cells the oxygen and nutrients needed for muscular activity and transports waste products from cells.
cardiovascular disease Any disease affecting the heart and blood vessels.
caries see **dental caries**
carnitine A molecule synthesized in the body that is needed to transport fatty acids and some amino acids into the mitochondria for metabolism. Supplements of carnitine are marketed to athletes to enhance performance.
carotenoids Natural pigments synthesized by plants and many microorganisms. They give yellow and red-orange fruits and vegetables their color.
carpal tunnel syndrome Numbness, tingling, weakness, and pain in the hand caused by pressure on the nerves.
case-control study A type of observational study that compares individuals with a particular condition under study with individuals of the same age, sex, and background who do not have the condition.
casein The predominant protein in cow's milk.
cash crop A crop that is grown to be sold for monetary return rather than as food for food the local population.
cassava A starchy root that is the staple of the diet in many parts of Africa.
catabolic pathways The biochemical reactions by which substances are broken down into simpler molecules releasing energy.
catabolism The processes by which substances are broken down into simpler molecules releasing energy.
catalase An iron-containing enzyme that destroys peroxides.
cataracts A disease of the eye that results in cloudy spots on the lens (and sometimes the cornea) that obscure vision.
celiac disease A disorder that causes damage to the intestines when the protein gluten is eaten.
cell differentiation Structural and functional changes that cause cells to mature into specialized cells.
cell membrane The membrane that surrounds the cell contents.
cells The basic structural and functional units of plant and animal life.
cellular respiration The reactions that break down glucose, fatty acids, and amino acids in the presence of oxygen to produce carbon dioxide, water, and energy in the form of ATP.
cellulite Subcutaneous fat that has a lumpy appearance because strands of connective tissue connect it to underlying structures.
cellulose An insoluble fiber that is the most prevalent structural material of plant cell walls.
cephalic response The stimulation of gastric secretion by the sight, smell, and sound of food preparation.
certified food color A food color that has been tested and certified for safety, quality, consistency, and strength of color.
ceruloplasmin The major copper-carrying protein in the blood.
cesarean section The surgical removal of the fetus from the uterus.
chemical or **amino acid score** A measure of protein quality determined by comparing the essential amino acid content of the protein in a food with that in a reference protein. The lowest amino acid ratio calculated is the chemical score.
chemical bond The force that holds atoms together.
Chinese restaurant syndrome See MSG symptom complex.
cholecalciferol The chemical name for vitamin D_3. It can be formed in the skin of animals by the action of sunlight on a form of cholesterol called 7-dehydrocholesterol.
cholecystokinin (CCK) A hormone released by the duodenum that stimulates the release of pancreatic juice rich in digestive enzymes and causes the gallbladder to contract and release bile into the duodenum.
cholesterol A lipid that consists of multiple chemical rings and is made only by animal cells.
cholic acid A bile acid.
choline A compound needed for the synthesis of the phospholipid phosphatidylcholine (lecithin) and the neurotransmitter acetylcholine. It is important for a number of biochemical reactions, and there is evidence that it is essential in the diet during certain stages of life.
chromium picolinate A form of chromium sold as a supplement promoted to change body composition. Chromium is involved in insulin action, and supplements claim to increase lean body mass, decrease body fat, and delay fatigue. There is little evidence of their effectiveness as an ergogenic aid.
chronic Refers to effects that develop slowly over a long period.
chylomicron A lipoprotein that transports lipid from the mucosal cells of the small intestine and delivers triglycerides to other body cells.
chyme A mixture of partially digested food and stomach secretions.
chymotrypsin A protein-digesting enzyme produced in an inactive form in the pancreas

and activated in the small intestine, where it aids digestion.

circulatory system The organ system consisting of the heart, blood, and blood vessels, that transports material to and from cells.

cirrhosis Chronic liver disease characterized by the loss of functioning liver cells and the accumulation of fibrous connective tissue.

citric acid cycle Also known as the Krebs cycle or the tricarboxylic acid cycle, this is the stage of cellular respiration in which two carbons from acetyl-CoA are oxidized, producing two molecules of carbon dioxide.

clinical trial see **human intervention study**

clones Copies that are identical to the original.

Clostridium botulinum A bacterium that produces a deadly toxin and grows in a low-acid, low-oxygen environment, such as inside certain canned goods.

Clostridium perfringens A bacterium found in meat and poultry that can cause food-borne illness.

coagulation The process of blood clotting.

cobalamin The chemical term for vitamin B_{12}.

coenzyme A small nonprotein organic molecule that acts as a carrier of electrons or atoms in metabolic reactions and is necessary for the proper functioning of many enzymes.

coenzyme Q or **ubiquinone** A coenzyme that transfers electrons from one molecule to another in the electron transport chain of cellular respiration.

cofactor An inorganic ion or coenzyme required for enzyme activity.

colic Inconsolable crying that is believed to be due to pain from gas buildup in the gastrointestinal tract or immaturity of the central nervous system.

collagen The major protein in connective tissue.

colon The largest portion of the large intestine.

colostrum The first milk, which is secreted in late pregnancy and up to a week after birth. It is rich in protein and immune factors.

complete dietary protein Protein that provides essential amino acids in the proportions needed to support protein synthesis.

complex carbohydrates Carbohydrates composed of sugar molecules linked together in straight or branching chains. They include glycogen, starches, and fibers.

compression of morbidity The postponement of the onset of chronic disease so that disability occupies a smaller and smaller proportion of the life span.

concentration gradient A condition that exists when the amount of a dissolved substance is greater in one area than it is in another.

conception The union of sperm and egg (ovum) that results in pregnancy.

condensation reaction Chemical reaction that joins two molecules together. Hydrogen and oxygen are lost from the two molecules to form water. Also called a dehydration reaction.

conditionally essential amino acid An amino acid that is essential in the diet only under certain conditions or at certain times of life.

connective tissue One of the four human tissue types; includes cartilage, bone, blood, adipose tissue, and the coverings of some organs.

constipation Infrequent or difficult defecation.

control group In a scientific experiment, the group of participants used as a basis of comparison. They are similar to the participants in the experimental group but do not receive the treatment being tested.

cornea The clear, transparent fibrous outer surface of the eye.

coronary heart disease A disease of the heart and blood vessels that supply blood to the heart.

correlation Two or more factors occurring together.

cortical or **compact bone** Dense, compact bone that makes up the sturdy outer surface layer of bones.

creatine A compound that can be converted into creatine phosphate, which replenishes muscle ATP during short bursts of activity. Creatine is a dietary supplement used by athletes to increase muscle mass and delay fatigue during short intense exercise.

creatine phosphate or **phosphocreatine** A compound found in muscle that can be broken down quickly to make ATP.

cretinism A condition resulting from poor maternal iodine intake during pregnancy that causes stunted growth and poor mental development in offspring.

critical control points Possible points in food production, manufacturing, and transportation at which controls can be applied and contamination be prevented or eliminated.

critical periods Times in growth and development when an organism is more susceptible to harm from poor nutrition or other environmental factors.

cross-contamination The transfer of contaminants from one food or object to another.

cross-sectional data Information obtained by a single broad sampling of many different individuals in a population.

cruciferous A group of vegetables (also called crucifers) named for the cross shape of their four-petal flowers. They include broccoli, Brussels sprouts, cabbage, cauliflower, kale, kohlrabi, mustard greens, rutabagas, and turnips. Their consumption is linked with lower rates of cancer.

cyclamate An alternative sweetener that was common in the United States in the 1960s; banned after it was found to cause cancer in laboratory animals.

cycle of malnutrition A cycle in which malnutrition is perpetuated by an inability to meet nutrient needs at all life stages.

cysteine A conditionally essential sulfur-containing amino acid; when methionine is available in sufficient quantities, cysteine is not essential in the diet.

cytoplasm The cellular material outside the nucleus that is contained by the cell membrane.

cytosol The liquid found within cells.

D

Daily Reference Values (DRVs) Reference values established for protein and seven nutrients for which no original RDAs were established. The values are based on dietary recommendations for reducing the risk of chronic disease.

Daily Value A nutrient reference value used on food labels to help consumers see how foods fit into their overall diets.

DASH diet A dietary pattern that is plentiful in fruits and vegetables and low-fat dairy products and therefore high in potassium, magnesium, calcium, and fiber, and low in saturated fat and cholesterol.

deamination The removal of the amino group from an amino acid.

decarboxylation The removal of a carboxyl group (COOH) from a chemical compound.

dehydration A condition that results when not enough water is present to meet the body's needs.

Delaney Clause A clause added to the 1958 Food Additives Amendment of the Pure Food and Drug Act that prohibits the intentional addition to foods of any compound that has been shown to induce cancer in animals or humans at any dose.

dementia A deterioration of mental state resulting in impaired memory, thinking, and/or judgment.

denaturation The alteration of a protein's three-dimensional structure.

dental caries The decay and deterioration of teeth caused by acid produced when bacteria on the teeth metabolize carbohydrate.

deoxyribose The 5-carbon sugar that is part of DNA.

depletion-repletion study A study that feeds a diet devoid of a nutrient until signs of deficiency appear, and then adds the nutrient back to the diet to a level at which symptoms disappear.

dermatitis An inflammation of the skin.

DHA See Docosahexaenoic acid.

DHEA (dehydroepiandosterone) A precursor of the sex hormones testosterone, estrogen, and progesterone; a banned substance used to slow aging and to increase muscle mass.

diabetes mellitus A disease caused by either insufficient insulin production or decreased sensitivity of cells to insulin. It results in elevated blood glucose levels.

diacylglycerol see **diglyceride**

diaphragm A muscular wall separating the abdomen from the thoracic cavity containing the heart and lungs.

diarrhea An intestinal disorder characterized by frequent or watery stools.

dicumarol An anticoagulant first isolated from moldy clover hay.

diet history Information about dietary habits and patterns. It may include a 24-hour recall, a food record, or a food frequency questionnaire to provide information about current intake patterns.

diet-induced thermogenesis See Thermic effect of food (TEF).

dietary antioxidant A substance in food that significantly decreases the adverse effects of reactive species on normal physiological function in humans.

dietary folate equivalents (DFEs) The unit used to express folate recommendations. One DFE is equivalent to 1 μg of folate naturally occurring in food, 0.6 μg of synthetic folic acid from fortified food or supplements consumed with food, or 0.5 μg of synthetic folic acid consumed on an empty stomach.

Dietary Guidelines for Americans A set of nutrition recommendations designed to promote population-wide dietary changes to reduce the incidence of nutrition-related chronic disease.

dietary reference intakes (DRIs) A set of reference values for the intake of energy, nutrients, and food components that can be used for planning and assessing the diets of healthy people in the United States and Canada.

dietary supplement A product intended for ingestion in the diet that contains one or more of the following: vitamins, minerals, plant-derived substances, amino acids, or concentrates or extracts.

diffusion The movement of molecules from an area of higher concentration to an area of lower concentration without the expenditure of energy.

digestion The process of breaking food into components small enough to be absorbed into the body.

digestive system The organ system responsible for the ingestion of food, the digestion and absorption of nutrients and the elimination of food residues; includes the gastrointestinal tract as well as a number of accessory organs.

digestive tract See Gastrointestinal tract.

diglyceride or **diacylglycerol** A molecule of glycerol with two fatty acids attached.

dipeptide Two amino acids linked by a peptide bond.

direct calorimetry A method of determining energy use that measures the amount of heat produced.

direct food additive A substance that is intentionally added to food. Direct food additives are regulated by the FDA.

disaccharide A carbohydrate made up of two sugar units.

dissociate To separate two charged ions.

diuretic A drug that promotes fluid excretion.

diverticula Sacs or pouches that protrude from the wall of the large intestine.

diverticulitis A condition in which diverticula in the large intestine become inflamed.

diverticulosis A condition in which outpouchings (or sacs) form in the wall of the large intestine.

DNA (deoxyribonucleic acid) The genetic material found in the nucleus that codes for the synthesis of proteins.

docosahexaenoic acid (DHA) A 22-carbon omega-3 polyunsaturated fatty acid found in fish that may be needed in the diet of newborns. It can be synthesized from α-linolenic acid.

double-blind study An experiment in which neither the study participants nor the researchers know which participants are in a control or an experimental group.

doubly labeled water technique A method for measuring energy expenditure based on measuring the disappearance of isotopes of hydrogen and oxygen in body fluids after consumption of a defined amount of water labeled with both isotopes.

Down syndrome A disorder caused by extra genetic material that results in mental retardation, distinctive facial characteristics, and other abnormalities.

duodenum The upper segment of the small intestine that connects to the stomach.

E

eating disorder A persistent disturbance in eating behavior or other behaviors intended to control weight that affects physical health and psychosocial functioning.

eating disorders not otherwise specified (EDNOS) A category of eating disorders that includes abnormal eating behaviors that don't fit into the anorexia or bulimia nervosa categories.

eclampsia A hypertensive disorder of pregnancy characterized by high blood pressure, protein in the urine, convulsions, and coma.

edema Swelling due to the buildup of extracellular fluid in the tissues or body cavities.

eicosanoids Regulatory molecules, including prostaglandins and related compounds, that can be synthesized from omega-3 and omega-6 fatty acids.

eicosapentaenoic acid (EPA) A 20-carbon omega-3 polyunsaturated fatty acid found in fish that can be synthesized from α-linolenic acid but may be essential in humans under some conditions.

electrolytes Positively and negatively charged ions that conduct an electrical current in solution. Commonly refers to sodium, potassium, and chloride.

electron High-energy particle carrying a negative charge that orbits the nucleus of an atom.

electron transport chain The final stage of cellular respiration in which electrons are passed down a chain of molecules to oxygen to form water and produce ATP.

element A substance that cannot be broken down into products with different properties.

elimination diet and **food challenge** A regimen that eliminates potential allergy-causing foods from an individual's diet and then systematically adds them back to identify any foods that cause an allergic reaction.

embryo The developing human from two to eight weeks after fertilization. All organ systems are formed during this time.

empty calories Calories from solid fats and/or added sugars, which add calories to the food, but few nutrients.

emulsifier A substance that allows water and fat to mix by breaking large fat globules into smaller ones.

endocrine system Organ system composed of cells, tissues, and organs that secrete hormones to help control body functions.

endoplasmic reticulum A cellular organelle involved in the synthesis of proteins and lipids and composed of a system of membranous tubules, channels, and sacs in the cytosol; rough endoplasmic reticulum has ribosomes on its outside surface.

endorphins Compounds that cause a natural euphoria and reduce the perception of pain under certain stressful conditions.

endosperm The largest portion of a kernel of grain. It is primarily starch and serves as a food supply for the sprouting seed.

energy The capacity to do work.

energy balance The amount of energy consumed in the diet compared with the amount expended by the body over a given period.

energy-yielding nutrient A nutrient that can be metabolized to provide energy in the body.

enriched Refers to a food that has had nutrients added to restore those lost in processing to a level equal to or higher than originally present.

enriched grain Grain to which specific amounts of thiamin, riboflavin, niacin, and iron have been added. Since 1998, folic acid has also been added to enriched grains.

enteral or **tube-feeding** A method of feeding by providing a liquid diet directly to the stomach or intestine through a tube placed down the throat or through the wall of the GI tract.

Environmental Protection Agency (EPA) U.S. government agency responsible for determining acceptable levels of environmental contaminants in the food supply and for establishing water quality standards.

enzymes Protein molecules that accelerate the rate of specific chemical reactions without being changed themselves.

EPA See Environmental Protection Agency

epidemiology The study of the interrelationships between health and disease and other factors in the environment or lifestyle of different populations.

epiglottis A piece of elastic connective tissue at the back of the throat that covers the opening of the passageway to the lungs during swallowing.

epinephrine A hormone secreted by the adrenal gland in response to stress that causes changes, such as an increase in heart rate, in preparation for "fight or flight"; also called adrenaline.

epithelial tissue One of the four human tissue types; includes the cells that cover external body surfaces and line internal cavities and tubes.

ergogenic aid Anything designed to increase work or improve performance.

erythropoietin A protein hormone made by the kidneys that stimulates bone marrow stem cells to produce red blood cells. It can be manufactured using genetic engineering and is injected by endurance athletes as an ergogenic aid known as EPO.

esophagus A portion of the GI tract that extends from the pharynx to the stomach.

essential amino acid or **indispensable amino acid** An amino acid that cannot be synthesized by the human body in sufficient amounts to meet needs and therefore must be included in the diet.

essential fatty acid A fatty acid that must be consumed in the diet because it cannot be made by the body or cannot be made in sufficient quantities to meet needs.

essential fatty acid deficiency A condition characterized by dry scaly skin and poor growth that results when the diet does not supply sufficient amounts of the essential fatty acids.

essential hypertension High blood pressure that has no obvious external cause.

essential nutrient A nutrient that must be provided in the diet because the body either cannot make it or cannot make it in sufficient quantities to satisfy its needs.

estimated average requirements (EARs) Intakes, established by the DRIs, that meet the estimated nutrient needs of 50% of individuals in a gender and life-stage group.

estimated energy requirements (EERs) The amount of energy recommended by the DRIs to maintain body weight in a healthy person based on age, gender, size, and activity level.

estrogen A steroid hormone secreted by the ovaries and by the placenta that is involved in the maintenance of pregnancy and the maintenance and development of female sex organs and secondary sex characteristics.

ethanol The type of alcohol in alcoholic beverages. It is produced by yeast fermentation of sugar.

Exchange Lists A food group system that groups foods according to energy and macronutrient content. It is used extensively in planning diabetic and weight loss diets.

excretory system Organ system involved in the elimination of metabolic waste products; includes the lungs, skin, and kidneys.

experimental controls Factors included in an experimental design that limit the number of variables, allowing an investigator to examine the effect of only the parameters of interest.

experimental group In a scientific experiment, the group of participants who undergo the treatment being tested.

extracellular fluid The fluid located outside cells. It includes fluid found in the blood plasma, lymph, gastrointestinal tract, spinal column, eyes, joints, and that found between cells.

extreme obesity See Morbid obesity.

F

facilitated diffusion The movement of substances across a cell membrane from an area of higher concentration to an area of lower concentration with the aid of a carrier molecule. No energy is required.

failure to thrive The inability of a child's growth to keep up with normal growth curves.

fallopian tubes or **oviducts** Narrow ducts leading from the ovaries to the uterus.

famine A widespread lack of access to food due to a disaster that causes a collapse in the food production and marketing systems.

fasting hypoglycemia Low blood sugar that is not related to food intake; often caused by an insulin-secreting tumor.

fat A lipid that is solid at room temperature; commonly used to refer to all lipids or specifically to triglycerides.

fat-free mass Body mass composed of all tissue except adipose tissue.

fat-soluble vitamins Vitamins that dissolve in fat.

fatigue The inability to continue an activity at an optimal level.

fatty acid An organic molecule made up of a chain of carbons linked to hydrogen atoms with an acid group at one end.

fatty liver The accumulation of fat in the liver.

fatty streak A cholesterol deposit in the artery wall.

FDA See Food and Drug Administration.

feces Body waste, including unabsorbed food residue, bacteria, mucus, and dead cells, which is excreted from the gastrointestinal tract by passing through the anus.

female athlete triad The combination of disordered eating, amenorrhea, and osteoporosis that occurs in some female athletes, particularly those involved in sports in which low body weight and appearance are important.

fermentation A process in which microorganisms metabolize components of a food and thus change the composition, taste, and storage properties of the food.

ferritin The major iron storage protein.

fertilization The union of sperm and egg (ovum).

fetal alcohol spectrum disorders (FASDs) A continuum of permanent birth defects in a child due to maternal alcohol consumption during pregnancy.

fetal alcohol syndrome A characteristic group of physical and mental abnormalities in an infant resulting from maternal alcohol consumption during pregnancy.

fetus The developing human from the ninth week to birth. Growth and refinement of structures occur during this time.

fiber A mixture of indigestible carbohydrates and lignin that is found in plants.

fitness The ability to perform routine physical activity without undue fatigue.

flavin adenine dinucleotide (FAD) and **flavin mononucleotide (FMN)** The active coenzyme forms of riboflavin. The structure of these molecules allows them to pick up and donate hydrogen ions and electrons in chemical reactions.

flexibility Range of motion.

fluorosis A condition caused by chronic overconsumption of fluoride, characterized by black and brown stains and cracking and pitting of the teeth.

foam cell A cholesterol-filled white blood cell.

folate and **folacin** General terms for the many forms of this B vitamin. The majority of the folate found naturally in foods is in the polyglutamate form, which contains a string of glutamate molecules that must be removed before it can be absorbed.

folic acid The monoglutamate form of folate, which is used in fortified foods and supplements.

food additive A substance that is intentionally added to or can reasonably be expected to become a component of a food during processing.

Food and Drug Administration (FDA) U.S. government agency responsible for the safety and wholesomeness of all food except red meat, poultry, and eggs; also sets standards and enforces regulations for food labeling and for food and color additives.

food challenge The introduction of foods into the diet one at a time to determine if allergy symptoms occur. It is used in combination with an elimination diet to identify foods that cause an allergic reaction.

Food Code A set of recommendations published by the FDA for the handling and service of food sold in restaurants and other establishments that serve food.

food desert An area that lacks access to affordable fruits, vegetables, whole grains, low-fat milk, and other foods that make up a healthy diet.

food diary A method of assessing dietary intake that involves an individual keeping a written record of all food and drink consumed during a defined period.

food disappearance surveys Surveys that estimate the food use of a population by monitoring the amount of food that leaves the marketplace.

food frequency questionnaire A method of assessing dietary intake that gathers information about how often certain categories of food are consumed.

food insecurity A situation in which people lack adequate physical, social, or economic access to sufficient, safe, nutritious food that meets their dietary needs and food preferences to lead an active and healthy life.

food intolerance An adverse reaction to a food that does not involve antibody production by the immune system.

food jag When a child will only eat one food item meal after meal.

food processing Any alteration of food from the way it is found in nature.

food security Access by all people at all times to sufficient food required for a healthy and active life.

food self-sufficiency The ability of an area to produce enough food to feed its population.
food shortage Insufficient food to feed a population.
food-borne illness An illness caused by consumption of food containing a toxin or disease-causing microorganism.
food-borne infection Illness produced by the ingestion of food containing microorganisms that can multiply inside the body and cause injurious effects.
food-borne intoxication Illness caused by consuming a food containing a toxin.
fortification A general term used to describe the process of adding nutrients to foods.
fortified food Food to which one or more nutrients have been added.
free radical An atom or group of atoms that has at least one unpaired electron and is therefore unstable and highly reactive and can cause cellular damage.
fructose A monosaccharide that is the primary form of carbohydrate found in fruit.
functional foods Foods that provide a potential benefit to health beyond that attributed to the nutrients they contain.

G

galactose A monosaccharide that combines with glucose to form lactose or milk sugar.
gallbladder An organ of the digestive system that stores bile, which is produced by the liver.
gastric banding A surgical procedure in which an adjustable band is placed around the upper portion of the stomach to limit the volume that the stomach can hold and the rate of stomach emptying.
gastric bypass A surgical procedure to treat morbid obesity that both reduces the size of the stomach and bypasses a portion of the small intestine.
gastric juice A collection of secretions including hydrochloric acid and pepsinogen, released by the gastric glands into the stomach.
gastric sleeve A surgical procedure to treat obesity that involves removing a large part of the stomach so the remainder resembles a tube or sleeve in order to limit the volume that the stomach can hold.
gastrin A hormone secreted by the stomach mucosa that stimulates the secretion of gastric juice.
gastroesophageal reflux disease (GERD) A chronic condition in which acidic stomach contents leak back up into the esophagus causing pain and damaging the esophagus.
gastroesophageal sphincter The muscular valve at the top of the stomach that allows food to enter the stomach from the esophagus and then keeps the food and digestive juice from leaking back up into the esophagus.
gastrointestinal tract A hollow tube consisting of the mouth, pharynx, esophagus, stomach, small intestine, large intestine, and anus, in which digestion and absorption of nutrients occur.
gelatin A protein derived from collagen that is deficient in the amino acid tryptophan.
gene A length of DNA containing the information needed to synthesize RNA or a polypeptide chain. Genes are responsible for inherited traits.
gene expression The events of protein synthesis in which the information coded in a gene is used to synthesize a product, either a protein or a molecule of RNA.
generally recognized as safe (GRAS) A group of chemical additives that are generally recognized as safe based on their long-standing presence in the food supply without obvious harmful effects.
genetic engineering A set of techniques used to manipulate DNA for the purpose of changing the characteristics of an organism or creating a new product.
genetically modified (GM) An organism whose genetic material has been altered using genetic engineering techniques; a product containing genetically modified ingredients or produced by genetic modification.
germ The embryo or sprouting portion of a kernel of grain. It contains vegetable oil, protein, fiber, and vitamins.
gestation The time between conception and birth, which lasts about nine months (or about 40 weeks) in humans.
gestational diabetes or **gestational diabetes mellitus** A form of diabetes that occurs during pregnancy and resolves after the baby is born.
gestational hypertension The development of hypertension after the twentieth week of pregnancy.
ghrelin A hormone produced by the stomach that stimulates food intake.
ginseng An herb used in traditional Chinese medicine; claims are made that it improves athletic performance and increases sexual potency.
gleaning The act of collecting the crops left in farmers' fields after they have been commercially harvested or collecting those that were not harvested for economic reasons.
glomerulus A ball of capillaries in the nephron that filters blood during urine formation.
glucagon A hormone made in the pancreas that stimulates the breakdown of liver glycogen and the synthesis of glucose to increase blood sugar.
glucogenic amino acid An amino acid that can be used to synthesize glucose through gluconeogenesis.
gluconeogenesis The synthesis of glucose from simple noncarbohydrate molecules. Amino acids from protein are the primary source of carbons for glucose synthesis.
glucose A monosaccharide that is the primary form of carbohydrate used to provide energy in the body. It is the sugar referred to as blood sugar.
glutamic acid A nonessential amino acid that is found in protein and in monosodium glutamate (MSG).
glutathione peroxidase A selenium-containing enzyme that protects cells from oxidative damage by neutralizing peroxides.
glycemic index A ranking of the effect on blood glucose of a food of a certain carbohydrate content relative to an equal amount of carbohydrate from a reference food such as white bread or glucose.
glycemic load An index of the glycemic response that occurs after eating specific foods. It is calculated by multiplying a food's glycemic index by the amount of available carbohydrate in a serving of the food.
glycemic response The rate, magnitude, and duration of the rise in blood glucose that occurs after a particular food or meal is consumed.
glycerol A 3-carbon molecule that forms the backbone of triglycerides and phosphoglycerides; also used as a humectant in food.
glycogen A carbohydrate made of many glucose molecules linked together in a highly branched structure. It is the storage form of carbohydrate in animals.
glycogen supercompensation or **carbohydrate loading** A regimen of diet and exercise training designed to maximize muscle glycogen stores before an athletic event.
glycogenesis The conversion of glucose to glycogen for storage.
glycolysis (also called **anaerobic metabolism**) Metabolic reactions in the cytosol of the cell that split glucose into two three-carbon pyruvate molecules, yielding two ATP molecules.
goiter An enlargement of the thyroid gland caused by a deficiency of iodine.
goitrogens Substances that interfere with the utilization of iodine or the function of the thyroid gland.
gout Disorder in which crystals of uric acid deposit in and around joints, especially of the big toe and foot, causing pain and arthritis.
GRAS See Generally recognized as safe.
growth hormone A hormone secreted by the pituitary gland that stimulates growth.

H

Hazard Analysis Critical Control Point (HACCP) A system for promoting food safety that focuses on identifying, preventing, and eliminating hazards that could cause food-borne illness.
health claim A statement made about the relationship between a nutrient or food and a disease or health condition.
Healthy Eating Index A system developed to evaluate the adequacy of the American diet. It is based on how well a person's diet meets current dietary recommendations.
Healthy People Initiative A set of national health promotion and disease prevention objectives for the U.S. population.

heart attack A condition in which an artery supplying blood to the heart becomes blocked, cutting off blood flow and hence oxygen and nutrients to a segment of the heart muscle, resulting in tissue death.

heartburn A burning sensation in the chest caused when acidic stomach contents leak into the esophagus through the gastroesophageal sphincter.

heat cramp A muscle cramp caused by an imbalance of sodium and potassium at the muscle cell membrane as a result of excessive exercise without adequate fluid and electrolyte replacement.

heat exhaustion Low blood pressure, rapid pulse, fainting, and sweating caused when dehydration decreases blood volume so much that blood can no longer both cool the body and provide oxygen to the muscles.

heat stroke Elevated body temperature as a result of fluid loss and the failure of the temperature regulatory center of the brain.

heat-related illness Conditions, including heat cramps, heat exhaustion, and heat stroke, that can occur due to an unfavorable combination of exercise, hydration status, and climatic conditions.

heavy drinker Someone who consumes five or more drinks per occasion on at least five different days.

Heimlich maneuver A procedure used to dislodge an object blocking an air passage; involves the application of sharp, firm pressure to the abdomen just below the rib cage.

heme iron A readily absorbed form of iron found in animal products that is chemically associated with proteins such as hemoglobin and myoglobin.

hemochromatosis An inherited condition that results in increased iron absorption.

hemoglobin An iron-containing protein in red blood cells that binds oxygen and transports it through the bloodstream to cells.

hemolytic anemia Anemia that results when red blood cells break open.

hemorrhoids Swollen veins in the anal or rectal area.

hemosiderin An insoluble iron storage compound that stores iron when the amount of iron in the body exceeds the storage capacity of ferritin.

hepatic portal circulation The system of blood vessels that collects nutrient-laden blood from the digestive organs and delivers it to the liver.

hepatic portal vein The vein that transports blood from the GI tract to the liver.

hepatitis Inflammation of the liver.

hepsidin A peptide hormone produced by the liver that regulates the absorption of iron across the intestinal mucosa.

herb The leaves, flowers, stems, roots, seeds, or any other part of a nonwoody seed-bearing plant that dies down to the ground after flowering.

herbicide An agent that kills weeds.

heterocyclic amines (HCAs) A class of mutagenic substances produced when there is incomplete combustion of amino acids during the cooking of meats—such as when meat is charred.

hiatal hernia When the upper part of the stomach bulges through the opening in the diaphragm into the chest cavity.

high-density lipoprotein (HDL) A lipoprotein that picks up cholesterol from cells and transports it to the liver so that it can be eliminated from the body. A high level of HDL decreases the risk of cardiovascular disease.

high-fructose corn syrup A sweetener made from corn syrup that is composed of approximately half fructose and half glucose.

high-quality protein or **complete dietary protein** Protein that provides essential amino acids in the proportions needed to support protein synthesis.

histamine A substance produced by cells of the immune system as part of a nonspecific response that leads to inflammation.

HIV (Human Immunodeficiency Virus) A virus that infects cells of the immune system and eventually leads to AIDS.

homeostasis A physiological state in which a stable internal body environment is maintained.

homocysteine A sulfur-containing amino acid that is produced from the metabolism of methionine. Elevated blood levels increase the risk of cardiovascular disease.

hormones Chemical messengers that are produced in one location, released into the blood, and elicit responses at other locations in the body.

hormone-sensitive lipase An enzyme present in adipose cells that responds to chemical signals by breaking down triglycerides into fatty acids and glycerol for release into the bloodstream.

human intervention study or **clinical trial** A study of a population in which there is an experimental manipulation of some members of the population; observations and measurements are made to determine the effects of this manipulation.

humectant A substance added to foods to retain moisture.

hunger A desire to consume food that is triggered by internal physiological signals. Also refers to the recurrent involuntary lack of food that over time may lead to malnutrition.

hybridization The process of cross-fertilizing two related plants with the goal of producing an offspring that has the desirable characteristics of both parent plants.

hydrochloric acid An acid secreted by the gastric glands of the stomach to aid in digestion.

hydrogen peroxide A reactive oxygen-containing compound that can form free radicals and cause oxidative damage. It can be eliminated by the selenium-containing enzyme glutathione peroxidase.

hydrogenation The process whereby hydrogen atoms are added to the carbon-carbon double bonds of unsaturated fatty acids, making them more saturated.

hydrolysis reaction Chemical reaction that breaks large molecules into smaller ones by adding water.

hydrolyzed protein See Protein hydrolysate.

hydroxyapatite A crystalline compound composed of calcium and phosphorus that is deposited in the protein matrix of bone to give it strength and rigidity.

hyperactivity Overactive, excitable, distractible behavior that is characteristic of attention deficit hyperactivity disorder.

hypercarotenemia A condition in which carotenoids accumulate in the adipose tissue, causing the skin to appear yellow-orange.

hyperemesis gravidarum A severe and intractable form of nausea and vomiting that affects some women during pregnancy.

hypertension Blood pressure that is consistently elevated to 140/90 mm Hg or greater.

hypertensive disorders of pregnancy High blood pressure during pregnancy that is due to chronic hypertension, gestational hypertension, preeclampsia-eclampsia, or preeclampsia superimposed on chronic hypertension.

hypertrophy An increase in the size of a muscle or organ.

hypoglycemia A low blood glucose level, usually below 40 to 50 mg of glucose per 100 mL of blood.

hyponatremia Low blood sodium concentration.

hypothalamus The region of the brain that monitors and regulates conditions and activities in the body, including food intake and energy expenditure.

hypothermia A condition in which body temperature drops below normal. Hypothermia depresses the central nervous system, resulting in the inability to shiver, sleepiness, and eventually coma.

hypothesis An educated guess made to explain an observation or to answer a question.

I

ileocecal valve The structure that separates the ileum of the small intestine from the large intestine.

ileum The 11-foot segment of the small intestine that connects the jejunum with the large intestine.

immunization An injection of a killed or inactivated organism into the body to stimulate the immune system to develop antibodies against the active disease-causing organism.

impaired glucose tolerance or **pre-diabetes** A fasting blood sugar level above the normal range but not high enough to be classified as diabetes (100–125 mg/dL).

implantation The process by which the developing embryo embeds in the uterine lining.

incomplete dietary protein Protein that is deficient in one or more essential amino acids relative to body needs.

indirect calorimetry A method of estimating energy use that compares the amount of oxygen consumed to the amount of carbon dioxide expired.

indirect food additive A substance that is expected to enter food unintentionally during manufacturing or from packaging. Indirect food additives are regulated by the FDA.

infant mortality rate The number of deaths during the first year of life per 1000 live births.

inflammation The response of a part of the body to injury that increases blood flow with an influx of white blood cells and other chemical substances to facilitate healing. It is characterized by swelling, heat, redness, pain, and loss of function.

inorganic molecules Those containing no carbon hydrogen bonds.

insensible losses Fluid losses that are not perceived by the senses, such as evaporation of water through the skin and lungs.

insoluble fiber Fiber that, for the most part, does not dissolve in water and cannot be broken down by bacteria in the large intestine. It includes cellulose, some hemicelluloses, and lignin.

insulin A hormone made in the pancreas that allows the uptake of glucose by body cells and has other metabolic effects such as stimulating protein and fat synthesis and the synthesis of glycogen in liver and muscle.

insulin-dependent diabetes See Type 1 diabetes.

insulin resistance A condition in which body cells have a reduced response to insulin.

integrated pest management (IPM) A method of agricultural pest control that integrates nonchemical and chemical techniques.

intermediate-density lipoprotein (IDL) A lipoprotein produced by the removal of triglycerides from VLDLs, most of which are then transformed to LDLs.

International Unit (IU) A unit of measure that has been used to express requirements of some vitamins.

interstitial fluid The portion of the extracellular fluid located in the spaces between the cells of body tissues.

intervention study See Clinical trial.

intestinal microflora Microorganisms that inhabit the large intestine.

intracellular fluid The fluid located inside cells.

intrinsic factor A protein produced in the stomach that is needed for the absorption of adequate amounts of vitamin B_{12}.

ion An atom or group of atoms that carries an electrical charge.

iron deficiency anemia An iron deficiency disease that occurs when the oxygen-carrying capacity of the blood is decreased because there is insufficient iron to make hemoglobin.

irradiation A process, also called cold pasteurization, that exposes foods to radiation to kill contaminating organisms and retard ripening and spoilage of fruits and vegetables.

isomers Molecules with the same molecular formula but a different arrangement of the atoms.

isotopes Alternative forms of an element that have different atomic masses, which may or may not be radioactive.

J

jejunum The 8-foot-long section of the small intestine lying between the duodenum and the ileum.

juvenile-onset diabetes See Type 1 diabetes.

K

kcalorie An abbreviation for kilocalorie, the unit of heat that is used to express the amount of energy provided by food. The term *calorie* is commonly used to refer to kcalorie.

keratin A hard protein that makes up hair and nails.

keratomalacia Softening and drying and ulceration of the cornea resulting from vitamin A deficiency.

Keshan disease A type of heart disease that occurs in areas of China where the soil is very low in selenium. It is believed to be caused by a combination of viral infection and selenium deficiency.

ketoacidosis A life-threatening increase in the acidity of the blood due to high ketone levels.

ketogenic amino acid An amino acid that breaks down to form acetyl-CoA and thus contributes to ketone synthesis.

ketones or **ketone bodies** Molecules formed in the liver when there is not sufficient carbohydrate to completely metabolize the two-carbon units produced from fat breakdown.

ketosis High levels of ketones in the blood.

kilocalorie (kcalorie, kcal) The unit of heat used to express the amount of energy provided by foods. It is the amount of heat required to raise the temperature of 1 kg of water 1 degree Celsius (1 kcalorie = 4.18 kjoules).

kilojoule (kjoule, kJ) A unit of work that can be used to express energy intake and energy output. It is the amount of work required to move an object weighing 1 kg a distance of 1 m under the force of gravity (4.18 kjoules = 1 kcalorie).

kinky hair disease An inherited defect in copper transport that results in copper deficiency. Symptoms include poor growth, neurological symptoms, and coarse, kinky, colorless, brittle hair. Also called Menkes disease.

kwashiorkor A form of protein-energy malnutrition in which only protein is deficient.

L

lactase An enzyme located in the microvilli of the small intestine that breaks the disaccharide lactose into glucose and galactose.

lactation Milk production and secretion.

lacteal A small lymph vessel in the intestine that absorbs and transports the products of fat digestion.

lactic acid A compound produced from the breakdown of glucose in the absence of oxygen.

lacto-ovo vegetarian One who eats no animal flesh but eats eggs and dairy products such as milk and cheese.

lacto-vegetarian One who eats no animal flesh or eggs but eats dairy products.

lactose A disaccharide that is formed by linking galactose and glucose. It is commonly known as milk sugar.

lactose intolerance The inability to digest lactose because of a reduction in the levels of the enzyme lactase. It causes symptoms including intestinal gas and bloating after dairy products are consumed.

large for gestational age An infant weighing more than 4 kg (8.8 lbs) at birth.

large intestine The portion of the gastrointestinal tract that includes the colon and rectum; some water and vitamins are absorbed, and bacteria act on food residues here.

laxative A substance that eases the excretion of feces.

LDL receptor A protein on the surface of cells that binds to LDL particles and allows their contents to be taken up for use by the cell.

lean body mass Body mass attributed to nonfat body components such as bone, muscle, and internal organs. It is also called *fat-free mass*.

leavening agent A substance added to food that causes the production of gas, resulting in an increase in volume.

lecithin A phosphoglyceride composed of a glycerol backbone, two fatty acids, a phosphate group, and a molecule of choline.

legumes Plants in the pea or bean family, which produce an elongated pod containing large starchy seeds. Examples include green peas, kidney beans, and peanuts.

leptin A protein hormone produced by adipocytes that signals information about the amount of body fat.

leptin receptors Proteins that bind the hormone leptin. In response to this binding, they trigger events that cause changes in food intake and energy expenditure.

letdown A hormonal reflex triggered by the infant's suckling that causes milk to be released from the milk glands and flow through the duct system to the nipple.

life expectancy The average length of life for a population of individuals.

life span The maximum age to which members of a species can live.

life-stage groups Groupings of individuals used in the DRIs based on stages of growth and development, pregnancy, and lactation that have similar nutrient needs.

lignin An insoluble fiber responsible for the hard woody nature of plant stems.

limiting amino acid The essential amino acid that is available in the lowest concentration in relation to the body's needs.

linoleic acid An essential omega-6 essential fatty acid with 18 carbons and 2 double bonds.
lipase A fat-digesting enzyme.
lipid bilayer Two layers of phosphoglyceride molecules oriented so that the fat-soluble fatty acid tails are sandwiched between the water-soluble phosphate-containing heads.
lipids A group of organic molecules, most of which do not dissolve in water. They include fatty acids, triglycerides, phospholipids, and sterols.
lipoprotein lipase An enzyme that breaks down triglycerides into fatty acids and glycerol; attached to the cell membranes of cells that line the blood vessels.
lipoproteins Particles containing a core of triglycerides and cholesterol surrounded by a shell of protein, phospholipids, and cholesterol that transport lipids in blood and lymph.
liposuction A procedure that suctions out adipose tissue from under the skin; used to decrease the size of local fat deposits such as on the abdomen or hips.
longevity The duration of an individual's life.
longitudinal data Information obtained by repeatedly sampling the same individuals in a population over time.
low birth weight A birth weight less than 2.5 kg (5.5 lbs).
low-density lipoprotein (LDL) A lipoprotein that transports cholesterol to cells. Elevated LDL cholesterol increases the risk of cardiovascular disease.
lumen The inside cavity of a tube, such as the gastrointestinal tract.
lutein A carotenoid found in corn and green peppers that has no vitamin A activity but provides some protection against macular degeneration.
lycopene A carotenoid that gives the red color to tomatoes. It cannot be converted to vitamin A.
lymphatic system The system of vessels, organs, and tissues that drains excess fluid from the spaces between cells, transports fat-soluble substances from the digestive tract, and contributes to immune function.
lymph vessel A tubular component of the lymphatic system that carries fluid away from body tissues. Lymph vessels in the intestine are known as lacteals and can transport large particles such as the products of fat digestion.
lysozyme An enzyme in saliva, tears, and sweat that is capable of destroying certain types of bacteria.

M

macrocytes Larger-than-normal mature red blood cells that have a shortened life span.
macrocytic anemia See megaloblastic anemia.
macronutrient A nutrient needed by the body in large amounts. These include water and the energy-yielding nutrients; carbohydrates, lipids, and proteins.
macular degeneration Degeneration of a portion of the retina that results in a loss of visual detail and blindness.
mad cow disease The common name for bovine spongiform encephalaopathy, a cattle disease that causes the brain to degenerate.
major mineral A mineral needed in the diet in an amount greater than 100 mg per day or present in the body in an amount greater than 0.01% of body weight.
malignancy A mass of cells showing uncontrolled growth, a tendency to invade and damage surrounding tissues, and an ability to seed daughter growths to sites remote from the original growth.
malnutrition Any condition resulting from an energy or nutrient intake either above or below that which is optimal.
maltase An enzyme found in the microvilli of the small intestine that breaks maltose into two molecules of glucose.
maltose A disaccharide made up of two molecules of glucose. It is formed in the intestines during starch digestion.
marasmus A form of protein-energy malnutrition in which a deficiency of energy in the diet causes severe body wasting.
maturity-onset diabetes See Type 2 diabetes.
maximal oxygen consumption or **VO_2 max** The maximum amount of oxygen that can be consumed by the tissues during exercise. Also called aerobic capacity.
maximum heart rate The maximum number of beats per minute that the heart can attain. It declines with age.
megaloblastic or **macrocytic anemia** A condition in which there are abnormally large immature and mature red blood cells in the bloodstream and a reduction in the total number of red blood cells and hence in the oxygen-carrying capacity of the blood.
megaloblasts Large, immature red blood cells that are formed when developing red blood cells are unable to divide normally.
melatonin A hormone involved in regulating the body's cycles of sleep and wakefulness. Levels decline with age. Supplements are claimed to boost antioxidant defenses, improve immune function, and slow aging.
menaquinones The forms of vitamin K synthesized by bacteria and found in animals.
menarche The onset of menstruation, which occurs normally between the ages of 10 and 15 years.
Menkes disease See Kinky hair disease.
menopause The physiological changes that mark the end of a woman's capacity to bear children.
menstruation The cyclic discharge of the uterine lining that, in the absence of pregnancy, occurs about every four weeks during the reproductive years of female humans.
metabolic pathway A series of chemical reactions inside a living organism that results in the transformation of one molecule into another.
metabolic syndrome A collection of health risks, including high blood pressure, altered blood lipids, high blood glucose and a large waist circumference, that increases the chance of developing heart disease, stroke, and diabetes. The condition is also known by other names including Syndrome X, insulin resistance syndrome, and dysmetabolic syndrome.
metabolism The sum of all the chemical reactions that take place in a living organism.
metallothionein Refers to proteins that bind minerals. One such protein binds zinc and copper in intestinal cells, limiting their absorption into the blood.
methionine An essential sulfur-containing amino acid found in protein.
methyl group A chemical group consisting of a carbon atom bound to three hydrogen atoms.
micelles Particles formed in the small intestine when the products of fat digestion are surrounded by bile acids. They facilitate the absorption of fat.
microbe An organism too small to be seen without a microscope; also called a microorganism.
microflora See Intestinal microflora.
micronutrient A nutrient needed by the body in small amounts. These include vitamins and minerals.
microorganism An organism such as a bacterium, too small to be seen without a microscope.
microsomal ethanol-oxidizing system (MEOS) A liver enzyme system located in microsomes that converts alcohol to acetaldehyde. Activity increases with increases in alcohol consumption.
microvilli or **brush border** Minute brush-like projections on the mucosal cell membrane that increase the absorptive surface area in the small intestine.
mineral In nutrition, an element needed by the body in small amounts for structure and to regulate chemical reactions and body processes.
miscarriage or **spontaneous abortion** Interruption of pregnancy prior to the seventh month.
mitochondrion (mitochondria) The cellular organelle that is responsible for producing ATP via aerobic metabolism; the citric acid cycle and electron transport chain are located here.
modified atmosphere packaging (MAP) A preservation technique used to prolong the shelf life of processed or fresh food by changing the gases surrounding the food in the package.
modified starch or **modified food starch** Starch that has been treated to enhance its ability to thicken or form a gel.
mold Multicellular fungi that form a filamentous branching growth.
molecular biology The study of cellular function at the molecular level.
molecule A group of two or more atoms of the same or different elements bonded together.
monoglyceride or **monoacylglycerol** A molecule of glycerol with one fatty acid attached.

monosaccharide A carbohydrate made up of a single sugar unit.
monosodium glutamate (MSG) An additive used as a flavor enhancer, commonly used in Chinese food; made up of the amino acid glutamate bound to sodium.
monounsaturated fatty acid A fatty acid that contains one carbon-carbon double bond.
morbid obesity or **extreme obesity** A body mass index greater than 40 kg/m^2.
morbidity The incidence or state of disease or disability.
morning sickness Nausea and vomiting that affects many women during the first few months of pregnancy and in some women can continue throughout the pregnancy.
mRNA (messenger RNA) A molecule that carries the information in a gene to ribosomes in the cytosol so polypeptides can be synthesized.
MSG symptom complex Symptoms of headache, flushing, tingling, burning sensations, and chest pain reported by some individuals after consuming monosodium glutamate (MSG); commonly referred to as Chinese restaurant syndrome.
mucosa The layer of tissue lining the GI tract and other body cavities.
mucus A viscous fluid secreted by glands in the gastrointestinal tract and other parts of the body. It acts to lubricate, moisten, and protect cells from harsh environments.
muscle-strengthening exercise Activities that are specifically designed to increase muscle strength, endurance, and size; also called strength-training exercise or resistance exercise.
mutagen Any agent that causes a change in a cell's genetic material.
mutations Changes in DNA caused by chemical or physical agents.
myelin A soft, white fatty substance that covers nerve fibers and aids in nerve transmission.
myocardial infarction Heart attack.
myoglobin An iron-containing protein in muscle cells that binds oxygen.

N

National Health and **Nutrition Examination Survey (NHANES)** An ongoing set of surveys designed to monitor the overall nutritional status of the U.S. population; combines food consumption information with medical histories, physical examinations, and laboratory measurements.
nephron The functional unit of the kidney that performs the job of filtering the blood and maintaining fluid balance.
nervous system A system of nerve cells organized in message sending, message receiving, and information processing pathways.
net protein utilization A measure of protein quality determined by comparing the amount of nitrogen retained in the body with the amount eaten in the diet.
neural tube The portion of the embryo that develops into the brain and spinal cord.
neural tube closure A developmental event in which neural tissue forms a groove and the sides fold together to form a tube; is completed about 28 days after fertilization.
neural tube defect A defect in the formation of the neural tube that occurs early in development and results in defects of the brain and spinal cord such as anencephaly and spina bifida.
neurotransmitter A chemical substance produced by a nerve cell that can stimulate or inhibit another cell.
Niacin A B vitamin, also called vitamin B_3, that is needed in energy metabolism.
niacin equivalents (NEs) The measure used to express the amount of niacin present in food, including that which can be made from its precursor, tryptophan. One NE is equal to 1 mg of niacin or 60 mg of tryptophan.
nicotinamide A form of niacin.
nicotinamide adenine dinucleotide (NAD) and **nicotinamide adenine dinucleotide phosphate (NADP)** The active coenzyme forms of niacin that can pick up and donate hydrogens and electrons. They are important in the transfer of electrons to oxygen in cellular respiration and in many synthetic reactions.
nicotinic acid A form of niacin.
night blindness The inability of the eye to adapt to reduced light, causing poor vision in dim light.
nitrogen balance The amount of nitrogen consumed in the diet compared with the amount excreted by the body over a given period.
nitrosamines Carcinogenic compounds produced by reactions between nitrites and amino acids.
nonessential or **dispensable amino acid** An amino acid that can be synthesized by the human body in sufficient amounts to meet needs.
nonexercise activity thermogenesis (NEAT) The energy expended for everything we do other than sleeping, eating, or sports-like exercise.
nonheme iron A poorly absorbed form of iron found in both plant and animal foods that is not part of the iron complex found in hemoglobin and myoglobin.
noninsulin-dependent diabetes See Type 2 diabetes.
nucleus The central core of an atom, consisting of positively charged protons and electrically neutral neutrons. In cells, it is an organelle containing DNA.
nursing bottle syndrome Extreme tooth decay in the upper teeth resulting from putting a child to bed with a bottle containing milk or other sweet liquids.
nutrient density An evaluation of the nutrient content of a food in comparison to the calories it provides.
nutrients Chemical substances in foods that provide energy and structure and help regulate body processes.
nutrition A science that studies the interactions between living organisms and food.
nutrition transition A series of changes in diet, physical activity, health and nutrition that occurs as poor countries become more prosperous.
nutritional assessment An evaluation used to determine the nutritional status of individuals or groups for the purpose of identifying nutritional needs and planning personal healthcare or community programs to meet those needs.
nutritional genomics or **nutrigenomics** The study of how diet affects our genes and how individual genetic variation can affect the impact of nutrients or other food components on health.
nutritional status State of health as it is influenced by the intake and utilization of nutrients.

O

obese A condition characterized by excess body fat. Obesity is defined as a body mass index of 30 kg/m^2 or greater.
obesity genes Genes that code for proteins involved in the regulation of food intake, energy expenditure, or the deposition of body fat. When they are abnormal, the result is abnormal amounts of body fat.
oil A lipid that is liquid at room temperature.
oleic acid A monounsaturated fatty acid with 18 carbons.
Olestra (sucrose polyester) An artificial fat made of sucrose with fatty acids linked to it. It cannot be digested or absorbed. It has been approved by the FDA for use in certain snack foods.
oligosaccharides Short-chain carbohydrates containing three to 10 sugar units.
omega-3 fatty acid A fatty acid containing a carbon-carbon double bond between the third and fourth carbons from the omega end.
omega-6 fatty acid A fatty acid containing a carbon-carbon double bond between the sixth and seventh carbons from the omega end.
opsin A protein in the retina of the eye involved in the visual cycle.
organ system A group of cooperative organs.
organelles Cellular organs that carry out specific metabolic functions.
organic food Food that is produced, processed, and handled in accordance with the standards of the USDA National Organic Program.
organic molecule A molecule that contains carbon bonded to hydrogen.
organs Discrete structures composed of more than one tissue that perform a specialized function.
osmosis The passive movement of water across a semipermeable membrane in a direction that will equalize the concentration of dissolved substances on both sides.
osteoarthritis The form of arthritis common in the elderly that is characterized by a wearing down of the joint surfaces and pain when the joint is moved.
osteoblast A type of cell responsible for the deposition of bone.

osteoclast A type of cell responsible for bone breakdown.

osteomalacia A vitamin D deficiency disease in adults characterized by a loss of minerals from bones. It causes bone pain, muscle aches, and an increase in bone fractures.

osteoporosis A bone disorder characterized by a reduction in bone mass, increased bone fragility, and an increased risk of fractures.

overload principle The concept that the body will adapt to the stresses placed on it.

overnutrition Poor nutritional status resulting from an energy or nutrient intake in excess of that which is optimal for health.

overtraining syndrome A collection of emotional, behavioral, and physical symptoms that occurs when training without sufficient rest persists for weeks to months.

overweight Being too heavy for one's height. It is defined as having a body mass index (a ratio of weight to height squared) of 25 to 29.9 kg/m^2.

oviducts or **fallopian tubes** Narrow ducts leading from the ovaries to the uterus.

ovum The female reproductive cell.

oxalate An organic acid found in spinach and other leafy green vegetables that can bind minerals and decrease their absorption.

oxaloacetate A 4-carbon compound derived from carbohydrate that combines with acetyl-CoA in the first step of the citric acid cycle.

oxidation The loss of electrons.

oxidation-reduction reaction A reaction in which electrons are transferred from a donor molecule (the reducing agent) to an acceptor molecule (the oxidizing agent).

oxidative damage Damage caused by highly reactive oxygen molecules that steal electrons from other compounds, causing changes in structure and function.

oxidative stress A condition that occurs when there are more reactive oxygen molecules than can be neutralized by available antioxidant defenses. It occurs either because excessive amounts of reactive oxygen molecules are generated or because antioxidant defenses are deficient.

oxidized Refers to a compound that has lost an electron or undergone a chemical reaction with oxygen.

oxidized LDL cholesterol A substance formed when the cholesterol in LDL particles is oxidized by reactive oxygen molecules. It is key in the development of atherosclerosis because it contributes to the inflammatory process.

oxytocin A hormone produced by the posterior pituitary gland that acts on the uterus to cause uterine contractions and on the breast to cause the movement of milk into the secretory ducts that lead to the nipple.

P

PA (physical activity) value A numeric value associated with activity level that is a variable in the EER equations used to calculate energy needs.

palmitic acid A saturated fatty acid containing 16 carbons.

pancreas An organ that secretes digestive enzymes and bicarbonate ions into the small intestine during digestion.

pancreatic amylase Starch-digesting enzyme produced in the pancreas and released into the small intestine.

pancreatic juice The secretion of the pancreas containing bicarbonate to neutralize acid and enzymes for the digestion of carbohydrates, fats, and proteins.

pantothenic acid A B vitamin needed in energy metabolism.

papain A protein-digesting enzyme found in papaya.

parasites Organisms that live at the expense of others.

parathyroid hormone (PTH) A hormone secreted by the parathyroid gland that increases blood calcium levels.

parietal cells Large cells in the stomach lining that produce and secrete intrinsic factor and hydrochloric acid.

partially hydrogenated vegetable oil Vegetable oil that has been modified by hydrogenation to decrease the number of unsaturated bonds, therefore raising the melting point and improving the storage characteristics.

pasteurization The process of heating food products to kill disease-causing organisms.

pathogen A biological agent that causes disease.

peak bone mass The maximum bone density attained at any time in life, usually occurring in young adulthood.

pectin A soluble fiber found in plant cell walls that forms a gel when mixed with acid and sugar.

peer review Review of the design and validity of a research experiment by experts in the field of study who did not participate in the research.

pellagra The disease resulting from a deficiency of niacin.

pepsin A protein-digesting enzyme produced by the gastric glands. It is secreted in the gastric juice in an inactive form and activated by acid in the stomach.

pepsinogen An inactive protein-digesting enzyme produced by gastric glands and activated to pepsin by acid in the stomach.

peptic ulcer An open sore in the lining of the stomach, esophagus, or small intestine.

peptide Two or more amino acids joined by peptide bonds.

peptide bond A chemical linkage between the amino group of one amino acid and the acid group of another.

peptide YY An appetite-suppressing peptide hormone that is released from the gastrointestinal tract after a meal.

periodontal disease A degeneration of the area surrounding the teeth, specifically the gum and supporting bone.

peristalsis Coordinated muscular contractions that move food through the gastrointestinal tract.

pernicious anemia A macrocytic anemia resulting from vitamin B_{12} deficiency that occurs when dietary vitamin B_{12} cannot be absorbed due to a lack of intrinsic factor.

peroxide A reactive chemical that can form free radicals and cause cellular damage.

pesticide Any substance or mixture of substances intended to prevent, destroy, repel, or mitigate insect, animal, plant, fungal, or microbial pests.

pesticide tolerances The maximum amount of pesticide residues that may legally remain in food, set by the EPA.

pH A measure of acidity.

pharynx A funnel-shaped opening that connects the nasal passages and mouth to the respiratory passages and esophagus. It is a common passageway for food and air and is responsible for swallowing.

phenylalanine An essential amino acid found in protein that cannot be metabolized by individuals with phenylketonuria (PKU).

phenylketone The product of phenylalanine breakdown produced when phenylalanine cannot be converted to tyrosine; when blood levels get too high, brain damage may result.

phenylketonuria (PKU) An inherited disease in which the body cannot metabolize the amino acid phenylalanine. If the disease is untreated, toxic by-products called *phenylketones* accumulate in the blood and interfere with brain development.

phosphoglyceride A type of phospholipid consisting of a glycerol molecule, two fatty acids, and a phosphate group.

phospholipid A type of lipid containing phosphorus. The most common is a phosphoglyceride, which is composed of a glycerol backbone with two fatty acids and a phosphate group attached.

photosynthesis The metabolic process by which plants trap energy from the sun and use it to make sugars from carbon dioxide and water.

phylloquinone The form of vitamin K found in plants.

physical frailty Impairment in function and reduction in physiological reserves severe enough to cause limitations in the basic activities of daily living.

phytate or **phytic acid** A phosphorus storage compound found in seeds and grains that can bind minerals and decrease their absorption.

phytochemicals Substances found in plant foods (phyto- means plant) that are not essential nutrients but may have health-promoting properties.

phytoestrogen An estrogen-like molecule produced by plants.

phytosterol Compound produced by plants that has a structure similar to cholesterol.

pica An abnormal craving for and ingestion of unusual food and nonfood substances.

placebo A fake medicine or supplement that is indistinguishable in appearance from the real thing. It is used to disguise the control and experimental groups in an experiment.

placenta An organ produced from both maternal and embryonic tissues. It secretes hormones, transfers nutrients and oxygen from the mother's blood to the fetus, and removes wastes.

plant sterols and **stanols** Compounds found in plant cell membranes that resemble cholesterol in structure. They can lower blood cholesterol by competing with cholesterol for absorption in the gastrointestinal tract.

plaque The cholesterol-rich material that is deposited in the blood vessels of individuals with atherosclerosis. It consists of cholesterol, smooth muscle cells, fibrous tissue and, eventually, calcium. Also refers to deposits on the teeth composed of mucus and bacteria that contribute to dental caries.

plasma The liquid portion of the blood that remains when the blood cells are removed.

plasmid A loop of bacterial DNA that is independent of the bacterial chromosome.

platelet A cell fragment found in blood that is involved in blood clotting.

polar Used to describe a molecule that has a positive charge at one end and a negative charge at the other.

polychlorinated biphenyls (PCBs) Carcinogenic industrial compounds that have found their way into the environment and, subsequently, the food supply. Repeated ingestion causes them to accumulate in biological tissues over time.

polycyclic aromatic hydrocarbons (PAHs) A class of mutagenic substances produced during cooking when there is incomplete combustion of organic materials—such as when fat drips on a grill.

polypeptide A chain of three or more amino acids joined together by peptide bonds.

polysaccharide A carbohydrate made up of many sugar units linked together.

polyunsaturated fatty acid A fatty acid that contains two or more carbon-carbon double bonds.

postmenopausal bone loss Accelerated bone loss that occurs in women for about 5 to 10 years surrounding menopause.

prebiotic A substance that passes undigested into the colon and stimulates the growth and/or activity of certain types of bacteria.

precursor Inactive form of a substance that can be converted into the active form.

pre-diabetes or **impaired glucose tolerance** A fasting blood glucose level above the normal range but not high enough to be classified as diabetes.

preeclampsia A condition characterized by an increase in body weight, elevated blood pressure, protein in the urine, and edema. It can progress to eclampsia, which can be life-threatening to mother and fetus.

pregnancy-induced hypertension See hypertensive disorders of pregnancy.

premature or **preterm infant** An infant born before 37 weeks of gestation.

premenstrual syndrome (PMS) A syndrome of mood swings, food cravings, bloating, tension and depression, headaches, acne, and anxiety, among other symptoms, that results from the hormonal changes during the days prior to menstruation.

preservative A compound that extends the shelf life of a product by retarding chemical, physical, or microbiological changes.

preterm or **premature** An infant born before 37 weeks of gestation.

prion A pathogenic protein that is the cause of degenerative brain diseases called spongiform encephalopathies. Prion is short for proteinaceous infectious particle.

prior-sanctioned substances Refers to substances that the FDA or the USDA had determined were safe for use in a specific food prior to the 1958 Food Additives Amendment.

pro-oxidant A substance that promotes oxidative damage.

probiotic Live bacteria that when consumed live temporarily in the colon and confer health benefits on the host.

processed food A food that has been specially treated or changed from its natural state.

progesterone A female sex hormone needed for development and function of the uterus and mammary glands.

programmed cell death The death of cells at specific predictable times.

prolactin A hormone released by the anterior pituitary that acts on the milk-producing glands in the breast to stimulate and sustain milk production.

protease A protein-digesting enzyme.

protein An organic molecule made up of one or more intertwining chains of amino acids.

protein complementation The process of combining proteins from different sources so that they collectively provide the proportions of amino acids required to meet the body's needs.

protein digestibility-corrected amino acid score (PDCAAS) A measure of protein quality that reflects a protein's digestibility as well as the proportions of amino acids it provides.

protein hydrolysate or **hydrolyzed protein** A mixture of amino acids or amino acids and polypeptides that results when a protein is completely or partially broken down by treatment with acid or enzymes.

protein quality A measure of how efficiently a protein in the diet can be used to make body proteins.

protein turnover The continuous synthesis and breakdown of body proteins.

protein-energy malnutrition (PEM) A condition characterized by wasting and an increased susceptibility to infection that results from the long-term consumption of insufficient amounts of energy and protein to meet needs.

protein-sparing modified fast A very-low-calorie diet with a high proportion of protein designed to maximize the loss of fat and minimize the loss of protein from the body.

prothrombin A blood protein required for blood clotting.

provitamin or **vitamin precursor** A compound that can be converted into the active form of a vitamin in the body.

psyllium A plant product high in soluble fiber that is used in over-the-counter bulk-forming laxatives.

puberty A period of rapid growth and physical changes that ends in the attainment of sexual maturity.

purging Behaviors such as self-induced vomiting and misuse of laxatives, diuretics, or enemas to rid the body of calories.

pyloric sphincter A muscular valve that helps regulate the rate at which food leaves the stomach and enters the small intestine.

pyridoxal phosphate The major coenzyme form of vitamin B_6 that functions in more than 100 enzymatic reactions, many of which involve amino acid metabolism.

pyridoxine The chemical term for vitamin B_6. It is used to refer to a group of compounds, including pyridoxal, pyridoxine, and pyridoxamine.

pyruvate A 3-carbon molecule produced when glucose is broken down by glycolysis.

Q

qualified health claims Health claims on food labels that have been approved based on emerging but not well-established evidence for a relationship between a food, food component, or dietary supplement and reduced risk of a disease or health-related condition.

R

raffinose An oligosaccharide found in beans and other legumes that cannot be digested by human enzymes in the stomach and small intestine.

recombinant DNA DNA that has been formed by joining DNA from different sources.

recommended dietary allowances (RDAs) DRI values that recommend intakes that are sufficient to meet the nutrient needs of almost all healthy people in a specific life-stage and gender group.

rectum The portion of the large intestine that connects the colon and anus.

reduced Refers to a substance that has gained an electron.

Reference Daily Intakes (RDIs) Reference values established for vitamins and minerals that are based on the highest amount of each nutrient recommended for any adult age group by the 1968 RDAs.

refined Refers to foods that have undergone processes that change or remove various components of the original food.

renewable resources Resources that are restored and replaced by natural processes and that can therefore be used forever.

renin An enzyme produced by the kidneys that converts angiotensinogen to angiotensin I.

reserve capacity The amount of functional capacity that an organ has above and beyond what is needed to sustain life.

resistant starch Starch that escapes digestion in the small intestine of healthy people.

respiratory system Organ system that includes the lungs and air passageways involved in the exchange of oxygen from the environment with carbon dioxide waste from cells by way of the bloodstream.

resting energy expenditure (REE) Energy expenditure at rest. It is measured after 5 or 6 hours without food or exercise.

resting heart rate The number of times that the heart beats per minute while a person is at rest.

resting metabolic rate (RMR) An estimate of basal metabolic rate that is determined by measuring energy utilization after 5 to 6 hours without food or exercise.

restriction enzyme A bacterial enzyme used in genetic engineering that has the ability to cut DNA in a specific location.

retinal The aldehyde form of vitamin A, which is needed for the visual cycle.

retinoic acid The acid form of vitamin A, which is needed for cell differentiation, growth, and reproduction.

retinoids The chemical forms of preformed vitamin A: retinol, retinal, and retinoic acid.

retinol The alcohol form of vitamin A, which can be interconverted with retinal.

retinol activity equivalent (RAE) The amount of retinol, β-carotene, α-carotene, or β-cryptoxanthin that provides vitamin A activity equal to 1 μg of retinol.

retinol-binding protein A protein that is necessary to transport vitamin A from the liver to other tissues.

rhodopsin A light-absorbing compound found in the retina of the eye that is composed of the protein opsin loosely bound to retinal.

Riboflavin A B vitamin, also called vitamin B_2, that is needed in energy metabolism.

ribose The 5-carbon sugar that is part of RNA.

ribosome The cell organelle where protein synthesis occurs.

rickets A vitamin D deficiency disease in children that is characterized by poor bone development because of inadequate calcium absorption.

risk factor A characteristic or circumstance that is associated with the occurrence of a particular disease.

risk-benefit analysis The process of weighing the risk of a substance against the benefits it provides; if the risk is small and the benefits great, small amounts of this substance may be acceptable.

RNA (ribonucleic acid) A single-stranded nucleic acid. It carries information in DNA from the nucleus to the cytosol, is a component of ribosomes, and delivers amino acids for protein synthesis.

S

saccharin An alternative sweetener that contains no energy and is about 300 times sweeter than sugar.

saliva A watery fluid produced and secreted into the mouth by the salivary glands. It contains lubricants, enzymes, and other substances.

salivary amylase An enzyme secreted by the salivary glands that breaks down starch.

salivary glands The internal structures located at the sides of and below the face and in front of the ears; secrete saliva into the mouth.

Salmonella A bacterium that commonly causes food-borne illness.

sarcopenia Progressive decrease in skeletal muscle mass and strength that occurs with age.

satiety The feeling of fullness and satisfaction, caused by food consumption, that eliminates the desire to eat.

saturated fatty acid A fatty acid in which the carbon atoms are bound to as many hydrogen atoms as possible and that therefore contains no carbon-carbon double bonds.

scavenger receptor A protein on the surface of macrophages that binds to oxidized LDL cholesterol and allows it to be taken up by the cell.

scientific method The general approach of science that is used to explain observations about the world around us.

scurvy The vitamin C deficiency disease.

secondary lactose intolerance Lactase deficiency that occurs as a result of disease and may resolve after the disease has ended.

secretin A hormone released by the duodenum that signals the release of pancreatic juice rich in bicarbonate ions and stimulates the liver to secrete bile into the gallbladder.

segmentation Rhythmic local constrictions of the intestine that mix food with digestive juices and speed absorption by repeatedly moving the food mass over the intestinal wall.

selective breeding Techniques to selectively control mating in plants and animals for the purpose of producing organisms that better serve human needs.

selectively permeable Describes a membrane or barrier that will allow some substances to pass freely but will restrict the passage of others.

selenoproteins Proteins that contain selenium as a structural component of their amino acids. Selenium is most often found as selenocysteine, which contains an atom of selenium in place of the sulfur atom.

self-esteem The general attitude of approval or disapproval that people make and maintain about themselves.

semiessential amino acid See Conditionally essential amino acid.

semivegetarian One who avoids only certain types of meat, fish, or poultry; e.g., an individual who avoids all red meat but continues to consume poultry and fish.

serotonin A neurotransmitter that functions in the sleep center of the brain.

set point A level at which body fat or body weight seems to resist change despite changes in energy intake or output.

set point theory The theory that when people finish growing, they have a stable weight range, known as their set-point. Weight will tend to return to this set point despite periodic changes in energy intake or output.

sickle cell anemia An inherited disease in which hemoglobin structure is altered. Red blood cells containing the altered hemoglobin are sickle-shaped; rupture easily, causing anemia; and block small blood vessels, causing inflammation and pain.

significant scientific agreement Refers to a means of authorizing health claims on food labels that involves extensive review of the scientific evidence.

simple carbohydrate A class of carbohydrates, known as sugars, that includes monosaccharides and disaccharides.

simple diffusion The movement of substances from an area of higher concentration to an area of lower concentration. No energy is required.

simple sugar See Simple carbohydrates.

single-blind study An experiment in which either the study participants or the researchers are unaware of who is in a control or an experimental group.

skinfold thickness A measurement of subcutaneous fat used to estimate total body fat.

small for gestational age An infant born at term weighing less than 2.5 kg (5.5 lbs).

small intestine A tube-shaped organ of the digestive tract where digestion of ingested food is completed and most of the absorption occurs.

smooth muscle Involuntary muscles that cause constriction of the gastrointestinal tract, blood vessels, and glands.

sodium bicarbonate A compound that is part of an important buffer system in pancreatic juice and in the bloodstream.

sodium caseinate A form of the milk protein casein that is frequently used as a food additive.

sodium-potassium ATPase or **sodium-potassium pump** An energy-requiring protein pump in the cell membrane that pumps sodium out of the cell and potassium into the cell.

sodium-potassium pump see sodium-potassium ATPase.

soluble fiber Fiber that dissolves in water or absorbs water to form viscous solutions and can be broken down by the intestinal microflora. It includes pectins, gums, and some hemicelluloses.

solutes Dissolved substances.

solution A solvent containing a dissolved substance.

solvent A fluid in which one or more substances dissolve.

sorbitol A sugar alcohol formed from the sugar sorbose; used as a sweetener or humectant in food.

sperm The male reproductive cell.

sphincter A muscular valve that helps control the flow of materials in the GI tract.

spina bifida A birth defect resulting from the incorrect development of the spinal cord that can leave the spinal cord exposed.

spontaneous abortion or **miscarriage** Interruption of pregnancy prior to the seventh month.

spore A dormant state of some bacteria that is resistant to heat but can germinate and produce a new organism when environmental conditions are favorable.

sports anemia A temporary decrease in hemoglobin concentration that occurs during exercise training. It occurs as an adaptation to training in which expanded plasma volume dilutes red blood cells, but it does not impair delivery of oxygen to tissues.

standards of identity Regulations that define the allowable ingredients, composition, and other characteristics of foods.

starch A carbohydrate made of many glucose molecules linked in straight or branching chains. The bonds that hold the glucose molecules together can be broken by human digestive enzymes.

starchyose An oligosaccharide found in beans and other legumes that cannot be digested by human enzymes in the stomach and small intestine.

starvation A severe reduction in nutrient and energy intake that impairs health and eventually causes death. It is the most extreme form of malnutrition.

stearic acid An 18-carbon saturated fatty acid that, unlike other saturated fats, does not raise blood cholesterol levels.

steroid hormone A hormone that is made from cholesterol; includes the male and female sex hormones.

sterol A type of lipid with a structure composed of multiple chemical rings.

stomach A muscular pouchlike organ of the digestive tract that mixes food and secretes gastric juice into the lumen and the hormone gastrin into the blood.

stroke A blood clot or bleeding in the brain that causes brain tissue death.

stroke volume The volume of blood pumped by each beat of the heart.

structure/function claims Claims on food and supplement labels that describe the role of a nutrient or dietary ingredient in maintaining normal structure or function in humans.

stunting A decrease in linear growth rate that is an indicator of nutritional well-being in populations of children.

subcutaneous fat Adipose tissue located under the skin.

subsistence crop A crop that is grown as food for a farmer's family, with little or nothing left over to sell.

sucralose An alternative sweetener made from sucrose that is about 600 times sweeter than sucrose; trichlorogalactosucrose. It is heat stable, and so can be used in baked products.

sucrase An enzyme in the microvilli of the small intestine that breaks sucrose into glucose and fructose.

sucrose A disaccharide that is formed by linking fructose and glucose. It is commonly known as table sugar or white sugar.

sudden infant death syndrome (SIDS) or **crib death** The unexplained death of an infant, usually during sleep.

sugar alcohols Sweeteners that are structurally related to sugars but provide less energy than monosaccharides and disaccharides because they are not well absorbed.

sulfites Sulfur-containing compounds used as preservatives to prevent oxidation in dried fruits and vegetables and to prevent bacterial growth in wine.

superoxide dismutase (SOD) An enzyme that protects the cell by neutralizing damaging superoxide free radicals. One form of the enzyme requires zinc and copper for activity, and another form requires manganese.

superoxide radical A type of reactive oxygen molecule that can form free radicals leading to oxidative damage. They can be neutralized by the enzyme superoxide dismutase.

sustainable agriculture Agricultural methods that maintain soil productivity and a healthy ecological balance while having minimal long-term impacts.

T

tannin A substance found in tea and some grains that can bind minerals and decrease their absorption.

teratogen A substance that can cause birth defects.

testosterone A steroid hormone secreted by the testes that is involved in the maintenance and development of male sex organs and secondary sex characteristics.

theory An explanation based on scientific study and reasoning.

thermic effect of food (TEF) or **diet-induced thermogenesis** The energy required for the digestion of food and the absorption, metabolism, and storage of nutrients. It is equal to approximately 10% of daily energy intake.

Thiamin A B vitamin, also called vitamin B_1, that is needed in energy metabolism.

thiamin pyrophosphate The active coenzyme form of thiamin. It is the predominant form found inside cells, where it aids reactions in which a carbon-containing group is lost as CO_2.

threshold effect A reaction that occurs at a certain level of ingestion and increases as the dose increases. Below that level there is no reaction.

thyroid gland A gland located in the neck that produces thyroid hormones and calcitonin.

thyroid hormones Hormones produced by the thyroid gland that regulate metabolic rate.

thyroid-stimulating hormone A hormone that stimulates the synthesis and secretion of thyroid hormones from the thyroid gland.

tocopherol The chemical name for vitamin E.

Tolerable upper intake levels (ULs) DRI values that represent maximum daily intakes that are unlikely to pose a risk of adverse health effects to almost all individuals in the specified life-stage and gender group.

total energy expenditure (TEE) The sum of basal energy expenditure, the thermic effect of food, and the energy used in physical activity, regulation of body temperature, deposition of new tissue, and production of milk.

total parenteral nutrition (TPN) A technique for nourishing an individual by providing all needed nutrients directly into the circulatory system.

toxin A substance that can cause harm at some level of exposure.

trabecular or **spongy bone** The type of bone forming the inner spongy lattice that lines the bone marrow cavity and supports the cortical shell.

trace mineral or **trace element** A mineral required in the diet in an amount of 100 mg or less per day or present in the body in an amount of 0.01% of body weight or less.

***trans* fatty acid** An unsaturated fatty acid in which the hydrogen atoms are on opposite sides of the double bond.

transamination The process by which an amino group from one amino acid is transferred to a carbon compound to form a new amino acid.

transcription The process of copying the information in DNA to a molecule of mRNA.

transferrin An iron transport protein in the blood.

transgenic An organism with a gene or group of genes intentionally transferred from another species or breed.

transit time The time between the ingestion of food and the elimination of the solid waste from that food.

translation The process of translating the mRNA code into the amino acid sequence of a polypeptide chain.

treatment groups See Experimental groups.

triacylglycerol See Triglyceride.

triceps Region at the back of the upper arm that is a common site for measuring skinfold thickness.

trichinosis The disease caused by infection with the roundworm *Trichinella spiralis* after eating undercooked contaminated pork or game meats; the juvenile form of this roundworm migrates to the muscles and causes flu-like symptoms and muscle pain and weakness.

triglyceride (triacylglycerol) The major form of lipid in food and in the body. Each consists of three fatty acids attached to a glycerol molecule.

trimester A term used to describe each third, or three-month period, of a pregnancy.

tripeptide Three amino acids linked together by peptide bonds.

tropical oils A term used in the popular press to refer to the saturated oils—coconut, palm, and palm kernel oil—that are derived from plants grown in tropical regions.

trypsin A protein-digesting enzyme that is secreted from the pancreas in inactive form and activated in the small intestine.

tuber The starchy underground storage organ of plants.

tumor initiator A substance that causes mutations and therefore may predispose a cell to becoming cancerous.

tumor promoter A substance that contributes to cancer formation by stimulating cells to divide.

twenty-four-hour recall (24-hour recall) A method of assessing dietary intake in which a trained interviewer helps an individual remember what he or she ate during the previous day.

type 1 diabetes A form of diabetes that is caused by the autoimmune destruction of insulin-producing cells in the pancreas, usually leading to absolute insulin deficiency; previously known as insulin-dependent diabetes mellitus or juvenile-onset diabetes.

type 2 diabetes A form of diabetes characterized by insulin resistance and relative insulin deficiency; previously known as noninsulin-dependent diabetes mellitus or adult-onset diabetes.

U

ubiquinone A compound that transports electrons in the electron transport chain but that is not essential in the diet; also called coenzyme Q.

UL See Tolerable Upper Intake Level (UL).

ulcer An open sore.

undernutrition Any condition resulting from an energy or nutrient intake below that which meets nutritional needs.

underwater weighing A technique that uses the difference between body weight underwater and body weight on land to estimate body composition.

underweight A body mass index of less than 18.5 kg/m^2, or a body weight 10% or more below the desirable body weight standard.

unsaturated fatty acid A fatty acid that contains one or more carbon-carbon double bonds.

urea A nitrogen-containing waste product formed from the breakdown of amino acids that is excreted in the urine.

urine A fluid produced by the kidneys consisting of metabolic wastes, excess water, and dissolved substances.

U.S. Department of Agriculture (USDA) U.S. government agency responsible for monitoring the safety and wholesomeness of meat, poultry, and eggs.

uterus A female organ for containing and nourishing the embryo and fetus from the time of implantation to the time of birth.

V

variable A factor or condition that is changed in an experimental setting.

vegan diet A pattern of food intake that eliminates all animal products.

vegetarian diet A pattern of food intake that includes plant-based foods and eliminates some or all foods of animal origin.

veins Vessels that carry blood toward the heart.

venule A small vein that drains blood from capillaries and passes it to larger veins for return to the heart.

very low birth weight A birth weight less than 1.5 kg (3.3 lbs).

very-low-calorie diet A weight-loss diet that provides fewer than 800 kcalories per day.

very-low-density lipoprotein (VLDL) A lipoprotein assembled by the liver that carries lipids from the liver and delivers triglycerides to body cells.

villi (villus) Finger-like protrusions of the lining of the small intestine that participate in the digestion and absorption of nutrients.

virus An infective agent that typically consists of a nucleic acid molecule in a protein coat, is not visible under an ordinary microscope and depends on cells for its metabolic and reproductive needs.

visceral fat Adipose tissue that is located in the abdomen around the body's internal organs.

vitamin A A fat-soluble vitamin needed in cell differentiation, reproduction and vision.

vitamin B_6 One of the B vitamins, needed in protein metabolism.

vitamin B_{12} One of the B vitamins, only found in animal foods.

vitamin C A water-soluble vitamin needed for the maintenance of collagen.

vitamin D A fat-soluble vitamin needed for calcium absorption that can be made in the body when there is exposure to sunlight.

vitamin E A fat-soluble vitamin that functions as an antioxidant.

vitamin K A fat-soluble vitamin needed for blood clotting.

vitamins Organic compounds needed in the diet in small amounts to promote and regulate the chemical reactions and processes needed for growth, reproduction, and maintenance of health.

W

warfarin An anticoagulant drug that acts by inhibiting the action of vitamin K. It is a derivative of dicumarol; also used as rat poison.

water A molecule composed of two hydrogen atoms and one oxygen atom; essential nutrient needed by the human body in large amounts.

water intoxication A condition that occurs when a person drinks enough water to significantly lower the concentration of sodium in the blood.

water-soluble vitamins Vitamins that dissolve in water.

weight cycling or **yo yo dieting** The repeated loss and regain of body weight.

Wernicke-Korsakoff syndrome A form of thiamin deficiency associated with alcohol abuse that is characterized by mental confusion, disorientation, loss of memory, and a staggering gait.

whole grain The entire kernel of grain including the bran layers, the germ, and the endosperm.

whole-wheat flour A flour that contains all components of the wheat kernel: the bran, the germ, and the endosperm.

Wilson's disease An inherited defect in copper metabolism that leads to excessive accumulation of copper in the body and can cause neurological symptoms and liver disease.

X

xanthan gum A plant extract used as a stabilizer in processed foods.

xanthomas Cholesterol deposits that form under the skin.

xerophthalmia A spectrum of eye conditions resulting from vitamin A deficiency that may lead to blindness. An early symptom is night blindness, and as deficiency worsens, lack of mucus leaves the eye dry and vulnerable to cracking and infection.

xylitol The sugar alcohol formed from the sugar xylose; used in sugarless gum.

Y

yo-yo diet syndrome See Weight cycling.

Z

zeaxanthin A carotenoid found in corn and green peppers that has no vitamin A activity but provides some protection against macular degeneration.

zoochemicals Substances found in animal foods (*zoo-* means animal) that are not essential nutrients but may have health-promoting properties.

zygote The cell produced by the union of sperm and ovum during fertilization.

Index

B

C

D

F

G

H

I

N

O

Q

R

T

U

W

X

Y

Z

Dietary Reference Intakes: Tolerable Upper Intake Levels (UL[a]): Minerals

Life Stage Group	Arsenic[b]	Boron (mg/day)	Calcium (g/day)	Chromium	Copper (μg/day)	Fluoride (mg/day)	Iodine (μg/day)	Iron (mg/day)	Magnesium (mg/day)[c]	Manganese (mg/day)	Molybdenum (μg/day)	Nickel (mg/day)	Phosphorus (g/day)	Selenium (μg/day)	Silicon[d]	Vanadium (mg/day)[e]	Zinc (mg/day)	Sodium (g/day)	Chloride (g/day)	Potassium
Infants																				
0–6 mo	ND[f]	ND	1.0	ND	ND	0.7	ND	40	ND	ND	ND	ND	ND	45	ND	ND	4	ND	ND	ND
6–12 mo	ND	ND	1.5	ND	ND	0.9	ND	40	ND	ND	ND	ND	ND	60	ND	ND	5	ND	ND	ND
Children																				
1–3 y	ND	3	2.5	ND	1,000	1.3	200	40	65	2	300	0.2	3	90	ND	ND	7	1.5	2.3	ND
4–8 y	ND	6	2.5	ND	3,000	2.2	300	40	110	3	600	0.3	3	150	ND	ND	12	1.9	2.9	ND
Males, Females																				
9–13 y	ND	11	3.0	ND	5,000	10	600	40	350	6	1,100	0.6	4	280	ND	ND	23	2.2	3.4	ND
14–18 y	ND	17	3.0	ND	8,000	10	900	45	350	9	1,700	1.0	4	400	ND	ND	34	2.3	3.6	ND
19–30 y	ND	20	2.5	ND	10,000	10	1,100	45	350	11	2,000	1.0	4	400	ND	1.8	40	2.3	3.6	ND
31–50 y	ND	20	2.5	ND	10,000	10	1,100	45	350	11	2,000	1.0	4	400	ND	1.8	40	2.3	3.6	ND
51–70 y	ND	20	2.0	ND	10,000	10	1,100	45	350	11	2,000	1.0	4	400	ND	1.8	40	2.3	3.6	ND
>70 y	ND	20	2.0	ND	10,000	10	1,100	45	350	11	2,000	1.0	3	400	ND	1.8	40	2.3	3.6	ND
Pregnancy																				
14–18 y	ND	17	3.0	ND	8,000	10	900	45	350	9	1,700	1.0	3.5	400	ND	ND	34	2.3	3.6	ND
19–50 y	ND	20	2.5	ND	10,000	10	1,100	45	350	11	2,000	1.0	3.5	400	ND	ND	40	2.3	3.6	ND
Lactation																				
14–18 y	ND	17	3.0	ND	8,000	10	900	45	350	9	1,700	1.0	4	400	ND	ND	34	2.3	3.6	ND
19–50 y	ND	20	2.5	ND	10,000	10	1,100	45	350	11	2,000	1.0	4	400	ND	ND	40	2.3	3.6	ND

[a]UL = the maximum level of daily nutrient intake that is likely to pose no risk of adverse effects. Unless otherwise specified, the UL represents total intake from food, water, and supplements. Due to lack of suitable data, ULs could not be established for arsenic, chromimum, silicon, and potassium. In the absence of ULs, extra caution may be warranted in consuming levels above recommended intakes.

[b]Although the UL was not deternmined for arsenic, there is no justification for adding arsenic to food or supplements.

[c]The ULs for magnesium represent intake from a pharmacological agent only and do not indude intake from food and water.

[d]Although silicon has not been shown to cause adverse effects in humans, there is no justification for adding silicon to supplements.

[e]Although vanadium in food has not been shown to cause adverse effects in humans, there is no justification for adding vanadium to food and vanadium supplements should be used with caution. The UL is based on adverse effects in laboratory animals and this data could be used to set a UL for adults but not children and adolescents.

[f]ND = Not determinable due to lack of data of adverse effects in this age group and concern with regard to lack of ability to handle excess amounts. Source of intake should be from food only to prevent high levels of intake.

Source: Dietary Reference Intake Tables: The Complete Set. Institute of Medicine, National Academy of Sciences. Available online at www.nap.edu. Reprinted with permission from *Dietary Reference Intakes: The Essential Guide to Nutrient Requirements*, 2006, by the National Academy of Sciences, Washington, D.C. Institute of Medicine, Food and Nutrition Board. Dietary Reference Intakes for Calcium and Vitamin D (2011), National Academies Press, Washington, DC, 2011.

Dietary Reference Intakes: Tolerable Upper Intake Levels (UL[a]): Vitamins

Life Stage Group	Vitamin A (μg/day)[b]	Vitamin C (mg/day)	Vitamin D (μg/day)	Vitamin E (mg/day)[c,d]	Vitamin K	Thiamin	Riboflavin	Niacin (mg/day)[d]	Vitamin B_6 (mg/day)	Folate (μg/day)[d]	Vitamin B_{12}	Pantothenic Acid	Biotin	Choline (g/day)	Carotenoids[e]
Infants															
0–6 mo	600	ND[f]	25	ND	ND	ND	ND	ND	ND	ND	ND	ND	ND	ND	ND
6–12 mo	600	ND	38	ND	ND	ND	ND	ND	ND	ND	ND	ND	ND	ND	ND
Children															
1–3 y	600	400	63	200	ND	ND	ND	10	30	300	ND	ND	ND	1.0	ND
4–8 y	900	650	75	300	ND	ND	ND	15	40	400	ND	ND	ND	1.0	ND
Males, Females															
9–13 y	1,700	1,200	100	600	ND	ND	ND	20	60	600	ND	ND	ND	2.0	ND
14–18 y	2,800	1,800	100	800	ND	ND	ND	30	80	800	ND	ND	ND	3.0	ND
19–70y	3,000	2,000	100	1,000	ND	ND	ND	35	100	1,000	ND	ND	ND	3.5	ND
>70 y	3,000	2,000	100	1,000	ND	ND	ND	35	100	1,000	ND	ND	ND	3.5	ND
Pregnancy															
14–18 y	2,800	1,800	100	800	ND	ND	ND	30	80	800	ND	ND	ND	3.0	ND
19–50 y	3,000	2,000	100	1,000	ND	ND	ND	35	100	1,000	ND	ND	ND	3.5	ND
Lactation															
14–18 y	2,800	1,800	100	800	ND	ND	ND	30	80	800	ND	ND	ND	3.0	ND
19–50 y	3,000	2,000	100	1,000	ND	ND	ND	35	100	1,000	ND	ND	ND	3.5	ND

[a]UL = The maximum level of daily nutrient intake that is likely to pose no risk of adverse effects. Unless otherwise specified, the UL represents total intake from food, water, and supplements. Due to lack of suitable data, ULs could not be established for vitamin K, thiamin, riboflavin, vitamin B_{12}, pantothenic acid, biotin, or carotenoids. In the absence of ULs, extra caution may be warranted in consuming levels above recommended intakes.

[b]As preformed vitamin A only.

[c]As α-tocopherol; applies to any form of supplemental α-tocopherol.

[d] The ULs for vitamin E, niacin, and folate apply to synthetic forms obtained from supplements, fortified foods, or a combination of the two.

[e]β-Carotene supplements are advised only to serve as a provitamin A source for individuals at risk of vitamin A deficiency.

[f]ND = Not determinable due to lack of data of adverse effects in this age group and concern with regard to lack of ability to handle excess amounts. Source of intakes should be from food only to prevent high levels of intake.

Source: Dietary Reference Intake Tables: The Complete Set. Institute of Medicine, National Academy of Sciences. Available online at www.nap.edu. Reprinted with permission from *Dietary Reference Intakes: The Essential Guide to Nutrient Requirements,* 2006, by the National Academy of Sciences, Washington, D.C. Institute of Medicine, Food and Nutrition Board, Dietary Reference Intakes for Calcium and Vitamin D (2011), National Academies Press, Washington. DC, 2011.